HANDBOOK OF MICROEMULSION SCIENCE AND TECHNOLOGY

edited by

Promod Kumar

Gillette Research Institute
Gaithersburg, Maryland

K. L. Mittal

Hopewell Junction, New York

CRC Press
Taylor & Francis Group
Boca Raton London New York

CRC Press is an imprint of the
Taylor & Francis Group, an **informa** business

CRC Press
Taylor & Francis Group
6000 Broken Sound Parkway NW, Suite 300
Boca Raton, FL 33487-2742

First issued in paperback 2019

© 1999 by Taylor & Francis Group, LLC
CRC Press is an imprint of Taylor & Francis Group, an Informa business

No claim to original U.S. Government works

ISBN-13: 978-0-8247-1979-1 (hbk)
ISBN-13: 978-0-367-39959-7 (pbk)

Visit the Taylor & Francis Web site at
http://www.taylorandfrancis.com

and the CRC Press Web site at
http://www.crcpress.com

Preface

Microemulsions are thermodynamically stable isotropic dispersions of oil and water containing domains of nanometer dimensions stabilized by the interfacial film of surface active agent(s). The word "microemulsion" was originally proposed by Jack H. Schulman and coworkers in 1959 (although the first paper appeared on this subject back in 1943), who observed that on addition of short chain alcohol to a coarse macroemulsion (containing water, benzene, hexanol, and potassium oleate), the system became transparent. Between 1943 and 1965, Schulman and coworkers described how to prepare these transparent systems and investigated them by using an array of experimental techniques. Basically a macroemulsion was prepared which was then titrated to clarity by adding a cosurfactant, and concomitantly it became transparent spontaneously when the composition was right. It was a widely held myth that a combination of a surfactant and a cosurfactant was needed to obtain transparent systems until 1967, when Shinoda and coworkers showed that clear isotropic systems could be achieved by using nonionic surfactants without the addition of a cosurfactant. Friberg and coworkers initiated the Scandinavian approach to microemulsion systems in 1975 following the work of Ekwall et al. on the amphiphilic association structures. The first attempt to define all the contributing factors to the stability of these systems was by Ruckenstein and coworkers in 1975.

Microemulsion research has been stimulated by the great potential of microemulsions for practical applications primarily because their use for enhanced oil recovery seemed an attractive alternative due to the work of Shah and colleagues (1971), and over the years a wide spectrum of other applications — ranging from mundane to sophisticated — have emerged and many more are under active development. Moreover, as we learn further about the interesting characteristics of microemulsions, new vistas will emerge.

The literature on microemulsions is quite extensive and is spread over many journals, book chapters, and monographs. The enormous growth and potential applications of microemulsions necessitated an up-to-date unified source of information related to these systems. This book was conceived with the idea of providing a panoramic view of the wonderful world of microemulsions, covering both the fundamental and applied aspects. We were most gratified with the tremendous interest and enthusiasm evinced by the invitees to write chapters for this compendium. This was a testimonial to the high tempo of research activity in this field and to the need for and timeliness of this treatise.

This book is divided into five parts as follows: Part I: Historical Perspective; Part II: Structural Aspects and Characterization of Microemulsions; Part III: Reactions in Microemulsions; Part IV: Applications of Microemulsions; and Part V: Future Prospects. The book opens with the chapter on the historical development of microemulsion systems by two leading authorities (Lindman and Friberg) who have significantly contributed to the field of microemulsions. In the next two chapters J. Th. G. Overbeek (the doyen of colloid science) and coworkers and E. Ruckenstein advance different approaches to describe the thermodynamics of microemulsion systems. While a full description of microemulsion thermodynamics is far from complete, the droplet type model predicts the experimental observations quite well. A theory that predicts the global phase behavior and the detailed properties of the phases as a function of experimentally adjustable parameters is still under development.

Significant progress has been made in theoretical statistical-mechanical description of these systems. Two different classes of microscopic models have been employed. In the first class are the microscopic models in which an amphiphilic system is considered to be a ternary system with one kind of molecule having complicated interactions with others (or with themselves). In the second class of microscopic models it is assumed that essentially all the amphiphilic molecules in the system are part of a monolayer or bilayer. A comprehensive review of microscopic models of microemulsions is presented in Chapter 4 by Gompper and Schick. Monte Carlo simulation methods for modeling self-association structures in surfactant systems with and without the presence of a nonpolar phase (oil) are reviewed in Chapter 5 by Rajagopalan and coworkers. In Chapter 6 the phase behavior and critical phenomena in microemulsion systems are presented and the effect of salt and alcohol chain length on the phase behavior and stability of surfactant phases is discussed in detail by Bellocq. The association behavior of amphiphilic molecules in one-phase ternary systems (Winsor IV) is presented in Chapter 7 by Garti and coworkers.

The general features of nonionic, ionic and supercritical microemulsions are presented in Chapters 7–9. The significance of the Winsor R-value in relation to designing an appropriate ternary phase microemulsion system is discussed. A systems approach to achieving the desired properties in a ternary system by taking into account the properties of nonpolar phase (oil) and of surfactant is discussed in Chapter 8 by Salager and Antón. Salager and Antón also discuss the pH dependency, which normally is an important issue for formulators, and furthermore microemulsions based on anionic–cationic surfactant mixtures are discussed in Chapter 8. The development of supercritical fluid technology has been hampered because supercritical fluids with a low critical point do not have large solubilizing capability. The use of surfactants in supercritical fluid extraction, first demonstrated by R. D. Smith and coworkers in 1987, has greatly extended the utility of supercritical fluid technology. In Chapter 9 McFann and Johnston review the phase behavior of supercritical (W/O) microemulsions and their potential applications.

The enormous interest in microemulsion systems required the development of analytical techniques by which microstructures and local molecular arrangements along with interactions and dynamics could be probed in detail. Thus a section of Part II is devoted to the analytical techniques for analyzing these systems. In Chapter 10 Lindman and coworkers give a comprehensive description of NMR self-diffusion and relaxation methods for the microstructural characterization of these systems. The rheological properties of microemulsions are covered in Chapter 11 by Gradzielski and Hoffmann. Elastic and inelastic light scattering techniques are extensively used for the characterization of microemulsions to obtain information on droplet sizes. The underlying principles for using light scattering techniques to probe these systems along with the surface scattering technique to measure the ultralow interfacial tension encountered in these systems are outlined in Chapter 12 by Langevin and Rouch. Transmission electron microscopy (TEM) is an important technique for studying microstructures because it directly produces images at high resolution, but the high vacuum in the microscope will result in alteration of the microstructure of microemulsions. Thus, advanced sample preparation techniques are needed to analyze the samples by using TEM. Sample preparation techniques and TEM analysis are described in detail by Bellare et al. in Chapter 13. The physical properties of microemulsion systems are dependent on thermodynamic conditions; when a system's thermodynamic conditions are altered instantaneously, the response of the system to such alterations provides valuable information about the kinetic properties of the systems. A particularly important method in this regard is electrical birefringence, which is discussed by Schelly in Chapter 14.

Microemulsions are the only systems which can cosolubilize high concentrations of both water-soluble and water-insoluble reactants and thus render themselves as a novel medium for chemical synthesis; the structure of the microemulsions critically determines the reaction parameters. The reactivity in microemulsions is discussed in Chapter 15 by Bunton and Romsted. The utility of microemulsions in spectroscopic analysis is described in Chapter 16 by Guo and Zhu.

The design and synthesis of novel materials requires a precise control of the shape and size of the precursor materials. The nanometer size domains act as nanoreactors for particle synthesis reactions. A precise control of microemulsion nanodomains would result in a precise control of the shape and size of the resultant particles. Nagy discusses this unique process with mechanism for metal borides particle formation and growth in Chapter 17. In the following chapter, the synthesis of metal oxides, ranging from silica to magnetic oxide to superconductors, in microemulsion systems is described by Osseo-Asare. The exchange of material among droplets determines the reaction kinetics in microemulsion systems. Luminescence quenching in microemulsion systems is a novel method to understand exchange dynamics, which is described by Almgren and Mays in Chapter 19. A comprehensive review of microemulsion structure in near-critical and supercritical fluids and their application as a reaction medium is provided in Chapter 20 by Fulton. A detailed review of the application of electrochemical techniques for the characterization of microemulsions and their application in electrochemical reactions is presented in Chapter 21 by Qutubuddin.

Emulsion polymerization has become a standard technique for the production of industrial polymers in the form of colloids or latexes. The concept of polymerization in microemulsions appeared only in the 1980s. Microemulsions provide a unique microenvironment that can be harnessed to produce novel polymeric materials with interesting morphologies or with specific properties. The current status of polymerization in these systems is provided in Chapter 22 by Candau. In this chapter the formulation of a polymerizable microemulsion and the effect of monomers on microemulsion structure, along with the kinetic and mechanistic features of these systems, are discussed. In Chapter 23 Holmberg discusses applications of microemulsions to enzyme-catalyzed reactions. A most interesting feature of a microemulsion is that it provides a propitious environment for an enzyme in the presence of water-insoluble substrates. Here Holmberg provides a comprehensive review of micellar-enzymology, including enzymatic polymerization. The application of microemulsions for enhanced oil recovery is the topic of Chapter 24 by Shah and cohorts.

The thermodynamic stability of microemulsions and their ability to solubilize both polar and nonpolar active materials make these systems an attractive alternative for drug delivery vehicles. Applications of microemulsions in pharmaceuticals are discussed in

Chapter 25 by Malmsten. Microemulsions have also found a special application in cosmetic formulations to reduce VOCs (volatile organic compounds) to levels allowed by recent regulation changes. In addition, transparent appearance of a microemulsion gives the perception of a clean system, which for some consumers is an essential property; thus microemulsions have earned a place in the cosmetic industry. The cosmetic applications are reviewed by Aikens and Friberg in Chapter 26. Certain foods contain microemulsions naturally, or they can form in the intestine during digestion. The possibility of using microemulsions, lipid based, in food production is discussed by Engström and Larsson in Chapter 27. The development of efficient separation processes is of considerable industrial importance. Energy efficient separation processes with higher selectivity and increased separation rate are required. Concentrating solutes from dilute solutions is of critical importance. Microemulsion systems with a significantly large interfacial area and faster, almost instantaneous, separation make them an attractive alternative to emulsion liquid membranes. In Chapter 28 the applications of microemulsions as liquid membranes are presented by Wiencek.

Because of the high solubilization power of both hydrophilic and hydrophobic substances and the ultralow interfacial tension as their inherent property, microemulsions are an excellent medium for textile detergency. The structure of microemulsions and its effect on detergency are presented in Chapter 29 by Dörfler. A complete discussion of all such applications would render this volume prohibitively long, so only eclectic applications are included.

The final chapter concludes with comments by Sjöblom and Friberg on the future prospects of microemulsions both in academic research as model systems and in technological processes.

In essence this book represents the cumulative wisdom of many internationally renowned researchers and technologists in the arena of microemulsions. We certainly hope this volume will become a *vade mecum* in the field of microemulsions, and the bounty of information garnered here will serve as a fountainhead for further exciting developments in this field. The book should be of interest to everyone with core or tangential interest in microemulsions. It should appeal to the novice (as a Baedeker) as well as to the veteran researcher (as a commentary on contemporary R&D activity). The book is profusely illustrated and copiously referenced.

Now comes the pleasant task of thanking those who helped in many and different ways to bring this project to fruition. First, we are most thankful to the authors for their interest, enthusiasm, and contributions without which this book could not have materialized. Special thanks are due to the unsung heroes (reviewers) for their valuable comments, as the peer review is a desideratum to maintain the highest standard of publication. Our appreciation goes to Anita Lekhwani for her unwavering interest in this project, and to Joseph Stubenrauch (both of Marcel Dekker, Inc.) for the excellent job done on the production of this book.

Promod Kumar
K. L. Mittal

Contents

Contributors

Patricia A. Aikens   Uniqema, Wilmington, Delaware

Mats Almgren   Department of Physical Chemistry, Uppsala University, Uppsala, Sweden

Raquel E. Antón   School of Chemical Engineering, Universidad de Los Andes, Mérida, Venezuela

Abraham Aserin   Casali Institute of Applied Chemistry, The Hebrew University of Jerusalem, Jerusalem, Israel

Jayesh R. Bellare*   Microstructure Engineering and Ultramicroscopy Laboratory, Department of Chemical Engineering, Indian Institute of Technology–Bombay, Mumbai, India

A. M. Bellocq   Centre de Recherche Paul Pascal–CNRS, Pessac, France

Clifford A. Bunton   Department of Chemistry, University of California, Santa Barbara, Santa Barbara, California

Françoise Candau   Institut Charles Sadron (CRM-EAHP), CNRS-ULP, Strasbourg, France

Hans-Dieter Dörfler   Department of Physical Chemistry, Technical University of Dresden, Dresden, Germany

Sven Engström   Food Technology, Chemical Center, Lund University, Lund, Sweden

Shmaryahu Ezrahi   Casali Institute of Applied Chemistry, The Hebrew University of Jerusalem, Jerusalem, Israel

Stig E. Friberg   Department of Chemistry, Clarkson University, Potsdam, New York

---

*Formerly at the Department of Chemical Engineering and Materials Science, University of Minnesota, Minneapolis, Minnesota.

John L. Fulton Environmental and Energy Sciences Division, Pacific Northwest National Laboratory, Richland, Washington

Nissim Garti Casali Institute of Applied Chemistry, The Hebrew University of Jerusalem, Jerusalem, Israel

Gerhard Gompper Theory Group, Max-Planck-Institut für Kolloid- und Grenzflächenforschung, Teltow, Germany

Michael Gradzielski Lehrstuhl für Physikalische Chemie I, Universität Bayreuth, Bayreuth, Germany

Rong Guo Department of Chemistry, Yangzhou University, Yangzhou, People's Republic of China

Manoj M. Haridas Microstructure Engineering and Ultramicroscopy Laboratory, Department of Chemical Engineering, Indian Institute of Technology–Bombay, Mumbai, India

Heinz Hoffmann Lehrstuhl für Physikalische Chemie I, Universität Bayreuth, Bayreuth, Germany

Krister Holmberg Department of Applied Surface Chemistry, Chalmers University of Technology, Göteborg, Sweden

Keith P. Johnston Department of Chemical Engineering, University of Texas, Austin, Texas

James R. Kanicky Center for Surface Science & Engineering, Department of Chemical Engineering, University of Florida, Gainesville, Florida

Willem K. Kegel Van't Hoff Laboratory for Physical and Colloid Chemistry, Debye Research Institute, Utrecht University, Utrecht, The Netherlands

D. Langevin Laboratoire de Physique des Solides, Université Paris-Sud, Orsay, France

Kåre Larsson Food Technology, Chemical Center, Lund University, Lund, Sweden

Henk N. W. Lekkerkerker Van't Hoff Laboratory for Physical and Colloid Chemistry, Debye Research Institute, Utrecht University, Utrecht, The Netherlands

Xiangbing Jason Li* Department of Chemical Engineering and Materials Science, University of Minnesota, Minneapolis, Minnesota

Björn Lindman Physical Chemistry 1, Chemical Center, University of Lund, Lund, Sweden

---

Current affiliation: Applied Materials, Inc., Santa Clara, California.

Martin Malmsten Institute for Surface Chemistry, Stockholm, Sweden

Holger Mays Department of Physical Chemistry, Uppsala University, Uppsala, Sweden

Gregory J. McFann Unilever Research U.S., Edgewater, New Jersey

Janos B.Nagy Laboratoire de Résonance Magnétique Nucléaire, Facultés Universitaires Notre-Dame de la Paix, Namur, Belgium

Ulf Olsson Physical Chemistry 1, Chemical Center, University of Lund, Lund, Sweden

K. Osseo-Asare Department of Materials Science and Engineering, The Pennsylvania State University, University Park, Pennsylvania

J. Theo G. Overbeek Van't Hoff Laboratory for Physical and Colloid Chemistry, Debye Research Institute, Utrecht University, Utrecht, The Netherlands

Vinod Pillai[†] Center for Surface Science & Engineering, Department of Chemical Engineering, University of Florida, Gainesville, Florida

Syed Qutubuddin Department of Chemical Engineering, Case Western Reserve University, Cleveland, Ohio

Raj Rajagopalan Department of Chemical Engineering, University of Florida, Gainesville, Florida

L. A. Rodriguez-Guadarrama Department of Chemical Engineering, University of Houston, Houston, Texas

Laurence S. Romsted Department of Chemistry, Wright-Rieman Laboratories, Rutgers, The State University of New Jersey, New Brunswick, New Jersey

J. Rouch Centre de Physique Moléculaire Optique et Hertzienne, Université Bordeaux I, Talence, France

Eli Ruckenstein Department of Chemical Engineering, State University of New York at Buffalo, Buffalo, New York

Jean-Louis Salager School of Chemical Engineering, Universidad de Los Andes, Mérida, Venezuela

Zoltan A. Schelly Department of Chemistry and Biochemistry, The University of Texas at Arlington, Arlington, Texas

Michael Schick Department of Physics, University of Washington, Seattle, Washington

---

[†] Deceased.

Dinesh O. Shah   Center for Surface Science & Engineering, Departments of Chemical Engineering and Anesthesiology, University of Florida, Gainesville, Florida

Johan Sjöblom   Department of Chemistry, University of Bergen, Bergen, Norway

Olle Söderman   Physical Chemistry 1, Chemical Center, University of Lund, Lund, Sweden

Sameer K. Talsania   Department of Chemical Engineering, University of Houston, Houston, Texas

John M. Wiencek   Chemical & Biochemical Engineering Department, The University of Iowa, Iowa City, Iowa

Xiashi Zhu   Department of Chemistry, Yangzhou University, Yangzhou, People's Republic of China

1

Microemulsions—A Historical Overview

Björn Lindman
University of Lund, Lund, Sweden

Stig E. Friberg
Clarkson University, Potsdam, New York

I. EARLY STAGE—EMULSIONS VS. AMPHIPHILIC ASSOCIATION STRUCTURES

The interest in emulsions with very small drops is understandable, as the use of emulsions or coarse dispersions in paints, wax dispersions, and similar systems is accompanied by stability problems, caking and creaming being the obvious ones. These problems are alleviated if the particle size is made smaller, and patents to this effect were issues as early as the 1930s [1,2].

The truly outstanding pioneer in the scientific treatment of emulsions was Schulman (see, e.g., Hoar and Schulman [3]), who noticed that coarse macroemulsions stabilized by an ionic surfactant became transparent after addition of a medium chain length alcohol. Schulman and his collaborators were the first to investigate these transparent liquids using an array of experimental methods such as X-ray diffraction [4], ultracentrifugation [5], light scattering [6], electron microscopy [7], viscosimetry [8], and nuclear magnetic resonance (NMR) [9,10] to characterize the size, shape, and dynamics of the dispersed phase. Schulman named these droplets hydrophilic oleomicelles or oleophilic hydromicelles depending on whether the continuous medium was nonaqueous or aqueous, respectively.

The sudden change from a turbid emulsion to a transparent liquid was a conspicuous feature, and Schulman and his collaborators were occupied with this problem for several years. It was obvious that the larger emulsion drops were spontaneously divided into small droplets, implying a spontaneous increase in the extent of the interface. Schulman and coworkers interpreted this change as being due to a temporary negative interfacial free energy for large droplets, expressing this phenomenon as a surface pressure [11]. A problem remaining with this approach was the huge surface pressures (≈ 50 mN/m) required. This requirement was alleviated by the Schulman–Prince approach [12,13], which considered the oil/water (O/W) interface as a duplex film with two tensions at the surface, one toward the aqueous phase and one toward the oil phase. Bending of the surface to reach an optimal radius would relax both tensions, giving a reduction in interfacial energy. This approach in fact introduces two of the three components of the interfacial free energy [14] and should be considered as a zero-order attempt to clarify the thermodynamic stability of microemulsions.

The first attempt to compile all the factors contributing to the stability of a microemulsion is due to Ruckenstein and Chi [15], who summarized calculations of enthalpic components (van der Waals attractive potential, electrical double layer repulsive potential, and the interfacial stretching and bending free energy) and entropic contributions from the location of droplets. These calculations as well as those following [16] were useful because they revealed the importance of extremely low interfacial tension.

Another approach was developed in Scandinavia following the pioneering contributions of Ekwall et al. on amphiphilic association structures [17] and in Japan by Shinoda and coworkers [18–24], who investigated the structural changes in systems of water, hydrocarbon, and nonionic surfactants of the polyethylene glycol alkyl ether type.

The Scandinavian approach to microemulsions was initiated by Gillberg et al. [25], who used the classical phase diagrams of water, ionic surfactant, and medium chain length alcohol of Ekwall et al. [17]. The diagram (Fig. 1) contains two regions of isotropic liquids and two regions of liquid crystals. Region L_2 was described by Ekwall et al. as containing inverse micelles in a continuum of alcohol. Gillberg et al. [25] showed that adding hydrocarbon to this liquid gave isotropic liquid compositions corresponding to the Schulman W/O microemulsions, as was also shown by Sjöblom and Friberg [26] (Fig. 2).

It is interesting to note that this Scandinavian contribution was in agreement with Schulman's original concept of microemulsions as micellar solutions [3]; the name "microemulsions" was coined much later [27]. Following Gillberg et al., Rance and Friberg [28] demonstrated that the O/W microemulsion is a direct continuation of the well-known Ekwall et al. aqueous micellar solutions [17] (Fig. 3). It should be noted that the common presentation of these four-component phase diagrams as phase maps in one plane

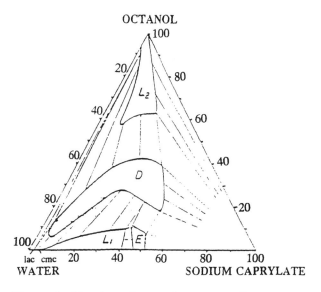

Figure 1 The phase diagram for water, sodium octanoate, and octanol according to Ekwall [17].

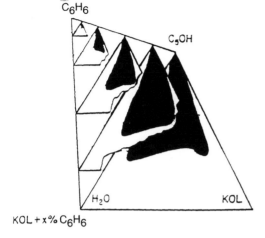

Figure 2 Water-in-oil microemulsion areas (dark) at different contents of benzene in the system water (H_2O)/potassium oleate (KOL)/pentanol (C_5OH)/benzene (C_6H_6). (According to Ref. 26.)

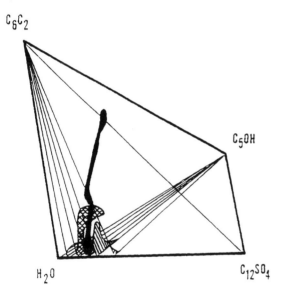

Figure 3 The O/W microemulsion regions (dark) in the system water (H_2O)/sodium dodecylsulfate ($C_{12}SO_4$)/pentanol (C_5OH)/p-xylene (C_6C_2).

(Fig. 4) may lead to a misunderstanding. The area of isotropic solutions in Fig. 4 may at first be considered as an isolated entity, but it is in fact part of a continuous isotropic liquid area emanating from the aqueous micellar solution (Fig. 5).

The third effort in the early development in microemulsion science originated with Saito and Shinoda [18,19] in Japan. In their studies on the temperature-dependent behavior of water–hydrocarbon–polyethylene glycol alkyl (aryl) ether systems, a relationship was observed between the cloud point of the surfactant and the solubilization of the hydrocarbon. For aliphatic hydrocarbons it was found that the solubilization of hydrocarbon

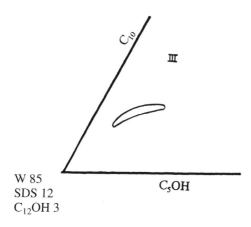

W 85
SDS 12
$C_{12}OH$ 3

Figure 4   Oil-in-water microemulsion area for aqueous solution in which the sodium dodecylsulfate was partially replaced by dodecanol. Compositions in weight percent. W = water; SDS = sodium dodecylsulfate; $C_{12}OH$ = dodecanol; C_5OH = pentanol; C_{10} = decane.

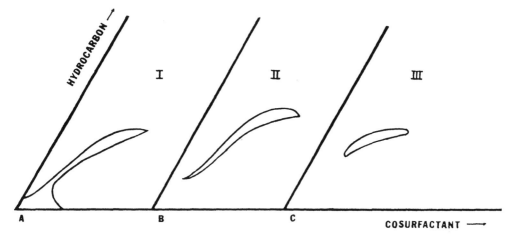

Figure 5   Oil-in-water microemulsion areas for aqueous solutions in which the sodium dodecylsulfate was partially replaced by dodecanol. Cosurfactant = pentanol; hydrocarbon = decane. Compositions in weight percent:

	Water	Sodium dodecylsulfate	Dodecanol
A	85	15	0
B	85	13.5	1.5
C	85	12	3

in water was strongly enhanced when an increase in temperature caused it to approach the cloud point of the surfactant. In the same manner, an increase in the water solubilization in hydrocarbon was found when the temperature was reduced to the haze point of the surfactant in the hydrocarbon in question. The results were presented in the form of temperature-dependent phase maps of water–hydrocarbon mixtures with a constant concentration of surfactant (Fig. 6). The region in the middle represented a three-phase region

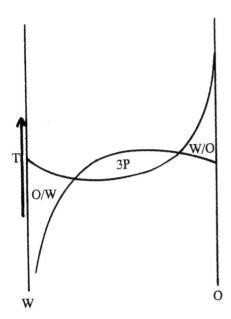

Figure 6 The principal phase diagram of water, hydrocarbon, and the nonionic surfactant polyethyleneglycol alkyl ether. W = water; O = hydrocarbon; O/W = oil-in-water microemulsion; W/O = water-in-oil microemulsion; 3P = three-phase region; T = temperature. (After Shinoda [18–24].)

in the early presentations [18–24]. The possibility of forming a microemulsion depends strongly on the balance between the surfactant's hydrophilic and lipophilic properties, determined not only by the surfactant's chemical structure but also by other factors such as temperature, salinity, and cosolvent or cosurfactant. Under balanced conditions there is a maximal simultaneous solubilization of oil and water.

Microemulsion research has since its inception been stimulated by the great potential for practical applications. In particular, considerable research interest has been invested in the possibility of using microemulsions for enhanced oil recovery (EOR). It was observed that surfactant formulations forming three-phase microemulsion systems, often termed Winsor III systems [29], in the oil well could increase the oil yield considerably. Important contributions to the understanding of the mechanisms involved were made by Shah and Hamlin [30] and the Austin group led by Schechter and Wade (see Bourrel et al. [31]).

II. LATER DEVELOPMENTS—MICROSTRUCTURE AND MODELING

These introductory and separate efforts were united and extended in the 1970s and since by experimental and theoretical work by several groups in France (de Gennes, Taupin, Langevin, Zana), Germany (Kahlweit, Strey), Australia (Israelachvili, Ninham), Sweden (Wennerström, Lindman), the United States (Scriven, Evans, Davis, Widom, Dawson, Friberg, Talmon, Prager), and elsewhere.

The most significant step in the development of the field of microemulsions was certainly the demonstration that they are thermodynamically stable phases [18–22,25,26]. This provided a bridge to other surfactant self-assembly systems, often well understood, and a basis for theoretical analyses of the thermodynamics. One consequence of recognizing microemulsions as equilibrium structures is clearly that previous approaches treating them as dispersions have to be reconsidered.

Abandoning the dispersion concept of microemulsions and realizing that they are closely related to micelles, liquid crystals, and other types of surfactant self-assemblies but distinctly different from (macro) emulsions clearly makes the term "microemulsion" less suitable. However, it has been kept for historical reasons. Confusion because of the wording continues, not so much concerning the thermodynamics but as regards microstructure; it seems that the term easily directs the mind toward a structure of discrete objects, "droplets," while we now know that this is not the typical situation.

The field of microemulsion microstructure started to develop forcefully in the 1970s. Realizing the thermodynamic stability and the relation to other surfactant self-assemblies, it became natural to assume that they were akin to micellar solutions, the most studied aspect of surfactant–water systems, and they were discussed in terms such as "swollen micelles."

However, even without structural studies, Friberg et al. [32], Shinoda [33], and others noted that the broad existence range with respect to the water/oil ratio could not be consistent with a micellar-only picture. Also, the rich polymorphism in general in surfactant systems made such a simplified picture unreasonable. It was natural to try to visualize microemulsions as disordered versions of the ordered liquid crystalline phases occurring under similar conditions, and the rods of hexagonal phases, the layered structure of lamellar phases, and the minimal surface structure of bicontinuous cubic phases formed a starting point. We now know that the minimal surfaces of zero or low mean curvature, as introduced in the field by Scriven [34], offer an excellent description of balanced microemulsions, i.e., microemulsions containing similar volumes of oil and water.

The problem of microstructure has been in focus for more than 20 years. Since microemulsions lack long-range order, the powerful diffraction techniques used for liquid crystalline phases are not applicable. Scattering techniques, which are powerful for dilute disperse systems, are not directly applicable to systems with oil and water domains occupying similar volumes. Furthermore, microemulsions, being equilibrium self-assembly structures, cannot be diluted without structural changes. While scattering studies were performed extensively and interpreted to give some apparent dimension, the analyses were based on an assumed general structure (such as micellar-like) that could not be tested easily.

In principle, we can distinguish (for surfactant self-assemblies in general) between a microstructure in which either oil or water forms discrete domains (droplets, micelles) and one in which both form domains that extend over macroscopic distances (Fig. 7a). It appears that there are few techniques that can distinguish between the two principal cases: uni- and bicontinuous. The first technique to prove bicontinuity was self-diffusion studies in which oil and water diffusion were monitored over macroscopic distances [35]. It appears that for most surfactant systems, microemulsions can be found where both oil and water diffusion are uninhibited and are only moderately reduced compared to the neat liquids. Quantitative agreement between experimental self-diffusion behavior and Scriven's suggestion of zero mean curvature surfactant monolayers has been demonstrated [36]. Independent experimental proof of bicontinuity has been obtained by cryo-electron microscopy, and neutron diffraction by contrast variation has demonstrated a low mean curvature surfactant film under balanced conditions. The bicontinuous microemulsion structure (Fig. 7b) has attracted considerable interest and has stimulated theoretical work strongly.

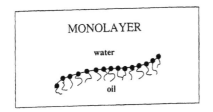

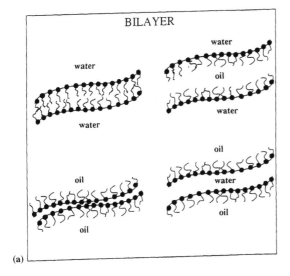

(a)

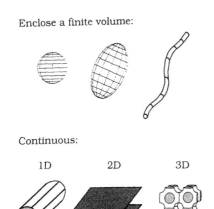

(b) cylinder   lamellae   "bicontinuous"

Figure 7 In surfactant–oil–water systems there is a segregation into oil and water domains and surfactant films. In (a) one can distinguish between cases of uncorrelated surfactant films (monolayers) and pairwise correlated films (bilayers). (b) Surfactant self-assembly can lead to discrete structures in which one of the solvents is enclosed or to structures that extend over macroscopic distances in one, two, or three dimensions. The "bicontinuous" structure, introduced by Scriven [34], in which both solvents form domains that are connected in three dimensions has stood in the foreground of microemulsion research. (Courtesy of Ulf Olsson.)

The thermodynamic modeling of microemulsions has taken various lines and gave conflicting results in the period before the thermodynamic stability and microstructure were established. It was early realized that a maximal solubilization of oil and water simultaneously could be discussed in terms of a balance between hydrophilic and lipophilic interactions; the surfactant (surfactant mixture) must be balanced. This can be expressed in terms of the HLB balance of Shinoda, Winsor's R value, and a critical packing parameter (or surfactant number), as introduced to microemulsions by Israelachvili et al. [37], Mitchell and Ninham [38], and others. The last has become very popular and useful for an understanding of surfactant aggregate structures in general.

However, these approaches do not give full insight into important characteristics of microemulsions. An important further concept is that of the spontaneous curvature of the surfactant layer. A bicontinuous microemulsion consists of a disordered, connected surfactant film separating the water and oil domains. The film has a spontaneous curvature that is close to zero, i.e., it has no tendency to curve toward either oil or water. There is a close analogy with the so-called sponge phase (L_3), a bilayer structure that also forms in binary surfactant–solvent systems. A characteristic feature of both bicontinuous microemulsions and the sponge phase is that they have only a narrow region of existence. An important aspect is the competition between a bicontinuous microemulsion and the lamellar liquid crystalline phase. An inspection of the phase diagrams shows that typically only slight changes in the conditions can lead to a transition from the microemulsion to the lamellar phase (see below); the latter typically shows an extensive swelling.

The modeling of microemulsions is still somewhat controversial, but a very powerful tool in establishing the relation between the geometrical description and the thermodynamics has been found in the flexible surface model of Helfrich [39]. The local energy is given by the local curvature, and the total curvature free energy is obtained by integrating over the surface. Talmon and Prager [40] pioneered the modeling of microemulsions and provided a thermodynamic model for the bicontinuous phase based on a subdivision of space into Voronoi polyhedra; they predicted a rich phase behavior, including the three-phase coexistence. Further important developments were due to de Gennes and coworkers [42,43]—who used a cubic lattice randomly filled with water and oil, with the lattice size chosen as the persistence length related to the bending energy of the surfactant film—and by Widom [44,45]. More recently, Wennerström and coworkers [46–48] demonstrated that with the flexible surface model a general form of the free energy of both balanced microemulsions and the sponge phase can be obtained. For the first time, the significant features of the stability of the balanced microemulsions with respect to other phases could be captured. In Fig. 8, we compare the theoretically predicted phase diagram with that of an experimental investigation [49]. Important characteristics such as the three-phase coexistence and the limited swelling of the bicontinuous microemulsion but not of the lamellar phase are nicely reproduced.

While microemulsions were introduced in four- and five-component systems, the simplification of working with three components, as is possible for the nonionic surfactants, is most significant both in experimental work of establishing the phase behavior and in theoretical modeling. Our understanding of the phase behavior or nonionic systems is to a great extent due to the groups of Shinoda and Kahlweit. To summarize our understanding of microemulsion structure, applicable, of course, to surfactant systems in general, we use a Shinoda cut; i.e., the surfactant concentration is kept constant and the temperature and the volume fractions of the solvents are varied. Figure 9 emphasizes a number of features important to capture in any theoretical modeling:

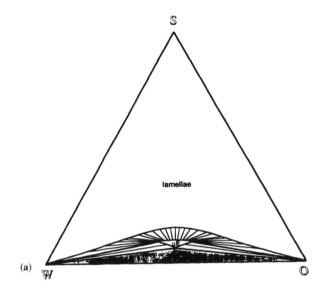

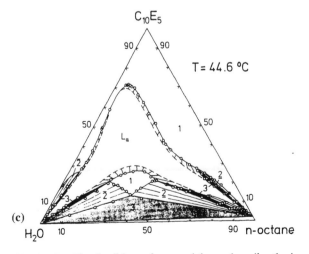

Figure 8 The flexible surface model can describe the important features of the thermodynamics of balanced microemulsions. (a, b) The calculated phase diagram of Daicic et al. [46]. (c) Experimental phase diagram of Strey [49]. μE = microemulsion.

1. The three-phase coexistence of a balanced microemulsion with oil and water
2. The narrowness (limited swelling) of both the microemulsion and the sponge phase
3. The closeness in stability ranges of the lamellar phase and the balanced microemulsion and the large swelling of the lamellar phase both for oil and for water
4. The continuous transition from sponge phase to bicontinuous microemulsion
5. The symmetry in the phase diagram around the coordinate of balanced temperature and 1 : 1 volume ratio of solvents

Figure 9 gives a generic picture, but the parameter controlling the spontaneous curvature is different for different surfactant systems. For nonionic surfactants, temperature controls the spontaneous curvature, changing it from positive (surfactant film curved toward oil) a low temperatures to negative at higher temperatures. For other systems, salinity (for ionics), surfactant composition in a mixture, and cosurfactant and cosolvent concentrations are controlling parameters. The generic structural change as a function of the spontaneous curvature is demonstrated in Fig. 9. At markedly positive spontaneous curvatures we have

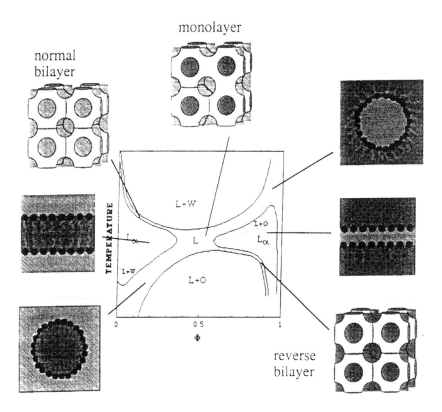

Figure 9 A schematic phase diagram of a cut at constant surfactant concentration through the temperature–composition phase prism of a ternary system with nonionic surfactant showing the characteristic X-like extension of the isotropic liquid phase L. (Φ is the volume fraction of oil in the solvent mixture.) Schematic drawings of the various microstructures are also shown. (Courtesy of Ulf Olsson.)

discrete oil-swollen micelles or oil droplets; at significant negative spontaneous curvatures there are discrete water-swollen reversed micelles or water droplets. At a zero spontaneous curvature we have a structure of minimal surface aggregates, which are disordered over larger distances. The exact nature and characteristics of the disorder are not fully characterized and furthermore the structures intermediate between discrete droplets and full bicontinuity require further study. However, we know that there is a continuous transition as the spontaneous curvature is varied: Droplets are growing and becoming connected, and the bicontinuous structure is disrupting into discrete particles.

III. CONCLUDING REMARKS

With a consensus on important issues regarding thermodynamic stability and microstructure, the microemulsion field has (again) become an integrated part of surfactant research in general. With a broadened scope there are rapid developments in theoretical modeling, in establishing microemulsions for new classes of amphiphiles such as lipids and amphiphilic polymers, and in technical applications.

REFERENCES

1. D. Bowden and J. Holmstine, U.S. Patent 2,045,455 (1936).
2. V. R. Kokatnur, U.S. Patent 2,111,100 (1935).
3. T. P. Hoar and J. H. Schulman, Nature 152: 102 (1943).
4. J. H. Schulman and D. P. Riley, J. Colloid Sci. 3: 383 (1948).
5. J. E. Bowcott and J. H. Schulman, Z. Electrochem. 11: 117 (1955).
6. J. H. Schulman and J. A. Friend, J. Colloid Sci. 4: 497 (1949).
7. W. Stoeckenius, J. H. Schulman, and L. M. Prince, Kolloid-Z. 169: 170 (1960).
8. C. E. Cooke and J. H. Schulman, in Surface Chemistry (P. Ekwall, ed.), Munksgaard, Copenhagen, Denmark, 1965, p. 231.
9. D. F. Sears and J. H. Schulman, J. Phys. Chem. 68: 3529 (1964).
10. I. A. Zlochower and J. H. Schulman, J. Colloid Interface Sci. 24: 115 (1967).
11. H. L. Rosano, H. Schiff, and J. H. Schulman, J. Phys. Chem. 66: 1928 (1962).
12. L. M. Prince, J. Colloid Interface Sci. 23: 165 (1967).
13. L. M. Prince, J. Colloid Interface Sci. 29: 216 (1969).
14. C. L. Murphy, Ph.D. Thesis, University of Minnesota, Minneapolis, MN, 1966.
15. E. Ruckenstein and J. C. Chi, J. Chem. Soc., Faraday Trans. II 71: 1690 (1975).
16. C. A. Miller and P. Neogi, AIChE 26: 212 (1980).
17. P. Ekwall, I. Danielsson, and L. Mandell, Kolloid-Z. 169: 113 (1960).
18. H. Saito and K. Shinoda, J. Colloid Interface Sci. 24: 10 (1967).
19. H. Saito and K. Shinoda, J. Colloid Interface Sci. 32: 647 (1970).
20. K. Shinoda, J. Colloid Interface Sci. 24: 4 (1967).
21. K. Shinoda, J. Colloid Interface Sci. 34: 278 (1970).
22. K. Shinoda and T. Ogawa, J. Colloid Interface Sci. 24: 56 (1967).
23. K. Shinoda and S. E. Friberg, Adv. Colloid Interface Sci. 4: 281 (1975).
24. K. Shinoda and H. Arai, J. Phys. Chem. 68: 3485 (1964).
25. G. Gillberg, H. Lehtinen, and S. E. Friberg, J. Colloid Interface Sci. 33: 40 (1970).
26. E. Sjöblom and S. E. Friberg, J. Colloid Interface Sci. 67: 16 (1978).
27. J. H. Schulman, W. Stoeckenius, and L. M. Prince, J. Phys. Chem. 63: 1677 (1959).
28. D. G. Rance and S. E. Friberg, J. Colloid Interface Sci. 60: 207 (1977).
29. P. A. Winsor, Solvent Properties of Amphiphilic Compounds, Butterworth, London, 1954.

30. D. O. Shah and R. H. Hamlin, Jr., Science 171: 483 (1971).
31. M. Bourel, C. Chambu, R. S. Schechter, and W. H. Wade, Soc. Pet. Eng. J. 22: 28 (1982).
32. S. Friberg, I. Lapczynska, and G. Gillberg, J. Colloid Interface Sci. 56: 19 (1976).
33. K. Shinoda, Prog. Colloid Polym. sci. 68: 1 (1983).
34. L. E. Scriven, Nature 263: 123 (1976).
35. B. Lindman, N. Kamenka, T.-M. Kathopoulis, B. Brun, and P.-G. Nilsson, J. Phys. Chem. 84: 2485 (1980).
36. D. M. Anderson and H. Wennerström, J. Phys. Chem. 94: 8683 (1990).
37. J. N. Israelachvili, D. J. Mitchell, and B. W. Ninham, J. Chem. Soc., Faraday Trans II I76:1525 (1976).
38. D. J. Mitchell and B. W. Ninham, J. Chem. Soc., Faraday Trans. II 77:601 (1981).
39. W. Helfrich, Z. Naturforsch. 28c:693 (1973).
40. Y. Talmon and S. Prager, J. Chem. Phys. 69:2984 (1978).
41. Y. Talmon and S. Prager, J. Chem. Phys. 76:1535 (1982).
42. P. G. de Gennes and C. Taupin, J. Phys. Chem. 86:2294 (1982).
43. J. Jouffroy, P. Levinson, and P. G. de Gennes, J. Phys. Fr. 43:1241 (1982).
44. B. J. Widom, J. Chem. Phys. 82:1030 (1984).
45. B. J. Widom, J. Chem. Phys. 84:6943 (1986).
46. J. Daicic, U. Olsson, and H. Wennerström, Langmuir I11:2451 (1995).
47. H. Wennerström and U. Olsson, Langmuir 9:365 (1993).
48. U. Olsson and H. Wennerström, Adv. Colloid Interface Sci. 49:113 (1994).
49. R. Strey, Ber. Bunsenges. Phys. Chem. 97:742 (1993).

2

Thermodynamics of Microemulsions I

Willem K. Kegel, J. Theo G. Overbeek, and Henk N. W. Lekkerkerker
Van't Hoff Laboratory for Physical and Colloid Chemistry, Debye Research Institute, Utrecht University, Utrecht, The Netherlands

I. INTRODUCTION

Microemulsions are thermodynamically stable mixtures of oil and water. The stability is due to the presence of fairly large amounts (several percent) of surfactants. Microemulsions are often transparent, but scattering of light, X-rays, etc. indicate that oil and water are not molecularly dispersed but are rather more coarsely mixed. By "coarse" in this case we mean that oil and water are present in domains of a few to over a hundred nanometers in size. Consequently, microemulsions contain huge oil/water interfacial areas, and to allow stability the interfacial tension must be quite low, usually $\ll 1$ mN/m. In that case the entropy of mixing, although small on account of the coarseness of the mixture, may be large enough to compensate for the positive interfacial free energy and to give the microemulsion a free energy lower than that of the unmixed components.

Microemulsions can have various textures, such as oil droplets in water, water droplets in oil, (random) bicontinuous mixtures, ordered droplets, or lamellar mixtures with a wide range of phase equilibria among them and with excess oil and/or water phases. This great variety is governed by variations in the composition of the whole system and in the structure of the interfacial layers.

Qualitatively the thermodynamics of microemulsions is well understood as the interplay between a small interfacial free energy and a small entropy of mixing. However, because of these contributions being small, other small effects, such as the influence of curvature on the interfacial tension and the influence of fluctuations, become important, and quantitative understanding still leaves a lot to be desired.

In Sec. II we discuss the mechanism by which the interfacial tension may become ultralow. After that, in Sec. III we mention curvature effects of the oil/water interface. Subsequently, a number of models for thermodynamic calculations are described (Sec. IV). In Secs. V–VII we discuss droplet-type microemulsions in some detail. Section V describes a thermodynamic formalism that incorporates the interfacial free energy (as influenced by the curvature) and the free energy of mixing of droplets and continuous medium and ultimately leads to equations for the size distribution of microemulsion droplets. This size distribution is important because measurable properties can be calculated once it is known. In Sec. VI we discuss the quantities that determine the size distribution of the droplets, and in Sec. VII the theory of droplet-type microemulsions is compared with experiments. Finally, in Sec. VIII we discuss the main results of the theory of droplet-type microemulsions and present a short discussion of bicontinuous microemulsions and lamellar structures.

II. ULTRALOW INTERFACIAL TENSION

A. Interfacial Tension and Micelle Formation

The O/W interfacial tension (O = oil, W = water or aqueous solution) is lowered by the addition and adsorption of surfactants. The higher the concentration of the surfactant, the lower the interfacial tension, until at the cmc (critical micelle concentration) micelles are formed. Micelles are aggregates of large numbers (often 50 or more) of surfactant molecules, structured in such a way that, for a water-soluble surfactant, the hydrophobic parts of the surfactant are in the interior of the micelle, and the hydrophilic parts form its skin. As a simple consequence of mass action, nearly all surfactant added beyond the cmc forms micelles with hardly any increase in the concentration of the single molecules, and thus of their activity, which governs the adsorption and therefore the interfacial tension.

A great variety of surfactants have been used in microemulsion formation. These include common soaps, other anionic and cationic surfactants, nonionic surfactants of the polyethylene oxide type, and other structures. The hydrophobic part contains one or two linear or branched hydrocarbon chains containing about 8–18 carbon atoms. Quite often microemulsions require, in addition to oil, water, and surfactant, the presence of simple electrolytes, alcohols, and/or other weakly surface-active substances.

B. Ionic Surfactants

In many cases the interfacial tension is not yet ultralow when the cmc is reached. Schulman and collaborators [1–3], who first raised scientific interest in microemulsions, realized that the addition of a cosurfactant (e.g., a medium-sized aliphatic alcohol or amine) to an ionic surfactant (in their case soap) solved this problem by lowering the interfacial tension to virtually zero.

A good example is found in the interfacial tension between aqueous solutions of SDS (sodium dodecyl sulfate) containing 0.30 M NaCl and pentanol solutions in cyclohexane containing 0–20% pentanol [4,5] as shown in Fig. 1.

We draw attention to two aspects of these curves. At 5% pentanol the interfacial tension already reaches very low values (<0.05 mN/m), and at 20% pentanol the curve seems to dip under the $\gamma = 0$ level. Furthermore, in the lower parts of the curves the points nearly follow a straight line, indicating, according to the Gibbs adsorption equation, Eq. (1), that the adsorption is constant ("saturation adsorption").

$$\left(\frac{\partial \gamma}{kT \, \partial \ln c_{sa}} \right) = -\Gamma_{sa} \tag{1}$$

In this equation γ is the interfacial tension of the (macroscopic) oil/water interface, c_{sa} is the concentration of SDS in the aqueous phase (in molecules per unit volume), Γ_{sa} its adsorption (in molecules per unit area), and k and T have their usual meaning.

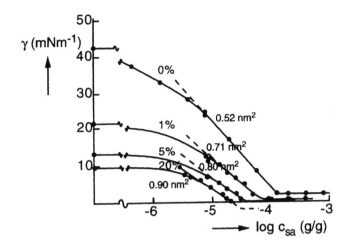

Figure 1 Interfacial tension γ as a function of the surfactant concentration c_{sa} (in g/g) between aqueous solutions of sodium dodecyl sulfate (SDS) containing 0.3 M NaCl and solutions of 0–20 wt% n-pentanol in cyclohexane. Without pentanol the cmc is found at about 0.0004 M SDS, and above the cmc, $\gamma = 2.42$ mN/m. Pentanol decreases the cmc and the interfacial tension above the cmc until, at 5% pentanol, γ above the cmc becomes as low as 0.036 mN/m. The area per molecule of SDS at saturation adsorption increases from 0.52 nm^2 in the absence of pentanol to about 0.9 nm^2 at 20% pentanol. Obviously the pentanol, which is adsorbed to the extent of two to three molecules per molecule of SDS, drives some of the SDS out of the interface. The experiments were carried out at 25°C.

C. Nonionic Surfactants

With nonionic surfactants of the PEO [poly(ethylene oxide)] type, the temperature is an important variable. They are water-soluble at lower temperatures and oil-soluble at high temperatures. In the narrow temperature range where the solubility changes, called PIT (phase inversion temperature [6]), the interfacial tension becomes extremely low, as sketched in Fig. 2. Below the PIT an O/W (oil-in-water) microemulsion is formed, above it a W/O (water-in-oil) microemulsion, with a continuous transition between them, possibly a bicontinuous mixture of oil and water, which at low surfactant concentrations may show a three-phase equilibrium with excess oil and water. Such equilibria are designated Winsor I (O/W + O), Winsor II (W/O + W), and Winsor III [(bicontinuous? O + W) + O + W] after P. A. Winsor [7–9], who studied phase equilibria in (mostly ionic) microemulsion systems extensively and at an early date.

D. Ionic Surfactants with Two Hydrophobic Chains

It should be mentioned here that ionic surfactants with two hydrocarbon chains do not need a cosurfactant to enable them to form microemulsions. Aerosol OT (AOT), which is sodium diethylhexylsulfosuccinate, has been known and used for a long time. In 1983 Angel et al. [10] remarked that substituted ammonium salts with two long chains, e.g., didodecyldimethylammonium bromide, would also form microemulsions without the use of a cosurfactant.

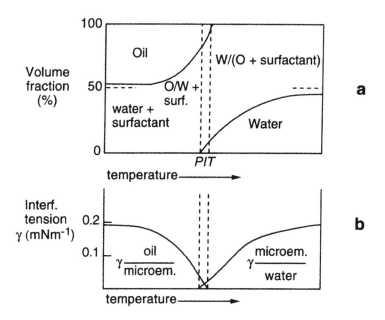

Figure 2 (a) Schematic diagram of volume fractions in a system of equal volumes of oil and water with, e.g., 5% of a PEO-type nonionic surfactant versus temperature. (b) Interfacial tensions in the same system. The line corresponding to the interfacial tension of the oil/microemulsion interface and the one corresponding to the microemulsion/water interface, are indicated. (After Ref. 6.)

E. Negative Interfacial Tension and Spontaneous Emulsification

The 20% pentanol line in Fig. 1 clearly suggests that the interfacial tension becomes negative if the surfactant concentration is increased above the concentration where $\gamma = 0$. This, however, would lead to an unstable situation, since a negative interfacial tension leads to a spontaneous increase in the area of the interface. When this happens, more surfactant will be adsorbed, thus lowering the surfactant concentration until γ is at least slightly positive. The large interfacial area thus formed may divide itself into a large number of closed shells around small droplets of either oil in water or water in oil and further decrease the free energy of the system as a direct consequence of the relatively large entropy of mixing of droplets and continuous medium. These considerations may serve as an explanation for the spontaneous formation of microemulsions that is often observed. The essential feature is that the composition of the mixture reaches $\gamma = 0$ before micelles are formed. To make this argument more quantitative we remark that the $\gamma = 0$ situation is often found at a surfactant concentration on the order of 0.1–1 wt%. Thus most of the surfactant in the mixtures will serve to make an interfacial area on the order of 1 nm^2 per surfactant molecule (see Sec. VI for a discussion of the order of magnitude of this molecular area), that is, on the order of 2000 m^2 per gram of surfactant, which is enough to stabilize 10 g of water or oil droplets with a radius of about 15 nm. So, although the surfactant is evidently the most important additive in the mixture and its amount determines the total interfacial area, it does not determine the structure of the microemulsion, not even whether it is O/W or W/O. That depends on curvature effects as discussed in the next section.

III. HOW TO AFFECT THE CURVATURE OF THE INTERFACE AND THE TYPE OF MICROEMULSION EQUILIBRIUM

Not only do surfactants and cosurfactants lower the interfacial tension, but also their molecular structures affect the curvature of the interface as shown schematically in Fig. 3. The hydrocarbon chains are rather closely packed (about 0.25 nm^2 per chain); they repel one another sideways and as a result have a tendency to bend the interface around the water side. The counterions of the ionic headgroups also repel one another sideways and thus tend to curve the interface around the oil side. The bulky polar groups of nonionic surfactants have a similar effect. So we understand qualitatively that more cosurfactant promotes W/O rather than O/W microemulsions. More electrolyte compresses the double layer, diminishes the sideways pressure of the double layer, and also promotes W/O microemulsions. The polar groups of PEO nonionics become more compact (less soluble) at higher temperatures, and so with this type of surfactants high temperature leads to W/O microemulsions.

From the above reasoning we expect that each composition of the interface has its own curvature at which the interface forms most easily and thus has the lowest interfacial tension (this interfacial tension, σ, of the droplet interface should not be confused with that of the macroscopic interface, γ). This consideration was made more quantitative by Helfrich [13], who presented an expression for the curvature free energy,

$$F_c = \int_A \left[\frac{K}{2}(c_1 + c_2 - 2c_0)^2 + \overline{K}c_1 c_2 \right] dA \tag{2}$$

as the free energy contribution due to bending of a flat interfacial area A. c_1 and c_2 are the local principal curvatures, c_0 is the spontaneous (or preferred) curvature, K is the

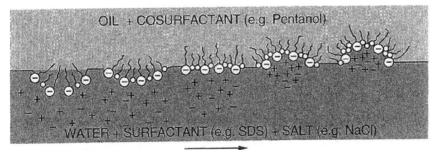

CONC. of SALT or CONC. of COSURFACTANT, or TEMPERATURE

Figure 3 Schematic representation of the influence of surfactants on the curvature of the interface. The hydrophilic parts of the surfactant molecules repel each other sideways, tend to curve the interface around the oil side, and promote O/W emulsions. This effect is most pronounced with long and/or bulky nonionic polar groups or, in the case of ionic surfactants, at low electrolyte content, where the double layers are extended. On the other hand, the mutual repulsion of the hydrophobic parts of the surfactants tend to curve the interface around the water side and promote W/O emulsions. Here, long hydrocarbon tails and/or close packing, as in combinations with cosurfactants, make the effects more pronounced. (After Refs. 11 and 12.)

bending elastic modulus, and $\overline{K}$ is the Gaussian or saddle splay modulus. $(c_1 + c_2)/2$ is referred to as "mean curvature," while the product $c_1 c_2$ is designated "Gaussian curvature."

For a single droplet of radius R we have

$$F_c = 8\pi K(1 - c_0 R)^2 + 4\pi \overline{K} \qquad \text{(single droplet)} \tag{3}$$

where $4\pi\overline{K}$ can be seen as the free energy needed (or furnished if $\overline{K}$ is negative) to detach a suitable area of flat interface and close it around one droplet.

In a Winsor I or Winsor II system (generally called saturated droplet type microemulsions), the interfacial tension, σ, of the droplets can be related to the (measurable) macroscopic interfacial tension, γ, of the flat interface separating the microemulsion and the excess phase by

$$\sigma(R) = \gamma + \int_0^{2/R} \frac{\partial\sigma(R)}{\partial(2/R)} \, d\left(\frac{2}{R}\right) = \gamma + \int_0^{2/R} \frac{\partial^2 F_c}{\partial A \, \partial(2/R)} \, d\left(\frac{2}{R}\right)$$
$$= \gamma - \frac{4Kc_0}{R} + \frac{2K + \overline{K}}{R^2} \tag{4}$$

We note that Eq. (4) does not contain finite size effects, which are expected to be present in the small systems we are dealing with here. An inclusion of such effects is postponed to Sec. V.

Figure 4 gives a schematic picture of Eq. (4), that is, the interfacial tension of the droplet interface as a function of the inverse radius of curvature, $1/R$. Curvature and radius of curvature are arbitrarily set positive for curvature around the water side (W/O) and negative for the O/W type.

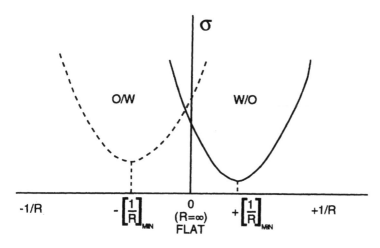

Figure 4 Tension of the droplet interface σ versus the inverse radius of curvature for two systems. In one (— — —) the repulsion between the polar groups wins and brings the "preferred" radius onto the O/W side; in the other (———) there is strong hydrophobic repulsion, and the "preferred" radius is on the W/O side.

From Eq. (4) it follows that the minimum of the interfacial tension of a droplet (the minimum in Fig. 4) is found at

$$\left(\frac{1}{R}\right)_{\mathrm{MIN}} = \frac{2Kc_0}{2K + \overline{K}} \tag{5}$$

$(1/R)_{\mathrm{MIN}}$ differs from c_0 due to the fact that according to Eq. (3), a term $4\pi\overline{K}$ must be added to the curvature free energy to form a droplet.

The bending elastic quantities being on the order of kT (as we show in Sec. VI), the system may increase its configurational entropy without much of a curvature free energy penalty by exploring a distribution of sizes around the one given by Eq. (5) (see Fig. 4). The entropy of mixing tends to maximize the number of droplets, and its effect is expected to result in a smaller average value of $|R|$ than the one that minimizes the curvature free energy, Eq. (5). In the case of saturated droplet-type microemulsions we may safely say that if the absolute value of the preferred curvature, c_0, is significantly different from 0 (if not, the very presence of a microemulsion with a droplet structure may become doubtful), its sign determines whether O/W or W/O microemulsions are found. Now, as already mentioned in a more qualitative way at the beginning of this section, the sign and magnitude of c_0 are determined by two competing forces at the droplet interface: The hydrocarbon chains that tend to bend the interface around the water side contribute positively to c_0 (in our definition of the sign of the curvature), whereas the counterions of the ionic headgroups contribute negatively. At low salt concentration, for example, the contribution of the electrical double layer to the preferred curvature dominates, and an O/W system is found (Winsor I). On increasing the salt concentration, the preferred curvature becomes less and less negative and the average droplet size is found to increase. At a certain salt concentration the contributions of the electrical double layer and the hydrocarbon chains to c_0 cancel out, and a structure with a zero mean curvature is expected. In fact, many structures have zero mean curvature, not only (on average) flat monolayers but also cubic phases where (again on average) $c_1 = -c_2$ (the so-called plumber's nightmare [14]), and it is a nontrivial task to determine the stable one, especially since fluctuations around the (mean) curvature of the monolayers are expected to be present. It turns out that at low surfactant concentrations, a system consisting of a surfactant-rich middle phase coexisting with excess oil and water phases is found. The structure of the middle phase is probably bicontinuous, although it has been argued [15] that the structure might switch over rapidly from water droplets in oil to oil droplets in water. Upon increasing the salt concentration even further, c_0 becomes positive and a W/O microemulsion becomes stable, the average size of the droplets decreasing with increasing salt concentration. A similar phase inversion is found in systems where a cosurfactant is present [16]. In that case, increasing the cosurfactant concentration in the system leads to a slightly increased adsorption of cosurfactant molecules at the droplet interface, resulting in an increased surface pressure, hence a positive contribution to c_0.

In order to proceed further, a thermodynamic formalism is needed that quantitatively relates the bending elastic quantities mentioned above to measurable properties of microemulsion systems (such as average droplet size or interfacial tension of the flat interface between the microemulsion phase and the excess oil or water phase). As a second step it is necessary to relate experimentally adjustable quantities such as the salt concentration and the nature of the surfactant to the bending elastic properties by means of molecular models. The state of the art is presented in the following sections.

IV. MODELS FOR THERMODYNAMIC CALCULATIONS

In the course of the years, several types of models have been proposed and worked out. In the early years (1940–1970) a microemulsion structure was postulated, e.g., that of isodispersed droplets of oil in water or water in oil, and the free energy of such a system was formulated from known or estimated interfacial tension and bending free energy of the monolayer as related to the overall composition of the system.

Interfacial tension and interfacial composition can be measured by performing experiments at low surfactant concentrations [see Eq. (1) and Fig. 1 for an example]. But bending free energies are much more difficult to come by, and even the entropy of mixing has posed some problems, as will be mentioned in Sec. V. In later years, starting with a paper by Talmon and Prager [17] in 1977, other models were proposed and worked out. These models are of three types:

1. Divide the total available volume into small volume elements (Voronoi polyhedra; later on, more simply, cubes were used by de Gennes and coworkers [18,19]), distribute oil and water (random distribution gives easy access to the entropy of mixing) among these volume elements, and place surfactant film at all oil/water boundaries. Choose the linear size of the volume elements equal to the persistence length [the length over which a (single) flexible film remains essentially flat]. Find the contribution of the film to the free energy from the interfacial tension and the bending free energy (caused by the difference between its average curvature as found from the model and the preferred curvature), and determine which monophasic or multiphasic situation gives the lowest value of free energy at a given total composition. An extension of this approach, where the lattice parameter is related to the ratio of the dispersed volume to the total interfacial area of the microemulsion, was put forward by Andelman et al. [20].

2. Construct an imaginary lattice (e.g., simple cubic lattice), and place O, W, and surfactant molecules on the lattice in such a way that a surfactant molecule occupies two neighboring sites—one with its polar group, the other with its hydrophobic part—and require that each lattice site contain only polar parts of molecules or hydrophobic parts of molecules. A simple example was proposed as early as 1968 by Wheeler and Widom [21], who assumed that, just as the surfactant occupies two neighboring sites, oil and water each also occupy two neighboring sites. Extending the system by adding suitable interaction energies between "oil" sites and "water" sites and making these energies dependent on the number of surfactant molecules meeting at one "oil" or one "water" site results in a model that represents the phase behavior of microemulsions in a rather realistic way as shown by several recent studies [22–24].

3. Finally, the Landau–Ginzburg approach can be applied to microemulsions. In this approach the thermodynamic behavior of a system in the neighborhood of critical points is studied. For microemulsions this approach uses the expansion of the free energy in (spatially varying) order parameter fields that are identified as the local concentration differences of oil, water, and surfactant. We will not elaborate this approach but refer to a recent review by Gompper [25].

These three theories predict the global phase behavior of microemulsion systems in terms of bending elastic properties (first approach), molecular interaction parameters (second approach), and expansion coefficients of order parameter fields (third approach). In this review, we try to fill the gap between the "early years" approach and item 1, above.

Whereas in approach 1 lattice models are used, we will work in the "continuum," making extensive use of interface thermodynamics. The advantage of such an approach, as it turns out, is that detailed properties such as the size distribution of microemulsion droplets and the interfacial tension of a flat monolayer separating a microemulsion and an excess phase can be predicted. On the other hand, the lattice approaches as summarized in item 1 predict global phase behavior, which is not (yet) possible with the thermodynamic formalism reviewed in the following section. The reason is that currently a realistic model for the middle phase is lacking. A more detailed discussion regarding this issue is presented in Sec. VIII.

V. THERMODYNAMIC FORMALISM OF DROPLET-TYPE MICROEMULSIONS

A. Free Energy of Polydisperse Droplet-Type Microemulsions

We briefly review the formalism given in Ref. 26, to which the reader interested in the details of the derivation is referred. We do, however, use a different expression for the entropy of mixing. We note that the result [cf. Eq. (7)] can also be derived starting with the droplet partition function [27]. The Gibbs free energy of a droplet-type microemulsion can immediately be written in two equivalent ways as

$$G = \sum_i N_i \mu_i = \sum_i N_{im} \mu_i + \sum_j N_{dj} \mu_{dj} \tag{6}$$

where N_i and μ_i are the number of molecules and the chemical potential of the components i making up the microemulsion. In the second step in Eq. (6) we have implicitly defined droplets of category j, where j refers to the number of surfactant molecules at the interface between a droplet and the continuous medium. Of course, other definitions of droplet categories are possible. The chosen definition assumes rapid exchange of oil and water molecules but a relatively slow exchange of the surfactant molecules. The subscript im in Eq. (6) refers to component i in the continuous medium phase, while the subscript dj refers to a droplet of category j. n_{dj} and μ_{dj} are the number of droplets of category j and their chemical potential, respectively.

The droplets are considered to also include the adsorption layers. For a detailed discussion in terms of the Gibbs surface between the droplets and the medium, the reader is referred to Ref. 28.

The Gibbs free energy can also be written by first considering the system in a reference state where the microemulsion droplets are hanging on microsyringes. The free energy of the system then equals the sum of the bulk free energies of the two different "phases," i.e., a "phase" of droplets and a "phase" of continuous medium, plus the interfacial free energy. An additional contribution to the free energy of the system arises once the droplets are allowed to move freely in the continuous medium (and the assignment "phase" as used above for the droplets and medium ceases to have meaning). This term is the free energy of mixing, which quantifies the number of additional configurations of the system due to the release of the droplets from the microsyringes. So we finally have

$$G = \sum_i N_{im} \mu_{im}^* + \sum_j \sum_i N_{ij} \mu_{ij}^* - \sum_j \Delta p_j V_{dj} + \sum_j \sigma_j A_j - TS_{\text{mix}} \tag{7}$$

where the μ^*'s are the chemical potentials of the different components at the pressures inside the two different phases. The subscript im again refers to the medium, and the subscript ij refers to component i inside all droplets of category j. In the present formulation we have assumed implicitly that all adsorbed molecules belong to the droplet. Δp_j is the difference between the pressure in the interior of a droplet of category j and the outer pressure, V_{dj} is the volume of all droplets of category j, and σ_j and A_j are the interfacial tension and interfacial area, respectively, of (all) droplets of category j. The last term (negative sign included) in Eq. (7) is the free energy of mixing, and S_{mix} the entropy of mixing. Equation (7) is most easily derived by first defining the Helmholtz free energy, since the two phases are at different pressures.

It is worth paying a little attention to the last term in Eq. (7). Neglecting excluded volume effects, the number of additional configurations of the droplets upon their release from their fixed positions is

$$\Omega = \prod_j \frac{1}{n_{dj}!} \left(\frac{V}{l^3}\right)^{n_{dj}}$$

with V the volume of the microemulsion phase and l the length over which a droplet should be translated in order to be counted as a new configuration. This is equivalent to the collection of droplets residing on a cubic lattice with lattice parameter l. Using the definition $S_{\text{mix}} = k \ln \Omega$ and taking the thermodynamic limit, the entropy of mixing becomes

$$S_{\text{mix}} = -k \sum_j n_{dj}[\ln(\rho_j l^3) - 1] = -k \sum_j n_{dj}\left[\ln \phi_j - 1 - \ln\left(\frac{v_{dj}}{l^3}\right)\right] \tag{8}$$

with ρ_j the number density of droplets of category j, ϕ_j their volume fraction, and v_{dj} the volume of a single droplet of category j.

The first equality of Eq. (8) is of the same form as the negative free energy of an ideal gas (where the summation is over only a single category) divided by temperature, with the de Broglie wavelength replaced by the length l. The fundamental difference between these two lengths is that the first is a momentum space quantity, while the second is to be evaluated in configuration space only. This evaluation is quite a subtle problem, as may be judged from the many choices that are (usually implicitly) made for this quantity, i.e., from the de Broglie wavelength of a drop [29] to the full size of a drop [20,30]. The treatment in Ref. 31 is based on the idea that the length l as defined in configuration space should be consistent with the full entropy of the system as defined in phase space. In other words, one must be careful not to either overcount or underestimate the number of additional configurations of the system that occur when the droplets are released from their fixed positions. In Ref. 31 an estimate of l is made by analyzing models that can be evaluated in full phase space and subsequently comparing the result to the definition of mixing entropy given by Eq. (7) for the same model. It turns out that the length l as defined by Eqs. (7) and (8) is on the order of the cube root of a molecular volume. This value is consistent with the one estimated much earlier [32] for microemulsion systems that were assumed to be monodisperse. In that estimate, l^3 was found to be equal to the molecular volume of the continuous phase.

We note that the calculation of mixing entropy transcends the microemulsion field and is relevant also to phenomenological theories of nucleation. In that field the length l appears in the "replacement free energy problem" (see Ref. 33 and, more general, Ref. 34).

It was stressed in the Introduction that microemulsions can exist because of the compensation for a small positive interfacial free energy by a small negative free energy of mixing. Here we are at a point where we can give some idea of how small is small. Taking droplets with an average radius of 10 nm at a volume fraction of 1%, assuming that most of the volume is spread among droplets between 8 nm ($j \approx 800$) and 12 nm ($j \approx 1800$), and assuming that $l = 0.5$ nm, we find a free energy of mixing per droplet of about 20 kT. This requires the interfacial tension σ to remain below 0.06 mN/m (say below 0.1 mN/m) to satisfy the relation $\sigma A < TS_{\text{mix}}$.

In Eqs. (7) and (8), droplet categories are defined by their number of adsorbed surfactant molecules. The only constraint in our model is that the droplets are spherical *on average*. Shape fluctuations are, at least in part, taken into account by the interfacial tension of the droplets. The analog of shape fluctuations on a flat interface is the capillary wave spectrum, the free energy of which is part of the interfacial free energy. Therefore, treating all configurations corresponding to different shapes of the droplets (i.e., counting all spherical harmonic modes) explicitly in our model leads to redundancies. However, there may be finite size effects due to the fact that not all (spherical analogs of) capillary wave modes are available for the droplets. In the next section we discuss how to implement this effect.

B. Size Distribution

The size distribution of the droplets [in the form of values of the individual ϕ_j's; cf. Eq. (19)] is obtained by minimizing the free energy, Eqs. (6)–(8). This will be done by applying the law of mass action. The chemical potential of a droplet of category j is calculated from Eq. (7), where use is made of the generalized Laplace equation. This chemical potential is subsequently equated to the chemical potential following from mass action; i.e., it should equal the sum of the products of the number of molecules and their respective chemical potentials for all components making up a droplet. In order to find the chemical potential of the droplet categories we first need to specify the pressure differences between the droplets and the outer pressure as well as the chemical potentials of the components. The first is given by the generalized Laplace equation,

$$\Delta p_j = \frac{2\sigma_j}{R_j} - \frac{2c_j}{R_j^2} \tag{9}$$

where R_j is the radius of a drop of category j and the bending force c_j is

$$c_j = \left(\frac{\partial \sigma}{\partial (2/R)} \right)_{R=R_j} = -2Kc_0 + \frac{2K + \overline{K}}{R_j} \tag{10}$$

where we have used Eq. (4). A straightforward derivation of Eq. (9) from mechanical equilibrium is given in Ref. 28. Before proceeding further, it is useful to specify the total interfacial area of the system as

$$\sum_j A_j = \sum_j n_{dj} 4\pi R_j^2 \tag{11}$$

and the total volume of the droplets of category j as

$$V_{dj} = n_{dj} \frac{4}{3} \pi R_j^3 = \sum_i N_{ij} v_i \tag{12}$$

where v_i is the molecular volume of component i. The volume fraction of droplets of category j is given as

$$\phi_j = \frac{4}{3} \pi R_j^3 \frac{n_{dj}}{V} = \frac{V_{dj}}{\sum_j V_{dj} + \sum_i N_{im} v_i} \tag{13}$$

The chemical potentials of the components in the continuous medium now follow from Eq. (7) with Eqs. (11)–(13).

$$\mu_i = \left(\frac{\partial G}{\partial N_{im}}\right)_{T,p,N_{jm \neq im}} = \mu_{im}^* - kT \frac{\sum_j n_{dj}}{V} v_i \tag{14}$$

so that

$$\sum_{im} N_{im}\mu_i = \sum_i N_{im}\mu_{im}^* - kT \sum_j n_{dj} \tag{15}$$

where a term linear in the total volume fraction of the droplets has been dropped. From this equation together with Eq. (7) it follows that, at constant n_{dk} ($k \neq j$),

$$\mu_{dj} = \frac{\partial G}{\partial n_{dj}} = \sum_i \frac{N_{ij}\mu_{ij}^*}{n_{dj}} + \frac{4}{3}\pi R_j^2\left(\sigma_j + \frac{2c_j}{R_j}\right) + kT\left[\ln \phi_j - \ln\left(\frac{v_{dj}}{l^3}\right)\right] \tag{16}$$

For the components making up the droplets we have

$$\mu_{ij}^* = \mu_i' + \int_p^{p+\Delta p_j} \frac{\partial \mu_i}{\partial p}\, dp = \mu_i' + \Delta p_j v_i = \mu_i' + \left(\frac{2\sigma_j}{R_j} - \frac{2c_j}{R_j^2}\right)v_i \tag{17}$$

where we have used Eq. (9). μ_i' is the chemical potential of the droplet components at pressure p. Putting μ_{dj} from Eq. (16) equal to the chemical potential of the droplets following from mass action, that is,

$$\mu_{dj} = \sum_i \frac{N_{ij}}{n_{dj}} \mu_i \tag{18}$$

eliminating μ_{ij}^* with Eq. (17) and making use of Eq. (12) for the droplet components, we arrive at the size distribution

$$\phi_j = \frac{v_{dj}}{l^3} \exp\left[-\frac{(4/3)\pi R_j^3 \Delta\bar{\mu} + 4\pi R_j^2 \sigma_j}{kT}\right] \tag{19}$$

with $\Delta\bar{\mu} = (\mu_i' - \mu_i)/v_i$, where the numerator is the difference between the chemical potentials of the components inside the droplets and that in a noncolloidal phase of these components, both at pressure p. It was shown in Ref. 28 that, at least for monodisperse systems, $(\mu_i' - \mu_i)/v_i$ does not depend on i. We assume that this remains true for the polydisperse microemulsion systems considered here. This requires that the compositions of the solutions in all the droplets and in a non-colloidal phase of these components be identical and that $\Delta\bar{\mu} = 0$ in order to obtain a microemulsion in equilibrium with an excess phase of the same composition as the droplet interiors. We are particularly interested in these kinds of systems, as in that situation the amount of dispersed phase and the average size of the globules are "chosen" by the system, i.e., not determined by the total volume of available dispersed phase, at least when this volume is large enough.

In Sec. III it was shown that if it is assumed that the interfacial tension of the droplets depends on curvature as prescribed by Eq. (4), then some essential features of microemulsion systems can already be qualitatively understood. In Ref. 27 the opposite route was followed. It was assumed that the interfacial tension does *not* depend on curvature, and in that case severe inconsistencies between theory and experiments were observed. As also mentioned in Sec. III, Eq. (4) does not account for finite size effects. We take into account both curvature dependence and finite size effects by using the Ansatz for the interfacial tension,

$$\sigma(R_j) = \gamma - \frac{4Kc_0}{R_j} + \frac{2K + \overline{K}}{R_j^2} + z'kT\frac{\ln(j)}{4\pi R_j^2} \tag{20}$$

The first three terms on the right follow from Helfrich's free energy expansion in the curvature [13] as discussed in Sec. III of this chapter and are identical to the right-hand side of Eq. (4). The last term quantifies the finite size effect as mentioned at the end of Sec. V.A. It was introduced by Fisher [35] in his treatment of condensation and is widely used in phenomenological theories of nucleation (see, e.g., Refs. 36 and 37). z' has an estimated value on the order of 1. We note that the calculation of z' from a model is far from trivial; see, for example, Ref. 38 for a discussion of the relatively simple case of on average flat interfaces.

Combining Eqs. (19) and (20) and setting $\Delta\bar{\mu} = 0$, the size distribution for a saturated droplet-type microemulsion can be written as

$$\phi_j = Cj^z \exp(\tilde{c}_0 j^{1/2} - \tilde{\gamma}j) \tag{21}$$

with

$$C = \frac{4\pi}{3l^3}\left(\frac{\sigma_s}{4\pi}\right)^{3/2}\exp\left(\frac{-4\pi(2K + \overline{K})}{kT}\right) \tag{22a}$$

$$\tilde{\gamma} = \frac{\gamma\sigma_s}{kT}; \qquad \tilde{c}_0 = \frac{16\pi Kc_0}{kT}\left(\frac{\sigma_s}{4\pi}\right)^{1/2}; \qquad z = \frac{3}{2} - z' \tag{22b}$$

where σ_s is the average area occupied by a surfactant molecule. The droplet radius R_j is defined by

$$\sigma_s j = 4\pi R_j^2 \tag{23}$$

and v_{dj} occurring in Eq. (19) has been replaced by

$$\frac{4}{3}\pi R_j^3 = \frac{4}{3}\pi\left(\frac{\sigma_s j}{4\pi}\right)^{3/2}$$

We need the values of z, $\tilde{c}_0$, C, and $\tilde{\gamma}$ for further calculations with the size distribution. The first three quantities should (ultimately) follow from a molecular model (see Sec. VI), but we still need to determine $\tilde{\gamma}$. This quantity plays the role of a Lagrange multiplier coupled to the total interfacial area in the system. Noting that the total number of surfactant molecules in the system is conserved, and assuming that the overwhelming majority of surfactant molecules reside at the droplet interface, we may write

$$c_{sa} = \frac{N_s}{V} = \sum_{j=1}^{\infty}\frac{jn_{dj}}{V} = \sum_{j=1}^{\infty}\frac{j\phi_j}{v_{dj}} \tag{24}$$

with c_{sa} the (known) number density of surfactant molecules in the microemulsion. Equations (24) and (21) together provide a relation for $\tilde{\gamma}$, which adjusts itself automatically to the required value by a slight adaptation of the free surfactant concentration. Note that a second Lagrange multiplier, $\Delta\bar{\mu}$ in Eq. (19) would appear in Eq. (21) if there were no excess phase present in the system. This multiplier couples to the total amount of dispersed volume in the system.

With Eqs. (21)–(24) it is, in principle, possible to obtain measurable properties of saturated droplet-type microemulsions: the total volume fraction of the droplets; the interfacial tension between the microemulsion phase and the excess phase; the average size of the droplets, i.e., moments of the size distribution; and the polydispersity. In order to obtain numbers one needs to know the values of five independent parameters [see Eqs. (21)–(24)], $(2K + \overline{K})$, Kc_0, σ_s, l, and z. These values should be obtained from molecular models or from independent experiments. A discussion of possible procedures is presented in the next section.

Before closing this section, we note that in the treatment of droplet-type microemulsions by Borkovec [39] and by Eriksson and Ljunggren [40], fluctuations in the number of molecules inside a droplet as well as fluctuations around the average area occupied by a surfactant molecule are explicitly taken into account. Their treatment leads to a size distribution that can be written in the same form as Eq. (21) but with an additional (weak) dependence of the "constant" C on droplet size. The major effect of treating the fluctuations mentioned above is an effectively different value of z from the one obtained here (but still on the order of 1).

VI. QUANTITIES THAT DETERMINE THE SIZE DISTRIBUTION OF SATURATED DROPLET-TYPE MICROEMULSIONS

A. Bending Elastic Quantities

As mentioned in Sec. III, the effect of an adsorbed surfactant monolayer at the oil/water interface on the bending elastic quantities can be split into two contributions.

1. The contribution of the "brush" formed by the tails of the surfactants (chain contribution)
2. The contribution of the electrical double layer formed by the charged heads of the surfactant molecules, the counterions, and the ions of added salt in the aqueous solution (electrical double layer contribution)

Before presenting results for these quantities, let us first consider how these contributions to the bending elastic quantities can be calculated.

The free energy per unit area of a uniformly bent interface can be written as

$$f_c = f_0 - 2Kc_0(c_1 + c_2) + \frac{1}{2}K(c_1 + c_2)^2 + \overline{K}c_1 c_2 \tag{25}$$

where f_0 is the free energy per unit area of flat interface. This equation is not simply F_c/A with F_c given by Eq. (2) but contains a constant contribution f_0 (containing $2Kc_0^2$) from the flat interface. For a sphere, $c_1 = c_2 = 1/R$, and Eq. (25) takes the form

$$f_c = f_0 - \frac{4Kc_0}{R} + \frac{2K + \overline{K}}{R^2} \quad \text{(sphere)} \tag{26}$$

[Note the equivalence of Eqs. (26) and (4) for the interfacial tension, $\sigma(R)$.] For a cylinder, $c_1 = 1/R$ and $c_2 = 0$, so that

$$f_c = f_0 - \frac{2Kc_0}{R} + \frac{K}{2R^2} \quad \text{(cylinder)} \tag{27}$$

If we now compute the free energy of an interface bent into a sphere and that of one bent into a cylinder, we can extract $K, \overline{K}$, and c_0 by comparing these free energies with the phenomenological expressions Eqs. (26) and (27). Such calculations have been carried out both for the bent brush part of the ionic surfactant monolayer and for the bent electrical double layer part of the adsorbed ionic surfactant monolayer.

Using the analogy that can be drawn between the system of end-grafted polymer chains and the chain part of a surfactant monolayer, Milner and Witten [41] obtained simple analytical expressions for the bending elastic quantities. For the case of a "melt brush," i.e., one in which the density is forced to be uniform, they obtain

$$K_{ch} = \frac{3}{2} f_{0,ch} h_0^2 \tag{28}$$

$$\overline{K}_{ch} = -\frac{2}{5} f_{0,ch} h_0^2 \tag{29}$$

$$(Kc_0)_{ch} = \frac{3}{8} f_{0,ch} h_0 \tag{30}$$

The subscript ch refers to the chain contribution, and here $f_{0,ch}$ is the free energy per unit area of a flat brush, for which Milner and Witten obtained the expression

$$f_{0,ch} = \frac{\pi^2 N \omega^3}{24 \rho_m^2} \left(\frac{3}{\ell^2 C_\infty} \right) kT \tag{31}$$

and $h_0 = N\omega/\rho_m$ is the height of the brush. In the above expressions N is the number of monomers per chain, ω is the number of chains per unit area, ρ_m is the number of monomers per unit volume, ℓ is the bond length, and C_∞ is the characteristic ratio relating the mean square end-to-end distance R_e^2 to the number of monomers per chain and the bond length ($R_e^2 = N\ell^2 C_\infty$ in the melt).

Using typical values for polyethylene chains ($\omega = 4$ nm^{-2}, $\rho_m = 34.6$ nm^{-3}, $\ell = 0.153$ nm, $C_\infty = 6$), one obtains

$$f_{0,ch} = 0.47 NkT \text{ nm}^{-2} \tag{32}$$

and

$$h_0 = 0.12 N \text{ nm} \tag{33}$$

The method used by Milner and Witten is valid only in the limit of very long hydrocarbon tails on the surfactant molecules. From the calculation presented above it appears that even for relatively short chain surfactants it yields commonly accepted values for the bending elastic quantities, i.e., K, $|\overline{K}|$ are on the order of kT and Kc_0 is on the order of $1kT$ nm$^{-1} \cong 10^{-12}$ N.

Starting from the Poisson–Boltzmann equation it is possible to calculate the free energy of a curved double layer in terms of an expansion in $1/\kappa R$, where κ is the inverse Debye length [$\kappa = (8\pi Q n_{el})^{1/2}$, with Q defined by Eq. (38) below and n_{el} the number of molecules of 1:1 electrolyte per unit volume] and R is the radius of curvature of the charged interface. Obviously this expansion is valid only for $\kappa R \gg 1$, i.e., for situations where the radius of curvature is (much) larger than the Debye length [28,42–44]. Using this method (the derivations are given in detail in Refs. 43 and 44), one obtains for the electrical contributions to the bending elastic quantities

$$K_{el} = \frac{kT}{2\pi Q\kappa} \left(\frac{(q-1)(q+2)}{q(q+1)} \right) \tag{34}$$

$$\overline{K} = \frac{-kT}{\pi Q\kappa} \int_{z=2/(1+q)}^{1} \frac{\ln(z)\,dz}{z-1} \tag{35}$$

$$(Kc_0)_{el} = \frac{-kT}{2\pi Q} \ln\left(\frac{q+1}{2} \right) \tag{36}$$

where the quantities p and q are related to the surface charge density σ_{el},

$$p = 2\pi Q \left| \frac{\sigma_{el}}{e} \right| \kappa^{-1}; \qquad q = (p^2+1)^{1/2} \tag{37}$$

where e is the elementary charge and Q is the Bjerrum length,

$$Q = \frac{e^2}{4\pi\varepsilon_0\varepsilon_r kT} \quad (= 0.7 \text{ nm in water at 298 K}) \tag{38}$$

In this equation, ε_r is the dielectric constant of the solution and ε_0 is the permittivity of vacuum. The negative sign in $(Kc_0)_{el}$, Eq. (36), is due to the fact that the electrical double layer tends to bend the interface around the oil side ($c_{0,el} < 0$).

In writing down Eq. (37) it is assumed that the surface charge density does not depend on curvature. In Ref. 28 arguments are given that the surface charge density does depend on the curvature. There, for an interface bent around the water side, it is assumed that

$$p_{curved} = p_{flat} \left(1 + \frac{\xi}{R} \right)^2$$

with ξ between 0.2 and 0.4 nm. Even such small values of ξ have a relatively large influence on the free energy of the bent interface.

It is interesting to consider two limiting cases.

1.   Low Surface Charge Densities.   In this limit $p \ll 1$, and therefore $q = 1 + p^2/2$. Substituting these results into the expressions above, one obtains

$$K_{el} = \frac{3\pi kTQ(\sigma_{el}/e)^2}{2\kappa^3} \tag{39}$$

$$\overline{K}_{el} = \frac{-\pi kTQ(\sigma_{el}/e)^2}{\kappa^3} \tag{40}$$

$$(Kc_0)_{el} = \frac{-kT\pi Q(\sigma_{el}/e)^2}{2\kappa^2} \tag{41}$$

Interestingly, the form of these expressions is similar to that of the expressions for the bent brush, that is, $f_0 h_0^2$ for K and $\overline{K}$ and $f_0 h_0$ for Kc_0. Indeed, the free energy per unit area of a flat double layer is, in the limit of small surface charge density, given by

$$f_{0,el} = \frac{2\pi kTQ(\sigma_{el}/e)^2}{\kappa} \tag{42}$$

while h_0 can be taken equal to the Debye length κ^{-1}. Using the above results we can write

$$K_{el} = \frac{3}{4} f_{0,el} h_0^2 \tag{43}$$

$$\overline{K}_{el} = -\frac{1}{2} f_{0,el} h_0^2 \tag{44}$$

$$(Kc_0)_{el} = -\frac{1}{4} f_{0,el} h_0 \tag{45}$$

2. High Surface Charge Densities. In this limit, $p = q \gg 1$, and we obtain

$$K_{el} = \frac{kT}{2\pi Q \kappa} \tag{46}$$

$$\overline{K}_{el} = \frac{-\pi kT}{6 Q \kappa} \tag{47}$$

$$(Kc_0)_{el} = \frac{-kT}{2\pi Q} \ln\left(\frac{p}{2}\right) \tag{48}$$

Again we find that for typical values of surface charge densities $|\sigma_{el}/e| \cong 1$ nm^{-2} and for salt concentrations between 0.1 and 0.5 M (of a 1: 1 electrolyte), K_{el} and $\overline{K}_{el}$ are on the order of kT, and $|(Kc_0)_{el}|$ is on the order of 10^{-12} N. Relatively small changes in salt concentration may lead to such a variation in the bending elastic quantities that it induces an inversion of the microemulsion, which, as discussed in Sec. III, is indeed experimentally observed.

Combining, for example, Eq. (30) for the chains [with the numerical values of Eqs. (32) and (33) and choosing $N = 6$ for the chain length] with Eq. (48) for the electrical double layer (with $|\sigma_{el}/e| = 1$ nm^{-2} and 0.1 M NaCl ($\kappa^{-1} = 1$ nm) in the aqueous solution), we find

$$Kc_0 = (Kc_0)_{ch} + (Kc_0)_{el}$$
$$= \frac{3}{8} f_{0,ch} h_0 - \frac{kT}{2\pi Q} \ln\left(\frac{p}{2}\right) = (0.76 - 0.18)kT \text{ nm}^{-1}$$
$$= 0.58 kT \text{ nm}^{-1} = 2.38 \times 10^{-12} \text{ N}$$

This (a quantity without any subscript symbolizes the total contribution) is of the expected order of magnitude and with a positive sign (i.e., curved as for water droplets in oil). Experiments show [16,28] that for SDS (C_{12} chain) as the surfactant combined with pentanol as the cosurfactant (20% w/w in cyclohexane), c_0 is zero at 0.15 M NaCl, and thus at 0.1 M NaCl the system forms an O/W microemulsion with Kc_0 negative.

The choice of $N = 6$ in the above numerical example is reasonable for comparison with the experiments, since according to Ref. 41 a mixture in the interface of about 25% C_{12} chains and 75% C_5 chains behaves in the bending as if all the chains had a length somewhat greater than C_5, the extra length of the C_{12} chains sticking out of the closely packed C_5 layer offering very little resistance to bending.

Our conclusion is that the theory, as described above, fits reasonably well to experiments, although the contribution of the chains to Kc_0 may be somewhat too large and/or the double layer contribution somewhat too small. The double layer contribution would become larger if we were to take account that the surface charge density increases on bending around the water side (see Ref. 28), and the chain contribution may have been overvalued by applying a theory developed for long chains (see Ref. 41) to rather short chains.

In general it seems a good idea to check theories at the condition where c_0 becomes zero and to remember that at the present stage the theories, both for the electrical effects and for the chains, may need quantitative adaptations on the order of several tens of percent.

Experimentally, the bending elastic modulus K as measured by ellipsometry is indeed found to be on the order of $(0.1–1)kT$ for quite a wide range of microemulsion systems; see, e.g., Ref. 45 for a system with a single-chain ionic surfactant plus cosurfactant, Ref. 46 for a system with a double-chain ionic surfactant in which the chain length of the oil (linear alkane) is varied, and Ref. 47 for a system with a nonionic surfactant in which the chain length of the surfactant is varied.

In ellipsometry the fluctuations in the interface position (capillary waves) manifest themselves, and when the interfacial tension γ is sufficiently small, the influence of K on the interfacial waves can be measured.

As follows easily from Eq. (3) for the bending free energy of a single drop, the term with the Gaussian modulus is proportional to the number of droplets in a system, just as the mixing entropy is. For this reason, the Gaussian modulus can be obtained from experiments only by using an expression for the mixing entropy. Several assumptions in the literature [27,48–52] point to values roughly in between $-kT$ and $+kT$ for $\overline{K}$.

Just like the Gaussian bending elastic modulus, the quantity Kc_0 can, in general, be obtained only by assuming a model for the mixing entropy. However, one of us [53] showed recently that when the average size of the droplets is large (say, several tens of nanometers), the quantity Kc_0 can, to a good approximation, be obtained from an average radius and the interfacial tension γ between the microemulsion phase and the excess phase without the need for numerical values of $(2K + \overline{K})$, σ_s, l, and z (see Sec. V, just before the last paragraph) via

$$\gamma = 2Kc_0/R_{ik}^{1/(i-k)} \qquad \text{(large droplets)} \tag{49}$$

In this equation, the average droplet radius is effectively a ratio of moments of the size distribution, Eq. (21), i.e.,

$$
\begin{aligned}
R_{ik} &= \frac{\langle R^i \rangle}{\langle R^k \rangle} = \frac{\sum_{j=1}^{\infty} R_j^i \rho_j}{\sum_{j=1}^{\infty} R_j^k \rho_j} = \frac{\sum_{j=1}^{\infty} R_j^i [\phi_j/(4/3)\pi R_j^3]}{\sum_{j=1}^{\infty} R_j^k [\phi_j/(4/3)\pi R_j^3]} \\
&= \frac{\sum_{j=1}^{\infty} R_j^i j^{z-3/2} \exp(-\tilde{\gamma} j + \tilde{c}_0 j^{1/2})}{\sum_{j=1}^{\infty} R_j^k j^{z-3/2} \exp(-\tilde{\gamma} j + \tilde{c}_0 j^{1/2})}
\end{aligned}
\tag{50}
$$

where R_j is defined by Eq. (23). Equation (49) can be obtained from Eq. (50) by replacing the summations by integrations and assuming that $\tilde{c}_0 \gg \tilde{\gamma}$. This derivation is given in some detail in Ref. 53. Note, however, that in this reference a slightly different definition of R_{ik} was used. i and k depend on the technique used to measure the droplet size; e.g., if the volume of the dispersed phase is measured and the interfacial area is known, one obtains R_{ik} with $i = 3$, $k = 2$, whereas by measuring the radius of gyration one has approximately $i = 8$, $k = 6$. Equation (49) is a generalization (to polydisperse systems) of the result of de Gennes and Taupin [18] using ideas from Robbins [11] for strictly monodisperse systems. We emphasize that, strictly speaking, Eq. (49) has only limited applicability, i.e., its right-hand side is the first term in a series in $1/R_{ik}^{1/(i-k)}$ of which convergence is questionable in situations where the average size of the droplets is smaller than about 10 nm. Once Eq. (49) is valid, however—that is, at large average radius and therefore at small γ—the quantity Kc_0 can be obtained *directly* from experiments without making assumptions for the mixing entropy and the value of z.

Incidentally, for the interfacial tension γ^* and the radius R^*, belonging to the maximum volume fraction ϕ_j (this R^* must be close to the average radius $R_{ik}^{1/(i-k)}$), an equation similar to Eq. (49) can be derived very simply from $\partial \phi_j / \partial j = 0$ [see Eqs. (21)–(24)]. This leads exactly to

$$\gamma^* = \frac{2Kc_0}{R^*} + \frac{zkT}{4\pi(R^*)^2} \tag{51}$$

The second term on the right in Eq. (51) is on the order of 10% of the first term.

In the next section we consider experimentally obtained data sets from which Kc_0 is extracted in the way described above. These results confirm the expectation that the spontaneous curvature, c_0, is on the order of the inverse average droplet radius in a saturated microemulsion system. This expectation follows from the relation, Eq. (5), between c_0 and R_{MIN} and the fact that the entropy of mixing makes the average droplet radii somewhat smaller than R_{MIN}.

B. Molecular Area of Surfactant(s)

The average area occupied by a surfactant molecule, σ_s, is obtained by measuring the tension of a macroscopic oil/water interface as a function of the surfactant concentration below the cmc and applying the Gibbs adsorption equation. This area turns out to be fairly independent of the concentrations of the constituent parts of a given microemulsion (i.e., salt and/or cosurfactant concentration) [5,54]. It varies between approximately 0.6 nm^2 for AOT systems [54] and about 1 nm^2 for SDS–cosurfactant (linear alcohol) systems [5]. In the latter situation, this area is shared with approximately two to three cosurfactant molecules. It is assumed here that the area per surfactant molecule at the droplet interface does not differ from the one at the macroscopic oil/water interface [see, however, the discussion following Eq. (38)]. The only experimental indication that we are aware of that indeed points to the validity of this assumption comes from small-angle X-ray scattering (SAXS) measurements [55].

C. Length Scale for Configurational Entropy

As already indicated, the length l cannot be measured directly. It is sensitive to the model employed in the description of a microemulsion. For the phenomenological model as described in this chapter, it is found to be on the order of the cube root of a molecular volume in the liquid state [31], that is, approximately 0.5 nm.

D. Value of z

In Ref. 27, Kegel and Reiss estimated z of the basis of experiments on microemulsions. Assuming that the average radius of a drop never exceeds $2/c_0$, they found the inequality (independent of their somewhat different definition of R_{ik})

$$3/4 < z < 3/2 \tag{52}$$

VII. GENERAL PROPERTIES OF THE SIZE DISTRIBUTION AND EXPERIMENTAL TEST OF THE THEORY

A. Order of Magnitude of Measurable Properties of Saturated Droplet-Type Microemulsions

We summarize the estimated ranges of the quantities discussed in the previous section:

$$0 < (2K + \overline{K})/kT < 3 \tag{53a}$$

$$0.01 < |Kc_0|/(kT \text{ nm}^{-1}) < 1 \tag{53b}$$

$$0.5 < \sigma_s \text{ nm}^{-2} < 1 \tag{53c}$$

$$l^3 \approx 0.1 \text{ nm}^3 \tag{53d}$$

$$3/4 < z < 3/2 \tag{53e}$$

As mentioned in the discussion around Eqs. (21)–(24), we need values of the five parameters listed in Eqs. (53) and of the surfactant concentration, c_{sa}, to be able to calculate values of various properties of microemulsions (O/W + O or W/O + W). Of these five parameters the area per surfactant molecule, σ_s, can be determined accurately from the Gibbs adsorption equation, Eq. (1), and there is only a small uncertainty regarding the values of l and z.

We therefore choose $\sigma_s = 1$ nm^2 in the following calculations and take $l^3 = 0.1$ nm^3 and $z = 3/2$ or $3/4$. We choose 10 mM for the surfactant concentration in Table 1 (and 50, 16.7, and 8.3 mM, respectively, in Fig. 5), assuming that all of it is adsorbed and the amount of dissolved surfactant is negligible. A number of values of $2K + \overline{K}$ and of Kc_0 are taken within the ranges indicated in Eqs. (53a) and (53b).

Then we (iteratively) calculate γ from Eqs. (21)–(24). This results in the size distribution $\{\phi_j\}$, from which the average size of the droplets follows using Eq. (50) and the total volume fraction of the droplets, ϕ, by summing Eq. (21) over all categories. In Fig. 5 we show a set of typical size distributions obtained in this way.

Table 1 Properties of the Size Distribution as a Function of the Input Parameters Kc_0 and $2K + \overline{K}$[a]

Kc_0 (kT nm^{-1})	$2K + \overline{K}$ (kT)	γ (kT nm^{-2})	R_{32} (nm)	ϕ	ε
0.01	1	0.1	1.3	0.0028	0.46
	2	0.0021	14.0	0.03	0.34
	3	0.00048	47.1	0.12	0.20
0.1	1	0.43	0.8	0.0017	0.36
	2	0.065	3.6	0.0072	0.23
	3	0.0285	7.6	0.0172	0.16
0.5	1	>2	<1	—	—
	2	1.0	1.1	0.0023	0.19
	3	0.545	2.0	0.004	0.14
1	1	>2	<1	—	—
	2	>2	<1	—	—
	3	1.97	1.1	0.0028	0.14

[a] The values of the other parameters were fixed at $c_{sa} = 10$ mM, $l^3 = 0.1$ nm^3, $z = 3/2$, $\sigma_s = 1$ nm^2. "Droplets" are considered to be of molecular size if $R_{32} < 1$ nm and $\gamma > 2kT$ nm^{-2}.

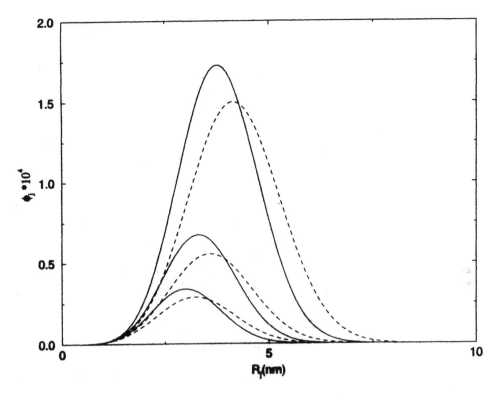

Figure 5 Size distributions calculated for $Kc_0 = 0.05kT$ nm^{-1}. The solid curves were calculated using $z = 3/2$; for the dashed ones, $z = 1$ was used. Going from the upper to the lower curves, the surfactant concentration was fixed at 50, 16.7, and 8.3 mM, respectively. The values of the other parameters are $l^3 = 0.1$ nm^3, $\sigma_s = 1$ nm^2, $2K + \bar{K} = 3kT/2$. (From Ref. 27.)

The polydispersity ε is obtained via

$$\varepsilon = \left(\frac{\langle R^2 \rangle}{\langle R \rangle^2} - 1 \right)^{1/2} \tag{54}$$

From experimental investigations reported in the literature in which the interfacial tension, the average droplet size, and/or the polydispersity of Winsor II systems is measured (see, e.g., Refs. 28, 46–48, 50, and 54–56), we estimate that at $c_{sa} \approx 10$ mM the average size of the droplets is in the range $R_{32} = 2$–20 nm, the macroscopic interfacial tension $\gamma = 0.001$–0.1 mN/m, the volume fraction of the droplets $\phi \approx 0.01$–0.1, and the polydispersity is roughly $0.1 \leq \varepsilon \leq 0.4$. We raise the question of whether the orders of magnitude of this set of data can be reproduced by the theory using the values of the parameters in the ranges given in Eqs. (53).

Using the values of the parameters as mentioned above, it is found that the combination $2K + \bar{K}$ should be larger than zero in order for a microemulsion with average droplet size on the order of 1 nm (and interfacial tension smaller than order kT nm^{-2}) to be stable. The results are summarized in Table 1.

First of all we can check that except for the upper two rows in Table 1, Eq. (49) describes the relation between γ, Kc_0, and R_{32} surprisingly well, even for very small R_{32}. It follows from Table 1 that when Kc_0 is increased, $2K + \bar{K}$ should be increasingly larger in order for drops of average sizes greater than 1 nm to be stable. If the value of z is set equal to 3/4, a larger average drop size is found (approximately 10–20% larger for the relevant values of the other parameters) as well as a slightly smaller polydispersity (not shown in Table 1). Generally, with the values of the parameters in the ranges given in Eqs. (53), the average droplet size varies from molecular lengths to approximately 50 nm, the volume fraction from about 0.002 to about 0.12, the interfacial tension from about 0.002 to 8 mN/m (at $T = 298$ K), and the polydispersity from approximately 14% to 46%.

We conclude from this section that values of the parameters indicated in Eqs. (53) estimated from experiments give predictions of measurable properties of droplet-type microemulsions that are of the same order of magnitude as those found in experiments. No additional concepts are required to force this agreement.

B. Quantitative Comparison with Experiments

In this section, measured average droplet sizes, droplet volume fractions, and interfacial tensions are compared with the thermodynamic formalism as presented in Sec. V. These quantities can be calculated by fixing values of l, z, σ_s, Kc_0, and $2K + \overline{K}$, numerically solving Eq. (24) with Eqs. (21)–(23) for $\tilde{\gamma}$ (with fixed c_{sa}), and subsequently calculating the average size of the droplets by using Eq. (50). In practice, not all values of these quantities are known. For l, z, and σ_s, one can make reasonable estimates, but in order to obtain the bending elastic quantities Kc_0 and $2K + \overline{K}$, one needs, at the current state of the art, input from independent experiments. At the moment there are no microemulsion systems for which both Kc_0 and $2K + \overline{K}$ are known without some theory (of droplet-type microemulsions) being used. Therefore the whole calculation as outlined above should be done iteratively, using trial values of one or more of the bending elastic quantities and varying them (within certain windows) until convergence is reached between theory and at least one measured quantity. It is therefore convenient to simplify the equations that interrelate these quantities. By replacing the summations by integrations in Eq. (50), substituting $x = j^{1/2}$, and using the method of steepest descent, the average radius can be approximated as (see Refs. 27 and 53 for details)

$$R_{ik} = \left(\frac{\sigma_s}{4\pi}\right)^{(i-k)/2} \exp[f(x_i^*) - f(x_k^*)] \tag{55}$$

and the total volume fraction of the droplets may be written as

$$\phi = 2C\left(\frac{\pi}{\tilde{\gamma}}\right)^{1/2} e^{f(x_3^*)} \tag{56}$$

with

$$f(x_i^*) = -\tilde{\gamma}(x_i^*)^2 + \tilde{c}_0 x_i^* + [2(z-1) + i] \ln(x_i^*) \tag{57}$$

and

$$x_i^* = \frac{\tilde{c}_0 + \{\tilde{c}_0^2 + 8[2(z-1) + i]\tilde{\gamma}\}^{1/2}}{4\tilde{\gamma}} \tag{58}$$

x_i^* is the maximum of the function $x^{2(z-1)+i} \exp(-\tilde{\gamma}x^2 + \tilde{c}_0 x) = \exp\{-\tilde{\gamma}x^2 + \tilde{c}_0 x + [2(z-1) + i] \ln x\}$. The subscript i of x_i^* refers to the value of i in Eqs. (57) and (58). Note that these expressions differ from the ones given in Refs. 27 and 53 owing to the slightly different definition of R_{ik} used in this work. Equation (58) is [by the definitions given in

Eqs. (22) and (23)] analogous to Eq. (51). Now the measurable properties of interest in satu-rated droplet-type microemulsions can be calculated without having to perform extensive computations.

1. Average Droplet Size and Macroscopic Interfacial Tension as a Function of Droplet Volume Fraction

As far as we are aware, the only system for which the volume fraction of the droplets, the average radius (R_{32}), and the tension of the macroscopic oil/water interface have been measured is a Winsor II system composed of SDS, pentanol, cyclohexane, and 0.2 M NaCl with equal volumes of water and oil phases. This system was studied in Ref. 55. The bending elastic modulus of this system was measured by ellipsometry [50]. The results of Sec. V imply that when l and z are fixed, there are still two unknown parameters: the Gaussian bending elastic modulus and the preferred curvature. Therefore, we choose to test the theory on consistency, that is, we fix z and l and fit the (R_{32}, ϕ) data with Eqs. (55)–(58)—we choose Kc_0 and $2K + \overline{K}$, determine the values of $\tilde{\gamma}$ that generate the experimental values of ϕ, and compare the calculated droplet radii with the experimental ones. This procedure is iterated until maximum agreement with the experimental droplet radii is obtained. Subsequently, we compare the values of the interfacial tension γ from $\tilde{\gamma}$ [Eq. (22)] that were necessary to generate these (R_{32}, ϕ) data with the experimentally measured values. Note that in this treatment the volume fraction of the droplets rather than the surfactant con-centration conveniently determines the Lagrange multiplier $\tilde{\gamma}$. We emphasize that the last mentioned comparison requires no adjustable parameters. Figure 6a shows the experimen-tally determined (R_{32}, ϕ) data together with the best fit results for constant $z = 3/2$ and c_0 and different values of $\overline{K}$. Figure 6b shows the set of values of the interfacial tension that were used to generate these data, again together with the experimental results from Ref. 55.

We note, as already mentioned, that the definition of R_{ik} that was used in Ref. 27 is different from the definition given in Eq. (50); that is, in Ref. 27 a volume fraction averaged radius was calculated instead of the number density averaged radius, Eq. (50). For this reason the values of Kc_0 and $2K + \overline{K}$ are slightly different from the ones found in Ref. 27.

Looking at Fig. 6, we note that the agreement is excellent. Comparable agreement is found over the whole range of z as given by Eq. (53e). The value of c_0 obtained in this way increases somewhat with decreasing z, whereas the value of $\overline{K}$ decreases and even becomes negative. This analysis proves that the theory is consistent for this aspect; i.e., fitting the theory to experimentally obtained (R_{32}, ϕ) data generates the variation of interfacial tension with volume fraction of the droplets without any additional adjustable parameters.

The value of c_0 can also be obtained independently from the work reported in Ref. 28. In this work, among other things, the droplet radii and the interfacial tensions as a function of the cosurfactant concentration for the same system as used in Ref. 55 were measured. For the system containing 19% pentanol, which is closest to the 20% used by van Aken [55], we estimate $\gamma = 0.02$ mN/m $= 0.00487kT$ nm^{-2}, and $R_{32} = 12$ nm. Substituting these values into Eq. (49) and using the fact that $K = 1kT$ for this system, we obtain $c_0 = 0.029kT$ nm^{-1}, which is remarkably close to the value of $0.035kT$ nm^{-1} used in Fig. 6. (Moreover, it is expected that c_0 indeed increases with increasing cosurfactant con-centration.)

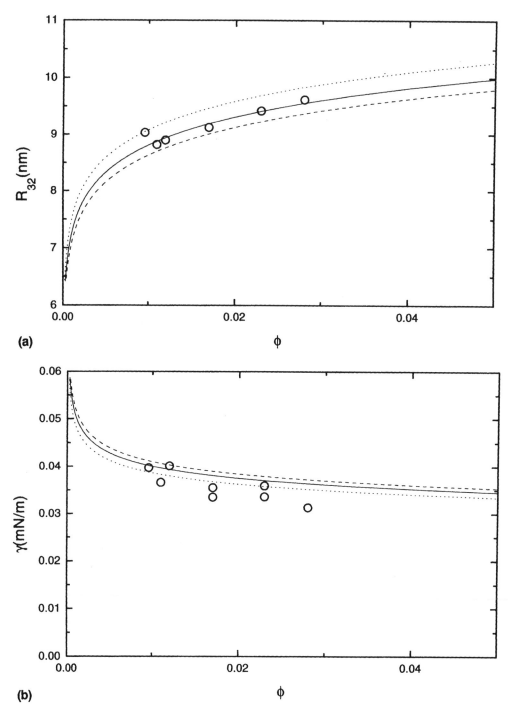

Figure 6 Experimental data of van Aken [55], experimental droplet radius R_{32} versus volume fraction ϕ of the droplets together with theoretical lines calculated from Eqs. (55)–(58) using $K = kT$ [50], $l^3 = 0.1$ nm^3, $\sigma_s = 1$ nm^2, $z = 3/2$, $c_0 = 0.035$ nm^{-1}, and $\bar{K}/kT = 0.25$ (dotted line), 0.22 (solid line), and 0.20 (dashed line). In (a) data were fitted using c_0 and $\bar{K}$ as free parameters. In (b) the variation of the interfacial tension γ necessary to produce the lines in (a) are plotted together with the experimental points. There is no additional free parameter.

2. Macroscopic Interfacial Tension and Average Droplet Sizes as a Function of Salt Concentration

In Sec. VI expressions were given for the chain contributions and the electrical double layer contributions to the bending elastic quantities. There it was argued that the theoretical results fit reasonably well to experiments but not well enough to give accurate values for all parameters involved.

With this in mind, we start by writing the preferred curvature as

$$c_0 = \frac{(c_0 K)_{\text{ch}} + (c_0 K)_{\text{el}}}{K} \tag{59}$$

where the subscript ch again refers to the contribution of the surfactant chains. The electrical part of the preferred curvature is given by Eq. (36). The chain contribution can in principle be obtained from Eq. (59) if the salt concentration at which $c_0 = 0$ (or, of course, any other known value) is known. We prefer this approach rather than using an estimate for the chain contribution based on Eq. (30) as suggested earlier following Eq. (48).

In Ref. 46, interfacial tension was measured as a function of the salt concentration for systems composed of AOT, brine, and linear alkanes of varying chain length (C_8, C_{10}, C_{12}, C_{14}). In this work, the bending elastic moduli of these systems were determined by ellipsometry. The droplet radii of these systems are reported elsewhere [48]. The interfacial tension as reported in Ref. 46 first decreases, goes through a minimum, and subsequently increases with increasing salt concentration. The minimum (and its immediate neighborhood) corresponds to an (actually quite special kind of) Winsor III system: a microemulsion phase coexisting with excess oil and brine. The preferred curvature should be zero at this interfacial tension minimum. Therefore we obtain $(c_0 K)_{\text{ch}} = -(c_0 K)_{\text{el}}$ at the salt concentration at which the interfacial tension is at a maximum. Note that in the Winsor III region we are not dealing with droplets but (most probably) with a structure with zero mean curvature. In that case we expect, contrary to the situation with droplets discussed in Sec. III [see in particular Eq. (5)], that the interfacial tension is at a minimum when $c_0 = 0$. Very similar ideas were used in Ref. 28 to give a quantitative explanation of the interfacial tension γ and the droplet radius R as a function of the salt concentration on W/O + W microemulsions stabilized with SDS and pentanol.

The constant value of the chain contribution may be considered as a zeroth-order approximation, where it is assumed that the density of the chains is independent of the salt concentration. A motivation for this assumption is that the molecular area of the surfactant as a function of salt concentration in the systems used in Ref. 46 was found [54] to be constant within the range of experimental error.

The interfacial tension was calculated from surfactant conservation by simplifying Eq. (24) in the same way as the expressions for the volume fraction and average radii were simplified by Eqs. (55)–(58). This expression reads

$$c_{sa} = 2\left(\frac{\pi}{\tilde{\gamma}}\right)^{1/2} l^{-3} \exp\left(\frac{-4\pi(2K + \overline{K})}{kT}\right) e^{f(x_2^*)} \tag{60}$$

with, from Eqs. (57) and (58),

$$f(x_2^*) = -\tilde{\gamma} x_2^{*2} + \tilde{c}_0 x_2^* + 2z \ln(x_2^*) \tag{61a}$$

where

$$x_2^* = \frac{\tilde{c}_0 + (\tilde{c}_0^2 + 16z\tilde{\gamma})^{1/2}}{4\tilde{\gamma}} \tag{61b}$$

It can be seen from Eqs. (60) and (61) that when K, c_{sa}, and Kc_0 are known and z and l are fixed, the interfacial tension still depends on a single variable, the Gaussian bending elastic modulus. The results together with the experimental data are shown in Fig. 7. The fixed independent parameters that were used to calculate these curves are listed in Table 2. Due to the fact that in Ref. 27 an erroneous expression for $(Kc_0)_{el}$ was used, the values of $(Kc_0)_{ch}$ in Table 2 are different from the ones used in Ref. 27. We note that the values of $(Kc_0)_{ch}$ listed in Table 2 are indeed of the same order of magnitude as the ones predicted by Eqs. (28)–(31) for polyethylene chains (with small N, say $N < 10$). The decrease in $(Kc_0)_{ch}$ with increasing chain length of the oil may be understood qualitatively as follows. Assuming that in going from C_8 to C_{12}, fewer and fewer oil molecules penetrate into the surfactant brush (and there are indeed strong experimental indications pointing to the validity of this approach; see Ref. 57), then effectively the number of monomers per unit volume, ρ_m, decreases and therefore, from Eqs. (28)–(31), $(Kc_0)_{ch}$ decreases.

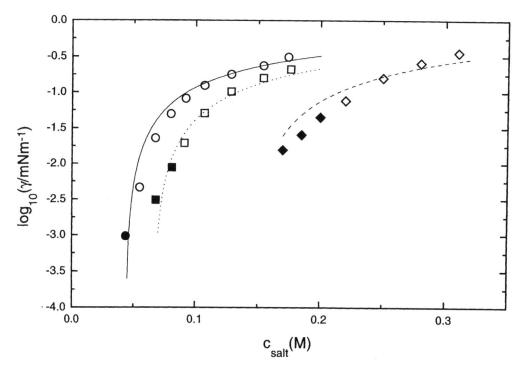

Figure 7 Interfacial tension of the planar interface between the microemulsion and the excess phase as a function of the salt concentration for systems composed of AOT (sodium diethylhexylsulfosuccinate), brine, and linear alkanes of varying chain length. Points are experimental data obtained for C_8 ($\bigcirc$, $\bullet$), C_{10} ($\square$, $\blacksquare$), and C_{12} ($\diamond$, $\blacklozenge$) linear alkanes making up the oil phase. Open symbols refer to Winsor II systems; corresponding filled symbols indicate the Winsor III region where the theory is no longer valid. The lines were calculated according to Eqs. (60) and (61), the fixed parameters listed in Table 2, and suitably chosen values of $\bar{K}/kT = 0.8$ (C_8 systems), 0.39 (C_{10} systems), and 1.2 (C_{12} systems). (Experimental data from Ref. 46.)

The theoretical lines in Fig. 7 correspond to suitably chosen values of the Gaussian bending elastic modulus $\overline{K}$. Other values of this parameter only translate the curves through the (γ, c_{salt}) plane without altering their shapes. It should be emphasized that the curves are all unique; that is, there are no values of $\overline{K}$ that allow the C_{10} or C_{12} data to be described by the parameters corresponding to the C_8 data. It is clear from Fig. 7 that the relevant experimental data (i.e., the ones corresponding to Winsor II systems) are described rather well by the theory. Even the agreement with the C_{12} data is reasonable, which is somewhat unexpected because of the small value of K for this system ($0.15kT$). For these small values of K, the *Ansatz* for the interfacial tension, $\sigma(R_j)$, Eq. (20) (i.e., up to second order in the inverse droplet radius), is no longer believed to be a good approximation.

Now, using the parameters listed in Table 2 and the values of $\overline{K}$ that were used to calculate the theoretical lines in Fig. 7, the droplet radii were calculated by using Eq. (55) with $i = 3$, $k = 2$. The results are shown in Fig. 8, together with the experimentally determined radii reported in Ref. 48. It is emphasized that the theoretical lines in Fig. 8

Table 2 Fixed Parameters Used to Calculate the Dependence of Interfacial Tension on Salt Concentration[a]

System (oil phase)	K (kT)	$(Kc_0)_{ch}$ $(kT\ nm^{-1})$	σ_s (nm^2)
Octane (C_8)	0.85	0.365	0.71
Decane (C_{10})	0.95	0.334	0.67
Dodecane (C_{12})	0.15	0.255	0.64

[a] A surfactant concentration of 10 mM in the microemulsion phase was used in all cases.
Source: Refs. 46 and 54.

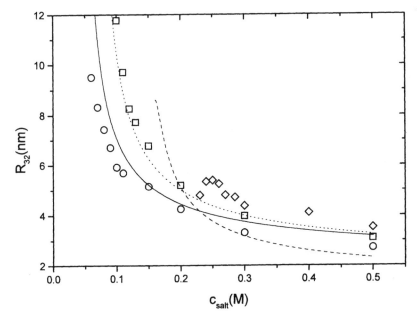

Figure 8 Variation of average droplet radius (R_{32}) as a function of salt concentration for the same systems as in Fig. 7. Theoretical lines were calculated using Eqs. (60) and (61) with Eq. (55) using the same parameters as in Fig. 7. The surfactant concentration was fixed at $c_{sa} = 60$ mM, as in Ref. 48. (Experimental points from Ref. 48.)

are calculated without any adjustable parameters.

As can be seen from Fig. 8, the agreement between theory and experiments is excellent for the C_{10} systems and quite good for the C_8 systems, although in the latter case it seems as if the difference between theory and experiments is a systematic shift parallel to the salt concentration axis. A possible explanation is that different batches of AOT might have been used in the work reported in Refs. 46 and 48. Very small amounts of adsorbed contaminants and/or hydrolysis products are expected to have a significant influence on (the chain contribution to) the preferred curvature. Indeed, if a slightly larger value of $(Kc_0)_{ch}$ than the one listed in Table 2 is used for the C_8 systems (i.e., $0.380kT$ nm^{-1} instead of $0.365kT$ nm^{-1}), then excellent agreement is observed between theoretical and experimental droplet radii as a function of salt concentration. Clearly the theory predicts values of the radii that are too small in the case of the C_{12} system.

We note that for the AOT systems the values of the Gaussian bending elastic moduli are found to be significantly greater than the one that was found for the SDS plus pentanol system (see Sec. VII.B.1). This difference might be due to the difference in molecular geometry between AOT and SDS/pentanol. We feel that more experimental work on other systems should be done and that sophisticated models for surfactant monolayers at oil/brine interfaces should be developed in order to be able to gain some understanding of this peculiar behavior of the Gaussian bending elastic modulus.

VIII. CONCLUDING REMARKS

A. What Can Be Understood?

The concept of a competing droplet size–dependent interfacial tension and entropy of mixing gives a good qualitative understanding of the behavior of saturated droplet-type microemulsions (variation of average droplet size and interfacial tension with salt concentration). As, at a certain point, the preferred curvature changes sign, an inversion from an O/W to a W/O system is found as well.

Quantitatively the thermodynamic formalism as worked out in Sec. V leads to good agreement with experiments. In Sec. VII.B.1 it was shown that the variation of average droplet radius with the volume fraction of the droplets can be described quantitatively by the theory presented in Sec. V using the same set of "adjustable" parameters [Kc_0 and $(2K + \overline{K})$] that quantitatively describes the variation of the interfacial tension γ with droplet volume fraction. The value of Kc_0 that has been used to describe these data could also be verified, at least approximately, with an independent experiment using Eq. (49).

There is still good agreement between theory and experiment even when all but one parameter are fixed (Sec. VII.B.2). In that comparison, a molecular model for the variation of the preferred curvature with salt concentration was used. It was shown that the variation in the interfacial tension γ with salt concentration can be described well by the theory using the Gaussian bending elastic modulus as a free parameter. Using the same values of the Gaussian bending elastic modulus and keeping all other parameters fixed, the variation of average droplet radius with salt concentration is also described quite well by the theory.

Interestingly, both in Sec. VII.B.1, where a system composed of SDS, pentanol, cyclohexane, and water plus salt was compared to the theory, and in Sec. VII.B.2, where a system containing AOT, water plus salt, and linear alkanes of different chain lengths was evaluated, the Gaussian bending elastic modulus was found to be greater than zero. Although in the first situation this positive sign is hardly significant (due to the experimental error in K of at least 10%), it definitely is in the second situation. This positive sign is not to be expected for surfactant monolayers. We believe that this observation is a challenging motive for further experimental work on droplet-type microemulsions and theoretical work on surfactant monolayers at oil/water interfaces.

B. What Is Not Understood or Is Only Partly Understood?

A theory that predicts both global phase behavior and detailed properties of the phases as a function of experimentally adjustable quantities without any adjustable parameters still seems far away. Below we list a number of, in our opinion, important problems that need to be solved in order to be able to proceed further.

First of all, there still are a few subtleties in the theory of droplet-type microemulsions:

1. As discussed in Sec. VI following Eq. (38), it is to be expected that the surface charge density at the droplet interface depends on curvature owing to the fact that the Gibbs surface does not necessarily coincide with the surface of closest packing of the surfactant molecules. Indeed, strong indications for this effect were found [28] for systems with SDS as a surfactant and pentanol as a cosurfactant. However, for the AOT systems analyzed in Sec. VII.B.2, it was not necessary to invoke such an effect. We conclude, therefore, that the magnitude of this effect [that is, the value of ξ as defined following Eq. (38)] depends on the type (geometry) of surfactant being used and is (to a good approximation) absent from the type of AOT systems studied in Sec. VII.B.2. This suggests another challenging issue in a theoretical treatment of surfactant monolayers at oil/water interfaces, in addition to the relation between the type (geometry) of surfactant and the value of $\overline{K}$ as discussed above and at the end of Sec. VII.B.2.

2. A consistent treatment of finite size effects and fluctuations of the volume of the droplets (beyond capillary waves) is needed. Ideally such a theory should come up with a value of z'. In the comparison between theory and experiments (see Sec. VII), we find no indications for a value of z' being different from 0.

C. Questions Regarding Non-Droplet Phases

The emphasis in this work was on droplet-type microemulsions. As promised in Sec. IV, we now present a short discussion on non-droplet phases.

At the end of Sec. III we mentioned that at zero preferred curvature and low surfactant concentrations, a system consisting of a surfactant-rich middle phase coexisting with excess oil and water phases (Winsor III) is found. If the preferred curvature is altered (by temperature, salt, or cosurfactant), a continuous transition to a droplet-type microemulsion coexisting with an excess phase is observed.

If more and more surfactant is added to a Winsor III system, the surfactant-rich phase swells at the expense of the excess oil and water phases. From a certain point, a single surfactant-rich phase is found. Upon increasing the amount of surfactant even further, a first-order transition to a lamellar phase may be observed. In a special case, it has been shown that the coexistence region between the (bicontinuous) microemulsion phase and a lamellar phase was extended into the region where the surfactant-rich phase coexists with excess oil and water, leading to a four-phase equilibrium: water–lamellar phase–microemulsion phase–oil [58]. In Ref. 46, even a three-phase equilibrium, water–lamellar phase–oil, was observed, the bicontinuous microemulsion phase apparently being absent.

In order to be able to extend the approach described in this contribution beyond microemulsions with a droplet structure, the (in our opinion) most important question to be resolved is: What is the structure of the middle-phase microemulsion, and how does it contribute to the mixing entropy of the system? An answer to this question may also provide a (qualitative) explanation regarding the limited swellability of a middle-phase microemulsion, that is, the observation that the surfactant-rich phase remains in between excess oil and water in spite of the fact that the middle phase is probably bicontinuous.

Strey and coworkers [59] showed evidence that the structure of such a middle phase in a nonionic surfactant system resembles a "molten cubic" structure. This is a multiply connected structure, and this may indeed provide an answer to the question as to why a middle phase does not swell. It is not clear yet whether this structure is the same for all middle-phase microemulsions.

In a rather successful phenomenological (lattice) theory put forward by Andelman and coworkers [20], the middle-phase microemulsion is regarded as an ensemble of noninteracting monolayers. These monolayers are characterized by their persistence length [18], which in turn depends exponentially on the bending elastic modulus (in contrast to polymers, where this dependence is linear). The collective character of a middle-phase microemulsion is then taken into account by invoking an "entropy of mixing" comparable to the Bragg–Williams approach in Ising models, also referred to as "random mixing" (i.e., the probabilities of neighboring lattice sites containing water or oil are uncorrelated). As a result, the characteristic length in the system (the so-called dispersion size) is predicted to be somewhat smaller than the persistence length. In this language, at constant interfacial area, the swellability of a middle-phase microemulsion is determined by the dispersion size.

A related question is: Why is a lamellar phase also sometimes found to exhibit a limited swellability [46,58]? Is this limited swellability caused by defects? Or is it necessary to invoke attractions between the surfactant monolayers? If attractions are responsible, classical van der Waals attraction is not a very good candidate, as it is found that the thickness of the water and oil layers is so great (several tens of nanometers) and the Hamaker constant for O/W layers is so small that van der Waals attraction seems to be too weak.

ACKNOWLEDGMENTS

The contribution of WKK to the work presented here was made possible by a fellowship of the Royal Netherlands Academy of Arts and Sciences.

REFERENCES

1. J. H. Schulman and E. G. Cockbain, Trans. Faraday Soc. 36:551 (1940).
2. T. P. Hoar and J. H. Schulman, Nature 152:102 (1943).
3. J. E. Bowcott and J. H. Schulman, Z. Electrochem. 59:283 (1955).
4. G. J. Verhoeckx, P. L. de Bruyn, and J. Th. G. Overbeek, J. Colloid Interface Sci. 119:409 (1987).
5. W. K. Kegel, G. A. van Aken, M. N. Bouts, H. N. W. Lekkerkerker, J. Th. G. Overbeek, and P. L. de Bruyn, Langmuir 9:252 (1993).
6. H. Saito and K. Shinoda, J. Colloid Interface Sci. 32:647 (1970).
7. P. A. Winsor, Trans. Faraday Soc. 44:376 (1948).

8. P. A. Winsor, Solvent Properties of Amphiphilic Compounds, Butterworth, London, 1954, pp. 7, 57–60, 68–71, 190.
9. P. A. Winsor, Chem. Rev. 68:1 (1968).
10. L. R. Angel, D. F. Evans, and B. W. Ninham, J. Phys. Chem. 87:538 (1983).
11. M. L. Robbins, in Micellization, Solubilization and Microemulsions, Vol. 2, (K. L. Mittal, ed.), Plenum, New York, 1977, p. 713.
12. J. Th. G. Overbeek, Kon. Ned. Akad. Wetenschap. Proc. B 89:66 (1986).
13. W. Helfrich, Z. Naturforsch. 28c:693 (1973).
14. D. A. Huse and S. Leibler, J. Phys. (France) 49:605 (1988).
15. J. Th. G. Overbeek in Surfactants in Solution, Vol. 11 (K. L. Mittal and D. O. Shah, eds.), Plenum, New York, 1991, p. 3.
16. P. L. de Bruyn, J. Th. G. Overbeek, and G. J. Verhoeckx, J. Colloid Interface Sci. 127:244 (1989).
17. Y. Talmon and S. Prager, Nature 267:333 (1977).
18. P. G. de Gennes and C. Taupin, J. Phys. Chem. 86:2294 (1982).
19. J. Jouffroy, P. Levinson, and P. G. de Gennes, J. Phys. (France) 43:1241 (1982).
20. D. Andelman, M. E. Cates, D. Roux, and S. A. Safran, J. Chem. Phys. 87:7229 (1987).
21. J. C. Wheeler and B. Widom, J. Am. Chem. Soc. 90:3064 (1968).
22. B. Widom, J. Chem. Phys. 84:6943 (1986).
23. Y. Levin and K. A. Dawson, Phys. Rev. A 42:1976 (1990).
24. B. Widom, Ber. Bunsenges. Phys. Chem. 100:242 (1996).
25. G. Gompper, Ber. Bunsenges. Phys. Chem. 100:264 (1996).
26. J. Th. G. Overbeek, Prog. Colloid Polym. Sci. 83:1 (1990).
27. W. K. Kegel and H. Reiss, Ber. Bunsenges. Phys. Chem. 100:300 (1996). Erratum: 101:1963 (1997).
28. J. Th. G. Overbeek, G. J. Verhoeckx, P. L. de Bruyn, and H. N. W. Lekkerkerker, J. Colloid Interface Sci. 119:422 (1987).
29. D. C. Morse and S. T. Milner, Europhys. Lett. 26:565 (1994).
30. K. M. Palmer and D. C. Morse, J. Chem. Phys. 105:11147 (1996).
31. H. Reiss, W. K. Kegel, and J. Groenewold, Ber. Bunsenges. Phys. Chem. 100:279 (1996).
32. J. Th. G. Overbeek, Faraday Disc. Chem. Soc. 65:7 (1978).
33. H. Reiss and W. K. Kegel, J. Phys. Chem. 100:10428 (1996).
34. H. Reiss, W. K. Kegel, and J. L. Katz, Phys. Rev. Lett. 78:4506 (1997).
35. M. E. Fisher, Physics 3:255 (1967).
36. A. Dillman and G. E. A. Meier, J. Chem. Phys. 94:3872 (1991).
37. V. I. Kalikmanov and M. E. H. van Dongen, Phys. Rev. E 51:4391 (1995).
38. W. Cai, T. C. Lubenski, P. Nelson, and T. Powers, J. Phys. II (France) 4:931 (1994).
39. M. Borkovec, J. Chem. Phys. 91:6268 (1989).
40. J. C. Eriksson and S. Ljunggren, Prog. Colloid Polym. Sci. 83:41 (1990).
41. S. T. Milner and T. A. Witten, J. Phys. (France) 49:1951 (1988).
42. A. N. Stokes, J. Chem. Phys. 65:261 (1976).
43. H. N. W. Lekkerkerker, Physica A 159:319 (1989).
44. H. N. W. Lekkerkerker, Physica A 167:384 (1990).
45. J. Meunier, J. Phys. Lett. (France) 46:L-1005 (1985).
46. B. P. Binks, H. Kelay, and J. Meunier, Europhys. Lett. 16:53 (1991).
47. L. T. Lee, D. Langevin, J. Meunier, K. Wong, and B. Cabane, Prog. Colloid Polym. Sci. 81:209 (1990).
48. H. Kellay, J. Meunier, and B. P. Binks, Phys. Rev. Lett. 70:1485 (1993).
49. F. Sicoli and D. Langevin, J. Chem. Phys. 99:4759 (1993).
50. W. K. Kegel, I. Bodnar, and H. N. W. Lekkerkerker, J. Phys. Chem. 99:3272 (1995).
51. J. C. Eriksson and S. Ljunggren, Langmuir 11:1145 (1995).
52. M. Gradzielski, D. Langevin, and B. Farago, Phys. Rev. E 53:3900 (1996).
53. W. K. Kegel, Langmuir 13:873 (1997).
54. R. Aveyard, B. P. Binks, and J. Mead, J. Chem. Soc., Faraday Trans. 82:1755 (1986).

55. G. A. van Aken, A study of Winsor II microemulsion equilibria, Doctor's Thesis, University of Utrecht, 1990.
56. M. Almgren, R. Johansson, and J. C. Eriksson, J. Phys. Chem. 97:1993 (1993).
57. H. Kellay, Y. Hendrikx, J. Meunier, and B. P. Binks. J. Phys. II 3:1747 (1993).
58. W. K. Kegel and H. N. W. Lekkerkerker, J. Phys. Chem. 97:11124 (1993).
59. R. Strey, J. Winkler, and L. Magid, J. Phys. Chem. 95:7502 (1991).

3

Thermodynamics of Microemulsions II

Eli Ruckenstein
State University of New York at Buffalo, Buffalo, New York

I. INTRODUCTION

A. General Introduction

The goal of this chapter is to demonstrate that the thermodynamics of microemulsions presented by Ruckenstein in 1982 at the Lund meeting on surfactants in solution and published in Refs. 1 and 2, can provide insight with respect to some of the relevant experimental results gathered regarding microemulsions. The following experimental results are particularly meaningful in the present context:

1. Schulman and coworkers [3–7] observed that isotropic and optically transparent dispersions of oil in water or water in oil are formed spontaneously in the presence of a surfactant and a cosurfactant such as an alkyl alcohol. Using low-angle X-rays, light scattering, ultracentrifugation, electron microscopy, and viscosity measurements, they concluded that the dispersed phase consists of almost uniform spherical droplets with diameters ranging between 8 and 80 nm.

2. Winsor [8,9] found that in ternary nonionic surfactant–oil–water phase diagrams there are three types of systems. In Winsor's type I systems, an oil-in-water (O/W) microemulsion coexists with excess oil, while in Winsor II systems a water-in-oil (W/O) microemulsion coexists with excess water. In both types of systems, almost spherical globules are dispersed in the continuous phase. In Winsor III systems, a microemulsion coexists with both excess phases, and one can no longer identify dispersed and continuous phases. For given amounts of the components, the type I system is present at relatively low temperatures; at intermediate temperatures, the type III system is generated; and, finally, at relatively high temperatures, the type II system is formed. Similar observations have been made with more complex systems that contained oil, water, surfactant, cosurfactant, and salt [10,11]. Indeed, if the temperature, the cosurfactant content, or the ionic strength increases, the system passes from an O/W microemulsion in equilibrium with excess oil to a middle-phase microemulsion in equilibrium with both excess phases and finally to a W/O microemulsion in equilibrium with excess water.

Numerous attempts have been made to explain the above observations, and before presenting our 1982 thermodynamic treatment, it is appropriate to review them. Ruckenstein and Chi [12] suggested a drop model for the microemulsion and derived an expression for its free energy that contained the interfacial free energy of the droplets, the entropy of dispersion of the globules in the continuous phase and the double layer, and van der Waals interactions between globules. A lattice model was used to derive approximate expressions for the entropy of dispersion, in which the size difference between globules and the molecules of the continuous phase was taken into account. They could thus predict the existence of a radius of the globules for which the free energy is minimum.

Similar attempts were made by Likhtman et al. [13] and Reiss [14]. Reference 13 employed the ideal mixture expression for the entropy and Ref. 14 an expression derived previously by Reiss in his nucleation theory. These authors added the interfacial free energy contribution to the entropic contribution. However, the free energy expressions of Refs. 13 and 14 do not provide a radius for which the free energy is minimum. An improved thermodynamic treatment was developed by Ruckenstein [15,16] and Overbeek [17] that included the chemical potentials in the expression of the free energy, since those potentials depend on the distribution of the surfactant and cosurfactant among the continuous, dispersed, and interfacial regions of the microemulsion. Ruckenstein and Krishnan [18] could explain, on the basis of the treatment in Refs. 15 and 16, the phase behavior of a three-component oil–water–nonionic surfactant system reported by Shinoda and Saito [19].

A useful suggestion was made by Robins [20] and Miller and Neogi [21]. Robins explained qualitatively the behavior of microemulsions in terms of the bending stress at the interface of the globules, and Miller and Neogi included the bending stress in the thermodynamic treatment of Refs. 15–18. A different kind of droplet model was developed by Safran and his colleagues [22–25], one in which the bending free energy of the interface and the entropy of dispersion of the globules in the continuous phase were assumed to be the main contributions to the free energy. For the bending free energy, they used an expression suggested by Helfrich [26], while for the entropy they employed the Flory–Huggins expression. Of particular interest are their studies of the effect of fluctuations. Regarding the middle phase, Scriven [27] suggested a rigid bicontinuous representation, while Talmon and Prager [28] described it as a random interspersion of oil and water domains generated by a Voronoi tessellation and derived an expression for the entropy of the system. Using a somewhat simpler representation than that of Talmon and Prager and introducing additional terms in the free energy, Jouffroy et al. [29] examined the phase behavior.

Finally, one word about the lattice theories of microemulsions [30–36]. In these models the space is divided into cells in which either water or oil can be found. This reduces the problem to a kind of lattice gas, for which there is a rich literature in statistical mechanics that could be extended to microemulsions. A predictive treatment of both droplet and bicontinuous microemulsions was developed recently by Nagarajan and Ruckenstein [37], which, in contrast to the previous theoretical approaches, takes into account the molecular structures of the surfactant, cosurfactant, and hydrocarbon molecules. The treatment is similar to that employed by Nagarajan and Ruckenstein for solubilization [38].

The emphasis of the present chapter is on the droplet model, and the methodology used in the treatment is a thermodynamic one. While the information obtained is incomplete, it will be shown that thermodynamics can provide understanding regarding the phase behavior of microemulsions, with respect to both the equilibrium between a microemulsion and an excess dispersed phase and the transition to a middle-phase microemulsion in equilibrium with both excess phases.

B. Thermodynamic Description of a Microemulsion

From the point of view of traditional thermodynamics, a microemulsion is a multicomponent mixture formed of oil, water, surfactant, cosurfactant, and electrolyte. There is, however, a major difference between a conventional mixture and a microemulsion. In the former case, the components are mixed on a molecular scale, while in the latter, oil or water is dispersed as globules on the order of 10–100 nm in diameter in water or oil. The surfactant and cosurfactant are mostly located at the interface between the two phases but are also distributed at equilibrium between the two media. In conventional mixtures, the sizes of the component species are fixed. In the case of microemulsions, the sizes of the globules are not given but are provided by the condition of thermodynamic equilibrium.

The thermodynamics of microemulsions has to account for the entropy of dispersion of the globules in the continuous medium, for the free energy of formation of the interface between the two media of the microemulsion, for the interactions among globules, and for the distribution at equilibrium of the species between the two media of the microemulsion and their interface. It is convenient to split the Helmholtz free energy F of the entire system into two terms. One of them, F_0, is the free energy of a dispersion of fixed globules without interactions among them (called in what follows a frozen noninteracting system), and the other, ΔF, is the free energy due to the entropy of dispersion of the globules in the continuous phase and to their interactions. This decomposition allows us to employ the Gibbs thermodynamics for the first contribution and conventional statistical mechanics for the second contribution.

II. SINGLE-PHASE MICROEMULSIONS

A. Basic Equations

Let us consider that the microemulsion contains spherical globules of uniform size. Their dispersion in the continuous medium is accompanied by an increase in the entropy of the system. As noted in Sec. I, the Helmholtz free energy f per unit volume of microemulsion ($f=F/V$, where V is the volume of the microemulsion) will be written as the sum between a frozen noninteracting free energy f_0 and a free energy Δf due to the entropy of dispersion of the globules in the continuous phase and to their interactions:

$$f = f_0 + \Delta f \tag{1}$$

Gibbs thermodynamics provides the following expression for df_0:

$$df_0 = \gamma \, dA + C_1 \, dc_1 + C_2 \, dc_2 + \sum \mu_i \, dn_i - p_2 \, d\phi - p_1 \, d(1 - \phi) \tag{2}$$

where γ is the interfacial tension; C_1 and C_2 are bending stresses associated with the curvatures c_1 and c_2, respectively; A is the interfacial area per unit volume between the two media of the microemulsion; μ_i and n_i are the chemical potential (at pressure p_1 for the species present in the continuous phase and at pressure p_2 for those present in the dispersed phase) and the number of molecules of species i per unit volume respectively; and ϕ is the volume fraction of the dispersed phase (which includes the surfactant and cosurfactant molecules present at the interface of the globules). It should be noted that p_1 represents the pressure in the continuous phase of the frozen noninteracting system, which differs from the pressure to which the continuous phase of the microemulsion is subjected.

For spherical globules, $c_1 = c_2 = 1/r$ and $C_1 = C_2 = C/2$, and Eqs. (1) and (2) lead to

$$df = \gamma \, dA + C \, d\left(\frac{1}{r}\right) + \sum \mu_i \, dn_i - p_2 \, d\phi - p_1 \, d(1 - \phi) + d\Delta f \tag{3}$$

where r is the radius of the globule (which includes the surfactant and cosurfactant layer adsorbed on its surface).

The equilibrium state of a microemulsion is completely determined by n_i, the temperature T, and pressure p. Therefore, the values of r and ϕ emerge from the condition that f be a minimum with respect to r and ϕ. Since for spherical globules

$$A = 3\phi/r \tag{4}$$

one obtains

$$\gamma = \frac{r^2}{3\phi} \left(\frac{\partial \Delta f}{\partial r}\right)_\phi - \frac{C}{3\phi} \tag{5}$$

and

$$p_2 - p_1 = \left(\frac{\partial \Delta f}{\partial \phi}\right)_r + \frac{r}{\phi}\left(\frac{\partial \Delta f}{\partial r}\right)_\phi - \frac{C}{r\phi} \tag{6}$$

An additional relationship between p_2 and p_1 is provided by the mechanical equilibrium condition between microemulsion and environment. Considering a variation dV of the volume V of the microemulsion at constant $N_i(= n_i V)$ and T, one can write

$$\gamma \, d(AV) + VCd(1/r) - p_2 \, d(V\phi) - p_1 \, d(V(1 - \phi)) + d(V\Delta f) - pdV_e = 0 \tag{7}$$

Since the variation in the volume of the environment dV_e is equal to $-dV$, combining Eqs. (5)–(7), one obtains

$$\frac{3\phi}{r}\gamma - (p_2 - p_1)\phi + (p - p_1) + \Delta f = 0 \tag{8}$$

Equations (6) and (8) yield

$$p_2 = p + \Delta f + (1 - \phi)\left(\frac{\partial \Delta f}{\partial \phi}\right)_r - \frac{C}{\phi r} + \frac{r}{\phi}\left(\frac{\partial \Delta f}{\partial r}\right)_\phi \tag{9}$$

and

$$p_1 = p + \Delta f - \phi\left(\frac{\partial \Delta f}{\partial \phi}\right)_r \tag{10}$$

Equation (10) demonstrates that the pressure p_1 in the continuous medium of the frozen noninteracting system (the system of free energy f_0) differs from the pressure p, and that the pressure p is equal to the sum between the pressure p_1 and the osmotic contribution due to the free energy Δf. The pressure p acts both on the microemulsion and on the continuous medium of the microemulsion. A globule senses in its immediate vicinity the pressure p_1 of the frozen noninteracting system plus the osmotic pressure due to the free energy Δf. While occurring at a scale large compared to the size of a globule, the latter contribution is felt also at the scale of a globule because of the continuity of the continuous medium. My previous assertion [1,2], that the globules sense only the pressure p_1 must be revised.

It is also important to note that the conventional Laplace equation or the generalized one, which includes the bending stress, is no longer valid for a microemulsion. Equation (9) replaces the Laplace equation. It can be rewritten in a more revealing form by combining Eqs. (5) and (9):

$$p_2 - p = \frac{2\gamma}{r} - \frac{C}{3\phi r} + \Delta f + (1 - \phi)\left(\frac{\partial \Delta f}{\partial \phi}\right)_r + \frac{r}{3\phi}\left(\frac{\partial \Delta f}{\partial r}\right)_\phi \qquad (11)$$

In addition to the first two terms on the right-hand side, which are present in the case of a single droplet in a continuum, Eq. (11) contains terms that are due to the entropy of dispersion of the globules in the continuous medium and to the energy of interactions among globules.

B. An Alternative Derivation of the Basic Equations

From the point of view of traditional thermodynamics, a microemulsion is a multicomponent mixture in thermodynamic equilibrium. The change at constant temperature of the Helmholtz free energy of such a system can therefore be written as

$$dF = \sum \mu_i^* dN_i - p\, dV \qquad (12)$$

where μ_i^* and N_i are the chemical potential in microemulsion and the number of moles of species i, respectively, and V is the volume of the microemulsion.

Let us now represent a microemulsion as a dispersion of globules of water in oil or of oil in water, with the surfactant and cosurfactant distributed at equilibrium among the dispersed and continuous media of the microemulsion and their interface. Assuming the globules to be spherical and of uniform radius, one can write the following expression for dF:

$$dF = \gamma\, d(AV) + CV\, d\left(\frac{1}{r}\right) + \sum \mu_i\, dN_i - p_2\, d(V\phi) - p_1\, d[V(1 - \phi)] + d(V\Delta f) \qquad (13)$$

Since Δf depends only on r and ϕ,

$$d\Delta f = \left(\frac{\partial \Delta f}{\partial r}\right)_\phi dr + \left(\frac{\partial \Delta f}{\partial \phi}\right)_r d\phi \qquad (14)$$

and Eq. (13) can be rewritten in the form

$$dF = \left(\frac{3\phi\gamma}{r} - p_1(1 - \phi) - p_2\phi + \Delta f\right) dV + \left[\frac{3\gamma V}{r} + p_1 V - p_2 V + V\left(\frac{\partial \Delta f}{\partial \phi}\right)_r\right] d\phi$$
$$+ \left[-\frac{3\phi\gamma V}{r^2} - \frac{CV}{r^2} + V\left(\frac{\partial \Delta f}{\partial r}\right)_\phi\right] dr + \sum \mu_i\, dN_i \qquad (15)$$

Equation (15), which is based on a model, must, however, be equivalent to Eq. (12), which is based on the traditional thermodynamics of a multicomponent mixture. For the free energy changes given by Eqs. (12) and (15) to be the same for arbitrary changes in the independent variables V, ϕ, r, and N_i, the respective coefficients multiplying dV, $d\phi$, dr, and dN_i must be equal. It should be emphasized, however, that ϕ, r, V, and N_i are independent quantities only if the pressure p_1 is not imposed. Indeed ϕ depends on the distribution at equilibrium of the moles N_i of species i between the two media of the microemulsion and their interface, hence on $N_i s$, V, the radius r, and the pressures p_1 and p_2. If pressure p_1 is imposed, then, because of the mechanical equilibrium condition [Eq. (16)], p_2 is a function of ϕ and r, and the N_i, V, ϕ, and r are not independent quantities. However, in the present case, p_1 is not imposed but is provided by the thermodynamic equilibrium condition (the minimum of the free energy F with respect to ϕ). Hence, ϕ, r, V, and N_i are independent variables. Consequently,

$$\frac{3\phi\gamma}{r} - p_1(1 - \phi) - p_2\phi + \Delta f = -p \tag{16}$$

$$\frac{3\gamma}{r} + (p_1 - p_2) + \left(\frac{\partial\Delta f}{\partial\phi}\right)_r = 0 \tag{17}$$

$$\frac{-3\phi\gamma}{r^2} - \frac{C}{r^2} + \left(\frac{\partial\Delta f}{\partial r}\right)_\phi = 0 \tag{18}$$

and

$$\mu_i(\text{at } p_1 \text{ or } p_2) = \mu_i^*(p) \tag{19}$$

Equations (16), (17), and (18) can be rearranged in the form of Eqs. (5), (9), and (10), respectively. It is clear that the treatments in Secs. II.A and II.B lead to the same results. Equation (19) indicates that the chemical potential in the microemulsion at pressure p is equal to the chemical potential of the corresponding component in the frozen noninteracting system.

C. A Relation Between γ and C

Equation (5) can provide the radius of a microemulsion for a given ϕ, while Eqs. (9) and (10) allow us to calculate p_2 and p_1. However, to perform such calculations, expressions for Δf, γ, and C are needed. Traditional thermodynamics can provide only some information about C, which can be obtained as follows. Integrating Eq. (3) at constant $1/r$, γ, μ_i, p_2, and p_1 yields

$$f = A\gamma + \sum n_i\mu_i - p_2\phi - p_1(1 - \phi) + \Delta f \tag{20}$$

which when differentiated leads to

$$df = A\,d\gamma + \gamma\,dA + \sum n_i\,d\mu_i + \sum \mu_i\,dn_i - p_2\,d\phi - p_1\,d(1 - \phi) - \phi \\ \times dp_2 - (1 - \phi)\,dp_1 + d(\Delta f) \tag{21}$$

The Gibbs–Duhem equations for the dispersed and continuous phases have the forms

$$\sum n_i^d\,d\mu_i - \phi\,dp_2 = 0 \tag{22}$$

and

$$\sum n_i^c\,d\mu_i - (1 - \phi)\,dp_1 = 0 \tag{23}$$

where the superscripts d and c stand for the dispersed and continuous media of the microemulsion, respectively.

Subtracting Eqs. (22) and (23) from Eq. (21) and using Eq. (3) yields

$$d\gamma + \sum \Gamma_i \, d\mu_i = \frac{rC}{3\phi} \, d\left(\frac{1}{r}\right) \equiv -\frac{C}{3\phi r} \, dr \tag{24}$$

where

$$\Gamma_i = \frac{n_i - (n_i^d + n_i^c)}{A}$$

is the surface excess of component i. Equation (24) leads to the expressions

$$C = \frac{3\phi}{r} \left(\frac{\partial \gamma}{\partial(1/r)}\right)_{\mu_i} \equiv -3\phi r \frac{\partial \gamma}{\partial r} \tag{25a}$$

and

$$\frac{\partial \Gamma_i}{\partial r} = \frac{1}{3\phi r} \frac{\partial C}{\partial \mu_i} \tag{25b}$$

The more compact arrangement of surfactant molecules on a surface of lower curvature suggests that $\partial \gamma / \partial r < 0$ and hence that C is a positive quantity. In addition, the surface excess is expected to increase with increasing radius, because the molecules can more easily pack on a surface of lower curvature. This means that $\partial C / \partial \mu_i$ is also a positive quantity.

III. MICROEMULSION COEXISTING WITH AN EXCESS DISPERSED PHASE

A. Phase Equilibria

The chemical potential of species i in the microemulsion can be defined as the partial derivative of F with respect to N_i at constant volume and temperature:

$$\mu_i^* = \left(\frac{\partial F}{\partial N_i}\right)_{V,T} \tag{26}$$

Equation (13), which is used to calculate μ_i^*, shows that F depends not only on N_i but also on r and ϕ. However, since $\partial F / \partial r = 0$ and $\partial F / \partial \phi = 0$, Eq. (26) can be replaced by

$$\mu_i^* = \left(\frac{\partial F}{\partial N_i}\right)_{V,T,r,\phi} \tag{27}$$

and Eq. (13) leads to

$$\mu_i^*(p) = \mu_i(\text{at } p_1 \text{ or } p_2) \tag{28}$$

Hence the chemical potential of any species of the microemulsion is equal to its chemical potential in the frozen noninteracting system, and this occurs because $\partial F / \partial r = 0$ and $\partial F / \partial \phi = 0$.

Since the composition of the excess dispersed phase is expected to be the same as that of the dispersed phase, the equality of their chemical potentials implies the equality of their pressures. The chemical potential in the dispersed phase is expressed in Eq. (3) at pressure p_2, while the pressure in the excess dispersed phase is equal to the external pressure p to which both the microemulsion and the excess phase are subjected. Consequently, at equilibrium,

$$p_2 = p \tag{29}$$

which, combined with Eq. (9), yields

$$\Delta f + (1 - \phi)\left(\frac{\partial \Delta f}{\partial \phi}\right)_r - \frac{C}{\phi r} + \frac{r}{\phi}\left(\frac{\partial \Delta f}{\partial r}\right)_\phi = 0 \tag{30}$$

Equations (5) and (30) can provide the values of r and ϕ for a microemulsion in equilibrium with an excess dispersed phase. Of course, expressions for γ, C, and Δf are needed to carry out such calculations.

Δf is calculated in what follows by neglecting the interactions among globules. Expressions for the entropy of dispersion of the globules in the continuous phase were derived by Ruckenstein and Chi [12] on the basis of a lattice model, assuming (as is usually done in this kind of model) that the volume of a site is equal to the volume v_c of a molecule of the continuous phase. Only upper and lower bounds could be obtained for the entropy of dispersion of spherical globules of radius r and volume fraction ϕ in the continuous phase. For the free energy Δf corresponding to the upper bound of the entropy, it was found that

$$\Delta f = -\frac{3\phi kT}{4\pi r^3}\left\{-\frac{1}{\phi}\ln(1 - \phi) + \ln\left[\frac{4\pi r^3}{3v_c}\left(\frac{1-\phi}{\phi}\right)\right]\right\} \tag{31}$$

and for that corresponding to the lower bound,

$$\Delta f = -\frac{3\phi kT}{4\pi r^3}\ln\left\{\frac{4\pi r^3}{3v_c}\left[\left(\frac{0.74}{\phi}\right)^{1/3} - 1\right]^3\right\} \tag{32}$$

where k is the Boltzmann constant.

The following empirical formula of Carnaham and Starling [39] for hard spheres, which is based on molecular dynamic simulations, will be used in this chapter:

$$\Delta f = -\frac{3\phi kT}{4\pi r^3}\left[1 - \ln\phi - \phi\frac{4 - 3\phi}{(1 - \phi)^2} + \ln\left(\frac{4\pi r^3}{3v_c}\right)\right] \tag{33}$$

Equation (33) provides values between the above two bounds (nearer to the upper bound).

Combining Eqs. (5) and (30) with Eq. (33) yields

$$\gamma = \frac{kT}{4\pi r^2}\left[\ln\left(\frac{4\pi r^3}{3v_c\phi}\right) - \frac{8\phi - 5\phi^2}{(1 - \phi)^2} + \phi\right] \tag{34}$$

and

$$C = \frac{3\phi kT}{4\pi r^2}\left[2\ln\left(\frac{4\pi r^3}{3v_c\phi}\right) + \frac{6\phi^2 - 5\phi - \phi^3}{(1 - \phi)^2}\right] \tag{35}$$

Equations (34) and (35) provide some information about γ and C. They are plotted in Figs. 1 and 2, which show that γ is very small and that the ratio $C/6\phi\gamma$ [in Eq. (11), γ appears as $2\gamma/r$ and C as $C/3\phi r$] is on the order of unity. The conclusion is that 2γ cannot be neglected compared to $C/3\phi$. Figures 1 and 2 also show that both γ and C become very small with increasing values of r and tend to zero for $r \to \infty$.

It is of interest to notice that for values of ϕ greater than about 0.5, the interfacial tension has negative values, at least in some range of radii. For values greater than 0.6, it becomes negative over the entire range of radii. Figure 3 shows a clearer dependence of γ on ϕ and r than Fig. 1. Since, as shown later (see Sec. III.B), the interfacial tension γ_∞ between the microemulsion and dispersed excess phase is smaller than γ, γ_∞ will also be negative when the former is negative. For this reason, the interface between the

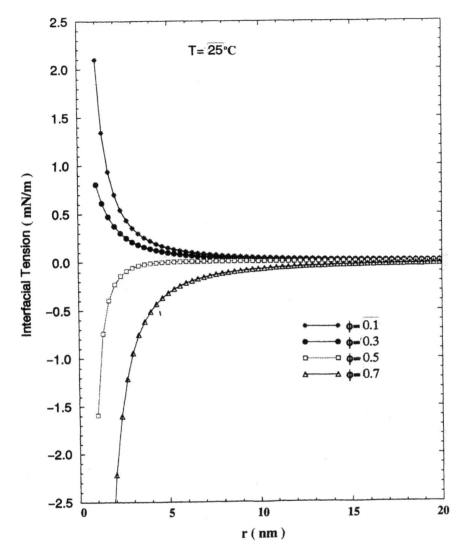

Figure 1 Interfacial tension γ versus radius r when a microemulsion coexists with an excess dispersed phase.

microemulsion and excess dispersed phase becomes unstable to thermal perturbations, spontaneous emulsification occurs, and the corresponding microemulsion cannot have physical reality. This may explain why for values of ϕ around 0.5 there is a transition to a middle-phase microemulsion that coexists with both excess phases.

The prediction of r and ϕ as a function of the natures of surfactant, cosurfactant, and hydrophobic phase requires the derivation of expressions for γ and C that should account for, among other things, their dependence on the curvature. Such expressions are not yet available.

It is of interest to note that, because at equilibrium $p_2 = p$, in a single-phase microemulsion the pressure p_2 in the globules is expected to be lower than that in the

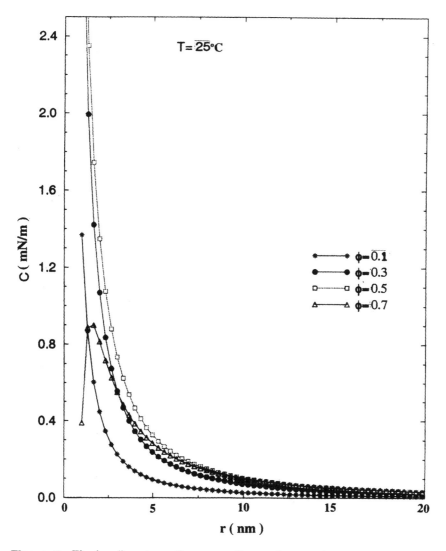

Figure 2 The bending stress C versus radius r when a microemulsion coexists with an excess dispersed phase.

continuous phase. Combining Eqs. (5) and (9) yields

$$p_2 - p = \frac{3\gamma}{r} + \Delta f + (1 - \phi)\left(\frac{\partial \Delta f}{\partial \phi}\right)_r$$

Because γ must be positive, $p_2 - p$ can become negative only if $\Delta f + (1 - \phi)\,\partial \Delta f/\partial \phi$ is sufficiently negative. The entropy of dispersion provides a negative contribution to Δf. When Δf is dominated by the entropy of dispersion, $\partial \Delta f/\partial \phi$ is expected to be negative for not too large values of ϕ, because the number of globules increases with increasing ϕ, and this increases the entropy of dispersion. At large values of ϕ, the excluded volume increases, which decreases the entropy. As long as the effect of the number of globules provides the main effect, $\partial \Delta f/\partial \phi < 0$.

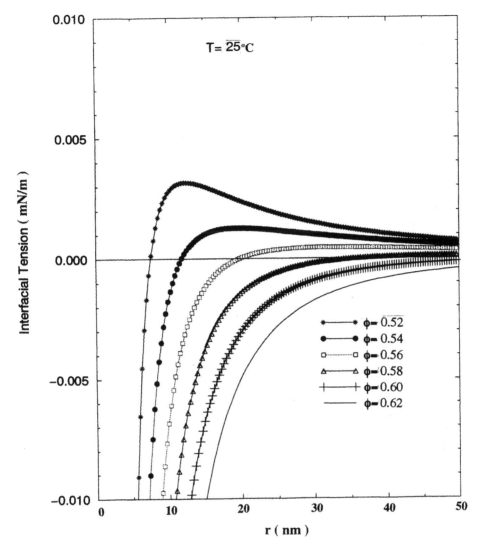

Figure 3 Interfacial tension γ versus radius r when a microemulsion coexists with an excess dispersed phase, for ϕ between 0.5 and 0.6.

B. Interfacial Tension Between Microemulsion and Excess Dispersed Phase

Integrating Eq. (24) at constant chemical potentials, one obtains

$$\gamma_\infty - \gamma = \int_r^\infty \frac{rC}{3\phi} \, d\left(\frac{1}{r}\right) \equiv - \int_r^\infty \frac{C}{3\phi r} \, dr \tag{36}$$

where, because the chemical potentials of the surfactants are the same at the surface of the globules and the interface between microemulsion and excess phase, γ_∞ can be considered the interfacial tension between microemulsion and excess dispersed phase. As already noted, C is a positive quantity. For this reason, $\gamma > \gamma_\infty$.

In Eq. (36), the quantity γ represents the interfacial tension at the surface of globules of radius r and is given by Eq. (34). Equation (35) cannot, however, be used in Eq. (36) for the bending stress C, because it does not involve in its derivation the restriction of constant chemical potentials on which Eq. (36) is based.

IV. MICROEMULSION COEXISTING WITH BOTH EXCESS PHASES

A microemulsion can coexist with an excess continuous phase when the chemical potential of the continuous phase of the microemulsion becomes equal to the chemical potential of the excess continuous phase. Since the chemical potential in the continuous phase of the microemulsion is equal to that of the frozen noninteracting system, which is expressed at pressure p_1, and that of the excess continuous phase is expressed at the external pressure p, the equality of the chemical potentials implies that

$$p_1 = p \tag{37}$$

Combining Eqs. (1) and (37) yields

$$\Delta f - \phi\left(\frac{\partial \Delta f}{\partial \phi}\right)_r = 0 \tag{38}$$

Using Eq. (33) for Δf, one can conclude that the radius of the globules diverges when an excess continuous phase coexists with the microemulsion. In addition, using Eq. (33) for Δf and Eq. (35) for C, Eq. (6) leads to

$$p_2 - p_1 = \frac{3\phi kT}{4\pi r^3}\left(\frac{1 + \phi + \phi^2 - \phi^3}{(1 - \phi)^3}\right) \tag{39}$$

Equation (39) shows that for $r \to \infty$,

$$p_2 - p_1 = 0 \tag{40}$$

In other words, the excess continuous phase cannot coexist alone with the microemulsion, since the equality $p_1 - p = 0$ implies that $p_2 = p_1 = p$, hence the coexistence of the microemulsion with both excess phases.

The preceding condition of thermodynamic equilibrium implies that the curvature of the dispersed phase becomes zero at the transition from two to three phases. In the middle-phase microemulsion, the pressures p_1 and p_2 fluctuate in time and space because of the instability of the interface between the two media (see below), and intuition suggests that Eqs. (37) and (40) be replaced with their average, hence that the condition of zero curvature be replaced with the condition of mean (with respect to time) average curvature.

Two structures are compatible with a zero average curvature. One of them is planar, and the other is bicontinuous [40,41]. A rigid bicontinuous structure for the middle phase was postulated by Scriven [27]. In contrast, in the present treatment, the zero average curvature is almost a natural result of the condition of thermodynamic equilibrium. Because of the extremely low interfacial tension, (Eq. (34) indeed suggests that the interfacial tension is extremely low in this case), the interface between the two media is extremely unstable to thermal perturbations; consequently, the interface cannot be rigid but probably has a complex oscillating, turbulent behavior accompanied by breakup and coalescence. The present considerations explain the large fluctuations observed experimentally [42] near this point and why this behavior is similar to that existing near a critical point.

V. CONCLUSIONS

1. General equations have been derived that can be used for single-phase microemulsions to calculate the dependence of the radius of the globules on the volume fraction of the dispersed phase, and for two-phase systems to calculate the values of both r and ϕ at equilibrium. Of course, thermodynamics provides only general equations among various quantities and is incomplete. Expressions for the interfacial tension and bending stress at the surface of the globules are needed to make predictions.
2. The phase equilibria between a microemulsion and the excess dispersed phase or both excess phases can be explained in terms of pressures.
3. The Laplace equation is no longer valid for a microemulsion. The new equation contains terms that are due to the entropy of dispersion of the globules in the continuous phase and to the energy of interaction among globules.
4. The curvature becomes zero at the transition to a microemulsion in equilibrium with both excess phases. This result is compatible with a planar or bicontinuous structure with a nonrigid interface because of its instability to thermal perturbations.

REFERENCES

1. E. Ruckenstein, Ann. N.Y. Acad. Sci. *404*:224 (1983).
2. E. Ruckenstein, Fluid Phase Equilib. *20*:189 (1985).
3. T. P. Hoar and J. H. Schulman, Nature *152*:102 (1943).
4. J. H. Schulman and D. R. Riley, J. Colloid Sci. *3*:383 (1948).
5. W. Stoeckenius, J. H. Schulman, and L. M. Prince, Kolloid Z. *169*:170 (1960).
6. J. E. L. Bowcott and J. H. Schulman, Z. Electrochem. *59*:283 (1955).
7. J. H. Schulman and J. A. Friend, J. Colloid Sci. *4*:1988 (1971).
8. P. A. Winsor, *Solvent Properties of Amphiphilic Compounds*, Butterworth, London, 1954.
9. P. A. Winsor, Chem. Rev. *68*:1 (1968).
10. R. N. Healy, R. L. Reed, and D. G. Stenmark, Soc. Petrol. Eng. J. Trans. AIME *261*:147 (1976).
11. M. Bourel, C. Konkounis, R. Schecter, and W. Wade, Dispersion Sci. Tech. *1*:13 (1980).
12. E. Ruckenstein and J. C. Chi, Faraday Trans. II *71*:1690 (1975).
13. V. I. Likhtman, E. D. Shchukin, and P. A. Rehbinder, *Physico-Chemical Mechanics of Metals*, Izdatelstvo Akad. Nauk SSSR, Moscow, p. 182. English translation by Israel Program for Scientific Translations, 1964.
14. H. Reiss, J. Colloid Interface Sci. *53*:61 (1975).
15. E. Ruckenstein, Chem. Phys. Lett. *56*:518 (1978).
16. E. Ruckenstein, Faraday Disc. Chem. Soc. *65*:141 (1978).
17. J. Th. G. Overbeek, Faraday Disc. Chem. Soc. *65*:7 (1978).
18. E. Ruckenstein and R. Krishman, J. Colloid Interface Sci. *71*:321 (1979).
19. K. Shinoda and H. Saito, J. Colloid Interface Sci. *26*:70 (1968); *30*:258 (1969).
20. M. Robbins, in *Micellization, Solubilization and Microemulsions*, Vol. 2 (K. L. Mittal, ed.), Plenum, New York, 1977, p. 713.
21. C. A. Miller and P. Neogi, AIChE J. *26*:212 (1980).
22. S. A. Safran, J. Chem. Phys. *78*:2073 (1981).
23. S. A. Safran and L. A. Turkevich, Phys. Rev. Lett. *50*:1930 (1983).
24. S. A. Safran, L. A. Turkevich, and P. A. Pincus, J. Phys. Lett. *45*:L69 (1984).
25. S. A. Safran, I. Webman, and G. S. Grest, Phys. Rev. A *32*:506 (1985).
26. W. Helfrich, Z. Naturforsch. *38*:6693 (1973).

27. L. E. Scriven, in *Micellization, Solubilization, and Microemulsions*, Vol. 2 (K. L. Mittal, ed.), Plenum, New York, 1977, p. 877.
28. Y. Talmon and S. Prager, J. Chem. Phys. *69*:2984 (1978).
29. J. Jouffroy, P. Levinson, and P. G. de Gennes, J. Phys. (Paris) *43*:1241 (1982).
30. B. Widom, J. Phys. Chem. *88*:6508 (1984).
31. B. Widom, J. Chem. Phys. *84*:6943 (1986).
32. M. Schick and W. H. Shih, Phys. Rev. *B34*:1797 (1986).
33. M. Schick and W. H. Shih, Phys. Rev. Lett. *59*:1205 (1987).
34. J. W. Halley and A. J. Kolan, J. Chem. Phys. *88*:3313 (1988).
35. R. G. Larson, J. Chem. Phys. *91*:2479 (1989).
36. G. Gompper and M. Schick, Phys. Rev. *A42*:2137 (1990).
37. R. Nagarajan and E. Ruckenstein, in *Equations of State for Fluids and Fluid Mixtures* (I.V. Sengers, ed.), (in press).
38. R. Nagarajan and E. Ruckenstein, Langmuir *7*:2934 (1991).
39. N. F. Carnaham and K. E. Starling, J. Chem. Phys. *51*:635 (1969).
40. E. R. Neovius, *Minimalflacken*, Frenkel, Helsingfors, Denmark, 1883.
41. M. A. Schwartz, *Gesammtte Mathematische Abhandlung*, Springer, Berlin, 1890.
42. A. M. Cazabat, D. Langevin, J. Meunier, and A. Pouchelon, Adv. Colloid Interface Sci. *16*:175 (1982).

4

Microscopic Models of Microemulsions

Gerhard Gompper
Max-Planck-Institut für Kolloid- und Grenzflächenforschung, Teltow, Germany

Michael Schick
University of Washington, Seattle, Washington

I. INTRODUCTION

The middle phase of ternary amphiphilic systems is quite intriguing because it is difficult to understand how the addition of a small amount of amphiphile to two immiscible fluids like oil and water can cause them to mix, producing a single isotropic phase. It is even more difficult to fashion a theory that can describe what the structure of this phase might be.

The theoretical description of microemulsions started in 1961 with the work of Schulman and Montagne [1], who showed that the interfacial tension of a flat oil/water interface decreases with increasing amphiphile density and would vanish if a saturated state of the amphiphile layer could be reached. In the late 1970s, it was realized that in addition to the small interfacial tension, entropy effects play an important role in the stabilization of microemulsions [2–4]. At about the same time, Scriven [5] suggested that microemulsions containing comparable amounts of oil and water could be bicontinuous, with two irregular networks of oil and water channels, separated by an amphiphilic monolayer, percolating through the whole system.

Much progress in the theoretical understanding of self-assembly, structure, and phase behavior of amphiphilic systems has been made since (for reviews, see Refs. 6–8). To a large extent, this is due to the careful analysis of statistical-mechanical models of microemulsions. Here, two different classes of models are used.

In the first class are microscopic models, in which the amphiphilic system is considered to be a ternary mixture, with one kind of molecule, the amphiphile, having complicated interactions with the others (and themselves). This does not mean that these models take into account the detailed chemical structure of an amphiphile, or even that of a water molecule, or their microscopic interactions. Rather, the focus is on generic features, which are common to many different species of amphiphiles and solvents. Therefore, the models are often very simplified. This has the advantage that they can be studied in detail by the established methods of statistical mechanics.

In the second class are interfacial models, in which it is assumed that essentially all the amphiphiles in the system are part of a monolayer or a bilayer. These models are restricted to the description of systems containing long-chain amphiphiles. Since they have only the local positions of the interfaces as the degrees of freedom, interfacial models are easier to analyze when the typical length scale of the oil and water regions gets very large, as it does for long-chain amphiphiles.

The two classes of models should be considered to be complementary descriptions of amphiphilic systems. Microscopic models can describe self-assembly and explicate the differences between weak and strong amphiphiles. Interfacial models describe very strong amphiphiles and emphasize their universal behavior.

Many of the phenomena in ternary amphiphilic systems also occur in mixtures of an AB diblock copolymer with A and B homopolymers (for a review, see Ref. 9). We emphasize the similarity of these systems by showing that the same models can be used to describe their properties.

II. MODELS

A large variety of "microscopic" models have been proposed to study the properties of ternary amphiphilic systems. None of these models is really microscopic, as they do not consider the detailed structure of the molecules involved or the full interactions between them. Rather, the molecules are taken as mostly structureless units. The interactions between molecules are also modeled by simple, short-range interaction potentials. Still they catch the most important part of the interaction in amphiphilic systems; i.e., the amphiphile reduces its energy by placing itself between oil and water. This conflicts with the tendency of oil and water to each aggregate separately and brings about competing interactions on different length scales.

A. Microscopic Lattice models

1. Widom–Wheeler Model

Probably the simplest of all lattice models of microemulsions is the Widom–Wheeler model [10,11]. In addition to its simplicity, its appeal lies in the fact that it is isomorphic to the Ising model of magnetism, a model well studied and readily simulated.

Consider any configuration of Ising spins on a cubic lattice. The configuration is mapped to one of the water, oil, and amphiphile molecules as follows. Two nearest-neighbor up spins σ_i, σ_j are interpreted as the presence of a water molecule on the bond ij connecting them. Similarly, two nearest-neighbor down spins are interpreted as an oil molecule on the bond. Finally, an up spin next to a down spin is an amphiphile whose hydrophilic head is the up spin and whose hydrophobic tail is the down spin. A spin on a given lattice site is part of the six different molecules that emanate from that lattice site. One easily sees that in such a mapping all amphiphiles are correctly oriented; e.g., there are no isolated amphiphiles completely surrounded by either oil or water.

Because each molecule is represented by two spins and neighboring molecules share a spin, interactions between two neighboring molecules translate into two-spin and three-spin interactions. Thus one finds an Hamiltonian of the form

$$H_W = -h \sum_i \sigma_i - J \sum_{\langle ij \rangle} \sigma_i \sigma_j - M \sum_{\langle ik \rangle}' \sigma_i \sigma_k - 2M \sum_{\langle ik \rangle}'' \sigma_i \sigma_k - L \sum_{\langle ijk \rangle} \sigma_i \sigma_j \sigma_k \qquad (1)$$

where the second sum is over all distinct pairs of nearest-neighbor spins, the third is over all district pairs of spins a distance of two lattice constants apart, the fourth is over all distinct pairs a distance $\sqrt{2}$ lattice constants apart, and the final sum is over all distinct contiguous triplets of spins. The interactions in this magnetic Hamiltonian are easily related to those between the oil, the water, and the head and tail of the amphiphile. One expects that $J > 0$, so that water prefers water and oil prefers oils, and that $M < 0$. This combination favors configurations like $+ + - - + + - -$, which represents a sequence of water, amphiphile, oil, amphiphile, water, etc. Without any calculation whatsoever, one knows from analogous models of magnetism that the Widom–Wheeler model will certainly be able to describe an oil-rich phase, a water-rich phase, and a disordered fluid phase as well as a lyotropic long-period lamellar phase.

2. Three-Component Model

The Widom–Wheeler model of water–oil–and amphiphile systems consists of only two kinds of variables, a water-like variable and an oil-like one. A water molecule consists of two water-like variables; an oil molecule, of two oil-like variables; and an amphiphile, of one of each. This guarantees that all amphiphiles are located at oil/water interfaces and are oriented correctly. In contrast, in the three-component model [12], there are three kinds of variables representing the three components, with the consequence that the amphiphiles may or may not be found at oil/water interfaces, depending on the relative energy gain to do so versus the entropy loss.

Because there are three independent components, the lattice version of this model is isomorphic to a spin 1 model of magnetism. If one considers any configuration of a spin 1 system, one interprets a spin up, or $S_i = 1$, as equivalent to the presence of water at site i, a spin down, or $S_i = -1$, as the presence of oil at that site, and $S_i = 0$ as the presence of an amphiphile. There is a molecule of some kind at every site. The Hamiltonian that describes the system with all possible nearest-neighbor pairwise interactions is

$$H_{\text{beg}} = - \sum_{\langle ij \rangle} [JS_iS_j + KS_i^2S_j^2 + C(S_i^2S_j + S_iS_j^2)] - \sum_i (HS_i - \Delta s_i^2) \tag{2}$$

which is the well-studied Blume–Emery–Griffiths model [13]. This Hamiltonian describes any three-component mixture and lacks any term indicating that the amphiphile is constructed so that it has a distinct head and tail that causes it to prefer to sit between oil and water so as to lower its energy. In this model, the amphiphile does not have a distinct head and tail. Nevertheless, the effect of having these two parts can be imitated by adding to the Hamiltonian a term that favors the location of an amphiphile between oil and water. A simple form of such a term is

$$H_{\text{amp}} = -L \sum_{\langle ijk \rangle} S_i(1 - S_j^2)S_k \tag{3}$$

where the sum is over all sets of three lattice sites in a row, and $L < 0$. One sees that this term reduces the energy of a configuration in which an amphiphile sits between an oil and a water molecule by the amount $-|L|$ and increases the energy of configurations in which an amphiphile sits between two water molecules or two oil molecules by $|L|$. The Hamiltonian of the three-component model is the sum of the two parts above,

$$H_{\text{three}} = H_{\text{beg}} + H_{\text{amp}} \tag{4}$$

Again, without calculation, one knows that the model will be able to describe water-rich, oil-rich, and long-period lamellar phases as in the Widom–Wheeler model. However, the additional interaction parameter L should make it possible to describe the variation in the behavior of amphiphiles from those that are strong, i.e., capable in low concentrations of solubilizing oil and water and markedly reducing the interfacial tension between them, and those that are weak. It turns out that this possibility is fully realized, and the description of the passage from weak to strong amphiphiles is one of the advantages of this model [14–16].

3. Vector Models

Like the three-component model above, vector models also begin with the generic Hamiltonian describing a mixture of three components. In magnetic language, this is simply H_{beg} of Eq. (2). But instead of simulating the effect of the interactions of oil and water with the head and tail of the amphiphile by means of a three-particle interaction, as in Eq. (3), vector models include orientational properties of the amphiphile explicitly by introducing an additional vector variable τ [17–22]. For simplicity, this vector is often restricted to point only along the directions of the underlying lattice. If one further restricts interactions to be between nearest neighbors and imposes symmetry under the simultaneous interchange of oil and water and of the two ends of the amphiphile, then the Hamiltonian is [21]

$$H = H_{\text{beg}} - \sum_{\langle ij \rangle} \sum_{v=2}^{4} [J_v Q_{v,ij} + K_v Q_{v,ij}^2] \tag{5}$$

where the interaction between the amphiphile and oil and water is governed by

$$Q_{2,ij} = S_i(1 - S_j^2)(\tau_j \cdot \mathbf{r}_{ji}) + (1 - S_i^2)S_j(\tau_i \cdot \mathbf{r}_{ij}) \tag{6}$$

with $\mathbf{r}_{ij}$ the lattice vector from site i to neighboring site j. Similarly, the interaction between amphiphiles parallel or antiparallel to one another is given by

$$Q_{3,ij} = (1 - S_i^2)(1 - S_j^2)(\tau_i \cdot \mathbf{r}_{ij})(\tau_j \cdot \mathbf{r}_{ji}) \tag{7}$$

and that between amphiphiles perpendicular to the vector joining them by

$$Q_{4,ij} = -(1 - S_i^2)(1 - S_j^2)(\tau_i \times \mathbf{r}_{ij}) \cdot (\tau_j \times \mathbf{r}_{ji}) \tag{8}$$

4. Polymer Chain Models

A further step in the direction of a more detailed and realistic description of the amphiphile is taken by models in which the amphiphile is modeled by a polymer-like chain of subunits that are water-like at one end and oil-like at the other end [23]. Such models have been studied both in the continuum [24–26] and on a lattice [27–31]. The advantage of these models is that they allow a calculation of the effect of the amphiphile chain length on the properties of microemulsions. These models have been studied mostly by computer simulations.

5. Charge-Frustrated Ising Model

The amphiphile chain length can also be taken into account on a somewhat more coarse-grained level by an effective interaction that binds head groups and tail groups together. Consider the head groups and the tail groups as independent molecules, A_n and B_m, respectively, that interact via some interaction potential $V_s(\mathbf{r})$. Obviously, the interaction has to be strong enough that the average separation of head and tail of a molecule is just the size l of the total amphiphilic molecule. It has been shown that this constraint of stoichiometry enforces an interaction potential of the form [32,33]

$$V_s(\mathbf{k}) = \frac{z_A z_B}{k^2} \qquad \text{for small } k \tag{9}$$

in Fourier space, where the "charges" z_A and z_B are given by

$$z_A = (3k_B T/4\pi \rho_s l^2)^{1/2} = -z_B \tag{10}$$

with ρ_s being the average amphiphile concentration. This form of the potential is valid to leading order in an expansion in powers of the wave vector $\mathbf{k}$. The potential (9) can be easily recognized as the Fourier transform of the familiar Coulomb potential.

The relation of a constraint of stoichiometry and the Coulomb form of the potential can be seen as follows [33]. Consider the spatial distributions $\rho_A(\mathbf{r})$ and $\rho_B(\mathbf{r})$ of heads and tails, respectively. Due to the constraint of stoichiometry, all possible configurations $\rho_A(\mathbf{r})$ and $\rho_B(\mathbf{r})$ must be identical on length scales $r \gg l$. This implies in Fourier space that

$$G_{AB}(k) \equiv V^{-1}\langle \rho_A(\mathbf{k})\rho_B(-\mathbf{k})\rangle \to 0 \qquad \text{for } k \ll l^{-1} \tag{11}$$

Thus, to leading order in a power series in the wave vector k,

$$G_{AB}(k) \sim k^2 \qquad \text{for small } k \tag{12}$$

On the other hand, for small deviations of the concentrations from a homogeneous state, $G_{AB}(k)$ is the inverse of the interaction term in the free energy functional; this leads immediately to the interaction potential (9). It is interesting to note the strong similarity with electrostatics: Accessible fluctuations of the densities of heads and tails are those that maintain stoichiometry on scales larger than the length l of the amphiphilic molecule. Similarly, accessible fluctuations of charged particles are those that maintain electroneutrality on length scales larger than the Debye screening length.

Stillinger [32] and Wu et al. [33] now argue that Eq. (9) contains the most relevant generic part of the intramolecular interaction of the amphiphile and therefore ignore all higher powers of k^2 in $V_s(\mathbf{k})$. Similar models have been studied in the context of diblock copolymer melts [34,35].

The lattice model for ternary amphiphilic systems can now be constructed [33,36]. A binary variable $s_i = \pm 1$ indicates whether a lattice site i is occupied by a polar or hydrophobic species, and a second variable $t_i = 1, 0$ indicates whether that species is part of a surfactant molecule or not. The Hamiltonian for these variables is

$$\mathcal{H}[s_i, t_i] = -\frac{1}{2}\sum_{ij} J_{ij} s_i s_j - \mu \sum_i s_i + \frac{z^2}{2}\sum_{i \neq j} V_{ij}(s_i t_i)(s_j t_j) - \mu_s \sum_i t_i \tag{13}$$

where $z \equiv z_A = -z_B$, μ_s is the chemical potential of the amphiphile, and $V_{ij} = V_s(\mathbf{r}_{ij})$ is the Fourier transform of the potential (9). The nearest-neighbor interaction $J_{ij} \geq 0$ expresses the preference of oil and water to phase separate.

Note that since the average amphiphile density ρ_s influences the strength z of the charges, but $\rho_s = \langle t_i \rangle$ in turn depends on the parameters μ_s and z of the model, a self-consistent procedure is unavoidable in a grand canonical ensemble to determine z, ρ_s, and μ_s at every point in the phase diagram [37]. Alternatively, a canonical ensemble can be used in which the total number of charges is fixed [33].

A similar, somewhat more complicated continuum version of the model has also been studied [32,38,39].

B. Ginzburg–Landau Models

The origin of the Ginzburg–Landau approach lies in the study of the thermal behavior near critical points, which is characterized by a set of universal critical exponents. One of the advantages of this approach is that many techniques that have been developed in this context can be applied to Ginzburg–Landau models of ternary amphiphilic systems as well.

1. Single-Order Parameter Model

The first step toward a Ginzburg–Landau description of oil–water–amphiphile mixtures was taken by Teubner and Strey [40], who noted that the scattering intensity in bulk contrast,

$$S_{\text{bulk}}(q) \sim [c_0 q^4 + g_0 q^2 + a_0]^{-1} \tag{14}$$

with $g_0 < 0$, describes the experimentally observed scattering data in the microemulsion phase extremely well. Furthermore, it can be calculated in the Ornstein–Zernike approximation from the free energy functional

$$\mathcal{F}_0[\Phi] = \int d^3 r [c_0 (\nabla^2 \Phi)^2 + g_0 (\nabla \Phi)^2 + a_0 \Phi^2] \tag{15}$$

The functional (15) determines via the Boltzmann weight $\exp(-\mathcal{F}_0[\Phi])$ the probability of each configuration of the order parameter field $\Phi(\mathbf{r})$ to occur in thermal equilibrium. In the ternary system under consideration, the order parameter field $\Phi(\mathbf{r})$ is identified with the local concentration difference of oil and water. The local amphiphile concentration is considered to be integrated out in this model [41].

Model (15) is not yet capable of describing any phase transitions, interfaces between different phases, or spatially inhomogeneous phases. Therefore, anharmonic terms have to be included in the free energy functional, which can then be written in the general form [42]

$$\mathcal{F}[\Phi] = \int d^3 r [c(\Phi)(\nabla^2 \Phi)^2 + g(\Phi)(\nabla \Phi)^2 + f(\Phi) - \mu \Phi] \tag{16}$$

The general form of the functions $f(\Phi)$, $g(\Phi)$, and $c(\Phi)$ can be inferred from the phase behavior and from the scattering intensities in the homogeneous phases. Since an oil-rich, a water-rich, and a microemulsion phase can coexist over a certain range of temperatures, $f(\Phi)$ has to have three local minima. The locations of these minima (Φ_o, Φ_m, Φ_w) are determined by the average oil and water concentrations in the three coexisting phases. Furthermore, since the scattering intensity in both the oil-rich and water-rich phases usually has a peak at wave vector $q = 0$, $g(\Phi)$ has to be positive for $\Phi \simeq \Phi_o$ and $\Phi \simeq \Phi_w$, while $g(\Phi_m) < 0$ is needed to give a peak at nonzero wave vector in the microemulsion phase. Finally, $c(\Phi)$ has to be nonnegative for all Φ and is often taken to be constant.

The amphiphile concentration seems not to appear explicitly in the free energy functional (16). This is not the case, in fact, as the functions f, g, and c depend parametrically on the average amphiphile concentration ρ_s. This dependence can be obtained by a derivation of the free energy functional from a more microscopic model.

2. Relation to Microscopic Lattice Models

Ginzburg–Landau models can be derived in a straightforward way from all microscopic lattice models of microemulsions. This has been done explicitly for the Widom model [43], for the three-component model [44], for vector models [45], and for the charge-frustrated Ising model [37]. In the case of the three-component model of Eqs. (2) and (3), the derivation shows, for example, that

$$g_0 = \frac{3}{2} J - 6\rho_s|L| \qquad \text{and} \qquad c_0 = -J/8 + 2\rho_s|L| \tag{17}$$

where $|L|$ is the amphiphile strength and ρ_s is again the amphiphile concentration. Thus, the parameter $g_0 = g(0)$ in Eqs. (15) and (16) decreases with increasing amphiphile strength and concentration.

The simplest results for the functions $f(\Phi)$, $g(\Phi)$, and $c(\Phi)$ have been obtained from the charge-frustrated Ising model [37], where one finds

$$c(\Phi) = c_0 \left(1 - \frac{1}{2} \Phi^2 \right)^2 \tag{18a}$$

$$g(\Phi) = g_0 + \frac{z_s l^2}{12} \Phi^2 \tag{18b}$$

$$f(\Phi) = a_0 \Phi^2 + \frac{1}{6} \Phi^4 + O(\Phi^6) \tag{18c}$$

where the coefficients are given by

$$c_0 = \frac{z_s l^4}{36}, \qquad g_0 = \beta J a^2 - \frac{z_s l^2}{6}, \qquad a_0 = 1 + z_s + 6\beta J \tag{19}$$

The parameter z_s in these equations is the fugacity of the amphiphiles, $z_s = \exp(\beta \mu_s)$. For balanced systems, $z_s = 2\rho_s$. With these results, the dependence of the parameters in the Ginzburg–Landau model on the experimental variables such as amphiphile concentration ρ_s and chain length l is now explicit.

Note that the dependence of the parameters g_0 and c_0 on the amphiphile concentration in Eqs. (17) and (19) is the same. The qualitative behavior as a function of the amphiphile chain length also agrees very well.

Until now, most of the work on Ginzburg–Landau models has been aimed at exploring their parameter space and obtaining an overview of their general properties.

3. Models with Several-Order Parameters

In some instances, the single-order parameter model (16) is not sufficient. This is the case when fluctuations of the amphiphile concentration play an important role, such as in the calculation of the scattering intensity in film contrast. A second scalar order parameter field, $\rho(\mathbf{r})$, which describes the deviation of the local amphiphile concentration from the average ρ_s, has to be included in the model. For a balanced system, the free energy functional then becomes [46–48]

$$\mathcal{F}[\Phi, \rho] = \mathcal{F}[\Phi] + \int d^3r[\alpha(\nabla\rho)^2 + \beta\rho^2] + \int d^3r[\gamma_1\rho\Phi^2 + \gamma_2\rho(\nabla\Phi)^2 + \gamma_3\rho\Phi\nabla^2\Phi] \quad (20)$$

where $\mathcal{F}[\Phi]$ is again the free energy functional (16). The terms in (20) that couple the two order-parameter fields favor a large concentration of amphiphile at the interface between oil and water (when the sign of the coupling constants γ is chosen correctly).

Ginzburg–Landau models that in addition to the scalar order parameter for the concentrations of oil, water, and amphiphile contain a vector order parameter for the amphiphile orientation have also been studied [41, 45, 49–53].

C. Other Microscopic Models

1. Hard-Sphere Models

In recent years, much progress has been made in understanding the thermal properties of simple liquids and colloidal systems [54]. Studies of these systems are often based on the density functional approach, in which the free energy is written as a sum of two contributions, one describing the hard-core repulsion and thus the steric hindrance of the particles, and a second, the long-range and possibly orientation-dependent interactions between them.

A similar approach can be used to construct a molecular model for microemulsions [55–57]. All three components—oil, water, and amphiphile—are taken to be hard spheres. The interactions, $W(r)$, between oil molecules and between water molecules are taken to be isotropic and attractive, and the interactive between oil and water, to be isotropic and repulsive. A Lennard-Jones or Yukawa form has been used for $W(r)$. The interaction between a surfactant molecule located at $\mathbf{r}_1$ and water molecule at $\mathbf{r}_2$ on the other hand, is anisotropic, with the dependence

$$V_{sw}(\mathbf{r}_{12}, \mathbf{n}) = W(r)\,\mathbf{n}\cdot\mathbf{r}_{12}/r_{12} \quad (21)$$

where $\mathbf{r}_{12} = \mathbf{r}_1 - \mathbf{r}_2$, and $\mathbf{n}$ is a unit vector that defines the orientation of the surfactant molecule. The interaction of a surfactant and an oil molecule is similarly $V_{so}(\mathbf{r}_{12}, \mathbf{n}) = -V_{sw}(\mathbf{r}_{12}, \mathbf{n})$. When the interactions are truncated at this level, amphiphile bilayers are found to be unstable [57]. To correct this problem, a quadrupolar interaction of the amphiphiles with water can be added to the potential.

Other models for ternary amphiphilic systems are based on mixtures of hard spheres and ellipsoids with Lennard-Jones interactions [58] or on mixtures of hard spheres and diatomic hard-sphere molecules [59]. Such models have been studied by molecular dynamics simulations.

2. Hybrid Model

A model that is somewhere in between the "hard-sphere model," where all three components are described as individual particles, and the Ginzburg–Landau approach, where all components are described by density fields, is the "hybrid model," where the oil and water molecules are described by density fields but the discrete nature of the surfactant molecules is retained [60].

To derive the effective Hamiltonian for this model, it is assumed that the interaction between the head of amphiphile molecule j and the water can be written as

$$\int d^3 r Q V_{ww}\left[\mathbf{r} - \left(\mathbf{r}_j + \frac{l}{2}\,\mathbf{n}_j\right)\right]\rho_w(\mathbf{r}) \tag{22}$$

where the center of mass of the amphiphilic molecule is located at $\mathbf{r}_j$, $\mathbf{n}_j$ is a unit vector describing the orientation of the molecule, l is its length, and Q measures the strength of the head–water interaction in units of the interaction between two water molecules. The interaction potential $V_{ww}(r)$ is assumed to be isotropic and of the exponential Yukawa form [60]. Similar expressions describe the interactions of the tails with water [with interaction potential $V_{ow}(r)$], of the heads with oil, and of the tails with oil [with interaction potential $V_{oo}(r)$] and also those between heads and heads, heads and tails, and tails and tails. All of these contributions are then expanded in a power series in the amphiphile length l, and the terms up to order l^2 are retained. This corresponds to the dipole approximation of electromagnetism. We want to consider here only the simplest case, where there is only one type of interaction,

$$V_{ww}(r) = V_{oo}(r) = -V_{ow}(r) \equiv V(r) \tag{23}$$

In this case, one finds the total Hamiltonian

$$\mathcal{H} = \mathcal{H}_{\text{bin}} + \mathcal{H}_{\text{A}} + \mathcal{H}_{\text{AA}} \tag{24}$$

where the first term describes the binary oil–water system, the second term the interactions of the amphiphile with oil and water, and the last term the direct interactions between amphiphiles. The binary oil–water system is described by the usual Ginzburg–Landau model (cf. Sec. II.B.1). The other contributions read

$$\mathcal{H}_{\text{A}} = Ql \sum_i \int d^3 r\, W(\mathbf{r} - \mathbf{r}_i)\mathbf{n}_i \cdot \nabla\phi(\mathbf{r}) \tag{25}$$

and

$$\mathcal{H}_{\text{AA}} = (Ql)^2 \sum_{i \neq j} (\mathbf{n}_i \cdot \nabla)(\mathbf{n}_j \cdot \nabla) W(\mathbf{r}_i - \mathbf{r}_j)$$
$$= (Ql)^2 \sum_{i \neq j} \left\{ \left[W''(r_{ij}) - \frac{1}{r} W'(r_{ij}) \right](\mathbf{n}_i \cdot \hat{\mathbf{r}}_{ij})(\mathbf{n}_i \cdot \hat{\mathbf{r}}_{ij}) + \frac{1}{r} W'(r_{ij})(\mathbf{n}_i \cdot \mathbf{n}_j) \right\} \tag{26}$$

where $W'(r) = dW(r)/dr$. Note that $\mathcal{H}_{\text{AA}}$ is again of the familiar form of the dipole–dipole interaction.

In the more general case of different attractive and repulsive interactions, the term $\mathcal{H}_{\text{AA}}$ looks more complicated [60] and is thus more difficult to interpret.

These expressions show the strong similarities between the vector models of Sec. II.A.3, the hard-sphere models of Sec. II.C.1, and the hybrid model. The main difference is whether the molecules are described as individual particles, which can move on a lattice or in the continuum, or by a density field.

3. Models of Surfactant Mono- and Bilayers

We have already mentioned in the Introduction that microemulsions containing long-chain amphiphiles can be described by interfacial models. These models are based on the curvature elasticity of the amphiphilic monolayer and thus contain as material parameters the bending rigidity and the saddle-splay modulus. These parameters have to be calculated from a more microscopic model. A somewhat similar problem occurs in the formation of micelles. In systems containing long-chain amphiphiles, the critical micelle concentration (cmc) is very small, so that the microscopic models described above cannot be expected to make reliable predictions. In both cases, chain packing effects play an important role.

Molecular models of chain packing [61–66] consider the amphiphile tails to be flexible polymer chains consisting of N monomers, with N varying typically from 2 to 50. The head groups are placed on a planar or curved surface with some prescribed area density. The allowed configurations of the chains have to satisfy the constraint that the segment density is constant in micelles or bilayers or does not exceed a maximum density for monolayers. By enumerating all possible configurations of the chains, the free energy can be calculated. Quantities like the interfacial tension or the bending rigidity then follow from the dependence of the free energy on the area density of the head groups or the radius of curvature of the head group surface.

Similar models have been studied for diblock copolymers [67–69], i.e., in the limit of large N.

The statistical mechanics of the surfactants within a monolayer or micelle is considerably simpler than the statistical mechanics of microemulsions, since thermal fluctuations are important only on the scale of a few molecules. Thus, models of monolayers or micelles can be very realistic [26] and have indeed been constructed for real surfactant and lipid molecules (see, e.g., Refs. 70–72).

III. RESULTS I: STATIC BEHAVIOR

A. Phase Behavior

Much of the phase behavior of binary and ternary amphiphilic systems has been reproduced by the models that we have described. Distinguishing which behavior is characteristic of generic multicomponent systems and which is particular to amphiphilic systems has been one of the successes of the microscopic approach, which is able to describe both kinds of systems.

For example, one of the behaviors that attracted much experimental attention is the characteristic progression from two-phase to three-phase to two-phase coexistence, i.e., the existence of a triple line that terminates at each end in a critical endpoint. This line is traversed in systems of nonionic amphiphiles by varying the temperature and in systems of ionic amphiphiles by varying the salt concentration. It is possible to decrease the length of the triple line by adding a fourth component [73] and thereby to approach a tricritical point.

The three-component model immediately makes clear that this behavior is generic [15,74]. What is unusual in, and specific to, the nonionic systems is that the triple line is traversed by varying the temperature. This is a consequence of the lower critical point that exists in these systems in addition to the usual upper critical point. That there is a lower critical point is related to hydrogen bonding, and a realistic phase diagram can be calculated only by taking this into account [75] (see Fig. 1).

The variation in phase behavior as the strength of the amphiphile is increased can also be followed in the three-component model [44]. One finds that for weak amphiphiles, characterized by a small negative value of L in Eq. (3) of Sec. II.A.2, there is a continuous transition to oil–water coexistence from a disordered phase that is not structured. As

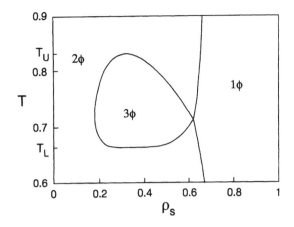

Figure 1 A vertical slice through the phase prism for equal oil and water concentrations. The three-phase region exists in the temperature interval T_L to T_U. (From Ref. 75.)

the strength of the amphiphile is increased, i.e., L made more negative, the disordered phase becomes structured as first the disorder line is crossed and then the Lifshitz line (compare Sec. III.C.1). The transition to oil–water coexistence becomes first-order, a tricritical point having been passed. Hence there is now three-phase coexistence between oil, water, and a structured disordered phase—a microemulsion. As the amphiphile is made stronger still, a lamellar phase appears and competes with the microemulsion in the phase space. Thermal fluctuations favor the disordered microemulsion over the more ordered lamellar phase, an effect shown most clearly in the simulations of the Ginzburg–Landau models [76]. However, the lamellar phase eventually pre-exempts the existence of the microemulsion. All of this is in complete accord with experiment.

While microscopic models do describe weak and moderately strong amphiphiles very well, they do not describe very strong ones, which are nearly insoluble in water. That is, the volume fraction of amphiphile in the microemulsion is rarely calculated to be less than 10% in these models. This is adequate for the short-chain nonionic amphiphiles C_iE_j but not for lipids.

The series of lyotropic phases that are characteristic of the binary water–amphiphile system are well described for the most part by Ginzburg–Landau models [50,51] and polymer chain models [29] (see Fig. 2). The most complicated lyotropic phase, the cubic gyroid phase, was the last phase to be determined theoretically, and it was found first [77] in models of long flexible diblock copolymers, as described in Sec. V.C.1. Quite recently, it has also been obtained in simulations of the short polymer chain model [78].

The spectacular swelling of the lamellar phase in some binary amphiphile–water systems [79] such as $C_{12}E_5$ has been reproduced in related diblock polymer systems [80–82], which are discussed in Sec. V.C.2.

In summary, the general trends of the phase behavior are given rather well by microscopic models. Although they describe generic amphiphiles for the most part, progress is being made in determining the effect of differences in the amphiphiles on phase behavior, differences such as in the number of head and tail groups [37]. One can foresee that the ability to predict the phase behavior of specific systems will eventually be achieved.

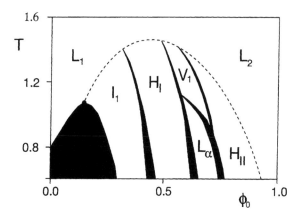

Figure 2 Phase diagram of a binary amphiphile–water mixture obtained from a Ginzburg–Landau model with a vector order parameter for the amphiphile orientation [50,51]. The phases L_1 and L_2 are micellar liquids, L_α is a lamellar phase, H_I and H_{II} denote hexagonal and inverse hexagonal phases, respectively, I_1 is an fcc crystal of spherical micelles, and V_1 is a simple cubic bicontinuous phase. (From Ref. 51.)

B. Structure and Topology of Microemulsions Between Lower and Upper Critical Endpoints

For systems containing oil, water, and a nonionic amphiphile, three-phase coexistence of the microemulsion with an oil-rich and a water-rich phase is possible only in a temperature interval $T_l < T < T_u$. At the lower critical endpoint T_l, the microemulsion and the water-rich phase become identical, while at the upper critical endpoint T_u, the microemulsion merges with the oil-rich phase [83,84].

Let us consider first how the approach to the critical endpoint (CEP) is described in the mean-field approximation, where the phase behavior is obtained simply by considering the minima of the free energy density $f(\Phi)$. Since we want to describe a system at three-phase coexistence, $f(\Phi)$ has to have three minima of equal depth. As the upper CEP is approached, the two minima describing the microemulsion and the oil-rich phase approach each other and merge at the CEP temperature T_u. Similarly, the minima of the microemulsion and the water-rich phase merge at the lower CEP T_l. Thus, in a series expansion in powers of Φ, the free energy density has to have the form

$$f(\Phi) = w(\Phi - \Phi_w)^2 \Phi^2 (\Phi - \Phi_o)^2 + \mu_s \Phi^2 \tag{27}$$

with $\mu_s = 0$. For simplicity, the location of the microemulsion minimum in Eq. (27) is assumed to remain unchanged by the oil–water asymmetry. To be specific, let us consider the approach to the upper CEP. The asymmetry is measured in this case by the dimensionless parameter $m_0 = |\Phi_o/\Phi_w|$. Note that $m_0 = 1$ for balanced systems with oil–water symmetry.

In the mean-field approximation, the CEP occurs at $m_0 = 0$. Since thermal fluctuations play an important role both in the microemulsion and near a critical point, Monte Carlo (MC) simulations and a variational approach have been used to study the variation of the microemulsion structure as a function of m_0 [85]. The calculation is carried out for a value of $g_0 \equiv g(0) = -1.2$, which corresponds to a system with strong (medium-

or long-chain) amphiphiles. In the variational calculation, the CEP is found to be located at $m_0 = 0.63$, in the MC simulations at $m_0 \simeq 0.75$; thus, in both cases thermal fluctuations cause a large shift of the CEP [85].

In order to characterize the microemulsion, the percolation threshold of the minority component (oil) and the geometry of the $\Phi = 0$ level surfaces, which can be identified with the amphiphile film separating the oil and water domains*, have been calculated. The percolation threshold is found in the MC simulations to occur at $m_0 = 0.765$, just before the CEP is reached along the triple line. This result is in good agreement with lattice model calculations [75] and with experiments on the system water–n-octane–$C_{12}E_5$ [86]. The geometry of the amphiphile film is characterized by the average mean curvature $\langle H \rangle$, the average of the mean curvature squared $\langle H^2 \rangle$, and the average Gaussian curvature $\langle K \rangle$. It is one of the advantages of (continuum) Ginzburg–Landau models compared to lattice models that quantities like the mean curvature squared are well-defined. The results are shown in Fig. 3. The data indicate that (1) there is an almost linear dependence of $\langle H \rangle$ on m_0 and (2) the Gaussian curvature $\langle K \rangle$ is negative for a balanced microemulsion (i.e., one with equal volume fractions of oil and water)—corresponding to a surface that is dominated by saddle-shaped configurations, but increases as the CEP is approached and changes sign at $m_0 \simeq 0.78$, again very close to the CEP. This behavior is in good agreement with experimental results for the system water–n-octane–$C_{12}E_5$ [86], where the data were found to be consistent with a linear temperature dependence of the two principal curvatures,

$$\langle c_1 \rangle \sim (T - T_u), \qquad \langle c_2 \rangle \sim (T - T_l) \tag{28}$$

This implies that $\langle H \rangle = \langle (c_1 + c_2)/2 \rangle \sim (T - \bar{T})$, where $\bar{T} = (T_u + T_l)/2$, and $\langle K \rangle = \langle c_1 c_2 \rangle \simeq \langle c_1 \rangle \langle c_2 \rangle \sim (T - T_l)(T - T_u)$. The linear temperature dependence of $\langle H \rangle$, with $\langle H \rangle = 0$ for balanced systems, and the quadratic temperature dependence of $\langle K \rangle$, with $\langle K \rangle = 0$ at both critical endpoints, is quite consistent with the Monte Carlo results if m_0 is assumed to vary linearly with temperature. The situation for $\langle H^2 \rangle$ is somewhat more complicated; the approximation $\langle (c_1 + c_2)^2 \rangle \simeq (\langle c_1 \rangle + \langle c_2 \rangle)^2 \sim (T - \bar{T})^2$ breaks down for small mean curvatures $H \simeq 0$, i.e., for $T \simeq \bar{T}$, since thermal fluctuations dominate in this regime. Therefore, one would expect $\langle H^2 \rangle$ to be roughly independent of m_0 for nearly balanced systems, again consistent with Fig. 3.

The breakup of the two percolating bicontinuous networks of oil and water regions (with an amphiphile film that is dominated by saddle-shaped configurations) into disconnected droplets near the critical endpoint can be seen very clearly in a sequence of typical configurations for different "temperatures" m_0, as shown in Fig. 4. Note that $\langle K \rangle \simeq 0$ does not necessarily imply the predominance of cylindrical configurations.

The value of the Gaussian curvature $\langle K \rangle$ is not a good measure of the connectivity of the amphiphile film, since a dilution of a system by a factor λ, with an unchanged topology of the amphiphile film, will reduce $\langle K \rangle$ by a factor λ^2. Thus, instead of $\langle K \rangle$ itself, the scaled quantity $\langle K \rangle V^2 \langle S \rangle^{-2}$ should be used to characterize the connectivity [76]. Here, V is the volume of the sample and $\langle S \rangle$ is the average area of the amphiphile film. This can be seen most easily from the following simple argument. The Gaussian curvature $\langle K \rangle$ has the dimension (length)$^{-2}$; therefore, it has to be multiplied by another length squared in order to obtain a dimensionless number. The relevant length scale is here the average domain

---

* For unbalanced systems, the choice of the level value, which determines the location of the amphiphile film, is not unique. See, e.g., Ref. [89] for a more detailed discussion.

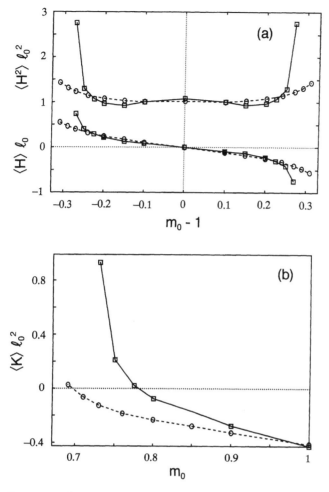

Figure 3 Averages of (a) the mean curvature H and the mean curvature squared H^2, and (b) the Gaussian curvature K, as a function of m_0 along the triple line. The length scale l_0 is on the order of the width of the oil/water interface. The squares connected by solid lines are the Monte Carlo data; the circles connected by dashed lines are the variational results. (From Ref. 85.)

size, $V/\langle S \rangle$. In systems containing long-chain amphiphiles of length , where essentially all amphiphile molecules are located at the microscopic oil/water interface, $l\langle S \rangle = \rho_s V$. Exact results [87] for level surfaces of Gaussian random fields* imply that

$$\langle K \rangle V^2 \langle S \rangle^{-2} = \frac{\pi^2}{8} \Theta \left(-\frac{\langle H \rangle^2}{\langle K \rangle} \right) \tag{29}$$

where $\Theta(x)$ is a universal scaling function, with $\Theta(0) = 1$. In Monte Carlo simulation, it is easier to calculate the Euler characteristic χ_E rather than the Gaussian curvature. These

* A Gaussian random field $\Phi(\mathbf{r})$ is defined by a probability distribution $\exp(-\mathcal{F}_0[\Phi])$ with a quadratic free energy functional $\mathcal{F}_0[\Phi]$. This applies in particular to the functional defined in Eq. (15).

two quantities are intimately related due to the Gauss–Bonnet theorem,

$$2\pi\langle\chi_E\rangle = \left\langle \int dS\ K \right\rangle \tag{30}$$

If the area fluctuations are sufficiently small, $2\pi\langle\chi_E\rangle \simeq \langle K\rangle\langle S\rangle$, so the calculation of $\langle\chi_E\rangle$ is equivalent to determining $\langle K\rangle$ in this case. A comparison of the Mote Carlo data for model (16) and the scaling function Θ for Gaussian random fields is shown in Fig. 5. The CEP of model (16) is located near $\langle H\rangle^2\langle S\rangle/2\pi\langle\chi_E\rangle \simeq -1.4$. It is also interesting to compare the value of $\Gamma \equiv \langle\chi_E\rangle V^2\langle S\rangle^{-3}$ for balanced microemulsions ($\Gamma \simeq 0.2$) with ordered bicontinuous ("plumber's nightmare") phases [76], such as the simple cubic P surface ($\Gamma = 0.314$) or the D surface with a diamond lattice structure ($\Gamma \simeq 0.284$). These numbers for Γ show that triple-periodic minimal surfaces have a considerably higher connectivity than balanced microemulsions.

In this context, it is interesting to note that it has been argued recently [88] that the $\Phi(\mathbf{r}) = 0$ surfaces of order-parameter configurations, which minimize the free energy functional (16), are (often) periodic minimal surfaces. The four best known triply-periodic minimal surfaces G, D, P, and $I-WP$ have been reproduced [88,89], and new candidates for minimal surfaces of high genus have been found [88].

C. Scattering Intensities

1. Scattering Intensity in Bulk Contrast

The scattering intensity in bulk contrast can be calculated easily in the Ornstein–Zernike approximation for all lattice [15, 90–92] and Ginzburg–Landau models. In the limit of wave vector $q \ll q_c = \pi/a$, one obtains in all cases the Teubner–Strey form

$$S_{\text{bulk}}(q) \sim [c_0 q^4 + g_0 q^2 + a_0]^{-1} \tag{31}$$

A Fourier transform of Eq. (31) for $g_0^2 < 4c_0 a_0$ gives the correlation function [40]

$$G(r) \sim e^{-r/\xi}\ \frac{\sin(q_0 r)}{q_0 r} \tag{32}$$

with the wave vector

$$q_0 = \left(\frac{a_0}{4c_0}\right)^{1/4}(1-\gamma)^{1/2} \tag{33}$$

and the correlation length

$$\xi = \left(\frac{4c_0}{a_0}\right)^{1/4}\frac{1}{(1+\gamma)^{1/2}} \tag{34}$$

where

$$\gamma \equiv g_0/(4a_0 c_0)^{1/2} \tag{35}$$

On the other hand, for $g_0 > (4c_0 a_0)^{1/2}$, i.e., for $\gamma > 1$, the correlation function is simply the sum of two monotonically decaying exponential functions. Thus, the structure of a bicontinuous microemulsion is characterized by *two* length scales: the domain size q_0^{-1} of oil or water domains and the correlation length ξ.

We can immediately conclude from Eqs. (31) and (32) that there are two lines within the region of stability of a microemulsion phase where the behavior of the scattering intensity or of the correlation function changes qualitatively. The first of these lines is the Lifshitz line

(a)

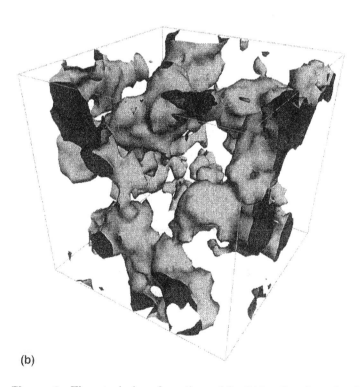

(b)

Figure 4 Three typical configurations of the $\Phi(\mathbf{r}) = 0$ surfaces for different "temperatures" m_0 near the critical endpoint. (a) $m_0 = 0.80$; (b) $m_0 = 0.775$; (c) $m_0 = 0.75$. The size of the smallest droplets in (c) is determined by a small-distance cutoff. (From Ref. 85.)

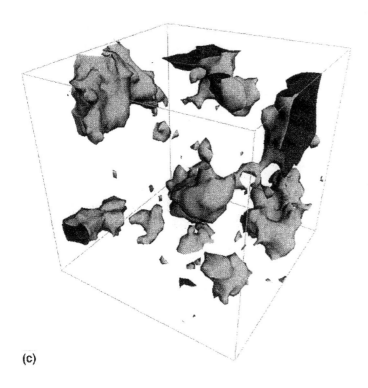

(c)

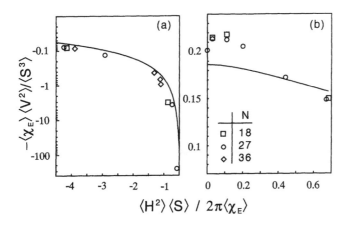

Figure 5 Scaled Euler characteristic $\langle \chi_E \rangle V^2 \langle S \rangle^{-3}$ as a function of the scaled mean curvature $\langle H \rangle^2 \langle S \rangle / 2\pi \langle \chi_E \rangle$. The full lines are the results for Gaussian random fields; the symbols represent Monte Carlo data for different system sizes N. (a) $\langle \chi_E \rangle > 0$, (b) $\langle \chi_E \rangle < 0$. (From Ref. 85.)

(LL), $g_0 = 0$, where the peak of the scattering intensity at nonzero wave vector q first appears; the second is the disorder line (DOL), $g_0 = (4c_0a_0)^{1/2}$, where the correlation function changes from a purely exponential decay to damped oscillations. We will consider only fluids on the "structured" side of the DOL to be microemulsions.

The interesting question is now that of how the two lengths (33) and (34) depend on amphiphile concentration and amphiphile strength. We have shown in Sec. II.B.2 that the coefficient g_0 in the scattering intensity, Eq. (31), becomes increasingly negative with increasing amphiphile concentration and strength. This leads to a shift of the scattering peak to larger values of the wave vector q. This behavior is consistent with experimental observation [40, 93, 94].

For a more quantitative prediction of the scattering behavior, the simple Ornstein–Zernike approximation is too simple, because it completely neglects the effect of fluctuations on the parameters c_0, g_0, and a_0. It has been shown [36, 43] that fluctuations indeed have a pronounced effect. The result that can most easily be compared with experimental observations has been obtained from the charge-frustrated Ising model, where one finds [36]

$$q_0 = \frac{\sqrt{3}}{l}\left(1 - \frac{3\beta\bar{\sigma}_{ow}a^3}{\theta_s l}\right)^{1/2} \tag{36}$$

where $\bar{\sigma}_{ow} = 2Ja^{-2}$ is the tension of the oil/water interface (without any amphiphile present) and $\theta_s = 2\rho_s a^2 l$ is the volume fraction of the amphiphile. A comparison of Eq. (36) with the experimental data for water–octane–C_4E_1 mixtures [94] gives indeed a nice confirmation of this result (see Fig. 6).

It should be noted, however, that Eq. (36) probably becomes less reliable for longer chain amphiphiles. This can be seen, for example, in a systematic variation of the values of $\bar{\sigma}_{ow}$ and l, which are needed to fit the experimental scattering data [36].

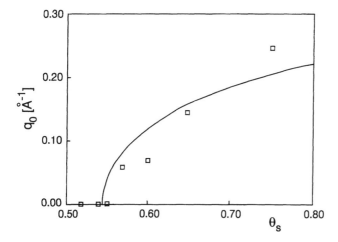

Figure 6   The position of the scattering peak as a function of the surfactant volume fraction θ_s at constant $l = 4.4$ Å, $\beta\bar{\sigma}_{ow} = 0.1$ Å^{-2}, and $a = 2$ Å. The Lifshitz line is at $\theta_s = 0.545$. The data points are for the water–octane–C_4E_1 systems of Ref. 94. (Redrawn from Ref. 36.)

2. Scattering Intensity in Film Contrast

While the scattering intensity in bulk contrast in systems containing medium- or long-chain amphiphiles shows a pronounced peak at nonzero wave vector $q = q_0$, the scattering intensity in film contrast, S_{film}, looks rather boring at first sight, with a peak at $q = 0$ and only a small peak or shoulder at wave vector $q \simeq 2q_0$ [95]. However, a more detailed look reveals a very interesting q dependence. A perturbative calculation based on model (20) gives an analytical expression for the film scattering intensity [48]. In the range $q \ll q_0$, the scattering intensity shows the q dependence [46, 48]

$$S_{film}(q) \sim (q\xi)^{-1} \arctan(q\xi/2) \tag{37}$$

Thus, for $\xi^{-1} \ll q \ll q_0$, the film scattering intensity decays as $1/q$, which differs significantly from the usual Ornstein–Zernike behavior. The $1/q$ behavior can be observed in its pure form only for systems with $q_0\xi \gg 1$, i.e., for strongly swollen microemulsions. In all other cases, the full expression for the scattering intensity has to be used. A fit to the scattering data [93] of the water–n-octane–C_8E_3 system is shown in Fig. 7. The agreement over the whole range of wave vectors is quite remarkable.

D. Interfaces and Capillary Waves

One of the motivations for introducing the anharmonic Ginzburg–Landau model, Eq. (16), was that it should be possible to describe interfaces between different phases. The simplest case is the interface between an oil-rich and a water-rich phase. A mean-field calculation [42] shows that the interfacial free energy is given by

$$\sigma_{ow} = \int dz \, p_s(z) \tag{38}$$

with

$$p_s(z) = 2g(\bar{\Phi})(\bar{\Phi}')^2 + 4c(\bar{\Phi})(\bar{\Phi}'')^2 \tag{39}$$

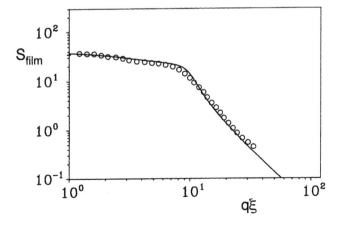

Figure 7 Amphiphile scattering intensity S_{film} in the microemulsion phase. The circles are experimental data for the system D_2O–C_8D_{18}–C_8E_3, taken from Ref. 93. The full line is the theoretical result (with $q_0\xi = 5.0$); see Ref. 48 for details.

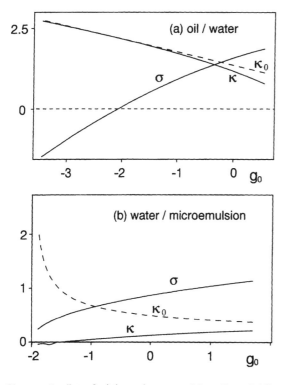

Figure 8 Interfacial tension σ and bending rigidity κ of (a) the oil/water interface and (b) the water/microemulsion interface, as a function of the parameter g_0, which decreases with increasing strength of the amphiphile. The dashed line is the approximation, Eq. (42), for the bending rigidity. For details see Ref. 96.

where z is the coordinate perpendicular to the interface, $\bar{\Phi}' = d\bar{\Phi}/dz$, and $\bar{\Phi}(z)$ is the density profile that minimizes the free energy functional, Eq. (16), with the boundary conditions $\lim_{z \to +\infty} \bar{\Phi}(z) = \Phi_o$ and $\lim_{z \to -\infty} \bar{\Phi}(z) = \Phi_w$. The function $p_s(z)$ can be interpreted as the stress profile across the interface. The result, Eqs. (38) and (39), indicates that when the function $g(\Phi)$ is sufficiently negative the interfacial free energy can become very small or even negative, as shown in Fig. 8. In the latter case, the oil–water coexistence is a metastable state; the system would rather like to have a large amount of internal oil/water interfacial area as in a microemulsion or a lamellar phase.

At any finite temperature, the oil/water interface is not flat but is roughened due to the presence of thermally excited capillary waves. The spectrum of capillary waves can be calculated by considering small fluctuations of the order parameter field, $\Phi(\mathbf{r}) = \bar{\Phi}(z) + \eta(\mathbf{r})$, around the planar interface. By an expansion of the free energy functional to second order in η, one finds the energy [42,96]

$$E(q) = \frac{1}{2}\sigma q^2 + \frac{1}{2}\kappa q^4 + O(q^6) \tag{40}$$

of a capillary wave with wave vector q. Here, σ is the interfacial tension, which turns out to be identical to the interfacial free energy, Eq. (38), as it should. A comparison of the spectrum of

Eq. (40) with the spectrum of Helfrich's curvature Hamiltonian for membranes [97],

$$\mathcal{H}_{\mathrm{curv}} = \int dS[\sigma + 2\kappa(H - H_0)^2 + \bar{\kappa}K] \tag{41}$$

where H and K are the mean and Gaussian curvatures as before, shows that the coefficient κ in Eq. (40) can be identified with the bending rigidity of the interface. In the limit of strong amphiphiles, the bending rigidity is given by [96]

$$\kappa \simeq \kappa_0 \equiv 2 \int dz \; c(\bar{\Phi})(\bar{\Phi}')^2 \tag{42}$$

This expression can also be obtained from a calculation of the free energies of spherical and cylindrical oil droplets in water (or vice versa) [93,98].* This calculation can also be used to extract the saddle-splay modulus $\bar{\kappa}$ of Eq. (41), with the result

$$\bar{\kappa} + 2\kappa_0 = \int dz \; z^2 p_s(z) \tag{43}$$

The approximation (42) becomes less reliable with decreasing amphiphile strength, as shown in Fig. 8.

Expression (42) can now be used to calculate the dependence of the bending rigidity as a function of the amphiphile chain length. A variational ansatz of the form $\bar{\Phi}(z) = \Phi_0 \tanh(z/\xi)$, where Φ_0 is the value of the order parameter of the oil-rich or water-rich phase, shows that for long-chain amphiphiles, $\xi \sim l$, as should be expected. Since Φ_0 is roughly independent of the chain length for large l and the fugacity z_s approaches a constant for saturated monolayers,[†] Eqs. (19) and (42) imply immediately that

$$\kappa_0 \sim c_0/\xi \sim l^3 \tag{44}$$

This is exactly the same l dependence that one gets for the bending rigidity of a thin solid sheet of thickness l [99]. This result is also consistent with a calculation of bending rigidity from a molecular model of monolayers [62,63,66].[‡] For l smaller than about 20 Å, where $\beta\kappa \simeq 1$, the bending rigidity is found to decrease more rapidly than l^3 [37].

The interfacial tension and the bending rigidity have also been calculated for the water/microemulsion interface. The results shown in Fig. 8 demonstrate that the bending rigidity is much smaller in this case than for the oil/water interface and that approximation (42) fails completely. This is consistent with the intuitive picture that the water/microemulsion interface is not a simple monolayer of amphiphile but consists of many channels and passages.

The result, Eq. (43), can also be used to calculate the elastic constants of interfaces in ternary diblock–copolymer systems [100]. The saddle-splay modulus is found to be always positive, which favors the formation of ordered bicontinuous structures, as observed experimentally [9] and theoretically [77,80] in diblock–copolymer systems. In contrast, molecular models for diblock–copolymer monolayers [68,69], which are applicable to the strong-segregation limit, always give a negative value of $\bar{\kappa}$. This result can be understood intuitively [68], as the volume of a saddle-shaped film of constant thickness is smaller than

---

* The result of Eq. (42) is a generalization of the result obtained in Ref. 96 for the case of $c(\Phi) = \mathrm{const.}$
† A saturated monolayer is defined here as an interface of vanishing interfacial tension.
‡ A fit to the data of Ref. 65 gives a dependence of the form $\kappa \sim N^2$, where N is the polymerization index.

that of a flat film and the chains must stretch upon bending, which costs free energy. The same argument should hold for other molecular models that assume a constant monomer density (of diblock copolymers or amphiphiles) in the film.

E.   Ultralow Interfacial Tensions

We address two questions in this section. First, Can the models that we have discussed reproduce the reduction in interfacial tensions observed experimentally? Second, Can some understanding be gained as to the origin of these great reductions?

   As to the first question, lattice models do exhibit oil/water interfacial tensions that are reduced to various degrees from the value in the absence of amphiphile. For example, in the three-component model solved within mean-field theory, a reduction on the order of 30 was found in the oil/water interfacial tension at three-phase coexistence with the microemulsion [101]. When simulated so that fluctuations were included [102], the reduction increased to about a factor of 100, which is characteristic of a weak amphiphile. Other lattice models [103] have obtained reductions as large as a factor of 800, larger than that provided by even the strong amphiphile C_6E_3 [104].

   The question as to why the interfacial tensions are so low is an extremely interesting one, to which several answers have been given. It was once thought that the answer lay in the fact that the system is near a critical point [105]. That this is clearly not the case appears when one notes that the compositions of the oil-rich and water-rich phases that exhibit such low tensions are not at all similar in composition as they would be near a tricritical point. It has also been argued [84] that the interfacial tension is so low because the upper and lower critical endpoints are so close to one another in temperature. It is true that the tension of the oil/microemulsion interface, σ_{om}, must vanish at one critical endpoint, and the tension of the water/microemulsion interface, σ_{wm}, must vanish at the other. Furthermore, the oil/water interfacial tension, σ_{ow}, satisfies the inequality

$$\sigma_{ow} \leq \sigma_{om} + \sigma_{wm} \tag{45}$$

If the endpoints are close to one another, then neither of the terms on the right-hand side of the above inequality can become very large. Hence the oil/water interfacial tension must also be small. The argument is correct in that a small oil/water tension will surely follow if the two critical endpoints are close together, but there seems no reason a priori why this should be so. Further it is easy to imagine a situation in which the endpoints are not close, so that the right-hand side of the above inequality is not small, yet the oil/water tension would be small.

   More recently it has been argued [6,106] that the small interfacial tensions are a consequence of a complete unbinding transition of the lamellar phase, i.e., a continuous increase *without limit* of the wavelength of the lamellar phase with the decrease in amphiphile concentration. The argument proceeds as follows. Suppose that the lamellar phase does not unbind but comes to three-phase coexistence with the water- and oil-rich phases, and let the wavelength of the lamellar phase be equal to some finite value 2L, with L large. In this case the free energy density of the lamellar phase differs from that of the oil-rich or water-rich phase by the amount $(\sigma_{ow} - B)/L$, where $B > 0$ is the binding energy of the internal interfaces. This statement merely notes that the free energy to create an internal interface between oil and water regions of thickness L is simply the free energy cost to create an interface between regions of infinite thickness, σ_{ow}, minus the free energy one gets back due to the attractive interaction between internal interfaces. If this net cost, $\sigma_{ow} - B$, were positive, such internal interfaces would be unfavorable, and the lamellar phase would be unstable with respect to the oil and water phases. Thus the system would not be in three-phase coexistence as assumed. Similarly, if this net cost were negative, such interfaces would be favorable, and the system would create more of them. Consequently, the homogeneous oil and water phases would be unstable to the lamellar phase, so that again the system would not be at three-phase coexistence. Hence along the triple line, this net cost must vanish. Therefore at three-phase coexistence with a lamellar phase of wavelength 2L, the oil/water interfacial tension must be $\sigma_{ow}(L) = B(L)$. Because this binding energy must decrease with increasing L, the interfacial tensions must also decrease as the period of the lamellar phase increases. Finally, if L were to actually diverge, as at a complete unbinding transition, then the tension between the coexisting oil and water phases would vanish.

In this scenario the low oil/water interfacial tension results from the proximity to a complete unbinding transition rather than the tricritical transition envisaged by Fleming and Vinatieri [105]. A definite prediction can be made from this picture if one further assumes that the binding energy between the internal interfaces is due to the van der Waals interaction between oil and water regions. It follows that the binding energy decreases as L^{-2} and therefore that $\sigma_{ow}(L) \propto L^{-2}$. That is, given a sequence of systems that exhibit three-phase coexistence with lamellar phases of different wavelengths 2L, the oil/water interfacial tension in those systems at three-phase coexistence will vary as L^{-2}. It is reasonable to believe that this will also be true when the structured phase is a microemulsion of characteristic wavelength $2L$ rather than a lamellar phase. Fortunately there are measurements of the oil/water interfacial tensions at three-phase coexistence in many different systems [84]. These tensions do in fact decrease as L^{-2} as the wavelength L increases [106].

Finally, as the density of amphiphile clearly decreases with increasing L, this argument explains the correction between low interfacial tensions and strong amphiphiles, ones that can, at very low volume fraction, solubilize both oil and water.

F. Wetting Behavior

It has long been known that one of the principal distinctions between the middle phase brought about by a weak amphiphile and the microemulsion brought about by a strong one is that the latter will not wet the interface between the coexisting oil and water phases whereas the former will [107]. One of the successes of the Ginzburg–Landau approach is in explaining this phenomenon in terms of the difference between the bulk structure of the microemulsion and that of an ordinary fluid [42]. This difference is manifest in the correlation function of Eq. (32), Sec. III.C.1,

$$G(r) \sim e^{-r/\xi} \frac{\sin(q_0 r)}{q_0 r} \tag{46}$$

Microemulsions exist for values of the parameter γ [in q_0 and ξ, see Eqs. (33) and (34)] less than 1 and greater than -1, with more negative values associated with more structure. As can be seen, the correlation function is an exponentially decaying oscillatory function of the separation r. On the other hand, for values of $\gamma > 1$, the Fourier transform is simply a sum of two monotonically decaying exponential functions, and the liquid is unstructured. It is this difference in bulk behavior that proves crucial to the interfacial wetting behavior.

We are quite familiar with the fact that the bulk properties of ordinary coexisting fluid phases determine the structure of the interface between them. For example, the thickness of the interfacial profile is set by the scale of the bulk correlation length. Similarly, the properties of the interface between the middle phase and the water-rich or oil-rich phase reflect the structure of the middle phase. Thus it was predicted [42] that at such an interface the density profile would oscillate just as the correlation function $G(\mathbf{r})$ does, a prediction later confirmed by studies of the microemulsion/air interface [108].

Now consider three-phase coexistence between the water-rich, oil-rich, and middle phases. To determine whether the middle phase wets the interface between the other two, we calculate the excess interfacial free energy of an oil/water interface as a function of the thickness of the layer of middle phase between them. An absolute minimum of this function at infinite thickness of this layer indicates wetting by the middle phase, whereas a minimum at any finite thickness indicates nonwetting. We find that the functional form of this excess free energy is again almost the same as the bulk correlation function, which from Eq. (46) has its absolute minimum at a finite value of its argument. Hence middle phases characterized by values of $\gamma \leq 1$, that is middle phases containing an amphiphile of sufficient strength to bring about a certain level of organization within it, will not wet the oil/water interface.

The calculation described above ignores both long-range interactions and fluctuations. If we include the effect of van der Waals forces while still ignoring fluctuations, then we find that the wetting transition is no longer predicted to occur at the disorder line but rather on the microemulsion side of it [42]. Fluctuations can be expected to have a similar effect and are likely to be quite important because the oil/water interfacial tension is so low [109]. To understand why they will have a similar effect, we picture the excess free energy of the oil/water interface, which is a function of the thickness of the layer of middle phase between them, as a potential energy for the configuration of the two interfaces that defines the layer of middle phase. The minimum of this potential gives the equilibrium thickness of the layer. As noted earlier, this function is similar to the bulk correlation function, being an exponentially damped oscillatory function of the layer thickness. Without fluctuations, this minimum moves continuously to infinity as the microemulsion is made weaker and γ approaches unity from below. Hence there is a continuous wetting transition that occurs at the disorder line. When fluctuations are added, there is an additional increase in free energy at finite thickness due to the entropy lost when the two interfaces collide with one another. This will tend to raise the minima at finite values of thickness compared to the minima at infinity. The unbinding of the two interfaces, which is the wetting transition, will now occur when an oscillatory potential still exists but is too weak to bind the fluctuating interfaces. That there is still an oscillatory potential at the wetting transition means that the fluid is still structured there. That the thickness of the intermediate film of the middle phase jumps discontinuously from a finite to an infinite value means that the wetting transition is first order.

A Monte Carlo simulation [102] of a system with short-range forces confirmed these notions. The correlation function clearly exhibited exponentially damped oscillations. From the ratio of the wavelength and correlation length, the value of γ characterizing the system could be obtained from Eq. (35), and it was found that $1 > \gamma > 0$, indicating that the microemulsion was structured but weakly so. Within the mean-field calculation, however, this is still strong enough that the middle phase should not wet the oil/water interface. However, measurement of all three interfacial tensions within the simulation revealed that Antonow's rule was obeyed, so that the interface was indeed wetted by the middle phase, an effect clearly attributable to the fluctuations included in the simulation.

The effect of fluctuations to quadratic order are easily calculated analytically [110]. It was found that for a balanced system, one in which the microemulsion contains equal volume fractions of oil and water, the wetting transition should be first-order and, for reasonable values of the parameters of the calculation, should occur in the vicinity of the Lifshitz line, $\gamma = 0$, as opposed to the mean-field prediction of a continuous transition occurring at the disorder line, $\gamma = 1$.

The calculations are in satisfactory agreement with experiment. Schubert and Strey [93] measured the structure functions of several balanced microemulsions containing n-alkyl polyglycol ethers (C_iE_j), which could be made weaker by the addition of formamide. They found that the wetting transition in their water–formamide–octane–C_iE_j systems occurred at $-0.53 < \gamma < -0.11$. This interval was further narrowed by Schubert et al. [94], who studied the system D_2O–n-alkane–C_6E_2 and induced the wetting transition by varying the alkane rather than adding formamide. They found the wetting transition at $-0.39 < \gamma < -0.22$ for such balanced mixtures.

In the case of a wetting transition brought about by driving the system toward an upper, or lower, critical endpoint so that it is not balanced at the wetting transition, an interesting result was obtained [110]. It was found that the closer the wetting transition occurred to the critical endpoint, the more structured the microemulsion was at the wetting transition, i.e., the more negative γ would be there. The reason is that as the critical endpoint is approached, the effect of fluctuations becomes very large, and only a very strong oscillatory potential can bind the interfaces. Therefore, if the wetting transition occurs close to the critical endpoint, then the fluid must be strongly structured, characterized by a negative γ of large magnitude.

Abillon et al. [111] measured the structure function of an unbalanced system driven to a wetting transition. They found that at that point the peak in the structure function was still at nonzero wave vector, not only confirming that the transition occurred on the microemulsion side of the disorder line but also showing that it occurred on the more ordered side of the Lifshitz line. Systems that were driven to their wetting transition by increasing the oil fraction from its balanced value were also examined by Schubert et al. [94], and the wetting transition located within the interval $-0.37 < \gamma < -0.33$. In a very recent experiment on D_2O–octane–C_8E_3 mixtures in which the wetting transition occurs quite close to the critical endpoint, Gradzielski et al. [112] found that at the transition, $\gamma \leq -0.5$, a value whose magnitude is indeed larger than that characterizing the wetting transition of weaker systems, in agreement with prediction.

There is another experimental observation related to the unbalanced systems that is noteworthy. Suppose one begins with a balanced system in which the microemulsion does not wet the oil/water interface. A droplet of middle phase will be characterized by some nonzero contact angle Θ_c with the oil/water interface. If the temperature is now either increased or decreased so as to approach one of the critical endpoints, a wetting transition occurs before the endpoint is reached as predicted by Cahn [113]. At the transition, $\Theta_c = 0$, of course. It might be expected that the contact angle would decrease uniformly as the endpoint was approached. Indeed, that is the case [104,114] in the octane–water–C_5E_2 system. Surprisingly, it is not so when the amphiphile is stronger, e.g., C_6E_3 or C_8E_4. In these cases, the contact angle actually increases before decreasing to zero at the wetting transition. Furthermore, as the amphiphile strength increases, the temperature of the wetting transition approaches that of the critical endpoint. This correlates nicely with the results of Gradzielski et al. [112] discussed earlier. They found that as the amphiphile was strengthened and the wetting temperature approached that of the critical endpoint, the value of $\gamma = g_0/(4a_0c_0)^{1/2}$ in the middle phase became more negative. As such behavior is correlated with a more structured middle phase, and it is just this structure that favors the nonwetting of the middle phase, it is reasonable that the contact angle would become larger, as observed.

It is not a trivial task to obtain this behavior from theory, because the contact angle depends on all three interfacial tensions in the system. Furthermore, near a critical endpoint, say the one at which the water and middle phases are becoming critical, one of these tensions, the one between the water and middle phases in this case, is very small, as is the difference between the other two tensions, $\sigma_{ow} - \sigma_{om}$. Yet the contact angle is essentially the ratio of these two small quantities,

$$\cos \Theta_c \approx \frac{\sigma_{ow} - \sigma_{om}}{\sigma_{mw}} \tag{47}$$

Hence the calculation must be rather precise to reproduce this behavior. Nonetheless, this variation of contact angle with temperature and amphiphile strength has been obtained from Ginzburg–Landau theories in the absence of fluctuations [115–117] as well as in the presence of them [110], as shown in Fig. 9. Under some conditions, it is predicted [117] that the contact angle can even increase to 180°. This indicates a transition quite different from the usual one in which the middle phase wets the oil/water interface. Approaching the endpoint at which the middle and water-rich phases are becoming critical in the presence of the oil-rich phase, the transition is one in which the water-rich phase wets the interface between the oil-rich and middle phases. Just such a transition has been reported by Chen and coworkers [118,119]. This is most interesting, for in order for there to be an ordinary wetting transition at one endpoint and an abnormal one at the other endpoint, as observed, there must be a change in sign of the Hamaker constant characterizing the van der Waals interactions of the middle phase between the oil and water phases. Just such a change in sign is expected at a critical wetting transition [120]. As such a transition has not been observed in any amphiphilic system, it would be of great interest to determine the order of this one.

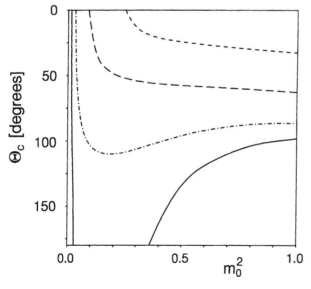

Figure 9 Behavior of the contact angle Θ_c versus reduced temperature $(T - T_{CEP})/T_{CEP} \sim m_0^2$. The four curves (from top to bottom) correspond to a sequence of systems that approaches the Lifshitz CEP. Note that for the bottom curve the water-rich phase wets the interface between the oil-rich and microemulsion phases at $m_0^2 \simeq 0.37$. (From Ref. 117.)

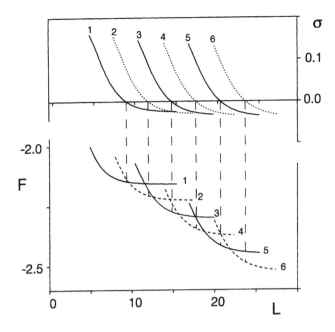

Figure 10 The lamellar phase between two walls, as a function of the wall separation L. The lower part shows the L dependence of the free energy F for various numbers of monolayers, as indicated. The upper part shows the interfacial tension σ. Note that $\sigma = 0$ occurs always when the configuration is stable. For details see Ref. 96.

G. The Lamellar Phase

Expression (38) for interfacial tension determines not only the coefficient of the q^2 contribution to the spectrum of capillary waves at the free oil/water interface but also the same coefficient of the spectrum in the lamellar phase—as long as all interfaces in the stack fluctuate in phase—except that $\bar{\Phi}(z)$ is now the mean-field density profile of the lamellar phase and the integral over z extends over only one lattice spacing. It now turns out that the "interfacial tension" σ is a function of the lattice spacing d of the lamellar phase [96]. The stable lamellar phase has a lattice spacing d^*. However, smaller or larger values of d are possible as metastable states if the stack is confined between two walls parallel to the interfaces so that an adiabatic change of the distance L between these walls leads to an expansion or compression of the stack. If the distance between neighboring membranes becomes too small or too large, the stack will respond be decreasing or increasing the number of interfaces. This is shown in Fig. 10 for a stack with a rather small number of interfaces. It can now be shown *exactly*, independent of the form of the functions $f(\Phi)$ and $g(\Phi)$, that the "interfacial tension" σ vanishes identically for $d = d^*$ (for an infinite stack) [96].* Furthermore, the tension depends exponentially on the intermembrane separation d,

$$\sigma(d) = \sigma_{ow} + \bar{\sigma}e^{-d/\xi} \tag{48}$$

---

* The same conclusion was reached by Golubovič and Lubensky [121], based on an interfacial model of lamellar phases.

as shown in Fig. 10. Here, σ_{ow} is the (negative) tension of the free oil/water interface and ξ is the width of its mean-field density profile.

It is important to note that the lamellar phase is thus stabilized by the balance of a negative interfacial tension (of the free oil/water interface covered by an amphiphilic monolayer), which tends to increase the internal area, and a repulsive interaction between interfaces.* The result, Eq. (48), indicates that the scattering intensity in a lamellar phase, with wave vector **q** parallel to the membranes, should have a peak at nonzero q for $d > d^*$ due to the negative coefficient of the q^2 term in the spectrum of Eq. (40), just as in the microemulsion phase. This effect should be very small for strongly swollen lamellar phases (in coexistence with excess oil and excess water), as both $|\sigma_{ow}|$ and $\exp(-d^*/\xi)$ are very small [96]. Very similar behavior has been observed in smectic liquid crystals (Helfrich–Hurault effect) [122]. Experimentally, the lamellar phase under an external tension can be studied with the surface-force apparatus [123,124]; simultaneous scattering experiments have to be performed to detect the undulation modes.

IV. RESULTS II: DYNAMICAL BEHAVIOR

The rich phase behavior of amphiphilic systems and the complicated structure of many phases on mesoscopic length scales indicate interesting dynamical behavior of these systems. Three principle cases of the dynamical behavior of many-body systems can be distinguished in general: the dynamics in the equilibrium state, the dynamics in a stationary state out of equilibrium, and the relaxation from a nonequilibrium state toward equilibrium. All three situations have been studied in some detail recently for amphiphilic systems.

The dynamical behavior of Ginzburg–Landau models is described by Langevin equations. In the simplest case, the equation of motion for a conserved order parameter field $\Phi(\mathbf{r}, t)$, which now depends on time t in addition to $\mathbf{r}$, reads

$$\frac{\partial \Phi}{\partial t} = \Gamma_\Phi \nabla^2 \frac{\partial \mathcal{F}}{\partial \Phi} + \zeta_\Phi \tag{49}$$

This equation can be obtained as follows. Since the order parameter is conserved, it obeys a conservation law, $\partial \Phi / \partial t + \nabla \cdot \mathbf{J} = 0$. The diffusion current $\mathbf{J}$ is given by $\mathbf{J} = -\Gamma_\Phi \nabla \mu$, where μ is the chemical potential difference between oil and water. With the relation $\mu = \partial \mathcal{F} / \partial \Phi$, which follows from standard thermodynamics, we arrive at the first part of Eq. (49). The second part, a random (Gaussian) noise source, is necessary to describe thermal fluctuations. It can be shown that the order parameter field has the correct Boltzmann statistics if the noise has the correlations

$$\langle \zeta_\Phi(\mathbf{r}, t) \zeta_\Phi(\mathbf{r}', t') \rangle = -2\Gamma_\Phi \nabla^2 \delta(\mathbf{r} - \mathbf{r}') \delta(t - t') \tag{50}$$

To describe the dynamical properties of ternary amphiphilic *liquids*, not only the order parameter field $\Phi(\mathbf{r}, t)$ [and the amphiphile concentration $\rho(\mathbf{r}, t)$] but also the hydrodynamic flow of the fluid, which is described by a velocity field $\mathbf{v}(\mathbf{r}, t)$, and the pressure field $p(\mathbf{r}, t)$ have to be taken into account. The dynamics of these five fields is then determined by a set of Langevin equations.

---

* In the vicinity of a first-order transition to oil–water coexistence, the long-range attractive part of the interaction dominates, and the interfacial tension σ_{ow} must be positive; compare Sec. III.E.

In many situations, such as the calculation of the frequency-dependent viscosity of a microemulsion, not all of these fields are necessary. In the simplest case, only the order parameter Φ and the transverse component $\mathbf{v}_T$ of the velocity field have to be included in the model. This description assumes that the diffusion of the amphiphile is fast compared to other transport processes and that the fluid is essentially incompressible, so that $\text{div}\,\mathbf{v} = 0$. In this case, the dynamics is governed by the Langevin equations [125,126]

$$\frac{\partial \Phi}{\partial t} = \Gamma_\Phi \nabla^2 \frac{\delta \mathcal{F}}{\delta \Phi} + g_0 \nabla(\Phi \mathbf{v}_T) + \zeta_\Phi \tag{51}$$

and

$$\frac{\partial \mathbf{v}_T}{\partial t} = \Gamma_T \nabla^2 \mathbf{v}_T + g_0 \left(\Phi \nabla \frac{\delta \mathcal{F}}{\delta \Phi} \right)_T + \zeta_T \tag{52}$$

where the Langevin forces $\zeta_\Phi(\mathbf{r}, t)$ and $\zeta_T(\mathbf{r}, t)$, which describe thermal fluctuations, have the usual white noise correlations [compare Eq. (50)]. The first part of Eq. (51) describes the diffusion of oil or water molecules in the fluid; the second part describes the transport of these molecules due to fluid flow. The first part of Eq. (52) is the linearized Navier–Stokes equation; the second part is a force due to the microemulsion structure.

A. Equilibrium Dynamics

Transport coefficients like the diffusion constant, the viscosity or the sound velocity, and dynamic light scattering or inelastic neutron scattering intensities are all dynamical properties of a system in thermal equilibrium. We concentrate here on the viscosity and on sound velocity and attenuation.

The frequency-dependent viscosity $\eta(\omega)$ is determined by the behavior of the velocity–velocity correlation function in the hydrodynamic limit $q \to 0$,

$$G_{v_T v_T}(\mathbf{q}, \omega) \sim \frac{1}{|i\omega - \eta(\omega)q^2|^2} \tag{53}$$

A typical result of a calculation [127] of the complex viscosity $\eta(\omega)$ is shown in Fig. 11. The real part of the viscosity, $\eta'(\omega)$, which describes the dissipation of energy when the fluid is sheared, is approximately frequency-independent for small ω, i.e., the fluid behaves as a Newtonian fluid. There is a characteristic frequency ω^* where $\eta'(\omega)$ drops rapidly. The imaginary part of the viscosity, $\eta''(\omega)$, which describes the elastic response of the fluid to an external perturbation, increases linearly for small ω and reaches a maximum at $\omega = \omega^*$. This behavior is not specific to microemulsions but has been observed in other complex fluids as well, such as in suspensions of spherical colloidal particles [128,129] and in dilute polymer solutions [130].

The most interesting aspect of the viscosity of microemulsion and sponge phases is therefore its dependence on the structural parameters q_0 and ξ. For a detailed analysis, the zero-shear viscosity $\eta(0)$ [131], the storage and loss moduli $G'(\omega) = \omega \, \text{Im}[\eta(\omega)]$ and $G''(\omega) = \omega \, \text{Re}[\eta(\omega)]$ in the terminal regime ($\omega \to 0$), and the characteristic frequency ω^* are obvious quantities to study. For the frequency ω^*, for example, one finds the scaling form [127]

$$\omega^* \sim \xi^{-6}\Omega(q_0\xi) \tag{54}$$

with a scaling function $\Omega(x)$, which approaches a constant for $x \to 0$ and diverges as x^4 for large x. Thus, the structural relaxation time $1/\omega^*$ increases with increasing correlation

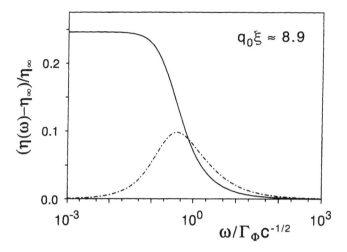

Figure 11 Complex viscosity $\eta(\omega)$ of a microemulsion as a function of the scaled frequency $\omega/\Gamma_\Phi c^{-1/2}$. The full line is the real part and the dash-dotted line the imaginary part of the viscosity. (From Ref. 127.)

length ξ and with increasing domain size $2\pi/q_0$. This qualitative behavior can be easily understood intuitively, for the larger ξ is, the longer it takes for regions of size ξ to rearrange their structure. For the same reason, the zero-shear viscosity increases with increasing ξ.

Similarly, the storage and loss moduli have a scaling behavior of the form [132]

$$G'(\omega) \sim \xi^9 \Gamma'(q_0\xi)\omega^2, \qquad G''(\omega) \sim \xi^3 \Gamma''(q_0\xi)\omega \tag{55}$$

to leading order in ω. The scaling functions Γ' and Γ'' approach a constant for $q_0\xi \to 0$, while

$$\Gamma'(x) \sim x^{-4}, \qquad \Gamma''(x) \simeq \text{const.} \qquad \text{for} \quad x \to \infty \tag{56}$$

Thus, as the transition to the lamellar phase is approached, G' diverges strongly as ξ^5, whereas G'' shows a weaker ξ^3 divergence. This asymptotic behavior was predicted first, in fact, for diblock copolymer melts (where $\xi \sim |T - T_c|^{-1/2}$) by Fredrickson and Larson [133], who studied a more detailed model of block copolymers (compare Sec. V.B) and used a different dynamic formalism.* This agreement demonstrates that the Ginzburg–Landau model is well suited for the calculation of generic dynamical behavior in complex fluids.

To calculate the sound velocity and damping, the simple model (51), (52) is not sufficient. Since sound is a longitudinal wave of the mass density, both the longitudinal component of the velocity, $v_L(\mathbf{r}, t)$, and the pressure field $p(\mathbf{r}, t)$ have to be taken into account in this case in addition to $\Phi(\mathbf{r}, t)$ and $v_T(\mathbf{r}, t)$ [132,135,136]. The frequency-dependent sound velocity $c(\omega)$ and damping $D(\omega)$ can then be obtained from the pole structure of the pressure correlation function in the hydrodynamic limit $q \to 0$, where

$$G_{pp}(\mathbf{q}, \omega) \sim \frac{1}{|\omega^2 - c^2(\omega)q^2 + i\omega D(\omega)q^2|^2} \tag{57}$$

---

* The results seem to be consistent with experiments for diblock copolymer melts by Bates [134].

An important result of this calculation [132] is that the quantities

$$-\gamma_L'(\omega) \equiv D(\omega) - D_\infty \tag{58}$$

and

$$-\gamma_L''(\omega) \equiv \frac{c^2(\omega) - c^2(0)}{\omega} \tag{59}$$

behave very much like the real and imaginary parts of the viscosity. A plot of $\gamma_L'(\omega)$ and $\gamma_L''(\omega)$ therefore looks very much like Fig. 11. In particular, for strongly structured microemulsions (with $q_0\xi \gg 1$), the same characteristic frequency ω^* that was observed in the viscosity curves also dominates sound velocity and damping [132].

The model given by the dynamic equations (51), (52) and the free energy functional (16) has also been used to calculate the dynamic scattering intensity in bulk contrast [137]. Related models have been employed to determine the dynamic scattering intensity in film contrast [132,138,139].

B. Stationary Shear Flow

The imposition of shear flow can have quite dramatic consequences on the structure and phase behavior of complex fluids. Steady shearing of binary amphiphilic systems can lead to a completely new phase of densely packed onionlike vesicles [140]. Shear flow also strongly affects the stability of the lamellar phase [141–145]. We want to discuss here the role of shear in the microemulsion-to-lamellar transition.

For a flow in the x direction, with a gradient of the average flow velocity in the y direction, the average flow field is given by $\langle \mathbf{v} \rangle = y\mathbf{e}_x$. The term in Eq. (51) that describes the hydrodynamic transport of order parameter therefore gives an additional contribution of the form $\nabla(\Phi\langle \mathbf{v} \rangle) = y\,\partial\Phi/\partial x$. The Langevin equation (49) then reads [146]

$$\frac{\partial\Phi}{\partial t} = \Gamma_\Phi\nabla^2\frac{\delta\mathcal{F}}{\delta\Phi} - Dy\frac{\partial\Phi}{\partial x} + \zeta_\Phi \tag{60}$$

where D is the shear rate. The same term has to be added to model (51), (52), which includes fluctuations of the hydrodynamic flow velocities.

As a result of the shear flow, order parameter fluctuations in the microemulsion phase are suppressed [142]. This destabilizes the microemulsion with respect to a lamellar phase, so that for a certain temperature range the lamellar phase can be induced by applying shear. Furthermore, fluctuations in the microemulsion become very anisotropic in shear flow. In particular, the lamellar fluctuations, which appear as the transition is approached, have wave vectors concentrated near $k_{max}\mathbf{e}_z$ transverse to both the flow velocity and its gradient. Therefore, a shear-induced lamellar phase is expected to occur preferentially in this orientation. A more detailed analysis [142]* based on model (60) shows that for small D the shift of the transition temperature, $T^*(D)$, is given by

$$T^*(D) - T^*(0) \sim D^2 \tag{61}$$

---

* The free-energy functional used in Ref. [142] has the form $\mathcal{F} = \int_q [\tau + (q - q_0)^2]\phi(\mathbf{q})\phi(-\mathbf{q})$ $+ (\lambda/4!) \int_q \int_{q'} \int_{q''} \int_{q'''} \phi(\mathbf{q})\phi(\mathbf{q}')\phi(\mathbf{q}'')\phi(\mathbf{q}''')\delta(\mathbf{q} + \mathbf{q}' + \mathbf{q}'' + \mathbf{q}''')$.

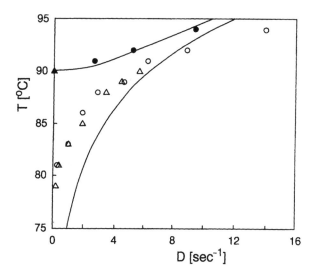

Figure 12 Equilibrium order–disorder transition and spinodal of the disordered state as a function of shear rate D, as predicted by Eqs. (61) and (62), respectively. The data points are experimental results for a nearly symmetrical diblock copolymer melt [147]. Data points given by open and full circles were obtained at fixed temperature T by varying D, while those represented by open and full triangles were determined by varying temperature at fixed share rate. (Redrawn from Ref. 147.)

while the shift of the spinodal temperature, $T_s(D)$, is found to be

$$T_s(D) - T_s(\infty) \sim -D^{-1/3} \tag{62}$$

Predictions (61) and (62) have been qualitatively confirmed in experiments for a symmetrical diblock copolymer melt [147] (see Fig. 12). We show in Sec. V.B that these systems are indeed closely related to ternary amphiphilic systems, and, in the weak segregation regime, can be described by the same Ginzburg–Landau models.

C. Phase Separation

Phase separation in binary alloys, polymer blends, and fluid mixtures has been studied intensively for many years [148]. It takes place when a two-component mixture is quenched from a disordered state into a two-phase coexistence region. After such a quench, composition fluctuations are created that grow and form domain structures. In the late stages, the domain structure coarsens in time, a process that is driven by the interfacial tension σ between the two coexisting phases. The growth of the characteristic length scale $R(t)$ follows a power-law behavior,

$$R(t) \sim t^{1/3} \tag{63}$$

[with a crossover to $R(t) \sim t$ for fluid mixtures due to hydrodynamics at very late stages [49]]. This behavior can be obtained heuristically from a simple scaling argument [150]. If we assume that there is only one length scale $R(t)$ present at long times, it will determine both the average distance between the domain walls and the thickness of the walls. Thus, the diffusion current can be estimated to be $\mathbf{J} = -\Gamma \nabla \mu \sim \Delta \mu / R(t)$, where $\Delta \mu$ is the change in μ across the wall. The curvature of the domain walls, also given by $R(t)$, leads to a Laplace "pressure" $\Delta \mu \sim \sigma / R(t)$, so that $J \sim \sigma R^{-2}$. On the other hand, the diffusion current is also

$J \sim dR/dt$, which finally yields

$$\frac{dR}{dt} \sim \frac{\sigma}{R(t)^2} \tag{64}$$

The solution of this equation is the power law, Eq. (63).

The temporal evolution of the domain structure can be monitored by studying the time dependence of the equal-time correlation function

$$S(\mathbf{k}, t) = \int d^3 r e^{-i\mathbf{k}\cdot\mathbf{r}} \langle \Phi(\mathbf{r}, t)\Phi(0, t) \rangle \tag{65}$$

which can be measured by light-scattering experiments. This function vanishes for small wave vectors k due to the conservation of the order parameter Φ. It has a peak at some wave vector k_{max}, which is proportional to the inverse domain size, i.e., $k_{max} \sim R(t)^{-1}$.

The question that now arises in amphiphilic systems is, Does the presence of amphiphilic molecules change this picture, and if so, how? Computer simulations of a Ginzburg–Landau model, without [151,152] or with [153] hydrodynamics, of the hybrid model [60,154–156], and of a hard-sphere model with diatomic surfactants [59] show that after a *quench into the coexistence region of an oil-rich and a water-rich phase* there is a power-law growth of the average domain size of the form (63) for intermediate time scales, followed by a much slower growth for longer times. In the simulations of Refs. 151 and 152, the long-time behavior was found to follow a logarithmic time dependence,

$$R(t) \sim \ln(t) \tag{66}$$

The slowing down of the domain growth is due to the assembly of amphiphilic molecules at the domain walls. When the domain walls are formed at the beginning of the phase separation process, their area is very large. Thus, for a sufficiently small amphiphile concentration, the density of amphiphiles in the domain walls must be very small. The interfacial tension is therefore very close to the tension σ_0 of an oil/water interface in the absence of amphiphile. As the domain structure coarsens, the area of the domain walls decreases and the density of amphiphile in the interfaces increases, thereby reducing the interfacial tension. For long times t, the domain wall approaches the state of a saturated amphiphilic monolayer, which can have a very small interfacial tension—if the amphiphile concentration exceeds the critical micelle concentration (cmc). It has been suggested [154,157] that the main effect of the amphiphile on the coarsening dynamics is an interfacial tension in Eq. (64), which also depends on the length scale $R(t)$. A possible ansatz for the interfacial tension is [154]

$$\sigma(R(t)) = \sigma_0 \exp[-R(t)/R_0] \tag{67}$$

where R_0 is the (vaguely defined) length where the density of the amphiphilic monolayer comes close to the saturation density. The solution of Eq. (64) with the tension, Eq. (67), indeed reproduces the logarithmic growth law (66) for $R(t) \gg R_0$.

It is most likely that the logarithmic behavior, Eq. (66), observed in Refs. 151 and 152 is a crossover phenomenon [59,154]. Indeed, a state of vanishing interfacial tension is possible only in the limit of very long chain lengths, where the three-phase triangle between the oil-rich, water-rich, and microemulsion phases degenerates into a line (compare Sec. III.E). If the equilibrium state of the system is a two-phase coexistence between an oil-rich and a water-rich phase, then the interfacial tension of the amphiphilic monolayer may be small, but it is finite. Thus, we expect a crossover from a $t^{1/3}$ behavior with a large prefactor, which is determined by the tension σ_0 of the bare oil/water interface, at intermediate times to another $t^{1/3}$ behavior with a small prefactor, which is determined by the tension σ_{ow} of an oil/water interface completely covered by an amphiphile layer, at very long times.

Finally, it is important to note that in balanced systems the intersection of the cmc surfaces with the oil–water coexistence region nearly coincides with the base of the three-phase triangle [84]. Therefore, a "saturated" monolayer is possible only at three-phase coexistence.

If the amphiphile concentration is too large, the system does not phase separate at all. The equilibrium state in this case is not a two-phase coexistence of an oil-rich and a water-rich phase, but a microemulsion. Thus, the length scale grows only until it reaches the characteristic domain size R_0 of a microemulsion and then stops [59]. This case can be described by a surface tension of the form

$$\sigma(R(t)) = \sigma_0 \left(1 - \frac{R(t)}{R_0} \right) \tag{68}$$

The solution of Eq. (64) is then given by

$$\frac{R(t) - R_0}{R_0} \sim \exp\left[-\frac{t}{t_0} \right] \tag{69}$$

where $t_0 \sim R_0^3/\sigma_0$, so that $R(t)$ should approach its equilibrium value exponentially for long times.

Another interesting case is a *quench into the coexistence region of a microemulsion and an unstructured fluid*. Interest has concentrated here on the early stages of phase separation [158]. In simple, unstructured binary fluids, fluctuations with small wave vectors k begin to grow exponentially in time when the system is quenched into the spinodal decomposition regime, which is separated by the classical spinodal line from the nucleation regime [148,150]. The situation is different for structured fluids like microemulsions. Here, modes with *finite* wave vector k_{max} become unstable *before* the classical spinodal is reached [158]. Furthermore, these modes grow much more quickly because their growth requires diffusion of molecules only over a distance on the order of π/k_{max}. This can be seen rather easily from the Cahn–Hilliard equation [159] for the structure function,

$$\frac{\partial}{\partial t} S(\mathbf{k}, t) = -2\Gamma k^2 A(\mathbf{k}) S(\mathbf{k}, t) + 2\Gamma k^2 \tag{70}$$

with

$$A(\mathbf{k}) = 2ck^4 + 2g(\langle \Phi \rangle)k^2 + f''(\langle \Phi \rangle) \tag{71}$$

Thus, for strongly structured microemulsions, the nucleation regime becomes very narrow, and the spinodal and binodal nearly coincide. No such effect is expected on the simple-fluid side of the phase diagram.

Finally, a *quench into the one-phase region of the microemulsion* has been investigated. An analysis based on a Ginzburg–Landau model for a single, conserved order parameter predicts [160] that the equal-time structure factor, Eq. (65), approaches its equilibrium form $S(\mathbf{k})$ algebraically for long times t.

V.   RELATED SYSTEMS: BLOCK COPOLYMERS

A.   Introduction

Diblock copolymers consist of a linear strand of repeating units, or monomers, of one kind, A, that is chemically bonded to another such strand of monomers of a different kind, B. The total number of units is referred to as the index of polymerization, N, and can range from as little as 10 to as large as 10^6. Each monomer prefers to be surrounded by its own kind, so there is an incompatibility, or effective repulsive interaction, between monomers of different kinds that is measured by a strength χ. Because of this incompatibility, the different mono-mers tend to self-assemble into structures in which these unfavorable contacts are localized at widely spaced internal interfaces separating regions of relatively pure A or B monomers, much as in systems of water and oil. In fact, AB diblock are used as compatibilizers in mixtures containing A homopolymers and B homopolymers in much the same way that amphiphiles are used as solubilizers in mixtures of water and oil.

The tendency to assemble is opposed not only by the loss of the entropy of mixing, as in small-molecule systems, but also by the loss of entropy in stretching the polymers to keep the interfaces widely spaced. Just what ordered structure forms depends on the relative fraction f of the A block in the molecule and the relative flexibility of the two blocks.

Even though this specific mechanism of entropy loss is particular to polymers, the patterns of self-assembly are almost exactly the same as those seen in water–oil–amphiphile systems. In particular, the structures and progression of structures seen in systems of AB diblocks in a solvent A homopolymer are quite analogous to those seen in amphiphile–water mixtures. That is, one observes lamellar phases, hexagonally packed cylinders, a body-centered cubic packing of spheres, and a bicontinuous cubic phase with $Ia\bar{3}d$ symmetry, often referred to as the gyroid [161,162]. A perforated lamellar phase, in which the lamellae of the minority component are pierced by a hexagonal array of holes filled by the majority component, has also been observed [163]. Furthermore, as one adds homopolymer solvent, the order of phases observed is the same as when water is added to amphiphile: lamellar to bicontinuous cubic to hexagonally packed cylinders. The one main difference is that diblock copolymers are sufficiently large that they can form ordered phases by themselves, whereas systems of pure short-chain amphiphiles do not.

B. Self-Consistent Field Approximation and Ginzburg–Landau Models

A few generic model Hamiltonians have been very successful in describing the properties of polymers. The configuration of the αth polymer is described by a space curve $r_\alpha(s)$, where $s = 0$ at the A monomer end, $s = f$ at the junction, and $s = 1$ at the B monomer end. With the volume of all monomers assumed to be equal and of a value $1/\rho$, and with each of the n polymers consisting of N monomers, the dimensionless density of A monomers at position $\mathbf{r}$ is

$$\hat{\Phi}_A(\mathbf{r}, \{r_\alpha(\cdot)\}) = \frac{N}{\rho} \sum_{\alpha=1}^{n} \int_0^f ds \ \delta(\mathbf{r} - \mathbf{r}_\alpha(s)) \tag{72}$$

and similarly for the local density of B monomers, $\hat{\Phi}_B(\mathbf{r})$. In these variables, one of the most common Hamiltonians used to describe a polymer melt (i.e., a system consisting only of polymers and no solvent) is

$$H_g\{r_\alpha(\cdot)\} = \frac{3k_BT}{2Na^2} \sum_{\alpha=1}^{n} \int_0^1 ds \left| \frac{d}{ds} r_\alpha(s) \right|^2 + k_BT\chi\rho \int d^3r \ \hat{\Phi}_A\hat{\Phi}_B \tag{73}$$

where k_B is the Boltzmann constant, T is the temperature, and a is called the statistical segment length. The second term is simply the usual repulsive interaction term between unlike monomers, with a strength taken to be χk_BT. The temperature appears here only to make the parameter χ dimensionless. The first term is simply the elastic energy of a series of discrete Hookean springs described in the continuum limit. The factor $3k_BT/Na^2$ is the spring constant, and the presence of k_BT here indicates that this Hookean behavior actually is entropic in origin; that is, any stretching of the polymer reduces the number of configurations that it can take. Hard-core interactions are accounted for by requiring that the ensemble-averaged concentration of monomers be uniform, so that the partition function is

$$
Z = \int \prod_{\alpha=1}^{n} \mathcal{D}\mathbf{r}_\alpha \delta[1 - \hat{\Phi}_A - \hat{\Phi}_B] \exp\left[-\frac{H_g\{\mathbf{r}_\alpha(\cdot)\}}{k_B T}\right]
\tag{74}
$$

where the integrals are over all polymer configurations.

The effects of fluctuations in such systems are small because each molecule contains so many monomers. Hence the mean-field, or self-consistent field, theory of such systems is expected to be quite good over a large part of the parameter space. The self-consistent theory was introduced by Edwards [164] and further developed by Helfand and coworkers [165]. It is easily derived as follows.

We begin by noting that what makes the partition function so difficult to carry out is that each density in the product $\hat{\Phi}_A \hat{\Phi}_B$ that appears in the Hamiltonian of Eq. (73) depends explicitly on the polymer configurations. To make the partition function somewhat more tractable, we insert a functional integral $1 = \int \mathcal{D}\Phi_A \mathcal{D}\Phi_B \delta[\Phi_A - \hat{\Phi}_A]\delta[\Phi_B - \hat{\Phi}_B]$, which permits replacement of the densities $\hat{\Phi}_A$ and $\hat{\Phi}_B$ by the functions Φ_A and Φ_B, which do not depend on the polymer configurations. Inserting representations for the delta functions in terms of Laplace transforms and Bromwich integrals, using the incompressibility constraint, and carrying out a Gaussian functional integral over Φ_A, we obtain

$$
Z = \mathcal{N} \int \mathcal{D}w_A \mathcal{D}w_B \exp\left[-\frac{F[w_A, w_B]}{k_B T}\right]
\tag{75}
$$

where $\mathcal{N}$ is a normalization constant and

$$
\frac{F[w_A, w_B]}{nk_B T} = -\ln Q + V^{-1} \int d^3r \left\{ \frac{\chi N}{4}\left[\frac{w_A(\mathbf{r}) - w_B(\mathbf{r})}{\chi N} - 1\right]^2 - w_B(\mathbf{r}) \right\}
\tag{76}
$$

with V the system volume and Q the partition function of a copolymer subjected to external fields w_A and w_B;

$$
Q \equiv \int \mathcal{D}\mathbf{r}_\alpha \exp\left[-\frac{3}{2Na^2}\int_0^1 ds\left(\frac{d}{ds}\mathbf{r}_\alpha\right)^2\right] \exp\left[-\int_0^f ds\ w_A(\mathbf{r}_\alpha(s)) - \int_f^1 ds\ w_B(\mathbf{r}_\alpha(s))\right]
\tag{77}
$$

Equation (75) expresses the partition function of the many-polymer system in terms of the partition functions of single polymers subjected to external fluctuating fields. The self-consistent field theory approximates this functional integral over the fields by the value of the integrand evaluated at those values of the fields, W_A and W_B, that minimize the functional $F[w_A, w_B]$. From the definition of F it follows that these functions satisfy the self-consistent equations

$$
W_A(\mathbf{r}) - W_B(\mathbf{r}) = \chi N[1 - 2\phi_A(\mathbf{r})]
\tag{78}
$$

$$
\phi_A(\mathbf{r}) + \phi_B(\mathbf{r}) = 1
\tag{79}
$$

where

$$\phi_A(\mathbf{r}) \equiv -\frac{V}{Q}\left(\frac{\mathcal{D}Q}{\mathcal{D}W_A(\mathbf{r})}\right) \tag{80}$$

$$\phi_B(\mathbf{r}) \equiv -\frac{V}{Q}\left(\frac{\mathcal{D}Q}{\mathcal{D}W_B(\mathbf{r})}\right) \tag{81}$$

Only the partition functions of single polymers in external fields appear in these equations, and these are readily calculated [164,165].

While the theory is simple to solve for spatially uniform phases, it has been extremely difficult to obtain solutions with a given desired symmetry such as that of the hexagonally packed cylinders. What has been done in the past is to further approximate the mean-field free energy of the system by expanding it in terms of the Fourier components of the local A and B densities, $\phi_A(\mathbf{k}) - \phi_B(\mathbf{k}) \equiv \Phi(\mathbf{k})$, an expansion that is precisely the Landau–Ginzburg free energy of the system.

For the system of n symmetric ($f = 0.5$) diblocks of polymerization index N, the first term in the expansion of the free energy is [34]

$$F = \frac{nk_BT}{2}\sum_q V_{\text{eff}}(\mathbf{q})\Phi(\mathbf{q})\Phi(-\mathbf{q}) \tag{82}$$

where the effective interaction

$$V_{\text{eff}}(\mathbf{q}) = I(x) - 2\chi N \tag{83}$$

with

$$I(x) = \frac{4}{4g(1/2, x) - g(1, x)} \tag{84}$$

and the Debye function

$$g(a, x) \equiv 2[ax + \exp(-ax) - 1]/x^2, \tag{85}$$

where $x = (qR)^2$ and R is the radius of gyration of an ideal chain of N monomers. It is worthwhile to note that in the limit of small q, the effective interaction becomes $V_{\text{eff}} \to 24/q^2R^2$. As noted earlier, this Coulombic form reflects the fact that the two parts of the diblock cannot be separated over large distances. This form of the effective interaction serves as the basis of the charge-frustrated Ising models of Sec. II.A.5.

The effective interaction has a minimum at q^* where $x^* \equiv (q^*R)^2 \approx 4$, and when this minimum value is reduced to zero, which occurs at the value $2\chi_c N \equiv I(x^*)$, the transition from the disordered phase to the lamellar phase with wave vector q^* occurs. It is instructive to expand the effective potential about q^*,

$$V_{\text{eff}}(\mathbf{q}) \approx I(x^*) - 2\chi N + (1/2)I''(x^*)[q^2 - (q^*)^2]^2 R^4 \tag{86}$$

to insert this approximation into the free energy, Eq. (82), and then to express this free energy in terms of the real-space order parameter $\Phi(\mathbf{r})$. We obtain

$$F = \frac{\rho k_BT}{2N}\int d^3r\{a_0\Phi^2(\mathbf{r}) + g_0(\nabla\Phi(\mathbf{r}))^2 + c_0[\nabla^2\Phi(\mathbf{r})]^2\} \tag{87}$$

where ρ is the monomer density. Note that this is precisely the form of the Ginzburg–Landau free energy used to describe the amphiphilic system in Eq. (16) of Sec. II.B.1. But in this case, the coefficients have been derived from the underlying microscopic model. They are

$$a_0 = \frac{(x^*)^2 I''(x^*)}{2}\left[1 + \frac{4(\chi_c N - \chi N)}{(x^*)^2 I''(x^*)}\right] \tag{88}$$

$$g_0 = -x^* I''(x^*) R^2 \tag{89}$$

and

$$c_0 = \frac{I''(x^*) R^4}{2} \tag{90}$$

From these, we calculate the ratio $\gamma = g_0/(4a_0 c_0)^{1/2}$ and obtain

$$\gamma = -\left[1 + \frac{4(\chi_c N - \chi N)}{(x^*)^2 I''(x^*)}\right]^{-1/2} \tag{91}$$

We see that in the disordered phase, for which $(\chi_c N - \chi N) > 0$, the parameter γ is negative and greater than -1. At the transition to the lamellar phase, $\gamma = -1$.

That this microscopic calculation produces a Ginzburg–Landau free energy of the form of Eq. (87) shows that this form indeed describes phases with structure characterized by some nonzero length. The major, and important, difference between the length scale in this problem of a diblock melt and that of an amphiphile in oil and water is that in the diblock the characteristic length is provided by the extended structure of the diblock itself so that the radius of gyration R sets the scale of all wave vectors. In the ternary amphiphile system, the characteristic length is much larger than the size of the amphiphile.

Also instructive is the form of the Landau free energy one obtains for the ternary system of A and B homopolymers and symmetrical AB diblocks, all of index N. For the case in which the homopolymers each have volume fraction $(1 - \theta)/2$ and the diblock has volume fraction θ that is not too large, $\theta \leq 2/3$, the free energy in terms of the real-space order parameter $\Phi(\mathbf{r})$ is [166]

$$F = \frac{k_B T \rho}{2N}\int d^3r\{c_0[\nabla^2\Phi(\mathbf{r})]^2 + g_0[\nabla\Phi(\mathbf{r})]^2 + a_0\Phi^2(\mathbf{r}) + \mathcal{B}\Phi^4(\mathbf{r}) + \mathcal{C}\Phi^6(\mathbf{r})\} \tag{92}$$

where ρ is again the density of monomers and the coefficients in the bulk free energy density are

$$a_0 = \frac{1}{1 - \theta} - \frac{\chi N}{2} \tag{93}$$

$$\mathcal{B} = \frac{2 - 3\theta}{12(1 - \theta)^3} \tag{94}$$

$$\mathcal{C} = \frac{8 - 25\theta + 20\theta^2}{120(1 - \theta)^5} \tag{95}$$

Finally the coefficients of the gradient terms are

$$g_0 = \frac{R^2}{3}\left[\frac{1}{1 - \theta} - \frac{(\chi N)^2\theta}{8}\right] \tag{96}$$

$$c_0 = \frac{R^4}{36}\left[\frac{1}{1 - \theta} + \frac{9(\chi N)^2\theta}{16} - \frac{(\chi N)^3\theta^2}{8}\right] \tag{97}$$

The free energy is again of the form of the Ginzburg–Landau free energy used to describe the amphiphilic system in Eq. (16) of Sec. II.B.1. From the bulk coefficients, we see that the consolute line obtained from $a_0 = 0$ is given by $(\chi N)^{-1} = (1 - \theta)/2$. As χ is inversely proportional to T, the transition temperature to A–B coexistence decreases linearly with the volume fraction of diblock. Furthermore, there is a tricritical point, given by $a_0 = \mathcal{B} = 0$, at $\chi N = 6$, $\theta = 2/3$. It is also very interesting to note that the coefficient g_0 becomes less positive as the concentration of diblock increases, just as one expects from the discussion of Sec. II.B.2. In fact, the Lifshitz line given by $g_0 = 0$ intersects the tricritical point, as is easily verified, indicating that this is a Lifshitz tricritical point [166,167].

Again, the form of the Ginzburg–Landau free energy reflects the fact that the phases it describes can be structured with some characteristic length. In the ternary polymer problem, we must determine whether this characteristic length is simply being provided by the radius of gyration R of the polymers or by the amphiphilic nature of the diblock. That is, if we examine the correlation function of all A and B monomers, there is always significant correlation at distances R because many of the monomers are part of diblocks. To investigate the efficacy of the diblock to solubilize the homopolymers, we must examine the correlations of only those monomers that belong to the homopolymers. When this is done [166], we find that the diblock behaves as a weak amphiphile.

The process of generating the Landau expansion becomes more cumbersome as more terms in the expansion are kept, particularly if we are interested in describing the lyotropic phases that are periodic. In practice, one rarely goes beyond fourth-order terms in the order-parameter components or considers more than one set of wave vectors that are related to one another by the symmetry operations of the phase considered, i.e., two independent wave vectors for the lamellar phase, three for the hexagonal packing of cylinders, and four for the bcc packing of spheres. Using such simplifying restrictions, Leibler [34] obtained an approximate phase diagram for a pure diblock copolymer system. Because of the approximations involved, it is expected to be accurate only near the transition from the disordered to the ordered phases. This use of the Landau–Ginzburg free energy is completely equivalent to the generation of phase diagrams for the small amphiphile–water system of Gompper and Klein [50].

The cumbersome procedure of generating the Landau expansion from the underlying mean-field theory has recently been made unnecessary because the problem of obtaining a solution to the mean-field equations with any desired symmetry and without further approximation was solved by Matsen and Schick [77]. All functions of position are expanded in a complete set of states that possess the desired symmetry, so that one is left with the equivalent self-consistent equations expressed in terms of the coefficients in the expansion. These equations can be solved numerically to whatever accuracy is possible with the guarantee that the solution has the desired symmetry. In this way solutions with up to 450 different wave vectors have been obtained, providing accuracy to very large incompatibilities χ or, equivalently, to very low temperatures.

C. Results

1. Lyotropic Phases in Binary Systems

Matsen [80,81] investigated the phase diagrams of several binary mixtures of AB diblocks and A homopolymers. Diblocks of varying A fraction f and fixed polymerization index N were considered and blended with homopolymers of varying polymerization index αN. Not surprisingly we find that if the copolymer consists mostly of B monomers (f small), the A homopolymers do not blend with it, and phase separation occurs. However, as f increases, the solubility of the system increases, and one finds several lyotropic phases. There is a normal lamellar phase, and a lamellar phase in which the minority B lamellae are pierced by a hexagonal array of holes containing A monomers. This is denoted "hexagonal perforated lamellar." There is also a bicontinuous cubic gyroid phase of symmetry group $Ia\bar{3}d$, a phase of hexagonally packed cylinders, and one of a body-centered cubic packing of spheres. Of interest here is that there are many systems (i.e., of given f and α) for which we observe with increasing AB diblock content the same progression from the hexagonal array of cylinders to the cubic $Ia\bar{3}d$ to the lamellar phase as is observed in amphiphilic systems. While this progression has been found in numerous theoretical treatments previously, they are all of a phenomenological character using as a starting point either a membrane Hamiltonian [168], which depends on unknown elastic constants, or a Ginzburg–Landau Hamiltonian, which also depends on unknown coefficients [44]. Matsen's calculations are the first that begin with a simple, well-characterized microscopic Hamiltonian of an amphiphilic system, obtain the free energy within mean-field theory, and show that these lyotropic phases appear, and in the order in which they are observed. Recently [78] an extensive Monte Carlo simulation of a binary system employing the polymer chain model described in Sec. II.A.4 produced the observed sequence of lyotropic phases. In addition, the transition temperatures of the various phases relative to one another and the amphiphile concentrations at which the phases appear are given rather well. This work demonstrates impressively that simple microscopic models can reproduce the observed sequence of lyotropic phases.

2. Swelling and Unbinding Transitions

One of the most interesting experimental phenomena that have been reproduced by calculations like those above is that of the unbinding of the lamellar phase. A spectacular example occurs in the system of $C_{12}E_5$ and water [79]. At temperatures of less than 20°C, a lamellar phase is observed at large amphiphile concentrations. As the amphiphile concentration is reduced, there is a transition to a cubic phase, which is most probably a gyroid phase. At even lower concentrations, the cubic phase undergoes a transition to a hexagonally packed phase of cylinders. This is just the sequence noted previously. Above 20°C, the lamellar phase is the only ordered phase observed. Between 20°C and about 54°C, it contains at most 50% water and the wavelength is no larger than 70 Å. Additional water destabilizes the phase, producing a disordered one. However, at 54°C a remarkable transition occurs. Additional water swells the lamellar phase almost indefinitely without destroying the order. The wavelength increases to approximately 3000 Å! An increase in wavelength without limit as the volume fraction of water is increased is denoted "complete unbinding" [169], and the change in the maximum wavelength of the lamellar phase from a finite value to an infinite one is denoted an unbinding transition. It is this unusual transition that has been reproduced for the first time from a microscopic calculation of the properties of an AB diblock in an A homopolymer solvent [80,81].

The question arises as to the nature of the unbinding transition. There are two possibilities. As the temperature is raised, the wavelength of the lamellar phases increases continuously and finally diverges at some temperature, or the wavelength jumps discontinuously to infinity at the transition. The former is denoted critical unbinding [170], the latter, first-order unbinding. In principle, either can occur [171]. This question was studied [82] for a particular system of a symmetrical ($f = 1/2$) AB diblock copolymer in a solvent of A homopolymer of equal polymerization index. The result is that the unbinding transition is first-order. It was shown that a consequence of the first-order nature is that at temperatures just above the unbinding transition, two lamellar phases of different wavelengths can coexist. It would be very interesting to look for such a coexistence of lamellar phases in the $C_{12}E_5$-water system.

3. Ternary Mixtures

Given the knowledge of the binary AB diblock–A homopolymer system, it is clearly interesting to see what changes are brought about by adding the B homopolymer. The simple lamellar phases have been investigated in such a system by Janert and Schick [172] using the self-consistent mean-field theory. In particular, the case of a symmetrical diblock, $f = 0.5$, and equal lengths of all polymers was considered. At low temperatures, there is a large three-phase coexistence triangle between a symmetrical lamellar phase of short period and A-rich and B-rich homopolymer fluids. As the temperature increases, a transition occurs above which the symmetrical lamellar phase of short period now coexists with an A-rich lamellar phase of bilayers and a B-rich lamellar phase of bilayers. The bilayer lamellar phases swell and unbind on, or near, the binary sides of the Gibbs triangle. At even higher temperatures, three-phase coexistence vanishes at a tricritical point, above which the symmetrical lamellar phase, when diluted by A and B homopolymers, becomes unstable to coexisting A-rich and B-rich lamellar phases.

In summary, upon dilution, the symmetrical lamellar phase in this system does not swell indefinitely and unbind. Instead it always becomes unstable to the asymmetrical lamellar phases, which do swell and unbind. The phase diagram is shown in Fig. 13. This

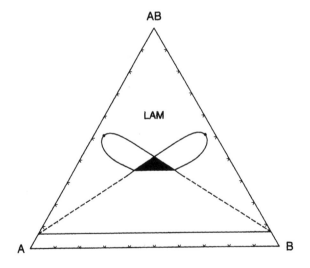

Figure 13   Phase diagram of a ternary mixture of AB diblock and A and B homopolymers, all of the same index of polymerization, N, at $\chi N = 11$. The lamellar phase (LAM) occupies the sole one-phase region and extends far down the binary sides where A- and B-rich lamellae undergo a complete unbinding transition. The symmetrical lamellar phase does not, becoming unstable to the asymmetrical lamellar phases. The shaded triangle is a three-phase coexistence region between the symmetrical and two asymmetrical lamellar phases. Two-phase coexistence regions emanate from it. The upper two end in critical points, shown by dots. The largest such region, shown with two of its sides in dashed lines, is between the asymmetrical lamellar phases. At AB volume fractions less than about 10%, there is two-phase coexistence between disordered A and B regions. (From Ref. 172.)

behavior is similar to that exhibited by oil–water–small amphiphile systems. An example is provided by the phase diagram of $C_{10}E_5$–water–octane [173] in which the lamellar phase almost unbinds completely along both binary sides, with amphiphile content barely a few percent in the bilayer phases. Yet the symmetrical lamellar phase becomes unstable when the amphiphile content is reduced to about 30%. The one significant difference between the calculated and measured phase diagrams is that in the former, the symmetrical lamellar phase is unstable to the asymmetrical lamellar phases, while in the latter it is unstable to a microemulsion.

We want to mention parenthetically that the characteristic shape of the one-phase region of the lamellar phase in the $C_{10}E_5$–water–octane phase diagram can also be understood from a calculation based on an interfacial model of membranes [174]. In this model, the undulation modes are responsible for the swelling of the lamellar phase; a disordered phase (which is modeled as a dilute droplet phase for simplicity) competes with the symmetric lamellar phase in the lower part of the phase triangle.

VI. SUMMARY AND OUTLOOK

We have shown in this review of microscopic theories of microemulsions that the theoretical description of amphiphilic systems has made enormous progress during the last decade. Much of the structure and phase behavior of these system is now well understood qualitatively. In some cases, it has even been possible to obtain quantitative agreement with experiment. A state of theoretical understanding has been reached in which an equally good description of specific systems seems possible.

We have also seen that there are strong similarities between systems containing small amphiphilic molecules on the one hand and diblock copolymers on the other. In both cases, the amphiphiles contain within themselves the properties of the components of two mutually insoluble liquids. The theory of diblock copolymers is more advanced than that of the small molecular systems not only because a simple microscopic model describes most properties of the polymers very well, but also because these properties depend on large-scale behavior of the chains, not small-scale behavior of the monomers. Furthermore, the large polymerization index guarantees that thermal fluctuations are less important than in small molecular systems, so that mean-field theories give very reliable results.

It is interesting to note that in biological systems both types of amphiphiles are present. Bilayers that establish a cell's perimeter consist of small amphiphilic lipids with hydrophilic heads and hydrophobic tails, whereas the proteins that undergrid the membrane or are embedded in it are simply polymers that consist of hydrophobic or hydrophilic amino acids. It is therefore not surprising that mixtures of polymers and small amphiphiles have received much attention recently. It is likely that the interest in biologically compatible and biologically inspired materials will bring together physicists, chemists, biologists, and materials scientists. The interdisciplinary aspect is one of the many attractions of this exciting field.

ACKNOWLEDGMENTS

We have been fortunate to have many stimulating collaborators, and we wish to acknowledge our debt to them: Jürgen Goos, Michael Hennes, Philipp Janert, Martin Kraus, Mark Matsen, Marcus Müller, Roland Netz, Friederike Schmid, and Ulrich Schwarz. Michael

Schick would like to thank the National Science Foundation, which has supported this work under past grants and the present one, DMR-9531161. Gerhard Gompper acknowledges the support of the work described in this review by the Deutsche Forschungsgemeinschaft through Sonderforschungsbereich 266.

REFERENCES

1. J. H. Schulman and J. B. Montagne, Ann. N.Y. Acad. Sci. *92*:366 (1961).
2. E. Ruckenstein and J. Chi, J. Chem. Soc., Faraday Trans. 2 *71*:1690 (1975).
3. Y. Talmon and S. Prager, J. Chem. Phys. *69*:2984 (1978).
4. P.-G. de Gennes and C. Taupin, J. Phys. Chem. *86*:2294 (1982).
5. L. E. Scriven, Nature *263*:123 (1976).
6. G. Gompper and M. Schick, in *Phase Transitions and Critical Phenomena*, Vol. 16 (C. Domb and J. Lebowitz, eds.), Academic, London, 1994, pp. 1176.
7. S. A. Safran, *Statistical Thermodynamics of Surfaces, Interfaces, and Membranes*, Addison-Wesley, Reading, MA, 1994.
8. W. M. Gelbart, A. Ben-Shaul, and D. Roux (eds.), *Micelles, Membranes, Microemulsions, and Monolayers*, Springer-Verlag, Berlin, 1995.
9. F. S. Bates, Science *251*:898 (1991).
10. J. C. Wheeler and B. Widom, J. Am. Chem. Soc. *90*:3064 (1968).
11. B. Widom, J. Chem. Phys. *84*:6943 (1986).
12. M. Schick and W.-H. Shih, Phys. Rev. Lett. *59*:1205 (1987).
13. M. Blume, V. Emery, and R. B. Griffiths, Phys. Rev. A *4*:1071 (1971).
14. G. Gompper and M. Schick, Phys. Rev. Lett. *62*:1647 (1989).
15. G. Gompper and M. Schick, Phys. Rev. B *41*:9148 (1990).
16. G. Gompper and M. Schick, Phys. Rev. A *42*:2137 (1990).
17. J. W. Halley and A. J. Kolan, J. Chem. Phys. *88*:3313 (1988).
18. A. Ciach, J. S. Høye, and G. Stell, J. Chem. Phys. *90*:1214 (1989).
19. A. Ciach, J. S. Høye, and G. Stell, J. Chem. Phys. *90*:1222 (1989).
20. G. Gompper and M. Schick, Chem. Phys. Lett. *163*:475 (1989).
21. M. W. Matsen and D. E. Sullivan, Phys. Rev. A *41*:2021 (1990).
22. K. A. Dawson and Z. Kurtović, J. Chem. Phys. *92*:5473 (1990).
23. R. G. Larson, L. E. Scriven, and H. T. Davis, J. Chem. Phys. *83*:2411 (1985).
24. B. Smit, P. A. J. Hilbers, K. Esselink, L. A. M. Rupert, N. M. van Os, and A. G. Schlijper, Nature *348*:624 (1990).
25. B. Smit, P. A. J. Hilbers, K. Esselink, L. A. M. Rupert, N. M. van Os, and A. G. Schlijper, J. Phys. Chem. *95*:6361 (1991).
26. B. Smit, K. Esselink, P. A. J. Hilbers, N. M. van Os, L. A. M. Rupert, and I. Szleifer, Langmuir *9*:9 (1993).
27. R. G. Larson, J. Chem. Phys. *89*:1642 (1988).
28. R. G. Larson, J. Chem. Phys. *91*:2479 (1989).
29. R. G. Larson, J. Chem. Phys. *96*:7904 (1992).
30. D. Stauffer, N. Jan, and R. Pandey, Physica A *198*:401 (1993).
31. D. Stauffer, N. Jan, Y. He, R. B. Pandey, D. G. Marangoni, and T. Smith-Palmer, J. Chem. Phys. *100*:6934 (1994).
32. F. H. Stillinger, J. Chem. Phys. *78*:4654 (1983).
33. D. Wu, D. Chandler, and B. Smit, J. Phys. Chem. *96*:4077 (1992).
34. L. Leibler, Macromolecules *13*:1602 (1980).
35. T. Ohta and K. Kawasaki, Macromolecules *19*:2621 (1986).
36. H.-J. Woo, C. Carraro, and D. Chandler, Phys. Rev. E *52*:6497 (1995).
37. H.-J. Woo, C. Carraro, and D. Chandler, Phys. Rev. E *53*:R41 (1996).
38. M. W. Deem and D. Chandler, Phys. Rev. E *49*:4268 (1994).

39. M. W. Deem and D. Chandler, Phys. Rev. E *49*:4276 (1994).
40. M. Teubner and R. Strey, J. Chem. Phys. *87*:3195 (1987).
41. K. Chen, C. Jayaprakash, R. Pandit, and W. Wenzel, Phys. Rev. Lett. *65*:2736 (1990).
42. G. Gompper and M. Schick, Phys. Rev. Lett. *65*:1116 (1990).
43. Y. Levin, C. Mundy, and K. Dawson, Phys. Rev. A *45*:7309 (1992).
44. J. Lerczak, M. Schick, and G. Gompper, Phys. Rev. A *46*:985 (1992).
45. A. Ciach, J. Chem. Phys. *104*:2376 (1996).
46. D. Roux, M. E. Cates, U. Olsson, R. C. Ball, F. Nallet, and A. M. Bellocq, Europhys. Lett. *11*:229 (1990).
47. D. Roux, C. Coulon, and M. Cates, J. Phys. Chem. *96*:4174 (1992).
48. G. Gompper and M. Schick, Phys. Rev. E *49*:1478 (1994).
49. K. Kawasaki and T. Kawakatsu, Physica A *164*:549 (1990).
50. G. Gompper and S. Klein, J. Phys. II (France) *2*:1725 (1992).
51. G. Gompper and U. Schwarz, Z. Phys. B *97*:233 (1995).
52. M. A. Anisimov, E. E. Gorodetsky, A. J. Davydov, and S. Kurliandsky, Liquid Cryst. *11*:941 (1992).
53. A. Ciach and A. Poniewierski, Phys. Rev. E *52*:596 (1995).
54. H. Löwen, Phys. Rep. *237*:249 (1994).
55. M. M. Telo da Gama and K. E. Gubbins, Mol. Phys. *59*:227 (1986).
56. M. M. Telo da Gama, Mol. Phys. *62*:585 (1987).
57. A. Somoza, E. Chacon, L. Mederos, and P. Tarazona, J. Phys. Condens. Matter *7*:5753 (1995).
58. J. R. Gunn and K. A. Dawson, J. Chem. Phys. *91*:6393 (1989).
59. M. Laradji, O. G. Mouritsen, S. Toxvaerd, and M. J. Zuckermann, Phys. Rev. E *50*:1243 (1994).
60. K. Kawasaki and T. Kawakatsu, Physica A *167*:690 (1990).
61. I. Szleifer, A. Ben-Shaul, and W. Gelbart, J. Chem. Phys. *86*:7094 (1987).
62. I. Szleifer, D. Kramer, A. Ben-Shaul, D. Roux, and W. M. Gelbart, Phys. Rev. Lett. *60*:1966 (1988).
63. I. Szleifer, D. Kramer, A. Ben-Shaul, W. M. Gelbart, and S. A. Safran, J. Chem. Phys. *92*:6800 (1990).
64. R. S. Cantor and P. M. McIlroy, J. Chem. Phys. *91*:416 (1989).
65. R. S. Cantor, J. Chem. Phys. *99*:7124 (1993).
66. A. Ben-Shaul and W. M. Gelbart, In *Micelles, Membranes, Microemulsions, and Monolayers* (W. M. Gelbart, A. Ben-Shaul, and D. Roux, eds.), Springer-Verlag, New York, 1994, pp. 1–104.
67. A. N. Semenov, Sov. Phys. JETP *61*:733 (1985).
68. S. T. Milner and T. A. Witten, J. Phys. (France) *49*:1951 (1988).
69. Z.-G. Wang and S. A. Safran, J. Chem. Phys. *94*:679 (1991).
70. M. C. Woods, J. M. Haile, and J. P. O'Connell, J. Phys. Chem. *90*:1875 (1986).
71. K. Watanabe, M. Ferrario, and M. Klein, J. Phys. Chem. *92*:819 (1988).
72. P. Ahlström and H. J. C. Berendsen, J. Chem. Phys. *97*:13691 (1993).
73. M. Kahlweit, R. Strey, M. Aratono, G. Busse, J. Jen, and K. V. Schubert, J. Chem. Phys. *95*:2842 (1991).
74. D. Furman, S. Dattagupta, and R. B. Griffiths, Phys. Rev. B *15*:441 (1977).
75. M. W. Matsen, M. Schick, and D. E. Sullivan, J. Chem. Phys. *98*:2341 (1993).
76. G. Gompper and M. Kraus, Phys. Rev. E *47*:4301 (1993).
77. M. W. Matsen, and M. Schick, Phys. Rev. Lett. *72*:2660 (1994).
78. R. G. Larson, J. Phys. II (France), *6*:1441 (1996).
79. R. Strey, R. Schomäcker, D. Roux, F. Nallet, and U. Olsson, J. Chem. Soc., Faraday Trans. *86*:2253 (1990).
80. M. W. Matsen, Phys. Rev. Lett. *74*:4225 (1995).
81. M. W. Matsen, Macromolecules *28*:5765 (1995).
82. P. K. Janert and M. Schick, Phys. Rev. E *54*:R33 (1996).
83. M. Kahlweit, R. Strey, and P. Firman, J. Phys. Chem. *90*:671 (1986).
84. M. Kahlweit, R. Strey, and G. Busse, Phys. Rev. E *47*:4197 (1993).

85. G. Gompper and J. Goos, in *Annual Reviews of Computational Physics*, Vol. II (D. Stauffer, ed.), World Scientific, Singapore, 1995, pp. 101–136.
86. R. Strey, Colloid Polym. Sci. *272*:1005 (1994).
87. M. Teubner, Europhys. Lett. *14*:403 (1991).
88. W. Gozdz and R. Holyst, Phys. Rev. Lett. *76*:2726 (1996).
89. G. Gompper and S. Zschocke, Phys. Rev. A *46*:4836 (1992).
90. B. Widom, J. Chem. Phys. *90*:2437 (1989).
91. M. Skaf and G. Stell, J. Chem. Phys. *97*:7699 (1992).
92. A. Ciach and A. Poniewierski, J. Chem. Phys. *100*:8315 (1994).
93. K.-V. Schubert and R. Strey, J. Chem. Phys. *95*:8532 (1991).
94. K.-V. Schubert, R. Strey, S. R. Kline, and E. W. Kaler, J. Chem. Phys. *101*:5343 (1994).
95. L. Auvray, J.-P. Cotton, R. Ober, and C. Taupin, J. Phys. Chem. *88*:4586 (1984).
96. G. Gompper and M. Kraus, Phys. Rev. E *47*:4289 (1993).
97. W. Helfrich, Z. Naturforsch. *28c*:693 (1973).
98. G. Gompper and S. Zschocke, Europhys. Lett. *16*:731 (1991).
99. L. D. Landau and E. M. Lifshitz, *Theory of Elasticity*, Addison-Wesley, Reading, MA, 1959.
100. M. W. Matsen and M. Schick, Macromolecules *26*:3878 (1993).
101. J. Lerczak, M. Schick, and G. Gompper, Phys. Rev. A *46*:985 (1992).
102. F. Schmid and M. Schick, Phys. Rev. E *49*:494 (1994).
103. T. P. Stockfisch and J. C. Wheeler, J. Phys. Chem. *99*:6155 (1993).
104. M. Aratono and M. Kahlweit, J. Chem. Phys. *95*:8578 (1991).
105. P. D. Fleming III and J. E. Vinatieri, J. Colloid Interface Sci. *81*:319 (1981).
106. M. Schick, Ber. Bunsenges. Phys. Chem. *100*:272 (1996).
107. B. Widom, Langmuir *3*:12 (1987).
108. X.-L. Zhou, L.-T. Lee, S.-H. Chen, and R. Strey, Phys. Rev. A *46*:6479 (1992).
109. R. Lipowsky and M. E. Fisher, Phys. Rev. B *36*:2126 (1987).
110. F. Schmid and M. Schick, J. Chem. Phys. *102*:7197 (1995).
111. O. Abillon, L. T. Lee, D. Langevin, and K. Wong, Physica A *172*:209 (1991).
112. M. Gradzielski, D. Langevin, T. Sottmann, and R. Strey, J. Chem. Phys. *104*:3782 (1996).
113. J. W. Cahn, J. Chem. Phys. *66*:3667 (1977).
114. M. Aratono and M. Kahlweit, J. Chem. Phys. *97*:5932 (1992).
115. J. Putz, R. Holyst, and M. Schick, Phys. Rev. A *46*:3369 (1992).
116. J. Putz, R. Holyst, and M. Schick, Phys. Rev. E *48*:635 (1993).
117. G. Gompper and M. Hennes, J. Chem. Phys. *102*:2871 (1995).
118. L.-J. Chen and W.-J. Yan, J. Chem. Phys. *98*:4830 (1993).
119. L.-J. Chen, W.-J. Yan, M.-C. Hsu, and D. L. Tyan, J. Chem. Phys. *98*:1910 (1994).
120. S. Dietrich and M. Schick, Phys. Rev. B *31*:4718 (1985).
121. L. Golubovic and T. C. Lubensky, Phys. Rev. B *39*:12110 (1989).
122. P.-G. de Gennes and J. Prost, *The Physics of Liquid Crystals*, Clarendon Press, Oxford, 1995.
123. P. Kekicheff and H. K. Christenson, Phys. Rev. Lett. *63*:2823 (1989).
124. O. Abillon and E. Perez, J. Phys. (France) *51*:2543 (1990).
125. E. Siggia, B. Halperin, and P. Hohenberg, Phys. Rev. B *13*:2110 (1976).
126. P. Hohenberg and B. Halperin, Rev. Mod. Phys. *49*:435 (1977).
127. G. Gompper and M. Hennes, in *Short and Long Chains at Interfaces* (J. Daillant, P. Guenoun, C. Marques, P. Muller, and J. Tran Thanh Van, eds.), Editions Frontières, Gif-sur-Yvette, 1995, pp. 385–390.
128. J. van der Werff, C. de Kruif, C. Blom, and J. Mellema, Phys. Rev. A *39*:795 (1989).
129. T. S. Chow, Phys. Rev. E *50*:1274 (1994).
130. J. D. Ferry, *Viscoelastic Properties of Polymers*, Wiley, New York, 1980.
131. C. Mundy, Y. Levin, and K. Dawson, J. Chem. Phys. *97*:7695 (1992).
132. M. Hennes and G. Gompper, Phys. Rev. E *54*:3811 (1996).
133. G. H. Fredrickson and R. G. Larson, J. Chem. Phys. *86*:1553 (1987).
134. F. S. Bates, Macromolecules *17*:2607 (1984).

135. D. M. Kroll and J. M. Ruhland, Phys. Rev. A *23*:371 (1981).

136. G. Gompper and M. Hennes, Europhys. Lett. *25*:193 (1994).

137. G. Gompper and M. Hennes, J. Phys. II (France): *4*:1375 (1994).

138. R. Granek, M. E. Cates, and S. Ramaswamy, Europhys. Lett. *19*:499 (1992).

139. R. Granek and M. E. Cates, Phys. Rev. A *46*:3319 (1992).

140. O. Diat and D. Roux, J. Phys. II (France) *3*:9 (1993).

141. G. H. Fredrickson, J. Chem. Phys. *85*:5306 (1986).

142. M. E. Cates and S. T. Milner, Phys. Rev. Lett. *62*:1856 (1989).

143. R. Bruinsma and Y. Rabin, Phys. Rev. A *45*:994 (1992).

144. S. Ramaswamy, Phys. Rev. Lett. *69*:112 (1992).

145. O. Diat, D. Roux, and F. Nallet, J. Phys. II (France) *3*:1427 (1993).

146. A. Onuki and K. Kawasaki, Ann. Phys. *121*:456 (1979).

147. K. Koppi, M. Tirrell, and F. Bates, Phys. Rev. Lett. *70*:1449 (1993).

148. J. D. Gunton, M. San Miguel, and P. S. Sahni, in *Phase Transitions and Critical Phenomena*, Vol. 8 (C. Domb and J. Lebowitz, eds.), Academic, London, 1983, pp. 267–466.

149. E. Siggia, Phys. Rev. A *20*:595 (1979).

150. P. M. Chaikin and T. C. Lubensky (eds.), *Principles of Condensed Matter Physics*, Cambridge Univ. Press, Cambridge, UK, 1995.

151. M. Laradji, H. Guo, M. Grant, and M. J. Zuckermann, J. Phys. A *24*:L629 (1991).

152. M. Laradji, H. Guo, M. Grant, and M. J. Zuckermann, J. Phys. Condens. Matter *4*:6715 (1992).

153. G. Pätzold and K. Dawson, Phys. Rev. E *52*:6908 (1995).

154. T. Kawakatsu, K. Kawasaki, M. Furusaka, H. Okabayashi, and T. Kanaya, J. Chem. Phys. *99*:8200 (1993).

155. T. Kawakatsu, K. Kawasaki, M. Furusaka, H. Okabayashi, and T. Kanaya, J. Phys. Condens. Matter *6*:6385 (1994).

156. T. Kawakatsu, K. Kawasaki, M. Furusaka, H. Okabayashi, and T. Kanaya, J. Chem. Phys. *102*:2247 (1995).

157. J. Yao and M. Laradji, Phys. Rev. E *47*:2695 (1993).

158. F. Schmid and R. Blossey, J. Phys. II (France) *4*1195 (1994).

159. J. W. Cahn and J. E. Hilliard, J. Chem. Phys. *28*258 (1958).

160. Y. Levin, C. Mundy, and K. Dawson, Physica A *196*:173 (1993).

161. D. A. Hajduk, P. E. Harper, S. M. Gruner, C. C. Honeker, G. Kim, E. L. Thomas, and L. Fetters, Macromolecules *27*:4063 (1994).

162. M. F. Schulz, F. S. Bates, K. Almdal, and K. Mortensen, Phys. Rev. Lett. *73*:86 (1994).

163. I. W. Hamley, K. Koppi, J. H. Rosedale, F. S. Bates, K. Almdal, and K. Mortensen, Macromolecules *26*:5959 (1993).

164. S. F. Edwards, Proc. Phys. Soc. *85*:613 (1965).

165. E. Helfand and Y. Tagami, J. Chem. Phys. *56*:3592 (1972).

166. R. Holyst and M. Schick, J. Chem. Phys. *96*:7728 (1992).

167. D. Broseta and G. H. Fredrickson, J. Chem. Phys. *95*:8532 (1991).

168. S. A. Safran, L. A. Turkevich, and P. Pincus, J. Phys. (Paris) Lett. *45*:L69 (1984).

169. S. Leibler and R. Lipowsky, Phys. Rev. Lett. *58*:1796 (1987).

170. R. Lipowsky and S. Leibler, Phys. Rev. Lett. *56*:2541 (1986).

171. R. Lipowsky, J. Phys. II (France) *4*:1755 (1994).

172. P. K. Janert and M. Schick, Macromolecules *30*:3916 (1997).

173. R. Strey, Ber. Bunsenges. Phys. Chem. *97*:742 (1993).

174. U.S. Schwarz, K. Swamy, and G. Gompper, Europhys. Lett. *36*:117 (1996).

5

Lattice Monte Carlo Simulations of Micellar and Microemulsion Systems

Raj Rajagopalan
University of Florida, Gainesville, Florida

L. A. Rodriguez-Guadarrama and Sameer K. Talsania
University of Houston, Houston, Texas

I. INTRODUCTION

The complexity of the equilibrium phases and nonequilibrium phenomena exhibited by multicomponent oil–water–surfactant systems is amply demonstrated in numerous contributions in this volume. Therefore, the need for theoretical (and computational) methods that make the interpretation of experimental observations easier and serve as predictive tools is readily apparent. Excellent treatments of the current status of theoretical advances in dealing with microemulsions are available in recent monographs and compendia (see, e.g., Refs. 1–3 and references therein). These references deal with systems consisting of significant fractions of oil and water and focus on the different phases and intricate microstructures that develop in such systems as the surfactant and salt concentrations are varied. In contrast, the present chapter focuses exclusively on simulations, particularly on a first level introduction to the use of lattice Monte Carlo methods for modeling self-association and phase equilibria in surfactant solutions with and without an oil phase. Although results on phase equilibria are presented, we spend a substantial portion of the review on micellization in surfactant–water mixtures, as this forms the necessary first step in the eventual identification of the most essential parameters needed in computer models of surfactant–water–oil systems.

In developing any theoretical method, however, a number of decisions must be made in advance. These include, in addition to a reasonable idea of what specific descriptions and predictions will be sought from the theories or models, a decision on what level of microscopic details will be incorporated into the model. Such a decision is dictated by the current limitations of the theoretical tools (e.g., classical or statistical thermodynamic theories) or computational resources. For example, microscopic models of micellization and solubilization can, in principle, be approached at the molecular level with a detailed structural representation of the various components along with their energetic interactions. Our current understanding of molecular dynamics is sufficiently comprehensive and well established to permit such a detailed approach to the evolution of mesoscopic and macroscopic structures and phenomena in surfactant–oil–water systems. However, the computational task such an approach requires is so demanding that such an ambitious program is beyond the reach of currently available computational resources. This implies that a certain level of spatial and temporal coarse-graining is inevitable.

This chapter will focus on a simpler version of such a spatially coarse-grained model applied to micellization in binary (surfactant–solvent) systems and to phase behavior in three-component solutions containing an oil phase. The use of simulations for studying solubilization and phase separation in surfactant–oil–water systems is relatively recent, and only limited results are available in the literature. We consider a few major studies from among those available. Although the bulk of this chapter focuses on lattice Monte Carlo (MC) simulations,* we begin with some observations based on molecular dynamics (MD) simulations of micellization. In the case of MC simulations, studies of both micellization and microemulsion phase behavior are presented. (Readers unfamiliar with details of Monte Carlo and molecular dynamics methods may consult standard references such as Refs. 5–8 for background.)

II. MOLECULAR DYNAMICS (MD) SIMULATIONS

Molecular dynamics simulations using atomistic models, and hence relatively detailed potentials of interatomic interactions, of surfactants and solvent molecules [9,10] have been attempted for studying surfactant assemblies. However, as mentioned earlier, detailed atomistic modeling approaches demand intensive computations and as a result require drastic simplifications that prevent examination of certain aspects of structural or temporal features of the system. One such example is an a priori selection of the structure of the micelle itself in the simulations; this clearly precludes the potential use of the simulations to examine the self-assembly process.

An alternative approach is to use a coarse-grained structural model for the surfactant and the other components so that the self-assembly of the surfactants can be monitored in the simulations [11,12]. The early studies of Karaborni, Smit, and coworkers [13–15] are such an example. These studies are mostly qualitative and primarily demonstrate that the basic features of surfactant self-assembly can be modeled by simple MD simulations. However, the more recent work of Esselink et al. [16] takes advantage of parallel processing techniques and faster computers to provide some insights into the forces of importance for surfactant aggregation and the effects of surfactant structure on the size and shape of the micelle. Moreover, Esselink et al. [16] also present a preliminary study of the mechanism of oil solubilization inside micelles. We restrict ourselves here to some of the details of the simulation and the results presented in their paper.

A. Details of the MD Algorithm and Model

In the simplest representation, the structural components of the molecular species in the system are reduced to four types of "beads," namely, solvent bead, represented by s; solute bead, c; and bead h and bead t representing the basic elements of the headgroup and tail of the surfactant. These beads are assembled appropriately to make up the three different

---

* A preliminary, two-dimensional off-lattice Monte Carlo simulation of the self-assembly of neutral and ionic surfactants has been presented recently by Bhattacharya et al. [4], but the results reported are essentially what has been found through lattice Monte Carlo systems.

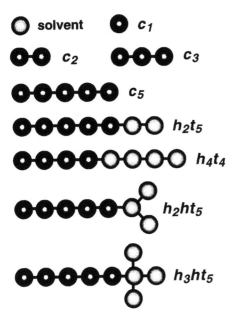

○ solvent ● c_1

c_2 c_3

c_5

h_2t_5

h_4t_4

h_2ht_5

h_3ht_5

Figure 1 Schematic drawing of the various types of surfactants used in the molecular dynamics (MD) simulations discussed in this chapter. The lighter spheres represent solvophilic beads (solvent bead or head bead), and the darker spheres are the solvophobic beads (tail bead or oil bead).

types of molecules present in the system; the solvent molecule, which is represented by just a single solvent bead; the oil (or solute) molecule, which may contain one or more connected beads; and the surfactant molecule, which consists of a combination of both head and tail beads. Figure 1 illustrates the different molecules used in the simulations [16].

The coarse-graining of the molecular structure also leads to a corresponding coarse-graining of the intramolecular and the intermolecular interactions. A number of options and variations exist for this purpose (see, e.g., Ref. 17 for details, discussed in the context of models used in polymer simulations). Esselink et al. [16] restrict the interactions to relatively simple expressions. For example, the bonds between the beads on the *same* molecule are represented by harmonic springs of length σ, the Lennard-Jones (LJ) size parameter, as

$$u_{bond}(r) = \frac{1}{2}\kappa(r - \sigma)^2 \tag{1}$$

where r is the distance between the beads. The magnitude of the spring constant κ is chosen such that 98% of the connected beads are within 2% of σ at all times.

The molecules interact according to a cut-and-shifted Lennard-Jones potential, $u_{S-LJ}(r)$, written in terms of the intermolecular distance r between two beads on two different molecules as

$$u_{S-LJ}(r) = \begin{cases} 4\epsilon[(\sigma/r)^{12} - (\sigma/r)^6] + |u_{LJ}(R_c)| & \text{if } r < R_c \geq 1.12\sigma \\ 0 & \text{otherwise} \end{cases} \tag{2}$$

where ϵ is the minimum in the usual Lennard-Jones potential, and

$$u_{LJ}(R_c) = 4\epsilon\left[\left(\frac{\sigma}{R_c}\right)^{12} - \left(\frac{\sigma}{R_c}\right)^6\right]$$

The choice of the cutoff distance R_c in Eq. (2) serves to differentiate the different types of interactions (e.g., solute–solute, solvent–solvent, solute–solvent). For example, the solute–solute and solvent–solvent interactions are truncated at $R_c = 2.5\sigma$; this cutoff length is large enough to include both excluded-volume (repulsive) and attractive forces [note that Eq. (2) is repulsive for $r \leq 1.12\sigma$]. The solute–solvent interactions, however, are cut off at $R_c = 1.12\sigma$ so as to include only the repulsive part of the potential. For most surfactants, the head and tail beads are considered equivalent to solvent and solute beads, respectively, for the purpose of calculating interactions. However, the head–head interaction can also be cut off at 1.12σ in order to study the effects of headgroup repulsion. The distances and the energies are usually scaled with respect to σ and ϵ, respectively. In the results reported by Esselink et al. [16], the interaction between surfactants is represented by the full Lennard-Jones potential except in the case of the bulky headgroups on branched h_3ht_5 surfactants, whose interactions are assumed to be purely repulsive.

The equations of motion can be solved numerically using the Verlet integration scheme [18]. Esselink et al. [16] use a time step of $\delta t = 0.005\tau_0$, where $\tau_0 = \sigma(m/\epsilon)^{1/2}$ and m is the mass of the bead. The temperature is kept constant in the canonical ensemble by scaling the velocities every 200 time steps. The results discussed below correspond to a density $\rho = 0.7\sigma^{-3}$, with the simulations done on a $30.4\sigma \times 30.4\sigma \times 60.8\sigma$ box (corresponding to 39,304 beads). The concentration of surfactant ranges from 0.75 to 3.0% by volume, with the remainder of the sites being assigned as solvent. Any two surfactants whose minimum separation distance is no more than 1.5σ are considered part of the same micelle.

B. Results of MD Simulations

Despite the considerable amount of simplification, the above approach still requires too great a computational effort to allow one to follow the macroscopic properties and phase behavior. However, the model described here is sufficient to examine the link between the coarse geometric features of the surfactants and their effects on the structure of the aggregates. We focus on a binary (surfactant–solvent) system to illustrate this point. For example, the relation between the shape of the micelles and the geometrical parameters such as optimal headgroup area a_0, critical tail length l_c, and the volume of the tail v, is well known (see, e.g., Refs. 19–21). The critical packing parameter p, defined in terms of the above geometrical parameters as $p = v/a_0l_c$, can be used as a rough predictor of the shape of the micelle. In general, a magnitude of $p \leq 1/3$ corresponds to spherical micelles, whereas bilayers are preferred for $p = 1$.

The results of Esselink et al. [16] are generally consistent with the above prescriptions. For instance, the surfactant h_2t_5 initially forms spherical structures, which gradually change to disks and then to bilayers due to collisions. The critical packing parameter p for this surfactant is approximately equal to unity, which indicates that bilayers are indeed the expected structure. The linear surfactant h_4t_4, however, forms sphere-like micelles with aggregation numbers ranging from 15 to 80, contrary to the expectation based on the magnitude of the packing parameter, which corresponds to aggregates with low curvature. The source of this discrepancy becomes evident when one examines some snapshots from the simulation, which show that the headgroups are not fully extended into the solvent but instead occupy a greater area at the interface. Hence, the actual optimal headgroup area is larger than expected and decreases the value of the packing parameter, thus causing the formation of spherical micelles. The branched surfactant h_2ht_5 forms oblate micelles,

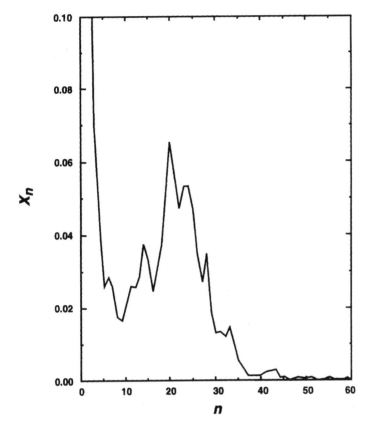

Figure 2 Micellar size distribution for h_3ht_5 as determined from MD simulations. (From Ref. 16.)

and the h_3ht_5 micelles are spherical, as expected from packing considerations. The micellar size distribution for h_3ht_5, shown in Fig. 2, is typical of ionic surfactants with large headgroups.

The results of Esselink et al. [16] also shed some light on the solubilization of oils in micelles and indicate three mechanisms of transfer of solute molecules from an "oil phase" into micelles. In one, solubilization results from the adsorption of surfactants on oil drops, which lowers the interfacial tension and causes the breakup of the oil drop into smaller droplets. In the second, the solute molecules that leave the drops are trapped by micelles in the immediate neighborhood of the drops. The last mechanism represents the exchange of surfactant and solute during a micelle–oil drop collision. The results of Esselink et al. also demonstrate quantitatively that the shorter solute molecule c_2 is solubilized to a greater extent (due to its higher solubility in the solvent) than the longer c_3.

These results show that simplified molecular dynamics simulations can qualitatively account for micellization quite well. However, the computation time necessary for even such simple models is too great to allow the model to be useful for the calculation of other micellar properties or phase behavior or for an in-depth study of solubilization. Stochastic dynamics simulations, in which the solvent effects are accounted for through a mean-field stochastic term in the equations of motion, can also be used to study surfactant self-assembly [22], but the most efficient approach to date is still the one based on lattice Monte Carlo simulations, which are discussed next.

III. LATTICE MONTE CARLO SIMULATIONS

As noted in Sec. I, the primary focus of this chapter is on lattice Monte Carlo simulations of micellization and solubilization. As is well known from statistical thermodynamics and mechanics of lattice models of solutions and lattice-gas automata, lattice theories offer several advantages. First, the discretization of space clearly simplifies the computations. In fact, in sufficiently simple cases (e.g., where the molecules can be represented by single beads or bonds between the nodes in the lattice), closed-form representations of the partition functions are possible, so that analytical solutions for the thermodynamic properties may be obtained. The excellent reviews of Gompper and Schick [1] and Gelbart et al. [23] can be consulted for examples of lattice models of micellar solutions and microemulsions based on the Ising model and the Potts model. For more complex cases, Monte Carlo simulations can be used to obtain the microscopic structure and macroscopic properties, as illustrated in the present chapter. The fact that bulk and interfacial properties of lattice fluids have been studied successfully for many years offers encouragement for exploring the same in the case of surfactant solutions and surfactant–solvent–oil mixtures. Lattice models offer a convenient way to examine the mesoscale structure in such systems—analytically for simpler models of the "architecture" of the components, and computationally for more complex models. The most popular computational schemes are based on the canonical (i.e., NVT) ensemble as it is the simplest and the most successful (since it does not require difficult moves such as "particle" insertions or volume changes).

In this section we present the details of a simple approach to simulating surfactant and micellar solutions. A basic introduction to the method, the types of problems for which it has been used, the specifics of the simulation method (including the details on the lattice model, the Monte Carlo algorithm, and the characterization methods) are presented.

A. Lattice Model

In the lattice model described here, a three-dimensional simple cubic lattice of size $L \times L \times L$ represents the discretized space, where L is in units of bond length between the beads in the molecules. All lattice points are assumed to be occupied either by a solvent molecule, denoted here by s; by a solute bead, c, of a solute molecule, $c_k, k \geq 1$; or by one of the beads of a surfactant molecule. The surfactant molecules are represented by a chain of head and tail beads and are denoted by $h_i t_j$, with $i \geq 1$ and $j \geq 1$. The head and tail beads are solvophilic and solvophobic, respectively, and are distinguished through an appropriate assignment of the energies of interactions with the solvent molecules. A surfactant chain contains at least one head bead and usually several tail beads, with each bead occupying one lattice site. The excluded-volume condition requires that each site may be occupied by only one bead. Periodic boundary conditions have been assumed in all three directions to mimic bulk conditions. One assumes that each bead interacts with its neighbors through an energy ϵ_{lm}, where the indices l and m represent the various types of molecular subunits considered in the simulation, i.e., $l, m = s, c, h$ or t. The total energy of the system is given by

$$\beta \mathcal{E}_{\text{total}} = \sum_{(l,m)} \beta \epsilon_{lm} \tag{3}$$

with $\beta = (k_B T)^{-1}$, where k_B is the Boltzmann constant and T is the absolute temperature. In the above equation, only the nearest-neighbor interactions of beads that are not adjacent on the same chain are considered. The neighborhood is usually taken to be the von Neumann neighborhood, i.e., each bead interacts only with the six sites immediately adjacent to it, although one may include additional coordination shells in the calculation (as discussed in Sec. IV).

B. Metropolis Monte Carlo Algorithm

The Monte Carlo simulations discussed here are based on the Metropolis algorithm [24]. For any given concentrations of the solute and the surfactant, appropriate numbers of c_k and $h_i t_j$ chains are randomly placed on the lattice sites, and the resulting configuration defines the initial configuration of the system. The total energy of this configuration, denoted by $\mathcal{E}_{old}$, is then calculated according to preassigned energy parameters, ϵ_{lm}. This initial configuration is then modified by moving a randomly chosen surfactant or solute chain. The new energy, $\mathcal{E}_{new}$, of this trial configuration is calculated, and the trial configuration is accepted or rejected according to the probability P, where

$$P = \begin{cases} 1 & \text{if } \beta\Delta\mathcal{E} = \beta(\mathcal{E}_{new} - \mathcal{E}_{old}) \leq 0 \\ e^{-\beta\Delta\mathcal{E}} & \text{otherwise} \end{cases} \tag{4}$$

and this procedure is continued until equilibrium is reached.

The Monte Carlo technique allows one to estimate the thermodynamic properties of the system. Once equilibrium is reached, a large number of configurations are generated, and the ensemble average of any property is calculated using the average of the required property over the equilibrium configurations. Snapshots are taken at regular intervals, during which several system properties are calculated. Properties are ensemble-averaged over a sufficient number of (≈ 200) snapshots, which are spaced about 1000 MC cycles apart to avoid correlations among the samples. However, for the results to be accurate, true equilibrium—and not a collection of metastable configurations—must be achieved. For this reason, the total energy of the system is continuously monitored in the simulations to ensure that the energy is stable and that equilibrium has indeed been achieved. However, it is often difficult to determine if the energy has truly stabilized due to large fluctuations, and therefore other system properties, discussed in Sec. III. D, are also monitored with respect to Monte Carlo time.

C. Monte Carlo Moves

Each Monte Carlo step involves the movement of a specific solute or surfactant chain. At the beginning of each MC step, a chain is chosen at random. A trial move of a part (or all) of this chain to a new site (or sites) is selected, as discussed in the next paragraph. If the new site(s) is (are) not occupied by another chain, the energy of this new (trial) configuration, $\mathcal{E}_{new}$, is calculated. The new configuration is then accepted or rejected according to the probability specified in Eq. (4).

A number of possible moves are available for moving the surfactant chains (the options for the movement of single beads or very short chains are clearly limited). Binder [17] may be consulted for a detailed discussion of the various possible moves (and for a discussion of how such moves can be made to satisfy microscopic reversibility conditions—and ergodicity conditions in the case of dynamic Monte Carlo simulations), especially for long-chain molecules such as polymer chains. Here we consider only a few of the many options.

If the randomly selected chain is a surfactant, the resulting Monte Carlo (trial) move is composed of two types of chain motions: reptation and local motions. If a solute is selected, only reptation is used. *Local motions* involve the movement of one or two beads in the chain, whereas *reptation* is a snakelike movement where one end of the chain is moved to a new site and the other beads are moved to the sites vacated by their neighbors in the chain; the net effect is a slithering motion. Reptation is a very efficient mode of chain rearrangement, for each bead of a chain is moved to a new site, whereas in the local motions only part of a chain is moved. Therefore, during each step, the number of trial local motions performed is equal to the number of beads per chain (i.e., the length of the chain), while only one trial reptation is performed per step.

Other chain rearrangement methods are also available, such as the configurational bias move, which has been used by Mackie et al. [25]. In this scheme, instead of a local move, entire surfactant chains are removed from the lattice and regrown from a random site. A *Rosenbluth weight*, which is a measure of the number of vacant sites around a chain, is calculated for both the initial and final configurations. The ratio of these weights is used in the acceptance criteria for this "move" such that all configurations are sampled in proportion to their correct probability. The configurational bias scheme allows for faster rearrangement of molecules, as an entire surfactant chain can be moved at once, although with a lower probability of acceptance. It also allows for greater flexibility of the chains and makes it easier to completely sample the entire phase space. However, for the short length of molecules considered here, these methods would not provide a substantial improvement over reptation and local motions. The efficiencies of the chain movements used have been investigated and are presented elsewhere [26].

1. Reptation

The specific steps in performing a reptation move are as follows. First, one end of the chain, either the head or the tail, is chosen randomly to be the lead end of the reptation. A site is then chosen randomly from the five, potentially available, nearest neighbors of the lead end (the sixth site is occupied by the bead adjacent to the end and is not considered). If this site is vacant (i.e., not occupied by a bead from another chain), then the end is moved to this new site and all other beads of the chain follow the end; otherwise, no move is made. Either way, the reptation move is completed, the new configuration is temporarily recorded, and the Monte Carlo acceptance criterion described above is used to determine whether this move is to be kept or not.

2. Local Motions

The procedure for local motions is slightly more involved than the one for reptation. In the case of local motions, a bead from the initially selected chain is chosen randomly. What type of local motion is performed depends on (1) the position of the bead on the chain (i.e., whether it is at the end or middle of the chain) and (2) the orientation of the chain (i.e., the position of the other beads on the chain). These motions differ in only one major respect, namely, selection of the new site(s) to which the bead(s) will move. The final step is the same as for reptation; the move is made if and only if the selected site is vacant. Again, this is followed by the MC algorithm to determine whether to keep the move or to reject it.

If the selected bead lies at the end of the chain (see Fig. 3a) the local motion performed is called the *endflip*. The new site for the endflip is chosen randomly from the remaining four nearest neighbors of the second bead on the chain (i.e., excluding the sites already occupied by the first and third beads). If the bead chosen for the trial move is a middle bead, there are three different possibilities: (1) If the chain makes a 180° angle at the selected bead,

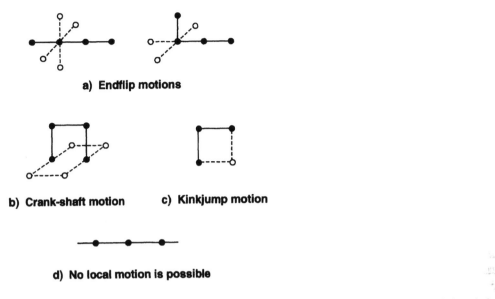

a) **Endflip motions**

b) **Crank-shaft motion**   c) **Kinkjump motion**

d) **No local motion is possible**

Figure 3 Possible local motions for a linear chain. The original conformation of the chain is indicated by solid lines and filled circles. The dashed lines and open circles show the new conformations allowed by the indicated local moves. (From Ref. 27.)

as in Fig. 3d, no move can be made; (2) if the chain makes a 90° angle at the bead, and the diagonally opposite lattice site is occupied by another bead on the same chain (see Fig. 3b), then the appropriate move is the *crankshaft* motion. In this motion, the selected bead and its adjacent bead are moved at a 90° angle to the adjacent plane; (3) If the diagonally opposite site is not occupied by the chain, as in Fig. 3c, then the move is a *kinkjump* [27].

D. Characterization Methods

Several properties can be calculated during the simulation to characterize the properties and structure of the system. These properties give information about the critical micelle concentration (cmc), the size and shape of the micelles, and the extent of solubilization of the solutes. Additionally, the number and type of contacts that the beads have inside the micelles are measured to give information about the micellar structure. All of these quantities are measured at each snapshot and are then averaged over several snapshots. Here, we consider only some of the simplest characterization methods, in particular, we do not discuss computation of radial distribution functions and the like.

Before we discuss the definitions of micellar properties, we first comment on how an aggregate is defined. Any surfactant chain with at least one tail segment that has a nearest-neighbor contact with a tail segment from another surfactant chain is considered to be part of an aggregate, which we will call a micelle. Nonaggregated surfactant chains are called free surfactants. Any solute chain that has at least one nearest-neighbor contact with a tail bead in a micelle is considered to be a solubilized or encapsulated solute. All other solute chains are considered to be unsolubilized solutes.

1. Micellar Molecular Weights

The micellar size is usually characterized by the aggregation number n, which is defined as the number of surfactant monomers that make up a micelle. However, since micelles are dynamic entities with surfactant monomers entering and leaving constantly, n is a fluctuating quantity. In general, the micelles are polydisperse and have a distribution of aggregation numbers. As a result, the size of a micelle is best represented by two quantities—n_{no}, the number-average aggregation number, and n_{wt}, the weight-average aggregation number—instead of a single aggregation number. These quantities are defined as

$$n_{no} = \frac{\sum_{m=1}^{M} n_m}{M} \tag{5}$$

and

$$n_{wt} = \frac{\sum_{m=1}^{M} n_m^2}{\sum_{m=1}^{M} n_m} \tag{6}$$

where M is the number of aggregates in a sufficiently large volume of the solution and n_m is the aggregation number of aggregate m. Note that free surfactants, which are not considered aggregates, are not used in this calculation. The polydispersity index, I_P, which is defined as n_{wt}/n_{no}, and an aggregate size distribution based only on the number of surfactants in an aggregate are also calculated. The aggregate size distribution is averaged over micelles of size $n \pm 2$, for $n \geq 10$, to reduce statistical fluctuation and to obtain a smooth distribution. Micelles of size $n < 10$, often referred to in the literature as submicellar aggregates, are not used in calculating certain micellar characteristics.

2. Micellar Shape and Structure

The shape of a micelle is hard to characterize, as there is no single well-defined numerical measure that represents all shapes. However, one can represent the micellar shape using the three principal moments of inertia, I_1, I_2, and I_3, where I_1 is the largest and I_3 the smallest. These moments are the eigenvalues of the matrix of radii of gyration, which are defined as [28]

$$R_{x_i, x_j}^2 = \frac{1}{N} \sum_{k=1}^{N} (x_{i,k} - x_{i,c.m.})(x_{j,k} - x_{j,c.m.}) \tag{7}$$

where x_i, $1 \leq i \leq 3$, represent the three lattice directions, and $x_{i,c.m.}$, the center of mass in direction i, is given by

$$x_{i,c.m.} = \frac{1}{N} \sum_{k=1}^{N} x_{i,k} \tag{8}$$

For a spherical molecule, all three principal moments would be equal, while for a cylinder with length greater than diameter, $I_1 = I_2 > I_3$. Thus, the ratio I_2/I_3 gives a rough estimate of whether a micelle is more spherical or more cylindrical.

Another quantity that is useful for characterizing the shape is the asphericity parameter A_s, defined by the equation [28]

$$A_s = \frac{\sum_{i>j=1}^{3}(I_i - I_j)^2}{2(\sum_{i=1}^{3} I_i)^2} \tag{9}$$

Table 1 Illustration of the Number of Solvent Contacts for Each Surfactant Bead of h_2t_4 (i.e., $h^{(1)}$–$t^{(4)}$) for Two Values N_s [a]

N_s	$h^{(1)}$	$h^{(2)}$	$t^{(1)}$	$t^{(2)}$	$t^{(3)}$	$t^{(4)}$
500	4.68	3.68	1.86	1.65	1.24	1.57
1000	4.66	3.66	1.80	1.55	1.16	1.42

[a] N_s = number of surfactants.

The asphericity parameter has a magnitude of 0 for spheres and 1 for an infinite cylinder, with the magnitudes in between representing ellipsoids ranging from sphere-like to elongated.

The micellar structure can also be inferred from the number of solvent contacts for each bead of the surfactant, from the first head bead, $h^{(1)}$, to the last tail bead, $t^{(4)}$. For a surfactant in a micelle, the average number of contacts with solvent, solute, and other head and tail beads can be measured for each bead starting with the first head bead, $h^{(1)}$, and ending with the last tail bead, $t^{(j)}$, and can be recorded for each bead individually. For example, the surfactant h_2t_4 would have six different measurements, one for each bead. The number of solvent contacts for each bead indicates the extent to which a solvent molecule penetrates into the micelle, and the number of solute contacts (see the following subsection) indicates whether the extent of solubilization is a function of depth inside the micelle. Also, this information can be used to calculate the internal and interfacial energies of each micelle. For example, Table 1, for a surfactant h_2t_4, shows the average number of solvent contacts for two concentrations of surfactant in the absence of solutes. The numbers of solvent contacts for the first and second head beads are almost at their maximum values—5 and 4, respectively—indicating that the headgroup mostly remains at the exterior of the micelle. Moreover, if one assumes that the extent of solvent penetration inside a micelle decreases with depth, then one can conclude from Table 1 that the tail beads in the interior of the micelle are disordered. That is, the fact that $t^{(4)}$ has more solvent contacts than $t^{(3)}$ shows that the tails of the micelle do not all point toward the center of the micelle but are disordered. The same result has also been obtained from neutron scattering experiments [29]. In addition to individual bead–bead contacts, one can also construct the radial distribution functions for the various types of beads. We shall not consider these here, however.

3. Partition Coefficient

In the case of low concentrations of the oil molecules, it is also useful to determine the extent of solubilization in the simulations, which can be represented in terms of the partition coefficient, K. If the micelles are considered to be a separate phase (as in the phase separation model of micellization [20,21]), then the concentrations of the solute in the aqueous and micellar phases can be related as

$$K = \frac{C_{mic}}{C_{aq}} \tag{10}$$

where

$$C_{mic} = \frac{\text{number of solubilized solutes}}{\text{volume of aggregates}} \tag{11}$$

and

$$C_{aq} = \frac{\text{number of unsolubilized solutes}}{\text{volume of aqueous solution}} \tag{12}$$

The volume of the aggregates is approximated by the volume occupied by the surfactants and solutes in the aggregates but does not include encapsulated solvent. Then the aqueous volume is just the total volume of the lattice minus the aggregate volumes. We do not consider the solubilization phenomena here in detail; some preliminary information on this is given by Talsania [26].

As noted in the previous subsection, several types of contacts inside a micelle can also be measured directly. For solubilized solutes, the average number of contacts per solute bead with solvent, tail, and head beads—n_{cs}, n_{ct}, and n_{ch}, respectively—are counted. These values give information on the locus of solubilization, as discussed later in the results.

IV. REVIEW OF MC SIMULATION RESULTS

Within the last decade a number of studies have appeared in print on lattice Monte Carlo simulations of surfactant–water–oil mixtures, many focusing on an examination of the simulation method using micellization of short-chain amphiphiles (or block copolymers) in solution. An almost equal number of studies focus on micellar and microemulsion phase behavior. In what follows, we include a brief discussion of a few papers on block copolymers as well, as these are conceptually similar to the systems of interest in this chapter and shed light on what could be expected in short-chain surfactant systems to some extent.

For convenience, we first divide the studies considered here into four groups, although in many cases they contain issues of common interest:

1. Micellization, which forms a major focus of this chapter, is the simplest class of phenomena accessible through simulations and, perhaps arguably, is the first step that needs to be examined before embarking on more complex problems. Both micellization [30,31] and micellar phase behavior [32] have been studied by Care and coworkers for a *binary* surfactant–solvent system using simple lattice models of the type we discuss here. The recent work of Talsania et al. [33] includes a preliminary examination of micellar solubilization, in addition to micellization in surfactant solutions. The latter is the first attempt to examine solubilization (i.e., encapsulation) at low solute (i.e., oil) concentration in a systematic manner.

2. *Ternary* surfactant–solute (oil)–solvent (water) systems based on a model along the lines described in the previous section have been studied extensively by Larson and coworkers [34–38]. These studies focus mostly on the phase behavior of micelles and microemulsions, with an emphasis on the microstructures formed by the surfactant aggregates and on the microstructural transitions at high surfactant concentrations (10–80% by volume). Larson has also reported results on micellization at relatively low surfactant concentrations in the absence of solutes [37], which is of particular interest here. In addition, variations of Larson's model have been presented by others [39–41].

3. Recently, the phase diagrams in various ternary surfactant–solute–solvent systems [25] and micellization in binary surfactant–solvent systems [42] based on the model of Larson et al. have also been presented using different simulation methodologies (i.e., Gibbs ensemble simulations and configurational bias NVT ensemble simulations). We shall also comment on these briefly.

4. Finally, Mattice and coworkers have used lattice Monte Carlo simulations for various studies of micellization of block copolymers in a solvent, including micellization of triblock copolymers [43], steric stabilization of polymer colloids by diblock copolymers [44], and the dynamics of chain interchange between micelles [45]. Their studies of the self-assembly of diblock copolymers [46–48] are roughly equivalent to those of surfactant micellization, as a surfactant can in essence be considered a short-chain diblock copolymer and vice versa. In fact, Wijmans and Linse [49,50] have also studied nonionic surfactant micelles using the same model that Mattice and coworkers used for a diblock copolymer. Thus, it is interesting to compare whether the micellization properties and theories of long-chain diblock copolymers also hold true for surfactants.

Although the basic methodology (i.e., the type of simulation and the simulation algorithms) discussed in detail in Sec. III is roughly the same for all of the above studies, each of these studies differs in specific parameters and techniques used in the respective simulations. These variations—particularly in the types of molecular interactions used and in the number of nearest neighbors considered—can have a strong effect on the final results. Hence, a brief description of each model is given next in Sec. IV. A before the results are reviewed. The results of these Monte Carlo studies are then separated into two categories, those that focus on micellization (Sec. IV. B. 1) and those that focus on micellar phase behavior (Sec. IV. B. 2).

A. Models and Their Scopes

Most of the above simulations are performed on three-dimensional simple cubic lattices with periodic boundary conditions in all directions. (Some of the early studies were based on two-dimensional square lattices but have since been updated.) Additionally, all of the works discussed in this section (except where noted otherwise) use the standard Metropolis Monte Carlo algorithm discussed in detail in Sec. III. B, but the major difference lies in the selection of which of the components contribute to the total energy of the system. Other differences include the lattice rearrangement methodology and parameters such as surfactant structure, temperature, composition, lattice size, and dimensionality. The specifics of each model are summarized below.

The model used by Care and coworkers [30–32] uses only nearest-neighbor interactions on a cubic lattice, i.e., the coordination number $z = 6$. Note that the numbers of all pair contacts in a lattice system can be specified using only three independent contact parameters, since there are only six possible bead–bead interactions among the three different beads. Desplat and Care [31] chose the three independent contact parameters to be tail–solvent, head–solvent, and head–head interactions. The total dimensionless energy of the system, $\beta\mathcal{E}$, can then be written as

$$\beta\mathcal{E} = \beta\epsilon_{ts}\left(n_{ts} + \frac{\epsilon_{hs}}{\epsilon_{ts}}n_{hs} + \frac{\epsilon_{hh}}{\epsilon_{ts}}n_{hh}\right) \tag{13}$$

where n_{ts}, n_{hs}, and n_{hh} are the numbers of tail–solvent, head–solvent, and head–head contacts, respectively, and ϵ_{ts}, ϵ_{hs}, and ϵ_{hh} are the corresponding energies. To model a typical micellar solution, ϵ_{ts} is taken to be positive (indicating tail–solvent repulsion), and ϵ_{hs} is chosen negative (indicating solvophilicity of the head segment) or zero. Desplat and Care [31] consider negative values of ϵ_{hs} and ϵ_{hh} and focus on $h_1 t_3$ for surfactant concentrations from 0.249–25.6 vol% on lattices of size ranging from $40 \times 40 \times 40$ to $184 \times 94 \times 93$. The chain motion is restricted to reptation.

The simulations of Larson use cubic lattices of various sizes with a coordination number of $z = 26$; i.e., each site has 26 adjacent sites with all nearest neighbors and diagonal nearest neighbors being equivalent. The model also allows for diagonal bonds. No distinction is made, with respect to interaction energies, between the two solvophilic units (head and solvent), denoted by A, or the two solvophobic units (tail and solute), denoted by B; hence, only three energies are needed: ϵ_{AB}, ϵ_{AA}, and ϵ_{BB}. Any change in energy of the system must be a multiple of

$$\epsilon = \epsilon_{AB} - \frac{1}{2}\epsilon_{AA} - \frac{1}{2}\epsilon_{BB}$$

and therefore only one energy parameter (or inverse temperature), ϵ (in units of $k_B T$), is needed to fully specify the system. Larson selects $\epsilon_{AA} = 0$ and $\epsilon_{BB} = 0$ and monitors only solvophobic–solvophilic pairs. A number of surfactant structures have been considered, although we restrict ourselves here to the results for $h_1 t_1$ to $h_4 t_4$. Three different methods are used for the rearrangement of molecules:

1. *Singlet interchange*: Two single-bead molecules (solute or solvent) are exchanged, regardless of the distance separating the two.
2. *Chain twisting*: A chain unit is exchanged with any adjacent unit, as long as no chain is severed during the exchange.
3. *Reptation*: This is the standard snakelike chain movement. Details of this motion are described in Sec. III. C.

Panagiotopoulos and coworkers [51] use the same parameters as Larson for the study of phase behavior, but with two different simulation methodologies. The first technique is the Gibbs ensemble method, in which each bulk phase is simulated in a separate cell and molecules are interchanged and volumes adjusted between the two for equilibration of the system [52]. The second is a standard canonical ensemble simulation, like Larson's, but employs the configurational bias Monte Carlo method. The configurational bias Monte Carlo method is much more efficient than the ones based on reptation and other local moves but is not useful if any dynamic information is sought from the simulations.

Mattice and coworkers performed their simulations of diblock copolymer micellization on a cubic lattice, typically of dimensions $44 \times 44 \times 44$ and with a coordination number of $z = 6$. As in the case of Larson's model, only one energy parameter, ϵ, is considered for all interactions, namely, the interaction between the tail and the solvophilic (head and solvent) beads. The base structure used for the copolymer is $h_{10} t_{10}$. The chains are rearranged using reptation and Verdier–Stockmayer [53] type local motions, both of which are discussed in detail in Sec. III. C. Wijmans and Linse [50] also based their simulations on this model and surfactant structure.

The above summary reveals the various combinations of parameters that are possible even in simple coarse-grained models, and no systematic examination of the effects of such variations in the parameters has been undertaken so far. For example, parameters that are a measure of the spatial structure and size–e.g., the lattice coordination number, dimensionality, and size—can have an effect on the results, especially on the size of fluctuations. Other model parameters such as energies of interactions are critical to micellization properties, as would be expected. As a result, the same surfactant can display a totally different behavior under two different models, and hence the results of the studies mentioned above cannot be compared with each other easily. Nevertheless, internal comparisons within models are still useful in understanding micellar behavior, although how best to choose a model surfactant (and its energies of interactions) to mimic a particular real surfactant of interest remains an open question [54].

B. Typical Results

Although the earliest attempts on simulating surfactant aggregation and microemulsions date back to the 1980s, it is still too early to develop a comprehensive picture of insights gained from simulations. This situation is not the result of any inadequacy of simulation methodologies; rather, it is a consequence of the relatively small number of studies that are available and the need for identifying and focusing on the crucial (architectural and interaction) parameters in the simulations. As mentioned earlier, the level of computational efforts needed even in the case of sufficiently coarse-grained models has also contributed to the present status.

For example, it is difficult to discuss the results of the studies cited in the previous section in a collective manner, because each study (or each group of studies) considered different types of surfactants and oils, different energy parameters, different numbers of neighbors, or different definitions of the cmc or what constitutes an aggregate. Therefore, we restrict ourselves here to a general overview of the results from the literature.

1. Micellization

As mentioned earlier, studies of simple linear surfactants in a solvent (i.e, those without any third component) allow one to examine the sufficiency of coarse-grained lattice models for predicting the aggregation behavior of micelles and to examine the limits of applicability of analytical lattice approximations such as quasi-chemical theory or self-consistent field theory (in the case of polymers). The results available from the simulations for the structure and shapes of micelles, the polydispersity, and the cmc show that the lattice approach can be used reliably to obtain such information qualitatively as well as quantitatively. The results are generally consistent with what one would expect from mass-action models and other theoretical techniques as well as from experiments. For example, Desplat and Care [31] report micellization results (the cmc and micellar size) for the surfactant $h_1 t_3$ (for a temperature of $\omega^{-1} = k_B T / \epsilon_{ts} = \beta^{-1} / \epsilon_{ts} = 1.18$ and a solvophilic strength of $\zeta = \epsilon_{hs} / \epsilon_{ts} = -2.0$). The cmc of the system can be determined from a plot of free surfactant concentration X_{fs} against the total surfactant concentration X_s (in mole fraction) and can be defined in a number of ways. Desplat and Care define the cmc as the intercept of a line of unit slope through the origin and another line through the high concentration data. For $h_1 t_3$, the cmc is approximately $X_{cmc} = 0.006$, as seen in Fig. 4. The free surfactant concentration decreases beyond the cmc, as has been observed in experiments, and the micellar size distribution, shown in Fig. 5, exhibits a maximum and a minimum for concentrations greater than the cmc as expected from mass-action models. The weight-average aggregation number,

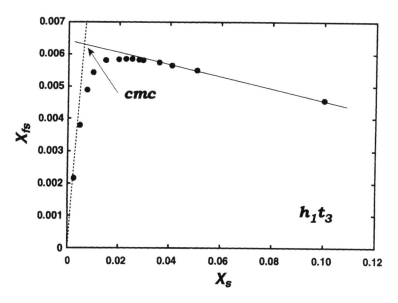

Figure 4   Free surfactant concentration X_{fs} as a function of total surfactant concentration X_s. The critical micellar concentration (cmc) is given by the intersection of the dashed line, which has a unit slope, and the line through the high concentration data. (From Ref. 31.)

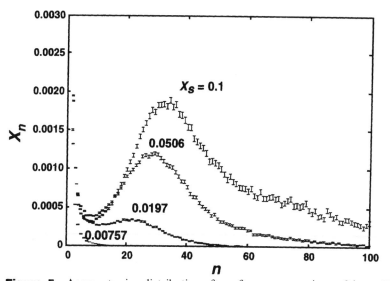

Figure 5   Aggregate size distributions for a few concentrations of h_1t_3. (From Ref. 31.)

n_{wt}, of the system also agrees well with Mukerjee's prediction [55] that

$$n_{\mathrm{wt}} \propto \left[\frac{X_s - X_1}{X_1}\right]^{1/2} \tag{14}$$

for concentrations above the cmc (Fig. 6).

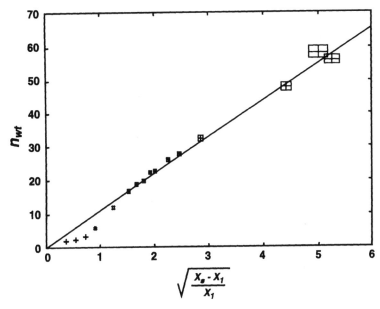

Figure 6   Weight-average aggregation number as a function of $[(X_s - X_1)/X_1]^{1/2}$. The best linear fit is also shown. (From Ref. 31.)

An examination of the effect of temperature ω^{-1} and solvophilic strength ζ on the system by Desplat and Care [31] also leads to physically reasonable results. For instance, an increase in temperature (from $\omega^{-1} = 1.08$ to $\omega^{-1} = 1.38$) leads to an increase in the cmc and a decrease in the characteristic micelle size, evidenced by Figs. 7 and 8, respectively. The results shown in Fig. 7 correspond to a linear increase in the cmc with respect to temperature in the range shown. However, the effect of solvophilic strength ζ is somewhat more complex. The model shows that micellar behavior can be expected only for a limited range of ζ (from -1.0 to -4.0). The cmc increases as ζ is decreased, but Figs. 9 and 10 show that the difference from $\zeta = -1.0$ to -2.0 is much greater than from $\zeta = -3.0$ to -4.0, indicating that the nearest neighbors of the head units become fully saturated by solvent at some sufficiently negative value of ζ.

Larson [37] also investigated micellization for various surfactant structures and concentrations in a binary surfactant–solvent system. For example, Fig. 11 shows that, for the surfactant h_2t_2, micelles do not form at an inverse temperature of $\epsilon = 0.1538$, but h_3t_3 does micellize at and above a surfactant concentration of $X_s = 0.04$, as seen by the peaks in Fig. 12. The surfactant h_1t_3 precipitates out of the solvent into a separate phase such that the heads are mixed in with the tails. If the heads were to repel each other, one would expect h_1t_3 to form micelles. Otherwise, the smallest surfactant that forms micelles is h_2t_3. These results cannot be compared directly with those of Care and coworkers because of the differences in the numbers of nearest neighbors and interaction energies used. Moreover, as noted earlier, the cmc can be defined in a number of different ways, and the exact magnitude obtained depends on the definition. For example, one can define the cmc as the concentration required to form a hump in the aggregate size distribution function (see, e.g., Fig. 5), i.e., when a significant volume fraction of surfactant resides in micelles instead of as monomers, or as the essentially constant singlet concentration beyond micellization. These definitions differ from the one used by Desplat and Care [31] illustrated in Fig. 4.

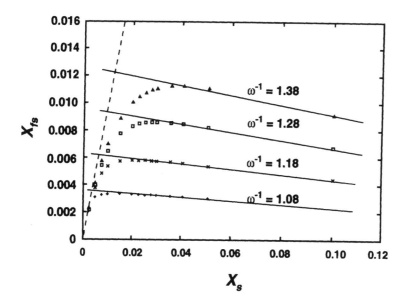

Figure 7 Temperature dependence of the free surfactant concentration as a function of the total surfactant concentration. A line of unit slope is shown for reference (dashed line). Lines through the high concentration regions are shown to guide the eyes. The points of intersection of these lines with the dashed line give the respective cmc's. Note that the temperature, $\omega^{-1} = \beta^{-1}/\epsilon_{ts}$, is in dimensionless units. (From Ref. 31.)

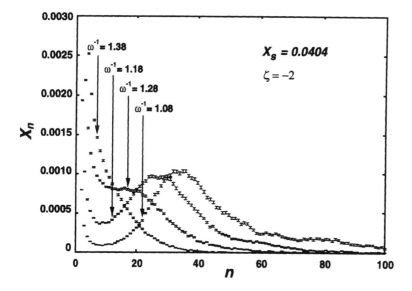

Figure 8 Temperature dependence of the aggregate size distribution for h_1t_3. Note that the temperature, $\omega^{-1} = \beta^{-1}/\epsilon_{ts}$, is in dimensionless units. (From Ref. 31.)

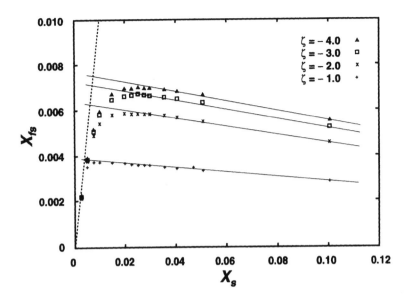

Figure 9 Free surfactant concentration versus total surfactant concentration for several values of the solvophilic strength ζ. The line of unit slope is shown for reference (dashed line). Lines through the high concentration regions are shown to guide the eyes. The points of intersection of these lines with the dashed line give the respective cmc's. (From Ref. 31.)

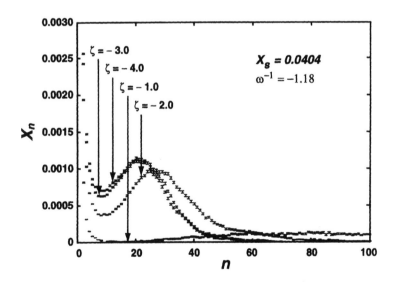

Figure 10 The dependence of the aggregate size distribution on solvophilic strength ζ for a solution of $h_1 t_3$. (From Ref. 31.)

Larson's studies show how lattice models can be used to investigate structural changes in surfactant systems. For example, Larson's results show that $h_4 t_4$ forms spherical micelles at low surfactant concentrations, up to $X_s = 0.2$. At these low concentrations, although the shape of the micelles remains the same, the micellar size grows with increasing

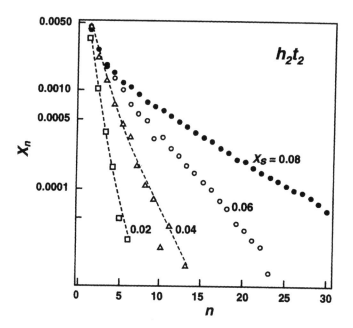

Figure 11 The aggregate size distribution for h_2t_2 for several values of the concentration X_s. (From Ref. 37.)

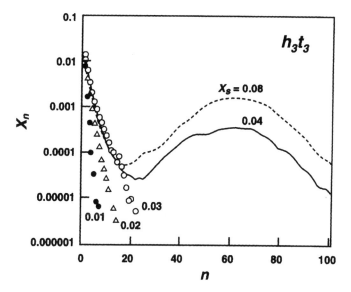

Figure 12 The aggregate size distribution for h_3t_3 for several values of the concentration X_s. (From Ref. 37.)

surfactant concentration X_s as seen by the shift in micelle peaks to the right in Fig. 13. At higher concentration, not only does the micellar size increase but the shape also changes. At about $X_s = 0.45$, the micelles become cylindrical, and if the concentration reaches about 0.7–0.8, the surfactants form a lamellar structure. During the transition from spherical to

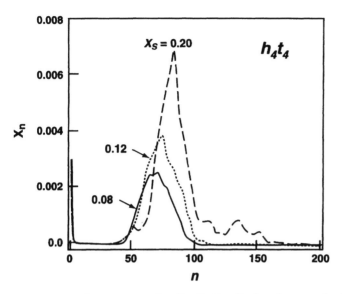

Figure 13 The aggregate size distribution for h_4t_4 for several values of the concentration X_s. (From Ref. 37.)

cylindrical micelles, which occurs around $X_s = 0.4$, spherical micelles coexist with oblong, cigar-shaped micelles of approximately equal diameters. The aggregation number for these oblong micelles, however, is much larger than that for the spherical ones. For an asymmetrical surfactant such as h_2t_6, a bicontinuous transitional structure can also form, in addition to spherical, cylindrical, and lamellar structures. For example, at $X_s = 0.2$, h_2t_6 forms spheres, and at 0.5 the aggregates become cylindrical, but at $X_s = 0.4$ the structure is irregular and bicontinuous. Interestingly, this structure is not seen in symmetric surfactants. However, the micellar shape is very sensitive to head/tail asymmetry, as the only slightly asymmetrical h_3t_4 forms significantly different structures than the symmetrical h_4t_4.

Larson's studies also illustrate the use of simulations to investigate how several other factors, in addition to the surfactant concentration, affect micellar properties such as the cmc and the size and shape of the micelles. In general, factors that tend to decrease the cmc tend to increase micellar size. These factors also tend to reduce the curvature in micellar shape from spherical to cylindrical to lamellar. As already shown in Fig. 13, increasing the total concentration of the surfactant leads to an increase in size and a reduction in curvature, as are expected from the typical surfactant phase diagram. Both lowering the temperature of the system and decreasing the surfactant head/tail ratio also increase the size, reduce the cmc, and decrease the curvature of the micelle. Figure 14 clearly shows that reducing the temperature (i.e., increasing ϵ from 0.1231 to 0.1538) increases the size of the micelles for h_3t_4. A comparison of Figs. 13 and 14—the micellar size distributions for h_4t_4 and h_3t_4, respectively—shows that removal of one head bead also has the same effect on micellar properties. Larson's work does not include head–head repulsion, but, as he suggested, the inclusion of head–head repulsion would be expected to cause a reverse trend in the size, shape, and cmc of the micelles; i.e., it would limit micellar growth, increase micellar curvature, and increase the cmc. A summary of these trends is given in Table 2.

As noted earlier, copolymer solutions are similar in many respects to short-chain amphiphilic solutions, and additional insights into the usefulness of lattice MC simulations for studying surfactant mixtures can be gathered from studies of copolymers. Mattice

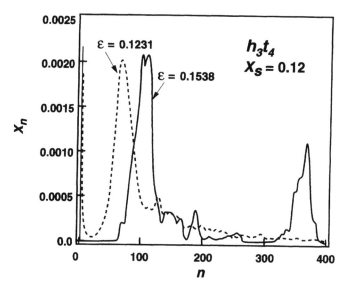

Figure 14 The aggregate size distribution for $h_3 t_4$ for a constant concentration $X_s = 0.12$ at two different inverse temperatures ϵ. (From Ref. 37.)

Table 2 Effect of Various System Parameters on Micellar Properties

Change in system parameter	n	cmc	Curvature of shape
Increasing surfactant concentration	↑	—	↓
Lowering system temperature	↑	↓	↓
Decreasing surfactant head/tail ratio	↑	↓	↓
Including surfactant head–head repulsion	↓	↑	↑

and coworkers (see, e.g., Wang et al. [46]) investigated parameters such as head size, tail size, and interaction energies for diblock copolymer micellization. They use the same definition of the cmc as Desplat and Care [31], discussed earlier in the context of short-chain surfactants. In a more recent work, Zhan and Mattice [48] used other definitions of the cmc, e.g., the concentration at which there is a breakpoint in the total number of aggregates in the system, N_{agg}. However, both definitions appear to give approximately the same results. In the case of the latter definition, the cmc is determined to be at a volume fraction of $X_{cmc} \approx 0.008$ for $h_{10} t_{10}$, as can be seen from Fig. 15. The results of Zhan and Mattice show that the cmc increases as the head size is increased and is inversely proportional to $j\epsilon$, where j is the length of the tail [46]. The polydispersity of the system also peaks at a concentration just above the cmc because of the presence of two types of aggregates in the system, namely, small aggregates and micelles. Below the cmc there are only small aggregates, whereas well above the cmc there are mostly larger micelles; hence, in both of these cases the polydispersity is small. However, near the cmc the polydispersity is high due to the presence of both types of aggregates, an observation that is also known for short-chain surfactants.

Finally, we comment on the lattice MC simulations of Wijmans and Linse [49], who investigated two cases similar to those examined by Mattice and coworkers using the same type of simulations. The objective here is to examine the origin of the differences between

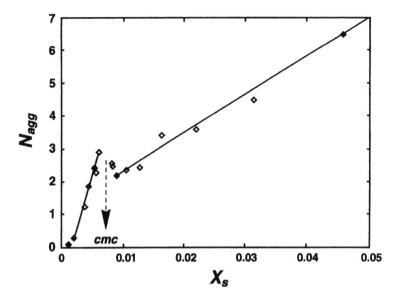

Figure 15   The total number of aggregates, N_{agg}, in the solution as a function of concentration for $h_{10}t_{10}$. (From Ref. 48.)

self-consistent field (SCF) calculations (see, e.g., Refs. 56–58) and simulations. In SCF calculations, the chains are modeled as self-intersecting random walks. Moreover, the conformations are weighted by a potential of mean force in each layer (of a prescribed geometry), and the fluctuations in the size and shape of the micelles are neglected. These drawbacks can be avoided in simulations. The two systems considered by Wijmans and Linse are 200 chains of $h_{10}t_{10}$ with $\epsilon = 0.45$ (system 1, equivalent to the system studied by Mattice and coworkers) and 200 chains of $h_{10}t_5$ with $\epsilon = 0.9$ (system 2). The results for system 1 indicate that at equilibrium the micellar solution is polydisperse with $n_{wt} = 20$ and $n_{no} = 4$ owing to a wide distribution of sizes with two major peaks—one representing monomers and the other, micelles of sizes 20–40 (see Fig. 16). However, there are also a small percentage of small aggregates of sizes 2–9 (near the minimum in the distribution) and a few very large aggregates up to size 60. For system 2, these very small and very large aggregates completely disappear, leaving only monomers and micelles, as seen in Fig. 17. The maximum in the distribution for system 2 shifts to 40 monomers, and the number of free chains for system 2 drops by a factor of 10 compared with system 1. Calculations of the principal moments of inertia show that micelles for system 2 are more spherical than those for system 1, as expected on the basis of geometrical considerations.

If the solutions are assumed to be ideal (i.e., the activity coefficients are equal to unity), then the difference in the standard chemical potential between a surfactant in an aggregate of size n and a free surfactant, $\mu_n^\circ - \mu_1^\circ$, can be written as

$$\beta(\mu_n^\circ - \mu_1^\circ) = \left[\ln X_1 - \frac{1}{n}\ln\left(\frac{X_n}{n}\right) \right] \qquad (15)$$

where X_n is the volume fraction of aggregates of size n. Figure 18, a plot of $\mu_n^\circ - \mu_1^\circ$ versus n for system 1, shows that $\mu_n^\circ - \mu_1^\circ$ decreases until the maximum in the size distribution (i.e., 30) is reached and then levels off. The difference in energies of the aggregate and the free surfactant, $\mathcal{E}_n^\circ - \mathcal{E}_1^\circ$, which is calculated by counting the total number of repulsive interac-

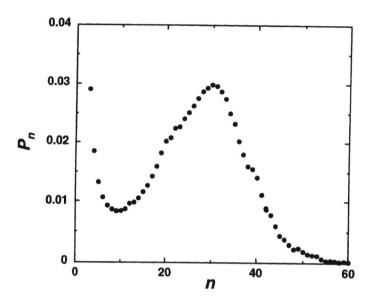

Figure 16 The aggregate size distribution for $h_{10}t_{10}$ for $\epsilon = 0.45$ considered by Wijmans and Linse [49]. See the text for other details. The normalized aggregate distribution P_n is defined as the volume of aggregates of a given aggregation number divided by the volumes of all aggregates; i.e.; if N_n is the number of aggregates with an aggregation number n, then $P_n = nN_n / \sum_{n=1}^{n_{max}} nN_n$, where n_{max} is the maximum aggregation number.

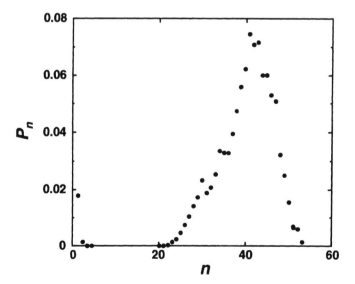

Figure 17 The aggregate size distribution for $h_{10}t_5$ for $\epsilon = 0.9$ considered by Wijmans and Linse [49]. See the text for other details. The normalized aggregate distribution P_n is defined in the legend for Fig. 16.

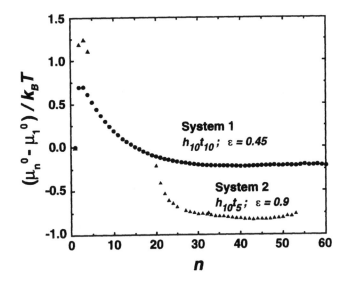

Figure 18 The difference in the standard chemical potentials, $\mu_n^{\circ} - \mu_1^{\circ}$, as a function of aggregation number n for the two systems of surfactants considered by Wijmans and Linse [49]. See the text for details.

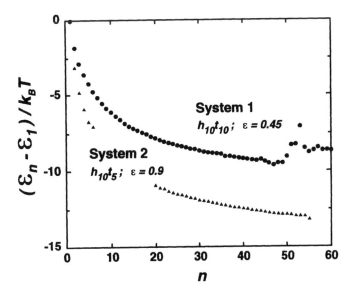

Figure 19 The difference in the average energy of free and aggregated surfactants, $\mathcal{E}_n - \mathcal{E}_1$, as a function of the aggregation number n for the two systems of surfactants shown in Fig. 18. (From Ref. 49.)

tions of an aggregate, continues to decrease at this point (Fig. 19). This indicates that the entropic contribution increases, making larger aggregates less favorable. Therefore, entropy is the dominant factor in limiting micelle growth—again a result that has been noted in the case of simple surfactants.

2. Microemulsion Phase Behavior

As noted earlier, the primary focus of Larson's studies is on the phase behavior in microemulsions; i.e., although the basic model and methodology are the same as the ones discussed earlier, the focus is quite different. Larson and coworkers [34–36] simulated three-component surfactant–solute–solvent systems, at surfactant concentrations much higher than in the micellization study, in order to examine the micellar phase diagrams and micellar microstructure. A general (qualitative) phase diagram of the sort that can be expected in surfactant–oil–water systems is shown in Fig. 20. A detailed construction of phase diagrams of this type using simulations is quite laborious and time-consuming but is not impossible in principle.

To determine the phase boundaries, one needs the Helmholtz free energy of the system, $\mathcal{F}$, which can be obtained from the average internal energy, $\langle \mathcal{E} \rangle$, calculated in the simulations. As noted by Larson et al. [34], one can calculate $\mathcal{F}$ by integrating $\langle \mathcal{E} \rangle$ over the temperature using the Gibbs–Helmholtz equation, as shown below. The Gibbs–Helmholtz equation is given by

$$\frac{\partial(\mathcal{F}/T)}{\partial T} = -\frac{\langle \mathcal{E} \rangle}{T^2} \tag{16}$$

which, on integration, leads to

$$\frac{\mathcal{F}_i}{T_i} - \frac{\mathcal{F}_{i-1}}{T_{i-1}} = \int_{1/T_i}^{1/T_{i-1}} \langle \mathcal{E} \rangle \, d\left(\frac{1}{T}\right) \tag{17}$$

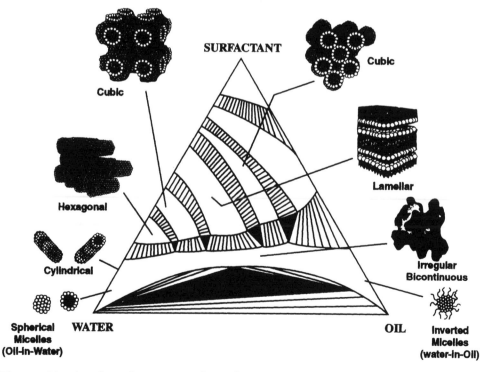

Figure 20 A schematic representation of a three-component phase diagram for a surfactant–oil–water system. (From Ref. 36.)

An approximation sufficient for $1/T_i$ close to $1/T_{i-1}$ is

$$\frac{\mathcal{F}_i}{T_i} - \frac{\mathcal{F}_{i-1}}{T_{i-1}} \cong \frac{1}{2}(\langle \mathcal{E} \rangle_i + \langle \mathcal{E} \rangle_{i-1})\left(\frac{1}{T_i} - \frac{1}{T_{i-1}}\right) \tag{18}$$

which can be written as

$$f_i \cong f_{i-1} + \frac{1}{2}\left(\frac{e_i}{\epsilon_i} + \frac{e_{i-1}}{\epsilon_{i-1}}\right)(\epsilon_i - \epsilon_{i-1}) \tag{19}$$

where f and e are the dimensionless Helmholtz energy and internal energy per lattice site, respectively, and ϵ is the dimensionless inverse temperature (i.e., interaction energy parameter). The initial values f_0 and e_0/ϵ_0 can be estimated from quasichemical theory. Phase separation occurs when a plot of Helmholtz energy f versus solute (i.e., oil) concentration (at constant surfactant concentration) has a region of zero slope. Figure 21, for a two-dimensional system, shows that $h_1 t_1$ does not form a separate phase at an areal fraction of $X_s = 0.25$ for solute concentrations up to $X_c = 0.40$, but, as seen in Fig. 22 (also for a two-dimensional system), a system of $X_s = 0.15$ $h_4 t_4$ does phase separate at a solute concentration of $X_c = 0.27$. This concentration represents a point on the phase envelope on a ternary phase diagram. Phase diagrams for $h_4 t_4$, as calculated from the simulations and from quasichemical and random solution theories, are shown in Fig. 23.

Mackie et al. [25] investigated the general phase behavior of Larson's model for a ternary surfactant–solute–solvent system. The Gibbs ensemble simulations were unsuccessful in modeling the phase behavior due to sampling problems. Therefore, the phase diagram was predicted using NVT simulations (canonical ensemble simulations) by separating the lattice into two bulk phases and physically measuring the composition of each phase. However, this method is subject to a large error because the fluctuations in the lattice density profile make the selection of the interface between the two phases on the lattice quite arbitrary. The phase diagrams of $h_1 t_3$, $h_2 t_4$, and $h_4 t_4$ as predicted from this method are shown in Figs. 24–26, respectively. The phase diagram for the symmetrical $h_4 t_4$ is logically symmetrical with respect to a solute/solvent ratio of unity. There is good agreement between the Monte Carlo simulations and quasichemical theory for the

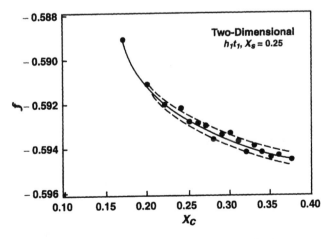

Figure 21 Helmholtz energy f as a function of solute concentration for a 0.25 areal fraction for a two-dimensional solution of $h_1 t_1$. (From Ref. 35.)

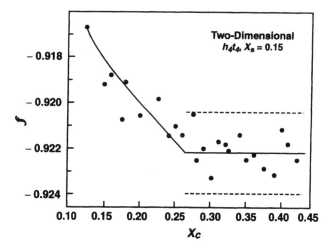

Figure 22   Helmholtz energy f as a function of solute concentration for a 0.15 areal fraction a two-dimensional solution of h_4t_4. (From Ref. 35.)

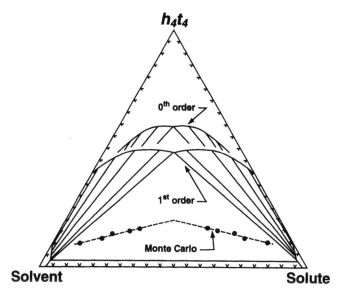

Figure 23   Phase diagram for the h_4t_4–solute–solvent system in two dimensions showing zeroth-order, first-order (quasi-chemical), and Monte Carlo results. (From Ref. 35.)

three-phase region of the phase diagram. Asymmetrical surfactants, such as h_1t_3 and h_2t_4, like to reside in an oil-rich phase; thus their solubility in the aqueous phase is very minimal. For h_1t_3, there is good agreement between the simulation and quasichemical data because h_1t_3 does not form any microstructures. However, h_2t_4 does form micelles, causing the quasi-chemical prediction to diverge from the simulation result, since quasichemical theory is a mean-field theory and cannot account for the presence of a nonuniform phase.

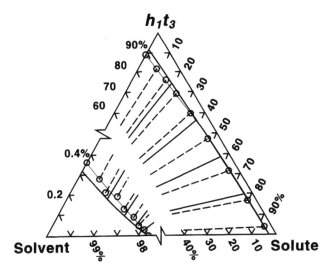

Figure 24 Phase diagram for the h_1t_3–solute–solvent system in three dimensions. The open circles represent Monte Carlo results, and the dashed lines are the corresponding tie-lines. (A thin line is drawn through the Monte Carlo data.) The quasi-chemical results are shown by solid lines. (From Ref. 25.)

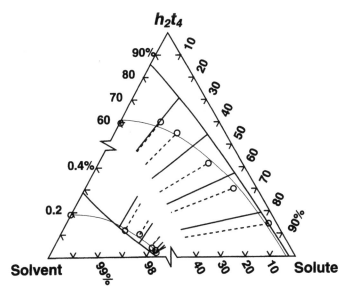

Figure 25 Phase diagram for the h_2t_4–solute–solvent system in three dimensions. The open circles are the data obtained from Monte Carlo simulations (a thin line is drawn through these points). The dashed lines are the tie-lines from the simulations. The quasi-chemical results are shown by the solid lines. (From Ref. 25.)

The present review is restricted to a few key studies because our objective here is merely to point out what kinds of issues can be examined using lattice simulations. Some related references on these topics can be found in an overview presented by Larson [59].

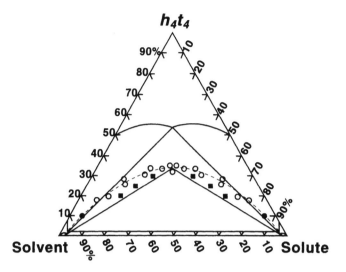

Figure 26 Phase diagram for the h_4t_4–solute–solvent system in three dimensions. The open circles are the data obtained from Monte Carlo simulations (a thin dashed line is drawn through these points to guide the eyes). The quasichemical results are shown by the solid lines. The filled squares are based on the thermodynamic integration technique, while the filled circles are plait point estimates. (From Ref. 25.)

V. SUMMARY AND CONCLUDING REMARKS

In this chapter we have reviewed several studies of micellar and microemulsion systems based on molecular dynamics and lattice Monte Carlo simulations. The major observations based on these studies can be summarized as follows.

Molecular dynamics simulations are consistent with calculations based on the critical packing parameter p, which indicate that the structure of the surfactant controls the shape of the micelle at the cmc. Esselink et al. [16] show that the surfactants h_2t_5, h_2ht_5, and h_3ht_5 form bilayers, cylindrical micelles, and spherical micelles, respectively, as expected. However, h_4t_4, expected to form micelles of low curvature based on p, instead forms sphere-like structures due to the coiling of the headgroup. If this increased effective headgroup area is accounted for in the calculation of the packing parameter, then a spherical shape is predicted, in agreement with the result of the simulations.

Similarly, Monte Carlo simulations show that the micelle shape, and also the micelle size and the cmc, are very sensitive to the sizes of the surfactant head and tails (in addition to other factors such as the topology of the lattice and the number of nearest neighbors considered). Larson [37] shows that as the head/tail ratio is increased, the aggregate shapes increase in curvature from bilayers to cylinders to spheres and the overall aggregate sizes decrease. An increase in the head/tail ratio also increases the cmc, which is evident as h_2t_3 micellizes at a finite surfactant concentration whereas h_2t_2 does not aggregate at all (i.e., it has an infinite cmc) and h_1t_3 associates into a separate phase (which can be thought of as having a cmc value of zero). Adriani et al. [47] also show that the cmc of h_it_{10} increases as i is increased and that the cmc has an inverse relationship with tail size.

Several different methods have been used to determine the cmc of a system from MC simulations—from aggregate size distributions (Larson [37]), from the free surfactant concentration (Desplat and Care [31]), from the total number of aggregates in the system (Zhan and Mattice [48]), etc. Nevertheless, each of these methods predicts a cmc that coincides with a breakpoint in various micellar properties, as expected from experimental observations. For example, the free surfactant concentration is found to increase until the cmc is reached and then to slightly decrease thereafter [31,47], while the polydispersity of the system is also found to peak at the cmc [47]. However, the micelle size continues to increase with increasing surfactant concentration after the cmc [37]; in fact, the number-average aggregation number is found to be proportional to the square root of the concentration [31], in agreement with Mukerjee's prediction [55]. Also, the micellar curvature decreases with increasing surfactant concentration [37].

In addition to the surfactant structure and concentration, the dependence of the micellar properties on other system parameters such as temperature or interaction energies can be predicted reliably from simulations. Increasing the temperature of the system causes an increase in the cmc [31,37,48], a decrease in micellar size [31,37], and a shift in micellar shape toward greater curvature [37]. Energies of interaction can also affect the cmc of the system, although the relationships are more complex. The critical micelle concentration is shown to have a nonlinear dependence on the solvophilic strength of the head [31]; the cmc decreases as the solvophilic strength is increased, although the increase is minimal beyond a certain level. This corresponds to a complete saturation of the head with solvent, after which a further increase in solvophilic strength has no effect. The solvophobicity of the tail has been determined to have an inverse relationship with the cmc [47].

Phase diagrams of several ternary surfactant–solute–solvent systems predicted using lattice MC simulations and quasichemical theory have been compared and shown to lead to qualitatively similar results. The two diagrams for h_1t_3 with c_1 solvent are in excellent agreement [25], as this surfactant does not aggregate to form any microstructures. However, similar comparisons for h_2t_4 [25] and h_4t_4 [25,36] are not as good, because quasichemical theory cannot account for the formation of micelles.

A comparison of all the results reported here reveals that all the models reviewed predict qualitatively similar micellar behavior. Although the magnitudes of quantities such as the cmc or average aggregation number may be quite different for the different models (e.g., h_1t_3 completely phase separates in Larson's model but forms well-behaved micelles in the simulations of Desplat and Care [31], because of the head–solvent attraction they used), each model predicts similar trends in these properties. This confirms the assumption that the solvophobic effect (i.e., the dislike of the solvophobic tail beads for the solvent) is the major driving force for micellization but also indicates that other forces are present that control specific micellar properties.

In closing, both molecular dynamics and Monte Carlo simulations can be used effectively to study micellar systems. The recent advances in the speed of computers and parallel processing enable the use of MD simulations for these studies, although they remain computationally expensive. Hence, MD simulations are still not effective for the study of equilibrium properties. Therefore, many studies have been published using MC simulations, evidenced by the number of works reviewed in Sec. IV. The studies cited in Secs. IV. B. 1 and IV. B. 2 are sufficient to illustrate the use of simulations to study micellization and overall phase behavior. However, none of the works reviewed here presents much information specifically on micellar solubilization or on phase behavior in the dilute solution region in which solubilization occurs. The first such studies have only now begun to appear in the literature [33,60]. Nevertheless, the works reviewed in this chapter provide a solid foundation for further, more targeted studies of micellization, micellar solubilization, and the formation of microemulsions.

It is important to note that certain crucial steps need to be taken first before lattice Monte Carlo simulations can be used as a guide for interpreting experimental data or to identify the types of experiments that can shed light on the phenomena of interest. For example:

1. An assessment needs to be made regarding the correspondence between real surfactants and their structural (bead–rod or bead–spring) and energetic representations in the simulations.
2. The sensitivity of the results of the simulation to the features of coarse-graining in the model (e.g., lattice topology and assumptions concerning nearest neighbors) also needs to be examined.
3. Dynamic versions of the simulations reviewed in this chapter can also be used to follow the kinetics of aggregate formation and breakup and the like, but the types of dynamic algorithms needed for such an effort need to examined as well.

Although much work needs to be done in these respects, the results reviewed in this chapter illustrate that efforts along the lines suggested are well worth the time.

REFERENCES

1. G. Gompper and M. Schick, *Self-Assembling Amphiphilic Systems*, Academic, London, 1994.
2. G. Gompper and M. Schick, in *Micelles, Membranes, Microemulsions, and Monolayers* (W. M. Gelbart, A. Ben-Shaul, and D. Roux, eds.), Springer-Verlag, New York, 1994, Chap. 8.
3. S.-H. Chen and R. Rajagopalan (eds.), *Micellar Solutions and Microemulsions: Structure, Dynamics and Statistical Thermodynamics*, Springer-Verlag, New York, 1990.
4. A. Bhattacharya, S. D. Mahanti, and A. Chakrabarti, J. Chem. Phys. *108*:10281 (1998).
5. D. W. Heermann, *Computer Simulation Methods in Theoretical Physics*, 2nd ed., Springer-Verlag, Berlin, 1990.
6. M. P. Allen and D. J. Tildesley, *Computer Simulation of Liquids*, Oxford Univ. Press, Oxford, 1987.
7. D. Frenkel and B. Smit, *Understanding Molecular Dynamics Simulations: From Algorithms to Applications*, Academic, San Diego, CA, 1996.
8. J. M. Haile, *Molecular Dynamics Simulation: Elementary Methods*, Wiley, New York, 1992.
9. J. Böcker, J. Brickmann, and P. Bopp, J. Phys. Chem. *92*:2881 (1988).
10. J. C. Shelley, M. Sprik, and M. Klein, Langmuir *9*:916 (1993).
11. J. R. Gunn and K. A. Dawson, J. Chem. Phys. *91*:6393 (1989).
12. D. R. Rector, F. van Swol, and J. R. Henderson, Mol. Phys. *82*:1009 (1994).
13. S. Karaborni, N. M. van Os, K. Esselink, and P. A. J. Hilbers, Langmuir *9*:1175 (1993).
14. B. Smit, K. Esselink, P. A. J. Hilbers, N. M. van Os, L. A. M. Rupert, and I. Szleifer, Langmuir *9*:9 (1993).
15. S. Karaborni, K. Esselink, P. A. J. Hilbers, and B. Smit, J. Phys. Condens. Matter *6*:351 (1994).
16. K. Esselink, P. A. J. Hilbers, N. M. van Os, B. Smit, and S. Karaborni, Colloids Surf. A *91*:155 (1994).

17. K. Binder, (ed.), *Monte Carlo and Molecular Dynamics Simulations in Polymer Science*, Oxford Univ. Press, New York, 1995.
18. L. Verlet, Phys. Rev. *159*:348 (1967).
19. C. Tanford, *The Hydrophobic Effect: Formation of Micelles and Biological Membranes*, Krieger, Malabar, FL, 1991.
20. P. C. Hiemenz and R. Rajagopalan, *Principles of Colloid and Surface Chemistry*, 3rd ed., Marcel Dekker, New York, 1997.
21. J. N. Israelachvili, *Intermolecular and Surface Forces*, 2nd ed., Academic, New York, 1991.
22. F. K. von Gottberg, K. A. Smith, and T. A. Hatton, J. Chem. Phys. *106*:9850 (1997).
23. W. M. Gelbart, D. Roux, and A. Ben-Shaul (eds.), *Micelles, Membranes, Microemulsions, and Monolayers*, Springer-Verlag, Berlin, 1995.
24. N. Metropolis, A. W. Rosenbluth, M. N. Rosenbluth, A. H. Teller, and E. Teller, J. Chem. Phys. *21*:1087 (1953).
25. A. D. Mackie, K. Onur, and A. Z. Panagiotopoulos, J. Chem. Phys. *104*:3718 (1996).
26. S. K. Talsania, Monte Carlo simulations of micellar system, Ph.D. Dissertation, University of Houston, Houston, TX, 1997.
27. M. T. Gurler, C. C. Crab, D. M. Dahlin, and J. Kovac, Macromolecules *16*:398 (1983).
28. A. E. van Giessen and I. Szleifer, J. Chem. Phys. *102*:9069 (1995).
29. R. Pynn, Los Alamos Sci. *19*:91 (1990).
30. C. M. Care, J. Chem. Soc., Faraday Trans. *83*:2905 (1987).
31. J. C. Desplat and C. M. Care, Mol. Phys. *87*:441 (1996).
32. D. Brindle and C. M. Care, J. Chem. Soc., Faraday Trans. *88*:2163 (1992).
33. S. K. Talsania, Y. Wang, R. Rajagopalan, and K. K. Mohanty, J. Colloid Interface Sci. *190*:92 (1997).
34. R. G. Larson, L. E. Scriven, and H. T. Davis, J. Chem. Phys. *83*:2411 (1985).
35. R. G. Larson, J. Chem. Phys. *89*:1642 (1988).
36. R. G. Larson, J. Chem. Phys. *91*:2479 (1989).
37. R. G. Larson, J. Chem. Phys. *96*:7904 (1992).
38. R. G. Larson, Chem. Eng. Sci. *49*:2833 (1994).
39. D. Stauffer, N. Jan, Y. He, and R. B. Pandey, Physica A *198*:401 (1993).
40. D. Stauffer, N. Jan, Y. He, R. B. Pandey, D. G. Marangoni, and T. Smith-Palmer, J. Chem. Phys. *100*:6934 (1994).
41. A. T. Bernardes, V. B. Henriques, and P. M. Bisch, J. Chem. Phys. *101*:645 (1994).
42. A. D. Mackie, A. Z. Panagiotopoulos, and I. Szleifer, Langmuir *13*:5022 (1997).
43. M. Nguyen-Misra and W. L. Mattice, Macromolecules *28*:1444 (1995).
44. K. Rodrigues and W. L. Mattice, J. Chem. Phys. *94*:761 (1991).
45. T. Haliloglu and W. L. Mattice, Chem. Eng. Sci. *49*:2851 (1994).
46. Y. Wang, W. L. Mattice, and D. H. Napper, Langmuir *9*:66 (1993).
47. P. Adriani, Y. Wang, and W. L. Mattice, J. Chem. Phys. *100*:7718 (1994).
48. Y. Zhan and W. L. Mattice, Macromolecules *27*:677 (1994).
49. C. M. Wijmans and P. Linse, Langmuir *11*:3748 (1995).
50. C. M. Wijmans and P. Linse, J. Chem. Phys. *106*:328 (1997).
51. A. D. Mackie, A. Z. Panagiotopoulos, and S. K. Kumar, J. Chem. Phys. *102*:1014 (1994).
52. A. Z. Panagiotopoulos, Mol. Phys. *61*:813 (1987).
53. P. H. Verdier and W. H. Stockmayer, J. Chem. Phys. *36*:227 (1962).
54. L. A. Rodriguez, S. K. Talsania, K. K. Mohanty, and R. Rajagopalan, Langmuir *15*: in press (1999).
55. P. Mukerjee, J. Phys. Chem. *76*:565 (1972).
56. F. A. M. Leermakers and J. M. H. M. Scheutjens, J. Phys. Chem. *93*: 7417 (1989).
57. F. A. M. Leermakers and J. M. H. M. Scheutjens, J. Colloid Interface Sci. *136*:231 (1990).
58. F. A. M. Leermakers, P. P. A. M. van der Schoot, J. M. H. M. Scheutjens, and J. Lyklema, in *Surfactants in Solution*, Vol. 7 (K. L. Mittal, ed.), Plenum, New York, 1990, p. 25.
59. R. G. Larson, Curr. Opinion Colloid Interface Sci. *2*:361 (1997).
60. S. K. Talsania, L. A. Rodriguez-Guadarrama, K. K. Mohanty, and R. Rajagopalan, Langmuir *14*:2684 (1998).

6

Effects of Alcohol Chain Length and Salt on Phase Behavior and Critical Phenomena in SDS Microemulsions

A. M. Bellocq
Centre de Recherche Paul Pascal–CNRS, Pessac, France

I. INTRODUCTION

Due to their amphiphilic nature, surfactants solubilize water and oil and form a wide variety of thermodynamically stable phases. Indeed, mixtures of amphiphiles, water, and oil may give rise to either isotropic phases (microemulsion, sponge) or lyotropic liquid crystalline phases (lamellar phase, hexagonal phase) [1–4]. The term "microemulsion" introduced in 1959 by Schulman et al. [5] in its most general use connotes a thermodynamically stable, fluid oil–water–surfactant mixture in which the oil and water regions are well separated and the surfactant molecules are organized as monolayers at the internal water/oil interfaces. Microemulsions often exist over a wide range of water and oil concentrations and can be formed with a few percent of surfactant. Many also contain alcohol and salt. The configuration of the oil and water domains varies with composition; for small fractions of oil in water (O/W) or of water in oil (W/O), the structure is that of swollen micelles. When the volume fractions of oil and water are comparable, a random bicontinuous structure is found. The first suggestion of this type of structure came from Scriven [6]. The oil regions have the structure of a connected random network, and the same property also holds for the water regions, the surfactant molecules making a continuous film between the oil and water domains with no long-range order such as that found in lyotropic liquid crystals.

One of the major steps to enhancing the understanding of surfactant association structures is to investigate the phase equilibria of water–oil–surfactant mixtures. Since the pioneering contributions of P. Ekwall in Scandinavia and K. Shinoda in Japan, who analyzed ternary systems containing ionic and nonionic surfactants, respectively, extensive experimental work has been devoted to different kinds of microemulsion systems [7–20]. During recent decades, detailed descriptions of the phase behavior of many ternary (water–oil–surfactant), quaternary (water–oil–surfactant–alcohol), and even quinary (water–oil–surfactant–alcohol–salt) systems have been presented (for example, see Refs. 21–36). These studies have provided evidence for several new phases where the surfactant creates surfaces. Thus, in addition to bicontinuous microemulsions, one finds dilute lamellar phases (L_α) [37–43] and liquid isotropic phases of randomly connected bilayers called sponge phases (phases L_3) [44–47]. In a few extreme cases, the smectic period of the L_α phases reaches values on the order of optical wavelengths, so Bragg peaks are observed with light scattering [40–43]. Recently, the existence of an extremely dilute crystalline phase exhibiting three-dimensional (3D) long-range order was also reported. In this phase, made of water–oil–surfactant and cosurfactant, the characteristic repeat distance can reach up to 220 nm [48]. Besides, these works have also established that microemulsion systems can give rise to critical points, critical endpoints, and even tricritical points [22–31,49–57]. Finally, a great deal of work has been devoted to understanding the structure and stability of all these phases [19]. From all the work that has been done on the states of surfactants in solution, there emerges a fundamental new concept: These phases are often better described as phases of fluctuating surfaces than as phases of particles. In this new approach the focus is on the interface between hydrophobic and hydrophilic regions, and a free energy is associated directly with this interface, featuring separately its bending (curvature) elasticity and its topological entropy. The more recent studies of microemulsions highlighted the relevance of flexible, thermally fluctuating surfactant sheets. In such systems thermal fluctuations have to be large, as they are in fact responsible for the very existence of the microemulsion structure. In practice, flexible films can be obtained by using nonionic surfactant or by adding to an ionic surfactant a cosurfactant, which is often an alcohol such as pentanol or hexanol.

In order to emphasize the role of the interfacial films and to highlight the most recent viewpoints on the stability of microemulsions, sponge phases, and dilute lamellar phases, some of the experimental facts about phase behavior of microemulsion systems containing alcohol are reviewed in this chapter. The systems investigated consist of water, oil, alcohol, and sodium dodecylsulfate (SDS). In the next section, the theoretical aspects of the stability of surfactant phases are briefly discussed. Then in Secs. III and IV the effects of varying alcohol and oil chain lengths and the addition of a water–soluble polymer are examined. The examination of multiphase regions provides the location of lines of critical points or critical endpoints. This chapter also deals with the study of several physical properties in the vicinity of critical points.

II. STABILITY OF MICROEMULSIONS: THEORETICAL ASPECTS

From a theoretical point of view the stability of the phases of surfactant results from the competition between entropic and elastic contributions to the free energy [19]. The ultimate structure of the aggregates existing in solution depends strongly on two intrinsic parameters of the interface: the mean bending modulus κ and the Gaussian bending modulus $\bar{\kappa}$. Within the framework of the membrane elasticity theory [58], the bending energy, F_b, of a piece of membrane is

$$F_b = \frac{1}{2}\kappa(C_1 + C_2 - 2C_0)^2 + \bar{\kappa}C_1C_2 \qquad (1)$$

where C_1 and C_2 are the local principal curvatures of the surfactant layer and C_0 the spontaneous curvature [59,60]. The contribution of F_b to the total free energy is crucial in determining the type and characteristic size of the structure. The first term, which characterizes the rigidity, represents the energy required to bend a unit area of interface by a unit amount of curvature. The second term plays an important role in the change of membrane topology and consequently in the phase transition where the genus of the surface is changed. This is a direct consequence of the Gauss–Bonnet theorem, which states that the integral of the Gaussian curvature over a given structure depends on its topological type. C_0 describes the tendency of the surfactant film to bend either toward the water or toward the oil. It arises from the competition in the packing of the polar heads and hydrocarbon tails of the surfactant molecules [61]. Roughly, one can say that if C_0 is sufficiently large, a droplet structure is formed, whereas when C_0 vanishes, a lamellar phase and/or a random bicontinuous structure is obtained.

Several theoretical models have been worked out in order to understand why a random structure of the bicontinuous type does not collapse into an ordered phase [59,60,62–66] (see also Ref. 19, p. 427). An answer has been suggested by de Gennes and coworkers [59,60], who first recognized that the stability of the microemulsion phase was controlled by the physics of the amphiphilic film. In the absence of thermal fluctuations a layer with a spontaneous curvature $C_0 = 0$ is assumed to be flat and de Gennes and Taupin introduced the concept of persistence length ξ_K of the interface; ξ_K is the distance over which the interface remains flat in the presence of thermal fluctuations. Then, below ξ_K the film is nearly flat and above ξ_K the film makes a two-dimensional random walk. The length ξ_K is similar to the persistence length of a linear polymer, but instead of being proportional to the bending constant κ of the film, it is an exponential function of it:

$$\xi_K = a \exp(2\pi\kappa/k_B T) \tag{2}$$

a is a molecular length, k_B the Boltzmann constant, and T the temperature. This behavior has some important effects. When κ is large compared to $k_B T$, ξ_K is macroscopic, meaning that the layers are flat over large distances; in this case the interfaces tend to stack, and a lamellar phase is obtained as soon as the distances between two films is smaller than ξ_K. In contrast, when κ is comparable to $k_B T$, the interface is extremely wrinkled, ξ_K is microscopic, the long-range order is lost, and either bicontinuous microemulsions or sponge phases appear. In the latter case, the theoretical models predict a competition between a bicontinuous microemulsion and a lamellar phase. Also, another main result of the de Gennes–Taupin model is that the minimum amount of surfactant needed to obtain a stable microemulsion phase with the same amount of water and oil, ϕ_s^{min}, is inversely proportional to ξ_K [59,60]. Since ξ_K is an exponential function of the bending elastic constant, κ, ϕ_s^{min} is extremely sensitive to the film properties.

The role of $\bar{\kappa}$ is also important because $\bar{\kappa}$ couples to the topology of the membrane. When $\bar{\kappa}$ is sufficiently negative, the surface forms many disconnected aggregates such as droplets or vesicles [66–70]; in contrast, when $\bar{\kappa}$ is positive, highly connected surfaces with many "handles" or "connections" are favored, and the bicontinuous microemulsions or sponge phases are stabilized [66–72].

In experimental terms, the elasticity parameters κ, $\bar{\kappa}$, and C_0 are expected to depend strongly on the particular surfactant selected, on the type and concentration of cosurfactant, on salt concentration, and also on temperature. Although the relationships involved are rather complex, qualitative comparisons with experiments are possible without knowing them in detail. It was early recognized that the elastic modulus κ was very sensitive to additives. Microemulsions are generally obtained by using mixtures of ionic surfactant and alcohol molecules. At least two effects of the cosurfactant can be related to the curvature energy: it changes C_0, and it decreases κ. A direct calculation of the effect of mixing surfactant and cosurfactant molecules has been published [73]. A very efficient lowering of the bending constant κ of the surfactant film is found by dilution with short chains. The fast decrease of κ with the short-chain fraction must be contrasted with its nearly linear variation with chain length in a pure monolayer. The elastic constant scales with the thickness L and with the area per polar headgroup, a, of the surfactant as

$$\kappa \propto \frac{L^n}{a^p}$$

where n and p are exponents ($n \sim 2$–3; $p \sim 5$) [73]. The reduction in κ obtained by the addition of alcohol molecules must be attributed to thinning of the interface [74]. Recently, an experimental study with deuterium solid-state NMR of the flexibility of molecular films composed of SDS and an alcohol confirmed this prediction. The results show that κ is sensitive to both the amount of alcohol in the film and the alcohol chain length, with typical values increasing from $1.3k_B T$ to $13k_B T$ from hexanol to decanol [75]. Thus in the following we examine the effect of the chain length of the alcohol on the competition between isotropic and ordered phases.

Moreover Porte et al. [52] have shown that in bilayer phases, $\bar{\kappa}$ is related to the spontaneous curvature of the monolayer and the elastic constants of the monolayer:

$$\bar{\kappa} = 2\bar{\kappa}_{\text{mono}} - 2LC_0\kappa_{\text{mono}}$$

Thus alcohol in reducing C_0 is expected to increase $\bar{\kappa}$.

III. PHASE DIAGRAMS

Very early, the Swedish school attempted to determine the extent and shape of the region of existence of microemulsions in quaternary systems [76–78]. By examination of sections of the phase diagram at several levels of oil, Friberg and coworkers established a direct connection between the microemulsion areas and the inverse micellar solutions described by Ekwall [1]. Thus, prior to describing the phase diagrams of the quaternary systems, those of ternary systems made of water, sodium dodecylsulfate (SDS), and an alcohol are first presented here.

A. Ternary Systems: Water–Surfactant–Alcohol

The phase diagrams of four ternary water–SDS–alcohol systems at 25°C are shown in Fig. 1. The alcohols used were pentanol, hexanol, octanol, and decanol. In the pentanol system, four one-phase regions are observed; three of them are mesophases: L_α, the lamellar phase; H_α; the hexagonal phase; and R, the rectangular phase [7]. These mesophases already exist in the binary system water–SDS but in very different ranges of temperature and concentrations. L_α and R are found only at very high temperatures ($T > 50°C$) and very high surfactant concentration (SDS $> 70\%$) [81]. In the ternary systems, the lamellar mesophase occurs at 25°C over a wide range of surfactant concentration. It extends down toward the water corner for up to 79% water, which corresponds to a maximum repeat distance of 10 nm. At 25°C, the pentanol system displays a continuous band of isotropic solution between pure water and pure pentanol. Provided that the surfactant concentration is adjusted, it is possible to go continuously from the water corner to the pentanol corner without any phase separation occurring. The main result of this continuity is the occurrence of a critical point P_C^A in the diagram. The accurate location of this point has been determined by phase composition analysis of several two-phase equilibria [82]. Significant changes in the phase equilibria occur on lowering the temperature. Below 20°C, the isotropic domain L is no longer continuous; it has split into two one-phase regions L_1 and L_2, which extend from the water and pentanol corners, respectively. These isotropic regions are separated by a

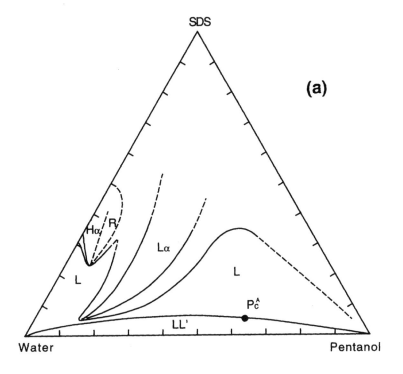

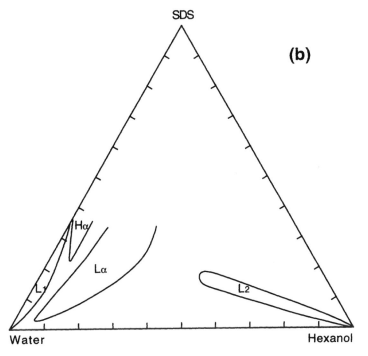

Figure 1 Phase diagrams at 25°C of four ternary systems. (a) Water–pentanol–SDS. (From Refs. 22 and 23.) (b) Water–hexanol–SDS. (From Refs. 22 and 23.) (c) Water–octanol–SDS. (From Ref. 79.) (d) Water–decanol–SDS. (From Ref. 80.) L_1, L_2, isotropic phases; H_α, hexagonal phase; R, rectangular phase; L_α, lamellar phase; L_4, vesicle phase; N, nematic phase. The dashed boundaries have not been determined accurately. P_C^A is a critical point.

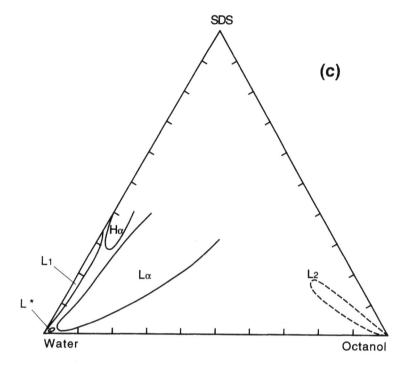

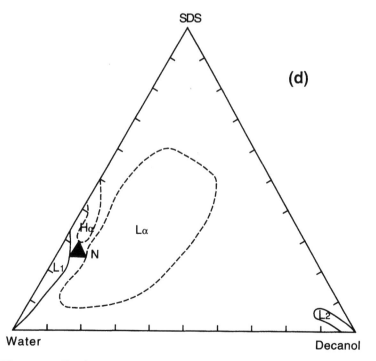

Figure 1 Continued.

complex multiphase region involving two three-phase domains, t_1 and t_2. These equilibria result in an indifferent state at 20°C [83]. At this temperature, the points representative of the three coexisting phases of both equilibria are located on the same straight line. Note that in an indifferent state the system does not show a particular thermodynamic behavior.

The ternary system with hexanol forms the same mesophases as pentanol. Replacement of pentanol with hexanol leads to a larger swelling of the lamellar phase, which can be prepared with 3.5% SDS instead of the 7.5% SDS used with pentanol. As a result, the isotropic domain is divided into two regions, L_1 and L_2, and the three phases L_1, L_2 and L_α are separated by one three-phase triangle and not two as in the pentanol case. Moreover, this system does not exhibit a plait point at 25°C. This situation resembles that described by Kunieda and Nakamura [84] for other water–ionic surfactant–alcohol systems. A further increase in alcohol chain length leads to a larger lamellar swelling of the L_α phase and the appearance of new phases: a vesicle phase L_4 in the case of octanol [79] and a nematic phase for decanol [80].

1. Effect of Salt

Significant changes in the phase behavior of water–alcohol–SDS systems occur with the addition of sodium chloride (Fig. 2) [69,79,85–88]. At a salinity of 20 g/L NaCl, the phase diagram of the hexanol–water–SDS system presents, in addition to L_1 and L_α, two new isotropic phases, L_3 and L_4, corresponding, respectively, to a sponge phase and a phase of vesicles. The L_1 and L_4 phases seem to merge at extremely low SDS concentration. At higher SDS concentration (above 1 wt%), when the alcohol concentration is increased, the following sequence of phases is found: L_1–L_4–L_α–L_3. A similar pattern of phase behavior has been observed in a few systems containing an ionic single-chain surfactant, alcohol, and brine [89–91] and also in systems made with a zwitterionic surfactant, alcohol, and water [92,93], but generally the L_4 phase is missing [89–91].

The macroscopic features of concentrated L_1 samples are modified by the addition of salt; they become very viscous and exhibit flow birefringence. Their structure has not yet been analyzed, but we may assume that it corresponds to living polymer-like structures. The L_3 phase is found to be stable over a large SDS concentration range and can be diluted to up to 0.4% SDS at 25°C. At higher dilution, it separates, with a dilute L_1 solution containing 0.05% SDS. As the temperature is lowered, the miscibility gap is reduced: at 20°C, the L_3 phase can be swollen up to 0.2% SDS. The dilute L_3 samples (~1% SDS) strongly scatter light and exhibit flow birefringence. Like the L_3 phase, the L_4 phase exists in an extremely narrow range of alcohol/SDS ratios R ($\Delta R \sim 2\%$), but it extends between almost 0 and 10% SDS. In the dilute part of the diagram, the width of the L_4 domain is typically one-fifth that of the sponge phase (alcohol width for the L_3 phase ~0.15%). Dilute L_4 samples are isotropic and of low viscosity; in the range 2–5% SDS they exhibit strong flow birefringence and become viscous, and finally above 5% SDS they are weakly birefringent even with no shear. In contrast to the L_3 phase, the birefringence appears as the surfactant concentration is increased. We must note that the boundary of the L_4 phase at high alcohol content is difficult to find especially because of the existence of a viscous, stable emulsion involving the L_4 and L_α phases (region S in Fig. 2b). The regions where the L_3, L_α, and L_4 phases exist are limited by straight lines along which the alcohol/surfactant ratio is constant.

In the region between the L_3 and L_4 phases, all the mixtures are birefringent. Examination of the textures of dilute samples (SDS < 10%) with a polarizing microscope provides evidence for the existence of three distinguishable regions labeled I, II, and S (Fig. 2).

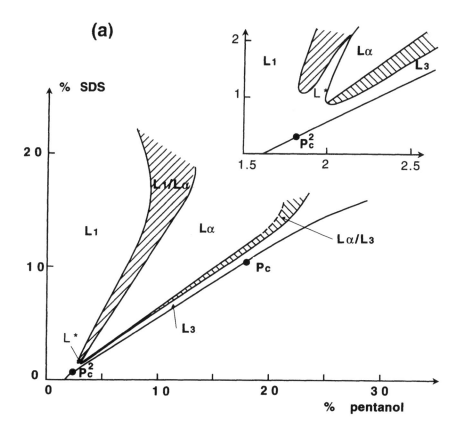

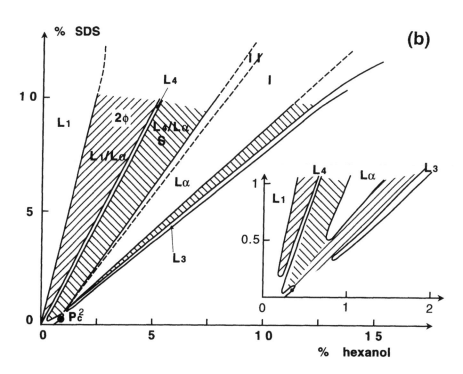

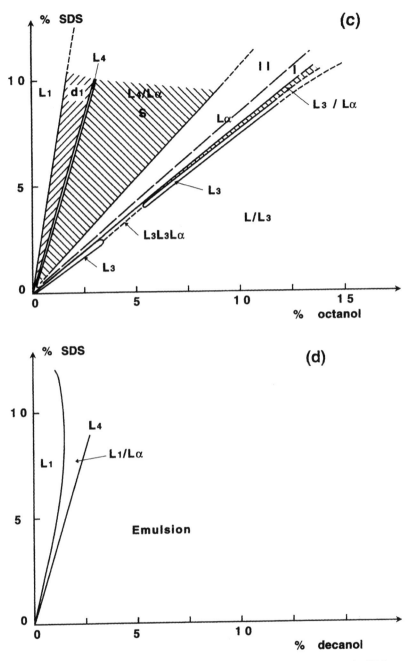

Figure 2 Partial phase diagrams of the systems (a) pentanol–SDS–water–NaCl; (b) hexanol–SDS–water–NaCl; (c) octanol–SDS–water–NaCl; (d) decanol–SDS–water–NaCl at $T = 25°C$. The concentration of NaCl in water is fixed at 20 g/L. L_1 is an isotropic phase. L_α is a lamellar phase. L_3 and L_4, both with a very narrow range in alcohol concentration, are, respectively, the sponge phase and the vesicle phase. In region I, the lamellar phase is made of a stack of flat parallel bilayers; in region II, spherulites form spontaneously. Region S is a two-phase region where phases L_4 and L_α coexist. The multiphase region at high alcohol content consists of two-phase and three-phase regions in the pentanol and hexanol systems. The dilute parts of the pentanol and hexanol systems are given in insets in Figs. 2a and 2b. (From Ref. 88.)

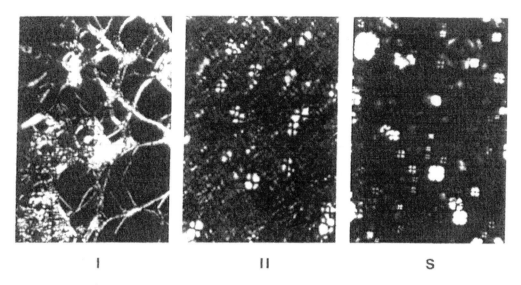

Figure 3 Optical textures observed in regions I, II, and S of the hexanol–SDS–water–NaCl (20 g/L) system. I: 5% SDS, 6% hexanol. II: 5% SDS, 4.5% hexanol; S: 5% SDS, 3% hexanol.

Mixtures in regions I and II are monophasic lamellar samples, while those in region S correspond to two-phase equilibria involving the lamellar phase L_α and the vesicle phase L_4. In photographs of the typical textures observed in the three zones one can see two types of defects characteristic of a lamellar phase (Fig. 3). Oily streaks associated with the usual focal domains with negative Gaussian curvature are seen at high alcohol content in region I, close to the L_3 phase boundary. In region II, at low alcohol content, one observes Maltese crosses dispersed in a birefringent matrix. This texture, found in the region close to the L_4 phase boundary, corresponds to a dispersion in the lamellar matrix of multilamellar vesicles with a positive Gaussian curvature, sometimes referred to as spherulites. As one heats very dilute samples located in region II, the spherulites and the lamellar matrix melt at the same temperature, and the spherulites do not reappear when the samples are slowly cooled. This observation indicates that the spherulites are textural defects of the L_α phase; as a consequence the lamellar phase L_α does indeed cover the two domains labeled I and II. Analogous textures were also found in other brine–alcohol–surfactant systems [79,86,87,89–95]. One must point out that the swelling of the lamellar phase increases with salinity. Very dilute lamellar phases containing only 0.5% SDS (instead of the 3.5% in pure water) exist at a salinity of 20 g/L NaCl.

Observation of textures in region S provides evidence for the coexistence of two phases in this third region; indeed, they exhibit large spherulites dispersed in an isotropic continuous medium (Fig. 3). The tendency for these two-phase equilibria to separate depends on the surfactant concentration. With the higher concentrations (SDS > 5%), no phase separation occurs even after several months. In contrast, for the dilute samples, a phase separation can be obtained; after a few days of storage one can see in a first step a gross separation between two dispersions, one rich in spherulites and the poor, and finally as the thermodynamic equilibrium is achieved one finds that the L_4 and L_α phases coexist. S samples with a low alcohol content are turbid; they become clearer with increasing alcohol content or an increase in temperature. When viewed through crossed polarizers they are seen to exhibit a slight cloudiness, and when disturbed regions of weak birefringence that relax slowly to the original cloudiness. These textures are characteristic of dilute lamellar dispersions and look very similar to those previously reported for other lamellar dispersions [79,89–93]. The weak birefringence is due to the existence of spherulites. Finally, during preparation of samples in region S it has been observed that their viscosity is considerably affected by the method of preparation. Solutions were obtained either by adding alcohol to the aqueous surfactant solutions or by adding brine to the surfactant–alcohol mixtures. The samples of the first type were found to be much more viscous than those of the second type. In the concentrated part of the phase diagram (for SDS > 10%), identification of the zones S and II by optical spectroscopy is made difficult because of the high concentration of spherulites. In summary, the region labeled S corresponds to an emulsion of the lamellar phase L_α in the L_4 phase, and regions I and II correspond to the lamellar phase L_α.

The phase diagrams obtained at a salinity of 20 g/L NaCl for other alcohols have been investigated [79,86,88], and those for pentanol, octanol, and decanol are presented in Fig. 2. The diagrams for octanol and hexanol are very similar; both exhibit the sequence of phases L_1–L_α–L_3 with increasing alcohol content. In the pentanol and decanol diagrams, only a few of these phases are stable. The short-chain alcohol forms phases L_1, and L_α, and L_3, and the long-chain alcohol, L_1 and L_4. Thus L_4 is not seen in the pentanol diagram, and L_α and L_3 are not present in the decanol one. For decanol, the lamellar phase exists only when the salinity is less than 10 g/L NaCl. In the pentanol system, the L_3 domain is continuous, connected to the micellar domain at high dilution. In addition, a new sequence of phases is found. One observes a large one-phase domain labeled L* on the pentanol-poor side of the lamellar phase instead of the L_4 phase. The L* domain is connected to the L_1 phase and also to the sponge domain L_3 through paths surrounding the lamellar phase. Finally, an apparently continuous evolution from L* to L_α is observed. This type of diagram is similar to that obtained for the system SDS–hexanol–glycerol–brine (10 g/L NaCL) [96,97]. Freeze-fracture electron microscopy and dielectric measurements provide evidence that L* is a sponge phase whose topology is different from the one in the concentrated part of the L_3 phase. These systems exhibit a re-entrant phase behavior. As the alcohol/SDS ratio is increased, the sequence of phases is sponge–lamellar–sponge instead of the vesicle–lamellar–sponge found in long-chain alcohol systems. In the systems investigated, the L_4 phase is stable over a narrow range of alcohol/SDS ratios, but it extends between 0 and 10 wt% surfactant for the alcohols hexanol to decanol. The width of the L_4 domain is typically one-half to one-fifth of that corresponding to the L_3 phases. Because of the existence of the emulsion, the boundary of the L_4 phase at high SDS concentration (above 2 wt%) is rather difficult to determine. Therefore, this limit has been established by using phase contrast optical microscopy, with which it is easy to detect the eventual presence of spherulites (L_α phase). The L_4 phase as well as the other phases, L_3 and $L\alpha$, are bounded by straight lines directed toward the brine corner for long-chain alcohols (C_8–C_{10}) and toward a point of the brine–alcohol axis for shorter chain alcohols. These features, similar to those previously described for hexanol, offer the possibility of investigating the properties of the L_4 phase along a dilution line at constant bilayer comparison. Figure 4 summarizes for several alcohols the domains of existence expressed as the molar ratio between alcohol and surfactant of the four phases L_1, L_4, L_α, and L_3. In the L_4 phases made with long-chain alcohols, the bilayer contains less than one molecule of alcohol per SDS molecule.

The structures of the phases L_4, L_α, and L_3 were characterized by means of several techniques and confirmed by freeze-fracture electron microscopy [47,79,86–88]. Typical electron micrographs showing the organization of the bilayers in phases L_3 and L_4 are shown in Fig. 5. These clearly show the continuity of the bilayer in the sponge phases and the existence

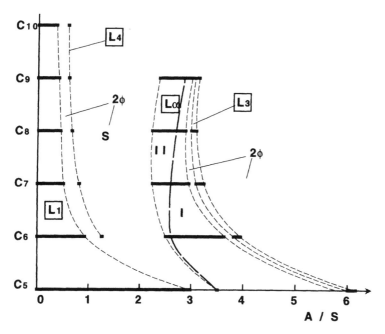

Figure 4 A/S locations of the L_1, L_3, L_4, and L_α phases for the various alcohols at $T = 25°$C. A/S is the molar alcohol/SDS ratio. I, II, and S have the same meaning as in Fig. 2. (From Ref. 88.)

of two types of aggregates in the L_4 phases. At low surfactant concentration, the L_4 phase consists of a dispersion of polydisperse unilamellar vesicles, while at high concentration the L_4 phase contains "onions" with a small number of layers ($N \sim 4$). In this concentrated regime the L_4 phase exhibits high viscosity and viscoelasticity [69,79].

As for hexanol, the lamellar phase is divided into two subregions identified by optical polarizing microscopy. In region I, at low alcohol content, large multilamellar vesicles are dispersed in the lamellar phase; in region II, the lamellar phase is free of these textural defects, which are replaced by more classical defects such as oily streaks. The extent of region I increases with the chain length of the alcohol. In the S domains, located between L_4 and L_α, an emulsion consisting of a dispersion of spherulites in the L_4 phase is found. The structures of the aggregates in zones I, II, and S were confirmed by freeze-fracture electron microscopy (Fig. 6) [79].

The various identified phases—L_1, L_3, L_4, and L_α—are separated by several complex multiphase equilibria including several three-phase regions. The tie-lines involving the L_4 phase have not been measured precisely. However, X-ray data obtained for octanol suggest that the tie-lines be drawn around the L_4 phase as shown in Fig. 7. Note that in region d_1 between the L_1 and L_4 phases, the macroscopic phase separation between an isotropic phase and the lamellar phase occurs after a few weeks.

Finally, one must note that the phase diagrams of the systems made with pentanol and hexanol present lines of critical points [85]. In the case of pentanol, the addition of salt generates a line of critical points that starts at P_C^A and terminates at the salinity of 3.5 g/L NaCl at a critical endpoint. A second line of critical points, P_C^2, exists in the water-rich region at very low SDS concentrations for salinities above 8.1 g/L NaCl. A line P_C^2 also occurs at very low SDS concentration ($\sim$0.1% SDS) for the hexanol system. In the pentanol

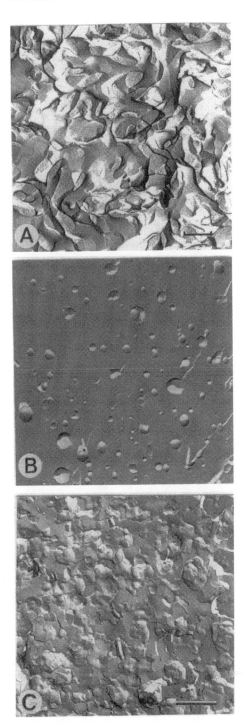

Figure 5 Freeze-fracture electron micrographs in L_3 and L_4 phases of nonanol–SDS–brine–glycerol 33% v/v, (A) L_3 phase (5% SDS) (bar = 250 nm); (B) Dilute L_4 phase (1% SDS); (C) concentrated L_4 phase (9% SDS). Note the existence of polydisperse unilamellar vesicles in (B) and small multilayer vesicles in (C), (B) and (C) are of the same magnification. (bar = 1 μm). (From Ref. 79.)

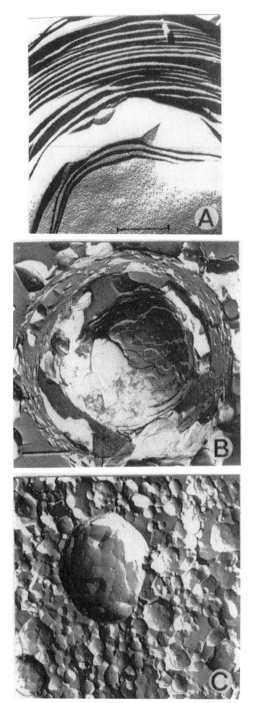

Figure 6 Freeze-fracture electron micrographs in the lamellar phase (regions I and II) and the two-phase region S of the brine–glycerol–octanol–SDS system. (A) L_α phase (region I) (7.5% SDS)l; (B) L_α phase (region II) (15% SDS); (C) region S (7.5% SDS). Notice the coexistence of small (L_4) and large (L_α) multilamellar vesicles. The bar represents 250 nm in (A) and 1 μm in (B). (B) and (C) are of the same magnification. (From Ref. 79.)

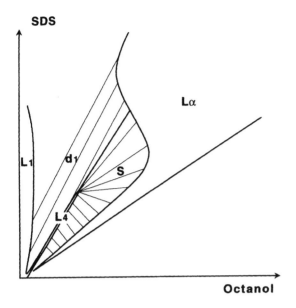

Figure 7 Schematic representation of the tie-lines involving the L_4 phase. The thick full line divides the two-phase region between domains where the samples separate (region d_1) and where they do not (region S). (From Ref. 79.)

system, around the points P_C^2, the mixtures separate between a symmetrical sponge phase and an asymmetrical sponge phase [98]. In the case of octanol, such a point does not exist at a salinity of 20 g/L NaCl.

In summary, in the ternary mixtures water–alcohol–SDS, the main effect of increasing the alcohol chain length from five to ten carbon atoms is to reduce the extent of the isotropic micellar domain to the benefit of the lamellar phase. This behavior reflects an increase in the elastic modulus κ as the alcohol chain length increases. The addition of salt to these ternary solutions produces very swollen lamellar phases, sponge phases, and vesicle phases. However, differences are observed when the alcohol chain length is varied; which also demonstrate the role of flexibility of the interface. Sponge phases do not occur with long-chain alcohols, whereas vesicle phases are not formed with short-chain alcohols.

One of the most striking properties of the lamellar phases encountered in the mixtures studied here is that the repeat distance between bilayers can be varied continuously by the addition of solvent (water or brine) from molecular sizes to extremely large values (10–100 nm). The ability to swell the smectic structure with water is associated with strong electrostatic repulsions. The even larger swellings obtained by dilution with brine originate from repulsive interactions between bilayers that have a nonelectrostatic origin. This interaction, known as the undulation interaction, arises in the high flexibility of the surfactant bilayers through a mechanism of entropy reduction proposed by Helfrich [99]. Under the influence of thermal fluctuations a free flexible membrane undulates. When the membrane is constrained in a lamellar phase it cannot explore the entire space around it. The restriction in its movement leads to an effective repulsive interaction, the amplitude of which is inversely proportional to the bending constant κ. This long-range repulsive interaction, which overcomes the van der Waals attraction at large distances d, is responsible for the very great dilution obtained in the systems described here [100]. The system containing decanol exhibits a particular behavior in comparison with the other alcohols because the lamellar phase does not exist at high salinity. The membranes made with SDS and decanol are rigid ($\kappa \sim 13k_B T$) [75], so one think that in this system the undulation interaction is too weak to stabilize both a swollen lamellar phase and a sponge phase.

Finally, in all the systems investigated here, the increase in cosurfactant concentration in the bilayer induces topological transitions from disconnected structures (vesicles) to connected ones (sponge). Thus, the sequence of phases $L_4 - L_\alpha - L_3$ allows us to conclude that the Gaussian modulus $\bar{\kappa}$ is governed by the composition of the bilayer and changes in sign from negative to positive as the alcohol/surfactant ratio increases. These results confirm previous predictions made by Porte et al. [71]. In nonionic surfactant–water mixtures, the same sequence of phases is produced by raising the temperature [101]. The modulus $\bar{\kappa}$ is controlled also by the alcohol chain length. An estimate of the Gaussian elastic constant $\bar{\kappa}$ was obtained from the size of the vesicles present in the L_4 phase; the value of $\bar{\kappa}$ increases from $-22k_B T$ to $-0.3k_B T$ as the alcohol chain length decreases from decanol to hexanol [88]. This trend suggests that $\bar{\kappa}$ is close to zero in the case of pentanol and could explain why the L_4 phase does not occur in the pentanol diagram.

B. Quaternary Systems: Water–Oil–Surfactant–Alcohol

The phase diagrams of two quaternary mixtures made of sodium dodecylsulfate (SDS)–water–dodecane and hexanol (system A) or pentanol (system B) have been investigated in detail [22,23]. In both cases, sections of the three-dimensional diagram with constant water/surfactant ratio have been examined. These cuts were chosen because they allow a good description of the oil region and also because the water/SDS ratio, termed X in the following, fixes the size of the droplets in the inverse microemulsion phase and the thickness of the bilayers in the oil-rich lamellar phase. In the description of the quaternary mixtures, we emphasize the details of the evolution of the phase equilibria as X is varied. We have focused our attention not only on the characterization and the location of the boundaries of the various phases but also on the equilibria between the phases.

1. System A: Water–Dodecane–SDS–Hexanol

Two typical pseudoternary diagrams are shown in Fig. 8. In each cut three one-phase regions L_{3-O}, L_α, and L_2 are observed. Regions L_2 and L_α correspond, respectively, to an inverse micellar phase and a lamellar phase. In the lamellar phase, the oil content may vary continuously between 0% and 85% (by volume). At high oil content, the lamellar phase consists of stacks of inverted bilayers (a water layer surrounded by two surfactant monolayers) separated by oil. The third phase, L_{3-O}, is a sponge phase (L_{3-O} designates an oil-rich sponge phase); in this phase, the inverted bilayers are connected. While motionless this phase is isotropic, but a flow transient birefringence appears in dilute samples as soon as any disturbance is created. In particular, flow birefringence is easily generated by shaking the sample tube. In addition, this phase scatters light. Minor changes are obtained as X is increased between 1.55 and 4.3 (by weight). In both cases, the microemulsion L_2 phase is bounded by a two-phase region where the L_2 phase coexists with the lamellar phase. The increase in X modifies the extents of the various regions but does not allow the oil-rich and water-rich regions to be connected.

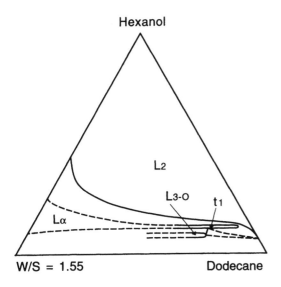

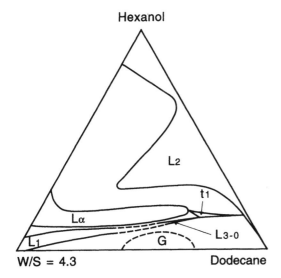

Figure 8   Sections at constant water/surfactant ratios $X = 1.55$ and $X = 4.3$ of the phase diagram of the water–dodecane–hexanol–SDS system (system A) at 21°C. L_1 and L_2 are isotropic phases. L_α is a lamellar phase, L_{3-O} is an oil-rich sponge phase. t_1 is a three-phase region. G is a gel phase. (From Refs. 22 and 23.)

2.   System B: Water–Dodecane–SDS–Pentanol

The pseudoternary phase diagrams corresponding to system B are strongly dependent on X (Fig. 9). In the planes below $X = 0.76$, only the microemulsion phase L_2 is detected in the oil-rich region. For $X = 0.76$, an oily swollen lamellar phase L_α appears. Both regions L_2 and L_α exist in all the diagrams for $X > 0.76$. In the sections corresponding to $X > 0.95$ two new phases appear: a sponge phase L_{3-O} and a hexagonal phase H_α. The amount of alcohol needed to form phases L_α and L_{3-O} increases with X for $X > 1$ (Figs. 10 and 11). Both phases may contain very large amounts of oil and alcohol, up to

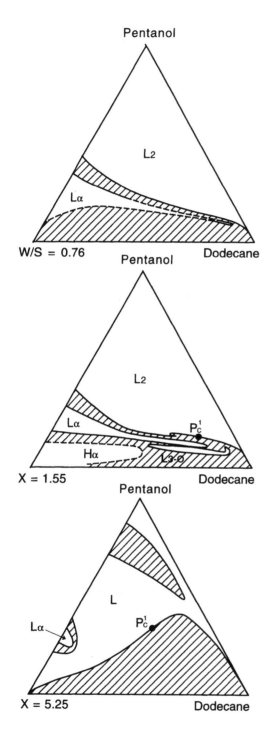

Figure 9   Sections at constant water/surfactant ratios $X = 0.76$, 1.55, and 5.25 of the phase diagram of the water–dodecane–pentanol–SDS system (system B) at 21°C. The hatched regions are the multiphase regions. P_C^1 is a critical point. L and L_2 are microemulsions; L_α is a lamellar phase; $L_{3\text{-}O}$ is an oil-rich sponge phase. (From Refs. 22 and 23.)

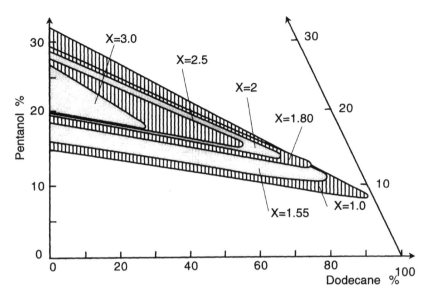

Figure 10 Effect of the water/surfactant ratio X on the location of the lamellar phase in the water–dodecane–pentanol–SDS system.

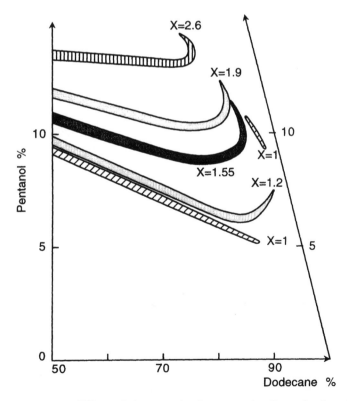

Figure 11 Effect of the water/surfactant ratio X on the location of the sponge phase in the water–dodecane–pentanol–SDS system.

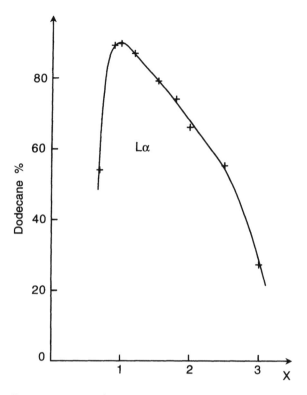

Figure 12 Maximum dodecane weight percentage contained in the lamellar phase of the water–dodecane–pentanol–SDS system versus the ratio X.

99 vol%. However, the swelling of the lamellar phase exhibits a maximum around $X = 1$ (Fig. 12). The sponge samples display flow birefringence when the oil content is greater than 75%. When X is greater than 3, the L_2 and L_{3-O} domains merge and form a continuous region L, which remains continuous up to $X = 5.3$. For the last value the amount of surfactant with respect to water is no longer sufficient to achieve the continuity of the single-phase domain L. For $X > 5.3$, L has split into two regions: one rich in oil and alcohol and the other rich in water and surfactant. It is worthwhile to note that for the section $X > 1$ small variations in the alcohol fractions (a few percent) lead to phase transformations and cause drastic changes in the properties of the oil-rich mixtures.

Moreover, one of the most interesting features occurring in any section where $X > 0.95$ is a critical point P_C^l along the coexistence boundary of the microemulsion L_2 region. A detailed analysis of the diagram shows that these points form a critical line P_C^l, which develops between P_C^A, located in the face of the water–pentanol–SDS tetrahedron at $X = 6.6$, and a critical endpoint P_{ce}^B, located in the oil-rich region at $X = 0.95$. The direct consequence of this critical line is the occurrence in each X section of a two-phase region (d_1) where two inverse microemulsions are in equilibrium (Fig. 13). In the sections where $X < 0.95$ and also far from the critical point, the microemulsion coexists with a lamellar phase. The appearance of new phases, on the one hand, and that of the critical point on the other, generates a very complex multiphase region in the sections where $X > 1$. As an example, Fig. 13 shows the multiphase regions corresponding to the section

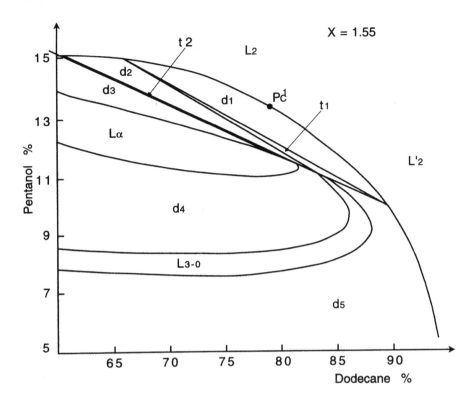

Figure 13 Magnification of the oil-rich corner of the section $X = 1.55$ of Fig. 8. d_1 is a two-phase region where two microemulsions, L_2 and L'_2, are in equilibrium. d_2–d_5 are two-phase regions. t_1 and t_2 are three-phase regions: $t_1 = L'_2 L_{3-O} L_2$. $t_2 = L_2$–L_α–L_{3-O}. (wt%). (From Refs. 22 and 23.)

$X = 1.55$. It includes five different two-phase regions (d_1–d_5) and two three-phase regions, t_1 and t_2; in region t_1 two microemulsion phases (L_2 and L'_2) coexist with the sponge phase (L_{3-O}); in region t_2, the middle phase is the lamellar phase and the upper and lower phases are the sponge and microemulsion phases, respectively. Both equilibria t_1 and t_2 disappear through a critical endpoint where $X \sim 3$ when the two isotropic phases L_2 and L_{3-O} merge [22,23].

The measurements of the phase composition of a large number of two- and three-phase equilibria t_1 and t_2 that involve both isotropic and mesomorphic phases provide evidence that X has the characteristic of a field variable in the oil-rich mixtures [22,23,102,103]; indeed this ratio takes the same value in all the coexisting phases. As an example, Fig. 14 shows compositions of conjugate phases of 10 samples occurring in region d_1 and two samples in regions t_1 and t_2. These data clearly show that X behaves as a chemical potential. The major consequence of this property is that the experimentally determined phase diagrams in which X is maintained constant are true pseudoternary diagrams. The tie-lines and tie-triangles are indeed located in the X section considered; however, for clarity, they are not shown in the figures. Moreover, since X behaves as a chemical potential, we used this variable to approach several critical points of the line P_C^1 at constant temperature [22,23,102,103].

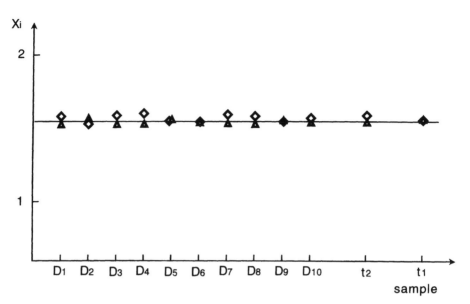

Figure 14 Values of water/surfactant ratio (X_i) measured in each phase i for 10 two-phase equilibria (termed D_1–D_{10}) lying in region d_1 and for three-phase equilibria in regions t_1 and t_2. For all these samples the global value of X is equal to 1.55 ($T = 21°C$). (From Refs. 22,23,102, and 103.)

3. Comparison of the Phase Diagrams of Systems A and B

From this brief description, it appears that the phase diagrams of systems A and B show several common points and also a few differences. Both systems form the same three phases in the oil-rich part of the diagram: a microemulsion phase L_2, a lamellar phase L_α, and a sponge phase L_{3-O}. The first two, L_2 and L_α, are the direct extensions of the micellar and lamellar phases found for the ternary mixtures in the absence of oil. The L_{3-O} phase is a new phase that is stable in a surfactant concentration range between a few percent and 20%. Several experimental facts based on neutron scattering and electrical conductivity measurements have demonstrated that in this phase the surfactant forms a continuous inverted bilayer slightly swollen with water [47]. In particular, the high conductivity measured all along the phase, even for the most dilute samples clearly indicates that the bilayers are connected and that continuous water paths through the membranes exist. In contrast to what is found in brine-rich SDS–pentanol systems, the transition from sponge to microemulsion is not continuous. However, in both systems the same sequence of phases, L_2–L_α–L_{3-O}, is obtained with a decrease in the alcohol concentration. This behavior is very similar to that previously described for brine rich alcohol–SDS mixtures. So, in this region too, the alcohol content in the bilayers controls the Gaussian elastic constant $\bar{\kappa}$. Finally, very oily swollen L_{3-O} and L_α phases, corresponding to distances between the bilayers larger than 100 nm, can be prepared. The large degree of swelling exhibited by both systems shows that long-range repulsive interactions between bilayers exist. Here again, as in salted systems, undulation interactions are responsible for the very existence of these extremely dilute phases [99].

Two main differences are found as one replaces hexanol with pentanol; the first is the existence of a critical line P_C^l, and the second is the possibility of going continuously from the water + surfactant vertex to the oil vertex when X is larger than 3. In the hexanol system, the L_2 and L_1 regions are never connected for any value of X. In this system the phase inversion L_1–L_2 occurs via the lamellar phase. Regarding the stability of the microemulsion phase, two characteristic behaviors are evidenced. In the first one no critical point occurs and the microemulsion region L_2 is bounded by a two-phase region where phase L_2 is in equilibrium with the lamellar phase L_α. The second type is characterized by the occurrence of a critical point and a two-phase region where two micellar phases coexist. This behavior is closely related to the intermicellar interactions, as we will see in Sec. IV.

From these phase diagrams, one important conclusion must be emphasized: A comparison of the water-rich side and the oil-rich side of the phase diagram shows similar properties in their phase behavior and structures. In each side as the alcohol concentration is varied, the system exhibits the sequence micelle (or vesicle)–lamellar–sponge. This reveals that although the experimental situation seems opposite, the physics is the same and can be described with the flexible surface model [19]. Symmetry properties of phase behavior were found also with nonionic surfactants systems [104].

C. Effect of Salt on Water–Oil–Surfactant–Alcohol Systems

It is worthwhile to point out that all the equilibria observed in these systems are different from the so-called Winsor equilibria where a microemulsion coexists with an excess oil phase (Winsor I equilibrium, WI), or an excess water phase (Winsor II equilibrium, WII), or both (Winsor III equilibrium, WIII). These equilibria are obtained only by the addition of salt.

1. Oil-Rich Regions

Figure 15 shows the effect of added salt on the stability of the three oil-rich phases L_2, L_α, and $L_{3\text{-}O}$ found in the section $X = 1.55$ of the phase diagram of the water–dodecane–SDS–pentanol system at low alcohol content. The three phases still exist at very high

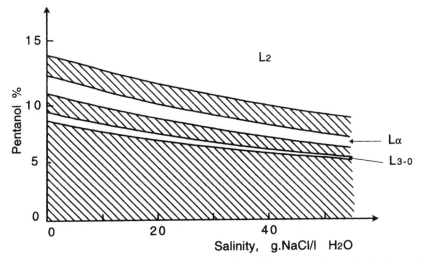

Figure 15   Water–NaCl–dodecane–pentanol–SDS system: effect of salinity on the microemulsion L_2, lamellar L_α, and sponge $L_{3\text{-}O}$. In the section considered, $X = 1.55$ and the dodecane content is 78 wt.%. The hatched regions represent multiphase regions. (From Ref. 105.)

salinities, but they are shifted to lower alcohol concentration. In fact, salt produces the same effect on the structures of the phases and on the topology of the interface as the alcohol does. Indeed, an increase in either salt or alcohol content produces the sequence of phases $L_3 \rightarrow L_\alpha \rightarrow L_2$ [105].

As one examines the entire diagram, one observes that the phase behavior is largely influenced by the addition of salt. Figure 16 compares three X sections of the phase diagrams obtained (Figs. 16a–c) in the absence of salt and (Figs. 16d–f) at a salinity of 20 g/L NaCl. At low X, salt causes the formation of Winsor II two-phase equilibria consisting of a microemulsion phase in equilibrium with an excess of brine and, consequently, a drastic reduction in the extent of the microemulsion domain. Upon increase X, the Winsor II region expands. This trend leads to the disappearance of oil-rich microemulsion phases, giving rise to the three-phase equilibria Winsor III.

2. Intermediate Region, Brine/Oil Ratio = 1

Winsor observed that a mixture of comparable amounts of oil and brine containing small quantities of both ionic surfactant and cosurfactant very often demixed into two or three phases [106]. One of these phases is a microemulsion presenting remarkable properties: low interfacial tension and maximal solubilization of both oil and brine for a given quantity of surfactant. The Winsor phase equilibria depend on the salinity as shown in Fig. 17. At low alcohol (A) + surfactant (S) concentrations, one observes the well-known progression of phase equilibria Winsor I → Winsor III → Winsor II as salinity increases. At low salinity an oil-in-water microemulsion coexists with an excess of oil (Winsor I); at high salinity a water-in-oil microemulsion coexists with an excess of brine (Winsor II); and in between the equilibrium is intermediate, i.e., a bicontinuous microemulsion coexists with both a water phase and an oil phase. Increasing the alcohol (A) + SDS (S) concentrations gives rise to an optimized, balanced microemulsion (points M in Fig. 17). At point M the amount of surfactant and cosurfactant needed to prepare a microemulsion with a water/oil ratio equal to 1 is minimized. At still higher A + S concentrations the phase behavior depends on the alcohol chain length. In the case of hexanol, a lamellar phase is observed between two microemulsions regions, while in the case of pentanol a continuous microemulsion domain is found.

Figure 17 shows that for a given alcohol/surfactant ratio (here A/S = 2), the increase in salinity generates a progressive inversion of the curvature of the interface. Another way to change the sign of the curvature is to maintain the salinity constant and vary the alcohol/surfactant ratio. The phase diagrams presented in Fig. 18 show that the same pattern of phases is obtained by increasing the alcohol concentration. For both alcohols, one observes at low SDS concentration the sequence Winsor I → Winsor III → Winsor II, and at high surfactant concentration the succession of phases microemulsion → lamellar phase → microemulsion is observed. In particular, with pentanol it appears that the lamellar phase is stable at alcohol/surfactant ratios lower than 1. Finally, these diagrams clearly show that the minimum amount of surfactant ϕ_s^m required to form dilute balanced microemulsion phases depends at a given salinity on both the alcohol and oil chain lengths. ϕ_s^m is smaller the longer the alcohol chain and shorter the oil chain. According to the results obtained by Binks et al. [107], the value of the bending elasticity of the SDS/alcohol monolayer can be expected to increase with decreasing alkane chain length. In the case of pentanol–dodecane, the lamellar phase is stable only for SDS concentrations above 15%, and the microemulsion phase exists over a large range of concentration, but ϕ_s^m is large ($\sim \phi_s^m = 6.5\%$ SDS). In contrast, in the case of hexanol–octane, both the lamellar and microemulsion domains

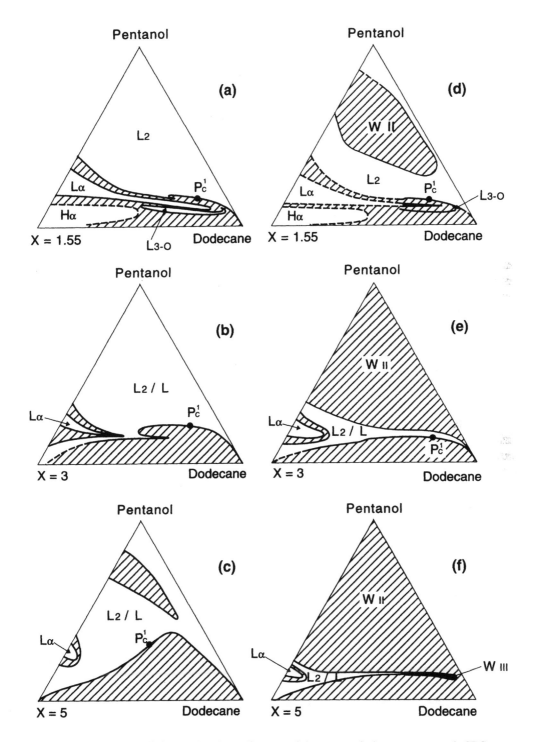

Figure 16 Effect of salinity on the phase diagram of the water–dodecane–pentanol–SDS system. (a,b,c) Diagrams for three X sections obtained without salt. (d,e,f) The same three sections obtained at the salinity of 20 g/L NaCl. W II, (Winsor II equilibria); W III, (Winsor III equilibria). (From Ref. 105.)

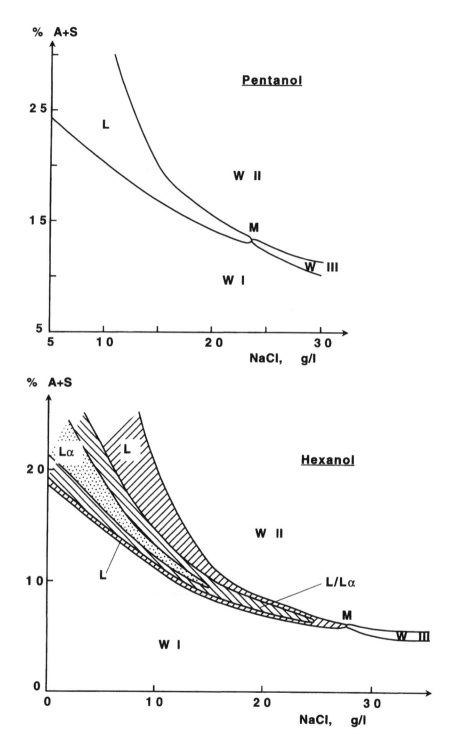

Figure 17 Phase behavior as a function of salinity for the water–octane–SDS–pentanol or hexanol systems. In the sections presented the water/oil ratio = 1 and the alcohol/SDS ratio (A/S) = 2. W I, W II, W III = Winsor equilibria; L = microemulsion phase; L_{α} = lamellar phase.

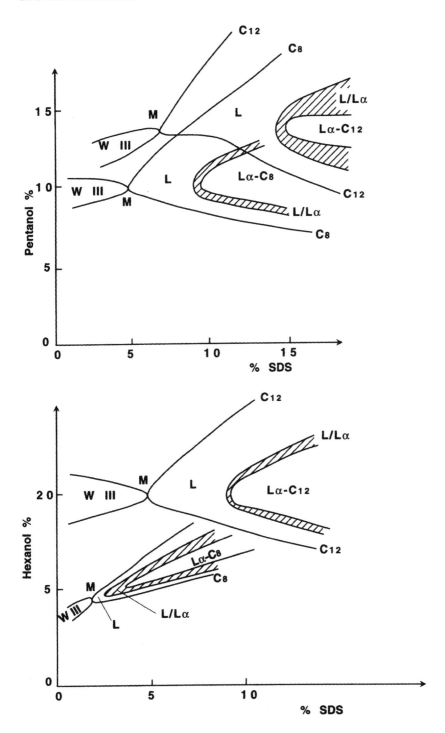

Figure 18 Phase behavior of four systems: brine (20 g/L NaCl)–octane (C₈)–pentanol–SDS; brine (20 g/L NaCl)–dodecane (C₁₂)–pentanol–SDS; brine (20 g/L NaCl)–octane (C₈)–hexanol–SDS; and brine (20 g/L NaCl)–dodecane (C₁₂)–hexanol–SDS in the alcohol–SDS concentration plane with equal amounts of brine and oil.

occur at lower ϕ_s and the concentration range over which the microemulsion is stable becomes significantly smaller. These trends are consistent with theoretical models for microemulsions that predict that the transition between a bicontinuous microemulsion and a lamellar phase is determined by the value of the bending constant κ [59,63,64]. Following these predictions and in agreement with experimental results [108], the recipe for making a microemulsion that is stable over a wide range of ϕ_s is to make κ small in order to push the lamellar phase to higher ϕ_s. However, in that case ϕ_s^m will be quite large and the microemulsion will use a considerable amount of surfactant. Consequently, the price to pay for having a microemulsion phase made with a very small amount of surfactant is to tolerate a narrow region of stability.

The elasticity parameters of flexible surfactant layers interacting with polymers may change. Recently, there have been some speculations on the effect the introduction of a neutral flexible polymer would have on a layered system. While some studies predict an increase in the rigidity or in the elastic bending modulus κ [109], some other theoretical studies predict κ to decrease and the Gaussian rigidity $\bar{\kappa}$ to increase [110]. One way to test these predictions is to examine the effect of the polymer on the quantity ϕ_s^m. The addition of polyethylene glycol (PEG, Mw 22,000) to the system described above leads to an increase in ϕ_s^m, suggesting that the elastic bending modulus κ is smaller than in samples without this polymer (Fig. 19).

3. Brine-Rich Region

The diagrams shown in Fig. 20 correspond to cuts of the water–salt–dodecane–SDA–pentanol quinary system obtained at fixed salinity (20 g/L NaCl) and fixed oil fraction. The addition of oil substantially alters the stability of the L_1, L_3, and L_α phases. When small fractions of dodecane are added to brine–pentanol–SDS mixtures, the L_1 and L_3 phases do not continue to form a single domain but split into two distinct regions. However, for oil fractions larger than 8%, the L_1 and L_3 phases merge and again form a single domain. Region L_1, which is the extension of the micellar phase found in the pseudoternary diagram, corresponds to oil-in-water microemeulsions. This phase is obtained only in the presence of alcohol. The addition of oil considerably reduces the swelling of both the lamellar phase L_α and the sponge phase L_3. Both regions are shifted to increasingly higher surfactant and pentanol concentrations as the oil content increases. These various phases are separated by complex multiphase regions including four three-phase regions denoted t_3, t_5, t_6, and W III. The three first regions exist only at low oil content (<9%) whereas the W III region is present up to high oil concentrations. In t_3 and t_5, the phases are isotropic ($t_3 = L_1, L_3, L_3' t_5 = L_1, L_3, L_1'$). In t_6 one of them is lamellar ($t_6 = L_1, L_3, L_\alpha$). Measurements of phase composition of three-phase equilibria t_3 and t_5 show that the ratio denoted Y between dodecane and SDS concentrations takes the same value in the upper and middle phases of those equilibria. This property is illustrated in Fig. 21, where the ratio Y_i measured in each phase i is plotted versus the value Y in the overall mixture. All the points fall along the bisecting line of the graph. The ratio Y has the characteristics of a chemical potential [111]. This behavior is fully identical to that of the ratio X (water/SDS ratio) found in the oil-rich region (Fig. 14).

4. Winsor III Equilibria–Modes of Appearance

The origin of the Winsor III three-phase equilibria where a microemulsion (m) coexists with both excess oil (u) and excess aqueous (l) phases has been discussed elsewhere [51–57, 112]. Several modes of appearance may be considered to account for the formation of a

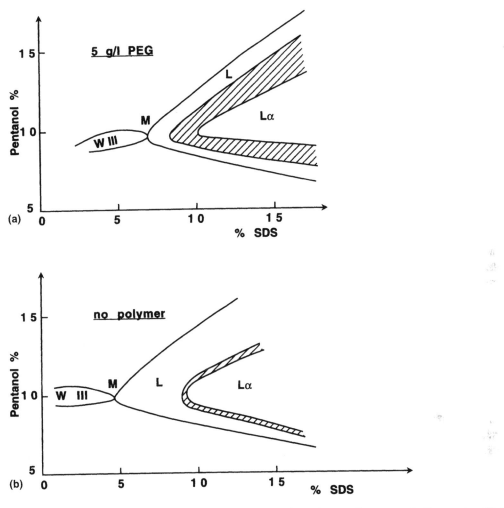

Figure 19 Effect of PEG (MW 22,600) on the phase diagram of the water–NaCl 20 g/L–dodecane–pentanol–SDS system. (a) With 5 g/L PEG. (b) With no PEG. In the section considered, brine/dodecane = 1; salinity = 20 g/L NaCl.

three-phase region in multicomponent systems. At constant temperature and pressure the three-phase region of a quaternary mixture is a volume made by a stack of tie-triangles whose vertices are located along three curves, Γ_u, Γ_m, and Γ_l, corresponding, respectively, to the upper, middle, and lower phases. Depending on the relative positions of these curves, the three-phase region may be generated in several different ways as illustrated in Fig. 22. One of these possibilities is the existence of two critical endpoints, P_{ce}^1, P_{ce}^2, where, respectively, the phases u and m and the phases m and l become identical. The critical endpoints P_{ce}^1 and P_{ce}^2 are the ends of two critical lines, P_C^1 and P_C^2. As the temperature changes, the three-phase region may shrink when a tricritical temperature is reached; in that case the three phases become identical, the critical endpoints merge, and the critical lines P_C^1 and P_C^2 form a connected single line (Fig. 22c). Two other origins can be considered;

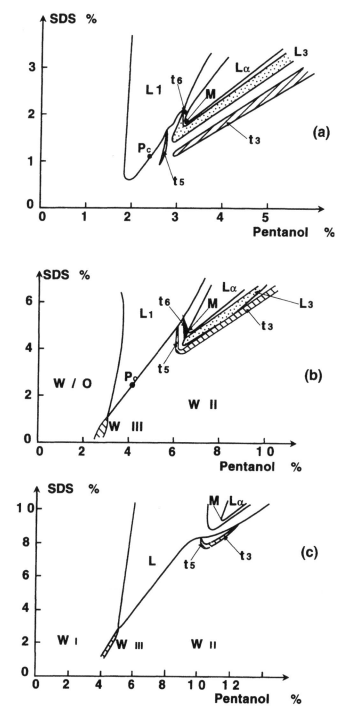

Figure 20 Three sections of the phase diagram of the quinary mixture water–NaCl–dodecane–SDS–pentanol ($T = 25°C$). In the three sections the concentration of salt in water is fixed at 20 g/L. In each section the dodecane weight fraction is constant; (a) 1%; (b) 4%; (c)8%. L, L_1, microemulsion; L_α, lamellar phase; L_3, sponge phase; t_3, t_5, t_6 are three-phase equilibria; W I, W II, W III are Winsor equilibria. (From Ref. 111.)

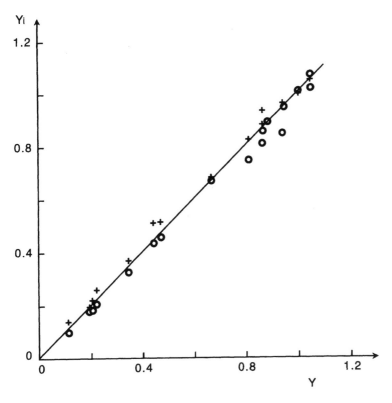

Figure 21 Values of the dodecane SDS ratio Y_i measured in the upper ($+$) and middle ($\circ$) phases of the three-phase equilibria t_3 as a function of the ratio Y in the overall mixture. (From Ref. 111.)

It is possible that the three-phase volume terminates at a four-phase region or arises by an indifferent state [83]. In this case, the points representative of the phases are along a straight line ABC (Fig. 22d).

The modes of appearance of a three-phase region for a quinary mixture are the same as those for a quaternary one, but the number of degrees of freedom of the corresponding states is increased by one. Consequently, at fixed P and T a three-phase region may result from one tricritical point; in that case it develops between two lines of critical endpoints L_{ce}^1 and L_{ce}^2. As in quaternary mixtures, three-phase regions in five-component mixtures may result in an indifferent state. The diagram of the quinary system water–salt–dodecane–pentanol–SDS can be represented by a series of tetrahedra, each corresponding to a constant brine salinity. In practice, one tetrahedron is constructed by examining a series of cuts in which the brine/surfactant ratio X is kept constant. This procedure allows one to describe completely the extent of the three-phase Winsor III (W III) region and to determine its mode of appearance (Fig. 23). The results of Bellocq and Gazeau show that this region, which starts to be observed at 4 g/L NaCl in water, arises from an indifferent state [57]. In such a state, the compositions of the three coexisting phases are located along a straight line ABC in the four-dimensional space of representation. At salinity above 5.5 g/L NaCl, the W III volume is bounded in the oil-rich region by a line of critical endpoints, L_{ce}^1. A second line of such points, L_{ce}^2, occurs in the water-rich region at salinities higher than 8.1 g/L NaCl. Then the three-phase region W III is located between two critical endpoints

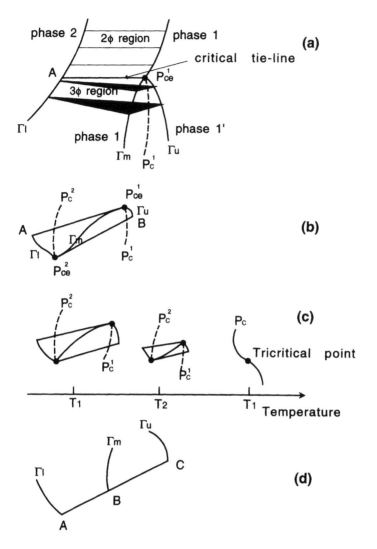

Figure 22   Modes of appearance of a three-phase region for a quaternary mixture. (a) Critical endpoint; (b) two critical endpoints; (c) one tricritical point; (d) one indifferent state. (From Ref. 57.)

only when salinity is higher than 8.1 g/L NaCl. The line L^1_{ce} is related to the critical line P^1_C found in the quaternary system water–dodecane–pentanol-SDS, the line along which the separated phases are inverse microemulsions. The line L^2_{ce} originates in the system without oil at the critical P^2_C; at this point the phase separation is between a sponge phase and an a asymmetrical phase (cf. Sec. III.A.1).

It should be mentioned that in the three-phase equilibria close to the limit ABC, the three phases strongly scatter light. This could indicate that the system is close to a tricritical point. It is likely that tricriticality, which would result from the connection of the two lines of critical endpoints L^1_{ce} and L^2_{ce}, could be achieved by raising the temperature. Tricritical points have been observed in some ionic and nonionic systems [51–56].

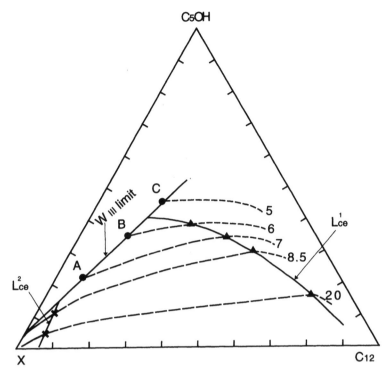

Figure 23 Projections on the planes X–pentanol–dodecane of the curves Γ_u (- - -), Γ_m (– – –), and $\Gamma\ell(2/m)$ observed at salinities of 5, 6, 7, 8.5, and 20 g/L NaCl. (▲, +) Critical endpoints P^1_{ce} and P^2_{ce}, respectively. (●) Limit of existence of the three-phase equilibria Winsor III. (From Ref. 57.)

IV. CRITICAL BEHAVIOR

As shown in the preceding sections, water–oil–surfactant mixtures exhibit a complex range of phase behavior and offer an interesting area for observing critical phenomena. Several detailed studies have established that ternary, quaternary, and quinary microemulsion systems give rise to critical points and have suggested that some exhibit tricritical points [19,49–56]. Critical points are observed in both the oil-rich and water-rich regions of the phase diagrams. For these multicomponent mixtures, phase separations are obtained either at constant temperature by varying the composition or at constant concentration by raising the temperature.

Over the last 10 years or so, a great deal of work has been devoted to the study of critical phenomena in binary micellar solutions and multicomponent microemulsions systems [19]. The aim of these investigations in surfactant solutions was to point out differences if they existed between these critical points and the liquid–gas critical points of a pure compound. The main questions to be considered were (1) Why did the observed critical exponents not always follow the universal behavior predicted by the renormalization group theory of critical phenomena? and (2) Was the order of magnitude of the critical amplitudes comparable to that found in mixtures of small molecules? The systems presented in this chapter exhibit several lines of critical points. Among them, one involves inverse microemulsions and another, sponge phases. The origin of these phase separations and their critical behavior are discussed next.

A Critical Behavior of Microemulsions

1. Critical Behavior and Intermicellar Interactions

In most cases, it appears possible to interpret the critical behavior of microemulsion mixtures as a liquid/gas-like critical point [113–116]. Several light- and neutron-scattering studies on oil-rich ternary and quaternary microemulsions have clearly demonstrated that the structure of these media can be described as a solution of interacting water-in-oil droplets. As first shown by Calje et al. [117], the droplets may behave essentially as hard spheres. However, in many systems an attractive contribution to the interactions exists. It has been established that the strength of attractions between W/O micelles is strongly dependent on the micellar size and on the chain lengths of both the alcohol and oil molecules. In particular, attractions have been found to increase when the micellar radius increases or the alcohol chain length decreases and the molecular volume of the oil increases [114, 115, 118-120].

By analogy with polymer solutions, the effect of oil can be easily understood as a change in the quality of the solvent for the alkyl chains of the surfactant surrounding the micelles. In the language of polymer physics, an increase in the chain length of oil leads to a bad solvent. A good solvent yields hard-sphere interactions, while a bad solvent yields an attractive interaction between the micelles. In a bad solvent the aliphatic chains of the surfactant molecules prefer to mix together rather than be penetrated by the solvent. Lemaire et al, [121] established that this attraction (which derives from van der Waals interactions) is due to the mutual interpenetration of the surfactants tails; the attraction increases with micellar radius and with the difference between surfactant and alcohol chain lengths. Later Auvray [122] proposed a mechanism of fusion of droplets induced by curvature to interpret the observations made for inverse micelles. A simplified calculation of the second viral coefficient of the osmotic pressure B, leads to an expression that depends on the shape of the dimer and on the elastic properties of the surfactant film. This mechanism also accounts for the variation of B with the alcohol chain length.

As the potential between inverse droplets becomes strongly attractive, a critical behavior is observed. The curves of reduced compressibility versus the micellar volume fraction Φ for three microemulsions formed with water, dodecane, and SDS (sodium dodecylsulfate) and heptanol or hexanol or pentanol, illustrate this effect [22,23]. The heptanol microemulsion is close to a hard-sphere system. In the case of pentanol, both $\partial \pi / \partial \phi$ and $\partial^2 \pi / \partial \phi^2$ are close to zero, which is the signature of a critical point. Thus for this system it appears that attractions between droplets are strong enough to induce a phase separation into a micelle-rich phase and a micelle-poor phase.

Comparison of the phase diagrams of the systems made with hexanol and pentanol reveals that they present two different types of phase behavior (Sec. II.2, Figs. 8 and 9). In the first type of diagram, that of hexanol, no critical point is observed, whereas the second type of diagram is characterized by the occurrence of a critical point. The critical point and the two-phase region where two microemulsions coexist found in the diagram of the pentanol system have been interpreted as a liquid–gas transition due to intermicellar interactions [113–115]. A quantitative interpretation of the diagram of the quaternary system with pentanol has been made possible by using the intermicellar potential developed by Lemaire et al. [121] A similar conclusion has been drawn for the ternary system AOT–water–decane [116,123]. In the hexanol system, intermicellar interactions remain moderately attractive; for this system the phase separation is not driven by interactions but is due to another mechanism involving probably the curvature energy of the oil/water interface [63,64].

2. Experimental Results on Critical Behavior of Microemulsion Systems

The description of thermodynamic anomalies observed near critical points has been presented in many books and reviews [124–135]. Sufficiently close to a critical point, thermodynamic properties A vary as simple power laws of the distance ε from the critical point.

$$A = A_0 \varepsilon^\lambda \qquad \text{with } \varepsilon = \frac{F - F_C}{F_C} \tag{3}$$

where λ is a critical exponent and the coefficient A_0 defines the amplitude of the divergence. The variable F is a field variable, and F_C is its value at the critical point. The fields are the variables such as temperature, pressure, or chemical potential of the species (μ_i), which have equal values in the coexisting phases. The exponents ν, γ, β, and μ, respectively, are characteristic of the correlation length, the osmotic compressibility χ, the coexistence curve, and the interfacial tension σ between phases in equilibrium.

The second major feature of critical phenomena is the concept of scaling. The critical exponents are not independent. There exist relations among the exponents called scaling relations. In particular,

$$\gamma = 2\nu \qquad \text{and} \qquad \mu + \nu = \gamma + 2\beta \tag{4}$$

Finally, the third important feature of critical phenomena highlighted by modern theories is the concept of universality. According to this, the critical exponents associated with the singularities are identical for all the systems within a given universality class. Pure fluids and fluid mixtures near normal critical points belong in the universality class of the three-dimensional Ising model. Renormalization group methods have yielded a detailed and accurate description of the critical thermodynamic behavior of such Ising-like systems. The Ising values for the exponents ν, γ, β, and μ are $\nu = 0.630 \pm 0.001$, $\gamma = 1.240 \pm 0.002$, $\beta = 0.325 \pm 0.001$, and $\mu = 1.260$.

Critical exponents are defined along precise paths; otherwise, the values of the exponents are renormalized [136–138]. Therefore, one of the main difficulties encountered in the study of a multicomponent mixture is that of finding an appropriate path to approach the critical point. This means that it is necessary to control, in addition to temperature and pressure, one or more chemical potentials. We have seen that in the oil-rich part of the phase diagram, the water/surfactant ratio X takes the same value in the coexisting phases. The major consequence of this remarkable property is that the section $X = $ constant of the phase diagram of the quaternary system can be considered to represent real ternary systems since the tie-lines and the tie-triangles corresponding to the two- and three-phase equilibria are lying in the X section. The association of X with a field variable in reducing the number of variables of the system leads to a situation equivalent to that of a ternary mixture. In a three-component system, the critical exponents are expected to be Fisher renormalized with respect to the respective exponents in a binary mixture if one varies temperature at a fixed composition to approach the critical point [139]. The theoretical renormalized values ν^*, γ^*, μ^*, β^* of the exponents are 10% higher than the Ising values. They are obtained by multiplying the Ising values by the coefficient $1/(1 - \alpha)$ with $\alpha = 0.11$ (α being the exponent of the specific heat). Such a renormalization has been shown experimentally in a number of different ordinary ternary mixtures [140–143], although γ^* and ν^* appear to take on values of $\gamma^* = 1.50$ and $\nu^* = 0.75$, which are different from the theoretically expected values, $\gamma^* = \gamma/(1 - \alpha) = 1.39$ and $\nu^* = \nu/(1 - \alpha) = 0.70$. The experimental situation with ternary microemulsions is far from clear. The exponent ν for the correlation length measured on microemulsions consisting of AOT, water, and decane appears to be rather close to the Fisher renormalized Ising exponent, whereas the exponent γ for the susceptibility seems to take the Ising value [144–148]. This is an unsatisfying situation, because either both or neither of the exponents should be renormalized. A recent very careful light-scattering experiment on a critical ternary microemulsion made of benzene, water, and BHDC was performed using temperature as the experimentally controlled variable. Contrary to the theoretical expectation, the experimental values of the exponents γ and ν indicate rather unrenormalized Ising-like behavior ($\gamma = 1.18 \pm 0.03$; $\nu = 0.60 \pm 0.02$) [149].

In the quaternary system water–dodecane–pentanol–SDS, the critical behavior is much more complex. As mentioned in Sec. II. B, at fixed temperature $T = 21°C$, this quaternary mixture presents a line of critical points that extends in the phase diagram between a critical point P_c^A belonging to the ternary mixture water–SDS–pentanol and a critical endpoint P_{ce}^B located in the oil-rich part of the diagram. The X values (expressed in weight) corresponding to P_{cc}^B and P_c^A are 0.95 and 6.6, respectively.

In the first place, Roux and Bellocq investigated the critical behavior at point P_c^A, using as variables either temperature (path I) or concentration at constant T (paths II and III). Paths II and III [82] are, respectively, perpendicular to the direction of the tie-lines and tangent to the coexistence curve at the critical point. The results obtained by light-scattering, turbidity, and density measurements yield values in good agreement with those expected for a ternary mixture [150]. The critical indices measured along the three different paths indicate an Ising-like behavior (Table 1). However, the question as to whether the Fisher renormalization is observed is open and requires a more accurate determination of the exponents.

In the quaternary system water–dodecane–pentanol–SDS, six different critical points of the line P_C^l defined by $X = 1.03, 1.207, 1.372, 1.59. 3.45,$ and 5.17 were approached, in the single-phase region, by using either temperature or X as the variable. In a first step, the critical behavior was characterized by divergences in the correlation length ξ and osmotic compressibility χ of the solutions [151]. Both quantities were extracted from light-scattering measurements. The scattered intensity as a function of the scattering vector q of a mixture in the critical region follows the well-known Ornstein–Zernike structure

Table 1 Critical Indices for the Ternary System Water–SDS–pentanol[a]

		Exponent	
Path	Variable	Theoretical	Experimental
I	T	$\gamma' = \dfrac{\gamma}{1-\alpha} = 1.39$	1.27 ± 0.1
		$\gamma' = \dfrac{\gamma}{1-\alpha} = 1.39$	1.45 ± 0.1
II	Concentration Y, x_y	$\nu' = \dfrac{\nu}{1-\alpha} = 0.71$	0.75 ± 0.05
		$\beta' = \dfrac{\beta}{1-\alpha} = 0.36$	0.38 ± 0.03
III	Concentration O, x_0	$\gamma/\beta = 3.81$	3.68 ± 0.2

[a] The direction Y is perpendicular to the tie-lines. The direction O is tangent to the existence curve at the critical point.
Source: Ref. 82.

factor [124]:

$$I(q) = \frac{I_{q=0}}{1 + q^2 \xi^2}$$

$I_{q=0}$ is the intensity at $q = 0$, and ξ is the correlation length. The osmotic compressibility at $q = 0$, $\chi_{q=0}$, is proportional to $I_{q=0}$ through the classical relationship, where k_B is the Boltzmann constant,

$$I_{q=0} = k_B T \chi_{q=0}$$

The experimental data obtained for the six critical points show typical features of critical phenomena. Both the correlation length ξ and the scattered intensity $I(0)$ are found to diverge at the critical points. Figure 24 shows log-log plots of ξ and $I(0)$ versus the reduced temperature $\varepsilon_T = (T - T_C)/T_C$. The data are well fitted by the power laws

$$\xi = \xi_0 \varepsilon_T^{-v_t} \qquad \text{and} \qquad I(0) = I_0 \varepsilon_T^{-\gamma_t}$$

In this analysis, the contribution to the scattering from the micelles in the mixtures was neglected, and the total scattered intensity was attributed to critical fluctuations over the entire ε range investigated. For each critical point, the values of v_t and γ_t obtained as described above are very close to those found in independent experiments by varying X at constant temperature [114]. These values are found to vary continuously from the Ising values ($\gamma_t = 1.21 \pm 0.06$; $v_t = 0.64 \pm 0.03$ for $X = 5.17$) to significantly smaller ones ($\gamma_t = 0.40 \pm 0.04$, $v_t = 0.21 \pm 0.02$ for $X = 1.03$) as the critical endpoint P_{ce}^B is approached. For the six critical points studied, the relation $\gamma = 2v$ is verified. The sharp decrease in the values of the exponents in the vicinity of P_{ce}^B is accompanied by a rapid increase in the correlation amplitude ξ_0 (3–30 nm).

In view of the anomalous critical behavior of the correlation length and the osmotic compressibility, it appeared of interest to characterize the behavior of other properties. Bellocq and Gazeau investigated how the interfacial tension between the coexisting phases on the one hand and the difference of density of these phases on the other hand vanished at various points of the critical line P_c^1 (Fig. 25) [152]. The aim of the experiments was to determine the associated critical exponents μ and β and check whether the scaling laws that relate v, γ, β, and μ were valid all along the critical line. Data obtained for two critical points defined by $X_C = 1.55$ and $X_C = 1.207$ indicate that the values of the critical exponents β and μ show an X dependence similar to that found for v and γ. Furthermore, within the experimental accuracy, the obtained values of v, γ, β, and μ are in reasonable agreement with the theoretical predictions $\mu + v = \gamma + 2\beta = 3v$ (Table 2).

Other properties of these critical mixtures have also been investigated. In particular, the refractive index [153], the thermal conductivity Λ_{th}, and the thermal diffusivity D_{th} [124135] have been measured. Near the critical endpoint P_{cc}^B, all these properties exhibit anomalous behavior as the temperature approaches the critical temperature.

Since the smallest values of the exponents were measured near the critical endpoint P_{cc}^B, measurements were performed in the plane $X = 1.034$ at three critical temperatures [154]. The new data show that the anomalous critical behavior is due, as a matter of fact, to a crossover of two critical phenomena. The temperature evolution of the phase diagram and the light-scattering data agree with this interpretation. The system exhibits a phase separation at low and high temperatures. The high temperature critical point P_c^1 corresponds to a phase separation between two inverted micellar solutions L_2 and L_2' driven by intermicellar interactions; this high temperature critical point is characterized by Ising exponents, and the amplitude for the correlation length ξ_0 is found to be very large,

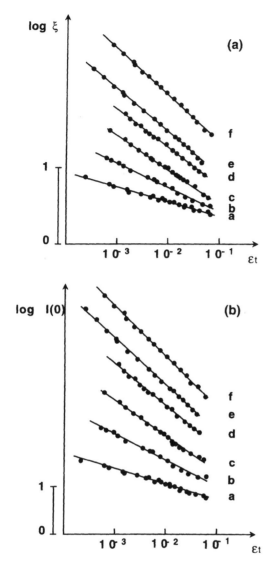

Figure 24 Log-log plots of (a) the correlation length ξ and (b) the total intensity $I(0)$ versus the reduced temperature for six critical mixtures: a, $X_c = 1.034$; b, $X_c = 1.207$; c, $X_c = 1.372$; d, $X_c = 1.552$; e, $X_c = 3.448$; f, $X_c = 5.172$. Each critical point has been approached in the single-phase region. (From Ref. 151.)

$\xi_0 = 2$ nm. The critical behavior found at low temperature is related to a transition from micellar L_2 to sponge phase L_{3-O}. This interpretation is based on theoretical arguments that state that the sponge phase has a particular symmetry resulting from the equivalence of the two continuous media (in and out) that form it. The sponge-to-micellar transition breaks this symmetry and corresponds to a second-order phase transition [68,155–157]. According to the analysis of Gazeau et al. [154] the size of the asymptotic critical domain where the behavior characteristic of the point P_c^1 is expected is very narrow and depends on the distance between the two critical points. Moreover, it appears that the competition

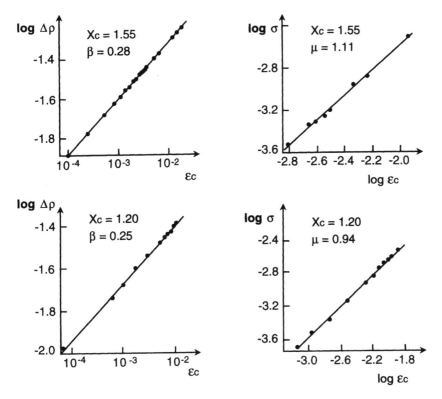

Figure 25 Log-log plots of the density difference $\Delta\rho$ (g/cm^3) and the interfacial tension σ (mN/m) versus a reduced composition variable ε_c for the critical points defined by $X_c = 1.55$ and $X_c = 1.20$. Each critical point has been approached in the two-phase region along a path perpendicular to the tie-lines. (From Ref. 152.)

Table 2 Values of the Critical Exponents γ, ν, β, μ, Measured for Two Critical Samples Defined by $X_C = 1.55$ and $X_C = 1.207$

	γ	ν	β	μ
Ising	1.240 ± 0.002	0.630 ± 0.01	0.325 ± 0.01	1.261 ± 0.1
$X_C = 1.55$	1.15 ± 0.05	0.58 ± 0.05	0.28 ± 0.03	1.11 ± 0.1
$X_C = 1.207$	0.77 ± 0.03	0.43 ± 0.03	0.25 ± 0.03	0.94 ± 0.03

between the two critical phenomena produces, within some accuracy, apparent power laws in the ε range investigated (i.e., in the range 10^{-3}–10^{-1}). Examination of the phase diagram shows that the distance between the two critical points increases with X. Consequently, the crossover should be less and less pronounced and effective exponents more and more Ising-like as X becomes larger. The X dependence of the exponents ν, γ, β, and μ is in agreement with this prediction.

Although the critical behavior in surfactant solutions is similar to that found in mixtures of simple molecules, one important difference has been noticed. It concerns the order of magnitude of the amplitudes. Indeed, the scale factors that control the amplitudes of the divergences are not universal. In the microemulsion systems the scaling factor for the interfacial tension is unusually small, over 100 times smaller than the corresponding numbers for simple binary mixtures. In contrast, the amplitude of the correlation is typically one order of magnitude larger than mixtures of small molecules. These results are consistent with the surface-active properties of the surfactant on the one hand and the fact that the phase separation is due to interactions between large aggregates on the other hand. Because of the two-scale factor universality principle, these unusual values of the scaling factors have several important consequences. In particular, the large value of ξ_0 means that some phenomena could not be observed whereas others will be more easily measured. For example, the magnitude of the critical heat capacity in microemulsions should be smaller by a factor of several thousands than those observed in binary mixtures, since $C_p^0 \xi_0^3 = $ constant. Such a heat capacity would be too small to be detected. This explains why no critical singularity of C_p is observed in microemulsions [158]. In contrast, the large value of ξ_0 found in microemulsions has allowed the spinodal decomposition of these media to be followed by light scattering [159], something that is essentially impossible in binary mixtures. Likewise, giant optical nonlinearities were observed in critical microemulsions [160]. Two mechanisms, electrostriction and thermodiffusion, have been shown to be at the origin of these nonlinearities, which are known to become progressively larger as the medium becomes more compressible [161].

B. Critical Behavior of Sponge Phases

In the phase diagrams presented above there exist, in addition to critical points involving micelles, critical points implying dilute sponge phases. Oil-rich and brine-rich sponges have been found to be stable over a large range of concentration. At high dilution, however, the L_{3-O} and L_{3-W} phases behave differently. The dilution of oil-rich phases L_{3-O} leads at some stage to a phase separation with a dilute microemulsion phase (Fig. 13, region d_5), whereas in the pseudoternary system made of salted water–pentanol–SDS the L_{3-W} domain is found to be continuous and connected to the micellar phase L_1. So, in that case, when solvent is added it is possible to go continuously from the concentrated sponge phase L_{3-W} to a dilute micellar solution L_1 with no phase separation. However, along such a dilution path a line denoted MTL (maximum turbidity line) is crossed where both the turbidity and the scattered intensity at a wave vector extrapolated to zero are maximum [98]. These experimental findings suggest a critical behavior along MTL. It should be noted that the critical point P_C^2 is located at approximately the end of this line.

In a sponge phase, surfactant molecules form a randomly connected surface separating the space into two statistically equivalent domains. Because of the equivalence of these two domains, both filled with the solvent, the bilayer system can exist in two different states: a symmetrical state S corresponding to perfect symmetry between the two domains and an asymmetrical state A where this symmetry is broken [66,155–157]. In the symmetrical phase the bilayer disconnects progressively, finally ending up in the form of a random dispersion of vesicles or micelles. From theoretical grounds, when diluting the symmetrical sponge phase, one expects a sponge-to-asymmetrical (S–A) phase transition. This transition can be either first- or second-order [66,156–157]. As discussed above, first-order transitions are observed in the oil-rich side of the diagram, and a line of continuous transition (MTL) in the water-rich side of salted systems. Therefore, the critical line MTL evidenced by light scattering has been identified to be the line of second-order S–A phase transition. However, as pointed out by Vinches et al. [162], in the vicinity of the MTL line the form factor of the bilayer dominates the scattered intensity so that the maximum of $I(0)$ could be a consequence of the maximum of the sponge size rather than the manifestation of a thermodynamic singularity. The very large values of the sponge sizes close to the MTL line and the limited $-q$ range accessible with most experimental light-scattering setups make the determination of the compressibility very difficult close to the transition. Consequently, a precise determination of a critical exponent approaching the second-order line is still an open problem.

In another salted system, a different scenario was proposed to explain the light-scattering behavior observed along the critical line MTL [163]. In that case the critical behavior was thought to be driven by a spontaneous tearing of the surfactant membrane as suggested by Huse and Liebler [164]. It should be noted that a recent freeze-fracture electron microscopic study of the very dilute sponge phase did not reveal the occurrence of hole defects, whereas conductivity results suggested their existence. Thus the critical behavior of sponge phases at high dilution remains a debated question.

V. CONCLUSION

We have seen that the water–(NaCl)–oil–SDS and alcohol mixtures exhibit a very rich phase behavior involving several phases—bicontinuous microemulsions, sponge phases, vesicle phases, and dilute lamellar phases—where the amphiphilic molecules self-assemble into sheetlike two-dimensional aggregates. The occurrence and the extents of these phases, which differ in the topology of the surfactant surface, depend on the alcohol chain length and salinity. Actually this phase behavior can be rationalized at least qualitatively in terms of the elastic energy expressed as a function of only three parameters: the mean (κ) and Gaussian ($\bar{\kappa}$) elastic constants and the spontaneous curvature C_0. κ controls the amplitude of the thermal undulation modes, while $\bar{\kappa}$ governs the topology of the interfaces. The relationships between the molecular parameters (alcohol chain length, area per polar head of the surfactant, alcohol/surfactant ratio, alkane chain length, salinity, etc.) and the elastic constants C_0, κ, and $\bar{\kappa}$ are now clear enough to know how to prepare bicontinuous microemulsions, dilute lamellar phases, sponge phases, or vesicle phases. In particular, the alcohol chain length controls both κ and $\bar{\kappa}$ while the alcohol concentration in the interface tune $\bar{\kappa}$. Starting from the lamellar phase, the sponge phase is obtained by the addition of alcohol ($\bar{\kappa}$ more positive), and the vesicle phase by tuning the alcohol content in the opposite direction ($\bar{\kappa}$ more negative).

Many industrial products use mixtures of both surfactant and polymer molecules or surfactant and colloid. Although the effects of polymer on the phase behavior and structure of surfactant phases have begun to be investigated in microemulsions, lamellar phases, and vesicle phases, further experimental work in mixed systems is necessary to understand how the polymer or the colloid modifies the elastic properties of the surfactant film.

Two lines of continuous phase transition implying inverse microemulsions and dilute sponge phases, respectively, have been observed in the phase diagrams presented in this chapter. The first phase separation is a liquid–gas-like critical transition driven by increasing attractions between inverse micelles, while the second critical phenomenon comes from a breaking of symmetry or a spontaneous tearing of the bilayers. All the experimental results obtained on the critical behavior in inverse microemulsions are successfully interpreted by the theory for simple molecules. In particular, the values of the critical exponents v and γ are very close to those measured for small molecules. It is important to note that the amplitudes of the numerical scale factors that depend on structure are significantly different from those of small molecules. In particular, ξ_0 is one order of magnitude larger than in the usual mixtures. This is consistent with the fact that the interacting units are large aggregates. Although up to now little consideration has been given to the amplitude ratios, one would expect the large value of ξ_0 to have interesting consequences. This should lead to exceptionally low or high values for some thermodynamic constants and offer opportunities for studying phenomena not accessible in the usual systems. It is worthwhile to point out also that the precise knowledge of the phase diagrams has allowed us to understand the origin of the anomalous behaviors observed in the quaternary system water–dodecane– pentanol–SDS.

The second-order phase transition between two isotropic phases, which is not a classical liquid–liquid phase separation, has been evidenced in brine-rich systems. One of these phases is the sponge phase S, and the second is an asymmetrical phase A of micelles or vesicles. The origin of this topological transition remains to be clarified, although some results are consistent with a symmetry-breaking mechanism and others with a tearing of the membranes. Theoretical considerations suggest that the special symmetry present in phases of connected membranes randomly distributed in space should allow various types of higher critical behavior to be observed [157]. Further experimental studies are required for a better understanding of the critical behavior in sponge phases.

ACKNOWLEDGMENTS

Parts of this work were carried out in collaboration with many colleagues. I specially thank D. Roux, F. Nallet, C. Coulon, T. Gulik, P. Honorat, D. Gazeau, F. Auguste, M. Maugey, and O. Babagbeto.

REFERENCES

1. P. Ekwall, in *Advances in Liquid Crystals*, Vol. 1 (G. H. Brown, ed.), Academic, New York, 1975, p. 1.
2. P. A. Winsor, Chem. Rev. *68*:1 (1968).
3. K. Shinoda and S. Friberg, Adv. Colloid Interface Sci. *4*:281 (1975).
4. K. Shinoda and S. E. Friberg, *Emulsions and Solubilization*, Wiley-Interscience New York, 1986.
5. J. H. Schulman, W. Stoeckenius, and L. M. Prince, J. Phys. Chem. *63*:1977 (1959).
6. L. E. Scriven, Nature *263*:123 (1976).
7. L. M. Prince (ed.), *Microemulsions: Theory and Practice*, Academic, New York, 1977.
8. K. L. Mittal (ed.), *Micellization, Solubilization and Microemulsions*, Vols. 1 and 2, Plenum, New York, 1977.
9. K. L. Mittal and B. Lindman (eds.) *Surfactants in Solution*, Vols. 1–3, Plenum, New York, Paris, 1984.
10. A. M. Bellocq, J. Biais, P. Bothorel, B. Clin, G. Fourche, P. Lalanne, B. Lemaire, B. Lemanceau, and D. Roux, Adv. Colloid Interface Sci. *20*:167 (1984).
11. A. M. Cazabat and M. Veyssie (eds.), *Colloides et Interfaces*, Les Editions de Physique, 1984.
12. M. Corti and V. Degiorgio (eds.), *Physics of Amphiphiles: Micelles, Vesicles and Microemulsions* North-Holland, Amsterdam, 1985.
13. K. L. Mittal and P. Bothorel (eds.), *Surfactants in Solution*, Vols. 4–6, Plenum, New York, 1986.
14. S. Martellucci and A. N. Chester (eds.), *Progress in Microemulsions*, Plenum, New York, 1986.
15. S. A. Safran and N. A. Clark, (eds.), *Physics of Complex and Supramolecular Fluids*, Wiley, New York, 1987.
16. H. L. Rosano and M. Clausse (eds.), *Microemulsion Systems*, (Surfact. Sci. Ser. Vol. 24), Marcel Dekker, New York, 1987.

17. J. Meunier, D. Langevin, and N. Boccara (eds.), *Physics of Amphiphilic Layers*, Springer, Heidelberg, 1987.
18. S. E. Friberg and P. Bothorel (eds.), *Microemulsions: Structure and Dynamics*, CRC, Boca Raton, FL, 1987.
19. W. M. Gelbart, A. Ben-Shaul, and D. Roux (eds.), *Micelles, Membranes, Microemulsions and Monolayers*, Springer-Verlag, New York, 1994.
20. J. Sjoblom, R. Lindberg, and S. E. Friberg, Adv. Colloid Interface Sci. *95*:125 (1996).
21. B. N. Knickerbocker, C.V. Pesheck, H. T. Davies, and L. E. Scriven, J. Phys. Chem. *83*:1984 (1979).
22. A. M. Bellocq and D. Roux, in *Microemulsions: Structure and Dynamics* (S. E. Friberg and P. Bothorel, eds.), CRC, Boca Raton, FL, 1987, p. 33.
23. A. M. Bellocq, in *Physics of Complex and Supramolecular Fluids* (S. A. Safran and N. Clark, eds.), Wiley, New York, 1987, p. 41.
24. M. Kahlweit, R. Strey, and G. Busse, J. Phys. Chem. *94*:3881 (1900).
25. M. Kahlweit, R. Strey, P. Firman, D. Haase, J. Jen, and R. Schomacker, Langmuir *4*:499 (1988).
26. M. Kahlweit, R. Strey, R. Schomacker, and D. Haase, Langmuir *5*:305 (1989).
27. R. Strey, Ber. Bunsenges. Phys. Chem. *97*:742 (1993).
28. K. Shinoda and H. Saito, J. Colloid Interface Sci. *26*:70 (1969).
29. K. Shinoda and H. Kunieda. J. Colloid Interface Sci. *75*:601 (1980).
30. H. Kunieda and Y. Sato, in *Organized Solutions* (S. E. Friberg and B. Lindman, eds.), Marcel Dekker, New York, 1992, p. 67.
31. H. Kunieda, K. Nakamura, and A. Uemoto, J. Colloid Interface Sci. *150*:235 (1992).
32. S. J. Chen, D. F. Evans, and B. W. Ninham, J. Phys. Chem. *88*:1631 (1984).
33. K. Fontell, A. Ceglie, B. Lindman, and B. W. Ninham, Acta Chem. Scand. A *40*:247 (1986).
34. U. Olsson, K. Nagai, and H. Wennerstrom, J. Phys. Chem. *92*:6675 (1988).
35. U. Olsson, K. Shinoda, and B. Lindman, J. Phys. Chem. *90*:4083 (1990).
36. U. Olsson, U. Wurtz and R. Strey, J. Phys. Chem. *97*:4535 (1993).
37. W. J. Benton and C. A. Miller, J. Phys. Chem. *87*:4981 (1983).
38. M. Dvolaitzky, R. Ober, J. Billard, C. Taupin, J. Charvolin and Y. Hendricks, C. R. Acad. II *45*:295 (1981).
39. M. Di Meglio, M. Dvolaitzky, and C. Taupin, J. Phys. Chem. *89*:871 (1985).
40. F. Larche, J. Appell, G. Porte, P. Bassereau, and J. Marignan, Phys. Rev. Lett. *56*:1700 (1985).
41. R. Strey, R. Schomacker, D. Roux, F. Nallet, and U. Olsson, J. Chem. Phys. Faraday Soc., Trans. *86*:2253 (1990).
42. N. Satoh and K. Tsujii, J. Phys. Chem. *91*:6629 (1987).
43. C. Thunig, H. Hoffmann, and G. Platz, Prog. Colloid Polym. Sci. *79*:297 (1989).
44. J. C. Lang and R. C. Morgan, J. Chem. Phys. *73*:5849 (1980).
45. W. J. Benton and C. Miller, in *Surfactants in Solution*, Vol. 1 (K. L. Mittal and B. Lindman, eds.), Plenum, New York, 1984, p. 205.
46. G. Porte, J. Marignan, P. Bassereau, and R. May, J. Phys. (France) *49*:511 (1988).
47. D. Gazeau, A. M. Bellocq, D. Roux, and T. Zemb, Europhys. Lett. *9*:447 (1989).
48. U. Peter, S. Konig, D. Roux, and A. M. Bellocq, Phys. Rev. Lett. *76*:3866 (1996).
49. B. Widom, J. Phys. Chem. *77*:2196 (1973).
50. J. C. Lang and B. Widom, Physica *81A*:190 (1975).
51. M. Kahlweit, R. Strey, P. Firman, and D. Haase, Langmuir *1*:281 (1995).
52. M. Kahlweit, R. Strey, and P. Firman, J. Phys. Chem. *90*:671 (1986).
53. M. Kahlweit and R. Strey, J. Phys. Chem. *90*:5239 (1986).
54. H. Kunieda and K. Shinoda, Bull. Chem. Soc. Jpn. *56*:980 (1983).
55. H. Kunieda, J. Colloid Interface Sci. *122*:138 (1988).
56. M. Yoshida and H. Kunieda, J. Colloid Interface Sci. *8*:273 (1990).
57. A. M. Bellocq and D. Gazeau, Prog. Colloid Polym. Sci. *76*:203 (1988).
58. W. Helfrich, Z. Naturforsch. C *28*:693 (1973).
59. P. G. de Gennes and C. Taupin, J. Phys. Chem. *86*:2294 (1982).
60. J. Jouffroy, P. Levinson, and P. G. de Gennes, J. Phys. (France) *43*:1241 (1982).

61. J. M. Israelachvili, D. J. Mitchell, and B. Ninham, J. Chem. Soc., Faraday Trans. 2 *72*:1525 (1976).
62. Y. Talmon and S. Prager, J. Chem. Phys. *69*:2984 (1978).
63. S. A. Safran, D. Roux, M. Cates, and D. Andelman, Phys. Rev. Lett. *57*:491 (1986); J. Chem. Phys. *87*:7229 (1987).
64. D. Andelman, S. A. Safran, D. Roux, and M. Castes, Langmuir *4*:802 (1988).
65. B. Widom, J. Chem. Phys. *81*:1030 (1984).
66. D. A. Huse and S. Leibler, J. Phys. (France) *49*:605 (1988).
67. D. C. Morse, Phys. Rev. E *50*:R 2423 (1994).
68. L. Golubovic and T. C. Lubensky, Phys. Rev. A *41*:343 (1990).
69. P. Herve, D. Roux, A. M. Bellocq, F. Nallet, and T. Gulik, J. Phys. II (France) *3*:1255 (1993).
70. B. D. Simons and M. E. Cates, J. Phys. II (France) *2*:439 (1992).
71. G. Porte, J. Appell, P. Bassereau, and J. Marignan, J. Phys. (France) *50*:1335 (1989).
72. H. Wennerström and U. Olsson, Langmuir *9*:365 (1993).
73. I. Szleifer, D. Kramer, A. Ben-Shaul, W. M. Gelbart, and S. A. Safran, J. Chem. Phys. *92*:6800 (1990).
74. C. R. Safinya, E. Sirota, D. Roux, and G. S. Smith, Phys. Rev. Lett. *62*:1134 (1989).
75. F. Auguste, P. Barois, L. Fredon, B. Clin, E. J. Dufourc, and A. M. Bellocq, J. Phys. II (France) *4*:2197 (1994).
76. G. Gillberg, H. Lehtinen, and S. Friberg, J. Colloid Interface Sci. *33*:40 (1970).
77. S. E. Ahmad, K. Shinoda, and S. Friberg, J. Colloid Interface Sci. *47*:32 (1974).
78. S. Friberg and I. Buraczewska, in *Micellization, Solubilization and Microemulsions*, Vol. 2 (K. L. Mittal, ed.), Plenum, New York, 1977, p. 791.
79. P. Herve, D. Roux, A. M. Bellocq, F. Nallet, and T. Gulik, (to be published).
80. L. Q. Amaral, M. E. M. Helène, D. R. Bittencourt, and R. Itri, J. Chem. Phys. *91*:5949 (1987).
81. K. Kekicheff and B. Cabane, J. Phys. (France) *48*:1571 (1987).
82. D. Roux and A. M. Bellocq, in *Micellar Solutions and Microemulsions* (S. H. Chen and R. Rajagopalan, eds.), Springer-Verlag, New York, 1990, p. 251.
83. I. Prigogine and R. Defay, *Thermodynamique Chimique*, 1950, p. 192.
84. H. Kunieda and K. Nakamura, J. Phys. Chem. *95*:1425 (1991).
85. G. Guerin and A. M. Bellocq, J. Phys. Chem. *92*:2550 (1988).
86. F. Auguste, Ph.D. Thesis, University Bordeaux I, 1993.
87. F. Auguste, A. M. Bellocq, F. Nallet, D. Roux, and T. Gulik, to be published.
88. F. Auguste, A. M. Bellocq, D. Roux, F. Nallet, and T. Gulik, Prog. Colloid Polym. Sci. *98*:276 (1995).
89. W. J. Benton, J. Natoli, S. Qutubuddin, C. A. Miller, and T. Fort, Jr., Soc. Petrol. Eng. J. [[]] 53 (1982).
90. W. J. Benton and C. A. Miller, J. Phys. Chem. *87*:4981 (1983).
91. R. Gomati, J. Appell, P. Bassereau, J. Marignan, and G. Porte, J. Phys. Chem. *91*:6203 (1987).
92. H. Hoffmann, C. Thuning, and U. Munkert, Langmuir *8*:2629 (1992).
93. H. Hoffmann, U. Munkert, C. Thuning, and A. Valiente, J. Colloid Interface Sci. *163*:217 (1994).
94. P. Boltenhagen, M. Kleman, and O. Lavrentovich, C. R. Acad. Sci. Paris II *315*:931 (1992).
95. M. Dubois, Th. Gulik-Krywicki, and B. Cabane, Langmuir, *9*:673 (1993).
96. I. Alibert, C. Coulon, and A. M. Bellocq, Europhys. Lett. *39*:563 (1997).
97. I. Alibert, Ph.D. Thesis, University Bordeaux I, 1997.
98. C. Coulon, D. Roux, and A. M. Bellocq, Phys. Rev. Lett. *66*:1709 (1991).
99. W. Helfrich, Z. Naturforsch. *33a*:305 (1978).
100. D. Roux, C. R. Safinya, and F. Nallet, in *Micelles, Membranes, Microemulsions and Monolayers* (W. M. Gelbart, A. Ben-Shaul, and D. Roux, eds.), Springer-Verlag, New York, 1994, p. 303.
101. M. Jonstromer and R. Strey, J. Phys. Chem. *96*:5993 (1992).
102. D. Roux and A. M. Bellocq, Phys. Rev. Lett. *52*:1895 (1984).
103. D. Roux and A. M. Bellocq, in *Surfactants in Solution*, Vol. 6 (K. L. Mittal and P. Bothorel, eds.), Plenum, New York, 1986, p. 1247.
104. U. Olsson and H. Wennerström, Adv. Colloid Interface Sci. *49*:113 (1994).

105. D. Gazeau, Ph.D. Thesis, University Bordeaux I, 1988.
106. P. A. Winsor, *Solvent Properties of Amphiphilic Compounds*, Butterworth, London, 1954.
107. B. P. Binks, H. Kellay, and J. Meunier, Europhys. Lett. *16*(1):53 (1991).
108. W. K. Kegel and H. W. Lekkerkerker, J. Phys. Chem. *97*:11124 (1993).
109. P. G. de Gennes, J. Phys. Chem. *94*:8407 (1990).
110. J. T. Brooks, C. M. Marques, and M. E. Cates, J. Phys. II (Paris) *1*:673 (1991).
111. G. Guerin, Thesis, University of Bordeaux I, 1986.
112. H. Kunieda, K. Nakamura, and A. Uemoto, J. Colloid Interface Sci. *163*:245 (1994).
113. D. Roux, A. M. Bellocq, and M. S. Leblanc, Chem. Phys. Lett. *94*:156 (1983).
114. D. Roux and A. M. Bellocq, in *Physics of Amphiphiles, Micelles, Vesicles and Microemulsions* (M. Corti and V. Degiorgio, eds.), North-Holland, Amsterdam, 1985, p. 842.
115. A. M. Bellocq and D. Roux, in *Marco and Microemulsions: Theory and Applications* (ACS Symp. Ser No. 272) (D. O. Shah, ed.), Am. Chem. Soc., Washington, DC, 1985, p. 105.
116. S. A. Safran and L. A. Turkevich, Phys. Rev. Lett. *50*:1930 (1983).
117. A. A. Calje, W. G. M. Agterof, and A. Vrij, in *Micellization, Solubilization and Microemulsions* Vol. 2 (K. L. Mittal, ed.), Plenum, New York, 1976, p. 779.
118. S. Brunetti, D. Roux, A. M. Bellocq, G. Fourche, and P. Bothorel, J. Phys. Chem. *87*:1029 (1983).
119. D. Roux, A. M. Bellocq, and P. Bothorel, in *Surfactants in Solution*, Vol. 3 (K. L. Mittal and B. Lindman, eds.), Plenum, New York, pp. 1843–1865.
120. A. M. Cazabat and D. Langevin, J. Chem. Phys. *74*:3148 (1981).
121. B. Lemaire, P. Bothorel, and D. Roux, J. Phys. Chem. *87*:1023 (1983).
122. L. Auvray, J. Phys. Lett. *46 L*:163 (1985).
123. S. H. Chen and M. Kotlarchyk, in *Physics of Amphiphiles, Micelles, Vesicles and Microemulsions* (M. Corti and V. Degiorgio, eds.), North-Holland, Amsterdam, 1985, p. 768.
124. H. E. Stanley, *Introduction to Phase Transitions and Critical Phenomena*, Oxford Univ. Press, Oxford, 1971.
125. C. Dom and M. S. Green, *Phase Transitions and Critical Phenomena*, Vols. 1–6, Academic, New York, 1972–1977.
126. J. M. H. Levelt-Sengers, R. Hocken, and J. V. Sengers, Phys. Today *30*:42 (1977).
127. M. Levy and J. Zinn-Justin (eds.), *Phase Transitions: Cargese 1980*, Plenum, New York, 1982.
128. B. Chu, in *Dynamics of Light Scattering* (R. Pecora, ed.), Plenum, New York, 1985.
129. S. C. Greer and M. R. Moldover, Ann. Rev. Phys. Chem. *32*:233 (1981).
130. J. V. Sengers and J. M. H. Levelt-Sengers, Ann. Rev. Phys. Chem. *37*:189 (1986).
131. A. Kumar, H. R. Krishnamurthy, and E. S. R. Gopal, Phys. Rep. *98*:57 (1983).
132. J. M. H. Levelt-Sengers, Pure Appl. Chem. *55*:437 (1983).
133. J. M. H. Levelt-Sengers, G. Morrison, and R. F. Chang, Fluid Phase Equilibria *14*:19 (1983).
134. D. Beysens and A. Bourgou, Phys. Rev. *A19*:2407 (1979).
135. D. Beysens, A. Bourgou, and P. Calmettes, Phys. Rev. *A26*:3589 (1982).
136. B. Widom, J. Chem. Phys. *46*:3324 (1967).
137. R. B. Griffiths and J. C. Wheeler, Phys. Rev. A *2*:1047 (1970).
138. J. V. Sengers and J. M. H. Levelt-Sengers, Ann. Rev. Phys. Chem. *37*:189 (1986).
139. M. E. Fisher, Phys. Rev. *176*:257 (1971).
140. C. S. Bak, W. I. Golburg, and P. N. Pusey, Phys. Rev. Lett. *25*:1420 (1970).
141. B. Chu and F. L. Lin, J. Chem. Phys. *61*:5132 (1975).
142. K. Ohbayashi and B. Chu, J. Chem. Phys. *68*:5066 (1978).
143. J. Rouch, A. Safouane, P. Tartaglia, and S. H. Chen, Phys. Rev. A *37*:4995 (1988).
144. J. S. Huang and M. W. Kim, Phys. Rev. Lett. *47*:1462 (1981).
145. M. Kotlarchyk, S. H. Chen, and J. S. Huang, Phys. Rev. A *28*:508 (1983).
146. P. Honorat, D. Roux, and A. M. Bellocq, J. Phys. Lett. *45*:L961 (1984).
147. M. W. Kim, J. Bock, and J. S. Huang, Phys. Rev. Lett. *54*:46 (1985).
148. Y. Jayalakshimi and D. Beysens, Phys. Rev. A *35*:8709 (1992).
149. R. Aschauer and D. Beysens, Phys. Rev. E *47*:1950 (1993).
150. J. C. Wheeler and B. Widom, J. Phys. Chem. *90*:3064 (1968).

151. A. M. Bellocq, P. Honorat, and D. Roux, J. Phys. (France) *46*:743 (1985).
152. A. M. Bellocq and D. Gazeau, J. Phys. Chem. *94*:8933 (1990).
153. N. Rebbouh, J. Buchert, and J. R. Lalanne, Europhys. Lett. *4*:447 (1987).
154. P. Dorion, J. R. Lalanne, B. Pouligny, S. Imaizumi, and C. W. Garland, J. Chem. Phys. *87*:578 (1987).
155. D. Gazeau, E. Freysz, and A. M. Bellocq, Europhys. Lett. *9*:833 (1989).
156. M. E. Cates, D. Roux, D. Andelman, S. T. Milner, and S. A. Safran, Europhys. Lett. *5*:733 (1988).
157. D. Roux, M. E. Cates, U. Olsson, R. C. Ball, F. Nallet, and A. M. Bellocq, Europhys. Lett. *11*:229 (1990).
158. D. Roux, C. Coulon, and M. E. Cates, J. Phys. Chem. *96*:1709 (1991).
159. D. Roux, J. Phys. (France) *47*:733 (1986).
160. E. Freysz, W. Claeys, A. Ducasse, and B. Pouligy, IEEE J. Quantum Electronics *QE-22*:1258 (1986).
161. B. Jean-Jean, E. Freysz, A. Ducasse, and B. Pouligny, Europhys. Lett. *7*:219 (1988).
162. C. Vinches, C. Coulon, and D. Roux, J. Phys. II (France) *4*:1165 (1994).
163. M. Filali, G. Porte, J. Appell, and P. Pfeuty, J. Phys. II (France) *4*:349 (1994).
164. D. Huse and S. Leibler, Phys. Rev. Lett. *66*:437 (1991).

7

Aggregation Behavior in One-Phase (Winsor IV) Microemulsion Systems

Shmaryahu Ezrahi, Abraham Aserin, and Nissim Garti
Casali Institute of Applied Chemistry, The Hebrew University of Jerusalem, Jerusalem, Israel

I. INTRODUCTION

One remarkable manifestation of the general phenomenon of molecular self-organization is the formation of well-defined microstructures in surfactant solutions [1]. This is the cornerstone for the widespread technological applications of amphiphiles and for the essential role they play in biology.

Surfactant molecules commonly self-assemble in water (or in oil). Even single-surfactant systems can display a quite remarkably rich variety of structures when parameters such as water content or temperature are varied. In dilute solution they form an isotropic solution phase consisting of micellar aggregates. At more concentrated surfactant–solvent systems, several isotropic and anisotropic liquid crystalline phases will be formed [2]. The phase behavior becomes even more intricate if an oil (such as an alkane or fluorinated hydrocarbon) is added to a water–surfactant binary system and the more so if other components (such as another surfactant or an alcohol) are also included [3]. In such systems, emulsions, microemulsions, and lyotropic mesophases with different geometries may be formed. Indeed, the ability to form such association colloids is the feature that singles out surfactants within the broader group of amphiphiles [4]. No wonder surfactants' phase behavior and microstructures have been the subject of intense and profound investigation over the course of recent decades.

Microemulsions were initially considered to be a special case of emulsions, i.e., kinetically stable dispersions of two mutually immiscible solvents, although of an unusually high stability. They are clear, seemingly one-phase systems and can be prepared without considerable input of mechanical energy. As they are thermodynamically stable, however, it is more plausible to conceive of microemulsions as being aggregated (micellar or swollen micellar) solutions [5].

II. PATTERNS OF PHASE BEHAVIOR IN AMPHIPHILIC SYSTEMS

Two issues of great importance for the study of microemulsions are their equilibrium phase behavior and microstructure characterization. It should be noted that the pioneers in surfactant solution chemistry have clung to a characterization by phase behavior alone and have virtually ignored microstructure within phases [6].

This chapter focuses mainly on the microstructures encountered in mixtures of oil and water with surfactants (mostly nonionic from the family of ethoxylated alcohols, designated as C_iE_j or $C_i(EO)_j$, where i and j are the number of carbon atoms in the alkyl chain and the number of ethylene oxide (EO) groups of the polar head, respectively, of the surfactant molecule) and sometimes in the presence of alcohols. However, a short survey concerning the investigation of the equilibrium phase behavior of these systems seems to be in order.

Historically, the first microemulsion systems were prepared by titrating mixtures of water, oil, and ionic surfactants with an alcohol [7–10]. Occasionally, an inorganic electrolyte was added to the water component [11]. The study of the phase diagrams of such multicomponent mixtures is essential in order both to define the realm of existence of microemulsions in these systems and to understand their nature. Microemulsions should not be treated as an isolated association phenomenon, but rather in the context of all interrelated, self-organized structures occurring in the systems [12]. Naturally, the inherent complexity of four- or five-component systems seemed to act as a deterrent to rationalization of their behavior. Eventually it was shown that suitable three-component systems also exhibited the salient features of microemulsion phase behavior [13]. McBain was apparently the first to study the equilibrium phase behavior at fixed temperature in mixtures of water, oil, and surfactant [14]. Three- and four-component systems, based mainly on ionic surfactants, were intensively investigated by research groups in Scandinavia in an attempt to determine the extent and shape of single- and multiphase regions of existence in the corresponding phase diagrams [15–20]. Systematic studies conducted by Shinoda [21–29] laid the basis for our current perception of the phase behavior of systems based on nonionic surfactants. In particular, the effect of temperature on such systems was established.

A microemulsion phase may be in equilibrium with excess oil (this type of phase diagram is called Winsor I), or water (Winsor II), or both oil and water (Winsor III). The microemulsion phase in this equilibrium is also known as "middle phase," "surfactant phase," or "midrange microemulsion." In addition to these two- or three-phase equilibria, phase diagrams of amphiphiles may also contain a single-phase microemulsion region (Winsor IV). These regions are called after the author who clearly catalogued them [30]. The Winsor I, II, and III phase equilibria are also designated by $\underline{2}$, 3, and $\overline{2}$, where the bar below or above the 2 shows that the microemulsion is the lower (water-rich) phase and the upper (oil-rich) phase, respectively [31,32] (Figs. 1 and 2). A lyotropic mesophase (usually lamellar) may be in equilibrium with a microemulsion phase at higher surfactant concentrations. The microemulsion composition corresponding to Winsor III is characterized by very low interfacial tension and maximal solubilization of oil and water for a given quantity of surfactant [35]. The system is said to be optimized or balanced [36]. Thus, the search for the three-phase region has been the subject of intensive investigation. The researchers primarily focused their interest on the factors influencing the transition Winsor I → Winsor II → Winsor III. One can achieve this transition by changing parameters such as temperature (first observed by Shinoda and Saito [25]), salinity, pH, water/oil ratio, chemical composition, and molecular geometry (e.g., branching or altering the number of carbons either in the alkane chain or in the headgroup of the surfactant) [37]. Many groups from all over the world have contributed findings on various aspects of phase behavior of multicomponent amphiphilic mixtures, and in view of the vast literature available it would be beyond the scope of this chapter to treat them all. We should, however, mention the work done by Schechter and Wade's group [37–39], the Minnesota group

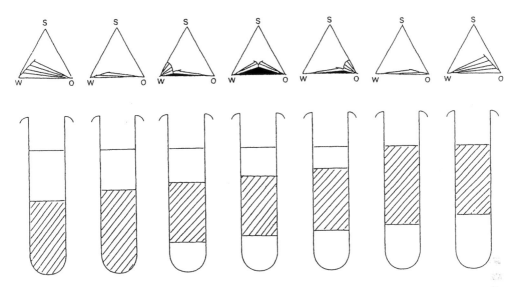

Figure 1 Phase changes of a system containing equal amounts of oil (O) and water (W) and a given (low) amount of surfactant (S). For a nonionic surfactant system, left-to-right transition may be induced by increasing temperature. (From Ref. 33.)

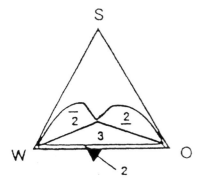

Figure 2 Schematic representation of Winsor-type equilibria in a ternary amphiphilic system. Between the water–oil side of the phase triangle and the base of the three-phase triangle there is a narrow two-phase zone at very low surfactant content that corresponds to the equilibrium between the excess phases that contain very little amounts of surfactant (about the critical micelle concentration). (From Ref. 34.)

[31,32,40–42], and, in particular, the incisive and exhaustive studies by the Göttingen group [43–63].

Two main conclusions may be drawn from the phase behavior studies:

1. The observed phase behavior patterns of mixtures of different amphiphiles were practically the same and, in fact, the macroscopic equilibrium properties (like the formation of a three-phase body within the phase diagram) are not peculiar properties of amphiphilic systems [64]. Rather, they can be observed in any mixture in which the third component changes its preferential solubility as a func-

tion of temperature or another field variable. For example, this common behavior was demonstrated for metal alloys [49]. The particular behavior of amphiphiles is revealed by the pronounced mutual solubility of water and oil: Ten molecules of a rather simple nonionic amphiphile may completely solubilize hundreds of water and oil molecules [57]. Another peculiarity is the appearance of ordered (usually lamellar) mesophases [65].

2. The incorporation of alcohol, salt, or another surfactant into a mixture of water, oil, and surfactant provides an additional degree of freedom that enables the adjustment of phase behavior [66]. A formulator may use such additives to control the phase sequence Winsor I → Winsor III → Winsor II and prepare microemulsions with predetermined properties [35,66].

A. Weakly to Strongly Structured Mixtures—Amphiphilicity Factor

The generic pattern of the macroscopically isotropic phases is the same for both long-chain (strong) and short-chain (weak) nonionic surfactants, except for the formation of lyotropic phases at high surfactant concentration [66], which may completely pre-empt the existence of the middle phase [65]. Thus, a phase inversion is observed as a function of temperature or alkane chain length, and the effect of additives (e.g., electrolytes or polar solvents) on the phase behavior is comparable in all mixtures [49]. Microscopically, a long-chain surfactant forms a monolayer of oriented molecules, separating water and oil domains [67]. In this way, the surfactant molecules order the domains into a variety of structures. Short-chain amphiphiles display a decreasing degree of ordering in the interface with decreasing chain length until the microstructure disappears and the components become molecularly dispersed with no interfacial film [68].

The degree of structuring in microemulsion phases was quantified by using an amphiphilicity factor, f_a, that determines the amphiphilic strength on a uniform system-independent scale [68–70]. This factor was defined in terms of fitting parameters derived from scattering experiments. Thus the scattering intensity distribution $I(q)$ is given by [71]

$$I(q) = (a_2 + c_1 q^2 + c_2 q^4)^{-1} \tag{1}$$

where a_2, c_1, and c_2 are fitting parameters. For negative c_1, this function yields a single maximum at nonzero wave vector q. At high q it decays in proportion to q^{-4}. Equation (1) corresponds to a real-space correlation function of the form

$$\gamma(r) = \frac{\sin kr}{kr} e^{-r/\xi} \tag{2}$$

where r is a real-space distance.

This correlation function describes a structure of periodicity $d (= 2\pi/k)$, i.e., a measure of the repeat distance between alternately arranged water and oil domains, dampened as a function of the correlation length ξ, which characterizes the decay of local order [65] and may also be interpreted [69] as a measure of the dispersion of d. The order parameter coefficients a_2, c_1, and c_2 can be grouped together to define an amphiphilicity factor $f_a \equiv c_1/(4a_2c_2)^{1/2}$. These parameters can also be used to express d and ξ. Figure 3 illustrates the use of this amphiphilicity scale [68].

For very strong amphiphiles (f_a is typically close to −1), the phase diagram is dominated by L_α (lamellar liquid crystal) [72]. Slightly less negative values of f_a characterize well-structured (so-called good) microemulsions. At three-phase coexistence, such a

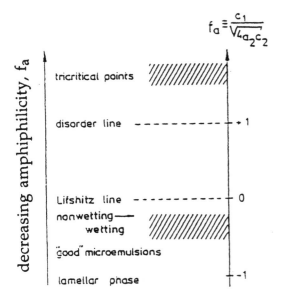

Figure 3 Schematic diagram of observations of microemulsion systems as a function of the amphiphilicity factor. (From Ref. 68.)

microemulsion does not wet the interface between the excess water phase and the excess oil phase in spite of the very low values of the water/microemulsion and oil/microemulsion interfacial tensions, which would tend to favor such wetting [56,73,74].

If the volume of the middle phase is reduced to a drop, this drop will contract to a lens floating at the water/oil interface [61,75]. This phenomenon is attributed to the large surface pressure of the surfactant monolayer coating the interface, which almost balances the bare oil/water interfacial tension. This layer stabilizes the microstructure of microemulsions. It does not form with small molecules [76,77]. On the other hand, if the amphiphile is sufficiently weakened (for example, by either varying temperature [78] or replacing part of the water with formamide [69,75]), such a drop will spread across the water/oil interface [61]. It was suggested [73] that microemulsions could be distinguished from less structured solutions by using this feature. Indeed, further decreasing the amphiphilicity (but with f_a still slightly negative) leads to this nonwetting → wetting transition. As was previously suggested, this transition may be associated with the disappearance of the surfactant monolayer and the microemulsion structure [79]. The decreasing structuring of weak amphiphiles may also be reflected in scattering behavior [80]. A peak at nonzero wave vector q indicates that there is a structure on a scale of $2\pi/q$. In the vicinity of the wetting transition there is a Lifshitz line [81]. This line is defined by $c_1 = 0$ (so that $f_a = 0$) and by $2\pi\xi/d = 1$. Since c_1 is negative for well-defined microemulsions characterized by correlated interfaces and becomes positive with increasing disorder [69], the Lifshitz line represents the locus at which the scattering peak is in the immediate vicinity of $q = 0$, and the correlation function is still dominated by the oscillatory behavior [65] [the sine term of Eq. (2)]. The relative distance on the amphiphilicity scale between the wetting transition and the Lifshitz line depends on the system investigated [68,70]. Upon further decreasing the amphiphilicity, one obtains the disorder line [82], where $f_a = 1$ (and $2\pi\xi/d = 0$). At this line, the solution loses its quasiperiodic order and the remains of the interface become uncorrelated [69].

Thus, the correlation function changes its behavior at the disorder line from oscillatory to monotonic (exponential decay) [65]. More positive values of f_a correspond to tricritical points (where three coexisting liquid phases become simultaneously identical) [83].

In conclusion, the amphiphilicity factor is important for the understanding of microemulsion structuring. It was also suggested as an additional means of classifying surfactants, together with the packing parameter or the hydrophilic–lipophilic–balance (HLB) scale [84].

III. STRUCTURES IN AMPHIPHILIC SYSTEMS

A. General Considerations

This chapter is intended to deal with the elucidation and characterization of structures occurring in monophasic Winsor IV equilibria. However, in many cases we use examples of binary amphiphilic systems whose isotropic solutions cannot, obviously, be considered microemulsions. This is because many of the microstructures found in three-component systems are simple extensions of their binary counterparts [16,27]. Furthermore, we included important results concerning liquid crystalline phases, as this information provides, we believe, the fundamental basis for more profound understanding of surfactant organization in multicomponent systems.

In this section we describe the more common structures observed in amphiphilic systems and try to obtain some physical insights into the basic mechanisms that are believed to be at the origin of these structures and their sequence of appearance in cases where surfactant concentration varies.

B. The Hydrophobic Effect

Surfactants self-organize spontaneously in water owing to local constraints originating from their having hydrophobic and hydrophilic molecular moieties linked together [85]. Therefore, they are opposite ends of a wandlike or straplike molecule [86]. Whereas oil and water, when mixed, form separate bulk phases in coexistence, surfactant molecules organize side by side into oriented finite aggregates (micelles). The two surfactants' tails are located on one side of the micelle, while their headgroups are located on the other side [86]. Thus, in micelles there exists a well-defined hydrophobic core consisting of the surfactants' "tails" in a liquidlike state rather than being pointed radially toward the micelle center. This misconceived "spokes of a wheel" picture should be rejected [87]. This core is covered by a hydrophilic mantle of hydrated headgroups. Very little water penetrates into the micelles, but it can be experimentally demonstrated that contacts occur between water and alkyl chains [87]. Luzzati, who introduced the concept of a variable polar–apolar partition [88,89], found that in the case of polyoxyethylene surfactants the nonpolar region could actually contain some fraction of the polar groups as well as the hydrocarbon chains. Thus, the water is mostly shielded from the hydrophobic moieties and the hydrophobic effect [90] is satisfied [91]. Such a microsegregation mechanism imposes strong restrictions on the shape of the aggregates formed, which are characterized by well-defined geometries [85].

One may observe monolayers, spherical (globular or rotund) micelles, cylindrical (tubelike or wormlike) micelles, and bilayers. (The more ordered liquid crystals are treated in the following.) The size and shape distributions of surfactant aggregates in solution depend on the chemical and physical properties of the surfactant, and, in contrast to those of ordinary solutions, they vary as a function of parameters such as total concentration, temperature, and ionic strength. Thus the average number of micelles increases with surfactant concentration [92]. The structure of surfactant aggregates at equilibrium is characterized by a minimum in free energy. The thermodynamically stable state may vary from a molecular or micellar solution via a liquid crystalline or crystalline phase to a multiphase system [86]. As the hydrophobic effect is not yet fully understood [93–95] and, moreover, surfactant aggregation may not always be driven solely by hydrophobic interactions [96], this picture is obviously too simplified.

C. Geometrical Considerations

We are not yet at the state of being able to determine the geometry assumed by any given surfactant in water from the underlying energetics, so it may be "explained" only a posteriori [92]. If the tendency to remove water from hydrophobic groups were the only factor in determining the aggregates' shape, spherical micelles (which have minimal water–hydrocarbon contact area for a given volume) [91] would be expected. Indeed, whenever it is sterically possible, spherical micelles (which also have the maximum gain in entropy of dispersion) [97] are preferred (Fig. 4).

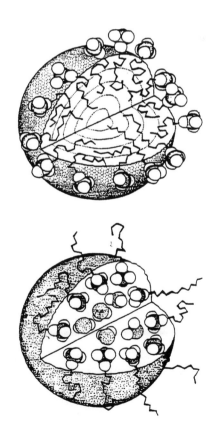

Figure 4 Schematic representation of a *normal micelle* (top) and a *reversed micelle* (bottom). (Drawn by Dr. J. N. Israelachvili from calculations by Dr. D. W. R. Gruen, Australian National University.)

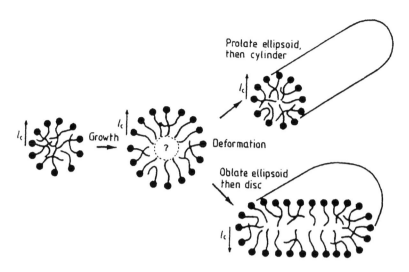

Figure 5 Limit of sphericity of a micelle as given by Tanford [90]. Large micelles cannot be spherical as a hole should occur at their center [91].

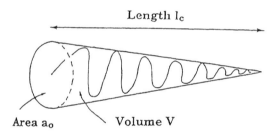

Figure 6 A surfactant molecule with headgroup area a_o, hydrocarbon critical chain length l_c, and hydrocarbon volume V. (After Ref. 33.)

However, the intramicellar free energy may be lower than for other geometrical shapes [97]. Now, large aggregate numbers can be reconciled only with nonspherical structures. Unlike the case of membrane droplets, the radius of a globular micelle (for binary systems) can never significantly exceed the fully extended chain length of a surfactant molecule (Fig. 5). This is because the hydrophobic interior must be filled uniformly with CH_2 groups [92], while the existence of voids within the hydrophobic region is physically impossible [97]. The same geometrical considerations apply to any defined aggregate shape. The surfactant (or packing) parameter p, which defines the spontaneous curvature of the interface, is fixed by geometrical packing and interfacial forces [98–100]. It is defined as $V/(a_o l_c)$, where V is the volume of the hydrocarbon chains per amphiphile; a_o is the optimal area per polar head and l_c is the length of the extended (all-trans) alkyl chain (Fig. 6). This parameter mar serve as a guide to the shape of the aggregates that will form spontaneously (see Fig. 7). For example, if $p < 1/3$, the formation of spherical micelles in water is predicted and experimentally observed [98–100]. Thus, one may distinguish, roughly speaking, between the classical structures of

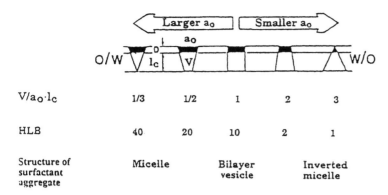

Figure 7 The packing parameter $V/a_o l_c$ governs the shape of the surfactant aggregate formed in solution. The figure also shows a correlation between packing parameter and HLB numbers. (From Ref. 33.)

amphiphilic assemblies, i.e., spheres (or polyhedra), cylinders, and planes. These aggregates can be spontaneously organized on a larger scale in the form of liquid crystalline phases (characterized by long-range order; Sec. IV. A-C.) or liquid isotropic phases (featured by only short-range correlations) [101]. A wide variety of surfactant molecules obey the geometrical rules embodied in the packing parameter [96]. For example, $C_{12}E_3$ forms vesicles, whereas $C_{12}E_7$, which has a larger headgroup, forms rodlike micelles [102]. Even the formation of supramolecular structures from tetracationic amphiphiles bearing a calix[4]arene core proceeds according to these rules [103]. However, it should be emphasized that since predictions based on p consider only geometrical arguments (ignoring, for instance, chain elasticity and entropy), they should be treated with caution [91]. Thus, p applies only to dilute micellar dispersions as interaggregate interactions are neglected [104]. Furthermore, the calculation of p for multicomponent systems is not simple: Oil may penetrate into the hydrophobic core and increase V; alcohol may adsorb at the interface and increase a_o. The area a_o is also sensitive to ionic strength for ionic surfactants and to temperature in the case of nonionic surfactants [105]; chain folding may shorten the alkyl chain length l_c.

The surfactant parameter was more recently rewritten in terms of the mean curvature H (also designated M) [106] and the Gaussian curvature K of the interface running between the headgroups and the alkyl chains of the surfactant [107]:

$$p \equiv V/a_o l_c = 1 + H l_c + (1/3) K l_c^2 \tag{3}$$

IV. STRUCTURES IN CONCENTRATED SOLUTIONS

In concentrated aqueous surfactant solutions, the sizes and shapes of the aggregates are also influenced by interaggregate forces. This leads to positionally ordered structures characterized by long-range orientational alignment and spatial periodicities that cannot be ascribed to spherical micelles [108]. Nevertheless, all three classical structural shapes—spheres, cylinders, and planes—are respectively revealed in hexagonal, discrete (globular) micellar,

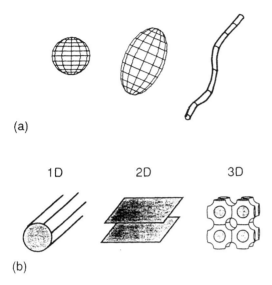

(a)

1D   2D   3D

(b)

Figure 8 Examples of (a) discrete and (b) continuous surfactant self-assembled structures. The latter can be extended in one (cylinders), two (lamellae), or three (bicontinuous bilayer) dimensions. (From Ref. 110.)

and lamellar mesophases [106]. These lyotropic mesophases cover all possible arrangements of repetitive surfaces in 3D space from infinite two-dimensional lamellae stacked on a one-dimensional lattice, via infinite cylinders stacked on a two-dimensional hexagonal lattice, to finite-size discontinuous inverse micelles packed on a three-dimensional cubic lattice [106,109] (Fig. 8), as will now be described.

A. The Lamellar Phase L_α

The lamellar phase (designated L_α) consists of alternating stacks of infinite planar bilayers (also designated as membranes) separated by intervening layers of solvent, usually water, and arranged parallel to one another [85,111]. The surfactants in the bilayers are organized in such a way that the hydrophobic tails of the surfactant molecules are at the center of the lamellae and the hydrophilic portions of the molecules are in contact with the solvent layer [112]. This phase exhibits quasi-long-range positional solid-like order (even at high dilution [108]) along the direction perpendicular to the layers. In the two other in-plane directions the system is liquidlike, i.e., the solvent and surfactant molecules are free to move in this plane [111] (Fig. 9). Stable lamellar phases are almost ubiquitous in the concentrated region of the phase diagrams of binary and multicomponent systems. However, their existence in very dilute solutions has only recently been observed and studied systematically [108].

Lamellar phases are known to exist in at least two configurations. In addition to a planar or continuous lamellar phase, where the bilayers are ordered in sheets [113], lamellar phases also exist with the bilayers ordered in closed concentric shells [114–116]. The structural units in the latter phase are often referred to as "onions", multilamellar vesicles (Fig.

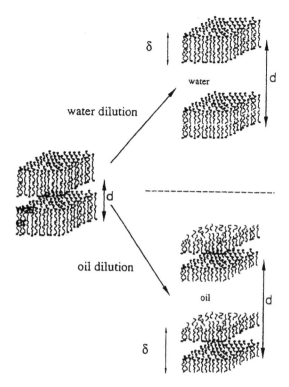

δ

water

d

water dilution

d

oil dilution

oil

d

δ

Figure 9 Schematic drawing of the membrane structure of lyotropic lamellar phases. The lamellar phase can be swollen either with water (hydrophilic solvent) or oil (hydrophobic solvent), leading to direct or inverted bilayers. The membrane thickness is δ, d is the smectic repeat distance. (From Ref. 111.)

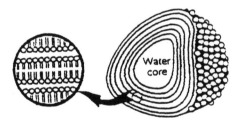

Water core

Figure 10 Schematic illustration of a liposome (vesicle). (From Ref. 117.)

10), lamellar droplets, or L_4 phase. It appears that the onions are stable even at low surfactant concentration when prepared from mixtures of nonionic and anionic surfactants at sufficiently high salinity [118].

The classical picture of the L_α phase was recently further elaborated by the addition of defective (solvent-filled lamellar phases, which will be treated in relation to certain phase transitions; see below) and strongly undulating bilayers [119]. Although these flexible bilayers are usually demonstrated in dilute solutions, they are treated here in order to present a more complete description of possible lamellar phases.

The swelling of lamellae by solvents depends on the elasticity of the bilayers. We can distinguish between rigid and flexible bilayers. In many binary surfactant–water systems the well-defined interface of the surfactant bilayers is insensitive to thermal fluctuations on a length scale comparable to the lamellar spacing. For rigid bilayers, $k \gg k_B T$, where k is the bending energy constant, k_B is the Boltzmann constant, and T is the absolute temperature. In this case, the membrane is virtually rigid, and thermal fluctuations have little effect on the shape of the membrane. For example, the system sodium laurate–water has a lamellar phase that exists only at high (>40%) surfactant concentrations, corresponding to rather small repeat distances [17]. On the other hand, for flexible bilayers, $k_B T \geq k$, and thus they exhibit a high configurational entropy state. In such a case the lamellar phases can swell considerably [112]. For example, in the systems $C_{12}E_5$-water [2,120] and AOT–water [121,122], the lamellar phases exist over a wide range of concentrations including very dilute regions. The corresponding repeat distances vary continuously from 4 to 300 nm in the $C_{12}E_5$-water system [2,120] and from 5 to 20 nm in the AOT–water system [121,122]. The existence of dilute lamellar phases may be conceived of in the following way. Two rigid membranes are subjected to a long-range attractive van der Waals force that tends to bring them together. This force may be balanced by a repulsive long-range electrostatic interaction that is evidently relevant only to charged membranes in polar solvents. At short distances (lengths smaller than 1 nm), hydration forces cause repulsion between the polar head regions of two membranes approaching each other. These forces are often attributed to the cost in free energy associated with the distortion of highly oriented water molecules near the membranes [123,124]. As the hydrophilic surfaces of the membranes approach each other, water molecules are squeezed out of the energetically favorable region, resulting in a repulsion [125]. Alternatively, it was claimed [126–128] that the origin of the hydration forces was the entropic repulsion of the surfaces interacting directly with each other, though the contribution due to the confinement of mobile surface groups [126–128] seems to be negligible [129].

When the membranes of a multilayer system are elastic, thermally excited undulations have to be included in the free energy. The undulations are confined in a smaller region of space as two such membranes approach each other to within a distance shorter than their persistence length ξ_k. ξ_k is the length scale over which thermal undulations led to a significant deviation from flatness and it is an exponential function of k. For lengths greater than ξ_k the membrane would become thermally crumpled [130]. This loss in configurational entropy leads to an effective long-range repulsive interaction inversely proportional to the bending constant k that can overcome the van der Waals attraction at large distances [131,132]. These thermal fluctuations are also thought to be the origin of membrane formation [130].

In summary, electrostatic repulsion stabilizes lamellar phases in ionic systems, whereas entropy reduction stabilizes lamellar phases in nonionic systems or in ionic systems in apolar solvents or in high ionic strength water. Also, the presence of suitable cosurfactants (generally alcohols), which increase the flexibility of the membranes, leads to the formation of dilute lamellar phases, for example, in the system brine–SDS–pentanol [133] or brine–SDS–pentanol–dodecane [134]. Recently, it was shown [135] that two distinct lamellar phases coexisted in the dilute region of the system cetylpyridinium chloride–hexanol–brine. The two phases differ in turbidity, viscosity, density, and some other physical properties. One of these lamellar phases is "classically" stabilized by the competition between van der Waals, hydration, and electrostatic forces. The other phase is entropically stabilized. The difference between electrostatically and sterically stabilized lamellar phases was demonstrated by transmission electron microscopy on thin vitrified

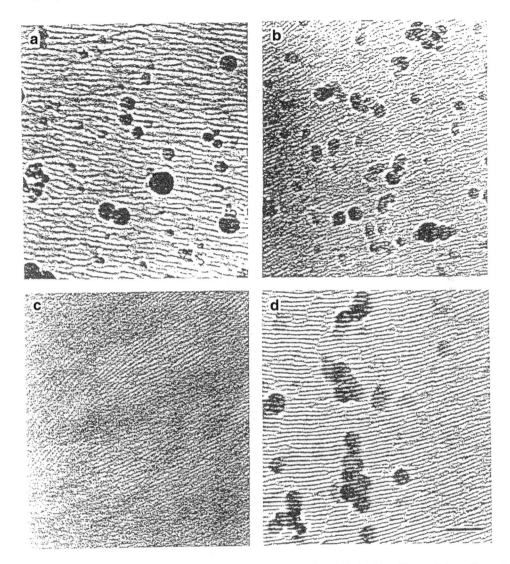

Figure 11 Cryo-transmission electron micrographs of vitrified thin films of lamellar phase (water–SDS 10%–pentanol 30%). Ordered dark lines correspond to the surfactant membranes. Dark roughly spherical objects of around 50 nm in size are frost particles. Bar = 100 nm. (From Ref. 136.)

liquid films [136]. Electrostatically stabilized lamellar phases consist of relatively rigid and well-ordered membranes (Fig. 11), whereas the sterically stabilized ones consist of soft and fluctuating membranes, as may be suggested by the strongly variable characteristic interlamellar distance [136] (Fig. 12).

B. The Hexagonal Phase

If the fluid surfactant aggregates consist of indefinitely long cylinders rather than bilayers, then two-dimensional fluid phases will be formed. The simplest and best established of these are the normal (H_I) and inverse (H_{II}) hexagonal phases. In the H_I phase, the surfactant

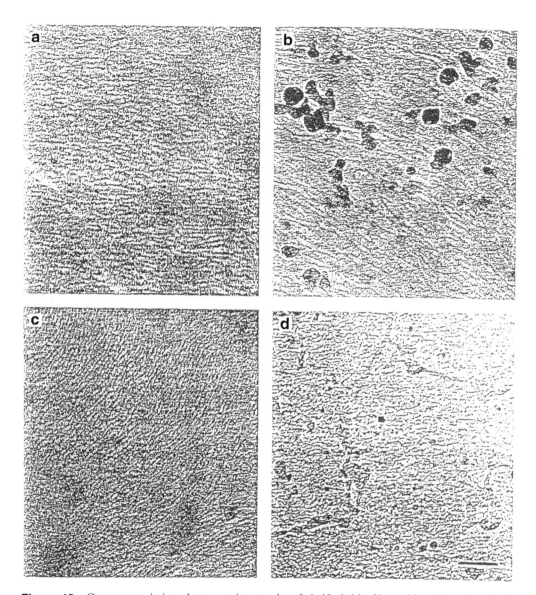

Figure 12 Cryo-transmission electron micrographs of vitrified thin films of lamellar phase (brine 0.5 M–SDS 9%–pentanol 7%). Dark lines correspond to the contrast produced by the surfactant membranes. Micrographs a–c display a lamellar order with a variable characteristic distance, which is interpreted as the sign of strong fluctuations of the membranes. Micrograph d displays a disorganized structure that could be due to a modification of the labile lamellar organization by shear or during specimen preparation. Bar = 100 nm. (From Ref. 136.)

molecules aggregate into circular cylindrical micelles that pack onto the hexagonal lattice, with a continuous water region filling the volume between the cylinders. In the H_{II} phase, the cylinders contain water cores surrounded by the surfactant polar headgroups, with the remaining volume completely filled by the fluidic chains at an essentially uniform liquid alkane density [137] (Fig. 13). It should be noted that the H_I phase can, in principle, swell

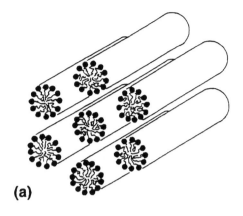

(a)

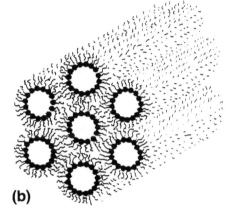

(b)

Figure 13 Topology of (a) normal (H$_I$) and (b) inverse (H$_{II}$) hexagonal phases. (From Ref. 138.)

without a significant change in the interfacial area per molecule, whereas swelling the H$_{II}$ phase lattice inevitably causes the interfacial area per molecule to increase. Also, for the H$_I$ phase, the water continuum is a true solvent, in the sense that although it is a structured fluid it is able to freely fill all the polar volume unoccupied by the surfactant headgroups. For the H$_{II}$ phase, this situation is not necessarily the case, since the hydrocarbon chains are pinned at one end of the polar interface by the headgroups, and the conformational state of the hydrocarbon chains, in part, determines whether the hydrophobic region can be uniformly filled and hence the H$_{II}$ phase allowed to form [138] (see Fig. 14). It has been suggested [139–141] that the free energy cost of some of the chains stretching to completely fill the hydrophobic volume can inhibit H$_{II}$ phase formation. Perhaps this is the reason for the absence of reversed micellar aggregates in binary nonionic surfactant–water mixtures [142].

Although most of the reported hexagonal phases are based on aggregates having a single curved surfactant layer (monolayers), a more complex type has been found in certain systems whose structure appears to be based on a hexagonal packing of cylinders formed by curved lipid bilayers that separate an inside and an outside of the same polarity [143].

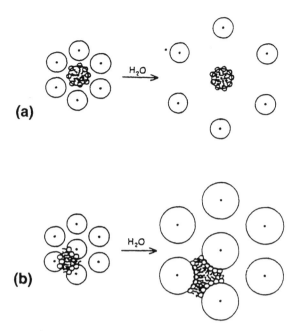

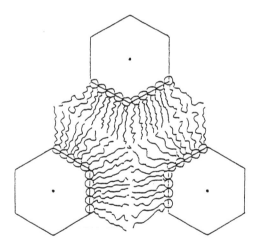

Figure 14 Swelling of (a) H_I and (b) H_{II} lattices with increasing water content. (From Ref. 138.)

Figure 15 Possible deviations of cylindrical aggregates from circular to hexagonal cross section in the H_{II} phase. (From Ref. 138.)

The assumption that the surfactant/water interface of the cylindrical aggregates of the H_{II} phase has a circular cross section is made purely as a convenient approximation to the true structure [143]. In principle, the cross section could, for example, become somewhat hexagonal in shape and still be compatible with a hexagonal packing [138] (see Fig. 15). It has been pointed out that a deformation of the interface toward a hexagonal cross section, although costing curvature elastic energy, would partially relieve the chain-packing stress of the H_{II} phase by bringing the interface closer to the center points [144,145] (see Fig. 15). For some systems the shape of the cylinders may deviate from being circular in cross section, leading to a packing into 2D phases of lower symmetry, not only in rectangular [146], oblique, or monoclinic [147] phases with ribbonlike aggregates, but also in hexagonal phases [148]. In the last case, deformation toward an elliptical cross section has been detected in the H_{II} phase of SDS–decanol–water ternary mixtures [130]. The explanation for this effect is that the SDS and the decanol have different tendencies for interfacial curvature, and as the decanol concentration increases, a partial lateral phase separation of the surfactant molecules occurs, leading to inhomogeneous interfacial curvature. Beyond a critical shape anisotropy of the aggregates, hexagonal packing is no longer possible, and transition to a less symmetrical rectangular phase occurs. In fact, the decanol component is not strictly necessary to bring about such deformations (although it may stabilize them); the purely binary system SDS–water adopts a series of intermediate phases between the H_I and L_α phases (Sec. V. F) [129,149]. As the water content is reduced, the cylinder radius of the H_I phase increase until it reaches a value of 1.83 nm, equal to the all-trans length of an SDS molecule. Beyond this point, the water layer thickness between cylinders would have to decrease rapidly if the cylinder cross section remained circular. However, this is resisted by repulsive electrostatic and hydrational forces between the aggregates. Instead, it costs less free energy for the cylinders to distort toward an elliptical shape, since in this way the opposing surfaces can maximize their separation. However, this then induces a phase transition to a less symmetrical oblique (2D monoclinic) phase. Although the SDS–water system is nominally a binary one, it may be regarded as containing three components: water, amphiphilic molecules bound to counterions, and dissociated molecules. Inhomogeneous interfacial curvature can then arise from a partial lateral phase separation of the charged and uncharged forms of the amphiphile [138].

In contrast to lamellar phases, swelling of hexagonal phases may a priori proceed via two different processes: either increasing the distance between adjacent cylindrical tubes or increasing the radius of the cylinders themselves [150]. The first possibility may be rejected on both theoretical [151,152] and experimental [153,154] considerations. This is because the cylinders may wander through the interstices of the hexagonal phase when the lattice parameter grows too large compared to their diameter [155]. The swelling of the cylindrical tubes themselves may be accomplished by tuning the spontaneous radius of curvature via the addition of salt, as was shown for the system SDS–brine–pentanol–cyclohexane [150].

Recently it was shown [156] that in the system $C_{12}E_8$–water, upon cooling samples containing less than 39% surfactant, a new type of nonbirefringent hexagonal micellar phase was observed. It had two quasi-spherical micelles per unit cell of hexagonal symmetry.

C. Cubic Phases

Lyotropic cubic phases have been the subject of many structural studies [138,157–159]. Their structure is more complicated and less readily visualized than that of other phases. Almost all 3D fluid phases so far observed are of cubic symmetry, although rhombohedral, tetragonal, and orthorhombic phases of inverse topology have been detected in a few systems (based, for example, on SDS or lipids) [160]. There are two distinct classes of cubic phases [110]:

1. The bicontinuous cubes (denoted by the symbol V) in which a single bilayer of surfactant divides the space into two interwoven continuous networks of water (for oil continuous, type II system, whereas for type I systems, the positions of the water and surfactant are reversed) and thus, the phase is continuous in water and in surfactant.
2. The micellar cubes (indicated by the symbol I), which consist of discrete micellar aggregates arranged on cubic lattices (Fig. 16).

Until now, the structures of only six or seven cubic phases have been firmly established [89]. Two of them are bicontinuous:

1. A cubic phase of space group $Ia3d$ (which was the first cubic structure to be solved [161] and is among the most commonly observed [162]). The structure of $Ia3d$ belongs to a body-centered space group of rods (which are essentially a surfactant bilayer with a circular cross section) connected 3×3 to generate two interwoven but unconnected 3D networks [162]. A chiral cubic phase of space group $P4_332$ has been observed so far in only one lipid–protein–water system [163]. Its proposed structure is similar to that of $Ia3d$: It has one water–lipid network interwoven with one network of quasi-spherical inverse micelles that encloses the protein molecules.
2. A cubic phase of space group $Pn3m$. It has a primitive cubic lattice formed by rods connected 4×4 at tetrahedral angles, forming two independent but interwoven diamond lattices [162] (see Fig. 16).

Direct evidence for the commonly accepted structure of the bicontinuous $Im3m$ cubic phase (orthogonal networks of water channels connected 6×6) [137] is still quite scant [89,164].

Bicontinuous cubic phases can be described by periodic minimal surfaces that are well-known in differential geometry [165]. In the case of inverted bicontinuous cubic phases, the periodic minimal surfaces lie along the middle of the bilayer. They are saddle surfaces with mean curvature zero everywhere (i.e., positive and negative curvatures of the bilayers forming the rods balance each other at every point [162] and with negative Gaussian curvature [138,157,159].

In contrast to the bicontinuous cubes, the less studied micellar cubic phases have positive Gaussian interfacial curvatures and are discontinuous, consisting of discrete micellar aggregates [164]. The structure of four micellar cubic phase has so far been established.

1. A cubic phase of space group $Fd3m$ is usually observed in type II systems based on a variety of hydrated lipid mixtures [163,166]. A type I $Fd3m$ phase with a very large lattice parameter has been claimed to form in a brine–butanol–SDS–toluene system [167]. The unit cell of the $Fd3m$ phase contains two quasi-spherical, differently sized types of inverse micelles [160]. There are eight of the larger and 16 of the smaller micelles per unit cell.
2. A cubic phase of space group $Pm3n$ is usually observed in type I systems [164]. Several structures have been suggested for the $Pm3n$ phase [168–171]. It is now agreed that it contains two types of micelles [172]: two quasi-spherical micelles packed on a body-centered cubic lattice and six slightly asymmetrical micelles arranged in parallel rows on opposite faces of the unit cell. The asymmetrical micelles are assumed to be disklike [173] or rodlike with rotational disorder around one of the short axes [174,175]. In order to pack space completely, each asymmetrical micelle, together with the water that surrounds it, takes the shape of a

Pn 3m

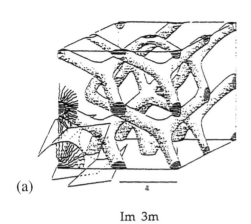

(a)

Im 3m

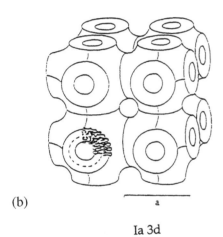

(b)

Ia 3d

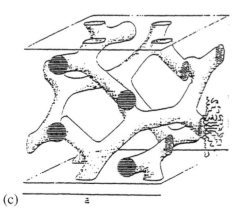

(c)

Figure 16 Examples of the structures of three-dimensional cubic mesophases. (a) *Pn3m*; (b) *Im3m*; (c) *Ia3d*. (From Ref. 138.)

tetrakaidecahedron, and each spherical micelle plus its associated water takes the shape of a dodecahedron [164]. The polyhedra fit together in space-filling packing [89].

3. A cubic phase of space group *Im3m* (which has an entirely different structure from the bicontinuous cubic phase with the same space group) [164]. It consists of a packing of quasi-spherical micelles on a body-centered cubic lattice. Each of these micelles with their associated water will on average have the shape of a tetrakaidecahedron. The polyhedra will assemble in space-filling packings.

4. A cubic phase of space group *Fm3m* that consists of a packing of quasi-spherical micelles on a face-centered cubic lattice. Each micelle with the water that surrounds it will on average have the shape of a dodecahedron. The dodecahedra pack together in space-filling aggregates [164].

The last two cubic phases have only recently been identified in systems based on gangliosides (a widespread class of natural glycolipids with unusual bulky polar headgroups) [176] and in the water–$C_{12}EO_{12}$ system [164].

D. Nematic Phases

In certain cases, nematic (orientational but not positional) order may be observed in phases of (relatively small) micellar rods (canonic) or disks (discotic) [177], which are associated with lower temperature hexagonal and lamellar phases, respectively [178]. (See Fig. 17.) The nematic isotropic phases were thought to be built up of discrete aggregates of different shapes, but the presence of continuous aggregates has also been recently suggested in this case [178]. Some systems (e.g., potassium laurate–decanol–water) form biaxial nematic phases. In this case, the micelles are believed to be neither rods nor disks but rather to

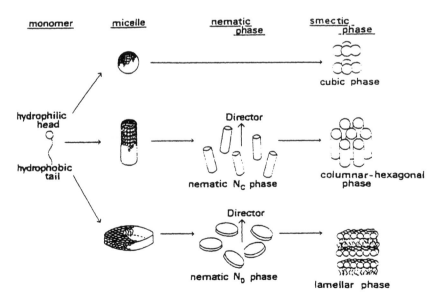

Figure 17 Schematic representation of possible structures of aggregates and their associated mesophases formed by soaplike surfactants in water. The concentration of surfactant should be read as increasing from left to right and from top to bottom. (From Ref. 177.)

have three axes of different lengths [179]. The nematic phase gives a very characteristic optical texture [180–182] and allows the LC director to be oriented both by external fields and by surface forces.

V. STRUCTURES IN DILUTE SOLUTIONS

The rich structural polymorphism that was long thought to be typical of concentrated amphiphilic systems can also be observed in very dilute dispersions [85,108]. The average interaggregate distance is relatively large in dilute systems, and thus the characterization of their structure by scattering methods is not easy. On the other hand, the weak interaggregate interactions can hardly affect the thermodynamic stability of the observed morphologies [108]. In the following, some typical structures encountered in dilute amphiphilic solutions are described.

A. "Giant" Micelles

Globular (mostly ionic) micelles may grow in a one-dimensional manner that leads to the formation of large, flexible wormlike aggregates referred to as "giant" or "threadlike" or "wormlike" micelles [183–186] (Fig. 18). Micellar elongation may be promoted by increasing the surfactant concentration [187]. More often, however, micellar growth is induced by adding salt [188,189] or alcohol [190]. One-dimensional growth was evidenced by using scattering data (q^{-1} dependence of the scattered intensity) [191] and corroborated by viscosity [190] and light-scattering [188] measurements.

›The major effect of the added salt or alcohol is to screen the electrostatic repulsion between the charged headgroups of the amphiphiles, thus lowering the optimal area per headgroup at the hydrocarbon/water interface of the micelle. This in turn implies an increase in the end cap energy, expressing the difference between the packing free energy in the highly curved (approximately hemispherical) end caps of the micelle and its (moderately curved) cylindrical body [92,97]. Consequently, the average micellar length, which increases exponentially with the end cap energy, can be dramatically enhanced, thereby reducing the spontaneous curvature [192] and leading to the formation of "giant"

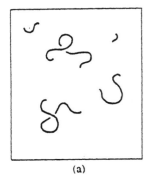

(a)   (b)

Figure 18 Intuitive representation of (a) the dilute regime and (b) the semidilute regime of a dispersion of flexible giant micelles. (From Ref. 108.)

micelles [193]. It should be noted that salts and alcohols are the very additives that are used to prepare nematic phases in concentrated solutions of surfactant systems [177]. It has been suggested [190] that at high salinity or alcohol concentration, the wormlike micelles could branch, leading at higher surfactant concentrations to the formation of a randomly cross-linked network of cylindrical micelles that either spans the whole system or separates as a densely connected phase coexisting with a very dilute phase of small micelles [186,194–200]. This network may serve as an intermediate structure between cylindrical micelles and flat bilayers [108]. The existence of branched micelles has been assumed on the basis of the unusual rheological behavior (decreased viscosity at high salinity) observed in certain semidilute solutions of wormlike micelles [186,194–200] through self-diffusion measurements [201] and direct imaging by transmission electron microscopy (TEM) [119,202,203]. Yet, some problems relating to sample preparation hamper the interpretation of cryoTEM data [192]. Thus, it is not entirely clear whether branched threadlike micelles are truly equilibrium entities or perhaps metastable transients induced by shear stresses in the course of sample preparation [193]. Recent calculations [193] showed that the formation of branched micelles (in aqueous solutions of a single-tailed surfactant component) is energetically unfavorable, but such micelles may appear as metastable structures or be entropically favored, as they provide an efficient alternative to entanglements and to high energy end caps.

The formation of nonspherical micelles can, in principle, occur either by one-dimensional extension of the nonpolar core of the individual micelles into an elongated (cylindrical or rodlike) shape or via clustering of the initial globular micelles [204]. However, the possibility [111] that long micelles can be formed by linear aggregation of spherical micelles (retaining their own shape) into a "pearl necklace" [108], "string of beads" [204,205], or "cross-linked micelles' [206] is usually considered unrealistic [108,110]. The arguments in favor of such linear aggregates could always be refuted [110,207]. In fact, this possibility has been ruled out, e.g., for the system SDS–brine, with the use of high neutron scattering data [208]. Nevertheless, clustering of spherical micelles has been alluded for the system C_8E_4–D_2O on the basis of NMR data [204].

Experimental evidence shows that quite often solutions of giant micelles posses static and dynamic properties analogous to those of macromolecules [108,192]. Their primary structure is both unidimensional and flexible. In the dilute regime, giant micelles are characterized by nonuniform size distribution, which is strongly influenced by experimental conditions [108]. At higher concentrations, in the semidilute regime, the main differences between micelles and polymer solutions vanish [108].

B. The Anomalous Isotropic (Sponge) L_3 Phase

The isotropic (sponge) L_3 phase was observed in nonionic [209] and ionic [210] surfactant–water binary systems as well as in nonionic surfactant–alcohol–water [211] and ionic surfactant–alcohol–brine ternary systems [212]. It presents no long-range order. Scattering methods [209,213], self-diffusion measurements [214,215], and freeze-fracture electron microscopic observations [216] have shown that it consists of an infinite bilayer bent everywhere with a saddle-like curvature so that it is multiconnected to itself (by many random "passages"—tubular connections or "handles" [85] in the sponge) isotropically in 3D space over macroscopic distances [85] and divides space into two independent solvent regions [217] (Fig. 19). This is in contrast to the lamellar L_α phase, where each membrane is singly connected to itself [85], and in contrast to the vesicle L_4 phase, where the bilayers

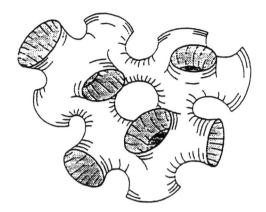

Figure 19 Schematic drawing of the randomly connected bilayer network proposed for the L_3 phase (From Ref. 216.)

form unilamellar or multilamellar spherical aggregates [218], but it is similar to cubic phases, which are also characterized by highly connected bilayers. Yet, as the L_3 phase lacks long-range order, it may be regarded as a disordered or melted cubic phase [217]. The L_3 phase has very low viscosity, and in the dilute regime it exhibits flow birefringence [217]. The respective stabilities of the L_α and L_3 phases, which are generally found in adjacent phase diagram regions, have been discussed in terms of the elasticity and fluctuations of their constituent membranes [219]. Such a membrane is usually envisaged as a two-dimensional incompressible liquid. This means that the molecules within the membrane have a high diffusion coefficient typical of fluids, while their interfacial area is virtually constant even when the surfactant aggregate is splayed [108]. The ultimate structure of the surfactant aggregate depends strongly on the parameters k and $\bar{k}$ [131,132]. k represents the energy cost to bend the surface film with a spherical splay deformation and is associated with changes in the mean curvature of the film. Thus, it characterizes the rigidity of the bilayer. $\bar{k}$ is related to the cost of saddle splay and is associated with the Gaussian curvature. (See Fig. 20.)

One theoretical treatment of the L_3 phase [220] considers k as the leading control parameter. The structure can be described as locally sheetlike sections of semiflexible surfactant bilayers joined at larger distances into a multiconnected random surface having a preferred structural length scale on the order of the persistence length ξ_k [134]. Upon dilution, when the diameter of a "passage" in the membrane exceeds ξ_k [219], the lamellar phase melts into the L_3 phase stabilized by entropy [134]. This behavior conforms to the general assumption [130] that a randomly oriented surface is favored when the typical radius of curvature is greater than ξ_k.

An alternative explanation for the stabilization of the sponge phase [221] is based on the phase diagrams of several brine–surfactant–alcohol systems, where the transition $L_\alpha \rightarrow L_3$ is brought about by the addition of alcohol. This transition is triggered by variations of $\bar{k}$, the parameter that determines the number of "passages" in the bilayer, so the higher the value of $\bar{k}$, the easier the formation of "passages" and hence the greater the stability of the L_3 phase. Conversely, lower values of $\bar{k}$ will favor the stability of the L_α phase [221]. The L_3 structure is then far from being a random surface, resembling rather a periodic minimal surface having lost its long-range cubic order [222]. The transition from an ordered (L_α)

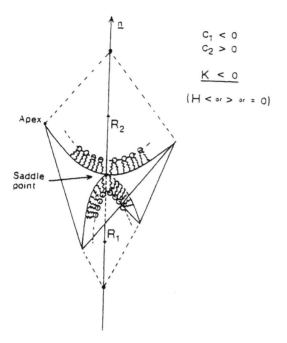

$$c_1 < 0$$
$$c_2 > 0$$

$$K < 0$$

$$(H < \text{ or } > \text{ or } = 0)$$

Figure 20 Saddle surface, of negative Gaussian curvature. In this figure K is related to $\bar{k}$ in the text. (From Ref 138.)

to a disordered (L_3) phase indicates the significance of curvature fluctuations at constant topology. Since $\bar{k}$ plays no role at constant topology, the amplitude of the thermal fluctuations is determined by k [85]. Large values of k are therefore expected to stabilize the L_α phase, while small values of k favor the formation of the L_3 phase [101,220].

C. Vesicles

Vesicles are closed bilayers that can be observed in two forms. At low surfactant concentration, the vesicles are unilamellar and behave like a colloidal suspension of polydisperse particles. At more concentrated surfactant solutions, small multilayered vesicles are formed [134]. Multilamellar vesicles (known also as spherulites) have also been observed in the lamellar phases of surfactant–brine (or even pure water–alcohol) systems [218]. The surfactant may be SDS [218,223] or DDAB (didodecyldimethylammonium bromide) [224]. In alcohol-containing systems the bilayer structural transformations are controlled by the alcohol/surfactant ratio [134]. Thus, in many SDS–brine (or water)–alcohol systems, a vesicle (L_4) phase is located between the micellar phase and the lamellar (L_α) phase. At fixed surfactant concentration, the sequence of phases L_4–L_α–L_3 (in water) is obtained by increasing the alcohol content, and the sequence L_2–L_α–L_3 (in oil) is obtained by decreasing the alcohol content [134].

A vesicle phase will form instead of the L_3 phase when the elastic constant k (which is controlled by the alkyl chain length of the alcohol) [134] and $\bar{k}$ (which is controlled by the alcohol concentration in the interface) [134] are chosen so that the curvature energy required to create a sphere is small [225]. The phases formed reflect the competition between entropic contributions (which favor smaller aggregates that allow for a more homogeneous distribution of the free counterions in space) [85] and elastic contributions to the free energy [223,225]. Decreasing surfactant concentration at constant k leads first to a highly swollen L_α

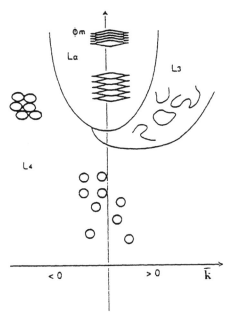

Figure 21 Schematic phase diagram of membranes. ϕ_m, membrane concentration; $\bar{k}$, Gaussian elastic constant. L_α, lamellar phases; L_3, sponge phase; L_4, vesicle phase. (From Ref. 134.)

phase due to thermal fluctuations, as we have seen. Further addition of solvent may induce the formation of either sponge phases (positive $\bar{k}$; leading to more "passages") or vesicle phases (negative $\bar{k}$; leading to more disconnected aggregates) [218]. Thus, increasing the alcohol/surfactant ratio induces a change in the sign of $\bar{k}$ from negative to positive [218]. A schematic phase diagram for L_α, L_3, and L_4 is shown in Fig. 21.

D. Phase Transitions

It seems quite remarkable that the structural diversity encountered in surfactant dispersions can be subsumed into an essentially universal phase behavior in water, common to all classical surfactants irrespective of their detailed chemical structure [177]. Thus, changing the structure of the polar headgroup or of the nonpolar tail (e.g., branching, moving from monoalkyl to dialkyl residue), replacing water by other polar solvents [226], and using copolymers instead of monomers [227–229]—all of these variations leave the structures and their sequence unchanged. In summary, transitions from one structure to the next occur along similar sequences when the dominant parameters of the phase diagrams are altered [230]. These transitions are accompanied by a prominent change in the topology of the aggregates (see Sec. V. E-F.) Thus, the interpretation of this universal phase behavior is based on an understanding of the factors leading to the changes in stability of the various aggregates with concentration [177]. This is now demonstrated using two recent models, the frustration relieving model [231] and the packing efficiency model [92,97].

1. Frustration-Relieving Model

"Frustration" is here defined as the conflict between antagonist physical forces acting on the amphiphilic systems. These forces (e.g., electrostatic, van der Waals, hydrophobic) have components parallel to the interface that determine the interfacial curvature and components

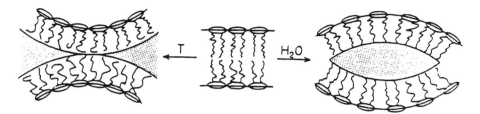

Figure 22 Frustration in bilayer packing with varying temperature or water content. (From Ref. 138.)

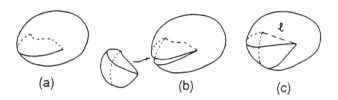

Figure 23 Creation of a defect in a piece of matter, following a Volterra process. (a) A cut is made; (b) the lips of the cut are separated, and extra matter is introduced in a manner respecting the symmetry of the matter; and (c) the matter relaxes around the line of defect ℓ. (After Ref. 230.)

normal to the interface that maintain the surfactant films at constant distances. This last requirement is tantamount to the packing constraint: the filling of the hydrophobic volume by hydrocarbon chains that are anchored at the interface [109]. For example, a bilayer with curved interfaces (which may occur when the headgroups occupy less space than the hydrocarbon tails) is in a frustrated state [230] because energetically expensive voids are formed in the core of the bilayer (Fig. 22). The system is then in a state of frustration as it is not able to satisfy the constraints of either uniform curvature or optimal packing [232].

Frustration is released by transferring the structure into a curved space in a geometrical procedure (Volterra process) similar to the mapping of a sphere onto a plane (Fig. 23). A sphere cannot be deformed into a plane without being torn unless matter is added during the process while retaining the symmetry of any structure that is present on the sphere. This can be done by creating disclinations (defects of rotation) in the structure. The idealized structures predicted by this approach to reconcile interfacial distances and curvatures are limited to a discrete number of points of real phase diagrams. Away from these points there is a buildup of frustration that must be resolved in other ways, for example by the occurrence of local fluctuations in the classical structures [233–236].

The need to add either salt or a long-chain alcohol in order to prepare nematic phases from classical surfactants (e.g., SDS–decanol–water [154]; decylammonium chloride–ammonium chloride–water [237]) may be understood in terms of this model. These additives shift the liquid crystalline phases away from their optimum concentration ranges and leading to a buildup of frustration. This frustration is resolved by the introduction of local structural fluctuations that destabilize the mesophases and lead to finite micelles of the appropriate sizes [177]. For instance, decanol limits the domains of existence of the mesophases for the system SDS–decanol–water [238]. The addition of NH_4Cl to the decylammonium chloride–water system suppresses the hexagonal phase and enhances the stability of the discotic phase [239].

2. Packing Efficiency Model

Repulsive (steric or electrostatic) intermicellar interactions favor larger, less curved aggregates because high curvature (e.g., spherical) surfactant molecules do not pack as efficiently as those of lower curvature (cylinders and sheets) [240]. This enables the micelles to pack more densely (for any given amount of surfactant) with less excluded volume [241]. The volume fractions of packed aggregates are 0.74, 0.91, and 1.00 for spheres (only in hexagonal close packing), (infinite) cylinders, and (infinite) sheets, respectively. A sphere-to-rod transition would then enhance packing energy and diminish repulsive energy (i.e., greater average distance between micelles). The transition will occur at sufficiently high concentrations where this free energy decrease can overcompensate for the self-energy (spherical curvature is locally preferred over cylindrical) and dispersion entropy (fewer cylinders than spheres) contributions. For the same reason, the short rods (which have many spherical end caps) will lengthen at higher surfactant concentrations, and while they lose the preferred spherical curvature they gain enhanced packing entropy (the fewer the rods, the larger the packing entropy) and reduce repulsion energy. A similar interpretation may be suggested for the hexagonal (cylinder) $\rightarrow$ lamellar (bilayer) transition. This transition proceeds via the formation of bilayers that are often riddled with curved holes or cracks (see Sec. V. F), thus satisfying more efficiently the hydrophobic effect, but concurrently they intensify the repulsive interaction between the bilayers. Creation of such a defect at a constant volume fraction necessitates the removal of surfactant molecules from the bilayer and their accommodation elsewhere in the same bilayer or the formation of new bilayers. Either way, this leads to smaller average interlamellar distance and reduces packing entropy. The characteristics of the defects depend on surfactant concentration. At low enough concentration, where interlamellar interactions are less significant, the larger numbers of defects facilitate transition to a more curved structure. At high concentrations, where interlamellar interactions are more intensive, the defects will gradually be expelled out of the lamellae, forming layers of fragmented bilayers [92] as an intermediate stage in the formation (at still higher concentration) of a "classical" lamellar phase.

E. Cubic Intermediate Phases

The study of the nature of the processes taking place as one structure transforms into another when the variables of the phase diagrams are changed appears much more complex in amphiphilic systems than in the classical case of atomic or molecular crystals. The topologies of liquid crystals (such as lamellar, hexagonal, and cubic phases) are so different that the structural transformations imply dramatic topological changes involving ruptures and fusions of the structural elements and films built by the amphiphilic molecules [242]. For example, the lamellar–hexagonal transition can be regarded as a progression in which the aggregate curvature decreases as a result of changes in internal interfaces arising from changes in surfactant concentration, temperature, or additives to the system (Fig. 24). Bicontinuous cubic phases [138,157,158] have been regarded as the topologically favored intermediate structure by which the hexagonal rods can undergo a morphological transition to a planar lamellar structure [236,242,243]. For medium alkyl chain lengths, in the concentrated regime, the classical phase sequence of hexagonal (H_I), bicontinuous cubic (V_1), and lamellar (L_α) phases is observed on increasing temperature or surfactant concentration [2]. It was suggested [236] that the lamellar–cubic and hexagonal–cubic transitions for the system $C_{12}E_6$–water proceed via cylinders arranged on a local hexagonal lattice. The lamellar phase, for instance, transforms itself into

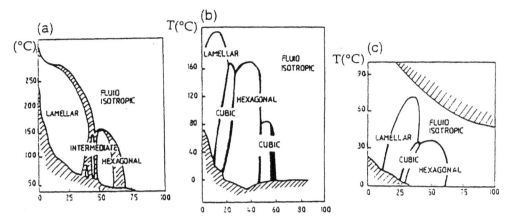

Figure 24 Schematic representations of the phase diagrams of the systems (a) $C_{12}Na-H_2O$, (b) $DTACl-H_2O$, and (c) $C_{12}EO_6-H_2O$. Broad lines and hatched regions represent polyphasic domains. Phase transitions along the temperature and (water) concentration axes are clearly seen. (From Ref. 230.)

the cubic phase, not directly by contacts and fusions of its lamellar planes but rather by contacts and fusions of the cylinders created by fragmentation of the bilayers. These cylinders are organized with a local hexagonal short order before the cylinders are deformed and transformed into the labyrinths of the cubic phase. This metastable hexagonal phase suggests that an important potential barrier (most likely related to the setting up of connections between the cylinders) separates the hexagonal and cubic structures [236]. The building of a cylindrical structure with homogeneous curvature rather than a cubic structure with inhomogeneous curvature is due to the limited number of degrees of freedom typical of nonionic molecules, whereas mixtures of ionic surfactant and cosurfactant [146] or ionic surfactant [147], whose molecules have different degrees of dissociation, can segregate along the interfaces and provide the inhomogeneous interfacial curvatures presented by the rectangular or cubic phases encountered in these systems [236].

Between micellar solution and H_I phase there is a micellar cubic phase with spherical or slightly anisotropic micelles. Between H_I and L_α or between L_α and H_{II} there are bicontinuous cubic phases as we have seen. Cubic phases based on the packing of inverse micelles between reverse micelle solution (L_2) and H_{II} are relatively rare, presumably owing to the low tendency of reverse micelles to organize in a regular manner [244], and have only quite recently been established [166,245]. Such a location of cubic phases might be anticipated for compounds having bulky hydrocarbon chains and small, weakly polar headgroups (ideally also with attractive lateral headgroup interactions such as hydrogen bonding). The *Fd3m* cubic phase has been discovered in fully hydrated phospholipids, mixtures of a strong polar lipid and a very weakly polar amphiphile [160,166,246]. In addition, it was shown that *Fd3m* cubic phases might be formed in purely binary glycolipid–water and nonionic surfactant–water systems [244]. It was suggested that the monolayer bending modulus in these cases was rather low, making the chain packing constraints less severe [244]. Also, a reverse (W/O) cubic phase was observed in a ternary system consisting of an amphiphilic diblock copolymer ($EO_{17}BO_{10}$, where

EO represents ethylene oxide and BO represents butylene oxide), water, and *p*-xylene [247]. Because of the weak (excluded volume) interactions between the reverse micelles, high micelle densities (and consequently high amphiphile concentration) would be needed to make a crystalline packing possible. Upon an increase in the amphiphile concentration it is more likely that the reverse micelles (L_2) elongate and then crystallize into a hexagonal array of crystals without going through a cubic ordering arrangement; this is usually the case with the typical surfactants. Apparently, in the case of macromolecular amphiphiles such as the block copolymer studied here, the micellar size and (steric) interactions between micelles are both large enough to facilitate the crystallization of the reverse micelles into a cubic lattice [247]. Along with *Fd3m*, a bicontinuous cubic phase *Ia3d* and a normal micellar cube of the space group *Im3m* were observed at the same temperature together with five more noncubic phases. In fact, this is the richest phase behavior reported to date in an amphiphile-containing system [248].

Transitions between lyotropic liquid crystalline phases might either take place randomly, through the destruction of the first phase and the building of the second, or occur in an organized way if the structural elements of the second structure grow from structural elements of the first. For example, at concentrations of $C_{12}E_6$ greater than 35 wt%, the system $C_{12}E_6$–water presents a cubic phase with space group *Ia3d* [249] at intermediate temperatures and falls between a lamellar (L_α) structure and a hexagonal (H_I) structure. It was shown that the transitions occurred in an organized manner, following well-defined epitaxial relations between particular reticular planes of the three phases.

F. Noncubic Intermediate Phases

Morphological transitions (such as the lamellar–hexagonal transition) may also proceed via noncubic intermediate mesophases [234,243,250–263]. Unlike the cubic phase, the intermediate phases are birefringent, showing that their structure is anisotropic [260]. There appear to be a wide variety of possible structures [143,257–259,261,264] that fall into one of three classes [142]. The first has a micellar structure similar to that of the H_I phase but in which the rod-shaped aggregates are distorted into biaxial ribbons and organized in a regular array [146,147,149,238,261], for example, the rhombohedral networks of connected structures [138,147,250,257] (designated R_α and shown in Fig. 25b). A second possibility is a "lamellar-like" structure of bilayer sheets stacked in one dimension but in which the planes are broken by irregular water-filled defects (such a defect is typical of L_α^H; see Fig. 26) that have no correlation between adjacent planes [147,149,233,235,250,257, 258,260,265–267]. Such structures have been given the generic name "mesh" phases, and they occupy large fractions of the phase diagrams, usually in solvent-rich regions, whereas other types of intermediate phases are restricted to narrow fields in surfactant-rich regions of the phase diagrams [119]. There can be a number of such phases depending on the type of interlayer correlation adopted by the water-filled defects. For example, there is a possibility of tetragonal phases (designated T_α) in which the lamellar planes are pierced by regular holes that are correlated from layer to layer [138,147,250,258,259] (Fig. 25a). Finally, a noncubic 3D bicontinuous structure has been proposed [267–270] (Fig. 27).

In some systems the cubic phase between the H_1 and L_α phases is replaced by one or more of these phases [260,262]. For example, four separate (intermediate) phases were observed in a narrow region of concentration in the binary system SDS–water [147,257,271]: (1) a two-dimensional monoclinic phase, (2) a three-dimensional rhombohedral or orthorhombic phase, (3) a three-dimensional cubic phase, and (4) a three-dimensional tetragonal phase. The 2D monoclinic phase can be seen as a deformed

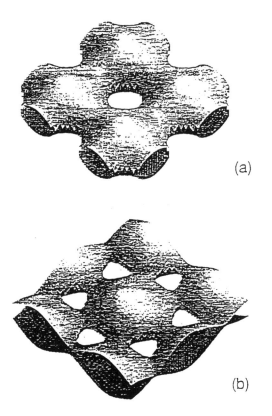

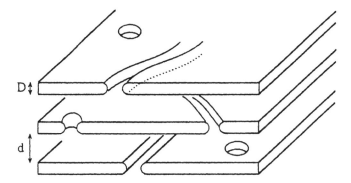

Figure 25 (a) One layer of a centered tetragonal mesh structure. (b) One layer of a rhombohedral mesh structure. (From Ref. 262.)

Figure 26 Schematic illustration of line and pore (curvature) defects in the lamellar phase. (From Ref. 97.)

hexagonal phase built up by rods with noncircular cross sections. Structures 2 and 4 can be seen as originating from cubic bicontinuous V_1 phases that have been deformed in one or two dimensions, while in other cases the cubic phase may exist in conjunction with these phases [147,243].

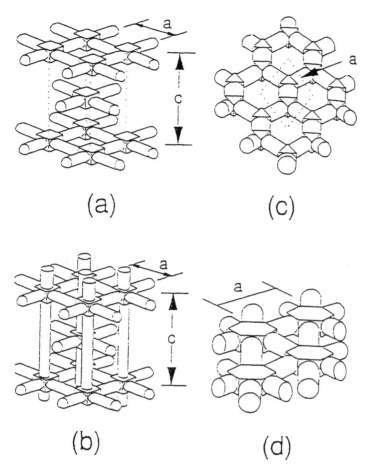

Figure 27 Structural models for the intermediate rods and boxes. The water/hydrocarbon interface is to be imagined as being located on the surface of the rods and boxes. (a) A centered tetragonal mesh; (b) a bicontinuous centered tetragonal structure; (c) one layer of a type a hexagonal mesh from which both a rhombohedral mesh and bicontinuous structures can be constructed; (d) as c, but for a type b hexagonal structure. (From Ref. 262.)

The occurrence of intermediate phases depends very much on the particular balance of forces between molecules. For example, it seems that noncubic intermediate phases are formed in preference to bicontinuous cubic phases when the alkyl chain portion of the surfactant molecule is more rigid, either because of fluorination [265,178,258] or because of increased strength [260,272], because in these cases the chains cannot pack in the cubic structure [243,263]. The effect of varying chain length was studied in the series of nonionic surfactants $C_n(EO)_m$. Burgoyne et al. [262] showed that the $C_{30}(EO)_9$–water system exhibited an extensive intermediate mesh phase between the H_1 and L_α phases. A recent small angle neutron scattering (SANS) study on aligned samples [273] identified the structure of the intermediate phase as a rhombohedral mesh phase of space group $R3m$ with a possible 6-coordinated structure [250]. In the system $C_{22}(EO)_6$ [260] the surfactant has a double bond at carbon 13, which reduces the packing efficiency in the crystalline state, enhancing the solubility. Although the bond is rigid, the chain does not show any preferred configuration. Because of the increased length of the chain and hence its decreased flexibility, intermediate phases are expected to be energetically more favored than cubic phases between the L_α and H_I phases [260]. Thus, there is no cubic phase at temperatures between which the hexagonal and lamellar phases are found, and only a defective lamellar phase (L_α^H) is observed. In this case, L_α^H is clearly a thermodynamically separate phase, being separated from the lamellar phase by a narrow two-phase region indicating a first-order phase transition.

Several possible models for the L_α^H phase have been discussed. These are disk micelles condensed onto planes (a water-continuous structure), a bilayer cut by irregular channels forming a ribbonlike structure (a bicontinuous structure), and a continuous bilayer structure pierced by holes (a surfactant-continuous structure). For the $C_{22}(EO)_6$ case, it was argued that the ribbon model was probably the most realistic and that the ribbons are arranged on a centered rectangular lattice. In the system $C_{16}(EO)_6$–water there is a very narrow cubic phase region and a metastable intermediate phase. At temperatures above this and below the lamellar phase there is an extensive region of L_α^H, which forms the conventional lamellar phase on heating without undergoing a first-order phase transition. Here the ribbons are irregular and connected within the lamellar planes. The defects are elongated rectangular water-filled holes of the type reported in another system [274]. The system forms the holes in the lamellae as a mechanism by which it can increase the surface curvature and hence the EO hydration without changing the surface area per molecule and without a major reorganization of the aggregate structure. Lamellae with elongated defects also act as a topological precursor to the lower temperature hexagonal phase.

The longer and hence more rigid alkyl chain in the system $C_{22}(EO)_6$-water plays an important role in stabilizing the intermediate phase and maintaining the surface area per headgroup constant, as it is not observed in its shorter chain homologs. It seems that the L_α^H phase becomes more stable as the surfactant alkyl chain length is increased. Under these conditions, the L_α^H–L_α transition becomes more prominent, with a two-phase coexistence region separating these microstructures; i.e., this is a first-order phase transition. In the system $C_{16}(EO)_6$–water [243] there is a region of the phase diagram where the phase behavior was reproducibly different on heating and cooling. In the region usually associated with the cubic or intermediate phases, the random mesh phase L_α^H, a more ordered rhombohedral intermediate phase, and a bicontinuous cubic phase of space group $Ia3d$ were all observed on cooling from the lamellar phase, whereas only the cubic and L_α^H phases are stable on heating from the two-phase region [142].

Short alkyl chain nonionic surfactants behave similarly to conventional ionic surfactants. For the $C_{12}(EO)_6$–water system, the transition from the L_α phase to the L_α^H phase appears to be a continuous process with no intervening two-phase region. The region between the L_α and H_I phases is occupied by a well-defined bicontinuous cubic phase [2,275] of space group $Ia3d$ [249]. On heating the hexagonal phase, micellar or cubic phases form before the lamellar phase does. In the adjoining lamellar and hexagonal phases there are structural fluctuations in the aggregate structure that act as precursors to the cubic phase [236,242]. The phase diagram of the system $C_{12}(EO)_2$–water lacks hexagonal phases [276], and on heating the lamellar phase transforms to the cubic phases $Ia3d$ and $Pn3m$ and then to the isotropic L_2 phase [277].

In summary, long-chain surfactants, be they nonionic [142,243,260,262], anionic [147,254,258,263,278], or cationic [278], lead to the formation of stable or metastable intermediate phases rather than cubic phases.

G. The Nature of the Defects

The L_α^H phase has been demonstrated to have a layerlike structure similar to that of the L_α phase but with the water-containing defected layers. The defects are thought to manifest as pores, which can be regularly or irregularly shaped, that pierce through the layers. It is not possible to be precise about the geometry of the defects except to state that they have broad interplanar reflections and that there is no interplanar correlation. The differences were detected with the use of electron spin resonance (ESR) spectroscopy [279], which reveals the existence of regions in the lamellae where the interfilm is highly curved; electron microscopy [280], which shows the existence of dislocation loops that cross the layers; or X-ray (or neutron) scattering [236]. Diffuse scattering, indicative of defects, has in fact been observed from a number of lamellar phases of binary or ternary surfactant systems. The existence of defects was also confirmed by birefringence and order parameter measurements [281] and rheological techniques [282]. The structure of the individual layers is thought to consist of arrays of small oblate (disklike) aggregates rather than bilayers [233,283]. In other systems the defects appear to consist of regions of correlated rodlike aggregates, either elliptical [234] or circular [236] in cross section. In addition to the bilayer-to-rod transition (which corresponds to an $L_\alpha \rightarrow L_1$ phase transition) the fragmentation of infinite bilayers may also lead to the creation of sites with increased curvature [119].

It was suggested [284] that the perforated lamellar phase may form via the growth of branched and multiconnected threadlike micelles. Interconnection of threadlike micelles reduces the overall curvature of the monolayer making up the micelles and thereby reflects a preference for microstructures of decreasing curvature [119]. The suggested morphological sequence for the system cetylpyridinium chloride–hexanol–brine is spheres, small disks, long capped cylinders, branched cylinders, perforated bilayers, smooth bilayers, loose network of connected bilayers (foamlike structure), and multiphasic domain [284].

H. Mechanisms for the Formation of Defective Phases

The variety of intermediate-phase behavior may be attributed to the delicate balance between the EO interactions and solvation and the alkyl chain interactions. The nonlamellar phases have curved surfactant/water (or lipid)/water interfaces. Generally, a surfactant layer will tend to curl if by doing so it relieves stress within the layer, which may arise from an asymmetry in the lateral interactions within the headgroups compared to the hydrocarbon regions. The phase changes seen in this system are the consequence of two mechanisms [262]. Increasing temperature causes a decreases in poly(ethylene oxide) chain hydration [285–288], which results in a decrease in the surface area per molecule at the surfactant/water interface and a reduction in the interfacial curvature, promoting phases with a flatter interface. Thus, the system moves from a highly curved H_1 phase at low temperature through a less curved intermediate phase to the zero mean curvature of the L_α phase at high temperature. On the other hand, the continuous increase in the interfacial curvature on decreasing temperature suggests that the defects may be initiated as "dimples" rather than full bilayer-crossing holes [142]. A similar suggestion for curvature inclusion in lamellar phases as precursor to a more ordered phase was proposed by Kékicheff et al. [149,259].

The dehydration of EO groups with increasing temperature also induces an attractive long-range force that leads to the cloud point and phase separation of dilute and concentrated surfactant phases [267]. The second process arises from the interaggregate interaction via the EO chains and mediated by the water. The main interaggregate repulsion forces clearly arise from the steric repulsion between solvated EO groups, which can be described either by a conventional exponential decay function as used for "hydration forces" or by a simple water-binding model [289,290]. The EO chains form "brushes" around each aggregate that tend to repel each other [259] and become compressed as the aggregates move together because of decreasing solvent concentration. This provides a repulsive pressure given by the Alexander–de Gennes expression [291,292] as the separation between the interfaces decreases. Essentially, this is the steric repulsion that is seen between colloidal particles stabilized by adsorbed poly(ethylene oxide) [127], the so-called headgroup overlap (HGO) interaction [292]. Liquid crystalline phases do not start to form as single phases until the mean water thickness between the rods has reached the value of twice the all-trans length of the EO chains. This represents the limits of the HGO interaction [142,262].

The question that still remains to be answered is, why does lengthening the alkyl chain (of ethoxylated nonionic surfactants) favor intermediate phases and disfavor cubic phases? Hyde [267,293] presented a pseudo-phase diagram derived from theoretical considerations of the global packing requirements of surfactant systems that shows the occurrence of phase types as a function of surfactant parameter (Fig. 28). Recently [106], the surfactant parameter was generalized to describe the domain shapes on both sides of any (homogeneous) hyperbolic interface of mean curvature M and Gaussian curvature K and chain volume fraction [see Eq. (3) in Sec. III.C]. For arbitrarily shaped aggregates of molecules with surfactant parameter s and an interaggregate spacing of d, the following scaling relationship holds [294]: $\phi_a \cong d^{-1/s}$, where ϕ_a is the volume fraction of the alkyl chain [295] (designated as ϕ in Fig. 28) allowing the determination of s from the calculated or measured values of ϕ_a and d. The H_1 phase occurs for all volume fractions for a surfactant parameter of

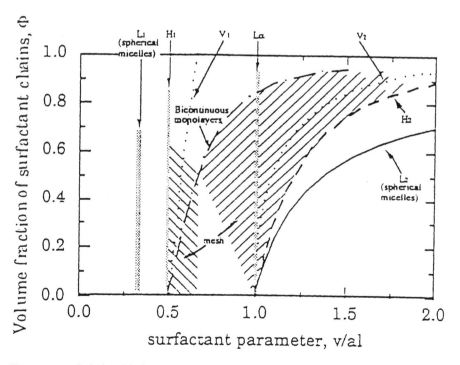

Figure 28 Relationship between volume fraction of surfactant chains, Φ and surfactant parameter (V/al) for a range of aggregation geometries. (From Ref. 267.)

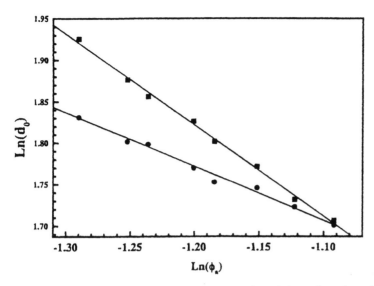

Figure 29   Plot of $\ln(d_0)$ versus $\ln(\phi_a)$, where ϕ_a and d_0 are the volume fraction of the alkyl chain and the first-order reflection, respectively, for the L_α phase at 50°C (■) and the phase at 35°C (●). The system was $C_{16}(EO)_6$–water. The gradient of the linear fit for the L_α phase is -1.01 (corresponding to $s = 1.0$ as expected for the lamellar phase) and at 35°C in the L_α^H phase it is -1.54 (giving $s = 0.65$, which is within the range predicted for a mesh structure). (From Ref. 142.)

$1/2$. Between $s = 1/2$ and $2/3$ are mesh phases that include regular tetragonal and hexagonal mesophases and also the defective lamellar phase [295]. For a surfactant parameter from $2/3$ to 1, continuous mesh phases or cubic phases are found. Finally, there is the L_α phase at a surfactant parameter of 1. Indeed, in the system $C_{16}(EO)_6$–water, the surfactant parameter is calculated [142] to be 0.65 (Fig. 29), as was found in the L_α^H phase of the cesium penta-decafluorooctanoate–water system [265]. In the system $C_{30}(EO)_9$–water, the surfactant parameter is calculated [262] to be 0.592.

The surfactant parameter approach is a simple formulation of the idea that interfacial curvature energy and its minimization determine the aggregate structure. Increasing the chain length of the surfactant moves the headgroup interfacial surface area away from the minimal surface. Ultimately, as the alkyl chain length increases, or as its rigidity increases, there must come a point where it is energetically more favorable for the system to form a structure that is not based on a minimal surface, and a mesh phase is formed [262].

I.   Effect of Oil and Alcohol on Phase Transitions

The effect of added oil on the phase behavior of the $C_{12}(EO)_n$–water system was investigated as a function of EO chain length at 25°C [296]. When decane is added to the system $C_{12}(EO)_3$–water, the lamellar L_α to reverse hexagonal H_{II} transition takes place, while in the system decane–$C_{12}(EO)_7$–water, the normal hexagonal H_I to hydrophilic discrete cubic I_I transition occurs. These phenomena may be interpreted by considering the two ways in which oil may affect the system [296]:

1.   Penetration effect.   Oil molecules penetrate into the surfactant palisade layer and expand the effective cross-sectional area, a_s, without increasing the hydrophobic volume of the aggregate. The increase in a_s due to oil penetration is minimized

via the transition to a more hydrophobic structure that would have negative or less positive (i.e., less convex toward water) spontaneous curvature. Thus, for the hydrophobic $C_{12}(EO)_3$ (whose curvature tends to be negative), the penetration effect will be dominant, leading to the L_α–H_{II} transition.

2. Swelling effect. Oil molecules are solubilized in the aggregate's core and expand its volume. In this case, a_s is almost constant. Since the volume of the surfactant's hydrophobic tail is increased, the surfactant mean curvature tends to be more positive (i.e., more convex toward water) in order to maintain a_s constant. Thus, for the hydrophilic $C_{12}(EO)_7$ (whose curvature tends to be positive), the swelling effect will be dominant, leading to the H_I–I_I transition. However, when an aromatic oil like *m*-xylene is used instead of decane, it penetrates into the surfactant palisade layer and makes the curvature negative, leading to the H_I–L_α transition [296].

For many years, short- and medium-chain alcohols were considered indispensable cosurfactants for the formation of microemulsions until Shinoda showed that microemulsions could be formed from just oil and noionic surfactants. Thus, the role of alcohol is to tune or adjust the phase behavior to bring the one-phase microemulsion region into the experimental window of composition and temperature [66]. Alcohol-containing nonionic microemulsion systems have been thoroughly investigated [62,296–306]. The addition of alcohol to nonionic microemulsions will generally increase the solubility of the surfactant in the aqueous phase [66] and enhance water solubilization [307]. For the system water–pentanol–dodecane–$C_{12}(EO)_8$ we have shown [308] that the addition of pentanol + dodecane (at 1 : 1 weight ratio) to mixtures of water and $C_{12}(EO)_8$ promotes the transition H_I–L_α (Fig. 30). It is known [2] that lamellar phases cannot form readily in $C_i(EO)_j$–water mixtures if the headgroup is too large ($j > 6$). Addition of alcohol + oil increases the effective volume of the hydrophobic alkyl chains, and the surfactant packing ratio [98,99] must approach closer to 1.

A similar volume increase was observed for the secondary alcohol ethoxylate *sec*-$C_{12}(EO)_8$, relative to its corresponding linear counterpart, and this gave rise to an extended region of lamellar phase [309]. In determining the packing ratio, however, headgroup hydration also plays a role: At 70 wt% surfactant, no additives are necessary to encourage lamellar geometry. As the water/surfactant ratio increases, the need for added oil + alcohol also increases in order to form a lamellar phase [308]. The growth of the hydrophobic core (i.e., increase of the nonpolar layer in the L_α phase and the radius of the nonpolar core in the H_I phase) and the interfacial surface area as a function of alcohol + oil content [308] indicates that the opposing trends (swelling vs. penetration) suggested by Kunieda et al. [296] may be used for the interpretation of the phase behavior in our system. It seems, however, that the dominant effect is the penetration of pentanol into the interface, as the observed H_I–L_α transition indicates that the interface curves increasingly toward the water [308]. Such a negative interfacial curvature induced by the presence of alcohol was also suggested for the system cetyltrimethylammonium bromide (CTAB)–pentanol–hexane–water [310].

J. Mesophase → Isotropic Solution Transitions

Many mesophase → isotropic solution transitions have been studied. Here we can only show a few examples.

(a) Dodecane (/Pentanol=1)

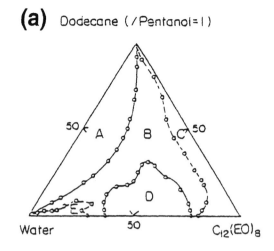

50   A   B   C   50

D

E

Water   50 $C_{12}(EO)_8$

(b) Dodecane /Pentanol = 1

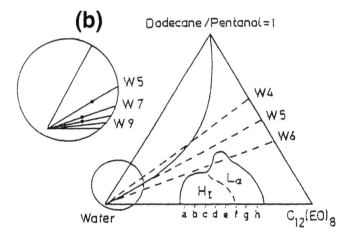

W5

W7

W9

W4

W5

W6

L_α

H_I

Water   a b c d e f g h $C_{12}(EO)_8$

Figure 30 (a) Superposition of phase diagrams at 27°C for the system water–n–dodecane–$C_{12}(EO)_8$ and water–n–dodecane–1–pentanol–$C_{12}(EO)_8$ (dodecane/pentanol weight ratio 1 : 1). The phase diagram for the pentanol-containing system consists of a two-phase region (A) and a liquid crystalline region (D), and the rest represents a one-phase region, each phase separated from its neighbor by a solid line. The phase diagram for the pentanol-free system consists of water-in-oil (C) and oil-in water (E) monophasic regions, with the rest (A + B + D) representing relatively large two-phase and mesophase regions. For the sake of clarity, the boundary between them is not shown. Those boundaries that are drawn are designated by dashed lines. The points (o) drawn on the phase diagram lie equidistant between consecutive experimental measurements on either side of the phase boundary. (b) The phase diagram for the water–dodecane–$C_{12}(EO)_8$–pentanol system, showing the water dilution lines WI (where $I \times 10\%$ is the initial weight fraction of surfactant). The boundary between the hexagonal (H$_I$) and the lamellar (L$_\alpha$) mesophases is represented by a dashed line and may be approximately described as $(N_P + N_D)/N_W = -2.19(N_S/N_W) + 0.175$, where N_P, N_D, N_W, and N_S are the mole numbers of pentanol, dodecane, water, and surfactant, respectively. The inset depicts (as dots) the compositions of six modeled O/W micelles. (From Ref. 308.)

The phase diagram of the $C_{12}(EO)_8$–water system shows the presence of a large micellar phase L_1 followed by a cubic phase I_1 at low temperature (below 15°C) and moderate concentration, between 20 and 40 wt%. This cubic phase is followed by a hexagonal phase H_I, which is present up to 58°C and between 35 and 70 wt% at 25°C [2]. Time-resolved fluorescence quenching (TRFQ) results have shown no discontinuity in micellar aggregation number when crossing the L_1/I_1 phase boundary and no large change in this number at approach to the I_1/H_I phase boundary. The aggregation number in both the L_1 and I_1 phases increases with temperature, the more so at higher surfactant concentration. However, cryo-TEM shows only spheroidal micelles even at temperatures quite close to the critical temperature, where anisotropic micelles should be present in the system, on the basis of the measured aggregation numbers [311]. This apparent inconsistency can be reconciled by assuming a partial mixing of alkyl chains and EO groups in $C_{12}(EO)_8$ micelles in a shell of thickness increasing with temperature. This enables $C_{12}(EO)_8$ and probably other similar ethoxylated nonionic surfactants to form spherical micelles much larger than anticipated on the basis of their alkyl chain length [311].

A mesophase → isotropic solution transition also occurs when appropriate amounts of pentanol + dodecane are added to the binary $C_{12}(EO)_8$-water liquid crystalline mixture [308]. Dodecane alone is unable to disrupt the mesophase, as is evident from the phase diagram of the ternary system water–$C_{12}(EO)_8$–dodecane [308] (Fig. 30). It has been suggested that addition of medium-chain alcohols to mixtures of water and long-chain surfactants makes the interfacial region thinner and more flexible, thereby promoting undulations that disrupt the mesophase order [312,313]. Yet the undulations in our system must be of small amplitude, since the periodicity d changes little across the phase boundary whereas the linewidth of the scattering peak (which is inversely related to the range of interparticle correlations or cluster size) [283] increases significantly. Thus, it seems reasonable to suggest that the pentanol promotes the liquid crystal-to-isotropic transition via mild local disruption of aggregate packing, which in turn results in a reduction in cluster size [307].

In the system $C_{16}(EO)_6$–water a nematic phase was observed for the first time in a nonionic system [243]. It was identified by optical microscopy as a narrow low viscosity birefringent phase. The fact that this phase is adjacent to a hexagonal phase is strong evidence that it is a nematic phase of rod-shaped microemulsions [182,314], the nematic phase acting as a precursor to the H_I phase. In the isotropic phase, near the isotropic-to-nematic transition the micelles are expected to have an elongated cylindrical shape, becoming orientationally ordered in the nematic phase. The H_I phase marks the onset of long-range positional order of the cylinders as the rods grow to be effectively infinite in length due to increased intermicellar interactions [243].

The suggested mechanism for the $H_I \rightarrow Im3m$ transition in the system $C_{12}(EO)_{12}$–water [164] is based on the desire for increased curvature due to the hydration of the hydrophilic headgroups as temperature is reduced. To accommodate this desire for greater curvature, the cylinders of the hexagonal phase start to undulate as if they were being "pinched" at regular intervals. This "pinching" continues until the narrowest points along the rods are closed off and discrete micelles form (Fig. 31). This suggested mechanism could also apply to the H_I–L_1 transition. Again, one would expect an increase in curvature to drive the regular pinching of the hexagonal phase and the formation of micelles. But, instead of arranging themselves in a stable body-centered lattice, as in the case of the $H_I \rightarrow Im3m$ transition, these micelles would then be randomly dispersed in solution [164].

The hexagonal-to-fluid isotropic phase transition was analyzed by Sallen et al. [315]. The exact nature of the defects observed on approaching the transition remains to be elucidated, since the structure of the isotropic phase above the hexagonal phase is itself

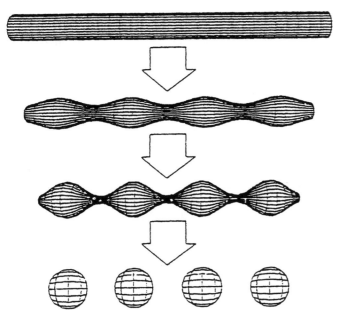

Figure 31 Suggested mechanism for the $H_I \rightarrow Im3m$ transition. Undulations form in the hexagonal phase rods, which are being "pinched" at regular intervals, and this continues until the narrowest points along the rods are pinched off and discrete micelles form. (From Ref. 164.)

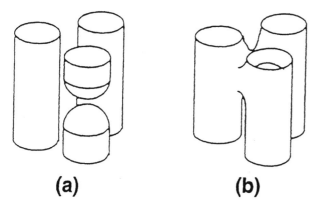

(a)   **(b)**

Figure 32 Schematic representation of a defect in the hexagonal phase. (a) A fragment of column ended by two spherical caps; (b) bridge connecting neighboring columns. (From Ref. 315.)

not completely understood [315]. The decrease in the hydration and hence the curvature of the aggregates with an increase in temperature would suggest that the fluid isotropic phase is made of disordered bilayers or branched interconnected cylinders rather than elongated micelles. Either the columns may spontaneously break up into shorter segments or bridges connecting close cylinders may develop (Fig. 32). In the case of fragmentation, two ends of a column are formed at each division. To avoid contact between the water and the aliphatic chains, the interface is spherical at each extremity, which increases the local curvature of the aggregate surface. In the second case, the curvature would decrease locally [315].

K. Sponge Phase-to-Isotropic Solution Transition

The sponge phase, like the lamellar phase, is usually found to be stable throughout the moderate concentration range [219]. At high dilution of the system, however, the phase behavior depends on the constituents of the studied systems. In some cases, the L_3 phase simply phase-separates at high dilution and expels a very dilute dispersion of micelles [316,317]. For other systems, the L_3 domain is continuously connected in the phase diagram to the domain of the micellar phase at high dilutions [101,318,319]. So, starting from the infinite multiconnected membrane in the L_3 domain, it is possible, with these systems, to move in the phase diagram toward the domain of dispersed disconnected micelles with no phase separation [219]. A boundary is crossed along such paths; light scattered in this boundary shows quite a sharp maximum, suggesting a critical behavior [101,319–322]. Two scenarios have been considered in the literature that can describe these experimental facts consistently. The first is based on considerations of the symmetry of the L_3 structure [101,318]. At moderate dilutions of the system, the membrane in the L_3 domain separates the phase into two statistically equivalent subvolumes; this simply reflects the intrinsic local symmetry of the membrane with respect to its midsurface. At high dilutions of the system, however, the global symmetry could well break spontaneously. Below the corresponding critical concentration of surfactant, the membrane disconnects progressively and finally ends up in the form of a random dispersion of vesicles and/or micelles [219].

The second scenario was motivated by consideration of the statistical distribution of defects such as holes that might form spontaneously in the bilayer [323]. It was suggested that "a sponge with free edges" will be formed when the free energy to create defects is low. Indeed, light-scattering and conductivity results suggest a spontaneous tearing of the membrane in the very dilute regime [319]. Conductivity and neutron-scattering results seem to be more compatible with the tearing scenario [194].

VI. MICROEMULSIONS

A. Spherical and Nonspherical Structures

Microemultions represent an important subject of self-organizing amphiphilic systems. They are thermodynamically stable, macroscopically homogeneous mixtures of at least three components: polar and nonpolar liquid phases (usually water and oil) and a surfactant that, on a microscopic level, forms a film separating the two incompatible liquids into two subphases. The microemusions form well-organized local structures like simple surfactant–water (or oil) binary systems. The distinctive feature of microemulsions, compared to micellar systems, is the presence of significant amounts of oil in the system. Also, the constraint of a maximum thickness for the hydrophobic medium in micellar systems is removed, since the hydrophobic region is now swollen with oil [108].

The diversity of microemulsion structures depends on the chemical composition, temperature, and concentrations of the constituents [324]. Different surfactants stabilize different microstructures. For example, the cationic surfactant DDAB (didodecyldimethylammonium bromide) is known to form both discrete spherical aggregates and interconnected bicontinuous structures [325], as does the nonionic surfactant $C_{12}(EO)_4$ [326,327], while a phospholipid soybean lecithin surfactant leads to the formation of W/O microemusions containing rod-shaped aggregates [328–330], and large entangled polymer-like reversed micellar aggregates are present in the L_2 phase, giving rise to high

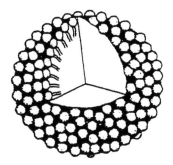

Figure 33 Schematic illustration of an O/W microemulsion. (From Ref. 117.)

viscosities [331]. Thus, on addition of water to a solution of the surfactant in isooctane, the viscosity increases dramatically, and at $W = 3$ (where W is the molar ratio of water to surfactant), it is a factor of 10^4 greater than that of the "dry" system [331].

Previously, it was thought that at a low volume fraction ($\phi \sim 0.05$), W/O microemulsions were systems containing discrete, spherical, surfactant-stabilized water droplets (Fig. 33). In many instances, the interdroplet interactions are purely repulsive, hard-sphere-like, but in certain cases, they exhibit an attractive component [332]. However, nonspherical structures may also be found. Reports of polymer-like and anisotropic W/O microemulsions aggregates have recently appeared with soybean lecithin [328–330], copper(II) dialkylcarboxylates [333], DDAB [334,335], metal bis(2-ethylhexyl) sulfosuccinate $[M^{n+}(AOT^-)_n]$ [336–339], and quaternary ammonium AOT [340] surfactants. Clearly, simple spherical structures are not a general feature of surfactant-based systems in nonaqueous media [341]. The changes in aggregate shape (such as from spheres to rods) have important consequences for the macroscopic physical properties of the system (e.g., viscosity). Small-angle neutron scattering (SANS), viscosity, and electrical conductivity measurements indicate that rigid monodisperse rod-shaped aggregates are formed at W values less than 5, with $[M^{n+}(AOT^-)_n]$ surfactants in alkane media [336,338,342] when the hydrated radius of the counterion is larger than 3 Å. Small radii lead to spherical aggregates [341]. This is because the larger ions are less effective at screening the repulsive negative ion–negative ion ($SO_3^- \to SO_3^-$) sulfonic interactions than the smaller metal ions. When water is added to the system (increasing W at constant surfactant concentration), the rod aggregates undergo a sharp change, and spherical droplets are favored. This contrasts with the case for Na(AOT), where the spherical curvature of the surfactant film persists over a wide range of composition, temperature, and pressure [343]. At the same time, a direct proportionality between the droplet radius and the composition parameter W was found [344,345].

The formation of extended bilayer structures (lamellar micelles) is believed to be energetically unfavorable due to edge effects [98]. Indeed, evidence for such structures in either the water (L_1) or nonpolar (L_2) media is scarce [346]. On the basis of SANS studies of the structure of hydrated micelles formed from the $C_n(EO)_m$ surfactants in oil media [347,348], it was concluded that the extent of solubilization of water (W_{max}) in the L_2 phase did not exceed the maximum hydration requirement of the hydrophilic EO groups. The scattering spectral analysis is also consistent with the formation of small bilayer structures that become thicker and more extended as W increases. At low W, the thickness of the bilayer micelles ("hanks") was found to be less than twice the fully extended conformation of the surfactant, suggesting that the EO chains of the surfactant were interdigitated (or coiled)

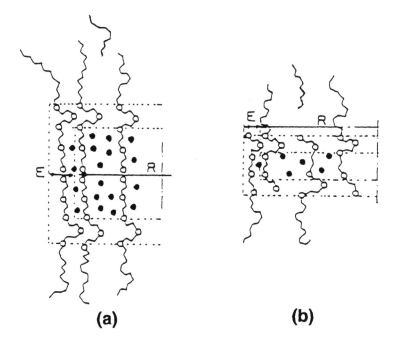

(a)   **(b)**

Figure 34 Schematic representation of the models for bilayered micelles with separate (a) and interdigitated (b) EO chains. (●) Water molecule. E = thickness of peripheral layer and R = radius of micellar core. (From Ref. 348.)

in the interior of the micelle (Fig. 34). As W increases, the bilayer thickness increases until at W_{max} the EO chains are fully hydrated and extended. It was concluded that the limit of stability was reached at the point when the surfactant became fully hydrated and it was not possible to form a discrete water "droplet" within the micelles. Similar conclusions can be drawn for the phase behavior of the system water–cyclohexane–L77-OH (ethoxylated polymethylsiloxane) [349]; i.e., hydrated reversed micelles are present rather than W/O microemulsion droplets. It is believed that liquid nonionic surfactants when weakly hydrated contain small, partially hydrated reversed micelles. For the L77-OH system, which forms a lamellar phase at higher W, such aggregates could conceivably be small bilayer "segments" of limited lateral extension (Fig. 35). The lamellar aggregates in the L_2 phase may be regular disks (the "edge effect" alone is presumably not sufficient to rule out the formation of disks), but a more realistic picture is probably one of irregularly shaped lamellar structures [349].

B. Theoretical Considerations

A theoretical basis for different shapes of microemulsions (even for small W/O or O/W volume fractions) has been established on the basis of the relationship between shape and interfacial curvature [350,351]. It is reasonable to expect that the relevant properties of the surfactant film are represented by a bending elasticity with a spontaneous curvature, c_o (as was demonstrated for binary systems). If the elastic modulii k, $\bar{k} \gg k_B T$, the fluctuations in curvature of the film are very small, and the entropy associated with them can be neglected. The actual morphology is the result of the competition between the tendency to minimize the bending free energy (which prefers spheres of optimal radius of curvature, $r_o = 1/c_o$) and the necessity to use up all of the water, oil, and surfactant

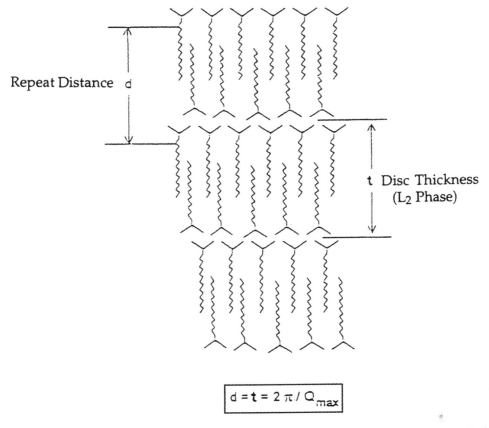

Repeat Distance d

t Disc Thickness
(L₂ Phase)

$$d = t = 2\pi / Q_{max}$$

Figure 35 Schematic representation of the lamellar phase showing the SANS contrast profile for the repeat distance d and disk thickness t in the L_2 phase. (From Ref. 349.)

molecules. This leads to a fixed total surface area of all of the interfaces in the system and a fixed total volume of all of the globules due to the incompressibility of the system components [351,352]. For a given positive value of c_o and for decreasing amounts of oil, one obtains the same sequence of shape changes as in classical micellar systems [350–352]. The optimum configuration corresponds to a monodisperse distribution of spheres having the optimal radius of curvature, r_o. If the volume fractions of the internal phase are increased, the globules fail to grow any further and reject the excess water or oil (whichever is internal to the globule)—the so-called emulsification failure [351]. When the content of the internal phase constituent becomes smaller at constant surfactant concentration, the area/volume ratio decreases as does the radius of the sphere. At some stage the film becomes too bent in the spherical configuration, and infinite cylinders best accommodate the volume and surface constraints but still maintain an average radius of curvature close to r_o. At small values of c_o, the same mechanism will favor the formation of flat lamellar structures as there is no energetic tendency to bend toward either the oil or the water [352].

DDAB is an example of a surfactant that forms rigid films. DDAB-based microemulsions are composed of initial globules opening progressively to form a network of branched cylinders [325,353,354], so that the curvature of the film remains constant over the whole range of composition variations [108].

The striking analogy between the shape sequences predicted in highly rigid microemulsions and those observed in micellar systems is explained in the following way [108]. Due to its high rigidity, the microemulsion's film tends to have everywhere a homogeneous constant curvature. The global ratio between the total area of the film and the total volume of the hydrophobic medium as fixed by the composition of the sample is therefore respected on a local scale. This constraint is thus parallel to the constraint imposed on the bilayer's or cylinder's thickness in micellar solutions [108].

Very different trends are expected in the low rigidity limit $k, \bar{k} \ll k_B T$. Large local deviations from the optimum curvature involve a moderate energetic price, and the entropy associated with the bending fluctuations of the film becomes the dominant contribution to the overall free energy of the mixture. Correspondingly, the appropriate approach is to maximize this entropy subject to the constraints of area and volume conservation. The main effect of the fluctuations is to smear out the well-defined morphologies characteristic of the high rigidity limit $k, \bar{k} \gg k_B T$ [108].

If the film of the microemulsions is flexible, one imagines easily that the radius of the droplets is not strictly fixed but is distributed around an average value (polydispersity). Moreover, the microemulsion droplets can be deformed by an external perturbation or by thermal fluctuations. The droplets may undergo attractive interactions that lead to aggregation between the droplets [324]. Two mechanisms have been invoked to explain the origin of the attractive interactions (which are too large to be due only to the pure van der Waals interactions between the cores of the droplets separated by the interfacial film): interpenetration of the tails of surfactant molecules residing on different droplets [355] or fusion of droplets, which lowers the curvature energy [356].

C. Bicontinuous Microemulsions

Bicontinuous structures are perhaps more frequently observed than discrete micellar aggregates. They are encountered in microemulsions, in mesophases, and even in relatively dilute surfactant solutions [5]. Furthermore, structures that cannot be described in terms of particles are also ubiquitous in nonamphiphilic systems [106]: zeolites [357], copolymer molecular melts [358], volcanic minerals, surfactant-templated synthetic mesoporous oxides [359–361], and composite media [362]. The structure of concentrated amphiphilic systems depends mainly on the properties of the interfacial film. Not only the flexibility but also the asymmetry of the amphiphilic films (with respect to the affinities for oil and water) play an important role. If a film is very asymmetrical with a strongly preferred (or spontaneous) curvature c_o, it will preferentially form droplets with radii close to $1/c_o$. If a film is flexible and symmetrical, or weakly asymmetrical, its local curvature may fluctuate strongly. In concentrated systems, this leads to fluid random structures, where the average curvature of the film is rather small. As first proposed by Scriven [363,364], there are no longer isolated droplets but rather a bicontinuous arrangement of oil and water channels. Most microemulsions that are bicontinuous contain comparable amounts of water and oil, and most interesting are those that are stabilized by only small amounts of surfactants [36]. However, as was mentioned before, systems also exist where bicontinuity appears at low water content [365–367]. The first experimental demonstration that microemulsions may indeed be bicontinuous came from NMR self-diffusion studies [368]. The relatively high diffusion coefficients of both water and oil [369–372] suggest that some continuity must exist between the water and oil domains [91]. The absence of birefringence or Bragg peaks indicates that the structure cannot be macroscopically ordered [91]. The demonstration of a microemulsion with a mean curvature on the average of zero is a strong indication of bicontinuity. This may be done by using the SANS contrast-variation technique [373]. If a surface is of zero or low mean curvature, due to the cancelation of oppositely signed principal curvatures, then this surface is predominantly saddle-like, which in mathematical terms means negative Gaussian curvature [362], and has a highly connected topology [67]. The first suggested structures were based on analogies with liquid crystalline structures [362]. Saito and Shinoda [374] suggested a structure in terms of thermally disrupted lamellar liquid crystal structure. Scriven [363,364] thought that it must be some kind of molten cubic phase. A further possible bicontinuous microstructure, considered in relation to water-poor systems, is similar to a hexagonal liquid crystal structure, where one of the liquids lies inside interconnected tubules or "conduits" [353] (this liquid as well as the other being continuous [67]).

The model structures that were proposed for the description of bicontinuous systems satisfy the mathematical constraint that the dividing surface between hydrophilic and hydrophobic domains is a surface of constant mean curvature [362]. The microstructure tends toward a state of homogeneous spontaneous mean curvature over the dividing surface (typically drawn at, or near, the surfactant headgroup) as a result of the balance of forces (steric, electrostatic, dispersion, etc.) on the two sides of the film [131]. As the microemulsion parameters, such as salinity in ionic surfactants or temperature in nonionic surfactant systems, are varied, this spontaneous mean curvature also varies, and there is an accompanying change in the oil/water ratio in the microemulsion phase. However, bicontinuity results from a particular spontaneous curvature of the surfactant films rather than a certain solvent volume fraction, which is a secondary factor in determining microstructure [5]. In microemulsions the structure is lacking any long-range order and the mean curvature over the dividing surface is not homogeneous at any given instant. However, we can visualize bicontinuous microemulsion structures using well-defined models (continuous paths between interconnected spheres, distorted lamellae, tubule structures with one of the solvents confined to branched tubes, and a minimal surface type of structure similar to that established for bicontinuous cubic phases) [5] but picture them as being thermally disrupted or "melted" [362].

The frequent description of bicontinuous structures in terms of a percolative process is misleading. Bicontinuity describes a situation with a dynamic equilibrium structure and not a temporary opening of extended pathways between droplets [5].

The large variety of structures ranging from interconnected spheres via a Scriven-type bicontinuity to interconnected tubules is qualitatively understood by the similarity to liquid crystal structures. However, there has been a lack of quantitative structural geometrical relations allowing a clear distinction between the intuitive concepts relating to bicontinuous structures [91].

Talmon and Prager were the first to provide a thermodynamic model for bicontinuous microemulsions [375]. In this model a random lattice is formed by a subdivision of space into Voronoi polyhedra (Voronoi tessellation), and its cells are randomly filled with water and oil with a probability proportional to the global volume fractions of water and oil (Fig. 36). The surfactant film volume fraction is considered negligible, and surfactant film is distributed only along the faces common to two adjacent cells filled with different phases, i.e., the surfactant is constrained to lie at the oil/water interface with constant area per molecule [375–377]. Although this model correctly predicts the average size of the resulting oil and water domains and the rich phase behavior for amphiphilic systems (including the existence of Winsor III equilibrium), it cannot interpret the nonmonotonic behavior of conductivity of microemulsions or, furthermore, the observed scattering peak [91]. Also, the way in which the bending energy was evaluated remained somewhat artificial [378]. The

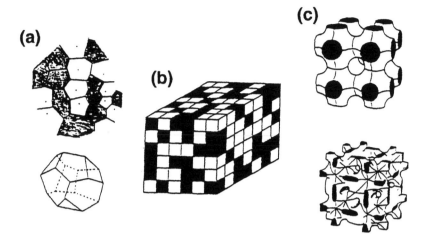

Figure 36 Schematic representations of bicontinuous microemulsions models [91] (a) Talmon and Prager [375]: random filling of Voronoi polyhedra with water or oil (top); unit water or oil cell (bottom). (b) De Gennes and Taupin [130]: random filling of cubes on a cubic lattice; the repetition distance equals the correlation length. (c) Scriven [364]: two examples of bicontinuous mesophase-like structures with minimal interfacial area. (From Ref. 91.)

cubic random cell (CRC) model [130,379] is an improved variation of the Talmon–Prager model [375,376]. The random polyhedra of the Talmon–Prager model are replaced by cubic cells filled at random with water and oil, according to the macroscopic volume fraction fixed by composition (Fig. 36a). The lattice size was chosen as the persistence length ξ_k, which is a measure of the order in membrane systems: Large values favor liquid crystal phases, and smaller values favor disordered phases such as bicontinuous microemulsions. ξ_k can be calculated [380] using the equation

$$\xi_k = \frac{a\phi_o\phi_w}{C_s\Sigma} \tag{4}$$

where ϕ_o and ϕ_w are the oil and water volume fractions (in the limit of negligible surfactant volume fraction), C_s is the number of surfactant molecules per unit volume, Σ is the area per surfactant molecule in the monolayer, and a is a numerical factor. The bicontinuous models differ only in the absolute numerical value of a [381]. For example, $a = 4$ [382], $a = 5.84$ [375], and $a = 6$ [130]. Recently, the value of a was experimentally determined to be $a = 7.16$, which is close to, but significantly different from, those used in theoretical models [381]. ξ increases with both ϕ_o and ϕ_w and reaches a maximum value, above which the oil and water are rejected as excess phases (emulsification failure). Experiments show that bicontinuous microemulsions are not completely random. Scattering peaks from bicontinuous microemulsions varied with ϕ_o (or ϕ_w) according to the random surface description [(Eq. (4)] [327,383], although random models do not necessarily predict a peak. Thus, there is a rather well defined characteristic distance between the oil and water microstructural elements [384]. This distance corresponds to the mean size in the space-filling models, as given by Eq. (4). It may be determined by using the Teubner–Strey method of fitting [71]. The CRC model also removed the interfacial area per molecule con-

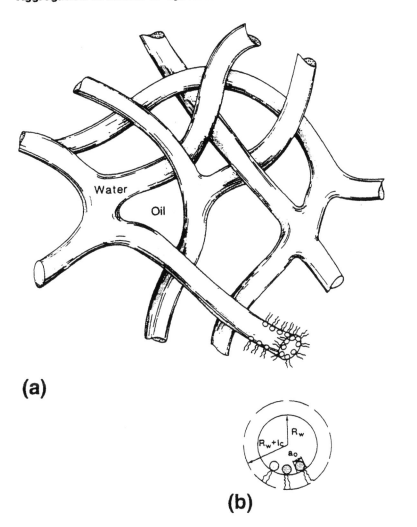

(a)

(b)

Figure 37 The interconnected conduits that comprise the bicontinuous structures in DDAB microemulsions according to the disordered open connected (DOC) model. Upon addition of water, the conduits disconnect and transform to water-in-oil droplets. (From Ref. 94.)

straint and showed that phase equilibria could occur at near-vanishing values of the interfacial tension. However, the Winsor III equilibrium could not be found with this model [378].

More advances were made, for example, by Widom [385,386], the Andelman–Cates–Roux–Safran (ACRS) model [387,388], the Ginzburg–Landau approach [65,389], and the theory of random interfaces [390,391]. Scattering spectra may be "interpreted" using parametric models such as those of Teubner and Strey [71] and Vonk et al. [392]. However, there is no guarantee that a given set of parameters (such as the spatial periodicity d, the correlation distance ξ, and the internal surface Σ of the Teubner–Strey model [71]) leads to a geometrically possible microstructure [393]. It is claimed that this difficulty is removed by the disordered open connected (DOC) model [91,353,354,394,395]. (A microstructure suggested according to this model is shown in Fig. 37.) The DOC model uses a Voronoi cell tessellation of space, leading to a macroscopically disordered structure. In the case of stiff interfaces, where bending energy dampens local curvature fluctuations, three constraints that govern microstructures have to be fulfilled simultaneously: the imposed volume fraction, the specific oil/water interface, and the minimum elastic energy with negligible entropic contributions. With this model, simple analytical expressions can be used to predict the characteristic size D^* and connectivity Z of the microstructure at any composition when the surfactant parameter p can be derived from the phase diagram shape. Once D^* and Z are known, the conductivity, scattering peak position, and phase boundary can be predicted from the spontaneous curvature alone [395].

For nonionic surfactant systems, the recent flexible surface model [378,396,397] has proved to be a useful tool in analyzing theoretically the microstructures in solution and predicting phase equilibria. The main aggregates described by means of the flexible surface model are ordinary and reversed micelles (O/W and W/O microemulsions, respectively), bicontinuous microemulsions, and lamellar, cubic, and sponge phases. The model estimates the mean curvature of the surfactant monolayer in different macroscopic phases with different aggregate geometries [12]. The curvature free energy density (G/V) is given by

$$G/V = a(\phi_s/l)^3 + b(\phi_s/l)^5 \tag{5}$$

where ϕ_s is the surfactant volume fraction, l is the film thickness, and a and b are coefficients whose values are $a = 2kH_0/l^2$ and $b = k/2l^3$ (where k is the bending modulus and H_0 is the spontaneous mean curvature). The G/V model describes the salient features of the phase behavior of surfactant systems in a simple and direct manner. For example, the observed phase behavior of balanced (i.e., having zero spontaneous curvature [378]) microemulsions can be understood if the sign of a is negative. This occurs when the bending modulus k is sufficiently small to allow for large fluctuations and the saddle-splaying constant $\bar{k}$ is not too negative, reducing the Gaussian curvature energy penalty to form the complex monolayer topology needed for bicontinuous microemulsions. This penalty must be overcome by the entropy contribution to the free energy. However, the microemulsion is entropically stabilized over only a very small range of compositions [397]. The reason for the existence of Winsor III equilibria is an anomalous entropy effect consisting of a decrease in entropy per surfactant molecule with decreasing concentration in contrast to normal ideal mixing behavior. This is because the structural length scale is inversely proportional to ϕ_s, making the system more effectively mixed, the higher the concentration. Thus, the microemulsion becomes destabilized upon dilution (at concentrations where the ϕ_s^3 term dominates the free energy) because the system will spontaneously expel excess liquid and contract to a higher concentration. The expelled liquid clearly separates into an oil phase and a water phase [396]. At the optimal concentration range, a positive value of b [Eq. (5)] also helps to stabilize the microemulsion due to contributions from surface forces and from higher order elastic terms [396]. At higher concentration, the microemulsion transforms into the more stable lamellar phase. It should be noted that this model is still controversial [398–400].

D. Local Lamellar Structure

By increasing the volume fraction of the dispersed phase, one often encounters aggregation phenomena associated with reorganizations of the interfacial film. The structures occurring at intermediate concentrations of the dispersed phase are the least well known and least studied [324].

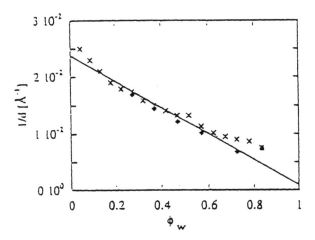

Figure 38 Dependence of the reciprocal of the microemulsion periodicity d on the water volume fraction ϕ_w for samples along the water dilution line W5 of Fig. 30b as measured by SAXS ($\times$) and SANS ($\blacklozenge$). (From Ref. 304.)

To account for structural results obtained by small angle x-ray scattering (SAXS) and for the transport properties of a concentrated microemulsion, a model based on a local lamellar structure was suggested [401]. By analogy with a liquid–gas or liquid–solid transition, this structure may be described as aggregates of stuck or fused droplets [324] that eventually lead to lamellae with very short correlation length [91]. In a similar model, the position of the scattering peak in Winsor III microemulsions was explained to a rough approximation as a molten cubic phase, thus indicating a local complexity in terms of assembly of "distorted minimal curvature" surfaces [91]. Vonk et al. [392] derived a general expression similar to the Teubner–Strey expression [71]. They assume, however, that any microemulsion structure is locally lamellar. The structure of the isotropic phase occurring in the ternary system AOT–water–n-heptane was interpreted as a local lamellar structure on the basis of SAXS data [402]. These data were recently reexamined, and according to the DOC model the isotropic phase should be considered an L_3 phase [395].

We have investigated some quaternary nonionic systems and found the microstructure within intermediate water concentrations to be locally lamellar [304,308,403]. This conclusion is based on SAXS data [one-dimensional swelling of $1/d$ (where d is the periodicity) with the water volume fraction ϕ_w leading to $1/d = 0$ at $\phi_w = 1$] (see Fig. 38) and on cryo-TEM micrographs, which were observed for the first time [for the system $C_{12}(EO)_8$–water–pentanol–dodecane] [304]. Compare the striated microstructure (typical of disordered lamellae structure) in Fig. 39 with the spotted structure (typical of globular micelles) in Fig. 40. The rather low values of the self-diffusion coefficients of both water and oil have led us to describe the structure as highly obstructed ordered bicontinuous [308]. This picture also fits quite well with conductivity measurements and the high value of the amphiphilicity factor [308]. It should be emphasized that as only one scattering peak was observed at any measured water concentration, and moreover the solution is nonviscous and nonbirefringent, we cannot envisage the local lamellar structure as distorted long-range ordered lamellae. Also, the possibility of interpreting our scattering data as relating to an L_3 (sponge) phase (as was recently suggested for the AOT–water–n-heptane system) [395] is ruled out as the system $C_{12}(EO)_8$–water–pentanol–dodecane does not show

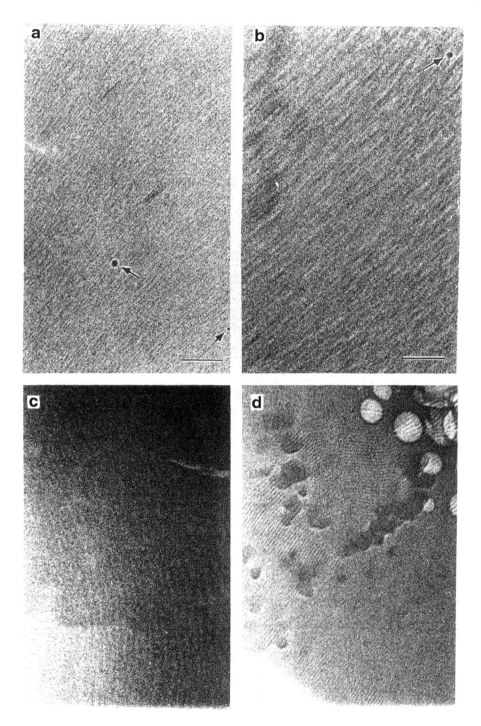

Figure 39 Microstructure in the intermediate region microemulsions, as observed by cryo-TEM. (a), (b) 40 wt% water; disordered lamellae were aligned during sample preparation; note unsmeared particles (arrows). (c), (d) 50 wt% water; ordered lamellae are seen. "Bubbles" in upper right corner of (d) are the result of electron beam damage. Bars = 100 nm. (From Ref. 304.)

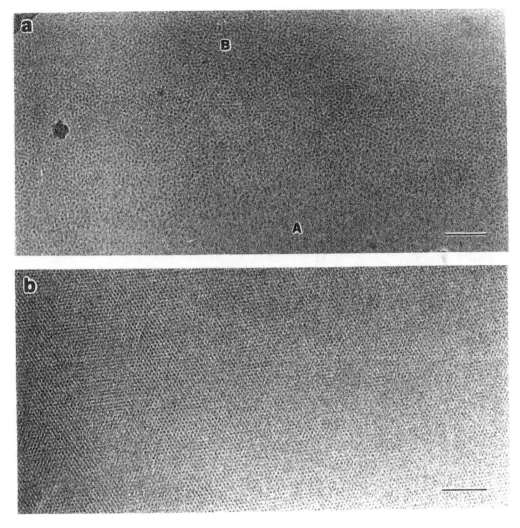

Figure 40   Cryo-TEM images of vitrified microemulsion samples. (a) 85 wt% water; note individual micelles in the thinner areas (A) and agglomerated micelles in the thicker areas (B). (b) 60 wt% water; the order is the result of micelle packing during specimen preparation. The image is a projection of several layers of ordered micelles. Bar = 100 nm. (From Ref. 304.)

streaming birefringence. A tentative model of local lamellar structure is shown in Fig. 41. In addition, a local lamellar structure was also shown to characterize aggregates of oligomeric surfactants with long hydrophilic headgroups (ethoxylated polydimethylsiloxanes) and strong interaction with the oily continuous phase [404]. However, it should be noted that in the oligomeric case no internal water core is formed, leading to different thermal behavior [404,405].

We suggest that the local lamellar structure is quite ubiquitous, but it will be observed only under specific experimental conditions and within restricted areas of the relevant phase diagrams.

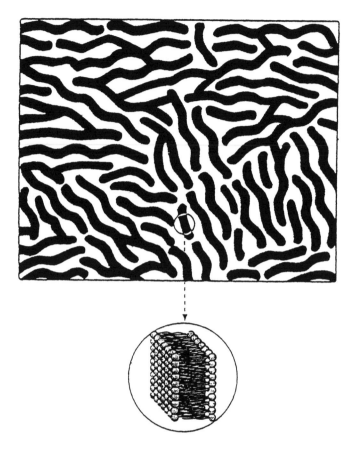

Figure 41 Cartoon of local lamellar structure. The black and white patches represent cross sections of the surfactant + alcohol + oil layers and the water layers, respectively. The relative thicknesses of the water- and non-water-filled regions were chosen approximately according to the values of d_0 and d for the 55 wt% water sample on the water dilution line W5 of Fig. 30b. The lamellae extend above and below the plane of the page. For simplicity, they are assumed to be perpendicular to the page, resulting in uniform cross-sectional widths. The inset shows a more detailed cartoon of the structure of the lamellar cross section as seen from above. (From Ref. 308.)

VII. CONCLUDING REMARKS

This review was an attempt to summarize the structural organization of surfactants in both concentrated and dilute solutions in areas where phase diagrams show thermodynamically stable isotropic solutions. The complexity of structures in concentrated systems is enormous, and transition sequences such as lamellar to normal hexagonal to cubic to reverse hexagonal can take place as a function of temperature, amphiphile concentration, and additives.

Theoretical considerations as well as experimental results show that such transitions are feasible and frequently occur. In dilute regimes of phase diagrams, structures that seem to be quite simple, such as normal micelles and reverse micelles, prove to be also of certain complexity, and transitions from swollen micelles into various other distorted and deformed structures can be found. For example, L_3 and L_4 phases as well as local lamellar structures have been discovered.

It seems that the structural aspects of aggregated surfactants in solution will be further elucidated in the future in view of the extensive work carried out in this area both by experimentalists and theoreticians.

Today's advanced analytical tools help to visualize some of these structures. Computer simulations facilitate modeling and visualizing three-dimensional structures and transformations. With the support of these facilities, the study of aggregation behavior of surfactants should lead to more realistic structures.

Note added in proof, recently, two review papers covering intermediate phases of surfactant-water mixture [406] and fluctuating Euler characteristics in lamellar and microemulsion phases [407] were published.

REFERENCES

1. Z. A. Schelly, J. Mol. Liq. *72*:3 (1997).
2. D. J. Mitchell, G. J. T. Tiddy, L. Waring, T. Bostock, and M. P. McDonald, J. Chem. Soc., Faraday Trans. I *79*:975 (1983).
3. A. Khan, Curr. Opinion Colloid Interface Sci. *1*:614 (1996).
4. R. G. Laughlin, in *Advances in Liquid Crystals*, Vol. 3 (G. H. Brown, ed.), Academic, New York, 1978, pp. 41–98.
5. B. Lindman and U. Olsson, Ber. Bunsenges. Phys. Chem. *100*:344 (1996).
6. B. W. Ninham, in *Structure and Reactivity in Reverse Micelles* (Stud. Phys. Theoret. Chem. Vol. 65) (M. P. Pileni, ed.), Elsevier, Amsterdam, 1989, pp. 3–12.
7. T. P. Hoar and J. H. Schulman, Nature *102*:152 (1943).
8. J. H. Schulman and D. P. Riley, J. Colloid Sci. *3*:383 (1948).
9. J. H. Schulman and J. A. Friend, J. Colloid Sci. *4*:497 (1949).
10. J. H. Schulman, W. Stoeckenius, and L. M. Prince, J. Phys. Chem. *63*:167 (1959).
11. P. A. Winsor, Chem. Rev. *68*:1 (1968).
12. J. Sjöblom, R. Lindberg, and S. E. Friberg, Adv. Colloid Interface Sci. *65*:125 (1996).
13. A. Kabalnov, B. Lindman, U. Olsson, L. Piculell, K. Thuresson, and H. Wennerström, Colloid Polym. Sci. *274*:297 (1996).
14. S. S. Marsden and J. W. McBain, J. Phys. Chem. *52*:110 (1948).
15. P. Ekwall, I. Danielsson, and L. Mandell, Kolloid Z. *169*:113 (1960).
16. P. Ekwall, L. Mandell, and L. Fontell, Mol. Cryst. Liq. Cryst. *8*:157 (1969).
17. P. Ekwall, in *Advances in Liquid Crystals*, Vol. 1 (G. H. Brown, ed.), Academic, London, 1975, pp. 1–142.
18. G. Gillberg, H. Lehtinen, and S. Friberg, J. Colloid Interface Sci. *30*:40 (1970).
19. S. E. Ahmad, K. Shinoda, and S. Friberg, J. Colloid Interface Sci. *47*:32 (1974).
20. S. Friberg and I. Buraczewska, Prog. Colloid Polym. Sci. *63*:1 (1978).
21. K. Shinoda and T. Ogawa, J. Colloid Interface Sci. *24*:56 (1967).
22. K. Shinoda, J. Colloid Interface Sci. *24*:4 (1967).
23. K. Shinoda and H. Arai, J. Phys. Chem. *68*:3485 (1964).
24. K. Shinoda and H. Arai, J. Colloid Interface Sci. *25*:429 (1967).
25. K. Shinoda and H. Saito, J. Colloid Interface Sci. *26*:70 (1968).
26. K. Shinoda and H. Takeda, J. Colloid Interface Sci. *32*:642 (1970).
27. K. Shinoda and S. Friberg, Adv. Colloid Interface Sci. *4*:281 (1975).
28. H. Kunieda and K. Shinoda, J. Dispersion Sci. Technol. *3*:233 (1982).
29. K. Shinoda and S. Friberg, *Emulsions and Solubilization*, Wiley, New York, 1986.
30. P. A. Winsor, Trans. Faraday Soc. *44*:376 (1948).
31. B. M. Knickerbocker, C. V. Peshek, L. E. Scriven, and H. T. Davis, J. Phys. Chem. *83*:1984 (1979).
32. B. M. Knickerbocker, C. V. Peshek, H. T. Davis, and L. E. Scriven, J. Phys. Chem. *86*:393 (1982).

33. K. Holmberg, in *Micelles, Microemulsions and Monolayers* (D. O. Shah, ed.), Marcel Dekker, New York, 1998, pp. 161–192.
34. J. L. Salager, in *Surfactants in Solution* (A. K. Chattpadhyay and K. L. Mittal, eds.), Marcel Dekker, New York, 1996, pp. 261–295.
35. R. Zana, Heterogene. Chem. Rev. *1*:145 (1994).
36. K. Shinoda and B. Lindman, Langmuir *3*:135 (1987).
37. M. Bourrel and R. S. Schechter, *Microemulsions and Related Systems* (Surfact. Sci. Ser. Vol. 30), Marcel Dekker, New York, 1988.
38. M. Bourrel, J. L. Salager, R. S. Schechter, and W. H. Wade, J. Colloid Interface Sci. *75*:451 (1980).
39. M. Abe, D. Schechter, R. S. Schechter, W. Wade, U. Veerasooriya, and S. Yiv, J. Colloid Interface Sci. *114*:342 (1986).
40. P. K. Kilpatrick, C. A. Gorman, H. T. Davis, L. E. Scriven, and W. G. Miller, J. Phys. Chem. *90*:5292 (1986).
41. P. K. Kilpatrick, H. T. Davis, L. E. Scriven, and W. G. Miller, J. Colloid Interface Sci. *118*:270 (1987).
42. H. T. Davis, J.-F. Bodet, L. E. Scriven, and W. G. Miller, in *Physics of Amphiphilic Layers* (Springer, Proc. Phys. Vol. 21) (J. Meunier, D. Langevin, and N. Boccara, eds.), Springer-Verlag, Berlin, 1987, pp. 310–327.
43. M. Kahlweit, J. Colloid Interface Sci. *90*:197 (1982).
44. M. Kahlweit, E. Lessner, and R. Strey, J. Phys. Chem. *87*:5032 (1983).
45. M. Kahlweit, E. Lessner, and R. Strey, J. Phys. Chem. *88*:1937 (1984).
46. M. Kahlweit, R. Strey, and D. Haase, J. Phys. Chem. *89*:163 (1985).
47. M. Kahlweit, R. Strey, P. Firman, and D. Haase, Langmuir *1*:281 (1985).
48. P. Firman, D. Haase, J. Jen, and M. Kahlweit, Langmuir *1*:718 (1985).
49. M. Kahlweit and R. Strey, Angew. Chem. Int. Ed. Engl. *24*:654 (1985).
50. M. Kahlweit, R. Strey, and P. Firman, J. Phys. Chem. *90*:671 (1986).
51. M. Kahlweit and R. Strey, J. Phys. Chem. *90*:5239 (1986).
52. M. Kahlweit and R. Strey, J. Phys. Chem. *91*:1553 (1987).
53. M. Kahlweit, R. Strey, D. Haase, H. Kunieda, T. Schmeling, B. Faulhaber, M. Borkovec, H.-F. Eicke, G. Busse, F. Eggers, Th. Funck, H. Richmann, L. Magid, O. Söderman, P. Stilbs, J. Winkler, A. Dittrich, and W. Jahn, J. Colloid Interface Sci. *118*:436 (1987).
54. M. Kahlweit and R. Strey, J. Phys. Chem. *92*:1557 (1988).
55. M. Kahlweit, R. Strey, P. Firman, D. Haase, J. Jen, and R. Schömacker, Langmuir *4*:499 (1988).
56. M. Kahlweit, R. Strey, D. Haase, and P. Firman, Langmuir *4*:785 (1988).
57. M. Kahlweit, Nature *240*:617 (1988).
58. M. Kahlweit, R. Strey, R. Schömacker, and D. Haase, Langmuir *5*:305 (1989).
59. M. Kahlweit, R. Strey, and G. Busse, J. Phys. Chem. *94*:3881 (1990).
60. K.-V. Schubert, R. Strey, and M. Kahlweit, J. Colloid Interface Sci. *141*:21 (1991).
61. M. Kahlweit, R. Strey, M. Aratono, G. Busse, J. Jen, and K.-V. Schubert, J. Chem. Phys. *95*:2842 (1991).
62. M. Kahlweit, R. Strey, and G. Busse, J. Phys. Chem. *95*:5344 (1991).
63. M. Kahlweit, B. Faulhaber, and G. Busse, Langmuir *10*:2528 (1994).
64. D. Furman, S. Dattagupta, and R. B. Griffith, Phys. Rev. *B15*:441 (1977).
65. G. Gompper and M. Schick, *Self-Assembling Amphiphilic Systems*, Academic, London, 1994.
66. K.-V. Schubert and E. W. Kaler, Ber. Bunsenges. Phys. Chem. *100*:190 (1996).
67. U. Olsson and B. Lindman, in *The Structure, Dynamics and Equilibrium Properties of Colloidal Systems* (D. M. Bloor and E. Wyn-Jones, eds.), Kluwer, Dordrecht, The Netherlands, 1990, pp. 233–242.
68. K.-V. Schubert, R. Strey, S. R. Kline, and E. W. Kaler, J. Chem. Phys. *101*:5343 (1994).
69. K.-V. Schubert and R. Strey, J. Chem. Phys. *95*:8532 (1991).
70. M. Gradzielski, D. Langevin, T. Sottmann, and R. Strey, J. Chem. Phys. *104*:3782 (1996).
71. M. Teubner and R. Strey, J. Chem. Phys. *87*:3195 (1987).
72. R. Strey, Ber. Bunsenges. Phys. Chem. *97*:742 (1993).

73. B. Widom, Langmuir *3*:12 (1987).
74. Y. Seeto, J. E. Puig, L. E. Scriven, and H. T. Davis, J. Colloid Interface Sci. *96*:360 (1983).
75. M. Aratono and M. Kahlweit, J. Chem. Phys. *95*:8578 (1991).
76. A. Pouchelon, D. Chatenay, J. Meunier, and D. Langevin, J. Colloid Interface Sci. *82*:418 (1981).
77. A. M. Cazabat, D. Langevin, J. Meunier, and A. Pouchelon, Adv. Colloid Interface Sci. *16*:175 (1982).
78. O. Abillon, L. T. Lee, D. Langevin, and K. Wong, Physica *A172*:209 (1991).
79. D. Langevin, Ber. Bunsenges. Phys. Chem. *100*:336 (1996).
80. G. Gompper and M. Schick, Phys. Rev. Lett. *65*:1116 (1990).
81. R. M. Hornreich, M. Luban, and S. Shtrikman, Phys. Rev. Lett. *35*:1678 (1975).
82. M. E. Fisher and B. Widom, J. Chem. Phys. *50*:3756 (1969).
83. B. Widom, J. Phys. Chem. *77*:2196 (1973).
84. P. D. I. Fletcher, Curr. Opinion Colloid Interface Sci. *1*:101 (1996).
85. G. Porte and J. Oberdisse, Colloids Surf. A *128*:101 (1997).
86. Z. Lin, H. T. Davis, and L. E. Scriven, Langmuir *12*:5489 (1996).
87. R. Zana, Colloids Surf. A *123/4*:27 (1997).
88. V. Luzzati, J. Phys. II *5*:1649 (1995).
89. V. Luzzati, H. Delacroix, and A. Gulik, J. Phys. II *6*:405 (1996).
90. C. Tanford, *The Hydrophobic Effect*, 2nd ed., Wiley, New York, 1980.
91. Y. Chevalier and T. Zemb, Rep. Progr. Phys. *53*:279 (1990).
92. W. M. Gelbart and A. Ben-Shaul, J. Phys. Chem. *100*:13169 (1996).
93. W. Blokzijl and J. B. F. N. Engberts, Angew. Chem. Int. Ed. Engl. *32*:1545 (1993).
94. D. F. Evans and H. Wennerström, *The Colloidal Domain: Where Chemistry, Physics, Biology and Technology Meet*, VCH, New York, 1994.
95. M. E. Paulaitis, S. Garde, and H. S. Ashbaugh, Curr. Opinion Colloid Interface Sci. *1*:376 (1996).
96. J. B. F. N. Engberts and J. Kevelam, Curr. Opinion Colloid Interface Sci. *1*:779 (1996).
97. A. Ben-Shaul and W. M. Gelbart, in *Micelles, Membranes, Microemulsions and Monolayers* (W. M. Gelbart, A. Ben-Shaul, and D. Roux, eds.), Springer-Verlag, New York, 1994, pp. 1–104.
98. J. N. Israelachvili, D. J. Mitchell, and B. W. Ninham, J. Chem. Soc., Faraday Trans. II *72*:1525 (1976).
99. J. N. Israelachvili, S. Marčelja, and R. G. Horn, Quart. Rev. Biophys. *13*:121 (1980).
100. D. J. Mitchell and B. W. Ninham, J. Chem. Soc., Faraday Trans. II *77*:601 (1981).
101. D. Roux, C. Coulon, and M. E. Cates, J. Phys. Chem. *96*:4174 (1992).
102. D. A. Van Hol, J. A. Bouwstra, A. Van Rensen, E. Jeremiasse, T. De Vringer, and H. E. Junginger, J. Colloid Interface Sci. *178*:263 (1996).
103. S. Arimori, T. Nagasaki, and S. Shinkai, J. Chem. Soc., Perkin Trans. 2:679 (1995).
104. D. Langevin, in *Structure and Reactivity in Reverse Micelles* (Stud. Phys. Theoret. Chem., Vol. 65) (M. P. Pileni, ed.), Elsevier, Amsterdam, 1989, pp. 13–43.
105. D. Langevin, Annu. Rev. Phys. Chem. *43*:341 (1992).
106. S. T. Hyde, Langmuir *13*:842 (1997).
107. S. T. Hyde, B. W. Ninham, and T. Zemb, J. Phys. Chem. *93*:1464 (1989).
108. G. Porte, in *Micelles, Membranes, Microemulsions and Monolayers* (W. M. Gelbart, A. Ben-Shaul, and D. Roux, eds.), Springer-Verlag, New York, 1994, pp. 105–151.
109. P. M. Deusing, R. H. Templer, and J. M. Seddon, Langmuir *13*:351 (1997).
110. B. Lindman, F. Tiberg, L. Picullel, U. Olsson, P. Alexandridis, and H. Wennerström, in *Micelles, Microemulsions and Monolayers* (D. O. Shah, ed.), Marcel Dekker, New York, 1998, pp. 101–126.
111. D. Roux, C. R. Safinia, and F. Nallet, in *Micelles, Membranes, Microemulsions and Monolayers* (W. M. Gelbart, A. Ben-Shaul, and D. Roux, eds.), Springer-Verlag, New York, 1994, pp. 303–346.
112. P. Versluis, J. C. van de Pas, and T. Mellema, Langmuir *13*:5732 (1997).

113. T. A. Bleasdale and G. J. T. Tiddy, in *The Structure, Dynamics and Equilibrium Properties of Colloidal Systems* (D. M. Bloor and E. Wyn-Jones, eds.), Kluwer, Dordrecht, The Netherlands, 1990, pp. 397–414.

114. A. Jurgens, Tenside, Surfact. Deterg. *26*:222 (1989).

115. J. C. van de Pas, Tenside, Surfact. Deterg. *28*:158 (1991).

116. H. Hoffmann, C. Thunig, P. Schmiedel, and U. Munkert, Langmuir *10*:3972 (1994).

117. S. Amselem, E. Zawoznik, A. Yogev, and D. Friedman, in *Handbook of Nonmedical Applications of Liposomes*, Vol. II (Y. Barenholz and D. D. Lasic, eds.), CRC, Boca Raton, FL, 1996, pp. 209–223.

118. A. Sein, J. B. F. N. Engberts, E. van Linden, and J. C. van de Pas, Langmuir *9*:1714 (1993).

119. J. Gustafsson, G. Orädd, G. Lindblom, U. Olsson, and M. Almgren, Langmuir *13*:852 (1997).

120. R. Strey, R. Schömacker, D. Roux, F. Nallet, and U. Olsson, J. Chem. Soc., Faraday Trans. *86*:2253 (1990).

121. P. Ekwall, L. Mandell, and K. Fontell, J. Colloid Interface Sci. *33*:215 (1970).

122. K. Fontell, J. Colloid Interface Sci. *44*:318 (1973).

123. S. Marčelja and N. Radić, Chem. Phys. Lett. *42*:129 (1976).

124. N. Ostrowsky and D. Sornette, Colloids Surf. *14*:231 (1985).

125. J. Forsman, C. E. Woodward, and B. Jönsson, Langmuir *13*:5459 (1997).

126. J. N. Israelachvili and H. Wennerström, Langmuir *6*:873 (1990).

127. J. N. Israelachvili and H. Wennerström, J. Phys. Chem. *96*:520 (1992).

128. J. N. Israelachvili and H. Wennerström, Nature *379*:219 (1996).

129. V. A. Parsegian and R. P. Rand, Langmuir *7*:1299 (1991).

130. P.-G. de Gennes and C. Taupin, J. Phys. Chem. *86*:2294 (1982).

131. W. Helfrich, Z. Naturforsch. *C28*:693 (1973).

132. W. Helfrich, Z. Naturforsch. *C33*:305 (1978).

133. G. Guérin and A.-M. Bellocq, J. Phys. Chem. *92*:2550 (1988).

134. A.-M. Bellocq, in *Emulsions and Emulsion Stability* (Surfact. Sci. Ser., Vol. 61) (J. Sjoblom, ed.), Marcel Dekker, New York, 1996, pp. 181–236.

135. K. M. McGrath, Langmuir *13*:1987 (1997).

136. V. Ponsinet and Y. Talmon, Langmuir *13*:7287 (1997).

137. J. M. Seddon and R. H. Templer, in *Handbook of Biological Physics* (R. Lipowski and E. Sackman, eds.), Elsevier, Amsterdam, 1995, pp. 97–160.

138. J. M. Seddon, Biochim. Biophys. Acta *1031*:1 (1990).

139. G. L. Kirk, S. M. Gruner, and D. L. Stein, Biochemistry *23*:1093 (1984).

140. J. Charvolin, J. Phys. *46*(C3):173 (1985).

141. S. M. Gruner, Proc. Natl. Acad. Sci. USA *82*:3665 (1985).

142. C. E. Fairhurst, M. C. Holmes, and M. S. Leaver, Langmuir *13*:4964 (1997).

143. V. Luzzati, in *Biological Membranes*, Vol. 1 (D. Chapman, ed.), Academic, London, 1968, pp. 71–123.

144. G. L. Kirk and S. M. Gruner, J. Phys. *46*:761 (1985).

145. S. M. Gruner, M. W. Tate, G. L. Kirk, P. T. C. So, D. C. Turner, D. T. Keane, C. P. S. Tilcock, and P. R. Cullis, Biochemistry *27*:2853 (1988).

146. S. Alpérine, Y. Hendrikx, and J. Charvolin, J. Phys. Lett. *46*:L27 (1985).

147. P. Kékicheff and B. Cabane, J. Phys. *48*:1571 (1987).

148. Y. Hendrikx and J. Charvolin, Liq. Cryst. *3*:265 (1988).

149. P. Kékicheff and B. Cabane, Acta Cryst. *B44*:395 (1988).

150. L. Ramos and P. Fabre, Langmuir *13*:862 (1997).

151. D. Roux and J. Coulon, J. Phys. *47*:1257 (1986).

152. M. P. Taylor and J. Herzfeld Phys. Rev. A *43*:1892 (1991).

153. F. Livolant and Y. Bouligand, J. Phys. *47*:1813 (1986).

154. Y. Hendrikx, J. Charvolin, M. Rawiso, L. Liébert, and M. C. Holmes, J. Phys. Chem. *87*:3991 (1983).

155. J. V. Selinger and R. F. Bruisma, Phys. Rev. A *43*:2922 (1991).

156. M. Clerc, J. Phys. II *6*:961 (1996).
157. G. Lindblom and L. Rilfors, Biochim. Biophys. Acta *988*:221 (1989).
158. K. Fontell, Adv. Colloid Interface Sci. *41*:127 (1992).
159. J. M. Seddon and R. H. Templer, Phil. Trans. Roy. Soc. Lond. *A344*:377 (1993).
160. V. Luzzati, R. Vargas, A. Gulik, P. Mariani, J. M. Seddon, and E. Rivas, Biochemistry *31*:279 (1992).
161. V. Luzzati and P. A. Spegt, Nature *215*:701 (1967).
162. S. S. Funari and G. Rapp, J. Phys. Chem. *B101*:732 (1997).
163. P. Mariani, H. Delacroix, and V. Luzzati, J. Mol. Biol. *204*:165 (1988).
164. P. Sakya, J. M. Seddon, R. H. Templer, R. J. Mirkin, and G. J. T. Tiddy, Langmuir *13*:3706 (1997).
165. H. A. Schwarz, Monatsber. Königl. Akad. Wiss. Berlin, 1865, pp. 149–153.
166. J. M. Seddon, Biochemistry *29*:7997 (1990).
167. A. de Geyer, Prog. Colloid Polym. Sci. *93*:76 (1993).
168. A. Tardieu and V. Luzzati, Biochim. Biophys. Acta *219*:11 (1970).
169. K. Fontell, K. K. Fox, and E. Hanson, Mol. Cryst. Liq. Cryst. Lett. *1*:9 (1985).
170. R. Vargas, P. Mariani, A. Gulik, and V. Luzzati, J. Mol. Biol. *225*:137 (1992).
171. H. Delacroix, T. Gulik-Krzywicki, P. Mariani, and V. Luzzati, J. Mol. Biol. *229*:526 (1993).
172. R.-O. Eriksson, G. Lindblom, and G. Arvidson, J. Phys. Chem. *89*:1050 (1985).
173. J. Charvolin and J. F. Sadoc, J. Phys. *49*:521 (1988).
174. P.-O. Eriksson, L. Rilfors, G. Lindblom, and G. Arvidson, Chem. Phys. Lipids *37*:357 (1985).
175. P.-O. Eriksson, G. Lindblom, and G. Arvidson, J. Phys. Chem. *91*:846 (1987).
176. A. Gulik, H. Delacroix, G. Kirschner, and V. Luzzati, J. Phys. II *5*:445 (1995).
177. N. Boden, in *Micelles, Membranes, Microemulsions and Monolayers* (W. M. Gelbart, A. Ben-Shaul, and D. Roux, eds.), Springer-Verlag, New York, 1994, pp. 153–217.
178. M. C. Holmes, M. S. Leaver, and A. M. Smith, Langmuir *11*:356 (1995).
179. V. Formoso, Y. Galerna, F. P. Nicoletta, G. Pepy, N. Picci, and R. Bartolino, J. Phys. IV *3*(Cl):271 (1993).
180. J. Charvolin, A. Levelut, and E. T. Samulsky, J. Phys. Lett. *40*:L587 (1979).
181. F. B. Rosevear, J. Soc. Cosmet. Chem. *1*:581 (1968).
182. M. C. Holmes, N. Boden, and K. Radley, Mol. Cryst. Liq. Cryst. *100*:93 (1983).
183. P. K. Vinson and Y. Talmon, Science *133*:288 (1990).
184. M. E. Cates and S. J. Candau, J. Phys. Condens. Matter *2*:6869 (1990).
185. K. Edwards and M. Almgren, J. Colloid Interface Sci. *147*:1 (1991).
186. S. J. Candau, A. Khatory, F. Lequeux, and F. Kern, J. Phys. IV *3*:197 (1993).
187. G. Porte, Y. Poggi, J. Appell, and G. Maret, J. Phys. Chem. *87*:1264 (1983).
188. E. W. Anacker and H. M. Ghose, J. Am. Chem. Soc. *90*:3161 (1968).
189. P. J. Missel, N. A. Mazer, G. B. Benedek, C. Y. Young, and M. Carey, J. Phys. Chem. *84*:1044 (1980).
190. R. Gomati, J. Appell, P. Bassereau, J. Marignan, and G. Porte, J. Phys. Chem. *91*:6203 (1987).
191. G. Porte, J. Marignan, P. Bassereau, and R. May, J. Phys. *49*:511 (1988).
192. P. Schurtenberger, Curr. Opinion Colloid Interface Sci. *1*:773 (1996).
193. S. May, Y. Bohbot, and A. Ben-Shaul, J. Phys. Chem. *B101*:8648 (1997).
194. S. J. Candau, E. Hirsch, and R. Zana, J. Colloid Interface Sci. *105*:521 (1985).
195. J. Appell, G. Porte, A. Khatory, F. Kern, and S. J. Candau, J. Phys. II *2*:1045 (1992).
196. A. Khatory, F. Lequeux, F. Kern, and S. J. Candau, Langmuir *9*:1456 (1993).
197. A. Khatory, F. Kern, F. Lequeux, J. Appell, G. Porte, N. Morie, A. Ott, and W. Urbach, Langmuir *9*:993 (1993).
198. T. J. Drye and M. E. Cates, J. Chem. Phys. *96*:1367 (1982).
199. K. Elleuch, F. Lequeux, and P. Pfeuty, J. Phys. I. *5*:465 (1995).
200. Y. Bohbot, A. Ben-Shaul, R. Granek, and W. M. Gelbart, J. Chem. Phys. *103*:8764 (1995).
201. M. Monduzzi, U. Olsson, and O. Söderman, Langmuir *9*:2914 (1993).
202. D. Danino, Y. Talmon, H. Levy, G. Beinart, and R. Zana, Science *269*:1420 (1995).

203. Z. Lin, Langmuir *12*:1279 (1996).
204. C. Stubenrauch, M. Nydén, G. H. Findenegg, and B. Lindman, J. Phys. Chem. *100*:17028 (1996).
205. C. Manohar, U. R. K. Rao, B. S. Valaulikar, and R. M. Iyer, J. Chem. Soc., Chem. Commun. *1986*:379 .
206. F. M. Menger and A. V. Eliseev, Langmuir *11*:1855 (1995).
207. F. A. L. Anet, J. Am. Chem. Soc, *108*:7102 (1986).
208. B. Cabane, R. Duplessix and T. Zemb, in *Surfactants in Solution* Vol. 1 (B. Lindman and K. L. Mittal, eds.), Plenum, New York, 1984, p. 373.
209. R. Strey, J. Winkler, and L. Magid, J. Physs. Chem. *95*:7502 (1991).
210. E. Radlinska, Z. Zemb, J.-P. Dalbiez, and B. Ninham, Langmuir *9*:2844 (1993).
211. H. Hoffmann, C. Thunig, and U. Munkert, Langmuir *8*:2629 (1992).
212. O. Ghosh and C. A. Miller, J. Phys. Chem. *91*:4528 (1987).
213. D. Gazeau, A.-M. Bellocq, D. Roux, and T. Zemb, Europhys. Lett. *9*:447 (1989).
214. B. Balinov, U. Olsson, and O. Söderman, J. Phys. Chem, *95*:5931 (1991).
215. A. Ott, W. Urbach, D. Langevin, and H. Hoffmann, Langmuir *8*:345 (1992).
216. R. Strey, W. Jahn, G. Porte, and P. Bassereau, Langmuir *6*:1635 (1990).
217. A. Maldonado, W. Urbach, and D. Langevin, J. Phys. Chem. B *101*:8069 (1997).
218. F. Auguste, J.-P. Douliez, A.-M. Bellocq, E. J. Dufourc, and T. Gulik-Krzywicki, Langmuir *13*:666 (1997).
219. G. Porte, Curr. Opinion Colloid Interface Sci. *1*:345 (1996).
220. M. E. Cates, D. Roux, D. Andelman, S. T. Milner, and S. A. Safran, Europhys. Lett. *5*:733 (1988).
221. G. Porte, J. Appell, P. Bassereau, and J. Marignan, J. Phys. *50*:1335 (1989).
222. D. Anderson, H. Wennerstrom, and U. Olsson, J. Phys. Chem. *93*:4243 (1989).
223. P. Hervé, D. Roux, A.-M. Bellocq, F. Nallet, and T. Gulik-Krzywicki, J. Phys. II *3*:1255 (1993).
224. M. Dubois, T. Gulik-Krzywicki, and B. Cabane, Langmuir *9*:673 (1993).
225. B. D. Simons and M. E. Cates, J. Phys. II *2*1439 (1992).
226. X. Auvray, C. Petipas, R. Anthore, I. Rico, and A. Lattes, J. Phys. Chem. *93*:7658 (1989).
227. A. Skoulius, Adv. Liq. Cryst. *1*:169 (1975).
228. E. L. Thomas, D. B. Alward, D. J. Kunning, D. C. Martin, D. J. Handlin, and L. J. Fetters, Macromolecules *19*:2197 (1986).
229. H. Hasegawa, H. Tanaka, K. Yamasaki, and T. Hashimoto, Macromolecules *20*:1651 (1987).
230. J. Charvolin and J.-F. Sadoc, in *Micelles, Membranes, Microemulsions and Monolayers* (W. M. Gelbart, A. Ben-Shaul, and D. Roux, eds.), Springer-Verlag, New York, 1994, pp. 219–249.
231. J. Charvolin and J.-F. Sadoc, J. Phys. Chem. *92*:37 (1988).
232. J. Erbes, C. Czeslik, W. Hahn, R. Winter, M. Rappolt, and G. Rapp, Ber. Bunsenges. Phys. Chem. *98*:1287 (1994).
233. M. C. Holmes and J. Charvolin, J. Phys. Chem. *88*:810 (1984).
234. P. Kékicheff, B. Cabane, and M. Rawiso, J. Phys. Lett. *45*:L813 (1984).
235. Y. Hendrikx, J. Charvolin, P. Kékicheff, and M. Roth, Liq. Cryst. *2*:677 (1987).
236. Y. Rançon and J. Charvolin, J. Phys. Chem. *92*:6339 (1988).
237. M. C. Holmes, J. Charvolin, and D. J. Reynolds, Liq. Cryst. *3*:1147 (1988).
238. Y. Hendrikx and J. Charvolin, J. Phys. *42*:1427 (1981).
239. M. R. Rizzatti and J. D. Gault, J. Colloid Interface Sci. *110*:258 (1986).
240. M. G. Noro and W. M. Gelbart, J. Phys. Chem. B *101*:8642 (1997).
241. C. Bagdassarian, W. M. Gelbart, and A. Ben-Shaul, J. Stat. Phys. *52*:1307 (1988).
242. Y. Rançon and J. Charvolin, J. Phys. Chem. *92*:2646 (1988).
243. S. S. Funari, M. C. Holmes, and G. J. T. Tiddy, J. Phys. Chem. *98*:3015 (1994).
244. J. M. Seddon, E. A. Bartell, and J. Mingins, J. Phys: Condens. Matter *2*:285 (1990).
245. J. M. Seddon, Z. Neelofar, R. H. Templer, R. N. McElhaney, and D. A. Mannock, Langmuir *12*:5250 (1996).
246. G. Orädd, G. Lindblom, G. Fontell, and H. Ljusberg-Wahren, Biophys. J. *68*:1856 (1995).
247. P. Alexandridis, U. Olsson, and B. Lindman, Langmuir *12*:1419 (1996).
248. P. Alexandridis, U. Olsson, and B. Lindman, Langmuir *13*:23 (1997).

249. Y. Rançon and J. Charvolin, J. Phys. *48*:1067 (1987).
250. V. Luzzati, A. Tardieu, and T. Gulik-Krzywicki, Nature *217*:1028 (1968).
251. F. Husson, H. Mustacchi, and V. Luzzati, Acta Cryst. *13*:668 (1960).
252. V. Luzzati, H. Mustacchi, A. Skoulios and F. Husson, Acta Cryst. *13*:660 (1960)
253. K. Rendall, G. J. T. Tiddy, and M. A. Trevethan, J. Chem. Soc., Faraday Trans, I *79*:2867 (1983).
254. K. Rendall and G. J. T. Tiddy, J. Chem. Soc., Faraday Trans. I *79*:637 (1983).
255. R. M. Wood and M. P. McDonald, J. Chem. Soc., Faraday Trans. I *83*:273 (1985).
256. G. Chidichimo, A. Golemme, and J. W. Doane, J. Chem. Phys. *82*:4369 (1985).
257. P. Kékicheff, J. Colloid Interface Sci. *131*:133 (1989).
258. P. Kékicheff and G. J. T. Tiddy, J. Phys. Chem. *93*:2520 (1989).
259. P. Kékicheff, Mol. Cryst. Liq. Cryst. *198*:131 (1991).
260. S. S. Funari, M. C. Holmes, and G. J. T. Tiddy, J. Phys. Chem. *96*:11029 (1992).
261. H. Hagslätt, O. Söderman, and B. Jönsson, Liq. Cryst. *12*:667 (1992).
262. J. Burgoyne, M. C. Holmes, and G. I. T. Tiddy, J. Phys. Chem. *99*:6054 (1995).
263. U. Henriksson, E. S. Blackmore, G. J. T. Tiddy, and O. Söderman, J. Phys. Chem. *96*:3894 (1992).
264. I. D. Leigh, M. P. McDonald, R. M. Wood, G. J. T. Tiddy, and M. A. Trevethan, J. Chem. Soc., Faraday Trans. I *77*:2867 (1981).
265. M. S. Leaver and M. C. Holmes, J. Phys. II *3*:105 (1993).
266. S. T. Hyde, J. Phys. Chem. *93*:1458 (1989).
267. S. T. Hyde, Pure Appl. Chem. *64*:1617 (1992).
268. D. M. Anderson and P. Ström, in *Polymer Association Structures: Microemulsions and Liquid Crystals* (ACS Symp. Ser. 384) (M. A. El-Nokaly, ed.), ACS, Washington, DC, 1989, pp. 204–224.
269. D. M. Anderson, Colloq. Phys. *51*:C7-1 (1990).
270. D. M. Anderson, H. T. Davis, L. E. Scriven, and J. C. C. Nitsche, Adv. Chem. Phys. *77*:337–396 (1990).
271. P. Kékicheff, C. Gabrielle-Madelmont, and M. Ollivon, J. Colloid Interface Sci. *131*:112 (1989).
272. C. Hall and G. J. T. Tiddy, in *Surfactants in Solution*, Vol. 8 (K. L. Mittal, ed.), Plenum, New York, 1989, pp. 9–23.
273. C. E. Fairhurst, M. C. Holmes, and M. S. Leaver, Langmuir *12*:6336 (1996).
274. M. C. Holmes, A. M. Smith, and M. S. Leaver, J. Phys. II *3*:1357 (1993).
275. K. Rendall and G. J. T. Tiddy, J. Chem. Soc., Faraday Trans. I *80*:3339 (1984).
276. J. P. Conroy, C. Hall, C. A. Lang, K. Rendall, G. J. T. Tiddy, J. Walsh, and G. Lindblom, Prog. Colloid Polym. Sci. *82*:253 (1990).
277. S. S. Funari and G. Rapp, J. Phys. Chem. B *101*:732 (1997).
278. E. S. Blackmore and G. J. T. Tiddy, J. Chem. Soc., Faraday Trans. II *84*:1115 (1984).
279. L. Paz, J. M. Di Meglio, M. Dvolaitzky, R. Ober, and C. Taupin, J. Phys. Chem. *88*:3415 (1984).
280. M. Allain, J. Phys. *46*:225 (1985).
281. P. Ostwald and M. Allain, J. Phys. *46*:831 (1985).
282. P. Ostwald and M. Allain, J. Colloid Interface Sci. *126*:45 (1988).
283. N. Boden, S. A. Corne, M. C. Holmes, P. H. Jackson, D. Parker, and K. W. Jolley, J. Phys. *47*:2135 (1986).
284. G. Porte, R. Gomati, O. El Haitamy, J. Appell, and J. Marignan, J. Phys. Chem. *90*:5746 (1986).
285. M. Anderson and G. Karlstrom J. Phys. Chem. *89*:4957 (1985).
286. G. Karlstrom, J. Phys. Chem. *89*:4962 (1985).
287. P. Nilsson and B. Lindman, J. Phys. Chem. *87*:4756 (1983).
288. T. Ahlnäs, G. Karlstrom, and B. Lindman, J. Phys. Chem. *91*:4030 (1987).
289. M. Carvell, D. G. Hall, I. G. Lyle, and G. J. T. Tiddy, Faraday Disc. Chem. Soc. *81*:223 (1986).
290. I. G. Lyle and G. J. T. Tiddy, Chem. Phys. Lett. *124*:432 (1986).
291. S. J. Alexander, J. Phys. *38*:983 (1977).
292. P.-G. de Gennes, Adv. Colloid Interface Sci. *27*:189 (1987).
293. S. T. Hyde, Prog. Colloid Polym. Sci. *82*:236 (1990).
294. S. T. Hyde, Colloids Surf. A *103*:227 (1995).
295. S. T. Hyde, J. Phys. *51*(C7):209 (1990).

296. H. Kunieda, K. Ozawa, and K. L. Huang, J. Phys. Chem. B *102*:831 (1998).
297. R. Strey and M. Jonströmer, J. Phys. Chem. *96*:4537 (1992).
298. M. Jonströmer and R. Strey, J. Phys. Chem. *96*:5993 (1992).
299. M. H. G. M. Penders and R. Strey, J. Phys. Chem. *99*:6091 (1995).
300. M. H. G. M. Penders and R. Strey, J. Phys, Chem, *99*:10313 (1995).
301. H. Kunieda, A, Nakano, and M. Akimaru, J. Colloid Interface Sci. *170*:78 (1995).
302. H. Kunieda, A. Nakano, and M. Angeles Pes, Langmuir *11*:3302 (1995).
303. S. Yamaguchi, Colloid Polym, Sci. *274*:1152 (1996).
304. O. Regev, S. Ezrahi, A. Aserin, N. Garti, E. Wachtel, E. W. Kaler, A. Khan, and Y. Talmon, Langmuir *12*:668 (1996).
305. M. Angeles Pes, K. Aramaki, N. Nakamura, and H. Kunieda, J. Colloid Interface Sci. *178*:666 (1996).
306. N. Nakamura, T. Tagawa, K. Kihara, I. Tobita, and H. Kunieda, Langmuir *13*:2001 (1997).
307. N. Garti, A. Aserin, S. Ezrahi, and E. Wachtel, J. Colloid Interface Sci. *169*:428 (1995).
308. S. Ezrahi, E. Wachtel, A. Aserin, and N, Garti, J. Colloid Interface Sci. *191*:277 (1997).
309. L. Thompson, J. M. Walsh, and G. J. T. Tiddy, Colloids Surf. A *106*:223 (1996).
310. G. Colafemmina, G. Palazzo, E. Balestrieri, M. Giomini, M. Giustini, and A. Ceglie, Prog. Colloid Polym. Sci. *105*:281 (1997).
311. D. Danino, Y. Talmon, and R. Zana, J. Colloid Interface Sci. *186*:170 (1997).
312. A. Martino and E. W. Kaler, Colloids Surf. A *99*:91 (1995).
313. S. A. Safran, *Statistical Thermodynamics of Surfaces, Interfaces and Membranes* Addison-Wesley, Reading, MA, 1994.
314. N. Boden, K. Radley, and M. C. Holmes, Mol. Phys. *42*:493 (1981).
315. L. Sallen, P. Sotta, and P. Ostwald, J. Phys. Chem. B*101*:4875 (1997).
316. J. Daicic, U. Olsson, H. Wennerström, G. Jerke, and P. Schurtenberger, J. Phys. II *5*:199 (1995).
317. J. Daicic, U. Olsson, H. Wennerström, G. Jerke, and P. Schurtenberger Phys. Rev. E *52*:3266 (1996).
318. D. Roux, M. E. Cates, U, Olsson, R. G. Ball, F. Nallet, and A. M. Bellocq, Europhys. Lett *11*:229 (1990).
319. M. Filali, G. Porte, J. Appell, and P. Pfeuty, J. Phys. II *4*:349 (1994).
320. C. Coulon, D. Roux, and A. M. Bellocq, Phys. Rev. Lett. *66*:1709 (1991).
321. C. Vinches, C. Coulon, and D. Roux, J. Phys, II *4*:1165 (1994).
322. M. Filali, J. Appell, and G. Porte, J. Phys, II *5*:657 (1995).
323. D. Huse and S. Leibler, Phys. Rev. Lett. *66*:437 (1991).
324. L. Auvray, in *Micelles, Microemulsions, Membranes and Monolayers* (A. Ben-Shaul, W. M. Gelbart, and D. Roux, eds.) Springer-Verlag, New York, 1994, pp. 347–393.
325. I. S. Barnes, S. T. Hyde, B. W. Ninham, P. J. Derian, M. Drifford, and T. N. Zemb, J. Phys. Chem. *92*:2286 (1988).
326. P. D. I. Fletcher, R. Aveyard, and B. P. Binks, Langmuir *5*:1210 (1989).
327. F. Lichterfeld, T. Schmeling, and R. Strey, J. Phys. Chem. *90*:5762 (1986).
328. P. Schurtenberger, R. Scartazzini, L. J. Magid, M. E. Leser, and P. L. Luisi, J. Phys. Chem. *94*:3695 (1990).
329. P. Schurtenberger, L. J. Magid, S. M. King, and P. Lindner, J. Phys. Chem. *95*:4173 (1991).
330. P. Schurtenberger and P. L. Luisi, Rheol. Acta *28*:372 (1989).
331. R. Scartazzani and P. L. Luisi, J. Phys. Chem. *92*:829 (1988).
332. W. G. M. Agterof, J. A. J. van Zomeren, and A. Vrij, Chem. Phys. Lett. *43*:363 (1976).
333. P. Terech, V. Schafhauser, P. Maldivi, and J. M. Geunet, Europhys. Lett. *17*:515 (1992).
334. J. Eastoe, Langmuir *8*:1503 (1992).
335. J. Eastoe, J. Dong, K. J. Hetherington, D, Steytler, and R. K. Heenan, J. Chem. Soc. Faraday Trans. *92*:65 (1996).
336. C. Petit, P. Lixon, and M. P. Pileni, Langmuir *7*:2620 (1991).
337. J. Eastoe, B. H. Robinson, G. Fragneto, T. F. Towey, R. K. Heenan, and F. J. Leng, J. Chem. Soc. Faraday Trans. I *86*:2883 (1990).

338. J. Eastoe, B. H. Robinson, G, Fragneto, R. K. Heenan and D. C. Steytler, Physica *B180/181*:555 (1992).

339. J. Eastoe, T. F. Towey, B. H. Robinson, J. Williams, and R. K. Heenan, Langmuir *9*:1459 (1993).

340. J. Eastoe, B. H. Robinson, and R. K. Heenan, Langmuir *9*:2820 (1993).

341. J. Eastoe, D. C. Steytler, B. H. Robinson, R. K. Heenan, A. N. North, and J. C. Dore, J. Chem. Soc., Faraday Trans, *90*:2497 (1994).

342. J. Eastoe, G. Fragneto, B. H. Robinson, T. F. Towey, R. K. Heenan, and F. J. Leng, J. Chem. Soc, Faraday Trans. *88*:461 (1992).

343. J. Eastoe, B. H. Robinson, and D. Thorn-Leeson, Adv. Colloid Interface Sci. *36*:1 (1991).

344. D. Langevin, Acc. Chem. Res. *21*:255 (1988).

345. B. H. Robinson, Chem Br. *26*:342 (1990).

346. N. A. Mazer, G. B. Benedek and M. C. Carey, Biochemistry *19*:601 (1980).

347. J. C. Ravey, M. Buzier, and C. Picot, J. Colloid Interface Sci. *97*:9 (1984).

348. J. C. Ravey and M. Buzier, J. Colloid Interface Sci. *116*:30 (1987).

349. D. C. Steytler, D. L. Sarageant, B. H. Robinson, J. Eastoe, and R. K. Heenan, Langmuir *10*:2213 (1994).

350. S. A. Safran, L. A. Turkevich, and P. Pincus, J. Phys. Lett. *45*:L69 (1984).

351. S. A. Safran, in *Micellar Solutions and Microemulsions. Structure, Dynamics and Statistical Thermodynamics* (S. H. Chen and R. Rajagopalan, eds.), Springer-Verlag,, New York, 1990, pp. 161–184.

352. S. A. Safran, in *Structure and Dynamics of Strongly Interacting Colloids and Supramolecular Aggregates in Solution* (S. H. Chen, J. S. Huang, and P. Tartaglia, eds.), Kluwer, Dordrecht, The Netherlands, 1992, pp. 237–263.

353. T. Zemb, S. T. Hyde, P. J. Derian, I. S. Barnes, and B. W. Ninham, J. Phys. Chem. *91*:3814 (1987).

354. B. W. Ninham, I. S. Barnes, S. T. Hyde, P. J. Derian, and T. N. Zemb, Europhys. Lett. *4*:561 (1987).

355. B. Lemaire, P. Bothorel, and D. Roux, J. Phys. Chem. *87*:1023 (1983).

356. S. A. Safran, J. Chem, Phys. *78*:2073 (1983).

357. S. Anderson, S. T. Hyde, and H. G. von Schnering, Z. Kristallogr, *168*:1 (1984).

358. D. A. Hajduk, P. E. Harper, S. M. Gruner, C. C. Honeker, G. Kin, E. L. Thomas, and L. J. Fetters, Macromolecules *27*:4063 (1994).

359. C. T. Kresge, M. E. Leonowicz, W. J. Roth, J. C. Vartuli, and J. S. Beck, Nature *359*:710 (1992).

360. Q. Huo, D. Margolese, U. Ciesla. P. Feng, T. Gier, P. Sieger, R. Leon, P. Petroff, F. Schüth, G. Stucky, Nature *368*:317 (1994).

361. S. T. Hyde, Phys. Chem. Miner. *20*:190 (1993).

362. B. Lindman, K. Shinoda, U. Olsson, D. Anderson, G. Karlström, and H. Wennerström, Colloids Surf. *38*:205 (1989).

363. L. E. Scriven, Nature *263*:123 (1976).

364. L. E. Scriven, in *Micellization, Solubilization and Microemulsions*, Vol. 2 (K. L. Mittal, ed.), Plenum, New York, 1977, pp. 877–893.

365. D. F. Evans, D. J. Mitchell, and B. W. Ninham, J. Phys. Chem. *90*:2817 (1986).

366. S. J. Chen, D. F. Evans, and B. W. Ninham, J. Phys. Chem. *88*:163 (1984).

367. K. Fontell, A. Ceglie, B. Lindman, and B. Ninham, Acta Chem, Scand. A *49*:247 (1986).

368. B. Lindman, N. Kamenka, T. M. Kathopoulis, B. Brun, and P. G. Nilsson, J. Phys, Chem, *84*:2485 (1980).

369. E. W. Kaler, K. E. Bennett, H. T. Davis, and L. E. Scriven, J. Chem. Phys. *79*:5673 (1983).

370. E. W. Kaler, H. T. Davis, and L. E. Scriven, J. Chem, Phys. *79*:5685 (1983).

371. P. Guéring and B. Lindman, Langmuir *1*:464 (1985).

372. B. Lindman and P. Stilbs, in *Microemulsions* (S. E. Friberg and P. Bothorel, eds.) CRC, Boca Raton, FL, 1987, pp. 119–152.

373. L. Auvray, J. Cotton, R. Ober, and C. Taupin, J. Phys, Chem. *88*:4586 (1984).

374. H. Saito and K. Shinoda, J. Colloid Interface Sci. *32*:647 (1970).

375. Y. Talmon and S. Prager, J. Chem. Phys. *69*:2984 (1978).

376. Y. Talmon and S. Prager, J Chem, Phys. *76*:1535 (1982).

377. E. W. Kaler and S. Prager, J. Colloid Interface Sci. *86*:359 (1982).
378. J. Daicic, U. Olsson, and H. Wennerström, Langmuir *11*:2451 (1995).
379. J. Jouffroy, P. Levinson, and P-G, de Gennes, J. Phys. *43*:1241 (1982).
380. D. Langevin, Adv. Colloid Interface Sci. *34*:583 (1991).
381. T. Sottmann, R. Strey, and S. H. Chen, J. Chem. Phys. *106*:6483 (1997).
382. P. Debye, H. R. Anderson, and H. Brumberger, J. Appl. Phys. *28*:679 (1957).
383. J. F. Billman and E. W. Kaler, Langmuir *6*:611 (1990).
384. D. Langevin, in *Structure and Dynamics of Strongly Interacting Colloids and Supramolecular Aggregates in Solutions* (S. H. Chen, J. S. Huang, and P. Tartaglia, eds.), Kluwer, Dordrecht, The Netherlands, 1992, pp. 325–349.
385. B. Widom, J. Chem. Phys. *82*:1030 (1984).
386. B. Widom, J. Chem. Phys. *84*:6943 (1986).
387. D. Andelman, M. E. Cates, D. Roux, and S. A. Safran, J. Chem. Phys. *87*:7229 (1987).
388. M. E. Cates, D. Andelman, S. A. Safran, and D. Roux, Langmuir *4*:802 (1988).
389. G. Gompper and J. Goos, Phys. Rev. E *50*:1325 (1994).
390. P. Pieruschka and S. Marčelja, J. Phys. II *2*:235 (1992).
391. P. Pieruschka and S. A. Safran, Europhys. Lett. *22*:625 (1993).
392. C. G. Vonk, J. F. Billman, and E. W. Kaler,. J. Chem. Phys. *88*:3970 (1988).
393. I. S. Barnes, P. J. Derian, S. T. Hyde, B. W. Ninham, and T. N. Zemb, J. Phys *51*:2605 (1990).
394. T. N. Zemb, I. S. Barnes, P. J. Derian, and B. W. Ninham, Prog. Colloid Polym. Sci. *81*:20 (1990).
395. T. N. Zemb, Colloids Surf. A *129–130*:435 (1997).
396. H. Wennerström and U. Olsson, Langmuir *9*:365 (1993).
397. H. Wennerström, J. Daicic, U. Olsson, G, Jerke, and P. Schurtenberger, J. Mol. Liq. *72*:15 (1997).
398. J. Daicic, U. Olsson, H. Wennerström, G. Jerke, and P. Schurtenberger, J. Phys. II *5*:199 (1995).
399. D. Roux, F. Nallet, C. Coulon, and M. E. Cates, J. Phys. II, *6*:93 (1996).
400. J. Daicic, U. Olsson, H. Wennerström, G. Jerke, and P. Schurtenberger, J. Phys. II *6*:95 (1996).
401. O. Parodi, in *Colloides et Interfaces* (A. M. Cazabat and M. Veyssie, eds.), Ecole d'Eté, Les Ulis, Aussois, France, 1983, pp. 355–370.
402. C. Cabos, P. Delord, and Marignan, Phys. Rev. B *37*:9796 (1988).
403. S. Ezrahi, A. Aserin, and N. Garti, J. Colloid Interface Sci. *202*:222 (1998).
404. N. Garti, A. Aserin, E. Wachtel, O. Gans, and Y. Shaul, Langmuir *submitted*.
405. N. Garti, A. Aserin, I. Tiunova, and S. Ezrahi, J. Thermal Anal. *51*:63 (1998).
406. M. C. Holmes, Curr. Opin. Colloid Interface Sci. *3*:485 (1998).
407. R. Hołyst, Curr. Opin. Colloid. Interface Sci. *3*:423 (1998).

8

Ionic Microemulsions

Jean-Louis Salager and Raquel E. Antón
Universidad de Los Andes, Mérida, Venezuela

I. INTRODUCTION

A typical computerized literature search using the key word "microemulsion" would return more than 1000 references per year in the past decade. It is thus obvious that everything known on microemulsions cannot be said within a reasonable number of pages and that a selection must be made according to a criterion that depends not only on the importance of the subject or the reviewed research but also on the authors' choice.

This chapter is intended to deal with structural and practical aspects of ionic microemulsions, particularly those that are related to the attainment of such systems from the formulator's point of view, since we have the feeling that these aspects have often been neglected in favor of theoretical developments but that they are quite important as far as practical problems and applications are concerned.

The text is kept as pedagogical and as simple as possible, and the literature review is not exhaustive, particularly on theoretical aspects, but rather selective of important fundamental developments, historical milestones, and technological breakthroughs. The scope is limited by editorial choice to systems containing ionic surfactants and some ionic–nonionic mixtures such as those involving pH-sensitive systems.

II. MICROEMULSION—A SELF-ASSEMBLY STRUCTURE

A. Microemulsion—A Controversial Label

The first microemulsions to be recognized as something different from other known structures were mentioned more than 50 years ago [1], though they were then labeled "oleopathic hydromicelles." These early authors were able to attain transparency by adding a medium-chain alcohol to a whitish macroemulsion, i.e., a two-phase system containing drops of oil dispersed in water or drops of water dispersed in oil. They interpreted the attainment of transparency as a drastic reduction of the drop size, much below the wavelengths of visible light, that is, in the colloidal range where fragments of dispersed matter are too large for the system to behave as a normal solution but too small to be sensitive to gravity and thus separate by settling.

The question that arises from this situation is whether such a system should be considered as a single phase like a colloid or as a two-phase dispersion, i.e., an emulsion with extremely small drops, deserving the label "microemulsion" as proposed in 1959 [2]. This is not merely a semantic problem, as will be discussed later, but something related to some important thermodynamic considerations.

Such systems had been mentioned and even used commercially much before they were referred to as microemulsions. As early as World War I, it was known that the addition of some natural fragrance oil such as eucalyptus, turpentine, or pine extract to a soap solution improved the detergency property of the system. In the 1920s, carnauba wax dispersions with a partially neutralized oleic acid soap were producing emulsions with extremely small droplets, a feature that results in self-shining when dry. In many perfumes, a considerable amount of ethanol is added to facilitate the dissolution of fragrance, e.g., terpenic oil components, as in Cologne water. Whenever the ethanol content has to be reduced to avoid toxicity or irritation, transparency can be maintained by designing a proper microemulsion structure [3,4].

Even today there is still some controversy about what system should be called a microemulsion. The point is that a microemulsion is not necessarily an emulsion with extremely small droplets. Actually, this may be the least interesting type of microemulsion, and many researchers think that the word "microemulsion" must be kept for those truly single-phase surfactant–oil–water (and often cosurfactant) mixtures in which sizable amounts of both oil and water are cosolubilized in a complex structure that has been described as percolated, bicontinuous, and transient. As will become clear later, from the point of view of applications these structures exhibit extremely interesting properties such as extremely low interfacial tension and high solubilization.

Thus, it can be said that in the definition of a microemulsion, some characteristics are accepted by (almost) everybody in the field, while others have been proposed by some and rejected by others with various kinds of arguments. Generally accepted features include thermodynamic stability as a single-phase system under natural gravity. This means that a microemulsion is a colloid in which the fragments of matter are small enough not to settle and are protected from agglomeration (even by Brownian motion), presumably by the surfactant adsorbed at the fragment surface. Note that if it is a single-phase system, then the word "interface" would be incorrect to describe the limit of the fragments. Nevertheless, if the dispersed phase fragments are too numerous, they may start agglomerating to form a separate phase referred to as an excess phase, which would stay in equilibrium with the microemulsion which can be viewed as a saturated colloid. In the most interesting case, referred to as optimum formulation, the microemulsion can be in equilibrium with two excess phases (oil and water), a definite proof that the microemulsion is something different from both oil and water since neither can be considered the dispersed phase or used as a dilution medium.

Many authors have defined a microemulsion as a transparent or translucent system, a characteristic that is linked to the difference in refractive index between the phases and mostly to the size of the matter fragments and their volume concentration. It can be said that transparency requires fragmentation below a 5 nm size whereas translucency with no opacity may be exhibited with up to 50 nm or even 100 nm (0.1 μm) dispersed objects. Some bicontinuous microemulsions can exhibit a rather strong light scattering—the transmitted light is typically reddish, an indication of a strong light–matter interaction, while the reflected light is whitish—almost as much as a macroemulsion.

It is known that Tyndall scattering is evidence of the presence of micelles in surfactant solutions, with a detectable structure size that can be as small as 5 nm, particularly if a laser beam is used to perform light scattering. Thus, transparency is not really a good criterion in defining a microemulsion, because it is not limiting and it can also be applied to other systems. Since the size is not a good criterion in defining a microemulsion, transparency is not a good one either.

A better criterion in defining a microemulsion could be its fluidity. In effect, the viscosity of (macro)emulsions and suspensions is known to increase as the fragment size decreases, and thus it is expected that an emulsion with extremely small drop size, an internal phase content greater than 20–30%, and a monodispersed distribution (as expected in a microemulsion) would be quite viscous. However, systems with such a high viscosity have been called gel emulsions or miniemulsions because the authors preferred to elude the label microemulsion in order to avoid confusion with single-phase microemulsions [5–7].

A microemulsion is thus anomalous, because it is relatively fluid and much less viscous than expected. This has been linked with the flexibility of the structure and its transient behavior, as early mentioned by Winsor and Shinoda for the surfactant phase, as they called it (they did not use the word microemulsion). This is particularly true in the case in which extremely low interfacial tension allows easy deformation and low interfacial curvature, a feature that can be attained by adjusting the physicochemical formulation.

Of course, the self-assembly of surfactant molecules, whether it happens in a single solvent phase or in the presence of both oil and water, can lead to solidlike organized structures called liquid crystals, which are nevertheless nonstoichiometric [8].

Most liquid crystal structures are based on lamellar or bent layers containing surfactant molecules in such an arrangement that their polar groups strongly interact with each other, thus forming some kind of solidlike crystal pattern, while the organization at the level of the hydrophobic tail, which is due to the weaker London interactions, is much looser and more liquidlike, from which the term liquid crystal originates. In some ways a microemulsion can be considered a disordered or molten state of a liquid crystal.

The overall rigidity of a liquid crystal can be altered by two effects. One is the disorder introduced by an increase in temperature and the corresponding increase in molecular motion that debilitates directional forces such as polar interactions. The other is the weakening of polar interactions (nonionic hydrophilic instead of ionic, or the introduction of alcohol cosurfactant) or the promotion of disparity in the molecular size and shape, by introducing double bonds, random tail length, tail branching, surfactant–cosurfactant mixtures, etc. All these alterations have been used to make microemulsions.

By the way, ionic surfactants generally do not lead to microemulsions at ambient temperature, but to liquid crystals, which are organized systems. To produce a microemulsion, it is often required to introduce disorder either by increasing temperature (up to 50°C) or by adding a cosurfactant, generally an alcohol [9,10].

B. Microemulsions vs. Mini- and Macroemulsions

What should be called a microemulsion is not important in itself, but since there are definition discrepancies in the publications of various authors it is important to make a point as to what will be called a microemulsion here and how to differentiate it from other self-assembly structures. The definition that is adopted here comes from the phase behavior of ternary systems that contain an amphiphile component and two immiscible so-called oil and water phases.

The term "water" refers to a polar phase that is generally an aqueous solution containing electrolytes and other additives, provided that they behave reasonably as a pseudophase. In extreme cases it can be quite different from water, e.g., ethylene glycol or formamide. In the practical cases that are discussed here, the polar phase would be plain water or sodium chloride brine, the concentration of which would be a characteristic of the polarity of water.

The word "oil" is used for an organic phase that is essentially immiscible with water and is thus relatively apolar. It can refer to a hydrocarbon, a partially or totally chlorinated or fluorinated hydrocarbon, single-chain alkane, cyclic or aromatic hydrocarbon, polar monoester such as ethyl oleate, long-chain alcohol such as dodecanol, triglyceride natural oil, or polycyclic cholesterol. The most typical oil phase is, of course, n-alkane, which would be characterized by its length or alkane carbon number (ACN).

The word "amphiphile" was coined by Winsor [11] from *amphi* (both sides, around) and *philos* (liking) to define substances with an affinity toward both polar and apolar phases. Surfactants are the most important amphiphiles, having a dual affinity that is strong enough for them to be driven to the interface so that the polar group is located in the polar phase and the apolar group is located in the oil phase. This means that the surfactant must exhibit a strong affinity for both phases and in some separated way. For instance, ethyl alcohol is an amphiphile, but its tail is so short that the corresponding hydrophobic effect is not strong enough to drive it out of the water phase. On the other hand, dodecyl alcohol is also an amphiphile, but the OH group is not polar enough to compensate for the long hydrocarbon chain, and it would stay mostly in the bulk of the oil phase.

In ionic surfactants the polar group is compact while the apolar group is a long hydrocarbon chain with typically 10 or more methylene groups. In effect, the lateral interaction between each surfactant methylene group and a similar group in the oil hydrocarbon chain is typically 10 times weaker than the polar group interaction, and thus a certain multiplicity of such groups is required to approximately reach a balance between the two tendencies (polar and apolar). This means that an ionic surfactant will exhibit some geometrical asymmetry in interactions, a situation that might not arise with nonionic surfactants. This is related, by the way, to the fact that a cosurfactant is in most cases required to form a microemulsion with ionic surfactants.

The phase behavior to be discussed in the following is the simplest ternary case presented by Winsor over 40 years ago [11], in which the amphiphile is either a pure surfactant or, as often in ionic systems, a mixture of surfactant and alcohol, which can be considered a pseudo-component, because the surfactant and alcohol behave in a collective way as far as their affinities of the oil and water phases are concerned. This is often the case with an intermediate alcohol such as *sec*-butanol.

Let us first analyze a typical phase behavior diagram such as the one in Fig. 1. On the left (AW) and right (AO) sides, the single-phase region extends from 100% amphiphile to 100% water or oil, which means that the amphiphile is completely soluble in both water and oil. In contrast there is a miscibility gap region near the OW side, since the O and W components are not compatible and their mixture results in a phase separation. The miscibility gap width decreases as more and more amphiphile is added to finally end up at the double line concentration of amphiphile, above which both the W and O phases are cosolubilized by the amphiphile in a single-phase region that would exhibit a large amount of both W and O phases and is likely to be a microemulsion or liquid crystal structure.

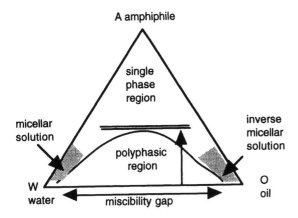

Figure 1 Typical amphiphile–oil–water ternary diagram.

On the left and right sides of the polyphasic region (near the OW side extremes), the single-phase region extends up to the pure W or O component. At the left corner, the amphiphile aqueous solution would contain micelles, which might eventually solubilize a small amount of oil (narrow band). At the right corner, inverse micelles in oil would eventually solubilize water. It is worth mentioning that in Fig. 1 the widths of these strips are grossly exaggerated for the sake of clarity, since in most practical cases the WO miscibility gap extends from less than 1% to more than 99% oil.

Note that since the miscibility gap decreases with increasing amounts of amphiphile, both micellar regions exhibit some wedge shape, i.e., their width increases as more and more amphiphile is added. This means that the solubilizing ability, e.g., the amount of oil solubilized in the aqueous micellar solution, increases. Since the amount of amphiphile is linked with the micellar surface area (where the amphiphile is located) and its solubilizing capacity is linked with the micellar core volume, it can be said that the solubilizing ability (as the ratio of solubilized substance to amphiphile) changes with the ratio of micelle volume to micelle surface, i.e., it is proportional to the micellar radius. Thus an increase in solubilization performance is necessarily associated with an increase in micellar size, which is in turn linked with curvature and bending considerations that are related to the concept of formulation.

As more and more (swollen) micelles are produced because of an increase of amphiphile amount, a point is reached where they start interacting according to the so-called percolation or clustering phenomenon, mostly studied for inverse micellar systems by conductivity [12–16]. This typically occurs when the micellar volume attains 20% of the overall volume but could take place even at a lower proportion if a charge fluctuation enhances the attraction of neighboring droplets [17], which could be promoted by other forces related to electric field or solute structure [18].

Above this 20% volume the structure of the dispersion is no longer that of independent fragments, and it is reasonable to use the term "percolated" or "bicontinuous" microemulsion.

Provided that there is no limitation from the standpoint of formulation, the microemulsion region thus extends above the polyphasic region from left to right, with the exception of two thin strips near the AW and AO sides, where the insufficient amount of solubilized either O or W would make the system binary rather than ternary. In this case, a dispersion of extremely fine and independent droplets would be referred to as an O/W or

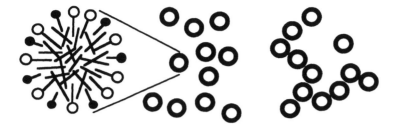

Figure 2 Oil-in-water microemulsions with dispersed structure (center) and percolated structure (right).

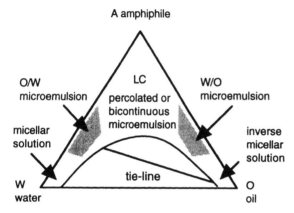

Figure 3 Regions where microemulsions are found.

W/O microemulsion, a good fit with the original labeling (see Fig. 2). Note that in such a case the amount of solubilized phase in the micellar core is less than the amount of surfactant, a situation associated with a low solubilization capacity.

Whenever a system has a composition that lies in the polyphasic region, it will generally separate (at equilibrium) into two phases, the representative points of which are located at the two extremes of the tie-line (see Fig. 3). In most cases the tie-line is inclined; i.e., one of the phases is rich in surfactant because it is located relatively near A or far from the OW side. If it is also located far from the AW and AO sides, then it contains both W and O in sizable amounts and fits the definition of a microemulsion (shaded region). Near the upper end of the tie-line in Fig. 3, it is an O/W type microemulsion. The other extreme of the tie-line is located near the OW side and near one of the component vertices (O in Fig. 3) and thus contains essentially one of the components. It is called an excess phase, in this case an oil excess phase. In most cases, particularly with ionic surfactant, the excess phase does contain a very small concentration of amphiphile, about the critical micelle concentration (cmc). In other words, the excess phase does not contain micelles, and as a consequence no micellar solubilization of the other phase can occur in the excess phase, an important feature when the mass balance is to be discussed.

Whenever a two-phase system is stirred, a macroemulsion is attained, either O/W or W/O depending on the dispersed phase, a situation that is related mainly to the formulation and to a lesser degree to the water/oil ratio. As a general rule, it can be said that the stirring of such polyphasic systems, even an extremely energetic stirring, cannot produce drops in the submicrometer range, whereas this fragmentation scale is spontaneously attained in microemulsions.

Nevertheless, some physicochemical tricks such as the triggering of instability in oversaturated systems can produce a very fine dispersion of one phase in the other, i.e., a macroemulsion made up of two immiscible phases but with extremely fine droplets, e.g., in the 10 nm range. Such extremely fine macroemulsions, which could be called miniemulsions, are not thermodynamically stable, but they can exhibit a stable appearance because of the large entropy term $(-T\Delta S)$ contribution that decreases their free energy and because of the presence of retardation phenomena in the kinetics of coalescence. It is worth remarking that a considerable number of such miniemulsions are called microemulsions in the scientific and technical literature. In any case, such miniemulsions can be diluted with their external phase, as is done with cutting oil concentrates, without losing their stability. On the other hand, microemulsions cannot be diluted indefinitely.

Coming back to the single-phase region at high amphiphile content, it is seen in Figs. 1 and 3 that there is a so-called bicontinuous microemulsion region from left to right, i.e. from O/W microemulsion to W/O microemulsion. In the central zone, i.e., in the nonshaded microemulsion area, the phase continuity must switch from W to O, and there is plenty of experimental evidence that some bicontinuity is exhibited, the characteristics of which depend on the formulation and the type of diagram, as will be discussed later. The flip from one continuity to the other can also occur through a rigid liquid crystalline phase, for instance a lamellar structure in which amphiphile bilayers alternate with water and oil layers.

The occurrence of lamellar liquid crystals or other types such as hexagonal or cubic depends on the relative amounts of O and W as well as on the formulation, i.e., the interaction balance between the surfactant and the O and W phases.

The actual structure also depends on the surfactant molecular structure. For instance, dual-tail amphiphiles such as sulfosuccinate surfactants are more likely to produce W/O-type miniemulsions and microemulsions with water core islands. If too much water is present, because of the inability of this surfactant to accommodate its branched double tails in an oily core, it would result in more complex structures such as vesicles, in which a surfactant bilayer closes on itself, as shown in Fig. 4.

The flexibility of the possible structure depends, in part, on the lateral interactions between neighboring amphiphile molecules. If the interaction is highly directional as with many ionic surfactants, an almost solid layer is formed that leads to liquid crystal or vesicle structures. To produce a liquidlike microemulsion, some disorder should be introduced. There are several ways to do so.

The first way to produce a microemulsion is to increase the temperature, which results in augmented molecular motion and debilitates the directional interactions. This method is used mostly in nonionic systems, but it works also with some ionic surfactants such as α-olefin sulfonates.

The most commonly used method is the addition of alcohol, which results in a double effect. First, the alcohol molecules get inserted in between surfactant molecules and push them apart, with a corresponding reduction in polar interactions and rigidity. Second, the alcohol also exhibits an overall lower interaction because of a shorter chain. The best alcohol to use as cosolvent is the one that adsorbs more at the interface or at the structure's palisade layer and has the same balance of affinity as the surfactant, i.e., C_3 or branched C_4-C_5 alcohols.

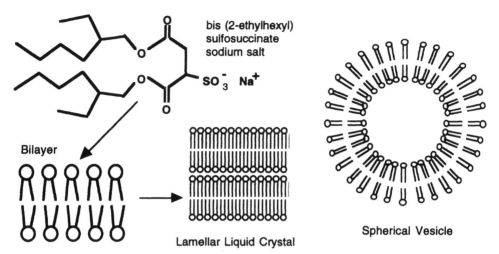

Figure 4 Structures occurring with double-tailed surfactants.

III. PHASE BEHAVIOR AND FORMULATION CONCEPTS

A. Winsor *R* Ratio Approach

In his pioneering work, Winsor associated the phase behavior of amphiphile–oil–water systems with the variables that are now recognized as those describing the physicochemical formulation, i.e., the nature of the different components that make up the systems. In the original series of eight papers [19] published in 1947, Winsor reported the phase behavior of systems containing an amphiphile such as alkyl sulfate or alkylammonium chloride, a hydrocarbon, an aqueous solution of electrolyte, and often a nonionic additive such as glycol, hexanol, or amine. By varying the nature of the different components and their respective amounts, he was able to determine the phase boundaries of completely solubilized systems and the nature of the phase equilibria that are still labeled according to his early definitions [11].

Figure 5 indicates the different cases of phase behavior and self-assembly structures according to Winsor notation, which has been followed by many other researchers who rediscovered Winsor's work during the enhanced oil recovery research drive of the 1970s [20–22].

In type I phase behavior, an S_1 type of aqueous micellar system (and its extension to an O/W microemulsion when swollen micelles occur or to a percolated microemulsion if a large amount of oil is solubilized in the micellar core) is in equilibrium with almost pure oil. This is the phase behavior exhibited in the polyphasic region of the so-called Winsor I ternary diagram. This phase behavior has also been labeled $\underline{2}$, since it appears as two phases with the surfactant-rich phase being the water or lower phase.

Conversely, a Winsor II diagram and phase behavior (also noted $\overline{2}$) corresponds to the opposite situation, in which the polyphasic equilibrium consists of an inverse micellar organic solution S_2 (that eventually solubilizes enough water to become a microemulsion) in equilibrium with an essentially pure aqueous phase.

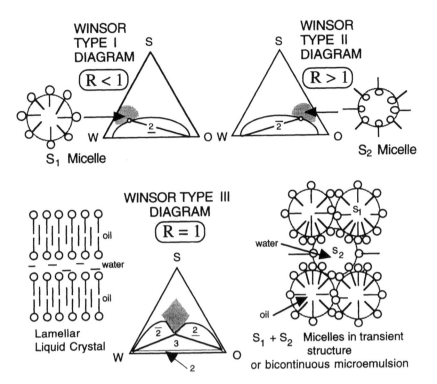

Figure 5 Phase behavior, R ratio, and types of structures according to Winsor notation. (From Ref. 11.)

It is worth noting that the tie-line slope in the polyphasic region clearly indicates the amphiphile partitioning, which is related to physicochemical formulation, for instance through the free energy of transfer of an amphiphile molecule from water to oil in a low surfactant concentration system, i.e., below the cmc [23].

In effect, the chemical potential of the surfactant in the water and oil phases can be written as

$$\mu^w = \mu^{*w} + RT \ \ln(C^w \phi^w) \tag{1}$$

$$\mu^o = \mu^{*o} + RT \ \ln(C^o \phi^o) \tag{2}$$

where the μ^*'s are the standard chemical potentials in some reference state, the C's are dimensionless concentrations, and the ϕ's are activity coefficients. Superscripts w and o refer to the excess water and oil phases, respectively. Since the water and oil phases are at equilibrium, their chemical potentials are equal. Since the surfactant concentration in the excess phases is low (at or below the cmc), it can be assumed that the activity coefficients are unity. As a consequence, the partition coefficient K of the surfactant between the water and oil phases can be written as

$$RT \ \ln \ K = RT \ \ln \frac{C^w}{C^o} = \mu^{*o} - \mu^{*w} = \Delta G_{w \to o} \tag{3}$$

Between the Winsor I and II cases, one could expect a situation in which the tie-lines would be horizontal as an indication that the amphiphile partitions equally in both phases. Such a case does occur with some amphiphiles such as alcohols [24], but not in general for surfactants, in which a type III diagram and phase behavior are exhibited instead [11]. The polyphasic region contains a three-phase zone surrounded by three two-phase zones. Systems whose compositions lie in the three-phase zone separate into a surfactant-rich phase that is in the middle of the diagram at the boundary of the microemulsion single-phase zone (shaded), and two excess phases, which are essentially pure aqueous phase and pure oil. This microemulsion phase has been called the middle phase because it appears between the oil and water phases in a test tube because of its intermediate density. Since it is at equilibrium with both excess phases, it cannot be diluted by either water or oil, and it is thus neither water nor oil-continuous but bicontinuous. These three-phase systems were the target of the enhanced oil recovery research drive in the 1970s, and a considerable amount of research has been carried out to determine both their conditions of occurrence and the properties of the associated systems.

In some cases Winsor [11] reported that the single-phase region (which he called type IV phase behavior) could also contain liquid crystalline phases with lamellar or other structure.

Based on the large number of experiments he carried out, Winsor was able to relate the phase behavior to the physicochemical situation at an interface and to propose an extremely pedagogical approach based on the ratio of the interaction energies at an interface. The Winsor notation is conserved in this text, as it is in most review work dealing with the subject [22].

Figure 6 represents the molecular population near the interfacial limit between the oil (O) and water (W) phases, where C represents the amphiphile or amphiphile mixture. The A's are the molecular interaction energies per unit area according to the regular solution theory. In Winsor notation,

$$A_{CW} = A_{HCW} + A_{LCW} \tag{4}$$

where A_{CW} represents the interaction between the amphiphile and the aqueous phase, A_{HCW} the interaction between the hydrophilic group of the amphiphile and the water, and A_{LCW} the interaction between the lipophilic group and the water, which is, by the way, probably much smaller than A_{HCW}. For a similar reason the interaction between the hydrophilic group of the surfactant and the oil, A_{HCO}, is probably much smaller than the interaction of the lipophilic group and oil noted as A_{LCO}.

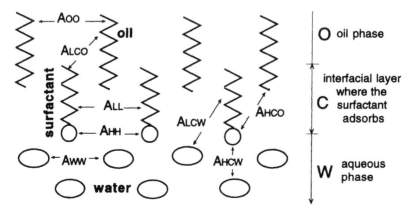

Figure 6   Molecular interactions near the interface according to Winsor notation.

According to Winsor, A_{CO} and A_{CW} promote miscibility, while A_{OO}, A_{CC}, and A_{WW} promote the segregation of the components in different phases or in regions of higher local concentration of these components. The cosolubilizing effect of the amphiphile C will be greater the higher both A_{CO} and A_{CW} are and the more nearly equal they are [25].

It was postulated by Winsor that the ratio of A_{CO} to A_{CW} would provide a way to determine the convexity of the C layer, that is, the phase behavior situation. The original R ratio was written as A_{CO}/A_{CW}, and it was suggested later that a better ratio would be

$$R = \frac{A_{CO} - A_{OO}}{A_{CW} - A_{WW}} \tag{5}$$

Today an overall expression,

$$R = \frac{A_{CO} - A_{OO} - A_{LL}}{A_{CW} - A_{WW} - A_{HH}} \tag{6}$$

which takes into account all the differences in interactions between the mixed system and the separated phase states, is generally preferred. In any case, it does not make that much difference in reasoning.

For a given O and W, the C region would become convex toward W if $R < 1$, and conversely. This is essentially equivalent to the wedge model in which each amphiphile molecule is associated with a certain amount of both O and W molecules, with the resulting formation of a wedge unit, which determines the bending according to its side-by-side stacking. Modern theory has provided the appropriate theoretical framework for such a simple figure that can be used to carry out straightforward reasoning [26].

It is worth noting that this description assumes implicitly that the $R = 1$ case is associated with a zero-curvature C layer, which could be provided either by a lamellar liquid crystal structure with alternating O and W flat layers or a zero curvature surface of the Schwartz type or as a transient and fluctuating combination of S_1 and S_2 structures. It is now well recognized that middle-phase microemulsions, which are in equilibrium with both oil and water excess phases, exhibit bicontinuous structures as shown in Fig. 7 [27] that are not far from the transient mixture of S_1 and S_2 swollen micelles predicted by Winsor.

Since the phase behavior could be readily determined by experience, and since it was possible to infer the qualitative effect of each formulation variable on R, Winsor was able to corroborate the physicochemical formulation influence as indicated in Table 1.

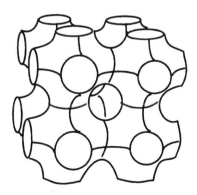

Figure 7 Shape of the amphiphile film in a bicontinuous structure model for microemulsion in a type III diagram.

Table 1 Influence of the Various Formulation Variables on the Interaction Energies and R

Variable	Interaction energy	R
Amphiphile hydrocarbon chain length increases	A_{CO} increases	Increases
Aqueous phase salinity increases	A_{CW} decreases	Increases
Aqueous phase electrolyte valence increases ($Na^+ \rightarrow Ca^{2+}$)	A_{CW} decreases	Increases
Oil phase length (hydrocarbon) increases	A_{CO} increases	
	A_{OO} increases more	
	$A_{CO} - A_{OO}$ decreases	Decreases
Temperature increases (ionic surfactant)	A_{CW} increases	Decreases

The first two cases are readily explained by the change in the numerator and denominator of R. The third case is related to the fact that with most anionic surfactants, sodium or potassium salts are more dissociated than their bivalent cation counterparts. As a consequence, calcium or magnesium salts are less hydrophilic.

The effect of the oil chain length or alkane carbon number (ACN) is slightly more complex to understand, since it alters two interactions with opposite effects. In order to make a qualitative interpretation it can be stated as a first approximation that these London-style interactions are proportional to the number of methylene groups in each interacting molecule [28]. Thus,

$$A_{CO} = \alpha(SACN)(ACN) \quad \text{and} \quad A_{OO} = \alpha(ACN)^2 \tag{7}$$

where SACN is the number of carbon atoms in the surfactant alkyl group. As the oil ACN increases, the second interaction increases faster (as the square) than the first one. As a consequence, $A_{CO} - A_{OO}$ decreases (provided that ACN is large enough). In other words, a higher ACN oil is a poorer solvent than a lower ACN oil because of its more favorable self-association energy.

B. Practical Experimental Techniques and Calculations

When R increases, the change in the phase behavior is Winsor I $\rightarrow$ Winsor III $\rightarrow$ Winsor II, and conversely. Since Winsor III occurrence precisely corresponds to both an experimentally observable situation and a fundamental concept, a considerable amount of work has been dedicated to it since Winsor's time.

The first technique is the so-called unidimensional formulation scan, in which a series of test tubes contain the same composition, i.e., the same amphiphile/oil/water proportions, but gradually changing formulation. For instance, the salinity of the aqueous phase is changed from one test tube to the next according to some progression (linear or geometric). It also could be the oil ACN or the surfactant tail length.

The aspect of the test tube sequence is depicted in Fig. 8, where the shaded area indicates the surfactant-rich phase. The composition of the system is typically 2% amphiphile and 49% each oil and water, so that the representative point is likely to be in the polyphasic region of the diagram.

At low salinity a type I phase behavior is observed, and at high salinity a type II. In the middle of the scan, type III phase behavior presents a surfactant-rich microemulsion middle phase in equilibrium with both water and oil excess phases. One of the type III tubes (here percent NaCl = 2.8) exhibits a microemulsion that contains exactly the same amounts of cosolubilized oil and water. This phase is represented by a point that is equidistant from

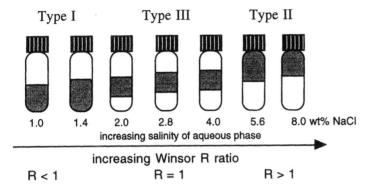

Figure 8 Unidimensional formulation scan. Typically observed phase behavior for a system containing 2% anionic surfactant, 49% brine, and 49% alkane, versus the brine salinity.

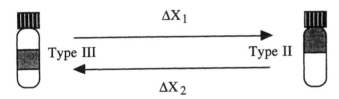

Figure 9 Double formulation change to produce compensating effects. Starting at an optimum formulation system (left), a formulation variable (X_1) is changed to attain an off-optimum system (right). Then a second formulation variable (X_2) is changed until three-phase behavior is restored.

the W and O vertices on a ternary diagram. This point of the formulation scan represents the so-called optimum formulation of the scan for the reasons discussed later. Such a phase behavior pattern takes place irrespective of the formulation variable (either in the increasing or decreasing direction of R, according to the scanned formulation variable).

The quantification of formulation effects requires a method to find the equivalence between any two changes in the physicochemical situation. The most common experimental technique has been the bidimensional scan, which is based on a double variation of competing effects and is carried out as illustrated in Fig. 9.

Starting at a type III physicochemical state ($R = 1$), a first formulation variable is changed, which results in a change in one (at least) of the interaction energies and also results in a change in the R value. For instance, suppose that the increase in variable X_1 produces a type III → type II transition, which means that R has increased to some value $R > 1$. Let us now select a second formulation variable X_2, and let us change it in the direction where the change in R would be opposite to the previous one, i.e., a decrease in R, until a type II → type III transition is attained again, back to $R = 1$. At this point it can be said that the two formulation changes ΔX_1 and ΔX_2 have produced exactly opposite effects on R and are thus equivalent from the physicochemical point of view but of opposite sign.

This technique of double change with compensating effects allows us to quantify the respective influence of the different formulation variables, and it has been widely used in performing a large amount of experimental work to quantify the Winsor concept. This was quite important because the Winsor R reasoning is fairly simple and is useful for

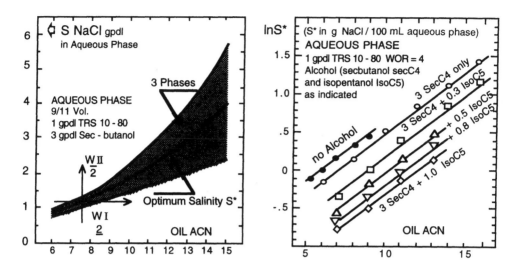

Figure 10 Phase behavior on a bidimensional scan map showing optimum formulation lines. Surfactant and alcohol concentrations are expressed in grams per deciliter (g/dL). (Adapted from Ref. 31.)

predicting general trends and carrying out qualitative reasoning, but it is not useful for making calculations because the computation of molecular interaction energies is still beyond supercomputer reach whenever more than 1000 molecules are considered, which is an extremely small amount of matter.

The formulation scan and dual compensating scan techniques were carried out on an experimental basis first with anionic surfactant systems such as alkyl aryl sulfonates, which were the primary candidates for enhanced oil recovery because of their low price and availability.

Extensive experimental work showed that a three-phase system with type III phase behavior was attained whenever a certain relationship, termed the correlation for optimum formulation, was satisfied. The term "optimum" came from the fact that this formulation was also associated with the lowest possible interfacial tension value in the scan, which led to the most reduction in capillary forces and maximum oil recovery. It is worth mentioning that type III systems are now recognized as corresponding to the bicontinuous microemulsions that exhibit the highest solubilization of oil and water per unit weight of amphiphile and thus also the optimum situation whenever the solubilizing ability is sought after, which is probably the most important application of microemulsions outside the petroleum industry.

As far as the experimental determination of the exact position of the optimum formulation of a scan is concerned, it should be remarked that three-phase behavior is generally exhibited over a certain range of the scanned variable. However, this is not a problem, because it has been shown that the optimum formulation can be taken as the center of this range in most cases [22,29,30].

Figure 10 shows the phase behavior and the optimum formulation line in the subspace S–ACN, i.e., with all other variables held constant. The arrows on the left graph indicate the two kinds of formulation scans, i.e., a salinity scan at constant oil nature (vertically) and an ACN scan at constant salinity (horizontally). It is seen that an increase in salinity produces a type I → type III → type II transition, whereas an increase in ACN results in the opposite.

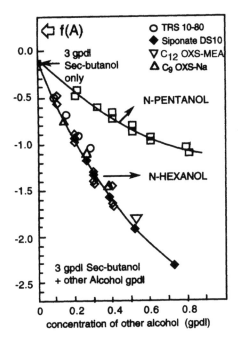

Figure 11 Alcohol effect as the shift of optimum formulation $f(A)$ versus alcohol concentration for different anionic surfactant systems. All systems contain 3 g/dL *sec*-butanol plus a certain concentration of other alcohol (*n*-pentanol or *n*-hexanol). (Adapted from Ref. 31.)

On the right-hand graph, the salinity scale is taken as the Naperian logarithm of salinity. It is seen that the centerline of the three-phase region becomes a straight line so that the expression ln S vs. ACN is linear, with a slope indicated as K. All the points along this line correspond to an optimum formulation where $R = 1$.

The correlation is verified when this dual scan is repeated with different kinds of formulation variable pairs, for instance, with different alcohol contents as in Fig. 10.

The first correlation for optimum formulation was found for anionic surfactant systems of the alkylbenzene sulfonate type, with sodium chloride brine and alkane oil and light alcohol as cosurfactant [31].

$$\ln\ S - K\,(\text{ACN}) - f(A) + \sigma - a_T \Delta T = 0 \tag{8}$$

where S is the salinity of the aqueous phase expressed as weight percent NaCl, ACN is the alkane carbon number, $f(A)$ is a function that takes into account both the type and concentration of the alcohol, σ is a parameter characteristic of the surfactant structure (both hydrophilic and lipophilic groups) that increases linearly with the hydrophobe tail length, and ΔT is the deviation from ambient standard temperature (25°C). K and a_T are numerical coefficients.

Figure 11 shows the alcohol effect as a function of the type and concentration of alcohol. It should be mentioned that this way of taking the alcohol into account, i.e., as a formulation "external" variable instead of considering it a part of the amphiphile, leads to a good approximation with most alcohols except *n*-butanol and *sec*-pentanol because the partitioning of these alcohols is affected by the type of oil and surfactant, so that the $f(A)$ value is not independent of the other variables. Figure 11 shows that this is not the case for n-pentanol and n-hexanol.

This kind of correlation was first attained from empirical experimentation, and a few years later it was recognized to be an expression of the "surfactant affinity difference" SAD (affinity is the negative of the chemical potential), i.e., the free energy of transfer of a surfactant molecule from the excess water to the excess oil in a three-phase system, which is related to the partitioning coefficient [32–34].

$$\frac{SAD}{RT} = \ln S - K \, ACN - f(A) + \sigma - a_T \Delta T \tag{9}$$

SAD values < 0, $= 0$, and > 0 correspond respectively to $R < 1$, $= 1$, and > 1 situations, but now the SAD numerical value and its changes can be readily calculated from the experimental data. Thus, SAD assigns a numerical value for the formulation concept, a quantitative improvement over Winsor's R.

Before discussing the respective numerical values and thus the relative importance of the different variables, it is worth noting the signs they bear. A positive sign means that an increase in the value of that variable would produce a type I → type III → type II transition (increase in R or in SAD), while a negative sign would correspond to the opposite transition. Results from Table 1 are thus corroborated; e.g., an increase in salinity or in the surfactant hydrophobe tail length (and thus σ) would produce a change opposite to the one resulting from an increase in oil length (ACN) or temperature.

K is found to be 0.16 ± 0.01 for alkyl aryl sulfonate sodium salts and 0.10 ± 0.02 for alkyl sulfates and fatty acid sodium salts. The temperature coefficient a_T is approximately 0.01 for all anionic surfactants; however, this value may be inaccurate ($\pm 20\%$), and a change in temperature over the whole 0–100°C range does not considerably affect the formulation [32].

For instance, a change in formulation corresponding to a doubling in aqueous phase salinity, for instance from 1 wt% to 2 wt% NaCl, i.e., a change $\Delta \ln S = 0.69$, can be compensated for by an increase in oil alkane carbon number $\Delta ACN = 0.69/0.16 = 4.3$ units. To attain the same compensation, the required increase in temperature should be $\Delta T = 0.69/0.01 = 69°C$.

In a similar fashion it would require $\Delta \sigma = -0.69$. Since σ/K is found to increase by 2.25 ± 0.05 units per additional methylene group or SACN (surfactant alkyl carbon number), as will be seen later, the same salinity increase can be compensated for by reducing the SACN by $0.69/(2.25 \times 0.16) = 1.9$, i.e., almost two methylene groups. Note that it is a reduction because the sign is positive in front of both $\ln S$ and σ and because σ increases with SACN.

The correlation allows us then to calculate the expected trade-off between formulation variables and to select the most suitable one for a given application, which is very useful technical and economical information in the practice of formulating bicontinuous microemulsions.

A recent study showed that the same correlation applies for cationic surfactant systems, although with a slight difference in the coefficient values [35,36].

$$\frac{SAD}{RT} = \ln S - K \, ACN - f(A) + \sigma - c_T \Delta T \tag{10}$$

The value of the constant K ranges from 0.15 to 0.20, which is essentially similar to the range found for sulfonate anionic surfactant systems. Although the available data are scarce, it seems that K is rather higher (0.19 ± 0.02) for quaternary compounds than for

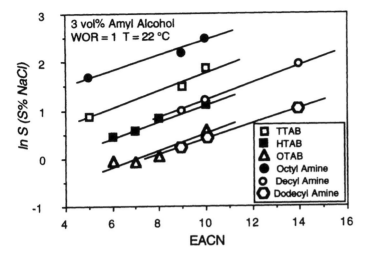

Figure 12 Optimum formulation of systems containing cationic surfactants (alkyltrimethyl-ammonium systems at pH 7 and amine systems at pH 3). (From Ref. 35.)

alkylammonium compounds at low pH ($K = 0.17 \pm 0.02$). The temperature coefficient c_T is about twice as large (0.02 ± 0.002) as in the case of anionic surfactant systems, an indication that cationic surfactants are more temperature-sensitive, as they also become more hydrophilic as temperature increases.

Figure 12 indicates the phase behavior of cationic systems as a function of two formulation variables, for both quaternary (tetra-, hexa-, and octadecyltrimethylammonium bromides) and amine salts at pH 3.

The effect of an alcohol is poorly documented. From the few data available it can be said that it is about the same for quaternary surfactants as for anionic ones, and 30% lower for alkylammonium salts.

Table 2 lists the characteristic parameter EPACNUS (σ/K) values for selected ionic surfactants. EPACNUS stands for extrapolated preferred alkane carbon number at unit salinity and no alcohol. It is the ACN value fround from equation (8) when $\ln S = 0$, $f(A) = 0$ and $\Delta T = 0$, and it is this characteristic of the surfactant. The EPACNUS parameter allows us to compare surfactant hydrophilicity against the ACN scale, which has exactly the same meaning in all correlations, whether for ionic or nonionic systems. The lower the value of EPACNUS, the higher the hydrophilicity of the surfactant.

From these data it can be readily stated that EPACNUS increases 2.25 units per additional methylene group in the surfactant lipophilic tail for all ionic surfactants, an indication that the transfer energy of an additional methylene group from water to oil is quite independent of the type of hydrophilic group, which is very consistent with the SAD concept and the independence of the different formulation variables in the correlations.

The data also show that the most hydrophilic polar group is the sulfate, very narrowly followed by the ionized carboxylic group (at high pH), i.e., as a soap. Another quite hydrophilic one is the quaternary ammonium, while it is seen that the benzene sulfonate and amine salt (at low pH) headgroups are much less hydrophilic.

For nonionic surfactants a similar correlation was found, but now the temperature coefficient is positive (thus an opposite effect) and is approximately 0.6, which is quite large compared to those of ionic surfactants. This antagonism can be useful in designing

TABLE 2 Characteristic Parameter Values for Selected Ionic Surfactants

Surfactant	EPACNUS (σ/K)	
C_9 alkylbenzene sulfonate Na salt	-10	
C_{12} alkylbenzene sulfonate Na salt	-3	
C_{15} alkylbenzene sulfonate Na salt	$+4$	
C_8 alkylammonium chloride (at low pH)	-15	
C_{14} alkylammonium chloride (at low pH)	0	
C_{14} alkyltrimethylammonium chloride	-14	
C_{18} alkyltrimethylammonium chloride	-4	
C_{14} alkylbenzene sulfonate Na salt	1.5	
C_{14} alkyl sulfate Na salt	-22	(± 3)
C_{14} fatty acid Na salt (at high pH)	-20	(± 3)
C_{14} alkylammonium chloride (at low pH)	0	
C_{14} alkyltrimethylammonium chloride	-14	

ionic–nonionic surfactant mixtures for making microemulsions and macroemulsions that are insensitive to temperature [37–43], a feature that is not easy to attain with single-surfactant systems [44]. On the other hand, the effect of the electrolyte is much less on nonionic surfactant systems than on ionic ones [45].

C. Advanced Features of Physicochemical Formulation

Since the first correlation was proposed for some anionic surfactant systems, its application has been extended to other systems.

The equivalent salinity produced by other electrolytes was analyzed [46]. The sodium salt anions were found to be equivalent on a normality basis (not on a weight percent basis) to sodium chloride according to the relationship

$$S_1 \text{ (in mol/L equivalent NaCl)} = \frac{2}{1 + Z_{\text{anion}}} S \text{ (in mol/L Na salt)} \tag{11}$$

or

$$S_2 \text{ (g-equiv NaCl/L, i.e., roughly wt\%)} = \frac{2}{1 + Z_{\text{anion}}} \left(\frac{58.5}{\text{MW}} \right) S \text{ (in g/L Na salt)} \tag{12}$$

where Z is the valence of the anion and MW is the molecular weight of the sodium salt. S_2 is the salinity value that is to be entered in the correlation for ionic surfactants, as $\ln S_2$ in Eq. (8), for polyvalent sodium salts.

The equivalent salinity for different sodium salt mixtures has been found to be easily computed by means of the equivalent valence of the mixture, Z_m, which is calculated from

$$\frac{2}{1 + Z_m} = \sum \frac{2x_i}{1 + Z_i} \tag{13}$$

where the x_i indicates the molar fraction of electrolyte i in the mixture.

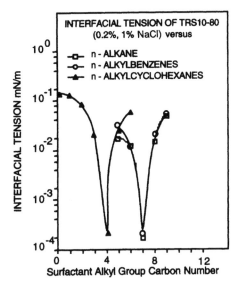

Figure 13 Change of interfacial tension along an ACN formulation scan showing the equivalence between the different oil families. (From Ref. 29.)

These results indicate that the higher the anion valence the less the ionic strength as far as equivalent salinity is concerned. This was readily translated by defining the valence activity factor (VAF) as follows, for both single sodium electrolyte and mixtures:

$$VAF = \frac{2}{1 + Z} \tag{14}$$

The effect of divalent cations is not amenable to empirical expression, probably because these cations influence the degree of dissociation of the salt, and thus the hydrophilicity of the polar group, in a complex way. It is found that the calcium and magnesium salts of most anionic surfactants are (much) less hydrophilic than their sodium counterparts as a general trend. Sulfonates are less sensitive than sulfates, which are less sensitive than carboxylates [47–49].

The correlations were also extended to oil phases other than n-alkanes. In each case an equivalent alkane carbon number (EACN) was found, which means that any oil or oil mixture behaves similarly to a certain alkane [50].

The experimental data on the attainment of the minimum interfacial tension, i.e., the same criterion as optimum formulation for the three-phase behavior at low surfactant concentration, showed the equivalence of linear and cyclic hydrocarbons [29]. Figure 13 shows that the EACN of alkylbenzene hydrocarbons is equal to the number of carbons of their alkyl group, while the EACN of alkyl cyclohexanes is equal to the number of carbon atoms in their alkyl group plus 3. Thus it can be said that a saturated cycle with six carbons is equivalent to a three-carbon linear chain, while the benzene ring does not bring any additional lipophilicity.

It has been found that the presence of a polar group in the oil molecule tends to drastically reduce its EACN. For instance, ethyl oleate EACN was found to be near 6, a considerable reduction from the expected value for the C_{18} chain [51,52]. Thus it can be said that the ester polar group results in a drastic reduction of the oil EACN, a characteristic parameter that is definitely linked to the oil-phase polarity. Other data with a natural triester such as soybean oil (a $C_{16}-C_{18}$ triglyceride) indicate an EACN around 6, i.e., a consistent value compared with ethyl oleate.

Chlorinated hydrocarbons were also found to exhibit a different EACN than their hydrocarbon counterparts [53–55]. The fact that these oils can form microemulsions make them retrievable in remedial cleaning processes, which have been developing at a fast pace in recent years [56,57].

Silicone oils with amino groups have also been solubilized in ionic surfactant microemulsions [58]. Organic bases such as pyridine, picoline, isoquinoline, and piperidine have been solubilized in anionic and cationic microemulsions [59], as well as dialkyl phthalate ester oil [60] or diesel oil [61], but no EACN value was reported.

Microemulsions have been achieved with supercritical fluids, particularly light hydrocarbons that can be solubilized in ionic systems [62,63], while carbon dioxide could be solubilized in only fluorocarbon surfactant and other nonionic systems [64].

Oil mixtures were found to exhibit an EACN that can be calculated according to a linear mixing rule on a mole fraction basis [50,65],

$$EACN_m = \sum x_i \, EACN_i \tag{15}$$

where x_i and $EACN_i$ are the mole fraction and EACN, respectively, of component i in the oil-phase mixture. According to this rule, even negative $EACN_i$ values have a physical meaning of reducing the EACN of the oil phase. For instance, it was found that the hydrogen sulfide effect in a liquid hydrocarbon mixture could be rendered by a negative value of its EACN.

Pressure is not expected to have a strong effect on a liquid system. This was corroborated in an experimental study showing that an increase in pressure produced the same transition as an increase in EACN, a change possibly due to an increase of the oil-phase density [66–68].

Other ionic surfactant systems were also studied, particularly isomerized α-olefin sulfonates [69] and other branched surfactants [70], polyalkyl ones such as dioleyl phosphates [71], and zwitterionic surfactants [72].

Another way to attain intermediate or different properties is to use surfactant mixtures. The characteristic parameter (σ or $\sigma/K = $ EPACNUS) of a surfactant mixture containing relatively similar substances can be calculated according to a linear mixing rule based on the molar fractions at the interface [73,74].

Since these proportions are not generally known, the molar fraction in the whole system is taken instead. This is a good approximation only if substantial fractionation does not take place, which is the case in many ionic surfactant mixtures of not-too-different species. Even petroleum sulfonates are found to follow that rule, and it can be said that most ionic surfactant mixtures of the same nature would exhibit almost linear behavior with respect to their characteristic parameter [37,73].

It is worth remarking that this is not the case for ethoxylated nonionic surfactants [75] or for some anionic ones such as disulfonates [76]. Sometimes surfactant mixing allows the cosurfactant requirements to be minimized [77].

A very particular mixture is the case of pH-sensitive surfactants, which is dealt with later. In such a case the pH controls the equilibrium between the dissociated and undissociated species, e.g., a fatty acid and its sodium salt that are both surfactants but exhibit very different hydrophilicity, since one is anionic and the other one nonionic, and are thus likely to fractionate.

Mixtures of anionic and nonionic surfactants were proposed to provide temperature-insensitive systems [37], a suggestion that has considerable practical interest not only for microemulsion systems but also in emulsion polymerization and enhanced oil recovery. It was recently shown that since both the anionic and nonionic surfactants can be selected, this double degree of freedom can be used to attain both temperature insensitivity and mixture composition insensitivity so that the formulation is a particularly robust one as far as the applications are concerned [39].

Ionic–nonionic intramolecular mixtures, i.e., with both groups included in the same molecule, have also been intended for different purposes, with the ionic part providing a good hydrophilic character that is rather insensitive to temperature whereas the polyethoxylated nonionic (short) chain ensures a certain tolerance to electrolytes. A proper combination of both ionic and nonionic characters can result in temperature insensitivity for ethoxylated sulfonates in enhanced oil recovery microemulsions [78,79]. Other types of desired performance such as salinity tolerance can be attained, as in the case of lauryl ethoxy sulfate foaming agent for shampoos and foam booster formulations. If the number of ethylene oxide groups is low, say one or two, then the surfactant behaves like an ionic one but with increased electrolyte tolerance, particularly to bivalent cations. The electrolyte sensitivity can be appropriately estimated from the decrease in the ln S–ACN slope [called K in Eq. (8)] as the number of ethylene oxide groups increases. When the number of ethylene oxide groups exceeds four, these ionic–nonionic surfactants behave as though they were nonionic. However, they are much less sensitive to temperature than ordinary nonionic surfactants.

These surfactants are found to exhibit a solubilization ability higher than those of their purely anionic and nonionic counterparts. It is found that for such surfactants the transition temperature from a liquid crystal mesophase to a (disordered) microemulsion structure is inversely related to the amount of alcohol cosurfactant, a quite useful feature in practice.

IV. PH-DEPENDENT SYSTEMS

Enhanced oil recovery by alkaline flooding was proposed some years ago as an inexpensive way to take advantage of the acid components that occur naturally in some crude oils [80,81]. The stabilization of oil-in-water emulsions can also be attained this way. In these cases the carboxylic acid contained in the crude oil adsorbs at the O/W interface, where it is neutralized into a carboxylic salt with surfactant properties such as interfacial tension lowering or emulsification. Fatty amines and their cationic counterparts at low pH are routinely used to stabilize asphalt emulsions for roads and pavement.

Other systems, particularly those dealing with biological aspects, often contain carboxylic acids or other pH-sensitive substances like fatty amines or alkylammonium bactericides.

Such pH-sensitive systems are best treated as containing a mixture of two amphiphilic species that occur in different proportions depending on the pH. For instance, the fatty acid–fatty soap equilibrium in water [82] can be expressed as

$$\text{AcH} \rightleftharpoons \text{Ac}^- + \text{H}^+, \qquad K_a = \frac{[\text{Ac}_w^-][\text{H}_w^+]}{[\text{AcH}_w]} \tag{16}$$

Type II   Type III   Type I   Type III Type II

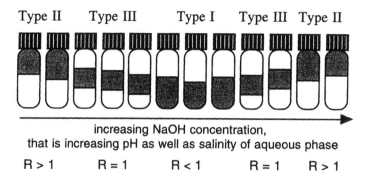

increasing NaOH concentration,
that is increasing pH as well as salinity of aqueous phase

R > 1   R = 1   R < 1   R = 1   R > 1

Figure 14 Phase behavior transitions as the concentration of sodium hydroxide is increased in a system containing a fatty acid, oil, and water.

Thus the relative amounts of the hydrophilic soap and the lipophilic carboxylic acid species depends on the pH according to

$$\frac{[Ac_w^-]}{[AcH_w]} = \frac{K_a}{[H^+]} \tag{17}$$

An increase in pH would favor the hydrophilic species, and thus the mixture would turn from lipophilic to hydrophilic at some point. As a consequence, the increase in pH is likely to produce an $R > 1 \rightarrow R = 1 \rightarrow R < 1$ transition as pointed out early by some authors [83].

However, it is worth noting that if the increase in pH is produced by adding a hydroxide such as NaOH, then the Na^+ salinity is also increased, and the reverse transition is likely to occur at some point when the second effect starts to dominate, as shown in Fig. 14. This is called a retrograde transition and has been found to happen whenever two opposite effects take place along a formulation scan.

To avoid this effect, some precaution has to be taken to deal only with the pH formulation effect. It is necessary to keep the electrolyte concentration constant while changing the pH, for instance by mixing NaCl and NaOH in different amounts in order to change the pH while keeping the whole sodium molar concentration constant so that the salinity stays unchanged.

Another kind of difficulty arises with pH-sensitive systems containing both an aqueous phase and an oil phase. It is due to the fact that the dissociated (ionic) and undissociated (nonionic) species are likely to selectively partition or fractionate between water and oil. As shown in Fig. 15, it can generally be assumed that the amount of dissociated salt in oil $[Ac_o^-]$ is negligible, whereas the amount of undissociated acid in water $[AcH_w]$ is to be taken into account since it appears in the dissociation equilibrium [84,85].

The partition coefficient of the undissociated acid between oil and water is defined by

$$P_a = \frac{[AcH_o]}{[AcH_w]} \tag{18}$$

For long-chain fatty acids, P_a can be in the 100–1000 range. In other words, neglecting the amount of acid present in the oil phase would be equivalent to neglecting most of it. It is not known for sure how the dissociation and partitioning equilibria are related to the composition of the interfacially adsorbed surfactant mixture layer, because such a system

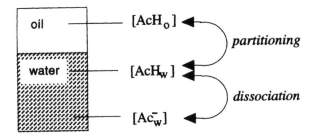

Figure 15 Acid–salt dissociation equilibrium in the aqueous phase and partitioning equilibrium of the undissociated species between the aqueous and oil phases.

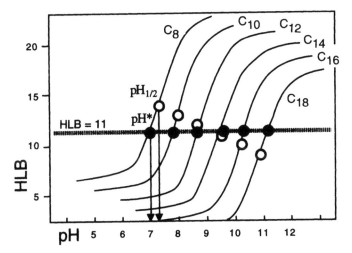

Figure 16 Optimum pH* for three-phase behavior and $pH_{1/2}$ at which half the acid is ionized (Adapted from Ref. 85.)

is more complex than the polyethoxylated nonionic case, the only one that has been elucidated. However, the partitioning effect can be qualitatively deduced from the optimum formulation shift (here the pH at which three-phase behavior is exhibited) with different water/oil ratios.

For a water/oil ratio equal to 1, it was found that $pH_{1/2}$, the pH at which half of the total acid in the system was dissociated, which is related [84,85] to the dissociation and partitioning constants by $pH_{1/2} = \log_{10}(P_a/K_a)$, was very near pH*, the pH at which three-phase behavior was exhibited.

Figure 16 illustrates this situation by assuming that the HLB of the interface is calculated by a linear mixing rule based on the total amounts of undissociated acid and salt with Davies [23] estimates for HLB versus chain length:

$$HLB_{mixture} = f_i \, HLB_{Ac^-} + (1 - f_i) \, HLB_{AcH} \qquad (19)$$

It is seen from Fig. 16 that the correspondence between $pH_{1/2}$ and pH* is essentially due to the shape of the HLB versus pH plot, which exhibits an almost vertical variation in the region where the dissociation takes place. This is why the actual pH* is not very much affected by the other formulation variables, i.e., the nature of the oil and water. In the present

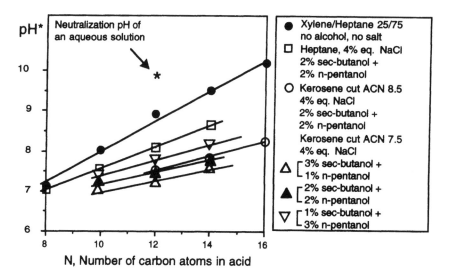

Figure 17 Optimum pH* for three-phase behavior for different fatty acid–oil–water–alcohol systems. (From Ref. 105.)

case, optimum formulation was taken to be at HLB = 11, but it is obvious that a ±1 unit departure from that value would not significantly change the location of the intersection with the sigmoid curves, i.e., the pH* values. On the contrary, it is seen that pH* is very dependent on the acid chain length.

Keeping in mind that HLB is essentially equivalent to an amphiphilic characteristic parameter such as σ, though not so accurate, the optimum formulation (as pH*) is found to vary linearly with the amphiphilic chain length, just as σ does. However, it is worth mentioning that pH* also depends on the other physicochemical formulation variables, as can be seen in Fig. 17, in which the slope of the linear variation changes with the type of oil phase, the salinity of the aqueous phase, and the alcohol content. This can be explained by the parallel changes between pH* and $pH_{1/2}$. In effect, the expression for $pH_{1/2}$ as $\log_{10}(P_a/K_a)$ shows that it changes linearly with the logarithm of the partitioning constant P_a, since K_a is essentially the same ($\cong 10^{-6}$) for all long-chain fatty acids.

A different slope in Fig. 17 thus indicates a change in the partitioning constant. It is worthwhile remarking that the highest slope is attained with a xylene–heptane oil mixture, which is the most aromatic oil in the Fig. 17 systems and also the one into which the acid partitioning would be the greatest.

This discussion indicates that pH-dependent systems are more complex to deal with than plain surfactant systems because of the eventual partitioning. This is also the case with many commercial surfactant mixtures, particularly those of ethoxylated nonionic surfactants, for which no simple and accurate rule can be used for formulation purposes [75,86].

The same kind of reasoning can be carried out for systems containing a fatty amine and a fatty amine salt as a function of pH, such as the ones that have been widely used in asphalt emulsions. At low pH (say pH 3), the amine salt behaves as a quaternary ammonium compound according to the ionic correlation for optimum formulation. However, a quaternary compound such as alkyltrimethylammonium is substantially more hydrophilic than the corresponding alkylammonium at low pH in spite of the three extra carbons from the methyl groups. For instance, it is seen in Table 2 that the C_{14} trimethylammonium salt has a σ/K characteristic parameter much lower than that of the C_{14} ammonium salt counterpart at pH 3 [35,87]. This is clear evidence that the alkylammonium group contribution is less hydrophilic at the interface than the fully ionized quaternary ammonium even at a pH at which full ionization is likely to take place in water. It can be conjectured that this is due either to the partitioning of the less hydrophilic species into the oil phase, where it remains as a nonionized amine, or to segregation of the adsorbed surfactant. In effect, even at low pH, there is still a small concentration of the nonionized alkylamine species in water, as indicated in the following equilibrium by [Am], while the ammonium salt concentration [AmH$^+$] is quite high.

$$Am(H_2O) \rightleftharpoons AmH^+ + OH^-, \qquad K_b = \frac{[AmH_w^+][OH^-]}{[Am_w]} \qquad (20)$$

The two species are competing for adsorption at the interface, and it is obvious that the nonionized amine exhibits a much stronger hydrophobic driving force to go to the interface. Thus, even if there is a very small proportion of nonionized amine in the aqueous bulk phase, there might be a substantial proportion of it at the interface. As a consequence, the interfacial formulation, which is, by the way, the true formulation, is less hydrophilic than expected from the overall or aqueous bulk-phase composition.

Such situations of interfacial segregation are not typical of pH-dependent systems, but rather they are inherent to mixtures that contain amphiphilic species with very different hydrophilic–lipophilic tendencies [75,86].

V. MICROEMULSIONS WITH ANIONIC–CATIONIC SURFACTANT MIXTURES

The mixtures of anionic and cationic surfactants are known to result in some association reaction that can be expressed as the equilibrium

$$A^- + C^+ \rightleftharpoons AC \qquad (21)$$

where A^- is an amphiphile anion and C^+ an amphiphile cation, while the counterions are not mentioned. If the anionic surfactant is some sodium sulfonate $R_1-SO_3^-Na^+$, and the cationic one is an alkyltrimethylammonium chloride $R_2(CH_3)_3N^+Cl^-$, then A^- stands for $R_1-SO_3^-$ while C^+ indicates $R_2(CH_3)_4N^+$. The AC association compound, a so-called catanionic structure, has quite interesting properties [88] and can exhibit some amphoteric or nonionic structure such as

$$R_1-SO_3^- {}^+N(CH_3)_3R_2 \qquad \text{or} \qquad R_1-SO_3-N(CH_3)_3R_2 \qquad (22)$$

This association equilibrium is used in the so-called two-phase titration of anionic surfactants by a cationic dyestuff in which the association compound is displaced in an oil phase and results in a coloration [89,90].

The AC catanionic compound is thus actually much less ionic than its separated components. Experimental evidence indicates that such a surfactant mixture in water results in the formation of either a precipitate, a liquid crystal, or a vesicle structure, with a molecular arrangement that is likely to be of the bilayer type, which is no wonder, as the association compound obviously looks like a double-tailed surfactant [91–97].

Very few studies have reported on the phase behavior and occurrence of microemulsions with anionic–cationic mixtures [98]. However, it was found that when one of the surfactant was in much lower proportion than the other (say, less than 10%

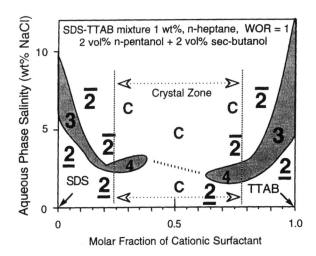

Figure 18 Phase behavior of systems containing a mixture of anionic and cationic surfactants with oil, water, and alcohol. (From Ref. 99.)

or 20%), then some kind of compatibility occurred [99]. This is exactly the situation in which a small amount of a cationic surfactant is added to an anionic detergent, a formulation trick that has been found to improve detergency and that is widely used by laundry powder manufacturers [100–104].

Figure 18 [99] shows the optimum formulation for three-phase behavior (as the optimum salinity) as a function of the composition of a mixture of anionic (sodium dodecyl sulfate) and cationic (tetradecyltrimethylammonium bromide) species loaded with a considerable amount of alcohol to avoid the formation of liquid crystals. The surfactant pair was selected so that both individual surfactants produced a three-phase microemulsion–oil–water behavior at about the same salinity, i.e., 5–10% NaCl. As some cationic surfactant is added to the anionic one, the shaded region that indicates the three-phase behavior goes down (from left to right). This downward displacement and the fact that three-phase behavior is still exhibited means that the addition of a small amount of the cationic surfactant to the anionic one results in a less hydrophilic surfactant mixture.

The same downward displacement (this time from right to left) occurs as some anionic surfactant is added to the cationic one, with the same conclusion with respect to the hydrophilicity of the mixture.

With further additions of one of the surfactants, the trend ends up at about 20%, and the three-phase system becomes coexistent with a liquid crystal phase (as indicated by the number 4). Further to the center of the diagram, the three-phase behavior disappears and a crystallized precipitate is exhibited instead (as indicated by the letter C).

These data indicate that microemulsions can be made with mixtures containing a small amount of cationic surfactant and a large amount of the anionic surfactant, or conversely.

As explained in detail elsewhere [99], it seems that the anionic–cationic surfactant mixtures would contain a certain amount of catanionic equimolecular compound, which is much less hydrophilic than its components, and an excess of either the anionic or cationic surfactant. In other words, a mixture containing 10 mol of cationic surfactant and 90 mol of anionic surfactant would really behave as a mixture of 80 mol of anionic surfactant with 10 mol of the catanionic equimolar component. There is evidence that this anionic–cationic mixture provides enhanced solubilization and an increase in the micellar aggregation number, which may be a good explanation for the improvement in detergency [105,106]. Another reason might be associated with the depression of the critical micelle concentration exhibited by such mixtures [107,108].

VI. APPLICATIONS OF IONIC MICROEMULSIONS

A. Bicontinuous Microemulsions

A sudden increase in research effort on microemulsion systems was driven by the economic impact of the oil embargo in the early 1970s and the development of the so-called enhanced oil recovery processes that followed. The plentiful research funding available from both industry and governmental agencies resulted in an unprecedented improvement in the basic and advanced knowledge of very complex phenomena, in particular the surfactant–oil–water phase behavior in all its intricacies. It was found that the interfacial tension could be lowered to an ultralow 0.001 mN/m in many systems provided that a particular physicochemical condition was attained. It turned out that this so-called first optimum salinity, and then optimum formulation, coincides with the occurrence of three-phase behavior in which a bicontinuous microemulsion is in equilibrium with oil and water excess phases, i.e., the Winsor III case [20,21,109,110].

Other petroleum applications such as emulsion breaking, particularly for crude oil dehydration, are of first importance. A by-product of enhanced oil recovery was a better understanding of the relationship between the phase behavior of surfactant–oil–water systems and the properties of the corresponding macroemulsions [111–118].

By the time the oil prices dropped in the early 1980s, many research groups had attained such proficiency that it was worthwhile for them to capitalize on this understanding and work on other potential applications.

The capacity of the middle-phase microemulsion to gather most of the surfactant was proposed as the basis for a purification technique [119,120].

Some reactions, e.g., the electrolytic dechlorination of DDT, can be carried out in bicontinuous ionic microemulsions in a less expensive and less toxic approach than the conventional process [121]. The kinetics rate of electrogenerated macrocyclics like vitamin B-12 can be controlled in a bicontinuous cationic microemulsion [122].

It was claimed that the spongelike bicontinuous microemulsion structures could be conserved as microporous (open-pore) polymeric materials with pore size depending on the surfactant concentration, by polymerizing methyl methacrylate and ethylene glycol dimethacrylate in cationic or zwitterionic surfactant systems [123–125].

Bicontinuous microemulsions have been formulated to solubilize polar oils, even of the triglyceride type, because of a new generation of the so-called extended surfactants, which contain a poly($\times$10)propylene oxide intermediate polarity chain in between the (ionic) hydrophilic and lipophilic groups [51,52,126].

Both bicontinuous and O/W microemulsions have been prepared with light hydrocarbons in a supercritical state [127–129], and even bicontinuous cationic surfactant structures have been obtained with near critical and supercritical propane [130].

B.   Microemulsions of the Swollen Micelle Type

The optimum formulation requires a proper balance of interactions to attain a physicochemical situation in which a bicontinuous spongelike microemulsion with high solubilization ability is formed. In some cases the required solubilization is quite low, and the attainment of microemulsions located near the ternary diagram sides, i.e., similar to swollen micellar systems, might be enough as far as the applications are concerned. It is worth noting that such microemulsions are very much like miniemulsions containing extremely fine droplets, both systems being stabilized by entropic effects and the main difference being the way they are produced [131], whether spontaneously (in a single-phase region) or by some physicochemical trick such as phase inversion transition (ending up in a two-phase region).

As a consequence, it is believed that some "microemulsions" in the literature are miniemulsions, and conversely. This point may not be important in practice, as the emulsions probably have very similar properties, although it does mean that some of these systems might lack thermodynamic stability, a particularly significant feature with respect to temperature variation or other cyclic changes.

Since the solubilization of oil phases other than alkanes in nonbicontinuous microemulsions is not very high (say, below 0.5 g of oil per gram of surfactant) unless fancy new molecules are used such as the so-called extended surfactants that are able to solubilize triglycerides [51,52], it is probable that many apparently solubilized systems referred to as O/W microemulsion are really miniemulsions produced by phase inversion [132].

In the same fashion as the early carnauba wax microemulsions, natural or synthetic wax micro- and miniemulsions stabilized by fatty soaps are being used for coating citrus fruit to avoid weight reduction and to lower internal carbon dioxide, with better performance than the conventional shellac or resin coatings [133]. Sometimes these are even more complex polyphasic systems involving both ionic and nonionic microemulsions in equilibrium with other phases [134,135].

In the case of cosmetics in which fragrances, flavors, or emollient oils are incorporated at very low concentrations solubilization may be ensured by micellar take-up. This also applies to alcoholic preparations such as mouthwash, cologne, or skin tonic. Nevertheless, a high solubilization, i.e., a bicontinuous system, would be required if the amount of alcohol were restricted by safety and toxicity constraints.

In most ionic surfactant studies concerning the formation of microemulsions, it is assumed that the brine is a pseudocomponent, one used solely to maintain the mass balance in the system. Recent evidence indicates that this may not be the case with some W/O microemulsions, in which the aqueous core of the W/O structure exhibits a lower salinity and may sometimes even be salt-free. Such segregation has been explained by negative adsorption of the coions according to the Gouy–Chapman double layer theory [136].

This may be quite important in processes in which the ionic strength is determinant such as sol-gel transitions or chemical reactions in microemulsions [137–141]. Double-tailed surfactants such as dioctyl sulfosuccinate or dialkylmethylammonium salts are likely to produce either vesicles (with excess water) or inverse W/O microemulsions with a polar core [142,143] that is used as a nanoreactor for a score of processes such as esterification or hydrolysis [144] in which enzymes are immobilized in an organogel [145]. Organogels can be made so that their structure depends on the composition of the microemulsion [146–148].

The formation of calcium carbonate particles can be carried out in W/O calcium sulfosuccinate microemulsions [149]. Hollow porous shells of calcium carbonate or reticulated calcium phosphate skeletons made under similar conditions are likely candidates for high technology applications such as the production of lightweight ceramics, catalysts, biomedical implants, and extremely strong membranes [150,151]. Other nanosized inorganic particles can be prepared in W/O microemulsions as well [152–154].

Among the reactions that are carried out in microemulsions, polymerization is probably the most studied, as can be found in recent reviews [155–157]. It takes place in a O/W or W/O microemulsion structure, in percolated microemulsion [158–160], and in bicontinuous systems [161,162].

The literature on microemulsions is still quite prolific, in spite of the disappearance of the original driving force of oil recovery. It is more difficult every day for the rookie investigator to start a learning process in this field because of the overwhelming number of references returned by computerized searches. It is thus almost compulsory to go through review books like those of the Marcel Dekker Surfactant Science Series or to review journals, particularly the recently established *Current Opinion in Colloid and Interface Science*, which carries a large number of different and noteworthy reviews on microemulsions and surfactant self-assembly topics.

VII. CONCLUDING REMARKS ON THEORETICAL APPROACHES AND MODELING

The purpose of this chapter was to present a consistent and simple view of what ionic microemulsions are and what their peculiarities are, particularly from practical and pedagogical points of view, without entering into complex theoretical aspects or overlapping on the topic of nonionic microemulsions, which is treated elsewhere in this book. However, it seems to us that it is worthwhile mentioning very briefly the main achievements in these fields as well as the investigators who have contributed to them, whose work can be traced in computerized searches and review articles.

Many theoretical and experimental research works have taken temperature as a special variable rather than a formulation variable like the other ones, as was done here. Most of these studies essentially deal with nonionic systems, which are beyond the scope of this text but are obviously related in many fundamental aspects. For instance, K. Shinoda, H. Kunieda, and their collaborators at Yokohoma University in Japan [6,38,43,58,135,163–169], as well as M. Kahlweit, R. Strey, G. Busse, and collaborators at Max Planck Institute in Göttingen, Germany [170–175], have extensively studied nonionic phase diagrams and many related theoretical, geometrical, and modeling aspects that are also quite useful in understanding ionic microemulsions such as the mapping of the three-phase region or the influence of the cosurfactant.

Even more theoretical work has been carried out by E. Ruckenstein [176–180] on the importance of dispersion entropy as the stabilizing mechanism and by P. G. de Gennes, C. Taupin, and collaborators at Collège de France in Paris [181–184] and C. Miller [185–187] and others [26,188–189] on the importance of interfacial tension, interfacial bending, and other stability mechanisms.

The many geometrical and dynamic structural aspects that make surfactant self-assembly such a fascinating subject have been discovered because of the development of sophisticated analytical techniques such as small-angle neutron or X-ray scattering and nuclear magnetic resonance [190].

REFERENCES

1. T. P. Hoar and J. H. Schulman, Nature *152*:102 (1943).
2. J. H. Schulman, W. Stoeckenius, and L. M. Prince, J. Phys. Chem. *63*:1677 (1959).
3. L. M. Prince, *Microemulsions—Theory and Practice*. Academic, New York, 1977.
4. I. D. Robb (ed.), *Microemulsions*, Plenum, New York, 1982.
5. R. Pons, C. Solans, and T. F. Tadros, Langmuir *11*:1966 (1995).
6. H. Kunieda, Y. Fukui, H. Uchiyama, and C. Solans, Langmuir *12*:2136 (1996).
7. M. Minana-Pérez, C. Gutron, C. Zundell, and J. L. Salager, in press J. Dispersion Sci. Technol. (1999).
8. R. Laughlin, *The Aqueous Phase Behavior of Surfactants*, Academic, London, 1994.
9. M. W. Matsen, M. Schick, and D. E. Sullivan, J. Chem. Phys. *98*:2341 (1993).
10. W. K. Kegel and H. N. Lekkerkerker, J. Phys. Chem. *97*:11124 (1993).
11. P. Winsor, *Solvent Properties of Amphiphilic Compounds*, Butterworth, London, 1954.
12. Y. Talmon and S. Prager, J. Chem. Phys. *69*:2984 (1978).
13. S. H. Chen, J. Rouch, F. Sciortino, and P. Tartaglia, J. Phys. Condens. Matter *6*:10855 (1994).
14. C. Boned, Z. Saidi, P. Xans, and J. Peyrelasse, Phys. Rev. E. *49*:5295 (1994).
15. D. Vollmer, J. Vollmer, and H. F. Eicke, Europhys. Lett. *26*:389 (1994).
16. W. Meier, Colloids Surf. A *94*:111 (1995).
17. H. F. Eicke, M. Borkovec, and D. DasGupta, J. Phys. Chem. *93*:314 (1989).
18. L. Schlicht, J. H. Spilgies, F. Runge, S. Lipgens, S. Boye, D. Schubel, and G. Ilgenfritz, Biophys. Chem. *58*:39 (1996).
19. P. A. Winsor, Trans. Faraday Soc. *43*:376, 382, 387, 390, 451, 455, 459, 463 (1947).
20. D. O. Shah and R. S. Schechter (eds.), *Improved Oil Recovery by Surfactant and Polymer Flooding*, Academic, New York, 1977.
21. D. O. Shah (ed.), *Surface Phenomena in Enhanced Oil Recovery*, Plenum, New York, 1981.
22. M. Bourrel and R. S. Schechter, *Microemulsions and Related Systems*, Marcel Dekker, New York, 1988.
23. J. T. Davies, in *Gas/Liquid and Liquid/Liquid interfaces*, Proc. 2nd Int. Congress Surface Activity, Vol. I. Butterworth, London, 1957, p. 426.
24. M. Bavière, R. S. Schechter, and W. H. Wade, J. Colloid Interface Sci. *81*:266 (1981).
25. P. A. Winsor, Trans. Faraday Soc. *44*:376 (1948).
26. A. Kalbanov and H. Wennerström, Langmuir *12*:276 (1996).
27. S. Scriven, in *Micellization, Solubilization and Microemulsions*, Vol. 2 (K. L. Mittal, ed.), Plenum, New York, 1977, p. 877.
28. M. Bourrel, J. Biais, P. Bothorel, B. Clin, and P. Lalanne, J. Dispersion Sci. Technol. *12*:531 (1991).
29. J. Morgan, R. S. Schechter, and W. H. Wade, in *Improved Oil Recovery by Surfactant and Polymer Flooding* (D. O. Shah and R. S. Schechter, eds.), Academic, New York, 1977, p. 101.
30. R. L. Reed and R. N. Healy, in *Improved Oil Recovery by Polymer and Surfactant Flooding* (D. O. Shah and R. S. Schechter, eds.), Academic, New York, 1977 p. 383.
31. J. L. Salager, J. Morgan, R. S. Schechter, W. Wade, and E. Vasquez, Soc. Petrol. Eng. J. *19*:107 (1979).
32. W. H. Wade, J. Morgan, J. Jacobson, J. L. Salager, and R. S. Schechter, Soc. Petrol. Eng. J. *18*:242 (1978).
33. J. L. Salager, Rev. Inst. Mex. Petrol. *11*:59 (1978).
34. J. L. Salager, in *Encyclopedia of Emulsion Technology*, Vol. 3 (P. Becher, ed.), Marcel Dekker, New York, 1988, p. 79.
35. R. E. Antón, N. Garcés, and A. Yajure, Proceedings IV Latin American Symp. Fluid Properties and Phase Equilibria for Process Design, EQUIFASES 95, Caracas, Venezuela, 1995, p. 571.
36. R. E. Antón, N. Garcés, and A. Yajure, J. Dispersion Sci. Technol. *18*:539 (1997).
37. J. L. Salager, M. Bourrel, R. S. Schechter, and W. H. Wade, Soc. Petrol. Eng. J. *19*:271 (1979).
38. H. Kunieda, K. Hanno, S. Yamaguchi, and K. Shinoda, J. Colloid Interface Sci. *107*:129 (1985).
39. R. E. Antón, J. L. Salager, A. Graciaa, and J. Lachaise, J. Dispersion Sci. Technol. *13*:565 (1992).

40. R. E. Antón, H. Rivas, and J. L. Salager, Proceedings World Congress on Emulsions, Paris, France, EDS Editeur, Paris, 1993, Paper 1-30-189.

41. R. E. Antón, F. Mosquera, and M. Oduber, Prog. Colloid Polym. Sci. *98*:85 (1995).

42. R. E. Antón, H. Rivas, and J. L. Salager, J. Dispersion Sci. Technol. *17*:553 (1996).

43. H. Kunieda and C. Solans, in *Industrial Applications of Microemulsions* (Surfact Sci. Ser. Vol. 66) (C. Solans and H. Kunieda, eds.), Marcel Dekker, New York, 1996,

44. K. H. Oh, J. R. Baran, W. H. Wade, and V. Weerasooriya, J. Dispersion Sci. Technol. *16*:165 (1995).

45. M. Bourrel, J. L. Salager, R. S. Schechter, and W. H. Wade, Colloq. Natl. CNRS *938*:337 (1978).

46. R. E. Antón, and J. L. Salager, J. Colloid Interface Sci. *140*:75 (1990).

47. V. K. Bansal and D. O. Shah, J. Am. Oil Chem. Soc. *55*:367 (1978).

48. M. C. Puerto and R. L. Reed, SPE Reservoir Eng. *5*:198 (1990).

49. H. Wennerström, A. Khan, and B. Lindman, Adv. Colloid Interface Sci. *34*:433 (1991).

50. J. L. Cayias, R. S. Schechter, and W. H. Wade, Soc. Petrol. Eng. J. *16*:351 (1976).

51. M. Minana-Pérez, A. Graciaa, J. Lachaise, and J. L. Salager, Prog. Colloid Polym. Sci. *98*:177 (1995).

52. M. Minana-Pérez, A. Graciaa, J. Lachaise, and J. L. Salager, Colloids Surf. A *100*:217 (1995).

53. J. R. Baran, G. Pope, W. H. Wade, V. Weerasooriya, and A. Yapa, Environ. Sci. Technol. *28*:1361 (1994).

54. J. R. Baran, G. Pope, W. H. Wade, V. Weerasooriya, and A. Yapa, J. Colloid Interface Sci. *168*:67 (1994).

55. J. R. Baran, G. Pope, W. H. Wade, and V. Weerasooriya, Langmuir *10*:1146 (1994).

56. G. Pope and W. H. Wade, in *Surfactant-Enhanced Subsurface Remediation: Emerging Technologies* (ACS Symp. Ser., Vol. 594) (D. Sabatini, R. Knoz and J. Harwell, eds.), Am. Chem. Soc., Washington, DC, 1995, p. 142.

57. P. Van der Meren and W. Verstraete, Curr. Opinion Colloid Interface Sci. *1*:624 (1996).

58. H. Katayama, T. Tagawa, and H. Kunieda, J. Colloid Interface Sci. *153*:429 (1992).

59. M. S. Bakri, Indian J. Chem. *A 34*:896 (1995).

60. R. Aveyard, B. P. Binks, P. D. Fletcher, P. A. Kingston, and A. R. Pitt, J. Chem. Soc., Faraday Trans. *90*:2743 (1994).

61. X. F. Li, G. H. Zhao, E. H. Lin, and S. Q. Qin, J. Dispersion Sci. Technol. *17*:111 (1996).

62. K. Bartscherer, M. Minier, and H. Renon, Fluid Phase Equilibria *107*:93 (1995).

63. R. D. Smith, J. L. Fulton, J. P. Blitz, and J. M. Tingey, J. Phys. Chem., *94*:781 (1990).

64. K. Harrison, J. Goveas, K. P. Johnson, and E. A. O'Rear, Langmuir *10*:3536 (1994).

65. R. Cash, J. L. Cayias, G. Fournier, D. McAllister, T. Shares, R. S. Schechter, and W. H. Wade, J. Colloid Interface Sci. *59*:39 (1977).

66. P. Fotland and A. Skange, J. Dispersion Sci. Technol. *7*:563 (1986).

67. M. W. Ki, W. Gallerghen, and J. Bock, J. Phys. Chem. *92*:1226 (1988).

68. A. Skange and P. Fotland, SPE Reservoir Eng. *5*:601 (1990).

69. R. D. Selliah, R. S. Schechter, W. H. Wade, and U. Weerasooriya, J. Dispersion Sci. Technol. *8*:75 (1987).

70. K. R. Wormuth and S. Zushma, Langmuir *7*:2048 (1991).

71. G. Nonaka, M. Harada, A. Shioi, M. Goto, and F. Nakashio, J. Colloid Interface Sci. *176*:1 (1995).

72. N. Kamenka, G. Haouche, B. Brun, and B. Lindman, Colloids Surf. A *25*:287 (1987).

73. W. H. Wade, J. Morgan, J. Jacobson, and R. S. Schechter, Soc. Petrol. Eng. J. *17*:122 (1977).

74. M. Puerto and W. W. Gale, Soc. Petrol. Eng. J. *17*:193 (1977).

75. A. Graciaa, J. Lachaise, J. G. Sayous, P. Grenier, S. Yiv, R. S. Schechter, and W. H. Wade, J. Colloid Interface Sci. *93*:474 (1983).

76. D. R. Zornes, G. P. Willhite, and M. J. Michnick, Symp. on Oilfield and Geothermal Chemistry, La Jolla, CA, 1977, Paper SPE 6595.

77. C. Lalanne-Cassou, I. Carmona, L. Fortney, A. Samil, R. S. Schechter, and W. H. Wade, J. Dispersion Sci. Technol. *8*:137 (1987).

78. V. K. Bansal and D. O. Shah, J. Colloid Interface Sci. *65*:451 (1978).

79. I. Carmona, R. S. Schechter, W. H. Wade, and U. Weerasooriga, Symp. on Oilfield and Geothermal Chemistry, Denver, CO, June 1–3, 1983, Paper SPE 11771.
80. N. Mungan, World Oil, June 1981, p. 209.
81. L. W. Holm and S. D. Robertson, J. Petrol. Technol. *33*:161 (1981).
82. P. Cratin, in *Chemistry and Physics of Interfaces*, Vol. 2 (S. Ross. ed.), American Chemical Society, Washington, DC., 1969, p. 37.
83. S. Qutubuddin, C. A. Miller, and T. Fort, J. Colloid Interface Sci. *101*:46 (1984).
84. R. E. Antón, A. Graciaa, J. Lachaise, and J. L. Salager, Proceedings 4th World Surfactants Congress, Vol. 2 (R. de Llúria, ed.), AEPSAT Barcelona, Spain, 1996, p. 244.
85. H. Rivas, X. Guttierrez, J. L. Ziritt, R. E. Anton, and J. L. Salager, in *Industrial Applications of Microemulsions* (Surfact. Sci. Ser. 67) (C. Solans and H. Kunieda, eds.), Marcel Dekker, New York, 1996, p. 305.
86. A. Graciaa, J. Lachaise, M. Bourrel, I. Osborne-Lee, R. S. Schechter, and W. H. Wade, Soc. Petrol. Eng. J. *24*:305 (1984).
87. A. Lipow, M. S. Thesis, Univ. Texas at Austin, 1979.
88. A. Khan and E. Marques, in *Speciality Surfactants* (I. D. Robb, ed.), Blackie, London, 1996.
89. G. F. Longman, *The Analysis of Detergents and Detergents Products*, Wiley, New York, 1975.
90. V. W. Reid, G. F. Egham, G. F. Longman, and E. Heinerth, Tenside Deterg. *4*:292 (1967).
91. J. Corkill, J. F. Goodman, S. P. Harrold, and J. R. Tate, Trans. Faraday Soc. *62*:247 (1967).
92. P. Jokela, B. Jönsson, and A. Khan, J. Phys. Chem. *91*:3291 (1987).
93. E. W. Kaler, A. K. Murthy, B. Rodriguez, and J. A. Zasadzinski, Science *245*:1371 (1989).
94. H. Fukuda, K. Kawata, H. Okuda, and S. L. Regen, J. Am. Chem. Soc. *67*:717 (1990).
95. A. S. Sadaghiani and A. Khan, J. Colloid Interface Sci. *146*:69 (1991).
96. E. Marques, A. Khan, M. G. Miguel, and B. Lindman, J. Phys. Chem. *97*:4729 (1993).
97. N. Filipovic-Vincekovic, M. Bujan, D. Dragcevic, and N. Nekic, Colloid Polym. Sci. *273*:182 (1995).
98. M. Bourrel, D. Bernard and A. Graciaa, Tenside Deterg. *21*:311 (1984).
99. R. E. Antón, D. Gómez, A. Graciaa, J. Lachaise, and J. L. Salager, J. Dispersion Sci. Technol. *14*:401 (1993).
100. Unilever, U. K. Patent 77/11,685 (1977).
101. Lion Corp., Jpn. Patent 55-129,497 (1980).
102. Colgate Palmolive, Eur. Patent 880,537 (1980).
103. Kao Soap, U. S. Patent 4,267,077 (1981).
104. Procter and Gamble, U. S. Patent 4,321,165 (1982).
105. R. E. Antón, Doctoral Dissertation, Univ. de Pau, France, 1992.
106. M. Chorro and N. Kamenka, J. Chim. Phys. *88*:515 (1991).
107. E. H. Lucassen-Reynders, J. Lucassen, and D. Giles, J. Colloid Interface Sci. *81*:150 (1981).
108. P. M. Holland, Colloids Surf. A *19*:171 (1983).
109. D. Wasan and A. Payatakes (eds.), *Interfacial Phenomena in Enhanced Oil Recovery*, AIChE Symp. Ser. 212, Vol. 78, Am. Chem. Soc., Washington, DC, 1982.
110. M. Bavière (ed.), *Basic Concepts in Enchanced Oil Recovery Processes*, Vol., 33, Elsevier Applied Science, London, 1991.
111. M. Bourrel, A. Graciaa, R. S. Schechter, and W. H. Wade, J. Colloid Interface Sci. *72*:161 (1979).
112. J. L. Salager, L. Quintero, E. Ramos, and J. Andérez, J. Colloid Interface Sci. *77*:288 (1980).
113. J. E. Viniatieri, Soc. Petrol. Eng. J. *20*:402 (1980).
114. J. L. Salager, I. Loaiza-Maldonado, M. Minana-Pérez, and F. Silva, J. Dispersion Sci. Technol. *3*:279 (1982).
115. R. E. Antón and J. L. Salager, J. Colloid Interface Sci. *111*:54 (1986).
116. J. L. Salager, Int. Chem. Eng. *30*:103 (1990).
117. J. L. Salager, M. Minana-Pérez, M. Pérez-Sanchez, M. Ramírez-Gouveia, and C. I. Rojas, J. Dispersion Sci. Technol. *4*:313 (1983).
118. J. L. Salager, in *Surfactants in Solution* (Surfact. Sci. Ser. No. 64) (A. Chattopadhyay and K. L. Mittal, eds.), Marcel Dekker, New York, 1996, p.261.

119. A. Graciaa, J. Lachaise, G. Marion, M. Bourrel, I. Rico, and A. Lattes, Tenside Deterg. *26*:384 (1989).

120. K. Shubert, R. Strey, and M. Kahlweit, Prog. Colloid Polym. Sci. *84*:103 (1991).

121. S. Schweizer, J. F. Rusling, and Q. Huang, Chemosphere *28*:961 (1994).

122. D. L. Zhou, J. X. Gao, and J. F. Rusling, J. Am. Chem. Soc. *117*:1127 (1995).

123. T. H. Chieng, L. M. Gan, C. H. Chew, L. Lee, S. C. Ng, K. L. Pey, and D. Grant, Langmuir *11*:3321 (1995).

124. T. D. Li, C. H. Chew, S. C. Ng. L. M. Gan, W. K. Teo, J. Y. Gu, and G. Y. Zhang, J. Macromol. Sci. Pure Appl. Chem. *32*:969 (1995).

125. L. M. Gan, T. D. Li, C. H. Chew, W. K. Teo, and L. H. Gan, Langmuir *11*:3316 (1995).

126. J. L. Salager, G. López-Castellanos, M. Minana-Pérez, C. Cucuphat-Lemercier, A. Graciaa, and J. Lachaise, J. Dispersion Sci. Technol. *12*:59 (1991).

127. R. W. Gale, J. L. Fulton, and R. D. Smith, J. Am. Chem. Soc. *109*:920 (1987).

128. R. D. Smith, J. L. Fulton, J. P. Blitz, and J. M. Tingey, J. Phys. Chem. *94*:781 (1990).

129. G. J. Fann and K. P. Johnson, J. Phys. Chem. *95*:4889 (1991).

130. J. M. Tingey, J. L. Fulton, D. W. Watson, and R. D. Smith, J. Phys. Chem. *95*:1445 (1991).

131. T. Foester, W. von Rybinski, and A. Wadle, Adv. Colloid Interface Sci. *58*:119 (1995).

132. J. H. Shittam, W. E. Gerbaccia, and H. Rosano, Cosmet. Technol. 35 (Oct. 1979).

133. R. D. Hagenmaier and R. A. Baker, J. Agric. Food Chem. *42*:899 (1994).

134. M. Nagai, H. Sagitani, and T. Takenpuchi, J. Cosmet. chem. Jpn. *14*:41 (1980).

135. K. Shinoda and H. Kunieda, in *Microemulsions* (L. M. Prince, ed.), Academic, New York, 1977, p.57.

136. G. A. Van Aken, J. Th. G. Overbeek, D. E. Bruijn, and H. N. Lekkerkerker, J. Colloid Interfac Sci. *157*:235 (1993).

137. J. Rivas and M. A. Lopez-Quintana, J. Colloid Interface Sci. *158*:446 (1993).

138. J. D. Ramsay, in *Controlled Particle, Droplet and Bubble Formation* (D. J. Wedlock, ed.), Butterworth-Heinemann, London, 1994, pp. 1–38.

139. Q. Huo, R. Leon, P. M. Petroff, and G. D. Stucky, Science *268*:1324 (1995).

140. P. Petroff and G. D. Stucky, Nature *368*:317 (1994).

141. J. H. Burban, M. He, and E. L. Cussler, AIChE J. *41*:159, 907 (1995).

142. P. Ekwall, L. Mandel, and K. Fontell, J. Colloid Interface Sci. *33*:215 (1970).

143. N. Mitra, L. Mukhopadhyay, P. Bhattacharya, and S. P. Moulik, J. Biochem. Biophys. *31*:115 (1994).

144. M. T. Patel, R. Nagarajan, and A. Kilara, Biotechnol. Appl. Biochem. *22*(1):1 (1995).

145. M. G. Nascimento, M. C. Rezenda, R. D.Vecchia, P. C. Jesus, and L. M. Aguiar, Tetrahedron Lett. *33*:5891 (1992).

146. P. J. Atkinson, B. H. Robinson, A. Howe, and A. P. Pitt, Colloids Surf. A *94*:231 (1995).

147. P. J. Atkinson, M. J. Grimson, R. K. Hennan, A. M. Howe, and B. H. Robinson, J. Chem. Soc. Commun. *1989*:1807 .

148. P. L. Luisi, R. Scartazzini, G. Haering, and P. Schurtenberger, Colloid Polym. Sci. *268*:356 (1990).

149. K. Kandori, K. Kon-no, and A. Kitahora, J. Colloid Interface Sci. *122*:78 (1988).

150. D. Walsh and S. Mann, Nature *377*:320 (1995).

151. D. Walsh, J. D. Hopwood, and S. Mann, Science *264*:1576 (1994).

152. P. Kumar, V. Pillai, S. R. Bates, and D. O. Shah, Mater. Lett. *16*:68 (1993).

153. P. Kumar, V. Pillai, and D. O. Shah, Appl. Phys. Lett. *62*:765 (1993).

154. F. J. Arriagada and K. Osseo-Asare, J. Colloid Interface Sci. *170*:8 (1995).

155. J. Eastoe and B. Warne, Curr. Opinion Colloid Interface Sci. *1*:800 (1996).

156. F. Candau, in *Polymerization in Organized Media* (C. M. Paleos, ed.), Gordon and Breach, London, 1992, p. 215–282.

157. M. Antonietti, R. Basten, and S. Lohman, Macromol. Chem. Phys. *196*:441 (1995).

158. J. O. Stoffer and T. Bone, J. Dispersion Sci. Technol. *1*:37 (1980).

159. S. S. Atik and J. K. Thomas, J. Am. Chem. Soc. *103*:4279 (1981).

160. F. Candau, Y. S. Leong, and R. M. Fitch, J. Polym. Sci. Polym. Chem. *23*:193 (1985).
161. F. Rabagliati, A. Falcon, D. Gonzalez, C. Martin, R. E. Antón, and J. L. Salager, J. Dispersion Sci. Technol. *7*:245 (1986).
162. W. R. P. Raj, M. Sasthav, and H. M. Cheung, J. Appl. Polym. Sci. *47*:499 (1993).
163. K. Shinoda (ed.), *Solvent Properties of Surfactant Solutions*, Marcel Dekker, New York, 1967.
164. K. Shinoda, J. Colloid Interface Sci. *24*:4 (1967).
165. K. Shinoda and H. Saito, J. Colloid Interface Sci. *26*:70 (1968).
166. K. Shinoda, *Proceedings International Congress Surface Activity 5th*, Barcelona, Spain, 1969, Vol. 2, p. 275.
167. K. Shinoda and B. Lindman, Langmuir *3*:135 (1987).
168. K. Shinoda and H. Kunieda, in *Encyclopedia of Emulsion Technology*, Vol. 1 (P. Becher, ed.), Marcel Dekker, New York, 1988, pp. 337–367.
169. H. Kunieda, A. Nakano, and M. Akimaru, J. Colloid Interface Sci. *170*:78 (1995).
170. M. Kahlweit, R. Strey, and G. Busse, J. Phys. Chem. *94*:3881 (1994).
171. M. Kahlweit, R. Strey, P. Firman, D. Haase, J. Jen, and R. Shomäcker, Langmuir *4*:499 (1988); *5*:305 (1989).
172. M. Kahlweit, R. Strey, and G. Busse, J. Phys. Chem. *94*:3881 (1990).
173. M. Kahlweit, R. Strey, and G. Busse, J. Phys. Chem. *95*:5344 (1991).
174. M. Kahlweit, J. Jen, and G. Busse, J. Chem. Phys. *97*:6917 (1992).
175. M. Kahlweit, R. Strey, and G. Busse, Phys. Rev. E *47*:4197 (1993).
176. E. Ruckenstein and J. C. Chi, J. Chem. Soc., Faraday Trans. II *71*:1690 (1975).
177. E. Ruckenstein, Chem. Phys. Lett. *57*:517 (1978).
178. E. Ruckenstein and R. Krishnan, J. Colloid Interface Sci. *71*:321 (1979); *75*:476 (1980).
179. E. Ruckenstein, Soc. Petrol. Eng. J. *21*:583 (1981).
180. E. Ruckenstein, in *Surfactants in Solution* (K. L. Mittal and B. Lindman, eds.), Plenum, New York, 1984, Vol. 3, pp. 1551–1582.
181. J. Jouffroy, P. Levinson, and P. G. de Gennes, J. Phys. (France) *43*:1241 (1982).
182. P. G. de Gennes and C. Taupin, J. Phys. Chem. *86*:2294 (1982).
183. L. Auvray, J. P. Cotton, R. Ober, and C. Taupin, J. Phys. Chem. *88*:4586 (1984).
184. J. M. Di Meglio, M. Dvolaitzky, and C. Taupin, J. Phys. Chem. *89*:871 (1985).
185. C. A. Miller, R. Hwan, W. J. Benton, and T. Fort, J. Colloid Interface Sci. *61*:554 (1977).
186. C. A. Miller and P. Neogi, AIChE J. *26*:212 (1980).
187. C. A. Miller, J. Dispersion Sci. Technol. *6*:159 (1985).
188. Y. Talmon and S. Prager, Nature *267*:333 (1977).
189. Y. Talmon and S. Prager, J. Chem. Phys. *69*:2984 (1978).
190. K. Kahlweit, R. Strey, D. Haase, H. Kunieda, T. Schmeling, B. Faulhaber, M. Borkovec, H. F. Eicke, G. Busse, F. Eggers, T. H. Funk, H. Ricmann, L. Magid, O. Söderman, P. Stilbs, J. Winkler, A. Dittrich, and W. Jahn, J. Colloid Interface Sci. *118*:436 (1987).

9

Supercritical Microemulsions

Gregory J. McFann
Unilever Research U.S., Edgewater, New Jersey

Keith P. Johnston
University of Texas, Austin, Texas

I. INTRODUCTION

A supercritical fluid is defined as one that is beyond its critical point and thus cannot be liquefied by a change in pressure. On a $P-T$ plot, the supercritical region lies at the end of the vapor–liquid coexistence curve and is bounded by the lines $P = P_c$ and $T = T_c$. In most applications, however, the compressed liquid region above the $P = P_c$ line and to the left of the $T = T_c$ line is also useful. Table 1 gives critical parameters for commonly used supercritical fluids.

Supercritical fluids offer a unique combination of properties. Compared to liquids, they have highly variable densities, much higher diffusivities, and much lower viscosities. Their solvent properties can be varied continuously from gaslike to nearly liquidlike by increasing pressure. Many applications have been either proposed or developed to take advantage of the adjustability of supercritical fluids. These include separations, reactions, materials processing, chromatography, and environmental remediation. Several reviews of supercritical fluid technology have been published in recent years [1–3] that complement earlier works [4,5].

The development of supercritical fluid technology has been hindered by the fact that those supercritical fluids that have reasonably low critical points do not have large solvent strengths, although their solvent strength is highly adjustable. Thus they are not effective solvents for large or hydrophilic molecules. Early supercritical fluid research focused on the enhancement of solvent strength by the addition of small polar cosolvents or complexing agents [6–8]. The use of surfactants in supercritical fluid systems was the next logical step in the search for greater solubilizing power. Reverse micelles or water-in-oil microemulsions in supercritical fluid solvents could act as hosts for hydrophilic molecules that would otherwise be insoluble [9–11]. Further studies were motivated by a desire to exploit the continuously variable solvent strength of supercritical fluids and compressed liquids to attain an enhanced understanding of solvent effects on surfactant phase behavior [12]. Initial work on microemulsions in supercritical fluids was surveyed in an earlier publication [13]. A comprehensive review paper with emphasis on experimental methods and interfacial thermodynamics has also been published [14].

Table 1 Thermodynamic Parameters of Common Supercritical Fluids

Compound	Boiling point (°C)	T_c (°C)	P_c (bar)	Crit. density g/cm^3
Ethylene	−103.7	9.2	50.3	0.22
Xenon	−108.1	16.6	58.4	0.11
Chlorotrifluoromethane	−81.4	28.9	38.7	0.58
Carbon dioxide	—	31.0	73.8	0.47
Ethane	−88.6	32.3	48.8	0.20
Nitrous oxide	−88.5	36.4	72.4	0.45
Monofluoromethane	−78.3	44.6	58.8	0.30
Propane	−42.0	96.7	42.5	0.22
Ammonia	−33.4	132.5	112.7	0.23
Water	100.0	374.1	220.9	0.32

Source: Ref. 59.

This brief survey begins in Sec. II with studies of the aggregation behavior of the anionic surfactant AOT (sodium bis-2-ethylhexyl sulfosuccinate) and of nonionic poly(ethylene oxide) alkyl ethers in supercritical fluid ethane and compressed liquid propane. One- and two-phase reverse micelle systems are formed in which the volume of the oil component greatly exceeds the volume of water. In Sec. III we continue with investigations into three-component systems of AOT, compressed liquid propane, and water. These microemulsion systems are of the classical Winsor type that contain water and oil in relatively equal amounts. We next examine the effect of the alkane carbon number of the oil on surfactant phase behavior in Sec. IV. "Unusual" reversals of phase behavior occur in alkanes lighter than hexane in both reverse micelle and Winsor systems. "Unusual" phase behavior, together with pressure-driven phase transitions, can be explained and modeled by a modest extension of existing theories of surfactant phase behavior. Finally, Sec. V describes efforts to create surfactants suitable for use in supercritical CO$_2$, and applications of surfactants in supercritical fluids are covered in Sec. VI.

II. REVERSE MICELLES IN ETHANE AND PROPANE

A. Initial Studies of AOT Aggregation

The anionic surfactant AOT was chosen for the initial experiments because its behavior in liquid solvents such as isooctane had been extensively studied [15]. AOT readily forms one-phase reverse micelle systems containing large amounts of solubilized water without the need for a cosurfactant. The added water gathers at the centers of the reverse micelles and forms water pools in which hydrophiles can be solubilized. When water and surfactant concentrations are high, AOT reverse micelle systems can cross a percolation boundary and behave as bicontinuous microemulsions [16].

Figure 1 shows the aggregation behavior of AOT in liquid cyclohexane and supercritical fluid ethane. The systems are one-phase without added water. Surfactant aggregation is indicated by the solvatochromic probe pyridine N-oxide. Pyridine N-oxide was used because of its small size and large dipole moment ($\mu = 4.3$ D), which allow it to partition to the center of reverse micelles instead of being trapped at the surfactant interface. This molecule is a "blue shift" indicator in that its UV absorption maximum shifts to lower

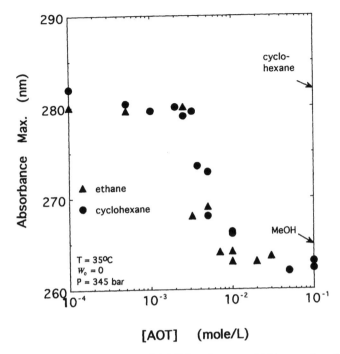

Figure 1 Aggregation of AOT in ethane at 345 bar and cyclohexane at ambient pressure in the one-phase region without added water, as indicated by the λ_{max} of pyridine N-oxide. The arrows indicate pyridine N-oxide λ_{max} in pure solvents. (From Ref. 20.)

wavelengths as the polarity of its surroundings increases. The λ_{max} is 281 nm in an alkane environment and 254 nm in an aqueous environment such as would be found in a reverse micelle water pool. It can be seen from Fig. 1 that the behavior indicated by the probe is the same in both solvents. AOT aggregation becomes prevalent at a concentration of about 0.002 M and is essentially fully developed at 0.01 M, which is consistent with other studies of AOT aggregation behavior [17,18].

The next step was to measure the ability of the reverse micelles to solubilize water. Known amounts of water were injected into one-phase AOT–fluid systems, and the location of the pressure boundary between the one-phase and two-phase regions was determined. It is customary in AOT studies to express the amount of solubilized water in terms of W_0, the molar water/surfactant ratio [19]. In liquid systems the W_0 value at which the system phase separates is known as W_0^{sat}. This number can be as high as 100 for AOT in solvents such as pentane [12]. In ethane and propane, however, W_0^{sat} is much smaller and varies with pressure, as shown in Fig. 2 [10,20].

A question that arises from W_0^{sat} data such as those in Fig. 2 concerns the mechanism by which the pressure-driven phase separation occurs. Two types of phase transitions are known from studies of liquid systems. At the solubilization phase boundary, reverse micelles expel excess water to form a second phase. Its location is determined by the natural curvature of the surfactant interface. The natural curvature is the preferred curvature of the interface when no interactions between droplets are present. At the haze point boundary, surfactant and water precipitate together to form a surfactant-rich second phase. This phase transition is driven by micelle–micelle interactions. It is analogous to the cloud point transition seen with increasing temperature in an aqueous micellar system.

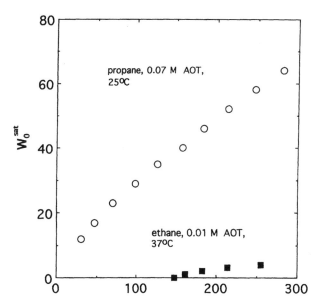

Figure 2 W_0^{sat} versus pressure for supercritical fluid ethane and compressed liquid propane. (From Ref. 20.)

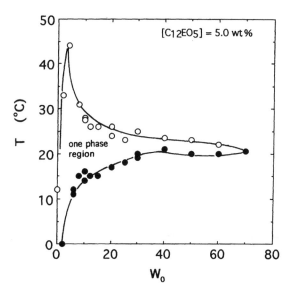

Figure 3 Phase behavior of a heptane–water–$C_{12}EO_5$ reverse micelle system. Open circles indicate solubilization boundary; closed circles indicate haze point boundary. (From Ref. 12.)

An example of surfactant behavior in conventional liquid solvents is shown in Fig. 3 for the nonionic $C_{12}EO_5$–water–heptane system [12]. The upper boundary of the narrow one-phase region, shown by open circles, is the solubilization boundary. The lower boundary, shown by filled circles, is the haze point curve. AOT systems have the same kind of phase behavior, but since AOT is anionic the relative positions of the solubilization and haze point boundaries are reversed. In the case of the pressure-driven transitions of Fig. 2, it has been shown that they occur by the solubilization mechanism, i.e., micelle–micelle interactions [12,20,21].

B. Pressure Effects on AOT Reverse Micelles in the One- and Two-Phase Regions

Pressure effects on AOT reverse micelles in ethane and propane have been studied by a variety of techniques, including phase behavior, light and neutron scattering, and solvatochromic and fluorescence probes [20]. All of the measurements show that a reverse micelle system containing solubilized water is insensitive to pressure changes in the one-phase region. This finding is consistent with studies in liquid alkanes, which show that the size of AOT reverse micelles is largely independent of the solvent but is strongly correlated with W_0 [17,19]. Changes in micellar size are seen only very near phase transition points, a result of increasing micelle–micelle interactions. The mechanism of micelle–micelle interactions in compressed solvents has been studied in detail with neutron scattering and lattice fluid self-consistent field theory [23–25].

In two-phase systems, however, where surfactant and water can partition between a fluid and a liquid phase, significant pressure effects occur. These effects were studied for AOT in ethane and propane by means of the absorption probe pyridine N-oxide and a fluorescence probe, ANS (8-anilino-1-naphthalenesulfonic acid) [20]. The UV absorbance of pyridine N-oxide is related to the interior polarity of reverse micelles, whereas the fluorescence behavior of ANS is an indicator of the freedom of motion of water molecules within reverse micelle water pools. In contrast to the blue-shift behavior of pyridine N-oxide, the emission maximum of ANS increases ("red shift") as polarity and water motion around the molecule increase. At low pressures the interior polarity, degree of water motion, and absorbance intensity are all low for AOT reverse micelles in the fluid phase because only small amounts of surfactant and water are in solution. As pressure increases, polarity, intensity, and water motion all increase rapidly as large amounts of surfactant and water partition to the fluid phase. The data indicate that the surfactant partitions ahead of the water; thus there is a constant increase in size and fluidity of the reverse micelle water pools up to the one-phase point. An example of such behavior is shown in Fig. 4 for AOT in propane with a total W_0 of 40. The change in the ANS emission maximum suggests a continuous increase in water mobility, which is due to increasing W_0 in the propane phase, up to the one-phase point at 200 bar.

C. Physical Properties of Water in AOT Reverse Micelle Water Pools

The nature of the water within AOT reverse micelles has been the subject of much study. Considered as a whole, the data tend to indicate the presence of three types of water in AOT reverse micelles, although it is difficult to establish precise W_0 boundaries between them [26]. "Bound" water is water of hydration that is tightly bound to individual AOT headgroups or shared between neighboring headgroups. It is the predominant type for W_0 below 8–10. "Trapped" water is not bound to any particular AOT molecule but is still confined to the immediate vicinity of the AOT headgroups. Water partitions between bound and trapped states for W_0 between 8 and 18. "Free" water begins to appear when W_0 is greater than 20. Free water has properties very close but not identical to normal water properties. The main difference appears to be a reduction in the freedom of motion for water molecules in the water pool [20,27].

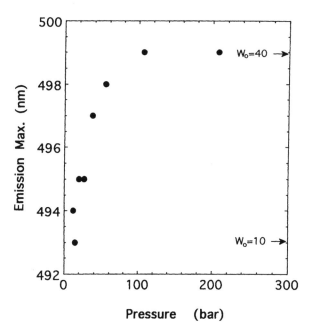

Figure 4 Partitioning of water and surfactant in the two-phase region, as indicated by the ANS emission maximum, for AOT (0.01 M) reverse micelles in liquid propane at 25°C with an overall W_0 of 40. Arrows indicate emission maxima for one-phase AOT systems. (From Ref. 20.)

AOT reverse micelle water pools in ethane, propane, and hexane were studied using the absorption probe pyridine N-oxide, which measures local polarity, and the fluorescence probe ANS, which indicates molecular dynamics. ANS fluorescence behavior was investigated by a combination of steady-state and time-resolved techniques [20,27,28]. It was found that the interior state of the reverse micelles was highly dependent on W_0 but independent of the solvent. It was also found that the microscopic environment in the water pools for one-phase AOT systems in ethane and propane was largely unaffected by pressure. An example of the spectroscopic data from these studies is shown in Fig. 5 for 0.01 M AOT at 25°C. The spectroscopic response, in this case the emission maximum of ANS, shows a typical pattern. There is a steep change in the maximum for W_0 from 0 to 6, which can be assigned to bound water, and a plateau region for W_0 greater than 15 that is characteristic of free water. The transition region between $W_0 = 6$ and $W_0 = 15$ can be attributed to the presence of trapped water. It is also noteworthy that the ANS emission maximum never reaches the value of 510 nm seen in bulk water, suggesting restricted freedom of motion of the molecules in the water pool.

D. Aggregation of Nonionic Surfactants in Ethane and Propane

Nonionic poly(ethylene oxide) alkyl ethers aggregate by means of hydrogen bonds and polar interactions between the poly(ethylene oxide) headgroups. Nonionic surfactant aggregation occurs by a stepwise mechanism, instead of by the critical micelle concentration (cmc) type of mechanism of AOT. The first step in the aggregation process is the establishment of multiple equilibria between free surfactant unimer and various *intra*molecular hydrogen-bonded species. As surfactant concentration increases, *inter*molecular hydrogen bonding becomes

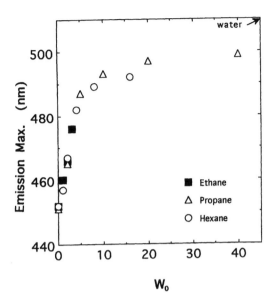

Figure 5 Degree of water motion about ANS as described by the emission maximum in AOT reverse micelles (0.01 M) in ($\triangle$) liquid propane at 25°C, 278 bar; ($\blacksquare$) supercritical fluid ethane at 37°C, 278 bar; and ($\bigcirc$) hexane at 25°C, 1 bar. (From Ref. 20.)

the predominant type [29]. In small nonpolar liquid solvents, it has been observed that the aggregation numbers are on the order of 5, in contrast to the aggregation number of approximately 21 seen for AOT reverse micelles [17]. When water is added to the system, large increases in reverse micelle size occur [30].

The aggregation of $C_{12}EO_3$ and $C_{12}EO_8$ was studied by Fourier transform infrared (FTIR) spectroscopy in supercritical fluid ethane [31]. The low solvent strength of ethane compared to a liquid alkane such as heptane discourages the solubilization of the surfactant as free unimers and encourages the formation of intramolecular hydrogen bonds at low surfactant concentration and of small aggregates as surfactant concentration increases. An average aggregation number of 4.5 was reported for $C_{12}EO_3$ in ethane at 40°C and 400 bar.

Although poly(ethylene oxide) alkyl ethers are readily soluble in ethane and propane, especially those having less than five ethylene oxide units in their headgroups, their solubility is reduced considerably when water is added to the system [31]. For the polydisperse surfactant $C_{11-14}EO_5$, the W_0^{sat} is 5 at 35°C and 200 bar [11], which is comparable to an AOT system at the same pressure and temperature [20]. At such low W_0^{sat} values it is likely that water hydrates the ethylene oxide units rather than forming water pools. In propane, higher W_0^{sat} values are possible, just as in the case of AOT. In $P-T$ space the W_0^{sat} boundaries curve back on themselves to form parabolic solubility loops, one for each value of W_0^{sat} [12]. The systems are one-phase inside the loops and two-phase outside. The loops are concave toward higher pressures and lower temperatures. The shape of the W_0^{sat} boundary is reminiscent of that seen in $\rho-T$ space for AOT in propane [32], but the curvature for the nonionic systems is much greater because of the greater temperature sensitivity of nonionic surfactants. Pressure effects on a mixed nonionic–cationic surfactant system in propane were studied by Klein and Prausnitz [33].

III. MICROEMULSIONS IN COMPRESSED LIQUID PROPANE

A. Introduction

A ternary system composed of oil, water, and surfactant can form a wide variety of aggregated structures. Two characteristic compositions are frequently studied: reverse micelle systems in which the amount of oil greatly exceeds the amount of water, and systems in which oil and water are present in relatively equal amounts (Winsor systems). Reverse micelle systems were discussed in the previous section; this section is devoted to Winsor systems having an oil phase composed of a supercritical fluid or compressed liquid alkane. It should be noted, however, that these two types of systems merely represent two specific regions in the space of ternary oil–water–surfactant compositions, and both are subject to the same thermodynamic considerations.

Systems containing equal amounts of oil and water often can be classified according to the four Winsor types [34]. These classical configurations are known as type I (normal micelles in equilibrium with excess oil, often designated $\underline{2}$, which means two phases with the surfactant in the lower phase), type II (reverse micelles in equilibrium with excess water, $\overline{2}$), type III (middle phase microemulsion with excess water and excess oil phases, 3), and type IV (one-phase microemulsion, 1).

Techniques for moving a system from one Winsor type to another are well known. Normally, either a $\underline{2}$–3–$\overline{2}$ or a $\overline{2}$–3–$\underline{2}$ transition occurs. For example, changing the oil phase to one of higher alkane carbon number (ACN) makes it a less favorable environment for the surfactant and promotes the $\overline{2}$–3–$\underline{2}$ transition for ACN > 6 [35,36]. Control of phase behavior via the hydrophile–lipophile balance (HLB) of the surfactant is a commonly practiced technique [37]. Surfactants having a low HLB number favor the $\overline{2}$ configuration, whereas surfactants of high HLB are hydrophilic and tend to form $\underline{2}$ systems. Temperature can also be an effective means of changing phase behavior, particularly in the case of nonionic surfactants. Increasing temperature causes a $\underline{2}$–3–$\overline{2}$ transition in nonionic surfactant systems, whereas a $\overline{2}$–3–$\underline{2}$ transition is normally seen in anionic surfactant systems. Other well-known techniques include the addition of electrolyte to force the surfactant out of the aqueous phase and change the interfacial curvature, and the use of cosurfactants such as medium-chain-length alcohols to reduce the interfacial tension [34].

Pressure effects on surfactant systems containing conventional liquid alkanes have not often been studied because of the very low compressibility of liquids. Conflicting results have been reported [38–40]. It is likely that the changes in cohesive energy density (solubility parameter) of the phases over the pressure ranges used were too low to produce definitive trends in phase behavior. The solubility parameter of compressed liquid propane, however, is moderately adjustable with pressure, and therefore a propane–brine–AOT system could be expected to show pressure-driven phase transitions [20,22,41].

B. Pressure, Salinity, and Temperature Effects on the Propane–Brine–AOT System

Experiments were carried out on a system composed of AOT, brine, and compressed liquid propane at 37°C [22]. The effect of pressure on the partitioning of AOT for a system containing 1.7 wt% NaCl in the aqueous phase is shown in Fig. 6, and the pressure effect on W_0 in the propane phase is shown in Fig. 7. The addition of NaCl to the system was necessary to prevent the formation of an AOT liquid crystal phase. These data were obtained by withdrawing small samples from each phase for analysis. Similar data have also been

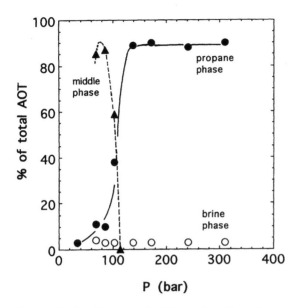

Figure 6 Partitioning of AOT as a function of pressure at 37°C and 1.7 wt% NaCl in the aqueous phase. Overall cell contents: 0.5 g AOT, 10 mL brine, 6.6 g propane. (From Ref. 22.)

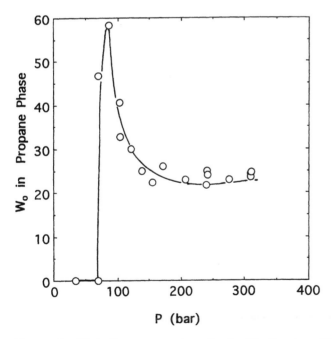

Figure 7 W_0 in the propane phase for the Fig. 6 system. (From Ref. 22.)

obtained in situ by FTIR spectroscopy [42]. At the highest pressure, 310 bar, a clear propane upper phase and a clear aqueous lower phase are observed. At this point, 90% of the AOT is in the upper phase, and the presence of water indicates the formation of AOT reverse micelles. This is the $\bar{2}$ configuration. The system initially changes very little in appearance as pressure is reduced. As pressure is reduced further, the propane phase becomes increasingly straw-colored. At 117 bar, a slight cloudiness appears in the propane phase, followed at 114 bar by a more distinct cloud point and the formation of a middle phase. The middle phase, characteristic of a 3 system, persists all the way to the propane vapor pressure of 13 bar.

The change in appearance of the propane phase as pressure is reduced toward the $\overline{2}$–3 phase boundary is a classical indication of a transition driven by micelle–micelle interactions [14]. The observed 3 configuration would therefore represent a split of the propane phase into a surfactant-rich phase and a surfactant-lean phase, both of which are propane continuous. This interpretation of the pressure-driven phase behavior is supported by the measured phase compositions in the 3 region [22] and also by theoretical modeling of the propane–brine–AOT system using interfacial thermodynamics [43].

Pressure also affects W_0 in the propane phase, as shown in Fig. 7. At pressures well above the phase transition pressure, W_0 changes only slightly with pressure. However, W_0 increases dramatically at the phase transition point. It keeps on increasing until the middle phase reaches its maximum extent, then decreases back to 0 as the last of the AOT leaves the propane phase. The simultaneous increase in W_0 with loss of AOT from the propane phase, as well as the known correlation of AOT W_0 with aggregation number, indicates that the reverse micelles are growing larger as the $\overline{2}$–3 phase transition point is approached, an indication of decreasing interfacial curvature, in accordance with theory [44].

For an ionic surfactant such as AOT, increasing salinity leads to a $\underline{2}$–3–$\overline{2}$; transition. The AOT–brine–propane system conforms to this classical behavior in several ways. For example, increases in the salinity of the aqueous phase increase the proportion of AOT in the propane phase. At high pressures, the size of the reverse micelles (as reflected by W_0) decreases as salinity increases. The pressure at which the $\overline{2}$–3 transition occurs decreases as salinity increases, an indication of increasing surfactant affinity for the propane phase. One interesting aspect of the AOT–brine–propane system is that the amount of NaCl required to effect phase changes appears to be higher than is the case in conventional liquid solvents. For propane at 310 bar and 37°C, about 1.1 wt% salt is required to drive AOT into the propane phase. For the heptane–brine–AOT system at the same concentration, only 0.5 wt% salt is required to achieve the same effect [36]. This result suggests that propane is a weaker solvent for AOT than heptane.

Temperature effects are more complicated. Existing theories and data predict that increasing temperature should move the AOT–brine–propane system from the $\overline{2}$ configuration toward the 3 configuration [45]. However, it was found that the small changes due to temperature were masked by large changes in propane density, which decreases its solubility parameter and also moves the system toward the 3 configuration [22].

C. "Fish" for Nonionic Surfactants in Propane

In the AOT–brine–propane system, $\overline{2}$–3 transitions occur readily. However, the structure of AOT is such that the 3–$\underline{2}$ transition is much more difficult. Furthermore, the phase behavior of AOT is not particularly temperature-sensitive. Nonionic surfactants, on the other hand, are very responsive to temperature and readily form both oil-in-water and water-in-oil phases when the surfactant size and HLB are chosen so as to be compatible with both water and oil. A plot of surfactant concentration vs. temperature has a characteristic fish shape, as seen in the work of Kahlweit et al. [46].

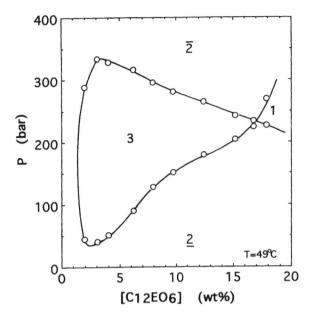

Figure 8 "Fish" for the $C_{12}EO_6$–propane–brine system generated by changing pressure. (From Ref. 12.)

"Fish" were created in propane at constant pressure using $C_{12}EO_6$ as the surfactant, with 5.0 wt% NaCl added to the aqueous phase to bring the fish down to a reasonable temperature range [12]. As expected, the system is in the $\underline{2}$ configuration at low temperatures and in the $\overline{2}$ configuration at high temperatures. The fish shift to higher temperature ranges as pressure decreases. The upper boundary of the fish moves upward in temperature much faster than the lower boundary, so the net effect is a large increase in the width of the fish. Since the upper boundary is the one associated with the compressible propane phase, its movement with pressure is an expected effect.

Additional experiments were conducted on the $C_{12}EO_6$–brine–propane system to determine whether a fish-shaped relationship could be created by means of pressure alone [12]. The result is shown in Fig. 8. This is believed to be the first example of a complete fish shape produced by pressure alone for nonionic surfactants and compressible oil phases. In this instance, both AOT and $C_{12}EO_6$ show the same $\underline{2}$–3–$\overline{2}$ transition with increasing pressure because of the enhanced ability of the oil phase to solubilize the surfactant as pressure increases.

D. Theoretical Discussion

The phase behavior observed in surfactant systems can be viewed from several theoretical perspectives. One is the qualitative Winsor R theory. In its simplest form, the Winsor R parameter is the ratio of the forces acting on the oil side of the surfactant interface to those on the water side. This definition has been extended to include other interactions that tend to oppose surfactant aggregation [34]. When $R > 1$, the forces on the oil side of the interface are the strongest, so the interface curves about water, resulting in a $\overline{2}$ system. When $R < 1$, the water-side forces are dominant, which causes the formation of a $\underline{2}$ system. When $R = 1$, the forces at the interface are balanced, which leads to a bicontinuous microemulsion of the 3 or 1 type. The Winsor theory affords a simple qualitative rationalization of many aspects of surfactant behavior. For example, oil chain length and surfactant HLB affect surfactant phase behavior through the oil–surfactant tail interaction. When this interaction is enhanced by the proper choice of surfactant and oil, R increases and the $\overline{2}$ configuration is favored.

In the phenomenological model of Kahlweit et al. [46], the behavior of a ternary oil–water–surfactant system can be described in terms of the miscibility gaps of the oil–surfactant and water–surfactant binary subsystems. Their locations are indicated by the upper critical solution temperature (UCST), T_α, of the oil–surfactant binary systems and the critical solution temperature of the water–surfactant binary systems. Nonionic surfactants in water normally have a lower critical solution temperature (LCST), T_β, for the temperature ranges encountered in surfactant phase studies. Ionic surfactants, on the other hand, have a UCST, T_δ. Kahlweit and coworkers have shown that techniques for altering surfactant phase behavior can be described in terms of their ability to change the miscibility gaps. One may note an analogy between this analysis and the Winsor analysis in that both involve a comparison of oil–surfactant and water–surfactant interactions.

The Kahlweit framework may be used to explain in a qualitative way the effects of pressure on surfactant phase behavior described above. In liquid–liquid systems both the water and oil phases are nearly incompressible, and so neither T_α nor T_β and T_δ change much with pressure. Therefore, the observed effects are variable and depend on whichever critical temperature happens to be more affected. This mechanism was documented in a study on the dodecane–water–C_7EO_5 system and its binary subsystems [40]. Both T_α and T_β increased with pressure, but T_α increased faster than T_β. The net effect according to Kahlweit's model would be to promote a $\overline{2}$–3–$\underline{2}$ transition, which was confirmed by the experimental observations of the ternary system. When the oil phase is a compressible one such as propane, however, the pressure effect is overwhelmingly greater on the oil–surfactant binary system than on the water–surfactant binary system. Increasing pressure increases the solubility parameter of the oil phase, reduces the oil–surfactant miscibility gap, and lowers T_α. This change moves the system toward the $\overline{2}$ configuration.

Quantitative predictions of surfactant phase behavior can be made by constructing a thermodynamic model. The classical expression for the free energy of a microemulsion is a function of the interfacial tension, bending moment, and micelle–micelle interactions [47]. Two quantitative models have been developed to describe supercritical microemulsions based on this concept. Here, the key challenge is to find accurate expressions for the oil–surfactant tail interactions and the tail–tail interactions. To do this, the first model uses a modified Flory–Krigbaum theory [43,44], and the second a lattice fluid self-consistent field (SCF) theory [25].

The first supercritical microemulsion model [43] was in good agreement with experimental data such as those of Figs. 6 and 7. The model also applies to the one- and two-phase reverse micelle systems of Sec. II. It was found that micelle–micelle interaction effects are dominant in reverse micelle systems where the water/oil ratio is small. However, in the Winsor microemulsion systems described in this section, which have a water/oil ratio near unity, the size of the reverse micelles in the oil phase is determined by natural curvature effects. Micelle–micelle interactions become important at phase transition points, as was observed experimentally in the AOT–brine–propane system [21,23]. The transition between the natural curvature and micelle–micelle interaction mechanisms can be understood in detail on a ternary phase diagram [43].

The model based on the lattice fluid SCF theory offers a means to calculate fundamental interfacial properties of microemulsions from pure component properties [25]. Because all of the relevant interfacial thermodynamic properties are calculated explicitly and the surfactant and oil molecular architectures are considered, the model is applicable to a wide range of microemulsion systems. The interfacial tension, bending moment, and interaction strength between the droplets can be calculated in a consistent manner and analyzed in terms of the detailed interfacial composition. The mechanism of the density effect on the natural curvature includes both an enthalpic and an entropic component. As density is decreased, the solvation of the surfactant tails is less favorable enthalpically, and the solvent is expelled from the interfacial region. Entropy also contributes to this oil expulsion due to the density difference between the interfacial region and the bulk. The oil expulsion and increased tail–tail interactions decrease the natural curvature.

IV. ALKANE CARBON NUMBER EFFECTS ON SURFACTANT PHASE BEHAVIOR

A. Introduction

Previous studies have shown that as the alkane carbon number (ACN) of the oil phase increases, it generally becomes a poorer solvent for the surfactant. In Winsor systems this effect leads to a $\overline{2}$–3–$\underline{2}$ transition [36]. In one-phase AOT reverse micelle systems, water solubilization at the phase boundary, W_0^{sat}, decreases as ACN increases [48]. None of these studies, however, used a solvent lighter than hexane.

Several objectives motivated the extension of ACN studies to light compressible solvents [12]. Initial studies of AOT in such solvents had demonstrated the possibility of intriguing solvent effects [20,21,32], which could be clarified by additional experiments. A second objective was to test the concepts generated from the thermodynamic models that were developed for the AOT–brine–propane system [25,44]. A final objective was to study the behavior of nonionic surfactant systems as a complement to AOT systems. Nonionic systems provide an enhanced opportunity to study temperature effects on surfactant phase behavior, as nonionic surfactants are much more responsive to temperature than the anionic surfactant AOT.

B. One-Phase AOT Reverse Micelle Systems

Water solubilization in AOT reverse micelles has been studied by measurements of W_0 at the phase boundary, W_0^{sat}, at a fixed temperature [16,20,21,48]. The data are in reasonable agreement for the heavy solvents octane, nonane, and decane and for the light solvents ethane, propane, and butane, but the results are not in agreement for the intermediate solvents pentane, hexane, and heptane [20]. To determine the location of W_0^{sat} for these solvents, phase boundary plots of the type shown in Fig. 3 were constructed, and a constant-temperature line was drawn across the plot to determine W_0^{sat}. This is a much more reliable technique than the alternative procedure of adding water aliquots to a reverse micelle solution at a fixed temperature, because the W_0^{sat} phase boundary can be difficult to observe in such an experiment [19].

When the new W_0^{sat} data for pentane through octane are combined with the results reported earlier [19], all at 25°C, a smooth trend in W_0^{sat} is obtained as a function of solvent density for solvents from ethane through decane, as shown in Fig. 9 [12]. A reversal in the

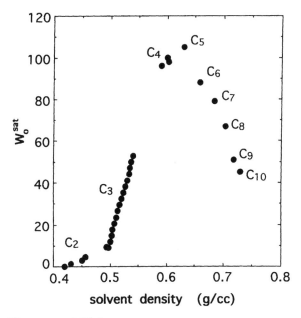

Figure 9 W_0^{sat} for 0.07 M AOT as a function of density for a series of n-alkanes. (From Ref. 12.)

W_0^{sat} trend is clearly visible in the neighborhood of pentane. This is an example of what has been called an "unusual" phase behavior trend [34]. The other important aspect of the Fig. 9 data is that all of the W_0^{sat} points are located on the haze point (micelle–micelle interaction) boundaries of their respective solvents [21]. An earlier discussion has suggested that these unusual effects mights be due to a transition from micelle–micelle interaction effects to natural curvature effects at the W_0^{sat} maximum [48]. For AOT at least, this is not the case. The thermodynamic model described in Sec. III indicates that they are in fact a result of enthalpic and entropic contributions to the solvent–surfactant tail interaction [43].

Figure 9 can be made more universal in character by replacing solvent density with solvent dielectric constant. In such a plot, not only branched alkanes but also the noble gases krypton and xenon form a continuous envelope of W_0^{sat} of the same general shape as that seen in Fig. 9. The dielectric constant at which the maximum W_0^{sat} is found corresponds to a minimum in the attractive dispersion interaction between microemulsion droplets as calculated from Lifshitz theory [21]. Therefore, reverse micelles would be most resistant to phase separation caused by micelle–micelle interactions at this point, and solubilization of water would reach a maximum. However, it seems unlikely that the dispersion interaction would be the sole contributor to W_0^{sat} behavior [14,43].

C. Nonionic Surfactant Reverse Micelle Systems

The behavior of the poly(ethylene oxide) alkyl ether surfactant $C_{12}EO_5$ was studied by constructing phase diagrams similar to Fig. 3 for a series of alkanes. From nonane through pentane the changes in the phase diagrams agree with the literature [49]. That is, the maximum W_0 of the one-phase region increases, and the one-phase region itself shifts to lower temperatures. For solvents lighter than pentane at constant pressure, "unusual" behavior is again seen, as the maximum W_0 decreases and the one-phase region moves back to higher temperatures in butane and propane [12].

An important issue in this type of behavior is the nature of the phase boundaries in the light solvents. In the case of the pentane and butane systems, the upper temperature phase boundary is marked by a gradual fogging of the solution due to precipitation of water, which is consistent with a natural curvature (solubilization) mechanism. The lower temperature boundary is marked by a sudden precipitation of both water and surfactant, which indicates a haze point (micelle–micelle interaction) transition. In the propane system it is difficult to determine the phase transition mechanism because of the low solubilization of water [12]. Nonetheless, it appears that the relative location of haze point and solubilization boundaries are the same in both light and heavy alkanes.

D. Winsor Systems

The commercial surfactant Igepal CO-520 was used to study ACN effects on ternary oil–water–nonionic surfactant systems. This surfactant is a polydisperse mixture whose main component is a poly(ethylene oxide) nonylphenol ether with five ethylene oxide units in its headgroup. Figure 10 shows the domains of the Winsor types for alkane–water–Igepal CO-520 systems as a function of temperature and ACN for a surfactant loading of 5 wt%. The right half of Fig. 10, from hexane through dodecane, shows a trend that is well known from previous work; i.e., the three-phase region grows wider and shifts to higher temperatures as ACN increases [34,50]. However, the left half of Fig. 10 is new and shows yet another unusual trend [12]. The same experiment was conducted using Igepal CO-530, whose main

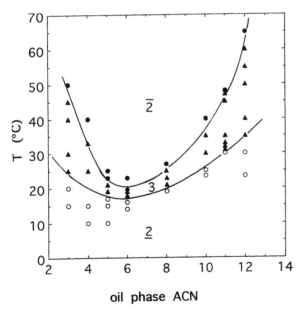

Figure 10   Microemulsion phase behavior for 5.0 wt% Igepal CO-520 in an oil–water–surfactant system. Butane and propane data are at 276 bar; others are at ambient pressure (○) $\underline{2}$ behavior observed; (●) $\overline{2}$ behavior observed; (▲) 3 behavior observed. (From Ref. 12.)

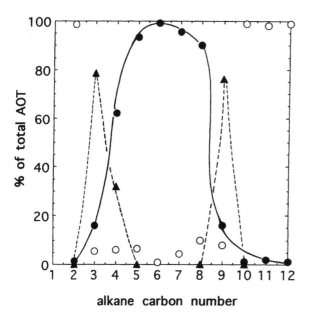

Figure 11 Partitioning of AOT as a function of the alkane carbon number in an oil–water–surfactant system. System composition: 0.15 g AOT, 10 mL brine, 10 mL alkane, 0.40 wt% NaCl in the brine. Ethane, propane, and butane at 310 bar. (●) Oil phase; (○) water phase; (▲) middle phase. (From Ref. 22.)

component contains six ethylene oxide units instead of five. The same pattern in the three-phase region was seen, but it was shifted upward by 15°C because of the increased hydrophilicity of Igepal CO-530 [12].

Fish-shaped diagrams (see Fig. 8) were constructed for Igepal CO-520 in a series of alkanes as a function of temperature and surfactant concentration [12]. The smallest fish shape is the one for hexane. As ACN increases, the fish shapes grow larger, and the optimum point of the system (where the "head" joins the "tail" of the fish) moves to higher temperature and higher surfactant concentration. This trend is in agreement with previous work [53]. For solvents lighter than hexane, an unusual trend is again seen in that the fish shapes get larger and the optimum point moves to higher temperatures and higher surfactant concentrations as ACN decreases.

The starting point for studies of solvent effects on AOT–alkane–brine systems was an earlier study using alkanes from hexane through dodecane [36]. Systems having the same composition as in the earlier work were made up, and the partitioning of AOT between the phases was determined [22]. The results are shown in Fig. 11. The data for hexane through dodecane match the earlier results and indicate a $\overline{2}$–3–$\underline{2}$ transition as AOT moves out of the oil phase and into the middle and lower phases. For lighter solvents at 25°C and 310 bar, an unusual $\overline{2}$–3–$\underline{2}$ transition is seen from pentane to ethane. The phase transitions are also indicated by the W_0 in the oil phase, as shown in Fig. 12. There is a local minimum in W_0 at hexane, where the oil–surfactant tail interaction is maximized and the surfactant interface is most strongly curved about water. For ACN larger and smaller than 6, the oil–surfactant tail interactions become weaker, so W_0 increases as interfacial cur-

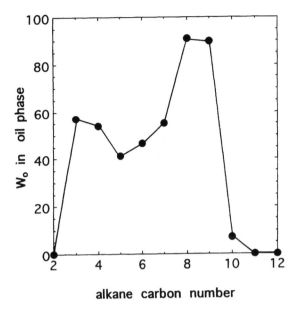

Figure 12 W_0 in the oil phase for the systems of Fig. 11. (From Ref. 22.)

vature decreases. At ACN values of 3 and 8, W_0 reaches an absolute maximum as the interfacial curvature goes to zero and the middle phase appears. For ACN less than 3 or greater than 9, W_0 goes to zero as the last of the AOT is expelled from the oil phase.

E. Theoretical Discussion

The trends in phase behavior as a function of oil-phase ACN indicate an optimum point in the neighborhood of pentane or hexane. This behavior is a consequence of the enthalpic and entropic components of the solvent–surfactant tail interaction [25,43]. Alkanes heavier than hexane are hindered in their ability to penetrate between the surfactant tails, which makes the combinatorial (chain length compatibility) effect the dominant one for these solvents. Alkanes lighter than hexane penetrate between surfactant tails very easily, but their enthalpic interaction with the tails is weak. At pentane and hexane, the combination of enthalpic and entropic effects is balanced. This balance favors the formation of the $\overline{2}$ configuration in Winsor systems. In reverse micelle systems, the optimum solvent–surfactant tail interaction stabilizes the reverse micelles against phase separation driven by micelle–micelle interactions. Also, there is a minimum in the attractive intermicellar dispersion interaction [13]. Therefore W_0^{sat} reaches a maximum.

V. MICROEMULSIONS IN SUPERCRITICAL FLUID CARBON DIOXIDE

A. Solvent Properties of CO_2

Liquid and supercritical fluid carbon dioxide are an attractive alternative to organic solvents because CO_2 is nontoxic, nonflammable, and inexpensive and has a reasonably accessible critical point ($T_c = 31°C$, $P_c = 73.8$ bar). It is an environmentally benign solvent that is being evaluated as a replacement for organic liquids in numerous applications. However,

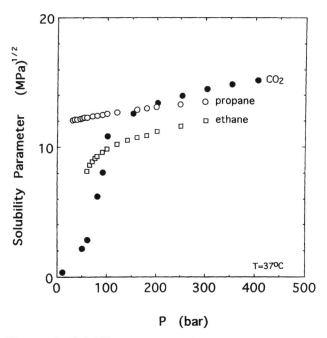

Figure 13 Solubility parameters of ethane, propane, and carbon dioxide at 37°C as a function of pressure. (From Ref. 52.)

the properties of CO_2 are very different from those of water or nonpolar organic solvents [51]. Unlike water, it has no dipole moment. Even when highly compressed, it has far weaker van der Waals forces than hydrocarbon solvents. Although the Hildebrand solubility parameter of CO_2 is larger than that of propane at pressures greater than 200 bar, as shown in Fig. 13, a substantial portion of it results from CO_2's large quadrupole moment [52]. The dispersion component of the solubility parameter, which is important for solubilizing surfactant tails, is actually less than that of ethane. Another measure of the strength of van der Waals forces, the polarizability per unit volume, is also very small for CO_2, below that of ethane at comparable conditions [53].

Like the alkane solvents discussed previously, CO_2 is a poor solvent for nonvolatile hydrophilic molecules. Some enhancement in hydrophile solvation was achieved by addition of cosolvents such as methanol and of complexing agents such as tri-*n*-butyl phosphate [8,54]. Chelating agents have been designed with tails that have highly favorable interactions with CO_2, such as fluoroethers and silicones [55]. A much wider range of hydrophilic compounds could be solubilized with surfactants that form reverse micelles and microemulsions in CO_2 with polar or aqueous cores. However, extremely few commercial surfactants have tails compatible with the weak van der Waals forces of CO_2. This limitation has made the formation of reverse micelles and microemulsions in CO_2 more difficult than was the case for the alkane solvents discussed in the previous sections.

B. Design of Surfactants for CO_2 Applications

At first, some preliminary evidence of aggregation in CO_2 was reported [56,57]. Next, Iezzi et al. [58] screened commercially available surfactants for their ability to raise the viscosity of CO_2 for improved oilfield flooding. None of the candidates selected for detailed study was found soluble enough in CO_2 to raise its viscosity by more than 10%, although this was a stringent requirement. Oates [59] studied the phase behavior of benzenesulfonates and naphthalenesulfonates with twin hydrocarbon tails. None was capable of transferring significant amounts of water to a CO_2 phase at pressures up to 83 bar. Consani and Smith [60] examined over 130 surfactants with a wide range of structures. Most ionic surfactants were insoluble in CO_2; some nonionic surfactants were soluble. Neither ionic nor nonionic surfactants would solubilize much water at pressures below 350 bar. The surfactants with perfluorinated or poly(dimethyl siloxane) tails were among the most soluble; this finding played a key role in later studies. McFann [61] evaluated 31 commercial and experimental surfactants at various conditions of temperature, pressure, and added water. In these studies, typical values of the number of water molecules per surfactant headgroup were less than 5. Aggregation of the poly(ethylene oxide) alkyl ethers $C_{12}EO_3$ and $C_{12}EO_8$ without added water was studied by FTIR [31]. For the small poly(ethylene oxide) alkyl ether C_8EO_5, W_0 values as high as 12 (above the baseline solubility of water in CO_2) were observed with the addition of significant amounts of n-pentanol as cosurfactant [52]. However, this corresponds to only about two water molecules per ethylene oxide unit.

Clearly, new concepts (and new surfactants) were needed to accommodate the unique solvent properties of CO_2. A key breakthrough in the design of surfactants for use in CO_2 was made by Hoefling et al. [62]. They synthesized a perfluorinated analog of AOT as well as fluoroalkyl and fluoroether carboxylate salts. Some of the fluoroalkyl and fluoroether compounds were converted to hydroxyaluminum soaps to produce a surfactant with a less polar headgroup. All of these compounds were found to be readily soluble in CO_2 at relatively modest conditions (40°C, 140 bar and below). The one-phase systems could dissolve the dye thymol blue, which is insoluble in CO_2. The AOT analogs and the hydroxyaluminum soaps were preferentially soluble in the CO_2 phase of a CO_2–water–surfactant system. However, the extent of aggregation and water solubilization in these systems was unknown, since the interfacial activity of thymol blue allows it to be solubilized in small aggregates without water pools as well as in fully developed reverse micelles.

The work of Hoefling et al. [62] and of Consani and Smith [60] shows that a surfactant tail group having favorable interactions with CO_2 is required for the formation of aggregated structures. The term "CO_2-philic" has been coined to denote this favorable interaction [63]. The effects of CO_2–philic interactions were also seen in the earlier surfactant screening studies [58–61], where surfactants whose tails had functional groups with low dispersion forces and polarizability per unit volume, such as fluorocarbons and silicones, tended to be the most soluble in CO_2. Based on phase behavior studies of surfactants, oligomers, and polymers in CO_2, the most promising candidates, in order of increasing interaction with CO_2, include siloxanes, fluorocarbons, fluoroethers, and fluoroacrylates [4,53,60,64–68]. Fluoroacrylates in particular are extremely CO_2-philic. Poly(1,1-dihydroperfluorooctyl acrylate) (PFOA) homopolymers and PFOA-based block copolymer surfactants have been synthesized for CO_2 applications [69,70]. Rapid progress has been made in dispersion polymerization in CO_2 using PFOA polymers as stabilizers, as will be discussed in the next section. Block and graft copolymers based on PFOA have been used to form micellar structures with nonpolar styrene [63,69] and polar poly(ethylene oxide) [71] cores. Because of the high CO_2-philicity of fluoroacrylates, they are particularly effective at pulling CO_2-phobic parts of surfactants into CO_2.

A W_0 of 32 was attained using a two-tailed hydrocarbon-fluorocarbon hybrid surfactant $C_7F_{15}CH(OSO_3^-Na^+)C_7H_{15}$ (designated F_7H_7) [53]. This high value of W_0 was for a one-phase system at 25°C and 231 bar, far in excess of any other surfactant

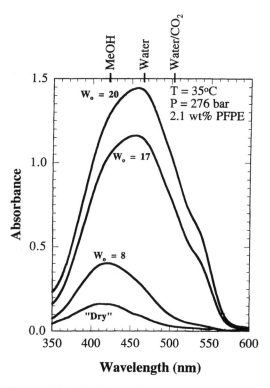

Figure 14 Absorbance of methyl orange in the PFPE + water + CO_2 system as a function of added water. (From Ref. 73.)

in CO_2. The key to the effectiveness of this surfactant is the combination of a hydrocarbon with a fluorocarbon tail. The fluorocarbon tail extends out into the CO_2 phase and provides a large solvent–tail interaction. Because the surfactant contains two tails, the natural curvature is about water, due to geometric packing constraints. It is likely that the hydrocarbon tails also form a palisade layer that shields the headgroups and water pool from the CO_2 phase. A limitation of this surfactant is that the molecule does not appear to be stable against slow hydrolysis over a 1–3 day period, which is facilitated by the acidity of CO_2. With small-angle neutron scattering (SANS) measurements, it was shown that spherical reverse micelles with a core radius around 2.5 nm were formed with a water/surfactant ratio of 33 [72].

In 1996 a stable microemulsion containing a waterlike core within a CO_2 continuous phase was reported for the first time [73]. The surfactant, an ammonium carboxylate perfluoropolyether $[CF_3O(CF_2CF(CF_3)O)_3CF_2COO^-NH_4^+]$ [74] that has a molecular weight of 740 and is commercially available in the COOH form, is similar to PFPE surfactants of molecular weights from 2500 to 7500 that were found in a previous study [66] to be soluble in CO_2. W_0 reached 17 (20 uncorrected) in the one-phase region. The polarity of the micelle cores was studied by means of the solvatochromic probe methyl orange. The spectra obtained at 276 bar are shown in Fig. 14 for several values of W_0 in the one-phase region. Benchmarks are placed for the λ_{max} of (surfactant-free) solvents, including methanol, water, and carbonic acid. As W_0 increases, there is a pronounced red shift in λ_{max} toward the value in bulk water, indicating that methyl orange is in a highly polar aqueous environment. The shoulder in the spectra at 540 nm suggests the presence of an acidic environment due to carbonic acid. The existence of bulklike water was confirmed by FTIR and electron paramagnetic resonance (EPR) spectroscopy [73].

C. Aggregates with Organic Cores in CO_2

Surfactants have been designed for CO_2/organic interfaces as well as CO_2/water interfaces. For example, semifluorinated alkanes have been shown to gel in liquid CO_2 [58]. PFOA homopolymers and PFOA block copolymer surfactants have been synthesized for forming micelles and stabilizing latexes in CO_2 [70,75]. Another type of organic core is poly(ethylene oxide). Large aggregates with diameters of 25 nm were formed from a graft copolymer, poly(1,1-dihydroperfluorooctylacrylate-g-ethylene oxide), having a molecular weight on the order of 100,000 [71]. This molecule has a CO_2-philic PFOA backbone to which are attached CO_2-phobic poly(ethylene oxide) grafts. At 60°C and 300 bar, micellar structures consisting of a PFOA shell surrounding a poly(ethylene oxide) core were observed by small-angle X-ray scattering (SAXS). The micellar core contained approximately 600 poly(ethylene oxide) grafts. The nominal ratio of water molecules to ethylene oxide units was 5.2. Because water can partition between the micellar cores and the external CO_2 phase, the true ratio in the cores was unknown but was probably about 2, based on the known solubility of water in CO_2. Water in these amounts would be expected to be dispersed among the poly(ethylene oxide) chains rather than collected into a water pool at the center of the core. However, the X-ray scattering technique did not provide information on this point. Spectroscopic studies using the probe methyl orange showed a relatively low polarity ($\lambda_{max} = 418$ nm) in the micellar core that was consistent with the small values of W_0 [61]. The λ_{max} of methyl orange in a reverse micelle having a hydrated poly(ethylene oxide) core was 420 nm [52,76], intermediate between a less polar organic core (in which methyl orange is insoluble) and a water core ($\lambda_{max} = 464$ nm).

The ability to design surfactants for the interfaces between organics and carbon dioxide and between water and carbon dioxide offers new opportunities in protein and polymer chemistry, separation science, reaction engineering, chemical waste minimization and treatment, and materials science. With the recent new developments described above, the field is poised for substantial growth.

VI. APPLICATIONS

A. Solubilization of Biomolecules and Other Hydrophilic Substances

Reverse micelles of AOT in liquid solvents such as isooctane have been the focus of applied studies designed to exploit their special interior features such as the water pool and the palisade region near the AOT headgroups. For example, reverse micelle extraction, especially of biomolecules, has been studied by several investigators [77,78]. The process consists of contacting an aqueous solution containing valuable biological products, such as a fermentation broth, with a nonpolar reverse micelle phase. The biomolecule of interest transfers to the interior of the reverse micelles, and the unwanted components remain in the aqueous phase. A second extraction is then carried out on the reverse micelle solution to recover the solubilized biomolecules.

A comprehensive set of studies were carried out on the solubilization of amino acids [79,80]. It was found that amino acids are solubilized into different regions of AOT reverse micelles according to their structure and polarity. The smallest and most hydrophilic amino acids, such as glycine and alanine, partition into the water pool. The other amino acids, especially those having nonpolar side chains, are preferentially solubilized in the palisade region or the surfactant interface. A theoretical model was developed to interpret the experimental results. Molecules solubilized near the surfactant interface can be recovered by increasing the interfacial curvature, a process known as "squeezing out." Normally this is done by contacting the reverse micelle phase with a brine phase. A clever alternative was proposed [81] in which the reverse micelle solution is cooled to near 0°C, then pressurized with methane to form clathrate hydrates in the AOT water pools. The clathrates precipitate, which produces a squeezing-out effect from the reduction of W_0. Another procedure makes use of molecular sieves to recover proteins from reverse micelles [82].

Solubilization of high molecular weight biomolecules such as cytochrome c and hemoglobin was observed in AOT reverse micelles in supercritical solvents [41]. Solubilization of amino acids into AOT reverse micelles in ethane and propane was investigated [83] as an extension of the earlier work in conventional liquids [79,80]. It was demonstrated that the selectivity for proline vs. tryptophan could be manipulated with pressure in ethane in the two-phase region. This is a direct result of changes in reverse micelle size and W_0 with pressure in the ethane phase, as discussed in Sec. II. Proline is preferentially solubilized in the water pools, whereas tryptophan resides in the interfacial region. Interfacial area increases as r^2, but water pool volume increases as r^3, so the increase in reverse micelle size and W_0 with pressure means that the number of proline solubilization sites increases faster than the number of tryptophan solubilization sites. Therefore, the selectivity for proline increases with pressure. The same manipulation of selectivity might be possible in the propane systems of Sec. III by adjusting the salinity of the aqueous phase to control the size and W_0 of reverse micelles in the propane phase.

The protein acrylodan-labeled bovine serum albumin (BSA-Ac) was solubilized into water-in-CO_2 microemulsions created with the PFPE surfactant mentioned in the previous section [73]. This particular protein was chosen because BSA is moderately large (67,000 Da) and the strong Ac fluorescence provides a measure of both BSA conformation and concentration in solution. Upon forming a microemulsion with PFPE, the BSA-Ac fluorescence is very strong and similar to that of native BSA-Ac in buffer at pH 7.0, as shown in Fig. 15. After recovery, the BSA-Ac is still recognized by the BSA antibody. These results show that BSA-Ac is solubilized within the aqueous microemulsion droplets in an environment similar to that of bulk buffered water and remains biologically active.

Water-in-supercritical fluid reverse micelles and microemulsions may be used to extract inorganic and organic wastes from aqueous phases. An alternative procedure is to extract solutes from normal micelles in water into a nonpolar supercritical fluid. Such a technique was recently studied as part of a proposed wastewater treatment process in which dilute organic contaminants are first concentrated by passage through a membrane into a micellar solution of poly(ethylene oxide)–poly(propylene oxide) block copolymer surfactant, which is then regenerated by extraction of the contaminant using supercritical CO_2. In the final process step the contaminant is recovered by depressurizing the CO_2. Thermodynamic analysis showed strong driving forces for solute transfer in both the membrane and supercritical fluid extraction steps [84,85].

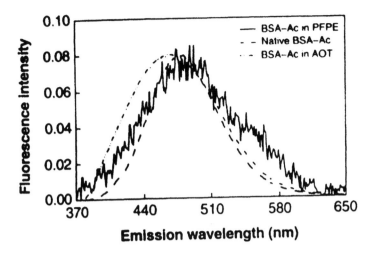

Figure 15 Emission of BSA-Ac encapsulated in a PFPE–stabilized water-in-CO_2 microemulsion. Also included are spectra for (- - -) native BSA-Ac and ($\cdot$ — $\cdot$) BSA-Ac encapsulated in an AOT water-in-oil microemulsion at $W_0 = 20$. (From Ref. 73.)

B. Polymerization Reactions

Acrylamide was successfully polymerized in a supercritical inverse emulsion composed of an ethane–propane mixture as the continuous phase, water and acrylamide as the dispersed phase, and a mixed nonionic surfactant system as the emulsifier [86]. AIBN [2,2'-azobis(isobutyronitrile)] was the initiator. The polymerization was subsequently repeated in supercritical CO_2 [87]. The CO_2-philic surfactant used to produce the inverse emulsion was an amide, end-capped poly(hexafluoropropylene oxide). The process yielded polymers of average molecular weights from 5×10^6 to 7×10^6.

The affinity of CO_2 for fluorinated compounds makes it an attractive choice as a medium for the synthesis of fluoropolymers, especially since certain halogenated organic solvents presently used are being phased out because of environmental concerns. Already several highly fluorinated monomers have been polymerized in supercritical CO_2 using free radical initiators [64,88]. CO_2 possesses additional advantages as a polymerization solvent, including pressure-adjustable solvent strength and ease of removal from the product, leaving the product itself in the desirable form of a loose powder. However, most polymers precipitate out of CO_2 before reaching any appreciable molecular weight because of solubility limitations. A remedy is to employ CO_2-philic surfactants to maintain the growing polymer chains in solution and sustain the free radical addition process. The surfactants prevent flocculation by steric stabilization, which has been described theoretically in supercritical fluids [24].

Poly(methyl methacrylate) (PMMA) and polystyrene (PS) have been synthesized by dispersion polymerization in supercritical CO_2 [70,75,89]. Dispersion polymerization, as distinguished from emulsion polymerization, is initiated homogeneously by the action of an initiator on free monomer in solution. The growing polymer then becomes a particle suspended in solution by the action of surface-active stabilizers. In this case the stabilizer was the poly(1,1-dihydroperfluorooctylacrylate) (PFOA) homopolymer mentioned in the previous section. AIBN was the initiator. The PMMA was recovered as a free-flowing

Table 2 Polymer Latexes Formed by Dispersion Polymerization in CO_2

Polymer	Stabilizer	Stabilizer conc. (w/v %)	Yield (%)	$\langle M_n \rangle$ ($\times 10^{-3}$) g/mol)	M_w/M_n	Avg. diam. (μm)
Poly(methyl	None		39	149	2.8	
methacrylate)	LMW PFOA	2	85	308	2.3	1.2
	HMW PFOA	2	92	315	2.1	1.7
Polystyrene	None		22	4	2.3	
	4.5K/25K PS-b-PFOA	2	98	22	3.1	0.2

Source: Refs. 72 and 87.

powder after the CO_2 was vented. This powder was composed of microspheres having sizes ranging from 0.9 to 2.7 μm, depending on reaction conditions, with a very tight size distribution of ± 0.2 μm. Molecular weights of up to 315,000 and conversions up to 92% were reported. The influence of the stabilizer on polymerization yields and molecular weights of PMMA and PS is shown in Table 2. Another advantage of PMMA polymerization in CO_2 is PMMA's highly plasticized and swollen state due to CO_2-induced reduction of its glass transition temperature [90]. This may allow for unique morphologies not accessible in conventional liquids. Low molecular weight fractions and other impurities can easily be removed by extraction with fresh CO_2 [91].

C. Other Applications

Many applications of reverse micelle technology have been studied in liquid systems but not yet explored in depth in compressible solvents. For example, reverse micelle water pools can serve as microreactors. One class of reactions that have been investigated is enzymatic catalysis [92–94]. The enzyme is located in the water pools, with reactants and products diffusing across the surfactant interface. Such processes have the potential for simultaneously achieving high reaction rates and preserving the biological activity of the enzyme. Another class of reactions are those leading to the formation of small particles. Here the narrow size range of water-swollen reverse micelles is exploited to produce particles of uniform shape and tight size distribution. Examples of such processes have been reported in conventional liquid solvents [95–98]. In a compressible solvent, pressure could be used to fine-tune the reverse micelle size, and thus the particle size, in the two-phase region. Submicrometer $Al(OH)_3$ particles have been formed in a propane microemulsion [99].

Yet another application would make use of systems consisting of CO_2, CO_2-philic surfactants, and possibly cosolvents as replacements for environmentally suspect chlorinated solvents in such cleaning applications as the degreasing of high value electronic components. Finally, a reverse micelle mobile phase could be used to extend supercritical fluid chromatography to nonvolatile or polar compounds. Such a technique would be analogous to gradient methods routinely used in liquid chromatography and gas chromatography. An initial demonstration of the method was published by Beckman et al. [13].

VII. CONCLUSIONS

Research on reverse micelle and microemulsion phases in supercritical fluids has progressed rapidly since initial observations in the late 1980s of AOT aggregation in supercritical ethane. It was discovered that pressure-driven transitions in compressible fluids, even unusual ones, could be explained by extending the existing theoretical framework to include the tunability of the solvent strength. The development of surfactants capable of forming reverse micelles and other types of aggregated structures in supercritical fluid carbon dioxide enhances the prospects of industrial applications, as CO_2 is a more benign solvent than ethane or propane. Experimental databases and theoretical models need to be developed for a wider range of surfactants, dispersed phases, and compressible solvents than is presently the case. The design of surfactants for particular applications, especially in CO_2 systems, is still in a primitive state, yet recent breakthroughs have occurred. High molecular weight latex particles have been synthesized in CO_2. Hydrophilic compounds, including proteins, have been solubilized in a CO_2-continuous phase containing microemulsion cores with bulklike water. The first applications to reach the commercialization stage may well represent extensions of existing processes, such as replacement of halogenated organic solvents in reaction, separation, and cleaning processes. Later, novel processes and products may appear in which the full potential of a supercritical medium is exploited.

REFERENCES

1. K. P. Johnston and J. M. L. Penninger (eds.), *Supercritical Fluid Science and Technology*, (ACS Symp Ser. No. 406), American Chemical Society, Washington, DC, 1989.
2. T. J. Bruno and J. F. Ely (eds.), *Supercritical Fluid Technology: Reviews in Modern Theory and Applications*, CRC, Boca Raton, FL, 1991.
3. K. W. Hutchenson and N. R. Foster (eds.), *Innovations in Supercritical Fluids: Science and Technology*, ACS Symp. Ser. No. 608, American Chemical Society, Washington, DC, 1995.
4. M. A. McHugh and V. J. Krukonis, *Supercritical Fluid Extraction: Principles and Practice*, 2nd ed., Butterworth-Heinemann, Boston, MA, 1994.
5. M. E. Paulaitis, J. M. L. Penninger, R. D. Gray, and P. Davidson (eds.), *Chemical Engineering at Supercritical Fluid Conditions,* Ann Arbor Science, Ann Arbor, MI, 1983.
6. S. Kim and K. P. Johnston, AIChE J. *33*:1603 (1987).
7. K. P. Johnston, S. Kim, and J. M. Wong, Fluid Phase Equilib. *38*:39 (1987).
8. R. M. Lemert and K. P. Johnston, Ind. Eng. Chem. Res. *30*:1222 (1991).
9. R. W. Gale, J. L. Fulton, and R. D. Smith, J. Am. Chem. Soc. *109*:920 (1987).
10. J. L. Fulton and R. D. Smith, J. Phys. Chem. *92*:2903 (1988).
11. K. P. Johnston, G. J. McFann, and R. M. Lemert, *Supercritical Fluid Science and Technology* (K. P. Johnston and J. M. L. Penninger, eds.), Am. Chem. Soc., Washington, DC, 1989, pp. 140–164.
12. G. J. McFann and K. P. Johnston, Langmuir *9*:2942 (1993).
13. E. J. Beckman, J. L. Fulton, and R. D. Smith, in *Supercritical Fluid Technology* (T. J. Bruno and J. F. Ely, eds.), CRC, Boca Raton, FL, 1991, pp. 405–449.
14. K. A. Bartscherer, M. Minier, and H. Renon, Fluid Phase Equilib. *107*:93 (1995).
15. P. L. Luisi and B. E. Straub (eds.), *Reverse Micelles; Biological and Technological Relevance of Amphiphilic Structures in Apolar Media*, Plenum, New York, 1984.
16. M. A. Middleton, R. S. Schechter, and K. P. Johnston, Langmuir *6*:920 (1990).
17. M. Kotlarchyk, J. S. Huang, and S. H. Cheng, J. Phys. Chem. *89*:4382 (1985).
18. S. V. Olesik and C. J. Miller, Langmuir *6*:183 (1990).
19. M. Zulauf and H. F. Eicke, J. Phys. Chem. *83*:480 (1979).
20. P. Yazdi, G. J. McFann, M. A. Fox, and K. P. Johnston, J. Phys. Chem. *94*:7224 (1990).

21. J. M. Tingey, J. L. Fulton, and R. D. Smith, J. Phys. Chem. *94*:1997 (1990).
22. G. J. McFann and K. P. Johnston, J. Phys. Chem. *95*:4889 (1991).
23. E. W. Kaler, J. F. Billman, J. L. Fulton, and R. D. Smith, J. Phys. Chem. *95*:458 (1991).
24. D. G. Peck and K. P. Johnston, Macromolecules *26*:1537 (1993).
25. D. G. Peck and K. P. Johnston, J. Phys. Chem. *97*:5661 (1993).
26. T. K. Jain, M. Varshney, and A. Maitra, J. Phys. Chem. *93*:7409 (1989).
27. J. Zhang and F. V. Bright, J. Phys. Chem. *96*:5633 (1992).
28. J. Zhang and F. V. Bright, J. Phys. Chem. *96*:9068 (1992).
29. W. Pacynko, J. Yarwood, and G. J. T. Tiddy, J. Chem. Soc., Faraday Trans. 1 *85*:1397 (1989).
30. J. C. Ravey, M. Buzier, and C. Picot, J. Colloid Interface Sci. *97*:9 (1984).
31. G. G. Yee, J. L. Fulton, and R. D. Smith, Langmuir *8*:377 (1992).
32. J. Eastoe, W. K. Young, B. H. Robinson, and D. C. Steytler, J. Chem. Soc. Faraday Trans. 1 *86*:2883 (1990).
33. T. Klein and J. M. Prausnitz, J. Phys. Chem. *94*:8811 (1990).
34. M. Bourrel and R. S. Schechter, *Microemulsions and Related Systems*, Marcel Dekker, New York, 1988.
35. K. R. Wormuth and S. Zushma, Langmuir *7*:2048 (1991).
36. R. Aveyard, B. P. Binks, and J. Mead, J. Chem. Soc., Faraday Trans. 1 *82*:1755 (1986).
37. K. Shinoda and S. Friberg, *Emulsions and Solubilization*, Wiley, New York, 1986.
38. J. P. O'Connell, J. D. Kim, P. T. Coram, R. J. Bragman, Prepr. Am. Chem. Soc. Div. Petrol. Chem. *26*:123 (1981).
39. P. Fotland and A. Skauge, J. Dispersion Sci. Technol. *7*:563 (1986).
40. C. L. Sassen, A. Gonzalez Casiellas, T. W. de Loos, and J. de Swaan Arons, Fluid Phase Equilib. *72*:173 (1992).
41. R. D. Smith, J. L. Fulton, J. P. Blitz, and J. M. Tingey, J. Phys. Chem. *94*:781 (1990).
42. G. G. Yee, J. L. Fulton, J. P. Blitz, and R. D. Smith, J. Phys. Chem. *95*:1403 (1991).
43. D. G. Peck and K. P. Johnston, J. Phys. Chem. *95*:9549 (1991).
44. D. G. Peck, R. S. Schechter, and K. P. Johnston, J. Phys. Chem. *95*:9541 (1991).
45. H. Kunieda and K. Shinoda, J. Colloid Interface Sci. *75*:601 (1980).
46. M. Kahlweit, R. Strey, and G. Busse, J. Phys. Chem. *94*:3881 (1990).
47. J. Th. G. Overbeek, G. J. Verhoeckx, P. L. de Bruyn, and H. N. W. Lekkerkerker, J. Colloid Interface Sci. *119*:422 (1987).
48. M. J. Hou and D. O. Shah, Langmuir *3*:1086 (1987).
49. R. Aveyard, B. P. Binks, and P. D. I. Fletcher, Langmuir *5*:1209 (1989).
50. M. Kahlweit, R. Strey, P. Firman, and D. Haase, Langmuir *1*:281 (1985).
51. J. A. Hyatt, J. Org. Chem. *49*:5097 (1984).
52. G. J. McFann, K. P. Johnston, and S. M. Howdle, AIChE J. *40*:543 (1994).
53. K. Harrison, J. Goveas, K. P. Johnston, and E. A. O'Rear III, Langmuir *10*:3536 (1994).
54. J. M. Dobbs, J. M. Wong, R. J. Lahiere, and K. P. Johnston, Ind. Eng. Chem. Res. *26*:56 (1987).
55. T. A. Hoefling, R. R. Beitle, R. M. Enick, and E. J. Beckman, Fluid Phase Equilib. *83*:203 (1993).
56. T. W. Randolph, D. S. Clark, H. W. Blanch, and J. M. Prausnitz, Science *239*:387 (1987).
57. J. M. Ritter and M. E. Paulaitis, Langmuir *6*:935 (1990).
58. A. Iezzi, R. Enick, and J. Brady, in *Supercritical Fluid Science and Technology* (K. P. Johnston and J. M. L. Penninger, eds.), (ACS Symp Ser. No. 406), American Chemical Society, Washington, DC, 1989. pp. 122–139.
59. J. Oates, Ph.D. dissertation, University of Texas, Austin, TX, 1989.
60. K. A. Consani and R. D. Smith, J. Supercrit. Fluid *3*:51 (1990).
61. G. J. McFann, Ph.D. Dissertation, Univ. Texas, Austin, TX, 1993.
62. T. A. Hoefling, R. M. Enick, and E. J. Beckman, J. Phys. Chem. *95*:7127 (1991).
63. E. E. Maury, H. J. Batten, S. K. Killian, Y. Z. Menceloglu, J. R. Combes, and J. M. DeSimone, Poly. Prepr. (Am. Chem. Soc. Div. Polym. Chem.) *43*:664 (1993).
64. J. M. DeSimone, Z. Guan, and C. S. Elsbernd, Science *257*:945 (1992).
65. A. Iezzi, P. Bendale, R. M. Enick, M. Turberg, and J. Brady, Fluid Phase Equilib. *52*:307 (1989).

66. D. A. Newman, T. A. Hoefling, R. R. Beitle, E. J. Beckman and R. M. Enick, J. Supercrit. Fl 6:205 (1993).
67. T. A. Hoefling, D. A. Newman, R. M. Enick, and E. J. Beckman, J. Supercrit. Fluids 6:165 (1993).
68. G. G. Yee, J. L. Fulton, and R. D. Smith, J. Phys. Chem. 96:6172 (1996).
69. Z. Guan and J. M. DeSimone, Macromolecules 27:5527 (1994).
70. Y. L. Hsiao, E. E. Maury, J. M. DeSimone, S. M. Mawson, and K. P. Johnston, Macromolecules 28:8159 (1995).
71. J. L. Fulton, D. M. Pfund, J. B. McClain, T. J. Romack, E. E. Maury, J. R. Combes, E. T. Samulski, J. M. DeSimone, and M. Capel, Langmuir 11:4241 (1995).
72. J. Eastoe, Z. Bayazit, S. Martel, D. C. Steytler, and R. K. Heenan, Langmuir 12:1423 (1996).
73. K. P. Johnston, K. L. Harrison, M. J. Clarke, S. M. Howdle, M. P. Heitz, F.V. Bright, C. Carlier, and T. W. Randolph, Science 271:624 (1996).
74. A. Chittofrati, D. Lenti, A. Sanguineti, M.Visca, C. M. C. Gambi, D. Senatra, and Z. Zhou, Prog. Colloid Polym. Sci. 79:218 (1989).
75. J. M. DeSimone, E. E. Maury, Y. Z. Menceloglu, J. B. McClain, T. J. Romack, and J. R. Combes, Science 265:356 (1994).
76. D. M. Zhu and Z. A. Schelly, Langmuir 8:48 (1993).
77. R. S. Rahaman, J. Y. Chee, J. M. S. Cabral, and T. A. Hatton, Biotech. Prog. 4:218 (1988).
78. M. Adachi, M. Haruda, A. Shioi, and Y. Sato, J. Phys. Chem. 95:7925 (1991).
79. E. B. Leodidis and T. A. Hatton, J. Phys. Chem. 95:5957 (1991).
80. E. B. Leodidis and T. A. Hatton, J. Colloid Interface Sci. 147:163 (1991).
81. H. Nguyen, V. T. John, and W. F. Reed, J. Phys. Chem. 95:1467 (1991).
82. R. B. Gupta and K. P. Johnston, Biotechnol. Bioeng. 44:830 (1994).
83. R. M. Lemert, R. A. Fuller, and K. P. Johnston, J. Phys. Chem. 94:6021 (1990).
84. G. J. McFann, K. P. Johnston, P. N. Hurter, and T. A. Hatton, Ind. Eng. Chem. Res. 32:2336 (1993).
85. P. N. Hurter and T. A. Hatton, Langmuir 8:1291 (1992).
86. E. J. Beckman and R. D. Smith, J. Phys. Chem. 94:345 (1990).
87. F. A. Adamsky and E. J. Beckman, Macromolecules 27:312, 5238 (1994).
88. K. A. Shaffer and J. M. DiSimone, Trends Polym. Sci. 3:146 (1995).
89. D. A. Canelas, D. E. Betts, and J. M. DeSimone, Macromolecules 29:2818 (1996).
90. P. D. Condo, I. C. Sanchez, C. G. Panayiotou, and K. P. Johnston, Macromolecules 25:6119 (1992).
91. J. Shim and K. P. Johnston, AIChE J. 37:607 (1991).
92. J.W. Shield, H. D. Fergusen, A. S. Bommarius, and T. A. Hatton, Ind. Eng. Chem. Fundam. 25:603 (1986).
93. D. K. Eggers and H. W. Blanch, Bioprocess Eng. 3:83 (1988).
94. P. L. Luisi, Angew. Chem. Int. Ed. (Eng.) 24:439 (1985).
95. P. Barnickel, A. Wokaun, W. Sager, and H. F. Eicke, J. Colloid Interface Sci. 148:80 (1992).
96. J. P. Roman, P. Hoornaert, D. Faure, C. Biver, F. Jacquet, and J. M. Martin, J. Colloid Interface Sci. 144:324 (1991).
97. C. A. Malbrel and P. Somasundaran, Langmuir 8:1285 (1992).
98. L. Motta, C. Petit, L. Boulanger, P. Lixon, and M. P. Pileni, Langmuir 8:1049 (1992).
99. D. W. Matson, J. L. Fulton, and R. D. Smith, Mater. Lett. 6:31 (1987).

10
Characterization of Microemulsions by NMR

Björn Lindman, Ulf Olsson, and Olle Söderman
University of Lund, Lund, Sweden

I. INTRODUCTION

A. Overview of the Chapter

Key ingredients in a physicochemical characterization of a microheterogeneous surfactant self-assembly system are (1) phase stability and phase behavior, (2) microstructure, and (3) local molecular arrangements, interactions, and dynamics. Microemulsions constitute but one type of surfactant self-assembly, and any attempt to understand or characterize microemulsions without a broad picture of surfactant systems will be rather meaningless. Microemulsions are, in contrast to (macro)emulsions, thermodynamically stable single-phase systems. They are isotropic solutions, like micellar solutions, with no long-range order. Microstructures of surfactant self-assembly systems show an enormous degree of polymorphism. We may classify structures in three ways (see Figs. 1 and 2):

1. Phases that possess long-range order and periodicity and those that do not.
2. Monolayer and bilayer structures.
3. Phases with discrete surfactant self-assemblies and phases that have self-assemblies that are infinite in one, two, or three dimensions.

Regarding local molecular dynamics, it is well established that all surfactant self-assemblies are characterized by very rapid short-range dynamics. The surfactant chains, as well as oil and water molecules, are in a "liquidlike" state. Local molecular reorientation occurs on time scales of 10^{-11} s, similar to pure water or hydrocarbon. On the other hand, due to the presence of the surfactant films, the reorientation is more or less anisotropic. Local translational motion is also quite fast, but long-range translation is strongly dependent on the presence of the surfactant films.

Spectroscopic techniques, in general, are very well suited for investigation into molecular aspects such as local molecular arrangements, molecular dynamics, and molecular interactions. However, for many microemulsions, in particular those that simultaneously contain large amounts of water and oil, these aspects are rather uninteresting. Local molecular dynamics are determined by short-range interactions and are essentially independent of the self-assembly structure. Nuclear magnetic resonance (NMR) is, however, particular as a spectroscopic technique in that it can also provide information on other central matters such as phase behavior and microstructure. In particular, NMR self-diffusion and, to some

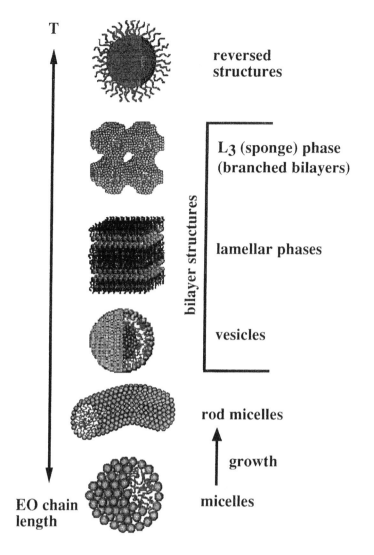

T

reversed
structures

L3 (sponge) phase
(branched bilayers)

bilayer structures

lamellar phases

vesicles

rod micelles

growth

EO chain
length

micelles

Figure 1 Polymorphism of surfactant self-assemblies, here illustrated in terms of the effect of temperature and ethylene oxide chain length for nonionic surfactants. (Not all possible structures are shown.)

extent, NMR relaxation constitute unique tools in providing microstructural information. Therefore, this chapter focuses on these approaches and leaves out a large number of NMR studies that have provided less unique original information. As argued above, microemulsions must be discussed with surfactant self-assembly in general as a background. The different NMR approaches mentioned here are therefore generally applicable and can sometimes be best exemplified by using systems other than microemulsions. We recently reviewed some of the central issues of the present topic. In particular we refer to two general reviews of NMR applied to surfactant systems [1,2] and one review concerned with NMR studies of microemulsion microstructure [3]. In many respects the present chapter is an elaboration of these previous publications.

Discrete particles:

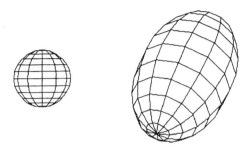

Connected structures:

1-D   **2-D**   **3-D**

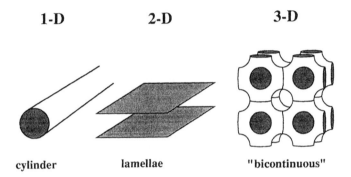

cylinder   lamellae   "bicontinuous"

Figure 2 Illustration of various surfactant self-assembly geometries classified according to the continuity or discontinuity of the surfactant film. The surfactant film may enclose a finite volume (top) or be continuous in one (1-D), two (2-D), or three (3-D) dimensions.

For two reasons NMR is an extremely versatile spectroscopic technique:

1. Since there are a large number of spectroscopic parameters, a comprehensive NMR investigation can provide a very detailed picture of static and dynamic molecular aspects of different types of systems. It is most powerful for the liquid state but is not limited to that.
2. There are a large number of atomic nuclei with nuclear spins, and for essentially any element it is possible to find at least one suitable nucleus. This has the important consequence that not only can different aspects be monitored but also different domains of a microheterogeneous system can be studied. In microemulsions, with local segregation into oil and water domains and surfactant films, we can obtain independent information on the different regions, often simultaneously in a single experiment.

This review deals with what can be and what has been learned about microemulsions from NMR studies. On the other hand, we do not provide any general introduction to NMR principles, theory, and methodology. This is a vast and still rapidly expanding field, well covered in textbooks and monographs [4,5]. We do, however, dwell on features, experimental design, theory, and evaluation, that are rather unique for microheterogeneous systems, not only because of their special relevance in the present context but also because they are not described in standard treatises on NMR.

B. Why and for What Should We Use NMR?

The success of NMR for studies of microemulsions has basically been that it can address one of the key questions, that of microheterogeneity. This is, for various reasons, very difficult to study by other physicochemical approaches and often informative regarding other types of complex formation, including certain aspects of surfactant self-assembly. To understand this we need to look at microemulsions in a broader context, and we start with micelles. The spherical micelle was the first firmly established surfactant self-assembly structure, and the spherical aggregate structure has penetrated thinking in this field of study to such an extent that almost every new phase or phenomenon in surfactant solutions has at some point been considered as based on spherical micellar units:

> Microemulsions were considered to be built up of either oil spheres in an aqueous continuum or water spheres in an oil continuum.
> Large micelles as well as the hexagonal phase based on rodlike aggregates have been discussed in terms of an array of spheres.
> Cubic liquid crystalline phases were considered to be built up of close-packed spherical micelles.

We now know that the spherical micelle is just one of many geometrical shapes of surfactant self-assemblies, and we have learned during the years that

1. Surfactant self-assemblies in microemulsions may depart strongly from a spherical shape.
2. Elongated micelles may have a connected hydrocarbon (or water) interior.
3. Cubic phases are built up of aggregates that depart from a spherical shape.

However, more significantly, we had to go one step further and admit more and more that the concept of discrete micellar aggregates had to be abandoned for a number of phases. In fact, we have accepted the ubiquity of structures that are connected in both the polar and nonpolar parts; we call them bicontinuous structures.

> Microemulsions typically contain both water and oil domains that extend over large distances.
> Many bicontinuous cubic liquid crystalline phases have been clearly identified.
> Recently we have learned that relatively dilute surfactant solutions may also be bicontinuous.

Above all we have learned how important it is to consider all types of surfactant aggregates in a unified picture. We have seen the confusion arising from considering microemulsion structure without reference to other structures and vice versa and how mistakes on micellar systems could have been avoided if microemulsions had been considered. This goes far beyond the simple bulk surfactant systems. For example, for interfaces the picture of connected structures, surfactant monolayers and bilayers, had to be abandoned in favor of a picture of discrete aggregates, with structures that, not unexpectedly in hindsight, parallel those of the bulk systems. Another area where our understanding has gained much from a more holistic approach is that of mixed polymer–surfactant solutions and surfactant self-assembly at a polymer chain.

The present review focuses on the study of microemulsion *structure* by NMR, since this is the main aspect about which NMR is capable of providing unique information. We mainly discuss NMR self-diffusion studies, as they have been the most informative and, in particular, self-diffusion work together with microscopy is probably the only direct way of probing the bicontinuity of microemulsions [6]. In fact, self-diffusion (tracer and NMR) studies gave the first direct evidence for a bicontinuous structure [7] and have continued to constitute the main approach to distinguish between discrete micellar, unicontinuous, and bicontinuous structures. Since this is the most fundamental issue in relation to microemulsion structure, we focus on this point. In developing the subject it is informative to broaden the view beyond microemulsions and consider the uni/bicontinuity issue also for some other types of surfactant systems, notably liquid crystalline phases and micellar solutions.

The alternative NMR approach that has provided information on microemulsions is relaxation. However, on the whole, relaxation has been less informative than anticipated from earlier studies of micellar solutions and has provided little unique information on microemulsion structure, although in the case of droplet structures it is probably the most reliable way of deducing any changes in droplet size and shape, particularly for concentrated systems. The reason for this is that NMR relaxation probes the rotational diffusion of droplets, which is relatively insensitive to interdroplet interactions. This is in contrast to, for example, translational collective and self-diffusion and viscosity which depend strongly on interactions. Furthermore, NMR relaxation is a useful technique for characterizing the local properties of the surfactant film.

C. Useful Isotopes and NMR Parameters

Most elements have NMR-active nuclei, i.e., nuclei possessing a magnetic moment. For surfactant systems, the situation is very good in that ^1H and ^{13}C nuclei, which occur in most surfactant molecules, have good NMR properties. For relaxation work, however, it is often advantageous to use ^2H NMR on selectively deuterated compounds. ^{19}F NMR is very useful for the study of fluorocarbon surfactants, ^{31}P NMR for phospholipids, and ^{29}Si NMR for silicon surfactants. For studies of water molecules we have a choice between three good alternatives: ^1H, ^2H, and ^{17}O NMR. Many common counterions, e.g., $^{23}Na, ^7Li, ^{87}Rb, ^{133}Cs, ^{19}F, ^{35}Cl, ^{81}Br,$ and ^{127}I, have highly sensitive nuclei, whereas others, such as $K^+, Mg^{2+}, Ca^{2+},$ and SO_4^{2-}, have lower sensitivity and are more difficult to study.

NMR parameters useful to investigate include chemical shifts, T_1, T_2, and other relaxation times, quadrupole splittings, shift anisotropy, and self-diffusion. The chemical shift defines the position or resonance frequency of the signal from a certain group or molecule. It changes with the intermolecular interactions and can be useful for studies of association phenomena in general; for alkyl chains it is generally dominated by conformational equilibria. The latter are also studied from spin–spin coupling constants. NMR relaxation is governed by time-modulated interactions involving the nuclear spin, and relaxation rates are determined by two factors: the interaction and the rate of fluctuations. Different interactions are significant for different nuclei: dipole–dipole, electric quadrupolar, chemical shift anisotropy, etc. The so-called static effects, dipolar couplings, quadrupole splittings, and chemical shift anisotropy convey information about phase anisotropy and are particularly useful for liquid crystalline phases and for studies of phase behavior. For the latter the general shape and fine structure of an NMR signal provide direct information on phase coexistence, even without macroscopic phase separation. Finally, a somewhat special use of NMR is to investigate the displacement in space of molecules. For the present chapter we are primarily concerned with NMR studies of self-diffusion but also, for example, those of macroscopic flow, including electrophoresis, which can be monitored in detail and with molecular differentiation.

In summary, NMR studies can deal with a wide range of problems in surfactant science. These include, e.g., molecular transport, phase diagrams, phase structure, self-association, micelle size and shape, counterion binding and hydration, solubilization, and polymer–micelle interactions. NMR is fruitfully applied to isotropic or liquid crystalline bulk phases, to dispersions (vesicles, emulsions, etc.), to polymer–surfactant mixtures, and to surfactant molecules at solid surfaces. In all cases NMR can provide information on molecular interactions and dynamics as well as on microstructure.

Reviews of NMR studies of surfactant systems include Refs. 1, 2, and 8–10. For microemulsions, several approaches have been used and a large number of studies have been published. However, information on microstructure can mainly be obtained in only two ways, by multicomponent self-diffusion and field-dependent relaxation.

Self-diffusion of the solvent molecules (oil and water) gives direct insight into connectivity, but it also reports on the size and shape of discrete particles and the type of bicontinuous structure. NMR relaxation of the surfactant molecules at different magnetic fields allows the extraction of dynamic parameters reporting on the curvature of surfactant films and thus on aggregate size and shape.

While self-diffusion allows a model-free and straightforward extraction of structural information, the interpretation of relaxation data requires the use of models. Self-diffusion is experimentally less demanding than relaxation studies, and self-diffusion studies can often be performed more rapidly. Because of its excellent sensitivity, self-diffusion is generally based on ^{1}H NMR. Under typical conditions, we can resolve different components of a microemulsion, and no isotopic labeling is required. Relaxation experiments require more care. Here, ^{2}H NMR studies of selectively deuterated surfactants is often the method of choice.

II. SELF-DIFFUSION AND MICROSTRUCTURE

A. Experimental Determination of Self-Diffusion Coefficients

Molecular self-diffusion coefficients are experimentally easily accessible and can be measured accurately with various techniques. However, the Fourier transform pulsed field gradient spin-echo NMR (FT PGSE NMR) approach has during the last decade proved to be superior to other approaches for several reasons. In this technique the displacement of nuclear spins in a controlled magnetic field gradient is monitored, and the contributions of different components are resolved by Fourier transformation of the NMR signal (spin echo).

Measurement of self-diffusion coefficients by means of PGSE techniques has evolved to become one of the most important tools in the characterization of surfactant systems. In particular, this is true of those surfactant systems that are isotropic liquid solutions such as micellar systems and microemulsions. The technique has been described in a number of review articles [9,11–13]. An account of the most recent developments of the method can be found in Ref. 9. We do not dwell on the technical aspects here but merely note that the technique requires no isotopic labeling (avoiding possible disturbances due to addition of probes); furthermore, it gives component-resolved diffusion coefficients with great precision in a minimum of measuring time.

The foundation for the use of NMR to monitor diffusion was laid in 1950 in the seminal paper by Hahn [14]. The next important development occurred around the middle of the 1960s, when Stejskal and Tanner (following a suggestion by Douglas and

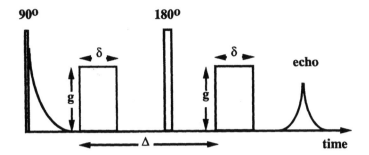

Figure 3 The pulsed gradient spin-echo pulse sequence. 90° and 180° are the two rf pulses. Two field gradient pulses of magnitude g, duration δ, and separation Δ are applied before and after the 180° refocusing pulse. The second half of the spin echo is Fourier transformed, and the relative intensities of the resolved absorption peaks are measured.

McCall [15]) demonstrated the use of pulsed gradients [16]. Finally, toward the end of the 1960s the FT approach was introduced [17], and thereby the method became component-resolved in that individual diffusion coefficients of different components of multicomponent systems could be measured simultaneously as long as they had resolved NMR signals.

In the simplest version, the PGSE experiment consists of two equal and rectangular gradient pulses of magnitude g and length δ, sandwiched on either side of the 180° rf pulse in a simple Hahn echo experiment; the PGSE sequence is schematized in Fig. 3. For molecules undergoing free diffusion characterized by a diffusion coefficient of magnitude D, the echo attenuation due to diffusion is given by [16,18]

$$I(\Delta, \delta, g) = I_0 \, \exp\left[-\gamma^2 g^2 \delta^2 \left(\Delta - \frac{\delta}{3}\right)D\right] \tag{1}$$

where Δ represents the distance between the leading edges of the two gradient pulses, γ is the magnetogyric ratio of the monitored spin, and I_0 denotes the echo intensity in the absence of any field gradient. By varying either $g\delta$, or Δ (while at the same time keeping the distance between the two rf pulses constant), D is obtained by fitting Eq. (1) to the observed intensities.

The diffusion coefficients that can be measured with the PGSE method cover the range from fast diffusion of small molecules in solutions with D values typically around 10^{-9} m^2/s to very slow diffusion of, for instance, polymers in the semidilute concentration regime, where D values down to 10^{-16} m^2/s can be measured [19]. Measurements of such very slow diffusion requires gradients of extreme magnitudes and places severe demands on the actual experimental setup [9]. What often limits the lowest value of D that can be measured is the value of spin–spin relaxation time, T_2. As a general rule, slow diffusion is often found in systems that also show rapid transverse relaxation. As a consequence, the echo intensity gets severely damped by T_2 relaxation in such systems. For microemulsion systems, such problems are virtually nonexistent for the solvents, while for the surfactant molecules the accuracy is often reduced because of T_2 effects.

One of the key properties of the PGSE diffusion experiment is the fact that the transport of molecules is measured over a time that we are free to choose at will in the range from around a few milliseconds to several seconds. If the molecules experience some sort

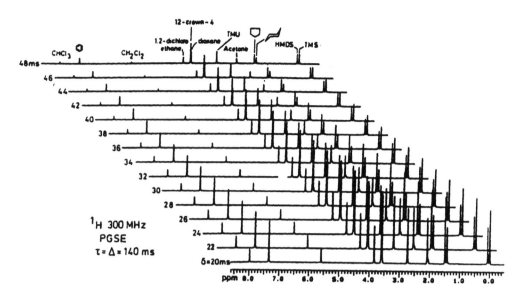

Figure 4 Fourier transform GSE spectra recorded for various values of the pulsed gradient duration δ. The sample is a 12-component liquid mixture, and the assignment of the various components are given on top. (Adopted from Ref. 12.)

of boundary with regard to their diffusion during this time, the outcome of the experiment becomes drastically changed [13,20,21], and Eq. (1) is no longer a valid description of the echo intensity. For microemulsions this simple equation has been found to be broadly applicable.

Through the spin echo experiment in the presence of (normally pulsed) magnetic field gradients, molecular transport over macroscopic distances (10^{-7}–10^{-4} m) can be monitored accurately and conveniently. For complex systems, Fourier transformation is essential and yields a very detailed picture of the molecular mobility in the system. Alternative ways of studying self-diffusion are generally based on using (most frequently radioactively or fluorescent) labeled compounds, as in the capillary tube technique. These techniques may be more sensitive and are sometimes more accurate, but NMR approaches have many advantages, such as no need for labeling, speed, and the possibility to study many compounds simultaneously in complex mixtures. The danger in adding foreign probes has been repeatedly demonstrated and should be avoided.

We note here also that, unlike collective diffusion coefficients, molecular self-diffusion coefficients are unaffected by critical effects, which is a significant advantage for systems with critical points, a typical situation in this context. We also note that self-diffusion studies are very general and that the NMR approach places little demand on the appearance of the sample (turbidity, color, rheology, etc.)

Figure 4, due to Stilbs [12], gives a good illustration of the performance of the FT PGSE technique. In a few minutes, the self-diffusion coefficients of all the components in a 12-component solvent mixture can be determined with an accuracy of 1%.

B. How to Obtain Information on Microstructure from Self-Diffusion

1. General Remarks

In a system of a surfactant mixed with hydrophilic and lipophilic components, there will be (depending on the amphiphilicity and amphiphobicity of the surfactant) segregation into "water" and "oil" domains. Surfactant systems can be classified into monolayer or bilayer systems depending on whether the surfactant films are uncorrelated or correlated. Here we are concerned with the former.

Depending on the size and shape of the domains, we can distinguish between structures of finite aggregates or particles and structures where the aggregates are infinite, extending over macroscopic distances in one, two, or three dimensions (Fig. 2).

Here we mainly address one problem of microemulsion microstructure, namely that of connectivity, in particular the distinction between uni- and bicontinuous structures. We also, to some extent, consider the problem of size and shape of aggregates for discrete particle structures and that of different bicontinuous structures. NMR can also shed light on other aspects of microstructure, such as the distribution of surfactant molecules between surfactant films and the oil and water domains and the local packing and ordering of surfactant molecules. Here we merely note that there is overwhelming evidence (mainly from NMR) that, locally, surfactant aggregates in different phases (micellar solutions, microemulsions, different liquid crystalline phases) show only quite minor differences.

We are here concerned with molecular self-diffusion over macroscopic distances, typically 10^{-6} m and above, i.e., much larger than the size of a surfactant molecular or the size of any discrete particle.

Let us start the description of the principles with a simple case. Assume a dispersion of one solvent in another. If the (discrete) drops of the dispersed phase are large, i.e., considerably larger than the distance over which diffusion is monitored, then the self-diffusion of both components will be unrestricted on the relevant time scale and we will observe high D values for both solvents. In fact, except for an obstruction correction, which may amount to at most about 30%, the D values will be the same as for the neat solvents. If the diffusion distance and the drop sizes match each other, the observed diffusion will be critically dependent on the (variable) diffusion time chosen. These cases are not applicable to microemulsions but give the basis for a very important general and noninvasive technique of monitoring drop sizes (and fusion processes) in (macro)emulsions [13,20–27].

For the case of an equilibrium surfactant self-assembly, as in microemulsions, discrete aggregates or droplets, if they exist, have extensions much smaller than the distance monitored in a diffusion experiment. Therefore, the experiment is not sensitive to the molecular displacement within droplets, and the only translation of the droplet component monitored is that of the entire droplet. This is much slower than the diffusion of the same component in the neat solvent or of the other solvent component of the microemulsion. For a droplet-type microemulsion, the diffusion coefficients of the two solvents differ dramatically, typically by two orders of magnitude and sometimes even more.

2. Obstruction Effects on Molecular Translation in the Continuous Medium

The molecules of the continuous medium experience an obstruction effect of the droplets that depends simply on droplet volume fraction and shape, and the obstruction effect may convey information concerning the shape and size of the colloidal particles in a broad range of systems. The relevant theory has been worked out by Jönsson et al. [28]. In short, the presence of spherical and rod-shaped aggregates gives rise to minor obstruction effects,

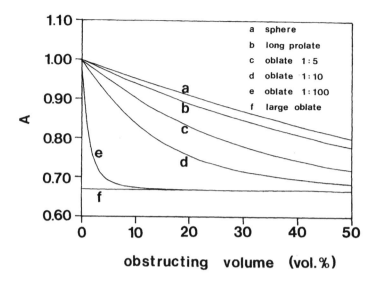

Figure 5 The variation of the obstruction factor A defined as the reduced self-diffusion coefficient D/D_0 of small solvent molecules in solutions as a function of the volume fraction of obstructing particles of different geometries. Curve a denotes spheres; b, long prolates; c, d, and e, oblates of axial ratios 1 : 5, 1 : 10, and 1 : 100, respectively; and f, large (i.e., infinite) oblates. (Redrawn from Ref. 28.)

while oblate or disk-shaped particles cause a larger effect. Given in Fig. 5 is the obstruction factor A, defined as the ratio of the diffusion coefficient in the presence and absence of obstructing particles, as a function of the particle volume fraction. The dependence of the obstruction effect for solvent molecules on the volume fraction of the obstructing particles provides a convenient way of qualitatively judging the geometry of the colloidal particles. A quantitative analysis is often hampered by the fact that the colloidal particles are always solvated to some extent. Solvation also reduces the observed diffusion coefficient as discussed below.

3.   Self-Diffusion of Colloidal Particles

An alternative approach to investigating particle size and shape is to monitor the self-diffusion of the colloidal particle itself. The self-diffusion coefficient is, in the absence of interactions (i.e., at infinite dilution), given by the Stokes–Einstein relation,

$$D_0 = \frac{k_B T}{6\pi\eta R_H} \tag{2}$$

where k_B represents the Boltzmann constant, T the absolute temperature, η the solvent viscosity, and R_H the hydrodynamic radius. At higher concentrations, interactions become important, and the self-diffusion coefficient decreases with increasing concentration [29]. The concentration dependence is often expressed in terms of an expansion,

$$D = D_0(1 - k_s\Phi + \cdots) \tag{3}$$

where Φ is the particle volume fraction and k_s a dimensionless constant that depends on the aggregate geometry and interactions. For hard-sphere systems [29,30], $k_s \approx 2.5$.

Strictly speaking, Eq. (2) is valid for spheres. However, it can also be taken as the definition of the hydrodynamic radius for particles of arbitrary geometry, and thus it can also be used to characterize particles of unknown shape.

Although the method outlined above is certainly useful, one should be aware of the limitations inherent in the approach. First, a direct application of Eq. (2) assumes infinite dilution. As surfactant aggregates are self-assembled units, one must be aware of the fact that dilution may change their properties. At higher concentrations, interactions become important [cf. Eq. (3)], and these must be taken into account [31,32]. In fact, the obstruction effects can be used to advantage, as information pertaining to the interparticle interactions can be derived from them.

4. Diffusion in Bicontinuous Microstructures

Anderson and Wennerström [33] calculated the geometrical obstruction factors of the self-diffusion of surfactant and solvent molecules in ordered bicontinuous microstructures, which serve as good approximations also for the disordered bicontinuous microemulsions and L_3 (sponge) phases. The geometrical obstruction factor is defined as the relative diffusion coefficient D/D_0, where D is the diffusion coefficient in the structured surfactant system and D_0 is the diffusion coefficient in the pure solvent. In a bicontinuous microemulsion the geometrical obstruction factor depends on the water/oil ratio. An expansion around the balanced (equal volumes of water and oil) state gives, to leading order,

$$D^w/D_0^w = 0.66 - \beta(\Phi_o - 1/2) \tag{4}$$

$$D^{\text{oil}}/D_0^{\text{oil}} = 0.66 + \beta(\Phi_o - 1/2) \tag{5}$$

where Φ_o is the oil volume fraction. The expansion coefficient β depends on the coordination number of the structure, i.e., on the family of constant mean curvature surfaces. For the P family (coordination number of 6), $\beta = 0.78$, while for the D family (coordination number of 4), $\beta = 0.54$.

The surfactant diffusion coefficient is maximum for the balanced state, and the corresponding expansion gives to leading order

$$\frac{D^s}{D_0^s} = \frac{2}{3} - \beta'\left(\Phi_o - \frac{1}{2}\right)^2 \tag{6}$$

Here D_0^s denotes the (two-dimensional) lateral diffusion coefficient of the surfactant within the surfactant film. The expansion coefficient β' has the value 1.8 for the P family of surfaces.

In the bilayer continuous structures occurring in the bicontinuous cubic and L_3 phases, the diffusion can be described by essentially the same equations. Finally, we note that Eqs. (4)–(6) have been applied to the analysis of self-diffusion data from a number of bicontinuous microemulsions, L_3 phases, and bicontinuous cubic phases [34–36].

5. Complications

One should note that the equations above describe only the effect of geometrical obstruction. In many cases, solvation effects (solvent molecules have a lower mobility in the vicinity of the surfactant film) may be of similar or even higher magnitude. The solvation effects increase with the surfactant concentration, and care has to be taken to separate solvation from obstruction effects [34].

Similarly, if segregation into domains is not complete, this affects the diffusion behavior. If some of the continuous medium molecules, for one of these reasons, diffuse with the droplets, this will slow down diffusion according to

$$D = p_f D_f + D_b p_b \tag{7}$$

Since the diffusion of the droplets, D_b, is typically much slower than the diffusion of the free solvent molecules, D_f, the solvent diffusion will be retarded in proportion to the fraction of solvent molecules, p_b, diffusing with the droplets. In the case of solvation this is typically proportional to the concentration of surfactant.

Diffusion with the droplets can thus occur due to incomplete segregation between water and oil and due to solvation of the surfactant molecules of the surfactant films. If some of the droplet solvent molecules are distributed in the continuous medium, this will accelerate diffusion, again in proportion to the fraction of molecules [Eq. (7)].

A further complication arises when the surfactant is not confined to the surfactant films but has significant solubility in one of the solvents. This retards solvent diffusion due to solvation. The effect occurs in general for weakly amphiphilic, mainly short alkyl chain, surfactants. Another example is that of nonionic ethylene oxide (EO) surfactants, which have a relatively high monomer solubility in oil, especially at higher temperatures.

These effects complicate the analysis and tend to make conclusions in terms of microstructure less certain. However, for reasonably amphiphilic surfactants and for typical oils this effect will be insignificant. For short-chain (say, C_6) surfactants, on the other hand, we have frequently run into problems of analyzing the microstructure because of an uncertainty of the distribution of surfactant between microdomains.

Therefore, for the interrelated cases of a high total surfactant concentration, high monomer solubility of the surfactant in one of the solvents, and incomplete solvent segregation, the structural deductions become, as with most other techniques, less unambiguous.

C. Connectivity of Solvent Domains from Oil and Water Self-Diffusion Coefficients

From the oil and water self-diffusion coefficients we can thus easily decide whether a given microemulsion is of the discrete oil-in-water (O/W), discrete water-in-oil (W/O), or bicontinuous type. Thus we have for the three cases (D_0 denoting the neat solvent value):

O/W: $D^{oil} \ll D^w$ and D^w close to D_0^w; typically D^{oil} on the order of 10^{-11} m²/s and D^w on the order of 10^{-9} m²/s.

W/O: $D^w \ll D^{oil}$ and D^{oil} close to D_0^{oil}; typically D^w on the order of 10^{-11} m²/s and D^{oil} on the order of 10^{-9} m²/s.

Bicontinuous: D^w and D^{oil} both high, on the order of 10^{-9} m²/s.

Finally, for a structureless or molecularly dispersed (i.e., simple solution) case we would also have high values of D^w and D^{oil}. It is normally trivial to distinguish between simple solutions and bicontinuous microemulsions (phase behavior, scattering, etc.), and a diffusion experiment will immediately tell, since for a simple solution the surfactant will show molecular, and thus fast, diffusion. For a bicontinuous microemulsion, surfactant diffusion is typically an order of magnitude lower.

The procedure of evaluating self-diffusion data in terms of microstructure is to calculate the reduced or normalized diffusion coefficient, D/D_0, for the two solvents, D_0 being the neat solvent value under the appropriate conditions. Here we also have to account for reductions in D resulting from factors other than microstructure, mainly solvation effects. As discussed above, solvation will lead to a reduction of solvent diffusion that is proportional to the surfactant concentration. Normally the correction has been empirical and based on diffusion studies for cases of established structure, notably micellar solutions. We need to distinguish between corrections due to polar head-water and alkyl chain–oil interactions. The latter have often been considered insignificant, but a closer analysis (either experimental or theoretical) is lacking. However, it is probably reasonable to assume, for example, that the resistance to translation is not very different in the lipophilic part of the surfactant film and in an alkane solution. (This is supported by observations of molecular mobilities of surfactant alkyl chains on the same order of magnitude as for a neat hydrocarbon.)

As for hydration, there are strong headgroup–water couplings for ionic heads, but the hydration numbers are relatively small due to the small headgroups. Therefore, corrections due to hydration are rather small. For the nonionic surfactants with large headgroups, in particular the ethoxylates, the correction becomes significant at moderately high surfactant concentrations. The water–EO interaction seems to be of general nature and quite independent of where the EO groups occur. Therefore, the correction can be easily obtained from measurements on an aqueous solution of poly(ethylene oxide) [37]. Hydration decreases with increasing temperature, so the correction is different at different temperatures.

In general, the correction for hydration is insignificant for an efficient surfactant, which forms microemulsions at a few weight percent of surfactant, but becomes significant above about 5 wt%. Up to surfactant concentrations of 15–20 wt%, the correction may be estimated with sufficiently high accuracy to make an analysis of diffusion data in terms of microstructure reliable. However, for weak surfactants, where the amount of surfactant required is well above this value, considerable caution must be exercised, as solvation effects on diffusion are similar in magnitude or larger than effects of microstructure.

In a typical experimental study, D values of oil and water are obtained over a range of conditions—temperature, salinity, cosurfactant or surfactant concentration, solvent composition, etc. From the deduced D/D_0 values, microstructures in certain ranges are normally directly obtained. If D/D_0 of water and oil differ by more than one order of magnitude, discrete particles of the slowly diffusing solvent are implied, whereas if they are of the same order of magnitude a bicontinuous structure is suggested.

For discrete micellar-type structures, we further want to establish the size and shape of the droplets. This is done by obtaining the hydrodynamic radius using the Stokes–Einstein relation [Eq. (2)], taking into consideration the effect of micelle–micelle interactions on micelle diffusion, normally by assuming hard-sphere interactions. Often the radius obtained is consistent with estimations for spheres based on the relation between the volume fraction of the droplet solvent and the total interfacial area of surfactant headgroups. If droplet sizes are considerably larger than those consistent with spheres, we can distinguish from the diffusion of the solvent of the continuous medium between elongated or prolate shape on the one hand and flattened out or oblate on the other hand. We recall that the obstruction effect is much larger for the latter case.

As regards bicontinuous structures, many cases may be possible, but in the literature mainly three types are discussed: a distorted lamellar type, a minimal surface type similar to that established for bicontinuous cubic phases, and a tubular type with one of the solvents confined to (branched) tubes. An ideal layered structure should give a D/D_0 value of $2/3$ for both solvents, because diffusion is hindered in one direction but free in the other two directions. The zero mean curvature structures give similar results (cf. above).

For a tubular structure where distances between branching points are significantly greater than the tube diameter, we expect D/D_0 of the solvent in the tubes to be close to $1/3$, because diffusion is allowed only along the axes of the tubules. The external solvent should have a D/D_0 value not much below 1. (The obstruction effect of cylinders is small.) More plausible is an interconnected rod model for which the diffusion was theoretically analyzed by Anderson and Wennerström [33].

D. Survey of Different Microemulsion Systems

In surveying the experimental studies of solvent diffusion in microemulsions we choose to focus on nonionic surfactants, one important reason being that these contain only three components. In particular, the phase behavior due to the pioneering work of Shinoda [38], Shinoda and Friberg [39], and later Kahlweit and coworkers [40,41] is very well characterized and understood. (The significance of accurate and careful phase diagram work cannot be overemphasized; much experimental work applying advanced physicochemical techniques on ill-defined, often multiphase, samples is reported in the literature.) A Shinoda diagram (i.e., a cut at constant surfactant concentration) is presented in Fig. 6. In a schematic way we also include here the information deduced on solution microstructure from NMR self-diffusion [42,43].

Nonionic surfactants show some other advantages: They are rather strongly amphiphilic, hydration effects can be simply accounted for, and the spontaneous curvature can be controlled simply by controlling temperature, i.e., without adding any component or changing the composition of the sample. By controlling temperature, microemulsions can, at a suitable surfactant concentration, be found at all mixing ratios between water and a hydrocarbon. The only disadvantage is the relatively high monomer solubility of surfactant in oil; in water it is very low.

Typically observed self-diffusion behavior [42,43] of nonionic surfactants is illustrated in Figs. 7–9. We note the following important observations:

1. D/D_0 values of oil and water vary dramatically over the microemulsion channel. At low temperatures water diffusion is one to two orders of magnitude faster than oil diffusion; at high temperatures we have the opposite situation.
2. The variation of solvent diffusion is mainly an effect of a change in curvature of the surfactant film. Thus diffusion changes strongly with temperature at constant solvent ratio (Fig. 8) but is independent of solvent ratio at constant temperature (Fig. 9).
3. At an intermediate temperature, D/D_0 of both solvents is above 0.6. Around this value, occurring for about equal volumes of the two solvents, the diffusion behavior is quite symmetric.

The diffusion behavior implies a rapid, but continuous, change in structure from discrete oil droplets to bicontinuous to discrete water droplets with increasing temperature. The bicontinuous structure appears to be well described by the constant mean curvature surface structures of low mean curvature rather than a tubular structure. The same applies to the bilayer phases, often denoted L_3, L_4, or sponge phases, also included in Fig. 6.

Completely analogous self-diffusion, and thus also structural, behavior is displayed by other, more complex, types of surfactant systems as exemplified by the following:

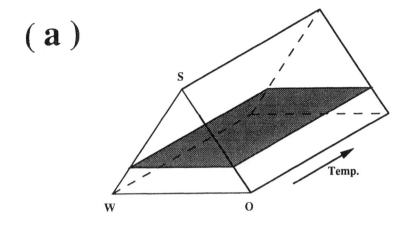

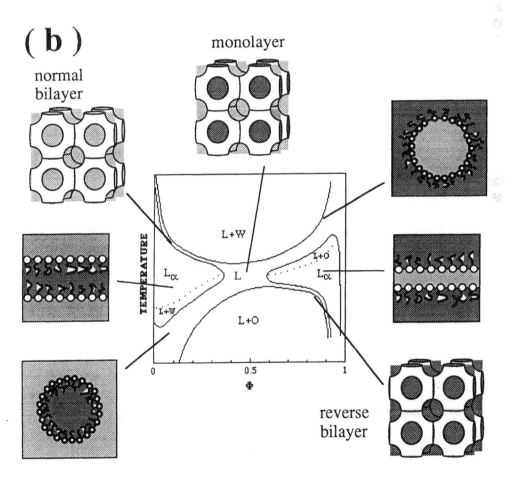

Figure 6 A schematic phase diagram cut at constant surfactant concentration through the temperature–composition phase prism of a ternary system with nonionic surfactant (Shinoda cut) showing the characteristic X-like extension of the isotropic liquid phase, L. Schematic drawings of the various microstructures are also shown.

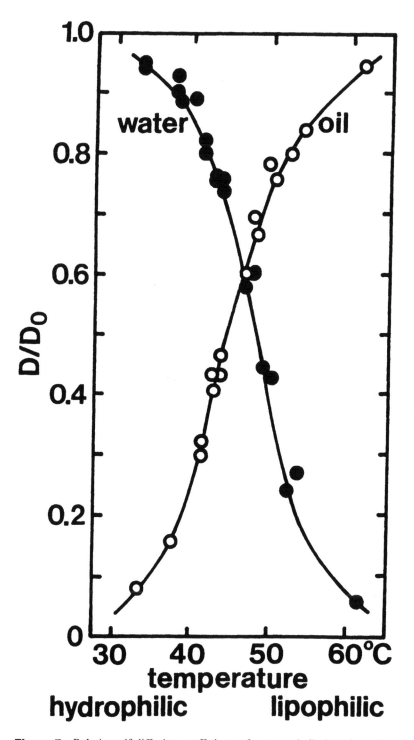

Figure 7   Relative self-diffusion coefficients of water and oil plotted as a function of temperature in a three-component system, $C_{12}E_5$–water–tetradecane, at constant (16.6 wt%) surfactant concentration. Note the symmetry around the balanced temperature (here approximately 47°C) corresponding to zero spontaneous curvature. (Data taken from Ref. 42.)

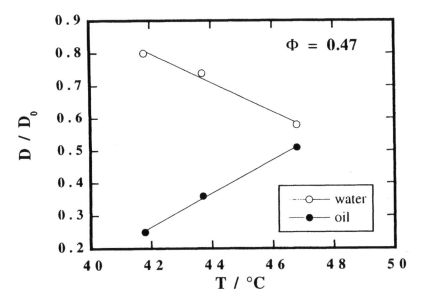

Figure 8 Relative self-diffusion coefficients of water and oil as a function of temperature for a sample containing 16.6 wt% $C_{12}E_5$ and roughly equal volumes of water and tetradecane. The opposite temperature dependences of water and oil clearly show that the structure evolves as a function of temperature at constant composition. The simultaneous increase in the oil and decrease in the water self-diffusion coefficients indicate a decrease in the interfacial film mean curvature with increasing temperature. (Data taken from Ref. 42.)

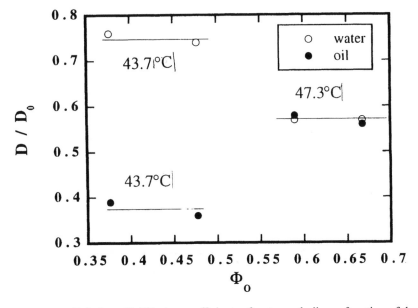

Figure 9 Relative self-diffusion coefficients of water and oil as a function of the oil volume fraction Φ_o at two temperatures, 43.7 and 47.3°C. The system is the same three-component system as in Figs. 7 and 8. Together with Fig. 8, this figure shows the striking behavior, that the self-diffusion coefficients and thus the microemulsion structure depend strongly on temperature while varying only weakly with the water/oil ratio.

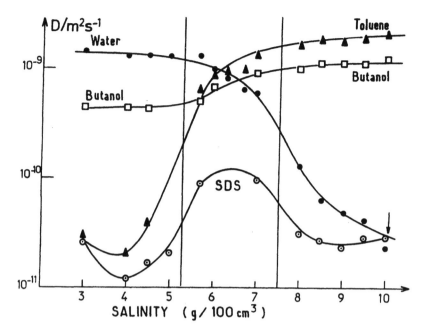

Figure 10 Variations of self-diffusion coefficients of water, oil, surfactant, and cosurfactant in a salinity scan for a five-component microemulsion, SDS–butanol–water–NaCl–toluene. The experiments were performed with excess solvent(s) in the so-called Winsor I–III–II sequence. The system is tuned by salinity at constant cosurfactant concentration. (Data taken from Ref. 44.)

Systems of a single-chain ionic surfactant and a cosurfactant, where the spontaneous curvature is controlled by controlling the salinity [44] (Fig. 10).

Systems of a double-chain ionic surfactant in the presence of electrolyte, where the spontaneous curvature is controlled by temperature [45] (Fig. 11).

Systems of a lipophilic lipid (lecithin), where the spontaneous curvature is controlled by controlling the polarity of the polar solvent [46] (Fig. 12).

Systems of one hydrophilic and one lipophilic surfactant, where the spontaneous curvature is controlled by controlling the surfactant composition [47] (Fig. 13).

All these systems are designed to allow the spontaneous curvature to be varied over wide ranges. In addition, many other types of microemulsion systems have been studied. With double-chain surfactants the microemulsion region is predominantly located toward the oil-rich part of the phase diagram. For Aerosol OT, water-in-oil droplets dominate over wide ranges of composition (Fig. 14), but a bicontinuous structure may be found at higher temperatures. (For ionic surfactants, the effect of temperature on spontaneous curvature is opposite that of nonionics surfactants.) For systems of a single-chain ionic surfactant and a long-chain alcohol, the oil-in-water droplet structure dominates (Fig. 15). For didodecyldimethylammonium bromide (DDAB) in dodecane–water mixtures, water-in-oil droplets are formed at high oil content, whereas a bicontinuous structure is formed as the DDAB + water concentration is increased [48] (Fig. 16).

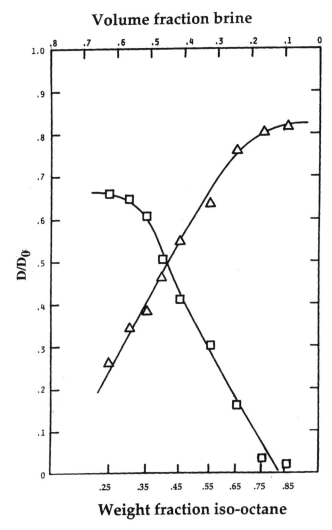

Volume fraction brine

Weight fraction iso-octane

Figure 11 Relative self-diffusion coefficients of (□) water and (△) oil as a function of the oil content in a four-component microemulsion, AOT–water–NaCl–isooctane. The system is tuned by temperature at constant salinity. (Data taken from Ref. 45.)

In many studies, much weaker variations are observed than those exemplified above. This refers, for example, to systems of short-chain surfactants and those of polar solvents other than water. Here the segregation of components between domains (oil, water, surfactant film) is weak, and distinct structures are not formed. This can be inferred from high values of the surfactant self-diffusion coefficient, which imply a considerable role of surfactant unimer translation. Such systems are intermediate between the organized microemulsions of strongly amphiphilic surfactants and simple molecular solutions. We note that, as expected, for nonassociating solvent mixtures, the self-diffusion coefficients vary little with composition, i.e., D/D_0 values throughout are not too different from unity.

"Percolation" is a term frequently used to describe microemulsions. It is often used to imply a bicontinuous structure. However, in our opinion the use of this term is unfortunate and can be misleading for the characterization of a bicontinuous structure. While

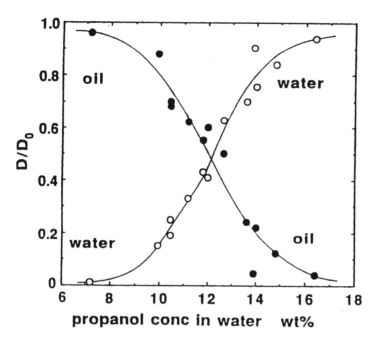

Figure 12 Relative self-diffusion coefficients of water and oil as a function of the aqueous concentration of propanol in a four-component microemulsion, lecithin–propanol–water–hexadecane. The system is tuned by the propanol concentration. (Data taken from Ref. 46.)

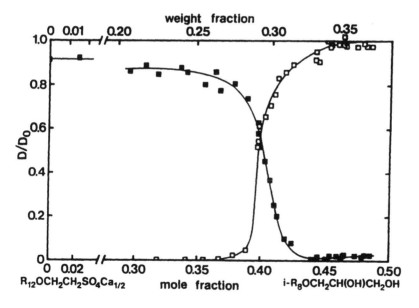

Figure 13 Relative self-diffusion coefficients of water and oil as a function of the surfactant mixing ratio in a five-component microemulsion consisting of $R_{12}OCH_2CH_2SO_4Ca_{1/2}$–$i$-$R_8OCH_2CH(OH)CH_2OH$–water with 8 wt% $CaCl_2$ and decane. Here R_{12} refers to a dodecyl ($C_{12}H_{25}$) chain and i-R_8 to an isooctyl $[(CH_3)_3C(CH_2)_4]$ chain. The system is tuned by the surfactant mixing ratio. (Data taken from Ref. 47.)

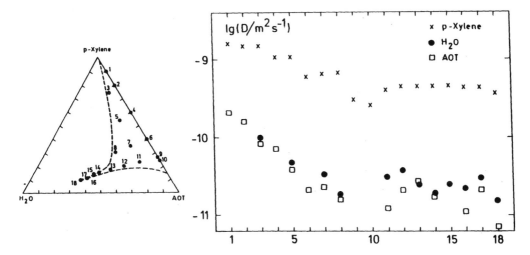

Figure 14 Self-diffusion coefficients of surfactant, water, and oil in a three-component microemulsion (L$_2$) phase with AOT, water, and p-xylene. The samples are labeled from 1 to 18, and the compositions are indicated in the phase diagram. Note the very similar diffusion coefficients of water and AOT over the full concentration range, showing that the structure is made up of discrete reverse micellar aggregates. The fact that the diffusion coefficient of the oil is high everywhere confirms that the structure is oil-continuous. (Data taken from Ref. 94.)

bicontinuity is taken as a structural concept, percolation is used to describe a process. Percolative behavior as generally understood can never give the very rapid simultaneous diffusion of oil and water observed with microemulsions.

We emphasize that bicontinuity describes a situation with (dynamic) equilibrium structure and not a temporary opening of extended pathways between droplets. The generally observed diffusion behavior cannot, furthermore, be described in terms of droplet inversion (e.g., in conjunction with the coexistence of water and oil droplets) or hopping or exchange between droplets (e.g., in transient merging).

We note that bicontinuity results from a particular spontaneous curvature of the surfactant films rather than from a certain solvent volume fraction, which is a secondary factor in determining microstructure. Note that for nonionic surfactants it was shown that the diffusion behavior was determined by temperature and not by solvent composition. For different systems at the same composition, we may have either water droplets, oil droplets, or a bicontinuous structure. An example is given in Fig. 17. Furthermore, one could argue that, to be consistent, all surfactant structures of infinite aggregates (including lamellar and hexagonal) should be described as percolated.

E. Self-Diffusion of "Double Solvent" Systems

Since self-diffusion coefficients are easily measured with high precision, there are numerous possibilities for probing deeper into microemulsion structure and the mechanisms of molecular transport by investigating specially designed systems including systems with added probe molecules. For our understanding of structural transitions, self-diffusion work where either the oil or the polar solvent is composed of two components has been highly informative [43,49].

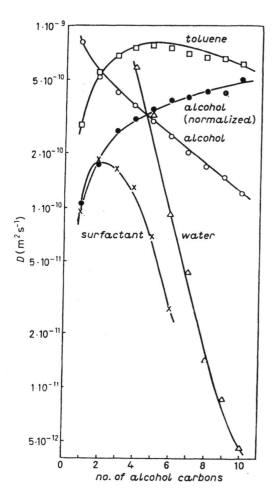

Figure 15 Self-diffusion data at 25°C in the toluene–water–alcohol–sodium dodecyl sulfate system (weight ratios 12.5 : 35.0 : 35.0 : 17.5) as a function of increasing chain length of the alcohol. The filled circles correspond to alcohol diffusion coefficients normalized by the diffusion coefficient of neat alcohol, which removes the trivial molecular weight dependence. Note that the self-diffusion coefficient of water decreases by more than two orders of magnitude as the cosurfactant is changed from butanol to decanol. (Data taken from Ref. 95.)

As illustrated in Fig. 18 for the "double oil" case ("double water" is completely analogous), two oils will have the same self-diffusion coefficient if they are confined to oil droplets. For a bicontinuous microemulsion we expect the ratio of the diffusion coefficients for the two oils to be identical to that in the surfactant-free solvent mixture. It is indeed observed for nonionic microemulsions that the ratio K between D values is unity at low temperatures but close to the bulk value at high temperature. This approach is most useful for monitoring microemulsion structure and, in particular, for quantifying contributions from coexisting droplet and bicontinuous structures to molecular transport. We note in Fig. 19 a very marked maximum in K as a function of temperature. This signifies that the change in microstructure starts with an association of droplets prior to growth. The

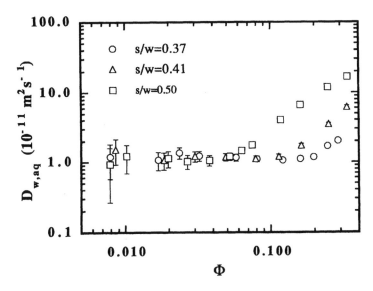

Figure 16 Variation of the water self-diffusion coefficient with the water + surfactant volume fraction Φ for three different dilution lines corresponding to different water/surfactant ratios in the ternary DDAB–water–dodecane system. Above a certain concentration, which depends on the water/surfactant ratio, the water diffusion coefficient increases dramatically, indicating a water droplet-to-bicontinuous transition. (Data from Ref. 48.)

transfer of oil molecules across the surfactant films in the microemulsions of associated droplets will be dependent on oil–headgroup interactions and is thus highly selective. A corresponding "doublewater" study is illustrated in Fig. 20.

F. Bicontinuous Structures and Structural Transitions

Calculations [33] of self-diffusion in ordered bicontinuous structures have, as illustrated in Fig. 21, reproduced the main features of the experimental studies for a large number of microemulsion systems. (For others, a quantitative comparison is difficult because of large influences of effects other than obstruction, such as a high surfactant film concentration or incomplete segregation between domains.) Furthermore, the symmetry of the self-diffusion pattern around the crossover of the oil and water curves implies a symmetry also in structural changes. This symmetry is easy to understand in terms of structures that have a constant mean curvature surface, where changes in spontaneous curvature away from zero in the two directions should be equivalent except for the two solvents changing place.

Alternative structures, termed tubular or interconnected rod models, seem not to explain these features well and are considered less probable for the case of maximal bicontinuity or in the vicinity of maximal bicontinuity. On the other hand, a branched tubular or interconnected rod structure is consistent with observations slightly away from the crossover point, assigned to zero mean curvature. In this region, D/D_0 of both oil and water are high but differ by a significant factor. The transition between discrete droplet and bicontinuous structure is more easily understood with the interconnected rod model.

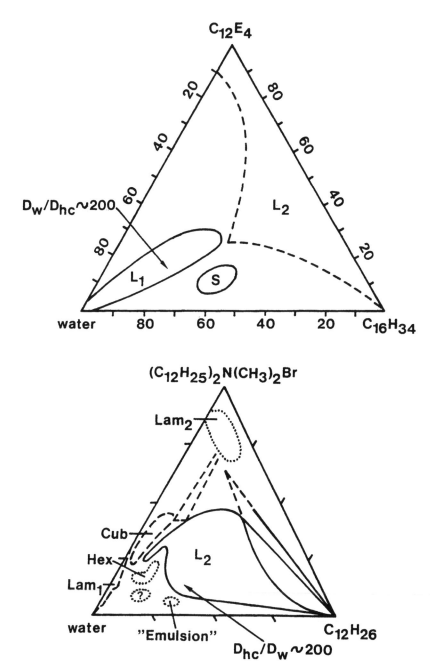

Figure 17 Illustration of the fact that microemulsion structure is not simply a function of composition. Shown are partial ternary phase diagrams with nonionic and cationic surfactants at room temperature. For a similar composition (approximately 15% surfactant, 65 wt% water, and 20 wt% oil), the microstructures of the two systems are widely different, as shown by the ratio of the water and oil diffusion coefficients, D_w/D_{hc}, where hc here denotes oil (hydrocarbon). The nonionic system has an oil-in-water structure ($D_w/D_{hc} = 200$), while the cationic system has a water-in-oil structure ($D_w/D_{hc} = 1/200$).

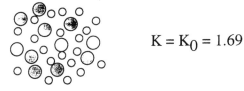

○ cyclohexane " 1 "

● hexadecane " 2 "

$$D_1/D_2 = K$$

NEAT BINARY LIQUID

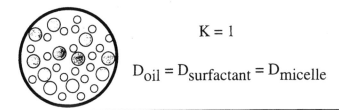

$$K = K_0 = 1.69$$

OIL SWOLLEN MICELLES

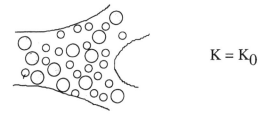

$$K = 1$$

$$D_{oil} = D_{surfactant} = D_{micelle}$$

BICONTINUOUS STRUCTURE

$$K = K_0$$

Figure 18 Illustration of the "double-oil" self-diffusion experiment with a cyclohexane–hexadecane mixture. $K = D_1/D_2$, where D_1 and D_2 are the cyclohexane and hexadecane diffusion coefficients, respectively. In the pure oil mixture the ratio of the two diffusion coefficients is $K = K_0 = 1.69$. For a water-in-oil droplet structure the two oil molecules have the same diffusion coefficient, that of the micelle, and the ratio K equals unity. In a bicontinuous structure, on the other hand, a molecular diffusion mechanism is dominating and the ratio K equals that of the pure oil mixture, K_0. By monitoring the diffusion coefficient ratio, the droplet-to-bicontinuous transition could be studied.

A summarizing picture consistent with the whole body of observations from a large positive value to a large negative value of the spontaneous curvature, H_0, can now be suggested. Starting with spherical oil droplets we have as a function of decreasing H_0 a progressive increase in micelle size to elongated, approximately prolate, micelles. In addition to the droplet growth there also appears to be some droplet–droplet association. At further lowering of H_0, further growth into large droplets is paralleled by a connection of rods (and a constant fusion/fission between droplets) and an onset of bicontinuity. Bicontinuity increases progressively with increasing number of connections and changes over to low

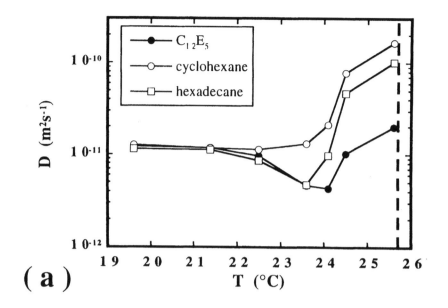

(a)

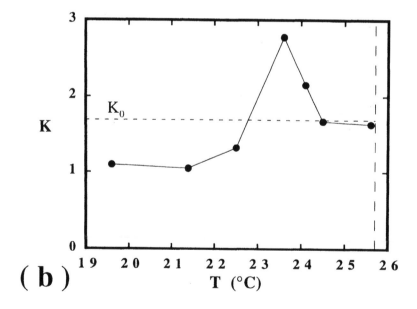

(b)

Figure 19 Double-oil diffusion experiment with nonionic surfactant. (a) Self-diffusion coefficients and (b) diffusion coefficient ratio K as a function of temperature in a water-rich microemulsion with nonionic surfactant. A transition from oil-in-water droplets to a bicontinuous microstructure occurs with increasing temperature (decreasing spontaneous curvature of the $C_{12}E_5$ surfactant film). The maximum in K indicates that an attractive interaction between the micelles is operating prior to the formation of a bicontinuous structure. $K_0 = 1.69$ is the diffusion coefficient ratio in the pure oil mixture and is indicated as a broken line in (b). Note that the initial decrease of the self-diffusion coefficients shows that the droplets grow in size before the bicontinuous transition. The phase boundary at 25.7°C is indicated as a vertical broken line. (Data from Ref. 43.)

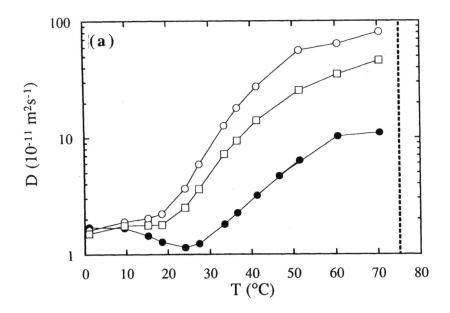

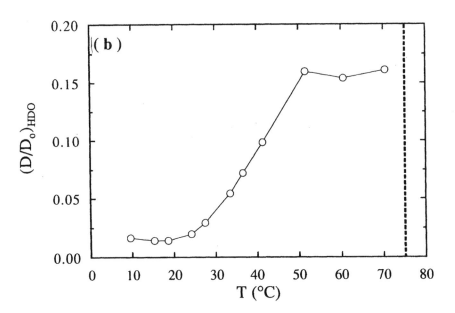

Figure 20 "Double-water" experiment, the aqueous analogy of the double-oil experiment, performed on an AOT microemulsion as a function of temperature. The polar solvent is a 5% *N*-methyl formamide (NMF) solution in heavy water (D$_2$O). The ratio of the water (here measured as trace "impurities" of HDO) and NMF diffusion coefficients is monitored as a function of temperature (c). Also shown as (a) the individual self-diffusion coefficients of water (○), NMF (□), and AOT (●) and (b) the relative diffusion coefficient of water. $K_0 = 1.73$ is the diffusion coefficient ratio in the pure water–NMF mixture and is indicated as a broken line in (c). The phase boundary at 75°C is indicated as a vertical broken line. The behavior with increasing temperature is completely analogous to that of the nonionic system (Fig. 19) and illustrates a transition from reverse micelles to a bicontinuous structure via growing droplets that become attractive. (Data from Ref. 49.)

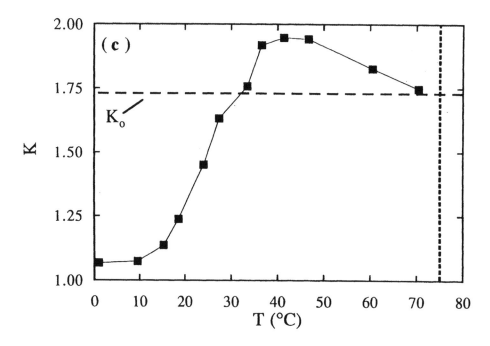

Figure 20 Continued.

mean curvature structure. The structural evolution is symmetrical, so starting from water droplets and increasing H_0 from a large negative value produces a completely analogous sequence of events.

G. Additional Information from Surfactant Self-Diffusion

Complementary information can be obtained from surfactant self-diffusion. In a typical NMR self-diffusion experiment, the value of the surfactant diffusion coefficient is obtained along with the values of the solvents. However, for two reasons the accuracy in self-diffusion coefficient of a surfactant is lower than that of a solvent. First, surfactant is typically present at a considerably lower concentration, and second, the transverse relaxation rate is higher and thus more unfavorable. Depending on the experimental conditions and the system, it may turn out to be impossible, to measure surfactant diffusion. However, by using stimulated echo techniques this problem can be diminished.

As for solvent diffusion, surfactant diffusion follows a rather general pattern for all microemulsion systems, and we illustrate it with an example (Fig. 10). As we vary the appropriate parameter (salinity in Fig. 10), changing the spontaneous curvature from positive to negative, we see at large positive H_0 values (low salinity) a slow surfactant diffusion that equals the oil diffusion value. Here, clearly, surfactant diffusion describes droplet diffusion, as also verified in measurements of droplet diffusion (by self-diffusion or

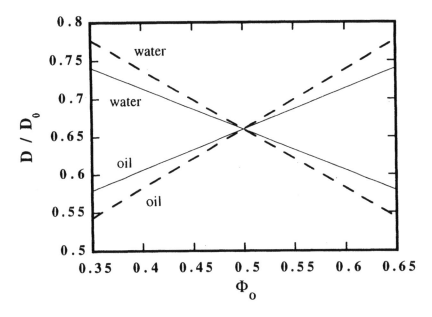

Figure 21 The variation of the reduced self-diffusion coefficients, D/D_0, of water and oil in model bicontinuous structures of constant mean curvature surfaces according to the results of Ref. 33. The results for two families of constant mean curvature surfaces, the D family (solid lines) and the P family (dashed lines), are given. There is a stronger variation of D/D_0 with Φ_0 for the P family, which has a coordination number of 6, compared to the D family which has a coordination number of 4.

dynamic light scattering). Decreasing H_0 by increasing the salinity leads to an accelerated surfactant diffusion, with a maximum for the situation with maximal bicontinuity. The increase is typically by an order of magnitude, and the maximal value is on the order of 10^{-10} m^2/s. This value is similar to the value found in bicontinuous cubic phases or the value for the lateral diffusion along the bilayers in a lamellar phase and thus supports a bicontinuous structure. On the transition from discrete droplet to bicontinuous structure, surfactant diffusion can occur freely along the infinite surfactant films. However, due to its confinement to the films, surfactant diffusion is generally lower than the diffusion of individual surfactant molecules in a molecularly dispersed solution.

Decreasing spontaneous curvature further to negative values (high salinity) leads to a progressive lowering of the surfactant self-diffusion coefficient to the value of water and to that of the droplets in the W/O microemulsions. (This decrease is not pronounced for nonionic surfactants because of the high solubility in oil and the concomitant contribution from single-molecule diffusion.)

In conclusion, surfactant self-diffusion clearly demonstrates the transitions between discrete aggregates and connected surfactant films.

III.   RELAXATION AND DISCRETE PARTICLE SIZE AND SHAPE

A.   Introduction

Nuclear magnetic resonance relaxation is a useful experimental technique to study surfactant aggregation in liquid solutions and liquid crystals [2,50,51]. It yields information on the local dynamics and the conformational state of the surfactant hydrocarbon chain and has, for example, demonstrated the liquidlike interior of surfactant micelles. However, the aim of NMR relaxation studies of microemulsions is often to study properties such as the surfactant aggregate (droplet) size.

The reorientation dynamics of aggregated surfactant molecules, which we discuss in more detail below, is characterized by a locally preferred orientation, in that they essentially form a monomolecular film of oriented molecules. Locally, surfactant molecules undergo rapid internal motions, such as trans–gauche isomerizations in the hydrocarbon chain. These motions are, due to the preferred orientation, slightly anisotropic. The situation can be pictured as the polar headgroups being anchored at the poplar/apolar interface, leaving room for dangling motion of the hydrocarbon chains, with certain restrictions on the number of conformations due to packing constraints in the film. The residual inter-action or anisotropy is, in isotropic solutions, further averaged by the thermal tumbling of the aggregates in combination with the diffusion of surfactant molecules within the curved surfactant film.

NMR spectra from surfactants in micellar solutions and microemulsions are generally in the motional narrowing regime, and the spin dynamics are characterized by well-defined relaxation rates. Exceptions can be found in solutions of very long (>100 nm) wormlike cylindrical micelles, where the large size in connection with a steric overlap of the micelles gives rise to a very slow ($\approx 10^{-4}$ s) aggregate reorientation [52,53]. On the other hand, the dynamics are typically outside the so-called extreme narrowing regime, and one often observes $R_2 \gg R_1$, where R_1 and R_2 are the longitudinal and transverse relaxation rates, respectively. Micellar aggregates reorient typically on the time scale of 10^{-9} s or slower. Thus, at conventional magnetic field strengths, the essential information on the surfactant aggregates is stored in the transverse relaxation rate, R_2.

Hydrocarbon, typically an alkyl chain of 12–16 carbons, constitutes the major of surfactant molecules, offering ^{1}H and ^{13}C nuclei for NMR studies. Being $I = 1/2$ nuclei, their relaxation is due mainly to dipole–dipole interactions. The easy access and the relatively high sensitivity (particularly important in early continuous-wave NMR studies) has made ^{1}H NMR (in particular bandwidth measurements) studies particular popular. Micellar growth can easily be studied in a qualitative manner by monitoring the bandwidths of aliphatic ($-CH_2-$) protons in the spectrum. A quantitative analysis is, however, complicated by the coupling of the various protons, with locally different motional characteristics, along the alkyl chain.

In an alkyl chain, the ^{13}C nuclei are relaxed mainly by the fluctuating dipole–dipole coupling to the directly bound protons. The various carbons along the chain are often resolved, allowing for individual characterizations of the different segments of the hydrocarbon chain. Usually one applies broadband proton decoupling, resulting in exponential longitudinal relaxation of the individual ^{13}C magnetizations, and it is possible to determine R_1 and the nuclear Overhauser enhancement (NOE) of the different methylene segments. On the other hand, it is difficult to measure R_2 to ^{13}C. Therefore ^{13}C is mainly useful for studying local chain dynamics and segmental order, since the aggregate dynamics occur on time scales that are normally significantly longer than the inverse ^{13}C resonance frequency.

Certain surfactants carry phosphate or ammonium polar headgroups, offering ^{31}P and ^{14}N nuclei, respectively, for relaxation studies. The phosphate ^{31}P ($I=1/2$) has a relatively large chemical shift anisotropy. ^{14}N is an $I=1$ nucleus, and its relaxation is dominated by the strong quadrupolar interaction and can be most useful for studies of microemulsion droplet size.

The most suitable nucleus for relaxation studies of surfactant systems is ^{2}H, which can be synthetically incorporated into the hydrocarbon chain of the surfactant molecule. Normally one selectively labels a single methylene segment, typically one adjacent to the polar headgroup, in order to avoid resonance overlap in the spectrum. ^{2}H is an $I=1$ nucleus, and the dominating interaction governing the relaxation is the strong quadrupolar interaction. For aliphatic (C–^{2}H) deuterons, the analysis is further simplified by the fact that the electric field gradient tensor has essentially cylindrical symmetry around the C–^{2}H bond direction, resulting in a vanishing asymmetry parameter. The ease with which both R_1 and R_2 can be accurately measured, the relatively simple interaction Hamiltonian, and the sufficiently high sensitivity to allow measurements at low resonance frequencies (down to 2 MHz using variable-field electromagnets), gives ^{2}H NMR relaxation a particularly important place among the experimental tools available to study surfactant systems.

To make the significance of the NMR technique as an experimental tool in surfactant science more apparent, it is important to compare the strengths and the weaknesses of the NMR relaxation technique in relation to other experimental techniques. In comparison with other experimental techniques to study, for example, microemulsion droplet size, the NMR relaxation technique has two major advantages, both of which are associated with the fact that it is reorientational motions that are measured. One is that the relaxation rate, i.e., R_2, is sensitive to small variations in micellar size. For example, in the case of a sphere, the rotational correlation time is proportional to the cube of the radius. This can be compared with the translational self-diffusion coefficient, which varies linearly with the radius. The second, and perhaps the most important, advantage is the fact that the rotational diffusion of particles in solution is essentially independent of interparticle interactions (electrostatic and hydrodynamic). This is in contrast to most other techniques available to study surfactant systems or colloidal systems in general, such as viscosity, collective and self-diffusion, and scattered light intensity. A weakness of the NMR relaxation approach to aggregate size determinations, compared with form factor determinations, would be the difficulties in absolute calibration, since the transformation from information on dynamics to information on structure must be performed by means of a motional model.

B. Time Scale Separation: The Two-Step Model

In this section we discuss a motional model [54–56] of aggregated surfactant molecules that involves a time scale separation of fast local and slow global motions. This model has been extensively applied to, and has been shown to rationalize, NMR relaxation data from aggregated surfactant systems. We discuss it in connection with ^{2}H relaxation experiments, since the most extensive experimental studies have been performed on ^{2}H-labeled surfactants. We begin by recalling that the longitudinal (R_1) and transverse (R_2) relaxation rates of an $I=1$ nucleus due to a quadrupolar interaction in an isotropic solution are given by [57]

$$R_1 = \frac{3\pi^2}{40} \chi^2 [2j(\omega_0) + 8j(2\omega_0)] \tag{8a}$$

$$R_2 = \frac{3\pi^2}{40} \chi^2 [3j(0) + 5j(\omega_0) + 2j(2\omega_0)] \tag{8b}$$

Here, χ is the quadrupolar coupling constant and $j(\omega_0)$ the reduced spectral density function evaluated at the Larmor frequency, ω_0.

On the basis of existing knowledge of surfactant aggregate structures, Wennerström and coworkers [54–56] suggested that the reorientational motion of surfactant molecules could be divided into (1) fast local motions (equilibration of internal modes), which are slightly anisotropic due to the preferred orientation of the surfactant molecules, and (2) slow isotropic motions associated with the aggregate tumbling and the lateral diffusion of surfactant molecules within the surfactant film. If these two motions occur on significantly different time scales, the fast motions generate a well-defined residual anisotropy to be averaged by the slow motions, and the spectral density can be written as a weighted sum of the fast and slow components:

$$j(\omega_0) = (1 - S^2)j_f(\omega_0) + S^2 j_s(\omega_0) \tag{9}$$

Here $j_f(\omega_0)$ and $j_s(\omega_0)$ represent the reduced spectral density functions describing the fast and slow motions, respectively. The parameter S is often referred to as an order parameter and is given by the average;

$$S = \tfrac{1}{2}\langle 3 \cos \theta_{MD} - 1 \rangle_f \tag{10}$$

where θ_{MD} is the angle between the axis of the maximum component of the electric field gradient tensor and the director axis.

The situation is illustrated in Fig. 22 for the case of a ^{2}H-labeled methylene segment of a surfactant molecule residing in a micelle. The director axis, Z_D, corresponds to the micellar surface normal. The field gradient tensor is cylindrically symmetrical around the C–^{2}H bond axis, with the maximum component in the bond axis direction, Z_M.

C. Interpreting the Spectral Densities

The fast motions described by the spectral density function $j_f(\omega_0)$ are expected to occur on a similar time scale ($\tau_f \approx 10^{-11}$ s) as in the corresponding liquid alkane. Thus this motion is in extreme narrowing ($\tau_f \ll 1/\omega_0$) in the accessible frequency range, and we may assume $j_f(2\omega_0) \approx j_f(\omega_0) \approx j_f(0) = 2\tau_f$, where τ_f should be considered an effective correlation time, representing the integral over a nonexponential correlation function.

The slow motion is a combination of droplet tumbling and the lateral diffusion of surfactant molecules within the surfactant film. For spherical aggregates, this motion is described by a Lorentzian spectral density function,

$$j_s(\omega_0) = \frac{2\tau_s}{1 + (\omega_0 \tau_s)^2} \tag{11}$$

where τ_s is the correlation time. The tumbling (subscript t) and the lateral diffusion (d) are expected to be statistically independent, and the slow motion correlation time can be written

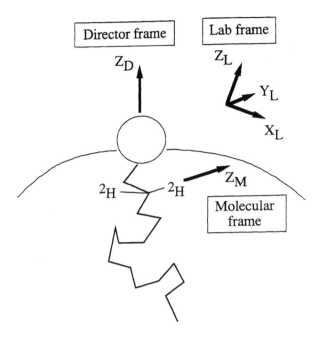

Figure 22 A schematic illustration of the various coordinate frames considered within the two-step model, for the case of a specifically deuterium-labeled methylene segment in the surfactant hydrocarbon chain. The laboratory frame (L) is set by the direction of the external magnetic field, where Z_L is the field direction. In this frame the nuclear quadrupolar moment tensor is diagonal. The director frame (D) is associated with the micellar aggregate where Z_D specifies the micellar surface normal. It is assumed that the fast local dynamics occur with an essentially cylindrical symmetry around Z_D. The molecular frame (M) corresponds to the principal axis of the electric field gradient tensor. For the case of a methylene segment, Z_M specifies the direction maximum component of the field gradient tensor, which is furthermore cylindrically symmetrical around Z_M.

as

$$\frac{1}{\tau_s} = \frac{1}{\tau_t} + \frac{1}{\tau_d} \tag{12}$$

The correlation time associated with the tumbling of a sphere of radius R is given by

$$\tau_t = \frac{4\pi\eta R^3}{3k_B T} \tag{13}$$

where η represents the solvent viscosity and $k_B T$ the thermal energy. The effect of lateral diffusion can be described in terms of diffusion on the surface of a sphere of radius R. In this case, the correlation time can be written as

$$\tau_d = \frac{R^2}{6D_{\text{lat}}} \tag{14}$$

where D_{lat} is the lateral diffusion coefficient.

With the assumptions of (1) extreme narrowing conditions for the fast motions and (2) a Lorentzian spectral density function of the slow motion, the expressions for the relaxation rates become

$$R_1 = \frac{3\pi^2}{40} \chi^2 \left[(1 - s^2)20\tau_f + S^2 \left(\frac{4\tau_s}{1 + (\omega_0\tau_s)^2} + \frac{16\tau_s}{1 + (2\omega_0\tau_s)^2} \right) \right] \tag{15a}$$

$$R_2 = \frac{3\pi^2}{40} \chi^2 \left[(1 - S^2)20\tau_f + S^2 \left(6\tau_s + \frac{10\tau_s}{1 + (\omega_0\tau_s)^2} + \frac{4\tau_s}{1 + (2\omega_0\tau_s)^2} \right) \right] \tag{15b}$$

Equations (15) have been applied in a large number of ^{2}H NMR relaxation studies of micellar systems. It turns out that frequency-dependent relaxation studies are particularly useful to study small spherical micelles. These have a typical radius of about 2 nm. Using the water viscosity we obtain with Eq. (7) $\tau_t = 7$ ns at room temperature. Including the effect of lateral diffusion, a typical value of the slow correlation time is $\tau_s = 5$ ns. Thus relaxation dispersion from slow motion occurs around $\nu_0 \approx (2\pi\tau_s)^{-1} \approx 30$ MHz, which is well within the accessible Larmor frequency range for ^{2}H. Examples of studies of micelle size and dynamics can be found in Refs. 58–61.

D. Examples of Applications

The small spherical micelles ($R \approx 2$ nm) can be seen as a limiting case, and in general surfactant aggregates can be much more extended. With an increase in the size of the aggregates τ_s increases, and the relevant relaxation dispersion moves out of the accessible frequency window. For most micellar and microemulsion systems, the motional narrowing condition still applies. In these cases, valuable information can also be obtained by measuring R_1 and R_2 for ^{2}H-labeled surfactants. For larger aggregates, the relevant information on the aggregate size is contained in the zero-frequency spectral density only. Assuming that the fast motions are in extreme narrowing, in which case they contribute to a constant offset to both R_1 and R_2, it is useful to consider the difference $\Delta R = R_2 - R_1$, which then depends only on the slow motions. For the case of very slow motions, we have also $j_s(0) \gg j_s(\omega_0) \approx j_s(2\omega_0)$ and ΔR reduces to the simple expression

$$\Delta R = \frac{9\pi^2}{40} (\chi S)^2 j_s(0) \tag{16}$$

where $j_s(0) = 2\tau_s$ in the case of a Lorentzian spectral density function [Eq. (11)]. Equation (16) has been used to study shape variations of surfactant aggregates in various microemulsion systems [48,58]. Here, the micellar aggregates are swollen by the addition of a third component (normal micelles in aqueous solution are swollen by oil, or reverse micelles in oil are swollen by water), where spherical aggregates can have radii on the order of 10 nm. Quantitative applications of Eq. (16) require accurate determinations of S from adjacent liquid crystalline phases. Furthermore, they require assumptions of the aggregate shape and a knowledge of D_{lat} to determine, for example, an aggregate radius of spherical aggregates. So, obviously, there are some uncertainties involved concerning the "absolute calibration" of the method.

The situation can, however, be optimized by noting that the aggregates are formed under the constraint of a constant ratio of average interfacial area to enclosed volume, set by the concentration ratio of surfactant to internal component. With this constraint, the minimum aggregate size corresponds to a spherical shape. Due to translational entropy, spherical aggregates, representing the minimum aggregate size, will always be stable at high dilutions, where the experiment can be calibrated. In many cases, D_{lat} is also known rather accurately from the various bicontinuous phases of the different systems.

Deviations from spherical shape can be modeled as a growth into prolate or oblate shapes. The area enclosed volume ratio constraint implies a constraint in the possible values of the two semiaxes describing the particle size. Halle [62] calculated the correlation functions for the combined particle tumbling and surface diffusion of prolate and oblate particles. His results have, for example, been applied to micro-emulsion systems [48,58], focusing on the ratio $j_{s,pr}(0)/j_{s,sph}(0)$, where the subscripts pr and sph refer to prolate and spherical particles, respectively. For a given ratio of interfacial area to enclosed volume, which specifies the radius R of the sphere, $j_{s,pr}(0)/j_{s,sph}(0)$ is a function of the prolate axial ratio, D_{lat}, and R. Knowing D_{lat} and R from other experiments, it is possible to determine the aggregate axial ratio from the relaxation experiment.

As mentioned above, the strength of the NMR relaxation technique in comparison to other (say, scattering) techniques for studying micelles and microemulsions is related to the fact that rotational dynamics are monitored. This has the advantage that the dynamics are (1) sensitive to small size variations and (2) essentially independent of interactions. In an extensive study of a three-component microemulsion system composed of the nonionic surfactant pentaethylene oxide dodecyl ether, water, and decane oil-swollen micelles dispersed in water were investigated over a large concentration range [58].

With nonionic surfactants, spherical aggregates are formed at lower temperatures, while the aggregates grow in size with increasing temperature [63]. In Fig. 23 we show the variation of ΔR with temperature for two compositions, $\Phi = 0.12$ and $\Phi = 0.23$, in the microemulsion phase [60]. Here, $\Phi = \Phi_s + \Phi_o$ is the total volume fraction of surfactant (Φ_s) and oil (Φ_o), and the ratio $\Phi_s/\Phi_0 = 0.815$ was kept constant.

ΔR decreases when the temperature is decreased in the microemulsion phase and levels off at lower temperatures at a minimum value, ΔR_{min}, when the aggregates have reached the limiting spherical shape corresponding to the minimum size. Note that the minimum ΔR value is the same for the two concentrations, as expected, since the radius, dictated by Φ_s/Φ_o, should be the same. By considering the reduced ΔR value, $\Delta R/\Delta R_{min} = j_{s,pr}(0)/j_{s,sph}(0)$, it was possible to calculate the increase in the axial ratio of the aggregates with increasing temperature, assuming a growth into prolate aggregates. The obtained variation of the axial ratio is shown in Fig. 24.

For polydisperse systems the measured relaxation rates correspond, in most cases, to a weighted average due to fast exchange of surfactant molecules between aggregates on the experimental time scale (ΔR_2). From measuring a single number only, for example the relaxation rate difference, of course only a single value can be determined. Widths and shapes of distribution functions remain undetermined. As a single-number measurement the information content is low; however, as outlined above, the technique is very sensitive to structural changes. The information content increases if the frequency dependence of relaxation rates can be measured as has been done for small micelles. For microemulsions, however, the interesting Larmor frequency regime is in the kilohertz region, which can be reached by field cycling techniques [64]. However, due to limited sensitivity, such experiments have not yet been performed on microemulsions where interest is focused on low surfactant concentrations.

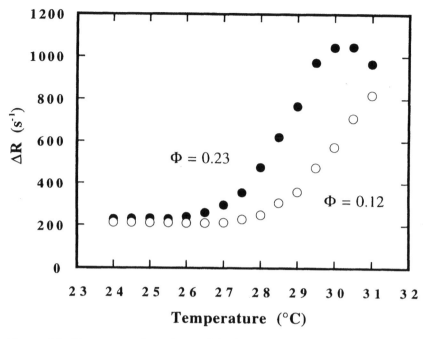

Figure 23 Temperature dependence of the deuterium relaxation rate difference, ΔR, measured with deuterium-labeled $C_{12}E_5$, in the ternary $C_{12}E_5$–water–decane system. The two volume fractions, $\Phi = 0.23$ and 0.12, respectively, refer to the total volume fraction of surfactant and oil for a constant surfactant/oil ratio $\phi_s/\phi_o = 0.815$. At lower temperatures the system forms spherical oil droplets in water and ΔR is the same for the two volume fractions. Above a certain temperature, the droplets grow in size as demonstrated by the increase in ΔR. (Data from Ref. 58.)

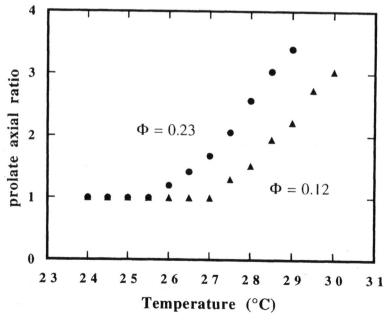

Figure 24 The variation of the axial ratio of the micelles with temperature evaluated from the ΔR values of Fig. 23. (▲) $\Phi = 0.12$ results, (●) $\Phi = 0.23$ results. (Data from Ref. 58.)

IV. MICROSTRUCTURE FROM SELF-DIFFUSION AND RELAXATION

In the foregoing we attempted both to analyze the type of information that NMR can provide on microemulsion structure and to present the current understanding that NMR approaches have provided on this problem. Only two types of NMR studies provide significant information on microstructure: multicomponent self-diffusion and surfactant relaxation.

Self-diffusion is the most powerful approach, and the following points can be made:

1. Self-diffusion appears to be the only technique, together with microscopy (cryo-TEM), that can demonstrate bicontinuity.
2. Self-diffusion provided the first proof of bicontinuity in microemulsions and has become a routine technique of microemulsion characterization used by many groups.
3. A simultaneously high D/D_0 value (above 0.6) for both solvents provides evidence for the highest conceivable degree of bicontinuity. This observation indicates a structure of close to zero mean curvature but appears to be inconsistent with a tubular structure.
4. Self-diffusion demonstrates a quite sharp transition from discrete micelles to a bicontinuous structure.
5. Structure is determined by spontaneous curvature rather than solvent composition.
6. Symmetry in diffusion behavior around $H_0 = 0$ demonstrates symmetry in structure, which favors constant mean curvature structures rather than connected rod structures.
7. Interpretation of diffusion data is straightforward and unambiguous for limiting structures (discrete droplets, bicontinuous structures) but is more difficult for intermediate situations, where there is a transition between structures. Here additional work is motivated.

On the other hand, NMR relaxation has provided less unique and original insight into microemulsion structure. Rather extensive efforts by several workers have been somewhat disappointing, in particular in relation to the successful studies of micellar solutions. Relaxation studies are mainly useful for cases of discrete droplets, for which they may provide direct and unique insight into size and shape changes. We may characterize the applicability of relaxation as follows:

1. Information on aggregate size from relaxation cannot be deduced directly; a model has to be used. Therefore, there is always some uncertainty in the exact values of droplet size, and some calibration must be employed.
2. On the other hand, relaxation is much more sensitive than other techniques (self-diffusion, scattering, viscosity, etc.) in picking up even quite minor changes in droplet size and/or shape. The reason is that relaxation depends on droplet radius to the second or third power.
3. Finally, relaxation is a reliable technique for monitoring changes in droplet size and shape, since it is to a good approximation independent of interdroplet interactions up to high droplet volume fractions. The reason for this is that relaxation arises from droplet reorientation and intradroplet surfactant diffusion, processes only weakly dependent on interdroplet interactions.

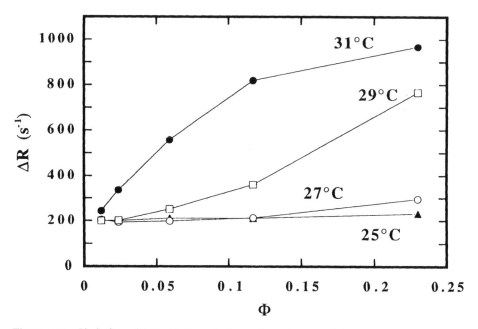

Figure 25   Variation of ΔR with the micellar volume fraction Φ from the same system as in Fig. 23. At 25°C ΔR is concentration-independent, demonstrating that the structure is constant (spherical droplets). At higher temperatures the droplets grow in size with increasing concentration as demonstrated by an increase in ΔR. (Data from Ref. 58.)

Figure 25 illustrates with the case of nonionic microemulsions at low temperatures that relaxation is constant up to high volume fractions for the case of constant-size microemulsions. Here, ΔR is plotted as a function of the droplet volume fraction Φ. At 25°C the droplet size and shape are independent of Φ, and ΔR is constant. At higher temperatures (29 and 31°C) the droplet size increases with increasing Φ as shown by the increase in ΔR.

V.   MISCELLANEOUS APPLICATIONS OF NMR FOR MICROEMULSIONS AND RELATED SURFACTANT SYSTEMS

A.   Phase Diagrams

Strikingly, NMR has become a standard technique for establishing phase diagrams of surfactant systems, with some very distinct advantages compared to alternative methods. Surfactant phase diagrams generally include both solution phases and liquid-crystalline phases [65–68]. Methods to elucidate phase behavior based on microscopic observation in polarized light, calorimetry, X-ray diffraction, and so on have been elaborated to great perfection, but various NMR observations are important complements, and it can in fact often be more practical to use NMR to determine entire phase diagrams. A significant advantage of the NMR method is that it is not necessary to achieve a macroscopic phase separation, and two- or three-phase character can be detected with single-phase domains on the micrometer scale.

Phase diagram studies by NMR are generally based on the fact that the rapid molecular dynamics causes an elimination of any spin interactions to an extent that is directly related to the degree of anisotropy of the structure. The most important use of NMR is to distinguish between phases that give residual dipolar or quadrupolar interactions and phases that do not; in general, this also marks the difference between optically anisotropic and optically isotropic phases. Whether residual couplings are seen depends on the extension of the anisotropic domains (microcrystallite sizes) in combination with the rate of molecular diffusion.

The presence of even very minute amounts of isotropic phase is generally visible in ^{1}H and ^{13}C spectra, whereas because of large signal width due to static dipolar couplings the signals from anisotropic liquid crystals are much more difficult to observe. Although ^{1}H and ^{13}C NMR can give some initial hints, they do not provide sufficiently detailed and unequivocal information on phase diagrams. It is much more useful and general to base an analysis on quadrupole splittings or chemical shift anisotropies.

For quadrupolar nuclei the simplest case is with $I = 1$ rather than $I = 3/2, 5/2$, or $7/2$, because any central signal peak can be referred to the presence of isotropic phase whereas a doublet signifies an anisotropic phase; for $I = 3/2, 5/2$, or $7/2$ nuclei, on the other hand, there is always a central peak, but otherwise the same considerations apply. Furthermore, the magnitude of the splitting depends on the degree of anisotropy of the phase, so that under identical local molecular conditions and composition a lamellar phase should give twice as large a splitting as a hexagonal phase. For nuclei with sizable shift anisotropies, the isotropic and anisotropic phases are again easily identified. Both the sign and the magnitude of the shift anisotropy are different for lamellar and hexagonal phases, so in this case the distinction is even more straightforward.

The quadrupole splitting method (generally using heavy water) has been used to establish entire phase diagrams for a number of two- and three-component systems [69–72].

For microemulsions, the main significance of, for example, studies of ^{2}H NMR of the water lies in the fact that a broad phase characterization is needed for any microemulsion system; leading proponents of this have been Friberg, Shinoda, and Kahlweit. However, sometimes a given system cannot be easily identified as a single-phase microemulsion, and then NMR work can be helpful. It should be noted, though, that while it is easy to obtain evidence for minute amounts of isotropic phase in a sample of mainly anisotropic liquid crystal, the converse is much more demanding.

B. Phase Structure

An initial aim is normally to obtain information on whether a phase is isotropic or anisotropic and on whether, in the latter case, it has a uniaxial or biaxial structure, as well as on the degree of anisotropy of an anisotropic phase. If a phase is isotropic on the relevant NMR time scale, static dipolar, quadrupolar, and shift anisotropy interactions are averaged to zero by molecular motion. If a phase is anisotropic, the spectrum should contain static interaction effects (e.g., quadrupole splittings). If the local order parameter can be predicted theoretically or if there are adequate reference data, one may decide from the magnitude of the splitting whether the phase is built up of rod aggregates with rapid diffusion around the rods or if it is built up of repeating planar aggregates.

If we turn to the isotropic phases, several possibilities exist with respect to both liquid solution phases and cubic liquid crystalline phases. Cubic phases, which have long-range order, can be distinguished from microemulsions in giving rise to low-angle X-ray diffraction patterns. A number of different cubic phase structures are possible, but the X-ray low-angle diffractograms normally do not give a sufficient number of reflections to permit a complete structural assignment [73]. However, here NMR self-diffusion studies [74–76] can very clearly distinguish between possible alternatives, as illustrated for the dodecyltrimethylammonium chloride–water system [74]. For the micellar phase the surfactant self-diffusion coefficient decreases strongly with increasing concentration; the micelles stay closely spherical, and the decrease at higher concentrations can be referred mainly to intermicellar interactions (repulsive electrostatic interactions). As one passes over into the first cubic liquid crystalline phase there is a regular change in the D value and the self-diffusion results require a structure based on discrete repulsive micelle-like units in a water continuum; later work showed that the micelles are short rods rather than spheres [59,77,78], and spheres are also inconsistent with the relatively rapid NMR relaxation. The cubic phase occurring at higher surfactant contents is characterized by a very much higher surfactant self-diffusion coefficient, and this phase cannot be discontinuous for the surfactant molecules. Instead these data are consistent with a bicontinuous structure with rodlike surfactant structures forming extended networks. Lindblom et al. [76] went a step further and made a quantitative comparison between self-diffusion data from cubic phases with self-diffusion along a surfactant bilayer (obtained from experiments on macroscopically oriented lamellar phases) to provide a more detailed structural characterization. This provided evidence for a bicontinuous cubic phase based on bilayer units. More recently detailed interpretation of self-diffusion data for several cubic phases has been successfully carried out in terms of minimal surface structures [34].

The work on microemulsions described above was developed on the basis of prior work on micelles and cubic phases.

C. Surfactant Self-Assembly

Several NMR parameters are markedly different in the unimeric and self-assembled states and are therefore possible candidates for characterization of surfactant self-association processes. NMR can be a convenient alternative for the determination of critical micelle concentrations (cmc's), but an analysis of variable-concentration NMR data generally also provides additional information. We here describe only how self-diffusion can be used to measure the free unimer concentration in simple and complex surfactant systems. In practice it is an alternative to surfactant-selective electrodes to measure surfactant activities. With a two-state assumption, we have, under the normal rapid exchange conditions [cf. Eq. (7)],

$$D = p_M D_M + p_{\text{free}} D_{\text{free}} \tag{17}$$

where p_M is the fraction of surfactant molecules existing in micelles and $p_{\text{free}} = 1 - p_M$ is the fraction of monomers. The phase-separation model gives

$$
\begin{aligned}
C < \text{cmc:} &\quad D = D_{\text{free}} \\
C > \text{cmc:} &\quad D = D_M - \frac{\text{cmc}}{C}(D_M - D_{\text{free}})
\end{aligned}
$$

The predicted behavior of two straight-line segments intersecting at the cmc is in rough agreement with observations and becomes a better approximation the longer the alkyl of a surfactant, as expected.

Surfactant self-diffusion coefficients can be analyzed directly assuming a two-state model,

$$D = \frac{C_M D_M + C_{\text{free}} D_{\text{free}}}{C} \tag{18}$$

to give the concentrations of free and self-assembled surfactant since D_{free} and D_M can be separately measured or estimated to a good approximation.

D. Micelle Size and Shape

Properties pertaining to micelle size or shape can be conveniently obtained from NMR relaxation data. The problem inherent in such a process is that one has to resort to modeling the relaxation parameters, as the information is not directly obtainable from the raw relaxation data. Needless to say, the models used must be as realistic as possible. One model that has proven to be useful in this regard is the so-called two-step model, described above.

To investigate the applicability of the two-step model of relaxation, Söderman et al. [59,60] studied aqueous micellar systems of alkyltrimethylammonium chloride surfactants of two different alkyl chain lengths, dodecyl and hexadecyl chains. The alkyl chains were ^{2}H-labeled in the α-methylene segment adjacent to the ammonium polar headgroup, and the relaxation rates, R_1 and R_2, were measured in the Larmor frequency range 2–55 MHz, to cover the dispersion of the slow motion. The lower frequencies were obtained by using a variable-field electromagnet. In Fig. 26 the results [60] are shown for the

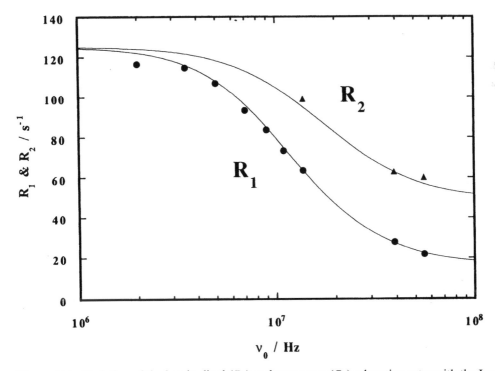

Figure 26 Variation of the longitudinal (R_1) and transverse (R_2) relaxation rates with the Larmor frequency for ^{2}H nuclei bound to the α-methylene segment of the surfactant hexadecyltrimethyl-ammonium chloride. The sample is an aqueous micellar solution containing 13 wt% of the surfactant. The experiments were performed at 27°C. (Data taken from Ref. 60.)

hexadecyltrimethylammonium chloride micelles (the concentration was 13 wt%). The solid lines are best fits of Eq. (15) to the whole data set, using τ_f, τ_s, and S as adjustable parameters. As is seen in Fig. 26, the data are well described by a Lorentzian spectral density function superimposed on a constant offset in the investigated frequency range, supporting the assumption of time scale separation. The fit yielded $\tau_f = 42 \pm 1$ ps, $\tau_s = 7.6 \pm 0.4$ ns, and $S = 0.186 \pm 0.003$, where the uncertainties correspond to an approximately 80% confidence interval taking only random errors into account, and the value $\chi = 167$ kHz was used for the quadrupolar coupling constant.

The slow motion correlation time was interpreted in terms of Eqs. (12)–(14), which contain two unknown parameters, R and D_{lat}. However, D_{lat} can be measured in phases of extended surfactant films, such as bicontinuous cubic phases, and R is expected to be close to the overall length of the surfactant molecule. Using $D_{\text{lat}} \approx 1 \times 10^{-11}$ m^2/s as measured in a bicontinuous cubic phase of the similar surfactant hexadecyltrimethylammonium fluoride and the radius $R = 2.4$ nm, a slow motion correlation time of approximately 10 ns is calculated from Eqs. (12)–(14), which is in reasonable agreement with the experimentally determined value.

The applicability of the two-step model receives further support when the S value obtained from the fit to the relaxation dispersion is compared with that measured from quadrupolar splittings in liquid crystalline phases forming at higher surfactant concentrations. In the liquid crystalline phases, extended surfactant aggregates are "locked" in a crystalline array. Two very common structures are the lamellar phase where surfactant bilayers are stacked with one-dimensional order and the hexagonal phase where cylindrical aggregates are packed into a two-dimensional hexagonal lattice. In these phases one has effectively frozen the slow motion, and the nuclei experience a static quadrupolar Hamiltonian, $\langle H_Q \rangle_f$, in addition to the Zeeman term, where the average is taken over the fast local motions. In the lamellar phase, where the surfactant film is planar, this results in a quadrupolar splitting, Δ_{lam}, of magnitude [55]

$$\Delta_{\text{lam}} = \tfrac{3}{4}\chi|S| \tag{19}$$

In the case of the hexagonal phase, the rapid surfactant diffusion around the cylinder axis results in an additional averaging of the interaction by a factor of 2, and the splitting is given by [55]

$$\Delta_{\text{hex}} = \tfrac{3}{8}\chi|S| \tag{20}$$

One may expect that the local motions, and hence the S value, should be similar in the liquid crystalline surfactant aggregates and in the micelles in the dilute solution phase. This is also found to be the case. As mentioned above, the S value obtained from the fit to the relaxation dispersion (Fig. 26) was found to be $S = 0.186$. A very similar value, 0.192, was measured in the hexagonal phase at higher concentrations [60].

In a modified form these principles have direct applications for the larger entities formed in microemulsions. This was described above.

Very large micelles may also form in binary surfactant systems. These are long wormlike micelles that become entangled at higher concentrations, giving rise to rheological properties similar to those in polymer solutions. Such systems have been examined by ^{1}H band shape analysis [52,53]. The protons of the surfactant hydrocarbon chain form a very large dipolar coupled spin system with an essentially continuous distribution of transverse relaxation rates. The distribution of relaxation rates is related to the distribution of order parameters [79], which can be obtained from analyzing the band shape in the liquid crystalline (lamellar or hexagonal) phases, where the band shape is given by the distribution of residual dipolar splittings [80].

Slow motions have also been studied by measuring the differential line broadening (DLB) of proton J-coupled ^{13}C nuclei [81]. Here the difference in bandwidth of the different resonances in a ^{13}C multiplet can be attributed to the interplay between the dipole–dipole and chemical shift anisotropy relaxation mechanisms, and information on the slow motion can be obtained by analyzing the difference.

E. Counterion Binding

In an ionic surfactant system there is an inhomogeneous distribution of counterions, with a higher concentration close to the charged microscopic interfaces than at large distances from them. A number of NMR parameters give insight into the counterion distribution as well as into specific ionic interactions. Following is a partial list of possible approaches.

1. The translational mobility of an ion is different by one to two orders of magnitude between moving freely in a solution and diffusing with an entity of colloidal dimensions such as a micelle. This allows counterion self-diffusion coefficients to be used for characterization of counterion distribution, for example, in terms of a counterion association degree.
2. Quadrupole relaxation reflects both the distribution of counterions and the dynamics of counterions associated with an aggregate. A full interpretation is typically difficult as information is not available on the quadrupole interaction of the associated state and is difficult to obtain independently because the relaxation dispersion is typically not observed.
3. Chemical shifts give a direct picture of the distribution of counterions between different environments.
4. For anisotropic liquid crystalline systems, both quadrupole relaxation and quadrupole splittings provide direct information into counterion interactions.

The two-site model leads to the definition of the degree of counterion binding, β, as the ratio of counterions to surfactant ions in a surfactant self-assembly. This is a useful but incomplete characterization of the counterion distribution. The value of β can be obtained directly from self-diffusion data because the self-diffusion coefficients of free ions are easily obtainable. For free counterion diffusion a correction is made for the obstruction effect. The micellar D value is obtained as described above or estimated; as $D_M \ll D_{free}$, an exact D_M value is not critical.

An example of a study of counterion binding by FT NMR self-diffusion is given in Refs. 82 and 83. The method is generally applicable to organic counterions, whereas only a few inorganic ions, such as Li^+ (see Ref. 84), F^-, Cs^+, and Be^{2+}, can be studied. A general observation for a large number of surfactant systems is a roughly concentration-independent counterion binding. This so-called ion condensation behavior has also been predicted theoretically in Poisson–Boltzmann calculations. A more satisfactory approach than using the two-site model would be to make use of the full theoretically deduced counterion distribution around micelles to calculate counterion self-diffusion coefficients. This is discussed in Ref. 85. The FT NMR self-diffusion method is very convenient for studying ion competition effects, as illustrated nicely in a study of the interaction of substituted acetate ions with cationic micelles [86].

The constancy of β to changes in micelle concentration is also directly borne out in studies of counterion quadrupole relaxation rates for a large number of systems using ^{23}Na, ^{35}Cl, ^{85}Rb, ^{133}Cs, and ^{81}Br NMR (reviewed in Ref. 87). These studies also demonstrate that the intrinsic relaxation rates of micellar counterions are nearly independent of micelle concentration and the alkyl chain length of the surfactant.

F. Hydration

Surfactant self-assembly is a delicate balance between hydrophobic and hydrophilic interactions, and the interactions between the headgroups and the solvent are decisive both for the onset of self-assembly and for the curvature of the surfactant films and thus for aggregate shape and phase behavior. ^{1}H, ^{2}H, and ^{17}O NMR have been used successfully to study the hydration of surfactant aggregates. The three by far most used approaches are ^{1}H (or ^{2}H) self-diffusion, ^{17}O quadrupole relaxation, and ^{2}H quadrupole splittings. We stress at the outset that a division into free and bound water molecules on which the concepts of hydration and hydration number are based is far from unambiguous, and furthermore this division is dependent on the physicochemical parameter monitored.

Water self-diffusion can be investigated using either ^{1}H or ^{2}H NMR. In studies of water diffusion as a function of surfactant concentration, one observes a change in behavior after the cmc. Thus the rate of decrease of D with surfactant concentration is relatively high below the cmc, while above the cmc it is much lower. Using a model with free water molecules and water molecules diffusing with monomeric and micellized surfactant molecules, one can directly conclude that the hydrocarbon–water contact is almost completely eliminated on micellization. For a number of ionic surfactants this type of investigation has yielded micellar hydration numbers in the range of 10–15 water molecules per surfactant molecule diffusing with the micelles [88,89]. Accounting for headgroup and bound counterion hydration, this suggests very little hydration of the alkyl chains. For nonionic surfactants of the ethylene oxide (EO) variety, self-diffusion work has demonstrated a decrease in hydration with increasing temperature, which is significant in discussions of the aggregation and phase behavior of these surfactants [90]. A striking observation is that at a given temperature, hydration is dependent only on the concentration of EO groups and not on whether the EO groups occur in spherical micelles, in rod micelles, or in poly(ethylene oxide).

Quadrupole effects in the NMR of water nuclei can be used to study hydration phenomena in exactly the same way as quadrupole effects of ions can be used to study counterion binding. ^{2}H and ^{17}O are rather unique in providing a detailed molecular picture of the hydration in these types of systems. The water exchange between different sites is (in all cases investigated so far) rapid, and often a simple two-site model is appropriate. The fact that the field gradients (being mainly of an intramolecular origin) are no influenced by changes in the system and the quadrupole coupling constants are known to a good approximation (0.222 MHz for ^{2}H and 6.7 MHz for ^{17}O) facilitates the analysis considerably.

Water NMR relaxation studies would be expected to be the richest source of information on hydration in colloidal and macromolecular systems in general. ^{1}H and ^{2}H NMR have been widely used for a long time, but unfortunately major problems of interpretation make many studies questionable [54]. The problems are associated with a complex relaxation mechanism of water protons and the very strong (often dominant) contributions from exchange with macromolecular ^{1}H or ^{2}H nuclei. These difficulties led Halle [91] to investigate in detail the possibilities of employing ^{17}O relaxation. Halle demonstrated that in interpreting multifield relaxation data using the two-step model of relaxation, it was indeed possible to map in detail the hydration of a large number of systems and that problems in previous work often were related to the neglect of the fast motions. The following are some general conclusions from water quadrupole relaxation and splitting studies:

1. Bound water molecules reorient quite rapidly, in fact less than a factor of 10 more slowly than in bulk water, and the degree of anisotropy of this motion is low.

2. One or two layers of water molecules are perturbed appreciably, and there are no indications of long-range hydration structures.

3. Fewer than two methylene groups in the alkyl chain are significantly exposed to water in surfactant micelles.

4. For lamellar phases there is often an "ideal swelling" behavior where water molecules in the vicinity of the surfaces have their orientation probability distribution almost constant on increasing the water content and where the addition of water only increases the amount of free water. There is no appreciable long-range ordering effect on the water molecules.

5. For liquid crystalline phases in general, even the water molecules in direct contact with the surfaces are very mobile and have quite low anisotropy in their motion.

G. Order–Disorder Transitions in Surfactant Continuous Microstructures

The formation of liquid crystalline phases at higher surfactant concentrations can often be considered as disorder–order phase transitions. For example, small micelles crystallize in a cubic lattice, and long cylindrical micelles are ordered in a two-dimensional hexagonal array. Analogously, the multiply connected bilayer structures, which is disordered in the liquid L_3 phase, crystallizes in some systems into a bicontinuous cubic phase at higher surfactant concentrations. This phase transition was studied by Balinov et al. [34] in the AOT–brine system. They measured the water and surfactant self-diffusion in the L_3 and cubic phases and found that the self-diffusion varied smoothly across the transition. The data are shown in Fig. 27. The results imply that the bilayer structures in the melted L_3 phase and the frozen cubic phase are very similar and that the phase transition can be regarded as a simple order–disorder transition. Analogous results have been obtained in a ternary system with nonionic surfactant [32], and the bilayer volume fraction at which the order–disorder transition occurs is thought to depend on the bilayer rigidity [92].

A similar transition was investigated in the cetylpyridinium chloride–sodium salicylate–water system, at a constant 1 : 1 molar ratio of surfactant to salt, by Monduzzi et al. [93]. In this system, which behaves very similarly to the binary cetylpyridinium salicylate–water system, the classical micellar L_1 phase is at higher concentrations and higher temperatures in equilibrium with the bicontinuous cubic V_1 phase. This unusual phase behavior results from an unusually low melting point of the hexagonal phase. The L_1–V_1 transition was studied by NMR self-diffusion and ^{14}N spin relaxation of the surfactant ion. Similar to the L_3–V_2 transition described above, the self-diffusion coefficients and the relaxation rates varied smoothly across the L_1–V_1 transition. From these results, including the relatively high value of the surfactant self-diffusion coefficient, it could be concluded that the aggregate structures in the two phases are similar and that the surfactant aggregate has to be multiply connected also in the L_1 phase. The interconnected cylinder structure, a description alternative to the minimal surface bilayer structure of the cubic phases (and microemulsions), provides a relevant description of the L_1 phase structure. The presence of a disordered interconnected cylinder network also appears to be quite general in binary nonionic surfactant–water systems.

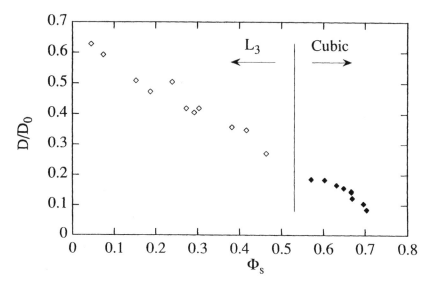

Figure 27   The relative self-diffusion coefficient (D/D_0) of water plotted as a function of the surfactant volume fraction in the L_3 and bicontinuous cubic phases of the AOT–water–NaCl system. Note the continuous variation of the self-diffusion coefficient across the L_3–cubic phase transition, demonstrating the structural similarity of the two phases. (Data taken from Ref. 34.)

ACKNOWLEDGMENTS

During our work on microemulsions, we have benefited greatly from interaction with several past and present members of the Division of Physical Chemistry 1 in Lund. In particular, Kozo Shinoda, Håkan Wennerström, Bengt Jönsson, and David Anderson have contributed much to our understanding of different aspects of the topic of the present review. We also acknowledge the important contributions of Peter Stilbs, who has been very generous in helping to develop our instrumentation.

REFERENCES

1. B. Lindman, U. Olsson, and O. Söderman, in *Dynamics of Solutions and Fluid Mixtures by NMR* (J.-J. Delpuech, ed.), Wiley, Chichester, 1995, pp. 346–392.
2. B. Lindman, O. Söderman, and H. Wennerström, in *Surfactant Solutions—New Methods of Investigation* (R. Zana, ed.), Marcel Dekker, New York, 1987, pp. 295–357.
3. B. Lindman and U. Olsson, Ber. Bunsenges. Phys. Chem. *100*:344 (1996).
4. R. R. Ernst, G. Bodenhausen, and A. Wokaun, *Principles of Nuclear Magnetic Resonance in One and Two Dimensions*, Clarendon Press, Oxford, 1986.
5. D. Canet, *Nuclear Magnetic Resonance*, Wiley, New York, 1996.
6. B. Lindman, K. Shinoda, U. Olsson, D. Anderson, G. Karlström, and H. Wennerström, Colloids Surf. *38*:205 (1989).
7. B. Lindman, N. Kamenka, T.-M. Kathopoulis, B. Brun, and P.-G. Nilsson, J. Phys. Chem. *84*:2485 (1980).
8. C. Chachaty, Prog. Nucl. Magn. Reson. Spectrosc. *19*:183 (1987).
9. O. Söderman and P. Stilbs. Prog. Nucl. Magn. Reson. Spectrosc. *26*:445 (1994).

10. O. Söderman and U. Olsson, in *Encyclopedia of Nuclear Magnetic Resonance* (D. M. Grant and R. K. Harris, eds.), Wiley, Chichester, 1996, pp. 3046–3057.
11. P. T. Callaghan, C. M. Trotter, and K. W. Jolley, J. Magn. Reson. *37*:247 (1980).
12. P. Stilbs, Prog. Nucl. Magn. Reson. Spectrosc. *19*:1 (1987).
13. P. T. Callaghan, *Principles of Nuclear Magnetic Resonance Microscopy*, Clarendon Press, Oxford, 1991.
14. E. L. Hahn, Phys. Rev. *80*:580 (1950).
15. D. C. Douglass and D. W. McCall, J. Phys. Chem. *62*:1102 (1958).
16. E. O. Stejskal and J. E. Tanner, J. Chem. Phys. *42*:288 (1965).
17. R. L. Vold, J. S. Waugh, M. P. Klein, and D. E. Phelps, J. Chem. Phys. *48*:3831 (1968).
18. J. E. Tanner, Thesis, Univ. of Wisconsin, Madison, WI, 1966.
19. P. Callaghan and A. Coy, Phys. Rev. Lett. *68*:3176 (1992).
20. P. T. Callaghan, A. Coy, T. P. J. Halpin, D. MacGowan, J. K. Packer, and F. O. Zelaya, J. Chem. Phys. *97*:651 (1992).
21. P. T. Callaghan and A. Coy, in *NMR Probes of Molecular Dynamics* (P. Tycko, ed.), Kluwer, Dordrecht, The Netherlands, 1993, pp. 489–523.
22. B. Balinov, B. Jönsson, P. Linse, and O. Söderman, J. Magn. Reson. A *104*:17 (1993).
23. B. Balinov, O. Urdahl, O. Söderman, and J. Sjöblom, Colloids Surf. *82*:173 (1994).
24. B. Balinov, O. Söderman, and T. Wärnheim, J. Am. Oil Chem. Soc. *71*:513 (1994).
25. B. Balinov, O. Söderman, and J. C. Ravey, J. Phys. Chem. *98*:393 (1994).
26. P. T. Callaghan, K. W. Jolley, and R. Humphrey, J. Colloid Interface Sci. *93*:521 (1983).
27. P. T. Callaghan, A. Coy, D. MacGowan, K. J. Packer, and F. O. Zelaya, Nature *351*:467 (1991).
28. B. Jönsson, H. Wennerström, P. Nilsson, and P. Linse, Colloid Polym. Sci. *264*:77 (1986).
29. P. N. Pusey, in *Liquids, Freezing and the Glass Transition (Les Houches Session LI)* (D. Levesque, J.-P. Hansen, and J. Zinn-Justin, eds.), Elsevier, Amsterdam, 1990, pp. 763–942.
30. M. Medina-Noyola, Phys. Rev. Lett. *60*:2705 (1988).
31. O. Söderman, E. Hansson, and M. Monduzzi, J. Colloid Interface Sci. *141*:512 (1991).
32. U. Olsson and P. Schurtenberger, Langmuir *9*:3389 (1993).
33. D. M. Anderson and H. Wennerström, J. Phys. Chem. *94*:8683 (1990).
34. B. Balinov, U. Olsson, and O. Söderman, J. Phys. Chem. *95*:5931 (1991).
35. P. Ström and D. M. Anderson, Langmuir *8*:691 (1992).
36. U. Olsson, B. Balinov, and O. Söderman, in *The Structure and Conformation of Amphiphilic Membranes* (Springer Proc. Phys., Vol. 66) (R. Lipowsky, and K. Kremer, eds.), Springer, Berlin, 1992, pp. 287–290.
37. M. Jonströmer, B. Jönsson, and B. Lindman, J. Phys. Chem. *95*:3293 (1991).
38. K. Shinoda, in *Solvent Properties of Surfactant Solutions* (K. Shinoda, ed.), Marcel Dekker, New York, 1967, pp. 27–63.
39. K. Shinoda and S. E. Friberg, *Emulsions and Solubilization*. Wiley-Interscience, New York, 1986.
40. M. Kahlweit and R. Strey, Angew. Chem., Int. Ed. Engl. *24*:654 (1985).
41. M. Kahlweit, R. Strey, D. Haase, H. Kunieda, T. Schmeling, B. Faulhaber, M. Borkovec, H.-F. Eicke, G. Busse, F. Eggers, T. Funck, H. Richmann, L. Magid, O. Söderman, P. Stilbs, J. Winkler, A. Dittrich, and W. Jahn, J. Colloid Interface Sci. *118*:436 (1987).
42. U. Olsson, K. Shinoda, and B. Lindman, J. Phys. Chem. *90*:4083 (1986).
43. U. Olsson, K. Nagai, and H. Wennerström, J. Phys. Chem. *92*:6675 (1988).
44. P. Guering and B. Lindman, Langmuir *1*:464 (1985).
45. J. O. Carnali, A. Ceglie, B. Lindman, and K. Shinoda, Langmuir *2*:417 (1986).
46. K. Shinoda, M. Araki, A. Sadaghiani, A. Khan, and B. Lindman, J. Phys. Chem. *95*:989 (1991).
47. B. Lindman, K. Shinoda, M. Jonströmer, and A. Shinohara, J. Phys. Chem. *92*:4702 (1988).
48. R. Skurtveit and U. Olsson, J. Phys. Chem. *96*:8640 (1992).
49. M. Jonströmer, U. Olsson and W. O. Parker, Langmuir *11*:61 (1995).
50. B. Lindman, O. Söderman, and H. Wennerström, Ann. Chim. (Rome) *77*:1 (1987).
51. B. Lindman, O. Söderman, and P. Stilbs, in *Surfactants in Solution*, (Vol. 7 K. L. Mittal, ed.), Plenum, New York, 1989, pp. 1–24.

52. J. Ulmius and H. Wennerström, J. Magn. Reson. *28*:309 (1977).

53. U. Olsson, O. Söderman, and P. Guéring, J. Phys. Chem. *90*:5223 (1986).

54. B. Halle and H. Wennerström, J. Chem. Phys. *75*:1928 (1981).

55. H. Wennerström, G. Lindblom, and B. Lindman, Chem. Scripta *6*:97 (1974).

56. H. Wennerström, B. Lindman, O. Söderman, T. Drakenberg, and J. B. Rosenholm, J. Am. Chem. Soc. *101*:6860 (1979).

57. A. Abragam, *The Principles of Nuclear Magnetism*, Clarendon Press, Oxford, UK, 1961.

58. M. S. Leaver, U. Olsson, H. Wennerström, and R. Strey, J. Phys. II *4*:515 (1994).

59. O. Söderman, H. Walderhaug, U. Henriksson, and P. Stilbs, J. Phys. Chem. *89*:3693 (1985).

60. O. Söderman, U. Henriksson, and U. Olsson, J. Phys. Chem. *91*:116 (1987).

61. O. Söderman, G. Carlström, U. Olsson, and T. C. Wong, J. Chem. Soc., Faraday Trans. 1 *84*:4475 (1988).

62. B. Halle, J. Chem. Phys. *94*:3150 (1991).

63. B. Lindman and H. Wennerström, J. Phys. Chem. *95*:6053 (1991).

64. F. Noack, Prog. Nucl. Magn. Reson. Spectrosc. *18*:171 (1986).

65. R. G. Laughlin, *The Aqueous Phase Behavior of Surfactants*, Academic, New York, 1994.

66. R. G. Laughlin, in *Surfactants* (T. Tadros, ed.), Academic, New York, 1984, pp. 53–82.

67. K. Shinoda, Prog. Colloid Polym. Sci. *68*:1 (1983).

68. G. J. T. Tiddy, Phys. Rep. *57*:1 (1980).

69. E. Marques, A. Khan, M. G. Miguel, and B. Lindman, J. Phys. Chem. *97*:4729 (1993).

70. G. T. Dimitrova, T. F. Tadros, P. F. Luckham, and M. R. Kipps, Langmuir *12*:315 (1996).

71. A. Khan, K. Fontell, G. Lindblom, and B. Lindman, J. Phys. Chem. *86*:4266 (1982).

72. N. O. Persson, K. Fontell, and B. Lindman, J. Colloid Interface Sci. *53*:461 (1975).

73. K. Fontell, Mol. Cryst. Liq. Cryst. *63*:59 (1981).

74. T. Bull and B. Lindman, Mol. Cryst. Liq. Cryst. *28*:155 (1974).

75. P. O. Eriksson, A. Khan, and G. Lindblom, J. Phys. Chem. *86*:387 (1982).

76. G. Lindblom, K. Larsson, L. Johansson, K. Fontell, and S. Forsén, J. Am. Chem. Soc. *101*:5465 (1979).

77. P.-O. Eriksson, G. Lindblom, and G. Arvidson, J. Phys. Chem. *89*:1050 (1985).

78. L. B.-Å. Johansson and O. Söderman, J. Phys. Chem. *91*:5275 (1987).

79. H. Wennerström and J. Ulmius, J. Magn. Reson. *23*:431 (1976).

80. H. Wennerström, Chem. Phys. Lett. *18*:41 (1973).

81. L.-P. Hwang, P.-L. Wang, and T. C. Wong, J. Phys. Chem. *92*:4753 (1988).

82. P. Stilbs and B. Lindman, J. Phys. Chem. *85*:2587 (1981).

83. B. Lindman and P. Stilbs, in *Proc. Int. Sch. Phys. "Enrico Fermi"* (V. Degiorgio and M. Corti, eds.), North-Holland, Amsterdam, 1985, pp. 94–121.

84. B. Lindman, M. C. Puyal, N. Kamenka, R. Rymdén, and P. Stilbs, J. Phys. Chem. *88*:5048 (1984).

85. D. Bratko and B. Lindman, J. Phys. Chem. *89*:1437 (1985).

86. M. Jansson and P. Stilbs, J. Phys. Chem. *89*:4868 (1985).

87. B. Lindman, in *NMR of Newly Accessible Nuclei* (P. Laszlo, ed.), Academic, New York, 1983, pp. 193–228.

88. B. Lindman, H. Wennerström, H. Gustavsson, N. Kamenka, and B. Brun, Pure Appl. Chem. *52*:1307 (1980).

89. B. Lindman, M. C. Puyal, N. Kamenka, R. Rymdén, and P. Stilbs, J. Phys. Chem. *88*:5048 (1984).

90. P. G. Nilsson, H. Wennerström, and B. Lindman, J. Phys. Chem. *87*:1377 (1983).

91. B. Halle, Thesis, Lund University, Lund, Sweden, 1981.

92. J. Bruinsma, J. Phys. II (France) *2*:425 (1992).

93. M. Monduzzi, U. Olsson, and O. Söderman, Langmuir *9*:2914 (1993).

94. P. Stilbs and B. Lindman, Prog. Colloid Polym. Sci. *69*:39 (1984).

95. B. Lindman, T. Ahlnäs, O. Söderman, H. Walderhaug, K. Rapacki, and P. Stilbs, Faraday Discuss. Chem. Soc. *76*:317 (1983).

11

Rheological Properties of Microemulsions

Michael Gradzielski and Heinz Hoffmann
Universität Bayreuth, Bayreuth, Germany

I. INTRODUCTION

One of the most easily observed macroscopic properties of colloidal systems is their flow behavior, and it may range anywhere between a low viscous fluid and a gel state. The rheological properties and, in particular, the viscosity of microemulsions are macroscopically observable parameters that characterize a given system. Of course, the viscosity is a relevant quantity for many practical applications of microemulsions. For instance, pumping such systems might be of interest in their application, and here viscosity plays an important role.

As an example the viscosity of microemulsions is an important factor in their ability to recover oil [1], which means that knowledge of it and, even more so, the capability to control it are important in the process of tertiary oil recovery. Therefore, it has been studied experimentally in some detail with respect to surfactant systems that are of interest for the oil recovery process [2–10]. Furthermore, there also exists theoretical work that models microemulsion viscosity as a function of the phase composition and phase type in order to predict properties of microemulsions under realistic conditions [11,12].

Another possible application where the typically low viscosity of microemulsions is very useful is in the preparation of alternative fuels for diesel engines. Such hybrid fuel microemulsions containing vegetable oil [13] and alcohols, with 1-butanol [14] or a lower trialkylamine [15] acting as surfactant attracted interest some years ago, and such systems were tested for their practical application [16]. Similar systems containing triglyceride, aqueous ethanol, and 1-butanol [17] or long-chain fatty alcohols [18] were also studied for the same purpose.

Moreover, viscosity is also an important property when W/O microemulsions are used for hydraulic fluids [19]. All this shows that rheological properties are an important physical quantity for a variety of applications of microemulsions.

II. SOME FUNDAMENTALS REGARDING RHEOLOGICAL PROPERTIES

A. Viscosity

All conventional fluids (excluding the superfluid state) will exhibit some friction during the flow process. This means that energy E is dissipated in the course of flow of a fluid, and its dissipation per unit volume V may be used to define a quantity—the viscosity η—that is specific for this fluid and characterizes this behavior:

$$\frac{d(dE/dV)}{dt} = \eta\left(\frac{dv}{dy}\right)^2 \tag{1}$$

where dv/dy is the velocity gradient or shear rate. The viscosity defined in this manner is also called the dynamic viscosity. It might be mentioned here that sometimes one also uses the kinematic viscosity v, which is related to the dynamic viscosity η simply by dividing η by the density ρ of the fluid:

$$v = \eta/\rho \tag{2}$$

Very frequently one will not use the viscosity itself, as defined above, but relate this quantity to the solvent viscosity and/or the concentration of the dispersed material. By doing so one obtains quantities that depend less on the particular system and the given experimental conditions; instead, more universal quantities are deduced that are more easily compared with other results. In the following, we briefly recall the most common definitions of such quantities.

In the case of dispersions or suspensions, the relative viscosity η_r, defined as the viscosity η of the system divided by the viscosity η_s of the pure solvent, is often used:

$$\eta_r = \eta/\eta_s \tag{3}$$

This is a quantity that is more specific for the dispersed material and the changes that occur in the viscosity of a given system (e.g., as a function of the concentration of the dispersed material). Furthermore, one may define the specific viscosity $\eta_{sp} = \eta_r - 1$, which gives the relative increase of viscosity due to the dispersed medium. Finally, the intrinsic viscosity $[\eta]$, given by

$$[\eta] = \lim_{c \to 0}\left(\frac{\eta_{sp}}{c}\right) \tag{4}$$

is another characteristic quantity that relates the relative increase of viscosity to the concentration c of the dispersed material. It is obtained by extrapolation to zero concentration, which means that interactions between particles can be neglected in the interpretation of this value. $[\eta]$ is very sensitive to the shape of the particles, and in the characterization of polymers in solution this quantity is used for a determination of shape and/or conformation of the chains of the polymer molecules [20–22]. However, for self-assembling systems (such as microemulsions) the extrapolation to zero concentration is not always straightforward, for here the shape might change as a function of the concentration and also care has to be taken to account properly for the cmc (critical micelle concentration).

B. Theoretical Expressions for the Viscosity

Viscosity, as it is observed for a given system, depends largely on the structure, i.e., the type of aggregates that are present, on their interactions, and on the concentration of the system. In principle, rheological measurements contain information regarding these parameters, and dynamic rheological experiments can also yield information on the dynamics of the given system.

As stated above, the viscosity of a system depends sensitively on the shape and interactions of the particles dispersed and can therefore be used to deduce information regarding these parameters or to monitor microstructural changes in a microemulsion system. In order to do so, one has to compare the experimental data to theoretical expressions that give the viscosity expected for certain model systems. Therefore, we give a short summary of some fundamental expressions that can be used to explain the viscosity of particles suspended in a fluid medium.

Let us first recall the general fact that whenever a second phase is dispersed in a solvent this should lead to an increase in the bulk viscosity. A solid theoretical description of this effect was first given by Einstein, who derived a formula that describes the increase in viscosity as a function of the volume fraction Φ of dispersed material for the case of compact spherical particles that have a sticky surface (no-slip condition) [23,24]:

$$\eta_r = 1 + 2.5\Phi \tag{5}$$

Here we have used the volume fraction Φ, but often it is more useful to generalize this equation [as well as the following ones, Eqs. (6)–(10)] and substitute Φ by $f\Phi$, where f is a factor that takes into account deviations from the ideal case that may occur. These deviations could arise for two different reasons: (1) The effective volume fraction might be larger than that of the "dry" aggregate because of solvent molecules that are relatively tightly bound to the aggregate (solvation shell) and (2) there may be deviations from the spherical shape—a question that we address in more detail somewhat later. In any case, if one wants to obtain a self-consistent picture of a given system, this factor f might be an important constant that has to be in agreement with other structural parameters as they are obtained from other experiments (such as scattering experiments and sedimentation experiments).

Einstein's formula is valid for hard spheres, but it is modified for the case of a dispersion of liquid spheres in another liquid medium. For that case Taylor obtained the expression [25]

$$\eta_r = 1 + \Phi\left(\frac{\eta_s + 2.5\eta_d}{\eta_s + \eta_d}\right) \tag{6}$$

where η_s is the viscosity of the solvent and η_d that of the dispersed phase. This expression reduces to Einstein's equation for the case of hard spheres (corresponding to $\eta_d \to \infty$) and will yield lower values of η_r for finite viscosities η_d (with a lower limit of a factor of 1 instead of 2.5 for the case where $\eta_s \gg \eta_d$).

It might be noted here that accordingly the intrinsic viscosity $[\eta]$ as defined in Eq. (4) will be given by $2.5/\rho$ for compact spheres, where ρ is the density of these spheres. However, Eq. (5) is valid only for relatively dilute systems (volume fraction < 0.02). For more concentrated systems this expression has to be augmented by higher order terms of the volume fraction, i.e., one has to expand Eq. (5) by higher orders of Φ, as

$$\eta_r = 1 + 2.5\Phi + B\Phi^2 + C\Phi^3 + \cdots \tag{7}$$

But this is not a simple task theoretically (especially since the hydrodynamic interactions that are important for higher concentrations are very difficult to account for), and even for the second-order term B quite different values can be found in the literature (e.g., 6.2 [26] and 14.1 [27]). However, for the description of concentrated dispersions, such a virial expression is not really very useful because it would require a fairly large number of coefficients that are theoretically not easily accessible. For spheres that are not correlated spatially the hydrodynamic interactions can be taken into consideration and yield the following expression due to Saito [28]:

$$\frac{\eta_r - 1}{\eta_r + 1.5} = \Phi \tag{8}$$

Because of the inherent disadvantage of the virial expression, more concentrated systems are customarily described by other functions that more appropriately take into account the steep increase in viscosity that occurs when the hard-sphere system approaches the concentrated regime, especially when it approaches the maximum packing condition [which is also not at all reflected in Eq. (8)]. One widely used expression of this type (that can be looked upon as a generalized version of the Einstein equation with the required divergence at the maximum packing volume fraction Φ_m, since at this point a solid is formed) is the Dougherty–Krieger formula [29,30]

$$\eta_r = \left(1 - \frac{\Phi}{\Phi_m}\right)^{-2.5\Phi_m} \tag{9}$$

A similar equation but with the exponent of -2 was advanced by Quemada [31]. For instance, this type of equation was shown to describe well the experimental data obtained on relatively monodisperse dispersions of hydrophobically modified silica particles suspended in cyclohexane [32], i.e., systems that can well be regarded as good model systems for a hard-sphere fluid. Typical values for Φ_m are 0.6–0.7 [32], i.e., values that correspond to the packing fraction of a cubic array of spheres.

There also exist expressions that describe the increased viscosity at high volume fractions in purely empirical terms. One such expression was advanced by Thomas [33]:

$$\eta_r = 1 + 2.5\Phi + 10.05\Phi^2 + 0.00273 \ \exp[16.6\Phi] \tag{10}$$

It was shown to describe well the viscosity of spherical objects such as glass beads and polystyrene latices [33] and was also applied successfully for the description of the viscosity of O/W microemulsion droplets [34].

Up to now we have always assumed the presence of spherical particles. However, microemulsions (like self-aggregating systems in general) may well contain anisometric particles, such as prolate, oblate, or rodlike structures. The viscosity of a dilute solution (here dilute means that the number density ^1N of the aggregates fulfills the condition $^1N < 1/L^3$ with L as the length of the rods) of rigid rods can be described by the expression of Doi and Edwards [35]:

$$\eta_r = 1 + {}^1NL^3 = 1 + \frac{4c_g L^2}{\pi d^2 \rho} \tag{11a}$$

where c_g is the concentration in mass per unit volume, ^1N the number density of the rods, L their length, and d the diameter of the rods.

Doi and Edwards [35] also treated the semidilute case, i.e., $1/L^3 <^1 N < 1/dL^2$, and for this situation they deduced the following expression for the relative viscosity:

$$\eta_r = \frac{96c_g^3 L^6}{5\pi^2 \rho^3 d^6 \ \ln(L/d)} \tag{11b}$$

Similarly, asymptotic expressions for rigid ellipsoids of revolution were derived that are valid for axial ratios larger than 10 and are given by

$$\eta_r = 1 + v\Phi \tag{12}$$

with

$$v = \frac{(a/b)^2}{5[\ln(2a/b) - 0.5]} + \frac{(a/b)^2}{15[\ln(2a/b) - 1.5]} + \frac{14}{15} \tag{12a}$$

for prolate ellipsoids and

$$v = \frac{(16/15)(a/b)}{\arctan(a/b)} \tag{12b}$$

for oblate ellipsoids, where a is the main axis and b the minor axis of the ellipsoid [36,37].

A very general way to account for the effects of either increased concentration or the particle shape is given by means of the Huggins equation [38],

$$\eta_{sp}/c = [\eta] + k_H[\eta]^2 c + \cdots \tag{13}$$

which is frequently used in the analysis of viscosity data from polymer solutions [3] but can, of course, as well be applied to all other colloidal systems. It is a way of determining the intrinsic viscosity $[\eta]$, and in addition one obtains the Huggins constant k_H [e.g., for random coils k_H is typically in the range 0.2–0.6; for hard spheres k_H is related to the coefficient B of Eq. (7) via $k_H = B/6.25$]. Both parameters yield information regarding the particle structure and can be compared to theoretical predictions for a given model [39].

A way to describe the relative viscosity of colloidal systems up to high concentrations is given by the Mooney equation [40]

$$\eta_r = \exp\frac{a\Phi}{1 - k\Phi} \tag{14}$$

Here a is a factor related to the intrinsic viscosity (typically in the range of 1–10) and k is a "crowding factor" (typically in the range 1.3–2) that takes into account the maximum packing fraction of the system.

From all this one may judge that viscosity measurement is an important tool for determining the structure of aggregates in solution, in particular the shape and elongation of anisometric particles.

The situation becomes more complicated if the systems considered are no longer composed of isolated aggregates but instead contain fairly large and interconnected structures. Such systems are discussed in some detail in Sec. III.C.

Microemulsions are usually Newtonian fluids; i.e., their viscosity is independent of the applied shear stress (or equivalently of the shear rate)—in other words the observed shear rate is proportional to the applied shear stress. For such a situation the applied shear stress does not affect the equilibrium structure, and for this case viscosity measurements can yield information regarding this equilibrium structure.

However, there might also exist more complicated situations when the microemulsion contains elongated structures that interact strongly. In this case non-Newtonian behavior might be observed that one can describe by introducing a complex viscosity η^* that contains a viscous part η' and an elastic part η'' (or equivalently a complex shear modulus G^*, where G' describes the elastic contribution and G' the viscous contribution). For details regarding the description of viscoelastic systems we refer to the specialized literature [41–43].

Deviations from Newtonian behavior mean that here the equilibrium structure becomes distorted. This is expected to happen for the case where the longest structural relaxation time τ becomes equal to or greater than $1/\dot{\gamma}$ ($\dot{\gamma}$ = shear rate). In this case deviations from Newtonian behavior are observed; i.e., the viscosity is no longer a constant but depends on other parameters (shear gradient, time, etc.). Some examples of such behavior are discussed in Secs. III.B and III.C. However, one can state quite generally that such behavior is the exception, whereas typical microemulsions show very simple (i.e., normally Newtonian) rheological properties.

III. EXPERIMENTAL RESULTS

In this section we consider some experimental results for the viscosity obtained for typical microemulsions. Here we mean microemulsions of the classical Winsor I (oil-in-water, O/W), II (water-in-oil, W/O), or III (bicontinuous, middle-phase microemulsion) types [44,45], where the O/W and W/O systems contain isolated aggregates. Rheological measurements can yield information on the structure of the microemulsion and in particular, if they are present, the shape and interactions of individual aggregates. In addition, it will be of interest to see how the microstructural transitions (from one Winsor type to another) influence the macroscopic viscosity. Such transitions can be observed either by changing the temperature (for nonionic surfactants) or the salinity (for ionic surfactants).

In the following, we first discuss systems composed of isolated aggregates and then bicontinuous microemulsions together with the changes observed as one crosses from one structural type of microemulsion to another. Finally, in Sec. III.D some examples of microemulsions are discussed whose rheological behavior is modified by the addition of polymeric compounds that can cross-link the microemulsion structure and thereby increase the viscosity considerably.

A. Oil-in-Water Microemulsions (Winsor I)

To begin with, we consider microemulsions that contain individual aggregates as they are typically observed in Winsor I and II systems. In principle, these systems can be considered as oil-swollen micelles (Winsor I) or water-swollen reverse micelles (Winsor II), where, of course, the distinction between a microemulsion droplet and a (reverse) micelle that contains solubilized material is somewhat arbitrary [46].

An example of such a simple O/W system was investigated with the nonionic surfactant pentaethylene glycol dodecyl ether ($C_{12}E_5$) together with hydrocarbon and water, i.e., the simplest possibility to form a microemulsion (surfactant, oil, water). This system had been studied in some detail before [47–49], and it is well established that it contains spherical aggregates close to the emulsification failure, i.e., when it is saturated with oil.

Viscosity measurements on this O/W microemulsion showed that the extrapolated intrinsic viscosity $[\eta]$ was somewhat higher than that expected for hard spheres but was still in agreement with spherical droplets if one assumed some hydration of the ethylene oxide headgroups, an effect that is to be expected because the water-soluble poly(ethylene oxide) is very hydrophilic [50]. For a given alkane/surfactant ratio, almost perfect hard-sphere behavior is obtained under this assumption close to the solubilization boundary

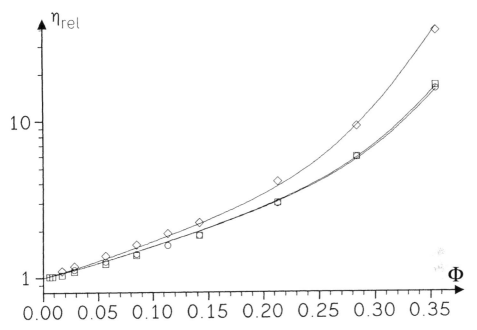

Figure 1 Relative viscosity η_r at 25°C as a function of the volume fraction Φ (of surfactant plus decane) for microemulsions of constant surfactant/decane molar ratio of 3.2 : 1. ($\square$) Pure TDMAO; ($\bigcirc$) 7.5 mol% TTABr; ($\diamondsuit$) 7.5 mol% SDS [solid lines fitted according to Eq. (10)]. (From Ref. 56.)

(from which one can move away by raising the temperature). At higher temperatures, i.e., by going away from the solubilization boundary, a moderate increase of the intrinsic viscosity is observed, whereas the slope of η_{sp}/c increases sharply. This has been interpreted as indicating that the spherical structure of the droplets remains unaltered but the attractive interactions between the droplets increase [50]. Again it should be noted here that viscosity measurements cannot distinguish such attractive interactions from an anisometric growth of the aggregates. In order to distinguish the different possibilities, additional experimental methods are required.

Further investigations [51] on a similar system containing about equal amounts of decane and surfactant were well in agreement with the hard-sphere model. However, owing to the relatively complex and bulky structure of the polyglycol headgroup, the authors found that at low concentrations the hydrodynamic radius of the aggregates and at higher concentrations their (somewhat smaller) hard-sphere radius were relevant to describe the viscosity; i.e., at higher concentrations the droplets become effectively smaller. In essence, similar results had been obtained for the microemulsion system tetradecyldimethyl amine oxide (TDMAO)–decane–water [34]. However, here the headgroup is less complex (in particular less bulky) and the viscous properties are characterized by a single radius (see Fig. 1).

Similar results, which showed behavior of the viscosity according to a hard-sphere model, have been observed for a variety of O/W microemulsion systems such as Brij 96–butanol–hexadecane–water [52], Tween 60–Span 80–glycerol–paraffin–water [53], and Brij 96–pentanol–hexadecane–water [54]. Again the extrapolated intrinsic viscosity is found to be about 60% greater than expected according to Einstein's equation and the Huggins coefficient k_H [cf. Eq. (13)] to be about 1.8 [52]. This shows that such behavior will quite generally be observed for droplet-type microemulsions irrespective of the inter-

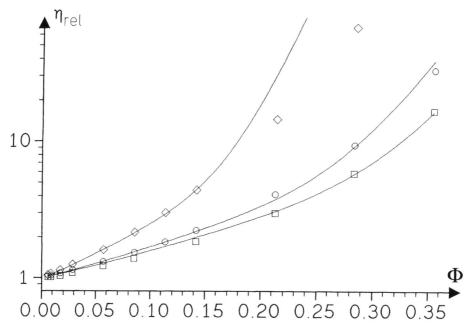

Figure 2 Relative viscosity η_r at 25°C as a function of the volume fraction Φ (of surfactant plus decane) for microemulsions of constant surfactant/decane molar ratio of 3.2 : 1. ($\square$) Pure TDMAO; ($\bigcirc$) 12.5 mol% Brij 76; ($\Diamond$) 7.5 mol% Brij 78 [solid lines fitted according to Eq. (10)]. (From Ref. 56.)

action potential that exists between the droplets (provided that it is not too strongly attractive, thereby leading to temporary association of the droplets). For instance, no difference has been observed for microemulsions that were originally uncharged and became charged by admixing ionic surfactants [55] as depicted in Fig. 1. One still observes normal hard-sphere behavior, and the differences that exist between the different systems are due to the different degrees of hydration of the headgroups of the different surfactants.

It should be remarked that typically the experimental viscosity data of O/W microemulsion droplets are well described by the corresponding theoretical expressions [in particular Eqs. (5)–(10)] given in Sec. II.B. The key parameter in these expressions is the hard-sphere volume fraction, which may be larger (and therefore also a higher viscosity will be observed) than the "real" volume fraction of the "dry" aggregates, since in the viscosity experiment one effectively probes the volume of the aggregates as they move around, which also includes a possible hydration shell. If one now fits a theoretical expression to the experimental data one obtains as a fit parameter a factor f [as already discussed in connection with Eq. (5)] that relates this effective (hard-sphere) volume fraction to the real volume fraction. This factor, therefore, contains quantitative information regarding the solvent molecules that are bound to the aggregates, i.e., viscosity measurements can yield a more detailed picture of the structure of the aggregates investigated.

Such an interpretation has, for instance, been applied to microemulsions based on the zwitterionic surfactant TDMAO, which was successively replaced by nonionic surfactants of the Brij type (polyglycol alkyl ethers). Here the effective volume fraction increases proportionally to the amount of polyglycol headgroups present, as these headgroups are fairly strongly hydrated. This is reflected in increasingly higher viscosities at a given volume fraction as demonstrated in Fig. 2, where Brij 76 and Brij 78 are oleyl moieties with 10 and 20 EO groups, respectively (the numbers of EO groups are average values) [56]. Furthermore it can be seen that the data of the Brij 78 system are no longer described well in the high concentration regime. To account for this deviation one would have to invoke a second particle radius (or equivalently a second factor f) that has to be used for the concentrated regime, as discussed above and applied in Ref. 51. Similar viscosity results were obtained for a mixture of Tween 60 and Span 80 (both nonionic surfactants with ethylene oxide headgroups), and the viscosity data conform well to hard-sphere behavior if one assumes a hydration of 3.6 water molecules per ethylene oxide group [53].

However, there is obviously a problem with automatically assigning an increased viscosity to an effective higher volume fraction, since it could also be due to the presence of anisometric aggregates. Because of this, such an interpretation should be supported by other experimental methods that yield information regarding the shape of the aggregates investigated (such as scattering experiments and electrical birefringence measurements). Provided such additional information has been acquired, viscosity measurements can yield valuable additional information regarding the state of solvation or possible attractive interactions that could lead to the formation of transient clusters (see also Sec. III.B).

B. Water-in-Oil Microemulsions (Winsor II)

All the systems discussed so far have been of the O/W type, but a similarly simple viscosity behavior of the solid-sphere type is normally also observed for W/O microemulsions. The simple type of microemulsion formed by a nonionic surfactant such as $C_{12}E_5$, hydrocarbon, and water (which has been described in Sec. III.A for the O/W case) was also investigated for the W/O situation. [It should be noted here that in such a system the inversion from O/W (Winsor I) to W/O (Winsor II) can simply be obtained by raising the temperature.] These investigations yielded results similar to those of the O/W case [57]; i.e., the intrinsic viscosity agrees well with spherical droplets that contain a solvation shell. Going away from the solubilization boundary (by lowering the temperature) changes the intrinsic viscosity only somewhat but significantly increases the viscosity at higher concentrations, which has been interpreted again as due to increasing attractive interactions [50].

The influence of the solvation shell on the viscosity data was also reported in a study on the system sodium alkylbenzenesulfonate (NaDBS)–hexanol–water–xylene. Here it was observed that the degree of solvation increased with increasing NaDBS/water ratio. This was attributed to the fact that with increasing ratio of surfactant layer thickness to droplet core the relative amount of bound solvent molecules increases, thereby increasing the hydrodynamic volume of the droplets [58]. This shows that viscosity measurements are very sensitive to the state of solvation of colloidal particles and can be used to extract information about this property.

Another very typical surfactant for the formation of W/O microemulsions is AOT [sodium bis(2-ethylhexyl) sulfosuccinate], and such systems have been intensively studied. For instance, viscosity of the hard-sphere type has recently been reported for a W/O microemulsion made up from water, AOT and hydrocarbon, and the relative viscosity equals those of latex or silica spheres of corresponding volume fraction [59]. However, here the situation is somewhat more complicated, since at higher concentrations and higher temperatures attractive interactions lead to a reversible aggregation of the droplets that has to be taken into account [60]. This clustering process (which has also been evidenced via other methods such as dielectric permittivity measurements [61] and dynamic Kerr effect experiments [62] leads to an increase in the relative viscosity with rising temperature.

The temperature dependence of the second coefficient in a virial expansion of the viscosity [Eq. (7)] can now be written in terms of the Gibbs free energy ΔG per mole of contact points. Doing so, one may write the relative viscosity as (again $f\Phi$ can be substituted for Φ as discussed in Sec. II.B) [60]

$$\eta_r = 1 + 2.5\Phi + 2.5(1 + a_2)\Phi^2 + O(\Phi^3) \tag{15}$$

with

$$a_2 = a_{2,0} + a_{2,1} \ \exp(-\Delta G/kT)$$

where the term $a_{2,0}$ is due to the hard-sphere interactions and the second term arises from the additional attractive interactions between the droplets that lead to temporary aggregation. Using $\Delta G = \Delta H - T \Delta S$ and assuming the temperature dependence of enthalpy ΔH and entropy ΔS to be negligible, it is evident that the entropy contribution can be included in the prefactor of the exponential and all the temperature dependence is then exclusively due to the enthalpy term. Via this approach, i.e., by carrying out viscosity measurements at different temperatures, the binding enthalpy for this clustering process in the ACT–water–isooctane system has been determined to be 89 and 121 kJ/mol for systems characterized by molar water/surfactant ratios of 25 and 35, respectively, i.e., the binding is stronger the larger the W/O droplets. This is, of course, equivalent to saying that the attractive interactions increase with increasing droplet size—a fact that was observed before by means of scattering experiments [63,64], and similar values for the binding enthalpy have been deduced from dielectric permittivity experiments on the identical system [61].

For higher volume fractions, the viscosity data were more recently interpreted by a similar model of aggregating droplets, and such a model is able to account for the experimental data up to volume fractions of 0.4 [65].

In general, reverse systems with AOT as surfactant have been studied intensively. For instance, in the system AOT–D_2O–n-decane it was found that the viscosity of the microemulsion first rapidly increases with increasing molar ratio $R = $[water]/[AOT], reaches a maximum around $R = 8$, and then decreases again upon further increase in R [66]. Here it is interesting to note that the temperature dependence of the maximum viscosity is reverse, i.e., it increases with increasing temperature. For a volume fraction of 0.4 the maximum relative viscosity was found to be close to 60, whereas for very low or high R values relative viscosities of less than 10 were observed, in agreement with a simple droplet picture. However, a dilution series showed that the Einstein coefficient [Eq. (5)] was always close to the expected 2.5 and slowly increased with increasing R [67]. This slow increase is due to the penetration of solvent molecules into the aggregates as has been observed before [68]. The large increase in viscosity in the intermediate R range for concentrated microemulsions is ascribed to increased attractive interactions due to a molecular rearrangement of the surfactant/water interface, which also explains the variation of the percolation transition with the molar ratio R.

In the AOT–water–oil system, when the temperature is increased sufficiently a cloud point (critical point) is reached. At temperatures well below this transition temperature the viscosity data are in agreement with a simple hard-sphere model. Upon approaching the critical point one first notices a relatively moderate increase of viscosity by about a factor of 4 followed by a critical divergence of viscosity very close to the cloud point. The critical divergence of such a system (with decane as oil) at the critical temperature was studied and was shown to scale almost Ising-like according to $\eta \sim [(T_c - T)/T_c]^{-\gamma_c}$, with a critical exponent of 0.03 [69].

As just discussed for AOT as surfactant, one observes spherical droplets in the W/O microemulsion. This situation can change if one changes the metal counterion of the surfactant, i.e., replaces the Na^+ ion with other metal ions. In such a study it was found that for identical water/surfactant molar ratios differently shaped aggregates are present. Whereas for Na^+, Mg^{2+}, and Ca^{2+}, spherical aggregates are found, for transition metal ions such as Co^{2+}, Ni^{2+}, Cu^{2+}, and Zn^{2+}, rodlike aggregates are formed. Viscosity measurements have been useful in differentiating these different aggregation forms [70]. For surfactants containing transition metal counterions, significantly higher intrinsic viscosities were observed that could be used to calculate the axial ratio according to Eq. (12a), since from SANS experiments it is evident that prolate aggregates are formed.

The change in the aggregate shape may not only be induced by the nature of the counterion of the ionic surfactant, it can also change with the water/surfactant molar ratio w, (See also the discussion on the percolation phenomenon in Sec. III.B.). Such a case has, for instance, been investigated in the case of didodecyldimethylammonium bromide DDAB–cyclohexane–water. For this system the viscosity shows markedly larger values for $w < 10$. If one extrapolates η_{sp}/Φ to zero volume fraction, one finds values of about 2.8 for $w = 10$ or 12, which indicates the presence of spherical aggregates [cf. Eq. (5), according to which one would expect 2.5 theoretically]. For samples with $w < 8$, values of 4–6 were observed, in agreement with the presence of rodlike aggregates, as also corroborated by SANS experiments [71].

Typically such W/O microemulsions (like their O/W counterparts) are Newtonian fluids up to shear rates of at least 10^4 s^{-1}. This has, for instance, been verified for the system ammonium heptadecylbenzene-p-sulfonate–cyclohexanol–water–decane [72]. This is a result of quite general validity; i.e., such systems are normally of low viscosity and behave as Newtonian fluids up to high shear rates or shear frequencies.

1. A Special Case of a Highly Viscous W/O Microemulsion

There also exist special situations of W/O systems of the microemulsion type that exhibit rheological behavior very different from that of typical microemulsions. An example of such a case is found in the system lecithin–water–oil. The addition of water to a low viscosity reverse micellar solution of lecithin leads to an increase in the viscosity of the solution by a factor of up to 10^6.

Despite the fact that here one has the typical composition of a microemulsion, i.e., surfactant–water–oil, one does not find a low viscosity microemulsion but instead a highly viscous system. The addition of water results in the formation of flexible cylindrical reverse micelles that form a transient network of entangled micelles and has been characterized by means of dynamic shear viscosity measurements [73,74]. Light scattering experiments on systems with cyclohexane as the oil have demonstrated that a water-induced micellar growth occurs and that these systems may be described analogously to semidilute polymer solutions [75–77].

Their rheological behavior is very similar to that of typical viscoelastic surfactant solutions made up of entangled rodlike micelles [78–81]. Thus these systems can be regarded as reverse analogs of the situation of "giant" micelles of the conventional type. Because of this we do not discuss their interesting rheological properties in the present context but instead refer to the general literature discussing the viscoelasticity of micellar solutions given above.

C. Bicontinuous Microemulsion (Winsor III) and the Transition from One Microemulsion Type to Another

The viscosity of bicontinuous microemulsions is less straightforward to describe theoretically than systems of isolated aggregates. In general, bicontinuous emulsions also exhibit low viscosity and simple Newtonian flow behavior, but it should be instructive to compare their rheological properties to that of O/W or W/O microemulsions, which in the phase diagram border on the bicontinuous microemulsion.

To see how the viscosity changes from one structural type to the other it might be particularly instructive to consider cases where all three different phases can be obtained in a similar system just by changing one physical parameter (e.g., temperature or salinity). The conventional phase sequence for these transitions is Winsor I → Winsor III → Winsor II or vice versa [45].

A well-known example for such a system are the polyglycol alkyl ethers C_iE_j. Together with oil and water one can obtain all three different structures simply by changing the temperature [47–49,82]. At low temperatures, O/W microemulsion structures are present, but upon raising the temperature the system becomes bicontinuous, and at still higher temperatures W/O systems are observed. This means that in this system one can observe the different microemulsions for a given surfactant concentration as a function of temperature. At the same time one can have a monophasic microemulsion by adjusting the surfactant concentration appropriately, i.e., an isotropic channel exists in the phase diagram [83].

For a particular system of such a nonionic surfactant ($C_{12}E_5$–n-octane–water) the viscosity was measured at constant surfactant concentration while always staying in the one-phase region (by properly adjusting the temperature according to the given oil/water ratio). It was found that the kinematic viscosity increased slightly upon approaching the bicontinuous system from either the O/W side or the W/O side [84]. However, in the range of the bicontinuous microemulsion itself, a minimum occurs. In the same study for a similar system ($C_{12}E_4$–n-tetradecane–water), the viscosity was measured as a function of temperature, with an O/W microemulsion present below 23.4°C, a W/O microemulsion above 28.4°C, and a bicontinuous microemulsion in between [84]. Here a multiphase equilibrium (two or three phases) was always present and the viscosity of the microemulsion phase was determined. Again it was found that the kinematic viscosity increased slightly upon approaching the bicontinuous system, yet for the bicontinuous microemulsion itself much lower viscosities (up to a factor 10 lower) were observed. Furthermore, it was found that the viscosity of the microemulsion correlated with the dynamics of phase separation in this multi-phase system, i.e., the lower the viscosity the faster the macroscopic phase separation. This means that the viscosity of the involved microemulsion phase can be a very important factor in separating two or three phases in a complex system, i.e., a type of process that is of large industrial importance.

The same system ($C_{12}E_5$–octane–water) was investigated in even more detail by combining NMR self-diffusion studies with rheological experiments in which the shear rate was varied [85]. Here a monophasic system was investigated at a constant water/octane ratio of 60 : 40 where upon changing the temperature (and correspondingly the surfactant concentration) one stays always in the one-phase region while the microemulsion structure is significantly altered. At low temperature isolated O/W aggregates are present, as shown by the low diffusivity of oil and surfactant (whereas the water diffusivity is close to that of bulk water, thereby proving a water-continuous structure). When the temperature is raised (at the same time the surfactant concentration is reduced) the system undergoes a structural

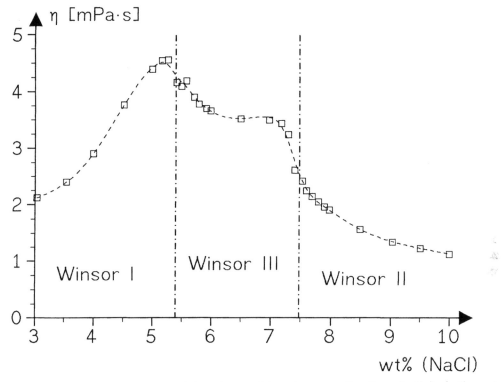

Figure 3 Zero shear viscosity of the microemulsion phase as a function of salinity in the system 46.8 wt% NaCl brine (the salinity of the brine is given on the horizontal axis), 47.2 wt% toluene, 2 wt% SDS, and 4 wt% butanol. The various microemulsion phases are marked. (Data taken from Ref. 86.)

transition that is marked by an increase in the self-diffusion coefficient of the oil by four orders of magnitude and that of the surfactant by three orders of magnitude. Concomitantly, the viscosity decreases by four orders of magnitude, reaching about 10 mPa · s for the bicontinuous system whereas the O/W microemulsion exhibits a viscosity of about 10^4 mPa · s [85].

Furthermore, in these experiments the shear rate was varied from 0.1 to 10 s^{-1} (in a cone-and-plate viscometer), and shear thinning behavior was observed for all the samples in this system; the effective viscosity decreased by about a factor of 3–5 in this range of shear rates. Such shear thinning behavior is more typically observed for bicontinuous microemulsions (as discussed in the following), but here it also occurs for the O/W microemulsion. This could be due to the fact that apparently this O/W microemulsion is not of the simple type of isolated aggregates (as evidenced by the much higher viscosities) but either contains very anisometric aggregates or, more likely, contains highly clustered aggregates.

Up to now we have considered systems where changes in the microemulsion structure are achieved by changing the temperature. However, similar transformations can be induced for systems containing ionic surfactant by varying the salinity, i.e., the ionic strength of the solution. One such system that was studied in some detail was made up of 46.8% NaCl brine, 47.2% toluene, 2% SDS, and 4% butanol [86]. Here a Winsor I → Winsor III transition occurs at 5.4% salinity and a Winsor III → Winsor II transition at 7.4%. A situation basically similar to the one discussed above for the same phase transitions is observed; the viscosity data as a function of salinity are given in Fig. 3. An increase in viscosity occurs upon approaching the Winsor III system (but still in the Winsor I range), and beyond a maximum the value remains relatively constant for the Winsor III phase. Upon approaching the Winsor II phase the viscosity again becomes much lower and readily approaches the value of the pure oil. This means that in the Winsor I and II systems, where isolated aggregates are present, the viscosity is fairly close to that of the continuous phase, whereas for the bicontinuous phase a somewhat larger value is observed.

The peak close to the Winsor I → Winsor III transition has been attributed to a clustering of the O/W droplets; i.e., the interaction between the droplets becomes more and more attractive, thereby leading to the formation of an increasing number of transient clusters (similar effects have been described more recently for other systems such as Brij 35–n-propanol–alkane–water or SDS–n-propanol–cyclohexane–brine [87] or Span80–Tween20 + propanol + oil (benzene, toluene, ethylbenzene, o-xylene [88]). A further increase in salinity then leads to coalescence of the aggregates in the clusters: the driving force for this process is the fact that at this increasing salinity the spontaneous curvature that the surfactant film would like to attain is decreasing. By coalescence the curvature of the film is decreased, which should render this situation energetically more favorable [86].

Similar behavior has been observed in viscosity measurements on microemulsions containing a commercial sulfonate surfactant [89]. With increasing salinity, they change from O/W to bicontinuous and then to W/O microemulsions, and again one observes a relatively sharp increase in viscosity upon approaching the bicontinuous microemulsion. This peak is much more pronounced on the O/W side, where the relative increase in viscosity is about a factor 4 (Fig. 4). It should be noted here that this maximum is always located inside the range where isolated structures are present (i.e., in either the Winsor I or II system).

Such behavior appears to be quite typical regardless of the type of surfactant present. For instance, in the system sodium bis(2-ethylhexyl)phosphate–n-heptane–water, the transition from a W/O to a bicontinuous and then to an O/W microemulsion can be induced by increasing the water/surfactant molar ratio. Again in the transition region higher viscosities are observed where this increase is significantly more pronounced at the O/W to bicontinuous transition [90].

1. The Percolation Process

The transition from a system of isolated droplets to an interconnected bicontinuous structure is often denoted as a percolation process. This phenomenon has been studied in much detail for the systems AOT–water–hydrocarbon [91–94] and didodecyldimethylammonium bromide–water–hydrocarbon [95]. The clustering of the droplets at the percolation threshold typically leads to an increase in viscosity [96] (as already discussed above). The percolation threshold is typically located at the point where in a logarithmic plot of viscosity versus volume fraction of droplets the slope attains a maximum. This means that a plot of $(1/\eta) (d\eta/d\Phi)$ versus Φ displays a maximum that coincides with the percolation threshold as determined by measurements of electrical conductivity. This has been confirmed for instance, for W/O microemulsions in the systems SDS–butanol–toluene–water [97] and AOT–n-heptane–glycerol [98]. Two examples of this behavior are given in Fig. 5. Accordingly, viscosity measurements can be used to determine the percolation threshold in microemulsions.

In this connection it might also be noted that the percolation process is hindered by the presence of electrolytes. In parallel to the percolation process, the viscosity decreases [99]. Finally, this behavior can be generalized for W/O microemulsions and is similarly observed

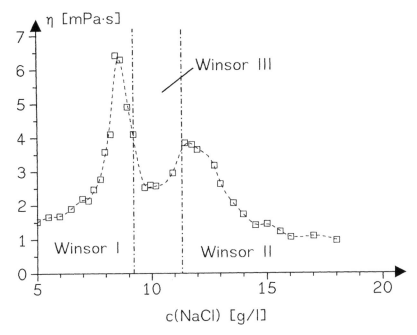

Figure 4 Viscosity of the microemulsion phase (made up with the commercial anionic surfactant mixture TRS 10-80, a sulfonate with an approximate molecular weight of 420) as a function of the salinity. (Data taken from Ref. 89.)

in systems where glycerol has been substituted for water. The viscosity of these waterless microemulsions shows behavior analogous to that of their aqueous equivalents [100,101]. For instance, the system glycerol–AOT–isooctane has been studied in some detail, and it has been verified that the viscosity data for this system are in good agreement with the dynamic theory of percolation. Here it has also been observed that the viscosity may increase or decrease with increasing temperature depending on the molar ratio of glycerol to AOT [102].

2. Rheological Behavior at High Shear Rates—Shear Thinning

Bicontinuous structures are also present in the well-investigated surfactant system didodecyldimethylammonium bromide (DDABr)–dodecane–water [95,103], where a percolation process occurs that is a function of the water content. Upon an increase in the water concentration, the interconnected water channels originally present (bicontinuous microemulsion) are transformed into globular water droplets (W/O microemulsion), and in this phase region the viscosity is somewhat lower than that of the bicontinuous microemulsion [104].

This system has also been intensively studied with respect to its rheological behavior, and these investigations showed that such a microemulsion always behaves like a Newtonian fluid, even up to shear rates of 3000–5000 s^{-1} [105]. This would have been expected for a microemulsion composed of isolated aggregates but might be somewhat surprising for the highly interconnected bicontinuous microemulsion. From a snapshot its structure

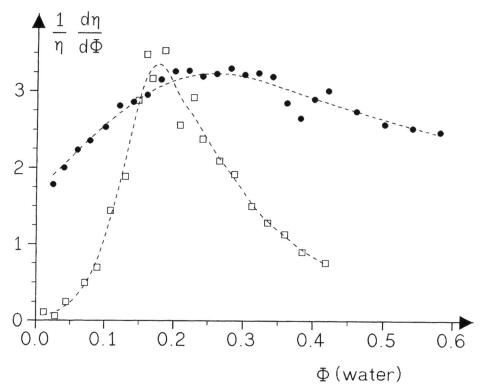

Figure 5 Plot of $(1/\eta)$ $(d\eta/d\Phi)$ as a function of the water volume fraction Φ for the systems AOT–undecane–water (●) and SDS–butanol–toluene–water (□). The maximum occurs at the percolation threshold. (Data taken from Ref. 97.)

resembles that of a cross-linked polymer network or that of a typical gel-like structure (e.g., that of a bicontinuous cubic phase, which is a solidlike substance). (Cubic phases and their relation to microemulsions are discussed in Sec. III.D). However, it is well known that such networks exhibit strongly viscoelastic properties. So why does a similar structure in a bicontinuous microemulsion show completely different rheological behavior? The answer to this question is that a microemulsion is a much more dynamic system, i.e., its structural relaxation time, which determines how fast structural rearrangements take place, is much shorter than that of typical gel-like structures. Therefore, in typical rheological experiments with shear rates of up to several thousand s^{-1}, the elastic contribution is simply not observed, because the system has sufficient time to relax the applied stress by means of structural changes. This means that its structural relaxation time clearly has to be shorter than ≈ 0.1–1 ms.

It should, however, be noted that there exist some hints that bicontinuous microemulsions behave elastically. This has been assumed to be due to the differences observed in measuring viscosities once in a Couette flow and in the other case by a capillary viscometer. Here it was observed that the values obtained with the capillary viscometer are markedly higher. It has been suggested that in capillary flow a component of elongational flow is observed and that in this type of flow elastic components can be observed much earlier than in shear flow [106,107].

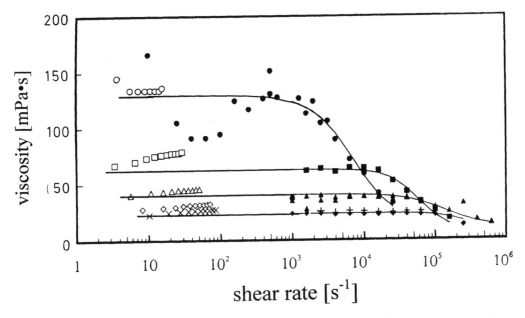

Figure 6 Viscosity as a function of the shear rate $\dot{\gamma}$ for microemulsions containing DDABr–water–dodecane. Samples have a constant dodecane/DDABr weight ratio of 4 : 6. Weight percentage of water: $(\times, +)$ 22%; $(\Diamond, \blacklozenge)$ 30%; $(\triangle, \blacktriangle)$ 40%; $(\Box, \blacksquare)$ 50%; $(\bigcirc, \bullet)$ 60%. Open symbols denote measurements in a Couette system, and closed symbols signify measurements taken with a Rheometrics RFX instrument. (From Ref. 110.)

More recent investigations studied the effect of high shear rates on both W/O and bicontinuous microemulsions containing DDAB, dodecane, and water [108]. Again these experiments showed that for both structural types of microemulsions Newtonian behavior was observed up to shear rates of 10^4 s^{-1}. This was the case for both shear and elongational flows, where the elongational viscosity was found to be three times as great as the shear viscosity, in agreement with Trouton's rule [109]. However, at still higher shear rates (determined by measuring the wall shear stress that develops by the Poiseuille flow between two concentric cylinders, using a Rheometrics RFX instrument, cf. Fig. 6), shear thinning was observed where the critical shear rate depended on whether a bicontinuous or droplet structure was present. The critical shear rate was about $0.3 \times 10^4 – 1 \times 10^4$ s^{-1} (and even somewhat lower) for the bicontinuous microemulsion, whereas it was about 10^5 s^{-1} for the droplet systems. In addition, the drop in viscosity was much more pronounced for the bicontinuous systems, as demonstrated in Fig. 6.

The drop in viscosity with increasing shear rate was described well [110] for all systems investigated by means of the Cross equation [Eq. (16)] that was derived for the flow of pseudoplastic systems [111].

$$\frac{\eta - \eta_\infty}{\eta_0 - \eta_\infty} = \frac{1}{1 + (\dot{\gamma}/\dot{\gamma}_c)^n} \tag{16}$$

where η_0 is the zero shear viscosity, η_∞ is the limiting high shear viscosity, $\dot{\gamma}_c$ the critical shear rate, and n a parameter characteristic of the steepness of the shear thinning. Here $\dot{\gamma}_c$ is an inverse measure of the structural relaxation time of the system.

All this indicates that the shortest structural relaxation time is significantly longer in the bicontinuous system than in the droplet microemulsion. However, it also demonstrates that even for the highly connective bicontinuous structure this relaxation time is still very short, since a snapshot of its structure would resemble that of highly cross-linked polymers, which have typical structural relaxation times orders of magnitude larger. This means that although the static picture is very similar for both systems, one observes strikingly different dynamic behavior, because the microemulsion structure is rapidly fluctuating [112], an effect that has been similarly observed in L_3 phases that structurally resemble a bicontinuous microemulsion [113].

3. Comparison of Bicontinuous Microemulsions with the L_3 Phase of Amphiphilic Systems

As just stated, the L_3 phase resembles structurally a bicontinuous microemulsion, which makes it interesting to compare their rheological properties. Viscosity measurements on an L_3 phase [in the system cetylpyridinium chloride–hexanol–brine (0.2 M NaCl)] showed Newtonian flow behavior for the range of shear rates of 0.1–100 s^{-1} [113]. The viscosity of this highly interconnected, spongelike system is always very low and close to the solvent viscosity, even for a volume fraction of 0.2 it is less than 10 mPa · s. (Similar viscosity values have been observed in the L_3 phase of the system tetradecyldimethylamine oxide–hexanol–water [114].) It increases linearly with the volume fraction of the amphiphilic material, where it is interesting to note that extrapolation to zero concentration does not yield the solvent viscosity but a value about three times as high. A similar value for the extrapolated viscosity was also reported more recently for another L_3 phase (in the system SDS pentanol–dodecane–water [115]), and it seems that this enhanced viscosity is a universal property connected to the structure of the L_3 phase.

Such behavior has also been explained from a theoretical point of view if one assumes that in a steady shear experiment the "passages" of the sponge structure are not forced to pass each other but that instead upon shearing the structure of the system can readjust by viscous flow of the surfactant in the amphiphilic film and a corresponding flow of the solvent through the passages [113]. Similar experiments on the viscosity change that occurs when a bicontinuous microemulsion is diluted have not, to our knowledge, been reported but should yield an interesting comparison. Of course, in many systems this is not possible because a significant surfactant concentration is often required to form such a bicontinuous microemulsion.

Another way to rationalize the small dependence of the viscosity of L_3 phases on the volume fraction Φ is to consider that the zero shear viscosity η_0 is related to the shear modulus G_0 and the structural relaxation time τ by

$$\eta_0 = G_0 \tau \tag{17}$$

However, the structural relaxation time τ, as determined by electrical birefringence measurements, varies with Φ^{-3} [114]. The shear modulus is not easily accessible but can be estimated according to

$$G_0 =^1 NkT \tag{18}$$

where ^1N is the number density of the elastic entities (for systems containing interacting aggregates, the number density of those aggregates). However, if one assumes that the characteristic length l of the sponge structure is proportional to the mean distance between the bilayers, i.e., $l \sim 1/\Phi$ (in agreement with experimental data), then the number density will be proportional to $1/l^3$, or equivalently $^1N \sim \Phi^3$ [114]. According to this approximation the viscosity should not change with the concentration and should remain low at all concentrations as observed experimentally, where typically an increase by only a factor of about 3–5 is found even for volume fractions of 0.3–0.4.

In a later study [115] it was also found that highly dilute samples showed shear thinning behavior. The reason for this is that here the characteristic length ξ of the sponge becomes very large (up to 600 nm), which renders the relaxation time of these systems sufficiently high to be observed in the rheological experiments. For samples of normal concentrations the critical shear rate for the observation of non-Newtonian behavior was estimated to be 10^3–10^6 s^{-1}, a range that is normally not accessed in rheological experiments [115]. With respect to the occurrence of shear thinning behavior at high shear rates, the properties of the L_3 phase and bicontinuous microemulsion are obviously quite similar, and this effect occurs only at very high shear rates as the typical relaxation times in these systems are very short.

4. Rheological Experiments at High Frequencies

Evidently the very short relaxation times in microemulsions require dynamic shear experiments at very high frequencies if one wants to observe experimentally the deviations from Newtonian flow behavior. Such measurements can be carried out, for instance, by means of torsion pendulum resonators [116] or a tube resonator [117]. These methods then allow for measurements in the range of 50–2500 Hz and 3.5–250 kHz, respectively. By means of such experiments, for instance, bicontinuous microemulsions in the systems SDS–n-heptane–brine [118] and nonylphenol (EO)$_{6.8}$ (NNP 7)–hexane–water [119] were investigated in some detail. Here it was observed that typically the elastic part of the viscosity (η'') becomes nonzero for frequencies above 10^4 Hz and at the same time the viscous part (η') is reduced. An example of such behavior is shown in Fig. 7 [118]. This means that such experiments now allow for a determination of the structural relaxation time in microemulsions and show that at sufficiently high frequencies the bicontinuous structure of Winsor III type microemulsions also exhibits elastic properties. A similar high frequency behavior of the complex viscosity is also expected for droplet microemulsions even at still higher frequencies. The corresponding rheological curves can be predicted theoretically from an emulsion droplet model that takes into account the thermal shape fluctuations of the droplets [118,120].

Another approach to determining the viscoelastic properties of dense microemulsions at high frequencies is to conduct ultrasonic absorption experiments. In such experiments it has been found that the percolation process is correlated to a shift of the ultrasonic dynamics from a single relaxation time to a distribution of relaxation times [121]. Other experiments showed an increase in the hypersonic velocity for samples at and beyond the percolation threshold. The complex longitudinal modulus deduced from such experiments is also correlated with the occurrence of the percolation phenomenon, which suggests that the velocity dispersion is clearly correlated with structural transformations [122].

D. Modifying the Viscosity of Microemulsions by Admixing Other Substances

For a number of applications it is important to modify the rheological properties of microemulsion systems. In general these systems are of low viscosity, so it is normally of interest to increase their viscosity in order to obtain more viscous, highly viscous, or

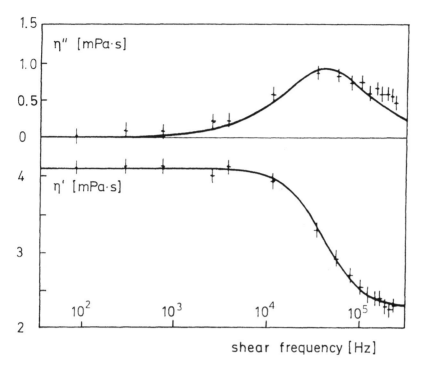

Figure 7 Elastic part η'' and viscous part η' of the complex shear viscosity η as a function of the shear frequency for a bicontinuous microemulsion containing SDS, 1-butanol, n-heptane, and NaCl brine. (From Ref. 118.)

even gel-like systems, as might be needed, for instance, for pharmaceutical applications or in cosmetics [123]. A modification of the rheological behavior is also of interest in the case of W/O microemulsions that are used as drilling fluids; by adding bentonite a shear-thinning system with a yield stress can be obtained [124].

1. Addition of Gellifiers

A good possibility to induce changes in rheological behavior is to add a gellifying compound to the microemulsion that is soluble in the continuous phase and is known to form a gel in this solvent. A well-known example of such a substance is gelatin (a polypeptide), which readily forms a gel in water even at low concentrations. However, gelatin can also be used to increase the viscosity of W/O microemulsions, and organogels can be produced via this route [125]. Here helical gelatin cross-links can be formed within the aqueous cores of the system (e.g., in AOT–water–isooctane [126]), and these interconnect the various aqueous droplets to form a network of connected microemulsion droplets. By doing so the low viscosity microemulsion is transformed into a transparent stable gel. NMR experiments showed that the individual components of the system—surfactant, oil, and water—retain their mobilities despite the fact that macroscopically the appearance of the systems changes considerably [126]. For sufficiently high gelatin concentrations the onset of interdroplet cross-linking is marked by a diverging viscosity [127,128]. This has been demonstrated, for instance, for the case of the W/O microemulsion of AOT–water–isooctane, where the addition of gelatin lowers the percolation threshold of the microemulsion at which an increase in viscosity occurs and at the same time this increase becomes much more pronounced with increasing gelatin concentration [127].

Another obvious way to obtain such rheologically modified microemulsions is to add a gellifying agent, typically a polymer that itself contains hydrophobic and hydrophilic parts and thereby interacts with the amphiphilic structures. The interaction between such polymers and amphiphilic systems (such as microemulsions) is a very broad field that has attracted considerable interest in recent years. Of course, it is beyond the scope of this chapter to cover this field extensively, and we wish to cover here only some of its aspects that are relevant to the modification of microemulsion rheology.

One straightforward way to obtain such microemulsion-mediated polymer networks is to add a triblock copolymer of type ABA to the microemulsion, where the A blocks are favorable to the dispersed phase and the B block to the continuous phase of the microemulsion. By doing so one may expect that the different A blocks will be located in different aggregates of the microemulsion and will thereby lead to a transient network treated theoretically in some detail in Refs. 129–133.

An example of such a system is the W/O microemulsion of AOT–water–isooctane to which the block copolymer poly(ethylene oxide) (PEO)–polyisoprene (PI)–poly(ethylene oxide) was added [128]. Here the hydrophilic PEO will reside in the aqueous cores of the microemulsion droplets, whereas the lipophilic PI will stay in the surrounding hydrocarbon. Depending on the molecular weight (correspondingly the length) of the PI block, either core–shell-type aggregates (with both PEO ends in the same droplet) are formed for long PI blocks or a bridging of the nanodroplets occurs for shorter PI blocks (with both PEO ends located in different droplets), since for short PI blocks a backfolding to the same droplet is conformationally constrained [134]. Thus an interconnected network is formed even at fairly low copolymer concentrations. Indeed, the addition of the block copolymer induces significant rheological changes, i.e., the originally Newtonian microemulsion is transformed into a viscoelastic fluid, which means that temporary network cross-links are formed in this system. The relaxation time for disentanglement of the PEO blocks and the droplets was shown to be about 10 ms at room temperature, i.e., in the range of the lifetime of the micellar aggregates [135].

The rheology of these so-called mesogels has been studied via oscillatory shear experiments in some detail, and by admixing the polymer (which is possible up to concentrations of more than 100 g/L) an increase in viscosity by a factor of more than 10^6 is observed. For this system three different regions (corresponding to dilute, semidilute, and concentrated) are observed in which the viscosity varies according to different scaling laws (Fig. 8) [136] (c_p = polymer concentration):

1. Up to 3 g/L: $\eta_{sp} \sim c_p^{1.5}$
2. 5–25 g/L: $\eta_{sp} \sim c_p^{3.6-4}$ (in agreement with theoretical predictions for entangled networks in the semidilute concentration regime [41,137])
3. More than 30 g/L: $\eta_{sp} \sim c_p^{5-5.6}$

For these transient networks formed by the interaction of an ABA triblock copolymer and a microemulsion it has been shown that their principal viscoelastic properties are not affected significantly by the chemical nature of the microemulsion, i.e., they are similar for systems with both nonionic and ionic surfactants. Also it should be noted that the phase behavior of the corresponding microemulsion is qualitatively preserved, i.e., the reversible aggregation of the nanodroplets and the phase transitions to lyotropic liquid crystalline phases remain essentially unchanged (although the concentrations at which they occur might

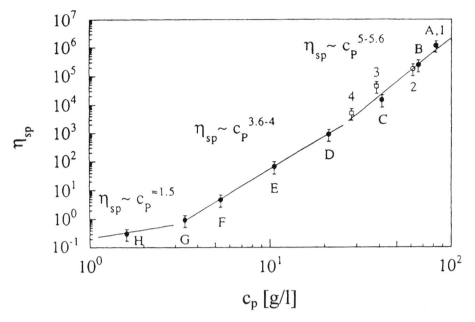

Figure 8 Specific viscosity η_{sp} at 25°C of microemulsion containing AOT, isooctane, and water (mass water/AOT ratio of 2.5) as a function of the amount of added triblock copolymer PEO–PI–PEO [PEO = poly(ethylene oxide); PI = polyisoprene]. (From Ref. 135.)

be changed somewhat [138]). The lifetimes of the network cross-links are related to the characteristic relaxation times derived from the mechanical spectra. The high-frequency shear modulus for samples that contain about 3–20 polymer molecules per nanodroplet is typically in the range of 100–1000 Pa [138].

Of course, the formation of such cross-linked microemulsion aggregates is also possible for the case of O/W microemulsions. Only here one has to use the opposite type of triblock structure, i.e., a hydrophilic central part and two hydrophobic ends. Such a triblock is given, for instance, by a poly(ethylene oxide) chain that is end-capped by two alkyl chains. This type of system was investigated for the case of EO_{55} that was end-capped by two oleyl chains (CL 428) in conjunction with the O/W microemulsion formed by tetradecyldimethylamine oxide (TDMAO), decane, and water [139]. For this system it was shown that for a microemulsion of a volume fraction of about 18% it is sufficient to add only 1.5% of the polymeric compound to increase the viscosity by more than three orders of magnitude (Fig. 9).

Of course, the viscosity also depends largely on the total concentration of the system as demonstrated in Fig. 10 for a system of constant relative composition. It can be seen that the viscosity increases greatly only for concentrations of more than 10 wt%. The reason for this is that at lower concentration the microemulsion droplets are on average too widely spaced to be bridged by the polymer molecule. Once the required distance becomes sufficiently small, a large increase in viscosity occurs that is also compared in Fig. 10 to the situation without added CL 428. The value of the shear modulus G_0 is in agreement with a simple network theory where it is assumed that each cross-linking polymer stores an elastic energy of kT [139,140]. Modifying the corresponding expression with the probability of finding the second hydrophobic end in another aggregate (assuming it to be simply given by the aggre-

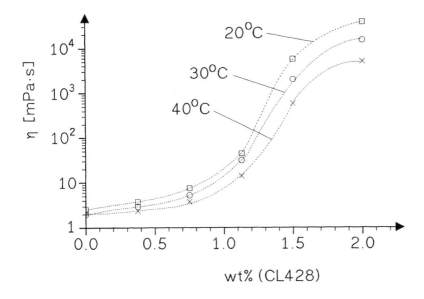

Figure 9 Viscosity as a function of the amount of added CL 428 (oleyl–EO_{55}–oleyl) at various temperatures. The total concentration of TDMAO+decane+CL 428 was kept constant at 16.5 wt%. (From Ref. 139.)

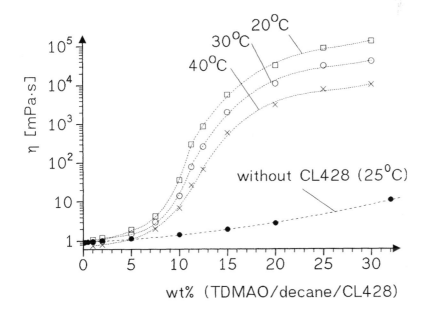

Figure 10 Viscosity as a function of the total concentration (TDMAO+CL 428+decane) at various temperatures for a system of constant composition [(mass (TDMAO)/mass (decane) = 5.4 : 1; mass (TDMAO+decane)/mass (CL 428) = 10 : 1]. For a comparison the viscosity of the corresponding system without CL 428 is also given. (From Ref. 139.)

gate volume fraction Φ) yields the expression

$$G_0 = {}^1N\Phi kT \qquad\qquad (19)$$

where ${}^1N$ is the number density of the polymer molecules. This simple relation is also able to at least semiquantitatively describe the experimental data obtained [139]. This investigation clearly shows that it is possible to transform an originally unconnected microemulsion system into a physical network, thereby greatly modifying the rheological properties of the given system.

In principle the addition of any kind of polymeric material to a microemulsion will influence its rheological behavior; however, the kind of changes observed will depend to a large degree on the nature of the polymer. For instance, a study on an O/W microemulsion composed of the nonionic Brij 96, butanol and hexadecane shows that, depending on the type of water-soluble polymer (of identical molecular weight of 300,000) added, different changes in the viscosity are observed [141]. The addition of poly(ethylene oxide) leads to a system where no specific interactions are present and where the viscosity of the mixtures can be described by a geometric mean Huggins coefficient. For the case where strong complexation takes place—as is the case for poly(acrylic acid)—much lower viscosities are observed; in this case one can reduce the viscosity of the polymeric solution by the addition of the microemulsion. This effect can be explained on the basis that the polymer chain collapses onto the surface of the droplets and wraps around them, thereby leading to a viscosity close to that of the unperturbed microemulsion. Finally, if weaker attractive interactions are present, as is the case for hydrolyzed polyacrylamide, a situation in between the two extrema is observed, i.e., the viscosity of a given system depends significantly on the strength of the polymer–microemulsion interaction [141]. This shows that very different situations can be encountered where, depending on the choice of the added polymer, one might not see any appreciable increase in viscosity or, on the other hand, gel-like systems may be formed (if the polymer is capable of cross-linking the microemulsion structure effectively).

2. Formation of Cubic Phases

Finally, another way to increase the viscosity of microemulsions in some cases is to increase the concentration of the dispersed phase. For instance, droplet systems will start to become densely packed, and typically for volume fractions of more than 0.5 the droplets become effectively frozen in. The systems undergo a phase transition, however, of such a type that their structural building units, the microemulsion droplets, remain unchanged. These gel-like systems have been known for quite a long time from cosmetics applications [142,143] and normally belong to the class of cubic phases (sometimes these systems have also been refered to as microemulsion gels because their composition is identical to that of the corresponding microemulsion with only the content of the dispersed phase being increased). The existence of a cubic phase in the phase diagram of a surfactant system is not necessarily linked to the presence of a microemulsion phase, but if it exists it may be found according to the recipe given above, i.e., by increasing the concentration of the dispersed phase while keeping the relative composition of the system constant. A typical phase diagram for such a surfactant system is given in Fig. 11.

Such a situation is, for instance, often observed in the case of nonionic surfactants of the type C_iE_j (alkyl polyglycol ethers) that form an O/W microemulsion with hydrocarbons that with increasing concentration is transformed into a cubic phase. This system has been

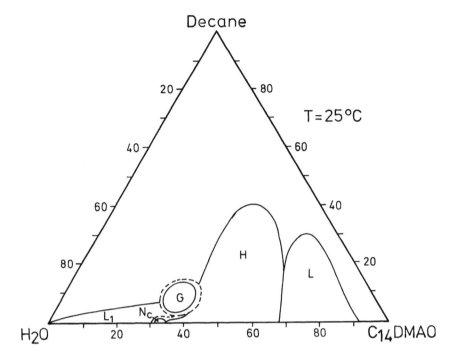

Figure 11 Phase diagram of the system tetradecyldimethylamine oxide (TDMAO)–decane–water at 25°C. Of interest here are the O/W microemulsion phase (L_1) and the cubic phase (G). (From Ref. 146.)

investigated to some extent with respect to its utility for pharmaceutical applications [144,145] because it allows for a very fine dispersion of a given hydrophobic substance in a system of creamlike consistency.

These systems are typically of very high viscosity and may even possess a yield stress [146]. They exhibit elastic properties, and their shear modulus is typically in the range of 10^4–10^6 Pa [56,146–149]. It should be noted here again that these cubic phases are formally no longer microemulsions, but they are structurally very closely related. Their microstructural units are almost identical to those in the corresponding microemulsion, but their macroscopic appearance is considerably different; i.e., whereas the microemulsion is a liquid of low viscosity, the cubic phase is a solidlike material.

Evidently the field of modifying the rheological properties of microemulsions is very broad, and in particular the addition of gellifiers leads to relatively complex systems that currently are a field of very active research. Coverage of this large subject is clearly beyond the scope of this chapter, but it should be kept in mind that it is possible to alter the rheological parameters of a given system by orders of magnitude via the addition of properly chosen polymer substances. The other way of increasing the viscous properties enormously is by changing from the microemulsion to its corresponding cubic phase (provided such a cubic phase exists in the phase diagram in the neighborhood of the microemulsion phase) where the structural units are preserved but the macroscopic appearance is largely altered and gel-like systems are obtained.

IV. SUMMARY

In summary, it may be stated that in general the rheological behavior of microemulsions is not very revealing. Typically they are of low viscosity and exhibit Newtonian flow behavior. Only at very high shear rates is shear-thinning behavior observed. Nonetheless, rheological experiments can yield valuable information regarding the structure of the corresponding microemulsion. For droplet-type microemulsions, normally hard-sphere behavior as a function of the volume fraction is observed for both O/W and W/O systems. Information regarding the state of hydration of the aggregates can be obtained from the analysis of these viscosity data.

For W/O microemulsions, deviations from hard-sphere behavior are sometimes observed. The reason for this is that here there is a stronger tendency for attractive interactions and the formation of anisometric shapes. Here viscosity data can be used to determine the shape of the corresponding aggregates or to extract information regarding the interaction potential that exists between the droplets. From temperature-dependent experiments the binding enthalpy of the droplets can be determined.

For microemulsions of bicontinuous structure also, very low viscosities are found despite the fact that they possess a highly interconnected structure. But at the same time these structures possess very short structural relaxation times of less than 1 ms, which explains why one normally observes simple Newtonian flow behavior. However, for bicontinuous microemulsions shear-thinning behavior is more easily observed than for their counterparts composed of isolated aggregates. For bicontinuous microemulsions, shear thinning can be observed at shear rates one or two orders of magnitude lower than for droplet microemulsions. To investigate this more interesting rheological behavior, dynamic experiments at high frequency (10^3–10^6 Hz) are necessary, a frequency range not easily accessible experimentally. Such experiments show that in this frequency range elastic properties of microemulsions can be observed.

It is interesting to note that in microemulsion systems where one can go from one structural type to another it has been observed that typically around the transition from a droplet structure to a bicontinuous structure a maximum in the viscosity occurs. The viscosity change can also be used to determine the percolation threshold in microemulsions, since here at this point the viscosity shows a maximum slope in a logarithmic plot.

Finally, the modification of the rheological properties, typically an increase of viscosity, is often desired for a microemulsion. Such changes can be achieved by admixing a gellifier with the system, where the gellifier is typically a polymeric compound. Its function is to form a highly interconnected network containing the microemulsion structure without altering it significantly. This way one is able to increase the viscosity by orders of magnitude or even form gels by adding relatively moderate amounts of gellifier. Another way to alter the rheological properties considerably is also possible for systems where the microemulsion borders on a cubic phase in the phase diagram. For such a situation one can cross into the cubic phase, which still contains basically identical structural entities but is typically a solidlike phase. These gel-like systems (obtained either with a gellifying agent or with the cubic phase) can be very interesting for applications in cosmetics, pharmaceutics, drilling fluids, etc.

REFERENCES

1. K. E. Bennett, J. C. Hatfield, H. T. Davis, C. W. Macosko, and L. E. Scriven, in *Microemulsions* (I. D. Robb, ed.), Plenum, New York, 1982, p. 65.
2. W. B. Gogarty and W. C. Tosch, Trans. Soc. Petr. Eng. AIME *243*:1407 (1968).
3. S. Vijayan, C. Ramachandran, H. Doshi, and D. O. Shah, in *Surface Phenomena in Enhanced Oil Recovery* (D. O. Shah, ed.), Plenum, New York, 1979, p. 327.
4. M. Bourrel, C. Chambu, R. S. Schechter, and W. H. Wade, Soc. Pet. Eng. J. *22*:28 (1982).
5. C. A. Miller, S. Mukherjee, W. J. Benton, J. Natoli, S. Qutubuddin, and T. Fort Jr., AIChE Symp. Ser. *78*:28 (1982).
6. J. Novosad, B. Maini, and Batycky, J. Am. Oil Chem. Soc. *59*:833 (1982).
7. B. Ohah and H. J. Neumann, Tenside Deterg. *20*:145 (1983).
8. M. C. Puerto and R. L. Reed, Soc. Pet. Eng. J. *23*:669 (1983).
9. H. Morita, Y. Kawada, J. Yamada, and T. Ukigai, Jpn. Patents 81-160296 811009 and 82-37186 820311 (1983).
10. B. W. Davis, in *Encyclopedia of Emulsion Technology*, Vol. 3 (P. Becher, ed.), Marcel Dekker, New York, 1988, p. 307.
11. J. R. Fanchi, Report DOE/BC/10033-9-Vol. 1; Order No. DE85000120, Energy Res. Abstr. No. 15664, 1985.
12. A. Lohne, Report RF-5/91 (1991); Order No. DE93752942; Energy Res. Abstr. No. 4985, 1993.
13. A. W. Schwab, M. O. Bagby, and B. Freedman, Fuel *66*:1372 (1987).
14. A. Schwab and E. H. Pryde, U.S. Patent 423402 (1983).
15. A. Schwab and E. H. Pryde, U.S. Patent 427229 (1983).
16. M. Ziejewski, K. R. Kaufman, A. W. Schwab, and E. H. Pryde, J. Am. Oil Chem. Soc *61*:1620 (1984).
17. A. W. Schwab, H. C. Nielsen, D. D. Brooks, and E. H. Pryde, J. Dispersion Sci. Technol. *4*:1 (1983).
18. R. O. Dunn, A. W. Schwab, and M. O. Bagby, J. Dispersion Sci. Technol. *13*:77 (1992).
19. N. Garti, R. Feldenkriez, A. Aserin, S. Ezrahi, and D. Shapira, Lubr. Eng. *49*:404 (1993).
20. H. Yamakawa, *Modern Theory of Polymer Solutions*, Harper and Row, New York, 1971.
21. M. Bohdanecky and J. Kovar, *Viscosity of Polymer Solutions*, Elsevier, Amsterdam, 1982.
22. H. G. Elias, *Makromoleküle*, Vol. 1, 5th ed., Hüthig & Wepf Verlag, Basel, 1990.
23. A. Einstein, Ann. Phys. *19*:289 (1906).
24. A. Einstein, Ann. Phys. *34*:591 (1911).
25. G. J. Taylor, Proc. Roy. Soc. Lond., Ser. A *138*:41 (1932).
26. G. K. Batchelor, J. Fluid Mech. *83*:97 (1977).
27. E. Guth and R. Simha, Kolloidn. Zh. *74*:266 (1966).
28. N. Saito, J. Phys. Soc. Jpn. *5*:4 (1950); *7*:447 (1952).
29. I. M. Krieger and T. J. Dougherty, Trans. Soc. Rheol. *3*:137 (1959).
30. R. Roscoe, Br. J. Appl. Phys. *3*:267 (1952).
31. D. E. Quemada, in *Lecture Notes in Physics: Stability of Thermodynamic Systems* (J. Cases-Vasquez and J. Lebon, eds.), Springer, Berlin, 1982, pp. 210–247.
32. J. C. van der Werff and C. G. de Kruif, J. Rheol. *33*:421 (1989).
33. D. G. Thomas J. Colloid Sci. *20*:267 (1965).
34. M. Gradzielski and H. Hoffmann, Adv. Colloid Interface Sci. *42*:149 (1992).
35. M. Doi and S. F. Edwards, J. Chem. Soc., Faraday Trans. II *74*:918 (1978).
36. R. Simha, J. Phys. Chem. *44*:25 (1940).
37. C. R. Cantor and P. R. Schimmel, *Biophysical Chemistry*, Part II, *Techniques for the Study of Biological Structure and Function*, Freeman, San Francisco, 1980.
38. M. L. Huggins, J. Am. Chem. Soc. *64*:2716 (1942).
39. C. Tanford, *Physical Chemistry of Macromolecules*, Wiley, New York, 1961.
40. M. J. Mooney, J. Colloid Sci. *6*:162 (1950).
41. J. D. Ferry, *Viscoelastic Properties of Polymers*, 3rd ed., Wiley, New York, 1980.

42. H. A. Barnes, J. F. Hutton, and K. Walters, *An Introduction to Rheology*, Elsevier, Amsterdam, 1989.
43. N. W. Tschoegl, *The Phenomenological Theory of Linear Viscoelastic Behaviour*, Springer-Verlag, Berlin, 1989.
44. A. M. Cazabat, D. Langevin, J. Meunier, and A. Pouchelon, Adv. Colloid Interface Sci. *16*:175 (1982).
45. D. Langevin, Mol. Cryst. Liq. Cryst. *138*:259 (1986).
46. D. Langevin, Ann. Rev. Phys. Chem. *43*:341 (1992).
47. K. Shinoda and S. Friberg, Adv. Colloid Interface Sci. *4*:281 (1975).
48. H. Kunieda and K. Shinoda, J. Dispersion Sci technol. *3*:233 (1982).
49. M. Kahlweit and R. Strey, Angew. Chem. *79*:655 (1985).
50. R. Aveyard, B. P. Binks, S. Clark, and P. D. I. Fletcher, Prog. Colloid Polym. Sci. *79*:202 (1989).
51. M. S. Leaver and U. Olsson, Langmuir *10*:3449 (1994).
52. D. B. Siano, J. Bock, P. Myer, and W. B. Russel, Colloids Surf. *26*:171 (1987).
53. D. Attwood, L. R. J. Currie, and P. H. Elworthy, J. Colloid Interface Sci. *46*:261 (1974).
54. D. B. Siano, J. Colloid Interface Sci. *93*:1 (1983).
55. M. Gradzielski and H. Hoffmann, J. Phys. Chem. *98*:2613 (1994).
56. M. Gradzielski, Dissertation, Universität Bayreuth, 1992.
57. R. Aveyard, B. P. Binks, and P. D. I. Fletcher, Langmuir *5*:1210 (1989).
58. R. C. Baker, A. T. Florence, R. H. Ottewill, and T. F. Tadros, J. Colloid Interface Sci. *100*:332 (1984).
59. S. Ray, S. Bisal, and S. P. Moulik, J. Surf. Sci. Technol. *8*:191 (1992).
60. J. Smeets, G. J. M. Koper, J. P. M. van der Ploeg, and D. Bedeaux, Langmuir *10*:1387 (1994).
61. M. A. van Dijk, J. G. H. Joosten, Y. K. Levine, and D. Bedeaux, J. Phys. Chem. *93*:2506 (1989).
62. E. van der Linden, D. Bedeaux, R. Hilfiker, and H.-F. Eicke, Ber. Bunsenges. Phys. Chem. *95*:876 (1991).
63. D. Roux, A. M. Bellocq, and M. S. Leblanc, Chem. Phys. Lett *94*:156 (1983).
64. W. D. Dozier, M. W. Kim, and R. Klein, J. Chem. Phys. *87*:1455 (1987).
65. G. J. M. Koper, W. F. C. Sager, J. Smeets, and D. Bedeaux, J. Phys. Chem. *99*:13291 (1995).
66. J. S. Huang, J. Surf. Sci. Technol. *5*:83 (1989).
67. J. Bergenholtz, A. A. Romagnoli, and N. J. Wagner, Langmuir *11*:1559 (1995).
68. J. B. Peri, J. Colloid Interface Sci. *29*:6 (1969).
69. R. F. Berg, M. R. Moldover, and J. S. Huang, J. Chem. Phys. *87*:3687 (1987).
70. J. Eastoe, G. Fragnetto, B. H. Robinson, T. F. Towey, R. K. Heenan, and F. J. Lang, J. Chem. Soc., Faraday Trans. *88*:461 (1992).
71. J. Eastoe and R. K. Heenan, J. Chem. Soc., Faraday Trans. *90*:487 (1994).
72. K. D. Dreher, W. B. Gogarty, and R. D. Sydansk, J. Colloid Interface Sci. *57*:379 (1976).
73. P. Schurtenberger, R. Scartazzini, and P. L. Luisi, Rhoel. Acta *28*:372 (1989).
74. P. L. Luisi, R. Scartazzini, G. Haering, and P. Schurtenberger, Colloid Polym. Sci. *268*:356 (1990).
75. P. Schurtenberger and C. Cavaco, Laugmuir *10*:100 (1994).
76. P. Schurtenberger and C. Cavaco, J. Phys. Chem. *98*:5481 (1994).
77. C. Cavaco and P. Schurtenberger, Helv. Phys. Acta *67*:227 (1994).
78. H. Rehage and H. Hoffmann, Faraday Discuss. Chem. Soc. *76*:363 (1983).
79. H. Rehage and H. Hoffmann, Mol. Phys. *74*:933 (1991).
80. F. Kern, P. Lemarechal, S. J. Candau, and M. E. Cates, Langmuir *8*:437 (1992).
81. J.-F. Berret, D. C. Roux, and G. Porte, J. Phys. II *4*:1261 (1994).
82. D. Anderson, H. Wennerström, and U. Olsson, J. Phys. Chem. *93*:4243 (1989).
83. R. Strey, Colloid Polym. Sci. *272*:1005 (1994).
84. M. Kahlweit, R. Strey, D. Haase, H. Kunieda, T. Schmeling, B. Faulhaber, M. Borkovec, H.-F. Eicke, G. Busse, F. Eggers, Th. Funck, H. Richmann, L. Magid, O. Söderman, P. Stilbs, J. Winkler, A. Dittrich, and W. Jahn, J. Colloid Interface Sci. *118*:436 (1987).
85. H. T. Davis, J. F. Bodet, L. E. Scriven, and W. G. Miller, Physica *A157*:470 (1989).

86. D. Quemada and D. Langevin, J. Theor. Appl. Mech. Num. Spec. 1985 p. 201.
87. S. Ajith, A. C. John, and A. K. Rakshit, Pure Appl. Chem. 66:509 (1994).
88. S. K. Mehta and K. Bala, Phys. Rev. E 51:5172 (1995).
89. E. W. Kaler, K. E. Bennett, H. T. Davis, and L. E. Scriven, J. Chem. Phys. 79:5673 (1983).
90. Z. J. Yu and R. D. Neuman, Langmuir 11:1081 (1995).
91. R. Hilfiker and H.-F. Eicke, J. Chem. Soc., Faraday Trans. I 83:1621 (1987).
92. J. Peyrelasse, M. Moha-Ouchane, and C. Boned, Phys. Rev. A 38:4155 (1988).
93. R. Hilfiker, H.-F. Eicke, and H. Thomas, NATO ASI Ser., Ser. B 258:531 (1991).
94. C. Boned and J. Peyrelasse, J. Surf. Sci. Technol. 7:1 (1991).
95. S. J. Chen, D. F. Evans, and B. W. Ninham, J. Phys. Chem. 88:1631 (1984).
96. D. Bedeaux, G. J. M. Koper, and J. Smeets, Physica A194:105 (1993).
97. C. Boned, J. Peyrelasse, and Z. Saidi, Phys. Rev. E 47:468 (1993).
98. C. Boned, Z. Saidi, P. Xans, and J. Peyrelasse, Phys. Rev. E 49:5295 (1994).
99. L. Garcia-Rio, J. R. Leis, J. C. Mejuto, M. E. Pena, and E. Iglesias, Langmuir 10:1676 (1994).
100. J. Peyrelasse, C. Boned, and Z. Saidi, Prog. Colloid Polym. Sci. 79:202 (1989).
101. S. Ray and S. P. Moulik, Langmuir 10:2511 (1994).
102. C. Mathew, Z. Saidi, J. Peyrelasse, and C. Boned, Phys. Rev. A 43:873 (1991).
103. K. Fontell, A. Ceglie, B. Lindman, and B. W. Ninham, Acta Chem. Chem. Scand. A 40:247 (1986).
104. U. Lenz, Dissertation, Universität Bayreuth, 1991.
105. C. M. Chen and G. G. Warr, J. Phys. Chem. 96:9492 (1992).
106. M. Yamamoto, J. Phys. Soc. Jpn. 12:1148 (1957).
107. D. W. Boger, Ann. Rev. Fluid Mech. 19:157 (1987).
108. G. C. Warr, Colloids Surf. A103:273 (1995).
109. R. B. Bird, R. C. Armstrong, and O. Hassager, *Dynamics of Polymeric Liquids*, Wiley, New York, 1987.
110. M. R. Anklam, R. K. Prud'homme, and G. G. Warr, AIChE J. 41:677 (1995).
111. M. M. Cross, J. Colloid. 20:417 (1965).
112. C. M. Chen and G. G. Warr, J. Phys. Chem. 96:9492 (1992).
113. P. Snabre and G. Porte, Europhys. Lett. 13:641 (1990).
114. C. A. Miller, M. Gradzielski, H. Hoffmann, U. Krämer, and C. Thunig, Colloid Polym. Sci. 268:1066 (1990).
115. C. Vinches, C. Coulon, and D. Roux, J. Phys. II (France) 2:453 (1992).
116. C. Blom and J. Mellema, Rheol. Acta 23:98 (1984).
117. M. Oosterbroek, H. A. Waterman, S. S. Wiseall, E. G. Altena, J. Mellema, and G. A. M. Kipp, Rheol. Acta 19:497 (1980).
118. C. Blom and J. Mellema, Prog. Colloid Polym. Sci. 76:228 (1988).
119. A. Eshuis and J. Mellema, Colloid Polym. Sci. 262:159 (1984).
120. J. Mellema, C. Blom, and J. Beekwilder, Rheol. Acta 26:418 (1987).
121. A. D'Aprano, G. D'Arrigo, M. Goffredi, A. Paparelli, and V. Turco-Liveri, J. Chem. Phys. 95:1304 (1991).
122. F. Mallamace, N. Micali, C. Vasi, and G. D'Arrigo, Phys. Rev. A 43:5710 (1991).
123. A. Y. Ozer, O. Cakoglu, B. Taylan, F. Mazda, and M. Summu, S.T.P. Pharm. Sci. 3:331 (1993).
124. J. B. Hayes, G. W. Haws, and W. B. Gogarty, U.S. Patent 4,012,329 (1978).
125. P. J. Atkinson, B. H. Robinson, A. M. Howe, and R. K. Heenan, J. Chem. Soc., Faraday Trans. 87:3389 (1991).
126. D. Capitani, A. L. Segre, G. Haering, and P. L. Luisi, J. Phys. Chem. 92:3500 (1988).
127. C. Quellet, H.-F. Eicke, and W. Sager, J. Phys. Chem. 95:5642 (1991).
128. H.-F. Eicke, U. Hofmeier, C. Quellet, and U. Zölzer, Prog. Colloid Polym. Sci. 90:165 (1992).
129. M. Yamamoto, J. Phys. Soc. Jpn. 11:433 (1956).
130. A. S. Lodge, Trans. Faraday Soc. 52:120 (1956).
131. L. G. Baxandall, Macromolecules 22:1982 (1989).
132. S. F. Edwards, Macromolecules 25:1516 (1992).

133. F. Tanaka and S. F. Edwards, J. Non-Newtonian Fluid Mech. *43*:247 (1992).
134. H.-F. Eicke, C. Quellet, and G. Xu, Colloids Surf. *36*:97 (1989).
135. C. Quellet, H.-F. Eicke, G. Xu, and Y. Hauger, Macromolecules *23*:3347 (1990).
136. U. Zölzer and H. F. Eicke, J. Phys. II *2*:2207 (1992).
137. P. G. de Gennes, Macromolecules *9*:587 (1976).
138. M. Odenwald, H. F. Eicke, and W. Meier, Macromolecules *28*:5069 (1995).
139. M. Gradzielski, A. Rauscher, and H. Hoffmann, J. Phys. IV, C1 *3*:65 (1993).
140. P. J. Flory, J. Chem. Phys. *18*:108 (1950).
141. D. B. Siano and J. Bock, Colloid Polym. Sci. *264*:197 (1986).
142. E. S. Lower and W. H. Harding, Manuf. Chem. *37*:52 (Sept. 1966).
143. F. B. Rosevear, J. Soc. Cosmet. Chem. *19*:581 (1968).
144. E. Nürnberger and W. Pohler, Prog. Colloid Polym. Sci. *69*:48 (1984).
145. E. Nürnberger and W. Pohler, Prog. Colloid Polym. Sci. *69*:64 (1984).
146. G. Oetter and H. Hoffmann, Colloids Surf. *38*:225 (1989).
147. L. Bohlin, H. Ljusberg-Wahren, and Y. Miezis, J. Colloid Interface Sci. *103*:294 (1985).
148. M. Gradzielski, H. Hoffmann, and G. Oetter, Colloid Polym. Sci. *268*:167 (1990).
149. M. Gradzielski, H. Hoffmann, J.-C. Panitz, and A. Wokaun, J. Colloid Interface Sci. *169*:103 (1995).

12

Light Scattering Studies of Microemulsion Systems

D. Langevin
Université Paris-Sud, Orsay, France

J. Rouch
Université Bordeaux I, Talence, France

I. INTRODUCTION

Elastic and inelastic light scattering are nowadays widely used techniques for the characterization of fluids. In particular, these techniques have been extensively and successfully used with microemulsion systems to obtain information about droplet sizes. Surface light scattering is a less common technique but has been used with microemulsion interfaces, in particular to measure the ultralow interfacial tensions found in these systems. In this chapter we discuss these aspects, first recalling the theoretical background and illustrating the potential of the techniques with experimental results.

II. BULK LIGHT SCATTERING

Bulk light scattering is a very popular method for the measurement of characteristic sizes in colloidal systems [1].

A. Light Scattering by Microemulsion Droplets

1. Monodisperse Spherical Microemulsion Droplets

In the elastic light scattering method, one measures the scattered intensity only. At a selected scattering angle θ, the intensity scattered in the θ direction for monodisperse spherical microemulsion droplets is given by

$$I(q) = kV\phi P(q)S(q) \tag{1}$$

where q is the scattering wave vector $q = (4\pi/\lambda) \sin(\theta/2)$, k a constant characteristic of the instrument, V the microemulsion droplet volume, ϕ the droplet volume fraction, P the form factor, and S the structure factor. Often, microemulsion droplets have radii R small compared to the wavelength of the light λ, and $qR \ll 1$. Then $P(q) \sim 1$ and

$S(q) \sim k_B T V/(\partial \Pi/\partial \phi)$, where T is the absolute temperature, k_B the Boltzmann constant, and Π the osmotic pressure of the droplets. The scattered intensity is then independent of q and is proportional to $\phi/(\partial \Pi/\partial \phi)$.

In the inelastic or quasi-elastic light scattering (QELS) method, the autocorrelation function of the scattered light is measured. Again for monodisperse spherical objects,

$$C(\tau) = \langle E(t)E(t+\tau) \rangle = E^2 \, \exp(-Dq^2\tau) \tag{2}$$

where E is the electric field of the light beam, the product inside the angular brackets is averaged over time t, and τ is the correlation time. One generally measures the correlation function of the scattered intensity:

$$C^2(\tau) = \langle I(t)I(t+\tau) \rangle = [C(\tau)]^2 + 1 = I^2 \, \exp(-2Dq^2\tau) + 1 \tag{3}$$

D is the diffusion coefficient of the droplets, given by

$$D(q) = D_0 \frac{H(q)}{S(q)}$$

$H(q)$ being the hydrodynamic factor and D_0 the Stokes diffusion coefficient, given by

$$D_0 = \frac{k_B T}{6\pi \eta_c R_H} \tag{4}$$

with η_c the viscosity of the continuous phase and R_H the hydrodynamic radius. In the limit $qR \ll 1$, $D(q)$ is independent of q and equal to

$$D = \frac{V \, \partial \Pi/\partial \phi}{f} \tag{5}$$

where f is the friction coefficient, which depends on ϕ, and is equal to $6\pi \eta_c R_H$ in the limit $\phi \to 0$.

In the same limit,

$$\left(\frac{\partial \Pi}{\partial \phi} \right)_{\phi \to 0} = \frac{k_B T}{V} \tag{6}$$

Since $V = 4\pi R^3/3$, one sees that *if the microemulsion can be diluted to very small volume fractions* (typically 1% or less), then R and R_H can be deduced from I and D. However, this is a very difficult task in many microemulsions, which are multicomponent mixtures. A dilution procedure was proposed early by Schulman et al. [2]. This allowed them to measure the microemulsion droplet radius using light scattering for the first time. The method was refined later by Graciaa et al. [3,4] and it became possible to use this method to obtain reliable radius values. Neutron and light scattering performed on the same systems showed that this dilution procedure was satisfactory [5]. Microemulsion phases are less frequent in ternary mixtures and are obtained only with particular surfactants, such as AOT [6–13]. In this case, dilution is straightforward. In theory, R and R_H are equal for droplet systems. In practice, R_H is frequently slightly larger in microemulsions because of solvent penetration in the surface layer.

2. Nonspherical Droplets

A similar difference between R and R_H can occur if the drops are not spherical. Microemulsion droplets are usually close to spherical, and it is very difficult to separate out the contributions of solvent penetration and droplet elongation from the light scattering data alone. A useful complementary experiment is the measurement of the solution viscosity η. For dilute enough systems,

$$\eta = \eta_c(1 + \beta\phi) \tag{7}$$

For spheres, β, the Einstein coefficient, is 2.5. For nonspherical particles, β is larger and strongly depends on the elongation ratio [14].

3. Virial Coefficients

It is not easy to predict the ϕ dependence of the osmotic pressure Π and the friction coefficient f. When ϕ is not too large, a virial expansion can be sufficient, i.e.,

$$\frac{\partial \Pi}{\partial \phi} = \frac{k_B T}{V}(1 + B\phi) \tag{8}$$

and

$$D = D_0(1 + \alpha\phi) \tag{9}$$

where α and B are virial coefficients. For hard spheres, $B = 8$ and $\alpha \sim 2$ [14]. The virial coefficients are smaller if there are supplementary attractive interactions and are larger in the case of additional repulsive interactions. It is also possible to relate B and α: $\alpha \sim 2 + (B - 8)/2$ [15]. Figure 1 shows an example of two microemulsions, one hard-sphere-like ($B = 7$, $R = 4.4$ nm), and the other with significant attractive interactions ($B = -12$, $R = 3.4$ nm) [16].

4. Polydisperse Droplets

The microemulsion droplets are generally quite monodisperse (typically the polydispersity p is around 10%). We see later (Sec. III) that this polydispersity depends on the bending elasticity of the surfactant layer. The correlation function is therefore not perfectly exponential. For small polydispersities, it can be shown that the correlation function can be approximated by a sum of two exponentials:

$$C(\tau) = A_1 \exp(-Dq^2\tau) + A_2 \exp(-D_s q^2\tau) \tag{10}$$

where D_s is the self-diffusion coefficient and A_1 and A_2 are constants. D_s describes the diffusive motion of a tracer object in a uniform medium, whereas D describes the diffusive motion of an object among other similar objects when the concentration is not uniform. In a light scattering experiment, D is associated to concentration fluctuations and goes to zero at a critical point; D_s is associated to size or connectivity fluctuations and goes to zero at a percolation point (formation of an infinite aggregate of droplets). One has for spheres

$$D_s = k_B T / f' \tag{11}$$

where f' is another friction coefficient. f and f' have the same limit for $\phi \to 0$. The ratio

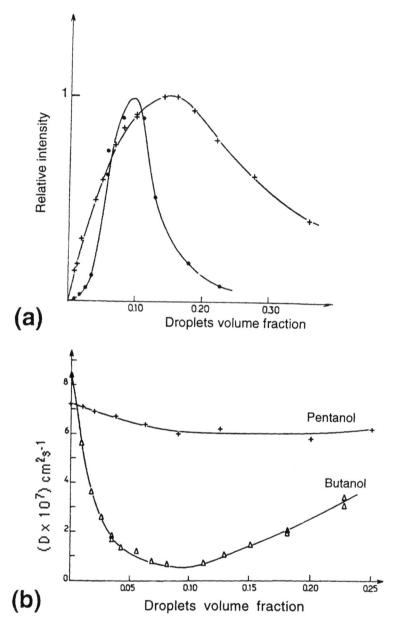

(a)

(b)

Figure 1   Scattered intensity (a) and diffusion coefficient (b) versus droplet volume fraction ϕ for two W/O microemulsions made with water, toluene, sodium dodecyl sulfate, and two different alcohols—butanol and pentanol. Data for pentanol microemulsions are represented with crosses. Although the compositions are similar, the interactions between droplets are very different: hard-sphere-like with pentanol and attractive with butanol. (Data from Ref. 16.)

A_1/A_2 is proportional to S/p. The deviation of $G(\tau)$ from a single exponential is therefore important if the polydispersity p is large or if the structure factor S is small (repulsive drops). This was checked with O/W drops in which repulsive electrostatic interactions were present: D and D_s had opposite ϕ dependences and a common limiting value for $\phi = 0$ (Fig.2) [17].

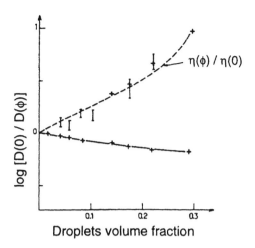

Droplets volume fraction

Figure 2 Plot of reduced viscosity η/η_c (dotted line) and inverse reduced diffusion coefficients D_0/D (lower crosses, from QELS experiments) and D_0/D_s (bars, from forced Rayleigh scattering experiments) versus droplet volume fraction for an O/W microemulsion. The upper crosses are the values of D_0/D_s as deduced from the fit of the correlation function of the scattered light with Eq. (10). (Data from Ref. 17.)

It is possible to create in the samples a refractive index gradient by using a light fringe pattern of spacing i. For instance, if photochromic probes are dissolved in the sample, the refractive index will vary in the light fringes. This is the principle of the forced Rayleigh scattering technique: This scattering produces a Bragg spot in the scattering angle associated with i. If the light fringes are suppressed, the refractive index gradient progressively disappears because of the diffusion of the photochromic probes. The time dependence of the intensity scattered in the Bragg direction allows determination of the diffusion coefficient. For instance, if the dye is in the microemulsion droplets (and stays there), then the diffusion coefficient measured is the droplet self-diffusion coefficient D_s [17]. Figure 2 shows the forced Rayleigh scattering determinations of D_s together with light scattering determinations of D and D_s with the same samples.

In some cases, the index of refraction of the spheres can be matched by the index of refraction of the solvent. The scattered intensity should vanish if the spheres are monodisperse, because the contrast would be identically zero at this point. The droplets being actually polydisperse, the residual intensity is a direct measure of the polydispersity. This procedure has been used in light scattering experiments on AOT microemulsions [18]. It can also be used in neutron scattering experiments, where the contrast matching is more easily adjusted by using mixtures of protonated and deuterated oils or water [19,20].

5. Concentrated Microemulsion Droplets

When ϕ is larger than typically a few percent, the above description no longer applies. In the case of hard spheres, the osmotic pressure is well represented by the Carnaham and Starling formula up to large volume fractions [21]. The friction coefficient is more difficult to evaluate because the hydrodynamic interactions are long-range. Microemulsion droplets behave as hard spheres in many circumstances. However, in W/O microemulsions, droplets frequently exhibit supplementary attractive interactions. It has been proposed that the osmotic pressure

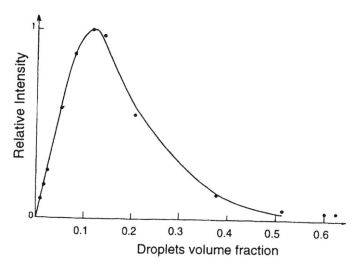

Figure 3   Relative scattered intensity for a W/O microemulsion made of SDS, pentanol, water, and cyclohexane, versus droplet volume fraction ϕ. The curve is the fit with Eqs. (1) and (12). (Data from Ref. 16.)

could be described by adding to the main hard-sphere contribution Π_{hs} a perturbation proportional to ϕ^2, i.e.,

$$\Pi = \Pi_{hs} + \frac{k_B T}{2V} A\phi^2 \tag{12}$$

where A is a constant. This expression leads to very good fits of the scattered intensity data up to large volume fractions (Fig. 3) [21]. It is more difficult to describe the volume fraction dependence of the diffusion coefficient. It should be noted, however, that even for small ϕ, the calculated virial coefficient α is larger than the one measured for attractive droplets [16]. This discrepancy may be related to the transient character of the droplets that exchange material with the continuous phase at rates faster than the correlation times in the light scattering experiment. Koper et al.[22] have proposed a theoretical analysis of the inelastic light scattering data taking into account these exchanges.

There is a good correlation between the strength of the attractive interactions and the amplitude of the electrical birefringence signals as well as the electrical conductivity variation around the percolation threshold [23]. This is because when attractive interactions are present, the formation of droplet aggregates is favored. These aggregates are anisotopic and are easily oriented by an electric field. They also promote percolation at lower volume fractions than for a hard sphere. This is discussed in more detail in Sec. II. B.

In the case of very concentrated systems ($\phi > 50\%$), the correlation function of the scattered light becomes nonexponential and is rather fitted by stretched exponentials. An analogy between these systems and gel systems has been proposed [24].

B.   Percolation and Critical Behavior

When the attractive interactions are strong enough, a phase separation between droplet-rich and droplet-poor phases occurs above a critical point. The light scattering behavior close to the critical point is similar to that of simpler fluids. The scattered intensity goes to infinity, and the diffusion coefficient goes to zero. The relation between the critical behavior and the percolation phenomenon is particular and is discussed below in some detail.

Close to the consolute critical point of a binary mixture a critical slowing down of dynamical properties associated with the order parameter are observed. These phenomena are medicated by the so-called long-range correlation length, which is a length characterizing the spatial decay of the static droplet correlation function $g(r)$. Close to the critical point and along the critical isochore, ξ is given by $\xi = \xi_0 \varepsilon^{-\nu}$, where ξ_0 is a constant and $\nu = 0.63$ is a universal exponent. From, Ornstein–Zernike (OZ) theory, the scattered intensity $I(q)$ at wave vector q, which is proportional to the Fourier transform of $g(r)$, can be written as

$$I(q) = I_0 \frac{\chi T}{1 + q^2 \xi^2} \tag{13}$$

where χ_T is the osmotic compressibility, I_0 a constant (at least close to T_c), and q the wave vector. In fact, some deviation from the OZ theory is observed, and the q-dependent intensity is usually fitted to experimental data by using the following static scaling function of the scaling variable $x = q\xi$:

$$g(q\xi)^{-1} = 1 + \Sigma_2 x^2 - \Sigma_4 x^4 \tag{14}$$

where Σ_2 and Σ_4 are constants. In the OZ theory, values of these constants are respectively 1 and 0 [25]. In simple fluids, a nonzero value of Σ_4 can usually be accounted for by the introduction of a small positive Fisher's critical exponent $\eta \simeq 0.03$ [26].

The decay rate (first cumulant) Γ of the time-dependent correlation function associated with the order parameter fluctuations can be measured by dynamic light scattering. Mode-coupling theories of critical phenomena including dynamical background contributions predict that is the sum of a critical part Γ_c and a background part Γ_B, i.e.,

$$\Gamma = \Gamma_c + \Gamma_B \tag{15}$$

The critical part Γ_c can be represented by

$$\Gamma_c = Rq^3 \frac{k_B T}{6\pi\eta(T)} \Omega(x) \tag{16}$$

where $R \simeq 1.03$ is the universal amplitude ratio and $\Omega(x)$ is the dynamical universal scaling function given by

$$\Omega(x) = (1/x^3)K(x)\sigma(x)^{x_\eta} \tag{17}$$

where $K(x)$, the Kawasaki function, is given by

$$K(x) = \frac{3}{4}\left[1 + x^2 + \left(x^3 - \frac{1}{x}\right)\tan^{-1} x\right] \tag{18}$$

and $\sigma(x)$, the correction to scaling, by

$$\sigma(x) = [1 + (b/a_0)^2 x^2]^{1/2} \tag{19}$$

The coefficient a_0 is related to the universal amplitude ratio R by $a_0 = R^{1/x_\eta}$, b is a system-dependent parameter, and x_η is the exponent describing the increase in the shear viscosity when approaching T_c. The theoretical expression for the shear viscosity $\eta(T)$ is

$$\eta(T) = \eta_B(Q_0\xi)^{x_\eta} = \eta_B(Q_0\xi_0)^{x_\eta}\varepsilon^{-\nu x_\eta} \tag{20}$$

in which η_B is the nonsingular part of the viscosity and Q_0 a cutoff parameter. A universal value of $x_\eta\nu \simeq 0.04$ has been found for many critical systems [27].

Analytical expressions for the background part Γ_B have been proposed both by Burstyn et al. [28] and by Oxtoby and Gelbart [29]. The Burstyn et al. expression reads

$$\Gamma_B = q^2\left(\frac{k_BT}{16\eta_B\xi}\right)\left(\frac{1+q^2\xi^2}{q_c\xi}\right) \tag{21}$$

where q_c is a Debye cutoff wavenumber. On the other hand, the one derived from Oxtoby and Gelbart [29] is

$$\Gamma_B = RCq^2\left(\frac{k_BT}{8\pi\eta_B\xi}\right)\left(\frac{1+q^2\xi^2}{q_c\xi}\right) \tag{22}$$

where $C=0.9$.

We now present experimental data on the extensively studied AOT–water–decane microemulsion system. At low or moderate volume fraction ϕ and not too high a temperature, one observes water-in-oil microemulsions with quasi-spherical and rather monodispersed droplets, whose sizes depends essentially on the molar ratio $X=[\text{water}]/[\text{AOT}]$. The radius of the water core of the microemulsion droplet is on the order of 5 nm at $X=40.8$. In the temperature–volume fraction phase diagram, these mixtures may show a cloud point curve with a lower critical point. At $X=40.8$, the coordinates of the critical point are $\phi_c = 0.098$ and $T_c = 39.9°\text{C}$. At the critical volume fraction, and for temperatures above T_c, the microemulsion phase separates. It has been shown by neutron scattering [30] that the two phases are produced with droplets having the same diameter but different volume fractions. Starting from the immediate vicinity of the critical point at low volume fraction and crossing all along the pass diagram toward high volume fraction, an electrical conductivity percolation line has been measured. In a narrow temperature range, when crossing the percolation line at constant ϕ, a very strong increase in the low frequency electrical conductivity σ is observed. The presence of a critical point and that of a percolation line are mediated by attractive interactions among droplets that do not behave as purely hard spheres. This is a very general observation for W/O microemulsion droplets, and the percolation and critical volume fractions are close in many systems [23].

The scattered light intensity correlation function $C^2(t)$ has been measured over a very wide range of temperature and wave vectors for various systems. Typical intensity correlation functions $C^2(t)$ are depicted in Fig. 4 for the sake of illustration. These graphs show systematic deviations from the usual exponential decay, which are also observed for most of the systems we report in this part. As a remark, nonexponential decays that are small at low concentration, close to the critical point, become large for dense systems. From the initial slope of the time-dependent intensity correlation function, one can deduce the first cumulant Γ, which is the relaxation rate of the order parameter fluctuations.

When the fits are performed with adjustable values of the background and exponents, the best fit of the data of Fig. 4 is obtained with $\gamma=1.40\pm0.04$, $\nu=0.70\pm0.03$, and $\xi=1.33\pm0.15$ nm. The exponents inferred from the fit are consistent with a Fisher-like

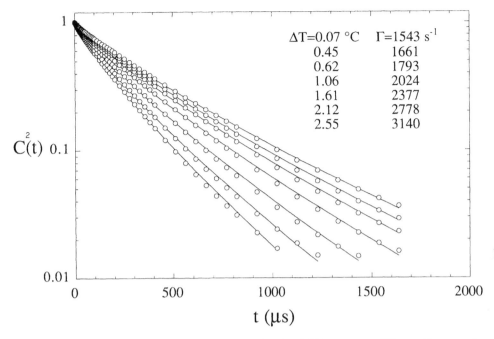

Figure 4 Scattered light correlation functions for the AOT system at different distances from the critical point, from 0.07° (top curve) to 2.55° (bottom curve). The curves are fits with the droplet model (see Appendix). The corresponding values of the first cumulant Γ are given after ΔT. Γ can be calculated by taking the $t = 0$ limit of the logarithmic derivative of $C(t)$. (Data from Refs. 31 and 32.)

renormalization of the critical indices. Applying Eq. (22), it can be shown that the inverse of the Debye cutoff is on the order of the diameter of the microemulsion droplet [31,32]. However, a discrepancy between the experimental results and the theoretical prediction is still observed in the hydrodynamic regime far from T_c. To account consistently for the background produced by the scattering of droplets, a complete model can be developed that incorporates the existence of droplet aggregates. Details can be found in the Appendix to this chapter.

For microemulsion systems with more than three components, it is more difficult to locate the critical point. Figure 5 shows the intensity and the diffusion coefficient for a near critical microemulsion. Both I and D follow the classical Orstein–Zernike and Kawasaki predictions. The correlation length can be determined from the q dependence of I and D and from $D(q = 0)$. The three determinations are consistent provided one is close to the critical point and that all the other light scattering contributions can be neglected. The critical behavior also produces increased light scattering from the interfaces between coexisting phases. This is discussed in Sec. III. E.

C. Other Microstructures

At large volume fractions ($\phi > 30\%$) for hard-sphere-like systems or above the percolation threshold for attractive spheres, the sphere description becomes unsatisfactory. This is probably because either the system is inverting from a W/O system to an O/W one or the droplets

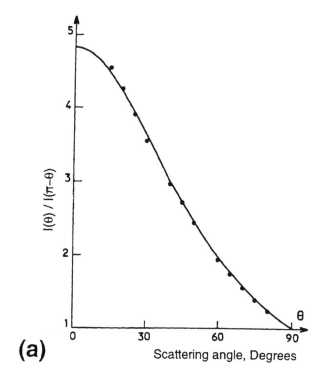

(a)

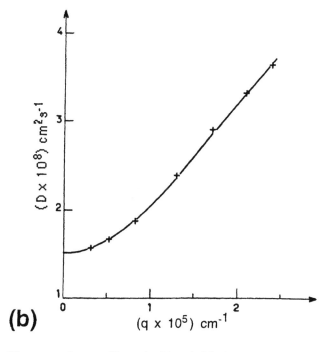

(b)

Figure 5 Scattered intensity (a) and diffusion coefficient (b) versus scattering angle and wave vector for a near-critical microemulsion made of water, sodium chloride, toluene, sodium dodecyl sulfate, and butanol. The lines are fits with Ornstein–Zernlike and Kawasaki formulas. (Data from Ref. 33.)

progressively evolve toward a bicontinuous structure. In the bicontinuous structure, there is still a well-defined characteristic size ξ_m, which is the average mean square distance between oil and water microdomains. Light scattering is unable to provide information about such a structure; dilution is no longer possible, and the light scattering technique does not provide enough information to separate the roles of the characteristic sizes, the characteristic shapes, and the interactions in the medium. Other methods such as X-ray or neutron scattering must be used. For bicontinuous structures, the correlation function of the scattered light is generally exponential, but if the size is extracted by using the Stokes formula, this size is typically one order of magnitude smaller than the actual characteristic size. This illustrates the influence of molecular exchanges, particularly important in the bicontinuous microemulsions. From a dynamics point of view, a bicontinuous microemulsion looks very much like a molecular solution.

III. SURFACE LIGHT SCATTERING

Light scattering by liquid surfaces is a powerful technique to investigate surface properties such as surface tension and surface viscoelasticity [34]. There are many methods available for the study of surface or interfacial tension. The light scattering method, however, is particularly useful when the interfacial tensions are low. Ultralow interfacial tensions cannot be measured with conventional devices such as Wilhelmy plates or drop shape methods, except for the so-called "spinning drop" method. The two- and three-phase equilibria involving microemulsions and excess phases are called Winsor equilibria [35], and the corresponding interfacial tensions γ are ultralow. As shown in the following, they are related to the microemulsion structure, i.e., $\gamma \sim k_B T / L^2$, L being the characteristic size of the microemulsion (R or ξ_m). The interfaces involving two microemulsions close to a critical point also have ultralow interfacial tensions, i.e., $\gamma \sim k_B T / \xi^2$.

A. Principles

Because the surface light scattering method is less well known than the bulk scattering one, we will give some details about its principle. In the surface scattering method, a laser beam impinges on the surface. Because of thermal motion, the surface is not perfectly flat and scatters light in all directions around the specularly reflected beam. The surface displacement ζ can be described as a superposition of sinusoidal deformations $\zeta = \sum_q \zeta_q(t) e^{i\mathbf{q}\mathbf{r}}$, where $\mathbf{r}$ is the position in the equilibrium plane of the surface and $\mathbf{q}$ is the wave vector. As for bulk scattering, each sinusoidal component scatters light in a direction related to $\mathbf{q}$ by $q = (\mathbf{k}_d - \mathbf{k}_r) \times n$, where $\mathbf{k}_d$ and $\mathbf{k}_r$ are the wave vectors of the scattered and reflected light, respectively, and $\mathbf{n}$ is the unit vector normal to the surface. The intensity rapidly decreases when q increases (as q^{-2}) and becomes smaller than the bulk scattering for scattering angles larger than a few degrees. One is therefore limited to small scattering angles and small wave vectors: $100 < q < 1000 \text{ cm}^{-1}$.

The time evolution of $\zeta_q(t)$ is described on average by the hydrodynamic equations subject to boundary conditions at the surface. The evolution is the same as for the capillary waves of wavelength $2\pi/q$, whose frequency and damping can be extracted from the dispersion equation. The light scattered by a sinusoidal component ζ_q is frequency-shifted due to the Doppler effect. The frequency shifts are small (in the kilohertz range or below), but they can be analyzed with optical mixing techniques. The power spectrum of the photomultiplier current is then generally a Lorentzian curve; its center frequency is the frequency of the capillary waves $\omega(q)$, and its half-width is the inverse damping time $\tau(q)$ of the waves.

In the case of microemulsion systems, the interfacial tensions are low and the capillary waves are overdamped, so the spectra are centered at zero frequency. They are usually Lorentzian, of half-width

$$\Delta\omega = \frac{\gamma + (\rho_2 - \rho_1)g/q}{2(\eta_1 + \eta_2) + \mu q} \tag{23}$$

where γ is the interfacial tension, ρ_2 and ρ_1 the densities and η_2 and η_1 the viscosities of the coexisting phases, μ the transverse shear viscosity, and q the scattering wave vector. When γ is ultralow and q is small, the gravity term $(\rho_2 - \rho_1)g/q$ might not be negligible. To determine γ and μ, ρ_1, ρ_2, η_1, and η_2 should be measured independently at the same temperature. Because the surfactant layers are very flexible in microemulsion systems, μ is negligible; in all the published studies it was found to be zero within instrumental accuracy.

The light scattering intensity can also be measured, but for the small scattering angles probed by the technique, it is usually mixed with stray light, and these intensity measurements might not be reliable.

B. Origin of the Low Interfacial Tensions

The surface light scattering method has been used to show that the low interfacial tensions in the Winsor I and II systems (O/W microemulsion in equilibrium with excess oil and W/O microemulsion in equilibrium with excess water, respectively) are due to the large surface pressure of the surfactant monolayer coating the interface, which almost balances the bare oil/water interfacial tension [36,37]. Schulman and Montagne [38] proposed early that the low interfacial tensions in microemulsion systems should be associated with these large surface pressures π, i.e., $\gamma = \gamma_0 - \pi \sim 0$. In other models, the origin of the low interfacial tensions was attributed to the vicinity of critical points [39,40].

To clarify the problem, Winsor systems composed of water, sodium chloride, toluene, butanol, and SDS (sodium dodecyl sulfate) were investigated [36,37]. The transitions Winsor I $\rightarrow$ Winsor III $\rightarrow$ Winsor II are obtained by increasing the water salinity S. By refining the Schulman et al. [2] and Graciaa et al. [3,4] dilution methods, it was possible to dilute the O/W and W/O microemulsions down to droplet volume fractions of less than 1%. The volumes of the dilute phases remain unchanged when they are placed in contact with the corresponding excess phases, and the interfacial tension between these two two-phase systems is about the same. This is very similar to the behavior observed in simpler micellar solutions, i.e., γ decreases when the surfactant concentration increases and saturates to a constant value above the critical micelle concentration (cmc). The value of the surface tension does not change appreciably between the cmc ($\sim 10^{-5}$ mN/m) and the high surfactant concentration of the microemulsion. This shows unambiguously that the low value of the surface tension has nothing to do with the presence of droplets. As in aqueous surfactant systems, the value of the surface tension is determined by the surface pressure of the monolayer.

Overbeek and coworkers [41] studied oil–water–alcohol–surfactant mixtures close to the cmc and showed that ultralow interfacial tensions could be obtained as well in their systems. They determined the monolayer composition from the $\gamma(c)$ curve below the cmc. Similar work on ternary oil–water–surfactant mixtures was reported by Aveyard et al. [42]. In these studies, γ was measured by using a spinning drop tensiometer.

C. Wetting Behavior

The largest interfacial tension in the Winsor III systems (bicontinuous microemulsions in equilibrium with both excess oil and water) is also equal to the interfacial tension between oil and water in the presence of a saturated surfactant monolayer, γ_{ow}, i.e.,

$$\gamma_{ow} = \sup(\gamma_{om}, \gamma_{wm}) \tag{24}$$

where the subscripts o, w, m stand for organic, aqueous, and microemulsion phases, respectively. Here, the strict Antonov inequality is verified, i.e., $\gamma_{ow} < \gamma_{om} + \gamma_{wm}$, the equality corresponding to the wetting case. Because of the fact that the interfacial tensions are ultralow, the microemulsion should not wet the oil/water interface, and this is observed experimentally [43,44].

In simple liquid systems, interfacial tensions are small only in the vicinity of a critical point. Let us call a, b, and c the three phases in equilibrium, and suppose that the critical interface is a/b and that phase b does not wet the a/c interface. Theory shows that γ_{ab} vanishes as $(\varepsilon - \varepsilon_c)^{\mu}$ and that $\gamma_{ac} - \gamma_{bc}$ vanishes as $(\varepsilon - \varepsilon_c)^{\beta'}$ when the critical point is approached [45]. Since the Antonov inequality is

$$\gamma_{ac} - \gamma_{bc} < \gamma_{ab}$$

it is seen that it cannot be satisfied very close to the critical point because γ_{ab} decreases much faster than $\gamma_{ac} - \gamma_{bc}$ when this point is approached ($\mu > \beta'$). Obviously, the distance at which wetting occurs depends on the system. In microemulsion systems, the critical points have not been approached close enough in the reported experiments, although the interfacial tensions were already ultralow. Again this points out that the main cause of the low interfacial tensions is the high surface pressure of the surfactant monolayers at the interface.

Experiments done in three-phase systems made with small molecules showed that the middle phase always wetted the interface between the upper and lower phases. This behavior was exhibited in particular by oil–water–short-chain nonionic surfactant mixtures [44]. It was therefore tempting to observe the behavior of intermediate-chain surfactants. As expected, a wetting transition was observed in the temperature range where the middle phase for the system C_6E_2–hexadecane–water existed [46,47]. The particular behavior of microemulsions is associated with the presence of monolayers, which do not form with small molecules. The microstructure of microemulsions is stabilized by similar monolayers. The wetting transition is then probably associated with the disappearance of the monolayers and of the microemulsion microstructure, leading to molecular mixtures of oil, water, and surfactant. As expected, this happens when the surfactant chain length is small and when the amphiphilicity of this molecule has been reduced compared to that of surfactants of longer chain length ("good" surfactants).

D. Relation Between Surface Tension and Surfactant Film Bending Elasticity

When the low interfacial tensions depend on the surfactant film properties, it is possible to relate these interfacial tensions to the film bending energy F, the expression for which has been given by Helfrich [48]:

$$F = \tfrac{1}{2}K(C_1 + C_2 - 2C_0)^2 + \overline{K}C_1C_2 \tag{25}$$

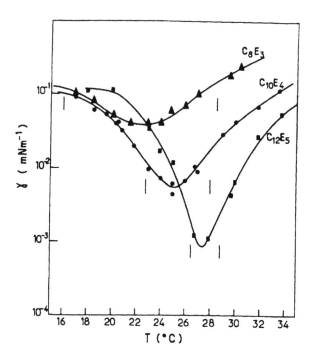

Figure 6 Interfacial tension between oil and water in the presence of a saturated surfactant monolayer versus temperature. The surfactants are alkyl polyoxyethylene glycol ethers: $C_{12}E_5$ with hexane, $C_{10}E_4$ with octane, and C_8E_3 with decane. The vertical bars indicate the transition temperatures between the different Winsor systems. (Data from Ref. 50.)

where C_1 and C_2 are the two principal curvatures of the surfactant layer, C_0 its spontaneous curvature, and K and $\overline{K}$ the mean and Gaussian bending elastic constants. The type of microstructure is closely related to the sign of the spontaneous curvature of the surfactant layer C_0; by convention, $C_0 > 0$ for aqueous dispersions and $C_0 < 0$ for reverse systems. One can show that the interfacial tension between a microemulsion made of oil or water droplets and excess oil or water is given by [49]

$$\gamma = \frac{2K + \overline{K}}{R^2} + \frac{k_B T}{4\pi R^2}(\ln \phi - 1) \tag{26}$$

Figure 6 shows the variation of γ for three nonionic surfactants. In the three-phase region, γ is the interfacial tension between the oil and water excess phases. As for other Winsor systems, the surface tension minimum occurs in the three-phase region where the characteristic size is maximum. Data analysis in the regions where the structure is made of droplets leads to values of $2K + \overline{K}$ that increase with surfactant chain length, as expected; i.e., theory predicts that the elastic constants should vary as l^3, l being the surfactant film thickness [51]. Ellipsometric determinations of K in the same systems are in agreement with these results [50].

The combined value of $2K + \overline{K}$ can also be deduced from the droplet polydispersity. Indeed, the surfactant film at the surface of the droplet will undulate due to thermal energy. As a result, the shape of a given droplet constantly fluctuates around an average spherical form with a mean radius R. Because this droplet can exchange material with surrounding

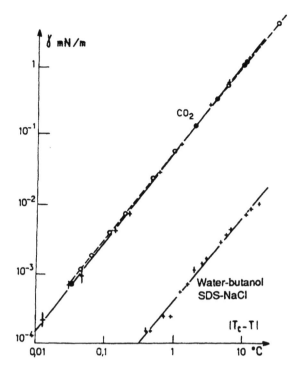

Figure 7 Interfacial tension versus distance to the critical point for CO_2 and a microemulsion system, showing that the scale factor γ_0 is smaller for the microemulsion. (Data from Ref. 33.)

droplets, the shape fluctuations are on average fluctuations at constant total surface area and constant total internal volume, and their amplitude is related to both K and $\bar{K}$ [52]. These shape fluctuations are equivalent to an average polydispersity, the nature of which is obviously very different from the polydispersity in more conventional colloidal systems such as dispersions of solid particles. It has been observed indeed that the polydispersity of microemulsion droplets seems to depend on the particular experiment carried out for the determination of this quantity; it is much smaller when measured with dynamic light scattering than with elastic light or neutron scattering (see, e.g., Refs. 16 and 19). This is simply because in dynamic measurements, averages are performed over times less than the time scale of the experiment, whereas in light scattering experiments, time scales are typically microseconds, i.e., comparable to or longer than the time for material exchange between the droplets and the surrounding medium. In static light scattering experiments, one measures average overall instantaneous droplet shapes, and the measurement probes the complete shape distribution. Polydispersity measurements performed on the systems of Fig. 7 give the same $2K + \bar{K}$ values as those of surface tension determinations [53].

E. Critical Behavior

The lowest interfacial tensions in Winsor systems have a different origin than that of the largest ones; they do not depend on the surfactant film properties and are due to the nearness of a critical point. Close to the boundaries Winsor I → Winsor III, $S = S_1$ and Winsor III → Winsor II, $S = S_2$, the corresponding excess phase becomes slightly turbid. This is reminiscent of the vicinity of a critical point. This particular type of critical point, where two phases become critical in the presence of a noncritical third phase, is called the critical end point [54]. The Winsor phase boundaries are generally not the critical end points, but they are close to these. Theory predicts that

$$\gamma = \gamma_0 \left(\frac{\varepsilon - \varepsilon_c}{\varepsilon_c} \right)^\mu \tag{27}$$

where μ is a critical exponent, ε a field variable, ε_c its critical value, and γ_0 a scaling factor.

Instead of measuring chemical potentials, a very difficult task in these multicomponent systems, it is easier to determine the critical behavior of a different property, the density difference between the phases, for instance:

$$\rho_2 - \rho_1 = \Delta\rho = \Delta\rho_0 \left(\frac{\varepsilon - \varepsilon_c}{\varepsilon_c} \right)^\beta \tag{28}$$

Eliminating $(\varepsilon - \varepsilon_c)/\varepsilon_c$ between Eqs. (27) and (28), one finds

$$\gamma = \gamma_0 \left(\frac{\Delta\rho}{\Delta\rho_0} \right)^{\mu/\beta} \tag{29}$$

Before the optimal salinity S^* for γ_{wm} and after S^* for γ_{om}, the variation of γ with $\Delta\rho$ is of the same form as predicted by Eq. (29), the optimal salinity being the salinity at which $\gamma_{wm} = \gamma_{om}$. It was found that $\mu/\beta = 4 \pm 0.4$ [55]. This result is in agreement with the 3D Ising values of the exponents $\mu = 2\nu$, $\nu = 2\beta$, where ν is the critical exponent for the correlation length ξ. It disagrees with mean-field exponents $\mu = 3$, $\beta = 1$. Similar results were found later by using a spinning drop technique [56] on other systems. The power law applies to only the smallest interfacial tensions. Above S^* for γ_{wm} and below S^* for γ_{om}, deviations clearly appear. These deviations are not surprising because the greater interfacial tensions are associated with monolayer properties and have nothing to do with critical behavior. Several authors extended the critical plots in the whole three-phase region and eventually in two-phase regions as well and concluded that all the interfacial tensions were low because of the vicinity of the critical end points [57–59]. This is because the accuracy of their measurements was not good enough to detect the deviations from critical behavior.

Turning back to the lowest interfacial tensions, and in order to fully confirm the critical behavior, the correlation length in the bulk microemulsion was measured by using elastic and inelastic bulk light scattering technique. ξ diverges as

$$\xi = \xi_0 \left(\frac{\varepsilon - \varepsilon_c}{\varepsilon_c} \right)^\nu \tag{30}$$

with $\nu = \mu/2$. The different ξ determinations [from $I(q)$, $D(q)$, and $D(q=0)$, see Sec. II. A. 5] are in agreement close to S_2; but close to S_1, I is q-dependent whereas D is not. The description in terms of individual droplets is more appropriate, at least for $S < S_1 : \xi \gg R$ only very close to S_2. The critical behavior is much less important for bulk properties than for surface tension. This is specific of microemulsions, for which the universal combination of scale factors $\gamma_0 \xi_0^2/k_B T$ is smaller than for simpler fluids (Fig. 7) [33,60]. This means that microemulsions may not belong to the same universality class as simple fluids.

In the SDS system, ξ does not become large enough to determine the ratio μ/ν accurately. This ratio was determined by Kim et al. [58,59] in another Winsor system where γ falls below 10^{-3} mN/m and ξ increases above 100 nm; they found $\nu/\mu = 0.52 \pm 0.04$. Critical behavior was also observed in a different type of three-phase microemulsion system: a W/O microemulsion (lower phase), a spongelike phase (middle phase), and excess oil [61]. The exponents measured for the interfacial tension were also in agreement with the Ising exponents.

IV. CONCLUSION

We have described bulk and surface light scattering techniques and some of their applications to microemulsion systems. These techniques are extremely powerful. The bulk light scattering limitations are associated with the fact that the light wavelength is much larger than characteristic sizes. However, we have seen that it is useful to combine this technique with neutrons or X-ray scattering.

APPENDIX

A. Dynamic Light Scattering from Critical Fractal Aggregates

A model to explain dynamic light scattering in terms of droplets was first proposed by Sorensen et al. [62] and Martin and coworkers [63,64]. A modified version of this dynamical droplet model making use of the Coniglio and Klein results [65] was proposed by Tartaglia and coworkers [66,67]. In this model it is assumed that the critical mixture is made of dilute polydispersed dynamical fractal clusters having a fractal dimension d_f and a polydispersity index τ. Starting from percolation theory [65], it is possible to reproduce the critical 3D Ising behavior by writing

$$d_f = d - \frac{\beta}{\nu} \tag{A.1}$$

and

$$\tau = 1 + \frac{d}{d_f} \tag{A.2}$$

where β and ν are the critical indices of the coexistence curve and of the correlation length. In three dimensions $d = 3$, and the above equations lead to $d_f = 2.5$ and $\tau = 2.2$ [68–71] when β and ν are assigned their universal values relevant for a 3D Ising system. The analogy between the physical clusters and critical fluctuations has been experimentally investigated by Guenoun et al. [72] and Beysens et al. [73]. These authors, by performing visual observations of the order parameter fluctuations near a critical point, were able to establish the self-similarity of the physical clusters. They deduced a fractal dimension $d_f \sim 2.8$, a value consistent with the theoretical estimate within the experimental error. As we said before, deviations from the static pair correlation function to the OZ behavior are accounted for by introducing Fisher's exponent η. When Fisher's correction is taken into account, one obtains

$$d_f = \frac{5 - \eta}{2} \tag{A.3}$$

and the index τ deduced from the hyperscaling relation is

$$\tau = \frac{11 - \eta}{5 - \eta} \tag{A.4}$$

With $\eta = 0.03$ we infer $d_f = 2.49$ and $\tau = 2.21$, values that are extremely close to those relevant for percolation. When comparing our experimental results to the modified version of the Martin and Ackerson [62–64] dynamical droplet model, which is described below, we use these numerical values as input parameters.

In this model it is assumed that the clusters are fractal and that the k cluster that contains k particles has a gyration radius given by

$$R_k = R_1 k^{1/d_f} \tag{A.5}$$

where R_1 is the average radius of a monomer. The spatial Fourier transform of the static correlation function of the clusters represents physically the q-dependent scattered intensity I and can be expressed as a function of $x = q\xi$. The droplet concentration in the sample is small, in any case less than 10%. Furthermore, close to the critical point, the density of clusters goes to zero. Thus it can be assumed that the critical fractal clusters are dilute and not interacting. With these hypotheses, $I(x)$ reads

$$I(x, x_1) = I_0 \left(\frac{x}{x_1}\right)^{d_f} (1 + x^2)^{d_f(\tau-3)/2} \frac{\Gamma(3 - \tau, (x_1/x)^{d_f}(1 + x^2)^{d_f/2})}{\Gamma(2 - \tau, (x_1/x)^{d_f})} \tag{A.6}$$

where $\Gamma(a, z)$ is the incomplete Euler gamma function, I_0 is almost independent of temperature close to T_c, and $x_1 = qR_1$ is the reduced size of the monomers. Thus in this equation a finite-size contribution associated with the droplets themselves is explicitly taken into account via x_1. When plotted as a function of x, and for realistic values of the parameters, the scattered intensity $I(x, x_1)$ reduces to the usual Ornstein–Zernike form, since $d_f(3 - \tau)/2 = 1 - \eta/2$. The situation is different when the intensity is plotted as a function of ε. In the so-called hydrodynamic regime, $x \ll 1$, the radius R_1 of the monomer plays a very important role. When R_1 is very small, as is typically the case for simple fluids or molecular binary mixtures, one observes the usual $\varepsilon^{-\gamma}$ behavior. However, when the size R_1 of the monomer is not very small compared to ξ, which is typically the case for solutions made of supramolecular aggregates, significant departures from the $\varepsilon^{-\gamma}$ law are observed. In particular, in the hydrodynamic regime, it is no longer possible to fit the scattered intensity to a power law over a large temperature range because an upward curvature of the plot is observed. Moreover, fitting intensity data to a power law in a restricted temperature domain would give an apparent value of the critical exponent γ smaller than the Ising value used as input parameter. Besides, this value would depend on the temperature domain in which the data were fitted.

We can also calculate the particle density–density time correlation function as

$$C(x, u, w) = \frac{1}{\Gamma(3 - \tau, u)} \int_u^{\infty} dz \, z^{2-\tau} \exp[-z - w z^{-1/d_f}] \tag{A.7}$$

where

$$w = D_1 R_1 q^3 t(1 + x^{-2})^{1/2} \tag{A.8}$$

is the reduced time. This quantity is also system-dependent via both R_1 and the translational diffusion coefficient of the monomer D_1. We express the latter as

$$D_1 = RK_BT/6\pi\eta R_1 \tag{A.9}$$

where R is the amplitude ratio. From these equations it is easy to infer the reduced first cumulant $\Gamma^*(x)$ as

$$\Gamma^*(x) = \Gamma/D_1 R_1 q^3 \tag{A.10}$$

or

$$\Gamma^* = \frac{3}{8\pi}\left[\frac{\Gamma(3-\tau, x_{1f}^d)\Gamma(3-\tau-1/d_f, u)}{\Gamma(3-\tau-1/d_f, x_{1f}^d)\Gamma(3-\tau, u)}\right]\left(1+\frac{1}{x^2}\right)^{1/2} \tag{A.11}$$

which explicitly depends on the droplet size. The relaxation rate we obtain exhibits striking differences from the prediction of the mode coupling or decoupling theories, since it contains a cutoff related to the size of the droplets. Nevertheless Γ^* reduces exactly to the Perl and Ferrel mode decoupling result [74,75] when the radius of the monomer $R_1 \to 0$. In this case, it is also numerically very close to Kawasaki's formula. However, when the radius of the monomer is not very small compared to ξ, significant deviations from generalized hydrodynamic models are observed.

B. Percolation

Percolating clusters can be treated much like critical clusters. However, in this case the assumption of scattering by dilute noninteracting clusters may not be valid as far as the scattered intensity is concerned. Indeed, no critical opalescence is observed upon crossing the percolation line. Therefore, we restrict ourselves to dynamical phenomena. The particle density–density time correlation function can be deduced from the formula as

$$C(q, t) = \frac{\sum_{k=1}^{\infty} k^2 N(k)S_k(q)\,\exp(-D_k q^2)}{\sum_{k=1}^{\infty} k^2 N(k)S_k(q)} \tag{A.12}$$

where S_k is the structure factor of the k cluster and D_k is the effective cluster diffusion constant including both the translational and rotational contributions. It can be approximately written as [28]

$$D_k = D_1 k^{-1/d_f}\left(1 + \frac{1}{2\rho^2}\right) \tag{A.13}$$

where D_1, the Stokes–Einstein translational diffusion coefficient for a single particle, is given by $D_1 = k_BT/6\pi\eta_s R_1$, where η_1 is the shear viscosity of the solvent and ρ is the ratio of the hydrodynamic radius to the radius of gyration of the cluster. The result is

$$\Gamma^*(x) = \frac{F(3-\tau-1/d_f, x)}{K(x, s)}(1+x^2)^{d_f(1-\tau-1/d_f)/2}$$
$$+ \left(1+\frac{1}{2\rho^2}\right)\left(\frac{G(2-\tau-1/d_f)}{K(x, s)}\right)\left(\frac{x}{h}\right)^{-d_f} \tag{A.14}$$

In the above equation, $K(x, s)$ is given by

$$K(x, s) = \frac{s}{\Gamma(2 - \tau)} \Gamma(3 - \tau)(1 + x^2)^{d_f(3-\tau)/2} \tag{A.15}$$

whereas $F(x, y)$ and $G(x, y)$ are, respectively,

$$F(a, x) = \Gamma(a) - \Gamma(a, u)$$

and

$$G(a, x) = \frac{\sin[(d_f - 1)\pi/2]}{d_f - 1} \Gamma\left(a, \frac{\sqrt{6}x}{[d_f(d_f + 1)]^{1/2}}\right)$$

with

$$u = \frac{d_f(d_f + 1)(1 + x^2)}{6x^2}$$

Two limiting cases can be identified. For $x \to 0$, $\Gamma^*(x)$ is inversely proportional to x:

$$\Gamma^*(x) \simeq \frac{[d_f(d_f + 1)]^{1/2}}{\sqrt{6}x} \left(\frac{\Gamma(3 - \tau - 1/d_f)}{\Gamma(3 - \tau)}\right) \tag{A.16}$$

but for large x limit,

$$\Gamma^*(x) \simeq \frac{[d_f(d_f + 1)]^{1/2}}{\sqrt{6}} \left(\frac{\Gamma(3 - \tau - 1/d_f, \infty)}{\Gamma(3 - \tau, \infty)}\right) \tag{A.17}$$

which is independent of x.

It is possible to cast the time correlation function $C(q, t)$ also into the scaled form, $C(x, v)$, where v is a natural nondimensional scaled time variable given by $v = D_1 q^2 t s^{-1/d_f}$. We note that in the long time limit the time correlation function approaches a stretched exponential form given by

$$C(v) \simeq \exp[-(d_f + 1)(v/d_f)^\beta] \tag{A.18}$$

where the stretch exponent β is universal and given by $\beta = d_f/(d_f + 1)$. Thus the droplet time correlation function starts off at short time as an exponential with a well-defined first cumulant and tends at long time to a stretched exponential having a universal stretch exponent. The crossover from the exponential to the stretched exponential occurs at $\Gamma \times t = 1$.

C. Analysis in terms of a Modified Version of the Droplet Model

Let us now compare the experimental results on the AOT system to the theoretical prediction deduced from the droplet model. In the hydrodynamic regime, significant deviation from the $\varepsilon^{-\gamma}$ behavior of the scattered intensity is deduced from the theory. The expected upward curvature in the intensity vs. temperature plot is in qualitative agreement with the theoretical predictions. Fits of the experimental data to the extended droplet model theory lead to results in quantitative agreement with theory in both the critical and hydrodynamics regimes. The reduced value of Γ, Γ^*, is plotted as a function of the inverse reduced wave vector, for four different temperatures, in Fig. 8. As far as the temperature dependence is concerned, it can be seen in the graphs that the overall behavior is similar to that observed for simple or molecular binary mixtures. Figure 8 shows in particular a leveling-off of Γ followed by a decrease in the linewidth upon a further increase in temperature.

A good quality fit to experimental data can be achieved by assuming polydisperse dynamical fractal clusters with a fractal dimension $d_f = 2.5$ and a polydispersity index $\tau = 2.2$. From the fits we can deduce the short-range correlation length ξ_0 and the radius of the cluster elements R_1. We obtain $\xi_0 = 1.3 \pm 0.1$ nm and $R_1 = 5 \pm 1$ nm. When none of the parameters is fixed, the best fit to the data leads to $d_f = 2.5 \pm 0.1$ and $\tau = 2.3 \pm 0.1$. These values are in good agreement with the theoretical estimates to within experimental error and lead to critical index values very close to the universal ones. The two other fitted parameters are $\xi_0 = 1.2 \pm 0.1$ nm and $R_1 = 6 \pm 1$ nm.

Using these values as input parameters, it is easy to account for the relaxation rate of the correlation function (first cumulant) not only in the critical regime but also in the hydrodynamic regime where the leveling-off of Γ is well reproduced. To obtain good agreement between theory and experiment we have to fix the coupling constant R to $R = 1.2 \pm 0.1$. At this point let us remember that in the previous section we analyzed the relaxation rate Γ as the sum of a critical part Γ_c, which can be calculated using the mode–mode coupling theory, and a system-dependent background Γ_B part, which depends explicitly on the Debye cutoff parameter q_c. We have shown phenomenologically that the dynamical background was roughly proportional to the size of the aggregates. In the present model the dynamical back-

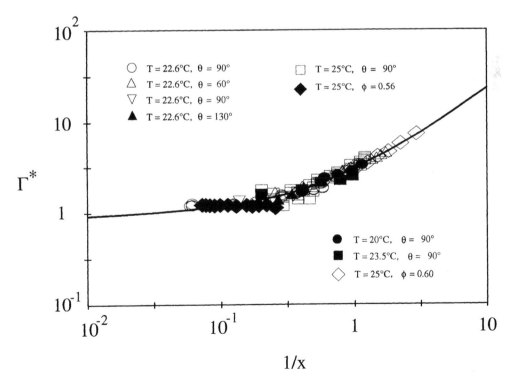

Figure 8 Reduced first cumulant Γ^* versus $1/x$ ($x = q\xi$) for the AOT microemulsions. θ is the scattering angle. The curve is the fit with the model of percolating clusters. (Data from Refs. 31 and 32.)

ground term enters in a natural way into the theory, and it is physically provided by the lower bound of the integral defining the correlation function. However, the price we have to pay is the introduction of a multiplicative constant R in the definition of D_1, the translational diffusion coefficient of the monomer, in a way similar to the renormalization group or mode coupling theories. In this latter case, however, the coupling coefficient R is universal. In our case, $R = 1.2$, a value close to the universal mode coupling prediction 1.03.

The intensity correlation function $C^2(t)$ is well fitted over more than three decades, from a delay time of 1 μs up to more than 1 ms with values of the parameters similar to those given above (Fig. 4). In particular, the deviations from the single exponential decay are well accounted for, at least in the time domain of interest. It is clear that a purely stretched exponential decay can be observed only at extremely long time where the amplitude of the correlation function is much less than the noise.

D. Analysis of the Experimental Results Close to the Percolation Line

At large volume fraction, $C^2(t)$ strongly deviates from a straight line. At long time, $C^2(t)$ can be well approximated by a stretched exponential, whereas at short time we observed a well-defined first cumulant that is also plotted in reduced units in Fig. 8. The theoretical prediction is compared to the experimental determination in the figure. From the fit we can deduce the value of the scaling variable x, which ranges from 1 for $\phi = 0.20$ up to 17 for $\phi = 0.60$. The geometrical correlation length can be fitted to a power law in terms of $(\phi - \phi_p)^{-\nu_p}$, where ϕ_p is the volume fraction at the percolation threshold and ν_p the characteristic exponent. We obtain $\nu_p \approx 0.9$, a value consistent with theory ($\nu_p = 0.88$) and a short-range geometrical length on the order of 20 nm, not too different from the diameter of the droplets.

ACKNOWLEDGMENTS

Part of the work described here was performed in collaboration with many colleagues at the Physics Laboratory of the Ecole Normale Supérieure. D.L. would like to thank particularly Anne-Marie Cazabat, Didier Chatenay, Jacques Meunier, Alain Pouchelon, Olivier Abillon, and Lay-Theng Lee. J.R. would like to thank his colleagues Professors S. H. Chen and P. Tartaglia for their very useful comments.

REFERENCES

1. B. Cabane, in *Surfactant Solutions: New Methods of Investigation* (R. Zana, ed.), Marcel Dekker, New York, 1987, p. 57.
2. J. H. Schulman, W. Stoeckenius, and L. M. Prince, J. Phys. Chem. *63*:1677 (1959).
3. A. Graciaa, J. Lachaise, A. Martinez, M. Bourrel, and C. Chambu, C. R. Acad. Sci. Ser. B *282*:547 (1976).
4. A. Garciaa, J. Lachaise, P. Chabrat, L. Letamendia, J. Rouch, C. Vaucamps, M. Bourrel, and C. Chambu, J. Phys. Lett. (France) *38*:253 (1977); *39*:235 (1978).
5. M. Dvolaitzky, M. Guyot, M. Lagues, J. P. Le Pesant, R. Ober, C. Sauterey, and C. Taupin, J. Chem. Phys. *69*:3279 (1978).
6. M. Kotlarchyk, J. S. Huang, and S. H. Chen, J. Phys. Chem. *89*:4382 (1985).
7. R. Aveyard, B. P. Binks, S. Clark, and J. Mead, J. Chem. Soc., Faraday Trans. I *82*:125 (1986).
8. C. Cabos and P. Delord, J. Phys. Lett. (France) *41*:L-455 (1980).

9. A. N. North, J. C. Dore, J. A. McDonald, B. H. Robinson, R. H. Heenan, and A. M. Howe, Colloids Surf. *19*:21 (1986).
10. M. Kotlarchyk, S. M. Chen, and J. S. Huang, J. Phys. Chem. *86*:3275 (1982).
11. M. Kotlarchyk, S. M. Chen, and J. S. Huang, Phys. Rev. A *28*:508 (1983).
12. A. M. Ganz and B. E. Boeger, J. Colloid Interface Sci. *109*:504 (1986).
13. W. M. Brouwer, E. A. Nieuwenhuis, and M. Kops-Werkhoven, J. Colloid Interface Sci. *92*:57 (1983).
14. C. Tanford, *Physical Chemistry of Macromolecules*, Wiley, New York, 1961.
15. B. U. Felderhof, J. Phys. A *11*:929 (1978).
16. A. M. Cazabat and D. Langevin, J. Chem. Phys. *74*:3148 (1981).
17. A. M. Cazabat, D. Chatenay, D. Langevin, J. Meunier, and L. Leger, in *Surfactants in Solution*, Vol. 3 (K. L. Mittal and B. Lindman, eds.), Plenum, New York, 1984, p. 1729.
18. HJ. Ricka, M. Borkovec, and U. Hofmeier, J. Chem. Phys. *94*:8503 (1991).
19. R. Ober and C. Taupin, J. Phys. Chem. *84*:2418 (1980).
20. F. Sicoli and D. Langevin, J. Phys. Chem. *99*:14819 (1995).
21. A. A. Calje, W. G. M. Atgerof, and A. Vrij, in *Micellization, Solubilization and Microemulsions*, Vol. 2 (K. L. Mittal, ed.), Plenum, New York, 1977, p. 779.
22. G. J. M. Koper, W. F. C. Sager, J. Smeets, and D. Bedaux, J. Phys. Chem. *99*:13291 (1995).
23. A. M. Cazabat, D. Chatenay, D. Langevin, and J. Meunier, Faraday Discuss. Chem. Soc. *76*:291 (1983).
24. E. Y. Sheu, S. H. Chen, J. S. Huang, and J. C. Sung, Phys. Rev. A *39*:5867 (1989).
25. C. A. Tracy and B. M. McCoy, Phys. Rev. B *12*:75 (1974).
26. M. E. Fisher, Physica *28*:172 (1962); J. Math. Phys. *5*:944 (1964).
27. J. V. Sengers, Int. J. Thermophys. *5*:805 (1985).
28. H. C. Burstyn, J. V. Sengers, J. K. Bhattacharjee, and R. A. Ferrell, Phys. Rev. A *28*:1567 (1983).
29. D. W. Oxtoby and W. M. Gelbart, J. Chem. Phys. *61*:3157 (1974).
30. M. Kotlarchyck, S. H. Chen, J. S. Huang, and M. W. Kim, Phys. Rev. A *28*:508 (1983); *29*:2054 (1984).
31. J. Rouch, P. Tartaglia, A. Safouane, and S. H. Chen, Phys. Rev. A *40*:2013 (1989).
32. J. Rouch, A. Safouane, P. Tartaglia, and S. H. Chen, J. Chem. Phys. *90*:3756 (1989).
33. O. Abillon, D. Chatenay, D. Langevin, and J. Meunier, J. Phys. Lett. (France) *45*:L223 (1984).
34. D. Langevin (ed.), *Light Scattering by Liquid Surfaces*, Marcel Dekker, New York, 1991.
35. P. A. Winsor, Trans. Faraday Soc. *44*:376 (1948).
36. A. Pouchelon, D. Chatenay, J. Meunier, and D. Langevin, J. Colloid Interface Sci. *82*:418 (1981).
37. A. M. Cazabat, D. Langevin, J. Meunier, and A. Pouchelon, Adv. Colloid Interface Sci. *16*:175 (1982).
38. J. H. Schulman and J. B. Montagne, Ann. N.Y. Acad. Sci. *92*:366 (1961).
39. H. T. Davis and L. E. Scriven, SPE Paper 9278, presented at the 55th Annual Fall Technical Conference on SPE-AIME in Dallas, TX, Sept. 21–24, 1980.
40. H. Kleinert, J. Chem. Phys. *84*:964 (1986).
41. G. J. Verhoeckx, P. L. de Bruyn, and J. Th. G. Overbeek, J. Colloid Interface Sci. *119*:409 (1987).
42. R. Aveyard, B. P. Binks, S. Clark, and J. Mead, J. Chem. Soc., Faraday Trans. 1 *82*:125 (1986).
43. D. Chatenay, O. Abillon, J. Meunier, D. Langevin, and A. M. Cabazat, in *Macro and Microemulsions* (D. O. Shah, ed.), American Chemical Society, Washington, DC, 1985, p. 119.
44. M. Kahlweit, R. Strey, D. Haase, H. Kunieda, T. Schmeling, B. Faulhaber, M. Borkovec, H.-F. Eicke, G. Busse, F. Eggers, Th. Funk, H. Richmann, L. Magid, O. Soderman, P. Stilbs, J. Winkler, A. Dittrich, and W. Jahn, J. Colloid Interface Sci. *118*:436 (1987).
45. J. Indekeu, in *Lectures on Thermodynamics and Statistical Mechanics* (M. Lopez de Haro and C. Varea, eds.), World Scientific, Singapore, 1991, p.175.
46. M. Robert and J. F. Jeng, J. Phys. (France) *49*:1821 (1988).
47. O. Abillon, L. T. Lee, D. Langevin, and K. Wong, Physica A *172*:209 (1991).
48. W. Helfrich, Z. Naturforsch. *28*:693 (1973).
49. B. P. Binks, J. Meunier, O. Abillon, and D. Langevin, Langmuir *5*:415 (1989).

50. L. T. Lee, D. Langevin, J. Meunier, K. Wong, and B. Cabane, Prog. Colloid Polym. Sci. *81*:209 (1990).
51. I. Szleifer, D. Kramer, A. Ben Shaul, D. Roux, and W. M. Gelbart, Phys. Rev. Lett. *60*:1966 (1988).
52. B. Farago, D. Richter, J. S. Huang, S. A. Safran, and S. T. Milner, Phys. Rev. Lett. *65*:3348 (1990).
53. M. Gradzielski, B. Farago, and D. Langevin, Phys. Rev. E *53*:3900 (1996).
54. B. Widom, in *Fundamental Problems in Statistical Mechanics*, Vol. 3 (E. D. G. Cohen, ed.), 1975.
55. A. M. Cazabat, D. Langevin, J. Meunier, and A. Pouchelon, J. Phys. Lett. (France) *43*:L89 (1982).
56. A. M. Bellocq, D. Bourbon, B. Lemanceau, and G. Fourche, J. Colloid Interface Sci. *89*:427 (1982).
57. P. D. Fleming, J. E. Vinatieri, and G. R. Glinsman, J. Phys. Chem. *84*:1526 (1980).
58. M. W. Kim, J. Bock, and J. S. Huang, *Waves on Fluid Interfaces*, Academic, San Diego, 1983, p. 151.
59. M. W. Kim, J. Bock, and J. S. Huang, SPE Paper 10788, 1982.
60. R. Moldover, Phys. Rev. A *31*:1022 (1995).
61. G. H. Findenegg, A. Hirtz, R. Rasch, and F. Sowa, J. Phys. Chem. *93*:4580 (1989).
62. C. M. Sorensen, B. J. Ackerson, R. C. Mockler, and W. J. O'Sullivan, Phys. Rev. A *13*:1593 (1976).
63. J. E. Martin and B. J. Ackerson, Phys. Rev. A *31*:1180 (1985).
64. J. E. Martin, J. Wilkonson, and D. Adolf, Phys. Rev. A *36*:1803 (1987).
65. A. Coniglio and W. Klein, J. Phys. A: Math. Gen. *13*:2775 (1980).
66. P. Tartaglia, J. Rouch, and S. H. Chen, Phys. Rev. A *45*:7275 (1992).
67. J. Rouch, P. Tartaglia, and S. H. Chen, Phys. Rev. Lett. *71*:1947 (1993).
68. D. Stauffer, in *On Growth and Form* (H. E. Stanley and N. Ostrowsky, eds.), Martinus Nijhoff, Dordrecht, 1986.
69. D. Stauffer and A. Aharony, *Introduction to Percolation Theory*, Taylor and Francis, London, 1992.
70. D. Stauffer, Phys. Rev. *54*:1 (1979).
71. Y. Gefen, A. Aharony, B. Mandelbrot, and S. Kirkpatrick, Phys. Rev. Lett. *47*:1771 (1981).
72. P. Guenoun, F. Perrot, and D. Beysens, Phys. Rev. Lett. *63*:1152 (1989).
73. D. Beysens, P. Guenoun, and F. Perrot, J. Phys.: Condens. Matter *2*:SA127 (1990).
74. R. Perl and R. A. Ferrel, Phys. Rev. Lett. *29*:51 (1972).
75. R. Perl and R. A. Ferrel, Phys. Rev. A *6*:2358 (1972).

13

Characterization of Microemulsions Using Fast Freeze-Fracture and Cryo-Electron Microscopy

Jayesh R. Bellare* and Manoj M. Haridas
Indian Institute of Technology–Bombay, Mumbai, India

Xiangbing Jason Li[†]
University of Minnesota, Minneapolis, Minnesota

I. ELECTRON MICROSCOPIC CHARACTERIZATION

Microemulsions and most surfactants in dilute solutions and dispersions self-assemble into a variety of microstructures: spherical or wormlike micelles, swollen micelles, vesicles, and liposomes. Such systems are of biological and technological importance, e.g., in detergency, drug delivery, catalysis, enhanced oil recovery, flammability control, and nanoscale particle production. The macroscopic properties—rheology, surface tension, and conductivity—of these systems depend on their microstructure. As these microstructures are small (1–1000 nm) and sometimes several microstructures can coexist in the same solution, it is difficult to determine their structure. Conventional techniques like radiation scattering, although useful, provide only indirect evidence of microstructures, and the structures deduced are model-dependent.

Transmission electron microscopy (TEM) is the most important technique for the study of microstructures because it directly produces images at high resolution and it can capture any coexistent structure and microstructural transitions. However, microemulsion microstructures cannot be easily imaged with TEM because (1) their high vapor pressures make them incompatible with the low pressures ($< 10^{-5}$ torr) in the microscope; (2) electrons may induce chemical reactions in microemulsion systems that can change the microstructure, especially for systems of high organic content; and (3) often there is insufficient contrast between the microstructures and their surroundings. Several sample preparation techniques have been used to overcome these problems, but the techniques themselves have introduced new problems. This is because sample preparation changes the structure of the sample, often radically enough to obscure the original microstructure. A new sample preparation technique that perturbs the sample so minimally

---

* Formerly at the Department of Chemical Engineering and Materials Science, University of Minnesota, Minneapolis, Minnesota.
† *Current affiliation*: Applied Materials, Inc., Santa Clara, California.

as to preserve the state of liquid water has overcome many of these limitations. This has opened a new window to study microemulsion microstructures, i.e., the ability to obtain direct, high resolution artifact-free images.

A. Introduction to Electron Microscopy

Electron microscopy is primarily concerned with the examination of specimen surfaces (scanning electron microscopy, SEM) or specimen bulks (TEM). Since the electron beam passes through the specimen in a TEM analysis, a TEM study necessitates a sample thickness of about 100 nm; thicker sections are impermeable to an electron beam. Also, overlapping of microstructures in a thicker sample will make TEM image interpretation a difficult task if it is even possible. Images are produced by both secondary and backscattered electrons, and information contained in either image is interpreted according to the sample being analyzed. The versatility of SEM for the study of specimens is derived in large measure from the wide variety of interactions that the primary beam undergoes with the specimen. The inelastic and elastic interactions lead to the generation of backscattered, secondary, and Auger electrons; X-ray photons; long-wavelength electromagnetic radiations in the visible, UV, and IR regions; phonons; and plasmons. In principle, all these interactions can be used to obtain information about the specimen that relates to its shape, composition, crystal structure, electronic structure, and internal electric or magnetic fields as well as cathodoluminescent properties, if any. Detailed information about electron optics and image formation in large depths of field in electron microscopy is available in standard texts [1–3].

Two important advances have permitted reliable, revealing, and direct visualization of microemulsion microstructures. The first is the ability to lower the vapor pressure of liquid water at ambient pressure without crystal formation. This is achieved by plunging thin (<1 μm) films of liquid water into cryogens such as ethane or propane at their freezing point [4]. The cooling rate is so fast [5], 10^4–10^5 K/s, that water vitrifies; water molecules do not rearrange into a crystalline form as they would at slower cooling rates. Vitrification of water is an important indicator of microstructural integrity. If small water molecules do not rearrange during cooling, then larger entities like microemulsions will not rearrange. This technique was used [6] to study droplet packing in water-diluted specimens of so-called glass-forming microemulsions and [7] to see microstructures in surfactant dispersions. However, Talmon [7] found that the thin samples prepared in an ambient atmosphere led to unavoidable evaporation and temperature changes and induced artifacts (microstructure not present in the original sample) in samples near a phase boundary. The second advance is the ability to prepare thin sample films in an environmental chamber and keep them under controlled temperature and chemical activities of the surrounding vapor until a few milliseconds before vitrifying them [8,9]. This advance prevents artifacts caused by changes in concentration or temperature. This technique was used to image vesicles and cylindrical micelles [8], to study spherical and cylindrical micelles in cetyltrimethylammonium bromide (CTAB) solutions [10], and to study polymer–surfactant interaction [11].

In this review, we discuss the development and application of cryogenic TEM techniques in the study of microemulsion microstructures. Our goal is to illustrate the power of the technique to directly image a variety of microemulsion microstructures that were once only figments of an artist's imagination. The chapter opens with a brief description of the sample preparation techniques, presents a series of micrographs obtained by investigators in some of the most comprehensive work in this field, a few of which have never been seen before, and closes with a discussion of future applications.

B. Sample Preparation Techniques: An Overview

1. Need for Sample Preparation

Fluid, labile systems such as colloidal dispersions, microstructured liquids, and living cells are a formidable challenge to the electron microscopist. Volatile constituents evaporate into the vacuum that surrounds the microscope stage, altering the composition of the specimen and often inducing phase changes with attendant artifacts. Low viscosity specimens can deform and flow under the subtle forces that develop during specimen preparation and imaging. Chemical reactions induced by the electron beam volatilize parts of a specimen or partially polymerize and cross-link other parts, causing stresses that lead to grossly altered microstructures. Moreover, contrast between different parts of the microstructure of aqueous and organic systems is low. Ever since the pioneering of transmission electron microscopy [12], various sample preparation techniques have been used to reduce vapor pressure and immobilize fluid, labile specimens; to reduce susceptibility to radiation damage by the electron beam; and, not infrequently, in conjunction with staining, to heighten contrast.

The two central features of sample preparation in TEM are (1) to produce specimens thin enough (100–200 nm) to transmit a substantial portion of the illuminating beam and (2) in the case of low temperature microscopy, to vitrify any aqueous phases in order to prevent ice crystal damage. In the case of bulk samples, it is necessary to cut frozen sections that are either dried before being examined at low temperatures or examined in the frozen hydrated state. If the objects are small enough—e.g., macromolecules, membrane fragments, viruses, particles, polymers—they may be incorporated into a thin liquid suspension that is rapidly frozen prior to being examined by TEM.

Sample preparation in electron microscopy is usually performed using chemical or thermal fixation techniques for sample immobilization. The evaluation and details of each preparation technique are presented below.

2. Chemical Fixation Techniques

Chemical fixation involves adding foreign chemical species to the sample, chemically altering the sample by reaction, and causing large compositional changes in the sample due to removal of one or more components. The technique broadly consists of one or more of the following: chemical stabilization, staining and drying, and/or polymerization stabilization. Samples are imaged in TEM by using conventional operating modes and stages.

(a) Chemical Stabilization. Chemical stabilization techniques consist of five basic steps with respect to sample preparation: fixing, dehydrating, embedding, microtoming, and staining. Several variations exist on the choice of chemicals and stabilization times, and cycles, and a comprehensive review on the experimental protocols is available in the literature [13].

Sample Fixation. In the fixation step, a chemical reagent is added to the sample and allowed to diffuse into and react with various components of the sample. The fixation step partially stabilizes the sample both chemically and structurally. Many different fixatives such as osmium tetroxide, glutaraldehyde, ruthenium red, and formaldehyde have been used, and each fixative has its own advantages and disadvantages, being suited to a specific sample; for instance, osmium tetroxide effectively fixes lipid membranes and also acts as a stain. Fixatives may be used in combinations, simultaneously or sequentially, and are added in a variety of ways such as vapor fixation for solid tissue chunks or immobilization of labile surfactant microstructures in agarose to provide a solid support that is fixed by immersion. An investigation into the experimental aspects of this technique has been reported [10].

Sample Dehydration.   Dehydration removes water and other liquids that remain in the sample after fixation. The sample is removed from the fixative and placed in a solvent that diffuses into the sample. The solvent must remove all sample fluids that are immiscible into the resin into which the sample is to be embedded so as to minimize the occurrence of gas or liquid pockets that make the sample unsuitable for microtomy and unstable under the electron beam. Typical dehydrating solvents are acetone, ethanol, dimethoxypropane, and water-soluble resins such as glycol methacrylate.

Dehydration can also be achieved through freeze-substitution, in which the fixed sample is rinsed in a buffer, quenched in a cryogen, and dehydrated over a period of days in cold solvent. This technique reduces the removal of elements from the sample that are soluble in the solvent at room temperature.

Sample Embedding.   After dehydration, the sample is embedded in resin so as to provide the mechanical stability necessary to withstand microtoming stresses. Embedding is done by infiltrating the sample with an exchange of resins until the dehydrating solvent is replaced by fresh resin. The sample is subsequently placed in an embedding mold that is filled with resin. The resin is polymerized to produce a solid block in which the sample is embedded. Most resins can be polymerized by ultraviolet light or by activation of a catalyst by heat.

Sample Microtoming.   When polymerization is complete, the resin block is removed from the mold and the block face is trimmed with a razor blade. The block should be trimmed so that the pressure exerted by the microtome knife on the block gradually increases during a cutting stroke. It serial sections are desired, the block face must be trimmed in a manner that allows a straight ribbon of sections to form. A trapezoidal block face mounted in the microtome with the parallel sides of the trapezoid oriented parallel to the knife edge allows cut sections to form a straight ribbon. Orienting the block with the shorter parallel side of the trapezoid contacting the knife first provides a gradual pressure increase during the cutting stroke. Once mounted into the microtome, the sample block is cut into thin sections about 70 nm thick. The cut sections are retrieved, mounted on a microgrid or on a continuous film support attached to a TEM grid, and allowed to dry. The dry sections are usually stained to enhance image contrast. Staining is the process of attaching heavy metal to certain components in the sample. Commonly used stains are uranyl acetate for carboxylic acids, lead citrate, potassium phototungstate for primary and secondary amines, and osmium tetroxide, which reacts with carbon double bonds.

Sample Staining and Drying.   Staining and drying techniques are relatively simple. The liquid sample is mixed with a drop of the stain solution, then placed on a film-covered grid and allowed to dry. The stains can be classified as positive or negative. A positive stain reacts with and attaches heavy metal atoms to the region of interest. Negative stains highlight the region of interest by reacting with adjacent areas, which makes the region of interest appear lighter than its surroundings.

(b) Polymerization Stabilization.   Polymerization stabilization, as applied to liquid microstructures, refers to cross-linking the molecules that compose a microstructure to produce a more stable arrangement that maintains the morphology and geometrical order of the

unpolymerized microstructure. This technique requires that the molecules in the microstructure contain reactive polymerizable groups. The microstructure is polymerized by ultraviolet light, gamma irradiation, or heat and a catalyst, depending on the reactive species present in the molecules.

(c) *Evaluation of Chemical Fixation Procedures.* Chemical fixation techniques are notoriously slow because the fix rates are governed by mass diffusivities and chemical kinetics. The fixation step can take from minutes to several weeks. Chemical stabilization appears to be successful in imaging relatively nonlabile liquid crystalline microstructures of low water content but offers little hope for the more mobile and labile microstructures that exist at higher water concentrations. Staining and drying techniques are unreliable, as they alter the system in several ways. The composition and electrostatic nature of the system are altered prior to drying through the addition of the heavy atom salt or stain. Drastic composition changes occur during the drying process due to the complete removal of volatile components. The drying process subjects the microstructures to powerful stresses not present in the bulk sample. Drying and staining techniques are reliable only when dealing with microstructures that are independent of composition, rigid, and mechanically stable. In the case of surfactant solutions, drying is equivalent to traversing the phase diagram in an uncontrolled manner.

3. Thermal Fixation

Thermal fixation techniques are preferred in the study of microstructured liquids, hydrated biological specimens, fluid-filled porous media, etc. Because thermal fixation is used for such diverse applications, there are a wide variety of fixation techniques and associated equipment that impart microscopic stability to the specimen.

Thermal fixation is the process of converting liquid specimens into solids by the rapid removal of heat. This process reduces the vapor pressure and arrests the internal motion of the sample, which allows the sample to be examined in the electron microscope. A sample is thermally fixed by contacting it with a cryogen or an object cooled to cryogenic temperatures. There are several advantages to using thermal fixation in the study of microstructured liquids. To begin with, thermal fixation eliminates the need to add foreign chemical species to the sample during the fixation process and is faster than chemical fixation because heat transfer is faster than the mass transfer and chemical reactions that limit the rate of fixation in chemical fixation techniques. In thermal fixation, the sample has the same chemical composition before and after fixing. Also, rapid fixation reduces the time in which phase separation processes such as nucleation and growth, spinodal decomposition, and solute translocation processes can occur. If thermal fixation occurs rapidly enough, liquid phases in a sample can be vitrified. Vitrification is the conversion of a fluid phase into a rigid phase without crystallization. The goal of thermal fixation is to vitrify the liquid phase and maintain the relative position of all the constituent atoms, molecules, and ions.

The essence of cryomicroscopy lies in the ability to vitrify the sample using the thermal fixation technique that alters the sample microstructure to the least extent. Cooling rate estimates of 10^5 K/s [14], 10^7 K/s [15], and 10^{10} K/s [16] are quoted as necessary to vitrify water or dilute aqueous suspensions. The ability to achieve the desired cooling rate depends on the sample, the cryogen, and the fixation technique. 10^5 K/s is probably the most accurate estimate. This evaluation is based on estimates of the cooling rate in freezing techniques that are known to produce vitreous specimens.

C. Sample Preparation for Cryomicroscopy

1. Factors Influencing Sample Preparation in Cryomicroscopy

Sample preparation techniques have been found to depend largely on the rate of cooling, the viscosity of the sample, its thickness, and the cryogen chosen for sample fixing.

(a) Choice of Cryogen. The essential requirements for a cryogen used in the thermal fixation of aqueous systems are a freezing point below the glass transition temperature of water, low viscosity, high density, relatively high boiling point, large heat capacity, and large thermal conductivity. The last three properties are important to reduce oxygen vaporization during vitrification, which produces a vapor film around the sample that reduces the rate of heat transfer. Table 1 lists several cryogens and their thermophysical properties at the melting temperatures.

It has been demonstrated [18] that liquid ethane gives a faster cooling rate than liquid propane for quenching 3 μL water drops, and before this report it was generally accepted that propane was the most efficient cryogen for rapid cooling. For practical use in cryo-electron microscopy, ethane is the best cryogen [10].

(b) Freezing Techniques and Cooling Rates. Quench cooling of an electron microscopic sample is a complicated heat transfer problem. High cooling rates can be achieved if the surface area/volume ratio of the sample is large, the cryogen temperature is low enough, and the heat transfer coefficient is optimized. As a result, samples are often prepared as thin films because this provides a large surface area/volume ratio. Comprehensive details on freezing techniques are available in specific literature on this subject [19].

Plunge Cooling. Plunge cooling refers to mechanical or manual immersion of a sample into a liquid cryogen. Several plunge cooling devices have been described [7–9,20]. A rotary, spring-propelled plunging device has been described [21]. Heat removal from thin film samples is by forced convection over the sample surface provided that excessive vapor formation does not occur in the vicinity of the sample. This sample precooling is avoided or reduced in designs that separate the sample preparation region from the cryogen [9] or those that very rapidly move samples from the bulk to the cryogen [21]. A cooling rate estimate of 2×10^4 K/s has been calculated [22] for 0.3–0.5 mm thick biological samples

Table 1 Thermophysical Properties of Cryogenic Liquids Near the Melting Point at Atmospheric Pressure

Liquid	Melting point (K)	Boiling point (K)	Density (g/cm^3)	Specific heat capacity [J/(g · K)]	Thermal conductivity [W/(cm · K) × 10^3]	Viscosity (P × 10^2)
Helium	2.2^a	4.2	0.147	4.0	0.181	0.0029
Hydrogen	14.1	20.4	0.077	7.3	1.08	0.022
Nitrogen	63.1	77.3	0.868	2.0	1.53	0.27
Propane	83.3	230.9	0.8	1.92	1.9	8.7
Ethane	89.7	184.4	0.655	2.27	2.4	0.9
Methane	90.5	109.0	0.451	3.29	2.23	0.18
Freon 22	113.0	232.5	1.72	1.07	1.52	2.15^b
Freon 12	115.0	243.2	1.77	0.83	1.38	2.38^b

^{a} Superfluid state.
^{b} Values estimated by intrapolation from ethane data.
Source: Ref. 17.

plunged into liquid propane. Cooling rates for thin films using liquid ethane are substantially higher [4].

Spray Cooling.   In spray cooling, the sample droplets, 10–20 μm in diameter, are sprayed into a cryogen. In an early attempt at fixing microemulsion samples [23], microemulsion droplets were sprayed into liquid nitrogen, but crystallization and excessive shear were found to alter the microstructure. Cooling rates on the order of 10^5 K/s have been estimated [2] for spray cooling techniques, and it is felt that cooling rates are probably much lower near the center of relatively large samples.

Jet Cooling.   In jet cooling, a jet of cryogen at several atmospheres of pressure is sprayed onto a specimen from one or both sides. This technique ensures rapid renewal of the cryogen in contact with the sample during the cooling and thereby rapid removal of heat. This technique is most often used with liquid specimens to prepare freeze-fracture replicas and other larger samples. Liquid propane is most often used as the cryogen. Cooling rates range from about 2×10^3 to 2.5×10^4 K/s depending on the pressure of the propane jet, the propane temperature, and the design of the sample holder [24].

Metal Block Cooling or Slam Cooling.   "Slam cooling" refers to the rapid pressing of a sample against a highly polished metal surface maintained at or near the temperature of liquid helium. This technique makes use of the high thermal conductivity of metals. For example, heat can be transferred through copper 10,000 times as fast as through liquid propane. Cooling rates are estimated at 10^5 K/s in the region of contact. This technique is used primarily for biological tissues because labile microstructures cannot withstand the impact from other techniques.

High Pressure Cooling.   The freezing point of water is reduced to $-22°$C at 2045 bars. It has been suggested [24] that high pressure suppresses nucleation and ice crystal growth, thereby lowering the cooling rate necessary to vitrify water. This idea was further exploited to build a high pressure cooling device in which specimens are subjected to a pressure of 2100 bars by applying a jet of liquid propanol to both sides of the specimen for a few milliseconds followed by a jet of liquid nitrogen. Vitrification in tissue samples up to 600 μm thick was demonstrated. For liquid samples, high pressure cooling is used mainly for preparing specimens for freeze-fracture replication.

II. PREPARATION OF MICROEMULSION SAMPLES FOR EM ANALYSIS

There are two variations of the TEM technique for fluid samples: (1) the cryo-TEM analyses in which samples are directly visualized after fast-freeze and freeze-fracture in the cold microscope and (2) the FFTEM technique in which a replica of the specimen is imaged under room temperature conditions.

A. Cryo-TEM Imaging of Microemulsions

Microemulsion microstructures are best imaged by cryo-TEM. A description of sample preparation methods with a view to obtaining direct and artifact-free images is described in this section as it is the key to microscopy.

For cryo-TEM analysis, about one drop of microemulsion sample is spread on a holey film-coated grid. After the droplet has reached equilibrium in a controlled environment at the desired temperature and chemical potential, it is thinned to the right thickness. After allowing it enough relaxation time, the thin film is fast frozen by being

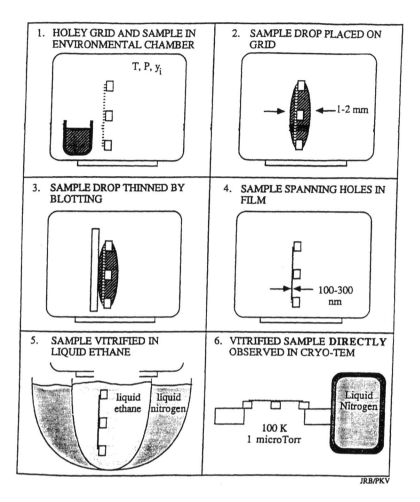

1. HOLEY GRID AND SAMPLE IN ENVIRONMENTAL CHAMBER

T, P, y_i

2. SAMPLE DROP PLACED ON GRID

1-2 mm

3. SAMPLE DROP THINNED BY BLOTTING

4. SAMPLE SPANNING HOLES IN FILM

100-300 nm

5. SAMPLE VITRIFIED IN LIQUID ETHANE

liquid ethane   liquid nitrogen

6. VITRIFIED SAMPLE DIRECTLY OBSERVED IN CRYO-TEM

Liquid Nitrogen

100 K
1 microTorr

JRB/PKV

Figure 1 Cryogenic transmission electron microscopy and optical microscopy with controlled environment vitrification. (From Ref. 9.)

rapidly plunged into a cryogen to achieve complete vitrification. The sample may be either plunge- or slam-cooled using a variety of cryogens such as liquid freon, liquid ethane, or liquid propane (Fig. 1). The cooling rates experienced by the sample are on the order of about 10^5 K/s, resulting in a rapidly solidified sample with a "liquidlike" microstructure entrapped within the sample bulk. The frozen microemulsion sample is amorphous as a result of the fast-freeze technique. The frozen sample is transferred to the cold storage of the electron microscope (maintained at liquid N_2 temperatures) and imaged.

Extensive documentation on the preparation of "holey" film substrates for cryo-TEM sample preparation is available in the literature. Holey films or nets have been produced [10,25–27] from a glycerol–water mixture in a solution of a mixture of formvar and triafol in dichloromethane spread on mica. After floating the thin film off the mica support, the films were treated with ethyl acetate to transform pseudoholes into holes or nets.

B. Freeze-Fracture TEM Imaging of Microemulsions

In freeze-fracture transmission electron microscopy (FFTEM), microemulsion samples are spread between two plates (Fig. 2). They are vitrified and transferred from liquid ethane into liquid nitrogen and mounted in a Balzer's freeze-etch device. The samples are held at a temperature below $-150°C$ and a pressure lower than 10^{-6} torr and are fractured by separating the two copper specimen support plates (Figs. 2 and 3). The fracture surfaces are replicated by shadowing them with platinum (by electron beam evaporation) and reinforcing them with carbon (by thermal evaporation). Fractured samples are seen to etch (sublime) when the partial pressure of water vapor in the replication system is below the partial pressure exerted by ice. Water vapor may also be found to recondense on the specimen surface and obliterate specimen detail or add artificial microstructure unless a cold trapping surface is kept near the sample. Hence, etching is reduced or avoided by fracturing the sample after platinum evaporation is started. Furthermore, the knife mount of the freeze-etch device is cooled with liquid nitrogen to act as a cold trap to prevent condensation of volatiles on the fractured surface. The replicas are floated onto nitric acid, washed with distilled water, air-dried, and examined in the conventional transmission mode of an electron microscope. Micrographs are usually recorded on Kodak 4489 film and developed for 4 min in Kodak D19 (1 : 2) developer. Comprehensive details are reported elsewhere [9].

The FFTEM technique, which has been the most widely used direct characterization tool in the imaging of microemulsions, has also been modified and improved to a certain extent through the design of a controlled environment vitrification system (CEVS). By rapidly cooling samples from a temperature and solvent-saturated controlled environment, the CEVS reduces or prevents artifacts during sample preparation prior to vitrification. The CEVS [8] is used with custom-designed tweezers [28] to which two thin copper plates (0.1 mm thick and about 4 mm square) are attached. About 2 µL of the microemulsion sample, previously equilibrated in the waver vapor and octane vapor saturated environmental chamber, held to within 0.1°C, are placed on one of the plates. The tweezers are then

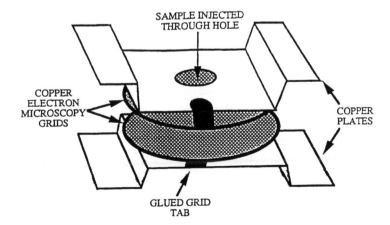

Figure 2 Sample holder for preparation of freeze-fracture replicas. Microemulsion is injected through the hole in the plate. The TEM grid tabs are folded over the copper plates and glued down. This ensures that the fracture propagates between the two grids and provides both cohesive and adhesive fracture surfaces. (From Ref. 10.)

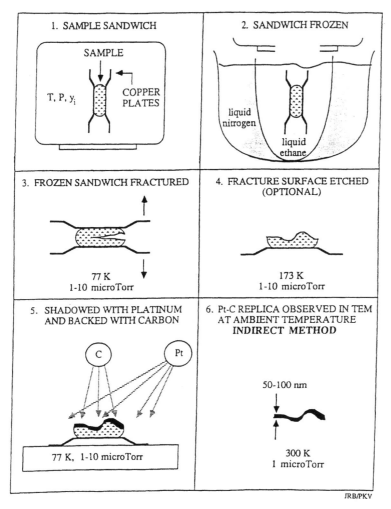

Figure 3 The freeze-fracture replication technique. (From Ref. 9.)

closed to squeeze the liquid specimen between the copper plates until only a thin layer of sample remains between them. The resulting sample is rapidly quenched by abruptly forcing the tweezers through a synchronously opened shutter in the environmental chamber and into a brass cup of liquid ethane that is freezing from the edges of the cup. The preparation technique ensures that all the sample components are kept at fixed chemical potentials until vitrification. Specimen cooling rates, estimated to be 2×10^4 K/s [5], are sufficient to vitrify the water in the samples [28]. Recent continuous improvements in the CEVS design, such as the flow-through controlled environment vitrification system (FT CEVS), which has a smaller chamber and more rapid and accurate temperature control with mixed humidified and dried flows, allow temperature jump and stability at relatively high temperatures during sample preparation [27].

Microemulsion samples can also be thermostated in mechanical plunging devices as described in Ref. 21. Specimens for fast freezing are transported by spring-driven mechanisms from such thermostated sample holders into cryogen and similarly into liquid nitrogen.

FFTEM samples can also be prepared [29] by a "double-replica" freeze-fracture technique, in which a drop of sample is deposited on a metal disk. A second disk is then placed on top, and the two disks are centered. A further drop is placed in the central bore of the specimen "sandwich," any excess sample being removed with filter paper. The sample is plunged into liquid cryogen and transferred to the cleavage unit of the microscope, also maintained under cold conditions. The sample is subsequently cleaved, etched, coated, and imaged.

C. Generation of Microscopic Contrast in the TEM

Contrast is the spatial variation in image intensity due to variations in microstructures. There are two important problems in imaging microemulsion systems by cryo-TEM: low contrast and radiation damage. Radiation damage is the microstructural change in a specimen induced by the electron beam due to heating, physical displacement of atoms, and chemical reactions, especially those induced by free radicals. Therefore, the electron exposure, i.e., the product of electron dose and exposure time, must be kept sufficiently small to prevent radiation damage artifacts. But short exposures yield images with small signal-to-noise ratio and therefore low contrast. The minimum contrast, C_m, detectable by the human eye is 0.05, i.e., $C_m \equiv (I_0 - I_b)/I_b > 0.05$, where I_0 is the intensity of the specimen and I_b that of the background. Exposures must be large enough to produce adequate contrast, yet small enough to prevent artifacts of radiation damage.

Images record intensity variations across a field of view. Intensity variation or contrast arises from changes in the amplitude of radiation traversing a specimen, which in turn arises from absorption of electrons within the specimen. Microemulsion systems contain molecules having relatively light atoms (e.g., carbon, hydrogen, oxygen, sodium) in a water or oil matrix that also has light atoms. The absorption of electrons by these atoms is small; furthermore, the absorption of the microemulsion structure does not differ appreciably from that of the matrix. Thus, there is negligible amplitude contrast in such systems. It is possible to increase amplitude contrast by staining—introducing heavy atoms (uranium or tungsten) into the specimen, typically by replacing a light counterion. However, this may not increase the contrast significantly, because of counterion dissociation. Moreover, staining may change the thermodynamic phase behavior of the system.

An incident electron beam that encounters the nucleus of an atom in the specimen is deflected through a relatively large angle without energy loss. The number of electrons deflected, or elastically scattered, in this manner increases with the specimen thickness and the scattering cross section of the atoms encountered. The scattering cross section is roughly proportional to the atomic number to the one-third power. If an aperture is inserted to prevent widely scattered electrons from contributing to the image, then the image will be dark in specimen regions of greater mass thickness. The correspondence of image intensity with mass thickness is called amplitude contrast. Amplitude contrast in the image arises from local heterogeneities in the sample. These local heterogeneities result in local variations in the number of electrons that leave the specimen and contribute to the image. Thus amplitude contrast depends on local variations in the number of elastically scattered electrons removed by the aperture and absorbed by the sample. Amplitude contrast can be increased by decreasing the size of the objective aperture.

While amplitude contrast can be partially attributed to variations in the number of elastically scattered electrons removed by the objective aperture, the elastically scattered electrons that pass through the objective aperture can also contribute to contrast in the image. Because the unscattered and elastically scatted electrons have not undergone energy exchange, both types of electrons will have the same well-defined wavelength. The elastically scattered electrons can be considered electron waves that can interfere constructively, or destructively, with waves of unscattered electrons. Contrast arising from such interference, called phase contrast, allows the visualization of specimen detail from materials of low atomic number that do not produce sufficient amplitude contrast to be imaged.

An ideal image is in the intensity distribution of an incident plane wave transmitted through the specimen and detected in a viewing plane. A basis $\mathbf{B}_v = [\mathbf{e}_x, \mathbf{e}_y, \mathbf{e}_z]$ with illumination along $\mathbf{e}_z$ and the viewing plane along $\mathbf{e}_x \mathbf{e}_y$ is a convenient choice of viewing frame of reference. In traversing the specimen, the incident wave undergoes scattering, absorption (multiple scattering), and phase shift, resulting in contrast, i.e., variations in image intensity. Therefore, the transmitted wave that has undergone absorption and phase shift can be expressed as

$$\psi(x, y; E, \alpha_0)_{id} = \psi_i \exp[-i\sigma\phi(x, y)t(x, y) - S_p(E, \alpha_0)\rho(x, y)t(x, y)] \tag{1}$$

This is also the amplitude of an ideal image of the specimen, because $|\psi|^2 = \psi\psi^*$ is the intensity distribution in the viewing plane $\mathbf{e}_x \mathbf{e}_y$. The object function is defined to be the normalized ideal image amplitude,

$$\psi_{id}/\psi_I \equiv \exp[-i\sigma\phi(x, y)t(x, y) - S_p(E, \alpha_0)\rho(x, y)t(x, y)] \tag{2}$$

It is the amplitude of the transmitted wave for a unit incident wave. It is assumed that $|\psi_I| = 1$ and therefore $\psi(x, y)$ is the object function.

Specimens are classified as phase objects if the term ϕ denotes the exponent and as mass-thickness or amplitude objects if the ρ term dominates. Furthermore, if $\phi t \ll 1$, the objects are known as weak-amplitude objects. For weak-phase and weak-amplitude objects,

$$\psi(x, y; E, \alpha_0)_{id} = 1 - i\sigma\phi(x, y)t(x, y) - S_p(E, \alpha_0)\rho(x, y) \tag{3}$$

where the exponent of the object function in Eq. (2) has been linearized. The first term on the right corresponds to the unscattered beam, and the remaining terms, to the scattered beam. Most thin specimens without heavy atoms, such as microemulsion systems, can be considered weak objects.

The phase contrast produced by an ideal-phase microscope is $C = 2\sigma\phi d$ [9], where $\sigma = 8.491 \times 10^{-3}$ nm^{-1} V^{-1}. The mean potential of water is $\phi_m = 5.8$ V, while that of hydrocarbon, ϕ_s, is 9.6 V. Hence the contrast between the aggregate and the matrix is $2\sigma(\phi_s - \phi_w)d = 64.53 \times 10^{-3}d$, and the resolution limit is 0.7 nm. However, a real microscope modifies the ideal image severely because it cannot produce the uniform phase shift of $\chi = \pi/2$ assumed in an ideal microscope. Nevertheless, the primary contrast in cryo-TEM of colloidal systems is phase contrast, which is strong enough to allow the detection of micelles and bilayers.

In FFTEM, the sample is a replica of the microstructure revealed by a fractured plane and possibly enhanced by shadowing. The contrast comes from mass-thickness variations almost exclusively, and interpretation is more straightforward. Also, there may be interfacial effects between the sample and the Pt/C coating material during sample preparation, thereby giving so-called phase detection effects [21] that permit identification of water and oil phases. However, in all replication techniques, the resolution is limited by the grain size of the replica as well as by inadvertent decoration effects that may result in loss of resolution.

III. MICROEMULSION SYSTEMS IMAGED USING CRYO-TEM/FFTEM TECHNIQUES

Freeze-fracture electron microscopic studies into microstructure have been carried out on several microemulsion systems. Some of the more extensively conducted studies on microemulsion systems are reviewed below.

A. Structural Studies in the Water–Pentanol–Potassium Oleate–Benzene System

In an early study on the structural determination of association structures in W/O microemulsions [30], the primary focus of characterization in water–alcohol–potassium oleate–benzene systems was through quasi-elastic light scattering observations, and cryo-TEM analyses were carried out mainly as corroborative studies. This study was prior to the fast-freeze sample preparation techniques currently accepted as the technique of choice. Nevertheless, this investigation into the development of structure with variations in each constituent established the importance of cryo-TEM analysis as a collaborative technique for a reliable structural interpretation. A conclusion of this study was that association conditions in typical W/O microemulsions are identical to those in inverse micellar solutions of water and soap in pentanol and the first association occurs when the number of water molecules per surfactant molecules exceeds eight. Association mechanisms leading to continuously evolving microstructures have been investigated in this study, and a few association microstructures are presented in Figs. 4a and 4b.

B. Direct Observation of Droplet Structure in a Vitrified Water–Propylene Glycol–Tween 80–CCl₄ System

Microstructural studies of glass-forming microemulsions, such as those in water–propylene glycol–Tween 80–CCl_4 systems, have been the focus of investigative research for quite some time now [29]. Microemulsion samples were prepared by titrating an initial mixture containing the aqueous phase and o-xylene in a volume ratio of 2 : 1 with a solution of Tween 80 and o-xylene in the volume ratio 2 : 1. The aqueous phase consisted of propylene glycol and water in the mole ratio 1 : 3. Compositions in the two-phase regions of the phase diagram were seen to form more or less stable emulsions that became optically transparent on addition of one or other of the components sufficient to reach the microemulsion region. Specimens were imaged by FFTEM using the double-replica freeze-fracture technique and the visualized microstructures, represented in Fig. 5, were found to show microemulsion droplets varying in size and shape. Droplet groups of 2–12 are seen to be more dominant than single droplets.

C. Imaging of Microemulsion Microstructures in the H₂O–n–Dodecane–DDAB and Related Systems

Microemulsion microstructures by cryo-TEM have been imaged for H_2O–n–octane–$C_{12}E_5$ systems for varying water/oil ratios at a constant amphiphilic concentration of 7 wt% [21], and on the basis of the micrographs, with supporting evidence from small-angle X-ray scattering (SAXS), small-angle neutron scattering (SANS), NMR, self-diffusion, and electrical conductivity measurements, a school of thought endorsing the notion of a bicontinuous network at comparable amounts of water and n-octane was proposed. The

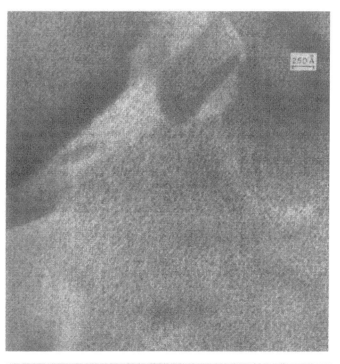

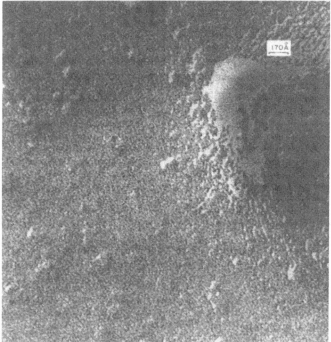

Figure 4 Continuously evolving association microstructures in water–alcohol–potassium oleate–benzene systems. (From Ref. 30.)

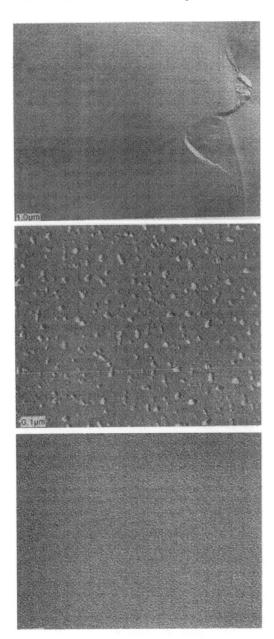

Figure 5 Cryo-TEM micrographs obtained by double-replica freeze fracture in the water–propylene glycol–Tween 80–CCl$_4$ system, showing microemulsion droplets varying in size and shape. (From Ref. 29.)

phase behavior of water–oil–nonionic amphiphile systems has been well studied. In the vicinity of the three-phase region, the landmark of these systems, the highest mutual solubility of water and oil occurs with the lowest amount of amphiphile along the line of points formed by the amphiphile-rich middle phase as it moves around the body of heterogeneous phases from the water-rich to the oil-rich side of the phase diagram. At slightly higher amphiphile concentrations than in the middle phase, one finds a one-phase channel connecting the water and oil corners of the Gibbs triangle. Along such a channel, the microstructure can be studied as it changes from an oil-in-water to a water-in-oil situation. Along this channel, the amphiphile concentration was kept constant at 7 wt% of $C_{12}E_5$. In the micrographs investigating such a research problem, water and oil were discriminated by a decoration effect of the shadow material on the oil fracture face. The results from the cryo-TEM characterization are presented in Fig. 6. The amphiphiles were seen to concentrate in the internal interface, fixing the size scale or the repeat distance to about 500–800 Å as independently verified by SAXS and SANS. Consequently, cryo-TEM imaging could yield an artifact-free direct imaging of vitrified microemulsion microstructures during their evolution.

In the same research attempt an investigation into microemulsion phases of the D_2O–decane–AOT system was also carried out. From prior work in this field, the investigators had concluded from the AOT concentration dependence of the scattering peak in SANS that a close packing of polydisperse spheres in a disordered cubic arrangement could explain the spatial ordering of microstructures in these systems. This conclusion was corroborated by cryo-TEM findings, presented in Fig. 7, in which the varying droplet fraction was decreased from 0.588 to 0.486 to 0.322.

The third system to be explored was the H_2O–n–dodecane–DDAB microemulsion system, and compared to the AOT system the droplets were seen to be larger due to lower amphiphilic concentration. The droplets showed a tendency to aggregate and flatten into polygon-type faces. The cryo-TEM results presented in Fig. 8 agree with theoretical predictions for this system made from model-dependent analytical techniques.

The H_2O–n–dodecane–1-hexanol–potassium oleate system has also aroused scientific curiosity. Previously reported SANS data indicate that this system manifests itself as a collection of water droplets in oil with a droplet size of about 100 Å. The results of a cryo-TEM analysis are presented in Fig. 9.

The conclusions from the cryo-TEM characterization, which have been validated for all the investigated systems, suggest that microstructures vary systematically with the oil/water ratio and, in particular, at comparable oil and water contents, both oil- and water-rich domains are mutually interwoven in an apparently bicontinuous network.

D. Fluid Microstructural Transition Studies from Globular to Bicontinuous Morphologies in Midrange Microemulsions

Freeze-fracture transmission electron microscopic studies on the $C_{12}E_5$–water–octane system to establish a morphological transition from globular to bicontinuous with temperature variations have been carried out [9]. The regions of adhesive and cohesive microstructural fractures in $C_{12}E_5$–H_2O–C_8 microemulsions are shown in the micrograph in Fig. 10. From earlier studies, bicontinuity was inferred from the fact that the self-diffusion coefficients of all components of the solution are large even though SANS indicates relatively large microstructures. Direct images obtained from FFTEM investigations establish a globular-to-bicontinuous transition with an increase in temperature as presented in Fig. 11. Independent confirmation of the morphological transition was obtained from a measurement of individual diffusion coefficients using PFGSE NMR.

A direct visualization of microstructural transition was attempted. Samples were prepared in a controlled environment vitrification system (CEVS) that reduces or prevents microstructural artifacts [9]. Direct images of a microemulsion sample are shown in Fig. 12. Micrograph 12A, taken at moderate electron exposure ($\sim 250 \, e^-/nm^2$) show

Figure 6 Sequence of images of freeze-fractured microemulsions of the H_2O–n-octane–$C_{12}E_5$ system with varying water/oil ratio. The shadow material may be observed to specifically decorate the oil fracture. Bar = 2000 Å. (From Ref. 21.)

Figure 7   Droplet structure of oil-continuous microemulsions of the D_2O–n-decane–AOT system with varying droplet volume fraction. Bar = 2000 Å. (From Ref. 21.)

Figure 8 Aggregated microstructure of deformed droplets in the H_2O–n-dodecane–DDAB system at equal volumes of water and oil. Bar = 2000 Å. (From Ref. 21.)

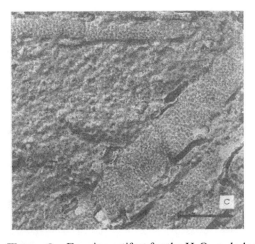

Figure 9 Freezing artifact for the H_2O–n-dodecane–DDAB system. Bar = 2000 Å. (From Ref. 21.)

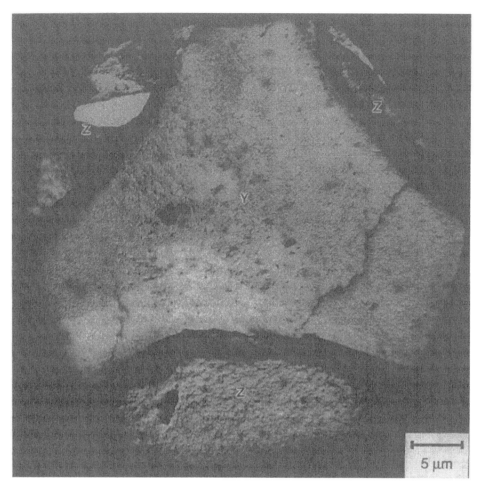

Figure 10 Low magnification of the replica showing an entire grid hexagon and regions corresponding to adhesive and cohesive fractures in $C_{12}E_5$–H_2O–C_8 systems. Bar = 5 μm. (From Ref. 10.)

network-like structures, indicating the possibility of a bicontinuous microemulsion. However, an independent PFGSE NMR diffusion coefficient measurement [9] indicated that the system is not bicontinuous but is water-continuous with discrete oil globules. This discrepancy was resolved by micrographs 12B and 12C, which were taken at low dose (~ 100 e$^-$/nm^2) and at a dose just high enough to expose the micrograph negative (~ 15 e$^-$/nm^2). As the electron exposure was decreased, the apparent microstructure seen was found to change. This is due to radiation damage, i.e., the introduction of structural or compositional changes due to the electron beam. Radiation damage is known to be a serious factor in limiting the amount of microstructural information that can be collected from an electron microscopic specimen, especially one that has water and hydrocarbons in close proximity [7]. For labile microemulsions, radiation damage at moderate and low exposures appears to obscure all original microstructure. This casts doubt on the interpretation of Fig. 12C, the micrograph taken at the lowest possible exposure required to record it on film.

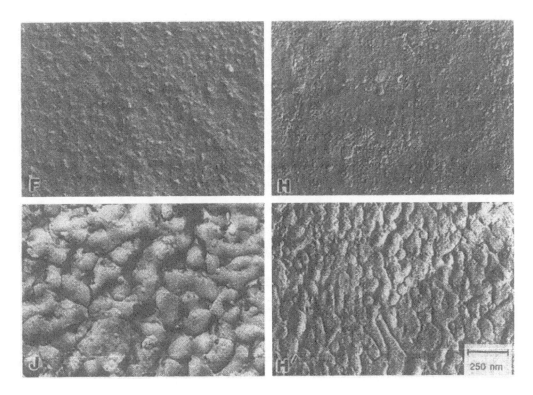

Figure 11 FFTEM micrographs depicting a globular-to-bicontinuous transition with an increase in temperature. Bar = 250 nm. (From Ref. 9.)

E. Microstructural Investigations in the $C_{12}E_5$–H_2O–C_8 System

Freeze-fracture TEM (FFTEM) and FFSEM studies were conducted on $C_{12}E_5$–H_2O–C_8 microemulsion systems [10] by using liquid ethane as a cryogen. Microemulsion systems were studied for shape, connectivity, periodicity, and size. Independent comparisons were also drawn from SAXS and SANS measurements. Adhesive and cohesive fractures were seen, and droplet sizes of about 250 nm were obtained. The FFTEM and FFSEM micrographs presented in Figs 13 and 14 demonstrate a connectivity in the microstructures. Structural connectivity could not be established from light or laser scattering but was independently corroborated from Fourier transform pulsed-field gradient spin-echo NMR (FT PGSE NMR) studies in which self-diffusion coefficients of the oil, water, and surfactant molecules were measured. This study was instrumental in establishing interconnectivity and continuous diffusivity in microemulsion systems.

F. Visualization of Hexagonal Phase Defects in SDS–Formamide Systems

The lyotropic H_α phase formed in an SDS–formamide system was studied [31] by TEM using a cryofracture method for sample fixing. Two morphologies were observed on cryo-TEM imaging: At low SDS concentration long straight rods grouped into large domains with classic liquid crystal defect, and at higher concentrations smaller domains and sinuous,

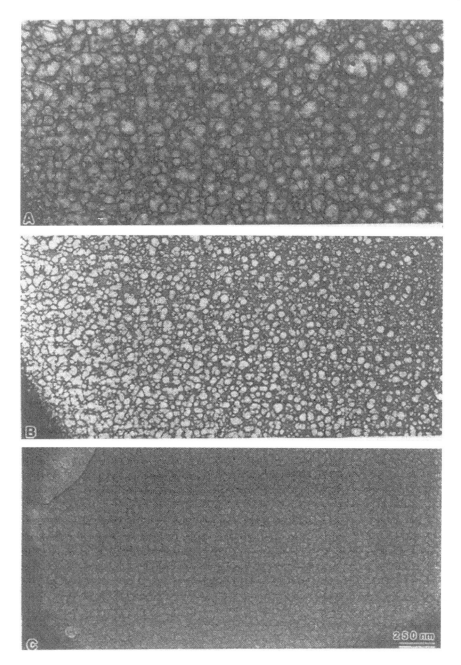

Figure 12 Micrographs from a series of control experiments to establish the globular-to-bicontinuous transition in microemulsion microstructures. Bar = 250 nm. (From Ref. 9.)

constitutive fibers, were seen. Additionally, as a result of cryo-EM observations, unusual defects were detected, which were then analyzed. An explanation of their origin was proposed by the investigators that sheds new light on the development of microstructure in such systems.

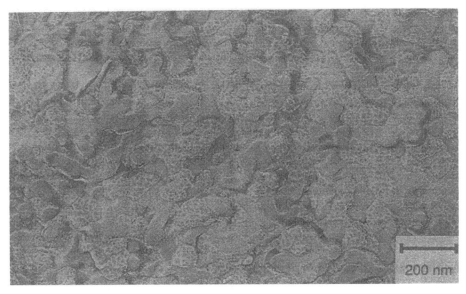

Figure 13 FFTEM micrograph from the $C_{12}E_5$–H_2O–C_8 microemulsion system. Bar $= 200$ nm. (From Ref. 21.)

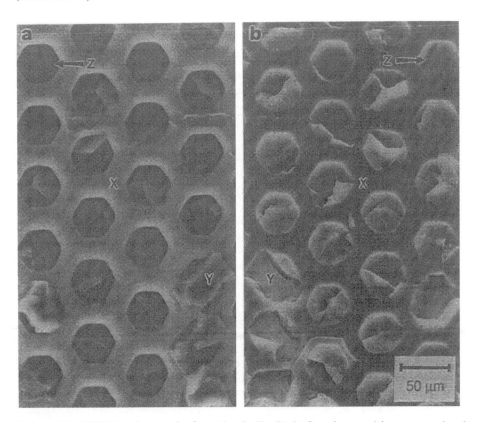

Figure 14 FFSEM micrographs from the $C_{12}E_5$–H_2O–C_8 microemulsion system showing complementary fracture surfaces. Adhesive (X, Y) and cohesive (Z) fractures are visible. Bar $= 50$ μm. (From Ref. 10.)

G. Microstructure Studies on Magnetic Particles Produced in Water-in-Oil Microemulsions

In a comprehensive attempt to evaluate processing–microstructure relationships in colloidal media, $Gd_{2-x}Ce_xCuO_4$, a typical magnetic material, was synthesized [32] using three means: the solid-state reaction, the sol-gel technique, and a water-in-oil microemulsion. The three techniques were chosen on the basis of their ability to yield particles of different sizes spanning the microscopic to macroscopic particle size range. Particles were characterized by XRD, QELS, and cryo-TEM. Magnetization and ac susceptibility measurements were performed on all samples. Analytical results were found to vary with a change in processing route, and there emerged a "graded" relationship between particle size and the magnetization curves for all three samples. Microstructural properties were hence optimized using cryo-TEM imaging for particle size and distribution analysis.

IV. SUMMARY

Freeze-fracture electron microscopy is a powerful tool for "directly" examining the microstructure of microemulsions. Samples are prepared by a fast freeze and a freeze-fracture process conducted in liquid cryogen to preserve the fluid, labile state of matter within a rapidly solidified sample. Samples are usually thermally fixed, and fracture exposes the fluid phase to the electron beam and makes direct viewing without the appearance of microscopic artifacts possible. Results obtained from this characterization technique are highly reproducible, and it has been demonstrated that a cryo-EM analysis provides new access to morphologies, microstructures, and microstructural evolution in microemulsions. FFTEM images of microemulsion microstructures provide conclusive evidence on the structure of midrange microemulsions, which can possess discontinuous globular or bicontinuous network-like structures.

Direct images of microemulsions using cryo-TEM are still elusive because of radiation damage induced by the electron beam and enhanced by the large interfacial area between oil and water phases in microemulsion systems. Despite attempts at lower dose microscopy, reliable results are yet to become accepted. Radiation damage in visualized microstructures can be reduced by lowering the electron exposure; this may be done by using the so-called low-dose attachments (or MDS, minimum dose system) on newer electron microscopes. Radiation damage can also be reduced by using higher voltage microscopes. Higher voltages increase electron energies and reduce their interaction with specimens, thereby reducing radiation damage but also contrast. Another way might be to increase the electron dose rate so that a very short exposure is sufficient to provide an image before damage becomes evident. Hydrocarbons in intimate contact with water damage easily; other systems, e.g., fluorocarbons, hydrosilanes, and fluorosilanes, may damage less and permit microscopy. Polymerizable systems may allow the removal of damage-inducing components; e.g., if the oil in a microemulsion is polymerized without structural reorganization, the water may be removed allowing the "dry" microemulsion to be imaged without fear of radiation damage. Even if the radiation damage problem were to be solved, the process of forming and the act of being a thin film, required for TEM, may induce microstructural reorganization, as has been shown [33] for thin phase-separated block-copolymer films. Newer forms of microscopy—scanning tunneling microscopy (STM), atomic force microscopy (AFM), X-ray microscopy, and magnetic resonance imaging (MRI)—may provide alternative approaches, as may novel detectors like charge-coupled detectors that may even replace conventional high quantum efficiency silver halide films on traditional microscopes.

ACKNOWLEDGMENTS

We acknowledge with gratitude the contributions of all members of the "Low Tension" Group, Department of Chemical Engineering and Materials Science, University of Minnesota, Minneapolis, in particular Prof. H. Ted Davis, Prof. L. E. Scriven, and Prof. W. G. Miller.

REFERENCES

1. L. Reimer, *Transmission Electron Microscopy: Physics of Image Formation and Microanalysis*, Springer-Verlag, Berlin, 1984.
2. D. E. Newbury, D. C. Joy, P. Echlin, C. E. Fiori, and J. I. Goldstein, *Advanced Scanning Electron Microscopy and X-Ray Microanalysis*, Plenum, New York, 1986, pp. 365–448.
3. J. I. Goldstein, D. E. Newbury, P. Echlin, D. C. Joy, C. E. Lyman, A. D. Romig, C. Fiori, and E. Lifshin, *Scanning Electron Microscopy and X-Ray Microanalysis*, Plenum, New York, 1992.
4. M. Adrian, J. Dubochet, J. Lepault, and A. W. McDowall, Nature *308*:32–36 (1984).
5. M. J. Costello, R. Fetter, and J. M. Corless, Optimum conditions for the plunge freezing of sandwiched samples, in *Science of Biological Specimen Preparation* (O. Johari, ed.), SEM Inc. AMF O'Hare, Chicago, pp. 105–115.
6. J. Dubochet, M. Adrian, J. Teixeira, C. M. Alba, R. K. Kadiyala, D. MacFarlane, and C. A. Angell, J. Phys. Chem. *88*:6727–6732 (1984).
7. Y. Talmon, Colloids Surf. *19*:237–248 (1986).
8. J. R. Bellare, H. T. Davis, L. E. Scriven, and Y. Talmon, An improved controlled-environment vitrification system (CEVS) for cryofixation of hydrated TEM samples, Proc. XIth Int. Cong. on Electron Microscopy, Kyoto, Japan, Aug. 31–Sept. 7, 1986; J. Electron Microsc. *35* (Suppl. 1):367–368 (1986).
9. J. R. Bellare, Cryo-electron and optical microscopy of surfactant microstructures, Doctoral Dissertation, University of Minnesota, 1988.
10. P. K. Vinson, Cryo-electron microscopy of microstructures in complex liquids, Doctoral Dissertation, University of Minnesota, 1991.
11. X. Li, Z. Lin, J. Cai, L. E. Scriven, and H. T. Davis, J. Phys. Chem. *99*:10865 (1995).
12. M. Knoll and E. Ruska, Z. Tech. Phys. *12*:389–399 (1931).
13. C. J. Dawes, *Biological Techniques for Transmission and Scanning Electron Microscopy*, 3rd ed., Ladd Research Industries, Burlington, VT, 1981.
14. J. Dubochet, F. P. Booy, R. Freeman, A. V. Jones, and C. A. Walter, Annu. Rev. Biophys Bioeng. *10*:133–149 (1981).
15. D. R. Uhlmann, J. Non-Cryst. Solids *7*:337–344 (1972).
16. F. Franks, Cryo-letters *2*:69–72 (1981).
17. W. B. Bald, J. Microsc. *134*:261–270 (1984).
18. N. R. Silvester, S. Marchese-Ragona, and D. N. Johnstone, J. Microsc. *128*:175–186 (1982).
19. E. B. Hunziker, W. Herrmann, R. K. Schenk, M. Müller, and H. Moor, J. Cell Biol. *98*:267–276 (1984).
20. M. H. Chestnut, D. P. Diegel, J. L. Burns, and T. Talmon, Microsc. Res. Tech. *20*:95 (1992).
21. W. Jahn and R. Strey, J. Phys. Chem. *92*:2294–2301 (1988).
22. H. Y. Elder, C. C. Gray, A. G. Jardine, J. N. Chapman, and W. H. Biddlecombe, J. Microsc. *126*:45, 221–229 (1982).

23. J. C. Hatfield, Freeze-fracture electron microscopy and electrical conductivity of microemulsions, Doctoral Dissertation, University of Minnesota, 1978.
24. M. Müller and H. Moore, Cryofixation of suspensions and tissues by propane-jet freezing and high-pressure freezing, Proc. 42nd Annual Meeting, Microscopy Society of America (G.W. Bailey, ed.), San Francisco Press, San Francisco, 1984.
25. S. Sheth and J. R. Bellare, Proc. 51st Annu. Meeting, Microscopy Society of America (G.W. Bailey and C. L. Rieder, eds.), San Francisco Press, San Francisco, 1993.
26. W. Jahn, J. Microsc. *179*:333–334 (1995).
27. X. Li, Phase behavior, microstructure and transitions in self-assembled colloidal systems, Doctoral Dissertation, University of Minnesota, 1996.
28. J. R. Bellare, J. K. Bailey, and M. Mecartney, Freezing dynamical sol-gel processes with the controlled environment vitrification system (CEVS), Proc. 45th Meeting: Electron Microscopy Society of America (G. W. Bailey, ed.), San Francisco Press, San Francisco, 1982, pp. 356–357.
29. E. A. Hildebrand, I. R. McKinnon, and D. R. MacFarlane, J. Phys. Chem. *90*:2784–2786 (1986).
30. E. Sjoblom and S. Friberg, J. Colloid Interface Sci. *67*:16–30 (1978).
31. M. Abiyaala and P. Duval, J. Phys. (France) II *4*:1687–1698 (1994).
32. J. Mahia, C. Vazquez-Vazquez, J. Mira, M. A. Lopez-Quintela, J. Rivas, T. E. Jones, and S. B. Oseroff, J. Appl. Phys. *75*:6757–6759 (1994).
33. C. S. Henkee, E. L. Thomas, and L. J. Fetters, J. Mater. Sci. *23*:1685–1694 (1988).

14

Characterization of Microemulsions by Electrical Birefringence

Zoltan A. Schelly
The University of Texas at Arlington, Arlington, Texas

I. INTRODUCTION

Similar to most chemical systems of interest, the characterization of colloidal solutions requires the determination of the size, shape, structure, and stability of the particles present. This information is especially important for the understanding and utilization of organized assemblies of surfactants, in particular microemulsions, because the physical properties of the particles usually depend strongly on the thermodynamic conditions such as overall composition, temperature, and external force fields. This dependence is mainly due to the sensitivity to conditions of the monomer–aggregate equilibrium of the surfactant, which is responsible for the existence of the particles, and to the delicate balance of forces that maintain their integrity. For microemulsions, an additional complication arises from the compartmentalization of the systems, which is a source of possible phase transitions but is also a reason for most of their practical applications.

The dependence of the physical properties of microemulsions on thermodynamic conditions also offers an avenue for their investigation. Namely, an intentional sudden change of a thermodynamic variable (such as temperature T, pressure p, electric field strength E) affects their properties, and the ensuing response of the system can be monitored as a function of time. In electrical birefringence studies the perturbation is achieved by exposing the system to an external high-voltage electric field E, and the relaxations are monitored through the induced birefringence of the solution. The primary data obtained are the sign and amplitude of the birefringence signals and their course as a function of time t, from which the relaxation times τ_i of the processes involved can be calculated. Under favorable circumstances, information about the size, shape, structure, and stability of the particles can be deduced directly from the birefringence data. More often, however, results of independent equilibrium measurements (dielectric, scattering, spectroscopic, osmometric, viscosimetric, conductimetric, etc.) must be invoked for the interpretation. In an ideal procedure, in general, the equilibrium properties and the dynamic behavior of a system are elucidated simultaneously through a synergistic modeling where all the available equilibrium and dynamic data are used [1].

Since the simplest oil-in-water (O/W) and water-in-oil (W/O) microemulsions are ternary systems in which the particles are swollen direct and reverse micelles, respectively, the examples given for the application of electrical birefringence will include both microemulsions and micelles. As the studies reveal, the experiments are usually carried out to find answers to specific questions instead of the complete physical characterization of the particles. Often, however, interesting additional information is derived such as the mechanism of phase separation or the elasticity constant of the monolayer in W/O microemulsions.

II. NATURAL AND INDUCED BIREFRINGENCE

A. Theoretical Background

Birefringence (or double refraction) is an optical property of structurally anisotropic crystals. Such materials transform an incident light beam into two perpendicularly linearly polarized rays, the ordinary (o) and extraordinary (e) rays, which propagate at different velocities in the medium. Unless the direction of incidence coincides with that of the optical axis of the (uniaxial) crystal, the e ray emerging from the crystal is displaced parallel to the o ray in the plane of the particular principal section of the crystal (see, for example, Ref. 2). The well-known double images that appear when an object is viewed through a polished calcite crystal are a manifestation of the phenomenon.

The electric field E of the o ray is normal to the principal section, and that of the e ray is parallel. Due to their different velocities, the waves of the two rays are out of phase by an optical retardation δ, and the medium exhibits different refractive indices, n_e and n_o, with respect to the two rays. The difference

$$\Delta n = n_e - n_o \tag{1}$$

is the measure of the birefringence. It is related to the retardation by

$$\delta = 2\pi l\, \Delta n/\lambda \tag{2}$$

where l is the path length of the sample and λ is the wavelength of the incident light in vacuo.

1. Electrical Birefringence

Structurally isotropic substances such as gases, most liquids (including colloidal solutions, e.g., microemulsions), and amorphous solids do not display birefringence. However, if exposed to an external force (mechanical stress, electric and magnetic fields, or a force due to a flow velocity gradient) of a strength sufficient to affect a bulk structural anisotropy, they may become birefringent. These artificial or induced birefringences are widely used in fundamental studies as well as for practical diagnostic purposes.

In electric-field-induced birefringence (or electrical birefringence, quadratic electro-optic effect, Kerr effect), an isotropic transparent substance becomes birefringent when placed in an electric field E. The sample assumes the characteristics of a uniaxial crystal, the optical axis of which is parallel to the direction of the applied field. When the sample is illuminated normal to E, the resulting two indices, $n_\parallel$ and $n_\perp$ (which can be thought of as n_e and n_o), are associated with the parallel and perpendicular orientations, respectively, relative to the external E, of the vibrations of the light wave. Their difference

$$\Delta n = n_\parallel - n_\perp \tag{3}$$

is the electrical birefringence, and it is found to be

$$\Delta n = \lambda K E^2 \tag{4}$$

(the Kerr law), where K is the Kerr constant. Since with increasing field strength the Kerr effect reaches saturation, the experimental Kerr constant, B,

$$B = \lim_{E \to 0} \left(\frac{\Delta n}{\lambda E^2} \right) \tag{5}$$

is also often determined and reported. Although K is most often positive, its sign is important, which is mentioned later.

Several factors contribute to the field-induced structural anisotropy that leads to optical anisotropy and hence to birefringence. All involve the particles' polarization by the field and the partial alignment of their resultant dipole moments parallel to E. The resultant dipole moment μ of a particle is the vector sum of its permanent and induced dipole moments. At the molecular level, electronic and atomic polarization occurs, the extent of which depends on the nature and symmetry of the molecule and on its polarizabilities ($\alpha_\parallel$ and $\alpha_\perp$) along the parallel and perpendicular directions relative to the electric field or, for cylindrical symmetry, along the molecular axes a and b (α_a and α_b). Naturally, the concept of the polarizability tensor is applicable to an assembly of molecules as a whole, e.g., a colloidal particle, as well. For such systems, and also for macromolecules and polyelectrolytes in an insulating medium, interfacial polarization may also have a major or even dominant contribution to the resultant dipole moment.

The induced birefringence is a function of the polarizabilities and the extent of alignment of the resultant dipole moments μ of the particles in the solution by the external field E. Under the assumption that interparticle interaction is negligible (dilute solutions) and the energy of interaction U between E and μ is less than the thermal energy kT, the following expressions can be derived [3] for the low-field limit:

$$\lim_{E \to 0} \left(\frac{\Delta n}{E^2} \right) = K_{sp} C_v n \tag{6}$$

where C_v is the volume fraction of the solute, n is the refractive index of the solution, and the specific Kerr constant K_{sp} is defined by

$$K_{sp} = \frac{2\pi(g_a - g_b)(P + Q)}{15n^2} \tag{7}$$

where the so-called intrinsic anisotropy factors g_i are related to the optical polarizabilities α_i^0 through $\alpha_i^0 = 4\pi\varepsilon_0 v g_i$, with ε_0 being the permittivity of free space and v the volume of the particle. In Eq. (7), $P = \mu_p^2/(kT)^2$ and $Q = (\alpha_a - \alpha_b)/kT$ represent the permanent and induced dipole terms, respectively, with μ_p denoting the permanent dipole moment of the particle.

2. Transient Electrical Birefringence

Upon the sudden application of an external electric field of constant strength E, the birefringence $\Delta n(t)$ asymptotically approaches a steady-state value, Δn_0 (static birefringence), and then Δn relaxes back to zero after the field is suddenly turned off. The time evolution of $\Delta n(t)$ reflects the rate and extent of field-induced processes that lead to the sample's optical anisotropy, i.e., of the polarization and reorientation processes. For liquid droplets, the different possible polarization processes (electronic, atomic, and interfacial) also result in an observable deformation of the particles present—for instance, the elongation of an originally spherical droplet in the direction of E [4–6]. For reverse micelles, in spite of the complexity of the systems and the polarization processes involved, the extent of elongation and the resulting birefringence can be estimated (in the limits of low volume fraction of water and infinite dilution) by using simplifying assumptions [7]. From the static birefringence (Δn_0), the magnitude of the bending modulus κ [8] (also called interfacial rigidity or bending, elasticity, and rigidity constant) of the surfactant monolayer can be deduced as well [7,9]. The significance of the bending modulus is due to the fact that given the vanishing interfacial tension in microemulsions, the elasticity of the interfacial layer plays a key role in determining the structure of the aggregates and the stability of the competing phases.

The usually spherical equilibrium shape of microemulsion droplets is the time-average (or ensemble-average) shape of undulating spherical shells. For small fluctuations $u(\theta, \varphi)$ from the mean spherical shape of radius r_0, the local radius $r(\theta, \varphi)$ of a droplet is

$$r(\theta, \varphi) = r_0[1 + u(\theta, \varphi)] \tag{8}$$

where u can be expanded into spherical harmonics [10]

$$u(\theta, \varphi) = \sum_{l, m} u_{lm} Y_{lm}(\theta, \varphi) \tag{9}$$

with $l = 0, 1, 2, \ldots$ and $-l \leq m \leq l$. The case $l = 0$ corresponds to a breathing fluctuation (change in size of a sphere), whereas $l = 1$ corresponds to translation. The first true shape fluctuations are the ellipsoidal modes with $l = 2$.

The shape fluctuation leads to an anisotropy of the polarizability that depends on the amplitudes of the fluctuation [11]. In addition, an asymmetric instantaneous structure may have a charge distribution that results in an instantaneous multipole moment [6]. Upon application of the electric field, the resultant of the instantaneous plus induced moments is partially aligned parallel to E. If the shape fluctuation is significantly slower than the rotational diffusion of the particle, the birefringence response of the particle will be equivalent to that of a rigid ellipsoid. In such a situation, provided all polarization mechanisms involved are faster than rotation, the relaxation time τ of the emerging birefringence $\Delta n(t)$ can be interpreted as the rotational relaxation time τ_r of the microemulsion particle. The rate of rotation, however, depends on the shape and size of the particle. In general,

$$\tau_r = 1/6\theta \tag{10}$$

where θ is the rotational diffusion coefficient. For a rigid, symmetrical ellipsoid rotating about a minor axis, θ is related to the major (a) and minor (b) semiaxes as [12]

$$\theta = \frac{3kT[2\ln(2a/b) - 1]}{16\pi\eta a^3} \tag{11}$$

where η is the viscosity of the solvent. Analogous equations for rigid rods [13] and disks [14] have also been derived, showing an inverse proportionality between θ and a^3 in each case. These equations are used instead of (11) if there is independent indication that the aggregates are rod- or disk-shaped (e.g., some aqueous micelles).

For a monodisperse size distribution of noninteracting ellipsoids, the relations between θ and the rise and decay of the electrical birefringence have been established [15–17]. At low field strength, for the time-dependent rise of the birefringence one finds

$$\frac{\Delta n(t)}{\Delta n_0} = 1 - \frac{3P/Q}{2(P/Q+1)} e^{-2\theta t} + \frac{P/Q-2}{2(P/Q+1)} e^{-6\theta t} \tag{12}$$

Examination of this equation shows that if the signs of P and Q are the same (i.e., if the permanent and induced dipole moments tend to orient the ellipsoid in the same direction), then the rate of reaching the steady state ($\Delta n \to \Delta n_0$) is determined by the magnitude of P/Q: The larger P/Q, the slower the process. If there is no permanent dipole (i.e., $P/Q = 0$, as expected for microemulsions) then Eq. (12) reduces to a simple exponential function,

$$\Delta n(t) = \Delta n_0 (1 - e^{-6\theta t}) \tag{13}$$

If P and Q have opposite signs, then the sign of birefringence changes during its evolution. If, in addition, $|P| = |Q|$, then Δn approaches zero at the steady state, and consequently there is no birefringence in the decay phase after the termination of the field.

After the sudden removal of the electric field, however, the orientational randomization of the ellipsoids results in the decay of the birefringence, which, in general, is described by

$$\Delta n(t) = \Delta n_0 e^{-6\theta t} \tag{14}$$

Thus, if the particle has no permanent dipole moment, the rise and decay signals of the birefringence are symmetrical [cf. Eqs. (13) and (14)]. The rotational diffusion coefficient θ is usually obtained directly from the decay curve, Eq. (14); and Eq. (11), together with the apparent hydrodynamic radius r_h determined in light scattering experiments, can be used to obtain information about the size and eccentricity a/b of the rotating ellipsoid.

If the rise and decay signals are found not to be symmetrical or if more than one relaxation is observed, it may indicate that the reorientation of permanent dipoles is involved and/or that the rate of induced polarization(s) is comparable to that of particle rotation. There is no comprehensive theory available that includes all the possible effects, and their analysis and interpretation have to be tailored to the particular system in question.

For further details on the theory of electrical birefringence, the reader is referred to more extensive treatments available on the subject [3,18–24].

B. Experimental Approach

1. Square-Pulse Perturbation

Although any functional form of the electric field $E(t)$ can be applied to induce the birefringence, the interpretation of the response function $\Delta n(t)$ is the simplest if a square-pulse perturbation is used, with rise and fall times significantly shorter than the relaxation times τ_i of the birefringence response. The corresponding equations in the preceding section are valid only for such conditions.

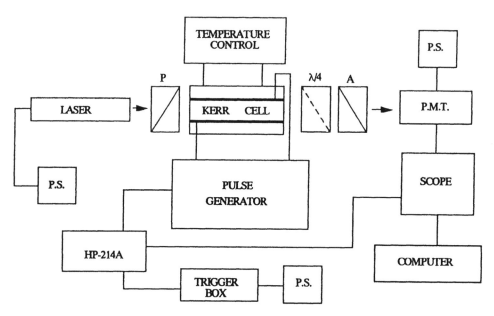

Figure 1 Block diagram of a typical electrical birefringence apparatus using either square-pulse (for transient electrical birefringence) or oscillating field (for dynamic electrical birefringence) perturbations. P, polarizer; A, analyzer; $\lambda/4$, quarter-wave retardation plate; P.S., power supply; P.M.T., photomultiplier tube; HP-214A, wave-form generator. (Reprinted with permission from Ref. 41. Copyright 1994 American Chemical Society.)

The block diagram of a typical transient electrical birefringence apparatus is shown in Fig. 1. The sample solution is located between the two electrodes (3 mm apart) of the Kerr cell, which is equipped with windows and has an optical path length $l = 5$ cm. The high voltage (up to a few kilovolts) square pulse is produced by a pulse generator (Cober 605P), with the rise and fall times of the pulse < 50 ns.

The birefringence detection system is mostly of the standard design [3]. A linearly polarized (extinction ratio 500 : 1) He-Ne laser (10 mW) is used as light source. Its beam (which should not be absorbed by the sample) passes through a polarizer, the Kerr cell, a quarter-wave retardation plate, and an analyzer before reaching the photomultiplier detector. The orientation of the optical components is schematically depicted in Fig. 2a. The planes of polarization of the laser and polarizer are set at $\pi/4$ to the direction of the applied electric field E with the slow axis of the quarter-wave plate at $3\pi/4$. The analyzer is set at a small angle $\pm \alpha$ from the crossed position. With such an arrangement, the intensity of light reaching the detector is $I_\delta(t)$ in the presence of the field and I_α in its absence. If both K and α are positive [cf. Eqs. (4) and (15) and Fig. 2a], then the induced birefringence effects an increase in the transmitted light intensity (positive birefringence). A negative K results in $I_\delta(t) < I_\alpha$ (negative birefringence). The magnitude of the change, $\Delta I(t) = I_\delta(t) - I_\alpha$, depends on the induced retardation $\delta(t)$ of the sample, and the relative change with respect to I_α is given by [3]

$$\frac{\Delta I_\delta}{I_\alpha} = \frac{\sin^2(\alpha + \delta/2) - \sin^2 \alpha}{\sin^2 \alpha} \tag{15}$$

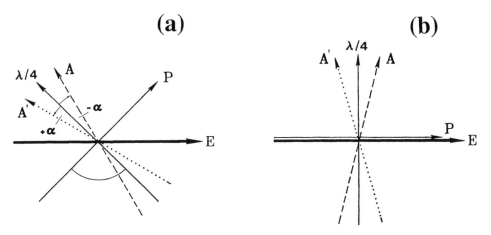

Figure 2 Schematic representation of the orientation of the optical components (as viewed from the photomultiplier). E, electric field; P, polarizer and laser; A and A', two alternative orientations ($\pm\alpha$) of the analyzer; $\lambda/4$, slow axis of quarter-wave plate. (a) Standard orientation: detects birefringence together with turbidity (if it occurs). (b) Detection of turbidity only. (Reprinted with permission from Ref. 6. Copyright 1989 American Chemical Society.)

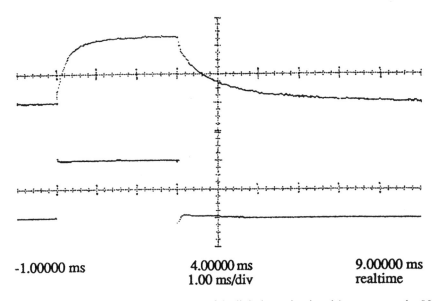

-1.00000 ms   4.00000 ms   9.00000 ms

1.00 ms/div   realtime

Figure 3 Typical oscilloscope traces of the light intensity signal (top trace; scale: 80 mV/1 ms) due to transient electrical birefringence and of the attenuated perturbation square pulse (bottom trace; scale: 400 mV/1 ms). Note that in this example the forward and reverse relaxation signals are not symmetrical.

The output of the photomultiplier, which must be linearly proportional to the incident light intensity, is recorded on a digital oscilloscope (Fig. 3). The data are transferred to a computer, where they may be smoothed, if necessary, by the method of zero determinants [25] or some other appropriate methods, and then $\Delta n(t)$ is computed. The resulting function is similar in general appearance to the light intensity signal, e.g., the one shown in Fig. 3. The rise (field on) and decay (field off) portions of $\Delta n(t)$ are individually analyzed. In the simple case, each of the forward and reverse relaxation curves can be represented by a multiple exponential function of the general form

$$\Delta n(t) = \sum_i A_i \exp\left(\frac{-t}{\tau_i}\right) \tag{16}$$

where A_i and τ_i are the relaxation amplitude and relaxation time, respectively, of the ith process contributing to the birefringence change. The number of relaxations involved and the values of A_i and τ_i can be determined by using the Z transform method with spike recovery [26] or some other appropriate fitting procedure [18–20]. An advantage of the Z transform method, however, is that it is specifically tailored for the analysis of functions with a set of equally spaced discrete data (as recorded by digital oscilloscopes or multichannel analyzers) and involves no numerical integration.

If, usually above a threshold field strength, the field-induced processes involve massive structural changes (beyond polarization, elongation, and reorientation) such as phase transition or percolation of the system, then the relaxation signals may become more complex, and $\Delta n(t)$ cannot be described by Eq. (16). Such behavior is easily recognized, for instance, from the nonmonotonic course of the forward relaxation or from the stretched exponential nature of the reverse relaxation. The analysis of such signals is, of course, more problematic, and the use of results of additional experiments is necessary for their interpretation.

Since the essence of the birefringence experiment is the measurement of transmitted light intensity change $\Delta I(t)$, any process in the sample that affects $I(t)$ will mimic an induced retardation $\pm \delta(t)$, i.e., contribute to or reduce the actual birefringence [6]. The most obvious such processes in microemulsions are the clustering of particles [27], phase separation [6], and percolation [28], each of which may cause transient turbidity $\Delta I_t(t)$, i.e., integrated loss of transmitted light intensity (over the full solid angle) due to increased light scattering. Obviously, it is mandatory to establish whether this occurs; if it does, $I_\delta(t)$ must be extracted from the observed light intensity $I(t)$ prior to the computation of the birefringence $\Delta n(t)$ [6,27]. Experimentally, the pure turbidity component $\Delta I_t(t)$ of the signal $I(t)$ can be obtained through a particular arrangement of the optical components of the detection system (Fig. 2b) that renders it blind to birefringence [6].

The birefringence $\Delta n(t)$ computed from the light intensity signal $\Delta I_t(t)$ is actually the sum of induced birefringences of several possible sources,

$$\Delta n(t) = \Delta n(t)_i + \Delta n(t)_f + \Delta n(t)_s + \Delta n(t)_w + \cdots \tag{17}$$

where $\Delta n(t)_i$ is the intrinsic birefringence of the oriented colloidal particles of interest, $\Delta n(t)_f$ is the form (textural or structural) birefringence [29,30] that arises when the refractive index of oriented isotropic rods or disks is different from that of the solvent, and $\Delta n(t)_s$ and $\Delta n(t)_w$ are the induced birefringences of the solvent and the windows, respectively. In practice, the last two terms are usually below the detection limit; otherwise, their magnitude must be determined and subtracted from $\Delta n(t)$. The contribution of form birefringence (positive for rods and negative for disks), if significant, can be estimated [3]. This seems to be the case especially for rod-shaped aqueous micelles of ionic surfactants where the form and intrinsic birefringences are found comparable in magnitude [31].

2. Reversing Pulse Technique

Although information about the electrical properties (permanent and induced dipole moments) of the particles can be obtained from the rise curve of the birefringence induced by a square pulse, the reversing pulse technique [32] is more accurate and convenient for this purpose. The reversing pulse (Fig. 4) is generated by two pulse generators connected in series but with opposite polarity, and the trailing edge of the first, positive pulse triggers the second, negative pulse. A fall time of about 0.5 μs for the polarity reversal can be relatively easily achieved.

Experimentally, the duration of the first pulse is adjusted to be long enough for the birefringence to reach steady state, but attention is focused on $\Delta n(t)$ subsequent to the polarity reversal. Theory [33] predicts that if there is no birefringence upon field reversal, then the orienting torque on the particle must be entirely due to a fast induced dipole moment. On the other hand, a change in birefringence (Fig. 5) indicates either some permanent dipole or slow induced dipole contribution to the orienting torque.

3. Laser-Induced Electrical Birefringence

Since birefringence is also induced by oscillating electric fields (dynamic electrical birefringence [18,34]), perturbation can also be achieved by exposing the sample to a high power linearly polarized laser step pulse (Fig. 6) in which, of course, E oscillates at optical

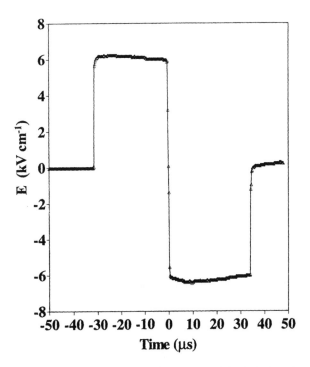

Figure 4 Image of a typical reversing (bipolar) electric perturbation pulse (100 ns per data point). The fall time of polarity reversal is < 0.5 μs. (Reprinted with permission from Ref. 40. Copyright 1995 American Chemical Society.)

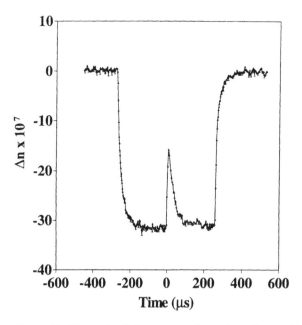

Figure 5 Example of the computed transient electrical birefringence $\Delta n(t)$ obtained in a reversing pulse experiment (2 μs per data point). The central spike coincides in time with the polarity reversal (Fig. 4). (Reprinted with permission from Ref. 40. Copyright 1995 American Chemical Society.)

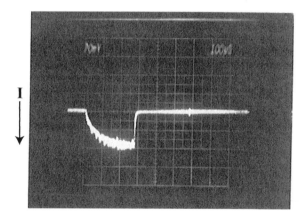

Figure 6 Oscilloscope trace of a linearly polarized laser step pulse used for perturbation in laser-induced electrical birefringence experiments. The pulse is generated by a Nd : YAG laser operated in the fixed-Q, TEM_{00} mode and terminated by the combined use of intra- and extracavity Pockels cells. I, light intensity; scale: 20 mV/100 μs. (Reprinted with permission from Ref. 27. Copyright 1995 American Chemical Society.)

frequency. The associated electric field strength E depends on beam geometry, power, and the intensity distribution within the laser beam's cross section. For instance, with a 2.5 kW pulse of a Nd : YAG laser, $E_0 = 6.5$ kV/cm or $E_{\text{rms}} = 4.6$ kV/cm can be obtained. The experimental setup [27] shown in Fig. 7 can be used for studying the dynamics of both

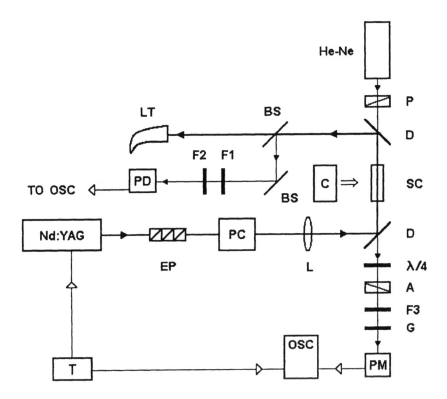

Figure 7 Schematic diagram of the laser-induced electrical birefringence apparatus. Path of Nd : YAG laser pulse: EP, extracavity polarizers; PC, Pockels cell; L, lens (focal length 63.5 cm, focal point about 20 cm past the sample cell); D, dichroic mirror; SC, sample cell; C, calorimeter; BS, beam splitter; F1, UV-Vis cutoff filter; F2, narrow-bandpass filter (1.060 μm); PD, photodiode; LT, light trap. Path of He-Ne monitoring beam: P, polarizer; $\lambda/4$, quarter-wave plate; A, analyzer; F3, narrow-bandpass filter (632.8 nm); G, glass plate diffuser; PM, photomultiplier. OSC, oscilloscope; T, timing and trigger. (Reprinted with permission from Ref. 27. Copyright 1995 American Chemical Society.)

optically induced birefringence (optical Kerr effect) [35] and optically induced light scattering [36]. Both effects are based on the nonlinear polarization of particles, and they appear simultaneously. The integrated loss (over the full solid angle) of scattered light intensity is detected as turbidity, which in microemulsions represents the dominant portion of the transmitted light intensity signal. The birefringence and turbidity components, however, can be separated [27].

The utility of laser-induced birefringence is based on the high frequency of the perturbation field used, which allows only electronic polarization to occur. If the duration of the laser pulse is sufficiently long, the induced dipoles are partially reoriented parallel to the E of the laser pulse. Comparison of the rate and amplitude of the resulting birefringence with that observed in square-pulse (dc) experiments permits elucidation of the contributions of different polarization mechanisms to the induced dipole moment of the particles [27].

III. EXAMPLES OF APPLICATIONS

For the quantitative interpretation of the usual (dc pulse) experimental results it is essential that the electric field strength be constant during the forward relaxation of the birefringence. For the unavoidable degradation of the field strength to be negligible, the electrical conductance of the solution must be as low as possible. Otherwise, due to the current flowing through the sample, Joule heating occurs and the field strength falls perceptibly.

   According to these requirements, clearly, ternary W/O microemulsions of a nonionic surfactant are ideal systems to be studied by transient electrical birefringence. The requirement is fulfilled to a progressively lesser degree as the microemulsion's conductance increases in the order O/W microemulsions of nonionic surfactants $<$ W/O ionic $<$ O/W ionic.

   Also, for W/O microemulsions, the conductance usually increases with the water content of the system, which is customarily expressed as the molar ratio of water to surfactant, symbolized by w_0 or R. Naturally, the situation is even worse for quaternary systems of any type (W/O or O/W) if the fourth component is an electrolyte. In this case, the birefringence experiment (E jump) is associated with a simultaneous temperature jump that complicates the interpretation of the data. Nevertheless, useful information has been obtained from such experiments.

A. Reverse Micelles and W/O Microemulsions

In a generic W/O microemulsion (L_2 phase), the monomeric surfactant and water in the continuous oil phase are in equilibrium with spherical droplets; however, the monomeric concentrations are very low (often neglected in calculations). Depending on temperature and volume fraction, the droplets are in equilibrium with clusters of droplets, mainly because of sticky collisions. Such is the constitution of the original equilibrium system that is perturbed by the electric field.

   Most studies have focused on ternary and quaternary systems (with an electrolyte as the fourth component or an alcohol as cosurfactant in the latter) of both ionic [6,37–41] and nonionic [28,42] surfactants. The shape of the transient birefringence signal and the number, amplitude, and rate of the relaxations typically depend on composition, temperature, and field strength. Since the thermodynamic conditions affect the aggregation number $\langle n \rangle$, size, and stability of the particles as well as the phase behavior of the system, the distance ($T_c - T$) from a critical temperature T_c and the distance from a critical composition C_c also have a major influence.

1. AOT Systems

The most detailed studies were carried out on ternary systems of the surfactant sodium bis(2-ethylhexyl) sulfosuccinate (aerosol-OT or AOT), in various nonpolar solvents [6,27,37,41]. Although the results cannot be generalized, the phenomena observed and the overall picture that has emerged [6,41] may be illustrative for many aspects of other systems.

   For AOT, birefringence can be observed above a threshold value of the applied field strength, which depends on the composition (especially the water content w_0) and temperature. At a low volume fraction of water (small droplets) the Kerr constant is negative

[43]. This is found to be a consequence of the optical anisotropy of the AOT molecules [7], which are all aligned orthogonal to the water/oil interface. Negative birefringence is observed also for small droplets stabilized by other amphiphiles [38,44].

For larger droplets the birefringence is positive and its rise is double-exponential, indicating that two detectable processes are involved in the evolution of the field-induced anisotropy. The relaxation times τ_f and τ_s of the fast and slow processes are on the order of 10 and 10^2 μs, respectively. The creation of the anisotropic structure can, in principle, occur through several possible processes. For the fast relaxation, the most likely possibilities are the following [41]:

1. The applied electric field interacts with the instantaneous dipole moments that may arise from uneven charge distribution due to the fluctuating shape of the individual microemulsion droplets. The ensuing reorientation of an instantaneous dipole by the field may involve the rotation of a droplet as a whole or a peristaltic type of rotation in which the instantaneous dipole moment vector rotates in (and the associated structural distortion propagates as a wave on the surface of) the droplet, which itself is either stationary or may also rotate but not necessarily in a correlated fashion. A peristaltic rotation of the instantaneous dipole may be the only mode of rotation on the observed time scale for droplets locked in clusters. If any of these descriptions applies, the time constant of the associated rise of the birefringence (τ_f) may be referred to as rotational relaxation time.

2. The electric field interacts with individual charges and as a consequence elongates the originally time-averaged spherical shape of the microemulsion droplet to a prolate ellipsoid with its major axis in the direction of the field E. This electromechanical distortion of the spherical droplet can also be viewed as the stabilization of shape fluctuations, preferentially in the direction of the applied field. If this mechanism is operative, the time constant for buildup of birefringence (τ_f) is related to the rate of elongation of the droplets, which is a function of, among other factors, the flexibility of the monolayer and the rate of displacement of Na^+ ions from their original equilibrium distribution in the aqueous pools during polarization by the dc field.

For AOT microemulsions with large aqueous pools, the forced vectorial migration of Na^+ ions may contribute significantly to the anisotropy [45]. In contrast, in dry systems ($w_0 \sim 0.1$), as found through laser E-jump perturbation, only structural changes due to electronic polarization can occur [27]. Although such experimentation does not allow directly for distinguishing the difference in mechanisms 1 and 2, they all lead to the elongation of originally spherical droplets, individually and as constituents of clusters.

The elongation of the droplets may also effect a shift to the right of the monomer–droplet equilibrium of the surfactant [41]. Namely, the deviation from sphericity of a droplet with an aqueous pool of constant volume results in an increase of its water/oil interfacial area. The instability that results may be reduced by the entrance of surfactant monomers from the bulk phase, leading to an increase in surfactant aggregation number of the microemulsion particle in the presence of the field.

In response to the dc [6,41] or ac [27,46] electric field, due to induced dipole–induced dipole interaction, the free polarized elongated droplets form linear clusters in the direction of E. The original, presumably globular, clusters are also reoriented and/or linearized to

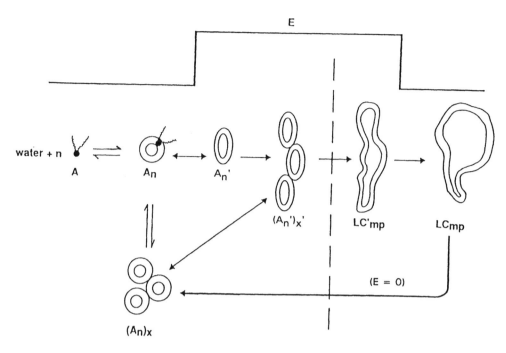

Figure 8 Schematic representation of the processes leading to birefringence (and turbidity) in a W/O microemulsion, in relation to an applied electric square pulse **E**. Below a (second) threshold value of the field strength and far from critical conditions, or under any conditions if the pulse is terminated at a time indicated by the dashed line, only birefringence is observed due to the formation of A'_n and $(A'_n)'_x$. Above the threshold of the field strength, close to critical conditions, and with a sufficiently long square pulse, turbidity contributes to the signal due to phase separation or/and percolation. The double wall of the particles symbolizes the water/oil interface. Symbols: A, surfactant monomer; A_n, microemulsion droplet; $(A_n)_x$, cluster; LC_{mp}, liquid-crystalline microphase or/and percolation structure. Primed symbols stand for polarized structures oriented parallel to **E**; ($\leftrightarrow$) reversible step with respect to turning the field on or off; ($\rightarrow$) irreversible step. (Reprinted with permission from Refs. 6 and 41. Copyright 1989 and 1994 American Chemical Society.)

strings of polarized microemulsion droplets (Fig. 8). These processes lead to an enhanced structural order that results in further increase of the anisotropy and thus the birefringence. The slow (τ_s) relaxation (usually of smaller amplitude) observed is ascribed to the formation of the aligned linear clusters.

   If the composition of the system is close to a phase boundary in the phase diagram or if $T_c - T$ is small and also the strength of the applied field is above a second threshold, phase separation [6] and/or percolation [28,47,48] occurs. These effects are manifested in transient turbidity [6], an increase in electrical conductivity (monitored through a current-viewing resistor in series with the Kerr cell) [28,48] or light scattering (monitored at 90°) [28]. They commence after an induction period (on the order of 10^2 μs) that depends on the applied field strength. Both phase separation and percolation seem to be initiated by fusion of the droplets in the linear clusters (Fig. 8). In phase separation ($L_2 \rightarrow L_2 + LC'_{mp}$), the formation of a liquid-crystalline microphase LC'_{mp} is indicated, which proceeds at a rate faster than linearly with time [6]. It may represent the precursor event for the subsequent percolation (observed at higher field strengths and in the presence of electrolytes) that occurs on the millisecond time scale. The current signal of percolation is sigmoid, suggesting a cooperative process [48].

Under conditions where no phase separation or percolation can occur, termination of the square pulse is followed by a double exponential decay (τ_{-f} and τ_{-s}) of the birefringence. The forward and reverse relaxations are found to be symmetrical, i.e., $\tau_i = \tau_{-i}$ and $\Delta n_i = \Delta n_{-i}$ within experimental error [41]. In the faster process the induced dipoles of the droplets (individually and as constituents of clusters) rapidly collapse, the shape of the droplets reverts to spherical, and/or the droplets randomize in their orientation. If the field-free fast relaxation is interpreted as the ellipsoid-to-sphere structural relaxation of the droplets, the bending modulus κ of the surfactant monolayer can be estimated from the measured τ_{-f} [49]. Depending on the polydispersity assumed, the values found in the range $\kappa = 0.4$–$1.0 \ kT$ are consistent with those obtained from the static birefringence Δn_0 [7,9].

At this point in the reverse relaxation, the only aligned species that remain are the linear clusters composed of nonpolarized droplets. Their slow orientational randomization or delinearization occurs at a rate characterized by τ_{-s} [41].

If the experimental conditions do allow for phase separation or percolation to occur, then either a third, slower process is observed on the time scale of seconds for phase separation [6] or the entire reverse relaxation signal is characterized by a stretched exponential function in the millisecond range for percolation [28,48]. This slow process represents the re-equilibration of the system in which the original species and their populations are reconstructed. Note, however, that the re-equilibration must occur via a route different from the forward path (Fig. 8), since the forward path involves states that are accessible only under an electric field [6,41].

2. NaDEHP Systems

Although the surfactant sodium bis(2-ethylhexyl) phosphate (or NaDEHP) differs from AOT only in its polar headgroup, the nature and behavior of its aggregates in nonpolar solvents are remarkably dissimilar. Since NaDEHP is only sparingly soluble under very dry conditions (up to $w_0 \sim 2$), it goes into solution as crystallites, i.e., about 10–100 nm long rods with a disordered ionic lattice core [50]. This basic structure is already present in the reverse hexagonal liquid crystalline solid state of the surfactant [51]. Although the aggregates in solution had been believed to be giant rodlike reverse micelles [52–54], square-pulse and reversing pulse electrical birefringence experiments were instrumental in establishing that they were actually brittle crystallites with permanent dipole moments [40] (Fig. 9a). From the negative birefringence observed (typical for particles in which the maximum optical polarizability is in the direction of a minor axis), a preferentially perpendicular orientation of the plane of the terminal $O-P-O^-$ of the phosphate to the crystallite's major axis can be concluded (Fig. 9c). The birefringence results also indicate that with increasing water content the crystallites progressively become shorter (dissolve), and only above a critical water content ($w_{0,c} \sim 3.0$) are nondipolar, nonspherical reverse micelles formed [40].

B. Aqueous Micelles

Numerous micellar systems of both ionic and nonionic surfactants in water have been investigated [55,56], occasionally with an added electrolyte or another amphiphile as a third component. The electrical birefringence dynamics observed reflect the significant differences in the structure and thermodynamic behavior of ionic and nonionic systems, while zwitterionic amphiphiles [57] show their ambivalent nature.

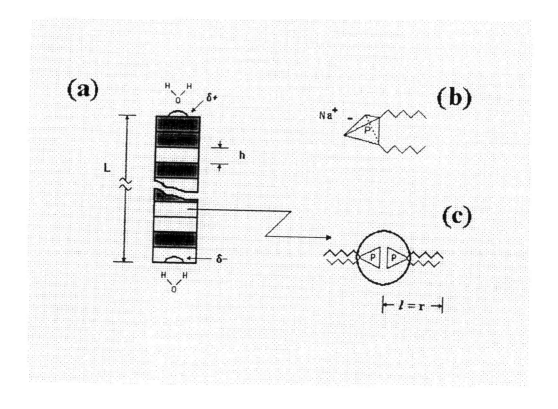

Figure 9 Cartoon of an NaDEHP rod (in the solid state or as a crystallite in solution) and its constituents. (a) Side view of the rod's crystalline core (of length L) consisting of a stack of disks (with average height h), each occupied by two (white) or three (shaded) $Na^+PO_4^-$ ion pairs. The protrusion ($\delta+$) and the recess ($\delta-$) symbolize the net overall axial displacement of the Na^+ ions relative to the PO_4^- ions, which results in a permanent dipole moment. The exposed polar ends of the rod are hydrated. (b) The PO_4^- tetrahedron of NaDEHP (showing only the main hexyl chains). (c) Top view of a disk, indicating the preferential orientation of the PO_4^- groups (Na^+ ions not shown). The length l of an NaDEHP molecule is the radius r of the crystallite. (Reprinted with permission from Ref. 50. Copyright 1995 American Chemical Society.)

1. Ionic Systems

Most ionic amphiphiles at high concentration or in the presence of an electrolyte form nonspherical (mainly rod-shaped) micelles, whereas at low concentration spherical aggregates are usually present. At equilibrium, surfactant monomers are rapidly exchanged between the bulk phase and the aggregates. Although a major fraction of the charges are neutralized by the counterions, the micelles are charged. Although Joule heating and electrode reactions are unavoidable under these conditions, they can be reduced by the use of short square pulses or an oscillating electric field. In spite of the potential complications, the birefringence dynamics of several ionic systems were thoroughly investigated—for instance, cetylpyridinium salicylate (CPS) and cetylpyridinium heptane sulfonate (CPHS) [58], tetraethylammonium perfluorooctane sulfonate (TEPOS) and tetramethylammonium perfluoronanoate (TMPN) [59], and hexadecyloctyldimethyl-ammonium bromide ($C_{16}C_8DMAB$) [60]—by the use of square-pulse and oscillating electric field perturbations. The birefringence signals observed exhibit a complex course where the sign of the birefringence may change during the relaxations and/or $\Delta n(t)$ may display extrema. The number of relaxations observed (up to four), the magnitude of the relaxation times τ_i and amplitudes, and the sign of the associated birefringence depend mainly on the surfactant concentration, the presence and concentration of electrolytes added, temperature, and the strength and frequency of the electric field applied. The interpretation that has evolved [60] is based on the recognition that for solutions of rodlike micelles the intrinsic birefringence and form birefringence [cf. Eq. (17)] may be comparable in magnitude but opposite in sign. The intrinsic birefringence is determined by the optical anisotropy of the alkyl chain(s), the headgroup, and the counterion as well as on the configuration of the chains and the orientation of the counterions at the micellar interface. Variation of the nature of the counterion may change the sign of the intrinsic birefringence [58].

Under an electric field, the processes and conditions responsible for the complex signals can generally be summarized as follows [60]:

1. In the dilute concentration range a single relaxation (τ_1), usually with negative birefringence, is observed due to the alignment (rotational relaxation) of the micelles.

2. In a semidilute concentration range where the mean distance between micelles is comparable with their length (overlap concentration), a second, slower relaxation (τ_2) with the opposite sign of birefringence becomes apparent. This part of the signal is due to the alignment of the rods perpendicular to E. Such an orientation is a consequence of a continuous three-dimensional network formed by the overlapping double layers of the rods that, in short times, permits the ion migration–induced buildup of an induced dipole moment only in the radial direction of the rods.

3. With further increase in the concentration, the enhanced repulsive interaction between rods seems to give rise to a third, even slower relaxation (τ_3). In this process the rods align again parallel to the field, and the sign of the birefringence is the same as for the first process. The amplitude of the associated birefringence increases rapidly with the surfactant concentration and ultimately dominates the overall signal.

4. In some cases an even slower fourth relaxation (τ_4) is observed that may be due to the global deformation of a temporary network of interacting rods.

All four effects also appear when rodlike micelles of nonionic surfactants are rendered charged by the addition of a small amount of an ionic cosurfactant. On the other hand, the addition of excess salt eliminates the second process (τ_2) by reducing the mutual repulsive interaction between micelles.

A convoluted birefringence behavior (including three distinct processes and change of sign of birefringence during relaxation) resembling the one described above was also observed at high concentrations of the nonionic polyoxyethylene surfactants [56]. Since ions are absent from these systems, several aspects of the interpretation offered for charged rodlike micelles [60] are not applicable for explaining the origin of their birefringence dynamics. Detailed studies, however, were done either at low concentrations or in the vicinity of the cloud point curve. The results are outlined in the next section.

2. Nonionic Systems

Aqueous solutions of many nonionic amphiphiles at low concentration become cloudy (phase separation) upon heating at a well-defined temperature that depends on the surfactant concentration. In the temperature–concentration plane, the cloud point curve is a lower consolution curve above which the solution separates into two isotropic micellar solutions of different concentrations. The coexistence curve exhibits a minimum at a critical temperature T_c and a critical concentration C_c. The value of T_c depends on the hydrophilic–lypophilic balance of the surfactant. A crucial point, however, is that near a cloud point transition, the properties of micellar solutions are similar to those of binary liquid mixtures in the vicinity of a critical consolution point, which are mainly governed by long-range concentration fluctuations [61].

At low concentration and significantly below T_c, the electrical birefringence of most nonionic micellar systems studied [62–65] is very small; however, the Kerr constant B increases considerably with concentration and when T_c is approached. A generic behavior emerges from experiments on polyoxyethylene (tetramethylbutyl)phenyl ether (or Triton X-100) [64] and several different polyoxyethylenes C_nE_m [63,65]. The rise and decay of the positive birefringence signal are asymmetric, each consisting of two relaxations:

1. Far from the critical temperature, the relaxations are exponential, and the fast relaxations ($\tau_{\pm 1}$) are due to the reorientation of individual nonspherical micelles [64].
2. In the critical range, however, where $T_c - T < 2$–5°C, the slower relaxations ($\tau_{\pm 2}$) become dominant and nonexponential.

Analysis of the slow forward relaxation (τ_2) reveals [63] that the associated Kerr constant follows a power law, i.e., B is proportional to the distance in temperature from T_c as $(1 - T_c/T)^{-\varphi}$, with $\varphi \sim 0.6$. This divergence of the static electrical birefringence is in accord with the droplet model [66,67] of critical binary mixtures. The central idea of the droplet model is that the electric field distorts (orients or vectorially amplifies) the spontaneous critical concentration fluctuations. The resulting anisotropic fluctuations then play the role of nonspherical particles in ordinary electrical birefringence. The magnitude of the concentration fluctuations rapidly increases as T_c is approached.

Since above a small cutoff radius (r_c) all sizes (r) of fluctuation contribute to the birefringence, in the dynamics of its rise,

$$\Delta n(t) = \Delta n_0[1 - f(t)] \tag{18}$$

the experimental $f(t)$ is not a simple exponential function. If $f(t)$ is derived by averaging the transient of single (fluctuation) droplets over the probability density of droplet sizes r scaled to the correlation length ξ of the fluctuations, then $f(t)$ defines a universal function of scaled time t^* [63]. The validity of this interpretation is confirmed by the congruence of the birefringence transients (observed at several different values of $T_c - T$) if displayed as a function of scaled time t^*.

ACKNOWLEDGMENT

My related work has been supported in part by the Welch Foundation, the National Science Foundation, the Petroleum Research Fund administered by the American Chemical Society, the research Corporation, and the Texas Advanced Research Program. Contributions by co-workers are gratefully acknowledged.

REFERENCES

1. Z. A. Schelly, Stud. Phys. Theor. Chem. *26*:140 (1983).
2. E. Hecht, *Optics*, 2nd ed., Addison-Wesley, Reading, MA, 1987.
3. E. Frederiq and C. Houssier, *Electric Dichroism and Electric Birefringence*, Clarendon Press, Oxford, 1973.
4. C. T. O'Konski and H. C. Thacher, Jr., J. Phys. Chem. *57*:955 (1953).
5. C. T. O'Konski and F. E. Harris, J. Phys. Chem. *61*:1172 (1957).
6. E. Tekle, M. Ueda, and Z. A. Schelly, J. Phys. Chem. *93*:5966 (1989).
7. E. van der Linden, S. Geiger, and D. Bedeaux, Physcia A *156*:130 (1989).
8. W. Helfrich, J. Phys. (Paris) *47*:321 (1986).
9. M. Borkovec and H.-F. Eicke, Chem, Phys. Lett. *157*:457 (1989).
10. S. A. Safran, J. Chem. Phys. *78*:2073 (1983).
11. M. Borkovec and H.-F. Eicke, Chem. Phys. Lett. *147*:195 (1988).
12. C. Tanford, *Physical Chemistry of Macromolecules*, Wiley, New York, 1961.
13. S. Broersma, J. Chem. Phys. *32*:1626 (1960).
14. V. J. Morris and B. R. Jennings, J. Colloid Interface Sci. *66*:313 (1978).
15. H. Benoit, J. Chim. Phys. *47*:719 (1950).
16. H. Benoit, Ann. Phys. *6*:561 (1951)
17. H. Benoit, J. Chim. Phys. *48*:612 (1951).
18. A. Peterlin and H. A. Stuart, Doppelbrechung insbesondere künstliche Dopplebrechung, in *Hand- und Jahrbuch der Chemischen Physik*, Vol. 8, Pt. 1B (A. Euken and K. L. Wolf, eds.), Akademische Verlagsgesellschaft, Leipzig 1943.
19. C. G. Le Fèvre and R. J.W. Le Fèvre, in *Techniques of Chemistry*, Vol. I, Pt. III (A. Weissberger and B. W. Rossiter, eds.), Wiley-Interscience, New York, 1972, p. 399.
20. C. T. O'Konski (ed.), *Molecular Elecro-Optics*, Parts I and II, Marcel Dekker, New York, 1976.
21. B. R. Jennings (ed.), *Electro-Optics and Dielectrics of Macromolecules and Colloids*, Plenum, New York, 1979.
22. S. Krause (ed.), *Molecular Electro-Optics: Electro-Optic Properties of Macromolecules and Colloids in Solution*, Plenum, New York, 1981.
23. H. Watanabe (ed.), *Dynamic Behavior of Macromolecules, Colloids, Liquid Crystals and Biological Systems by Optical and Electro-Optical Methods*, Hirokawa, Tokyo, 1989.
24. B. R. Jennings and S. P. Stoylov (eds.), *Colloid and Molecular Electro-Optics*, Institute of Physics Publishing, Bristol, UK, 1992.
25. J. Szamosi and Z. A. Schelly, J. Comput. Chem. *5*:182 (1984).
26. J. Szamosi and Z. A. Schelly, J. Phys. Chem. *88*:3197 (1984).
27. H. M. Chen and Z. A. Schelly, Langmuir *11*:758 (1995).
28. F. Runge, L. Schlicht, J.-H. Spilgies, and G. Ilgenfritz, Ber. Bunsenges. Phys. Chem. *98*:506 (1994).
29. O. Wiener, Abh. Sächs. Akad. Wiss. *33*:507 (1912).
30. A. Peterlin and H. A. Stuart, Z. Phys. *112*:129 (1939).
31. H. Hoffmann, in *Physics of Amphiphiles: Micelles, Vesicles and Microemulsions* (V. Degiorgio and M. Corti, eds.), North-Holland, Amsterdam, 1985, p. 160.
32. C. T. O'Konski and A. J. Haltner, J. Am. Chem. Soc. *79*:5634 (1957).
33. I. Tinoco, Jr. and K. Yamaoka, J. Phys. Chem. *63*:423 (1959).

34. G. B. Thurston and J. Bowling, J. Colloid Interface Sci. *30*:34 (1969).
35. A. D. Buckingham, Proc. Phys. Soc. (Lond.) *B69*:344 (1956).
36. S. Kielich, Acta Phys. Polon. *A37*:718 (1970).
37. H.-F. Eicke and Z. Markovic, J. Colloid Interface Sci. *85*:198 (1982).
38. P. Guering and A. M. Cazabat, J. Phys. (Paris) Lett. *44*:601 (1983).
39. A. M. Cazabat, Adv. Colloid Interface Sci. *38*:33 (1992).
40. K.-I Feng and Z. A. Schelly, J. Phys. Chem. *99*:17212 (1995).
41. E. Tekle and Z. A. Schelly, J. Phys. Chem. *98*:7657 (1994).
42. J. Gu and Z. A. Schelly, Langmuir *13*:4256 (1997).
43. R. Hilfiker, H.-F. Eicke, and H. Hammerich, Helv. Chem. Acta *70*:1531 (1987).
44. G. J. M. Koper, C. Ramán Vas, and E. van der Linden, J. Phys. II (France) *4*:163 (1994).
45. Z. Markovic, in *Reverse Micelles* (P. L. Luisi and B. E. Straub, eds.), Plenum, New York, 1984, p. 201.
46. H. M. Chen and Z. A. Schelly, Chem. Phys. Lett. *224*:61 (1994).
47. H.-F. Eicke, R. Hilfiker, and H. Thomas, Chem. Phys. Lett. *125*:106 (1987).
48. L. Schlicht, J.-H. Spilgies, F. Runge, S. Lipgens, S. Boye, D. Schübel, and G. Ilgenfritz, Biophys. Chem. *58*:39 (1996).
49. E. van der Linden, D. Bedeaux, R. Hilfiker, and H.-F. Eicke, Ber. Bunsenges. Phys. Chem. *95*:876 (1991).
50. K.-I Feng and Z. A. Schelly, J. Phys. Chem. *99*:17207 (1995).
51. A Faure, J. Lovera, P. Gregoire, and C. Chachaty, J. Chim. Phys. *82*:779 (1985).
52. Z.-J. Yu and R. D. Neuman, J. Am. Chem. Soc. *116*:4075 (1994).
53. Z.-J. Yu and R. D. Neuman Langmuir *10*:2553 (1994).
54. Z.-J. Yu and R. D. Neuman, Laugmuir *11*:1081 (1995).
55. D. F. Nicoli, J. G. Ellas, and D. Eden, J. Phys. Chem. *85*:2866 (1981).
56. M. E. Nash, B. R. Jennings, and G. J. T. Tiddy, J. Colloid Interface Sci. *120*:542 (1987).
57. H. Hoffmann, G. Oetter, and B. Schwandner, Prog. Colloid Polym. Sci *73*:95 (1987).
58. H. Hoffmann and W. Schorr, Ber. Bunsenges. Phys. Chem. *89*:538 (1985).
59. M. Angel, H. Hoffmann, U. Krämer, and H. Thurn, Ber. Bunsenges. Phys Chem. *93*:184 (1989).
60. H. Hoffmann, U. Krämer, and H. Thurn, J. Phys. Chem. *94*:2027 (1990).
61. V. Degiorgio, in *Physics of Amphiphiles: Micelles Vesicles and Microemulsions* (V. Degiorgio and M. Corti, eds.), North-Holland, Amsterdam, 1985, p. 303.
62. P. G. Neeson, B. R. Jennings, and G. J. T. Tiddy, Faraday Disc. Chem. Soc. *76*:353 (1983).
63. V. Degiorgio and R. Piazza, Phys. Rev. Lett. *55*:288 (1985).
64. Y. Dormoy, E. Hirsh, S. J. Candau, and R. Zana, Prog. Colloid Polym. Sci. *73*:81 (1987).
65. V. Degiorgio and R. Piazza, Prog. Colloid Polym. Sci. *73*:76 (1987).
66. J. Goulon, J. L. Greffe, and D. W. Oxtoby, J. Chem. Phys. *70*:4742 (1979).
67. J. S. Hoye and G. Stell, J. Chem. Phys. *81*:3200 (1981).

15
Organic Reactivity in Microemulsions

Clifford A. Bunton
University of California, Santa Barbara, Santa Barbara, California

Laurence S. Romsted
Rutgers, The State University of New Jersey, New Brunswick, New Jersey

I. INTRODUCTION

Surfactants (also called amphiphiles and sometimes detergents, emulsifiers, and now more rarely soaps) are surface-active agents that contain both apolar residues, typically linear alkyl groups, and polar or ionic headgroups [1–6]. Some examples are shown in Scheme 1. In water or in solvents such as sulfuric acid [7], hydrazine, diols, or triols [8,9] that associate in three dimensions, surfactants and other components aggregate to form association colloids such as micelles, microemulsions, and vesicles, whose size and shape depend on the structures and concentrations of components present (Fig. 1) [1–6]. These association colloids are composed of oil-like and solvent-like regions separated by an interfacial region that is the locus of chemical reactivity in aggregated systems. Association is driven by the solvophobic effect, which reduces unfavorable interactions of apolar groups with solvent. The term "hydrophobic" applies strictly to aqueous media but is often used more generally. Most kinetic work with normal micelles, microemulsions, vesicles, and polyelectrolytes is based on aqueous media, and we focus attention on water as solvent [6,10–12].

In dilute aqueous solution, micelles formed from surfactants with a single hydrophobic tail are approximately spherical aggregates with their apolar groups in the interior and polar or ionic groups at the surface in contact with water (Fig. 1) [1,6,13]. Micelles typically contain 50–150 surfactant monomers, depending on surfactant concentration and type, that are in dynamic equilibrium with monomers in the surrounding bulk aqueous phase. Addition of polar organic solvents that are readily miscible with water, e.g., short-chain primary alcohols and urea, shifts the monomer–micelle equilibrium in favor of monomer by disrupting the water structure or by destabilizing the micelles or both [14–16]. The position of equilibrium depends on the relative stabilities of the aggregates versus monomer in bulk solution, i.e., on the balance between hydrophobically induced association of apolar groups and interactions between ionic or polar headgroups, water, and other added solvents and solutes. The situation changes when moderately hydrophobic solvents, e.g., medium chain length alcohols, amides, or sulfoxides, are added to water [1–5,17,18]. These solvents act as cosurfactants and enter the micelles. Normal micelles in water do not tolerate large amounts of apolar solvents, e.g., alkanes, but oil-in-water (O/W) microemulsions formed

457

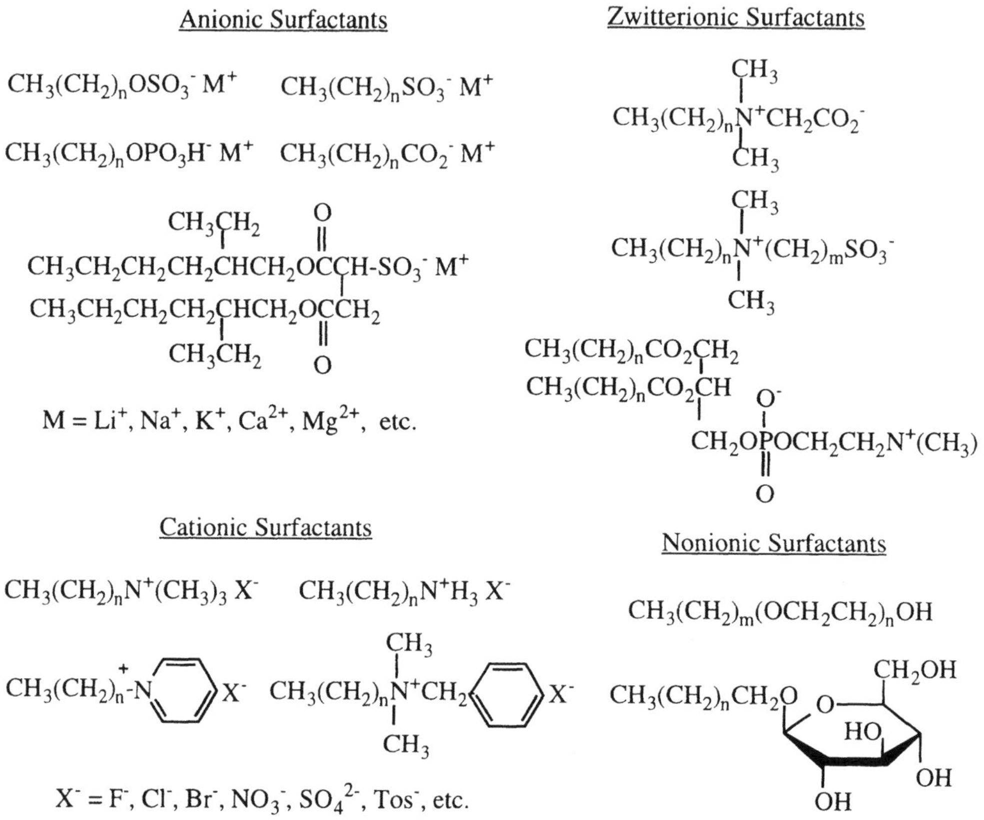

Anionic Surfactants

$CH_3(CH_2)_nOSO_3^- \ M^+$ $CH_3(CH_2)_nSO_3^- \ M^+$

$CH_3(CH_2)_nOPO_3H^- \ M^+$ $CH_3(CH_2)_nCO_2^- \ M^+$

$$CH_3CH_2CH_2CH_2\underset{\underset{CH_3CH_2}{|}}{\overset{\overset{CH_3CH_2}{|}}{C}}HCH_2O\overset{\overset{O}{\|}}{C}\underset{\underset{O}{\|}}{C}H\text{-}SO_3^- \ M^+$$

$$CH_3CH_2CH_2CH_2\underset{\underset{CH_3CH_2}{|}}{C}HCH_2O\overset{}{C}CH_2$$

$M = Li^+, Na^+, K^+, Ca^{2+}, Mg^{2+}, \text{ etc.}$

Cationic Surfactants

$CH_3(CH_2)_nN^+(CH_3)_3 \ X^-$ $CH_3(CH_2)_nN^+H_3 \ X^-$

$$CH_3(CH_2)_n\overset{+}{N}\bigcirc X^- \qquad CH_3(CH_2)_n\underset{\underset{CH_3}{|}}{\overset{\overset{CH_3}{|}}{N^+}}CH_2\bigcirc X^-$$

$X^- = F^-, Cl^-, Br^-, NO_3^-, SO_4^{2-}, Tos^-, \text{ etc.}$

Zwitterionic Surfactants

$$CH_3(CH_2)_n\underset{\underset{CH_3}{|}}{\overset{\overset{CH_3}{|}}{N^+}}CH_2CO_2^-$$

$$CH_3(CH_2)_n\underset{\underset{CH_3}{|}}{\overset{\overset{CH_3}{|}}{N^+}}(CH_2)_mSO_3^-$$

$$CH_3(CH_2)_nCO_2CH_2$$
$$CH_3(CH_2)_nCO_2CH$$
$$CH_2O\overset{\overset{O^-}{|}}{\underset{\underset{O}{\|}}{P}}OCH_2CH_2N^+(CH_3)$$

Nonionic Surfactants

$$CH_3(CH_2)_m(OCH_2CH_2)_nOH$$

$$CH_3(CH_2)_nCH_2O \ \begin{array}{c} CH_2OH \\ O \\ HO \\ OH \end{array} \ OH$$

Scheme 1

by the addition of cosurfactants are much better solubilizing agents than normal micelles, and this property is very important in industrial and medicinal chemistry [19,20]. Properties of microemulsions are similar to those of micelles made up of surfactant and cosurfactant, especially in terms of their effects on chemical reactivity.

Most ionic surfactants are insoluble in liquid hydrocarbon. AOT, di-2-ethylhexyl-sulfosuccinate (Scheme 1), is an exception to this generalization. However, they dissolve in slightly more polar solvents, e.g., halogenated compounds and long-chain alcohols [6]. If water is present, "water-pool" reverse micelles form with surfactant headgroups in the interior and apolar groups in contact with the organic solvent. These clear isotropic dispersions tolerate cosurfactants, generating water-in-oil (W/O) microemulsions. Conversion of O/W into W/O microemulsions often results in phase separation, but sometimes bicontinuous microemulsions are formed without phase separation [2,21,22].

This chapter is devoted to chemical reactivity, so structures of association colloids are taken as givens. The properties of solvent, surfactant, and cosurfactant that control aggregate structure are discussed in terms of theoretical models in Refs. 1–6, 18, and 23–27. Current interpretations of the effects of association colloids on chemical reactivity view aggregates as microreactors, i.e., reaction regions distinct from bulk solvent, but distributed throughout the solution (Fig. 1). In normal and cosurfactant-modified micelles and in O/W microemulsions, ionic or polar groups that are in contact with the bulk aqueous medium

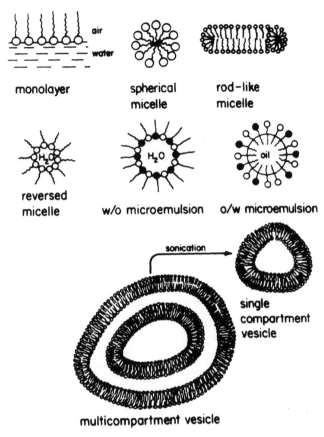

monolayer   spherical   rod-like
             micelle   micelle

reversed
micelle   w/o microemulsion   o/w microemulsion

single
compartment
vesicle

multicompartment vesicle

Figure 1 An oversimplified representation of some organized structures. (Adapted from Fig. 1 of Ref. 159.)

surround a hydrocarbon-like droplet containing surfactant and cosurfactant tails and any added oil. In reverse micelles and W/O microemulsions, a microdroplet of water interacts with the polar or ionic headgroups of the surfactant and cosurfactant, and their apolar residues are in contact with organic solvent. Most kinetic studies to date have been carried out in O/W microemulsions with water as bulk solvent, and quantitative treatments are generally adapted from theoretical models that were originally developed for normal micelles in water [6,10,11,28].

II. REACTIONS IN O/W MICROEMULSIONS AND COSURFACTANT MODIFIED MICELLES

Microemulsions offer the intriguing capability of cosolubilizing high concentrations of water-insoluble and water-soluble reactants, and many studies of chemical reactivity in microemulsions have focused on potential large-scale applications of these systems as reaction media. For example, the high solubilizing power of microemulsions has led several groups to use them for the destruction of toxic materials, e.g., the still dangerous and

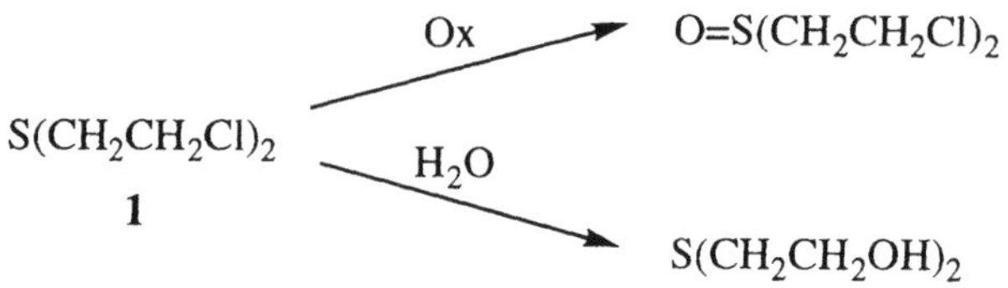

$$S(CH_2CH_2Cl)_2 \xrightarrow{\text{Ox}} O{=}S(CH_2CH_2Cl)_2$$

$$\mathbf{1} \xrightarrow{\text{H}_2\text{O}} S(CH_2CH_2OH)_2$$

unstable stockpiles of chemical weapons in Russia and the United States [29,30], because some chemical agents are only sparingly soluble in water, aqueous micelles, or aqueous organic solvents [20]. The highly toxic and very hydrophobic blister agent mustard gas (**1**) and less toxic model compounds are rapidly destroyed by hydrolysis or oxidation, but their solubilities in water are generally too low for rapid reaction under mild conditions. Their reactions are very rapid in microemulsions that contain hypochlorite ion and *tert*-butyl alcohol, which generates *tert*-butyl hypochlorite, the probable oxidant. The high solubility of very hydrophobic chlorosulfides in microemulsions is of key importance in this reaction [31,32].

Mackay and coworkers [33,34] and Moss and coworkers [35] showed that reactions of nucleophiles with phosphorus(V) esters, which are models for nerve agents, can be carried out effectively in microemulsions. The active agents often contain polymeric thickeners, and the solubilizing power of microemulsions is of crucial importance. Typically cationic surfactants are used with alcohols or alkylpyrollidinones as cosurfactants. Iodosobenzoate ions (**2**) are nucleophilic turnover catalysts toward fluorophosphonate nerve agents and phosphate esters, and Moss and coworkers used amphiphilic iodosobenzoate ions extensively in this work [36] (see Scheme 2).

The use of microemulsions or reverse micelles as media for chemical and enzymatic reactions has been reviewed in recent years [20,37,38]. Microemulsions, including those based on organogels, are also useful media for enzyme-catalyzed synthetic reactions [37,39–43] and for preparation of nanoparticles [44]. In a very different direction, Vanag and Hanazaki [45] showed that the ferroin-catalyzed Belousov–Zhabitinskii oscillatory reaction exhibits frequency-multiplying bifurcations in reverse AOT microemulsions in octane. A clear understanding of reactivity in microemulsions and insight into how to optimize the experimental conditions requires kinetic models with predictive power. We focus attention primarily on this problem.

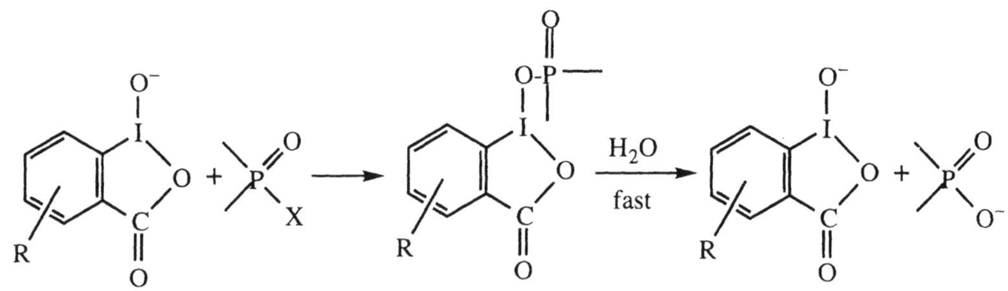

X = F, OAr

Scheme 2

A. Quantitative Treatments of Reactivity in Microemulsions and Cosurfactant-Modified Micelles

Reactivity in aqueous association colloids is generally analyzed with the assumption that colloids provide a discrete reaction medium distinct from bulk solvents. The colloidal microdroplets and the surrounding bulk solvent are treated as separate phases or, more correctly, pseudophases, and distributions of reactants are described by transfer equilibria between the aqueous and aggregate pseudophases [10,46–48]. This approach was originally developed to account for rates of reaction in aqueous micelles, but it has been extended to other colloidal systems, e.g., vesicles and microemulsions, and, because of similarities in the interfacial regions, to alcohol or other cosurfactant-modified micelles [1,10,11,28].

Much of the approach used in pseudophase models for treating reactivity in aqueous micelles is applicable to cosurfactant-modified micelles and microemulsions, but there are some limitations. Structures of aqueous micelles depend primarily on surfactant structure and, for ionic micelles, added electrolyte, in particular the counterion. The situation is more complicated for other association colloids. For example, with ionic O/W microemulsions and cosurfactant-modified micelles the interfacial region contains surfactant headgroups, associated counterions, cosurfactant, and water. In addition, the size of the microemulsion microdroplet depends on, among other things, the amount and type of oil, which may also penetrate the interfacial region [49,50]. Although reactivity in solutions of association colloids is not very sensitive to aggregate size and shape [10], treatments of the distributions of reactants, usually expressed as transfer equilibria between the aggregates and the bulk pseudophases, are more difficult because transfers of ions and cosurfactant from water to the microdroplet are mutually dependent.

The problem is even more difficult for systems containing mixtures of counterions. The presence of interacting solutes complicates measurement of local concentrations in the interfacial region, and some experimental methods are useful only under limited conditions. For this reason we give considerable emphasis to a recently developed chemical trapping method for estimating interfacial compositions (see Sec. III). It promises to be generally applicable to a variety of association colloids and to provide information on the compositions of microemulsions and other association colloids that are useful as reaction media.

B. Pseudophase Model of Reactions in Association Colloids

Reactants, which are usually polar organic molecules and organic or inorganic ions, partition between the bulk solvent and colloidal microdroplets. Transfers are generally orders of magnitude faster than most thermal reactions [6,10,11,28,51–54], and they are written as equilibria. This assumption may not be true for some very fast enzyme-catalyzed reactions in phospholipid micelles and vesicles [55] or for some photochemical and electron transfer reactions [6,56–58]. Bulk solvent and the colloidal droplets are treated as distinct reaction regions, and the overall, or observed, rate of reaction is assumed to be the sum of the rates of reaction in the pseudophases. Substantial evidence shows that reactions between aggregate-bound ionic or polar reactants occur in the interfacial region; in ionic micelles this region is often identified with the Stern layer [1,6,10,11]. These assumptions are used below to develop kinetic equations for first-order spontaneous reactions in micelles and adapt them to treat second-order reactions in micelles, microemulsions, and other cosurfactant-modified micelles.

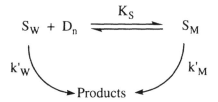

Scheme 3

The kinetic treatment for spontaneous reactions is straightforward; see Scheme 3 and Eq. (1).

$$\frac{d[S_T]}{dt} = k_\Psi[S_T] = k'_W[S_W] + k'_M[S_M] \tag{1}$$

$$K_S = \frac{[S_M]}{[S_W][D_n]} \tag{2}$$

Subscripts T, W, and M indicate the stoichiometric, aqueous, and micellar concentrations, respectively, of the organic reactant or substrate, S, and square brackets indicate concentration in moles per liter of total solution volume. Equation (1) defines the observed rate as the sum of the rates of reaction in the aqueous and micellar pseudophases where k_Ψ, k'_W, and k'_M are the observed, aqueous, and micellar first-order rate constants, respectively, typically expressed in units of inverse seconds (s^{-1}). Substrate is assumed to be in dynamic equilibrium between the two pseudophases throughout the time course of reaction, and its distribution depends on the binding or association constant, K_S, Eq. (2). $[D_n]$ denotes the concentration of micellized surfactant (detergent). In the pseudophase model, $[D_n] = [D] - cmc$, i.e., the concentration of micellized surfactant is the difference between the stoichiometric amphiphile concentration and the critical micelle concentration (cmc) under the experimental conditions. The first-order rate constant is given by

$$k_\Psi = \frac{k'_W + k'_M K_S([D] - cmc)}{1 + K_S([D] - cmc)} \tag{3}$$

Equation (3) predicts that the change in k_Ψ with added surfactant depends on the values of k'_W, k'_M, and K_S. Values of k_Ψ are independent of the surfactant concentration below the cmc, i.e., $[D] \leq cmc$ and $[S_W] = [S_T]$, and at elevated surfactant concentrations at which the substrate is completely micellar bound, i.e., $K_S([D] - cmc) \gg 1$ and $[S_M] = [S_T]$. Kinetic forms of spontaneous reactions, both accelerated and inhibited, fit Eq. (3) [6,10–12], and this treatment holds for various types of colloidal assemblies.

The situation is more complicated for nonsolvolytic, bimolecular, reactions, where the observed rate depends on the concentrations of both reactants in both pseudophases. Experiments are generally run under first-order conditions with the concentration of the second reactant in large excess over that of the substrate. In modeling colloidal effects on bimolecular reactions, first-order rate constants [Scheme 3 and Eq. (1)] are written in terms of second-order rate constants and concentrations of the second reactant in each pseudophase.

In the aqueous pseudophase,

$$k'_W = k_W[X_W] \tag{4}$$

where k_W is the second-order rate constant, typically in units of inverse mole-seconds ($M^{-1} s^{-1}$). $[X_W]$ is the concentration of the second reactant expressed as moles per liter of solution volume, i.e., molarity, because the micelles or other association colloids generally occupy a small fraction of the total solution.

In the micellar pseudophase,

$$k'_M = k_2^M X_M \tag{5}$$

where k_2^M is a second-order rate constant in units of local molarity ($M^{-1} s^{-1}$), which can be compared directly with second-order rate constants in bulk water, k_W ($M^{-1} s^{-1}$). The local molarity, X_M, of the second reactant is written in terms of the reaction volume within the micellar pseudophase, i.e., in units of moles per liter of reaction volume, which cannot be measured directly (see below). Alternatively, the concentration of the second reactant can be expressed as a mole ratio of micellar bound X to micellized surfactant:

$$k'_M = \frac{k_M[X_M]}{[D] - cmc} \tag{6}$$

where k_M is in units of s^{-1}. $[X_M]$ and $[D] - cmc$ can be estimated experimentally, but k_M and k_W have different units and cannot be compared directly.

The two micellar rate constants are related in terms of the molar volume, V_M, of the reaction region in the colloidal droplets (see below):

$$k_2^M = k_M V_M \tag{7}$$

First-order rate constants with respect to substrate are written in different forms depending on the concentration unit selected.

Molar ratio:

$$k_\Psi = \frac{k_W[X_W] + k_M K_S[X_M]}{1 + K_S([D] - cmc)} \tag{8}$$

Local molarity:

$$k_\Psi = \frac{k_W[X_W] + k_2^M K_S X_M([D] - cmc)}{1 + K_S([D] - cmc)} \tag{9}$$

For many bimolecular reactions, e.g., of organic nucleophiles or electrophiles, the distribution of the second reactant, X, between bulk solvent and the colloidal assemblies can be determined directly or calculated from a binding constant of the same form as that used for the substrate, Eq. (2). Binding constants can be determined by solubility, chromatography, or spectroscopy [6] or estimated from empirical relations [59,60]. Estimates of k_M are obtained as best fit values from simulations of k_Ψ versus surfactant concentration profiles by using Eq. (8) and independent estimates of k_W and the cmc. To estimate k_2^M from k_M by using Eq. (7), a value of V_M must be assumed. Values for V_M for bimolecular reactions are not known, because the reaction volume available to each reactant within the colloidal droplets cannot be measured [10]. Typically V_M is set equal to the molar volume of the aggregates [48] or the interfacial region [11,12], which for microemulsions involves both surfactant and cosurfactant.

Determination of the distribution of small ions is more difficult, although in favorable cases experimental methods such as electrochemistry [61,62], NMR spectrometry [63,64], or chemical trapping [65] can be used. However, there are no experimentally feasible methods for some chemically interesting ionic reagents, e.g., OH^-, and theoretical models must then be used. As with organic reactants, most experimental methods provide estimates of the fraction of reactant in the colloidal pseudophase, i.e., [X_M], rather than its local concentration in the aggregate, X_M. When the ionic reagent, X, is the counterion of the surfactant and no other counterions are present, X_M is related to the molar ratio of bound counterions and the degree of counterion binding, β, through V_M:

$$X_M = \frac{[X_M]}{([D] - cmc)V_M} = \frac{\beta}{V_M} \tag{10}$$

β is related to the fractional charge of an association colloid droplet containing an ionic surfactant, α ($\alpha = 1 - \beta$), which can be estimated by a variety of experimental methods and provides a method for estimating counterion concentration at an interface, Eq. (10) [6,10,11,66]. For aqueous ionic micelles, typical values of α range from 0.1 to 0.5 and are largest for small hydrophilic ions, showing that 50–90% of the ionic headgroups are neutralized by counterions. Cosurfactant-modified micelles and O/W microemulsion droplets have relatively high α values, because intercalation of a nonionic cosurfactant in the interfacial region decreases its affinity for counterions [61,62,65,67]. As with any reactant, V_M for counterions cannot be measured directly and is assumed to be equal to either the molar volume of the aggregates or that of the interfacial region. However, X_M can be estimated experimentally by chemical trapping [65] and, in principle, can be calculated by molecular dynamics simulations [68], thus avoiding the need to assume values of β and V_M (see below).

Many chemical reactions in aggregated systems are run in the presence of mixtures of reactive and inert ions, and their distributions can be treated by ion exchange [10,66,69,70]. Provided that α is small and insensitive to counterion type and concentration, ions are assumed to exchange on a one-to-one basis. Equation (11) describes exchange between a reactive ion X and an inert ion N:

$$K_X^N = \frac{[X_W]N_M}{X_M[N_W]} \tag{11}$$

If α is constant, Eq. (11) combined with the mass-balance relations for the two ions can be used to estimate the concentration of X in the interfacial region in terms of its molar volume V_m. Most estimates of K_X^N are for ions of like charge [66,71,72], but a few exchange constants are available for ions of unequal charge [73]. Trends in K_X^N for a number of different ions can be interpreted qualitatively in terms of a balance between entropic and enthalpic forces. Entropically ions distribute themselves over all regions in solution, i.e., bulk water and the interfacial region, except the apolar interior of an association colloid. Enthalpy depends on Coulombic interactions between ionic headgroups and counterions and on specific interactions that depend on hydration, dispersive, and hydrophobic forces and are largest for polarizable, low charge density ions. These specific interactions are primarily responsible for the differences in values of K_X^N. Affinity orders of alkali metals and halide ions toward

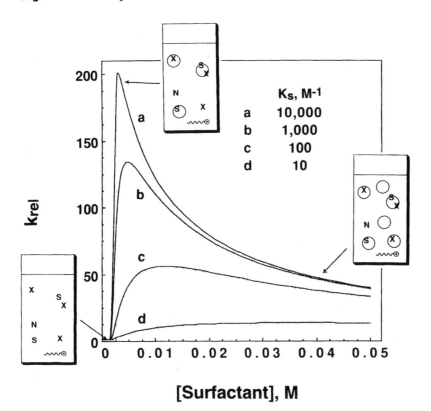

Figure 2 Effect of increasing substrate hydrophobicity, K_S, on k_{rel} as a function of surfactant concentration. Parameter values are $k_2^M/k_W = 1.0$, cmc $= 0.002$ M, $[X_T] = 0.01$ M, $K_X^N = 4$, $\beta = 0.7$, $V_M = 0.15$ M^{-1}. K_S values are listed in the figure.

anionic and cationic micelles, respectively, and their effects on aggregate properties such as cmc, size, and chemical reactivity [12,74] generally, but not always [75], correlate with ionic radius, i.e., they follow the Hofmeister series [76].

The combination of the pseudophase assumption with mass action binding constants of substrates and ion exchange of reactive and nonreactive counterions is called the pseudophase ion-exchange (PIE) model [10,48,66]. It successfully fits the kinetics of many bimolecular reactions and also shifts in apparent indicator equilibria in a variety of association colloids, especially reactions between organic substrates and inorganic ions in normal micelles over a range of surfactant and salt concentrations and types (up to about 0.2 M). It has also been successfully applied to cosurfactant-modified micelles [77,78], O/W microemulsions [79–81], and vesicles [82].

Increases in surfactant and salt concentrations have characteristic effects on rate constant–concentration profiles for bimolecular reactions. Figures 2 and 3 are theoretical profiles generated by using the PIE model for reaction between an uncharged organic substrate and a reactive ion of opposite charge to that of the micelle, i.e., a counterion. In both plots, k_{rel} is the observed first-order rate constant relative to the first-order rate constant in water. In the PIE model, changes in the observed rate brought about by aggregate formation and binding of the reactants are attributed to changes in the concentrations of reactants and the differences in the rate constants for reaction within each pseudophase,

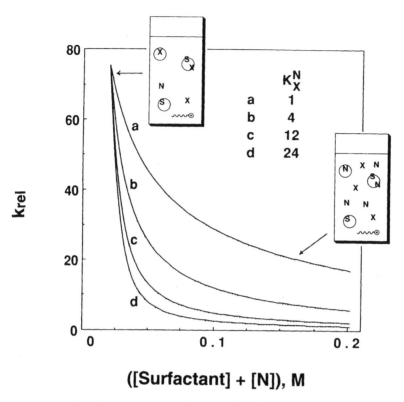

Figure 3 Specific salt effects on k_{rel} as a function of added NaX, [X], for typical K_X^N values (see figure), at surfactant concentration $= 0.02$ M and $K_S = 1000$ M^{-1}. Other parameters are as in Fig. 1. Surfactant and salt have the same counterion in curve b.

e.g., Eq. (9). To focus attention on concentration effects, we set the second-order rate constant in micelles equal to that in water, $k_2^M = k_W$, in Figs. 2 and 3 so that changes in k_{rel} are caused only by changes in the distributions of reactants between the two pseudophases and in their concentrations within each pseudophase as illustrated in the cartoon reaction flasks, e.g., cuvettes in a UV/Vis spectrophotometer. Note: The totality of the micelles in solution is generally a small fraction of the total solution, typically on the order of 0.1–1% by weight in most experiments, unlike the micellar images in the cuvettes. Figure 2 shows that increasing the surfactant concentration at constant total concentrations of the reactive ion, X_T, and inert ion, N_T, generates a maximum in k_{rel} that increases dramatically as K_S becomes larger. Once the surfactant concentration exceeds the cmc, k_{rel} increases rapidly, because binding (in the sense of association) of the reactants to the micelles concentrates them within the micellar pseudophase. Stronger binding gives greater rate enhancements at lower surfactant concentrations. K_S is increased experimentally by making the substrate more hydrophobic (e.g., by increasing the length of a hydrocarbon tail) without changing the reactive group. The rate maximum with increasing surfactant concentration occurs because the increase in the local concentration of reactants within the aggregate pseudophase caused by transfer of the reactants from the aqueous pseudophase into the aggregates is opposed by dilution of the reactants within the aggregate pseudophase. The dilution effect dominates at higher surfactant concentrations, and because $k_2^M = k_W$ in this simulation, k_{rel} will be the same for all substrates in this limit.

Figure 3 shows the effect of adding different nonreactive counterions, i.e., with different K_X^N, at constant surfactant concentration. The changes in k_{rel} illustrate the substantial and specific effects of added salts on k_Ψ that are commonly observed [66]. Specific salt effects follow the affinity order of the aggregates for the nonreactive counterion, X, i.e., the Hofmeister series [76]. Although the PIE model readily fits rate–surfactant profiles and inhibition by added electrolyte, it has limitations. It breaks down with high concentrations of ions and when competing ions have very different affinities for micelles or other association colloids with variable surface charge densities as indicated by significant differences in α. Under these conditions, the basic assumption of one-to-one exchange is suspect. Strongly interacting anions, e.g., Cl^- and Br^-, readily displace a weakly interacting anion, e.g., OH^-, from cationic micelles. However, even at high OH^- concentration, Cl^- and Br^- are not completely displaced [64]. Under some conditions, fitting of rate–surfactant profiles becomes indeterminate and K_X^N and k_2^M (or k_M) [see Eqs. (8), (9) and (11)] cannot be treated as independent parameters [83].

Results with various reactions show that if α is relatively large and/or the concentration of added counterions is high, i.e., > 0.2 M, then the basic assumption of one-to-one exchange (i.e., constant α for both counterions) is no longer valid [64] and the simple treatment fails. One solution [84–86] has been to model the interfacial concentration of a counterion as the sum of the initial concentration of counterions at the aggregate surface in the absence of added salt, as given by $1 - \alpha$, plus its concentration in aqueous bulk solvent,

$$X_M = \frac{1 - \alpha}{V_M} + [X_W]$$ (12)

At low concentrations of added salt ($[X_W] \leq 0.2$) the interfacial concentration is represented reasonably well by Eqs. (10) and (11), but in more concentrated electrolyte $[X_W]$ makes a significant contribution, and in the high concentration limit, $X_M \approx [X_W]$.

In principle, the Coulombic contribution to interactions between a colloidal macroion and mobile ions can be calculated from the surface electrical potential in terms of classical electrostatics [87–90]. The potential of an isolated colloidal macroion cannot be determined experimentally, but it can be calculated by solving the Poisson–Boltzmann equation (PBE) in the appropriate symmetry for a given surface charge density. Specific interactions between ions and association colloids are treated by including an empirical term for non-Coulombic interactions. This treatment has been used to fit kinetic data for various reactions of co- and counterions mediated by ionic micelles, and it leads to relations between k_Ψ and surfactant and electrolyte concentrations with forms similar to those in Figs. 2 and 3 [88,91–93]. The limited ability of very hydrophilic ions to displace less hydrophilic ions from cationic micelles is accounted for by assuming that very hydrophilic ions do not interact specifically [64]. The PBE model has been extended to reactions in alcohol-modified micelles with allowance for the effects of the alcohols on the geometry and surface charge density of the association colloid [94]. The PBE treatment depends on the surface charge density, which affects Coulombic interactions with ions, and therefore on the geometry of the colloidal microdroplet. It is difficult to know the effect of cosurfactant on the charge density and morphology of cosurfactant-modified micelles and microemulsion droplets. Therefore, treatments of ion binding and kinetic analyses are currently more reliable for aqueous micelles than for cosurfactant-modified micelles and O/W microemulsions.

Much of the experimental work on chemical reactivity involving small ions has been with counterions because their local concentrations are enhanced by ionic micelles. However, it is important to recognize that co-ions are not completely excluded from the interfacial regions of ionic association colloids. Their concentrations in this region are very low in dilute electrolyte but increase as electrolyte becomes more concentrated and the micellar surface potential decreases [48,84,88,95]. For example, Eq. (12) provided a good fit of acid-catalyzed ketal hydrolyses in cationic micelles at high concentrations of added NaBr and NaCl after correction for the salt effect on the activity of the acid [84]. Concentration gradients of ions between water and ionic association colloids decrease as electrolyte concentrations increase and become small with greater than 5 M electrolyte. As a result, ionic association colloids inhibit bimolecular reactions of dilute co-ions where substrate enters the colloids and the ionic reactant remains largely in the bulk solvent, but the inhibition decreases with increasing ionic concentration. Conversely, for reactions of counterions, observed rate constants decrease, relative to those in bulk solvent, with increasing concentration of the ionic reactant.

C. Examples of Quantitative Analyses of Rate Effects

Kinetic treatments are relatively simple for spontaneous reactions, both unimolecular and bimolecular water-catalyzed reactions, because only the transfer equilibrium of the substrate between solvent and the association colloid has to be considered, Eq. (3). For example, the rate of decarboxylation of 6-nitrobenzisoxazole-3-carboxylate ion (3) is a useful indicator of medium polarity, and reaction is inhibited by solvents that hydrogen bond to the carboxylate moiety [96]. The reaction is accelerated by a variety of colloidal species that incorporate 3, including ionic and zwitterionic micelles [97,98] and O/W microemulsions [99]. Reactions are slightly slower in microemulsions derived from cetyltrimethylammonium bromide (CTABr) than in the corresponding aqueous micelles, but changes in the alcohol cosurfactant or the hydrocarbon, or in their relative concentrations, do not have major rate effects, and it appears that these microemulsion droplets are similar to aqueous micelles as submicroscopic reaction media. These observations are consistent with estimates of surface polarities [99–101] determined with bound fluorescent probes [102].

Rates of spontaneous water-catalyzed hydrolyses are usually similar in water and at surfaces of aqueous ionic micelles. This generalization applies to the hydrolysis of bis(4-nitrophenyl)carbonate (4) in micelles and in O/W microemulsions of CTABr, t-amyl alcohol, and n-octane [99], indicating that the alcohol does not significantly exclude water from the interfacial region or markedly decrease its reactivity (see Sec. III.A).

The situation is more complicated for nonspontaneous bimolecular reactions involving a second reactant, whose distribution between the two pseudophases has to be considered. The simplest situation is that for reaction of a hydrophobic species whose solubility in water is sufficiently low that it is incorporated essentially quantitatively in the association colloid. For example, for reactions of nucleophilic amines in aqueous micelles, second-order rate constants in the micellar pseudophase calculated in terms of local concentrations are lower than in water [103,104], because these reactions are inhibited by a decrease in medium polarity and micelle/water interfaces are less polar than bulk water [59,60,99101]. Nonetheless, these bimolecular reactions are generally faster in micellar solutions than in water because the nucleophile is concentrated within the small volume of the micelles. Similar results were obtained for the reaction of 2,4-dinitrochlorobenzene (**5**) with the cosurfactant *n*-hexylamine in O/W microemulsions with CTABr and *n*-octane [99], again consistent with the postulated similarities in the interfacial regions of aqueous micelles and O/W microemulsions.

Transfer equilibria of halide ions between bulk water and association colloids have been followed electrochemically, e.g., by use of specific ion electrodes or conductimetrically [61,62], or by chemical trapping (Sec. III) [65]. Bromide ion is an effective nucleophile in S_N2 displacements at alkyl centers, and rate constants in aqueous and alcohol-modified micelles and in O/W microemulsions have been analyzed quantitatively in terms of local concentrations of substrate and Br^- in the interfacial region of the colloid microdroplets [99,105]. The local second-order rate constants are typically slightly lower in the colloidal pseudophases than in water but are similar for micelles and microemulsions prepared with CTABr, indicating that interfacial regions provide similar kinetic media for these S_N2 reactions. However, reactions with the same overall concentrations of Br^-, or other ionic reactant, are slower in microemulsions or alcohol-modified micelles than in normal micelles for two reasons: (1) The fractional ionization, α, is lower in the normal micelles and (2) the increased volume of the reaction region, due to the presence of cosurfactant, "dilutes" Br^- in the pseudophase provided by the association colloid [66,69,105].

Theoretical models similar to those applied to reactions in normal aqueous micelles have been used to simulate microemulsion effects on reactivity when the transfer equilibria of the nucleophiles could not be measured directly. The ion-exchange (PIE) model [Eq. (6)] was applied to the reaction of *p*-nitrophenyldiphenyl phosphate (**6**) with OH^- or F^- in microemulsions of CTABr with 1-butanol as cosurfactant [79]. The rate data could be fitted to this model, even though in solutions of OH^- the 1-butoxide ion formed by deprotonation of the alcohol is a competing nucleophile.

Nu = F, OH, 1-BuO

Similar observations were made on reactions of OH^- with 2,4-dinitrochloro- and 2,4-dinitrofluorobenzene in microemulsions and alcohol-modified micelles derived from CTABr. With primary alcohols, 1-butanol, or benzyl alcohol, extensive amounts of ethers were generated by attack of alkoxide ions, but this reaction became unimportant in micelles of CTABr and *tert*-amyl alcohol where OH^- was the only nucleophilic reagent and rate data were fitted by the PIE model, as for normal aqueous micelles [80].

The PIE model was also applied successfully to the chromic acid oxidation of alcohols in micelles of sodium dodecyl sulfate (SDS) [81]. In these systems the medium chain length alcohols act as cosurfactants. However, as noted above, key assumptions of the PIE treatment are the constancy of α and the one-to-one ion exchange, which are reasonably satisfactory for aqueous micelles but are less reliable for cosurfactant-modified micelles and O/W microemulsions, where α is relatively large and sensitive to both ionic and nonionic solutes.

The kinetics of acid hydrolysis of the *p*-methoxybenzaldehyde-*O*-acyloximes (**7**) in SDS micelles modified by BuOH has also been fitted to the PIE model [86]. The substrates differ only in their hydrophobicities, and while the acetyl derivative partitions between water and micelles, the octanoyl derivative is wholly micelle-bound. The simple PIE model fits rate data in dilute HCl [Eq. (6)], but it underpredicts observed rate constants in more concentrated acid. This increased rate was analyzed in terms of a model that does not involve a constant value of α but allows concentrations of reactive and inert ions, H^+ and Na^+, in the micellar pseudophase to increase, following Langmuir isotherms [106]. This model was reasonably satisfactory except at high 1-butanol concentration. Alternatively, the rate data in more concentrated acid can be fitted in terms of Eq. (12).

The alkaline fading of crystal violet ($4\text{-}Me_2NC_6H_4)_3C^+$, in micelles of CTABr modified by 1-hexanol or by 1-octanol has been treated quantitatively by estimating the local concentration of OH^- in the interfacial region by solving the Poisson–Boltzmann equation. The local second-order rate constants calculated by this method are very similar to those calculated for normal aqueous micelles of CTABr [107], but these rate constants are uncertain because the authors did not correct for the competing reaction with alkoxide ion formed by deprotonation of the alcohol by OH^- (cf. Refs. 79 and 80).

The synthesis of sodium decanesulfonate from the alkyl halide and Na_2SO_3 can be carried out very conveniently in microemulsions based on nonionic surfactants [108]. The rate data fit a pseudophase model, and the nonionic surfactants make this method preparatively very useful because phase separation with a change of temperature allows recovery of the surfactants. This approach is very useful because surfactants have high molecular weights and large amounts are needed in preparative reactions and the solubilizing ability of surfactants complicates product isolation. Therefore, more research is needed on methods for surfactant recovery, reuse, or destruction following the lead of Jaeger, who has developed a number of methods for the selective destruction of surfactants used as reaction media [109].

In most reactions it is reasonable to analyze rate data in terms of reactions in water or in the interfacial region of the association colloids. This simplification fails for reactions of very hydrophobic substrates in O/W microemulsions. These substrates may be solubilized in a region of the oil microdroplets that is inaccessible to ionic or polar reactants, and it is then necessary to consider the partitioning of substrates between aqueous, interfacial, and oil regions [110,111].

D. Reactions in W/O Microemulsions and Reverse Micelles

Structures of W/O microemulsions are similar to "water pool" reverse micelles, and ionic and hydrophilic reactants, which are excluded from the bulk solvent, are highly concentrated in the small microdroplets of water [6,19,20,37]. As a result, many reactions of water-soluble reactions are very rapid in these reverse micelles or W/O microemulsions. Much of the work on reactivity and acid–base equilibria has been reviewed [6,112], especially as regards effects of the size of the "water pool."

The alkaline fading of crystal violet $(4\text{-}Me_2NC_6H_4)_3C^+$ in cationic W/O microemulsions of CTABr–alcohol–cyclohexane occurs in the "water pools" [113]. The corresponding reactions of crystal violet and malachite green, $(4\text{-}Me_2NC_6H_4)_2C^+Ph$, in anionic W/O AOT–isooctane microemulsions also occur in the "water pools," and rate data were analyzed in terms of their size and the competing base-promoted hydrolysis of AOT [114]. This reaction with crystal violet is strongly inhibited by ClO_4^-, which pairs with the carbocation and blocks nucleophilic attack.

Marcia-Rio et al. used this W/O AOT microemulsion as the medium for nitroso transfer to secondary amines from N-methyl-N-nitroso-p-toluenesulfonamide (**8**). Their quantitative treatment, which includes consideration of reactant solubilities, shows that reaction occurs at the microemulsion interface, where it is slower than in water. This rate difference is understandable on the very reasonable assumption that the polarity of the microemulsion interface is lower than that of water [99–101]. These kinetic data indicate that the interfacial regions of the "water pool" microdroplets in O/W microemulsions and reverse micelles can be regarded as reaction media corresponding to descriptions applied to normal aqueous association colloids. This concept has also been applied to acid–base equilibria, especially by El Seoud and his group [112,116,117].

The "water pool" description of reverse micelles and O/W microemulsions is not appropriate if only small amounts of water are present. In that event the surfactants form ion pairs or small ion clusters with associated water [118]. These clusters are catalytically very effective in the decarboxylation of 6-nitrobenzisoxazole-3-carboxylate ion (**3**), and for solutions of cationic surfactants and hydrophobic ammonium ions rate constants of reaction in CH_2Cl_2 decrease significantly when there is sufficient water to generate "water pool" reverse micelles [119,120]. Similar results were obtained for the spontaneous hydrolysis

$$O_2N-\text{(ring)}-\overset{\overset{\displaystyle O}{\|}}{\underset{\underset{\displaystyle O^-}{|}}{P}}-O^- \xrightarrow{\;H_2O\;} O_2N-\text{(ring)}-O^- \;+\; HO-\overset{\overset{\displaystyle O}{\|}}{\underset{\underset{\displaystyle O^-}{|}}{P}}-O^-$$

9

of the 2,4-dinitrophenylphosphate dianion (**9**) in solutions of CTABr in moist CH_2Cl_2 [121]. In this reaction, with $w = [H_2O]/[CTABr] = 13$, first-order rate constants at the interior surface of a water microdroplet are very similar to those at the exterior surface of a normal aqueous micelle but increase sharply with decreasing w. This kinetic evidence for a change of the properties of water is consistent with ^1H NMR and other spectrophotometric evidence [119,120,122].

III. ESTIMATING AGGREGATE COMPOSITIONS IN MICROEMULSIONS BY CHEMICAL TRAPPING WITH ARENEDIAZONIUM IONS

Chemists studying aggregate effects on chemical reactivity have always been aware that changes in chemical reactivity should provide quantitative information about the compositions of the interfacial regions of aggregates. In principle, interfacial concentrations can be estimated from variations in k_Ψ (or from changes in apparent ionization constants of indicators) by using a pseudophase model. For example, when experimental conditions are such that substrate is fully micelle-bound and reaction occurs only in the micellar pseudophase, the relations between k_Ψ and the interfacial concentration of the second reactant in bimolecular reactions, as expressed in Eqs. (7)–(9), simplify to

$$k_\Psi = k_2^M X_M = \frac{k_2^M [X_M]}{V_M} \tag{13}$$

Kinetic data obtained under these conditions have been fitted by pseudophase models in terms of k_2^M by solving the PBE or in terms of k_2^M/V_M by using the PIE model [10]. However, because these treatments contain reasonable but unproven assumptions [64] and because values of parameters such as β, V_M, and K_X^N are only approximate, values of k_2^M may not be unique [123]. Extensive evidence shows that $k_2^M \approx k_W$ for many reactions of anionic nucleophiles, but this generalization does not hold for anionic electrophiles [83,124,125]. Therefore, although a great many kinetic data are consistent with the assumption that counterions concentrate at surfaces of ionic association colloids, it is difficult to obtain quantitative estimates of interfacial ion concentrations from measured rate constants.

   An alternative approach is to estimate interfacial compositions from product yields obtained by competitive chemical trapping of a very reactive intermediate that is strongly bound to aggregates with its reactive group in the interfacial region. Problems caused by uncertainties in k_2^M, β, and V_M disappear, because products form competitively in the same reaction medium and volume. Scheme 4 summarizes the dediazoniation mechanism for spontaneous, rate-limiting formation of a highly reactive and very unselective aryl cation intermediate, z-Ar^+, that is trapped by some weakly basic nucleophiles, water, alcohols, and anions. z-ArN_2^+ is prepared and stored as its crystalline tetrafluoroborate, which is stable for long periods. The long-tailed derivative, $z = 16$, is water-insoluble and

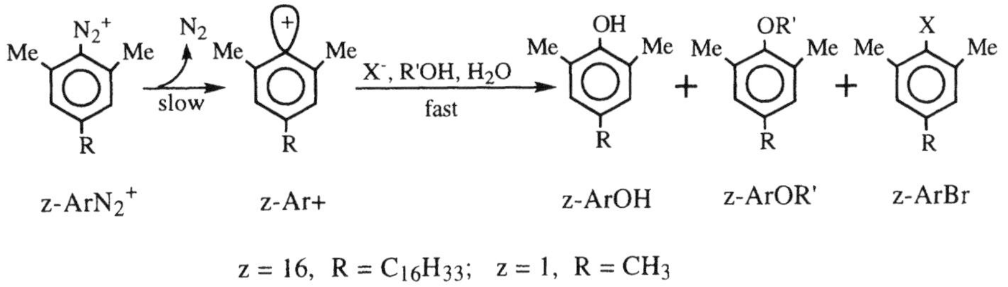

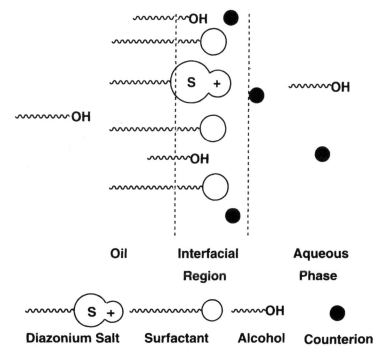

Scheme 4

The scheme shows:

z-ArN$_2^+$ → (slow) → z-Ar+ → (X$^-$, R'OH, H$_2$O, fast) → z-ArOH + z-ArOR' + z-ArBr

$$z = 16, \quad R = C_{16}H_{33}; \quad z = 1, \quad R = CH_3$$

is used for determining interfacial compositions of aggregates. The short-tailed, water-soluble analog, $z = 1$, is used to determine the selectivity of the arenediazonium ion toward different nucleophiles in bulk solution (see below).

The characteristics that make arenediazonium ions good probes of aggregate interfaces are discussed in detail elsewhere [65,126]. In brief, the arenediazonium ion is amphiphilic and acts as a pseudomonomer; the hexadecyl tail of 16-ArN$_2^+$ ensures strong binding to aggregates, and the cationic charge orients the diazonio group within the interfacial region (Fig. 4). z-ArN$_2^+$ traps many of the components, e.g., counterions and surfactant headgroups, commonly used in surfactant chemistry (see Sec. III.C). The selectivity of the aryl cation toward different nucleophiles (including anions competing with neutral nucleophiles for the aryl cation) is extremely low, and reasonable product yields

Oil   Interfacial   Aqueous

Region   Phase

Diazonium Salt   Surfactant   Alcohol   Counterion

Figure 4 Cartoon of the interfacial region of a four-component microemulsion composed of an ionic surfactant and its counterion, an alcohol, water, and oil. The amphiphilic arenediazonium ion probe, 16-ArN$_2^+$, is shown with its reactive headgroup located in the interfacial region.

are obtained simultaneously with mixtures of nucleophiles [65,127]. Product yields as low as 2%, obtained from initial 16-ArN$_2^+$ concentrations of $\leq 10^{-4}$ M, are detected with little or no workup and with excellent reproducibility by using an HPLC with an autosampler, a C-18 reversed-phase column, and a UV detector [65].

A. Applications of the Chemical Trapping Method

To date the chemical trapping method has been used to estimate interfacial concentrations of water, alcohols, and counterions in cationic micelles and microemulsions [65,128]; the affinity of cationic micelles toward Cl$^-$ versus Br$^-$, expressed as an ion-exchange constant [Eq. (11)] [129], and interfacial alcohol/water molar ratios in microemulsions; and distributions of 1-butanol and 1-hexanol between aggregate interfaces and bulk aqueous phases (O/W microemulsions) [130] and bulk oil phases (W/O microemulsions) over a range of alcohol and surfactant concentrations [131]. The focus here is on results in microemulsions.

1. Interfacial Concentrations

Relative product yields from dediazoniation of 1-ArN$_2^+$ in H$_2$O–R'OH solution are proportional to rates of formation of these products from reaction with the aryl cation intermediate, 1-Ar$^+$ in the fast step, Scheme 4. The equation

$$\frac{\%(1\text{-ArOR}')}{\%(1\text{-ArOH})} = \frac{k_A[\text{R'OH}]}{k_W[\text{H}_2\text{O}]} = S_W^A \frac{[\text{R'OH}]}{[\text{H}_2\text{O}]} \tag{14}$$

summarizes the relationships between product yield ratios, the rate constants of product formation, k_A and k_W, and concentrations of R'OH and H$_2$O in solution. We assume that the selectivity, S_W^A, of 16-Ar$^+$ toward H$_2$O and R'OH in the microemulsion interface is the same as that of 1-Ar$^+$ toward H$_2$O and R'OH in bulk solution,

$$S_W^A = \frac{[\text{H}_2\text{O}](\%1\text{-ArOR}')}{[\text{R'OH}](\%1\text{-ArOH})} = \frac{\text{H}_2\text{O}_M(\%16\text{-ArOR}')}{\text{R'OH}_M(\%16\text{-ArOH})} \tag{15}$$

In aqueous solutions of 1-butanol (BuOH), up to the solubility limit of BuOH (1 M), $S_W^A = 0.31$, and in 9 : 1 BuOH/H$_2$O, $S_W^A = 0.28$. The near identity of these two values and the fact that S_W^A is also constant for a wide range of compositions of water–polyethylene glycol mixtures [127] show that S_W^A for competitive reactions of nonionic nucleophiles is independent of solution composition. Thus, H$_2$O$_M$/R'OH$_M$ molar ratios can be estimated directly from %(16-ArOR')/%(16-ArOH) product yield ratios by assuming that the selectivity of the dediazoniation reaction in aggregates is the same as that in bulk solution. Equation (15) also shows that when the yield i.e., %(16-ArOR') in a microemulsion is the same as %(1-ArOR') in R'OH-H$_2$O, the concentration of alcohol at the interface, R'OH$_M$ is the same as R'OH in the bulk solution. Thus yield data are used to estimate interfacial concentrations, including interfacial concentrations of counterions [65].

Figure 5 shows the concentrations of two alcohols, BuOH and 1-hexanol (HexOH), Br$^-$, and water at the interfaces of three-component microemulsions composed of CTABr, R'OH, and H$_2$O as a function of the concentration of the alcohols up to their solubility limits. Adding either alcohol lowers the interfacial Br$^-$ concentration, but because HexOH binds more strongly than BuOH, much less HexOH is needed to displace an equivalent amount of Br$^-$. The decrease in interfacial concentration with added BuOH is similar to the increase in aqueous Br$^-$ concentration observed from conductivity [105,132]. Surprisingly, added alcohol has little effect on the concentration of interfacial H$_2$O (Fig.

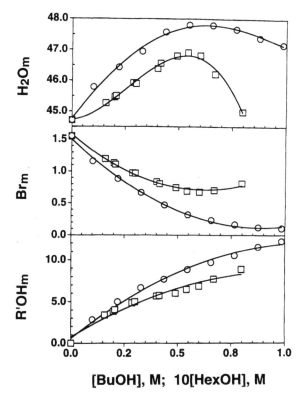

Figure 5 Effect of added BuOH (○) and HexOH (□) on interfacial concentrations (molarity) of water (H_2O_m), bromide ion Br_m, and alcohol ($R'OH_m$) in 0.01 M CTABr, 0.01 M HBr at $40 \pm 0.1°C$. Note the scale changes on the y axis and for HexOH on the x axis. (From Ref. 65.)

5), which varies only by about ±1.5 M (note the expanded scale for H_2O). These results show that when alcohols bind to cationic micelles, they do not displace interfacial water but swell the interface.

2. Alcohol Distributions

The determination of distributions of alcohols between the aqueous, interfacial, and oil regions of microemulsions is an active area of research, because the balance of forces controlling aggregate structure and stability depends on alcohol concentration within the aggregate, rather than on its stoichiometric concentration in solution [13,133,134]. Alcohol distributions have been estimated by a variety of physical methods, including solubility, changes in critical micelle concentrations and solution densities, excess enthalpies, heat capacities, ultrafiltration, vapor pressure, conductivity, small-angle neutron scattering, and fluorescence [135–144]. Alcohol distributions are generally reported as partition constants, i.e., the ratio of the mole fraction of alcohol in the aggregate divided by the mole fraction of the alcohol in the bulk phase. Partition constants for BuOH and HexOH in $CTABr-R'OH-H_2O$ microemulsions obtained by chemical trapping agree with those from solubility measurements within experimental error [145]. Chemical trapping experiments run over a range of alcohol concentrations in $CTABr-R'OH-H_2O$ microemulsions show that partition constants decrease with increasing alcohol concentration, consistent with results in other surfactant systems obtained by fluorescence quenching [146] and by theoretical treatments [147].

Partition constants report the equilibrium distribution of components in aggregated systems but do not represent the free energy of transfer of alcohols between aggregates and the bulk aqueous or oil phases, and their values need not be independent of solution composition, especially when the alcohol concentration in the aggregates or the bulk phase is high. However, alcohol distributions expressed as mass-action binding constants in aqueous three-component microemulsions [reaction (16)] and in W/O microemulsions [reaction (17)] are independent of alcohol concentration.

$$R'OH_W + D_n \overset{K'_A}{\rightleftharpoons} R'OH_M, \qquad K'_A = \frac{[R'OH_M]}{[R'OH_W][D_n]} \tag{16}$$

$$R'OH_O + D_n \overset{K'_a}{\rightleftharpoons} R'OH_M + oil, \qquad K'_a = \frac{[R'OH_M][oil]}{[R'OH_O][D_n]} \tag{17}$$

Table 1 summarizes values of K'_A and K'_a obtained in aqueous three-component (no oil) and W/O four-component R'OH–CTABr–oil–H$_2$O microemulsions obtained over a range of alcohol concentrations. These results illustrate several important features of the chemical trapping method and the properties of cationic microemulsions. First, the average deviation in the binding constants of BuOH and HexOH in hexadecane is within 4% over a range of alcohol concentrations [130,131]. Second, HexOH binds more tightly than BuOH to aqueous microemulsions, but the reverse is true in hexadecane W/O microemulsions, consistent with their relative hydrophobicities. Third, unlike partition constants, these mass-action binding constants are independent of R'OH and CTABr concentrations, which indicates that the primary driving force controlling alcohol distribution is the hydrophobic effect and that interactions between alcohol –OH groups and quaternary ammonium groups (and the Br⁻ counterion) have minimal effects on alcohol distribution. This need not be the case in anionic microemulsions where hydrogen bonding between alcohol –OH and the anionic headgroup of the surfactant may contribute significantly to alcohol binding, but mass-action binding constants for this system have not been estimated. Fourth, recent results indicate that values of K'_a are not very sensitive to oil type [148].

Table 1 Mass-Action Binding Constants for R'OH Estimated by Chemical Trapping from 16-ArOR' Product Yields in Aqueous Three-Component Microemulsions, K'_A and W/O Microemulsions with Different Oils, K'_a, at 40°C

Oil	R'OH	K'_A (M^{-1})	K'_a
Hexadecane	HexOH	32.0[a]	17.8[b]
Hexadecane	BuOH	5.99	23.1[b] (22.3[c])
Octane	BuOH		19.6[c]
Benzene	BuOH		23.6[c]
Tributyrin	BuOH		19.7[c]
Triolein	BuOH		13.6[c]

[a] From Ref. 65.
[b] From Ref. 131.
[c] Obtained by a modified treatment of the data of Ref. 148.

B. Advantages of the Chemical Trapping Method

Chemical trapping differs from other techniques used to probe the distributions of solutes at aggregate interfaces such as electrochemistry and spectrometry (NMR, UV/Vis, fluorescence, ESR, IR, and CD). The method reports simultaneously on the interfacial concentrations of all weakly basic nucleophiles within the interfacial region, it can be used over a very wide range of solution compositions, and many functional groups commonly used in basic research on aggregated systems and in commercial products are detected, including a number that are spectrophotometrically transparent or difficult to detect in the presence of water. The capability of reporting simultaneously on concentrations of water, in molarity or as hydration numbers, and on other nucleophiles is unique. The short-chain diazonium ion, $1\text{-}ArN_2^+$, can also be used to probe aggregated systems. Product yields from reaction of $1\text{-}ArN_2^+$ with Cl^- and Br^- in the aqueous pseudophase around cationic micelles were used to estimate α for each ion and their ion exchange constant [149].

C. New Directions for Chemical Trapping

Many important functional groups commonly found in association colloid research, biomembranes, and commercial products are weakly basic nucleophiles that trap the aryl cation (Scheme 4). For example, published and unpublished results indicate that the anions $X^- = Cl^-$, Br^-, I^-, SO_4^{2-}, HCO_2^-, $CH_3CO_2^-$, and $C_{10}H_{21}OPO_3H^-$ and the nonionic molecules $R'OH$ = water, alcohols, or nonionic surfactants all react by the dediazoniation mechanism. Surprisingly, amides such as urea, acetamide, and N-methyl acetamide also react with the aryl cation at both the amide nitrogen and the acyl oxygen. The photochemical reaction of $z\text{-}ArN_2^+$ is complete in about 2 min, and unpublished results indicate that product distributions from reactions with Cl^- and Br^- are very similar to those from the thermal reaction. Thus, interfacial compositions might be monitored over time by pulsed laser photolysis. Chemical trapping provides a new tool for exploring the relationship between interfacial composition and aggregate stability. Trapping of aryl cations by urea at interfaces should provide the first clear evidence about the distribution of urea in association colloids. Trapping by the functional groups of the peptide backbone and amino acid side chains in polypeptide–surfactant aggregates may also provide information on the binding, conformation, and orientation of the aggregated polypeptide. Trapping in membrane mimetic vesicles might be used to obtain information on the relationships between vesicle stability, hydration state, and local counteranion concentration. Chemical trapping is applicable to a very wide variety of colloidal systems, including, in principle, its use in estimating interfacial compositions and hydration states of opaque systems such as macroemulsions.

IV. CONCLUSIONS

Effects of association colloids on rates of reactions that are slow relative to diffusion are understandable on the assumption that these colloidal aggregates provide reaction media distinct from bulk solvent. Rate constants depend on the transfer equilibria and the individual rate constants in the colloidal aggregates and in bulk solvent. The quantitative treatments developed for reactions mediated by aqueous micelles are applicable to other association colloids including microemulsions. The phase volumes provided by O/W microemulsions contain contributions of surfactant, cosurfactant, and oil and are typically much larger than those provided by aqueous micelles. Bimolecular reactions in O/W microemulsions are therefore typically slower than in otherwise similar normal micelles, because bound reactants are more dilute when extracted out of the bulk solvent. However, the solubilizing power of these microemulsions is much greater than that of the corresponding micelles. The situation is different for spontaneous reactions, where there is no effect due to concentration in colloidal droplets, although for water-catalyzed reactions the availability of water in the interfacial region may be important, and these reactions are often sensitive to medium polarities, which depend on the composition of the association colloid. Chemical trapping of interfacial components with arenediazonium ions promises to provide new insight into the effects of changing interfacial compositions on the properties of multicomponent association colloids.

ACKNOWLEDGMENTS

We are grateful for support from the following agencies: the Chemical Dynamics and International Programs of the National Science Foundation, the U.S. Army Office of Research, and the Center for Advanced Food Technology at Rutgers University (Publication No. F-10535-1-96). We are also deeply indebted to the contributions of many colleagues, most of whom are noted in the references.

APPENDIX

Since completion of this chapter, some important papers have appeared that included quantitative treatments of chemical reactivity in microemulsions including comparisons of behaviors of micelles and microemulsions. They are briefly annotated here.

García-Río et al. [150] analyzed the kinetics of nitrosation of amines by alkyl nitrites in AOT–isooctane–water microemulsions, and Blagoeva et al. [151] used the pseudophase model to treat the effect of AOT microemulsions on the bromination of 1-octene. Reactions between multicharged metal ion complexes of the same charge as that of the surfactant occur in the aqueous pseudophase, and the kinetic salt effects of the micelles or microemulsions on these reactions can be treated with a combination of the Brønsted–Bjerrum equation and the extended Debyl–Hückel equation for estimating activity coefficients. This approach was used to fit observed rate constants of like charged ions in SDS W/O microemulsions [152] and the oxidation of $Fe(CN)_2(bpy)_2$ with $S_2O_8^{2-}$ in AOT micelles and microemulsions [153]. Holmberg and coworkers [154–156] describe the use of microemulsions as reaction media for amphiphile preparation, including product isolation and reuse of the amphiphile. Svensson et al. [157] showed that the lipase Humicola lanuginosa produces octyl decanoate from its alcohol and acid precursors in W/O microemulsions of egg phosphatidylcholine and heptane, and Yao and Romsted [158] used the chemical trapping method to estimate the effects of hydrocarbon and triglyceride oils on butanol distributions in W/O microemulsions.

REFERENCES

1. C. Tanford, *The Hydrophobic Effect: Formation of Micelles and Biological Membranes*, Wiley, New York, 1980.

2. R. Zana (ed.), *Surfactant Solutions: New Methods of Investigation*, Marcel Dekker; New York, 1985.
3. J. N. Israelachvili, D. J. Mitchell, and B. W. Ninham, J. Chem. Soc., Faraday Trans. 2 *72*:1525 (1976).
4. R. Nagarajan and E. Ruckenstein, J. Colloid Interface Sci. *71*:580 (1979).
5. R. Nagarajan and E. Ruckenstein, J. Colloid Interface Sci. *60*:221 (1977).
6. J. H. Fendler, *Membrane Mimetic Chemistry*, Wiley-Interscience, New York, 1982.
7. F. M. Menger and J. M. Jerkunica, J. Am. Chem. Soc. *101*:1896 (1979).
8. C. A. Bunton, L.-H. Gan, F. H. Hamed, and J. R. Moffatt, J. Phys. Chem. *87*:336 (1983).
9. D. F. Evans and D. D. Miller, in *Organized Solutions* (Surfact. Sci. Ser., Vol. 44) (S. E. Friberg and B. Lindman, eds.), Marcel Dekker, New York, 1992, p. 33.
10. C. A. Bunton, F. Nome, F. H. Quina, and L. S. Romsted, Acc. Chem. Res. *24*:357 (1991).
11. C. A. Bunton and G. Savelli, Adv. Phys. Org. Chem. *22*:213 (1986).
12. C. A. Bunton, in *Kinetics and Catalysis in Microheterogeneous Systems* (Surfact. Sci. Ser., Vol. 38) (M. Gratzel and K. Kalyanasundaram, eds.), Marcel Dekker, New York, 1991, p. 13.
13. D. M. Bloor and E. Wyn-Jones (eds.), *The Structure, Dynamics and Equilibrium Properties of Colloidal Systems*, Vol. 324, Kluwer, Boston, MA, 1990.
14. K. Shinoda, T. Nakagawa, B.-I. Tamamushi, and T. Isemura, *Colloidal Surfactants*, Academic, New York, 1963.
15. G. C. Krescheck, in *Water: A Comprehensive Treatise*, Vol. 4, *Aqueous Solutions of Amphiphiles and Macromolecules* (F. Franks, ed.), Plenum, New York, 1975, p. 95.
16. L. S. Romsted, Ph.D. Thesis, Indiana University, 1975.
17. K. L. Mittal and P. Bothorel (eds.), *Surfactants in Solution*, Plenum, New York, 1986, Vol. 6, Part VII.
18. R. Zana, Adv. Colloid Interface Sci. *57*:1 (1995).
19. M.-J. Schwuger, K. Stickdorn, and R. Schomaker, Chem. Rev. *95*:849 (1995).
20. K. Holmberg, Adv. Colloid Interface Sci. *51*:137 (1994).
21. K. Shinoda and B. Lindman, Langmuir *3*:135 (1987).
22. S.-H. Chen and R. Rajagopalan (eds.), *Micellar Solutions and Microemulsions: Structure, Dynamics, and Statistical Thermodynamics*, Springer-Verlag, New York, 1990.
23. A. Ben-Shaul and W. M. Gelbart, in *Micelles, Membranes, Microemulsions, and Monolayers* (W. M. Gelbart, A. Ben-Shaul, and D. Roux, eds.), Springer-Verlag, New York, 1994, p. 608.
24. R. Zana, Langmuir *11*:2314 (1995).
25. A. Ray and P. Mukerjee, J. Phys. Chem. *70*:2138 (1966).
26. M. J. Blandamer, P. M. Cullis, L. G. Soldi, J. B. F. N. Engberts, A. Kacperska, N. M. Van Os, and M. C. S. Subha, Adv. Colloid Interface Sci. *58*:171 (1995).
27. J. Daicic, U. Olsson, and H. Wennerstrom, Langmuir *11*:2451 (1995).
28. M. Gratzel and K. Kalyanasundaram (eds.), *Kinetics and Catalysis in Microheterogeneous Systems*, Vol. 38, Marcel Dekker, New York, 1991.
29. Y.-C. Yang, Chem. Ind.:337 (1995).
30. Y.-C. Yang, J. A. Baker, and J. R. Ward, Chem. Rev. *92*:1729 (1992).
31. F. M. Menger and A. R. Elrington, J. Am. Chem. Soc. *113*:9621 (1991).
32. F. M. Menger and H. Park, Rec. Trav. Chim. Pays-Bas *113*:176 (1994).
33. S. M. Garlick, H. D. Durst, R. A. Mackay, K. G. Haddaway, and F. R. Longo, J. Colloid Interface Sci. *135*:508 (1990).
34. R. A. Mackay, Adv. Colloid Interface Sci. *15*:131 (1981).
35. R. A. Moss, R. Fujiyama, H. Zhang, Y.-C. Chung, and K. McSorley, Langmuir *9*:2902 (1993).
36. R. A. Moss, K. Y. Kim, and S. Swarup, J. Am. Chem. Soc. *108*:788 (1986).
37. P. L. Luisi, in *Kinetics and Catalysis in Microheterogeneous Systems* (M. Gratzel and K. Kalyanasundaram, eds.), Marcel Dekker, New York, 1991, p. 115.
38. R. Strey, Curr. Opinion Colloid Interface Sci. *1*:402 (1996).
39. G. D. Rees, M. D. Nascimento, J. R. J. Jenta, and B. H. Robinson, Biochim. Biophys. Acta *1073*:493 (1991).

40. C. P. Singh and D. O. Shah, Colloids Surf. A: Physicochem. Eng. Aspects 77:219 (1993).
41. C. Oldfield, Biotechnol. Genet. Eng. Rev. 12:255 (1994).
42. S. Backlund, F. Eriksson, L. T. Kanerva, and M. Rantala, Colloids Surf. B: Biointerfaces 4:121 (1995).
43. N. L. Klyachko, A. V. Levashov, A. V. Kabanov, Y. L. Khmelnitsky, and K. Martinek, in *Kinetics and Catalysis in Microheterogeneous Systems* (M. Gratzel and K. Kalyanasundaram, eds.), Marcel Dekker, New York, 1991, p. 135.
44. M.-P. Pileni, Adv. Colloid Interface Sci. 46:139 (1993).
45. V. K. Vanag and I. Hanazaki, J. Phys. Chem. 99:6944 (1995).
46. E. J. R. Sudholter, G. B. van der Langkruis, and J. B. F. N. Engberts, Rec. Trav. Chim. Pays-Bas 99:73 (1980).
47. C. Minero and E. Pelizzetti, in *Organized Solutions (Surfact. Sci. Ser.*, Vol. 44) (S. E. Friberg and B. Lindman, eds.), Marcel Dekker, New York, 1992, p. 307.
48. H. Chaimovich, R. M. V. Aleixo, I. M. Cuccovia, D. Zanette, and F. H. Quina, in *Solution Behavior of Surfactants: Theoretical and Applied Aspects*, Vol. 2 (K. L. Mittal and E. J. Fendler, eds.), Plenum, New York, 1982, p. 949.
49. A. Ceglie, K. P. Das, and B. Lindman, J. Colloid Interface Sci. 115:115 (1987).
50. S. J. Chen, D. F. Evans, B. W. Ninham, D. J. Mitchell, F. D. Blum, and S. Pickup, J. Phys. Chem. 90:842 (1986).
51. M. Almgren, P. Linse, M. Van der Auweraer, F. C. De Schryver, E. Gelade, and Y. Croonen, J. Phys. Chem. 88:289 (1984).
52. J. Lang and R. Zana, in *Surfactant Solutions: New Methods of Investigation* (R. Zana, ed.), Marcel Dekker, New York, 1985, Chap. 8.
53. M. H. Gehlen, N. Boens, F. C. De Schryver, M. Van der Auweraer, and S. Reekmans, J. Phys. Chem. 96:5592 (1992).
54. E. A. G. Aniansson, S. N. Wall, M. Almgren, H. Hoffman, I. Kielmann, W. Ulbricht, R. Zana, J. Lang, and C. Tondre, J. Phys. Chem. 80:905 (1976).
55. M. H. Gelb, M. K. Jain, A. M. Hanel, and O. G. Berg, Annu. Rev. Biochem. 64:653 (1995).
56. R. G. Weiss, V. Ramamurthy, and G. S. Hammond, Acc. Chem. Res. 26:530 (1993).
57. H. F. Eicke, J. C. W. Shepherd, and A. Steinemann, J. Colloid Interface Sci. 56:168 (1976).
58. B. H. Robinson, D. C. Steyter, and R. D. Tack, J. Chem. Soc., Faraday Trans. 1 75:481 (1979).
59. M. H. Abraham, H. S. Chadha, J. P. Dixon, C. Fafols, and C. Treiner, J. Chem. Soc., Perkin Trans. 2:887 (1995).
60. F. H. Quina, E. O. Alonso, and J. P. S. Farah, J. Phys. Chem. 99:11708 (1995).
61. P. Lianos and R. Zana, J. Colloid Interface Sci. 88:594 (1982).
62. R. Zana, J. Colloid Interface Sci. 78:330 (1980).
63. E. Fabre, N. Kamenka, A. Khan, G. Lindblom, B. Lindman, and J. T. G. Tiddy, J. Phys. Chem. 84:3428 (1980).
64. A. Blasko, C. A. Bunton, G. Cerichelli, and D. C. McKenzie, J. Phys. Chem. 97: 11324 (1993).
65. A. Chaudhuri, J. A. Loughlin, L. S. Romsted, and J. Yao, J. Am. Chem. Soc. 115:8351 (1993).
66. L. S. Romsted, in *Surfactants in Solution*, Vol. 2 (K. L. Mittal and B. Lindman, eds.), Plenum, New York, 1984, p. 1015.
67. J. W. Larsen and L. B. Tepley, J. Colloid Interface Sci. 49:113 (1974).
68. J. Bocker, J. Brickmann, and P. Bopp, J. Phys. Chem. 98:712 (1994).
69. L. S. Romsted, in *Micellization, Solubilization and Microemulsions*, Vol. 2 (K. L. Mittal, ed.), Plenum, New York, 1977, p. 489.
70. F. H. Quina and H. Chaimovich, J. Phys. Chem. 83:1844 (1979).
71. J. D. Morgan, D. H. Napper, and G. G. Warr, J. Phys. Chem. 99:9458 (1995).
72. J. Morgan, Ph.D. Thesis, University of Sydney, Sydney, Australia, 1994.
73. E. A. Lissi, E. B. Abuin, L. Sepulveda, and F. H. Quina, J. Phys. Chem. 88:81 (1984).
74. B. Lindman and H. Wennerström, Top. Curr. Chem. 87:32 (1980).
75. L. S. Romsted and C.-O. Yoon, J. Am. Chem. Soc. 115:989 (1993).
76. K. D. Collins and M. W. Washabaugh, Quart. Rev. Biophys. 18:323 (1985).

77. E. Abuin and E. Lissi, J. Colloid Interface Sci. *151*:594 (1992).
78. C. R. A. Bertoncini, M. F. S. Neves, F. Nome, and C. A. Bunton, Langmuir *9*:1274 (1993).
79. R. A. Mackay, J. Phys. Chem. *86*:4756 (1982).
80. V. Athanassakis, C. A. Bunton, and F. de Buzzaccarini, J. Phys. Chem. *86*:5002 (1982).
81. E. Perez-Benito and E. Rodenas, Langmuir *7*:232 (1991).
82. M. K. Kawamauro, H. Chaimovich, E. B. Abuin, E. A. Lissi, and I. M. Cuccovia, J. Phys. Chem. *95*:1458 (1991).
83. A. Blasko, C. A. Bunton, and H. J. Foroudian, J. Colloid Interface Sci *175*:122 (1995).
84. Z.-M. He, J. A. Loughlin, and L. S. Romsted, Bol. Soc. Chil. Quim. *35*:43 (1990).
85. L. C. M. Ferreira, C. Zucco, D. Zanette, and F. Nome, J. Phys. Chem. *96*:9008 (1992).
86. D. A. R. Rubio, D. Zanette, F. Nome, and C. A. Bunton, Langmuir *10*:1155 (1994).
87. C. A. Bunton and J. R. Moffatt, Langmuir *8*:2130 (1992).
88. C. A. Bunton, M. M. Mhala, and J. R. Moffatt, J. Phys. Chem. *93*:7851 (1989).
89. C. A. Bunton and J. R. Moffatt, Ann. Chim. *77*:117 (1987).
90. G. Gunnarsson, B. Jonsson, and H. Wennerstrom, J. Phys. Chem. *84*:3114 (1980).
91. C. A. Bunton and J. R. Moffatt, J. Phys. Chem. *92*:2896 (1988).
92. C. A. Bunton and J. R. Moffatt, J. Phys. Chem. *90*:538 (1986).
93. F. Ortega and E. Rodenas, J. Phys. Chem. *91*:837 (1987).
94. E. Rodenas, C. Dolcet, and M. Valiente, J. Phys. Chem. *94*:1472 (1990).
95. A. Blasko, C. A. Bunton, C. Armstrong, W. Gotham, Z.-M. He, J. Nikles, and L. S. Romsted, J. Phys. Chem. *95*:6747 (1991).
96. D. S. Kemp and K. G. Paul, J. Am. Chem. Soc. *97*:7305 (1975).
97. C. A. Bunton, M. J. Minch, J. Hidalgo, and L. Sepulveda, J. Am. Chem. Soc. *95*:3262 (1973).
98. G. Cerichelli, G. Mancini, L. Luchetti, G. Savelli, and C. A. Bunton, J. Phys. Org. Chem. *4*:71 (1991).
99. C. A. Bunton and F. de Buzzaccarini, J. Phys. Chem. *85*:3139 (1981).
100. E. H. Cordes and C. Gitler, Prog. Bioorg. Chem. *2*:1 (1973).
101. C. Reichardt, *Solvents and Solvent Effects in Organic Chemistry*, 2nd ed., VCH, Weinheim, 1988.
102. E. M. Kosower, J. Am. Chem. Soc. *80*:3253 (1958).
103. K. Martinek, A. K. Yatsimirski, A. V. Levashov, and I. V. Berezin, in *Micellization, Solubilization, and Microemulsions*, Vol. 2 (K. L. Mittal, ed.), Plenum, New York, 1977, p. 489.
104. C. A. Bunton, G. Cerichelli, Y. Ihara, and L. Sepulveda, J. Am. Chem. Soc. *101*:2429 (1979).
105. C. R. A. Bertoncini, F. Nome, G. Cerichelli, and C. A. Bunton, J. Phys. Chem. *94*:5875 (1990).
106. C. A. Bunton, L.-H. Gan, J. H. Moffatt, and L. S. Romsted, J. Phys. Chem. *85*:4118 (1981).
107. M. Valiente and E. Rodenas, Langmuir *6*:775 (1990).
108. S. G. Oh, J. Kizling, and K. Holmberg, Colloids Surf. A: Physicochem. Eng. Aspects *97*:169, 217 (1995).
109. D. A. Jaeger, Supramol. Chem. *5*:27 (1995).
110. C. Minero, E. Pramauro, and E. Pelizzetti, Langmuir *4*:101 (1988).
111. R. da Rocha Pereira, D. Zanette, and F. Nome, J. Phys. Chem. *94*:356 (1990).
112. O. A. El Seoud, Adv. Colloid Interface Sci. *30*:1 (1989).
113. M. Valiente and E. Rodenas, Colloid Polym. Sci. *271*:494 (1993).
114. J. R. Leis, J. C. Mejuto, and M. E. Pena, Langmuir *9*:889 (1993).
115. L. Garcia-Rio, J. R. Leis, M. E. Pena, and E. Iglesias, J. Phys. Chem. *97*:3437 (1993).
116. A. M. Chinelatto, M. T. M. Fonseca, N. Z. Kiyan, and O. A. El Seoud, Ber. Bunsenges. Phys. Chem. *94*:882 (1990).
117. A. M. Chinelatto, L. T. Okomo, and O. A. El Seoud, Colloid Polym. Sci. *269*:264 (1991).
118. E. Sjoblom and S. J. Friberg, J. Colloid Interface Sci. *67*:16 (1978).
119. R. Germani, G. Savelli, G. Cerichelli, G. Mancini, L. Luchetti, P. P. Ponti, N. Spreti, and C. A. Bunton, J. Colloid Interface Sci. *147*:152 (1991).
120. R. Germani, P. P. Ponti, N. Spreti, G. Savelli, A. Cipiciani, G. Cerichelli, C. A. Bunton, and V. Si, J. Colloid Interface Sci. *138*:443 (1990).

121. F. del Rosso, A. Bartolletti, P. di Profio, R. Germani, G. Savelli, A. Blasko, and C. A. Bunton, J. Chem. Soc., Perkin Trans. 2: 643 (1995).

122. H. Kondo, I. Miwa, and J. Sunamoto, J. Phys. Chem. *86*:4826 (1982).

123. C. A. Bunton, Y.-S. Hong, and L. S. Romsted, in *Solution Behavior of Surfactants: Theoretical and Applied Aspects*, Vol. 2 (K. L. Mittal and E. J. Fendler, eds.), Plenum, New York, 1982, p. 1137.

124. A. Blasko, C. A. Bunton, and S. Wright, J. Phys. Org. Chem. *97*:5435 (1993).

125. G. Cerichelli, C. Grande, L. Luchetti, and G. Mancini, J. Org. Chem. *56*:3025 (1991).

126. A. Chaudhuri, J. A. Loughlin, and L. S. Romsted, in *Atualidades de Fisico-Quimica Organica-1991* (E. J. J. Humeres, ed.), Florianopolis, Brazil, 1991, p. 176.

127. L. S. Romsted and J. Yao, Langmuir *12*: 2425 (1996).

128. A. Chaudhuri and L. S. Romsted, J. Am. Chem. Soc. *113*: 5052 (1991).

129. J. A. Loughlin and L. S. Romsted, Colloids Surf. *48*:123 (1990).

130. A. Chaudhuri, L. S. Romsted, and J. Yao, J. Am. Chem. Soc. *115*:8362 (1993).

131. J. Yao and L. S. Romsted, J. Am. Chem. Soc. *116*:11779 (1994).

132. R. Zana, S. Yiv, C. Strazielle, and P. Lianos, J. Colloid Interface Sci. *80*:208 (1981).

133. S. E. Friberg and P. Bothorel (eds.), *Microemulsions: Structure and Dynamics*, CRC, Boca Raton, FL, 1987.

134. J. Israelachvili, *Intermolecular and Surface Forces*, Academic, London, 1991.

135. Y. Muto, K. Yoda, N. Yoshida, K. Esumi, K. Meguro, W. Binana-Limbele, and R. Zana, J. Colloid Interface Sci. *130*:165 (1989).

136. C. Treiner, J. Colloid Interface Sci. *93*:33 (1983).

137. R. De Lisi, S. Milioto, M. Castagnolo, and A. Inglese, J. Solution Chem. *19*:767 (1990).

138. R. De Lisi and S. Milioto, J. Solution Chem. *17*:245 (1988).

139. R. De Lisi and S. Milioto, J. Solution Chem. *16*:767 (1987).

140. C. Gamboa, A. Olea, H. Rios, and M. Henriquez, Langmuir *8*:23 (1992).

141. L. Damaszewski and R. A. Mackay, J. Colloid Interface Sci. *97*:166 (1983).

142. M. Manabe, H. Kawamura, S. Kondo, M. Kojima, and S. Tokunaga, Langmuir *6*:1596 (1990).

143. E. Caponetti, A. Lizzio, R. Triolo, W. L. Griffith, and J. S. Johnson, Jr., Langmuir *8*:1554 (1992).

144. E. A. Lissi and D. Engel, Langmuir *8*:452 (1992).

145. J. Gettins, D. Hall, P. L. Jobling, J. E. Rassing, and E. Wyn-Jones, J. Chem. Soc., Faraday Trans. *274*:1957 (1978).

146. E. B. Abuin and E. A. Lissi, J. Colloid Interface Sci. *95*:198 (1983).

147. I. V. Rao and E. Ruckenstein, J. Colloid Interface Sci. *113*:375 (1986).

148. J. Yao and L. S. Romsted, Colloids Surf. A: Physicochem. Eng. Aspects *123–124*:89 (1986).

149. I. M. Cuccovia, I. N. da Silva, H. Chaimovich, and L. S. Romsted, Langmuir *13*:647 (1997).

150. L. Garcia-Rio, J. R. Leis, and J. C. Mejuto, J. Phys. Chem. *100*:10981 (1996).

151. I. B. Blagoeva, P. Gray, and M.-F. Ruasse, J. Phys. Chem. *100*:12638 (1996).

152. P. Lopez, F. Sanchez, M. L. Moya, and R. Jimenez, J. Chem. Soc., Faraday Trans. 1 *92*:3381 (1996).

153. F. Sanchez, M. L. Moya, A. Rodriguez, R. Jimenez, C. Gomez-Herrera, C. Yanes, and P. Lopez-Cornejo, Langmuir *13*:3084 (1997).

154. K. Holmberg, S. G. Oh, and J. Kizling, Prog. Colloid Polym. Sci. *100*:281 (1996).

155. S. G. Oh, J. Kizling, and K. Holmberg, Colloids Surf. A: Physicochem. Eng. Aspects *104*:217 (1995).

156. S. G. Oh, J. Kizling, and K. Holmberg, Colloids Surf. A: Physicochem. Eng. Aspects *97*:169 (1995).

157. M. Svensson, G. D. Rees, B. H. Robinson, and G. R. Stephenson, Colloids Surf. B: Biointerfaces *8*:101 (1996).

158. J. Yao and L. S. Romsted, Colloids Surf. A: Physicochem. Eng. Aspects *123-124*:89 (1997).

159. J. H. Fendler, Pure Appl. Chem *54*:1809 (1982).

16

Application of Microemulsions in Enhancing Sensitivity of Reactions in Spectroscopic Analysis

Rong Guo and Xiashi Zhu
Yangzhou University, Yangzhou, People's Republic of China

I. INTRODUCTION

Micelles are dynamic aggregates of surfactant molecules [1,2]. Microemulsions are isotropic, thermodynamically stable dispersions of two immiscible liquids, generally oil and water that are interfacially stabilized by the surfactant and cosurfactant molecules [3–5]. These systems have important industrial [5–8] and biomedical [9–11] applications.

The utility of micelles in spectroscopic methods of analysis is perhaps the most popular and oldest micellar application in analytical chemistry. Hinze [12] first reviewed this subject in 1978. Since that time, the number of reports in this area has continued to grow at a fast rate.

The utility of micelles in spectroscopic measurements is derived from several possible effects. The well-documented effects of micellar systems on the acid–base chemistry of even slightly associated molecules can enhance (or degrade) the sensitivities of spectroscopic analysis [1,13–16]. In the field of metal ion complexation, much work has been carried out that suggests that the surfactant (within the micelle) takes part in the formation of a ternary complex with concomitant shifts in the wavelength of absorption [11,13,17]. Finally, as always, the ability of the micellar system to solubilize slightly insoluble or even very insoluble complexes and/or ligands has been used to enhance the sensitivities of spectroscopic analysis [12,13,18,19].

The first paper on the application of microemulsions in spectroscopic measurements was published by our group in 1987 [20]. Since that time, a series of studies have been reported, most of them published in the Chinese journals published by the Chinese Chemical Society [21–30].

In comparison with micelles, microemulsions used as a medium in spectroscopic measurements enhance considerably the sensitivity of reactions and simplify the experimental conditions. For example, a metal–PAN complex is not soluble in water but can be solubilized in the O/W microemulsion so that the O/W microemulsion can be used as a medium to determine the content of metal ion. Investigations of the mechanism show that the reason for the highest sensitivity of reactions in a microemulsion is that a microemulsion possesses a higher solubilization capacity than a micelle [31,32]. Without a doubt, the utility of microemulsions in spectroscopic measurements will continue to develop in the foreseeable future.

II. PHASE BEHAVIOR AND STRUCTURE OF MICROEMULSIONS

There are two kinds of phase diagrams for an ionic surfactant microemulsion system. The first [33] is shown in Fig. 1, in which the isotropic solution region connecting with the water corner is the O/W microemulsion region and that connecting with the alcohol (as cosurfactant) corner or oil corner is the W/O microemulsion region. In the second [34], an isotropic region exists continuously from the water corner to the alcohol corner (Fig. 2). Conductively measurements are usually employed to determine the structure of this kind

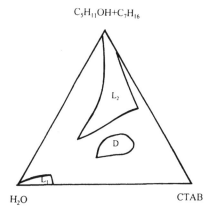

Figure 1   Part of the phase diagram of the system $CTAB-n-C_5H_{11}OH-n-C_7H_{16}OH-H_2O$. L_1, O/W region; L_2, W/O region; D, lamellar liquid crystal region. (From Ref. 33.)

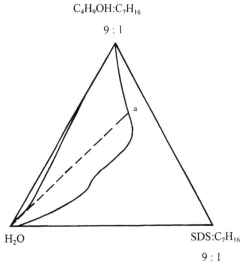

Figure 2   Part of the phase diagram of the system $SDS-n-C_4H_9OH-n-C_7H_{16}OH-H_2O$. Dotted line shows the results of conductivity measurements in Fig. 3. (From Ref. 34.)

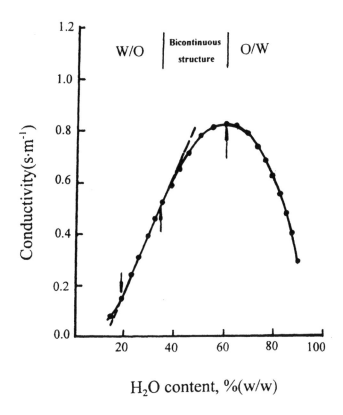

Figure 3 Conductivity vs. water content for the W/O, bicontinuous, and O/W microemulsion regions of the $SDS-n-C_4H_9OH-n-C_7H_{16}OH-H_2O$ system. The $SDS-n-C_4H_9OH:n-C_7H_{16}OH$ weight ratio $= 2.76 : 6.24 : 1.00$. (From Ref. 34.)

of microemulsion system [34,35]. The variation in the conductivity of such a micro-emulsion with water content is determined along the dashed line shown in Fig. 2, and the results are shown in Fig. 3 [34]. The conductivity rapidly increases linearly with water content ϕ up to $\phi = 34\%$ after an initial nonlinear increase from 13% to 18%. For values of ϕ greater than 34%, this linear region is followed by a nonlinear one until maximum conductivity is reached at $\phi = 60\%$. Above 60% it decreases continuously and nonlinearly. According to the investigations of Clausse et al. [36], such systems should show an O/W structure in the region $\phi > 60\%$, a W/O structure in the region $\phi < 34\%$, and a bicontinuous structure in the region $\phi = 34$–60%.

Figure 4 shows the typical phase diagram for a nonionic surfactant microemulsion system containing Triton X-100, decanol, and H_2O [37]. The isotropic region near the water corner shows the O/W structure, and an isotropic region on the right-hand side of the phase diagram shows the W/O structure.

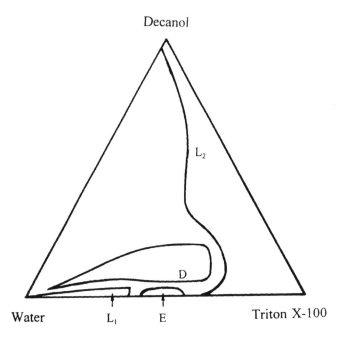

Figure 4 Part of the phase diagram of the system Triton X-100–decanol–H$_2$O. L$_1$, O/W region; L$_2$, W/O region; D, lamellar liquid crystal region; E, hexagonal liquid crystal region. (From Ref. 37.)

III. ANALYTICAL SENSITIVITY ENHANCEMENT IN MICROEMULSION MEDIA

A. W/O, O/W, and Bicontinuous Structure of an SDS–n–C$_4$H$_9$OH–n–C$_7$H$_{16}$–H$_2$O Microemulsion with an SDS: n-C$_4$H$_9$OH: n-C$_7$H$_{16}$ Weight Ratio of 0.27 : 0.63 : 0.1

1. The Ferric Thiocyanate Complex

The first report on the enhanced sensitivity of microemulsion analyzed spectroscopically was on the effect of the anionic microemulsion SDS–n–C$_4$H$_9$OH–n–C$_7$H$_{16}$–H$_2$O as a medium for ferric thiocyanate complexes [20] (Fig. 5). The results show that the maximum absorption wavelength λ_{max} exhibits a red shift and the analytical sensitivity is enhanced in comparison to water as the medium when the three kinds of microemulsions—O/W, W/O, and bicontinuous structure—are used as the medium, but the W/O microemulsion with 20% water content is the best one. λ_{max} shows red shifts of about 20 nm, and the analytical sensitivity is doubled in the W/O microemulsion with 20% water (Fig. 6).

This result has been used to determine the amount of iron in ferric ammonium citrate because the limiting content of citrate for the determination of iron increases greatly in the W/O microemulsion with 20% water [21]. The limiting content of citrate for the determination of iron is 1.0×10^{-6} g/25 mL in water but 3.5×10^{-4} g/25 mL in the W/O microemulsion.

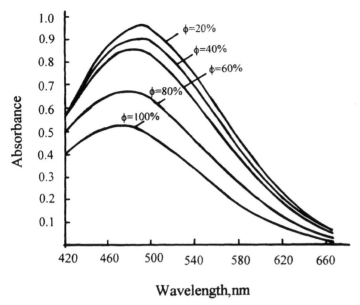

Wavelength,nm

Figure 5 Absorption spectra of Fe(III)–thiocyanate complex in a microemulsion medium of SDS–n-C$_4$H$_9$OH–n-C$_7$H$_{16}$–H$_2$O with different water contents ϕ. The SDS : n-C$_4$H$_9$OH : n-C$_7$H$_{16}$OH weight ratio = 2.76 : 6.24 : 1.00. (From Ref. 20.)

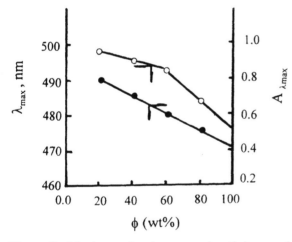

Figure 6 Maximum absorbance wavelength λ_{max}, and absorbance at λ_{max}, $A_{\lambda_{max}}$, versus water content ϕ in SDS–n-C$_4$H$_9$OH–n-C$_7$H$_{16}$–H$_2$O microemulsion. The SDS : n-C$_4$H$_9$OH : n-C$_7$H$_{16}$OH weight ratio = 2.76 : 6.24 : 1.00. (From Ref. 20.)

2. The Zinc(II)–1-(2-Pyridylazo)-2-Naphthol Complex

The apparent molar absorption coefficient, ε, of the Zinc–1-(2-pyridylazo)-2-naphthol (Zn-PAN) complex in an O/W microemulsion with 80% water is 5.15×10^4 L/(mol · cm) at 550 nm, which is higher than the 3.51×10^4 L/(mol · cm) in SDS micelles with the same water content (Fig. 7). On the other hand, some experimental conditions have been improved, such as reduction of the required volume of color agent with maximum absorption (Fig. 8).

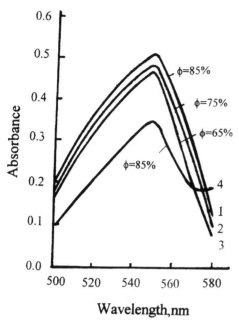

Figure 7 Absorption spectra of Zn(II)-PAN complex in SDS micelle and in O/W microemulsion of SDS–butanol–heptane–H$_2$O with different water contents ϕ. Media: Curves 1–3, O/W microemulsions with SDS : n-C$_4$H$_9$OH : n-C$_7$H$_{16}$OH weight ratio $= 2.76 : 6.24 : 1.00$; curve 4, micelle. (From Ref. 22.)

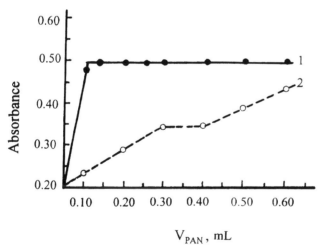

Figure 8 The effect of PAN volume on absorbance in different media. Media: Curve 1, O/W SDS–butanol–heptane–H$_2$O microemulsion with SDS : n-C$_4$H$_9$OH : n-C$_7$H$_{16}$OH weight ratio $= 2.76 : 6.24 : 1.00$; curve 2, SDS micelles. (From Ref. 22.)

3. The Mn(II)-PAN Complex

From the absorption spectra of Mn(II)-PAN [23] (Fig. 9), the maximum absorption wavelength is 510 nm in the O/W microemulsion medium and 500 nm in the micellar medium, and the absorbance A increases with increasing water content. At the same water content,

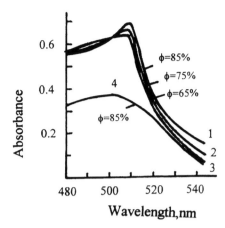

Figure 9 Absorption spectra of Mn(II)-PAN complex in SDS micelle and in O/W microemulsions of SDS–n-C$_4$H$_9$OH–n-C$_7$H$_{16}$OH–H$_2$O with different water contents ϕ. Media: Curves 1–3, O/W microemulsions with SDS : n-C$_4$H$_9$OH : n-C$_7$H$_{16}$OH weight ratio $= 2.76 : 6.24 : 1.00$; curve 4, micelles. (From Ref. 23.)

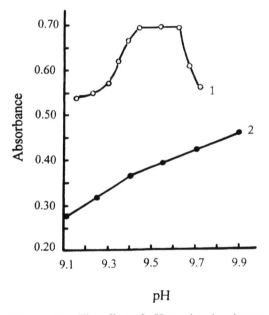

Figure 10 The effect of pH on the absorbance of Mn(II)-PAN complex. Media: curve 1, O/W SDS–butanol–heptane–H$_2$O microemulsion with SDS : n-C$_4$H$_9$OH : n-C$_7$H$_{16}$OH weight ratio $= 2.76 : 6.24 : 1.00$; curve 2, SDS micelles. (From Ref. 23.)

$\phi = 85\%$, $A_{max} = 0.69$, $\varepsilon = 3.79 \times 10^4$ L/(mol · cm) in the O/W microemulsion, and $A_{max} = 0.37$, $\varepsilon = 2.03 \times 10^4$ L/(mol · cm) in the micellar medium. Obviously, the maximum absorption wavelength shifts to the red by 10 nm and the sensitivity is doubled.

Microemulsions also improve some experimental conditions. For example, in Fig. 10 the absorbance of the Zn-PAN complex always increases with pH, but there is a plateau between pH 9.45 and 9.60 in the O/W microemulsion, which is very useful in spectroscopic analysis.

4. The Cu(II)-PAN Complex

For the same water content, $\phi = 65\%$, Fig. 11 shows $A_{max} = 0.425$, $\varepsilon = 2.86 \times 10^4$ L/(mol·cm) in the O/W microemulsion and $A_{max} = 0.38$, $\varepsilon = 2.42 \times 10^4$ L/(mol·cm) in the micellar medium. It is clear that analytical sensitivity is enhanced in the O/W microemulsion medium compared to that in the micellar medium.

5. The Cd(II)-PAN Complex

Using a micellar system with 75% water content as the medium (Fig. 12), $\lambda_{max} = 545$ nm, $A_{max} = 0.31$, $\varepsilon = 0.877 \times 10^4$ L/(mol·cm); but in the O/W microemulsion with the same water content $\lambda_{max} = 540$ nm, $A_{max} = 0.735$, $\varepsilon = 2.07 \times 10^4$ L/(mol·cm), the maximum absorption wavelength shifts by 5 nm, and the analytical sensitivity is enhanced about 2.5-fold.

6. The Aluminum Phenylfluorone Complex

In the spectroscopic measurements of aluminum–phenylfluorone (Al-PF) in Fig. 13, the analytical sensitivity is highest in the O/W microemulsion medium with a water content of 85%, and $\varepsilon = 8.09 \times 10^4$ L/(mol·cm), compared to $\varepsilon = 5.53 \times 10^4$ L/(mol·cm) in the SDS micellar medium and $\varepsilon = 4.72 \times 10^4$ L/(mol·cm) in water.

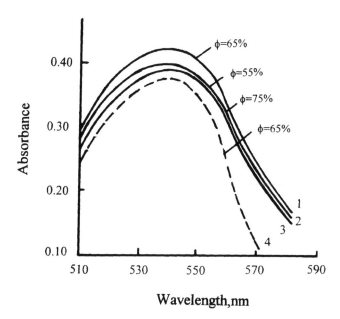

Figure 11 Absorption spectra of Cu(II)-PAN in SDS micelles and in O/W SDS–butanol–heptane–H$_2$O microemulsions with different water contents ϕ. Media: Curves 1–3, O/W microemulsions with SDS : n-C$_4$H$_9$OH : n-C$_7$H$_{16}$OH weight ratio = 2.76 : 6.24 : 1.00; curve 4, micelles. (From Ref. 24.)

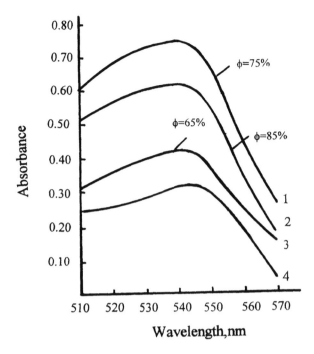

Figure 12 Absorption spectra of Cd(II)-PAN complex in SDS micelle and in O/W SDS–butanol–heptane–H_2O microemulsions with different water contents ϕ. Media: Curves 1–3, O/W microemulsions with SDS : n-C_4H_9OH : n-$C_7H_{16}OH$ weight ratio $= 2.76 : 6.24 : 1.00$; curve 4, micelles. (From Ref. 25.)

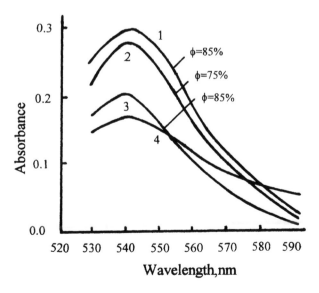

Figure 13 Absorption spectra of Al-PF complex in water, in SDS micelles, and in O/W SDS–butanol–heptane–H_2O microemulsions with different water contents. Curves 1 and 2, in O/W microemulsions with SDS : n-C_4H_9OH : n-$C_7H_{16}OH$ weight ratio $= 2.76 : 6.24 : 1.00$; curve 3, in micelle; curve 4, in water. (From Ref. 26.)

B. Oil-in-Water CTAB–n-C₅H₁₁OH–n-C₇H₁₆–H₂O Microemulsion

1. The Manganese(II)–2,6,7-Trihydroxy-9-Salicylfluorone Complex

The results in Fig. 14 show that the maximum absorption wavelength for manganese(II)–2,6,7-trihydroxy-9-salicylfluorone [Mn(II)-TSF] is about 575 nm, and the analytical sensitivity in a CTAB–n-C₅H₁₁OH–n-C₇H₁₆–H₂O O/W microemulsion medium is higher than in either CTAB micellar medium or water medium.

The CTAB : n-C₅H₁₁OH : n-C₇H₁₆ : H₂O weight ratio in O/W microemulsion is 1.30 : 0.36 : 0.04 : 98.30.

2. The Zinc(II)–4-(2-Pyridylazo)Resorcinol Complex

Figure 15 shows that the maximum absorption wavelength $\lambda_{max} = 502$ nm and $A_{max} = 0.630$, $\varepsilon = 8.6 \times 10^4$ L/(mol·cm) in the Zinc(II) 4-(2-pyridylazo)resorcinol [Zn(II)-PAR] O/W cationic microemulsion are higher than $A_{max} = 0.56$, $\varepsilon = 7.6 \times 10^4$ L/(mol·cm) in CTAB micelles at the same water content. The weight ratio CTAB : n-C₅H₁₁OH : n-C₇H₁₆ : H₂O in the O/W cationic microemulsion is 0.13 : 0.12 : 0.01 : 99.74.

C. Nonionic O/W Triton X-100–n-C₅H₁₁OH–n-C₉H₂₀–H₂O Microemulsion

1. The Copper(II)–Chrome Azurols Complex

In the copper (II)–chrome azurol [Cu(II)-CAS] water system in Fig. 16, $\lambda_{max} = 520$ nm, $A_{max} = 0.185$, $\varepsilon = 1.8 \times 10^4$ L/(mol·cm). However, $\lambda_{max} = 550$ nm, $A_{max} = 0.450$, $\varepsilon = 2.86 \times 10^4$ L/(mol·cm) in the Triton X-100 micellar system and $A_{max} = 0.550$, $\varepsilon = 3.52 \times 10^4$ L/(mol·cm) in the O/W microemulsion. Obviously, the analytical sensitivity is highest in the O/W microemulsion medium.

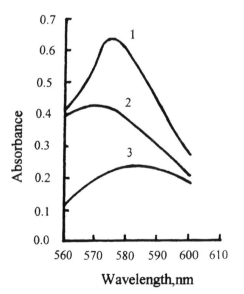

Figure 14 Absorption spectra of Mn(II)–2,6,7-trihydroxy-9-salicylfluorone complex in different media. Curve 1, in O/W CTAB–n-C₅H₁₁OH–n-C₇H₁₆–H₂O microemulsion with weight ratio 1.30 : 0.36 : 0.04 : 98.3; curve 2, in CTAB micelles; curve 3, in water. (From Ref. 27.)

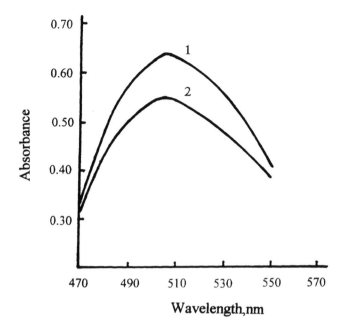

Figure 15 Absorption spectra of Zn(II)-PAN complex in different media. Curve 1, in O/W CTAB–n-C$_5$H$_{11}$OH–n-C$_7$H$_{16}$OH–H$_2$O microemulsion with weight ratio 0.13 : 0.12 : 0.01 : 99.73; curve 2, in CTAB micelles. (From Ref. 28.)

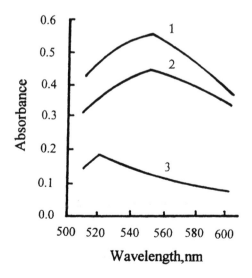

Figure 16 Absorption spectra of Cu(II)-CAS complex in different media. Curve 1, in O/W Triton X-100–n-C$_5$H$_{11}$OH–n-C$_9$H$_{20}$OH–H$_2$O microemulsion with weight ratio 0.03 : 0.02 : 0.005 : 0.945; curve 2, in Triton X-100 micelles; curve 3, in water. (From Ref. 29.)

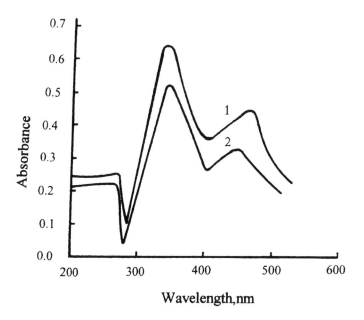

Figure 17 Absorption spectra of Fe(III)-BPHA complex in different media. Curve 1, in mixed O/W Triton X-100–SDBS–n-$C_5H_{11}OH$–H_2O microemulsion with weight ratio 1.2 : 0.12 : 0.72 : 97.96; curve 2, in mixed micelles of Triton X-100–SDBS with weight ratio 1.2 : 0.12. (From Ref. 30.)

The Triton X-100 : n-$C_5H_{11}OH$: n-C_9H_{20} : H_2O weight ratio in the O/W microemulsion is 0.03 : 0.02 : 0.005 : 0.945.

D. Mixed O/W Triton X-100–SDBS–$C_5H_{11}OH$–H_2O Microemulsion

1. The Iron(III)–N-Benzoyl-N-Phenylhydroxylamine Complex

The iron(III)–N-benzoyl-N-phenylhydroxylamine [Fe(III)-BPHA] complex is insoluble in water but can be solubilized in the mixed micelles of Triton X-100 and SDBS (sodium dodecylbenzene sulfonate) and in the mixed microemulsion of Triton X-100, SDBS, n-$C_5H_{11}OH$, and H_2O. In the mixed micellar medium in Fig. 17, $\lambda_{max} = 460$ nm, $\varepsilon = 4.00 \times 10^3$ L/(mol·cm), and in the mixed microemulsion, $\lambda_{max} = 460$ nm, $\varepsilon = 5.25 \times 10^3$ L/(mol·cm). So the analytical sensitivity is enhanced by about one-third in the mixed O/W microemulsion medium.

The Triton X-100 : SDBS : n-$C_5H_{11}OH$: H_2O weight ratio in the mixed O/W microemulsion is 1.2 : 0.12 : 0.72 : 97.96.

IV. INVESTIGATION OF MECHANISMS TO ENHANCE THE SENSITIVITY OF REACTIONS IN MICROEMULSIONS IN SPECTROSCOPIC ANALYSIS

For understanding the mechanism of enhancing the sensitivity of a reaction in a microemulsion, the effects of the O/W microemulsion on the apparent pK values of indicators and the relations between the pK values of the indicators in the microemulsion

Table 1 pK Values of Indicators in an SDS O/W Microemulsion[a] at 25°C

Indicators	pK_m	pK_i	$\dfrac{pK_m - pK_i}{pK_i}$ (%)	Buffer
Hydrophilic				
Cresol red	8.84	8.46	4.5	$HCl–Na_2B_4O_7$
Alizarin red S (20°C)	5.81	5.50	5.6	$HAc–NaAc$
Hydrophobic and ethanol-soluble				
Methyl red	6.44	4.97	29.5	$HAc–NaAc$
Bromothymol blue	8.59	7.3	17.5	$KH_2PO_4–Na_2B_4O_7$
Bromocresol green	6.18	4.90	26.1	$HAc–NaAc$
Hydrophilic acid ethanol-soluble				
Chromazurol S				
pK_1	5.28	4.90	7.8	$HAc–NaAc$
pK_2	12.07	11.50	5.0	$Na_2B_4O_7–NaOH$

$pK_m = pK$ value in O/W microemulsion; $pK_i = pK$ value in water.
[a] SDS 6.9%, C_4H_9OH 15.6%, C_7H_{16} 2.5%, H_2O 75%.
Source: Ref. 34.

and the hydrophilic character of the indicators were studied [34]. From a comparison of pK_m values (in microemulsion) with pK_i values (in water) (Table 1), it is clear that an O/W microemulsion enhances the effect of all kinds of indicators studied, but the difference between pK_m and pK_i is related to the hydrophilic character of the indicators. For hydrophobic and ethanol-soluble indicators, the increase in pK values is 15–30% in the O/W microemulsion. However, for hydrophilic and ethanol-soluble indicators, the increase in pK values is only 4–8%. The difference in the increase in pK values can be attributed to the difference in the location of the indicators in the microemulsion. The indicator molecules with hydrophobic character can enter deeply into the interphase of droplets of the O/W microemulsion where the electrostatic effect of the $-SO_4^-$ group of the SDS is very strong so that larger pK values appear. However, a hydrophilic indicator molecule can be solubilized in the water-continuous phase where the electrostatic effect of the $-SO_4^-$ group of the SDS is much weaker than that in the interphase of droplets of the O/W microemulsion, so only smaller pK values can be obtained.

The distribution coefficients of PAN and CAS indicators between microemulsion droplets and the water-continuous phase in O/W microemulsions constituted of anionic, cationic, and nonionic surfactants have been measured in order to investigate the mechanism of enhanced sensitivity of reactions in O/W microemulsion [31]. From Table 2 one can see that the distribution coefficients of PAN and CAS indicators in all O/W microemulsions are larger than those in micelles with the same surfactants. Thus, we can conclude from these results that the reason for higher sensitivity in microemulsions is that a microemulsion has greater solubilization capacity for indicators or complexes.

This conclusion has been confirmed by the measurements of the distribution coefficient of BPHA in the O/W microemulsion as shown in Table 2 [30].

Table 2 Distribution Coefficient K_D of Color Reagents

Color reagent	Medium	λ_w [a] (nm)	λ_m [b] (nm)	pH	K_D	Concn region of surfactant (10^{-3} mol/L)
PAN[c]	CTAB–H$_2$O	470	475	7.2–10.0	1440	1.5–2.0
	CTAB–C$_5$H$_{11}$OH–H$_2$O	470	475	7.2–10.0	2280	1.5–2.0
	CTAB–C$_5$H$_{11}$OH–C$_7$H$_{16}$–H$_2$O	470	475	7.2–10.0	2260	1.5–2.0
	SDS–H$_2$O	470	475	7.2–10.0	64	50–80
	SDS–C$_4$H$_9$OH–H$_2$O	470	475	7.2–10.0	15	48.64
	SDS–C$_4$H$_9$OH–C$_7$H$_{16}$–H$_2$O	470	475	7.2–10.0	73	70.100
	Triton X-100–H$_2$O	470	475	7.2–10.0	360	1.5–4.0
	Triton–C$_5$H$_{11}$OH–H$_2$O	470	475	7.2–10.0	4010	2.0–2.5
	Triton X-100–C$_5$H$_{11}$OH–C$_9$H$_{20}$–H$_2$O	470	475	7.2–10.0	1660	1.0–3.3
CAS[d]	CTAB–H$_2$O	429	425	5.6	2200	1.0–1.8
	CTAB–C$_5$H$_{11}$OH–H$_2$O	429	425	5.6	2310	0.6–1.0
	CTAB–C$_5$H$_{11}$OH–C$_7$H$_{16}$–H$_2$O	429	425	5.6	4500	0.5–1.6
	SDS–H$_2$O	429	435	5.6	374	10–16
	SDS–C$_4$H$_9$OH–H$_2$O	429	435	5.6	23	48–80
	SDS–C$_4$H$_9$OH–C$_7$H$_{16}$–H$_2$O	429	435	5.6	560	130–200
	Triton X-100–H$_2$O	429	425	5.6	4200	2.0–4.0
	Triton X-100–C$_5$H$_{11}$OH–H$_2$O	429	450	5.6	4610	2.0–4.0
	Triton X-100–C$_5$H$_{11}$OH–C$_9$H$_{20}$–H$_2$O	429	440	5.6	10500	2.5–3.5
BPHA[e]	Triton X-100–SDBS–C$_5$H$_{11}$OH–H$_2$O	253	254	5.5	2000	1.5–3.0
	Triton X-100–SDBS–H$_2$O	253	256	5.5	1300	1.5–3.0

[a] λ_{max} in water.
[b] λ_{max} in surfactant system.
[c] 1-(2-pyridylazo)-2-naphthol.
[d] Chrome azurols.
[e] N-Benzoyl-N-phenylhydroxylamine.
Source: Ref. 31.

V. SUMMARY

The analytical sensitivity of metal–ionic indicator complexes can be enhanced in an appropriate microemulsion medium because such complexes can be solubilized more in a microemulsion medium than in a micellar medium or an aqueous medium.

ACKNOWLEDGMENT

This work was supported by the China National Science Foundation (29733110).

REFERENCES

1. J. H. Fendler and E. J. Fendler, *Catalysis in Micellar and Macromolecular Systems*, Academic, New York, 1975.
2. C. Tanford, *The Hydrophobic Effect: Formation of Micelles and Biological Membranes*, 2nd ed., Wiley, New York, 1980.
3. T. P. Hoar and J. H. Schulman, Nature (Lond.) *152*:102 (1943).

4. R. N. Healy and R. L. Reed, Soc. Pet. Eng. J. *17*:129 (1977).
5. C. A. Miller and S. Qutubuddin, in *Interfacial Phenomena in Apolar Media* (H. F. Eicke and G. D. Parfitt, eds.), Marcel Dekker, New York, 1987, pp. 117–185.
6. K. Shinoda (ed.), *Solvent Properties of Surfactant Solutions*, Marcel Dekker, New York, 1967.
7. D. O. Shah (ed.), *Macro- and Microemulsions: Theory and Applications*, ACS Symp. Ser. No. 272, American Chemical Society, Washington, DC, 1985.
8. D. O. Shah and R. S. Schechter (eds.), *Improved Oil Recovery by Surfactant and Polymer Flooding*, Academic, New York, 1977.
9. M. Bourrel and R. S. Schechter, *Microemulsion and Related Systems*, Marcel Dekker, New York, 1988.
10. R. A. Mackay, Adv. Colloid Interface Sci. *15*:131 (1981).
11. D. Lichtenberg, R. J. Robson, and E. A. Dennis, Biochim. Biophys. Acta *737*:285 (1983).
12. W. L. Hinze, in *Solution Chemistry of Surfactants*, Vol. 1 (K. L. Mittal, ed.) Plenum, New York, 1979, p. 79.
13. E. Pelizzetti and E. Pramauro, Anal. Chim. Acta *169*:1 (1985).
14. M. J. Rosen, *Surfactants and Interfacial Phenomena*, Wiley, New York, 1978.
15. D. C. Hall, J. Phys. Chem. *91*:4287 (1987).
16. R. Staroscik, E. Maskiewicz, and F. Malcki, Fresenius Z. Anal. Chem. *329*:472 (1987).
17. I. Mori, Y. Fujita, K. Fujita, A. Usami, H. Kawabe, Y. Koshiyama, and T. Tanaka, Bull. Chem. Soc. Jpn. *59*:1623 (1986).
18. J. Hernandez-Mendez, B. Moreno-Cordero, J. L. Perez-Paron, and J. Cerda-Miralles, Inorg. Chem. Acta *140*:245 (1987).
19. M. Aihara, M. Arai, and T. Taketatsu, Analyst *111*:641 (1986).
20. R. Guo and X. Zhu, Chem. J. Chin. Univ. *8*:508 (1987).
21. C. Kang and R. Guo, Chin. J. Pharm. Anal. *12*:152 (1992).
22. X. Zhu and R. Guo, Chin. J. Anal. Chem. *20*:452 (1992).
23. X. Zhu and R. Guo, Phys. Test Chem. Anal. *27*:31 (1991) (in Chinese).
24. X. Zhu and R. Guo, Metall. Anal. *13*:51 (1993) (in Chinese).
25. X. Zhu, R. Guo, and Y. Liu, J. Yangzhou Univ. *11*:49 (1991) (in Chinese).
26. X. Zhu, R. Guo, P. Yan, and M. Shen, J. Yangzhou Univ. *12*:29 (1992) (in Chinese).
27. X. Zhu, R. Guo, and P. Yan, Chin. J. Anal. Chem. *22*:35 (1994).
28. X. Zhu, R. Guo, and C. Kang, Phys. Test Chem. Anal. *31*:335 (1995) (in Chinese).
29. X. Zhu, R. Guo, and M. L. Shi, Chin. J. Anal. Chem. *21*:1276 (1993).
30. X. Zhu, R. Guo, and X. Liu, Chin. J. Anal. Chem. *22*:860 (1994).
31. X. Zhu, R. Guo, and W. Qi, Chin. J. Anal. Chem. *23*:989 (1995).
32. X. Zhu, R. Guo, C. Kang, and M. Shen, Acta Chim. Sin. *53*:716 (1995) (in Chinese).
33. P. Yan, R. Guo, X. Zhu, and M. Shen, Acta Phys. Chim. Sin. *8*:690 (1992) (in Chinese).
34. R. Guo and X. Zhu, J. Surf. Sci. Technol. *1*:41 (1991).
35. R. Guo and G. Li, Acta Chim. Sin. *45*:55 (1987).
36. M. Clausse, A. Zradba, and L. Nicolas-Morgantini, in *Microemulsion systems* (H. L. Rosano and M. Clausse, eds.), Marcel Dekker, New York, 1987, p. 387.
37. P. Yan, R. Guo, Z. Liu, X. Zhu, and M. Shen, Acta Phys.-Chim. Sin. *10*:468 (1994) (in Chinese).

17

Preparation of Ultrafine Particles of Metals and Metal Borides in Microemulsions

Janos B. Nagy
Facultés Universitaires Notre-Dame de la Paix, Namur, Belgium

I. INTRODUCTION

In the mid-1970s S. Friberg and the late F. Gault proposed an original method using microemulsions to prepare monodisperse nanosized particles. These ideas were followed by a rapid increase in original research works related to the preparation of metal and metal boride nanoparticles [1–5].

A previous review dealt with a rather comprehensive literature survey up to 1985 [3] involving microemulsions, vesicles, polymer solutions, surfactant in water, sodium citrate in water, and general aqueous solutions as reaction media. It was followed by another that was based only on the use of microemulsions [4]. In 1996, a general overview was devoted to the results obtained in the Facultés Universitaires Notre-Dame de la Paix on various metals, metal borides, metal oxides, and silver halides [6].

However, the above-mentioned reaction media represent only a small part of a large variety of colloidal particle preparation. In particular, one could cite, among others, physical vapor deposition, chemical vapor deposition, Langmuir–Blodgett films, polymer films [6–9], zeolite-entrapped nanoparticles [10], and supported catalysts [11].

For a quantitative evaluation of the properties of colloidal dispersions, monodispersity of particles—uniformity in size and shape—is a prerequisite. The quantum size effects are particularly studied, since they lead to interesting mechanical, chemical, electrical, optical, magnetic, electro-optical, and magneto-optical properties that are quite different from those reported for bulk materials [8,10,12,13].

Instead of continuing the previous series of review papers [3,4], I place emphasis in this contribution on the fundamental aspects of monodisperse nanoparticle formation. We shall see how the inner water cores of the microemulsion systems work as microreactors. Moreover, the nucleation process is approached, and a minimum number of atoms forming the nuclei is proposed. The role of the surfactant and the cosurfactant are analyzed in the light of the formation of the first nuclei. Finally, the role of the adsorbed molecules in the monodisperse nature of the particles is examined. The different parts are illustrated taking into account the available literature.

A. Preparation of Nanoparticles Using Microemulsions

Table 1 lists the types of nanoparticles prepared in microemulsions together with the precursor, the reactant, and the microemulsion system.

1. Description of the Microemulsions

A water-in-oil microemulsion is a thermodynamically stable, optically transparent dispersion of two immiscible liquids stabilized by a surfactant. The important properties are

Table 1 Preparation of Monodisperse Nanoparticles from Microemulsions

Particle	Diameter (nm)	Precursor	Reactant	Microemulsion	Radius of inner water cores (nm)	Ref.
Ni_2B	3.0; 7.0	$NiCl_2$	$NaBH_4$	CTAB–hexanol–water	0.7–2.3	2–4, 14
Co_2B	2.5; 7.0	$CoCl_2$	$NaBH_4$	CTAB–hexanol–water	0.7–2.3	2–4, 15–19
Co_2B	4.0; 6.0	$CoCl_2$	$NaBH_4$	Triton X-100–decanol–water	0.6–1.4	20
Ni-Co-B	1.5; 3.0	$NiCl_2$, $CoCl_2$	$NaBH_4$	CTAB–hexanol–water	0.7–2.3	3, 4, 18, 19
FeB–Fe_2O_3	3.0; 8.0	$FeCl_3$	$NaBH_4$	CTAB–hexanol–water	0.7–2.3	2–4, 21, 22
Pt	2.0; 7.0	H_2PtCl_6	N_2H_4, $NaBH_4$	CTAB–hexanol–water	0.5–3.0	4
Pt	1.5; 13.0	K_2PtCl_4	N_2H_4	PEGDE–hexane–water	—	4, 23
Au	3.0; 4.5	$AuCl_3$	N_2H_4	PEGDE–hexane–water	—	24
Pt-Au	3.2; 11.0	$AuCl_3$, K_2PtCl_4	N_2H_4	PEGDE–hexane–water	—	4
ReO_2	2.0; 3.0	$NaReO_4$	N_2H_4	PEGDE–hexane–water	—	4, 23
Pt–ReO_2	1.5; 3.0	K_2PtCl_4, $NaReO_4$	N_2H_4	PEGDE–hexane–water	—	4, 23
Pt	4.0	H_2PtCl_6	N_2H_4	AOT–heptane–water	1.0–10.0	25
Pt	0.5	H_2PtCl_6	N_2H_4	AOT–heptane–glycerol	0.9	25
Pt	2.5; 4.0 2.3, 2.4	H_2PtCl_6	H_2, N_2H_4	CTAB–alcohol[a]–water PEGDE–hexane–water	4.0 6.0	26, 27
Pt/CdS	< 10.0	K_2PtCl_4	Uv-Vis irradiation	AOT–isooctane–water	15.0	28
Cu	20.0; 28.0	$Cu(AOT)_2$	$NaBH_4$	AOT–isooctane–water	9.0–30.0	29
Cu	2.0; 10.0	$Cu(AOT)_2$	N_2H_4	AOT–isooctane–water	9.0–22.5	29, 30
Cu	2.0; 10.0	$Cu(AOT)_2$	N_2H_4	AOT–isooctane–water	9.0–12.0	31
Pd	5.0	$PdCl_2$	H_2, N_2H_4	PEGDE–hexane–water	6.0	32–34
Pd	1.2; 1.9	K_2PdCl_4	H_2, N_2H_4	AOT–heptane–water	—	35, 36
Rh	0.8; 3.0	$RhCl_3$	H_2	PEGDE–hexane–water	6.0	32–34
Ir	2.5	$IrCl_3$	H_2 (2% Pt/Al_2O_3)	PEGDE–hexane–water	6.0	32, 33
Au	10.0; 60.0	$HAuCl_4$	Solvated electron	PEGDE–hexane–water	15.0; 22.0	37
Au	2.5; 4.2	$HAuCl_4$	N_2H_4	AOT–heptane–glycerol	0.9	35, 36
Ni_2B	150.0	—	$NaBH_4$	Ni decanoate–cyclohexane–water	—	38

[a] Alcohol: C_8OH–C_5OH, C_4OH.

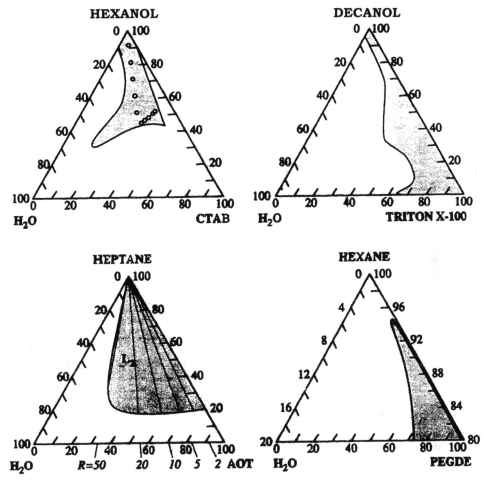

Figure 1 Microemulsion regions of various ternary systems.

governed mainly by the water-surfactant molar ratio ($R = [\text{H}_2\text{O}]/[\text{surfactant}]$). Cationic, anionic, and neutral surfactants have all been used for the preparation of nanoparticles.

The cationic cetyltrimethylammonium bromide–hexanol–water system contains hexanol, which forms the organic phase and plays the role of cosurfactant. It has been mapped by Ahmad and Friberg [39] (see Fig. 1). Although possessing a large L_2 domain, this microemulsion is capable of solubilizing only small amounts of water at low R values. The maximum water dispersion appears at 2.16 m CTABr in hexanol for an R value of 44.

The anionic AOT–heptane–water system is one of the best characterized microemulsions [40]. Up to 2 m AOT can be solubilized in heptane. The phase diagram was determined by Rouvière et al. [40]. It shows a particularly large L_2 region (Fig. 1) in which water can be dispersed, even at very low surfactant concentrations. It is the wedge-shaped structure of the surfactant that favors the reverse micellar association. The reverse micelles can take up to 60 molecules of water per surfactant headgroup.

The nonionic Triton X-100–decanol–water system was mapped by Ekwall et al. [41] (Fig. 1). The surfactant is completely soluble in decanol. Maximum water dispersion is obtained at a relatively high surfactant concentration: 1.54 m Triton X-100 in decanol.

The nonionic penta(ethylene glycol) dodecyl ether–hexane–water system (Fig. 1) was studied by Friberg and Lapczynska [42]. The reverse micellar droplets have a cylindrical shape in which the surfactant molecules stay parallel to each other, forming a bilayer impregnated with water.

2. Methods of Preparation of Nanoparticles

Two different methods were used for preparing the nonoparticles. The two reactants were dissolved in the same microemulsion system and then mixed under vigorous stirring (scheme II in Fig. 2). This method led to the smallest sizes and was used whenever the aqueous solutions of the reactants were stable enough. The nanoparticles of Pt, Pt-Au, ReO_2, Pt-ReO_2, Rh, and Pd were all prepared following scheme II (see Table 1).

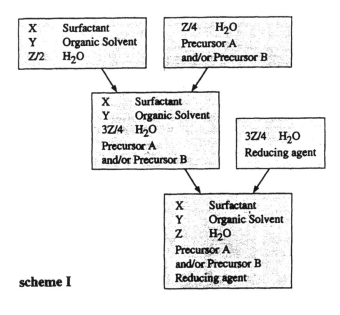

scheme I

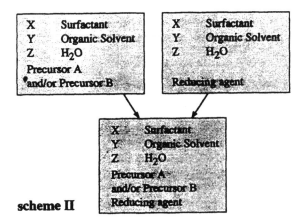

scheme II

Figure 2   Methods of preparation of monodisperse particles.

The Ni_2B, Co_2B, and Ni-Co-B particles were prepared by adding the aqueous solution of $NaBH_4$ to the microemulsion containing the metal salt (scheme I in Fig. 2). Moreover, these particles were prepared in a glove box under argon atmosphere by adding dropwise a threefold excess of aqueous $NaBH_4$ solution of 0°C under vigorous stirring. The expected microemulsion composition was achieved after complete mixing of the reactants. At the end of the reaction, the temperature was raised to room temperature until complete hydrolysis of excess $NaBH_4$ occurred.

II. RESULTS AND DISCUSSION

In order to understand why the nanoparticles are monodisperse, the microemulsion systems have to be well characterized. Important parameters are the radius of the inner water cores, which depends on R; the possible change in the microemulsion due to the presence of reactants in the inner water cores; a possible percolation phenomenon within the microemulsion; the localization and site of solvation of the precursor ion; the distribution of reactants in the inner water cores; and the rate of exchange of the content of the aqueous droplets.

The monodispersity depends on several parameters. The first to be emphasized is the differentiation between nucleation and growth of the particles [43]. Indeed, Zsigmondy [44] had already shown that by using seeding, which drastically separates the two phenomena, monodisperse colloidal gold particles could be obtained.

The second factor is the compartmentalization of the reaction medium where each water core is considered as a separate microreactor. The third factor concerns the role of adsorbed species—surfactant, ions, etc.—that could stabilize the colloidal particles at a certain size and prevent them from coagulating [45,46]. The existence of "magic numbers" was evoked to explain the higher stability of certain clusters [47]. This could be the fourth factor to be considered.

Finally, it could be considered that the spherical monodisperse particles are formed by secondary coagulation of the primary particles [46,48,49].

A. Physicochemical Characterization of Microemulsions

1. Nature of the Microemulsions

Three of the microemulsions studied were characterized by conductivity measurements: AOT–heptane–water, Triton X-100–decanol–water, and CTAB–hexanol–water. The last two were also characterized by cryofracture measurements.

The AOT–heptane–water microemulsion shows the lowest conductivity (2.7×10^{-5} S/m) at a water mole fraction of 0.1 (Table 2) [6].

The conductivity of the Triton X-100–decanol–water system is also quite low, 2.3×10^{-4} S/m, while the CTAB–hexanol–water microemulsion shows a rather high conductivity, 3.9×10^{-2} S/m [20]. The latter system can be considered as being in a percolated state, especially at high surfactant concentration.

The cryofracture measurements showed individual spherical forms for the Triton X-100–decanol–water microemulsion, whereas the spherical forms were tightly packed together for the CTAB–hexanol–water microemulsion (Table 3) [20].

Table 2 Comparison of Conductivity Values (κ, in S/m) at Constant Volume Mole Fraction of Water (0.1)

Microemulsion	Surfactant concn[a]	R[b]	κ (S/m)
AOT–heptane–water	0.29	3.5	2.7×10^{-5}
Triton X-100–decanol–water	1.54	8	2.3×10^{-4}
CTAB–hexanol–water	0.68	14	3.9×10^{-2}
	1.83	6	7.0×10^{-2}

[a] Molal vs. total microemulsion.
[b] $R = [H_2O]/[\text{surfactant}]$.

Table 3 Diameters of Spherical Objects Determined by Cryofracture and the Diameter of Inner Water Core Determined by ^{19}F NMR of Probe Molecules

Microemulsion	R	d_a (nm)	d_b (nm)	d_F (nm)	d_a/d_F	d_b/d_F
Triton X-100–decanol–water	13.1	10	7	2.9	3.4	2.4
CTAB–hexanol–water	5.0	8	7	2.4	3.4	2.9

d_a = diameter of the water core obtained by subtracting the interface layer thickness.
d_b = diameter of the water core obtained by subtracting the interface layer thickness increased by a layer of cosurfactant.
d_F = diameter of the water core determined by ^{19}F NMR of probe molecules.

The diameters determined by cryofracture are compared with those computed from an indirect determination method using ^{19}F NMR of probe molecules (see below). It is interesting to note that the diameters determined by the two methods are of the same order of magnitude. The values determined by cryofracture were computed from the observed diameter from which the thickness of the interface (d_a) and in addition the length of a layer of the cosurfactant (d_b) were subtracted. Note also that for bicontinuous systems such as sodium oleate–pentanol–water–benzene or sodium oleate–butanol–water, the ratio between the two types of diameters is equal to about 10 [20].

2. Site of Solubilization of the Metal Ions

It is generally known that anionic surfactants attract countercations. The opposite is true for cationic surfactants. Indeed, it was shown in the preparation of nanoparticles using LB films that particles could be made only from precursor ions attracted at the interface [8].

Therefore it can be stated that in the AOT–heptane–water microemulsion the Cu^{2+} ions are attracted preferentially to the interface, and this case is not treated further here [29–31]. I emphasize essentially the case of microemulsions containing either cationic or neutral surfactants.

(a) Solubilization of Paramagnetic Ions. The surfactant and cosurfactant molecules were characterized in the presence of shift reagents for the elucidation of the structure of the interface.

The use of shift reagents is rather well documented [50]. The unpaired electrons produce an additional magnetic field at the level of the nucleus, thus modifying the nuclear chemical shift. Two types of interactions can essentially be distinguished. The first is a through-space interaction of the dipolar type [50]:

$$\frac{\Delta v^{\text{dip}}}{v} = -D \frac{3 \cos^2 \theta - 1}{r^3} \tag{1}$$

where $\Delta v = v_M - v$ [chemical shifts in the presence (v_M) and in the absence (v) of paramagnetic ions]. The factor D depends on magnetic anisotropy, r is the distance between the two interacting nuclei, and θ is the angle between the principal axis of the paramagnetic entity (usually taken as the vector joining the paramagnetic atom and the first coordinating atom of the molecule) and the internuclear vector r. As can be seen, this dipolar contribution decreases rapidly with increasing r. Note that the simplified formula above is valid only for axially symmetrical systems [50].

The second type of interaction is the so-called Fermi contact interaction,

$$\frac{\Delta v^{con}}{v} = -A_i \frac{2\pi \, g_j \beta (g_j - 1) J(J + 1)}{\gamma_N \quad 3kT} \tag{2}$$

where A_i is the hyperfine coupling constant, γ_N is the nuclear magnetogyric ratio, β is the Bohr magneton, and g_j is the g factor depending on the quantum number J. In this case a coordinate bond exists between the paramagnetic ion and the molecule under study.

It is rather difficult to separate these two different contributions to the chemical shifts. Nevertheless, if one of them dominates the chemical shift, quite useful information can be obtained on the coordination of the paramagnetic ions.

For all the microemulsions studied, the presence of only one resonance line for each carbon atom is characteristic of a rapid exchange of molecules between several sites (interface, organic phase, vicinity of the paramagnetic ion, etc.) [2–4]. Figure 3 illustrates the variation of the chemical shifts ($\Delta\delta = \delta_M - \delta$, where δ_M and δ are the chemical shifts in the presence and absence of paramagnetic ions, respectively).

Three different behaviors can be clearly distinguished for Ni(II), Co(II), and Fe(III) chlorides dissolved in the microemulsion CTAB–hexanol–water.

In the system containing $CoCl_2$, the chemical shifts of the hexanol carbon atoms H_x vary linearly with the amount of ions, but in the system containing $NiCl_2$, an upward curvature is observed. The linear variation can be explained by a 1 : 1 cobalt(II)–hexanol association, while a stepwise replacement of water by hexanol in the coordination shell of Ni(II) ions could characterize the interaction of Ni(II) with hexanol molecules [2–4].

In all cases the variations ($\Delta\delta$) are quite linear as a function of the ion concentrations. In CTAB, the $N–CH_3$ carbon atom (C_0) is much more influenced than the $N–CH_2$ carbon atom (C_1). This suggests a dominant dipole–dipole interaction between the polar heads of the surfactant molecules and the paramagnetic ions and assumes that the paramagnetic ions are essentially located in the inner water cores of the microemulsions.

For the system containing Ni(II) or Co(II) ions, the chemical shift of the hexanol carbon atom in the β position (H_2) is more influenced than that of the carbon atom in the α position (H_1). This means that the interaction between the paramagnetic ions and the hexanol molecules includes an important contribution of the contact type. Therefore the hexanol molecules must participate in the first coordination shell of solubilized Co(II) or Ni(II) ions.

When dissolved in the same microemulsion, Fe(III) ions have a quite different influence on the carbon atoms of hexanol; they are only slightly shifted (-0.6 to 0.4 ppm). It follows that no direct contact occurs between the hexanol molecules and the Fe(III) ions that are solubilized in the water cores, where they are largely solvated by water molecules.

If the concentration of paramagnetic ions is normalized to the total amount of water, similar dependences are observed irrespective of the microemulsion system (Fig. 4). These relationships confirm that the paramagnetic ions are essentially dissolved in the inner water cores of the microemulsions. Note that these conclusions are also valid for the Triton X-100–decanol–water system containing $CoCl_2$ salt. In this case, both decanol and Triton

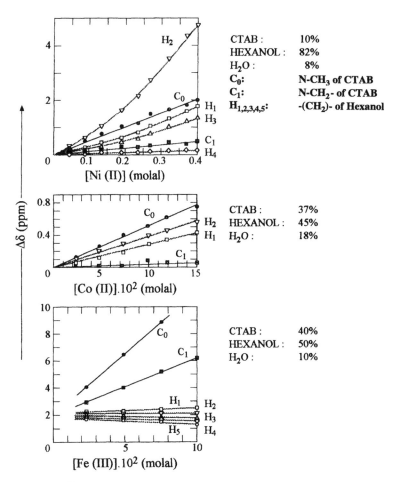

Figure 3 ^{13}C NMR chemical shift variation as a function of paramagnetic ion concentration.

X-100 molecules show a dominant contact type of interaction with the Co(II) ions, suggesting that both molecules enter the first coordination sphere of the paramagnetic ions. Nevertheless, in this case a non-negligible amount of Co(II) ion is also dissolved, probably in the organic phase formed by Triton X-100 and decanol molecules.

As was mentioned above, the upward curvature of the $\Delta\delta$ vs. Ni(II) concentration for the first three carbon atoms of hexanol molecules suggests a stepwise replacement of water by hexanol molecules in their first coordination shell. These observations can be expressed quantitatively by the equations

$$Ni^{2+} + 6H_2O \overset{K_A}{\rightleftharpoons} Ni(H_2O)_6^{2+} \tag{3}$$

$$Ni(H_2O)_6^{2+} + n_H C_6H_{13}OH \overset{K_B}{\rightleftharpoons} Ni(H_2O)_{6-n_H}(C_6H_{13}OH)_{n_H}^{2+} + n_H H_2O \tag{4}$$

$$Ni(H_2O)_{6-n_H}(C_6H_{13}OH)_{n_H}^{2+} + n_C CTAB \overset{K_C}{\rightleftharpoons} Ni(H_2O)_{6-n_H}(C_6H_{13}OH)_{n_H}^{2+} \cdots n_C CTAB \tag{5}$$

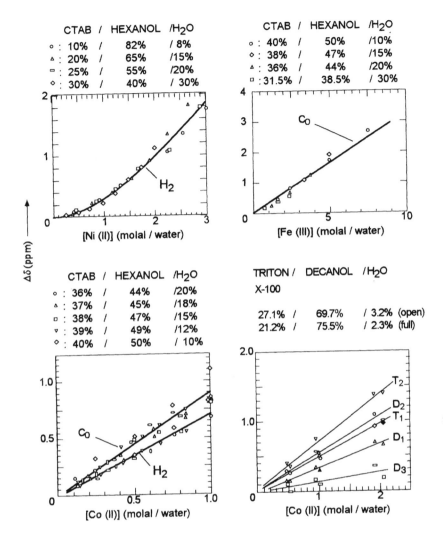

Figure 4 ^{13}C NMR chemical shift variations as a function of paramagnetic ion concentration versus water.

where n_H and n_C denote the numbers of hexanol and CTAB molecules, respectively. The most self-consistent results were obtained by simulating the NMR curve at low nickel concentration, assuming that $n_H = 3$ and $n_C = 1$. The computed values of the limiting chemical shifts $\Delta\delta_0$ of carbon atoms 1, 2, and 3 of hexanol were in good agreement with literature data [51]: C_1-20 (10.5); C_2-40 (40.7), and C_3-17 (10.4) ppm. Note that the maximum deviation on the product $K_B K_C = 12.5$ is about 60%. The limiting $\Delta\delta_0$ values for N–CH$_3$ and N–CH$_2$-carbon atoms are 3 and 0.7 ppm, respectively.

At higher nickel concentrations, the number of hexanol molecules can be computed if one considers a rapid exchange between the Ni(II) coordination sphere (δ_1) and the organic phase (δ_0):

$$\delta_{obs} = X_0\delta_0 + X_I\delta_I \tag{6}$$

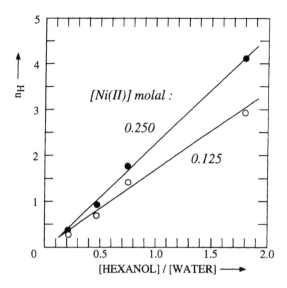

Figure 5 Variation of n_H, number of hexanol molecules in the first coordination shell of Ni(II), as a function of the ratio [hexanol]/[water].

where

$$X_I = \frac{\delta_{obs} - \delta_0}{\delta_I - \delta_0} = \frac{\Delta\delta}{\Delta\delta_0} \tag{7}$$

and also

$$X_I = \frac{n_H[Ni(II)]_{total}}{[hexanol]_{total}} \tag{8}$$

hence

$$n_H = \frac{[hexanol]_{total}}{[Ni(II)]_{total}} \frac{\Delta\delta}{\Delta\delta_0} \tag{9}$$

In this hypothesis, all the Ni(II) ions are at the interface. For $\Delta\delta_0$, the literature value of 10.5 ppm is taken for C_1 of hexanol [51].

For a given microemulsion, n_H increases with Ni(II) concentration. This corresponds quite well to a competition between water and hexanol. If [Ni(II)] increases, less water is available for the coordination of Ni(II) ions, because water molecules also need to solvate the Cl^- anions introduced in the water cores. This effect is the greater the lower the water concentration. [In a subsequent section of this chapter, the number of Ni(II) ions per microemulsion droplet is also computed.] The competition between water and hexanol molecules is nicely illustrated in Fig. 5. Linear correlations are obtained between n_H and the [hexanol]/[water] ratio.

Finally, it is also possible to compute the average conformation angle φ formed by the planes $Ni-O-C_\alpha$ and $O-C_\alpha-C_\beta$ in the complex Ni(II) and hexanol molecule from the chemical shift induced by the paramagnetic ions $\Delta\delta_0$:

$$\rho = \rho^T \cos^2\varphi \tag{10}$$

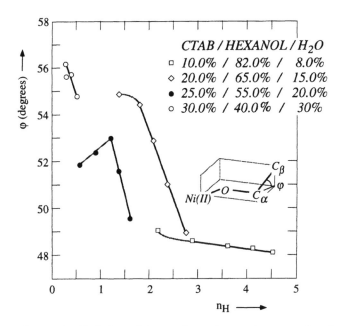

Figure 6 Variation of the conformation angle φ as a function of n_H.

where ρ and ρ^T are, respectively, the actual spin density and the one corresponding to the trans conformation. ρ^T is taken from Ref. 51, and ρ is computed from

$$\rho = k \, \Delta\delta_0^{C_2} \tag{11}$$

The k value is computed from data of the same reference [51].

The average φ angle decreases with increasing n_H value (Fig. 6). This suggests a steric constraint between the coordinating hexanol molecules, which tend to adopt a conformation where the C_α–C_β axis is an anticonformation with respect to the Ni–O bond. This also means that the interface becomes more rigid on average.

(b)   Nature of the Co(II) Complexes.   Because UV-Vis absorption energy is very sensitive to the number and the nature of ligands in the complexes of transition metal ions, UV-Vis spectroscopy is used to study the Co(II) complexes formed in microemulsions.

The spectrum of a CTAB–hexanol–water microemulsion containing Co(II) ions shows two different absorption bands: one at $\lambda_{max} = 688$ nm and another at 512 nm. The first one is attributed to a tetrahedral cobalt species having a molar absorption coefficient about 800 L/(mol · cm). The second band corresponds to an octahedral complex having a molar absorption coefficient of 5–5.8 L/(mol · cm) depending on the microemulsion composition. This is quite close to the value of 4.76 L/(mol · cm) obtained in aqueous solutions [52].

The concentrations of the two complexes increase with increasing Co(II) concentration, but at fixed [Co(II)] the absorbance due to the tetrahedral complex decreases quite rapidly with increasing water content [3] (Fig. 7). There exists an equilibrium between the octahedral and tetrahedral Co(II) species, and the octahedral complex is favored at high water concentration. This behavior can be easily understood if we assume that some of the water molecules are used to solvate the polar headgroups of CTAB molecules [2–4].

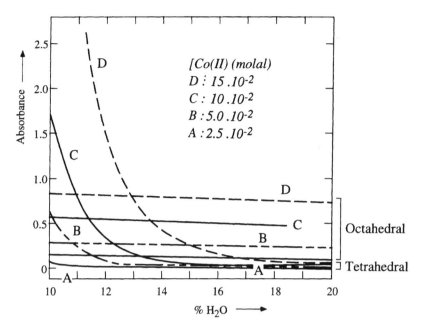

Figure 7 Variation of the absorbance of the Co(II) complexes as a function of the water content for different Co(II) concentrations.

In fact, in the case of low water content, there are not enough water molecules available to completely solvate the Co(II) ions. Consequently, some of the ions are transformed into tetrahedral complexes. With an increasing amount of water, the concentration of "free" water molecules also increases, which favors the formation of an octahedral complex [52]. The approximate relative amount of the tetrahedral complex is rather low: 0–2.5% of the total Co(II) amount. In water, the octahedral cobalt complex, $Co(H_2O)_6^{2+}$, absorbs at 510 nm, while in the microemulsion we found a band at 512 nm; this small difference is attributed to the coordination by hexanol molecules (shown also by ^{13}C NMR). Thus, the octahedral complex in the micellar water core can be of the form $(Co(H_2O)_{6-x}Hexanol_x)^{2+}$, where x is very small [2–4].

Tetrahedral complexes of the type $CoBr_4^{2-}$ or $CoCl_xBr_{4-x}^{2-}$ are formed in a micellar solution of the system CTAB–chloroform–water containing Co(II) ions. The maximum in the absorption band is at 719 nm [52]. On the other hand, in pure hexanol, the tetrahedral complex $CoCl_2Hexanol_2$ absorbs at 656 nm, and in HBr, $CoBr_4^{2-}$ absorbs at 710 nm. As in our microemulsions, the tetrahedral complex absorbs at 688 nm, we can assume that it has an intermediate composition, i.e., $CoCl_xBr_{3-x}Hexanol^-$ with $0 \leq x \leq 3$.

The tetrahedral complex is negatively charged, whereas the octahedral complex bears a positive charge. In the former case, the electrostatic interaction with the CTAB polar headgroup is favorable. It might seem surprising that the positively charged octahedral complex can also be found at the interface. The hexanol molecule included in the first coordination shell of Co(II) ions maintains this complex close to the interface. Consequently, the positively charged CTA^+ ion can interact with the Co(II) ions through dipolar interactions (see above). Similar conclusions have been arrived at for Ni(II) ions dissolved in the same reverse micellar system.

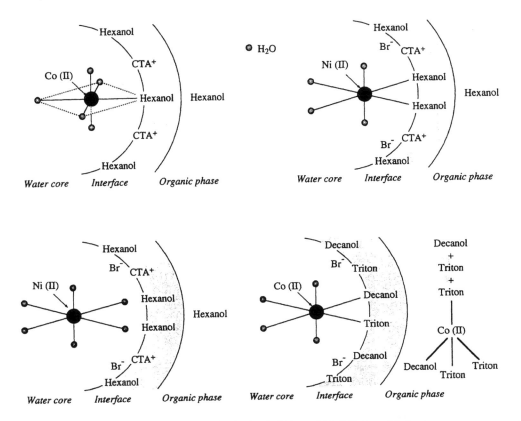

Figure 8 Models of solubilization of Ni(II), Co(II), and Fe(III) ions in the inner water core of microemulsions.

However, the system $CoCl_2$–Triton X-100–decanol–water shows quite different behavior. The tetrahedral complex has its absorption maximum at $\lambda = 669$ nm in the microemulsion. In addition, a tetrahedral complex is also formed in the organic medium, in pure decanol ($\lambda_{max} = 666$ nm), and in pure Triton X-100 ($\lambda_{max} = 673$ nm). The intermediate value in the microemulsion suggests that the Co(II) ions in the tetrahedral complex are solvated by both decanol and Triton X-100 molecules. The absorption maximum of the octahedral complex, $\lambda_{max} = 525$ nm, is quite different from the aqueous solution value, $\lambda_{max} = 510$ nm. This difference confirms the coordination of Co(II) ions by both Triton X-100 and decanol molecules (see also the ^{13}C NMR results). As the solubility of $CoCl_2$ is non-negligible in both pure decanol and Triton X-100, further work is necessary to analyze more quantitatively this rather complicated system.

Based on the ^{13}C NMR and UV-Vis spectrophotometric results, the following models can be proposed for the four systems investigated (Fig. 8). Both Co(II) and Ni(II) atoms are retained at the interface in the different systems. More than one hexanol molecule enters the first coordination shell of Ni(II) ions. Co(II) interacts with one hexanol molecule in the CTAB–hexanol–water microemulsions, whereas both decanol and Triton X-100 molecules enter its first coordination shell. The Fe(III) ions are strongly hydrated in the inner water cores, and no hexanol molecules are able to replace the strongly held water molecules with this highly charged ion. Finally, the CTA$^+$ ions interact indirectly only with the positively charged complexes.

(c)   Site of Solubilization of H_2PtCl_6.   Hexachloroplatinic acid (H_2PtCl_6) is soluble in both hexanol and the aqueous phase of the CTAB–hexanol–water microemulsion. Moreover, as the $PtBr_6^{2-}$ is much more stable than $PtCl_6^{2-}$ [53] and the available Br^- concentration is roughly 100–1000 times higher than that of Cl^-, $PtBr_6^{2-}$ ions are preferentially formed. The partition coefficient of H_2PtBr_6 between the two immiscible liquids, hexanol and water, was determined to be equal to 0.58 ± 0.04 [54].

A solution of H_2PtCl_6 in the CTAB–hexanol–water microemulsion shows the UV-Vis absorption bands of $PtBr_6^{2-}$ in both hexanol and water solutions. In addition, as the transformation of $PtCl_6^{2-}$ into $PtBr_6^{2-}$ is much slower in the hexanol phase [44], $PtCl_6^{2-}$ is also detected in that phase.

As a conclusion, the H_2PtBr_6 precursor to the Pt particle is partitioned between the organic and aqueous phases, as is the case for $CoCl_2$ in the Triton X-100–decanol–water system.

(d)   Site of Solubilization of K_2PtCl_4.   Potassium chloroplatinate is dissolved essentially in the inner water cores of the various microemulsions (Table 1). These data are essential for an understanding of monodisperse nanoparticle formation (see below).

3.   Nature of the Water in the Inner Water Cores

The reverse micelles, or the closely related water-in-oil (W/O) microemulsions, can solubilize a large number of polar and nonpolar hydrocarbons as well as inorganic and organic salts [9,55–58].

Nonpolar or weakly polar hydrocarbons such as benzene, cyclohexane, or hexanol are dissolved in the bulk hydrocarbon solvent or the interface [59], while more polar substances such as methanol, pyrazole, or dimethyl sulfoxide (DMSO) are preferentially attracted by the polar water core [55,60].

The polarity of the water core can be measured by relating the observed property (i.e., fluorescence maximum) of a probe molecule to solvent polarity parameters such as the Kosower's Z values [61] or the Dimroth–Reichardt E_T values [62]. The interesting conclusion is that the limiting polarity parameter in the inner water core remains quite below the polarity of pure water [63]. Eicke and Christen [64] also emphasized the different kinds of water molecules in the inner core: those solvating the polar headgroups of the surfactant molecules and the remaining water molecules in the interior of the water core. The solubilization of organic molecules in the normal micelles can also be decomposed into two parts: solubilization at the interface (adsorbed fraction) and solubilization in the hydrocarbon core (dissolved fraction) [65]. The polarity of the microenvironment is explored by UV-Vis measurements of paramagnetic nitroxides [66,67] as well as by EPR measurements determining the hyperfine coupling constants [68].

To avoid the rather large perturbation introduced by fluorescent probe molecules, the use of water-soluble and small ions is recommended to probe the polarity of the inner water cores [22,69–71].

The $n \rightarrow \pi^*$ absorption band of nitrate anion is used in this work as a polarity probe of the water pools of the microemulsions. The ground state of the nitrate anion, with an electron pair on the oxygen atom, can hydrogen bond with the solvent with a greater strength than the excited state, leading therefore to a blue shift of the absorption band [71]. In all the CTAB–hexanol–water microemulsions, with the water content varying from 10 to 30 wt%, the observed λ_{max} is 306 nm, corresponding to 93.76 kcal/mol. By comparing this value with the literature data [70], it can be inferred that the observed polarity of the inner water core has a value intermediate between that of isopropanol and *tert*-butanol. This value of polarity for such a high water content is still quite far from the value of pure water ($\lambda_{max} = 302$ nm, corresponding to 95.00 kcal/mol) [22].

Another valuable tool to probe the nature of the inner water core is ^{1}H NMR [71–73]. The study of water self-diffusion showed the presence of a non-negligible amount of water in the organic phase of the W/O microemulsion composed of sodium octanoate, decanol, and water [74].

In the Triton X-100–decanol–water microemulsion, the chemical shift of water increases with increasing water content [20]. At the same time, the ^{1}H NMR linewidth decreases and the T_1 values increase. The downfield shift observed for the water protons from 4.15 to 4.65 ppm upon increased water addition reflects the gradual buildup of bulk-type water in the inner water cores [20,71]. This change is accompanied by a higher mobility (decreased linewidth and increased T_1 value) of the water molecules. Nevertheless, the corresponding values for bulk water are not reached even at the highest water concentrations: $\delta_{water} = 5.00$ ppm [71] and $T_1 = 3.6$ s [20].

The use of the paramagnetic salt $CoCl_2$ sheds some more light on the behavior of water molecules in this microemulsion. The influence of the paramagnetic ions on the water chemical shift increases with increasing water content. This phenomenon can be understood if a non-negligible amount of water is solvating the polar headgroups of the Triton X-100 molecules, some of which are dissolved in the outer organic phase. In addition, the Co(II) ions adopt a tetrahedral configuration at low water content, while the dominant configuration becomes octahedral for high water content [20,71] (see above). At higher water concentrations, a larger amount of water can thus be influenced by the paramagnetic Co(II) ions, which are dissolved in both the inner water cores and the outer organic medium.

4. Size, Number, and Molar Composition of Microemulsions

Generally, the radius of the inner water cores is a linear function of R [25]. This relationship was used, where available, to compute the radii and hence the number of water cores in the microemulsion. Table 4 shows the data for the available microemulsions. Note that no data could be found for the PEGDE–hexane–water microemulsion.

For the CTAB–hexanol–water and Triton X-100–decanol–water microemulsions, the approximate sizes of the inner water cores were determined by means of ^{19}F NMR of 6-fluorohexanol and ω-fluoroundecanol. It is assumed that the probe molecules behave like

Table 4 Linear Correlations for the Size Dependence (r_w) on the Molar Ratio of Water to Surfactant, R^a

Microemulsion	[Surfactant] (m)	a	b
		Correlation coefficient	
CTAB–hexanol–water	2.20	0.131	0.054
			(0.9882)
Triton X-100–decanol–water	1.0	0.099	−0.022
			(0.9999)
AOT–heptane–water	0.14	0.128	0.083
			(0.9990)

^{a} $r_w = aR + b$.

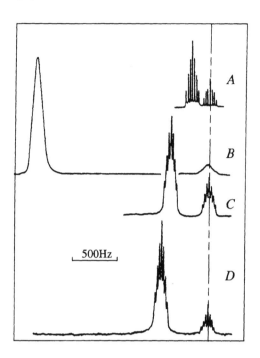

Figure 9 ^{19}F NMR of 6-fluoro-1-hexanol (2 wt%) in (A) hexanol, (B) normal micelle (CTAB 33%–H$_2$O 67%; [Ni(II)] = 0.5 m vs. water); (C) microemulsion (CTAB 31.6%–hexanol 52.6%–H$_2$O 15.8%; [Ni(II)] = 5 × 10^{-2} m); (D) microemulsion (CTAB 24.9%–hexanol 62.5%–H$_2$O 12.6%; [Ni(ii)] = 8 × 10^{-2} m); reference: 6-fluoro-1-hexanol (2 wt%) in CD$_3$O$_D$ in a sealed capillary tube.

the cosurfactant and/or the organic molecules in the organic phase [75,76]. This indirect method to measure the average radius of microemulsions was proposed by Nguyen and Ghaffarie [76]. It is based on the following hypothesis:

1. The water cores are assumed to be spherical [77].
2. The total amount of surfactant is at the interface. Only the cosurfactant is distributed between the interface and the continuous organic medium.
3. The total amount of water is in the inner water core. The relative amount dissolved in the organic phase is neglected.
4. The interfaces in the O/W and W/O microemulsions are of identical nature as far as the chemical shift of the probe molecules is concerned.

Advantage is taken of the great dependence of the ^{19}F NMR chemical environment experienced by the probe molecule at the interface and in the outer organic phase [2–4,45] (Fig. 9). This probe is distributed between the water core, the interface, and the bulk organic phase. Because of the fast exchange between these phases, the observed chemical shift (δ_{obs}) of the probe molecule is related to the amounts present in each phase by the equation

$$\delta_{obs} = X\delta_w + Y\delta_I + Z\delta_0 \tag{12}$$

where δ_w, δ_I, and δ_0 are the chemical shifts of the fluoro probe in water, interface, and organic phase, respectively, with the corresponding mole fractions X, Y, and Z.

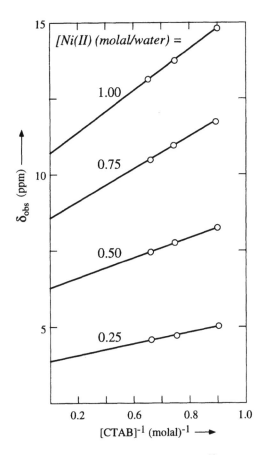

Figure 10 Variation of the observed ^{19}F NMR chemical shift of 6-fluoro-1-hexanol as a function of 1/[CTAB] for different Ni(II) concentrations in normal micelles.

Because of the very low solubility of the probe in water, $X \cong 0$. Moreover, the probe molecule in the organic phase is used as reference and δ_0 is therefore equal to zero. The relationship is then simplified to

$$\delta_{\text{obs}} = Y\delta_I \tag{13}$$

δ_I is determined from normal micelles formed by the surfactant in aqueous solutions with 2 wt% of the fluoro probe. The observed chemical shifts in solutions with different surfactant concentrations are related in the fast-exchange approximation to δ_I, δ_w, and the cmc of the surfactant according to Ref. 78:

$$\delta_{\text{obs}} = \delta_I + \frac{\text{cmc}}{[\text{surfactant}]}(\delta_W - \delta_I) \tag{14}$$

From the variation of δ_{obs} as a function of 1/[surfactant], the intercept of the straight line yields δ_I. As a function of the concentration of the paramagnetic ions, a linear increase is observed [2–4,45,79] (Fig. 10). From the latter, the following relationships are obtained for the CTAB normal micelles (Fig. 11).

$$\delta_I \text{ (ppm)} = 1.5(\pm 0.1) + 9.5(\pm 0.05) \text{ [Ni(II)]} \tag{15}$$

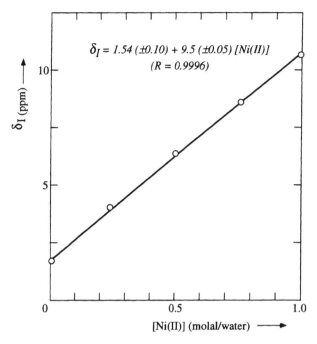

$\delta_I = 1.54\ (\pm 0.10) + 9.5\ (\pm 0.05)\ [Ni(II)]$
$(R = 0.9996)$

Figure 11 Variation of δ_I (chemical shift of 6-fluoro-1-hexanol at the interface) as a function of Ni(II) concentration.

The corresponding equations for Co(II) and Fe(III) ions are, respectively,

$$\delta_I\ (\text{ppm}) = 1.7(\pm 0.1) + 27.5(\pm 0.1)\ [Co(II)] \tag{16}$$

and

$$\delta_I\ (\text{ppm}) = 1.4(\pm 0.1) + 0.3(\pm 0.02)\ [Fe(III)] \tag{17}$$

For the Triton X-100 normal micelle, the following relationship was observed:

$$\delta_I\ (\text{ppm}) = 3.0(\pm 0.5) + 51.0(\pm 0.9)\ [Co(II)] \tag{18}$$

In the CTAB–hexanol–water microemulsion system, the Ni(II) and Co(II) ions are not dispersed in the water core; they are located in an envelope in the vicinity of the interface as was shown by ^{13}C and UV-Vis spectroscopic measurements. The Ni(II) local concentration is more important than the global concentration, with a correction factor given by [3]

$$F_{\text{corr}} = \frac{r_M^3}{r_M^3 - (r_M - r_{Ni})^3} \tag{19}$$

where r_M is the radius of the inner water core and r_{Ni}, the thickness of the envelope containing the Ni(II) ions, is taken as equal to 0.32 nm, on the basis of the coordination shell of the ions. This correction is needed only for Ni(II) and Co(II) ions. As Fe(III) ions are essentially solubilized in the interior of the water core, no such correction had to be carried out.

A knowledge of the δ_I value enables us to calculate, from the observed chemical shift, the mole fraction of probe molecules at the interface.

Assuming a monodisperse distribution of spherical water droplets, the average radius of the inner water cores is given by

$$r_M = \frac{3V}{S_T} \tag{20}$$

where V is the total volume of water in the microemulsion and S_T is the total area of the interface. S_T is the sum of the areas occupied by the fluoro probe molecules and the surfactant. Typical values of the polar head cross section and the molecular length are reported in Table 5.

If we assume that all the surfactant molecules are located at the interface, we can compute the total surface area of the interface as

$$S_T = n_{\text{cosurfactant}}^{\text{interface}} \sigma_{CS} + n_{\text{surfactant}} \sigma_S \tag{21}$$

where σ_{CS} and σ_S are the cross sections of the cosurfactant and surfactant molecules, respectively.

$$n_{\text{cosurfactant}}^{\text{interface}} = Y \frac{(W_{CS}/100) \times 6.02 \times 10^{23}}{M_{w,(CS)}} \tag{22}$$

$$n_{\text{surfactant}} = \frac{(W_S/100) \times 6.02 \times 10^{23}}{M_w, (W_S)} \tag{23}$$

with W_{CS} and W_S being the weight percent concentrations of the cosurfactant and surfactant, respectively.

Figures 12 and 13 show the average size of the inner water core as a function of the paramagnetic ion concentration and water content, respectively.

Table 5 Cross Section (σ_S) and Length of Surface-Active Molecules

Molecule	Cross section (nm^2)	Length (nm)
Triton X-100	0.65	3.4
$C_{12}E_5$	0.43	—
CTAB	0.34	2.5
AOT	0.41	1.1
Decanol	0.21	1.5
Hexanol	0.21	1.0

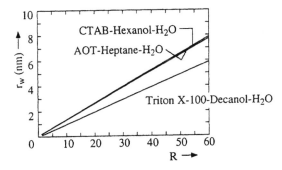

Figure 12 Variation in the radius of the inner water core as a function of $R = [H_2O]/[\text{surfactant}]$.

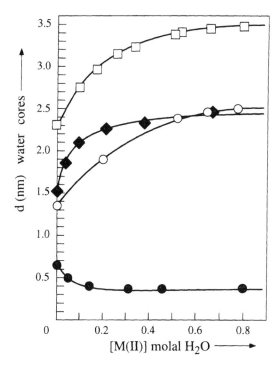

Figure 13 Variation in the diameter of the inner water core as a function of M(II) ion concentration. (●) Triton X-100–decanol–water/CoCl$_2$ ([Triton X-100] = 0.85 m vs. decanol); $R = 4$. (○) CTAB–hexanol–water/CoCl$_2$ ([CTAB] = 0.40 m vs. hexanol); $R = 10$. (□) CTAB–hexanol–water/NiCl$_2$ ([CTAB] = 1.65 m vs. hexanol); $R = 10$. (◆) CTAB–hexanol–water/CoCl$_2$ ([CTAB] = 2.60 m vs. hexanol); $R = 5$.

The average size of water cores increases with increasing M(II) ion concentration in the CTAB–hexanol–water microemulsion, whereas the opposite is true for the influence of Co(II) ions in the Triton X-100–decanol–water microemulsion.

The concentration is expressed with respect to water because the M(II) ions are essentially solubilized in the water cores. The increase in radius can be explained as being due to the increase in the interfacial free energy in the presence of Ni(II) ions in the water core. Indeed, the repulsion between the CTA$^+$ and Ni(II) ions is an unfavorable interaction, hence an increase in the free energy. Consequently, the total interfacial area has to decrease and the number of alcohol molecules at the interface also decreases. The water volume being constant, the average radius increases. This conclusion is also valid for Co(II) ions, because they also are essentially dissolved at the interface. In contrast, in the presence of Fe(III) ions the radius of the water core somewhat decreases [3,6].

The decrease of the radius of the inner water core in the Triton X-100–decanol–water microemulsion can be explained by the coordination of both Triton X-100 and decanol to the Co(II) ion. Similar decreases in droplet sizes in the presence of dissolved ions have also been reported [80,81]. For both systems the size variation in the presence of guest ions reaches a plateau at higher concentrations of the ions.

The most important effect on the average size of the water cores is due to the water content of the microemulsion (Fig. 12 and Table 4). The increase in the water radius with increasing water content (R value) is due to an increase in the total volume of water.

5. Kinetic Observations During the Formation of the Colloidal Particles

Two different methods are used for the preparation of colloidal nanoparticles. The first consists in mixing the two microemulsions containing the reactant precursors (Fig. 2, scheme II). The second, for practical reasons, introduces an aqueous solution with one reactant into the microemulsion containing the second reactant (Fig. 2, scheme I).

(a) Polarity and Dynamics of the Interface.   In the section on the preparation of colloidal metal boride particles, we shall see how the lability of the interface influences the number of nuclei formed, hence also the size of the metal boride particles. When the interface is more rigid, the number of particles is higher and their size is smaller. Under these conditions, the water cores are also better isolated and therefore limit the aggregation of metal boride particles, which should eventually yield smaller particles [3,6].

To investigate the dynamics of the interface, the usefulness of nitroxide spin probes [82] and the ^1H relaxation times T_1 is considered.

The hyperfine coupling constant between the electron spin ($S = 1/2$) and ^{14}N nuclear spin ($I = 1$) depends on the local polarity, whereas linewidths reflect the rate of molecular tumbling, which is influenced by the microviscosity of the medium. In general, the coupling constants decrease and the linewidths increase with decreasing water content, indicating clearly the diminution of local polarity [66,68] as well as the higher immobilization of the radical [83].

In the system dodecylammonium propionate–benzene–water, only the neutral spin probe is partitioned between benzene and water, whereas the anionic and cationic spin probes are essentially solubilized in water [84].

$I : CH_3-CH_2-C-(CH_2)_{14}-COOH$

$II : CH_3-(CH_2)_{12}-C-(CH_2)_3-COOH$

Spin probes I and II are studied in the system CTAB–hexanol–water [3,21]. Spin probe I, with its nitroxyl radical quite distant from the polar headgroup, probes the nonpolar region of the interface; whereas spin probe II, with the nitroxyl radical close to the polar headgroup, explores the polar region of the interface. From the heights (h) and the linewidths (ΔH) of the different multiplets (due to the electron–nitrogen hyperfine coupling), one can compute the rotational correlation times (τ_c) of these spin labels [84] from the relationship

$$\tau_c = A \, \Delta H_{(+1)} \left[\left(\frac{h_{(+1)}}{h_{(-1)}} \right)^{1/2} - 1 \right] \tag{24}$$

where $\Delta H_{(+1)}$ is the peak-to-peak width (in gauss) of the low-field resonance line and $h_{(+1)}$ and $h_{(-1)}$ are the peak-to-peak heights for the low and high field lines, respectively. A is a constant assumed equal to 6.6×10^{-10} s/G. Table 6 lists the τ_c values obtained for different water contents.

Table 6 Correlation Times τ_c (s) for Spin Probes I and II in CTAB–1-Hexanol–Water Microemulsions

	Composition (wt%)		$\dfrac{[H_2O]^a}{[CTAB]}$	$\dfrac{[CTAB]^b}{[HEX]_{int}}$	$\tau_{cI} \times 10^{10}$	$\tau_{cII} \times 10^{10}$
CTAB	1-Hexanol	Water			(s)	(s)
40	50	10	5.06	0.68	3.7	16.0
30	40	30	20.2	—	3.3	12.9
19	65	16	17.0	—	2.7	9.7
15	75	10	13.5	—	2.8	9.1
—	100	—	—	—	1.9	7.1
38	47	15	8.0	0.63	3.4	15.0
36	44	20	11.24	0.61	3.2	12.8
31.5	38.5	30	19.45	—	2.4	12.1
31.6	52.6	15.8	10.11	0.49	—	11.7
18.5	72.1	9.4	10.28	0.29	—	9.9
6.1	90.8	3.1	10.28	—	—	7.3

[a] $[H_2O]/[CTAB] = R$.
[b] $[CTAB]/[HEX]_{int}$ = ratio of CTAB to hexanol at the interface.

Table 7 ^{13}C NMR Study of Linewidths (ΔH in Hz) for CTAB (40 wt%–1-hexanol 50 wt%–Water 10 wt% Microemulsion with and Without Spin Probe II

	ΔH (Hz)		
Carbon	Without spin probe	With spin probe	$\Delta\Delta H$ (Hz)
C_0	80	133	53
C_1	64	106	42
C_m	80	160	80
H_1	64	64	0
H_2	53	53	0
H_3	53	59	6
H_4	53	59	6
H_5	53	59	6
H_6	59	59	0

C_0: N-CH$_3$ of CTAB; C_1: N-CH$_2$ of CTAB; Cm: -(CH$_2$)$_x$- of CTAB H$_{1,2,3,4,5,6}$: -(CH$_2$)$_x$- of Hexanol.

We have observed that the rotational correlation times (τ_c) depend strongly on the composition of the micellar solution and that $\tau_{cII} > \tau_{cI}$ for all the investigated microemulsion solutions. This shows clearly that the polar region of the interface is more rigid that the nonpolar part of the microemulsion. In addition, it is observed that the mobility of the interface (τ_c) is related linearly to the [CTAB]/[hexanol] ratio at the interface at constant [CTAB]/[water] ratio [3]. When the former ratio decreases, the rotational correlation time also decreases, indicating a "dilution" of the interface. This is turn leads to decreased interaction between the CTAB molecules and results in a more labile interface.

Indeed, these conclusions are valid only if the spin probes are well situated at the interface. The localization of the spin probe is possible if the influence of unpaired electrons on the ^{13}C NMR linewidth of the CTAB and hexanol molecules is studied. Table 7 shows that

Table 8 Electron–Nitrogen Hyperfine Coupling Constants for Spin Probes I and II Dissolved in CTAB–Hexanol–Water Microemulsion

| Composition (wt%) | | | | |
CTAB	Hexanol	Water	A_{N_I} (G)	$A_{N_{II}}$ (G)
40	50	10	14.47	14.85
38	47	15	14.56	14.86
36	44	20	14.60	14.14
31.5	38.5	30	14.58	14.59
—	100	—	14.58	14.50
—	—	100	15.93[a]	15.93[a]

[a] From Ref. 71.

only the CTAB carbon atoms are significantly influenced by the presence of spin probe II. As the total amount of CTAB is essentially at the interface, the spin probe molecules are also situated at the interface. [Of course, the influence on hexanol molecules is expected to be smaller because of their higher concentration. Nevertheless, the influence of Co(II) ions is very well detected, while the spin probe molecules have only quite a small effect.]

Finally, the nitrogen hyperfine coupling (A, in gauss) is highly dependent on the medium polarity. The A values can therefore be used to characterize the local polarity in the microemulsions [66,68]. The A values of spin probes I and II (Table 8) show clearly that they are close to the hexanol value and are quite different from that obtained in water. The interface can therefore be assumed to be nonpolar, and no water penetration can be detected in our system. Note that even the nitroxyl group of spin probe II, although close to the polar headgroup, is not in direct interaction with the more polar interface.

The ^1H T_1 relaxation times have been determined for the system Triton X-100–decanol–water. The T_1 values of the Triton X-100 molecules decrease with increasing water concentration [20], and the linewidths simultaneously increase. Both parameters show an increase in the correlation time of the surfactant molecules, showing continuous incorporation of Triton X-100 into the interface. On the other hand, the T_1 values that characterize the decanol molecules increase with increasing water content. This confirms the observation that the mobility of the interface decreases when the amount of water increases in the microemulsions.

(b)   Re-Equilibration of the System.   The first method of colloidal particle preparation (Fig. 2, scheme I) depends on the water dissolution rate, and both methods depends on the exchange rate between the water pools.

(c)   Water Dissolution Rate in the System.   It is almost obvious that the sudden addition of a reducing agent, NaBH$_4$ dissolved in one/fourth of the microemulsion's total water content, perturbs the microemulsion and leads to the dispersion of the reducing agent solution in a microemulsion of given composition.

Simple observation reveals that adding water to a microemulsion leads to the sudden appearance of turbidity. This turbidity is due to the momentary presence of another phase in the normally homogeneous microemulsion solution. Under the influence of stirring, this turbidity disappears rather quickly, depending mainly on the composition of the microemulsion sample. This could be observed by the naked eye during particle formation, but because the sample turns black during the particle formation it sometimes is not. The compositions of the microemulsion samples for particle preparation were chosen such that at the end they remained microemulsions within the single-phase domain.

To follow the water uptake separately from the particle formation, we measured the sudden increase in turbidity with the help of an experimental setup comparable to the one used in fast kinetics: the stopped-flow experiment. Changes were made by replacing the mixing chambers by a reaction vessel identical to the one used in the preparation of the particles. Instead of obtaining fast cross-flow mixing of the reactants, as in a classical stopped-flow experiment, a normal syringe as used in the particle preparation was used for the introduction of the reducing agent. To ensure mixing, a small mechanical stirrer of known frequency (10 Hz) was inserted in the vessel outside the optical axis of the stopped-flow apparatus.

We made an approximation, however, by admitting the hypothesis that the dissolution of pure water is similar to that of the reducing agent solution. Indeed, pure water was used in the experiments instead of a reducing agent solution to avoid the inconvenience caused by hydrogen evolution during the water uptake process in the microemulsion.

Theoretically the turbidity (τ) resulting from the injection of water into a microemulsion can be defined by the relation

$$I = I_0 e^{-\tau b} \tag{25}$$

where I_0 is the intensity of the incident light, I the intensity of the transmitted light, and b the optical path length of the cell.

When the droplets or the particles have a radius r smaller than $\lambda/20$, then the turbidity τ can be considered proportional to the concentration of the droplets (N_P) and the square of their volume (V_P^2). The concentration of the droplets is expressed as the number of droplets per milliliter of microemulsion. This proportionality is given by the Rayleigh equation

$$\tau = A N_P V_P^2 \tag{26}$$

with A an optical constant defined as

$$A = \frac{24\pi^3 n_0^4}{\lambda^4} \left(\frac{n^2 - n_0^2}{n^2 + 2n_0^2} \right)^2 \tag{27}$$

where n_0 is the refractive index of the solvent, n the refractive index of the droplets or particles, and λ the wavelength of the light at which the turbidity is measured.

Three types of experiments were carried out as part of this brief analysis:

First, during the water uptake the decrease in turbidity of the system immediately after the mixing was observed as a function of time for samples at different surfactant concentrations.

Second, the time process was also monitored separately for a microemulsion with a constant surfactant concentration but at different R values, adding a supplementary 25% water for the sample.

Third, the influence of successive additions of water to a microemulsion, thereby changing the total constant 25% water added, was examined.

Let us first consider the results obtained for the first type of experiment. Traces of the time process involving the decrease in turbidity from representative solutions of the Triton X-100–decanol–water microemulsion were recorded. The water content of each of them was increased by 33.33%. In these experiments it is assumed that the turbidity change measures the re-equilibration rate of the microemulsion. This overall rate of re-equilibrating

Table 9 Re-Equilibration of Microemulsion Samples During Water Addition

[Surf]	$R_1 - R_2$	l (s^{-1})	m (s^{-1})	A (a.u)	B (a.u)
a. Re-equilibration as a function of the surfactant concentration					
Triton X-100–decanol–water					
0.435	3.91–5.22	0.37	0.05	1.33	0.59
2.109	3.97–5.30	1.80	0.10	1.52	0.16
CTAB–hexanol–water					
0.686	5.00–6.67	3.21	0.31	1.02	0.47
1.829	5.00–6.67	2.71	0.42	0.70	0.33
b. Re-equilibration as a function of the water content					
Triton X-100–decanol–water					
3.608	2.75–3.67	3.55	0.18	0.82	0.17
3.611	9.13–12.16	1.30	0.19	1.47	0.94
c. Re-equilibration after successive additions until phase separation					
Triton X-100–decanol–water					
0.389	0–1.3	0.69	–	1.28	—
	5.6–6.9	0.43	0.04	1.05	0.73
	9.9–11.3	0.11	0.02	0.56	1.43
3.605	0–1.6	6.93	0.23	1.33	0.39
	6.5–7.3	2.53	0.20	1.16	0.47
	11.1–12.3	1.57	0.18	0.71	0.63

the microemulsion system did not seem to be monitored in a single step. Traces showed clearly the superposition of two different time processes. Hence they were deconvoluted, which revealed that the decrease in turbidity as a function of time, $\tau(t)$, could be recomposed by the sum of two decreasing exponentials:

$$\tau(t) = Ae^{-lt} + Be^{-mt} \tag{28}$$

with A, B ($\pm 10\%$) constants and l, m ($\pm 10\%$) the apparent rate constants of the water uptake. The values of these apparent rate constants and the compositions of the solutions analyzed are given in Table 9. Water uptake was measured at constant R value in increasingly concentrated samples (Table 9a) as well as for some microemulsion samples with increasing water content (Table 9b).

Finally, to complete the behavior of a microemulsion under the addition of water, the turbidity of successive additions of a constant amount of water (0.100 mL shots) was measured as a function of time. These experiments were carried out until turbidity persisted (Table 9c).

The following conclusions could be drawn from these results:

It is evident that during the dissolution of water in a microemulsion both the concentration (N_P) of the droplets and their volume (V_P) change as a function of time. Nevertheless, it is the decrease in volume of the droplets that predominates and leads to the decrease in turbidity.

The fast component (l) of the decrease in turbidity increases with the surfactant concentration but decreases with the water content. Successive addition of water reveals similar behavior as the R value increases. The slow component changes little, although a slight increase is observed in more concentrated microemulsion samples.

From these conclusions we can propose a mechanism for the water dissolution in a microemulsion. Two steps are revealed by the kinetics. The first step, the dispersion of the aqueous phase in emulsion form, is to be associated with the fast apparent rate constant. The second step, the progressive dispersion of water, is in competition with the original microemulsion droplet collisions, leading to a new equilibrium state for the microemulsion with larger sizes of its nanodroplets. This interaction can be characterized by the slow rate constant.

(d) *Exchange Rate Measurements.* After the water uptake rate, the pool communication rate within a microemulsion is the second basic parameter for a better understanding of particle formation. Various techniques such as fluorescence probing with Tb^{3+} [85] and fast indicator reactions [86,87] with continuous flow and integrated observation (CFMOI) [88–90] and stopped-flow techniques have been used to measure this value. We based ours on the analysis of fast reaction kinetics with the stopped-flow method.

The method used involves the fast complexation reaction between transition metal ions Me^{2+} [Ni(II), Co(II), Zn(II)] and murexide (ammonium purpurate or Mu^-). This reaction in an aqueous medium can be schematically represented by

$$Me^{2+} + Mu^- \rightleftharpoons MeMu^+ \qquad (29)$$

The reactants are sufficiently hydrophilic that no significant partitioning occurs between the different phases of the system. Hence the reaction can be considered to take place via the water droplets of the microemulsion.

Two mechanisms are proposed to explain the reaction that occurs in a microemulsion medium, considering that the reactants are each dissolved individually in a single water core of the microemulsion. In the first, the encounter with fusion of two reverse micellar aggregates permits the complexation reaction. The reaction can be represented in a two-step mechanism scheme considering each reactant to be dissolved in a reverse micelle:

$$Me^{++} \qquad Mu^- \qquad\qquad Me^{++}\ Mu^- \qquad MeMu^+ \qquad EM$$

with Me^{2+} and Mu^-, the two reactants, forming the encounter complex (EC) in the fused water cores of the micelles and EM being the empty micelles formed after reaction. The k_1 rate constant characterizes the faster step of the mechanism leading to the product formation.

In the second mechanism, the encounter permits the transfer of one of the reactants to the other water droplet, in which the reaction then occurs. This is schematically represented as

The efficiency of this transfer mechanism without fusion is linked to the lifetime of the encounter. Note that these processes must take place before the complexation reaction can occur. Also, it must be considered that not every encounter successfully leads to a fusion or to an exchange of droplet contents. However, there is a certain efficiency leading to the maximum rate possible for this hydrophilic reaction carried out in an oil-in-water microemulsion medium.

The recommendations to measure the intermicellar dynamics are (1) to use a fast reaction with reactant concentration such that only single-ion occupation appears and (2) to choose the droplet concentration of the microemulsion low enough to be in a diffusion-controlled state. Therefore, experiments were carried out in low surfactant concentration samples at low R values with single-ion occupation for the droplets. The reactants were solubilized in freshly prepared microemulsions of the same composition. For all the rate measurements the pseudo-first-order conditions were applied so that the metal ion concentration exceeded the murexide concentration by a factor of 10.

Equal-volume aliquots were mixed together in a small-volume stopped-flow instrument at the University of East Anglia at Norwich (UK). As the maximum absorption of the murexide is at 520 nm and that of the complex is at 450 nm, their visible light change is monitored by trace matching.

The observed rate constant is calculated from the half-reaction time on the digitized traces:

$$k_{obs} = \frac{\ln 2}{t_{1/2}} \tag{30}$$

which can further be expressed in a rate equation as

$$v = k_{obs}[Mu^-] \quad \text{with } k_{obs} = k_{fus}[Me^{2+}] \tag{31}$$

Measuring the rate at different metal ion concentrations enables us to determine the apparent reaction constant of the reaction.

Rates were measured in the presence of cobalt and nickel ions in Triton X-100–decanol–water microemulsion samples (Table 10).

Rate results for the two types of ions, cobalt(II) or nickel(II), were normally expected to be similar in a microemulsion solution despite the difference in their reaction rates in aqueous media, because of the reaction limitation by the droplet exchange. This was not observed for the Triton X-100–decanol–water system, suggesting that the droplet collision rate was not limiting the complexation rate and was consequently fast compared to the complexation reaction.

Table 10 Observed Rates k_{obs} (in s^{-1}) and Calculated Rates k_{fus} [in dm^3/(mol · s)] for the Complexation of Metal Ions [Co(II), Ni(II)] with Murexide in Triton X-100–Decanol–water[a]

Ion concn	R	k_{obs} (s^{-1})	k_{fus} [dm^3/(mol · s)]
$[Ni^{2+}] = 1.5 \times 10^{-2}$	2.5	0.52	31.7
	7.5	0.35	20.0
$[Co^{2+}] = 1.5 \times 10^{-2}$	2.5	0.90	56.9
	7.5	0.54	32.7

[a] [Triton X-100] = 0.312 m vs. decanol. [Mu$^-$] = 1.5×10^{-3} M vs. water.

Table 11   Kinetic Data Obtained int the CTAB 18%–Hexanol 70%–Water 12% Microemulsion at 25°C, pH 6.5, and $[Mu^-] = 2.5 \times 10^{-5}\ m$

Reactant	$[M^{2+}]$, $\times 10^3$	k_{obs} (s^{-1}) ± 0.05	k_{fus} [dm^3/(mol · s)] ± 0.2, $\times 10^{-2}$
Ni^{2+}	0.5	0.77	9.5
	2.0	2.20	
Co^{2+}	0.5	0.73	9.3
	2.0	2.15	
Zn^{2+}	0.5	0.73	9.3
	2.0	2.13	

Indeed, the complexation rate of the cobalt ion in water is some 100 times faster than for nickel. However, a slowing down of the rates for both ions is observed in the presence of microemulsions. Further experimental data would certainly be required to analyze the effect of the microemulsion on the murexide complexation reaction. However, we can conclude that the droplet collision is faster than expected.

The results obtained in the CTAB–hexanol–water microemulsion are reported in Table 11. The second-order rate constants are not dependent on either the nature or the concentration of the metal ions. It can thus be accepted that these values, $9.4\ (\pm 0.2) \times 10^2$ dm^3/(mol · s), represent the rate of rearrangement of the microemulsion droplets. These values are not very dependent on the R values. Only some 10% increase is observed from $R = 6$ to $R = 24$.

The rearrangement rate is quite slow with respect to other microemulsions. For example, for the AOT–heptane–water systems, the rate of fusion is determined to be equal to 10^6–10^7 dm^3/(mol·s). There is a difference of three to four orders of magnitude between the rates of fusion of the two microemulsion systems. The differences are even more astonishing if we recall the near-percolating state suggested by the conductivity measurements for the CTAB–hexanol–water microemulsion.

A more detailed study is necessary to assess the influence of the different components on the complexation rates of the metal ions by the Mu$^-$ ion.

B.   Preparation of Monodisperse Colloidal Particles

The different monodisperse nanoparticles shown in Table 1 were prepared following either scheme I or scheme II of Fig. 2. We first discuss the mechanism of formation following scheme I, where small amounts of aqueous solutions are added to the initial microemulsion.

From the known size of the water core, it is possible to calculate the number of aggregates in 1 kg of microemulsion, N_M, and the average composition of one aggregate. These results are reported in Tables 12a and 12b. The number of micellar water cores first decreases with increasing Ni(II) ion concentration, and it remains quasi-constant at higher Ni(II) concentration. It also decreases with increasing water content in the different reverse microemulsions (Table 12a). On the other hand, the average number of Ni(II) ions per micellar aggregate increases with increasing Ni(II) concentration and with increasing proportion of water in the solution (Table 12a). These results are used in the next paragraph to explain the size of nickel boride particles.

Table 12a Composition and Size of Microemulsion Aggregates in the CTAB–Hexanol–Water Microemulsion Containing $NiCl_2$

Ni(II)[a] (m vs. total)	Ni(II)[b] (m vs. water)	r_M (nm)[c]	$N_M^d \times 10^{-20}$ (per kg of soln)	n_{CTAB}^e	n_{Hex}^e	$n_{H_2O}^e$	$n_{Ni(II)}^e$	$\dfrac{n_{Ni(II)}^e \times 10^{-20}}{V}$
\multicolumn{9}{c}{1. CTAB 12.3%–hexanol 81.5%–H_2O 6.2%}
0.000	0.000	0.69	450	4	21	46	0.0	0.0
0.032	0.525	1.2	86	23	48	239	2.3	3.2
0.048	0.790	1.25	76	27	50	274	3.9	4.8
0.080	1.334	1.22	82	25	48	255	6.1	8.0
\multicolumn{9}{c}{2. CTAB 18.5%–hexanol 72.1%–H_2O 9.4%}
0.000	0.000	0.85	365	18	30	87	0.0	0.0
0.010	0.110	1.02	211	15	38	147	0.3	0.7
0.040	0.446	1.34	93	32	55	338	2.7	2.7
0.080	0.887	1.47	71	42	60	441	7.0	5.3
\multicolumn{9}{c}{3. CTAB 25%–hexanol 62.5%–H_2O 12.5%}
0.000	0.000	1.02	281	14	39	148	0.0	0.0
0.010	0.082	1.17	186	22	46	224	0.3	0.5
0.040	0.339	1.44	100	41	58	421	2.6	2.0
0.120	0.999	1.59	74	56	61	566	10.2	6.0
\multicolumn{9}{c}{4. CTAB 31.6%–hexanol 52.6%–H_2O 15.8%}
0.000	0.000	1.18	230	23	46	229	0.0	0.0
0.010	0.091	1.39	140	37	57	379	0.6	0.6
0.040	0.257	1.58	96	54	61	550	2.6	1.5
0.120	0.802	1.72	74	71	63	719	10.4	4.8
0.160	1.075	1.76	69	76	62	760	14.7	6.5

[a] Concentration with respect to the total weight of microemulsion.
[b] Concentration with respect to the weight of water.
[c] Radius of inner water core determined from ^{19}F NMR.
[d] Number of water droplets per kg solution: $N_M = V_{H_2O}/4/3\pi r_M^3$.
[e] n_{CTAB}, n_{Hex}, n_{H_2O}, $n_{Ni(II)}$, and $n_{Ni(II)}/V$ stand for the number of either CTAB, hexanol, water, Ni(II) ions per aggregate and Ni(II) ions in 1 cm^3 of water, respectively.

1. Size of Metal Boride Particles

Monodisperse colloidal nickel boride and cobalt boride particles are synthesized by reducing, with $NaBH_4$, the metallic ions solubilized in the water cores of the microemulsions. The $NaBH_4/MCl_2$ ratio was held equal to 3, because larger particles were obtained for a lower value, the particle size remaining constant above that ratio [2–4].

The composition of the particles was determined by XPS to be respectively Ni_2B and Co_2B. In every case, the size of particles (2.5–7.0 nm) is much smaller than that obtained by reduction of Ni(II) or Co(II) in water (300–400 nm) or in ethanol (250–300 nm), and the size distribution is quite narrow (± 0.5 nm).

Figure 14 shows the dependence of the nickel boride particle size on the water content in the microemulsion and on the Ni(II) ion concentration. The average size of the particles decreases with decreasing size of the inner water core (decreasing water content), while

Table 12b Composition and Size of Microemulsion Aggregates in the CTAB–Hexanol–Water Microemulsion Containing $CoCl_2$

$Co(II)^a$ (m vs. total)	$Co(II)^b$ (m vs. water)	r_M (nm)^c	$N_M^d \times 10^{-20}$ (per kg of soln)	n_{CTAB}^e	n_{Hex}^e	$n_{H_2O}^e$	$n_{Co(II)}^e$	$\dfrac{n_{Co(II)}^e \times 10^{-20}}{V}$
			1. CTAB 12.3%–hexanol 81.5%–H_2O 6.2%					
0.00	0.00	0.71	667	10	14.0	50	0.0	0.0
0.005	0.05	0.90	327	20	15.5	102	0.09	0.3
0.050	0.50	1.0	239	27	14.7	140	1.26	3.0
0.150	1.50	1.04	212	31	14.3	158	4.36	9.2
			2. CTAB 39%–hexanol 49%–H_2O 12%					
0.000	0.00	0.87	434	15	21.0	92	0.0	0.0
0.005	0.04	1.07	219	29	25.7	183	0.14	0.3
0.050	0.42	1.24	150	42	24.0	267	2.0	2.4
0.150	1.25	1.27	140	46	21.9	287	6.4	7.5
			3. CTAB 38%–hexanol 47%–H_2O 15%					
0.000	0.00	1.07	292	21	34	172	0.0	0.0
0.005	0.03	1.34	149	42	39	337	0.20	0.2
0.050	0.33	1.56	94	69	38.4	534	3.2	2.0
0.150	1.00	1.64	81	78	35.5	619	11.2	6.0
			4. CTAB 37%–hexanol 45%–H_2O 18%					
0.000	0.00	1.33	182	34	51	330	0.0	0.0
0.005	0.028	1.58	109	56	58	552	0.28	0.2
0.050	0.28	1.92	61	100	59	991	5.0	1.7
0.150	0.83	2.0	54	114	56	1120	16.8	5.0
			5. CTAB 36%–hexanol 44%–H_2O 20%					
0.000	0.00	1.49	144	41	66	464	0.0	0.0
0.005	0.025	1.85	75	79	77.4	886	0.40	0.2
0.050	0.25	2.16	47	125	76.8	1410	6.4	1.5
0.150	0.75	2.28	40	148	71.6	1660	22.4	4.5

For Footnotes a–e see Table 12a.

complex behavior is observed as a function of the Ni(II) ion concentration: A minimum is detected at approximately 5×10^{-2} m concentration. These observations can be understood if one analyzes the nucleation and the growth processes of the particles.

To form a stable nucleus a minimum number of atoms are required [91]. Thus, for nucleation several atoms must collide at the same time, and the probability of this phenomenon is much lower than the probability of collision between one atom and an already formed nucleus. The latter phenomenon is called the growth process. At the very beginning of the reduction, nucleation occurs only in those water cores that contain enough ions to form a nucleus. At this moment, the micellar aggregates act as "reaction cages" where the nuclei are formed. On the other hand, the microemulsion being dynamic, the water cores rapidly rearrange. The other ions brought into contact with the existing nuclei essentially participate in their growth process. The latter being faster than nucleation,

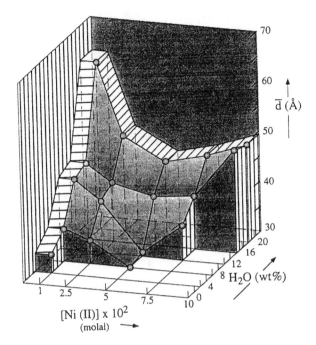

Figure 14 Variation of the average diameter (in Å) of the nickel boride particles as a function of water content and Ni(II) ion molal concentration.

no new nucleus is synthesized at this moment. As all the nuclei are formed at the same time and grow at the same rate, monodisperse particles are obtained. In summary, the particle size depends on the number of nuclei formed at the very beginning of the reduction, and this number is a function of the number of water cores, containing enough ions to form a stable nucleus, reached by the reducing agent before the rearrangement of the system. However, the stabilization of the nuclei by surfactants is probably one of the most important factors in explaining the monodispersity of the particles.

2. Quantitative Aspects of the Formation of Monodisperse Colloidal Particles

The first step in the determination of the essential parameters that control particle size is a study of the distribution of the ions in the microemulsion water cores.

Knowing the average radii of the microemulsion water cores (r_M) and the total volume of water (V_T) per kilogram of microemulsion, one can calculate the number of water cores per kilogram of reverse micelles (N_M), neglecting the solubility of water in the hexanol organic phase (Table 12).

$$N_M = \frac{V_T}{(4/3)\pi r_M^3} \tag{32}$$

The parameter N_M and the initial concentration of metal ions expressed in molality (see Table 12) allow us to determine the average number of ions per water core (n_{ions}):

$$n_{ions} = \frac{[ions] \times 6.023 \times 10^{23}}{N_M} \tag{33}$$

The ions are statistically distributed in the aggregates. To calculate this distribution, Poisson statistics are perfectly adequate [92]. This gives the probability (p_k) of having k ions per water core (k is an integer taking the values 0, 1, 2, 3, . . .), provided the average number of ions per water core ($\lambda = n_{ions}$) is known:

$$p_k = \frac{\lambda^k e^{-\lambda}}{k!} \tag{34}$$

The number of nuclei formed (N_n) when the ions solubilized in 1 kg of solution are reduced is proportional to the number of aggregates containing enough ions for nucleation. If the minimum number of ions required to obtain a nucleus is i, then N_n can be calculated from the relation

$$N_n = F N_M \sum_{k=i}^{\infty} p_k \tag{35}$$

where $\sum_{k=i}^{\infty} p_k$ is the probability of having i or more ions per aggregate; hence $N_M \sum_{k=i}^{\infty} p_k$ is the number of water cores containing i or more ions. F is a proportionality factor taking into account the proportion of aggregates reached by the reducing agent before rearrangement of the system can occur.

In Eq. (35) we do not know the value of i and F but we can calculate all the other parameters. Indeed, the number of nuclei (N_n) is the number of particles prepared, and the latter is given by

$$N_n = W_t / W \tag{36}$$

where W_t is the total weight of the particles prepared per kilogram of micellar solution. W is the weight of one particle, and

$$W_t = \frac{[ions] \times M_{particle}}{x} \tag{37}$$

where $M_{particle}$ is the molecular weight of the particles and x is the number of metal atoms per molecule of particle.

$$W = \frac{4}{3}\pi \left(\frac{d}{2}\right)^3 M_{v,particle} \tag{38}$$

where d is the diameter of the particles measured by electron microscopy and $M_{v,particle}$ is the volumetric mass of the particle.

All the experimental and computed data are reported in Tables 13a and 13b for Ni_2B and Co_2B particles, respectively.

The diameter of the particles is systematically higher than the diameter of the inner water cores. For all the particles synthesized, we calculated the proportionality factor F by systematically varying the value of the minimum number of ions required to form a nucleus (i). Only if i takes the value of 2 is the factor F reasonably constant (see Tables 13a and b). The order of magnitude of the factor F is always 10^{-3}. This means that at the very beginning of the reduction, i.e., when the nuclei are formed, only one aggregate per thousand leads to the formation of metal boride particles.

There is another indication that the nucleation occurs at the very beginning of the reduction. Indeed, the average radii of the water cores used for the calculation of the formation parameters of colloidal particles are measured for the system containing only

Table 13a Important Parameters for the Formation of Ni_2B Colloidal Particles

$[Ni(II)]$ $\times 10^{-2}$ (m)	$r_M{}^a$ (nm)	$N_M{}^{a,b}$ $\times 10^{-22}$	$n_{Ni(II)}{}^a$	d (nm)	$W_t{}^c$ (g)	w^d $\times 10^{19}$ (g)	$N_n{}^b$ $\times 10^{-18}$	$N_n/N_M{}^b$ $\times 10^{-18}$	$\sum_{k=2}^{\infty} p_k$	F^e $\times 10^3$
			CTAB 18%–hexanol 70%–water 12%							
1.00	1.02	2.11	0.29	4.4	0.64	3.52	1.82	8.63×10^{-5}	0.0347	2.5
2.50	1.19	1.33	1.13	3.6	1.60	1.93	8.29	6.23×10^{-4}	0.3119	2.0
7.50	1.46	0.72	6.24	3.7	4.81	2.09	23.01	3.19×10^{-3}	0.9859	3.2
			CTAB 24%–hexanol 60%–water 16%							
1.00	1.17	1.86	0.32	4.5	0.64	3.77	1.70	9.14×10^{-5}	0.0415	2.2
2.50	1.32	1.29	1.17	4.2	1.60	3.06	5.23	4.05×10^{-4}	0.3265	1.2
7.50	1.54	0.81	5.58	4.0	4.81	2.65	18.15	2.24×10^{-3}	0.9752	2.3
10.00	1.57	0.77	7.82	5.1	6.41	4.87	11.67	1.52×10^{-3}	0.9964	1.8
			CTAB 30%–hexanol 50%–water 20%							
1.00	1.34	1.54	0.39	6.7	0.64	12.44	0.51	3.31×10^{-5}	0.0589	0.6
2.50	1.48	1.16	1.30	4.9	1.60	4.87	3.28	2.83×10^{-4}	0.3732	0.8
7.50	1.68	0.79	5.72	4.6	4.81	4.03	11.93	1.51×10^{-3}	0.9780	1.5
10.00	1.72	0.74	8.14	4.9	6.41	4.87	13.16	1.78×10^{-3}	0.9973	1.8

Table 13b Important Parameters for the Formation of Co_2B Colloidal Particles

$[Co(II)]$ $\times 10^{-2}$ (m)	$r_M{}^a$ (nm)	$N_M{}^{a,b}$ $\times 10^{-22}$	$n_{Ni(II)}{}^a$	d (nm)	$W_t{}^c$ (g)	w^d $\times 10^{19}$ (g)	$N_n{}^b$ $\times 10^{-18}$	$N_n/N_M{}^b$ $\times 10^{-18}$	$\sum_{k=2}^{\infty} p_k$	F^e $\times 10^3$
			CTAB 39%–hexanol 49%–water 12%							
2.50	0.89	3.14	0.48	3.1	1.61	1.26	12.78	4.07×10^{-4}	0.0842	4.8
10.00	0.93	2.75	2.19	2.8	6.43	0.93	69.14	2.51×10^{-3}	0.6430	3.9
15.00	0.93	2.75	3.28	3.0	9.65	1.14	84.65	3.08×10^{-3}	0.8389	3.7
			CTAB 38%–hexanol 47%–water 15%							
0.50	1.04	2.48	0.12	5.9	0.32	8.71	0.37	1.49×10^{-5}	0.0066	2.3
2.50	1.15	1.83	0.82	4.9	1.61	4.99	3.23	1.76×10^{-4}	0.1984	0.9
10.00	1.23	1.50	4.01	3.4	6.43	1.67	38.50	2.57×10^{-3}	0.9091	2.8
15.00	1.24	1.46	6.19	3.8	9.65	2.33	41.42	2.84×10^{-3}	0.9853	2.9
			CTAB 37%–hexanol 45%–water 18%							
0.50	1.21	1.90	0.16	6.7	0.32	12.76	0.25	1.32×10^{-5}	0.0115	1.2
2.50	1.42	1.18	1.28	5.2	1.61	5.96	2.70	2.29×10^{-4}	0.3661	0.6
10.00	1.52	0.96	6.27	4.1	6.43	2.92	22.02	2.29×10^{-3}	0.9852	2.3
15.00	1.54	0.92	9.82	4.6	9.65	4.13	23.37	2.54×10^{-3}	0.9994	2.5
			CTAB 36%–hexanol 44%–water 20%							
0.50	1.37	1.47	0.20	7.1	0.32	15.18	0.21	1.43×10^{-5}	0.0175	0.8
2.50	1.59	0.94	1.60	5.9	1.61	8.71	1.85	1.97×10^{-4}	0.4751	0.4
5.00	1.66	0.82	3.67	4.8	3.22	4.69	6.87	8.38×10^{-4}	0.8810	1.0

^{a} Values given for the system containing three-fourths of the total amount of water.
^{b} Values given for 1 kg of solution.
^{c} W_t is calculated with $M(Ni_2B) = 128.23$ g/mol and $M(Co_2B) = 128.68$ g/mol.
^{d} w is calculated with $M_V(Ni_2B) = 7.9$ g/cm^3 and $M_V(Co_2B) = 8.1$ g/cm^3.
^{e} Correction factor from $N_n = FN_M \sum_{k=2}^{\infty} p_k$.

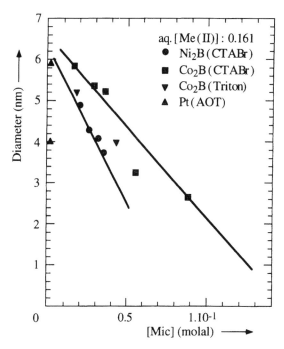

Figure 15 Sizes of nanoparticles prepared in various microemulsions as a function of micellar droplet concentrations.

three-fourths of the total amount of water, which is the composition of the solution before the addition of the reducing agent. If the final composition is used, however, no coherent results based on the above analysis can be obtained.

The order of magnitude of the factor F is constant, but its value decreases with the increase of water content in the microemulsion (see Tables 13a and b). This phenomenon can be easily understood because the rearrangement rate of the microemulsion decreases with the water amount and hence the number of aggregates reached by the reducing agent before rearrangement decreases. As the number of nuclei formed decreases, of a constant concentration of precursor ions, the particle size increases with the increase in the water content in the system.

The diameter of the particles is plotted as a function of micellar droplet concentration in Fig. 15. The values of the particle size are those obtained by interpolation of previous results in the presence of 0.161 m aqueous metal ion for the CTAB–hexanol–water systems. Particles prepared in the AOT–heptane–water system at much lower droplet concentration are included for comparison.

The size of the particles decreases linearly with the droplet concentration. This is a strong indication that the final size obtained for the particles is governed by the presence of reverse micellar aggregates. Indeed, if initial nucleation takes place in the water cores, then nucleation should be related to the droplet concentration of the system. Further, the greater the number of microemulsion droplets, the greater the number of nucleation sites possible (the aqueous metal ion concentration being obviously maintained constant).

The results of Tables 13a and b also allow us to explain the minimum in the particle size as a function of the concentration of ions (see Fig. 14).

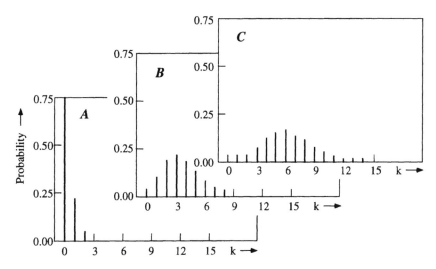

Figure 16 Variation in the probability of having k Ni(II) ions per aggregate for the microemulsion CTAB 18%–hexanol 70%–H_2O 12%. [Ni(II)] (m): A, 1×10^{-2}; B, 5×10^{-2}; C, 7.5×10^{-2}.

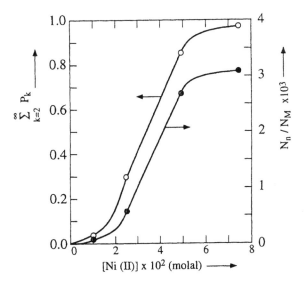

Figure 17 Variation in ($\bullet$) the number of nuclei formed per aggregate and ($\bigcirc$) the probability of having two or more ions per aggregate as a function of Ni(II) concentration in the microemulsion CTAB 18%–hexanol 70%–water 12%.

For a constant microemulsion composition, at low ion concentration, only a few water cores contain the minimum number of ions (two) required to form a nucleus; hence, few nuclei are formed at the very beginning of the reduction, and the metal boride particles are relatively large. When the ion concentration increases, the distribution of precursor ions in the microemulsion is very different (Fig. 16), and the number of nuclei obtained by reduction increases faster than the total number of ions (Fig. 17). This results in a decrease

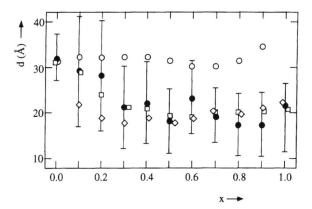

Figure 18 Variation of particle size as a function of the molar fraction in cobalt in the catalysts (x). (●) Experimental sizes. Hypothesis of a mechanical mixture of Ni_2B and Co_2B: (○) Sizes calculated for Ni_2B; (◇) sizes calculated for Co_2B; (□) weighted average sizes for $Ni_2B + Co_2B$.

in the particle size. When more than 80% of the water cores contain two or more ions, the number of nuclei formed remains quasi-constant with increasing ion concentration. Hence the size of the particles increases again.

Figure 14 also shows the particle size as a function of water content in the microemulsion for different Ni(II) concentrations. An increase in the average diameter is observed with increasing proportion of water. The decrease in the number of micellar aggregates (N_M) with water (Table 13a) is accompanied by an increase in their size. For the same Ni(II) concentration with respect to water (i.e., for the same probability of collision between the ions in the same water core), the total number of nuclei formed in the early stage of the reduction decreases with increasing water concentration, and more ions can participate in the growth process. This results in an increase in the particle size. One should also keep in mind that the total number of Ni(II) ions also increases with increasing water content. This is also shown if the size of the particles is plotted as a function of micellar droplet concentration (Fig. 15). For most of the systems studied a monotonous decrease in the size with increasing N_M is observed. These results reinforce the hypothesis leading to the computation of the number of nuclei and underline the importance of the water cores as reaction cages.

From the CTAB 18%–hexanol 70%–water 12% microemulsion and M(II), 5.00×10^{-2} m bimetallic particles of Ni-Co-B were also prepared.

The F values for nickel boride and cobalt boride particles are quite different. For the former the value obtained for F is equal to 3.2×10^{-3}, and for the latter, 17.4×10^{-3}. As for these experiments, the rearrangement rate of the microemulsion system is constant in the first approximation, the difference between the F values probably being due to the different solvation of the two types of ions at the interface. The Co(II) ions contain, on average, one hexanol molecule in their first coordination shell, while the Ni(II) ions are multiply coordinated with hexanol at the interface. The mobility of the latter is hence lower, and the probability of collision between the two reduced Ni atoms required to form a nucleus is also lower. In other words, the rate of nucleation is higher for cobalt boride than for nickel boride particles.

The average particle size and the width of the size distribution were measured by electron microscopy (Fig. 18). No consistent values are obtained for the factor F if the particles are considered to be homogeneous bimetallic particles. On the other hand, knowing the values of F for Ni_2B and Co_2B for this microemulsion, we calculated the expected sizes for the case where a mechanical mixture of separate particles of the monometallic borides is formed. These values are reported in Fig. 18 as well as the weighted average sizes for these two types of particles. Only the latter can be compared with the experimental results, which are average sizes. The average sizes so calculated are close to those measured experimentally. In most of the cases, the experimental size distributions are narrow, whereas a mechanical mixture of monometallic particles would result in a broad bimodal distribution. Hence, the particles are probably bimetallic but not completely homogeneous. The nucleation rate is higher for Co(II) ions than for Ni(II) ions (see above), the nuclei are formed preferentially from Co(II) ions, and the particles contain more nickel at the surface.

The essential parameters for the formation of the monodisperse colloidal particles are thus quantified. I have shown that two metal atoms are required to form a stable nucleus and that nucleation occurs only in those aggregates that are reached by the reducing agent before the rearrangement of the system can occur (one per 1000 aggregates). Figure 19 illustrates quite well the mechanism of reduction in a water-in-oil microemulsion.

After fast diffusion of the reducing agent, nucleation occurs in the water droplets where the above-mentioned conditions are satisfied. The nucleus is stabilized by the adsorbed surfactant molecules. The growth of the particles requires an exchange between different water cores. Finally, surfactant-protected monodisperse particles are formed that can be used directly or by being deposited on a support. This study also allowed us to gain some information about the composition of the bimetallic Ni-Co-B particles.

In the treatment shown above, the nucleation step could not yet be clearly described. The model is based on the presence of discrete water pools in the microemulsion, whereas recent conductivity measurements showed that percolation already occurs in these systems, favoring the exchange between water pool contents [20]. More experiments are needed to determine the formation of the first nuclei using fast kinetics measurements.

Nevertheless, the stabilization of the nuclei by the surfactant molecules at the interface could play a definite role in controlling their number formed at the very beginning of the reduction. The method of addition of the reducing agent in aqueous solution is indeed very important, because if higher amounts of microemulsion systems are used, larger particles are obtained.

Finally, I should mention the formation of monodisperse colloidal Co_2B particles in the microemulsion Triton X-100–decanol–water. The diameter of Co_2B particles is shown in Table 14 as a function of micellar composition.

Table 14 Size of Colloidal Co_2B Particles Obtained in Triton X-100–Decanol–Water Microemulsion

[CoCl$_2$] (M) vs. H$_2$O	[Triton X-100] (m)	[NaBH$_4$] (M) vs. H$_2$O	R [a]	r_w [b] (nm)	[Aggregate] [c] $\times 10^2$ (m)	N_M $\times 10^{22}$ [d]	d [e] (nm)
0.05	0.43	0.5	4.2	0.62	4.12	2.48	4.0
	3.59	0.5	13.1	1.42	2.67	1.61	4.9

[a] $R = [H_2O]/[Triton]$.
[b] r_w = radius of inner water cores.
[c] Concentration of micellar droplets.
[d] Number of micellar droplets per kilogram of solution.
[e] Diameter of Co_2B particles.

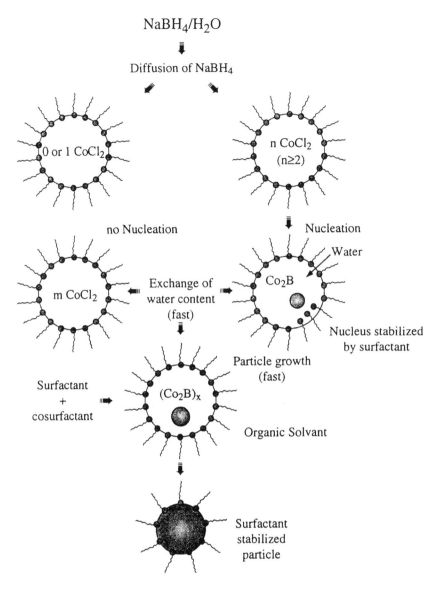

Figure 19   Model of the preparation of colloidal Co_2B particles from water-in-oil microemulsion.

The diameter of Co_2B increases with increasing water content. As the diffusion of the reactant through the heterogeneous system of the microemulsion is one of the important factors in controlling the size of the particles, fast kinetics measurements were carried out to follow the Co_2B formation in Triton X-100–decanol–water microemulsion. The turbidity of an aqueous, an alcoholic, and the microemulsion system was followed after having added to the initial solution containing the $CoCl_2$ precursors a small amount of $NaBH_4$ solution. All three traces show a small change in absorbance due to first increasing and then decreasing turbidity following the injection of the solution of the reducing agent (Fig. 20).

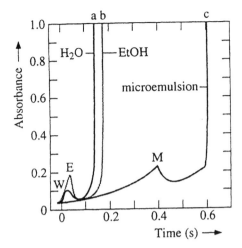

Figure 20 Three traces obtained during the formation of cobalt boride particles. (a) Absorption variation of an aqueous solution and (b) that of an ethanolic cobalt chloride solution (0.1 M) during the reduction of the cobalt ions. (c) Trace obtained for the same reduction but in a Triton X-100 microemulsion sample ([Triton X-100] = 1.5 M and $R = 13.3$). Note the small changes (W, E, and M) in absorbance due to addition of the reducing agent solution.

The small changes before the onset of particle formation are at 30, 35 and 400 ms for water, ethanol, and the microemulsion, respectively. Although the traces also represent a sudden increase in absorption due to particle formation, the trace in the microemulsion medium is different.

Particle preparation is achieved after 140, 178, and 610 ms for water, ethanol, and the Triton X-100–decanol–water system (Fig. 20). Indeed the fact that the particles appear later in the microemulsion solution presumably results from the nonimmediate dispersion of the reducing agent. This delayed particle appearance can be linked to the water uptake of the microemulsion (see above).

However, at this moment we can say that the difference between the homogeneous media (water and ethanol) and the microemulsion is that in the first case the aqueous reducing agent is dissolved straightforwardly because of the miscibility of both solutions, whereas in the microemulsion solution the aqueous phase of the reducing agent has to be taken up in the form of small droplets before efficient reaction can take place. It can be concluded that the dispersion of water in a microemulsion is a much slower phenomenon that the dissolution of water in normal homogeneous solutions.

Another point that is demonstrated by Fig. 20 is that the sudden increase in absorbance (the solution becomes black) is the fastest in aqueous solution, slowest in the microemulsion sample, and of intermediate rate in the ethanolic solution of cobalt chloride. This illustrates the two-step mechanism in particle formation, i.e., reactant dissolution or dispersion first and reduction reaction afterward.

Furthermore, for all traces at the onset of particle preparation, we observe a nonlinear increase in turbidity as particles of cobalt boride are formed. This increase can be fitted to an exponential law:

$$f(x) = ax^n + b \qquad (39)$$

From all traces characterizing the appearance of colloidal particles, we can say, with some caution, that the exponent n is close to 2 while b is equal to 0.37. Hence x^n is in agreement with Eq. (26) and can be considered as being the square of the volume of the particles, V_P^2. The exponent n for each of the traces calculated by curve fitting for the nonlinear section of the traces from the onset is 1.82 for Co_2B in Triton X-100–decanol–water microemulsion ($a = 10.2$), 2.02 in ethanol ($a = 22.7$), and 2.56 in water ($a = 165$).

The results indicate a difference between the traces obtained for generating colloidal boride particles in pure solvents (a single-component system) and those obtained in microemulsions.

In the single-component systems the reducing agent is dissolved from the place of injection throughout the whole solution (water or alcohol). The rate at which this occurs is similar to the diffusion rate of the reducing agent. Effective stirring is required to ensure complete immediate physical mixing of the two solutions containing the reactants, although the contribution of the physical mixing is small compared to the diffusion rate of the reactant in the homogeneous solution.

The situation is quite different when an aqueous solution is added to a microemulsion. In this case of mixing, the system evolves between nonmiscibility and miscibility, which is characterized by the water uptake rate.

Before going to further interpretation, I must state that all the reactants are predominantly water-soluble. This was demonstrated for the cobalt chloride in microemulsion. Following the solubility properties of the sodium borohydride there seems to be no substantial difference. Turbidity results indicate that the reducing agent seems first to be dispersed and become part of the macroemulsion before any particles appear.

The fact that no immediate particle formation is detected could lead to the erroneous conclusion that the reducing agent is not reacting at all before being entirely dispersed in the form of a microemulsion. There is no reason why no reduction should take place during the water uptake; however, we have to be cautious because we do not know how efficient the particle formation could be. This formation certainly depends on the dynamics of the reverse micelle. Further, the droplets of reducing agent dispersed in the microemulsion during the reduction are comparable to a separate phase in nonequilibrium with the original microemulsion. This means that on the microscopic scale, surfactant is transferred from the microemulsion to stabilize the interface of the reducing agent water droplets, and this through collisions with the original reverse micelles.

Therefore, there is good reason to think that the sodium borohydride and the cobalt chloride react really efficiently only after the sodium borohydride is encapsulated in the form of reverse micelles. The reduction could then be treated as an interface reaction, implicating droplet collisions. An argument for this could be that the reducing agent is fully dispersed, so the interface associated with it is much greater, making it much easier for the reducing agent to react with the cobalt chloride through this interface. Experimental evidence for this would require more results. Indeed, nuclei are small, and the onset of turbidity is difficult to detect on this time scale. Second, nuclei may have short lifetimes, and, third, their absorption properties can differ from those of the colloidal particles themselves.

Therefore the conclusions that we can draw at this stage from these experimental results on particle formation are very general and certainly need further investigation. To summarize, I can state that the mechanism of particle formation in microemulsions is certainly different from that in homogeneous media. (This difference is induced by the structures present in the microemulsion.) Taking into account the solubility properties of microemulsions, we can think of carrying out a reaction in little "individual" nanoscale reaction vessels. But it is precisely the nature of these reaction vessels that plays a major role in the size control of the colloidal particles obtained. This is emphasized in the following section.

3. Size of Platinum, Rhenium Dioxide, and Gold Particles

(a) CTAB–Hexanol–Water Microemulsion. Colloidal particles of Pt, ReO$_2$, and Au were prepared following scheme I of Fig. 2. The monodisperse Pt particles prepared from H$_2$PtCl$_6$ dissolved in the CTAB–hexanol–water microemulsion had an average diameter of 4.0 ± 0.5 nm, and their size was not dependent on the H$_2$PtCl$_6$ concentration (5×10^{-3}–2×10^{-2} m with respect to water) [54]. The aqueous solution of hydrazine containing a tenfold molar excess of hydrazine with respect to H$_2$PtCl$_6$ had an initial pH of 10. The metal particle precursor is soluble in both the dispersed inner water core and the continuous (or hexanol) phases. If it is assumed that the nucleation occurs in both phases, the particle size is dependent only on its stabilization by the adsorbed surfactant molecules [54].

It is interesting to note that particles of a similar size were obtained, independently of water and H$_2$PtCl$_6$ concentration, from the AOT–heptane–water microemulsion [25].

If K$_2$PtCl$_4$ is used instead as the particle precursor (for the same hydrazine to K$_2$PtCl$_4$ excess), complex behavior is observed as a function of pH. At low pH values ($1 <$ pH < 4), no Pt particles could be obtained. At $5 <$ pH < 8, dispersed Pt particles were formed, but the reduction was not complete even after 24 h of reaction. For high pH values (pH < 9), complete reduction of the Pt salt occurred, but the particles thus obtained were aggregated.

It is thus clear that the surface charge does influence the aggregation of the metal particles. In addition, the adsorption of the surfactant molecules, also pH-dependent, can also greatly influence the particle aggregation.

(b) PEGDE–Hexane–Water Microemulsion. Colloidal Pt particles were prepared following both schemes I and II of Fig. 2. To avoid particle aggregation, a neutral surfactant, PEGDE, was used to form a microemulsion of composition PEGDE 9.5%–hexane 90%–water 0.5%. Only K$_2$PtCl$_4$ was tested as precursor salt, however, because it is insoluble in the organic medium.

Table 15 and Fig. 21 show the variation in the size of the Pt particles obtained following scheme I as a function of initial K$_2$PtCl$_4$ concentration. The standard deviation was small in all cases studied.

Table 15 Analysis of Pt Particle Size as a Function of the Number of Nuclei N_n

[K$_2$PtCl$_4$] (m vs. H$_2$O)	$n_{PtCl_4^{2-}}$ [a,b]	d (nm)	$W_t \times 10^3$ (g) [c]	$w \times 10^{19}$ (g) [c,d]	$N_n \times 10^{-16}$
0.001	0.54	1.5 ± 0.3	0.98	0.38	2.6
0.01	5.4	2.5 ± 0.3	9.8	1.75	5.6
0.05	27.0	5.0 ± 0.3	49.0	14	3.5
0.1	54.0	9.0 ± 1.0	98.0	81.9	1.2
0.3	162.0	13.0 ± 1.5	294.0	247	1.2

[a] Number of PtCl$_4^{2-}$ ions per inner water core.
[b] N_M (number of inner water cores) $= 5.56 \times 10^{18}$ per kg solution; $r_M = 6.0$ nm.
[c] Values given for 1 kg of solution.
[d] M_V of Pt $= 21.45$ g/cm^3.

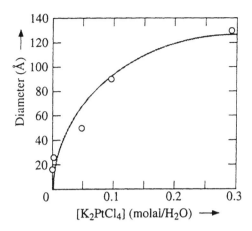

Figure 21 Variation of the Pt average diameter as a function of K_2PtCl_4 concentration with respect to water prepared according to scheme I of Fig. 2.

The particle diameter increases monotonously with increasing K_2PtCl_4 concentration and approaches a plateau at high concentration. This behavior seems to be different from those previously observed for the Pt particles using H_2PtCl_6 [25,74] and for the Ni_2B or Co_2B particles. In the first case, a constant particle size was obtained irrespective of the initial H_2PtCl_6 concentration, while in the second cases a minimum was observed in the particle size (Ni_2B or Co_2B) versus $NiCl_2$ or $CoCl_2$ concentration curve.

For low initial K_2PtCl_4 concentration (up to 0.01 m with respect to water), N_n increases as a function of Pt concentration. This behavior was found earlier for the case of Ni_2B and Co_2B particles. However, for higher K_2PtCl_4 concentrations, the N_n value decreases, leading to larger particles. The probable nucleus is a surfactant-stabilized Pt atom that is able to form the final Pt particle [23].

If the particles are prepared following scheme II, where the two microemulsions containing the precursor K_2PtCl_4 and the reducing agent N_2H_4, respectively, are mixed together, smaller sizes are obtained. Indeed, the Pt particles prepared from the microemulsion with [K_2PtCl_4] with respect to water have a diameter of 3.5 ± 0.5 nm, while the diameter is much greater (9.0 ± 1.0 nm) if scheme I is used (Table 15). Figure 21 illustrates the variation of the average diameter of the Pt particles as a function of the concentration of K_2PtCl_4 prepared by scheme I.

The larger size of the Pt particles obtained by the method of scheme I can be explained in a first approximation by the diffusion of the aqueous solution through the organic phase being slower than the exchange between the water cores. Although in the PEGDE–hexane–water microemulsion no separate spherical droplets are present, the water is probably the dispersed phase in the microemulsion. The structure of the microemulsion is better represented as a lamellar aggregate where the surfactant molecules are associated head-to-head along a cylinder.

(c)   Preparation of Monodisperse ReO_2 Particles.   Monodisperse ReO_2 particles were obtained by reducing $NaReO_4$ with hydrazine in the system PEGDE 9.5%–hexane 90%–water 0.5% following scheme I of Fig. 2. The presence of ReO_2 is confirmed by XPS experiments. However, the $NaReO_4$ is only partially reduced under these conditions.

Table 16 Variation in Monodisperse ReO_2 Particle Size as a Function of $NaReO_4$ Concentration

$[NaReO_4]$ $(m$ vs. $H_2O)]$	d (nm)
0.01	1.8 ± 0.2
0.05	2.7 ± 0.3
0.1	3.1 ± 0.4
0.3	4.2 ± 0.5
1.0	5.5 ± 0.6

Table 17 Variation in Monodisperse Pt-ReO_2 Particle Size as a Function of the Mole Fraction (x) of K_2PtCl_4[a,b]

Mole fraction x of K_2PtCl_4	d Pt $(nm)^c$	d ReO_2 $(nm)^c$	Pt-ReO_2
0	—	3.0	3.1 ± 0.3
0.16	3.0	≈ 2.9	2.5 ± 0.3
0.33	≈ 3.5	≈ 2.8	2.4 ± 0.3
0.5	5.0	≈ 2.7	2.7 ± 0.3
0.66	7.0	≈ 2.2	2.5 ± 0.2
0.8	≈ 8.0	≈ 2.0	3.8 ± 0.4
0.9	≈ 8.5	≈ 1.8	7.0 ± 0.5
1	9.0	—	9.0 ± 1.0

[a] PEGDE 9.5%–hexane 90%–H_2O 0.5%.
[b] $[K_2PtCl_4] + [NaReO_4] = 0.10$ m with respect to water.
[c] Hypothetical particle size estimated for the case where the systems would contain pure Pt or ReO_2 particles.

Table 16 illustrates the variation in particle size as a function of $NaReO_4$ concentration. Once again the size of monodisperse particles approaches a plateau for high $NaReO_4$ concentrations, and this behavior is quite similar to that of the Pt particles.

(d)   Preparation of Monodisperse Pt-ReO$_2$ Particles.   Monodisperse Pt-ReO_2 particles were prepared following scheme I from the PEGDE–hexane–water microemulsion using a total ion concentration $[K_2PtCl_4] + [NaReO_4] = 0.10$ m with respect to water. Table 17 and Fig. 22 show the variation in the particle size as a function of the mole fraction x of K_2PtCl_4.

It is surprising that up to $x = 0.7$, the diameter of the particles remains quasi-constant and is close to that of the pure ReO_2 particles. For higher initial $[K_2PtCl_4]$, the diameter of the particles increases monotonously to reach that of the pure Pt particles. The quasi-constancy of the particle diameter for low K_2PtCl_4 concentration suggests that in this region of concentration the Pt is dispersed on the ReO_2 particles. Indeed, the slight

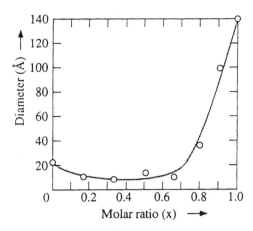

Figure 22 Variation in Pt-ReO$_2$ particle size as a function of ratio x of K$_2$PtCl$_4$ ([K$_2$PtCl$_4$] + [NaReO$_4$] = 0.10 m with respect to water).

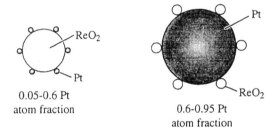

Figure 23 Models for Pt-ReO$_2$ particles.

decrease in the size could be due to the decrease in the size of the ReO$_2$ particles, as can be seen in Table 17. For high K$_2$PtCl$_4$ content, the reverse situation could occur, i.e., the dispersion of ReO$_2$ particles on the larger Pt particles (Fig. 23).

All these results are different from those one could expect on the basis of a mechanical mixture. Indeed, in that case a bimodal distribution is expected at least for $x \geq 0.5$, based on the different sizes of the separate Pt and ReO$_2$ particles. Table 17 also includes the particle sizes estimated for the hypothetical case of a system containing pure Pt and ReO$_2$ particles (see Tables 15 and 16).

(e) DOBANOL – Hexane – Water Microemulsion. Particles of Pt, Au, and Pt-Au have been prepared in a DOBANOL–hexane–water microemulsion following scheme II of Fig. 2. DOBANOL is a mixture of penta(ethylene glycol) undecyl (<1 wt%), dodecyl (41 wt%), tridecyl (58 wt%), and tetradecyl (<1 wt%) ethers. The microemulsion region is smaller than for the PEGDE system [24].

Table 18 illustrates the influence of DOBANOL and PEGDE surfactants. A difference is noted only for low K$_2$PtCl$_4$ concentrations, where larger Pt particles are formed with DOBANOL.

Table 18 Average Diameter of Pt Particles Synthesized from PEGDE 9.5%–Hexane 90%–Water 0.5% and DOBANOL 9.5%–Hexane 90%–Water 0.5% Microemulsions

	d (nm)	
[K$_2$PtCl$_4$] (m vs. H$_2$O)	PEGDE	DOBANOL
0.001	1.9 ± 0.2	2.3 ± 0.3
0.01	2.2 ± 0.3	2.7 ± 0.4
0.05	2.6 ± 0.3	2.7 ± 0.4
0.1	2.8 ± 0.3	2.8 ± 0.4
0.3	3.8 ± 0.4	3.9 ± 0.4

Table 19 Average Diameter of Au Particles Showing the Bidispersion in PEGDE–Hexane–Water Microemulsions

a. PEGDE 9.5%–hexane 90%–water 0.5%		
[AuCl$_3$] (m vs. H$_2$O)	d (nm)	d (nm)
0.001	—	2.9 ± 0.4
0.01	—	3.0 ± 0.3
0.05	8.2 ± 1.2	3.6 ± 0.4
0.1	11.0 ± 1.9	3.8 ± 0.5
0.3	13.9 ± 2.5	4.4 ± 0.5

b. DOBANOL 9.5%–hexane 90%–water 0.5%	
[AuCl$_3$] (m vs. H$_2$O)	d (nm)
0.001	3.3 ± 0.5
0.01	7.1 ± 1.1
0.05	9.7 ± 1.2
0.1	11.5 ± 1.4
0.3	13.2 ± 1.6

The Au particles were obtained from the precursor AuCl$_3$. In the PEGDE–hexane–water microemulsion, at low precursor concentrations (less than 0.05 m with respect to water), only small particles (about 3.0 nm diameter) were formed (Table 19a and Fig. 24), whereas both small and large (about 10 nm diameter) particles were formed at higher precursor concentrations.

In the DOBANOL–hexane–water microemulsion only one type of particle is obtained, the size of which increases with increasing precursor concentration (Table 19).

The Pt-Au particles were prepared in both microemulsion systems (Table 20). Both small (about 3.0 nm diameter) and big (about 12 nm diameter) particles were obtained in both systems. The size of the particles is not dependent on the composition of the precursor salts. The big particles are clearly formed by aggregation of the small particles. The nanoparticles are true mixed Pt-Au particles, as was shown by STEM/EDX measurements [24].

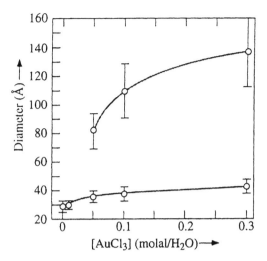

Figure 24 Variation in gold particle diameter as a function of precursor $AuCl_3$ concentration versus water synthesized in PEGDE 9.5%–hexane 90%–water 0.5% microemulsion. The presence of bigger particles shows particle aggregation.

Table 20 Average Diameter of Pt-Au Particles as a Function of Pt Mole Fraction x Showing the Bidispersion in Both Systems[a]

x	d (nm)	d (nm)
a. PEGDE 9.5%–hexane 90%–water 0.5%		
0.16	9.51 ± 1.8	3.4 ± 0.4
0.33	11.3 ± 2.0	3.2 ± 0.4
0.5	10.7 ± 1.9	3.7 ± 0.5
0.66	13.5 ± 2.4	2.9 ± 0.3
0.8	9.16 ± 1.3	3.4 ± 0.4
b. DOBANOL 9.5%–hexane 90%–water 0.5%		
0.16	11.2 ± 2.2	2.6 ± 0.4
0.33	12.5 ± 2.5	2.9 ± 0.4
0.5	14.6 ± 3.2	2.6 ± 0.4
0.66	12.7 ± 2.5	2.7 ± 0.5
0.8	12.1 ± 2.6	2.8 ± 0.4

[a] $[AuCl_3] + [K_2PtCl_4] = 0.1 \; m$ vs. H_2O.

III. CONCLUSION

One of the most important conditions for obtaining monodisperse colloidal particles from microemulsions is shown to be the differentiation between nucleation and growth. Detailed analysis showed that the formation of Ni_2B or Co_2B particles was governed by the initial composition of the microemulsion. In most cases the particles obtained are larger than the inner water core. All known cases emphasize the compartmentalization of the reaction medium, which is the determining factor during the nucleation step. Generally the rate of exchange between the water cores is higher than the rate of particle growth.

Finally, the stabilizing role of the surfactant molecules is also emphasized. Such stabilization is thought to be responsible for the constant particle size as a function of precursor concentration. Other examples taken from the literature show the formation of secondary monodisperse spherical particles by coagulation of the primary particles.

Despite the fact that many parameters influencing the size of nanoparticles could be rationalized and explained, a detailed understanding of the formation of monodisperse nanosized particles is still lacking. Subtle differences can change the systems entirely. One possible influence is the nature of the organic phase, the role of which was not explicitly considered in our simplified schemes.

ACKNOWLEDGMENTS

I thank Professors A. Persoons (Katolieke Universiteit Leuven, Belgium), B. H. Robinson (University of East-Anglia, Great Britain), and A. Lopez-Quintela (Universitad de Santiago de la Compostella, Spain) for the facilities offered in fast reaction techniques.

REFERENCES

1. M. Boutonnet, J. Kizling, P. Stenius, and G. Maire, Colloids Surf. *5*:209 (1982).
2. J. B. Nagy, A. Gourgue, and E. G. Derouane, Stud. Surf. Sci. Catal. *16*:193 (1983).
3. J. B. Nagy, E. G. Derouane, A. Gourgue, N. Lufimpadio, I. Ravet, and J.-P. Verfaillie, in *Surfactants in Solution, Modern Aspects*, Vol. 10 (K. L. Mittal, ed.), Plenum, New York, 1989, p. 1.
4. J. B. Nagy and A. Claerbout, in *Surfactants in Solution*, Vol. 11 (K. L. Mittal and D. O. Shah, eds.), Plenum, New York, 1991 p. 363.
5. J. H. Fendler, Chem. Rev. *87*:877 (1991).
6. J. B. Nagy, D. Barette, A. Fonseca, L. Jeunieau, Ph. Monnoyer, P. Piedigrosso, I. Ravet-Bodart, J.-P. Verfaillie, and A. Wathelet, in *Nanoparticles in Solids and Solutions* (J. H. Fendler and I. Dékány, eds.) Kluwer, Dordrecht, 1996, p. 71.
7. Faraday Symposia of the Chemical Society, Yonnay like to updated London, No. 14, 1980.
8. J. H. Fendler and F. C. Meldrum, Adv. Mater. *7*:607 (1995).
9. J. H. Fendler, *Adv. Polym Sci. 113*:236 (1994).
10. G. A. Ozin, A. Kuperman, and A. Stein, Angew. Chem., Int. Ed. Eng. *28*:359 (1989).
11. B. C. Gates, L. Guczi, and H. Knözinger (eds.), *Metal Clusters in Catalysis*, Stud. Surf. Sci. Catal. Vol. 29, 1986.
12. J. Belloni, M. Mostafavi, J.-L. Marignier, and J. Amblard, J. Imaging Sci. *35*:68 (1991).
13. A. Henglein, J. Phys. Chem. *97*:5457 (1993).
14. D. Rosier, J. L. Dallons, G. Jannes, and J. P. Puttemans, Acta Chim. Hung. *124*:57 (1987).
15. I. Ravet, N. B. Lufimpadio, A. Gourgue, and J. B. Nagy, Acta Chim. Hung. *119*:155 (1985).
16. I. Ravet, A. Gourgue, Z. Gabelica, and J. B. Nagy, Proc. 8th Int. Congress on Catalysis, Vol. IV, Verlag Chemie, Weinheim-Basel, West Berlin, p. 871.
17. I. Ravet, A. Gourgue, and J. B. Nagy, in *Surfactants in Solution* Vol. 5 (K. L. Mittal and P. Bothorel, eds.), Plenum, New York, 1986, p. 697.
18. I. Ravet, J. B. Nagy, and E. G. Derouane, Stud. Surf. Sci. Catal. *31*:505 (1987).
19. I. Bodart-Ravet, Ph.D. Thesis, Namur, Belgium, 1988.
20. J.-P Verfaillie, Ph.D. Thesis, Namur, Belgium, 1991.

21. N. B. Lufimpadio, J. B.Nagy, and E. G. Derouane, in *Surfactants in Solution*, Vol. 3 (K. L. Mittal and B. Lindman, eds.), Plenum, New York, 1984, p. 1983.
22. N. B. Lufimpadio, Ph.D. Thesis, Namur, Belgium, 1983.
23. A. Claerbout and J. B.Nagy, Stud. Surf. Sci. Catal. *63*:705 (1991).
24. D. Barette, Mémoire de Licence, FUNDP, Namur, Belgium, 1992.
25. A. Khan-Lodhi, B. H. Robinson, T. Towey, C. Herrmann, W. Knoche, and U. Thesing, in *The Structure, Dynamics and Equilibrium Properties of Colloidal Systems*, NATO ASI Ser. C 324 (D. M. Bloor and E. Wyn-Jones, eds.), Kluwer, Dordrecht, 1990, p. 373.
26. T. H. Hsieh and H. S. Fogler, Paper presented at the 5th Int. Congress of Surface and Colloid Science, Potsdam, New York, 1985.
27. M. Boutonnet, C. Andersson, and R. Larsson, Acta Chim. Scand. A *34*:639 (1980).
28. M. Meyer, C.Wallberg, K. Kurihara, and J. H. Fendler, J. Chem. Soc., Chem. Commun. 90 (1984).
29. I. Lisiecki and M. P. Pileni, J. Am. Chem. Soc. *115*:3887 (1993).
30. I. Lisiecki and M. P. Pileni, J. Phys. Chem. *99*:5077 (1995).
31. I. Lisiecki, M. Björling, L. Motte, B. Ninham, and M. P. Pileni, Langmuir *11*:2385 (1995).
32. P. Stenius, J. Kizling, and M. Boutonnet, U. S. Patent 4,425,261 (1984).
33. M. Boutonnet, J. Kizling, P. Stenius, and G. Maire, Colloids Surf. *5*:209 (1982).
34. M. Boutonnet, J. Kizling, V. Mintsa-Eya, A. Choplin, R. Touroude, G. Maire, and P. Stenius, J. Catal *103*:95 (1987).
35. A. Khan-Lodhi, Ph.D. Thesis, University of Kent, UK, 1988.
36. B. H. Robinson, A. N. Khan-Lodhi, and T. Towey, Stud. Phys. Theor. Chem. *65*:198 (1989).
37. K. Kurihara, J. Kizling, P. Stenius, and J. H. Fendler, J. Am. Chem. Soc. *105*:2574 (1983).
38. G. Jannes, J.-P. Puttemans, and P. Vanderwegen, Catal. Today *5*:265 (1989).
39. S. I. Ahmad and S. Friberg, J. Am. Chem. Soc. *94*:5196 (1972).
40. J. Rouvière, J.-M. Couret, M. Lindheimer, J.-L. Dejardin, and R. Marrony, J. Chim. Phys. *76*:289 (1979).
41. P. Ekwall, L. Mandell, and K. Fontell, Mol. Cryst. Liq. Cryst. *8*:157 (1969).
42. S. Friberg and I. Lapczynska, Prog. Colloid Polym. Sci. *56*:16 (1975).
43. V. K. LaMer and R. H. Dinegar, J. Am. Chem. Soc. *72*:4847 (1950).
44. R. Zsigmondy, Z. Anorg. Allgem. Chem. *99*:105 (1917).
45. J. B.Nagy, Colloids Surf. *35*:201 (1989).
46. T. Sugimoto, Adv. Colloid Interface Sci. *28*:65 (1987).
47. W. D. Knight, K. Clemenger, W. A. de Heer, W. A. Saunders, M. Y. Chou, and M. L. Cohen, Phys. Rev. Lett. *52*:2141 (1984).
48. L. Lerot, F. Legrand, and P. De Bruycker, J. Mater. Sci. *26*:2353 (1991).
49. T. Ogihara, N. Mizutani, and M. Kato, J. Am. Ceram. Soc. *72*:421 (1989).
50. G. N. La Mar, W. Horrock, Jr., and R. H. Holm (eds.), *NMR of Parmagnetic Molecules*; Academic, New York, 1973.
51. A. A. Obynochnyi, O. I. Bel'chenko, P.V. Schastnev, R. Z. Sagdeev, A.V. Dushkin, Yu. N. Molin, and A. I. Rezvukin, Zh. Strukt. Khim. *17*:620 (1976).
52. J. Sunamoto and T. Hamada, Bull. Chem. Soc. Jpn. *51*:3130 (1978).
53. A. Von Zelewsky, Helv. Chim. Acta *51*:803 (1968).
54. A. Wathelet, Mémoire de Licence, Namur, Belgium, 1984.
55. J. H. Fendler and E. J. Fendler, *Catalysis in Micellar and Macromolecular Systems*, Academic, New York, 1975.
56. A. S. Kertes and H. Gutman, in *Surface and Colloid Science*, Vol. 8 (E. Matijevic, ed.), Wiley, New York, 1976.
57. A. Kitahara, Adv. Colloid Interface Sci. *12*:109 (1980).
58. J. B.Nagy, in *Solution Behavior of Surfactant*, Vol. 2 (K. L. Mittal and E. J. Fendler, eds.), Plenum, New York, 1982, p. 743.
59. E. Sjöblom and S. Friberg, J. Colloid Interface Sci. *67*:16 (1978).
60. J. H. Fendler, Acc. Chem. Res. *9*:153 (1976).
61. E. M. Kosower, *An Introduction to Physical Organic Chemistry*, Wiley, New York, 1968.

62. C. Reichardt, *Solvent Effects in Organic Chemistry*, Verlag Chemie, Weinheim, 1979.
63. M. Wong, J. K. Thomas, and M. Grätzel, J. Am. Chem. Soc. *98*:2391 (1976).
64. H. F. Eicke and H. Christen, Helv. Chim. Acta *61*:2258 (1978).
65. R. A. Pyter, C. Ramachandran, and P. Mukerjee, J. Phys. Chem. *86*:3206 (1982).
66. P. Mukerjee, C. Ramachandran, and R. A. Pyter, J. Phys. Chem. *86*:3189 (1982).
67. C. Ramachandran, R. A. Pyter, and P. Mukerjee, J. Phys. Chem. *86*:3198 (1982).
68. R. Briere, H. Lemaire, and A. Rassat, Tetrahedron Lett. *27*:1775 (1964); *33*:2318 (1964).
69. D. Balasubramanian and C. Kumar, J. Indian Chem. Soc. *58*:633 (1981).
70. E. Rotlevi and A. Treinen, J. Phys. Chem. *69*:2645 (1965).
71. D. Balasubramanian and C. Kumar, in *Solution Behavior of Surfactants*, Vol. 2 (K. L. Mittal and E. Fendler, eds.) Plenum, New York, 1982, p. 1207.
72. T. P. Gentile, F. Ricci, F. Podo, and P. E. Gina, Gazz. Chim. Ital. *106*:423 (1976).
73. M. Wong, J. K. Thomas, and T. Nowak, J. Am. Chem. Soc. *99*:4730 (1977).
74. B. Lindman, and P. Stilbs, in *Surfactants in Solution*, Vol. 3 (K. L. Mittal and B. Lindman, eds.), Plenum, New York, 1984, p. 1657.
75. T. Nguyen and H. H. Ghaffarie, C. R. Acad. Sci. Paris, Ser. C. *290*:113 (1980).
76. T. Nguyen and H. H. Ghaffarie, J. Chim. Phys. *76*:513 (1979).
77. J. H. Fendler, *Membrane Mimetic Chemistry: Characterization and Applications of Micelles, Microemulsions, Monolayers, Bilayers, Vesicles, Host–Guest Systems and Polyanions*, Wiley-Interscience, New York, 1982.
78. B. Lindman and H. Wennerström, *Top. Curr. Chem.* *87*:1 (1980).
79. J. B. Nagy, I. Bodart-Ravet, E. G. Derouane, A. Gourgue, and J.-P. Verfaillie, Colloids Surf. *36*:229 (1989).
80. A. Pouchelon, D. Chaternay, J. Meunier, and D. Langevin, J. Colloid Interface Sci. *82*:418 (1981).
81. A. M. Cazabat, D. Langevin, and A. Pouchelon, J. Colloid Interface Sci. *73*:1 (1980).
82. G. I. Likhtenshtein, *Spin Labeling Methods in Molecular Biology*, Wiley-Interscience, New York, 1976.
83. F. M. Menger, G. Saito, G. V. Sanzero, and J. R. Dodd, J. Am. Chem. Soc. *97*:909 (1975).
84. Y. Y. Lim and J. H. Fendler, J. Am. Chem. Soc. *100*:7490 (1978).
85. H. F. Eicke, J. C. W. Shepherd, and A. Steinmann, J. Colloid Interface Sci. *56*:168 (1976).
86. J. M. Furois, P. Brochette, and M. P. Pileni, J. Colloid Interface Sci. *97*:552 (1984).
87. P. D. I. Fletcher and B. H. Robinson, Ber. Bunsenges. Phys. Chem. *85*:863 (1981).
88. J. F. Holzwarth, in *Techniques and Applications of Fast Reactions in Solution* (W. U. Gettins and E. Wyn-Jones, eds.), Springer-Verlag, Berlin, 1979, p. 509.
89. S. S. Atik and J. K. Thomas, Chem. Phys. Lett. *79*:351 (1981).
90. C. Oldfield, J. Chem. Soc., Faraday Trans. *87*:2607 (1991).
91. P. C. Hiemenz, *Principles of Colloid and Surface Chemistry*, Marcel Dekker, New York, 1977, p. 234.
92. R. D. Vold and M. J. Vold, *Colloid and Interface Chemistry* Addison-Wesley, London, 1983, p. 181.

18

Microemulsion-Mediated Synthesis of Nanosize Oxide Materials

K. Osseo-Asare
The Pennsylvania State University, University Park, Pennsylvania

I. INTRODUCTION

The core of the earth is composed of a mass of metals ($\sim$90% iron, 9% nickel, and 1% other metals), while the earth's crust—the outer shell extending to a depth of about 13 km—is dominated by oxygen (47%) [1]. Not surprisingly, metal oxides constitute the most abundant and most accessible of inorganic materials in nature [1,2]. The fact that the periodic table is dominated by metals means, further, that a wide variety of metal oxides can be formed, spanning a broad range of physical and chemical properties. These properties provide the basis for the many scientific and technological applications of metal oxides, some selected examples of which are collected in Table 1 [3].

Manufacturing processes based on metal oxides typically require that these materials be used in the form of fine particulates. In many cases, it is found that the smaller the particle size, the better the desired property, be it biological, chemical, electrical, electronic, magnetic, mechanical, optical, or thermal. Many new and unusual physical and chemical properties also arise as particles attain nanosize dimensions [3–8]. There is increasing recognition that aqueous synthesis offers growth control capabilities that can be conveniently exploited to prepare these desirable fine particles [5,6,9]. Compared to conventional solid-state reaction methods, solution-based synthesis results in higher levels of chemical homogeneity. Also, in solution systems, mixing of the starting materials is achieved at the molecular level, and this is especially important when multicomponent oxides are being prepared.

As a solution-based materials synthesis technique, the microemulsion-mediated method [10–18] offers the unique ability to effect particle synthesis and particle stabilization in one step. The solubilized water droplets serve as nanosize test tubes, thus limiting particle growth, while the associated surfactant films adsorb on the growing particles, thereby minimizing particle aggregation. The purpose of this chapter is to review the literature on the microemulsion-mediated synthesis of metal hydroxides and oxides; the definition of a metal is extended here to include the semimetal silicon. Since metal oxides are frequently produced by decomposing metal salts, aspects of the literature on microemulsion-derived metal salts are also considered. In principle, any previously established aqueous precipitation chemistry can be adapted to the microemulsion synthesis technique. Accordingly,

Table 1　Selected Industrial Applications of Metal Oxides

Type of function	Material	Property exploited	Application
Electrical/Electronic	Al_2O_3	Insulation	IC substrates
	BeO	Insulation	IC substrates
	Fe_3O_4	Magnetism	Magnetic core
	SnO_2	Semiconductivity	Gas sensors
	$ZnO\text{-}Bi_2O_3$	Semiconductivity	Varistors
	$\beta\text{-}Al_2O_3$	Ionic conductivity	NaS battery
	ZrO_2	Ionic conductivity	Oxide sensors
	$BaTiO_3$	Dielectricity	Capacitors
	$BaTiO_3$	Semiconductivity	Resistance junctions
	$Pb(Zr, Ti)O_3$ (PZT)	Piezoelectricity	Oscillators, ignition junctions
	$LiNbO_3$	Piezoelectricity	Piezoelectric oscillators
	$MgAl_2O_4$	Insulation	IC substrates
	MFe_2O_4	Magnetism	Magnetic recording
	$YBa_2Cu_3O_x$	Superconductivity	Superconductors
	$La_{2-x}Sr_xCuO_4$	Superconductivity	Superconductors
Mechanical	Al_2O_3	Wear resistance	Polishing materials, grindstones
	ZrO_2	Wear resistance	Polishing materials, grindstones
Optical	Al_2O_3	Transparency	Sodium lamp
	SiO_2	Light conductivity	Optical fibers
Thermal	Al_2O_3	Heat resistance	Structural refractories
	ZrO_2	Heat insulation	Heat-insulating materials
	BeO	Heat conductivity	Substrates
Nuclear	UO_2	Nuclear properties	Nuclear fuel
Chemical	TiO_2	Color, chemical inertness	Paint pigments
	Fe_2O_3	Color	Paint pigments
	Al_2O_3	Chemical inertness	Catalyst support
Biological	Al_2O_3	Biocompatibility, hardness	Artificial teeth and bones

Source: After Ichinose [3].

in the discussion that follows, a brief summary is first presented of the reaction paths available for the precipitation of (hydrous) metal oxides in aqueous solution. This is followed by a description of the synthesis protocols that have been developed for microemulsion-mediated materials synthesis. Next, the physicochemical factors that influence materials synthesis in microemulsion media are examined. A survey is then presented of the various oxide materials that have been prepared by the microemulsion method. Finally, some thoughts are offered concerning outstanding issues and future directions.

II. PRECIPITATION REACTION PATHS

A. Equilibrium Considerations

The solution chemical conditions that favor the formation of metal hydroxides and oxides can be identified readily by referring to aqueous stability diagrams, such as potential versus pH (Eh–pH or Pourbaix) diagrams [19–21] and log[M]–pH (solubility) diagrams [21,22].

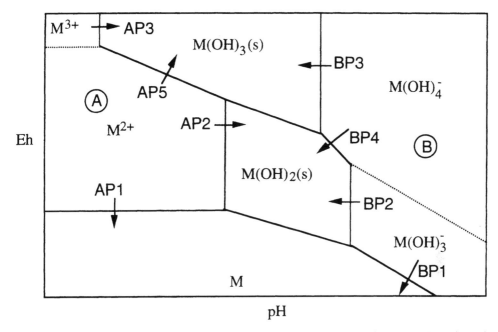

Figure 1 Schematic Eh–pH diagram illustrating precipitation reaction paths for a hypothetical M–H_2O system. M denotes a metal.

Table 2 Reaction Paths for Metal Hydroxide Precipitation

Reaction path	Chemical reaction	Process
AP2	$M^{2+} + 2OH^- \rightarrow M(OH)_2(s)$	Base addition
AP3	$M^{3+} + 3OH^- \rightarrow M(OH)_3(s)$	Base addition
AP5	$M^{2+} + 3OH^- \rightarrow M(OH)_3(s) + e^-$	Oxidative base addition
BP2	$M(OH)_3^- + H^+ \rightarrow M(OH)_2(s) + H_2O$	Acidification
BP3	$M(OH)_4^- + H^+ \rightarrow M(OH)_3(s) + H_2O$	Acidification
BP4	$M(OH)_4^- + e^- + 2H^+ \rightarrow M(OH)_2(s) + 2H_2O$	Reductive acidification

Figure 1 presents a schematic Eh–pH diagram for an arbitrary metal (M)–water system [20]. This diagram shows two aqueous stability regions, i.e., region A, the acidic aqueous stability field occupied by the metallic species M^{2+} and M^{3+}, and region B, the basic (i.e., alkaline) aqueous stability field dominated by the species $M(OH)_3^-$ and $M(OH)_4^-$. The arrows indicate the reaction paths that connect the dissolved metal to the relevant solid phases. The corresponding chemical equations are summarized in Table 2.

Figure 1 and Table 2 highlight the fact that three reaction paths are available for the precipitation of the M(II) hydroxide phase (i.e., paths AP2, BP2, and BP4) and three for the M(III) hydroxide (i.e., paths AP3, AP5, and BP3). Depending on the particular metal salt solution, hydroxide formation may be effected by base addition only (paths AP2, AP3), acidification only (paths BP2 and BP3), base addition accompanied by metal oxidation (path AP5), and acidification accompanied by metal reduction (path BP4).

B. Hydrolytic Reactions

In Fig. 1, only the predominant species are shown in the aqueous stability regions. These ions, however, are not necessarily the only kinetically active species with respect to solid formation. The key factors are, in fact, metal hydroxo complexes [22], sometimes containing bound anions such as SO_4^{2-}. The hydrolytic processes that precede precipitation proper involve hydroxylation [i.e., Eqs. (1) and (2)], olation [i.e., Eq. (3)], and oxolation [i.e., Eq. (4)] [9,22–24]:

$$M(H_2O)_6^{3+} \rightarrow M(OH)(H_2O)_5^{2+} + H^+ \tag{1}$$

$$M(OH)(H_2O)_5^{2+} \rightarrow M(OH)_2(H_2O)_4^+ + H^+ \tag{2}$$

$$M(OH)(H_2O)_5^+ + M(OH)(H_2O)_5^+ \rightarrow (H_2O)_4\,M\overset{\displaystyle \overset{H}{\underset{}{O}}}{\underset{\underset{H}{O}}{\diagup\diagdown}}M\,(H_2O)_4^{2+} + 2H_2O \tag{3}$$

$$-M-OH + HO-M- \rightarrow -M-O-M- + H_2O \tag{4}$$

In addition to the hydroxide (OH^-) and oxo (O^{2-}) ions, certain anions, such as the sulfate anion, SO_4^{2-}, can serve as bridging ligands:

$$MOH^{2+} + MOH^{2+} + SO_4^{2-} \rightarrow HO\overset{\displaystyle M-O}{\underset{\displaystyle M-O}{\diagdown\underset{OH}{}\diagup}}S\overset{\displaystyle O}{\underset{\displaystyle O}{\diagup\diagdown}} \qquad (M_2(OH)_2SO_4^{2+}) \tag{5}$$

The nature of the complex solutes can be manipulated by adjusting pH and temperature, by varying the metal ion concentration, and by changing the type and concentration of anion. The clustering of metal ions achieved through hydroxylation, oxolation, and anion bridging eventually results in the formation of critical nuclei, which then grow into the final solid product. Considerable controversy remains as to whether the particle growth occurs by monomer (i.e., discrete metal complex) addition (classical crystal growth mechanism) or by nuclei aggregation (colloidal interaction mechanism) [25–28]. Nonetheless, there is increasing recognition of a correlation between the precursors for the nucleating species and the basic units of the solid phase. Examples are cited below for selected $M-H_2O$ systems of relevance to microemulsion-mediated materials synthesis.

The overall reaction resulting in the formation of aluminium hydroxide may be expressed as

$$Al^{3+} + 3OH^- \rightarrow Al(OH)_3\,(s) \tag{6}$$

All three polymorphs of aluminium hydroxide (i.e., bayerite, gibbsite, and nordstrandite [29]) are built on hexameric structural units, and the evolution of the hydrolytic species with increasing OH/Al ratio was proposed by de Bruyn and coworkers [30–32] to proceed as illustrated in Fig. 2. As base is added to the Al(III) solution, cyclic structures grow, up to OH/Al ≈ 2.5. Beyond this, the additional OH^- ions contribute to charge neutralization by attaching themselves, through single bonds, to the Al^{3+} ions. In order for solid nucleation to occur, (1) the cyclic structures should grow to a sufficiently large size, (2) the electrostatic charge on the polynuclear complexes must be adequately neutralized, and (3) the hexameric sheets must be stacked appropriately along the crystallographic c axis [31].

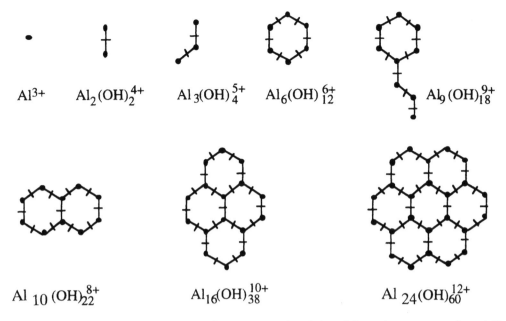

$$Al^{3+} \qquad Al_2(OH)_2^{4+} \qquad Al_3(OH)_4^{5+} \qquad Al_6(OH)_{12}^{6+} \qquad Al_9(OH)_{18}^{9+}$$

$$Al_{10}(OH)_{22}^{8+} \qquad\qquad Al_{16}(OH)_{38}^{10+} \qquad\qquad Al_{24}(OH)_{60}^{12+}$$

Figure 2 Schematic representation of aqueous species deduced from the structure of crystalline $Al(OH)_3$. (After Ref. 31.)

Reaction conditions influence the manner in which the aqueous cyclic structures are stacked, and this in turn affects the structure of the resulting crystalline phase. This situation is well illustrated by the $Zr(IV)$–H_2O system [33,34], where the tetrameric species $(Zr(OH)_2 \cdot 4H_2O)_4^{8+}$ has been identified in both the solid oxide phase and aqueous solutions. Polymerization occurs through the formation of hydroxyl bridges between two tetrameric species. In the pH 7–11 range, rapid precipitation yields monoclinic zirconia primarily, while slow precipitation favors the tetragonal phase. This behavior has been rationalized by Clearfield [33,34] in terms of two different pathways for formation of the hydroxyl bridges. Conditions that promote a random linkage of the tetramers (e.g., rapid base addition) produce the monoclinic phase, whereas conditions that permit the tetramers to self-organize into ordered arrays lead to the tetragonal phase.

In the case of the $Fe(III)$–H_2O system, the initial precipitation product is an amorphous ferric hydroxide, $Fe(OH)_3(s)$. It has been postulated by van der Woude and de Bruyn [35] that the formation of the corresponding supercritical nucleus may be viewed in terms of the reaction

$$Fe_9(OH)_{20}^{7+} + Fe_3(OH)_4^{5+} \rightarrow \text{supercritical cluster} \qquad (7)$$

The resulting colloidal ferric hydroxide particles are amorphous and grow to average particle diameters of 3–5 nm [35,36]. It is estimated that a typical primary particle ($\sim$3.8 nm diameter) contains $\sim$820 Fe atoms [35]. Crystalline phases such as goethite (α-FeOOH) and akagemite (β-FeOOH) develop from this initial material when the spherical amorphous colloidal particles link up to form rods, and the rods in turn aggregate to form rafts [35,36].

The formation of silica from silicate solution also proceeds through polymerization reactions that involve cyclic hydroxo structures [9,23,37]: Monomer$\rightarrow$dimer $[Si_2O_3(OH)_4^{2-}]$ $\rightarrow$cyclic tetramer $[Si_4O_6(OH)_6^{2-},\ Si_4O_8(OH)_4^{4-}]\rightarrow$particle. The polymerization occurs via proton-catalyzed [Eqs. (8) and (9)] and hydroxide-catalyzed [Eqs. (10) and (11)] condensation reactions [9,37]:

$$Si - OH + H_3O^+ \rightarrow Si - OH_2^+ + H_2O \tag{8}$$

$$Si - OH_2{}^+ + HO - Si \rightarrow Si - O - Si + H_3O^+ \tag{9}$$

$$Si - OH + OH^- \rightarrow Si - O^- + H_2O \tag{10}$$

$$Si - O^- + HO - Si \rightarrow Si - O - Si + OH^- \tag{11}$$

It is reasonable to expect that all the above processes would be influenced by the microemulsion environment, e.g., electrostatic and metal–ligand interactions through the surfactant polar groups, and size and stereochemical constraints imposed on the evolution of the cyclic structures by the geometrical dimensions of the microemulsion water pools. However, systematic studies of these effects are not yet available.

C. Homogeneous Precipitation

In a conventional precipitation process, a solution of the precipitant is added directly to a solution of the metal salt (direct strike method) or vice versa (reverse strike method). In either case, pockets of high local concentrations typically develop during mixing of the two solutions, and this results in particles with irregular shape and broad size distributions. To produce particles with desired physical and chemical characteristics, it is necessary to have good control over the chemical reactions associated with metal complexation, particle nucleation, and particle growth. The technique of homogeneous precipitation was introduced by analytical chemists who sought to improve the purity, crystallinity, and physical characteristics of precipitates prepared for gravimetric analysis [38]. In this technique, the precipitant is generated homogeneously in situ instead of being introduced externally. For example, the decomposition of acetamide (CH_3COONH_2) releases hydroxide ions:

$$CH_3COONH_2 + 2H_2O \rightarrow CH_3COOH + NH_4^+ + OH^- \tag{12}$$

From a separations standpoint, the slow release of precipitant species is desirable, since this avoids the generation of the high local supersaturations that produce slimy voluminous precipitates. At the same time, from a materials synthesis standpoint, the controlled release of reactant results in a single burst of nuclei, leading to monodisperse particles. Five main avenues are available for the homogeneous generation of precipitation reactants, as summarized in Table 3. Matijevic [25,39,40] exploited these homogeneous precipitation methods to prepare monodisperse colloidal particles of various shapes for a wide variety of hydrous metal oxides. In principle, all the methods in Table 3 can be adapted to microemulsion-mediated synthesis. It must be noted, though, that the microemulsion technique itself may be viewed as an example of "homogeneous precipitation." In fact, the intermicellar exchange that occurs when a metal-containing water pool collides with a hydroxide-containing water pool represents a "controlled release" of reactants, which, as noted above, should promote monodispersity. The resulting nanoscale mixing is more gentle and controlled than conventional precipitation.

Table 3 Homogeneous Precipitation Methods for Hydrous Metal Oxides

Method	Starting solution	Process description
Hydroxide ion generation by chemical reaction	Solution of a metal salt and the precursor anion/molecule	Hydrolytic decomposition of precursor releases the precipitating anion.
Forced hydrolysis	Acidic solution of a metal salt	Heating promotes the deprotonation of hydrous metal ions.
Thermolysis of metal complexes	Strongly alkaline solution of metal–ligand complex	Heating decomposes the metal–ligand complex; released metal ions react with the readily available OH$^-$ ions.
Cation generation by redox reaction	A solution of a metal salt and the corresponding reductant/oxidant	Reduction or oxidation of a multivalent cation releases the lower or higher valent ion.
Solution-mediated phase transformation	A preformed hydrous oxide dispersed in aqueous solution	A metastable (amorphous) hydrous oxide forms first and then dissolves to release metal ions needed to form new more stable crystalline phase.

D. Alkoxide Sol-Gel Process

Metal alkoxides [M(OR)$_n$] react with water to form metal oxides and hydroxides, as described by the overall reaction

$$M(OR)_n + nH_2O \rightarrow M(OH)_n + nROH \tag{13}$$

This alkoxide-to-hydroxide reaction may be viewed in terms of a two-step process, i.e., hydrolysis and condensation [9,24,37]. Hydrolysis involves the reaction of the alkoxide with water to generate M—OH groups:

$$M - OR + H - OH \rightarrow M - OH + R - OH \tag{14}$$

The hydroxo groups generated by the hydrolysis step are used in the condensation step, which involves the formation of M—O—M or M—OH—M bridges via olation [Eqs. (15) and (16)], oxolation [Eq. (17)], or alcoxolation [Eq. (18)]:

$$M - OH + M \leftarrow O{\overset{\displaystyle /\,H}{\underset{\displaystyle \backslash\,H}{}}} \rightarrow M \overset{H}{-} O - M + H_2O \tag{15}$$

$$M - OH + M \leftarrow O{\overset{\displaystyle /\,H}{\underset{\displaystyle \backslash\,R}{}}} \rightarrow M \overset{H}{-} O - M + ROH \tag{16}$$

$$M - OH + M - OH \rightarrow M - O - M + H_2O \tag{17}$$

$$M - OH + M - OR \rightarrow M - O - M + ROH \tag{18}$$

Table 4 Metal Alkoxides in Microemulsion-Mediated Materials Synthesis

Metal	Name	Chemical formula	Abbreviation
Si	Tetramethyl orthosilicate	$Si(OCH_3)_4$	TMOS
	Tetraethyl orthosilicate	$Si(OC_2H_5)_4$	TEOS
Ti	Titanium tetraethoxide	$Ti(OC_2H_5)_4$	TTEO
	Titanium tetraisopropoxide	$Ti(i\text{-}OC_3H_7)_4$	TIPO
	Titanium tetra-n-propoxide	$Ti(n\text{-}OC_3H_7)_4$	TNPO
	Titanium tetrabutoxide	$Ti(OC_4H_9)_4$	TTBO
Al	Aluminum tributoxide	$Al(OC_4H_9)_3$	ATBO
Sr	Strontium isopropoxide	$Sr(i\text{-}OC_3H_7)_2$	SIPO
Zr	Zirconium tetrapropoxide	$Zr(OC_3H_7)_4$	ZTPO
	Zirconium tetrabutoxide	$Zr(OC_4H_9)_4$	ZTBO
Ba	Barium isopropoxide	$Ba(i\text{-}OC_3H_7)_2$	BIPO

As in the case of inorganic silica systems [Eqs. (8)–(11)], both hydrolysis and condensation of alkoxides may be acid- or base-catalyzed [9,24,37]. The relative contributions of the four reactions—hydrolysis, olation, oxolation, and alcoxolation—determine the physical and chemical characteristics of the resulting oxide material. These contributions are in turn determined by the nature of the metal alkoxide (i.e., nature of the metal atom, nature of the alkyl groups) and the characteristics of the chosen experimental conditions (i.e., the water/alkoxide molar ratio, the nature and concentration of the catalyst, the nature of the solvent, and temperature). A major factor is the hydrolysis ratio, h, defined as the water/alkoxide molar ratio:

$$h = [H_2O]/[M(OR)_n] \tag{19}$$

By changing the hydrolysis ratio, the degree of condensation can be manipulated in interesting ways. When $h < 1$, condensation proceeds predominantly via olation and alcoxolation. Gelation or precipitation does not occur; instead, discrete oxo-alkoxide molecules form. When $1 \leq h \leq n$, the formation of chain polymers is promoted. Finally, with $h > n$, the tendency is to form three-dimensional polymers, particulate gels, or precipitates. Regarding the nature of the catalyst, it is known, for example, that the use of an acid catalyst promotes the formation of spinnable sols and monolithic gels [9,24]. On the other hand, the use of a base catalyst tends to produce particulate gels. Table 4 presents a list of alkoxide precursors that have been exploited in microemulsion-mediated materials synthesis. Of the abbreviations indicated in this table, only TMOS and TEOS may be considered to be well established [9,24,37]; the rest are provided here only for convenience.

III. MICROEMULSION-MEDIATED SYNTHESIS

A. Synthesis Protocols

Microemulsion-mediated materials synthesis employs three basic methods, as illustrated in Fig. 3 [17]. The microemulsion-plus-trigger method (Fig. 3, method a) is based on a single microemulsion. The fluid system is activated in some way in order to initiate the reactions that eventually lead to particle formation. Pulse radiolysis and laser photolysis have served as triggers for the preparation of nanosize gold particles [41]. In the case of metal oxides, temperature elevation can provide the needed trigger action; hydrated metal ions solubilized

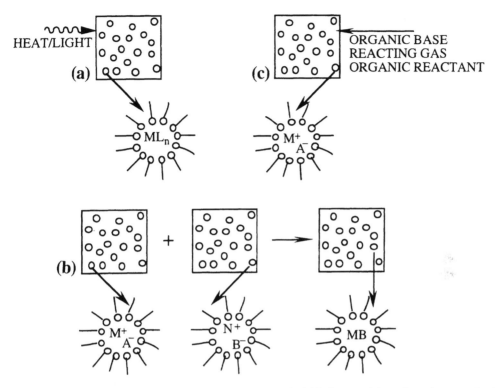

Figure 3 Microemulsion-mediated synthesis protocols. (a) Microemulsion-plus-trigger method; (b) two-microemulsion (or microemulsion-plus-microemulsion) method; (c) microemulsion-plus-reactant method. (From Ref. 17.)

in the aqueous pseudophase undergo forced hydrolysis (Table 3) when the temperature is raised. With the two-microemulsion (or microemulsion-plus-microemulsion) method (Fig. 3, method b), the metal ion and the precipitant are solubilized in different microemulsions; reaction takes place when the two microemulsions are mixed. In one scenario of the microemulsion-plus-reactant method (Fig. 3, method c), the metal ions are first solubilized in the water pools of a W/O microemulsion. Then the precipitant, in the form of an aqueous solution (e.g., aqueous NaOH) or a gas phase [e.g., $NH_3(g)$], is introduced into the microemulsion phase. In a second scenario, the precipitant is first solubilized in the polar core and the metal-containing solution (e.g., an alcoholic or hydrocarbon solution of an alkoxide) is subsequently added to the microemulsion. Using these techniques and microemulsions based on a wide range of surfactants (Table 5) [42,43], a variety of oxide materials have been prepared, as summarized in Table 6 [44–120].

A number of variations of the basic synthesis methods have been reported. In the works of Morooka and coworkers [104,105], the microemulsion was prepared by a solvent extraction process in which a bulk organic phase containing the surfactant dioleyl phosphoric acid (DOLPA) in isooctane was contacted with a bulk aqueous phase containing KCl and NH_3 (for pH adjustment). The extraction process can be expressed as

$$(C_{18}H_{35})_2POOH(org) + K^+(aq) = (C_{18}H_{35})_2POO^- \cdot K^+(org) + H^+(aq) \qquad (20)$$

Text continues on page 574.

Table 5   Surfactants Used in Microemulsion-Mediated Materials Synthesis

Type of surfactant	Name of surfactant	Chemical formula
Anionic	Aerosol OT (AOT); sodium bis(2-ethylhexyl) sulfosuccinate	$ROOC-CH_2$ $\|$ $ROOC-CHSO_3^-\,Na^+$ C_2H_5 $\|$ $(R = CH_2CH(CH_2)_3CH_3)$
	Potassium oleate	$CH_3(CH_2)_7CH{=}CH(CH_2)_7COO^-\,K^+$
	Sodium dodecyl sulfate (SDS)	$C_{12}H_{25}OSO_3^-\,Na^+$
	Dioleyl phosphoric acid (DOLPA)	$C_{18}H_{35}\diagdown \underset{C_{18}H_{35}\diagup}{\overset{O}{P}}{\diagdown OH}$
Cationic	Didodecyldimethylammonium bromide (DDAB)	$\left[\begin{smallmatrix}R_1 & R_2 \\ &N& \\ R_1 & R_2\end{smallmatrix}\right]^{+BR^-}$, $R_1 = C_{12}H_{25}$; $R_2 = CH_3$
	Cetyltrimethylammonium bromide (CTAB)	$\left[\begin{smallmatrix}R_2 & R_2 \\ &N& \\ R_1 & R_2\end{smallmatrix}\right]^{+BR^-}$, $R_1 = C_{16}H_{33}$; $R_2 = CH_3$
	Cetyldimethylbenzylammonium bromide (CDBA)	$\left[\begin{smallmatrix}R_1 & R_2 \\ &N& \\ R_3 & R_2\end{smallmatrix}\right]^{+Cl^-}$, $R_1 = C_{16}H_{33}$; $R_2 = CH_3$; $R_3 = {-}\langle\bigcirc\rangle{-}CH_2$
Nonionic	Polyoxyethylene-(n)-dodecyl ether ($C_{12}EO_n$) or polyethylene glycol ether; $n = 4$ (tetraoxyethylene dodecyl ether, $C_{12}EO_4$) or tetra(ethylene glycol) dodecyl ether (TEGDE) or polyoxyethylene-(4)-lauryl ether (POELE)	$C_{12}H_{25}-(OCH_2-CH_2)_nOH$
	Polyoxyethylene-(n)-nonylphenyl ether [NP-4, $n = 4$ (Triton N-42); NP-5, $n = 5$ (Triton N-57); NP-6, $n = 6$ (Triton N-60); NP-9, $n = 9{-}10$ (Triton N-101)]	$C_9H_{19}{-}\langle\bigcirc\rangle{-}(OCH_2-CH_2)_nOH$
	Polyoxyethylene-(n)-octylphenyl ether [OP-1, $n = 1$ (Triton X-15); OP-3, $n = 3$ (Triton X-35); OP-5, $n = 5$ (Triton X-45); OP-10, $n = 9{-}10$ (Triton X-100)]	$C_8H_{17}{-}\langle\bigcirc\rangle{-}(OCH_2-CH_2)_nOH$
	Polyoxyethylene-(n)-dodecylphenyl ether (DP-6, $n = 6$)	$C_{12}H_{25}{-}\langle\bigcirc\rangle{-}(OCH_2-CH_2)_nOH$
	Sorbitan monooleate (SPAN 80)	$\begin{smallmatrix}CH_2\\ \|\\ HCOH\\ \|\\ HOCH\\ \|\\ HC\\ \|\\ HCOH\\ \|\\ CH_2OOCR\end{smallmatrix}$ O , $R = (CH_2)_7CH{=}CH(CH_2)_7CH_3$

Table 6 Oxide Materials Synthesized in Microemulsions[a]

Material	Polar phase reactants	Microemulsion system	Synthesis route[b]	Comments
Al(OH)$_3$	Al(NO$_3$)$_3$ ([Al] = 0.01–0.1 M)	AOT/supercritical propane/water ([AOT] = 50 mM, R = 5)	C(NH$_3$ addition)	T = 110°C; 100–500 nm particles, partially agglomerated [44].
Al(OH)$_3$	Na aluminate solution ([Al] = 0.15 M, [NaOH] = 1.8 M)	AOT/isooctane/water (R = 8)	A (heating to 85°C, 24 h)	Agglomerated submicrometer particles (SEM; not shown in original paper) [45].
α-FeOOH	FeSO$_4$(NH$_4$)$_2$SO$_4$ [5×10^{-3}–0.01 M Fe(II)]	Potassium oleate/hexanol/hexane/water	C [O$_2$ (air) bubbling]	30–70°C; needle-shaped primary particles, aggregated into ~80 nm cubic clusters [46].
		AOT/hexane or heptane/water	C [O$_2$ (air) bubbling]	Data on particle characteristics not provided [46].
		AOT/cyclohexane/water	C [O$_2$ (air) bubbling]	Data on particle characteristics not provided [46].
β-FeOOH	FeCl$_3$, HCl (0.025 M FeCl$_3$ + 0.01 M HCl)	C$_{12}$EO$_4$/hexane/water (47% C$_{12}$EO$_4$/48% hexane/5% aq soln; 43% C$_{12}$EO$_4$/42% hexane/15% aq soln)	A (heating to 60,80°C)	Particle size (d_p) decreased with increase in heating time (t); e.g., d_p = 40–50 nm for t = 5 min; d_p = 2 nm for t = 4 h [47].
(α,γ)-Fe$_2$O$_3$	Fe(NO$_3$)$_3$	Sorbitan monooleate/2-ethylhexanol/water	C (NH$_4$OH addition)	Calcination at 250°C (15 min) yielded microcrystalline particles (5–80 nm); <30 nm: α-Fe$_2$O$_3$; >30 nm: γ-Fe$_2$O$_3$ [49].
γ-Fe$_2$O$_3$ (FeC$_2$O$_4$ · 2H$_2$O precursor)	FeSO$_4$(NH$_4$)$_2$SO$_4$ (μE 1), H$_2$C$_2$O$_4$ (μE 2) [0.3 M Fe(II); 0.3 M H$_2$C$_2$O$_4$]	Triton X-100/hexanol/cyclohexane/water (Triton X-100/hexanol molar ratio = 1 : 5; R = 30–60)	B (two μE)	Acicular ferrous oxalate particles; calcination (225–300°C): acicular γ-Fe$_2$O$_3$ particles (d_p ~7–8 nm) [50].

Table 6 Continued

Material	Polar phase reactants	Microemulsion system	Synthesis route[b]	Comments
Fe_3O_4	$FeCl_2 + FeCl_3$ (μE 1); NH_3 (μE 2) [1.4 M Fe(II), 1.4 M Fe(III), 15 M NH_3]	NP-6/cyclohexane/water; AOT/isooctane/water; AOT/cyclohexane/water	B (two μE); C ($FeCl_2$ addition)	Superparamagnetic particles; d_p ~5 nm [51–53].
Fe_3O_4	$FeCl_2 + FeCl_3$ (μE 1); NH_4OH (μE 2) [0.1 M Fe(II), 0.2 M Fe(III); 2 M NH_4OH]	AOT/heptane/water (R = 10, 15% aqueous phase)	B (two μE)	Superparamagnetic particles; $d_p \approx 10$ nm, assumed to correspond to droplet diameter [54].
		Labrasol/plurol/ethyl oleate/water (35% labrasol + plurol, 52% ethyloleate, 13% water)	B (two μE)	Particle size $d_p \approx 30$ nm, assumed to correspond to droplet diameter.
Fe_3O_4	$FeCl_3$ (μE 1), NaOH (μE 2) [0.1 M $FeCl_2$, 0.15 M $FeCl_3$, 1 M NaOH]	AOT/isooctane/water (R = 3.7–14.9)	B (two μE)	Particle size ~1 nm (DLS) in μE, 5–20 nm (TEM, dried by evaporation). Magnetic properties: 1 nm particles superparamagnetic, 5–20 nm particles ferromagnetic [55].
SiO_2	H_2O (+HCl, pH 2)	SDS/pentanol/water	C (TEOS addition)	Approximately triangular μE stability region (corner compositions: 100% C_5OH; 35% SDS, 45% C_5OH, 20% water; 5% SDS, 25% C_5OH, 70% water). 9 wt% TEOS gives transparent silica gel over nearly entire μE stability region; increase in TEOS shifts the transparent gel region toward the water-rich corner [57]

SiO$_2$	H$_2$O (+HCl, pH 2)	SDS/pentanol/water	C (TEOS addition)	Phase diagrams were determined for SDS/pentanol/water system with and without ethanol or TEOS addition [58].
SiO$_2$	H$_2$O(+HNO$_3$, pH 2)	SDS/pentanol/water	C (TEOS addition)	Viscoelectric properties were investigated. Gelation time 30–35 h [59].
SiO$_2$	H$_2$O (pH 1)	Ethanol/C$_n$OH ($n = 4$, 5, 6)/H$_2$O	C (TEOS addition)	Phase diagrams were determined with and without TEOS [60].
SiO$_2$	Silica sol (partially hydrolyzed TMOS); 0.4–10 wt% HF (Silica sol: 33.3 wt% 0.25 M H$_2$SO$_4$, 66.7 wt% TMOS)	DDAB/decane/water (48.7% DDAB, 19.5% decane, 31.8% aqueous silica sol)	C (HF addition)	Microporous silica gels; monodisperse pores (2 nm pore radius), large specific surface area (~103 m^2/g) [61].
SiO$_2$	Nonaqueous μE (formamide)	CTAB/decanol/decane/formamide	C [TEOS + H$_2$O (pH 1, HNO$_3$)] [TEOS/μE = 4 : 1 (w/w); TEOS/H$_2$O = 0.5 (molar ratio)]	Both decanol- and formamide-continuous μE's yielded gels [62].
SiO$_2$	Nonaqueous μE (glycerol)	AOT/decanol/glycerol	C [TEOS + H$_2$O (pH 1, HNO$_3$)]	Gels obtained only in glycerol-rich region of phase diagram [62].
SiO$_2$	Nonaqueous μE (glycerol)	AOT/decane/glycerol	C [TEOS + H$_2$O (pH 1, HNO$_3$)]	No gels were obtained [62].

Table 6 continued

Material	Polar phase reactants	Microemulsion system	Synthesis route[b]	Comments
SiO_2	Nonaqueous μE (formamide)	CTAB/decanol/decane/formamide (decanol/decane = 75/25 w/w)	C [TEOS + H_2O (pH 1)] (water/TEOS molar ratio = 2)	Viscoelastic properties of silica gels were investigated; gels based on formamide/(decane + decanol) molar ratio ~2.5–3 gave the highest elasticity [63].
SiO_2	Nonaqueous μE (formamide)	CTAB/decanol/decane/formamide	C [TEOS + H_2O (pH 1, HNO_3)] (μE/H_2O/TEOS = 5 : 1 : 5 w/w)	Condensation rate monitored with ^{29}Si NMR [64].
SiO_2	Nonaqueous μE (ethylene glycol)	SDS/octanol/toluene/ethylene glycol (weight ratio: SDS/octanol, toluene (50/50 w/w)/ethylene glycol = 13.1 : 29.8 : 29.0)	C [TEOS + H_2O (pH 2, HCl)]	Reaction rate monitored with ^{29}Si NMR [65].
SiO_2	H_2O + $Cu(NO_3)_2$ (48%) (+HNO_3, pH 1.25)	TEGDE/cyclohexane/water (25% TEGDE, 40% cyclohexane, 35% aq soln)	C (TEOS addition) (TEOS/μE = 1 : 1 w/w)	$Cu(NO_3)_2$-encapsulated silica gel product [66].
SiO_2	Nonaqueous μE (formamide)	CTAB/decanol/decane/formamide	C [TEOS + H_2O (pH 1, HNO_3; 10^{-2} M laser dye)]	Silica gels doped with laser dyes (rhodamine B, rhodamine 6G) gave fluorescence quantum yields, indicating promise as candidate solid-state laser dye materials [67].
SiO_2	Nonaqueous μE (glycerol)	AOT/decanol/glycerol	C [TEOS + H_2O (pH 1, HNO_3; 10^{-2} M laser dye)]	Silica gels doped with laser dyes (rhodamine B, rhodamine 6G) were synthesized [67].
SiO_2	Nonaqueous μE (formamide)	CTAB/decanol/decane/formamide	C [TEOS + H_2O (pH 1, HNO_3; coumarin 120 or coumarin 311)]	Preparation of silica gels doped with coumarin 120 and coumarin 311 [68].

SiO$_2$	HCl (0.1 M) or NaOH (0.1 M)	AOT/isooctane/gelatin/water [R = 30, 14% w/v gelatin (referred to the aqueous pseudophase)]	C (TMOS addition) (water/TMOS molar ratio = 4)	Product: silica-gelatin nanocomposites, ~30 nm silica particles [69].
SiO$_2$		AOT/isooctane/chitosan/water	C (TMOS addition)	Silica-chitosan nanocomposites [69].
SiO$_2$		AOT/isooctane/water	C (TMOS addition)	Silica nanoparticles (~30 nm) aggregate to form porous gel structure [69].
SiO$_2$	Na$_2$SiO$_3$ (0.2 M)	Triton X-100/hexanol/cyclohexane/water	C (acid addition?)	~30 nm diameter particles [70].
SiO$_2$	NH$_4$OH (0.7–3.6 M)	AOT/isooctane/water (R = 6–18)	C (TEOS addition) (water/TEOS molar ratio = 0.25–4)	Polydisperse, 15–70 nm porous particles; specific surface area = 100–300 m^2/g [71].
SiO$_2$	H$_2$O, NH$_3$	AOT/toluene/water (R = 1–6)	C (addition of TEOS and MPS) (water/TEOS molar ratio = 2.8–17)	Spherical, 20–70 nm particles; vinyl groups in surface via MPS [72].
SiO$_2$	H$_2$O, NH$_3$	POELE/cyclohexane/water (R = 5.5)	C (addition of TEOS and MPS)	Spherical particles, d_p~28 nm [72].
SiO$_2$	H$_2$O + NH$_3$ (30 wt%)	AOT/toluene/water/ammonia (R = 2.1–10)	C (TEOS + MPTMS addition) (water/TEOS molar ratio = 6–21)	Particle size 28–113 nm [73].
SiO$_2$	H$_2$O + NH$_3$ (30 wt%)	Isopropanol/toluene/water/ammonia	C (TEOS + MPTMS addition)	Highly unstable particle dispersion [73].

Table 6 continued

Material	Polar phase reactants	Microemulsion system	Synthesis route[b]	Comments
SiO_2	H_2O, NH_3 (29 wt%)	NP-4/heptane/water/ammonia ([NP-4] = 0.06–0.25 M; [H_2O] = 0.12–0.58 M; [NH_3]/[H_2O] = 0.086–2.0)	C (TEOS addition) ([TEOS] = 0.0357 M in µE–TEOS system; h = [H_2O]/[TEOS] = 3.36–16.2)	Silica particle size 26–43 nm, depending on water and surfactant concentrations; particle size distribution approximately constant during growth period [74].
SiO_2	H_2O, NH_3 (29 wt%)	NIS/heptane/water/ammonia (NIS = NP-4, NP-5, or DP-6; 0.04–0.26 M NIS, 0.174–1.10 M H_2O, 0.075 M NH_3)	C (TEOS addition) ([TEOS] = 0.018 M in µE–TEOS mixture)	Monodisperse spherical nanoparticles, d_p = 28–50 nm; particle size increased in the order NP-5 > NP-4 > DP-6 [75].
SiO_2	H_2O, NH_3 (29 wt%)	NP-5/oil/water/ammonia (oil = heptane, cyclohexane, or heptane/cyclohexane (50/50 v/o); 0.04–0.23 M NP-5, 0.174 M H_2O, 0.075 M NH_3)	C (TEOS addition) ([TEOS] = 0.018 M in µE–TEOS mixture)	Monodisperse spherical nanoparticles, d_p = 30–75 nm; particle size increased in the order heptane > heptane/cyclohexane > cyclohexane [75].
SiO_2	H_2O, NH_3 (29.6 wt%)	NP-5/cyclohexane/water ammonia ([NP-5] = 0.09–0.3 M; R = 0.7–2.3)	C (TEOS addition) ([TEOS] = 0.027 M in µE–TEOS mixture; h = [H_2O]/[TEOS] = 7.8)	Spherical monodisperse particles, d_p = 50–70 nm [76].
SiO_2	H_2O, NH_3 (29.6 wt%)	NP-5/cyclohexane/water/ammonia ([NP-5] = 0.056–0.277 M; R = 0.5–3.5)	C (TEOS addition) ([TEOS] = 0.023 M in µE–TEOS mixture; h = [H_2O]/[TEOS] = 7.8)	Spherical monodisperse particles, d_p = 35–68 nm, depending on R and time; minimum in d_p at ~R = 1.5–2 [77].

SiO$_2$	H$_2$, NH$_3$ (1.6–29.6 wt%)	NP-5/cyclohexane/water/ammonia ([NP-5] = 0.05–0.3 M; R = 0.5–6.8)	C (TEOS addition) ([TEOS] = 0.0082–0.102 M in μE–TEOS mixture; h = [H$_2$O]/[TEOS] = 1.6–19.9)	Spherical monodisperse particles, d_p = 32–76 nm; minimum in d_p at ~R = 2–3 [78,79].
SiO$_2$	H$_2$O, NH$_3$	NP-5/cyclohexane/water/ammonia (R = 0.7–5.4)	C (TEOS addition) ([TEOS] = 0.025 M; water/TEOS molar ratio (h) = 7.8)	Effect of R on the time evolution of particle size was investigated [80].
SiO$_2$	H$_2$O, NH$_3$ (29.6 wt%)	NP-5/cyclohexane/water/ammonia ([NP-5] = 0.05–0.3 M; R = 2.29–6.8)	C (TEOS addition) ([TEOS] = 0.0082–0.102 M in μE–TEOS mixture; h = [H$_2$O]/[TEOS] = 1.6–19.9)	Spherical particles, d_p = 30–70 nm; bimodal size distribution; phase separation during synthesis reaction [81].
SiO$_2$	H$_2$O, NH$_3$ (13.9 wt%)	AOT/decane/water/ammonia ([AOT] = 0.09–0.24 M; R = 2.0–9.5)	C (TEOS addition) (water/TEOS molar ratio (h) = 18.5 [TEOS] = 0.044 M)	No particles observed (TEM) for R < 4; d_p = 10–60 nm; d_p increased with R [82].
SiO$_2$	H$_2$O, NH$_3$ (13.9 wt%)	AOT/decane/benzyl alcohol (BA)/water (R = 6.8; BA/AOT molar ratio = 0–2.5)	C (TEOS addition) ([TEOS] = 0.044 M, water/TEOS molar ratio (h) = 18.5)	Microemulsions with BA/AOT > 1.5 became unstable during synthesis reaction; nearly spherical nanoparticles; maximum in particle size at BA/AOT = 1.5 [83].

Table 6 continued

Material	Polar phase reactants	Microemulsion system	Synthesis route[b]	Comments
TiO_2	H_2O	TX-100/decane/water ($R = 0.9-1.8$)	C (TTBO addition)	50–300 nm particles at sol-gel transition; size distribution broader as R increased; gelation time = 400 min ($R = 0.9$), 15 min ($R = 1.8$) [84].
TiO_2	H_2O	NIS/decane/water (NIS = TX-15, TX-35, or TX-45; $R = 0.7-2.0$)	C (TIPO or TTBO addition) ($h = 0.7-2.0$)	Gelation time ranged from minutes to hours, depending on the surfactant type, microemulsion composition, and alkoxide concentration [85].
TiO_2	H_2O, H_2O_2 [20–40% (w/v)]	TX-35/decane/water ($R = 1.1-1.7$)	C (TIPO addition) ([TIPO] = 0.8–1.0 mol/kg μE solution; $h = 1.7-2.3$)	Replacing H_2O with $H_2O_2-H_2O$ changed gelation time from days to hours [86].
TiO_2	H_2O	TX-100?/decane/water	C (TIPO addition)	Textural evolution of TiO_2 gels investigated via thermoporometry, dilatometry, and electron microscopy [87].
TiO_2	H_2O	TX-100?/decane/water ($R = 0.9$)	C (TIPO addition) ([TIPO] = 1.0 mol/kg μE solution; [H_2O]/[TIPO] = $h = 2$)	Gelation time = 15.5 h; solvent effects on textural evolution of TiO_2 gels investigated via thermoporometry [88].
TiO_2	H_2O	TX-35/decane/water ($R = 1.0-2.0$)	C (addition of TIPO or TTBO) ([Alkoxide] = 0.4–1.2 mol/kg μE solution)	Above a certain critical value of R (≤ 1.5) and below a certain initial value of alkoxide concentration, the gel structure changed from a narrow to a broader pore size distribution [89].

TiO_2	H_2O	TX-100?/decane/water ($R = 1.1$–1.5)	C (TIPO addition)	Rheological investigation of sol-to-gel transition [90].
TiO_2	H_2O	TX-100?/decane/water	C (TIPO addition)	X-ray and neutron scattering investigation of TiO_2 gel structure [91].
TiO_2	H_2O	TX-35/oil/water (oil = decane or cyclohexane, $R = 1$–2)	C (addition of TIPO) ([TIPO] = 0.1 mol/kg μE; $h = [H_2O]/[TIPO]$ = 2–3)	The fractal dimension of the gels (determined by small-angle X-ray scattering) varied between ~2.0 and ~2.5 [92].
TiO_2	H_2O	SDS/pentanol/cyclohexane/water (H_2O/SDS = 2.5 w/w; H_2O/SDS/TIPO = 0.4 : 0.01 : 0.2 M)	C (addition of TIPO)	Gelling occurs in hours [93].
TiO_2	H_2O	TX-100/cyclohexane/water (H_2O/TX-100/TIPO = 0.4 : 0.2 : 0.2 M)	C (addition of TIPO)	Gelling occurs in hours [93].
TiO_2	H_2O	SDS/pentanol/water (H_2O/SDS = 50 : 1 M)	C (addition of TIPO)	Gelling occurs in days [93].
TiO_2	H_2O	TX-35 or TX-45/decane/water	C (addition of TIPO and photoactive species, Eu^{3+}, Nd^{3+}, pyrene)	Preparation of photoactive microemulsion-derived titania gels [94].
TiO_2	H_2O	AOT/isooctane/water ($R = 10$; 0.1 M AOT, 1.0 M H_2O)	C (TIPO addition)	Ultrafine, ? nm particles [95].
TiO_2	H_2O	AOT/cyclohexane/water ($R = 4$–10)	C (addition of TTEO, TIPO, or TTBO) ([Alkoxide] = 3×10^{-2} M (in μE))	Discrete nanoparticles [15–30 nm] produced at short times, but flocculation and particle settling occurred at longer times [96].

Table 6 continued

Material	Polar phase reactants	Microemulsion system	Synthesis route[b]	Comments
TiO_2	H_2O	AOT/heptane/water ($R = 4$–10)	C (addition of TTEO, TIPO, or TTBO ([Alkoxide] $= 3 \times 10^{-2}$ M (in μE))	Discrete nanoparticles (15–30 nm) produced at short times, but flocculation and particle settling occurred at longer times [96].
TiO_2	H_2O	NP-5/cyclohexane/water ($R = 0.15$–5)	C (addition of TTEO, TIPO, or TTBO ([Alkoxide] $= 3 \times 10^{-2}$ M (in μE))	Discrete nanoparticles (15–30 nm) produced at short times, but flocculation and particle settling occurred at longer times [96].
TiO_2	Nonaqueous μE (glycerol)	AOT/cyclohexane/glycerol [0.4–4.3 wt% H_2O in glycerol–water mixture; glycerol/surfactant molar ratio (R_g) ≤ 2.5]	C (addition of TIPO or TTBO) ([Alkoxide] $= 3 \times 10^{-4}$–5×10^{-3} M)	$d_p \approx 20$–80 nm; increase in R_g > 0.8 gave gel-like structures [96].
TiO_2	Nonaqueous μE (glycerol)	AOT/heptane/glycerol [0.4–4.3 wt% H_2O in glycerol–water mixture; glycerol/surfactant molar ratio (R_g) ≤ 2.5]	C (addition of TIPO or TTBO) ([Alkoxide] $= 3 \times 10^{-4}$–5×10^{-3} M)	d_p 20–80 nm; increase in R_g > 0.8 gave gel-like structures [96].
TiO_2	H_2O	AOT/isooctane/water (0.05–0.1 M AOT, $R = 9$–30)	C (addition of TTBO–butanol solution) ([TTBO] $= 10^{-5}$–10^{-4} M)	Size of reverse micelles $= 9$–19.3 nm (DLS); $d_p \approx 3$ nm (DLS); quantitative model of particle formation process [97].
TiO_2	H_2O	AOT/cyclohexane/water ($R = 2$–20)	C (addition of TIPO–cyclohexane solution) ([TIPO] $= 5 \times 10^{-4}$–5 $\times 10^{-2}$ M)	Size of reverse micelles $= 1.6$–6.6 nm; nanoparticles, $d_p < 10$ nm; formed aggregates (20–200 nm) and gelatinous precipitates; crystalline phase: anatase for $R > 10$, amorphous particles for $R < 6$ [98].

Material	Precursor	Surfactant system	Method	Comments
TiO_2	H_2O	CDBA/benzene/water ([CDBA] = 0.04 M; R = 4.5–13.5)	C (addition of $TiCl_4$–benzene solutions; 0.625 M $TiCl_4$)	Typical TiO_2 concentration in μE: 6.25×10^{-3} M; size of reverse micelles = 2.5–5 nm; $d_p \approx 1$ nm [99].
TiO_2	$TiCl_4$ (0.3 M, μE 1), NH_3 (1.2 M, μE 2)	TX-100/hexanol/cyclohexane/water	B (two μE)	Calcined particles: aggregates of nanoparticles, d_p = 15–30 nm at 700°C; d_p = 40–60 nm at 1000°C [100].
V_2O_5	H_2O	AOT/toluene/water (10–20 wt% AOT; 2.6–5 wt% H_2O)	C (addition of VIPO–toluene solution) (~0.4 wt% VIPO)	Alkoxide hydrolysis time scale ~40 ms [101].
V_2O_5	H_2O	AOT/toluene/water (24–40 wt% AOT; ≥5 wt% water)	C (addition of VIPO–toluene solution) (0.37–0.53% VIPO in μE)	Small-angle X-ray scattering gave Porod slopes of −2.0 (5 min) and −2.5 (3 h); dynamic light scattering indicated lag time in particle growth [102].
ZnO (ZnCO₃ precursor)	$Zn(NO_3)_2$ (μE 1); $(NH_4)_2CO_3$ (μE 2) [0.1 M $Zn(NO_3)_2$, 0.1 M $(NH_4)_2CO_3$]	CTAB/butanol/octane/water (10 wt% CTAB, 10 wt% butanol, 44 wt% octane, 36 wt% aqueous phase)	B (two μE)	ZnO obtained by calcination of $ZnCO_3$; ZnO particle size 5–40 nm [103].
ZrO_2	H_2O	DOLPA/isooctane/water (0.02 M DOLPA in isooctane; [H_2O]/[DOLPA] = 15)	C (addition of ZTBO–butanol solution) (0.5×10^{-5}–5×10^{-5} M ZTBO)	Reverse micelles ~5 nm (DLS); particle size ~2 nm; as-prepared material, tetragonal zirconia [TEM-EDAX], calcined (673–873 K) particles—partial formation of zirconium phosphate [104,105].

Table 6 continued

Material	Polar phase reactants	Microemulsion system	Synthesis route[b]	Comments
ZrO_2	H_2O, H_2SO_4	NP-6/cyclohexane/water ([NP-6] = 200 mmol/kg; R = 2–40, R_s = 0–0.3)	C (addition of ZTBO) ([ZTBO] = 10 mmol/kg)	Solubilization of H_2SO_4 in NP-6/cyclohexane solution took ~7 days to reach equilibrium; no particles formed in the absence of H_2SO_4; spherical particles, d_p = 10.5–76 nm [106].
ZrO_2	$ZrO(NO_3)_2$ (μE 1); NH_3 (μE 2) [0.75 M $ZrO(NO_3)_2$, 2.0 M NH_3]	NP-5/NP-9/cyclohexane/water (35 wt% NP-5 + NP-9, NP-5/NP-9 = 1 : 1, 53 wt% cyclohexane, 12 wt% aqueous phase)	B (two μE)	As-prepared particles: spherical, d_p = 5–10 nm (TEM); calcination of $Zr(OH)_4$ product yielded ZrO_2 [107].
Aluminosilicate zeolite	H_2O, sodium aluminate sodium silicate [1.0 M Al(III) + 1.5 M NaOH, 1.0 M Si(IV) + 1.5 M NaOH]	AOT/isooctane/water (R = 4)	C (addition of sodium aluminate solution to a sodium silicate–containing μE) + A (heating to 80°C)	Zeolite crystals (1–2 μm) after 8 h at 80°C [45,108].
$BaFe_{12}O_{19}$ (barium ferrite; barium iron oxalate precursor)	Nonaqueous microemulsion (ethanol)	M-AOT/isooctane/ethanol (M = Fe^{3+}, Ba^{2+}; Ba^{2+}/Fe^{3+} = 1 : 12)	C (addition of oxalic acid in ethanol)	Barium iron oxalate coprecipitate, well dispersed, d_p = 10–20 nm; calcined (950°C) product ($BaFe_{12}O_{19}$), d_p = 60 nm [109].
$3Al_2O_3 \cdot 2SiO_2$ (mullite)	H_2O, $Al(NO_3)_3$ [10–30 wt% $Al(NO_3)_3$]	SDS/pentanol/water	C (addition of TEOS)	Synthesis not successful (no gelation) [110].
$BaTiO_3$ (barium titanyloxalate precursor)	$BaCl_2$, $TiCl_4$ (μE 1); $H_2C_2O_4$ (μE 2) (0.05 M $BaCl_2$, 0.05 M $TiCl_4$, 0.1 M $H_2C_2O_4$	GOX-30/decane/water (10 g GOX-30, 70 g decane, water/surfactant mass fraction = 0.2–0.75?)	B (two μE)	30 min reaction time; as-prepared particles: $d_p \sim$ 5 nm (TEM); XRD did not reveal Ba, Ti oxalate diffraction pattern; calcination (400–1200°C) did not give phase-pure (XRD) $BaTiO_3$ [111].

Compound	Precursor solution	Microemulsion system	Method	Remarks
$BaTiO_3$	H_2O	Brij 30/cyclohexane/water (6 g Brij 30, 36 mL cyclohexane, 50 μL water)	C (addition of μE to TNPO–BIPO–isopropanol solution)	Average crystallite size = 9 nm (XRD line broadening) [112].
$SrTiO_3$	H_2O	Brij 30/cyclohexane/water (6 g Brij 30, 36 mL cyclohexane, 50 μL water)	C (addition of μE to TNPO–SIPO–isopropanol solution)	As-produced material: $SrTiO_3$, $SrCO_3$, TiO_2 mixture; calcination (700°C, 1 h) product; phase-pure $SrTiO_3$, crystallite size=18 nm [112].
$BaZrO_3$	H_2O	Brij 30/cyclohexane/water (6 g Brij 30, 36 mL cyclohexane, 50 μL water)	C (addition of μE to ZTPO–BIPO–isopropanol solution)	As-produced material: $BaCO_3$–ZrO_2 mixture; calcination (800°C, 1 h) product: phase-pure $BaZrO_3$, crystallite size = 11 nm [112].
$SrZrO_3$	H_2O	Brij 30/cyclohexane/water (6 g Brij 30, 36 mL cyclohexane, 50 μL water)	C (addition of μE to ZTPO–SIPO–isopropanol solution)	As-produced material: $SrCO_3$–ZrO_2 mixture; calcination (900°C, 1 h) product; phase-pure $SrZrO_3$, crystallite size = 19 nm [112].
$BaPbO_3$ (mixed Ba, Pb oxalate precursor)	$Ba(NO_3)_2 + Pb(NO_3)_2$ (μE 1); $(NH_4)_2C_2O_4$ (μE 2) [0.1 M $Ba(NO_3)_2$, 0.1 M $Pb(NO_3)_2$; 0.2 M $(NH_4)_2C_2O_4$]	NP-5/octane/water [mixture of μE 1 plus μE 2: 25.5 wt% NP-5, 59.5 wt% octane (or 42.5 wt% NP-5, 42.5 wt% octane), 15 wt% aqueous solution]	B (two μE)	15–30% excess oxalate needed for complete coprecipitation. Colloidal stability: stable dispersion for first 2 h visible flocculation and precipitation subsequently. Ba-, Pb oxalate coprecipitate: spherical particle size ≈ 8 nm (TEM); calcination (600–650°C, 12 h) product: phase-pure $BaPbO_3$ [113].

Table 6 continued

Material	Polar phase reactants	Microemulsion system	Synthesis route[b]	Comments
La$_2$CuO$_4$ (mixed La, Cu oxalate precursor)	La(NO$_3$)$_3$ + Cu(NO$_3$)$_2$ (μE 1); H$_2$C$_2$O$_4$ (μE 2) [0.2 M La(NO$_3$)$_3$, 0.1 M Cu(NO$_3$)$_2$; 0.4 M H$_2$C$_2$O$_4$]	NP-5/petroleum ether/water 25.5 wt% NP-5, 59.5 wt% petroleum ether (or 42.5 wt% NP-5, 42.5 wt% petroleum ether), 15 wt% aqueous solution)	B (two μE)	15–30% excess oxalate needed for complete coprecipitation. Colloidal stability: stable dispersion for first 1 h, visible flocculation and precipitation subsequently: $d_p \sim$5 nm. Calcination (600–700°C, 2 h) product: single-phase La$_2$CuO$_4$ [113].
LaNiO$_3$ (mixed La, Ni oxalate precursor)	La(NO$_3$)$_3$ + Ni(NO$_3$)$_2$ (μE 1); H$_2$C$_2$O$_4$ (μE 2) [0.2 M La(NO$_3$)$_3$, 0.2 M Ni(NO$_3$)$_2$; 0.5 M H$_2$C$_2$O$_4$]	NP-5/petroleum ether/water (25.5 wt% NP-5, 59.5 wt% petroleum ether (or 42.5 wt% NP-5, 42.5 wt% petroleum ether), 15 wt% aqueous solution)	B (two μE)	15–30% excess oxalate needed for complete coprecipitation. Colloidal stability: stable dispersion for first 2 h, visible flocculation and precipitation subsequently. As-produced material: spherical, monodisperse, $d_p \sim$20 nm. Calcination (800°C, 20 h) product: single-phase LaNiO$_3$ [113].
Yittrium-iron garnet (YIG)	Not given	CTAB/butanol/octane/water	Not given	Experimental details for synthesis not provided; magnetic properties of annealed samples ($>$600°C) investigated [114].
Yttrium-iron garnet (YIG)	Not given	Igepal CA-570/heptane/water	Not given	Experimental details for synthesis not provided; magnetic properties of annealed samples ($>$600°C) investigated [114].

Material (precursor)	Composition	Microemulsion[a]	Synthesis route[b]	Results
Yttrium-iron garnet (YIG) (carbonate, hydroxide precursor)	$Fe(NO_3)_3 + Y(NO_3)_3$ (μE 1); NH_4OH or $(NH_4)_2CO_3$ (μE 2) [0.2 M $Fe(NO_3)_3$, 0.12 M $Y(NO_3)_3$; 1.5–2 M NH_4OH, 0.4–0.6 M $(NH_4)_2CO_3$]	Igepal CA-520/heptane/water (38.13% Igepal CA-520, 53.87% heptane, 8% aqueous phase)	B (two μE)	Carbonate, hydroxide coprecipitates, $d_p \sim 3$ nm; calcination ($>700°C$) to YIG [115].
$YBa_2Cu_3O_{7-x}$ (oxalate precursor)	$Y(NO_3)_3 + Ba(NO_3)_2 + Cu(NO_3)_2$ (μE 1); $(NH_4)_2C_2O_4$ (μE 2) [0.05 M $Y(NO_3)_3$, 0.1 M $Ba(NO_3)_3$, 0.15 M $Cu(NO_3)_3$; 0.45 N $(NH_4)_2C_2O_4$]	CTAB/butanol/octane/water (29.25%, CTAB + butanol (1 : 0.73 w/w), 59.42% octane, 11.33% aqueous solution)	B (two μE)	As-produced material, $d_p \sim 47$ nm; calcination (air, 820°C, 2 h) product: single-phase YBCO, $d_p \sim 275$ nm [116–118].
$YBa_2Cu_3O_{7-x}$ (oxalate precursor)	$Y(NO_3)_3 + Ba(NO_3)_2 + Cu(NO_3)_2$ (μE 1); $(NH_4)_2C_2O_4$ (μE 2) [0.05 M $Y(NO_3)_3$, 0.1 M $Ba(NO_3)_2$, 0.1 M $Cu(NO_3)_2$; 0.3 M $(NH_4)_2C_2O_4$]	Labrasol/isostearic plurol/ethyl oleate/water [35% labrasol/isostearic plurol (3 : 1), 52% ethyl oleate, 13% aqueous solution]	B (two μE)	Particle size: $d_p \approx 30$ nm (as prepared), 100 nm [after heat treatment: 825°C, air, 24 h, plus 450°C, oxygen, 6 h] [119].
Bi-Pb-Sr-Ca-Cu-O (2223) oxide superconductor (Bi-Pb-Sr-Ca-Cu oxalate precursor)	Bi(III), Pb(II), Sr(II), Ca(II), Cu(II) acetates [0.0368 M Bi^{3+}, Bi/Pb/Sr/Ca/Cu = 1.84 : 0.34 : 1.91 : 2.03 : 3.06; $H_2C_2O_4$ (10% excess)]	Igepal CO-430/cyclohexane/acetic acid–water (50 : 50 v/v)	B (two μE)	As-produced material: mainly monodisperse, $d_p = 2$–6 nm; calcination (air, 800°C, 12 h), sintering (840°C, 96 h): nearly phase-pure 2223 oxide [120].

[a] μE = microemulsion; GOX-30 = nonionic surfactant Genapol OX 30 (Hoechst); Igepal CA-570 = Ethylene glycol monobutyl ether; Igepal CA-520 = Penta (ethylene glycol) monoisononyl phenyl ether; Igepal CO-430 = Nonylphenoxypoly(ethyleneoxy) ethanol.
[b] For synthesis routes A–C, see Fig. 3.

The surfactant used was therefore the potassium salt of DOLPA. The resulting organic phase was a water-in-oil microemulsion since the potassium transfer was accompanied by the coextraction of water molecules. Following liquid–liquid separation of the bulk phases, the organic phase was used for material synthesis.

Dutta and Robbins [45,108] used a synthesis protocol that combined the microemulsion-plus-reactant route (C) with the microemulsion-plus-trigger method (A). An isooctane solution of AOT was first prepared, and an aqueous solution of sodium silicate was then added to convert the organic solution into a microemulsion. Following this, an aqueous solution of sodium aluminate was introduced, a procedure that corresponded to the microemulsion-plus-reactant synthesis, protocol C. After phase separation (apparently the final microemulsion could solubilize only 50% of the added aqueous phase), the temperature of the microemulsion was raised (synthesis protocol A) for further reaction to proceed.

In addition to the above variations, which involve the manipulation of reactant addition sequencing, further modifications of the three basic synthesis protocols have been achieved through (1) substitution of the aqueous pseudophase with a polar organic solvent to prepare nonaqueous microemulsions [62–65,96,109], (2) introduction of functional coupling agents that effect in situ surface modification of the synthesized particles [72,73], (3) synthesis of inorganic materials within the aqueous microdomains of microemulsion organogels [69], and (4) introduction of additives (e.g., metal salts and organic or inorganic photoactive species) that are encapsulated in the matrices of the synthesized materials [66,67,94].

Very little is currently available in terms of systematic comparisons of synthesis protocols. Liz et al. [54] criticized the microemulsion-plus-aqueous solution method used by Gobe et al. [51] and by Bandow et al. [53] to prepare magnetite. It was argued [54] that mixing the microemulsion fluid with the aqueous solution is likely to result in partial destruction of the water pools that serve as test tubes for the particle formation reactions. The importance of using identical microemulsions for the reactants in the two-microemulsion protocol was also stressed by Chhabra et al. [50].

B. Surfactants and Microemulsion Phase Behavior

It can be seen from Table 6 that anionic, cationic, and nonionic surfactants have all been exploited in formulating microemulsions for materials synthesis. Anionic and nonionic surfactants appear to be the most popular types of surfactants, with Aerosol OT (AOT) and the polyoxyethylated alkylphenyl ether surfactants (e.g., NP-5) leading. Part of the attraction of AOT and the NP surfactants is related to the fact that they permit microemulsion formulation without the need for cosurfactants. Also, a large body of information is already available on the phase behavior and structure of AOT microemulsions [121], and this makes it convenient to work with this anionic surfactant. A unique advantage of the nonionic surfactants is the fact that their use does not involve the introduction of (potentially undesirable) counterions. The ability to alter the size of the hydrophilic (oxyethylene) groups and/or the hydrophobic (alkyl) groups provides additional flexibility in surfactant selection.

In order for microemulsion-based materials synthesis to be feasible, surfactant/oil/water formulations that give stable microemulsions must be identified. Phase diagrams already available in the literature [122–124] provide a useful starting point. Frequently, however, these published diagrams do not extend to conditions directly relevant to materials synthesis, e.g., in terms of the specific metal salt, base, acid, and temperature. Of important consideration, therefore, are investigations into the effects of the reactants

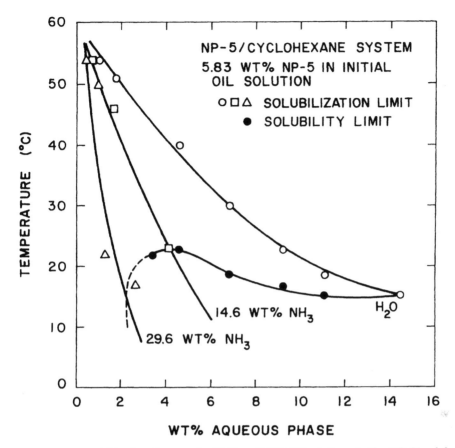

Figure 4 Solubilization diagram for H_2O and NH_4OH solutions in the NP-5/cyclohexane system; 5.83 wt% NP-5. (From Ref. 78.)

and products on the domain of stability of water-in-oil microemulsions. In the course of their work on microemulsion-mediated Fe(III) precipitation, O'Sullivan et al. [47] noted that the maximum Water uptake in the $C_{12}EO_4$/hexane/water microemulsion decreased from 36 wt% to 26 wt% when the aqueous pseudophase was changed from 100% H_2O to a 0.025 M $FeCl_3$ solution. Morooka and coworkers [104] reported that decreasing the pH from 6 to 3.5 lowered the water content of their DOLPA-based microemulsion by two orders of magnitude.

The solubilization behaviors of water and aqueous ammonia solutions in the NP5/cyclohexane/water microemulsion system are compared in Fig. 4 [78]. It can be seen that replacing water with ammonium hydroxide solution shifts the solubilization curve to lower temperatures. The observed ammonia effect is attributable to a competition between hydroxyl ions and the oxyethylene groups of the surfactant for interaction with water molecules. The microemulsion-based alkoxide sol-gel systems provide an interesting example of reactant effects on microemulsion phase stability. In this case, water serves multiple roles, including that of a reaction medium (i.e., microemulsion water pools) and a reactant [for alkoxide hydrolysis, Eq. (13)]. Thus, consumption of water via the alkoxide hydrolysis reaction can lead to destabilization of the microemulsions [57,77]. Also, the alcohol reaction product [Eq. (13)] can contribute to additional microemulsion destabilization [58,60,81].

The susceptibility of microemulsions to destabilization by electrolytes severely limits the highest metal concentrations that can be used for precipitation reactions. This, in turn, discourages the large-scale application of microemulsion-mediated materials synthesis. A possible approach to tackling this problem appears to lie in the judicious selection of cosurfactants for microemulsion formulations. Darab et al. [125] reported that addition of SDS to the AOT/isooctane/water microemulsion increased dramatically the tolerable concentration of metal salts in the water pools. According to Chhabra et al. [50], addition of n-hexanol to the Triton X-100/cyclohexane/water microemulsion led to a significant improvement in the water-solubilizing capacity.

C. Physicochemical Processes in Particle Formation

A central issue in the microemulsion-mediated synthesis of materials is the relationship between the properties of the microemulsion fluid phase and those of the ensuing particles. To address this question systematically, it is necessary to consider how the properties of the microemulsion influence each stage in the solid formation process, i.e., nucleation, growth of primary particles, ripening, and coagulation of primary particles. Among the key microemulsion properties of interest are the size of the water pools (radius $= r_w$), the surfactant aggregation number (N), the nature of the surfactant polar groups, the nature of the solubilized water, and the rate of intermicellar communication (rate constant k_{ex}). These properties, in turn, may be altered by varying the water/surfactant molar ratio (R), the nature of the continuous oil phase, the type of surfactant and/or cosurfactant, the nature and concentration of electrolyte in the water cores, and the temperature.

1. Nuclei Formation in Microemulsions

It is well established that in order to obtain a stable nucleus, a cluster containing a critical number of monomers (n_c) must form [126–128]. It follows, therefore, that for microemulsion-mediated synthesis, if the water pools are viewed as isolated microreactors, then a relevant question is, How many of the available water pools contain the minimum number of monomers needed for nucleation? Here the concept of occupancy number (n_{oc}) is helpful [13]. For a reactant solubilized in the reverse micellar pseudophase, the quantity n_{oc} represents the number of solubilisate molecules present in a given water pool.

It should be noted that the solubilisate will be distributed among the water pools according to the Poisson distribution law [129]. That is, the probability P_k of having k monomers per water pool such that the average occupancy number is n_{av} is given by

$$P_k = [(n_{av})^k \exp(-n_{av})]/k! \tag{21}$$

where k is an integer. As n_{av} increases, the probability that a water pool contains the critical number of monomers, and therefore the probability of nucleation increases. Figure 5a presents a schematic illustration of the expected trend; it can be seen that with an increase in n_{av}, the number of nuclei increases up to a constant value. In principle, when all the water pools satisfy the condition $n_{oc} \geq n_c$, no further increase in the number of nuclei is expected.

It would be expected that for a given concentration of water pools, the average occupancy number would increase with an increase in reactant concentration. On the other hand, for a given reactant concentration, an increase in the concentration of the water pools

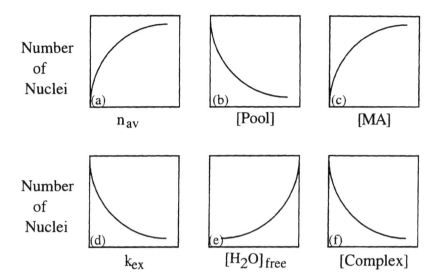

Number
of
Nuclei

(a) n_{av}

(b) [Pool]

(c) [MA]

Number
of
Nuclei

(d) k_{ex}

(e) $[H_2O]_{free}$

(f) [Complex]

Figure 5 Physicochemical factors in crystal nucleation in microemulsions.

would be expected to decrease the average occupancy number. Based on these considerations and Fig. 5a, it follows that with all things being equal, an increase in the pool concentration should decrease the number of nuclei, while an increase in the reactant (MA) concentration will have the opposite effect, as illustrated in Figs. 5b and 5c, respectively.

Nucleation is a rate phenomenon; therefore, even though the condition $n_{oc} \geq n_c$ is necessary, it is nonetheless insufficient to bring about nucleus formation. An additional requirement must be satisfied. That is, in order for cluster formation to proceed to completion, the monomers that represent potential candidates for a given critical nucleus must be retained for a sufficiently long period of time within a volume of molecular dimensions. If intermicellar exchange proceeds extremely rapidly, then the candidate monomers will be redistributed before nucleation can occur. Accordingly, it may be deduced that an increase in the communication rate will decrease the probability of nucleus formation. The expected trend of number of nuclei versus the rate of intermicellar exchange (k_{ex}) is illustrated schematically in Fig. 5d.

Since the water pool serves as the reaction medium, the characteristics of the solubilized water present within the polar core would be expected to influence the nucleation process. The properties of the water pool of relevance here include the proportion of free versus bound water molecules and the polarity of the solubilized water. When a volume of aqueous solution containing a given reactant is introduced into an initially dry surfactant–oil solution, a portion of the added water molecules will be immobilized through hydration of the surfactant polar groups. With increasing immobilization of the water molecules, reactions that require ionic dissociation (e.g., those involving weak acids and bases) will become less favorable. Thus, under such circumstances, an increase in the water content should enhance nucleation (see Fig. 5e). In the case of reactions involving hydrolysis, water serves as a reactant, and therefore a decrease in the availability of free water will lead to lowered nucleation rates (Fig. 5e).

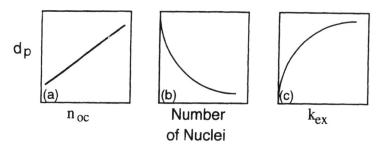

Figure 6 Physicochemical factors in crystal growth in microemulsions.

The nature of the surfactant polar groups can influence the nucleation process in two major ways. The polar groups may serve as sites for heterogeneous nucleation; this effect would be specific to particular surfactant–solid combinations. Adsorption and/or complexation of the reactant species by the polar groups will result in changes in the local concentrations at the interface and in the bulk of the water pool. Complexation will reduce the bulk concentration of the reactant species, and for nucleation within the water pool this will have the effect of decreasing the tendency toward nucleation (see Fig. 5f).

2. Crystal Growth in Microemulsions

Following the formation of nuclei, the monomers not incorporated into nuclei are used for particle growth. Thus, the size of the primary particles will be determined by the number of nuclei initially formed. That is, for a given amount of reactant, the greater the number of nuclei, the smaller the primary particles. As indicated above, the nucleation reaction takes place in the water pools via the combination of monomers. Accordingly, if the water pools are truly isolated from each other, then the final particle size will be determined by the average occupancy number at the time the reaction is initiated. That is, following the formation of the nucleus, the remaining $(n_{oc} - n_c)$ monomers in a given water pool will be used for crystal growth therein. Thus, smaller primary particles will be obtained by decreasing the occupancy number. Hence, by maintaining a relatively low number of monomers per water pool, and by isolating the water pools from each other, the system may be constrained to yield extremely small particles. Figure 6a illustrates the expected trend in primary particle size (d_p) versus occupancy number (n_{oc}).

In general, the water pools are not completely isolated, and therefore the effects of intermicellar communication on the growth process must be considered. Once nucleation takes place in the eligible (i.e., supersaturated) water pools, the monomers in the undersaturated (particle-free) pools will contribute to crystal growth through intermicellar matter exchange. In this case, the size of the primary particles will be determined by the number of nuclei initially formed. That is, for a given amount of reactant, the greater the number of nuclei, the smaller the number of free monomers available for growth and therefore the smaller the size of the primary particles. The resulting relationship between particle size and the number of nuclei is illustrated schematically in Fig. 6b. An increase in the intermicellar exchange rate (k_{ex}) increases the likelihood that the young nuclei in two different water pools will see one another and interact. It follows then that an increase in k_{ex} will result in larger particles (through colloidal aggregation), as illustrated schematically in Fig. 6c.

IV. MATERIALS SYSTEMS

A. Aluminum Oxide

Only limited work has been reported on microemulsion-mediated synthesis of aluminum hydroxide [44,45]. In the two publications available [44,45], AOT served as the surfactant. It is possible to form reverse micelles in supercritical fluid media [130], and Matson et al. [44] used such a medium and the microemulsion-plus-reactant technique to synthesize $Al(OH)_3$ particles at $110°C$. With supercritical propane as the continuous phase, anhydrous ammonia was injected into the reversed micellar solution containing solubilized Al^{3+} [as an aqueous $Al(NO_3)_3$ solution]. Referring to Fig. 1 and Table 2, the resulting precipitation process followed reaction path AP3; the added ammonia reacted with water molecules in the aqueous pseudophase of the microemulsion to generate hydroxide ions:

$$NH_3 + H_2O \rightarrow NH_4^+ + OH^- \tag{22}$$

An agglomerated product of spherical particles with a mean particle size on the order of 0.5 μm was obtained when the polar core contained 0.1 M $Al(NO_3)_3$. However, Matson et al. [44] did not identify the $Al(OH)_3$ phase obtained in their experiment. Decreasing the aluminum salt concentration resulted in smaller but more highly agglomerated particles. The observed decrease in primary particle size with decrease in Al^{3+} concentration is suggestive of the trend depicted in Fig. 6a.

Aluminum hydroxide was synthesized by Dutta and Robbins [45] using the microemulsion-plus-trigger protocol. A sodium aluminate solution was solubilized in an AOT/isooctane/water microemulsion. Heating this solution at $85°C$ for 24 h yielded a fine powder product that consisted of agglomerated submicrometer particles, as revealed by transmission electron microscopy (the authors, unfortunately, did not provide the corresponding SEM micrographs). It was concluded, on the basis of X-ray diffraction and Raman spectroscopy, that this material was gibbsite. In an extensive study of $Al(OH)_3$ precipitation from supersaturated aluminate solutions, de Bruyn and coworkers [131,132] identified the following precipitation sequence: amorphous phase→pseudo-boehmite→bayerite→gibbsite. It may be that systematic investigation of temperature and time effects would also reveal a similar sequence in microemulsion-mediated synthesis. It is interesting to note that previous work in conventional aqueous systems revealed that alkali metal ions can have dramatic effects on the rate of precipitation as well as the nature of the solid product. In the case of Li^+, a lithium aluminate phase ($LiAl_2(OH)_7 \cdot 2H_2O$) was identified [132]. Similar effects may be possible in the constrained environment of AOT water pools, due to the high local concentrations of sodium ions associated with the dissociation of AOT [133].

B. Iron Oxide

1. Hydrous Ferric Oxide

The first report of microemulsion-mediated synthesis of an iron oxide phase was by Inouye et al. [46]. The microemulsion-plus-reactant method (C of Fig. 3) was used by bubbling molecular oxygen into a ferrous ion–containing microemulsion. Microemulsions based on either AOT (i.e., AOT/hydrocarbon/water, where the hydrocarbons were cyclohexane, hexane, or heptane) or potassium oleate (i.e., potassium oleate/n-hexane/n-hexanol/water) were used; the aqueous pseudophase consisted of a ferrous ammonium sulfate solution (5×10^{-3} and 10^{-2} M). Electron microscopy revealed needle-shaped primary particles that aggregated into cubic clusters ($\sim$80 nm). It was concluded from the particle shape and electron diffraction data that the oxide phase was α-FeOOH. The observation of aggregated nanosize primary particles in microemulsion-mediated synthesis is reminiscent of the previous findings that in homogeneous solution spherical amorphous particles aggregate to form rods and rafts [35,36].

The Fe(II)–O_2 reaction is an example of the homogeneous precipitation method of "cation generation by redox reaction" (Table 3). The oxidation of Fe^{2+} by O_2 generates Fe^{3+} ions homogeneously:

$$2Fe^{2+} + (1/2)O_2 + 2H^+ \rightarrow 2Fe^{3+} + H_2O \tag{23}$$

It is noted from Eq. (23) that the oxidation reaction raises the pH (via consumption of H^+). Therefore, referring to Fig. 1 and Table 2 (path AP3), precipitation should occur eventually.

Inouye et al. [46] investigated the effect of the microemulsion environment on the rate of the Fe(II)–O_2 reaction. The oxidation of Fe(II) was found to obey a first-order rate law up to the point where the microemulsion became turbid (i.e., visual observation of precipitation became possible). The observed rate constants were two to three orders of magnitude greater than the corresponding values for bulk aqueous phase reaction. The rate enhancement was attributed to two main factors. First, there is a concentrative effect due to electrostatic attraction of Fe^{2+} ions to the anionic polar groups of the surfactants (sulfonate, oleate). According to this view, O_2 molecules approaching from the continuous organic phase encounter a high local concentration of Fe^{2+} ions at the surfactant/aqueous core interface. The second effect is related to the fact that both the polar groups of the surfactant molecule and the Na^+ and K^+ ions released by the partially dissociated surfactant molecules compete with the Fe(II) ions for hydration by water molecules. This results in incompletely hydrated (and therefore more reactive) Fe(II) ions [46,134].

It is possible that the polar groups play an additional role, i.e., they may assist in orienting the Fe^{2+} ions for reaction with O_2 molecules. According to Astanina and Rudenko [135], the first step in the Fe(II)–O_2 reaction mechanism involves the formation of a ferrous ion–oxygen complex:

$$Fe(OH_2)_6^{2+} + O_2 \rightleftharpoons \left[(H_2O)_5Fe(II) \cdots O_2 \begin{matrix} H \cdots O \\ | \\ H \cdots O \end{matrix} \right]^{2+} \tag{24}$$

This step is followed by formation of a binuclear aquo-complex with an oxygen bridge:

$$\left[(H_2O)_5Fe(II) \cdots O_2 \begin{matrix} H \cdots O \\ | \\ H \cdots O \end{matrix} \right]^{2+} + Fe(OH_2)_6^{2+} \rightleftharpoons$$

$$[(H_2O)_5Fe(II) \cdots O_2 \begin{matrix} H \cdots O \cdots H \\ \\ H \cdots O \cdots H \end{matrix} O_2 \cdots Fe(II)(OH_2)_5]^{4+} \tag{25}$$

The adsorption of Fe^{2+} to the polar groups may be viewed as a surface complexation reaction [136]:

$$Fe(OH_2)_6^{2+} + RX^- \rightleftharpoons RX\ Fe(OH_2)_5^+ + H_2O \tag{26}$$

With the Fe^{2+} ion thus immobilized, the formation of the Fe(II)–O_2 complex via Eq. (24) should be facilitated.

The microemulsion-plus-trigger method, based on forced hydrolysis (Table 3) was used by O'Sullivan et al. [47,48] to prepare ferric oxyhydroxide from ferric chloride. The microemulsions were formulated with mixtures of the commercial nonionic surfactants, Neodol 25-3 and Neodol 25-7 (Shell Chemical Co.), corresponding to an average of four oxyethylene groups per surfactant molecule; the mixtures were therefore designated as $C_{12}EO_4$. The iron oxide phase was established as β-FeOOH by X-ray diffraction. The particle size decreased with increase in aging time. For example, material produced at 333 K with a reaction time of 5 min had a particle size of 40–50 nm; after 4 h, the particle size had decreased to $\sim$2 nm. X-ray diffractograms indicated that there was no change in the crystalline phase, thus implying redissolution of the particles. The dissolution was attributed to the complexation of Fe(III) ions by the surfactant molecules, a view that was believed to be supported by NMR spectral evidence of a strong interaction between Fe^{3+} ions and the surfactant polar groups. It should be noted, however, that a β-FeOOH-to-α-Fe_2O_3 transformation in homogeneous aqueous Fe(III) chloride solution is known where the phase transformation is mediated through the aqueous phase [26]. The initially formed β-FeOOH dissolves to supply Fe^{3+} ions for the nucleation and growth of α-Fe_2O_3. It is possible that the results of O'Sullivan et al. [47] therefore represent a stage in this transformation process, where the α-Fe_2O_3 nuclei were too small to be responsive to X-ray diffraction.

2. Maghemite

The microemulsion technique formed the basis for the study of Ayyub et al. [49] on size-induced structural phase transitions of microcrystalline Fe_2O_3. The synthesis experiments were conducted in a three-component microemulsion consisting of sorbitan monooleate as the surfactant, 2-ethylhexanol as the oil, and an aqueous $Fe(NO_3)_3$ solution as the water pool. Precipitation was effected by adding NH_4OH to the microemulsion (path AP3, Fig. 1), and the resulting amorphous precipitate was transformed to Fe_2O_3 by calcining at 250°C. It was found that increasing the $Fe(NO_3)_3$ concentration in the aqueous precursor solution from 0.312% to 20% resulted in an increase in the particle size of the final product from 5 to 80 nm; this is consistent with Fig. 6a. Further, the nature of the solid product varied with ferric nitrate concentration in the starting microemulsion, with γ-Fe_2O_3 and α-Fe_2O_3, respectively, being the predominant phases at low and high ferric ion concentrations. Two possible interpretations of these results were offered [49]. In the first case, the difference in the calcination products was attributed to the formation of different solid products during the initial precipitation step, i.e., γ-FeOOH and amorphous Fe_2O_3, respectively, at low and high $Fe(NO_3)_3$ concentrations. An alternative interpretation considered that decreasing the particle size subjects the bulk material to a negative effective pressure. This second interpretation was preferred by the authors because they noted that similar size-dependent phase transitions had been observed in Fe_2O_3 prepared by other methods and also in different materials such as Al_2O_3 and ZrO_2.

Chhabra et al. [50] prepared γ-Fe_2O_3 by thermochemical decomposition of ferrous oxalate prepared in a microemulsion medium. The two-microemulsion method was used with the system Triton X-100/hexanol/cyclohexane/water. Ferrous oxalate ($FeC_2O_4 \cdot 2H_2O$) particles were obtained by mixing a ferrous ion–containing microemulsion and an oxalate-containing microemulsion. The resulting ferrous oxalate particles decomposed in the temperature range 225–300°C (15 min). The calcination was conducted in a stream of moist air for the first 30 s, followed by dry nitrogen gas. A low water/surfactant molar ratio R (and therefore small droplet size) and low calcination temperatures ($T \leq 225°C$) favored the formation of γ-Fe_2O_3; otherwise the calcination yielded mixtures of γ-Fe_2O_3

and α-Fe_2O_3. Based on the average number of Fe ions in the water pools used to synthesize the precursor ferrous oxalate, Chhabra et al. [50] estimated the corresponding diameter of γ-Fe_2O_3 particles (assuming complete conversion). Noting that the value of this diameter coincided with the equivalent spherical diameter (d_{esd}; based on surface area determination by the BET method) of the γ-Fe_2O_3 particles (7–8 nm), it was concluded that a given microemulsion droplet produced only one γ-Fe_2O_3 particle.

3. Magnetite

Synthesis of nanosize magnetite (Fe_2O_4) via microemulsion-mediated techniques was investigated by Kitahara and coworkers [51–53]. Two different synthesis protocols were used. In one approach based on AOT, a $FeCl_3$ microemulsion was added to an NH_3 microemulsion, and an aqueous solution of $FeCl_2$ was added to the resulting mixture with vigorous agitation. Apparently, the limited solubility of $FeCl_2$ in AOT-based microemulsions precluded the direct use of a $FeCl_2$ microemulsion. In a second method, a $FeCl_2$/NP-6 microemulsion was added to a $FeCl_3$/NH_3/AOT microemulsion. The sol particles were identified as magnetite by Mössbauer spectroscopy, electron diffraction, X-ray diffraction, and chemical analysis. The particles exhibited superparamagnetic behavior [52,53] and followed a log-normal size distribution, with a mean diameter of 3.6 nm and a standard deviation of 1.28 [53]. The colloidal dispersion of magnetite nanoparticles was extremely stable, as demonstrated by the fact that a sample stored for over 2 years showed no signs of stability breakdown [53]. Based on a theoretical consideration of the relevant attractive and repulsive energies (i.e., magnetic, van der Waals, electrostatic, and steric), the marked dispersion stability was attributed to steric (entropic) interactions of the surfactant chains [51,52].

Liz et al. [54] preferred the microemulsion-plus-microemulsion method. Two different microemulsion systems were used; namely, AOT/heptane/water and labrasol/isostearic plurol/ethyl oleate/water. Labrasol, a nonionic surfactant, consists of glycerides and polyglycerides of fatty acids; isostearic plurol, the cosurfactant, is polyglycerol isostearate; ethyl oleate represented the oil. Thus, using the same surfactant, the aqueous pseudophase of one microemulsion consisted of a solution of ferric and ferrous chlorides ($[Fe^{2+}]/[Fe^{3+}] = 1 : 2$), while the second microemulsion was based on an aqueous pseudophase of NH_4OH solution. The particles produced were identified as Fe_3O_4 by X-ray diffraction. Magnetic measurements revealed the particles to be superparamagnetic. It was found that an increase in reactant concentration resulted in a decrease in the dispersion stability. Successive additions of fresh reactants to particle-containing microemulsions led to larger particles. The labrosol-based microemulsion gave larger particles (as deduced from magnetization data), and this was attributed to the presence of larger water pools and greater intermicellar interaction (see Fig 6c). Lee et al. [55] also synthesized magnetite by the two-microemulsion method. Using the microemulsion system AOT/isooctane/water, they obtained particle sizes on the order of 1 nm. Samples dried by evaporation yielded larger particle sizes, indicating that the drying process resulted in particle aggregation.

C. Silica

The available reports on microemulsion-mediated synthesis of silica [56–83] are based almost exclusively on the alkoxide sol-gel method; only one exception has been found [70]. The first attempt to extend the alkoxide sol-gel process to microemulsion systems was reported by Yanagi et al. in 1986 [56]. Since then, additional papers have appeared from different research groups. In most cases the microemulsion-plus-reactant protocol was used;

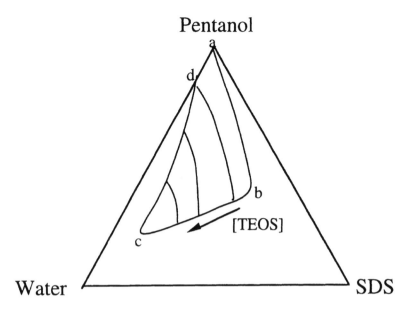

Figure 7 Schematic ternary phase diagram for the SDS/pentanol/water system, illustrating the effect of TEOS concentration on the microemulsion stability region that produces a transparent silica gel. (After Ref. 57.)

that is, the oil-soluble metal alkoxide was added to a microemulsion containing solubilized water. In order for reaction to occur, the alkoxide must diffuse through the continuous oil phase to the reverse micelle; it must then cross the surfactant film before reaching the water pool where the hydrolysis and condensation reactions [Eqs. (13)–(18)] take place. A variety of microemulsion systems have been used, including SDS/pentanol/water [57–59], polyoxyethylene alkylphenyl ether/alkane/water [74–81], AOT/alkane/water [69,71,82,83], AOT/decane/ glycerol [62], and CTAB/decanol/decane/formamide [62–64,67,68]. Compared with the materials systems previously considered above (i.e., aluminum and iron oxides), microemulsion sol-gel processing differs in two major ways. First, water is an active reactant and is consumed in the particle formation reaction [Eq. (13)]. Second, the synthesis reaction may result in a gel product [57–69] or in discrete particles [70–83], depending on whether acid or base catalysis is used.

1. Silica Gels in Aqueous Microemulsions

Friberg et al. [57–60] prepared silica gel in SDS/pentanol/water microemulsions, with tetraethoxysilane (TEOS) as the alkoxide precursor. The alkoxide was introduced into an acidic aqueous pseudophase (pH 2), and viscosity measurement indicated a gelation time of 30–35 h [59]. The composition of the organic-aqueous solution had a profound influence on the region of the phase diagram in which a stable transparent silica gel could be prepared [57–59], as illustrated schematically in the ternary phase diagram presented in Fig. 7. In the absence of TEOS, the microemulsion stability domain is approximately a triangle, with one apex at the pentanol corner, i.e., the region *abcd*. With the addition of increasing amounts of TEOS, the microemulsion stability region that produces a transparent silica gel shrinks toward the water-rich corner. This behavior may be rationalized by recognizing that in the microemulsion sol-gel process, water plays multiple roles; i.e., water is needed to (1) form a stable microemulsion before and after TEOS addition (this requires that the surfactant polar groups be adequately hydrated), (2) hydrolyze TEOS molecules, and (3) solvate the ethanol molecules produced by the TEOS hydrolysis reaction [Eq. (13)]. The microemulsion stability field is constrained on the right-hand side by the need to adequately hydrate the surfactant (SDS) polar groups. Therefore, the observed leftward shift of this boundary with increasing addition of TEOS is a reflection of the corresponding decrease in available water caused by the consumption of water molecules in the TEOS hydrolysis reaction.

Cussler and coworkers [61] prepared high surface area silica gels by polymerizing partially hydrolyzed TMOS within the aqueous pseudophase of a bicontinuous microemulsion based on didodecyldimethylammonium bromide (DDAB). In contrast to the work of Friberg et al. [57–59], the hydrolysis of the alkoxide was initiated outside of the microemulsion. That is, first a sol derived from partially hydrolyzed TMOS was prepared by adding the alkoxide to a dilute aqueous solution of H_2SO_4. Apparently the methanol reaction product acted as a cosolvent, transforming the initially biphase system into a homogeneous solution. Next, the silica-containing solution was added to a DDAB/decane solution to form the microemulsion. Aqueous HF, a polymerization catalyst, was then introduced. Polymerization converted the previously clear bicontinuous microemulsion into a transparent solid, without phase separation. The as-produced silica gel was treated with ethanol to remove the surfactant, decane, and the residual alkoxide. Vacuum or supercritical drying of the microemulsion-derived silica gels yielded specific surface areas that were greater than those obtained with conventional vacuum-dried gels.

2. Silica Gels in Nonaqueous Microemulsions

Polar organic solvents may be substituted for water to prepare nonaqueous microemulsions [137–139]. Friberg and coworkers [62–65] investigated the feasibility of using such solutions for the microemulsion gel method. The polar organic solvents considered included formamide [62–64, 67,68], glycerol [62,67], and ethylene glycol [65]. Nonaqueous solvents have been exploited in the conventional sol-gel process as drying control additives that serve to minimize cracking during the drying of silica gels [140]. Thus, it was of interest to see if the replacement of the usual aqueous pseudophase with a polar organic solvent would provide some benefits to the microemulsion sol-gel process.

Friberg and coworkers [62] prepared silica gel in the system CTAB/decane/decanol/formamide. The nonaqueous microemulsion was first prepared, and then the stoichiometric amounts of water and TEOS were added sequentially. Gelation time (as signified by visual observation of solidification) ranged from 1 to 7 days, depending on the microemulsion composition. In general, the gelation time decreased with increase in CTAB concentration. An increase in the decane/decanol ratio resulted in enhanced cracking of the gels following evaporation. The systems AOT/decanol/glycerol and AOT/decane/glycerol were also investigated [62]. The decanol system gave gels, but no gelation was obtained when the decane system was used. A viscoelastic investigation of the silica gels derived from CTAB/decane/decanol/formamide microemulsions indicated that with a decanol/decane weight ratio of 75/25, solutions with the molar ratio formamide/(decane + decanol) $\sim$2.5–3 gave the highest elastic behavior [63]. It was argued that this microemulsion composition gave a "balanced mixture of polar and nonpolar entities within the microemulsion. As a consequence, the final gel had an optimal mixture of cluster–cluster bonds and cluster–linear chain–cluster bonds.

The rates of the hydrolysis and condensation reactions of TEOS in formamide microemulsions were investigated by Jones and Friberg [64] with the aid of ^{29}Si NMR. It was observed that the initial stages of the TEOS–water reaction were dominated by the cyclic tetramer. A decrease in the solvent polarity (as reflected by an increase in the decane/decanol ratio) increased the rate of the early stages of the condensation (as demonstrated by the faster disappearance of the NMR signal associated with the cyclic tetramer). This behavior was attributed to the fact that the relatively polar hydrolyzed and partially condensed intermediate products would be confined to the microemulsion polar regions, where the resulting proximity promotes additional condensation. The gelatin time, on the other hand, increased with a decrease in the solvent polarity. This observation was rationalized by arguing that the faster initial condensation gave rise to larger SiO$_2$ particles, which subsequently condensed relatively slowly to give gels. Silicon NMR spectroscopy was also used to monitor the gelatin reaction in the microemulsion system SDS/octanol/toluene/ethylene glycol [65]. Samples taken at timed intervals ranging from 1 h to 2 days indicated the presence of only TEOS and SiO$_2$.

3. Doped Silica Gels

The alkoxide sol-gel process provides a convenient method of preparing metal-containing silicate coatings [141,142]. Friberg and coworkers [66] explored the feasibility of incorporating copper nitrate into microemulsion-gel glass. The microemulsion system was TEGDE/cyclohexane/aqueous solution [48% Cu(NO$_3$)$_2$, pH 1.25 with HNO$_3$]. The prepared silica gel samples were stored for 30 days and then characterized by UV-Vis spectroscopy, X-ray diffraction, SEM, and optical microscopy. Copper nitrate crystals were identified within the silicate network. The size and size distribution of the particles could be altered by heating (100°C) or by subjecting the samples to Soxhlet extraction with cyclohexane (70°C, 30 h).

The microheterogeneous nature of microemulsions, i.e., the presence of both polar and nonpolar pseudophases, facilitates the solubilization of organic molecules. Friberg and Jones [67] exploited this to prepare microemulsion silica gels doped with laser dyes. Both formamide- and glycerol-based nonaqueous microemulsions were investigated, and the microemulsion-plus-reactant method was used in which water containing dissolved dye (rhodamine B or rhodamine 6G) and TEOS were added sequentially. The doped silica gels gave fluorescence quantum yields that indicated that the microemulsion gel method is a promising technique for preparing solid-state laser dye materials. The effectiveness of relatively polar dyes (coumarins) was also examined [68]. However, the results were not as encouraging as those obtained with the rhodamine dyes.

Watzke and Dieschbourg [69] prepared silica–biopolymer nanocomposites by synthesizing silica within the aqueous microdomains of a microemulsion organogel. A microemulsion based on AOT/isooctane/water/gelatin was first formed, and the resulting gelatin organogel was melted by heating. The alkoxide (TMOS) was added, and then the solution was gelled by cooling to room temperature. Scanning electron microscopy of samples subjected to critical point drying showed that the pH of the aqueous pseudophase had a significant effect on the structure of the composite gel. With neutral pH, a porous structure was obtained, characterized by pores with a tubular appearance; the pore walls were thick and consisted of small silica particles packed into a dense body. In contrast, when the aqueous pseudophase was acidic, a highly porous structure was obtained.

4. Silica Nanoparticles

Silica is highly insoluble in an acidic solution [23]. Thus, a convenient reaction path for silica synthesis in homogeneous aqueous solution involves the acidification of soluble silicate solutions, i.e., path BP3 of Fig. 1:

$$SiO(OH)_3^- + H^+ \rightarrow Si(OH)_4^0 \rightarrow SiO_2 + 2H_2O \tag{27}$$

Wang et al. [70] used this method to prepare silica nanoparticles in a Triton X-100 microemulsion. Particles with a mean size of 31.8 nm and a standard deviation of 11.5% were obtained.

The base-catalyzed alkoxide sol gel route has been exploited by a number of authors to synthesize nanosize silica [71–83]. The work of Yamauchi et al. [71] was based on the TEOS/AOT/isooctane/water/ammonia system. The resulting silica particles were characterized with a variety of techniques, including nitrogen and water adsorption, TEM, SEM, and IR spectroscopy. By varying the water/surfactant molar ratio (R) in the range 5.7–15.8, the water/TEOS molar ratio (h) in the range 0.25–4, and ammonia concentration from 0.7 to 3.6 M (based on the aqueous phase volume), small spherical particles were obtained, with average diameters in the 15–70 nm range and surface areas in the 100–300 m^2/g range. The nitrogen and water adsorption data were interpreted to indicate the presence of microporosity in the particles. The TEM micrographs showed the presence of approximately 1 nm black spots within the 15–70 nm particles, and these darker regions were taken to represent primary oligomeric particles. The TEM micrographs indicated a uniform distribution of the black spots within each particle, and it was therefore concluded that the spheres did not have hollow cores.

Using a variation of the microemulsion-mediated sol-gel process, Espiard et al. [72] prepared hydrophobic silica particles coated with vinyl groups by cohydrolyzing and cocondensing TEOS and trimethoxysilylpropyl methacrylate (MPS). Ammonia served as the catalyst, and two different microemulsion systems were investigated: AOT/toluene/water/ammonia and polyoxyethylene-(4)-lauryl ether/cyclohexane/water/ammonia. An increase in the water/surfactant molar ratio (R) increased the particle size. Particles characterized by a narrow size distribution and a size range of 20–70 nm were obtained. With the AOT-based microemulsions, the silica particle size decreased with decreases in the concentration of water and with increases in the surfactant concentration. The results of Espiard et al. [72] also demonstrated the important role played by the nature of the surfactant. Thus, for the same water/surfactant molar ratio ($R = 5.5$–6) and water/alkoxide molar ratio ($h = 17$), the AOT-based microemulsion produced larger silica particles ($d_p = 65$ nm) than did the polyoxyethylene-based microemulsion ($d_p = 28$ nm). In a later publication, Espiard et al. [73] extended this work by using the surface-functionalized microemulsion-derived silica particles for inverse emulsion polymerization and also for conventional emulsion polymerization.

To establish clear relationships between synthesis conditions and the particle formation process, it is necessary to alter the composition variables in a systematic manner. Investigations motivated by this consideration are beginning to appear [74–83]. Using the NP-4/heptane/water/ammonia system, Chang and Fogler [74] studied the effects of the microemulsion environment on the rates of TEOS hydrolysis and silica particle growth.

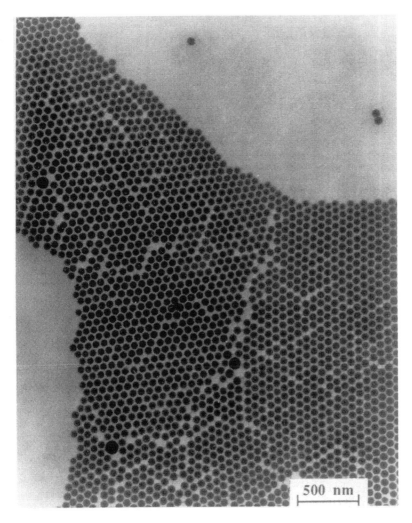

Figure 8 TEM micrograph of SiO_2 particles obtained by hydrolysis of TEOS in the NP-5/cyclohexane/water/ammonia microemulsion system; $R = 1.0$, $[TEOS] = 0.027$ M, $h = [H_2O]/[TEOS] = 7.78$. (From Ref. 76.)

For a given surfactant concentration, an increase in the concentration of water led to a decrease in particle size; this trend is consistent with Figs. 5e and 6b. On the other hand, when the water concentration was kept constant, the particle size went through a minimum (at $R = [H_2O]/[NP-4] = 1.9$) with an increase in surfactant concentration. The effects of surfactant molecular structure and type of oil were also investigated by Chang and Fogler [75]. Three different polyoxyethylene-type surfactants were used, i.e., NP-4, NP-5, and DP-6. The particle size of the silica particles followed the order NP-5 > NP-4 > DP-6. The particle size was found to be sensitive to the type of oil, with the size decreasing in the order heptane > heptane/cyclohexane (50/50 v/o) > cyclohexane.

Osseo-Asare and Arriagada [17,76] reported on the preparation of nanosize monodisperse silica particles by the controlled hydrolysis of TEOS in the NP-5/cyclohexane/water/ammonia water system. Particles in the range of 50–70 nm were produced with standard deviations below 8.5% around the mean diameters. Figure 8 [76]

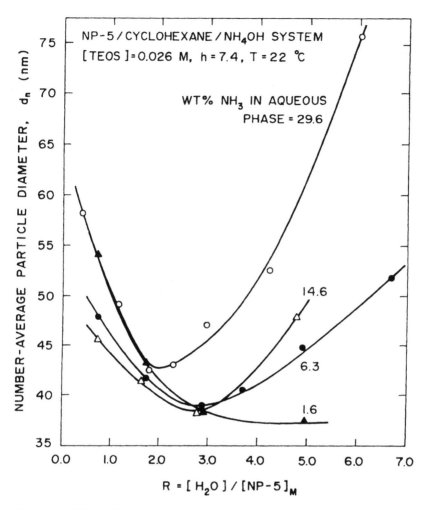

Figure 9 Effect of the water/surfactant molar ratio (R) on the mean diameter of SiO_2 particles prepared with different ammonia concentrations. (From Ref. 78.)

presents a typical TEM micrograph of the resulting silica particles. As the water/surfactant molar ratio (R) increased from 0.7 to 2.3, the particle size decreased and the size distribution became narrower. Further work showed that the dependence of particle size on R became more complicated as a wider range of R values was used and the ammonia concentration was varied [77,78]. Figure 9 [78] shows particle size versus R data for several constant ammonia concentrations. Referring to the data for concentrated ammonium hydroxide (29.6%), it can be seen that as R increases the particle size goes through a minimum. A similar but less dramatic trend is observed for intermediate ammonia concentrations (i.e., 6.3% and 14.6%). In the most dilute ammonia solution used (1.6% NH_3), no minimum is observed within the R range investigated, and the particle size decreases continuously.

To rationalize the above results, it is necessary to recall that changes in the water content result in significant changes in several properties of the microemulsion. Thus, with an increase in R, the surfactant aggregation number (N), the concentration of "free water," and the intermicellar communication rate (k_{ex}) all increase [143,144]. For a given surfactant concentration, an increase in N means a decrease in the concentration of water pools, which in turn should result in an increase in the average number of TEOS molecules per water pool (i.e., the average TEOS occupancy number, n_{av}). As already noted above (Fig. 5a), an increase in n_{av} is expected to increase the number of nuclei. Since water is an active participant in the hydrolysis reaction [Eq. (13)], the increase in the number of free water molecules is also expected to increase the number of nuclei (Fig. 5e). Thus, it may be argued that the decrease in particle size with R observed in Fig. 9 is attributable to the formation of an increasing number of nuclei (Fig. 6b). The subsequent rise in particle size at relatively high R values may be the combined result of several effects. As more and more of the water pools acquire TEOS occupancy numbers in excess of the critical number (n_c) needed for nucleation, a stage is eventually reached where any additional increase in n_{av} results mainly in growth rather than in generation of new nuclei. In this case particle size would be expected to increase with the TEOS occupancy number (n_{oc}) and therefore with R (Fig. 6a). The increase in k_{ex} that accompanies an increase in R is also expected to result in enhanced growth and therefore larger particles (Fig. 6c).

However, the fact that the minimum in the particle size versus R plot is dependent on the ammonium hydroxide concentration implies that additional factors are at work. It is known that silica is relatively unstable in highly basic solutions, dissolving to give anionic species such as $SiO(OH)_3^-$ [9,22,23]. Thus, it is likely that at higher ammonia concentrations, larger particles are observed at high R values because a high fraction of the nuclei redissolve. The particle size data obtained by transmission electron microscopy [78] have been analyzed in terms of a statistical nucleation model based on the reverse micellar populations, the partition of TEOS molecules between the reverse micellar pseudophase and the bulk oil phase, and the Poisson distribution of TEOS molecules and hydroxyl ions among the reverse micelles [79]. The growth kinetics of silica in the NP-5/cyclohexane/water/ammonia microemulsion has been analyzed in terms of a reverse micellar pseudophase model [80].

It was observed in certain experiments that even though particle synthesis was initiated in a clear one-phase microemulsion, the fluid phase became unstable during the reaction and phase separation occurred [81]. The continuing nucleation and growth of silica in the resulting biphase system resulted in a bimodal size distribution. With the aid of phase diagrams of temperature versus weight percent aqueous phase and temperature versus the ethanol/H_2O mole ratio, the phase separation was traced to microemulsion destabilization via H_2O depletion and ethanol release [81].

The effect of R on particle size was also investigated for silica nanoparticles synthesized in AOT/decane/water/ammonia microemulsions [82]. At R values below about 4, no particles were produced. On the other hand, as R increased in the range of 5–9.5, the diameter of the resulting particles increased and the size distribution narrowed. The fluorescence spectra of pyrene tetrasulfonic acid trapped in the water pools indicated that free water pools did not become available until the water/surfactant ratio exceeded about 10 [82]. The finding that particle formation is completely inhibited below $R = 4$ is therefore attributable to the fact that under these conditions the water molecules are strongly bound to the surfactant polar groups and the sodium counterions; the net result is to inhibit TEOS hydrolysis. The increase in particle size with R may be rationalized in terms of occupancy effects, as previously discussed in relation to Fig. 6a. That is, particle nucleation in the AOT system is primarily an intramicellar process in which the water core plays the role of a microreactor. AOT reverse microemulsions are characterized by a relatively rigid oil/water interfacial surfactant layer, so the rates of intermicellar exchange in this system are the lowest among most reverse microemulsion systems [145,146].

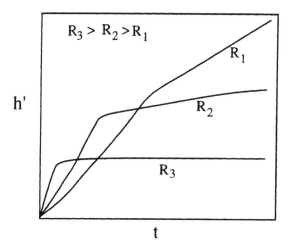

Figure 10 Schematic illustration of the effect of R on the rate of release of alcohol during microemulsion-mediated alkoxide hydrolysis. (After Ref. 84.)

It is interesting to note that the final particle sizes were significantly greater than the corresponding water pool diameters [82]. This suggests that further particle growth relied on intermicellar matter exchange. In order to explore the role of the exchange process in particle synthesis, AOT microemulsions were formulated with different amounts of benzyl alcohol (BA) [83], a reagent that is known to significantly increase the rate of intermicellar communication in AOT reverse microemulsions [129,146,147]. It was found that benzyl alcohol had a significant effect on the rate of particle growth. For BA/AOT ratios below 1.5, the final particle size and the rate of growth increased markedly as the BA/AOT ratio increased, and furthermore the corresponding size distribution became narrower. For BA/AOT ratios higher than 1.5, on the other hand, the growth rate decreased again and was similar to that obtained in the absence of the additive. This latter effect was attributed to the formation of an excess aqueous phase in the course of the reaction (phase instability), so that the growing particles (being located in this aqueous phase) were less accessible to the unreacted TEOS mainly solubilized in the coexisting W/O microemulsion [83].

D. Titania

1. Alkoxide-Derived Titania Gels

Guizard et al. [84,85] studied the acid-catalyzed hydrolysis and condensation of titanium alkoxides [i.e., titanium isopropoxide (TIPO) and titanium tetrabutoxide (TTBO)] in polyoxyethylene-(n)-octylphenyl ether/decane/water microemulsions. The nonionic surfactants used were Triton X-15 (TX-15), TX-35, and TX-45. Dynamic light scattering [85] showed that the hydrodynamic radii of droplets in the TX-35/decane/water system increased from about 2.2 nm in the absence of water to about 4 nm at water/surfactant molar ratio $R = 2$. The radius remained approximately constant up to about $R = 4$, above which phase separation occurred. The constant droplet size observed above $R = 2$ suggests that 2 mol H_2O per mole of surfactant is needed for the hydration of the polar groups.

Titania gels were synthesized by hydrolyzing the titanium alkoxides in the nonionic surfactant-based microemulsions. Gas chromatography was used to monitor the rate at which alcohol was released according to Eqs. (14) and (17). Figure 10 presents a schematic illustration of the manner in which R affected the observed trends; here the parameter h' represents the total moles of alcohol produced per mole of initial alkoxide. It can be seen that each curve is characterized by an initial period of rapid release of alcohol followed by a later stage of slower kinetics. The first part of a given curve represents the hydrolysis step, while the subsequent portion reflects the much slower alcohol-forming condensation reaction. According to Fig. 10, an increase in R enhances the hydrolysis rate. For relatively high R values, the plateau observed in the second part of the h' vs. t plot suggests that the alcohol-forming condensation reaction is unimportant. Hence, under these circumstances, condensation will proceed primarily via the water-forming reaction [Eq. (17)]. The nature of the alkoxide was also found to affect the kinetics of both the hydrolysis and condensation reactions. Thus, it was observed that the hydrolysis rates followed the order TIPO > TTBO [84,85].

It was found that the gelation time decreased with increases in the water/surfactant molar ratio (R). This behavior was attributed to effects associated with the local concentration of free water. At relatively high R, more free water is available; this promotes the hydrolysis step, and a relatively high concentration of hydroxo groups is generated. Accordingly, the relatively fast oxolation reaction [Eq. (17)] dominates network formation, and shorter gelation times are obtained. In contrast, when the water/surfactant molar ratio is relatively low, only limited free water is available. Hydrolysis is therefore not favored, and condensation proceeds primarily via the slower alcoxolation pathway [Eq. (18)]; hence, longer gelation times are obtained.

The gelation time (t_g) was found to decrease exponentially with the water/surfactant molar ratio [84,85]:

$$t_g = A \exp[-kR] \tag{28}$$

The A parameter was found to be characteristic of the particular surfactant–alkoxide couple, whereas the parameter k was dependent solely on the nature of the surfactant. For the hydrolysis of TIPO, the values of both the A and k parameters for TX-45 were higher than the corresponding values for TX-35. It is noteworthy that the difference between the molecular structures of the surfactants TX-45 and TX-35 is that TX-45 has an average of five oxyethylene groups compared with three oxyethylene groups for TX-35 (Table 5). Presumably, for the same water/surfactant molar ratio, TX-35 (with fewer oxyethylene groups to bind the available water molecules) has more free water molecules, and thus the hydrolysis and condensation proceed more rapidly in this system. The gelation time for TIPO was smaller than that for TTBO, a reflection of the lower reactivity of the bulkier butyl group.

Using a microemulsion system based on TX-35/decane/water, Peres-Durand et al. [86] found that the presence of hydrogen peroxide in the aqueous pseudophase resulted in a drastic decrease in the gelation time. This effect was attributed to the ability of the peroxo group to complex the cation:

$$Ti(OR)_4 + H_2O_2 \rightarrow TiO_2(OR)_2 + 2ROH \tag{29}$$

Apparently, the presence of the O_2^{2-} ligand increases the positive partial charge on the titanium atom, thereby enhancing the reactivity of the precursor [24].

The structure and textural evolution of the gels were investigated with a number of techniques, including thermoporometry [85,87–89], dilatometry [85,87], electron microscopy [85,87], rheology [90], and X-ray and neutron scattering [91,92]. It was observed that the gels swelled upon aging in decane or in water [85,87–89]. Investigations into the effects of the solvent type on the aging of the gels revealed that the pore volume of the aged gels increased with increasing polarizability of the solvent [88]. The scattering results indicated that a surfactant film covers the surface of the TiO_2 material and that washing the gel removes this film. The scattering results also revealed a fractal gel structure, with the backbone of this structure constituted by a three-dimensional network of rods. The rods were polydisperse and had a radius of gyration of about 0.7 nm. At scales greater than 3 nm, the gels were characterized by a fractal dimension of about 2.08.

Papoutsi et al. [93] prepared TiO_2 gels with microemulsions based on the systems SDS/pentanol/cyclohexane/water and Triton-100/cyclohexane/water; TIPO served as the alkoxide precursor. By visual inspection, it was found that a few hours sufficed to gel the SDS- and Triton-based microemulsions. In contrast, SDS/pentanol/water dispersions and ethanol–water solutions required several days to achieve gelation. These results were rationalized by arguing that in the microemulsion systems the hydrolysis reaction had to compete with the surfactant polar groups for water molecules. The resulting incomplete hydrolysis favored polycondensation, leading to relatively rapid gelation. It must be noted, however, that this reasoning is counter to that posited by Guizard et al. [84,85] to rationalize their observation of a decrease in gelation time with a decrease in R.

The potential of microemulsion-derived titania gels as host matrices for photoactive ions and molecules was explored by Guizard et al. [94] with the dopants Eu^{3+}, Nd^{3+}, and pyrene. It was concluded that the gel matrix presented three types of host sites associated with the microporosity, the mesoporosity, and the macroporosity.

2. Alkoxide-Derived Nanoparticles

The microemulsion-mediated alkoxide sol-gel method was used by Fendler and coworkers [95] to prepare TiO_2 nanoparticles. For the AOT/isooctane/water system with 0.1 M AOT, 1.0 M H_2O, and 2×10^{-4} M TiO_2, an optically clear dispersion was obtained whose colloidal stability lasted for several days. However, the stability decreased with an increase in TiO_2 concentration. Arriagada and Osseo-Asare [96] synthesized nanosize TiO_2 using both aqueous and nonaqueous reverse microemulsion systems. The aqueous microemulsions were based on either AOT or NP-5, and the alkoxides used were TTEO, TIPO, and TTBO. Using short reaction times (≤ 40 min) gave discrete nanosize particles (15–30 nm). However, with longer times, flocculation occurred and the solid material settled out of the microemulsion solution. Substitution of the solubilized aqueous phase with a glycerol–water mixture led to the formation of small polydisperse spherical particles in the size range of 20–80 nm. The properties of the solid product as well as the colloidal stability of the solid particles were strongly influenced by the glycerol/surfactant molar ratio (R_g). When R_g was below about 0.8, discrete and approximately spherical particles resulted and the particle–fluid system was stable. On the other hand, above $R_g \approx 0.8$, gel-like structures were obtained. It is likely that metal-organic polymeric networks formed under these circumstances [148,149].

Hirai et al. [97] used the AOT/isooctane/water system and titanium tetrabutoxide (TTBO). Particle diameters on the order of 3 nm were obtained by dynamic light scattering, a dimension that is smaller than the diameters of the reverse micelles (9–19.3 nm). Particle formation was strongly influenced by the water/surfactant molar ratio (R) and by the alkoxide concentration. The reaction kinetics was followed with UV-Vis absorption spectrophotometry. The absorbance of the reaction system increased monotonically with time for $R = 9$, whereas it went through a peak (after ~ 400 min) for $R = 30$. These results were rationalized in terms of differences in availability of reactant water molecules. For

$R = 9$, the solubilized water molecules are mostly bound to the polar groups on the AOT molecules. Thus, hydrolysis is incomplete and condensation (leading eventually to particle formation) is inhibited. In the case of $R = 30$, free water molecules are available, and therefore extensive hydrolysis, followed by condensation, can occur. The peak in the absorbance and the subsequent decline in the absorbance mark the removal of the hydrolyzed monomers for the condensation reaction.

Moran et al. [98] investigated the effects of solution composition and reactant concentrations on the sol-gel synthesis of TiO_2 in the microemulsion system AOT/cyclohexane/H_2O. Their results further demonstrate that the nature of the solid product can be altered drastically by manipulating the water/surfactant molar ratio. Using $R = 1$–6 resulted in the initial production of opaque sols that subsequently transformed into gelatinous translucent flocs upon standing. On the other hand, with $R \geq 8$, a dense but colloidally unstable precipitate was produced. Initially all the sols were amorphous. However, after aging for 2 weeks in their corresponding reaction media, the particles prepared with $R \geq 6$ became crystalline, and the major phase was identified as anatase (via Raman spectroscopy).

3. Titanium Tetrachloride–Derived Nanoparticles

Working with the CDBA/benzene/water microemulsion system, Joselevich and Willner [99] used $TiCl_4$ as the titanium source, the overall reaction being

$$TiCl_4 + 2H_2O \rightarrow TiO_2 + 4H^+ + 4Cl^- \tag{30}$$

The UV absorption spectra of the solubilized TiO_2 particles showed an absorption onset at $\lambda = 345$ nm, i.e., bandgap energy $E_g = 3.6$ eV. (In comparison, for bulk TiO_2, $E_g = 3.2$ eV for anatase and $E_g = 3.0$ eV for rutile.) On the basis of a mathematical analysis of the bandgap shift, the particle diameter was estimated to be 1.2–1.5 nm. Introduction of a hydrophilic photosensitizer, ruthenium(II) tris(bipyridine), $Ru(bpy)_3^{2+}$, into the TiO_2-containing microemulsion water pools revealed that an increase in the TiO_2 concentration caused a decrease in the luminescence intensity of $Ru(bpy)_3^{2+}$. A mathematical analysis of the emission decay curves gave the mean agglomeration number of the TiO_2 particles as 11 ± 2. Assuming a density of 4 g/cm^3 for TiO_2, an estimate of the particle diameter was then obtained as 0.9 nm, which is in fair agreement with the value of 1.2–1.5 nm given by the bandgap shift. The diameter of the TiO_2 particles increased from 0.7 nm to 0.9 nm as the water/surfactant molar ratio increased from 4.5 to 13.5.

The $TiCl_4$ hydrolysis reaction was also used by Chhabra et al. [100] to synthesize TiO_2. In this case, the microemulsion was based on the nonionic surfactant TX-100. Differential thermal analysis (DTA), thermogravimetric analysis (TGA), and X-ray diffraction indicated the sequence of phase transitions as

$$TiO_2 \text{ (amorphous)} \xrightarrow{530°C} TiO_2 \text{ (anatase)} \xrightarrow{890°C} TiO_2 \text{ (rutile)} \tag{31}$$

TEM micrographs indicated that the calcined material consisted of aggregated nanoparticles; $d_p = 15$–30 nm at a calcination temperature of 700°C and, $d_p = 40$–60 nm at a calcination temperature of 1000°C.

E. Vanadium Oxide

Desai and Cussler [101,102] exploited the microemulsion-mediated synthesis route to prepare vanadium oxide of enhanced microporosity for possible application as an intercalation cathode for lithium batteries. A vanadyl isopropoxide (VIPO)–toluene solution was added to an AOT-based microemulsion. Material collected after the initial few minutes showed a ribbonlike structure. The kinetics of the process was investigated with the aid of stopped-flow spectrophotometry, dynamic light scattering, and small-angle X-ray scattering. Three stages in the synthesis process were identified: (1) hydrolysis of the alkoxide to produce aqueous soluble polyions (about the first 100 ms), (2) reaction of the polyions to give polymer ribbons (~10 min), and (3) combination of the ribbons into large clusters (≥1 h).

F. Zinc Oxide

Only one study was found on microemulsion-based preparation of zinc oxide [103]. The two-microemulsion synthesis protocol was used by Hingorani et al. [103] to prepare zinc carbonate that was then calcined (~220°C) to produce zinc oxide. Working with the CTAB/butanol/octane/water microemulsion system and the two-microemulsion protocol, one aqueous pseudophase contained zinc nitrate while the other contained ammonium carbonate. X-ray diffraction identified the resulting calcined particles as ZnO with an average particle size of 14 nm.

G. Zirconia

Morooka and coworkers [104,105] and Kawai et al. [106] prepared zirconia nanoparticles by the microemulsion-mediated alkoxide sol-gel method. In all cases, zirconium tetrabutoxide (ZTBO) served as the alkoxide precursor. In the work of Morooka and coworkers [104,105], the surfactant was dioleyl phosphoric acid (DOLPA) and the microemulsion was prepared by a solvent extraction process [Eq. (20)]. After phase separation, zirconia particles were produced by introducing a butanol solution of ZTBO into the microemulsion solution. The resulting zirconia particles had an average size of 2 nm.

The work of Kawai et al. [106] was based on the NP-6/cyclohexane/water/sulfuric acid system, and ZTBO was added as its methanolic solution. The water/surfactant molar ratio (R) and the sulfuric acid/surfactant molar ratio (R_s) were varied in the ranges of $2 \leq R \leq 40$ and $0.05 < R_s < 0.30$. The resulting spherical particles ranged in size between 10.5 and 76 nm. Kawai et al. [106] investigated the effect of the state of the solubilized water on the particle formation process and on particle characteristics. Among the characterization techniques used were H NMR near-infrared spectroscopy, and light scattering. Three states of water were identified, depending on the water/surfactant molar ratio. For $R \leq 5$ (regime I), the water molecules were bound directly to the oxyethylene groups of the surfactant molecules (reverse micelles); for $5 \leq R \leq 9$ (regime II), the additional water molecules were bound to the hydrated oxyethylene groups through hydrogen bonds (swollen micelles); for $R > 20$ (regime III), bulklike water was present (microemulsions). For all three regimes, spherical particles were obtained. However, the particles in regimes I and II (reverse and swollen micelles) had a narrow size distribution, in contrast to the relatively broad size distribution found in regime III (W/O microemulsions). The broader size distribution observed for regime III was attributed to a less controlled hydrolysis/condensation associated with the greater availability of reactant water molecules.

Wang et al. [107] synthesized zirconium hydroxide particles with the two-micro-emulsion method. The microemulsion system adopted was NP-5/NP-9/cyclohexane/water, and a Zr-containing microemulsion was mixed with an NH_3-containing microemulsion (path AP3, Fig. 1). The as-produced zirconium hydroxide particles were spherical, and particle size was in the 5–40 nm range, as determined by TEM. The extracted particles, dried at 100°C, were X-ray amorphous. Thermogravimetric analysis (TGA) revealed two main weight loss steps, i.e., from room temperature to 120°C and from 120°C to 410°C. The first weight loss was attributed to the removal of residual water and solvent, while the second and more drastic weight loss was assigned to the transformation of zirconium hydroxide to zirconia, with additional contributions from solvent and surfactant removal. Differential thermal analysis (DTA) of the particles gave a major exotherm at 362°C, followed by a minor peak at 453°C. On the basis of X-ray diffraction data, the 362°C exotherm was attributed to the hydroxide-to-oxide transformation, while the 453°C exotherm was attributed to secondary crystallization. The primary crystallization temperature of 362°C was about 100°C lower than the corresponding temperature for conventional bulk aqueous phase synthesis.

H. Complex Oxides

Extension of the microemulsion reaction method to complex oxides has been explored for a wide variety of materials, including aluminosilicate zeolite [45,108] ferrites [109], mullite [110], barium titanate [111,112], and various perovskite-type mixed oxides [112,113], yttrium-iron garnets (YIGs) [114,115], and high T_c superconducting oxides [116–120].

1. Aluminosilicate Zeolite

Dutta and Robbins [45,108] used the AOT/isooctane/water system in a synthesis protocol that combined the microemulsion-plus-reactant route (C, Fig. 3) with the microemulsion-plus-trigger route (A, Fig. 3). Cloudiness was observed within an hour after reaction was initiated by heating, and a white precipitate collected at the bottom of the reactor. This initial material was amorphous (XRD); however, crystals in the 1–2 μm size range were identified by SEM after 8 h heating.

The role of the microemulsion environment in this zeolite synthesis process is undoubtedly complex. Presumably, addition of the sodium aluminate solution to the silicate-containing microemulsion first results in the formation of an amorphous aluminosilicate precipitate. Apparently this material does not remain as dispersed particles in the microemulsion fluid phase. One possible reason is that the stability field of the one-phase microemulsion regions shrinks with an increase in temperature. The ability of NaOH to promote the hydrolytic decomposition of AOT [150,151] will also contribute to the destabilization of the surfactant aggregates. It is likely that the adsorbed AOT molecules associated with the amorphous precipitate play some role in the subsequent transformation to crystalline zeolite. However, the details are yet to be determined.

2. Ferrite

In the aqueous synthesis of multicomponent oxides, a common approach is to first prepare a coprecipitate that can then be calcined to produce the desired complex oxide as the final product. When the component metal ions exhibit widely different aqueous solubilities, effecting coprecipitation becomes a challenge. Chhabra et al. [109] overcome this difficulty in the case of the ferrite $BaFe_{12}O_{19}$ by using a nonaqueous microemulsion system based on ethanol as the dispersed phase. The barium and ferric forms of AOT were first synthesized. Isooctane solutions of the Ba-AOT and Fe-AOT salts were then prepared ($Ba^{2+}/Fe^{3+} = 1 : 12$), followed by ethanol addition. An ethanolic solution of oxalic acid was introduced into the microemulsion to produce a barium iron oxalate coprecipitate. Calcination of this material yielded $BaFe_{12}O_{19}$ at 950°C, which was 250°C lower than the corresponding temperature for conventional solid-state synthesis.

3. Mullite

Klassen and Fischman [110] attempted to synthesize mullite ($3Al_2O_3 \cdot 2SiO_2$) by extending the method used by Friberg and coworkers [57–59] to prepare silica gel. Using the SDS/pentanol/water microemulsion system, the approach was to dissolve aluminum nitrate in the water pools, followed by addition of the stoichiometric amount of TEOS. The synthesis was not successful, however. A number of problems were encountered; for example, the presence of $Al(NO_3)_3$ drastically destabilized the microemulsion. A limited region of microemulsion stabilization was obtained when the aqueous pool contained 10 wt% $Al(NO_3)_3$, but then the corresponding stoichiometric amount of TEOS was too low to cause gelation.

4. Perovskites

The complex oxides with the group name perovskite have the general formula ABO_3, where the ionic charges on the metals can assume the forms $A^+B^{5+}O_3$ (e.g., $KNbO_3$), $A^{2+}B^{4+}O_3$ (e.g., $MTiO_3$, M = Ca, Sr, Ba, Pb), and $A^{3+}B^{3+}O_3$ (e.g., $LaMnO_3$, $LaCrO_3$) [3]. The first reported attempt to synthesize a perovskite by the microemulsion method was by Schlag et al. [111], who used the two-microemulsion method and a microemulsion system based on a commercial (Hoechst) nonionic surfactant, i.e., Genapol OX 30/decane/water. A microemulsion containing $BaCl_2$ and $TiCl_4$ in the dispersed phase was mixed with an oxalic acid ($H_2C_2O_4$)-containing microemulsion. Reaction gave a stable colloidal dispersion, where the as-prepared particles had an average particle size of 5 nm as determined by TEM. However, X-ray diffraction could not confirm the presence of barium titanyl oxalate. Calcination of the recovered solid material in the temperature range of 400–1200°C did not yield phase-pure (XRD) $BaTiO_3$.

The efforts of Herrig and Hempelmann [112] were more successful. These investigators also used a nonionic surfactant, i.e., polyoxyethylene-(4)-laury ether (Brij 30, ICI); however, alkoxide precursors and the microemulsion-plus-reactant protocol were used. Successful synthesis was reported of barium and strontium titanates ($BaTiO_3$, $SrTiO_3$) as well as barium and strontium zirconates ($BaZrO_3$, $SrZrO_3$). In the case of $BaTiO_3$, the as-prepared particles gave the perovskite XRD pattern. With $SrTiO_3$, the as-prepared material contained $SrCO_3$ and TiO_2 as impurities; thermal treatment (700°C, 1 h) led to phase-pure titanate. Barium and strontium zirconates did not form directly from the microemulsion reaction. Instead, zirconium oxide and barium (or strontium) carbonate formed, which, when subjected to calcination, underwent solid-state reaction to give the corresponding zirconates.

Gan et al. [113] combined the oxalate precipitation method with the two-microemulsion protocol to prepare $BaPbO_3$, La_2CuO_4, and $LaNiO_3$. The NP-5–octane (or petroleum ether–water) system was used, with one microemulsion solution containing the nitrate salts of the metals [i.e., $Ba(NO_3)_2$-$Pb(NO_3)_2$, $La(NO_3)_3$-$Cu(NO_3)_2$, $La(NO_3)_3$-$Ni(NO_3)_2$] and the other microemulsion containing the oxalate precipitant as ammonium oxalate or oxalic acid. Metal oxalate coprecipitates were formed that lost their colloidal stability after about 2 h, as manifested by the observation of flocculated materials.

Calcination of the oxalate coprecipitates readily yielded phase-pure perovskite-type complex metal oxides. The required calcinations for the microemulsion-derived mixed oxalates were 100–250°C below the temperatures used for oxalates prepared in homogeneous aqueous solutions.

5. Yttrium-Iron Garnet

Teijeiro et al. [114] reported on the synthesis of yttrium iron garnet (YIG), $Y_3Fe_5O_{12}$. The CTAB/butanol/octane/water system was used, but little was given of the experimental details. In a more extensive investigation by Vaqueiro et al. [115], a microemulsion containing solubilized Fe(III) and Y(III) nitrates was added to a second microemulsion containing solubilized NH_3 or $(NH_4)_2CO_3$. The hydrodynamic radius of the electrolyte-free microemulsion droplets was obtained as 7 nm, while the as-produced precipitates consisted of 3 nm primary particles. Washing (ethanol) followed by drying (80°C) and calcination ($> 700°C$) produced crystalline YIG. Below 700°C, significant amounts of other phases [e.g., maghemite (γ-Fe_2O_3), hematite (α-Fe_2O_3), perovskite ($YFeO_3$)] were also obtained.

6. High T_c Superconducting Oxides

Ayyub and coworkers [116,117] used the two-microemulsion technique and the system CTAB/butanol/octanol/water [118] to prepare nanosize oxalate coprecipitates of yttrium, barium, and copper. Calcination of the oxalate product at ~820°C for 2 h gave single-phase $YBa_2Cu_3O_{7-x}$ (YBCO). Dynamic light scattering gave the hydrodynamic diameter of the microemulsion droplets as ~13.0 nm. On the other hand, the dried oxalate precipitate and the calcined product had equivalent spherical diameters of 47.4 nm and 274.8 nm, respectively. The comparable diameters for materials produced by conventional precipitation were given as 280.6 nm for the dried oxalate precipitate and 626.6 nm for the calcined product. Thus, significantly smaller particles were obtained with the microemulsion method. It was also reported by Ayyub et al. [116] that the microemulsion route was less pH-sensitive than conventional aqueous precipitation. In the conventional process, the oxalate precipitate attained the stoichiometric Y:Ba:Cu value of 1:2:3 only at pH 2.4. In contrast, microemulsion-based precipitates with initial pH values in the broader range of 3.5–4.0 gave calcined materials of single-phase YBCO.

López-Quintela and Rivas [119] also exploited the two-microemulsion method to prepare Y-Ba-Cu oxalate coprecipitates as precursors for YBCO. The initial oxalate precipitate had a particle size of about 30 nm. Calcination (825°C in air) produced an oxide material with a particle size of ~100 nm. The small geometrical dimensions of microemulsion polar domains serve to limit the spatial propagation of chemical inhomogeneities. Kumar et al. [120] took advantage of this to prepare a mixed oxalate of Bi-Pb-Sr-Ca-Cu in the microemulsion system Igepal CO-430/cyclohexane/acetic acid/water. The resulting nanosize oxalate precipitate had a particle size in the 2–6 nm range. Calcination followed by sintering yielded nearly phase-pure Bi-Pb-Sr-Ca-Cu-O (2223) oxide superconductor.

V. SUMMARY AND CONCLUSIONS

Microemulsion-mediated synthesis has attracted the attention of many researchers [10–18, 44–120]. In the specific case of oxide materials, the clearest achievement of the work to date is the demonstration that a wide variety of nanosize metal oxides—binary oxides from alumina to zirconia, as well as complex metal oxides—can be prepared by exploiting the ability of microemulsions to solubilize, compartmentalize, and concentrate reactants and products. The major challenges ahead include the need to (1) undertake more systematic investigations that probe the mechanistic aspects of microemulsion-mediated materials synthesis (2) further broaden the range of oxide materials prepared via the microemulsion method, and (3) develop strategies that will facilitate the practical applications of microemulsion-derived materials.

Much of the focus in the past has been on the successful synthesis of specific metal oxide compounds as nanoparticles. Thus, only limited information is available on the influence of important process variables, such as microemulsion composition (e.g., type of surfactant, oil, and cosurfactant, and the proportion of water), the nature of the reactant, reactant concentration, temperature, and synthesis protocol (i.e., two-microemulsion, microemulsion-plus-reactant, microemulsion-plus-trigger) on the nature of the resulting particles. In order to exploit microemulsions more effectively for materials synthesis, it is important to recognize the multiple roles of surfactant and water molecules. The particle formation process involves several different physicochemical processes, e.g., solvation/desolvation, ligand exchange (complexation), nucleation, particle growth by monomer addition, particle growth by aggregation, and particle dissolution. Surfactant molecules may influence one or more of these processes. Unfortunately, there is a tendency to view the structured microdomains within microemulsions merely as indifferent templates whose main function is the mechanical transfer of nanodimensionality to the synthesized materials. In fact, surfactants are not only needed to stabilize the polar microdomains; they also serve to stabilize the product particles and may also act as crystal growth modifiers. The experimental results examined in this chapter further reveal that water molecules are needed not only to serve as reaction media but also to (1) hydrate the polar groups of the surfactant molecules, (2) hydrate metal ions and the associated anions in the water pools, (3) serve as reactants (e.g., for alkoxide hydrolysis), and (4) hydrate the product solids (e.g., hydrated metal oxalates). There is clearly a need for systematic investigations that will further clarify the manner in which the solution behavior, the microstructure, and the dynamics of microemulsion systems control particle characteristics and dispersion stability. The hope is that it may be possible to exploit the insights accrued from the fundamental investigations in designing novel materials at the molecular level.

Besides nanodimensionality and particle size uniformity, a number of interesting and desirable particle characteristics have been attributed to the unique microenvironment of the microemulsion fluid phase. Thus, it is generally accepted that the spatial propagation of chemical inhomogeneities is suppressed by the small geometrical dimensions of the microemulsion "nanoreactors," i.e., the polar microdomains. A lowering of phase transformation temperatures for microemulsion-derived materials has been reported by some investigators [107,109,113]. The bicontinuous structure of microemulsions has been employed as a template for preparing highly porous materials [61]. It has also been suggested that microemulsion-mediated synthesis results in a more ordered solid-phase structure; this has been attributed to the relatively gentle and controlled mixing mediated through the collision and fusion of the polar phase droplets [107]. Many of the above claims and observations, however, are based on limited research by only a few investigators and are generally focused on only a single metal oxide material. If these are not to remain merely anecdotal claims, then more extensive studies must be conducted, involving more investigators and different metal oxides. Examination of Table 6 reveals that most of the entries pertain to three metals—Si, Ti, and Fe—with ~39%, 24%, and 12%, respectively, of the entries. The remaining materials have at most three entries, individually; several have only one. Clearly, the possibilities are far from exhausted, in terms of material–protocol (Fig. 3), material–surfactant (Table 5), and material–reaction path (Fig. 1, Table 2, and Table 3) combinations.

Many potential applications of microemulsion derived materials have been mentioned in the literature; the wish list includes catalysts, nanoporous membranes, nanocomposites, and precursor powders for functional ceramics [10–18,44–120,153]. However, before microemulsion-derived materials can become more than laboratory curiosities, two key issues need to be addressed: product recovery and product yield. Materials generated in microemulsion media can be used in two ways: as prepared in the fluid phase or as recovered solids. An example of the first case is the infiltration technique adopted by Morooka and coworkers [104,105] to prepare nanoporous inorganic membranes. Where solid–liquid separation is required, a major challenge is how to harvest nanoparticles without promoting excessive particle aggregation and how to remove solvents and surfactants from gels without adversely changing the internal structures of the materials. A common practice found in the literature involves phase destabilization via the addition of polar solvents such as ethanol and acetone. This approach works well as a laboratory technique that permits particles to be recovered for later characterization. However, it is unlikely that this could be an economically viable approach for a practical processing system, given the eventual need to separate the added solvents in order to reformulate the original microemulsion.

As already noted above, the susceptibility of microemulsions to destabilization by electrolytes severely limits the highest metals concentrations that can be used for precipitation reactions; this has an adverse impact on prospects for large-scale processing. The feasibility of microemulsion-based synthesis depends on the availability of surfactant/oil/water formulations that give stable microemulsions before, during, and after precipitation reactions. Therefore, investigations that emphasize the effects of reactants and products on the stability domains of microemulsions will fill a significant need. Hopefully, such investigations will lead to guidelines for formulating microemulsion compositions that are compatible with large-scale materials synthesis.

REFERENCES

1. E. G. Kelly and D. J. Spottiswood, *Introduction to Mineral Processing*, Wiley, New York, 1982, p. 5.
2. N. V. Sidgwick, *The Chemical Elements and their Compounds*, Oxford University Press, New York, 1962.
3. N. Ichinose (ed.), *Introduction to Fine Ceramics*, Wiley, New York, 1987.
4. N. Ichinose, Y. Ozaki, and S. Kashu, *Superfine Particle Technology*, Springer-Verlag, New York, 1988.
5. E. Matijevic, MRS Bull., December, 1989, pp. 18–20.
6. M. Haruta and B. Delmon, J. Chim. Phys. *83*:859–868 (1986).
7. C. Hayashi, Physics Today, December 1989, pp. 44–51.
8. R. P. Andres, R. S. Averback, W. L. Brown, L. E. Brus, W. A. Goddard III, A. Kaldor, S. G. Louie, M. Moscovits, P. S. Peercy, S. J. Riley, R. W. Siegel, F. Spaepen, and Y. Wang, J. Mater. Res. *4*:704–736 (1989).
9. C. J. Brinker and G. W. Scherer, *Sol-Gel Science*, Academic Press, San Diego, 1990.
10. M. Boutonnet, J. Kizling, P. Stenius, and G. Maire, Colloids Surf. *5*:209–225 (1982).
11. K. Kandori, N. Shizuka, M. Gobe, K. Kon-no, and A. Kitahara, J. Dispersion Sci. Technol. *8*:477–491 (1987).
12. J. H. Fendler, Chem. Rev. *87*:877–899 (1987).
13. J. B. Nagy, Colloids Surf., *35*:201–220 (1989).

14. A. J. I. Ward and S. E. Friberg, MRS Bull., December 1989, pp. 41–46.

15. M. J. Hou and D. O. Shah, in *Interfacial Phenomena in Biotechnology and Materials Processing* (Y. A. Attia, B. M. Moudgil, and S. Chander, eds.), Elsevier, Amsterdam, 1988, pp. 443–458.

16. A. Khan-Lodhi, B. H. Robinson, T. Towey, C. Hermann, W. Knoche, and U. Thesing, in *The Structure, Dynamics and Equilibrium Properties of Colloidal Systems* (D. M. Bloor and E. Wyn-Jones, eds.), Kluwer, Dordrecht, 1990, pp. 373–383.

17. K. Osseo-Asare and F. J. Arriagada, in *Ceramic Powder Science III* (G. L. Messing, S. Hirano, and H. Hausner, eds.), American Ceramic Society, Westerville, OH, 1990, pp. 3–16.

18. M. P. Pileni, J. Phys. Chem. *97*:6961–6973 (1993).

19. M. Pourbaix, *Atlas of Electrochemical Equilibria in Aqueous Solutions*, National Association of Corrosion Engineers, Houston, TX, 1974.

20. K. Osseo-Asare, in *Challenges in Mineral Processing* (K.V. S. Sastry and M. C. Fuerstenau, eds.), SME, Littleton, CO, 1989, pp. 585–604.

21. K. Osseo-Asare and K. K. Mishra, J. Electronic Mater. *25*:1599–1607 (1996).

22. C. F. Baes, Jr. and R. E. Mesmer, *The Hydrolysis of Cations*, Wiley, New York, 1976.

23. R. K. Iler, *The Chemistry of Silica*, Wiley, New York, 1979.

24. J. Livage, M. Henry, and C. Sanchez, Prog. Solid State Chem. *18*:259–341 (1988).

25. E. Matijevic, Chem. Mater. *5*:412–426 (1993).

26. T. Sugimoto, J. Colloid Interface Sci. *184*:626–638 (1996).

27. P. Calvert, Nature *367*:119–120 (1994).

28. G. H. Bogush and C. F. Zukoski, J. Colloid Interface Sci. *142*:19–34 (1991).

29. K. Wefers and C. Misra, *Oxides and Hydroxides of Aluminium*, Technical Paper 19, Alcoa Research Laboratories, Alcoa Center, PA, 1987.

30. A. C. Vermeulen, J. W. Geus, R. J. Stol, and P. L. de Bruyn, J. Colloid Interface Sci. *51*:449–458 (1975).

31. R. J. Stol, A. K. van Helden, and P. L. de Bruyn, J. Colloid Interface Sci. *57*:115–131 (1976).

32. H. de Hek, R. J. Stol, and P. L. de Bruyn, J. Colloid Interface Sci. *64*:72–89 (1978).

33. A. Clearfield, J. Mater. Res. *5*:161–162 (1990).

34. A. Clearfield, Rev. Pure Appl. Chem. *14*:91–108 (1964).

35. J. H. A. van der Woude and P. L. de Bruyn, Colloids Surf. *8*:55–78 (1983).

36. P. J. Murphy, A. M. Posner, and J. P. Quirk, J. Colloid Interface Sci. *56*:312–320 (1975).

37. C. J. Brinker, in *The Colloid Chemistry of Silica*, Adv. Chem. Ser. Vol. 234 (H. Bergna, ed.), ACS, Washington, DC, 1994, pp. 361–402.

38. L. Gordon, M. L. Salutsky, and H. H. Willard, *Precipitation from Homogeneous Solution*, Wiley, New York, 1958.

39. E. Matijevic, Pure Appl. Chem. *50*:1193–1210 (1978).

40. E. Matijevic, Ann. Rev. Mater. Sci. *15*:483–516 (1985).

41. K. Kurihara, J. Kizling, P. Stenius, and J. H. Fendler, J. Am. Chem. Soc. *105*:2574–2579 (1983).

42. R. J. Hunter, *Foundations of Colloid Science*, Vol. II, Oxford Univ. Press, New York, 1989, pp. 914–915.

43. M. J. Schick (ed.), *Nonionic Surfactants*, Marcel Dekker, New York, 1967.

44. D. W. Matson, J. L. Fulton, and R. D. Smith, Mater. Lett. *6*:31–33 (1987).

45. P. K. Dutta and D. Robbins, ACS Preprints, Div. Petroleum Chem. *34*(3):461–464 (1989).

46. K. Inouye, R. Endo, Y. Otsuka, K. Miyashiro, K. Kaneko, and T. Ishikawa, J. Phys. Chem. *86*:1465–1469 (1982).

47. E. C. O'Sullivan, A. J. I. Ward, and T. Budd, Langmuir *10*:2985–2992 (1994).

48. E. C. O'Sullivan, R. C. Patel, and A. J. I. Ward, J. Colloid Interface Sci. *146*:582–585 (1991).

49. P. Ayyub, M. Multani, M. Barma, V. R. Palkar, and R. Vijayaraghavan, J. Phys. C *21*:2229–2245 (1988).

50. V. Chhabra, P. Ayyub, S. Chattopadhyay, and A. N. Maitra, Mater. Lett. *26*:21–26 (1996).

51. M. Gobe, K. Kandori, K. Kon-no, and A. Kitahara, J. Colloid Interface Sci. *93*:293–295 (1983).

52. M. Gobe, K. Kandori, K. Kon-no, and A. Kitahara, J. Jpn Soc. Color Mater. *57*:380–385 (1984).

53. S. Bandow, K. Kimura, K. Kon-no, and A. Kitahara, Jpn. J. Appl. Phys. *26*:713–717 (1987).

54. L. Liz, M. A. López Quintela, J. Mira, and J. Rivas, J. Mater. Sci. 29:3797–3801 (1994).
55. K. M. Lee, C. M. Sorensen, K. J. Klabunde, and G. C. Hadjipanayis, IEEE Trans. Magn. 28:3180–3182 (1992).
56. M. Yanagi, Y. Asano, K. Kandori, and K. Kon-no, Abstr. 39th Symp. Div. Colloid Interface Chem., Chem. Soc. Jpn. 1986, p. 396.
57. S. E. Friberg and C. C. Yang, in *Innovations in Materials Processing Using Aqueous, Colloid and Surface Chemistry* (F. M. Doyle, S. Raghavan, P. Somasundaran, and G. W. Warren, eds.), The Minerals, Metals & Materials Society, Warrendale, PA, 1988, pp. 181–191.
58. S. E. Friberg, C. C. Yang, and J. Sjöblom, Langmuir 8:372–376 (1992).
59. S. E. Friberg, A. U. Ahmed, C. C. Yang, S. Ahuja, and S. S. Bodesha, J. Mater. Chem. 2:257–258 (1992).
60. S. E. Friberg, S. M. Jones, and C. C. Yang, J. Dispersion Sci. Technol. 13:65–75 (1992).
61. J. H. Burban, M. He, and E. L. Cussler, AIChE J. 41:159–165 (1995).
62. S. E. Friberg, S. M. Jones, and C. C. Yang, J. Dispersion Sci. Technol. 13:45–63 (1992).
63. S. E. Friberg, S. M. Jones, A. Motyka, and G. Broze, J. Mater. Sci. 29:1753–1757 (1994).
64. S. M. Jones and S. E. Friberg, J. Dispersion Sci. Technol. 13:669–696 (1992).
65. S. E. Friberg and Z. Ma, J. Non-Cryst. Solids 147/148:30–35 (1992).
66. S. M. Jones, A. Amran, and S. E. Friberg, J. Dispersion Sci. Technol. 15:513–542 (1994).
67. S. E. Friberg, S. M. Jones, and J. Sjoblom, J. Mater Synth. Process. 2:29–44 (1994).
68. S. M. Jones and S. E. Friberg, J. Non-Cryst. Solids 181:39–48 (1995).
69. H. J. Watzke and C. Dieschbourg, Adv. Colloid Interface Sci. 50:1 (1994).
70. W. Wang, X. Fu, J. Tang, and L. Jiang, Colloids Surf. 81:177–180 (1993).
71. H. Yamauchi, T. Ishikawa, and S. Kondo, Colloids Surf. 37:71–80 (1989).
72. P. Espiard, J. E. Mark, and A. Guyot, Polym. Bull. 24:173–179 (1990).
73. P. Espiard, A. Guyot, and J. E. Mark, J. Inorg. Organometal. Polym. 5:391–407 (1995).
74. C. L. Chang and H. S. Fogler, AIChE J. 42:3153–3163 (1996).
75. C. L. Chang and H. S. Fogler, Langmuir 13:3295–3307 (1997).
76. K. Osseo-Asare and F. J. Arriagada, Colloids Surf. 50:321–339 (1990).
77. F. J. Arriagada and K. Osseo-Asare, in *The Colloid Chemistry of Silica*, Adv. Chem. Ser., Vol. 234 (H. E. Bergna, ed.), American Chemical Society, Washington, DC, 1994, pp. 113–128.
78. F. J. Arriagada and K. Osseo-Asare, J. Colloid Interface Sci., in press.
79. F. J. Arriagada and K. Osseo-Asare, Colloids Surf., in press.
80. K. Osseo-Asare and F. J. Arriagada, J. Colloid Interface Sci., in press.
81. F. J. Arriagada and K. Osseo-Asare, Colloids Surf. 69:105–115 (1992).
82. F. J. Arriagada and K. Osseo-Asare, J. Colloid Interface Sci. 170:8–17 (1995).
83. F. J. Arriagada and K. Osseo-Asare, J. Dispersion Sci. Technol. 15:59–71 (1994).
84. C. Guizard, A. Stitou, A. Larbot, L. Cot, and J. Rouviere, in *Better Ceramics Through Chemistry III*, Mater. Res. Soc. Symp. Proc., Vol. 121 (C. J. Brinker, D. E. Clark, and D. R. Ulrich, eds.) Materials Research Society, Pittsburgh, 1988, pp. 115–120.
85. C. Guizard, A. Larbot, L. Cot, S. Perez, and J. Rouviere, J. Chim. Phys. 87:1901–1922 (1990).
86. S. Peres-Durand, J. Rouviere, and C. Guizard, Colloids Surf. 98:251–270 (1995).
87. J. F. Quinson, N. Tchipkam, J. Dumas, C. Bovier, J. Serughetti, C. Guizard, A. Larbot, and L. Cot, J. Non-Cryst. Solids 99:151–159 (1988).
88. J. F. Quinson, J. Dumas, M. Chatelut, J. Serughetti, C. Guizard, A. Larbot, and L. Cot, J. Non-Cryst. Solids 113:14–20 (1989).
89. J. F. Quinson, M. Chatelut, C. Guizard, A. Larbot, and L. Cot, J. Non-Cryst. Solids 121:72–75 (1990).
90. C. Guizard, J. Achddou, A. Larbot, and L. Cot, J. Non-Cryst. Solids 147/148:681–685 (1992).
91. J. Marignan, C. Guizard, and A. Larbot, Europhys. Lett. 8:691–696 (1989).
92. F. Molino, J. M. Barthez, A. Ayral, C. Guizard, R. Jullien, and J. Marignan, Phys. Rev. E 53:921–925 (1996).
93. D. Papoutsi, P. Lianos, P. Yianoulis, and P. Koutsoukos, Langmuir 10:1684–1689 (1994).

94. C. Guizard, J. C. Achddou, A. Larbot, L. Cot, G. Le Flem, C. Parent, and C. Lurin, SPIE Proc. *1328*:208–219 (1990).
95. Y.-M. Tricot, R. Rafaeloff, A. Emeren, and J. H. Fendler, ACS Symp. Ser. *278*:99–111 (1985).
96. F. J. Arriagada and K. Osseo-Asare, in *Refractory Metals: Extraction, Processing and Applications* (K. Liddell, D. R. Sadoway, and R. G. Bautista, eds.), TMS, Warrendale, PA, 1991, pp. 259–269.
97. T. Hirai, H. Sato, and I. Komasawa, Ind. Eng. Chem. Res. *32*:3014–3019 (1993).
98. P. D. Moran, J. R. Bartlett, J. L. Woolfrey, G. A. Bowmaker, and R. P. Cooney, J. Sol-Gel Sci. Technol. *8*:65–69 (1997).
99. E. Joselevich and I. Willner, J. Phys. Chem. *98*:7628–7635 (1994).
100. V. Chhabra, V. Pillai, B. K. Mishra, A. Morrone, and D. O. Shah, Langmuir *11*:3307–3311 (1995).
101. S. D. Desai and E. L. Cussler, Langmuir *13*:1496–1500 (1997).
102. S. D. Desai and E. L. Cussler, Langmuir *14*:277–282 (1998).
103. S. Hingorani, V. Pillai, P. Kumar, M. S. Multani, and D. O. Shah, Mater. Res. Bull. *28*:1303–1310 (1993).
104. K. Kusakabe, T. Yamaki, H. Maeda, and S. Morooka, ACS Prepr. *38*(1):352–357(1993).
105. T. Yamaki, H. Maeda, K. Kusakabe, and S. Morooka, J. Membrane Sci. *85*:167–173 (1993).
106. T. Kawai, A. Fujino, and K. Kon-no, Colloids Surf. *109*:245–253 (1996).
107. J. Wang, L. S. Ee, S. C. Ng, C. H. Chew, and L. M. Gan, Mater. Lett. *30*:119–124 (1997).
108. P. K. Dutta and D. Robbins, Langmuir *7*:1048–1050 (1991).
109. V. Chhabra, M. Lal, A. N. Maitra, and P. Ayyub, J. Mater. Res. *10*:2689–2692 (1995).
110. G. Klassen and G. Fischman, Ceram. Eng. Sci. Proc., September/October 1992, pp. 1089–1093.
111. S. Schlag, H.-F. Eicke, D. Mathys, and R. Guggenheim, Langmuir *10*:3357–3361 (1994).
112. H. Herrig and R. Hempelmann, Mater. Lett. *27*:287–292 (1996).
113. L. M. Gan, L. H. Zhang, H. S. O. Chan, C. H. Chew, and B. H. Loo, J. Mater. Sci. *31*:1071–1079 (1996).
114. A. G. Teijeiro, D. Baldomir, J. Rivas, S. Paz, P. Vaqueiro, and A. López, Quintela, J. Magn. Magnetic Mater. *140–144*:2129–2130 (1995).
115. P. Vaqueiro, M. A. López-Quintela, and J. Rivas, J. Mater. Chem. *7*:501–504 (1997).
116. P. Ayyub, A. N. Maitra, and D. O. Shah, Physica C *168*:571–579 (1990).
117. P. Ayyub and M. S. Multani, Mater. Lett. *10*:431–436 (1991).
118. P. Ayyub, A. Maitra, and D. O. Shah, J. Chem. Soc. Faraday Trans. *89*:3585–3589 (1993).
119. M. A. López-Quintela and J. Rivas, J. Colloid Interface Sci. *158*:446–451 (1993).
120. P. Kumar, V. Pillai, and D. O. Shah, Appl. Phys. Lett. *62*:765–767 (1993).
121. J. Eastoe, B. H. Robinson, D. C. Steytler, and D. Thorn-Leeson, Adv. Colloid Interface Sci., *36*:1–31 (1991).
122. J. Sjoblom, R. Lindberg, and S. E. Friberg, Adv. Colloid Interface Sci., *95*:125–287 (1996).
123. K. Shinoda and B. Lindman, Langmuir *3*:135–149 (1987).
124. K. Kon-no, Surf. Colloid Sci. *15*:125–151 (1993).
125. J. G. Darab, D. M. Pfund, J. L. Fulton, J. C. Linehan, M. Capel, and Y. Ma, Langmuir *10*:135–141 (1994).
126. A. E. Nielsen, *Kinetics of Precipitation*, Macmillan, New York, 1964.
127. J. A. Dirksen and T. A. Ring, Chem. Eng. Sci. *46*:2389–2427 (1991).
128. J. Nyvlt, O. Sohnel, M. Matuchova, and M. Browl, *The Kinetics of Industrial Crystallization*, Elsevier, New York, 1985.
129. S. S. Atik and J. K. Thomas, J. Am. Chem. Soc. *103*:3543–3550 (1981).
130. R. W. Gale, J. L. Fulton, and R. D. Smith, J. Am. Chem. Soc. *109*:920–921 (1987).
131. H. A. van Straten and P. L. de Bruyn, J. Colloid Interface Sci. *102*:260–277 (1984).
132. H. A. van Straten, M. A. A. Schoonen, and P. L. de Bruyn, J. Colloid Interface Sci. *103*:493–507 (1985).
133. M. Wong, J. K. Thomas, and T. Nowak, J. Am. Chem. Soc. *99*:4730–4736 (1977).
134. J. Sunamoto, H. Kondo, and K. Akimaru, Chem. Lett. 821–824 (1978).
135. A. N. Astanina and A. P. Rudenko, Russ. J. Phys. Chem. *45*:191–194 (1971).

136. D. A. Dzombak and F. M. M. Morel, *Surface Complexation Modeling—Hydrous Ferric Oxide*, Wiley, New York, 1990.
137. I. Rico and A. Lattes, Nouv. J. Chim. *8*:429–431 (1984).
138. S. E. Friberg and M. Podzimek, Colloid Polym. Sci. *262*:252–253 (1984).
139. P. D. I. Fletcher, M. F. Galal, and B. H. Robinson, J. Chem. Soc. Faraday Trans. I *80*:3307–3314 (1984).
140. L. L. Hench and J. K. West, Chem. Rev. *90*:33–72 (1990).
141. F. Orgaz and H. Rawson, J. Non-Cryst. Solids *82*:378–390 (1986).
142. N. D. S. Mohallem and M. A. Aegerter, J. Non-Cryst. Solids *100*:526–530 (1988).
143. S. Clark, P. D. I. Fletcher, and X. Ye, Langmuir *6*:1301–1309 (1990).
144. A. S. Bommarius, J. F. Holzwarth, D. I. C. Wang, and T. A. Hatton, J. Phys. Chem. *94*:7232–7239 (1990).
145. R. Zana and J. Lang, in *Microemulsions: Structure and Dynamics* (S. E. Friberg and P. Bothorel, eds.), CRC Press, Boca Raton, FL, 1987, p. 153.
146. P. D. I. Fletcher, A. M. Howe, and B. H. Robinson, J. Chem. Soc. Faraday Trans. I *83*:985–1006 (1987).
147. S. S. Atik and J. K. Thomas, J. Am. Chem. Soc. *103*:3543–3550 (1981).
148. E. Dutkiewicz and B. H. Robinson, J. Electroanal. Chem. *251*:11–20 (1988).
149. H. Schmidt, J. Non-Cryst. Solids *73*:681–691 (1985).
150. P. Judeinstein, J. Livage, A. Zarudiansky, and R. Rose, Solid State Ionics *28–30*:1722–1725 (1988).
151. P. Delord and F. C. Larche, J. Colloid Interface Sci. *98*:277–278 (1984).
152. P. D. I. Fletcher, N. M. Perrins, B. H. Robinson, and C. Toprakcioglu, in *Reverse Micelles* (P. L. Luisi and B. E. Straub, eds.), Plenum, New York, 1984, p. 69.
153. J. H. Adair, T. Li, T. Kido, K. Havey, J. Moon, J. Mecholsky, A. Morrone, D. R. Talham, M. H. Ludwig, and L. Wang, Mater. Sci. Eng. *R23*:139–242 (1998).

19

Time-Resolved Luminescence Quenching in Microemulsions

Mats Almgren and Holger Mays
Uppsala University, Uppsala, Sweden

I. INTRODUCTION

Emitted light can be measured with extremely high sensitivity, so background luminescence from various sources sets the detection limit of luminescent molecules. High sensitivity is usually not required in microemulsion systems, since the volume fractions of the aqueous, oil, and interfacial domains are typically so large that the amount of probe required to allow measurements with rather insensitive methods still corresponds to such low concentrations that the perturbation is minimal.

The most important applications of luminescence probing in microemulsions involve the deactivation dynamics or excitation energy transfer properties of the excited states. With a brief flash of light a population of excited species is created in the sample, and the subsequent deactivation is observed over time. The decay of the excited probe, and the fluorescence spectrum, may depend on the interactions with the environment, which reveal useful information. In time-resolved luminescence quenching (TRLQ),* however, it is the interaction of the probe with another added component, a quencher, that is studied. This method is dealt with here. For micellar systems, several publications have already discussed it in both experimental and theoretical detail [1–6].

The rate of quenching depends on how the quenchers are distributed within the environment and in general also on the rate of the diffusive motion of both the excited state and the quencher, leading to an encounter between them. In the case of excitation energy transfer, the interaction may occur over a considerable distance, the Förster radius, which may exceed 4 nm for many interesting donor–acceptor pairs. A donor with a short excited state lifetime can be employed, so that the transfer occurs with little change in the donor and acceptor positions. Under such circumstances the fluorescence decay can provide information on the distribution of acceptors around the donor. In principle, it should be poss-

---

* This name is preferred to time-resolved fluorescence quenching, because phosphorescence is sometimes involved.

ible to observe a difference between acceptors distributed over flat, cylindrical, and spherical surfaces with a radius of curvature comparable to the Förster radius [7,8]. To our knowledge such studies have not yet been done.

Luminescence quenching by mechanisms other than Förster energy transfer is often close to being diffusion-controlled; i.e., it is highly probable that quenching occurs before the reactants separate after a diffusive encounter. This type of nearly diffusion-controlled quenching has been used to obtain information about various microheterogeneous systems. A model for fluorescence quenching in spherical micelles that allows the migration of quenchers between the micelles during the lifetime of the excited states was suggested as early as 1974 by Infelta et al. [9], and later a clear mathematical derivation was given by Tachiya [10]. The Infelta–Tachiya model (or more properly the IGTT model, since Infelta's coauthors should also be mentioned) is still the basis for the interpretation of most TRLQ experiments in micellar media. Originally it was presented to explain the quenching kinetics and not proposed as a method to determine aggregation numbers. Turro and Yekta [11] were the first to use fluorescence quenching in this way (although in their case in static measurements); however, already in 1964 Förster and Selinger [12] had shown that pyrene excimer emission could be used as a qualitative indicator of the presence of micelles.

The IGTT model and its many elaborations have been widely used in studies of microheterogeneous systems. The model is based on the stochastic distribution of probes and quenchers over the confinements. Before discussing it, however, we approach the problem from the aspect of diffusion-limited reactions and consider how a change from a homogeneous three-dimensional (3-D) solution into effectively 2-D, 1-D, and 0-D systems (with 0-D we refer to a system limited in all three dimensions such as a spherical micelle) will affect the diffusion-controlled deactivation process. The stochastic methods apply only to the zero-dimensional systems; we present some of the elaborations of the IGTT model with particular relevance to microemulsion systems and the complications that arise therein. We then review and discuss some of the experimental studies. It appears as if much more could be done with microemulsions, but the standard methods from studies of normal micelles have to be used with utmost care.

II. BASIS OF QUENCHING EXPERIMENTS IN MICROEMULSIONS

A. Luminescence Quenching in Spatially Unobstructed Media

If an ensemble of luminescent probe molecules in a homogeneous solution is excited by a δ pulse of light, the excited states decay in a first-order process with a characteristic rate constant k_0, which is the inverse of the natural lifetime τ_0. The decay rate is determined by the competing processes of radiationless deactivation and light emission. Further pathways for deactivation are offered by quenching reactions. Such processes employed in the study of microemulsions are usually dynamic: The reactants meet by diffusion, transfer the stored energy, and separate again, often without undergoing any net chemical change (upper reaction in Fig. 1). This mechanism is also denoted collisional quenching with the corresponding rate constant k_0, in distinction to static quenching, in which the ground state of the probe molecule forms a nonluminescent complex with the quencher (lower reaction in Fig. 1). Static quenching is not considered here, since dynamic quenching is essential to obtaining information on microemulsions.

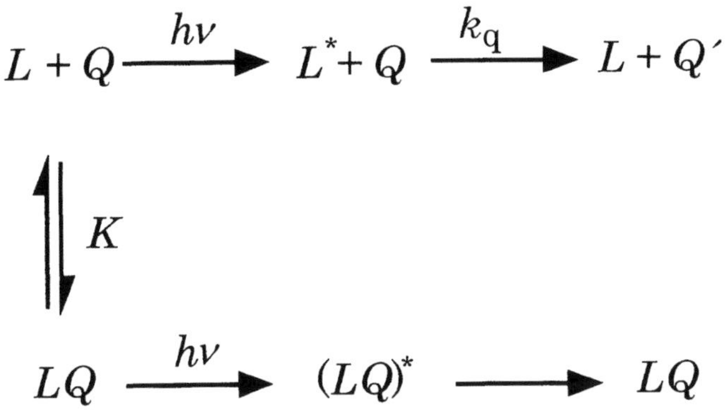

$$L + Q \xrightarrow{\ h\nu\ } L^* + Q \xrightarrow{\ k_q\ } L + Q'$$

$$\Big\updownarrow K$$

$$LQ \xrightarrow{\ h\nu\ } (LQ)^* \longrightarrow LQ$$

Figure 1 Simplified reaction scheme for dynamic and static luminescence quenching. L denotes the probe molecule in the electronically ground state, L* is the excited luminescer, and Q is a quencher molecule.

Neglecting fast transient effects, the quenching reaction in a homogeneous liquid phase follows quasi-first-order kinetics:

$$F(t) = F_0 \exp(-k_0 t - k_q C_q t) \tag{1}$$

where $F(t)$ is the intensity at time t, proportional to the concentration of excited lumophore; F_0 the initial intensity; C_q the (constant) quencher concentration; and k_q the second-order quenching rate constant. According to Eq. (1) all logarithmic decay curves of a certain system at different quencher concentrations, including the quencher-free solution, meet on extrapolation to $t = 0$ at one point, F_0. This is the normal behavior in homogeneous solutions; in microheterogeneous systems it is different, as we see below.

Although quenching reactions in general can be described by Eq. (1) with a characteristic rate constant k_q for a given probe–quencher pair in a particular solvent, there are many different quenching mechanisms. The fluorescence emission of organic molecules such as pyrene or its water-soluble sulfonate derivatives can be suppressed by increasing the rate of singlet–triplet transitions through the addition of heavy ions, e.g., Cu^{2+} or I^-, which enhance the spin–orbit coupling. The luminescence probes $Ru(bpy)_3^{2+}$ and $Tb(pda)_3^{3-}$ can easily be oxidized in the excited state; hence electron acceptors such as Cu^{2+}, methyl viologen MV^{2+}, or $Fe(CN)_6^{3-}$ reduce the probe lifetime and are suitable quenchers. Other quasi-first-order quenching reactions are excimer and exciplex formation, whereas triplet–triplet annihilation reactions, which have also been used to probe microemulsion systems, are of second order. Both singlet and triplet energy transfer reactions requiring molecular contacts are very suitable for dynamic experiments, whereas typical Förster energy transfer between dye molecules, which may occur over distances approaching 10 nm, is better suited for situations with statically localized dyes. The latter has been suggested recently for experiments to distinguish between different geometries on this length scale [8].

The rate constants for many quenching reactions between energy donors and acceptors are often very large, in the diffusion-controlled limit, corresponding to a situation where the species meet and react without appreciable probability of separating without having reacted. With few exceptions—notably Förster energy transfer—the quenching mechanisms involve close contact between the excited state and the quencher, and the quenching step occurs in a transient encounter complex. Following Weller [13], the quenching rate constant can be expressed as

$$k_q = \gamma k_{dc}, \qquad \gamma = \frac{k_2}{k_2 + k_{-dc} + k_0 + (\tau_0/\tau_q)k_{dc}C_q} \tag{2}$$

where γ is the quenching efficiency of the quenching reaction, k_{dc} and k_{-dc} are the diffusion-limited rate constants for encounter complex formation and separation, k_2 is the rate constant of the reaction in the encounter complex, and τ_q is the quenched luminescence lifetime. The rate constant k_{dc} depends on the charges in the involved species, their diffusion coefficients, and the reaction distance in the encounter complex and lies normally in the range of 10^8–10^{10} dm^3/mol·s). Note that when k_2 is not large compared to the other terms in the denominator of Eq. (2) the quenching reaction will be slower than diffusion-controlled and accompanied by an apparent activation energy. This situation was found in the quenching of Tb(pda)$_3^{3-}$ with MV^{2+}, where E_a is about 14 kJ mol [14]. Some consequences of these rate limitations in microemulsion systems are discussed later.

Since the determination of surfactant aggregation numbers and compartment sizes is based on the distinction between compartments occupied by both excited probe and quenchers from those occupied by only probe, a high quenching efficiency is required in such measurements.

B. Diffusion-Controlled Deactivation in Three to Zero Dimensions

Let us now consider in general terms the effects of the confinement of a probe and quencher in a space that is finite in at least one dimension. Let us suppose that a relatively long lived luminescence probe has been excited with a brief pulse of light in a fluid solution containing an effective quencher. Neglecting transient effects, as is normally possible, the fluorescence decay will be exponential, and if we choose to represent it as in Fig. 3, i.e., on a logarithmic scale and with the intensities corrected for the natural decay by multiplication with a factor of exp(k_0t), then a constant intensity level is obtained without quenchers, and a linear decay is obtained when quenchers are present.

Imagine that the volume containing the excited states and quenchers, initially randomly distributed, is subdivided into thin slabs of thickness a, assumed to be a few nanometers, and surrounded by impenetrable walls (see Fig. 2). Now the quenching reaction can occur only between excited states and quenchers within the same slab. The initial part of the decay will not be affected by this confinement, since at times shorter than $a^2/2D$, where D is the diffusion coefficient, the displacements of the reactants are smaller than the width of the slab. At longer times, however, there will be less quenching than in the 3-D system, since only quenchers within the same slab as the excited state may take part. The frequency of encounters, and hence the rate of quenching, then depends on the diffusion in the two unlimited dimensions.

A further subdivision of the volume into long columns reduces the quenching at long times even more. The quenching rate becomes limited by one-dimensional diffusion along the axis of the column. Finally, by cutting the volume in the third direction also so that small cubes are formed, we arrive at the quasi-0-D case. Quenching will then occur only in those small volumes that contain both the excited probe and at least one quencher. In these confinements the reaction will occur within a time of about $a^2/2D$; all excited states surviving appreciably longer must originally have been created in confinements without quenchers.

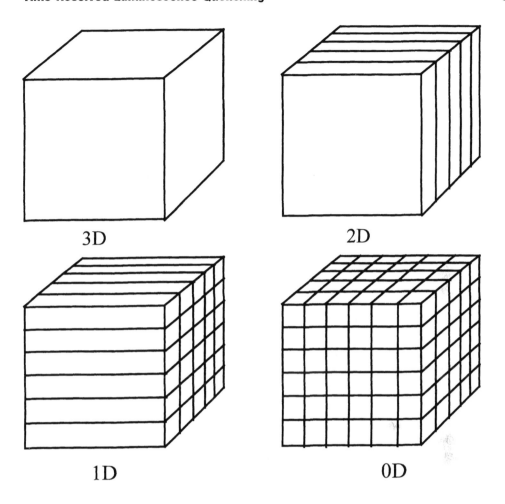

3D   2D

1D   0D

Figure 2 Illustration of restrictions in space for an excited probe. The first sketch represents a 3-D system, and by dividing the space along the three axes the reaction space is reduced to two, one, and zero dimensions.

The expected type of behavior is illustrated in Fig. 3 with simulated fluorescence decay data. Results of this type have been experimentally observed for micellar systems, and equations are available that describe the resulting decays [15].

It is possible to draw some general conclusions that are valid irrespective of the specific model used to describe the deactivation in detail. It is obvious that if a constant final level in the representation of Fig. 3 (implying a final exponential decay with the same decay constant as in the absence of quenchers) is reached, then a certain fraction of the excited states are inaccessible for quenching; and this is the case whatever the mechanism of the quenching is and no matter how the quenchers are distributed over the micelles. This fraction, present in confinements without quenchers, is what is primarily measured in the TRLQ method to determine micellar aggregation numbers. It is important to note that this interpretation is general and, in particular, independent of how the quenching proceeds. However, the calculation of the number of confinements and their average size from the primarily deter-

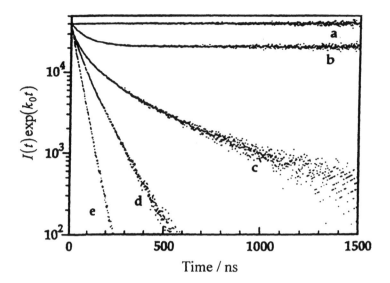

Figure 3 Simulated fluorescence decay data illustrating the changes in quenching behavior with the confinement structure. The influence of natural decay has been eliminated by multiplication with the factor $\exp(k_0 t)$. Curve a, without quencher; b–e, quenching in zero to three dimensions (spherical micelles, rods, bilayers, neat surfactant). The parameters were chosen to mimic $C_{12}E_6$ in water, with pyrene as probe and dimethylbenzophenone as quencher. (From Ref. 15.)

mined fraction of quencher-free confinements requires knowledge of how the quenchers and surfactant molecules are distributed over the confinements. Usually a random distribution over equal-sized micelles is assumed and represented by the Poisson distribution. If it holds, the same aggregation number should be obtained at different quenchers/surfactant ratios, whereas a variation of the apparent aggregation number with quencher concentration signifies a non-Poissonian distribution. A possible reason for such deviations is that the micelles are polydisperse in size. It has been shown that under some assumptions polydisperse micelles give an apparent aggregation number that decreases from the weight-average aggregation number at zero quencher concentration, with an initial slope proportional to the width of the distribution and a curvature depending on the skewness [16,17].

Specific forces, e.g., electrostatic attractions, between the quenchers, probes, and surfactants may give non-Poissonian distributions even if the micelles are of equal size. Such effects have been assessed in just a few cases but found to be important [14,18–20], so more work has to be done in order to learn how to handle them properly. Depending on the type of interaction, the apparent aggregation number may either increase or decrease with the quencher concentration. As long as the nonideal mixing affects mainly the quencher distribution and not the probes, so that by excitation a representative sample of micelles is still selected, the extrapolation to zero quencher concentration still gives a correct aggregation number (which is the weight-average aggregation number if the micelles are polydisperse).

The effect of the growth of the micelles (or other domains confining probe and quenchers) on the luminescence decays is worth some general comments. In many cases the growth process is accompanied by an increased domain size polydispersity, which makes the construction of a detailed model of the decay somewhat complex. One way to build such a model would be to add contributions from each size class. The resulting sum over a large number of exponential terms cannot normally be reduced to a simple expression [16], and an additional complexity is the fact that the decay function will be nonexponential even for an excited probe and a quencher in an oblong or flat domain. From the perspective of diffusion-controlled reactions in confinements, it is obvious that as long as one dimension of the confinement is larger than $(2Dt_m)^{1/2}$ (where t_m is the time window of the measurements, at most a few natural half-lives), the reaction will occur as if the space were unlimited in that dimension. Conversely, in rodlike micelles with a finite length l, the decay function at times much shorter than $l^2/2D$ will be the same as for infinite rods but will be unaffected by quenching for times much longer than that.

In typical microemulsion systems, a prominent composition region shows bicontinuous structures. With probes and quenchers confined to either oil or water, the domains in the bicontinuous region may be so large in all dimensions that normal exponential decays are observed. Only in the region with discrete droplets, and in a transition region where droplets cluster and merge, can the micellar type of quenching be expected. However, if the amphiphilic probes and quenchers are bound to the interfacial surfactant film in the bicontinuous microemulsions, one would expect 2-D behavior.

At long times—how long depends on the system—the effects of the confinement will eventually disappear. Through various mechanisms, excited probes and quenchers from different confinements will meet and react, resulting in a decay that is exponential at long times but with decay constant larger than for the unquenched decay. A mechanism that has been invoked and studied in microemulsions is solubilisate exchange between discrete droplets in "sticky" collisions, with partial or complete merging of two droplets into an unstable larger one [21,22]. With stronger attractive interactions between the droplets, clusters may form, possibly as the first stage in a transition from the discrete to the bicontinous state. Large, highly ramified clusters should be present at the percolation point. Jumps of quenchers from droplet to droplet, either through narrow (transient) channels or by fusion–fission events, would result in a quenching that can be looked upon as a random walk over a percolation cluster [23]. A nonexponential decay would be expected, but on a long time scale the dynamic reshaping of the cluster itself, mediated by cluster collisions or fluctuations [14,24], would add further transport possibilities, again resulting in a final exponential decay.

C. Micellar Systems with Reduced Dimensionality—The 1-D Case

The shape and size of self-assembled micellar systems depend on the conditions for a given system. Changes can be induced, e.g., by addition of cosurfactant or salt or by high surfactant concentrations. An aqueous solution of cetyltrimethylammonium chloride (CTAC) exhibits a transition from spherical micelles to elongated cylinders upon addition of chlorate anions, which is shown, for instance, by a strong increase in viscosity [25]. This system was experimentally studied by TRLQ, with results in good agreement with theory, as shown in Fig. 4.

The probe and quencher were in this case hydrophobic enough to be confined within the aggregates. By choosing a low probe concentration and low intensity of the excitation flash, it was ensured that no aggregate contained more than one excited lumophore. A characteristic feature of the decay curve family in Fig. 4 is that all curves start in the same nonexponential way but separate and become linear and parallel to the unquenched exponential curve after a certain time, which is longer the larger the micelles. There is no difference between the curves for the two highest concentrations of chlorate, in spite

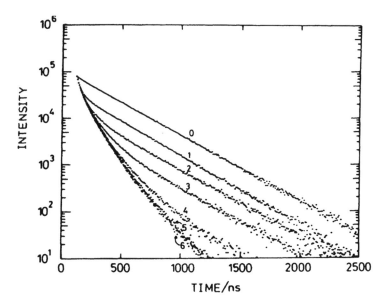

Figure 4 Experimental decay profiles of pyrene quenched by dimethyl benzophenone in micellar solutions of CTAC containing various concentrations of NaClO₃. The top curve (0) is without quencher; for the others, $C_q = 4.5 \times 10^{-3}$ mol/dm³. The chlorate concentration is increased from 0 (curve 1) to 0.101 mol/dm³ (curve 6). (From Ref. 25.)

of the fact that the viscosity indicates a strong increase in this concentration region. Already at this relatively short length, the TRLQ measurements give results as for an infinite rod, which means that the rod length is longer than $(2Dt_m)^{1/2}$. In the measurements of Fig. 4, t_m was about 2 μs and $D \approx 9 \times 10^{-11}$ m⁻² s⁻¹, so that "infinite" in this case means longer than 20 nm.

Two ways to describe quenching in rods theoretically have been advanced. Van der Auweraer et al. [26,27] discussed the decay kinetics of an immobile probe localized in the middle part of a finite cylinder (or in a torus) containing a mobile quencher. Almgren et al. [25], on the other hand, proposed a model in which an excited probe was localized in an infinite rod of radius R containing diffusing quenchers. A reaction zone was introduced to account for the finite rate of the quenching reaction between the excited probe and a quencher in close proximity; even if both are at the same axial coordinate, they are not necessarily in contact. Within the reaction zone the quenching reaction was assumed to occur with the first-order rate constant k_{qm}. The length of the reaction zone was originally chosen to give the same volume as a sphere with the same radius as the rod, but later the length was simply taken as the radius of the rod; the different assumptions affect only the magnitude of k_{qm}. The resulting decay equation is

$$\ln \frac{F(t)}{F_0} = -k_0 t - 4\pi R^3 c_q \left\{ \sqrt{\frac{t}{\pi \tau_q}} - \frac{3}{4hR} + \frac{3}{4hR} \left[\exp\left(\frac{4h^2 R^2}{9\tau_q} t\right) \operatorname{erfc}\left(\frac{2hR}{3}\sqrt{\frac{t}{\tau_q}}\right) \right] \right\} \quad (3)$$

where c_q is the number density of quencher per unit volume. The parameters h and τ_q are related to D, the sum of the diffusion coefficients of quencher and excited probe:

$$h = \frac{k_{qm}\tau_q}{DR}, \qquad \tau_q = \frac{R^2}{D} \tag{4}$$

The experimental decays from the "infinite" rods at high salt concentrations in Fig. 4 fit well to the 1-D model. The same model has been used in several investigations of the sphere-to-rod transition and to determine the diffusion coefficients of the probe and quencher in rodlike micelles. However, equations based on other assumptions could fit the experimental data about equally well; there is no distinctive feature in the decay that points to the 1-D case. Often one can reject alternative models due to unreasonable values for estimated parameters. A 2-D model, for example, would probably also fit well but would result in too low a value of the diffusion coefficient. Numerical studies demonstrated that 2-D and (with consideration of transient effects) 3-D models generally fit well to 1-D data but with unreasonable values suggested for the parameters, whereas a decay curve from a 2-D structure does not fit to a 1-D model [15].

Also strongly polydisperse samples can give decays similar to 1-D or 2-D decays, in particular if some of the confinements are so large that the final exponential tail is not clearly defined within the time window available for measured. If, in addition, exchange processes occur and increase the decay rate in the limiting exponential tail, then TRLQ measurements alone are not sufficient to distinguish between 1-D and 2-D quenching, polydispersity, and migration between compartments.

These difficulties are particularly crucial in microemulsions, which are more ambiguous and complex than micellar solutions. In microemulsions a change of the conditions can drive an evolution from closed droplets, via clustering and increasing size and shape polydispersity, to bicontinuous structures. The first part of this change is accompanied by an increase in the solute exchange rate between the compartments. A further effect not fully considered in the literature is the limited lifetime of the aggregates. This factor would limit the time window over which a model, e.g., for quenching in an infinite rod, will apply; at long times an exponential decay is likely to appear.

D. Quenching in Microemulsion Droplets—The Limiting 0-D Case

The equation most often used to analyze quenching experiments in microheterogeneous systems is the IGTT relationship, originally derived for quenching in micelles [9,10]. The time-dependent luminescence intensity (cf. Fig. 5) in logarithmic form is given by

$$\ln \frac{I(t)}{I_0} = -A_2 t + A_3[\exp(-A_4 t) - 1] \tag{5}$$

with the parameters $A_2 - A_4$ related to the natural probe lifetime τ_0 (equal to $1/k_0$, the reciprocal of the natural decay constant), to the first-order intramicellar quenching rate constant, k_{qm}, to the rate constant of quencher escape into the bulk phase, k_-, and to the average quencher occupancy number, n:

$$A_2 = k_0 + \frac{k_{qm}k_-}{k_{qm} + k_-}n, \qquad A_3 = \frac{k_{qm}^2}{(k_{qm} + k_-)^2}n, \qquad A_4 = k_{qm} + k_- \tag{6}$$

Important assumptions behind Eq. (5) are that (1) the micelles are monodisperse, (2) the quenchers are distributed according to a Poisson distribution, (3) the excited probe remains in the same micelle during its lifetime, and (4) the intramicellar quenching rate is proportional to the number of quenchers in the micelle. If no migration between the

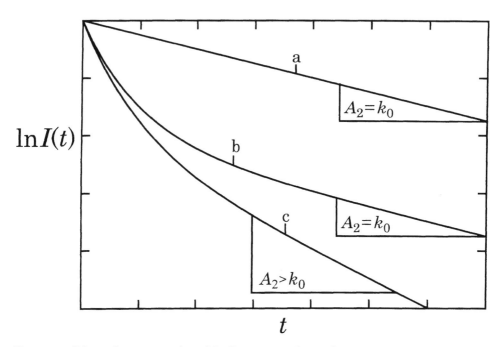

Figure 5 Schematic representation of the fluorescence decay after the IGTT model, Eq. (5). Curve a is without quencher, curve b with quencher but without solubilizate exchange, and in curve c the slope of the final decay is changed due to exchange of probe or quencher.

micelles occurs, i.e., $k_- = 0$, then Eq. (5) reduces to

$$\ln \frac{I(t)}{I_0} = -k_0 t - n + n \exp(-k_{qm} t) \tag{7}$$

which is the ideal case for the determination of surfactant aggregation numbers, N_{agg}. Since n is related to the micelle concentration C_m and the quencher concentration C_q by $n = C_q/C_m$, we have

$$N_{agg} = \frac{C_s - C_{s,f}}{C_q} n \tag{8}$$

where C_s is the total surfactant concentration and $C_{s,f}$ is the concentration of free surfactant, often taken as the critical micelle concentration. In W/O microemulsions the number of water molecules per droplet can be obtained in a similar way. If spherical micelles are assumed, the average water core radius of the aggregates can be calculated from a simple geometric model;

$$r = \left(\frac{N_{agg} A_s}{4\pi}\right)^{1/2} \tag{9}$$

with A_s the area occupied by a surfactant molecule in the interfacial layer. Another possibility is to calculate the total volume of the aqueous core, using the molar volume of water and an assigned volume for the surfactant polar head.

The quenched decay curves b and c in Fig. 5 are clearly different from those in a homogeneous solution; due to the fast intramicellar quenching, the extrapolations of the final exponential decays to zero time do not meet the quencher-free curve at a single point, but instead an initial drop appears. If exchange of probe and/or quencher occurs between the compartments, the slope of the logarithmic tail is different than in a quencher-free microemulsion, but the quenching is still slower than in a corresponding homogenous solution.

In static luminescence measurements, all the measured intensity stems from the fraction of the micelles with excited lumophore but without quencher, provided that k_{qm} is much larger than k_0 and k_-. If also the Poisson distribution holds, then the fraction is given by $\exp(-n)$. This leads directly to the Turro–Yekta equation [11],

$$\ln \frac{I_0}{I} = n \tag{10}$$

where I_0 and I are the emission intensities without and with quencher, respectively.

III. EXPERIMENTAL APPLICATIONS

Much work applying luminescence quenching methods to microemulsions has involved the anionic surfactant Aerosol OT (AOT) in W/O systems [14,23,24,28–35] and also in O/W microemulsions [36,37]. Other surfactants investigated include sodium dodecyl sulfate (SDS) [38–41], cationic surfactants such as quaternary ammonium surfactants [33,41–46], and nonionic surfactants such as the alkyl oligo(ethylene oxides) C_iE_j [14,43,47–49]. Both ternary systems of surfactant–water–oil and quaternary systems with an additional cosurfactant, usually an alcohol, have been examined. In the extensive studies of Lang and coworkers [29,33,42–45,50], many useful correlations were found concerning the effects on droplet size and exchange rates of parameters such as oil (and cosurfactant) chain length, water/surfactant ratio, concentration of droplets, and temperature. Most remarkable is the finding that the solute exchange rate increases in parallel with the electrical conductivity of the microemulsion, so that the nominal second-order rate constant always reaches a value of $(1–2) \times 10^9$ dm³/(mol · s) at the percolation threshold. Furthermore, these transport properties are mainly determined by the attractive interactions between the droplets, which in turn increase with the droplet size. Some of these findings are discussed below.

There are also other luminescence studies, for instance with microemulsions as reaction media, where the primary aim was to investigate how reaction rates were affected by the partitioning of the reactant molecules between the oil, water, and interfacial domains in bicontinuous systems [51], e.g., with AOT and C_iE_j as surfactants.

The first application of the IGTT model to quenching reactions in microemulsions was presented by Atik and Thomas in 1981 [28], with the parameters A_2–A_4 of Eq. (5) interpreted as

$$A_2 = k_0 + \frac{k_{ex}k_{qm}}{k_{qm} + k_{ex}C_m}C_q, \qquad A_3 = \left(\frac{k_{qm}}{k_{qm} + k_{ex}C_m}\right)^2 \left(\frac{C_q}{C_m}\right), \qquad A_4 = k_{qm} + k_{ex}C_m \tag{11}$$

The second order rate constant k_{ex} was introduced to account for the observed slow exchange, assumed to occur on droplet collisions.

A. Surfactant Aggregation Numbers and Size of Closed Droplets

The TRLQ method can show the presence of closed aggregates, distinguish between clustering of small units and coalescence into large connected aggregates, and pick up even small changes in the size of closed aggregates brought about by variations in composition or other conditions [32]. It also gives an absolute measure of the size, which, however, is often rather uncertain. The determination of aggregation numbers in microemulsions is not so easy as may appear from Eqs. (5) and (11). The TRLQ method often works well, but one has to be careful in the choice of probe–quencher pair and must consider the interactions between the reactants and the surfactant. Some studies have compared the droplet size determined by fluorescence quenching using different probe–quencher pairs with results from other methods [31,32]. The values obtained with different methods, or with different probe–quencher pairs, often deviate considerably, and the deviations tend to increase with droplet size.

Under optimum conditions a droplet size of about 3.6 nm, corresponding to a W_0 value of 20 (where W_0 is the number of water molecules per surfactant molecule), seems to be close to the upper limit for the determination of aggregation numbers by the luminescence method in AOT microemulsions. The main limiting factor is the competition between intradroplet quenching and exchange. The rate of the former decreases with increasing size, since k_{qm} is a measure of the encounter frequency of a reactant pair within the droplet, whereas the latter rate often increases [29,32,52]. In other words, the condition $k_{qm} \gg k_{ex}C_m$ does not hold for large droplets and/or slow quenching systems. One may ask why this condition must be fulfilled. The reason is that with rapid exchange of probe and quenchers between the droplets, each probe senses an average concentration of quencher, and the stochastic feature, which is the very basis of the method, is lost.

A rapid exchange process can be recognized by a change in the slope of the final part of the decay compared to the corresponding quencher-free microemulsion; i.e., parameter A_2 is larger than k_0 in Eqs. (6) and (11), as shown in Fig. 5. Further disturbances may arise from the spatial distributions of the probe and quencher, depending on their charges and the charge of the surfactant, or from amphiphilic properties of the probe or quencher itself. It is advantageous to use a probe and quencher that both have the same charge sign as the interface. In addition, bulky probes and quenchers may perturb small droplets or the distribution of the reactants over the droplets, which was suggested as the reason for the incorrect aggregation numbers obtained using $Ru(bpy)_3^{2+}$ and $Fe(CN)_4^{3-}$ in small AOT droplets [31].

Only a few claims of the discovery of nonspherical structures in microemulsions by TRLQ have been published as yet. The quenching of $Ru(bpy)_3^{2+}$ by MV^{2+} in water-in-oil microemulsions of the nonionic amphiphile $C_{12}E_4$ in decane indicates the presence of nonglobular structures at high water/surfactant ratios. This comes mainly from the fact that if spherical structures are assumed, the aggregation numbers estimated by TRLQ would imply unreasonably small areas per surfactant at the interface, and simultaneously the radius of the aqueous droplet that would far exceed the length of the polar chain (four ethylene oxides) of the surfactant. A pool of pure water would thus be present in the middle of the droplet, and the EO tails would be compressed close to the interface. It appears more likely that the micelles take on a nonspherical shape, and this would furthermore be compatible with the observed decay curves [47].

Luminescence studies of bicontinuous systems, in particular of systems undergoing a transition from discrete droplets to bicontinuous structures, have an unexploited potential to give more detailed information about the structures and mechanisms involved. New experiments can be tried, e.g., using Förster energy transfer as mentioned above. The experimental studies have to be accompanied by theoretical developments, however, if reliable interpretations of the experimental data from such complex systems is to be obtained.

B. Exchange Between Microemulsion Droplets and Clusters

In many luminescence deactivation studies the interdroplet quenching has been used to study the transport or exchange of solutes between microemulsion droplets. In contrast to normal micelle solutions, where an exchange, usually of the quencher only, often occurs by diffusion through the intermicellar aqueous solution, the normal mechanism in microemulsions is considered to be the exchange between colliding droplets. By this mechanism both the probe and quencher may be exchanged equally fast. This case was not covered by the original IGTT model, but a generalization was made later [53]. However, the data analysis with the general exchange mechanism is more complex and not straightforward, and since the deviations are typically less than 20% [29], the parameters as expressed by Eq. (11) are usually used in the data analysis. For all exchange mechanisms the most obvious measurable effect is a change in the rate of the final exponential decay, which can be determined in a relatively simple measurement. It is often more difficult to make sure that the process really is an exchange between closed droplets.

In the early work on exchange in microemulsions, the conditions were chosen so that the exchange between separated droplets was relatively slow, appearing in the millisecond time range, and with a second-order rate constant of about 10^7 dm^3/(mol · s) or less [28,29,32]. The systems studied were AOT microemulsions with short-chain alkane oils such as hexane or heptane. These systems have high upper phase separation temperatures, and at the commonly used experimental temperatures the microemulsions are far from their haze points. Previously it had been shown that an exchange of strongly hydrophilic solubilisate occurred exclusively by collision and transient merging of the surfactant-coated water pools (the fusion–fission mechanism, pictured in Fig. 6), since the insertion of a dialysis membrane between two microemulsions prevented exchange [21]. From the low polydispersity it was also concluded that droplet fragmentation and coalescence played no important role with respect to the exchange in ionic W/O microemulsions [29].

In 1987 Howe et al. [32] found that the exchange was facilitated when a critical point was approached, in parallel with a clustering of the droplets. The formation of dimers and larger clusters in droplet microemulsions has long been claimed to occur. An account of the prehistory is given in a more recent paper by Koper et al. [54], who also present a simple quantitative model of cluster formation.

A similar observation of increasing k_{ex} when the temperature is increased toward the haze point was made in a comprehensive study by Lang et al. [29]. The results were later correlated with the percolation phenomenon of microemulsions as revealed by conductivity measurements [33]. Surprisingly high values of the second-order exchange rate constant

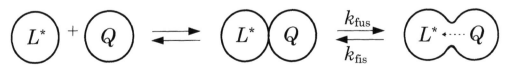

Figure 6 The reaction steps of the fusion–fission exchange mechanism between microemulsion droplets. In the first step an encounter pair is built up by sticky collision, followed by either temporary droplet merging (k_{fus}) or dissociation. L* represents an excited lumophore and Q a quencher molecule.

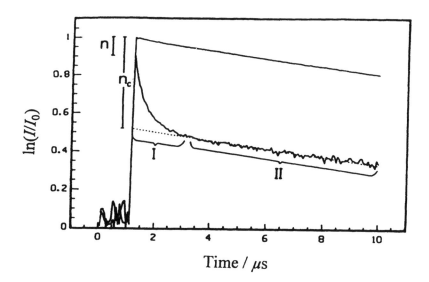

Figure 7 Quenched phosphorescence decay in a clustered droplet system using the probe $Cr(bpy)_3^{3+}$ with I^-. The initial drop n corresponds to the intramicellar quenching. Stage I is the time-resolved intercluster quenching, with the average cluster quencher occupation number n_c as amplitude. In stage II the natural decay of the probe is the dominant deactivation mechanism. (From Ref. 30.)

were reported in these studies and also by Verbeeck and De Schryver [31]. The latter authors found, for instance, that k_{ex} reached 6.5×10^{10} $dm^3/(mol \cdot s)$ in an AOT–hexane system, with pyrene tetrasulfonate (PTSA) as probe and thiocyanate (SCN^-) as quencher. High values were reported also for somewhat larger droplets when I^- was employed as quencher (the temperature of the measurements was not mentioned but was probably not above 25°C). These values exceed the diffusion-controlled rate constant for uncharged species in the same solvent by more than a factor of 2. The reason for such results must be carefully analyzed.

Using the same microemulsion and under closely similar conditions, Fletcher et al. [52] determined solubilisate exchange rates on the order of 4×10^6 $dm^3/(mol \cdot s)$ from stopped-flow experiments. When the discrepancy is a factor of more than 10^4, it is of little concern that the TRLQ experiments were analyzed with expressions from Eqs. (11) for the parameters $A_2 - A_4$, which are not strictly valid if both quencher and probe are exchanged [53,55,56]. Obviously, it cannot be the same process that is measured in the two types of experiments.

An explanation of the discrepancies was offered by the results from some experiments made by Johannsson et al. [30], employing a more long lived probe, $Cr(bpy)_3^{3+}$ ($\tau \approx 25$ μs), in the AOT–alkane–water system. From the observation of two decay processes, well separated in time as shown in Fig. 7, the authors concluded that small clusters of reverse micelles were present in the microemulsions. The initial fast process, the intramicellar quenching, occurs on the submicrosecond time scale and appears only as an initial drop since it is not resolved on the time scale used with $Cr(bpy)_3^{3+}$. It is this part of the deactivation that is possible to monitor in normal TRLQ measurements with short-lived probes. The initial drop is followed by a second decay with a characteristic time of a few microseconds before the final, very slow deactivation occurs. The results suggest that the fast exchange

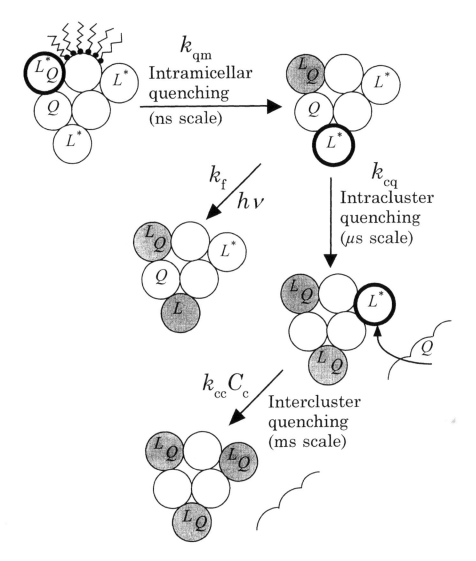

Figure 8 The three different quenching pathways within droplet clusters in microemulsions. L* denotes an excited lumophore and Q a quencher molecule. Although in reality the different reaction pathways compete, for illustration purposes they are pictured here as subsequent processes. After a quenching reaction has finished, the considered droplet is shaded.

process observed in earlier TRLQ experiments was in reality an exchange within small clusters, whereas the stopped-flow experiments monitored the much slower exchange requiring droplet (or cluster) collisions. Figure 8 visualizes the different stages in the deactivation process in microemulsions.

The second nonexponential process is not covered by the IGTT formalism, and an extended equation for deactivation in small clusters was proposed [30]:

$$\ln\frac{I(t)}{I_0} = -k_0 t + n[\exp(-k_{qm}t) - 1] + (n_c - n)[\exp(-k_{cq}t) - 1] \tag{12}$$

where the additional intracluster deactivation process is described by the first-order rate constant k_{cq}, with an amplitude determined by the average number of quenchers per cluster, n_c. In this model an excited lumophore in a droplet without quenchers can be quenched after solubilisate exchange between the droplets in the same cluster.

From the magnitude of n_c the average number of aggregated droplets was found to be about 8 in an AOT–dodecane W/O system at a surfactant concentration of 0.2 mol/dm^3. The average cluster size decreased both on dilution with alkane (to 5.6 at 0.1 mol/dm^3 surfactant concentration) and when a shorter alkane oil was employed (2.5 droplets per cluster in isooctane). It should be noted that at 25°C the compositions studied are rather far from the haze limit. The first-order rate constant k_{cq} was on the order of 10^6 s^{-1}, also decreasing with the cluster size. The results seem reasonable, since long alkane oils are known to enhance droplet aggregation, and the intracluster process should require longer time in larger clusters, just as intradroplet quenching does in large droplets.

Gehlen et al. [57] showed that similar clustering occurred in some quaternary cationic microemulsions. They used an advanced model that permitted both quencher and probe exchange. Apparent second-order solute exchange rate constants with values exceeding the diffusion-controlled limit were obtained and could be reinterpreted in terms of cluster size and intracluster exchange in a consistent way.

The temperature dependence of k_{cq} was determined using Cr(bpy)$_3^{3+}$ as a phosphorescent probe in AOT microemulsions [34]. The rate constant k_{cq} obeys an Arrhenius relationship that yields an activation energy of intracluster exchange. The activation energy depends on the droplet size but is independent of the oil. The determined activation energies were clearly smaller than those obtained from stopped-flow measurements. For example, in a system with $W_0 = 20$, phosphorescence quenching gave $E_a = 33$ kJ/mol, whereas stopped-flow measurements gave 87 kJ/mol [52]. These differences and the difference with respect to the oil dependence indicate again that different processes are measured in the two methods. It has been proposed [24,34] that the differences in E_a may exist because the interior monolayers separating the droplets in a cluster, and thus involved in the exchange within a cluster, are less penetrated by alkane oils than the outer monolayers, which are involved in exchange between clusters.

It is also important that the interaction between the droplets becomes more attractive with increasing temperature (approaching the haze point), so that the cluster lifetime should increase. Three factors determining the overall rate of solute exchange between clusters—namely, the collision rate, the lifetime of the cluster, and the exchange within the cluster—would then all increase with temperature and contribute to the activation energy determined in the stopped-flow experiments [34].

The last stage of the deactivation in W/O microemulsions, the slow decay following the intracluster exchange, was investigated in detail through measurements on the millisecond time scale using the probe Tb(pda)$_3^{3-}$ with a natural lifetime of about 2 ms [14,21]. This final exponential decay was affected by quenchers, indicating an exchange process involving exchange between clusters or between clusters and droplets. At this point the importance of freely moving quenchers became apparent. Methyl viologen, which is adsorbed strongly at negatively charged interfaces, showed, in contrast to anionic quenchers, no or only very slow exchange and quenching in a system built up with AOT, whereas it migrates much faster in a nonionic microemulsion. To account for the intercluster exchange, the IGTT equation—or more precisely Eq. (12)—was extended by introducing a term containing the second-order rate constant k_{cc}. In the long-time approximation (which is indicated in the following equation by the abbreviation lt) the intracluster quenching is apparent only as an initial drop of amplitude n_c:

$$\frac{I(t)}{I_0}\bigg|_{lt} = -k_0 t - n_c - k_{cc} C_q t \tag{13}$$

As mentioned above, the exchange rate constant k_{ex}, as determined on the microsecond time scale using $Ru(bpy)_3^{2+}$, was found to reach $(1-2) \times 10^{10}$ $dm^3/(mol \cdot s)$ at the percolation threshold. The intercluster exchange rate constant, k_{cc}, increased also when the percolation region was approached but remained much lower than k_{ex}, with a common value for a series of alkane oils of $k_{cc} \approx 4 \times 10^8$ $dm^3/(mol \cdot s)$ at the percolation threshold. This difference can again be attributed to contributions from fast intracluster quenching to k_{ex}. It is interesting to note that the activation energy of the slow intercluster quenching process increases with the chain length of the oil used as solvent, as can be seen later from Fig. 10, from 137 kJ/mol for n-heptane to 188 kJ/mol for n-undecane [24], and that the extrapolation of E_a to infinitely short alkanes, i.e., $N_{alk} = 0$, gives a value in agreement with the oil-independent activation energy of the intracluster quenching at the same W_0 value. It is also noted that the activation energy determined by Fletcher et al. [52] in stopped-flow experiments agrees with results obtained by Mays [58] from long-lived phosphorescence. The same slow exchange process seems to be monitored in both cases.

Comparing values at a constant droplet size corresponding to $W_0 = 25$, the slow process (in n-heptane) gave $E_a = 100$ kJ/mol [58], and the intracluster exchange, 40 kJ/mol (no oil dependence) [34]. With good agreement between the results from three different techniques, the enthalpy $\Delta H°$ of cluster formation (in isooctane) was reported to be about 80 kJ/mol [54]. (However, a much lower value for $\Delta H°$ has also been reported [59].) If the exchange within the cluster is slow compared to cluster dissociation, there is a simple relationship between these values. The apparent activation energy of the intercluster exchange process should be given by the sum of the cluster formation enthalpy and the activation energy of the intracluster exchange. Even if there may be some difference between isooctane and n-heptane, the sum of the latter two values seems to be at least as large as the first value, indicating that the cluster dissociation is so slow that the reaction occurs in a large fraction to the formed clusters. At the other extreme, the diffusion-controlled encounter rate of clusters or of a droplet and a cluster would limit the exchange rate, contributing an activation energy of about 13 kJ/mol as for a diffusion-controlled process, similar to the "activation energy" against the viscosity in an alkane solvent. With this value, the sum of about 53 kJ/mol is evidently too low.

Contrary to expectations, on the millisecond time scale the initial drop n_c in the luminescence studies of Mays and Ilgenfritz [24] did not increase significantly with the temperature-induced cluster growth but remained constant even when an infinite percolation cluster was present. Furthermore, the observed decays were always exponential (Fig. 9). Evidently, the initial drop no longer reflects the cluster size. The process responsible for it is over within 50 μs and should perhaps rather be looked upon similarly to the transient "active sphere" part of normal diffusion-controlled decay. The diffusion in this case is a random walk performed by the quencher on a (percolation) cluster. A stretched exponential decay would be expected for a random walk deactivation on a static cluster, as was observed close to the percolation threshold in earlier studies [23,24]. Those measurements were performed over a time window of about 50 μs, which is close to the reported value of the cluster lifetime from electrical birefringence measurements [60]. It is very likely that

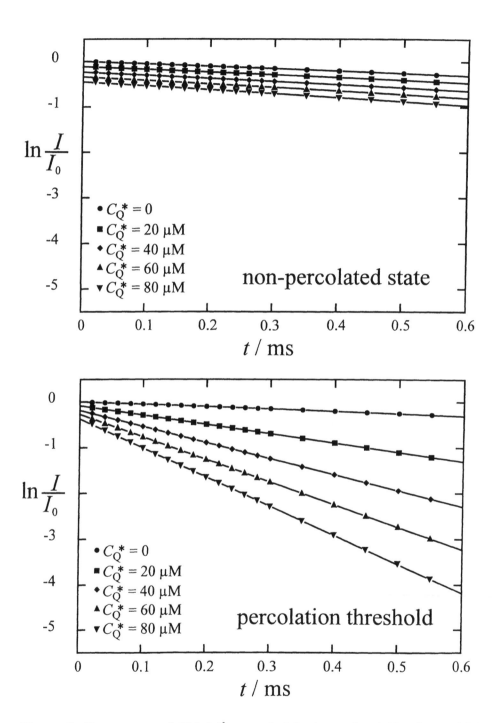

Figure 9 Decay curves of $Tb(pda)_3^{3-}$ quenched by bromophenol blue in an AOT–octane microemulsion at $W_0 = 46$. The upper diagram was measured at 20°C, slightly above the solubilization phase boundary, and the lower diagram at 35°C, just at the percolation threshold. The quencher concentrations are given as concentrations in the water domains of the microemulsion. Part of the initial drop is due to inner filter effects of the quenching dye molecule. (From Ref. 24.)

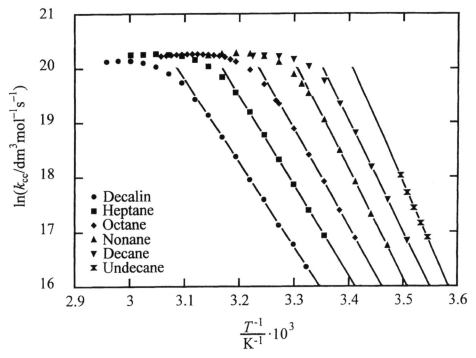

Figure 10 Arrhenius plot of the intercluster quenching rate constant k_{cc} for an oil series at constant droplet size and droplet concentration. The strong deviation from linearity is probably due to the finite reaction rate between the probe $Tb(pda)_3^{3-}$ and the quencher employed, bromophenol blue. (From Ref. 24.)

the reshuffling occurring when the highly dynamic clusters break and re-form delimits the random walk process and provides the additional transport mechanism (k_{cc}) that is responsible for the exponential decay on longer time scales.

The long time scale available with $Tb(pda)_3^{3-}$ as probe is accompanied by the disadvantage of relatively slow quenching reactions; k_q with bromophenol blue in water is only about 6×10^7 dm^3/(mol · s). A consequence of using a slow reaction is that if the microemulsion dynamics becomes too fast compared to the limited reaction rate, the effects of compartmentalization disappear. This leads, for instance, to strong deviations in the Arrhenius plots in Fig. 10 [24], similar to the deviations expected from a percolation transition or from a transition to a bicontinuous state. However, in the activation-controlled regime, time-resolved luminescence quenching with very long lived probes gives results similar to those of stopped-flow measurements but is easier to perform and gives higher experimental precision.

Summarizing, it should be pointed out that with the choice of the luminescent probe and its particular lifetime, different processes are selected for study. Generally it can be stated that the exchange dynamics between the compartments depends strongly on the temperature and on the droplet size as well as on additives such as salt and cosurfactants. Although the reason for the influence of the droplet size is not clear, it seems that factors decreasing the monolayer stiffness and factors increasing the attraction between the droplets result in a faster solubilisate exchange. Obviously, the dynamic properties obtained by TRLQ measurements are closely related to the phase behavior of the microemulsion. This is reflected in the fact that the temperature dependence of exchange rates of nonionic W/O microemulsions is opposite that of the exchange rates of ionic W/O systems [14], yielding formally negative activation energies; and in nonionic O/W microemulsions the situation is reversed again; here a temperature increase facilitates exchange, as in ionic W/O systems [37].

C. Microemulsions Containing Polymers

In the 1990s the TRLQ has been applied to investigate the effects of additives such as enzymes [61–63] or polymers [35,41,64–67] on microstructure, compartment size, and exchange dynamics of microemulsions. Building on their earlier studies of ternary and quaternary microemulsions, Lang and coworkers [35,41] added polymers to the W/O-systems, both water-soluble poly(ethylene glycol) (PEG) and poly(vinyl alcohol) and oil-soluble polybutadiene, as well as poly(propylene glycol), which is distributed between water and oil. The water-soluble polymers decrease the surfactant aggregation number and the rate of exchange in all the systems studied: ternary, quaternary, and systems with different surfactants. The oil-soluble polymer, on the other hand, increases N_{agg} and k_{ex} [35,41]. The percolation threshold, as determined both from electrical conductivities and from the point where k_{ex} reaches a value of $(1–2) \times 10^{10}$ dm^3/(mol · s), is displaced toward higher droplet concentration by a water-soluble polymer and toward lower droplet concentration by the oil-soluble polymer. Lang et al. [35,41] suggested that the primary effect of the polymer was to change the droplet size and that the effect of this on the interactions was the cause for the other changes. They also presented interesting thoughts about the reason for the increased interactions and exchange rates when the droplet size increases and about why the water-soluble polymers reduce the droplet size.

Lianos and coworkers [64–67] studied PEG in SDS-based quaternary microemulsions by TRLQ and dynamic light scattering (DLS). Unfortunately, most of the TRLQ results were analyzed using a "fractal" model, a mathematical model containing two stretched exponential terms. Since this description lacks a clear physical background, we discuss only the results that were also analyzed with the conventional IGTT model [65]. These results indicate that the polymer addition increased the number of droplets and thus decreased the average size. The DLS results in the same systems showed that the apparent hydrodynamic radius increased in the presence of polymer. The paradox was resolved by measurements using polymers of different sizes [67]. Small PEGs, up to a molecular weight of 400, either did not affect the droplet size or decreased it slightly; an $M_w = 11,000$ almost doubled the hydrodynamic radius, and at $M_w = 35,000$ and above there were two diffusion modes, one corresponding to a hydrodynamic radius as without polymer and the other very large. An obvious explanation would be that small polymers are dissolved in the core of the droplets, with only a small effect on the droplet size and structure, whereas the very long polymers connect several droplets into larger clusters; the two modes observed should then correspond to bound and free droplets. The last point was corroborated by the observation that the amplitude of the slow relaxation mode (large R_h) increased and that of the fast relaxation mode decreased when the polymer concentration was increased. It would be interesting to perform TRLQ measurements with long-lived probes, to compare the size and intracluster exchange in these structures with that in percolation clusters.

D. Further Applications of Luminescence in Microemulsion Studies

In this section some other luminescence techniques are mentioned briefly, with examples, to show the possibilities and what has been tried. A few examples of mainly analytical methods, using the microemulsion only as a convenient medium, are given at the end of the section.

As already mentioned in the beginning, triplet–triplet annihilation reactions as well as delayed fluorescence of amphiphilic pyrene derivatives were used to probe the microemulsion dynamics. [68]. The fact that the interacting species were associated with the surfactant monolayer made it possible to probe the dynamics in an unusual way. It was found that at low occupation numbers, the annihilation between triplets associated with different droplets occurred at a rate close to diffusion-controlled. Evidently, this reaction did not require a fusion or exchange between the aqueous compartments of the droplets, which is the process we discussed above and which would be slower than diffusion-controlled by two or three orders of magnitude.

Steady-state and time-resolved fluorescence anisotropy have been used to determine the hydrodynamic radius of relatively small (up to 3 nm [69]) microemulsion droplets. In small droplets, the rapid molecular rotation of an excited probe in the aqueous core or at the interface will not fully randomize the orientations, but a part of the anisotropy will decay with the rotation of the whole droplet (or by translational diffusion within it). The two relaxation modes can be distinguished, and the correlation times have been determined. It is generally found that the molecular rotational correlation time is longer than that for the probe in pure water, approaching the value for the free probe with increasing radius of the droplets. The rotational correlation time of the droplet is related to its hydro-dynamic volume through the Stokes–Einstein equation. For small droplets, the size determined in this way is in good agreement with other results, whereas in larger droplets the internal motion of the probe gets increasingly important and prevents the size determination [69–72].

Within its limitations, this method is both convenient and precise and has found an interesting application in the study of AOT microemulsions in supercritical and near-critical alkanes [72]; fluorescence probes have also found other uses in supercritical microemulsions [73]. Visser et al. [71] compared water and glycerol as the polar cores of microemulsion droplets. The high viscosity of glycerol is, of course, particularly beneficial in these experiments, as the internal motion of the probe is almost inhibited. An interesting aspect of the results is that rotational correlation times obtained for glycerol droplets in AOT–heptane were in good agreement with the known size of the droplets, whereas considerably longer correlation times were observed in dodecane. This may be an effect from the stronger attractive interactions between droplets in dodecane, inducing some clustering.

Chen et al. [74] successfully used fluorescence polarization of an amphiphilic and an oil-soluble probe to monitor the changes of monolayer curvature within the microemulsion phase in didodecyldimethylammonium bromide (DDAB)–alkane–water.

By analyzing the fast portion of the anisotropy decay it is possible to obtain information about the microviscosity at the location of the probe, and for molecules incorporated at the interface the fluidity of the interfacial layer itself will influence the anisotropy relaxation rate [75]. Another way to obtain information on the water pool microviscosity is given by the application of 8-anilino-1-naphthalenesulfonic acid (ANS) or Auramin O [76]. For these molecules the quantum yield of fluorescence increases with the microviscosity of the environment.

The fluorescence emission spectrum of ANS depends strongly on the polarity of the environment, with a red shift of the emission maximum in more polar media. Therefore this probe has been used to obtain information on the shell-like structuring in the interior of microemulsion water droplets [73]. Fluorescence spectroscopy can often be used to obtain information on the location of additives in microemulsions [70]. A change in the emission intensity or in the wavelength of the emission maximum indicates interactions of the probe with the interfacial layer, either direct or indirect. From a comparative analysis of the fluorescence spectra of labeled enzyme it was concluded that hydrophobically modified probes were forced toward the interface [77].

The efficiency of Förster energy transfer between a water-soluble donor, an enzyme, and an oil-soluble acceptor has been used to discuss the localization of the enzyme at the interface in a microemulsion droplet [78]. And by investigating the exciplex formation of a pyrene derivative bound to the interface in the presence of N-dimethylaniline (DMA) via fluorescence spectra and quantum yields, the surfactant layer has been probed directly [79]. However, it is difficult to quantify the effects, so such studies are mostly highly qualitative and sometimes speculative.

Fluorescence recovery after photobleaching (FRAP) is a technique that allows the diffusion coefficients of a fluorescent probe to be measured. With a highly intense laser pulse all fluorophores in a selected spot are destroyed irreversibly, and the subsequent diffusion of fresh probe molecules into the area is followed by an increase in the fluorescence intensity. To give an example, gelatine-based organogels in microemulsions were investigated with this method [80].

Finally, in analytical applications the luminescence is often used only to detect specific components of interest. The microemulsion is then employed to extract these components, e.g., polycyclic aromatic hydrocarbons [81], or to separate neutral metal ion complexes by electrophoresis with charged microemulsion droplets [82]. But it must not always be luminescence induced by external light; chemiluminescence in microemulsions was also reported [83] to give an appreciable increase in the intensity compared to homogeneous solutions.

REFERENCES

1. G. R. Ramos, M. C. G. Alvarez-Coque, A. Berthod, and J. D. Winefordner, Anal. Chim. Acta *208*:1 (1988).
2. R. Zana and J. Lang, Colloids Surf. *48*:153 (1990).
3. M. Almgren, Adv. Colloid Interface Sci. *41*:9 (1992).
4. M. H. Gehlen and F. C. De Schryver, Chem. Rev. *93*:199 (1993).
5. A. V. Barzykin and M. Tachiya, Heterogen. Chem. Rev. *3*:105 (1996).
6. M. Tachiya, in *Kinetics of Nonhomogeneous Processes* (G. R. Freeman, ed.), Wiley, New York, 1987, p. 575.
7. A. V. Barzykin and M. Tachiya, J. Chem. Phys. *102*:3146 (1995).
8. A. V. Barzykin, Chem. Phys. *163*:63 (1992).
9. P. P. Infelta, M. Grätzel, and J. K. Thomas, J. Phys. Chem. *78*:190 (1976).
10. M. Tachiya, Chem. Phys. Lett. *33*:289 (1975).
11. N. J. Turro and A. Yekta, J. Am. Chem. Soc. *100*:5951 (1978).
12. T. Förster and B. K. Selinger, Z. Naturforsch. A *19*:38 (1964).
13. A. Weller, Prog. React. Kinet. *1*:187 (1961).
14. H. Mays, J. Pochert, and G. Ilgenfritz, Langmuir *11*:4347 (1995).
15. B. Medhage, M. Almgren, and J. Alsins, J Phys. Chem. *97*:7753 (1993).

16. M. Almgren and J. E. Löfroth, J. Chem. Phys. *76*:2734 (1982).
17. G. G. Warr and F. Grieser, J. Chem. Soc., Faraday Trans. 1 *82*:1813 (1986).
18. B. L. Bales and C. Stenland, J. Phys. Chem. *97*:3418 (1993).
19. B. L. Bales and M. Almgren, J. Phys. Chem. *99*:15153 (1995).
20. M. Almgren, P. Hansson, and K. Wang, Langmuir *12*:3855 (1996).
21. H. F. Eicke, J. C. W. Shepherd, and A. Steinemann, J. Colloid Interface Sci. *56*:168 (1976).
22. P. D. I. Fletcher and B. H. Robinson, Ber Bunsenges. Phys. Chem. *85*:863 (1981).
23. M. Almgren and R. Johannsson, J. Phys. Chem. *96*:9512 (1992).
24. H. Mays and G. Ilgenfritz, J. Chem. Soc., Faraday Trans. *92*:3145 (1996).
25. M. Almgren, J. Alsins, J. van Stam, and E. Mukhtar, Prog. Colloid Polym. Sci. *92*:4479 (1988).
26. M. van der Auweraer and F. C. De Schryver, Chem. Phys. *111*:105 (1987).
27. M. van der Auweraer, S. Reekmans, N. Boens, and F. C. De Schryver, Chem. Phys. *132*:91 (1989).
28. S. S. Atik and J. K. Thomas, J. Am. Chem. Soc. *103*:3543 (1981).
29. J. Lang, A. Jada, and A. Malliaris, J. Phys. Chem. *92*:1946 (1988).
30. R. Johannsson, M. Almgren, and J. Alsins, J. Phys. Chem. *95*:3819 (1991).
31. A. Verbeeck and F. C. De Schryver, Langmuir *3*:494 (1987).
32. A. M. Howe, J. A. McDonald, and B. H. Robinson, J. Chem. Soc., Faraday Trans. 1 *83*:1007 (1987).
33. A. Jada, J. Lang, and R. Zana, J. Phys. Chem. *93*:10 (1989).
34. R. Johannsson and M. Almgren, Langmuir *9*:2879 (1993).
35. M. J. Suarez, H. Levy, and J. Lang, J. Phys. Chem. *97*:9808 (1993).
36. P. D. I. Fletcher, J. Chem. Soc., Faraday Trans. *83*:1493 (1987).
37. P. D. I. Fletcher and R. Johannsson, J. Chem. Soc., Faraday Trans. *90*:3567 (1994).
38. P. Lianos and S. Modes, J. Phys. Chem. *91*:6088 (1987).
39. P. Lianos, J. Chem. Phys. *89*:5237 (1988).
40. S. Modes, P. Lianos, and A. Xenakis, J. Phys. Chem. *94*:3363 (1990).
41. M. J. Suarez and J. Lang, J. Phys. Chem. *99*:4626 (1995).
42. A. Jada, J. Lang, R. Zana, R. Makhloufi, E. Hirsch, and S. J. Candau, J. Phys. Chem. *94*:387 (1990).
43. J. Lang, G. Mascolo, R. Zana, and P. L. Luisi, J. Phys. Chem. *94*:3069 (1990).
44. J. Lang, N. Lalem, and R. Zana, J. Phys. Chem. *95*:9533 (1991).
45. J. Lang, N. Lalem, and R. Zana, J. Phys. Chem. *96*:4667 (1992).
46. J. Zhang, J. L. Fulton, and R. D. Smith, J. Phys. Chem. *97*:12331 (1993).
47. M. Vasilescu, A. Caragheorgheopol, M. Almgren, W. Brown, J. Alsins, and R. Johannsson, Langmuir *11*:2893 (1995).
48. S. Clark, P. D. I. Fletcher, and X. Ye, Langmuir *6*:1301 (1990).
49. P. D. I. Fletcher and D. I. Horsup, J. Chem. Soc., Faraday Trans. *88*:855 (1992).
50. A. Jada, J. Lang, and R. Zana, J. Phys. Chem. *94*:381 (1990).
51. R. Johannsson, M. Almgren, and R. Schomäcker, Langmuir *9*:1269 (1993).
52. P. D. I. Fletcher, A. M. Howe, and B. H. Robinson, J. Chem. Soc., Faraday Trans. 1 *83*:985 (1987).
53. M. Almgren, J. E. Löfroth, and J. van Stam, J. Phys. Chem. *90*:4431 (1986).
54. G. J. M. Koper, W. F. C. Sager, J. Smeets, and D. Bedeaux, J. Phys. Chem. *99*:13291 (1995).
55. M. H. Gehlen, M. van der Auweraer, S. Reekmans, M. G. Neumann, and F. C. De Schryver, J. Phys. Chem. *95*:5684 (1991).
56. M. H. Gehlen, M. van der Auweraer, and F. C. De Schryver, Langmuir *8*:64 (1992).
57. M. H. Gehlen, F. C. De Schryver, G. Bhaskar Dutt, J. van Stam, N. Boens, and M. van der Auweraer, J. Phys. Chem. *99*:14407 (1995).
58. H. Mays, Ph.D. Thesis, University of Cologne, 1996.
59. D. Vollmer, J. Vollmer, and H. F. Eicke, Europhys Lett. *26*:389 (1994).
60. P. Guering, A. M. Cazabat, and M. Pailette, Europhys. Lett. *2*:953 (1986).
61. V. Papadimitriou, A. Xenakis, and P. Lianos, Langmuir *9*:912 (1993).
62. A. Xenakis, H. Stamatis, A. Malliaris and F. N. Kolisis, Prog. Colloid Polym. Sci. *93*:373 (1993).
63. A. Xenakis, V. Papadimitriou, and P. Lianos, Prog. Colloid Polym. Sci. *93*:370 (1993).
64. P. Lianos, S. Modes, G. Staikos, and W. Brown, Langmuir *8*:1054 (1992).
65. D. Papoutsi, P. Lianos, and W. Brown, Langmuir *9*:663 (1993).

66. D. Papoutsi, W. Brown, and P. Lianos, Prog. Colloid Polym. Sci. *97*:243 (1994).

67. D. Papoutsi, P. Lianos, and W. Brown, Langmuir *10*:3402 (1994).

68. C. Bohne, E. B. Abuin, and J. C. Scaiano, Langmuir *8*:469 (1992).

69. E. Keh and B. Valeur, J. Colloid Interface Sci. *79*:465 (1981).

70. A. J. W. G. Visser, J. S. Santema, and A. van Hoek, Photochem. Photobiol. *39*:11 (1984).

71. A. J. W. G. Visser, K. Vos, A. van Hoek, and J. S. Santema, J. Phys. Chem. *92*:759 (1988).

72. J. Eastoe, B. H. Robinson, A. J. W. G. Visser, and D. C. Steytler, J. Chem. Soc., Faraday Trans. *87*:1899 (1991).

73. P. Yazdi, G. J. McFann, M. A. Fox, and K. P. Johnston, J. Phys. Chem. *94*:7224 (1990).

74. V. Chen, G. G. Warr, D. F. Evans, and F. G. Prendergast, J. Phys. Chem. *92*:768 (1988).

75. M. Aoudia, M. A. J. Rodgers, and W. H. Wade, J. Colloid Interface Sci. *144*:353 (1991).

76. M. Hasegawa, T. Sugimura, Y. Shindo, and A. Kitahara, J. Phys. Chem. *98*:2120 (1994).

77. F. Pitre, C. Regnaut, and M. P. Pileni, Langmuir *9*:2855 (1993).

78. H. Stamatis, A. Xenakis, F. N. Kolisis, and A. Malliaris, Prog. Colloid Polym. Sci. *97*:253 (1994).

79. C. D. Borsarelli, J. J. Cosa, and C. M. Previtali, Langmuir *9*:2895 (1993).

80. P. J. Atkinson, D. C. Clark, A. M. Howe, R. K. Heenan, and B. H. Robinson, Prog. Colloid Polym. Sci. *84*:129 (1991).

81. N. B. Elliott, A. J. Prenni, T. T. Ndou, and I. M. Warner, J. Colloid Interface Sci. *156*:359 (1993).

82. H. Watarai and K. Ogawa, Anal. Chim. Acta *277*:73 (1993).

83. N. Dan and M. L. Grayeski, Langmuir *10*:447 (1994).

20

Structure and Reactions in Microemulsions Formed in Near-Critical and Supercritical Fluids

John L. Fulton
Pacific Northwest National Laboratory,* Richland, Washington

I. INTRODUCTION

Since the discovery of microemulsion phases in supercritical fluids in the mid-1980s [1] and their subsequent characterization [2–16], there has been much interest in exploiting the unusual properties of the supercritical fluid phase in applications of these systems. One such application is as a new type of solvent for chemical reactions. In the following sections, I discuss the properties of these systems for reactions, review the progress so far, and analyze the future potential. As a prelude to these discussions, I begin with a brief overview of what is known about the molecular structure of microemulsions in near-critical and supercritical fluids. The details of the primary and secondary molecular structures of various types of microemulsion phases can dramatically affect the reactivity in these systems.

A supercritical fluid is in a thermodynamic region where the substance's temperature ($T > T_c$) and pressure ($P > P_c$) are above those at the critical point (see Fig. 1). The critical temperature (T_c) is the highest temperature at which a gas can still be condensed to a liquid. As one approaches the critical point of the substance starting from the vapor/liquid region, the density of the vapor phase becomes equal to that of the liquid phase at which point the interface between the two phases disappears. The properties of the fluid in the supercritical region are uniquely different from those of either the conventional gas or liquid states [17–21]. In addition, the near-critical liquid region, where the temperature is just slightly below the critical point temperature ($1.0 > T/T_c > 0.75$), is also of considerable interest for microemulsion formation because these liquids are still quite compressible due to their proximity to the critical point. Table 1 gives the critical pressures, temperatures, and densities of several fluids whose moderate critical temperatures make them attractive candidates for microemulsion formation in the near-critical or supercritical states.

---

* Operated by Battelle Memorial Institute.

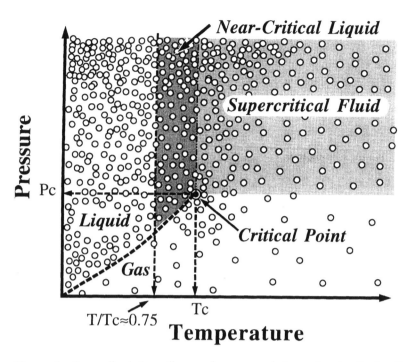

Figure 1 Generalized phase diagram for a pure substance showing the locations of the supercritical and near-critical regions.

Table 1 Critical Parameters of Some Near-Critical and Supercritical Fluids That Have Moderate Critical Points Suitable for Microemulsion Formation

Compound	Critical temperature (°C)	Critical pressure (bar)[a]	Critical density (g/mL)
CO_2	31.3	73.8	0.468
Ethane	32.3	48.8	0.203
Ethylene	9.2	50.4	0.218
Propane	96.7	42.5	0.217
Propylene	91.8	46.2	0.232
n-Butane	152.0	37.5	0.228
n-Pentane	196.6	33.3	0.232
Xenon	16.7	58.3	1.112

[a] 10 bar = 1 MPa.

A generalized phase diagram for a pure substance is shown in Fig. 1. For the simplest supercritical fluid systems, the density is continuous on going from high density to low, but this is not the case below the critical point where a density discontinuity occurs at the liquid–vapor transition. By manipulating the pressure at a constant temperature above T_c, we can choose a continuum of densities from gaslike to liquidlike. A classic demonstration of the "tunability" of supercritical fluids is shown in Fig. 2, which gives the solubility of naphthalene in carbon dioxide (CO_2) as a function of fluid density [22]. Small changes in fluid density in response to modest changes in the system pressure at constant temperature can alter the solubility by several orders of magnitude. Another important aspect of

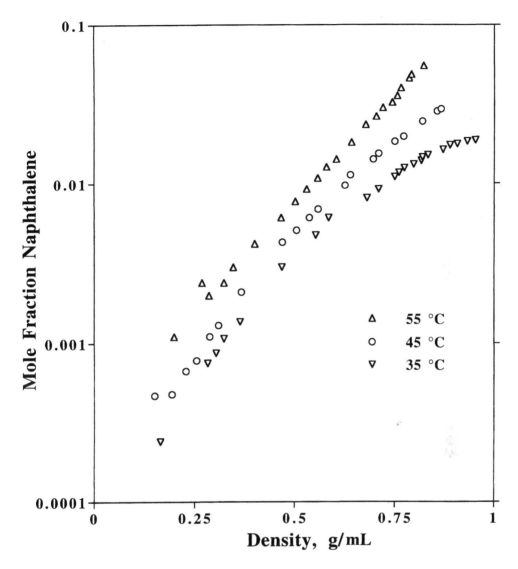

Figure 2 Solubility of naphthalene in supercritical carbon dioxide as a function of the fluid density. (From Ref. 22.)

supercritical fluids is that solute diffusion coefficients are 10–100 times higher in supercritical fluids than in liquids, which is a significant advantage for applications (e.g., reactions) that depend heavily on mass transport mechanisms. These unique properties of supercritical fluids offer great latitude for controlling reaction rates and pathways in supercritical microemulsions.

Recent advances in the development of reverse micelle and microemulsion phases in CO_2 will provide an important new solvent system for reactions [12,23–33]. Although the principles are mostly the same as for conventional reverse-micelle systems, the surfactants required for CO_2 must contain "CO_2-philic" tails compatible with the unique aspects of the CO_2-continuous phase. A CO_2-based microemulsion would have minimal environmental impact, since at this scale of use the solvent is environmentally benign.

In addition, the CO_2 continuous phase exhibits solvent properties typical of hydrocarbon solvents such as pentane or hexane [34–36] that have a very high capacity for lower polarity organic solutes. More details about the solvent properties and thermodynamics of these and other supercritical systems are given in a companion chapter by McFann and Johnston in this book (Chapter 9).

II. Microemulsion Structure in Near-Critical and Supercritical Fluids

Knowing a solution's microstructure is of utmost importance in understanding the rates and pathways of chemical reactions in microemulsions. Dynamic light scattering (DLS) as well as small-angle X-ray scattering (SAXS) and small-angle neutron scattering (SANS) have been used for microstructure determination. A summary of these types of studies of near-critical and supercritical microemulsions is given in Table 2. As described in the following sections, a variety of microstructures have been identified. In addition, changes in the fluid density have been shown to affect not only the primary structure but also the secondary structure involving the spatial distribution of microemulsion droplets in the continuous-phase solvent. This can have dramatic effects on the reaction rates and pathways.

A. Dynamic Light Scattering

Light scattering [37,38] was one of the first methods used to derive microemulsion structure in near-critical and supercritical fluids [2,4,39–42]. This technique provided the first direct evidence that microemulsion droplets existed in a supercritical fluid.

The DLS method or photon correlation spectroscopy directly measures the translation diffusion coefficient for those microemulsion droplets that are noninteracting. Further, the droplet hydrodynamic diameter (or alternatively the correlation length) can then be calculated from the Stokes–Einstein equation under the assumption of noninteracting droplets. The micelle diffusion coefficients for a wide range of liquid and supercritical alkanes are shown in Fig. 3 [40]. The diffusion coefficients of the microemulsion droplets (at $W = [H_2O]/[\text{surfactant}] = 5$) in near-critical propane (25°C) and supercritical ethane are significantly higher than in the larger alkanes, which is to be expected given the significantly lower viscosities of these lower alkanes. For instance, the diffusion coefficients of micelles in supercritical ethane are four to five times higher than in liquid isooctane. Interestingly, supercritical ethane microemulsion droplets have diffusion coefficients that *increase* with pressure. These results are attributed to formation of micelle–micelle clusters that collectively have lower diffusion coefficients [4,39]. At higher pressures the clusters are dispersed, and the high diffusion coefficients of the individual micelles are recovered. This suggests that significantly improved mass-transport properties for these systems may often be obtained at higher pressures, even though the viscosity of the fluid-continuous phase is higher. However, for reactions that require exchange of reactants in the micelle cores, the reaction rates under conditions of micelle clustering are found to be much higher due to the more effective exchange that occurs for micelles in the close proximity of a cluster [43]. This effect is described in more detail in Sec. IV.

The apparent hydrodynamic diameters of the droplets (or the correlation length), as calculated using the Stokes–Einstein equation for a number of different systems, are given in Table 2. These early findings showed that the micelle sizes measured in near-critical and supercritical solutions were similar to those found for conventional water-in-oil microemulsions in liquid alkane. At lower fluid densities, DLS probes the combined effect of the collective diffusion coefficient of the micelle cluster and that of the individual micelles.

Table 2 Structure of Microemulsions in Compressible Fluids from Scattering Techniques Including DLS, SAXS, and SANS

Continuous-phase fluid	Surfactant	Temp/pressure	W^a	Geometry	Diam. (nm)	Method	Ref.
Ethane	AOT[b]	25, 37°C/ 224–550 bar	1, 5	Spherical	3–6[c]	DLS	2,39
Xenon	AOT	25°C/ 150–550 bar	1, 5	Spherical	3–6[c]	DLS	2
Propane	AOT	25°C/ 90–700 bar	30	Spherical	8–32[c]	DLS	4
Ethane/ propane	$C_{12}E_4^d$ with acrylamide	65°C/ 325–475 bar	3.5	Spherical	34–60[c]	DLS	41,42
Propane	AOT	25°C/ 90–500 bar	20, 30	Spherical	7.6, 10 (water core)	SANS	4,47
Propane	AOT	25°C/ 24–472 bar	15	Spherical	6.2 (water core)	SANS	8
Propane	AOT	110°C/ 250–400 bar	12	Spherical	3.2	SAXS	77
Propane	DDAB[e]	25°C/ 35–400 bar	20	Spherical	7.6–8.4 (water core)	SANS	47
Propane	$C_{12}E_5$ [f]	55°C/ 120–400 bar	12	Spherical	10.8–14.0 (water core)	SANS	47
Xenon	AOT	25°C/ 400 bar	1	Spherical	3.4	SAXS	78
Propylene	Lecithin[g]	30°C/ 200 bar	14	Extended rod	6.1	SAXS	78
CO_2	PFOA-*g*-PEO[h]	60°C/ 255–470 bar	0.32 (w/w)[i]	Spherical	25.3	SAXS	32
CO_2	Zonyl FSO-100[j]	65°C/ 530 bar	0	Spherical polydisperse	16.7	SAXS	32
CO_2	Fluorinated diblock[k]	65°C/ 220–400 bar	0	Very small aggregates	1.4	SAXS	32
CO_2	PFOA-*g*-PEO[h]	65°C/ 333 bar	0, 0.32 (w/w)	Spherical	11.2, 25.6	SANS	52
CO_2	Dichain hybrid[l]	25°C/ 500 bar	33	Spherical	5.0 (water core)	SANS	54
CO_2	PS-*b*-PFOA[m]	65°C/ 340 bar	0	Spherical	13.6–32.4	SANS	53
Water/ xenon	SDS[n]	70°C/ 400 bar	—	Normal micelle	5.7	SAXS	This chapter and Ref. 58

[a] [H_2O]/[surfactant].
[b] Bis-2-ethylhexylsulfosuccinate sodium salt.
[c] Apparent hydrodynamic diameter or correlation length.
[d] Alkyl poly(ethylene glycol) ethers.
[e] Didodecyldimethylammonium bromide.
[f] Dodecyl penta(ethylene glycol) ether, nonionic.
[g] L-α-Phosphatidylcholine.
[h] Multiple 5 kilodalton poly(ethylene oxide) grafts onto a backbone of poly(1,1-dihydroperfluorooctyl acrylate).
[i] Weight H_2O/weight surfactant.
[j] $F(CF_2)_{6-10}CH_2CH_2O(CH_2CH_2O)_{3-8}H$.
[k] $F(CF_2)_{10}(CH_2)_{10}H$.
[l] $(C_7H_{15})C_7F_{15})CHSO_4^-Na^+$.
[m] Polystyrene-*block*-poly(1,1-dihydroperfluorooctyl acrylate).
[n] Sodium dodecyl sulfate.

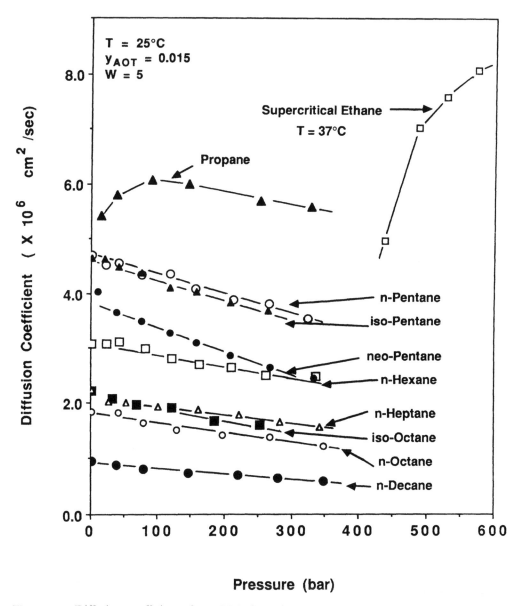

Figure 3 Diffusion coefficients from DLS for microemulsion droplets in various liquid and supercritical alkanes as a function of solution pressure (10 bar = 1 MPa). $T = 25°C$, $W = 5$, and mole fraction, y_{AOT}, of AOT = 0.015. (From Ref. 40.)

For those systems near a phase transition, the apparent hydrodynamic diameter of the droplets (or the correlation length), as calculated using the Stokes–Einstein equation, appears to decrease as pressure increases [2,4,39]. For example, the apparent hydrodynamic diameter of a microemulsion droplet (for [surfactant] = 150 mM and $W = 5$) in supercritical xenon [2] decreases from 6.5 to 4.5 nm as pressure is increased from 350 to 550 bar (10 bar = 1 MPa). This effect is due to the change in the extent of micelle clustering rather than an actual change in the micelle size.

After the initial DLS studies were complete, it became apparent that the very strong interdroplet attractive interactions in near-critical and supercritical fluids limited the standard DLS technique to systems of higher dilution or to high fluid densities. Thus, small-angle scattering techniques were later used to better resolve the full dimensional scale range of these microemulsions over a wider range of conditions.

B. Small-Angle X-Ray and Neutron Scattering

A SAXS or SANS study is one of the most powerful methods of resolving the microstructure of a microemulsion phase. The dimensional scale probed by these methods can resolve the spacing between individual surfactant molecules in the interfacial region, the geometry and size of individual microemulsion droplets, and the larger scale secondary macromolecular structure or micelle clustering. Results for numerous near-critical and supercritical systems are given in Table 2. These studies showed that a variety of microstructures exist in compressible fluids and that changes in the density of the fluid can affect both the primary structure and the secondary structure (including the spatial distribution of microemulsion droplets).

In the method of small-angle scattering, the radiation is scattered to low angles because of inhomogeneities in the properties of the micelles or particles relative to those of the solvent [44–46]. In a simplified form of the general theory, for systems containing microemulsion droplets, the scattering data contain contributions to the total excess X-ray scattering, $I(q)$, arising from two sources: (1) the particle contributions, $P(q)$, which depend solely on the size and shape of the particle and (2) an appropriate structure factor, $S(q)$, which accounts for attractive or repulsive interactions between particles. Assuming that the solvent is structureless, the total scattering is simply

$$I(q) = \rho S(q)P(q) \tag{1}$$

where ρ is the particle number density $q = (4\pi\lambda)\sin\theta$ and θ is the scattering angle. When the system is sufficiently dilute, interparticle interactions are significantly reduced, yielding $S(q) = 1$.

An example of $P(q)$ for a monodisperse system of homogeneous spheres of radius r is given by

$$P(q) = P(0)\left[3\frac{\sin(qr) - qr\cos(qr)}{(qr)^3}\right]^2 \tag{2}$$

where $P(0)$ is the scattered intensity at $q = 0$.

For SAXS and SANS, the theoretical methods required to interpret the scattering information are virtually identical except for the basis of the scattering contrast: For SAXS, inhomogeneities in the electron densities give rise to coherent, excess X-ray scattering, whereas in SANS, inhomogeneities in the nuclei's scattering cross section give rise to excess neutron scattering. Both methods require a high-pressure cell with beam-transparent windows. For examining small molecular clusters, neutron scattering perhaps holds an advantage since the scattering contrast can be finely controlled by changing the percent deuteration of the solute or solvent. (Protons and deuterons have very different scattering cross sections.) For studies of organic liquid solutions, SAXS is often not the preferred method because of the poor electron density contrast between the solute and solvent. For supercritical fluids, however, this problem is eliminated because the lower density supercritical fluid phase has good scattering contrast (i.e., a large difference in the electron densities) with the solute having higher electron density. Further, for systems containing fluorocarbon surfactants (very

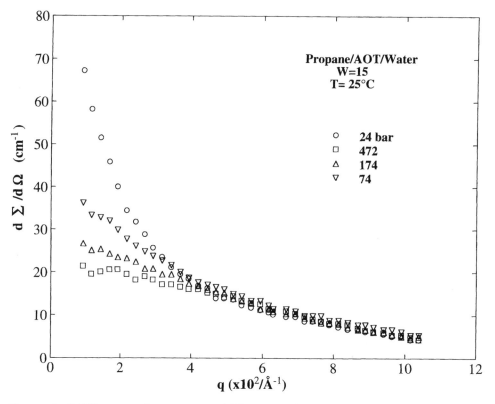

Figure 4 SANS spectra for a propane–AOT–water microemulsion (water-in-oil). (From Ref. 8.)

high electron density) in supercritical CO_2 (very low electron density), the scattering contrast is one of the highest available by either SAXS or SANS. Clearly, both SAXS and SANS are powerful techniques for probing structure in supercritical fluid solutions, and each has strengths to explore different regions of the microemulsion structure.

In two early SANS studies [4,8], the structure in a near-critical propane microemulsion formed with sodium bis(2-ethylhexyl)sulfosuccinate (AOT) was explored. These studies conclusively demonstrated that as the continuous-phase density was adjusted the micelle size remained constant for systems containing ionic surfactants in the single-phase region under low to moderate volume fractions of the dispersed phase. Figure 4 shows the SANS spectra for this propane microemulsion. In the region above $q = 0.04$ Å^{-1}, where the scattering is dominated by the particle size and shape ($P(q)$), the pressure does not affect the scattering. However, in the low-q region, the sharply rising scattering at low pressure indicates micelle clustering. Because of changes in this secondary structure of the microemulsion measured by SANS [8], the interdroplet attractive potential could be directly determined. Using a square-well model for the attractive potential, the micelles were found to interact strongly at short range, corresponding to overlap and intersolvation of the tails of surfactants on adjoining micelles.

The SANS method has also been applied to microemulsions formed by using cationic and nonionic surfactants in near-critical propane [47]. A propane microemulsion formed with the cationic surfactant didodecyldimethylammonium bromide (DDAB) exhibits a percolation of the electrical conductivity as the pressure is reduced toward the phase boundary [48].

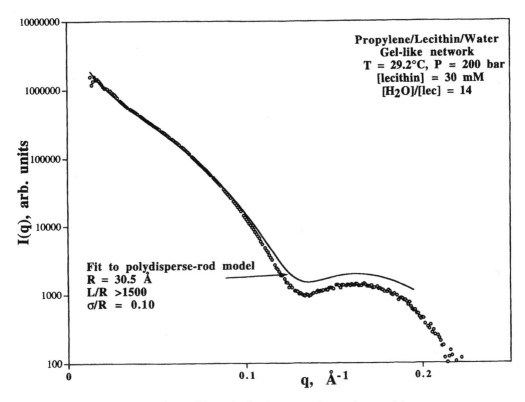

Figure 5 SAXS spectrum for rodlike micelles in a propylene microemulsion.

The SANS study [47] confirmed early findings by fluorescence quenching [43] that this transition was related to clustering microemulsion droplets rather than a structural change to a bicontinuous microemulsion (an earlier hypothesis in Ref. [48]). The SANS result also confirmed that the size of the micelle was mostly unchanged as continuous-phase density was adjusted in accordance with the early results for the anionic surfactant AOT. The nonionic surfactant, triethylene glycol dodecyl ether ($C_{12}E_3$), whose aggregation is driven by hydrogen bonding rather than the stronger electrostatic interactions, forms reverse micelles whose primary structure does depend on the fluid density. In the SANS study by Eastoe et al. [47] of a system that was earlier described by McFann and Johnston [13], the micelle size was found to decrease from 14 nm diameter at 400 bar to 10.8 nm at 120 bar. This effect was ascribed to a partitioning of the surfactant monomer, dissolved in the propane-continuous phase at high pressure, into the micellar interface at lower pressure, thereby reducing the effective W.

The shape and size of surfactant aggregates formed in fluids depend strongly on the physical properties of the surfactant molecule. The zwitterionic surfactant lecithin (L-α-phosphatidylcholine) forms very long, rodlike micelles in near-critical propylene under certain conditions of temperature, pressure, and water concentration. Figure 5 shows the SAXS data for such a system and the corresponding $P(q)$ fit to the data using a model of long rods having a small amount of polydispersity in diameter. Further support for the proposed rodlike structure is obtained from the observation that the viscosity of this solution is gel-like at low pressures, increasing by several orders of magnitude above that of pure propylene [49]. These effects are also observed under relatively low volume fractions of the surfactant phase, $\phi_s = 0.06$. Again, a dramatic effect of changing the fluid density

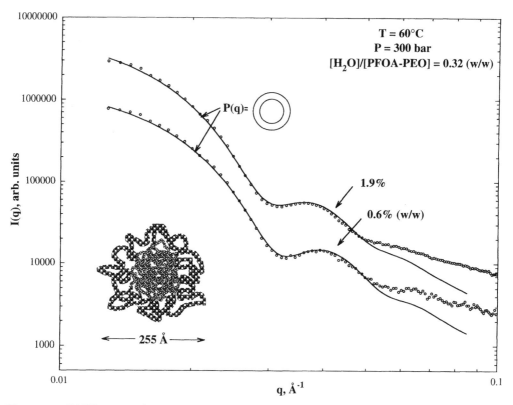

Figure 6 SAXS spectra for a CO_2–PFOA-g-PEO–water microemulsion. (From Ref. 32.)

was manifested in this system, since a continuous change from liquid to gel-like viscosities could be induced by altering the density of the solution by only 4%. The changes in fluid viscosity are due to increased chain entanglement. Chain entanglement is induced at lower pressures by the increased interchain attractive interactions or by growth and stiffening of the rods. Further the SAXS results showed that the viscosity transition was not the result of a sphere to rod structural transition.

More recent scattering studies are of reverse micelles and microemulsions formed in CO_2. In the SAXS study by Fulton et al. [32], the aggregation of three different amphiphiles was explored. Of significance, the results showed that a surfactant based on the graft copolymer poly(ethylene oxide) *graft*-poly(1,1-dihydroperfluorooctyl acrylate)(PEO-g-PFOA) [28,31,50,51] formed large aggregates in supercritical CO_2. The SAXS spectra of this system are shown in Fig. 6. The diameter of these highly monodisperse structures is quite large, about 25 nm, and before this discovery it was not known that structures of this size could be stably suspended in CO_2. The overall structure of these microemulsion droplets was later confirmed in a SANS study [52]. The most likely structure is a core-and-shell micelle with the fluorinated portion of the polymer decorating the outer shell and acting as a "CO_2-philic" transition to the hydrophilic core. This core contains the PEO portion of the molecule and appreciable amounts of water (32 wt% with respect to surfactant). In a related system McClain et al. [53] synthesized a nonionic surfactant based on the block copolymer poly-styrene-*block*-poly(1,1-dihydroperfluorooctyl acrylate) that disperses otherwise insoluble

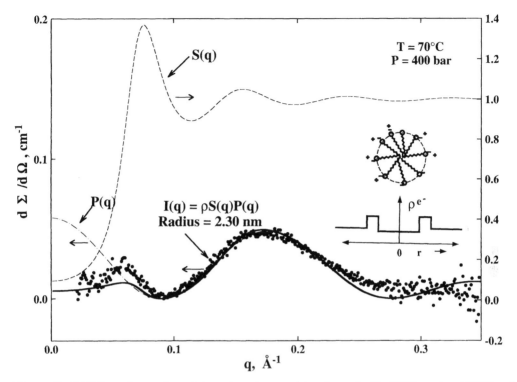

Figure 7 SAXS spectrum of an oil-in-water microemulsion formed with SDS in water at 70°C and 400 bar.

material into CO_2. SANS studies of these systems have shown highly monodisperse spheres with outer diameters of 14–32 nm. For another CO_2–surfactant system [32] containing the fluorinated nonionic surfactant polytetrafluoroethylene (PTFE)-*block*-PEO or the analog of $C_{7-11}E_{4-9}$, the polydisperse micelles were found to have a diameter of about 17 nm. The behaviour of this surfactant system was previously described by Consani and Smith [23].

Eastoe et al. [54], following the initial work of Harrison et al. [29], used SANS to study the structure of another CO_2–surfactant system. In this study the fluorinated hybrid of AOT was used to produce spherical microemulsion droplets having a water core diameter of 5 nm at a relatively high $W = 35$.

The systems described thus far have been water-in-"oil" microemulsions. It is also possible to form an "oil"-in-water microemulsion that incorporates an appreciable amount of a supercritical fluid. For the classic normal-micelle system consisting of sodium dodecyl sulfate (SDS) in water, organic solutes are readily solubilized in the lipophilic regions of micelle cores. The micelle core is also a solvent environment favorable for partitioning of gases, giving overall solubilities that are much higher than for pure water [55–57]. It was subsequently demonstrated [58] that this core region is also an environment favorable for the partitioning of appreciable amounts of a supercritical fluid phase. The uptake of supercritical xenon in these normal-micelle cores is vividly illustrated by SAXS studies. (To avoid the xenon/water gas hydrate, these experiments were conducted at 70°C.)

Figures 7 and 8, show the SAXS spectra for SDS micelles without and with xenon in the cores, respectively. For the SDS micelle in the absence of xenon, the origin of the scattering is the slightly higher electron density of the interfacial layer, as the core and

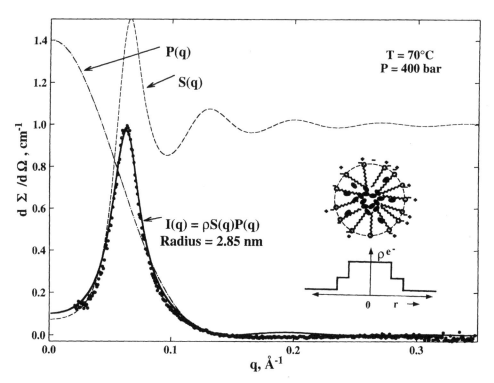

Figure 8 SAXS spectrum of an oil-in-water microemulsion formed with SDS in water at 70°C and 400 bar containing supercritical xenon in the cores of the micelles.

the water-continuous phase have nearly identical electron densities. The scattered intensity increases dramatically after the incorporation of electron-rich xenon into the micelle core. In treating these scattering curves, it is important to include the strong intermicellar electrostatic repulsions. The structure factor $S(q)$ shown in Figs. 7 and 8 for this micelle system is that given by Hayter et al. [59–61], which was derived from an analytical solution of the rescaled mean spherical approximation (MSA) for solution consisting of screened charged particles. The effect of temperature, pressure, and micelle size on the degree of micelle ionization was determined from the theory of "dressed micelles" of Evans et al. [62]. Over the experimental range of these studies, only very small changes occurred in the degree of micelle ionization.

Because of the excellent contrast available for a relatively electron-rich micelle in an electron-poor solvent, we are able to determine the extent of supercritical solvent penetration into the hydrocarbon tail region of the micelle. These studies also provide an interesting result relative to previous SAXS and SANS studies of these systems [63]. Previous SAXS studies also show the weak scattering peak at $q = 0.18$ Å^{-1}, whereas SANS studies of SDS in D$_2$O show a strong scattering peak at about $q = 0.06$ Å^{-1}. The SAXS studies with xenon clearly show the transition from shell contrast to core contrast and unify the results for the two techniques relative to both $P(q)$ and $S(q)$.

As shown in Fig. 9, the micelle size can be increased in a continuous fashion from its "empty" state (Fig. 7), which has an outer radius of 2.3 nm, up to a radius of 2.85 nm at a pressure of 400 bar. Thus, the total volume of the micelle doubles with the incorporation

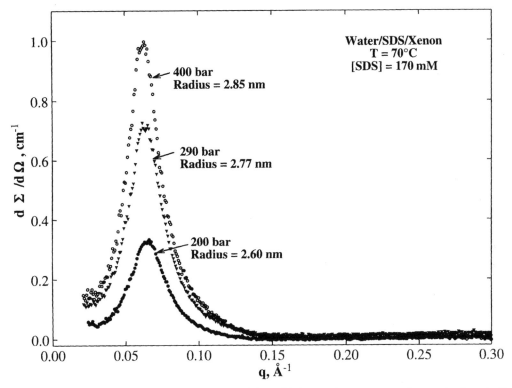

Figure 9 SAXS spectra of an SDS microemulsion with increasing amounts of xenon in the micelle core.

of xenon. In addition, the largest growth in the micelle size occurs in a pressure range where the largest changes in density of the pure fluid occurs. Above a pressure of about 400 bar, the SDS micelles are essentially saturated with xenon. This system is an example of a case where the primary structure of the microemulsion droplet can be altered by the density of the supercritical fluid.

III. PROPERTIES OF SUPERCRITICAL MICROEMULSION REACTION MEDIA

Our knowledge of these systems has progressed rapidly. Since the discovery in the mid-1980s that micelles and microemulsion phases could be stabilized in supercritical fluid solutions. The combination of the unique properties of supercritical fluids (viscosity, diffusion rates, solvent properties, etc.) with those of a dispersed microemulsion (or reverse micelle) phase creates a whole new class of solvents. The microemulsion phase adds to the properties of the supercritical fluid phase what amounts to a second solvent environment, which is highly polar and may also be manipulated using pressure. Although pure supercritical fluids have pressure- or density-dependent properties that are very attractive for reaction processes, they are limited at moderate temperatures and pressures by their inability to appreciably solvate most moderately polar solutes and nearly all ionic materials. Adding a dispersed droplet phase (forming a microemulsion) provides a convenient means of solubilizing highly polar or ionic reactive or catalytic species into the low-polarity environment of the supercritical fluid phase. Hence, the combination of supercritical solvents with microemulsion structures provides a new type of solvent with some unusual and important properties of potential interest for chemical reactions.

Supercritical microemulsions represent a radically different type of reaction media. Whereas the aqueous microdomains of these systems are much like their analogs in liquid systems, the interfacial region and the continuous-phase solvent have unusual and potentially advantageous properties. The specific benefits include:

1. Control of the reaction rates and pathways by changing the density of the continuous phase solvent
2. Control of the partitioning of the reactant into and out of the interface or core region by changing density
3. A simple product and catalyst recovery scheme that uses the unique phase behavior of these systems
4. Control of the reaction by altering the primary or secondary structure of the microemulsions through changes in the fluid density
5. Greatly improved mass transport properties
6. An enivronmentally benign solvent for CO_2-based systems

In this section, these unique aspects of supercritical microemulsions are discussed.

The high diffusivities and the low viscosities of supercritical and near-critical fluids are well established. Studies have shown that the favorable mass transport properties of dense gases are retained in microemulsions formed in these systems, and multiple benefits are expected for chemical reactions. The diffusion rates of microemulsion droplets in near-critical and supercritical fluids are up to 10 times greater than in liquids. Conventional liquid microemulsion systems often contain nanometer-sized droplets whose diffusion rates are 10–100 times lower than those of molecularly dispersed species. The higher diffusion rates of these droplets in near-critical and supercritical fluids offset the transport limiting effect due to large droplet size.

The three sites of solvation of the reactants, products, and catalyst have many different permutations. These sites include the near- or supercritical continuous phase for nonpolar or low-polarity solutes, the interfacial region for amphiphilic solutes, and the core for the highly polar solutes. The surfactant interfacial region provides a dramatic transition from the nonpolar continuous-phase solvent to the highly polar aqueous core. This region represents a unique type of solvent environment where amphiphilic solutes can reside. Such amphiphilic species will be strongly oriented in the interfacial film so that their polar ends are in the core of the microemulsion droplet and the nonpolar end is pointed toward or dissolved in the continuous-phase solvent.

Changes in the near-critical or supercritical fluid density can be used to affect the partitioning of the reactants into different microdomains of the microemulsion droplet. Fluids such as ethane or propane are very "nonpolar" and have dielectric constants (ϵ) ranging from $\epsilon = 1.55$ to 1.8 (visible frequencies). In contrast, the typical organic solvents have much higher dielectric constants, for example, $\epsilon = 1.88$ or 1.98, for hexane or isooctane, respectively. The hydrocarbon tail region of the microemulsion droplet thus provides the first part of a transition from the low dielectric constant environment of the supercritical phase to the high dielectric constant of the aqueous core. The surfactant alignment and strong interactions between the surfactant headgroup counterion and aqueous core can strongly affect molecular orientation, a significant advantage for a catalyst contained in the aqueous core. Many potential reactants have slightly polar functional groups, such as aromatic or alcohol moieties, which will be preferentially solvated in this interface region. By lowering the dielectric constant of the continuous-phase solvent (i.e., by reducing pressure), a concentrating effect will be observed where the reactant is appreciably enhanced in the interfacial region. This is a highly favorable factor affecting the kinetics of the catalytic reaction. The increased concentration of the reactant at the interface would result in reaction rate increases for those reactions in which the catalyst is contained within the droplet core. This also means that the reactants must be highly aligned to reach reactive sites within the aggregate core. Hence the selectivity can be controlled.

The ability to dissolve highly polar or ionic species in the polar microdomains of the microemulsion depends to a large extent on the amount of water incorporated into the microemulsion. The molar water/surfactant ratio (W) is a convenient measure of the polar solvent strength of the microemulsion. For ionic surfactants, the microemulsion will have a solvent power for polar species approaching that of bulk water when W is above approximately 10. At lower W values, the water acts to hydrate the headgroups and counterions of the surfactant, yielding a solvent microenvironment that has a dielectric constant appreciably lower than that of bulk water and hence a much reduced capacity for dissolving ionic or highly polar species. Manipulating the W value of the microemulsion can be viewed as a means to adjust the polarity of the microemulsion for reactions. For supercritical microemulsions, the W value can be adjusted through controlled addition of water to the solution or by manipulating the pressure or temperature of the solution. The degree of polarity of the core region affects the types of reactions that can be conducted. In addition, the ability to dissolve catalytic species in the core depends strongly on the properties of the surfactant headgroup. A more detailed discussion of the solvent environment of the fluid, interface, and core regions is covered in a companion chapter on supercritical microemulsions by McFann and Johnston in this volume (Chapter 9).

Another advantage of microemulsions having supercritical fluid continuous phases is that the phase behavior of the system can be readily controlled by manipulating the system pressure because the density of supercritical fluids depends strongly on the pressure. This creates some unusual and potentially advantageous opportunities for microemulsions formed in such systems that can be used to significant advantage in reactions. The phase behavior of microemulsions formed in near-critical liquids also depends strongly on the pressure of the system [64–66], much more so than in conventional liquids, because the magnitude of the pressure effect on phase behavior is related to the fluid compressibility. A simple means of recovering the high-value catalyst following the reaction is possible, as small changes in the pressure of the system can be used to create or destroy the microemulsion phase.

Changes in the hydrostatic pressure affect the reaction rates for molecularly dispersed species. In addition, pressure may affect the structure of the microemulsion in a way that significantly affects the rates of reaction. Changes in the fluid pressure can cause changes in the microstructure of the fluid, and the accessibility of the reactant to a catalyst will be altered. As indicated in previous sections, several different studies have shown that the density of the continuous-phase solvent can be used to induce structural changes in both the primary and secondary structures of the microemulsion. These structural changes can dramatically alter the reaction kinetics.

Several classes of chemical reactions are possible in microemulsions formed in supercritical fluids. Catalytic hydrogenation or oxidation reactions using molecular hydrogen and oxygen as reactants are particularly well suited for these studies as both are very soluble in supercritical fluid solvents. A potentially useful role for these oxidation reactions is the destruction of hazardous chemical wastes or contaminated materials.

Nucleophilic substitution reactions are another class of reactions that are easily conducted in these microemulsions. Finally, enzymatic reactions using supercritical CO_2 microemulsions may be of high interest because of the general biocompatibility of the enzymes with the two primary solvents, water and CO_2.

A reverse micelle or microemulsion system of particular significance is one based on CO_2, as it would have minimal environmental impact at this scale of use. Since water and CO_2 are the two most abundant, inexpensive, and environmentally compatible solvents, the application of such a system could have tremendous implications for the chemical industries of the 21st century. Many advances have been made recently [12,23,24,26–33,53] in designing and synthesizing surfactants for CO_2, and we expect much development in this area in the future.

IV. CHARACTERISTIC REACTIONS

Few results have been published in this rich area, but as the advantages and properties of these systems become known we expect to see more work in this area. A fundamental study of reaction mechanisms in a near-critical microemulsion was carried out by Zhang et al. [43]. Further, microemulsions in near-critical and supercritical fluids have been used in particle synthesis, hydrogenation, and polymerization reactions [41,67,68]. These diverse areas are discussed in this section.

A. Dynamic Luminescence Quenching

The method of dynamic luminescence quenching provides a highly controlled system for exploring many aspects of microemulsion reaction kinetics. Many unexpected consequences of fluid density on reaction kinetics have been shown to occur in near-critical microemulsions by using this technique. It has been used to explore the exchange kinetics between microemulsion droplets formed in near-critical propane using the surfactant didodecyldimethylammonium bromide (DDAB) [43,69]. It is useful to describe these results in some detail, because they highlight some of the dominant mechanisms for reactions in these classes of systems. In this DDAB–propane study, the ruthenium tris(bipyridyl) ion $Ru(bpy)_3^{2+}$ was used as the luminescent ionic probe and methyl viologen, MV^{2+}, was used as the excited-state quencher. The rate of radiative decay of the luminescence probe $[Ru(bpy)_3^{2+}]$ from the excited state depends strongly on the concentration of the luminescence quencher. The original kinetics model [70,71] for these types of micellar reactions was later refined for reverse micelle systems by Atik and Thomas [72]. The reaction kinetics model is simplified somewhat by using low reactant concentrations where the distribution of probe and quencher among the reverse micelles can be well described by Poisson statistics.

The results of Zhang et al. [43] confirmed earlier studies for liquid systems, that the collisional exchange between the droplets was not very efficient. For the DDAB–propane system, only about one in 500 intermicellar collisions results in the exchange of the micellar cationic reactants. The micelle interfacial layer creates a substantial barrier to exchange of the micelle contents upon collision. Of course, the rates of reaction are much higher in the near-critical propane than in normal liquids, as the micellar collision rates are much higher. This is predicted from the Smoluchowski and Stokes–Einstein relationships [43,73]. The viscosity of propane is about five times lower than that of normal liquids. It is also possible that at higher temperatures the micelle interfacial region may become much more liquidlike, thus substantially improving the exchange process.

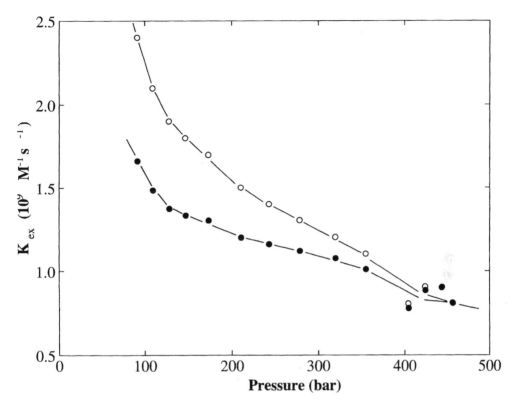

Figure 10 The intermicellar exchange rate constant (○) for the system of 10 μm Ru(bpy)$_3^{2+}$ in a DDAB–water ($W=24$)–propane microemulsion. The reduced rate constant (●) corrected for the propane viscosity is also shown. (From Ref. 43.)

Another interesting finding by Zhang et al. provides insight into the effect of pressure on the electrical conductivity percolation threshold. At higher micelle concentrations, an electrical percolation phenomenon is observed where the electrical conductivity of the solution increases dramatically at some threshold of the continuous-phase density (or system pressure) [43,48]. The results indicate that the size and shape of the microemulsion droplets are unchanged by the density of the continuous phase. However, strong interdroplet attractive interactions occur for most of the near-critical and supercritical microemulsion systems studied so far, and the magnitude of the attractive forces depends strongly on the density of this continuous-phase solvent. These strong attractive forces drive the formation of large micellar clusters. It is believed that electrical conductivity percolation occurs because of enhanced charge transport between micelles in a cluster.

In the DDAB–propane microemulsion, the intermicellar exchange rate increases as the density is reduced (Fig. 10). The increase in the intermicellar exchange rate at lower densities is much greater than would be expected from reduced viscosity effects alone. Thus, the formation of large micellar clusters significantly increases the exchange rates. At the electrical conductivity percolation transition, even higher rates are observed. For example, the exchange rate increases by a factor of 2.5 when the system pressure is decreased from 400 to 100 bar, a position that is well into the percolation region. These changes in the secondary structure lead to important and unexpected consequences for conducting chemical reactions in supercritical and near-critical microemulsions.

Figure 11 A scanning electron micrograph of Al(OH)$_3$ particles formed in a supercritical propane microemulsion at 110°C and 200 bar using the surfactant AOT at $W = 5$. (From Ref. 67.)

B. Particle Synthesis and Catalytic Reactions

Supercritical microemulsions have been used to generate inorganic particles in the cores of the microemulsion droplets. In the study by Matson et al. [67], 5 μm diameter Al(OH)$_3$ particles were produced in the cores of sodium bis(2-ethylhexyl)sulfosuccinate (AOT) micelles formed in supercritical propane at 110°C and 200 bar. A scanning electron micrograph of these particles is shown in Fig. 11. Aqueous Al(NO$_3$)$_3$ in the core of the dispersed droplet phase reacted with NH$_3$ dissolved in the supercritical propane continuous phase to produce a particle in the core of a microemulsion droplet. In this case, the NH$_3$ is highly soluble in the supercritical propane phase, where it diffuses to the micelle and probably first reacts with the water to form NH$_4^+$ before reacting with the Al(NO$_3$)$_3$ to form the aluminum hydroxide particles. In this study, the particle size was reported to depend strongly on the concentration of the starting reactants. When the concentration was reduced from 0.1 M to 0.05 M, the particle sizes were reduced from 5 μm to less than 0.1 μm. Clearly, the 5 μm material is of a size that is much larger than an individual microemulsion droplet, as the droplet size is typically much less than 0.5 μm diameter for $W = 5$. Matson et al. [67] observed that as the reaction proceeded for the higher concentration studies, the solutions became turbid after about 1 min. This is consistent with the formation of these larger particles that result from the agglomeration of many smaller particles in the micelle cores. The supercritical fluid offers many mechanisms for much finer control over the particle morphology than is possible with conventional liquid systems.

In a small exploratory study, Zemanian et al. [68] synthesized a variety of catalyst materials in supercritical butane microemulsions and subsequently used these to explore the hydrogenation reactions of naphthalene to tetralin using molecular hydrogen dissolved in the continuous-phase fluid. Ultrasmall catalyst particles of molybdenum(VI) oxide (MoO_3), molybdenum(IV) sulfide (MoS_2), iron(II) sulfide (FeS), and the cocatalyst CoMoO$_3$ were prepared in situ by reaction of appropriate precursor salts with either CO_2 or Na_2S. The molybdenum-containing catalyst particles appeared to be stably suspended in the cores of the microemulsion droplets, whereas the FeS catalyst material agglomerated and precipitated from the solution after 1 h at reaction conditions. The reaction conditions for this study were 135–155°C and 300–500 bar. Interestingly, under these conditions the surfactant (DDAB) appears to be stable. In this reaction, both of the reactants, naphthalene and hydrogen, are dissolved in the supercritical butane-continuous phase. The naphthalene hydrogenation reaction showed slight conversion (0.25%) after 4 h using the CoMoO$_3$ cocatalyst, implying that some partitioning of the reactants into the core regions of the micelle occurs to allow the reaction to proceed. The very high surface areas of the ultrasmall particles in the micelle cores may improve the apparent activity; in contrast, commercially available CoMoO$_3$ on a solid support gave no conversion under identical conditions. The results emphasized one of the interesting applications of these systems for hydrogenation or oxidation reactions: The supercritical phase is an excellent solvent for these small gases, and hence high concentrations can be used to promote the reactions. These preliminary reaction results highlight many interesting aspects of this potentially fruitful area.

C. Polymerization Reactions

Another promising application of supercritical microemulsions is in the broad area of polymerization reactions. An example of one such system is in the work by Beckman et al. [41,74]. In this study, acrylamide monomer dissolved in the surfactant interfacial region of a supercritical ethane–propane microemulsion was catalyzed by azobis(isobutylnitrile) (AIBN) dissolved in the reverse micelle core to produce the micelle-soluble polyacrylamide.

Beckman et al. observed an effect of the secondary microemulsion structure on the molecular weight and yield of the polymer. Under conditions where extensive micelle–micelle clustering occurred, at lower fluid density the molecular weight of the polymer was as much as two times higher. Thus, the density of the supercritical phase could be used to control the polymer morphology. Beckman and Smith also completed an extensive study [74] of the effect that acrylamide, surfactant, and water concentrations as well as the pressure and temperature had on the phase stability of the microemulsions. The phase behavior of these systems depends on the choice of operating parameters, and this behavior can be exploited to optimize the properties of the polymer.

A major new development in a related area is the work of DeSimone et al. [26,31,50,51,75,76], who conducted dispersion polymerizations in supercritical CO_2. In the early stages of the dispersion–polymerization reaction, the solutions are homogenous microemulsions containing surface-active polymers with CO_2-philic moieties. The monomer is soluble in the continuous phase. As the polymer grows, its solubility rapidly diminishes to form precipitated polymer particles that are stabilized by the surface-active polymer. This approach has been expanded to several different polymer systems [50].

V. CONCLUSIONS

Microemulsions tremendously expand the potential applications of supercritical fluids as reaction media for chemical reactions. By themselves, near-critical and supercritical fluids are much weaker solvents than the typical organic liquid. However, microemulsions create a highly polar region that is capable of solvating polar catalysts, reactants, or products. The unique aspects of the near-critical or supercritical continuous phase offer many advantages over their liquid-phase analogs.

The microstructure is wide and varied, and it appears that the fascinating variety of geometries found in liquid-phase microemulsions will be available for supercritical microemulsions. Thus far, several different water-in-"oil" microemulsions containing spherical droplets have been identified in a variety of fluids using cationic, anionic, and nonionic surfactants. Further, long, extended-rod structures have been formed with a zwitterionic surfactant in near-critical propylene. Finally, an "oil"-in-water microemulsion can be formed in which cores of the spherical droplets are filled with a near-critical or supercritical fluid. New surfactants are appearing for use in CO_2. These CO_2 systems have exciting potential for "solvent-free" organic synthesis reactions.

Supercritical microemulsions have thus far been used for particle synthesis, polymerizations, and hydrogenation reactions. These types of systems represent a potentially rich area of high industrial importance. Studies of these systems should expand in the coming years.

ACKNOWLEDGMENT

This research was supported by the Director, Office of Energy Research, Office of Basic Energy Sciences, Chemical Sciences Division of the U.S. Department of Energy under Contract DE-AC06-76RLO 1830. It was performed at Pacific Northwest National Laboratory, which is operated by Battelle Memorial Institute for the U.S. Department of Energy.

REFERENCES

1. R. W. Gale, J. L. Fulton, and R. D. Smith, J. Phys. Chem. *109*:920–921 (1987).
2. J. L. Fulton, J. P. Blitz, J. M. Tingey, and R. D. Smith, J. Phys. Chem. *93*:4198–4204 (1989).
3. K. P. Johnston, G. J. McFann, and R. M. Lemert, in *Supercritical Fluid Science and Technology* (K. P. Johnston and J. L. M. Penninger, eds.), American Chemical Society, Washington, DC, 1989, 140–164.
4. J. Eastoe, W. K. Young, B. H. Robinson, and D. C. Steytler, J. Chem. Soc., Faraday Trans. *86*:2883–2889 (1990).
5. T. Klein and J. M. Prausnitz, J. Phys. Chem. *94*:8811–8816 (1990).
6. R. D. Smith, J. L. Fulton, J. P. Blitz, and J. M. Tingey, J. Phys. Chem. *94*:781–787 (1990).
7. J. M. Tingey, J. L. Fulton, and R. D. Smith, J. Phys. Chem. *94*:1997–2004 (1990).
8. E. W. Kaler, J. F. Billman, J. L. Fulton, and R. D. Smith, J. Phys. Chem. *95*:458–462 (1991).
9. E. J. Beckman, J. L. Fulton, and R. D. Smith, in *Supercritical Fluid Technology: Reviews in Modern Theory and Applications* (T. J. Bruno and J. F. Ely, eds.), CRC Press, Boca Raton, FL, 1991, pp. 405–449.
10. D. G. Peck, R. S. Schechter, and K. P. Johnston, J. Phys. Chem. *95*:9541–9549 (1991).
11. J. Zhang and F. V. Bright, J. Phys. Chem. *96*:9068–9073 (1992).

12. G. G. Yee, J. L. Fulton, and R. D. Smith, Langmuir *8*:377–384 (1992).
13. G. J. McFann and K. P. Johnston, Langmuir *9*:2942–2948 (1993).
14. K. A. Bartscherer, M. Minier, and H. Renon, Fluid Phase Equilibria *107*:93–150 (1995).
15. P. Yazdi, G. J. McFann, M. A. Fox, and K. P. Johnston, J. Phys. Chem. *94*:7224–7232 (1990).
16. D. G. Peck and K. P. Johnston, J. Phys. Chem. *97*:5661–5667 (1993).
17. M. McHugh and V. Krukonis, Supercritical Fluid Extraction, Butterworths, Boston, 1986.
18. M. E. Paulaitis, J. M. L. Penninger, R. D. Gray, and P. Davidson, *Chemical Engineering at Supercritical Fluid Conditions*, Ann Arbor Science, Ann Arbor, MI, 1983.
19. T. J. Bruno and J. F. Ely (eds.), *Supercritical Fluid Technology: Reviews in Modern Theory and Applications*, CRC Press, Boca Raton, FL, 1991.
20. E. Kiran and J. M. H. Levelt Sengers (eds.), *Supercritical Fluids: Fundamentals and Applications*, Kluwer, Boston, 1994.
21. K. W. Hutchenson and N. R. Foster (eds.), *Innovations in Supercritical Fluids: Science and Technology*, American Chemical Society, Washington, DC, 1995.
22. Y. V. Tsekhanskaya and M. B. Iomtev, Russ. J. Phys. Chem. *38*:1173–1176 (1964).
23. K. A. Consani and R. D. Smith, J. Supercrit. Fluids *3*:51–65 (1990).
24. T. A. Hoefling, R. M. Enick, and E. J. Beckman, J. Phys. Chem. *95*:7127–7129 (1991).
25. G. G. Yee, J. L. Fulton, and R. D. Smith, J. Phys. Chem. *96*:6172–6181 (1992).
26. J. M. DeSimone, Z. Guan, and C. S. Elsbernd, Science *257*:945–947 (1992).
27. T. A. Hoefling, R. M. Enick, and E. J. Beckman, *Proceedings of 6th FPECPD*, Cortina, 1992, pp. 23–27.
28. E. E. Maury, H. J. Batten, S. K. Killian, Y. Z. Menceloglu, J. R. Combes, et al., Polym. Prepr. (Am. Chem. Society, Div. Polym. Chem.) *43*:664 (1993).
29. K. Harrison, J. Goveas, K. P. Johnston, and E. A. O'Rear, Langmuir *10*:3536–3541 (1994).
30. G. J. McFann, K. P. Johnston, and S. M. Howdle, AIChE *40*:543–555 (1994).
31. J. M. DeSimone, E. E. Maury, Y. Z. Menceloglu, J. B. McClain, T. J. Romack, et al., Science *265*:356–359 (1994).
32. J. L. Fulton, D. M. Pfund, J. B. McClain, T. J. Romack, E. E. Maury, et al., Langmuir *11*:4241–4249 (1995).
33. K. P. Johnston, K. L. Harrison, M. J. Clarke, S. M. Howdle, M. P. Heitz, et al., Science *271*:624–626 (1996).
34. A. W. Francis, J. Phys. Chem. *58*:1099–1114 (1954).
35. M. E. Paulaitis, V. J. Krukonis, R. T. Kurnik, and R. C. Reid, Rev. Chem. Eng. *1*:179–250 (1983).
36. K. D. Bartle, A. A. Clifford, S. A. Jafar, and G. F. Shilstone, J. Phys. Chem. Ref. Data *20*:713–756 (1991).
37. R. Pecora, *Dynamic Light Scattering*, Plenum, New York, 1985.
38. M. Kerker, *The Scattering of Light and Other Electromagnetic Radiation*, Academic, New York, 1969.
39. J. P. Blitz, J. L. Fulton, and R. D. Smith, J. Phys. Chem. *92*:2707–2710 (1988).
40. R. D. Smith, J. P. Blitz, and J. L. Fulton, in *Supercritical Fluid Science and Technology* (K. P. Johnston and J. L. M. Penninger, eds.), American Chemical Society, Washington, DC, 1989, pp. 165–183.
41. E. J. Beckman, J. L. Fulton, D. W. Matson, and R. D. Smith, in *Supercritical Fluid Science and Technology* (K. P. Johnston and J. M. L. Penninger, eds.), American Chemical Society, Washington, DC, 1989, pp. 184–206.
42. E. J. Beckman and R. D. Smith, J. Phys. Chem. *94*:3729–3734 (1990).
43. J. Zhang, J. L. Fulton, and R. D. Smith, J. Phys. Chem. *97*:12331–12338 (1993).
44. O. Glatter and O. Kratky, *Small Angle X-Ray Scattering*, Academic, New York, 1982.
45. L. A. Feigin and D. I. Svergun, *Structure Analysis by Small Angle X-Ray and Neutron Scattering*, Plenum, New York, 1987.
46. L. J. Magid, in *Nonionic Surfactants* (M. J. Schick, ed.), Marcel Dekker, New York, 1987, 677-752.
47. J. Eastoe, D. C. Steytler, B. H. Robinson, and R. K. Heenan, J. Chem. Soc., Faraday Trans. *90*:3121–3127 (1994).

48. J. M. Tingey, J. L. Fulton, D. W. Matson, and R. D. Smith, J. Phys. Chem. *95*:1445–1448 (1991).
49. J. Zhang, G. L. White, and J. L. Fulton, J. Phys. Chem. *99*:5540–5547 (1995).
50. K. A. Shaffer and J. M. DeSimone, Trends Polym. Sci. *3*:146–153 (1995).
51. Z. Guan and J. M. DeSimone, Macromolecules *27*:5527–5532 (1994).
52. D. Chillura-Martino, R. Triolo, J. B. McClain, J. R. Combes, D. E. Betts, et al., J. Mol. Struct. *383*:3–10 (1996).
53. J. B. McClain, D. E. Betts, D. A. Canelas, E. T. Samulski, J. M. DeSimone, et al., Science *274*:2049–2051 (1996).
54. J. Eastoe, Z. Bayazit, S. Martel, D. C. Steytler, and R. K. Heenan, Langmuir *12*:1423–1424 (1996).
55. P. L. Bolden, J. C. Hoskins, and A. D. King, J. Colloid Interface Sci. *91*:454–463 (1983).
56. D. W. Ownby, W. Prapaitrakul, and A. D. King, J. Colloid Interface Sci. *125*:526–533 (1988).
57. W. Prapaitrakul and A. D. King, J. Colloid Interface Sci. *112*:387–395 (1986).
58. J. Zhang and J. L. Fulton, in *Innovations in Supercritical Fluids: Science and Technology* (K. W. Hutchenson and N. R. Foster, eds.), American Chemical Society, Washington, DC, 1995, pp. 111–125.
59. J. B. Hayter and J. Penfold, Mol. Phys. *42*:109–118 (1981).
60. J. B. Hayter and J. Penfold, J. Chem. Soc., Faraday Trans. I *77*:1851–1863 (1981).
61. J. P. Hansen and J. B. Hayter, J. Mol. Phys. *46*:651–656 (1982).
62. D. F. Evans, D. J. Mitchell, and B. W. Ninham, J. Phys. Chem. *88*:6344–6348 (1984).
63. T. Zemb and P. Charpin, in *Surfactants in Solution* (K. L. Mittal and P. Bothorel, eds.), Plenum, New York, 1986, pp. 141–154.
64. J. L. Fulton and R. D. Smith, J. Phys. Chem. *92*:2903–2907 (1988).
65. G. J. McFann and K. P. Johnston, J. Phys. Chem. *95*:4889–4896 (1991).
66. D. G. Peck and K. P. Johnston, J. Phys. Chem. *95*:9549–9556 (1991).
67. D. W. Matson, J. L. Fulton, and R. D. Smith, Mater. Lett. *6*:31–33 (1987).
68. T. S. Zemanian, R. M. Bean, J. L. Fulton, J. C. Linehan, and R. D. Smith, in *Proceedings of the 2nd International Symposium on Supercritical Fluids* (M. A. McHugh, ed.), Johns Hopkins Univ. Press, Baltimore, MD, 1991, pp. 193–195.
69. T. S. Zemanian, J. L. Fulton, and R. D. Smith, in *Supercritical Fluid Engineering Science* (E. Kiran and J. F. Brennecke, eds.), Americal Chemical Society, Washington, DC, 1993, pp. 258–270.
70. P. P. Infelta, M. Grätzel, and J. K. Thomas, J. Phys. Chem. *78*:190 (1974).
71. M. Tachiya, Chem. Phys. Lett. *33*:289 (1975).
72. S. S. Atik and J. K. Thomas, J. Am. Chem. Soc. *103*:3543 (1981).
73. J. Lang, A. Jada, and A. Malliaris, J. Phys. Chem. *92*:1946 (1988).
74. E. J. Beckman and R. D. Smith, J. Phys. Chem. *94*:345 (1990).
75. K. A. Shaffer, T. A. Jones, D. A. Canelas, and J. M. DeSimone, Macromolecules *29*:2704–2706 (1996).
76. D. A. Canelas, D. E. Betts, and J. M. DeSimone, Macromolecules *29*:2818–2821 (1996).
77. D. M. Pfund and J. L. Fulton, in *Proceedings of the Third International Symposium on Supercritical Fluids* (M. Perrut, ed.), Institut National Polytechnique de Lorraine (Vandoeuvre), Strasbourg, France, 1994, pp. 235–240.
78. J. L. Fulton and D. M. Pfund, in *Proceedings of Third International Symposium on Supercritical Fluids* (M. Perrut, ed.), Institut National Polytechnique de Lorraine (Vandoeuvre), Strasbourg, France, 1994, pp. 391–396.

21

Electrochemical Studies in Microemulsions

Syed Qutubuddin
Case Western Reserve University, Cleveland, Ohio

I. SCOPE OF THE REVIEW

The use of surfactants or amphiphilic molecules in electrochemistry dates back over four decades [1,2]. Extensive research on electrochemistry in surfactant systems has been reported primarily in the last 20 years. Surfactant systems are ubiquitous. The aggregation of surfactant molecules may produce a variety of systems including micelles, monolayers and bilayers, vesicles, lipid films, emulsions, foams, and microemulsions. Developments in the area of electrochemistry in such association colloids and dispersions have been documented by Mackay and Texter [3]. Mackay [4] reviewed the developments in association colloids, particularly micelles and microemulsions. Rusling [5,6] also reviewed electrochemistry in micelles, microemulsions, and related organized media. This chapter focuses on microemulsions and does not deal with micelles, monolayers, emulsions, and other surfactant systems per se.

II. BRIEF INTRODUCTION TO MICROEMULSIONS

The purpose of this section is to briefly introduce the main features of microemulsions that are relevant to electrochemical investigations. Detailed descriptions of the properties of microemulsions are available in various chapters of this book. A microemulsion is generally defined as a thermodynamically stable isotropic solution of two immiscible fluids, commonly oil and water, containing one or more surface-active species. The surface-active species or surfactant molecules are mostly located at the interface between the domains of polar and nonpolar fluids. Historically, Schulman and coworkers [7,8] first described microemulsions as transparent or translucent systems formed spontaneously by mixing oil and water with relatively large amounts of an ionic surfactant together with a cosurfactant. Microemulsions need not be transparent and are not required to contain cosurfactants or cosolvents and electrolytes [9]. The surfactant can be ionic or nonionic. Many systems of practical interest do contain a cosolvent such as a medium-chain alcohol and also electrolyte [10]. The above definition is broad and includes among microemulsions systems with cosolvents, reverse micellar solutions, and "swollen" or "solubilized" micellar solutions. This definition also does not identify a specific structure for microemulsion aggregates. The use of microemulsions preceded their introduction into the scientific community by Schulman and coworkers [7,8]. For instance, microemulsions formulated with eucalyptus oil, water, soap, and white spirits were used more than a century ago to wash wool efficiently. Several industrial products such as lubricating oils, liquid waxes, and detergent formulations were patented during the 1930s [11].

Microemulsions have received great attention during the last quarter-century because of their interesting thermodynamics, intricate physicochemical behavior, and increasing number of applications. The importance of microemulsions for enhanced oil recovery [12,13] has played a key role in the fostering of extensive theoretical and experimental research programs worldwide. The salient features of microemulsions include (1) spontaneous formation and thermodynamic stability, (2) intricate phase behavior, (3) large interfacial area per unit volume, (4) a wide range of interfacial curvature, (5) ultralow interfacial tensions and critical phenomena, and (6) large solubilization of both organic and aqueous phases. The special features of microemulsions make them tunable for a wide variety of applications. Commercial applications are increasing, and the scope is tremendous, as is evident from the increasing number of papers and patents on microemulsions.

Physicochemical aspects of the phase behavior of microemulsions continue to be extensively studied, as is made evident in several chapters on this topic in this book. The phase behavior exhibited by microemulsions is very rich. Besides single phases, microemulsions can exist in equilibrium with excess oil or water or both. Winsor [14] referred to these equilibria as types I, II, and III, respectively. Such transitions occur when an appropriate variable such as salinity is changed. Microemulsions containing a few percent of an anionic surfactant (e.g., petroleum sulfonates), a cosurfactant such as a short-chain alcohol, and equal volumes of oil and NaCl brine have been extensively studied [13]. The phase behavior is termed "simple" [15] when the microemulsions behave as though composed of three pure components and can be easily represented in a ternary phase diagram. At low salinity there is a two-phase region in which microemulsions along the binodal curve exist in equilibrium with oil containing molecularly dispersed surfactant. The surfactant is partitioned predominantly into the water-continuous microemulsion phase. This oil-in-water (O/W) microemulsion system, which corresponds to Winsor's type I, is often called a "lower" phase. At high salinity there is another two-phase region in which microemulsions exist in equilibrium with excess brine, i.e., Winsor's type II. Such microemulsions are commonly named "upper" phase and are oil-continuous. Three phases exist in equilibrium at intermediate salinity: a Winsor type III microemulsion, excess brine, and excess oil. Type III microemulsions have been designated as "middle" phase [15] or "surfactant" phase [16] and are bicontinuous [17] in microstructure. It is also possible to have four coexisting phases, two of which may be microemulsions in equilibrium with one another or a microemulsion in equilibrium with a liquid crystalline phase. A liquid crystalline phase may also exist in equilibrium with only brine and oil [18]. The pseudoternary diagrams are slices at constant salinity of the tetrahedron necessary to display the phase boundaries of a four-component system.

Parameters that may be used to tune the phase behavior of microemulsions include salinity, surfactant type and concentration, cosolvent type and concentration, pH, oil composition, temperature, and pressure. As salinity increases, there is a steady progression from lower phase to middle phase to upper phase microemulsions. This reflects a continuous evolution of the preferred curvature of the surfactant film and corresponds to an increase in hydrophobicity with added electrolyte such as NaCl. At low salinity the droplet size in the water-continuous lower phase increases with increasing salinity. This corresponds to an increase in the solubilization of oil and is reflected in increased light scattering. As salinity increases further, the middle phase appears and is initially water-continuous.

Table 1 Variables for Tuning the Phase Behavior of Ionic Microemulsions

Variable	Phase transition
Salinity ↑	O/W→bicontinuous→W/O
Oil alkane carbon number ↑	W/O→bicontinuous→O/W
Oil aromaticity ↑	O/W→bicontinuous→W/O
Alcohol concentration ↑ (low MW)	W/O→bicontinuous→O/W
Alcohol concentration ↑ (high MW)	O/W→bicontinuous→W/O
Surfactant hydrophobic chain length ↑	O/W→bicontinuous→W/O
Temperature ↑	W/O→bicontinuous→O/W
pH ↑ (for carboxylic acids, amines, and other pH-sensitive surfactants)	W/O→bicontinuous→O/W

The microemulsion is bicontinuous over an intermediate salinity range. The middle phase becomes oil-continuous at a higher salinity. After the transition from three to two phases, the microemulsion remains oil-continuous, with drop size decreasing with increasing salinity. The droplet diameter in microemulsions is in the range of 5–100 nm.

Table 1 summarizes the qualitative changes in the phase behavior of microemulsions containing ionic surfactants. Some details of the effects of different variables are available in Ref. 13 and various chapters in this book. The phase transitions are generally understood in terms of relative strengths of hydrophilic and hydrophobic properties of the surfactant film in the microemulsion. The phase behavior depends strongly on the type and structure of the surfactant. For example, microemulsions containing nonionic surfactants are less sensitive to salinity but are more sensitive to temperature than those with ionic surfactants. The partitioning of cosolvents such as alcohols between the surfactant film, the organic phase, and the aqueous phase also affects the phase behavior. Microemulsions can be tailored for specific applications by adjusting an appropriate variable. For example, as indicated in Table 1, the effect of salinity on the phase behavior can be counterbalanced by an increase in the pH of an appropriate microemulsion [18,19].

There are two bulk interfaces in middle phase microemulsions and one in lower or upper phase microemulsions. Thus, one or three values of interfacial tension (IFT) may be measured depending on system composition: (1) γ_{mo} between microemulsion and excess oil phase, (2) γ_{mw} between microemulsion and excess brine phase, and (3) γ_{ow} between excess oil and brine phases. Phase volumes and consequently the volumes of oil (V_o) and brine (V_w) solubilized in the microemulsion depend on the variables that control the phase behavior. The solubilization parameters are defined as V_o/V_s and V_w/V_s, where V_s is the volume of the surfactant in the microemulsion phase. These parameters are easily determined from phase volume measurements if all the surfactant is assumed to be in the microemulsion phase. The magnitude of γ_{mo} decreases as V_o/V_s increases, i.e., as more oil is solubilized. Similarly, the magnitude of γ_{mw} decreases as V_o/V_s increases. The salinity at which the values of γ_{mo} and γ_{mw} are equal is known as the optimal salinity based on IFT. Similarly, the intersection of V_o/V_s and V_w/V_s defines the optimal salinity based on phase behavior. The optimal salinity concept is very important for enhanced oil recovery.

III. MICROEMULSIONS AS REACTION MEDIA: GENERAL FEATURES

A rapidly growing field of application for microemulsions is as media for a variety of chemical reactions including electrochemical, photochemical, enzymatic, and polymerization reactions. The existence of microdomains or droplets with large interfacial area per unit volume allows the possibility of controlling the reaction rates, pathways, and stereochemistry as well as the morphology of the precipitation products. Catalysis of various chemical reactions in microemulsions has a great potential for industrial applications [20,21]. One factor in the rate enhancement is the tremendous increase of contact surface between the reactant molecules that may be adsorbed at the interfacial film or solubilized inside the dispersed phase. The rate enhancement may far exceed the value predicted by simple partitioning of substrates between the dispersed phase and the continuous phase. The pathway of a reaction system and the stereochemistry may be controlled by the local environment of the microdomains. For instance, in the case of enzymatic reactions, the microenvironment may mimic the active sites, and certain configurations or orientations could be favorable for the bound substrate.

Some factors to be considered in using microemulsions as reaction media include the dynamics of the surfactant molecules, collisions between droplets, flexibility of the interfacial film, and partitioning of the reactants. The average lifetime of a surfactant molecule in the interfacial film is on the order of microseconds, while the exchange time between droplets is on the order of milliseconds [22]. The number of "sticky" collisions that allow exchange between droplets decreases with increasing rigidity of the interfacial film. When the reaction rates are faster than the exchange times, as in electron and proton transfers, the droplets may be treated as virtually isolated or "frozen." At the other extreme, when the rates are very slow, the dispersed phase may be regarded as "continuous" from the reaction kinetics viewpoint, allowing regular rate expressions to apply. For the intermediate case, statistical distributions and interdroplet exchange will affect the reaction rate [23]. This review of electrochemistry in microemulsions is approached from two different, though interrelated, perspectives, (1) electrochemistry as a characterization tool, and (2) microemulsions as media to conduct electrochemical reactions. The next section provides an introduction to electrochemical tools for the characterization of microemulsions.

IV. ELECTROCHEMICAL TECHNIQUES FOR CHARACTERIZATION OF MICROEMULSIONS

A battery of techniques are available to characterize the microstructure and also understand the interactions in microemulsions. Tools widely used include dynamic light scattering, nuclear magnetic resonance, neutron scattering, small-angle X-ray scattering, and fluorescence spectroscopy. Some of these topics appear elsewhere in this book. Electrochemical techniques provide a complementary approach that is simple, fast, and inexpensive for characterization of microemulsions [24]. Any electrochemical technique that allows the determination of the diffusion coefficient of an electroactive substance can, in principle, be used for measuring diffusion in micellar and microemulsion systems [24–27]. A predetermined concentration of an electroactive probe is dissolved in the surfactant system, and an apparent diffusion coefficient of the aggregate (micelle or microemulsion droplet) is measured. The information obtained depends on the nature of the electroactive probe, its relative partitioning between the continuous and discontinuous pseudophases, and interactions. Electrochemical techniques used for such studies include polarography, cyclic voltammetry (CV), rotating disk voltammetry (RDV), chronoamperometry/ chronocoulometry, and chronopotentiometry. The current–diffusion coefficient relationships that are applicable for each of these techniques are given below [28].

Cyclic voltammetry (Randles–Sevcik equation):

$$i_p = 0.4463 \frac{n^{3/2}F^{3/2}}{R^{1/2}T^{1/2}} ACD^{1/2}V^{1/2} \tag{1}$$

Polarography (Ilkovic equation):

$$i_d = 708nCD^{1/2}m^{2/3}t^{1/6} \tag{2}$$

Rotating disk voltammetry (Levich equation):

$$i_l = 0.62nFACD^{2/3}v^{-1/6}\omega^{1/2} \tag{3}$$

Chronoamperometry (Cottrell equation):

$$i(t) = nFACD^{1/2}\pi^{-1/2}t^{-1/2} \tag{4}$$

Chronopotentiometry (Sand equation):

$$i(\tau) = 1/2nFACD^{1/2}\pi^{-1/2}\tau^{-1/2} \tag{5}$$

The symbols in the above mean the following: i_d is the polarographic diffusion-limited current (A), n is the number of electrons transferred, C is the concentration of the electroactive probe (mol/cm^3), D is the diffusion coefficient of the electroactive probe (cm^2/s), m is the mass flow rate of mercury at the dropping electrode (mg/s), t is the drop time (s), i_p is the peak current (A), F is Faraday's constant (coulombs/mol), A is the area of the electrode (cm^2), R is the gas constant (J/mol), T is the temperature (K), V is voltage scan rate (V/s), i_1 is the limiting current (A), v is the kinematic viscosity of the solution (cm^2/s), ω is the angular velocity of the rotating disk electrode (rad/s), $i(t)$ is the diffusion current (A) at time t (s), and τ is the transition time (s).

The electrochemical experiments are typically conducted in a three-electrode cell with working, counter, and reference electrodes. The current flows between the working and counter electrodes. The potential of the working electrode is recorded with respect to the reference electrode. Typical reference electrodes used in studying surfactant systems include the saturated calomel electrode (SCE) and the Ag/AgCl (saturated KCl) electrode. Carbon (glassy or pyrolytic), platinum, and mercury are generally used as the working electrode. A dropping mercury electrode (DME) or a static mercury drop electrode is common for polarography. Both solid electrodes and hanging mercury drop electrodes are used in voltammetry. The geometric surface area of the electrode varies from 1 to 10 mm^2.

A cyclic voltammogram is obtained by measuring the current at the working electrode during a scan of the potential. The potential applied to the working electrode is scanned linearly from an initial value $E(i)$ to a predetermined limit $E(f)$ called the switching potential, where the direction of the scan is reversed. From the magnitude of the anodic peak current (I_{pa}) and cathodic peak current (I_{pc}), and from the potentials at which the peak currents are observed, E_{pa} and E_{pc}, respectively, information can be obtained on reversibility or irreversibility of the system and on chemical reactions accompanying electrode processes [29–32].

One drawback of electrochemical investigations of surfactant systems used to be that the measurements could be made only in systems that are fairly conducting. Usually results are not reliable due to charging current and iR drop contributions in systems without sufficient indifferent electrolyte. Addition of an electrolyte very often changes the microstructure of the surfactant system compared to the extant microstructure without the electrolyte. Electrochemical measurements were usually restricted to surfactant systems containing added electrolytes. Recent developments have made it possible to apply voltammetry to highly resistive systems by using ultramicroelectrodes [33]. The limiting current

(i_1) is given by

$$i_1 = 4nFDCr \tag{6}$$

where r is the radius of the ultramicroelectrode (cm) and the other terms are as defined above. A few investigators have applied this technique for diffusion measurements in surfactant systems [34–37].

V. MICROEMULSION CHARACTERIZATION: DIFFUSION COEFFICIENT MEASUREMENTS USING ELECTROCHEMICAL TECHNIQUES

Mackay and coworkers [38,39] were the first to measure diffusion coefficients in microemulsions using an electrochemical technique. They used polarography to determine the diffusion coefficients of Cd(II) in a mineral oil–water–1-pentanol–Tween 40 [polyoxyethylene(20) sorbitan monopalmitate] microemulsion over a wide range of water content, 25–85 wt%. The diffusion coefficient D of Cd(II) located in the aqueous phase decreased with increasing dispersed phase volume fraction, ϕ. However, there was no apparent discontinuity in the value of D over the entire range of water content. Similar results were obtained with Tl(I) and ferricyanide and ferrocyanide ions [40]. Mackay [41] used a surface-active probe, 1-dodecyl-4-cyanopyridinium ($C_{12}CP$), to investigate Tween 40 or Tween 60–1-pentanol–hexadecane and sodium cetyl sulfate (SCS)–1-pentanol–mineral oil microemulsions. The diffusion coefficients in microemulsions were found to obey the following relationships based on the "obstruction" effect [42]:

$$D = D_w (1 - \phi)^p \tag{7}$$

where D_w is the diffusion coefficient of the ion in water.

The theoretical value of p is 1.5 in Eq. (7), which was derived assuming spherical droplets. This relationship is valid for both ionic and nonionic systems provided the electroactive probe is solubilized only in water. The value of p observed by Mackay and coworkers was higher than the expected value of 1.5. The disagreement was accounted for by the rather large amount of bound water in the droplet or nonaqueous pseudophase [43–45].

Mackay et al. [25] used $C_{12}CP$ to determine droplet diffusion coefficients in an octane–1-butanol–CTAB microemulsion containing 60% water. The polarography results were in agreement with quasi-elastic light scattering (QELS) results. Hexadecane–1-butanol–cetyltrimethylammonium bromide (CTAB) and mineral oil–1-pentanol–SCS systems were also investigated. Alkyl pyridinium and viologen probes were used in the CTAB and SCS microemulsions containing 60% water. As illustrated in Fig. 1 [25], the small-chain-length species partition between the water and the organic phase, while the probe is completely associated with the droplet (or interface) for chain lengths of 8 or higher. For the same microemulsion systems, the diffusion coefficient was found to be relatively constant over the temperature range 18–45°C [46]. Polarography of Cd(II) in hexadecane–1-butanol–CTAB microemulsions gave a diffusion coefficient of 7×10^{-7} cm²/s independent of water content in the range of 35–70% [47]. It was suggested that Cd(II) was bound to the surfactant film and diffused with the oil droplet.

Georges and Berthod [26,48] investigated microemulsions containing methylene chloride, sodium p-octylbenzenesulfonate, 1-pentanol, and water. The apparent diffusion coefficient of a hydrophobic probe, 10-methylphenothiazine, increased whereas the diffusion coefficient of hydrophilic hydroquinone decreased with increasing oil content. The diffusion

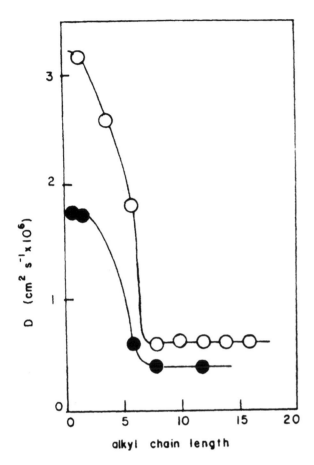

Figure 1 Polarographic diffusion coefficient D versus alkyl chain length at 25°C for (○) 60% water CTAB microemulsion, 1-alkyl-4-cyanopyridinium ion (1–3 mM), 0.10 M KCL; (●) 60% water SCS microemulsion, alkyl viologen (0.8 mM), 0.05 M KCl. (From Ref. 25.)

coefficients were generally too large, indicating a bicontinuous rather than a droplet microstructure for most of the compositions studied. Electrochemical measurements on the reduction of oxygen and hydrogen peroxide in the same microemulsion system led to similar conclusions. Georges and coworkers [49,50] also measured diffusion coefficients in dodecane–SDS–brine microemulsions with 1-pentanol or 1-heptanol as cosurfactant. The electrochemical data were consistent with conductivity measurements. Microviscosity and polarity were also estimated using a fluorescence probe [49]. The combined experimental approach was useful in interpreting the phase diagram of SDS microemulsions [50].

Chokshi et al. [24] described the use of two electrochemical techniques, cyclic voltammetry (CV) and rotating disk voltammetry (RDV), to characterize oil-in-water microemulsions containing a cationic surfactant CTAB, 1-butanol as cosurfactant, n-octane, and water (in the range of 90–95%). NaBr was used as the electrolyte. Diffusion coefficients of microemulsions droplets were determined using ferrocene as a hydrophobic electroactive probe. Typical voltammograms and a corresponding plot of anodic peak current versus square root of scan rate are shown in Fig. 2 [24]. The diffusion coefficient is calculated from

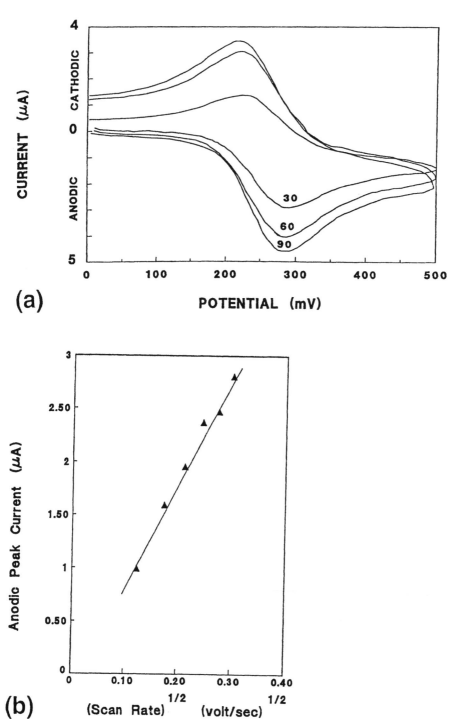

Figure 2   (a) Typical voltammograms for ferrocene in a CTAB microemulsion for three scan rates: 30, 60, and 90 mV/s. (b) A plot of peak current (anodic) versus square root of scan rate. The composition of the microemulsion is 2.3% CTAB, 1.3% NaBr, 4.6% *n*-butanol, and 2.3% *n*-octane. (From Ref. 24.)

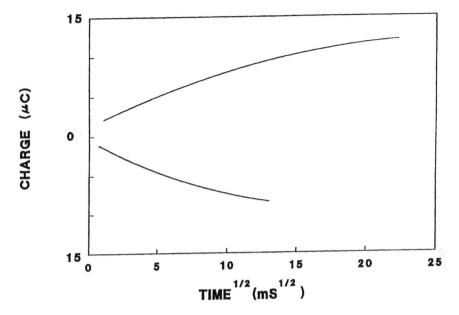

Figure 3 Typical plot of charge versus square root of time for chronocoulometric experiment in a CTAB microemulsion system. (From Ref. 24.)

the slope using Eq. (1). One possible concern about electrochemical measurements is the adsorption of the electroactive probe on the working electrode. Chronocoulometric experiments are useful in evaluating the adsorption behavior. Typical results obtained with CTAB microemulsion are shown in Fig. 3 [24]. In this example, the amount of probe adsorbed, obtained from the difference of intercepts, was negligible compared to the amount of probe added. Furthermore, light scattering experiments were performed to determine whether the addition of ferrocene affected the diffusion coefficient. No significant difference was observed when the diffusion coefficient in the CTAB microemulsion was measured by dynamic light scattering with and without the probe. Cyclic voltammetric and rotating disk voltammetric techniques gave diffusion coefficients that were in good agreement for the microemulsion compositions studied. The results were also compared with QELS measurements [24].

A few typical results obtained in CTAB microemulsions are shown in Fig. 4 [51]. The amount of oil was varied, keeping all other components constant. While the CV and RDV measurements are within experimental error, there are significant differences between electrochemical and QELS values. The discrepancy is due to the fact that electrochemical techniques yield values of the self-diffusion coefficients, whereas light scattering techniques yield mutual-diffusion coefficients [24]. The light scattering diffusion coefficients are strongly affected by both thermodynamic and hydrodynamic interactions between the microemulsion droplets and also by contributions due to the scattering of monomers.

Diffusion coefficients are often related to particle size by use of the Stokes–Einstein equation

$$D = kT/6\pi\eta a \tag{8}$$

where k is the Boltzmann constant, η is viscosity, and a is the particle radius. For a given surfactant concentration, the average size of the microemulsion droplet increases with

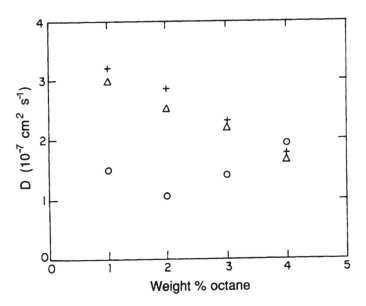

Figure 4 Comparison of diffusion coefficient data obtained by CV (+), RDV (△), and QELS (○). The microemulsion comprised CTAB 2 wt%, butanol (4 wt%), aqueous NaBr (2 wt%) solution, and octane. (From Ref. 51.)

increasing amount of oil. Thus, the effective charge density of the oil droplet decreases, and the repulsive electrostatic interactions decrease. The interactions between the surfactant aggregates are lower at higher oil concentrations, leading to better agreement between electrochemical and QELS measurements as shown in Fig. 4.

Diffusion coefficients were also reported for microemulsions containing the anionic surfactant SDS, n-butanol as cosurfactant, toluene as the oil, and NaCl as the electrolyte in the aqueous phase [51]. The measurements were performed as a function of salinity. Reliable NMR data are available for this system [52,53]. As discussed in a previous section, the microemulsion structure changes from oil-in-water (lower phase) to bicontinuous (middle) and then to water-in-oil (upper phase) as the salt concentration is increased. Ferrocene was used as a hydrophobic probe for lower and middle phase microemulsions. Potassium ferrocyanide was used as a hydrophilic probe for measurements in the middle phase. The RDV measurements with ferrocene are compared in Fig. 5 with the NMR diffusion coefficients of toluene reported by two different groups [52,53]. The NMR diffusion coefficients of toluene in the lower phase should equal the diffusion coefficients of the microemulsion droplets. The diffusion coefficient of toluene measured by NMR in the middle phase represents the diffusion coefficient of toluene in the interconnected hydrophobic domains. The electrochemical diffusion coefficient of ferrocene in the lower phase corresponds to that of the oil droplet. The electrochemical diffusion coefficient of ferrocene in the middle phase corresponds to the diffusion coefficient of ferrocene in the interconnected toluene domains. The diffusion coefficients of ferrocene and toluene are expected to be similar in magnitude, as their sizes are comparable. The electrochemical and NMR values in Fig. 5 are in very good agreement. The data in middle phase microemulsions suggest that ferrocene is essentially diffusing in a continuous toluene medium, indirectly supporting the bicontinuous structure. The electrochemical diffusion

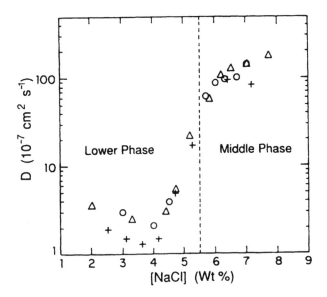

Figure 5 SDS microemulsion. Comparison of diffusion coefficient data obtained by RDV (+) and by NMR [(○) Ref. 52; (△) Ref. 53]. The illustrated microemulsion formulation was 1.99 wt% SDS, 47.25 wt% toluene, 3.96 wt% butanol, and 46.80 wt% brine.

coefficients obtained using potassium ferrocyanide are comparable to the NMR diffusion coefficients of water in middle phase microemulsions [51]. The agreement between electrochemical and NMR values supports the notion that both techniques measure self-diffusion coefficients.

Compared to widely used dynamic light scattering, the electrochemical approach provides a faster and less expensive tool for characterizing microemulsions. Electrochemical techniques do not require any prior knowledge of physical properties except viscosity and are also applicable to opaque systems. However, caution is recommended in interpreting electrochemical diffusion coefficients, as discussed in the following section.

VI. ANALYSIS OF DIFFUSION COEFFICIENTS MEASURED ELECTROCHEMICALLY

Entrance–exit kinetics for probe partitioning has been ignored in numerous treatments of electrochemically determined diffusion coefficients. It is generally accepted that an ideal electrochemical probe should partition completely in the micelle or microemulsion droplet and be insoluble in the continuous phase [24]. Under such conditions, the need for considering entrance–exit kinetics does not arise. This "complete partitioning" limit is the simplest case and allows one to directly use the current–concentration equations described earlier, wherein the diffusion coefficient is that of the droplet and the parameters n and C refer to the probe.

The zero-kinetics limit has been applied to the situation where re-equilibration in concentration gradients near the electrode surface is neglected [54,55]. Thus, the exit rate for a partitioned probe is zero, and the distribution of probe far from the electrode is maintained (in absolute amount) in the diffusion layer at the electrode surface. In the zero-kinetics limit, the contributions of current from the probe in the continuous phase and from the probe in the dispersed phase (droplets) are treated as additive. The term CD^z in the different current–concentration equations of Sec. IV is given by

$$CD_a^z = C_c D_c^z + C_d D_d^z \tag{9}$$

where the subscript a denotes the apparent diffusion coefficient, subscript c refers to the continuous-phase probe concentration and diffusion coefficient, and subscript d refers to the discontinuous- or droplet-phase probe concentration and diffusion coefficient. The superscript z is the exponent of the diffusion coefficient in the original current–concentration relationships. Equation (9) may be rearranged as

$$D_a = [xD_c^z + (1 - x)D_d^z]^{1/z} \tag{10}$$

where x represents the mole fraction of probe in the continuous phase and $1-x$ is the mole fraction in the discontinuous phase or pseudophase.

The entrance–exit rates for probe partitioning are rapid in the fast-kinetics limit. This implies that the probe exchange occurs on a time scale that is faster than otherwise unassisted diffusion or convection (in the case of hydrodynamic electrodes). Although this limit has not been widely investigated theoretically, the analysis of Evans [56] provides a basis for the limiting expression

$$D_a = xD_c + (1 - x)D_d \tag{11}$$

The apparent diffusion coefficient, D_a in Eq. (11) is a mole fraction–weighted average of the probe diffusion coefficient in the continuous phase and the microemulsion (or micelle) diffusion coefficient. It replaces D in the current–concentration relationships where total probe concentration is used. Both the zero-kinetics and fast-kinetics expressions require knowledge of the partition coefficient and the continuous-phase diffusion coefficient for the probe. Texter et al. [57] showed that finite exchange kinetics for electroactive probes results in zero-kinetics estimates of partitioning equilibrium constants that are lower bounds to the actual equilibrium constants. The fast-kinetics limit and Eq. (11) have generally been considered as a consequence of a "local equilibrium" assumption. This use is more or less axiomatic, since existing analytical derivations of effective diffusion coefficients from reaction-diffusion equations are approximate.

Alternatively, the kinetics considerations may be treated in terms of a two-state model, where the probe is assumed to be "free" in the continuous phase or uniformly "bound" with monodisperse aggregates, either micelles or microemulsion droplets. For instance, according to the so-called pseudophase model, the following relationship is obtained [5]:

$$D_a = \frac{D_c}{1 + K_D} + \frac{D_d K_D}{1 + K_D} \tag{12}$$

where K_D is an apparent distribution coefficient, i.e., ratio of bound to free probe. It is possible to account for nonuniform binding by using a number of binding constants and for polydispersity by using a number of diffusion coefficients. Three-state models were used by Rusling et al. [58] to account for polydispersity in SDS and CTAB systems and by Texter et al. [57] to describe polydispersity in CTAB–NaCl systems.

Equation (12) does not predict any dependence of the apparent diffusion coefficient D_a on the probe concentration C_x. Assuming that the probe is almost entirely bound to the particle (droplet or micelle), the following approximate model for D_a was obtained by

Table 2 Apparent Diffusion Coefficients (D_a) in Micelles and Microemulsions

System	Diffusion coefficient (10^{-7} cm^2/s)		
	Electrochemical		Light scattering
	Ferrocene	MV^{2+}	
Aqueous solution (D_c)	—	59	—
(1.0% NaBr)			
CTAB microemulsion	—	74	—
(2.3% CTAB+1.3% NaBr+			
2.3% n-octane+4.6% 1-butanol)			
TX100 microemulsion	—	53	—
(10% Triton X-100+			
1.0% NaBr+1.4% n-octane)			
SDS micelles	9.6	9.4	14.1
(2.0% SDS+1.0% NaBr)			
SDS–1-pentanol micelles	7.8	19	10.4
(2.0% SDS+1.0% NaBr			
+3% 1-pentanol)			
SDS microemulsion 1	7.2	8.2	4.6
(5.5% SDS+1.0% NaBr			
+5.1% dodecane			
+10.3% 1-pentanol)			
SDS microemulsion 2	2.0	14	4.1
(2.0% SDS+1.0% NaBr			
+2.0% dodecane			
+5.5% 1-pentanol)			

Source: Ref. 51.

Rusling et al. [5,58]:

$$D_a = \frac{D_c}{1 + C_d K^m C_x^{m-1}} + \frac{D_d C_d K^m C_x^{m-1}}{1 + C_d K^m C_x^{m-1}} \tag{13}$$

where m is the number of probe molecules bound to the droplet, C_d is the total droplet concentration, and K^m denotes an "apparent" binding constant [5,58].

The zero-kinetics limiting equation has been developed for partitioning solutes [54]. When RDV is used as the electrochemical technique, the expression for D_a is given as

$$D_d = \left[D_a^{2/3} \frac{C}{C_d} - D_c^{2/3} \frac{C_c}{C_d} \right]^{3/2} \tag{14}$$

where C, C_c, and C_d represent, respectively, the total (volume) concentration of probe, the concentration of probe in the aqueous (continuous) phase, and the concentration of probe in the droplet (discontinuous) phase.

Table 2 lists some measurements of diffusion coefficients in microemulsions of different surfactants [51]. Some data on SDS micellar solutions are included for comparison. The electrochemical measurements in all SDS systems were performed using ferrocene as probe to obtain the apparent diffusion coefficients. These values are compared to QELS

Table 3 Apparent Diffusion Coefficients (D_a) Measured in Microemulsions Using Different Probes

System	Diffusion coefficient (10^{-7} cm^2/s)			
	Ferrocene	PPD1	PPD2	$K_4(Fe(CN)_6)$
Aqueous buffer				
(5.46% K_2CO_3+3.15% $KHCO_3$)				
25°C (D_c)	—	—	36.9	—
40°C (D_c)	—	50.0	47.0	58.0
CTAB microemulsion	6.6	22.0	11.0	54.0
(3.00% CTAB, 3.22% n-butanol,				
2.20% n-octane, 95.58% aqueous buffer,				
40°C)				
C_{18}DMB microemulsion	2.3	37.0	11.0	—
(1.60% C_{18}DMB, 1.88% n-butanol,				
3.00% dodecane, 98.72%				
aqueous buffer 40°C)				
Triton X-100 microemulsions				
(8% Triton X-100, pH 10 buffer,				
dodecane, 25°C)				
0.5% dodecane	1.6 (1.0)[a,b]	—	6.2	—
1.0% dodecane	2.6 (1.5)[a,c]	—	6.7	—
1.5% dodecane	2.4 (1.4)[a,d]	—	6.4	—

[a] D_d values in parentheses calculated from Eq. (14).
[b] 97.6% ferrocene in droplet.
[c] 96.1% ferrocene in droplet.
[d] 96.4% ferrocene in droplet.
Source: Ref. 51.

values. Although the trends with changing composition are similar, the differences are significant and arise due to the effect of interactions. As mentioned previously, QELS measures mutual-diffusion coefficients, whereas electrochemical techniques provide self-diffusion coefficients. The diffusion coefficients obtained with methyl viologen (MV) as a probe [59] are also included in Table 2 to illustrate the role of probe partitioning. The electrochemistry of methyl viologen in microemulsions is discussed in the next section. The dictation MV^{2+} is primarily water-soluble and is not expected to partition significantly into the oil phase. The diffusion coefficients in cationic and nonionic microemulsions are high and of the same order of magnitude as in the aqueous phase as expected. However, the MV^{2+} diffusion coefficients in anionic micelles and microemulsions are much smaller than the aqueous diffusion coefficient. This behavior is caused by the partitioning of MV^{2+} due to electrostatic attraction between the anionic surfactant and the cationic MV^{2+}.

Diffusion coefficients measured in different microemulsions using various probes are depicted in Table 3 [51,52]. The apparent diffusion coefficients obtained with ferrocene are lowest in value, consistent with its partitioning into the droplets. The ferrocyanide, PPD1 [4-amino-3-methyl-N-ethyl-N-(β-sulfoethyl)aniline], and PPD2 [4-amino-3-methyl-N,N-diethylaniline] have similar molecular diffusivities in aqueous buffer, in the range of $(47-58) \times 10^{-6}$ cm^2/s. The cationic CTAB system clearly demonstrates the wide range of interactions shown by this group of probes. The diffusion coefficient of ferrocyanide anion is similar to the value in aqueous buffer, indicating limited interaction with the microemulsion droplet. The ionic strength of the buffer affects the range of interaction of the droplet double layer. The diffusion of anionic PPD1 is further reduced. However, it is present predominantly in the aqueous phase. In contrast, uncharged PPD2 is relatively hydrophobic [60] and partitions significantly into the oil droplet. The apparent diffusion coefficient D_a is therefore significantly larger than the actual droplet coefficient D_d. Ferrocene has a very low solubility in water and exhibits the lowest D_a. The solubility and the diffusion coefficient of ferrocene in 0.1 M NaCl have been reported to be 5×10^{-5} M and 6.7×10^{-6} cm^2/s, respectively [61]. The effect of probe type on the D_a values shows a similar pattern in the zwitterionic octadecyldimethylbetaine (C$_{18}$DMB) system. The electrostatic interactions of PPD1 with C$_{18}$DMB are less, leading to a larger value of D_a than in the CTAB microemulsion. The order of binding is ferrocyanide $\ll$ PPD1 $\ll$ PPD2 in the CTAB system, with the same binding order of PPD1 and PPD2 in the C$_{18}$DMB microemulsion.

Table 3 also shows data obtained in Triton X-100 nonionic microemulsions with low oil content (swollen micelles). Based on the analysis presented earlier in this section, corrections can be made to obtain a better estimate of the actual diffusion coefficient from knowledge of the relative concentration of the probe in the aqueous phase and the droplet. Differences between the measured (D_a) and corrected (D_d) values for various micellar solutions and microemulsions are discussed by Dayalan et al. [51]. Corrected values of diffusion coefficients obtained using Eq. (14) are shown in Table 3 for Triton X-100 microemulsions. The corrected diffusion coefficients can also be calculated by using the fast-kinetics approximation, such as Eq. (13). In general, the results from the two models are of the same magnitude and within experimental error [5,51].

Mackay et al. [62] also used a variety of probes with different solubility behavior to investigate diffusion coefficients in a CTAB–1-butanol–hexadecane–water microemulsion system. The probes used are ferricyanide, ferrocene, diferrocenylethane, acetyl ferrocene, and methyl viologen. The value of D_a for ferricyanide was found to be on the order of 10^{-6} cm^2/s for 19–60% water content, while D_a for ferrocene decreased from 4×10^{-6} to 0.6×10^{-6} cm^2/s when the water content was increased from 20% to 90%. The ferrocene data suggest a transition from a bicontinuous to a water-continuous microemulsion. The diffusion coefficients measured for diferrocenylethane indicated that it was strongly bound to the oil and apparently not accessible to the electrode. The D_a values for polar acetyl ferrocene and methyl viologen were found to be larger, indicating partitioning into the aqueous phase.

Recently, Mackay et al. [63] reported a similar study using different probes to investigate diffusion coefficients in an SDS–1-pentanol–dodecane–water microemulsion system with NaCl as electrolyte. Ferrocene, ferricyanide, and methyl viologen were used as probes to study the effects of different microstructures and interactions on the apparent diffusion coefficients. Diffusion coefficients for the reduced and oxidized forms of the probes were calculated from chronocoulometric results. The compositions examined correspond to a relatively low oil content (<10%) and a wide range of water (34–89%) SDS (3.3–20%), and 1-pentanol (6.7–40%) concentrations. The formal potentials of ferrocene and methyl viologen in the SDS microemulsions were observed to be significantly different from their potentials in aqueous solutions. This suggests that the electron exchange may not be fast [56,64]. However, considering the residence time of a probe in a microemulsion to be shorter than the time scale of chronocoulometry, Mackay et al. [63] used the fast-kinetics approach. Using Eq. (13), the authors concluded that the concentration dependence of D_a of ferrocene was significantly less in SDS microemulsions [63] than in CTAB microemulsions [62]. The diffusion coefficient measurements as a function of water content demonstrate the role of probe partitioning and microstructure of the microemulsion, as discussed previously.

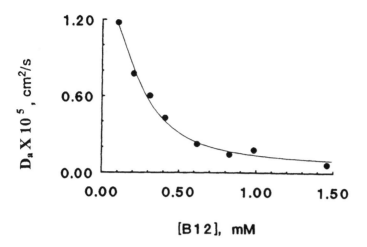

Figure 6 Influence of concentration of vitamin B_{12r} on apparent diffusion coefficient in microemulsion of 0.2 M AOT, 4 M water at pH 3, and isooctane. Circles are experimental data; line is the best fit by nonlinear regression onto Eq. (13) for $m=3$. (From Ref. 35.)

A similar microemulsion system containing SDS, 1-pentanol, dodecane, and water with NaBr as electrolyte was studied by Shah et al. [65]. Cyclic voltammetry and chronocoulometry were used to measure diffusion coefficients and half-wave potentials ($E_{1/2}$) of a series of amphiphilic cobalt complexes and ferrocene derivatives. The diffusion coefficients of the cobalt(III) complexes were essentially independent of microemulsion composition varying from 35 to 87% water. Such behavior is consistent with the association of the complexes with the anionic surfactant film. However, the smaller diffusion coefficients observed with the methyl- and phenyl-substituted complexes indicate some partitioning into the oil phase. In contrast to the cobalt(III) complexes, higher diffusion coefficients were observed with the ferrocene derivatives at low water content. The dependence of D_a of ferrocene derivatives on the oil content was related to variations in the hydrodynamic radii of oil droplets. The droplet size measurements are consistent with electrochemical [59] and light scattering [66] results in similar SDS microemulsions. The binding constants and partition coefficient for the ferrocene derivatives were also calculated [65]. In yet another recent study of the SDS–1-pentanol–dodecane–water–NaBr microemulsion, Santhanalakshmi and Vijayalakshmi [67] determined the diffusion coefficients using cobalticenium hexafluorophosphate as the electrochemical probe.

The diffusion studies described in the above sections pertain to water-continuous and bicontinuous microemulsions. Chen and Georges [34] were the first to study diffusion in oil-continuous microemulsions using steady-state microelectrode voltammetry. Ferrocene was used to probe diffusion in an SDS–dodecane–1-heptanol–water system. The diffusion coefficient of the hydrophobic probe indicated the microviscosity of the oil rather than the bulk viscosity of the microemulsion. Owlia et al. [36] reported diffusion coefficient measurements of water droplets in an Aerosol OT [AOT, bis(2-ethylhexyl)sulfosuccinate] microemulsion using a microelectrode. Water-soluble cobalt(II) corrin complex (vitamin B_{12r}) was used in an oil-continuous microemulsion containing 0.2 M AOT, 4 M water buffered at pH 3, and isooctane. The apparent diffusion coefficient decreased with the probe concentration in accordance with Eq. (13) as shown in Fig. 6 [36]. The water droplet size was estimated to be 7.5 nm, larger than the size of 5 nm observed in solute-free microemulsion [36]. Similar size changes in water droplets were caused by vitamin B_{12} in the dodecylammonium propionate–benzene–water system [68].

VII. MICROEMULSIONS AS MEDIA FOR ELECTROCHEMISTRY

A. Water-Continuous Microemulsions

The previous sections dealt with using electrochemistry as a tool to characterize microemulsions. The remainder of this chapter deals with examples of electrochemistry in microemulsions as tunable media. The electrochemistry of methyl viologen (MV) as a model system illustrates the usefulness of microemulsions as media for fundamental electrochemical studies [59]. This reaction involves a water-soluble reactant, a sparingly water-soluble intermediate, and a water-insoluble product. Microemulsions are very suitable for conducting electrochemical investigations of such systems since the three different types of species can be solubilized in the same medium. Methyl viologen has been widely investigated to understand the electron transfer mechanism between the various redox forms, namely, the dication MV^{2+}, the cation radical, $MV^{\cdot+}$, and the neutral form MV. An understanding of the electrochemistry of methyl viologen is important because of its frequent use as an electron acceptor in photochemical energy conversion and electrochemical display devices, as a mediator in electron transfer in biological studies, and in reducing electrochemically inactive compounds.

The electrochemical behavior of methyl viologen in aqueous media (Fig. 7) differs significantly from its behavior in microemulsions (Fig. 8). The cyclic voltammogram (CV) for 1 mM MV^{2+} in NaBr solution at 100 mV/s scan rate (Fig. 7a) shows two peaks for the anodic reaction corresponding to the second step of methyl viologen reduction. One peak is at -1.0 mV and the other at -0.875 mV. This implies that two forms of the product of the second reduction step (MV) are present, one easily oxidizable compared to the other. The peak positions change with scan rate. At 20 mV/s, only the peak at -0.875 V remains (Fig. 7b). However, at 500 mV/s (Fig. 7c) the peak at -1.0 V is predominant, while a very small shoulder is left at 0.875 V. The hydrophobic nature of MV leads to the adsorption peak. The reduction of MV^{2+} in three different types of microemulsions (Fig. 8) takes place in two reversible steps [60]. The complications observed in aqueous solutions due to the adsorption and rearrangement of MV on the electrode surface are eliminated in microemulsion media. Microemulsions provide a nonpolar environment for the preferential solubilization of the hydrophobic product, thereby preventing its adsorption on the electrode surface. The half-wave potentials for the reduction steps and the diffusion currents depend on the surfactant type and the composition of the microemulsion [59].

An extensive study of the electrochemistry of ferrocene, methyl viologen, and ferricyanide was conducted in SDS–dodecane–1-pentanol–0.1 M NaCl microemulsion [69]. The ferricyanide was confined to the aqueous phase, while the methyl viologen was partitioned. The shifts in the half-wave potentials were analyzed in terms of the model of Ohsawa et al. [70]. Mackay and coworkers [71–75] investigated microemulsions as media for the degradation of the nerve agent simulant p-nitrophenyldiphenyl phosphate (PNDP) using iodosobenzoate as catalyst. Luminescence and electrochemical studies were conducted in SDS, CTAB, and cetyltrimethylammonoum chloride (CTAC) microemulsions used

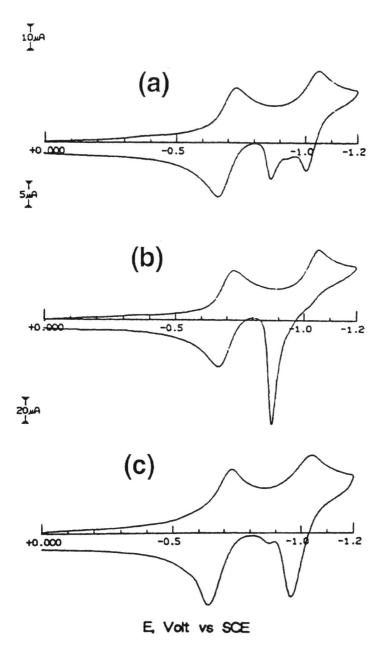

Figure 7 Cyclic voltammograms of 1 mM methyl viologen in aqueous 100 mM NaBr. Sweep rate (mV/s): (a) 100; (b) 20; (c) 500. (From Ref. 59.)

for the degradation kinetics of PNDP [76]. The authors concluded that there was no simple correlation between the catalyzed hydrolysis of phosphate esters and aggregation number, although the reactivity enhancement of the microemulsions was due to the reagent-concentrating effect of the interfacial area.

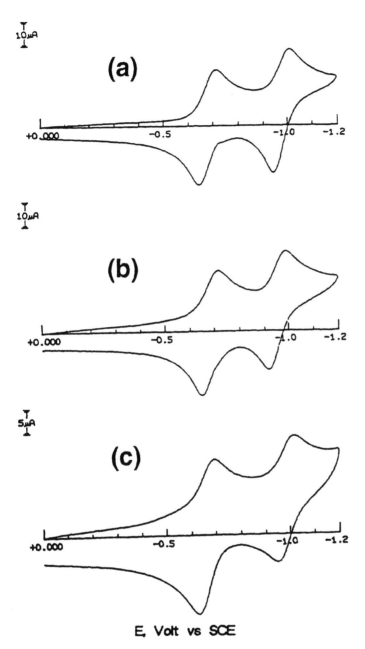

Figure 8 Cyclic voltammograms of 1 mM methyl viologen in microemulsions of (a) cationic CTAB, (b) nonionic Triton X-100, (c) anionic SDS. Sweep rate 100 mV/s. (From Ref. 59.)

B. Oil-Continuous Microemulsions

It is rather difficult to conduct electrochemistry in oil-continuous or W/O microemulsions because of the low conductivity, as stated earlier. Chen and Georges [34] used carbon fiber ultramicroelectrodes to study SDS–1-heptanol–dodecane microemulsions, using ferrocene as the probe. The limiting current exhibited steady-state behavior at low scan rates. Wang et al. [35] reported that adsorption of MV^{2+} on carbon microdisk electrodes led to incorrect diffusion coefficients in AOT–isooctane–water microemulsions. Owlia et al. [36] investigated the reduction of the central Co(II) atom of vitamin B_{12a} in AOT–isooctane–water microemulsions. The aqueous phase was buffered with phosphate at a nominal pH 3.0. The electrochemistry of vitamin B_{12} is strongly pH-dependent due to its basic benzimidazole side chain, which can act as an axial ligand. Protonation of this side chain (L) yields "base-off" forms that are reduced more rapidly at electrodes. The formal potentials and heterogeneous rate constants of the Co(III)/Co(II) and Co(II)/Co(I) reactions in the AOT microemulsions were compared with the values in buffered homogeneous solutions. The reactions were reported as quasi-reversible in both microemulsions and aqueous media [36].

Garcia et al. [77,78] reported an electron transfer percolation threshold in highly resistive oil-continuous microemulsions. The Faradaic electron transfer is modulated by the amount of cosurfactant present in AOT–toluene–water microemulsions. Below a certain threshold concentration of the cosurfactant, the electron transfer between electroactive solutes in the water droplets and ultramicroelectrode is retarded or blocked. Electron transfer becomes facilitated, and a sharp increase in Faradaic current is observed above the threshold concentration. This effect was demonstrated for ruthenium hexamine reduction [77,78], ferrocyanide oxidation [77,78], acrylamide oxidation [77], and alkylamide oxidation [77,79] with acrylamide, alkylamides, and acetonitrile as cosurfactants in AOT microemulsions. NMR results [80] suggest that there is an interfacial packing transition of the surfactant (AOT) at about the same cosurfactant concentration as the threshold transition observed electrochemically.

C. Bicontinuous Microemulsions

Iwunze et al. [81] examined the behavior of several redox couples in bicontinuous microemulsions containing DDAB, dodecane, and water. Hydrophilic ruthenium(III) hexamine, ferrocyanide, and vitamin B_{12} and hydrophobic ferrocene and polycyclic aromatic hydrocarbons (PAHs) were studied using cyclic voltammetry. The quasi-reversible reactions observed in the bicontinuous microemulsions could be simply described by voltammetric theory developed for homogeneous solutions [81]. Although the bulk microemulsion viscosity is high, the nonpolar molecules and ions diffused as if in oil or water, respectively. In fact, the apparent diffusion coefficients for ruthenium hexamine and ferrocyanide ions are decreased respectively to 0.2 and 0.14 times the value in pure water. Table 4 [81] lists the diffusion coefficients, standard heterogeneous rate constants (k^0), and formal potentials ($E^{0'}$) for the various probes. Unlike in water-continuous and oil-continuous microemulsions or micellar solutions, there was apparently no need to consider coupled diffusion and dynamic binding equilibria or partitioning in bicontinuous microemulsions. Georges and Berthod [26] studied the stability of electrochemically generated N-methylphenothiazine cation radical in sodium p-octylbenzenesulfonate– methylene chloride–1-pentanol microemulsion. They reported a drastic decrease in the half-life of the cation radical in the microemulsion compared to aqueous and SDS micellar solutions. This was due to the presence of alcohol.

The reactions of polyaromatic hydrocarbons (PAHs) in DDAB bicontinuous microemulsions followed the ECE (electrochemical–chemical–electrochemical) mechanism at low scan rates [81,82]. The three-step ECE process consists of electron transfer to the PAH to form an anion radical, protonation of the anion radical to yield a neutral radical,

Table 4 Apparent Electrochemical Parameters at Glassy Carbon Electrodes in Bicontinuous Microemulsions

Species	% DDAB[a]	$10^6 D$ (cm²/s)	k^0 (cm/s)	$E^{0'}$ (V vs. SCE)
Ferrocyanide	21	1.0	0.027	0.027
	13	0.79	0.017	0.022
Ru(II) hexamine	21	0.68	0.016	−0.213
Co(II) alamine	21	0.3	0.0002	−0.87
Ferrocene	21	6.3	0.010	0.34
	13	5.6	0.004	0.38
Perylene	21	5	0.2	−1.64
Pyrene	21	6	0.1	−2.06
9-PA	21	8	0.2	−1.95

[a] Compositions of microemulsions: 21% DDAB–40% dodecane–39% water and 13% DDBA–59% dodecane–28% water.
Source: Ref. 81.

and a second electron transfer to the neutral radical. The mechanism is similar to that observed in an organic solvent containing proton donors. The electrochemistry of perylene was investigated in DDAB bicontinuous microemulsions containing cyclohexane [82]. Cyclic voltammetry in microemulsions containing 19–39% water showed transitions from a two-electron process at low scan rates to a one-electron process at higher scan rates. A second-order rate constant of 3.7 ± 0.7 M^{-1}/s was observed for reaction of perylene anion radical with water [82]. The stability of the anion radical improved with decreasing water content.

The electrochemical oxidation of ferrocene and amphiphilic ferrocenes, 2- and 5-(ferrocenylcarboxy)dodecyltrimethylammonium nitrates (2-Fc and 5-Fc, respectively), were investigated in a bicontinuous microemulsion containing CTAC, *n*-tetradecane, pentanol, and water [83]. The electron transfer rates for ferrocene, 2-Fc, and 5-Fc in microemulsions were an order of magnitude slower than in acetonitrile, possibly due to partial inhibition by microemulsion components adsorbed on the electrode [5,6]. The rates of electron transfer of 2-Fc and 5-Fc were only two-fold smaller than that of ferrocene, in contrast to 10-fold (2-Fc) and 100-fold (5-Fc) smaller in CTAB micelles [84]. These results indicate an increased disorder and mobility at the electrode/fluid interface in the CTAC microemulsion compared to CTAB micellar system [83].

D. Electrocatalysis in Microemulsions

Electrochemical catalysis constitutes a general synthesis route that is amenable to rate control and enhancement in microemulsions [5,6]. Owlia et al. [36] were the first to investigate electrochemical catalysis in oil-continuous microemulsions. The kinetics of reduction of several alkyl vicinal dibromides was studied in the presence of vitamin B_{12} as catalyst. The following reaction mechanism was suggested:

$$Co(II) + e \rightleftharpoons Co(I) \quad \text{(at electrode)} \tag{15a}$$

$$Co(I) + RX_2 \xrightarrow{k_1} RX^{\cdot} + X^- + Co(II) \tag{15b}$$

$$RX^{\cdot} + Co(I) \rightarrow \text{alkene} + X^- + CO(II) \tag{15c}$$

Table 5 Apparent Rate Constants[a] for Catalytic Reduction of Alkyl Vicinal Dibromides by Vitamin B_{12}

	0.2 M AOT–4 M water–isooctane[b]		pH 2.3 acetonitrile–water[c]	
Substrate	$10^{-3}k_{obs}$ $(M^{-1} s^{-1})$	Rel k	$10^{-6}k_{obs}$ $(M^{-1} s^{-1})$	Rel k
DBB	0.31±0.19	1	1.3±0.2	1
EDB	1.34±0.17	4.3	2.6±0.5	2
t-DBCH	6.2+1.4	20	5.1±1.3	3.9

[a] Data obtained at 10 mV/s at carbon microdisk (r=6 μm) electrodes.
[b] 17 mM phosphate (pH 3) in water pools; total concentrations: 0.5 mM vitamin B_{12}, 2 mM substrate.
[c] 17 mM phosphate buffer, 0.5 mM vitamin B_{12}, 5 mM substrate. DBB=1,2-dibromobutate; EDB-ethylene dibromide; 1,2-DBCH-*trans*-1,2-dibromocyclohexane.
Source: Ref. 36.

The rate-determining step was identified as an inner-sphere electron transfer between $B_{12}Co(I)$ and the alkyl dibromide, Eq. (15b). This electron transfer may alternatively be due to a concerted E_2 mechanism. Ultramicroelectrode voltammetry was used to determine the kinetics in AOT–isooctane–water microemulsions. The polar vitamin B_{12} resides entirely in the water pools, whereas the substrates are present primarily in the continuous oil phase. The rate constant k_1 was determined from the increase of the limiting current caused by the addition of the substrates [ethylene dibromide (EDB), 1,2-dibromobutane (DBB), and *trans*-1,2-dibromocyclohexane (*t*-DBCH)] to the AOT microemulsion. As shown in Table 5 [36], the apparent rate constants were three orders of magnitude lower in the microemulsion than in the acetonitrile–water system. The rate decrease was attributed to the spatial separation of the catalyst and the substrate [36].

The ratio of the rate constants for DBB : EDB : *t*-DBCH was 1 : 4 : 20 in the microemulsion and 1 : 2 : 4 in the acetonitrile–water system. The higher reactivity of *t*-DBCH in the microemulsion may be due to specific interactions between the substrate and the surfactant. This AOT microemulsion system with vitamin B_{12} was employed to evaluate EDB content in leaded and unleaded gasoline [85].

Bicontinuous DDAB microemulsions were found to increase the rate of bulk electrolysis of 4,4'-dichlorobiphenyl (4,4'-DCB) by an order of magnitude compared to CTAB micellar solution [86,87]. The dechlorination of 4,4'-DCB was conducted in DDAB microemulsions containing 21% surfactant, 22% dodecane, and 57% water using zinc phthalocyanine as catalyst [88]. A lead cathode was found to be more efficient than carbon felt. DDAB–dodecane–water (21 : 39 : 40) bicontinuous microemulsions were also used to study the dehalogenation of alkyl vicinal dibromides (DBB and *t*-DBCH) and trichloroacetic acid (TCA) [89]. Water-soluble nickel(II) phthalocyanine tetrasulfonate and copper(II) phthalocyanine tetrasulfonate were used as catalysts. The reactions resemble Eqs. (15a)–(15c), producing alkenes from the vicinal dibromides and acetic acid from TCA. Cyclic and square-wave voltammetric experiments with glassy carbon electrodes revealed much higher efficiencies for DBB and *t*-DBCH in the microemulsions than in acetonitrile–water solutions. The rate enhancement was attributed to the coadsorption of catalyst, surfactant, and hydrophobic substrate (DBB and *t*-DBCH) as in micellar solutions [86]. In contrast, the catalytic efficiency for water-soluble TCA was lower in the microemulsion than in acetonitrile–water solution [89]. In a follow-up study, Kamau and Rusling [90] compared the apparent pseudo-first-order rate constants for the rate-determining step involving reduction of t-DBCH, DBB, and TCA in bicontinuous DDAB microemulsion and acetonitrile–water solution. Water-insoluble copper and nickel phthalocyanines were used as catalysts in addition to their tetrasulfonates. The higher rates for t-DBCH and DBB in microemulsions than in homogeneous solvent were explained in terms of their preconcentration into hydrophobic regions of an adsorbed film on the surfactant-catalyst-coated electrode [90].

Zhou et al. [91] studied the catalytic reduction of $trans$-1,2-dibromocyclohexane (t-DBCH) and nucleophilic substitution (S_N2) reactions of n-alkyl bromides with electrochemically generated Co(I) complexes. DDAB–dodecane–water bicontinuous microemulsions and N,N-dimethylformamide (DMF) were used as media. Macrocyclic complexes vitamin B_{12} and Co(salen) partitioned into the water domains, and the alkyl bromides were present in the oil domains of the microemulsion. Rates of the bimolecular reactions were similar in bicontinuous microemulsions and homogeneous solutions. The reduction of t-DBCH was 40 times faster in bicontinuous medium than in oil-continuous microemulsions, apparently due to larger interfacial area [91]. The bicontinuous microemulsion behaves as a homogeneous solvent with a proton donor, in agreement with previous reports [81,82]. For the alkyl bromides, the activation free energies determining the rate constants were controlled primarily by the formal potential of the Co(II)/Co(I) redox couple rather than by distribution of reactants between phases. In general, the formal potential depends on specific interactions between the microemulsion components, as reported by various investigators.

Zhou et al. [92] also examined the catalytic reduction of benzyl bromide with macrocyclic cobalt complexes in a bicontinuous DDAB microemulsion and homogeneous solvent. The reaction sequence involves electrochemical generation of a Co(I) complex, Co^IL, followed by oxidative addition of benzyl bromide to give benzyl-$Co^{III}L$. Reductive cleavage of the Co–C bond yields benzyl radicals or anions depending on the potential of reduction of benzyl-$Co^{III}L$. The rate of the initial coupling of Co^IL with benzyl bromide depends largely on the activation free energy governed by the standard potential of the Co(II)/Co(I) couple. The one-electron catalyst, vitamin B_{12}, has a relatively positive redox potential and a smaller rate of initial coupling than the two-electron catalyst, Co(salen). Vitamin B_{12} facilitated a fast radical pathway yielding bibenzyl, whereas Co(salen) facilitated a less efficient anionic pathway producing toluene. Thus, a radical or anionic pathway for benzyl reduction may be achieved by controlling the redox potential of the catalyst and the reactant–catalyst adduct.

Recently, Gao et al. [93] reported carbon–carbon bond formation by electrochemical catalysis in conductive microemulsions. Cobalt complexes react with alkyl halides to give alkyl cobalt complexes as mentioned above. The carbon–cobalt bonds can be cleaved by electrolysis, visible light, or reducing agents to yield carbon-centered radicals that can react in situ with activated alkenes to form carbon–carbon bonds. Conjugated additions of primary alkyl iodides to 2-cyclohexen-1-one to produce 3-alkyl cyclohexanones and cyclization of 2-(4-bromobutyl)-2-cyclohexen-1-one to 1-decalone were compared in bicontinuous CTAB and SDS microemulsions and a homogeneous solvent, DMF. Comparable and high yields were obtained in the two microemulsions as well as in DMF. However, excellent stereoselectivity for the trans isomer of 1-decalone was observed only in CTAB microemulsion. SDS microemulsion and DMF did not exhibit much stereoselectivity [93]. The results suggest a possible role of electrostatic interactions, as was also observed with CTAB in the case of methyl viologen [59].

More recently, Gao and Rusling [94] reported another example of electrochemical catalysis and reaction control in microemulsion media, a 5-*endo*-trig cyclization. Such cyclization is not favored, requiring large distortions of bond angles and lengths to obtain the necessary reaction geometry. A significant increase in the yield of unfavorable 5-*endo*-trig cyclization product 4-hydrindanone was achieved in SDS and CTAB microemulsions [94], illustrating the potential of microemulsions for electrochemical synthesis of fused five-membered rings under mild conditions.

An attempt was made to model the interfacial chemistry in microemulsions by using the interface between two immiscible electrolyte solutions [95]. The reaction between the electrochemically generated Co(I) form of vitamin B_{12} in the aqueous phase and *t*-DBCH in benzonitrile was probed directly at the interface by using scanning electrochemical microscopy. The kinetics of *t*-DBCH reduction by vitamin B_{12} was observed to be more complex at the liquid/liquid interface than in a homogeneous solution [95].

The investigations of electrochemical catalysis in microemulsions by Rusling and coworkers as summarized above used catalysts that are solubilized in the medium or adsorbed on the electrode surface. The adsorbed films are relatively unstable. Onuoha and Rusling [96] investigated films cast from DDAB microemulsions containing the protein myoglobin (Mb) onto pyrolytic graphite electrodes. Direct electron transfer was observed between the electrode and the Fe(III)/Fe(II) redox couple of Mb with these films when they were used in the same microemulsions. The Mb-DDAB films were less than a micrometer thick and were usable for nearly a week in an unstirred DDAB–water–dodecane (13 : 28 : 59) microemulsion. However, the protein in the film may be partly denatured. The Mb-DDAB films were used to mediate redox reactions of polar (trichloroacetic acid) and nonpolar (oxygen) solutes in DDAB microemulsions [96].

Onuoha et al. [97] reported that ferrylmyoglobin species, which are active oxidant forms of myoglobin, can be prepared by electrochemical reduction of metmyoglobin in the presence of oxygen in aqueous neutral buffer and in microemulsions. Mb facilitates the reduction of oxygen to hydrogen peroxide, which oxidizes nearly all of the Mb in solution to ferrylmyoglobin. Oxidation reactions with ferrylmyoglobin were demonstrated [97]. Fiftyfold higher yields of styrene oxide and benzaldehyde were obtained in a microemulsion compared to electrochemical or chemical oxidation of styrene in pH 7.4 buffer.

E.   Electrochemistry and Microemulsion Polymerization

Garcia et al. [98] conducted coulometric initiation of acrylamide polymerization in oil-continuous AOT–toluene–water microemulsions using platinum/Nafion solid polymer electrodes (SPEs). The SPE served to separate the microemulsion from an aqueous electrolyte phase. Polymerization was initiated at room temperature by constant-potential electrolytic reduction of potassium persulfate initiator solubilized in the microemulsion droplets. Acrylamide monomer behaved as a cosurfactant and was required for the redox process. Latex particles and solid polyacrylamide were obtained. The kinetics of electroinitiated polymerization was slower than observed with UV or thermal initiation. Latex stability results suggest that coalescence is the primary mechanism for particle growth.

Phani et al. [99] reported the electrodeposition of polyparaphenylene (PPP) films on an indium-tin oxide glass and platinum electrode from a water-continuous SDS microemulsion containing benzene and concentrated sulfuric acid. It is difficult to obtain homogeneous coherent films of high molecular weight PPP via electrodeposition. The PPP films deposited from microemulsion were smooth and showed good redox activity in sulfuric acid solutions. The film morphology and crystallinity were found to depend on the cycling potential, surfactant, and film thickness [100].

Novel conductive composite films have been developed by Kaplin and Qutubuddin [101] using a two-step process: microemulsion polymerization to form a porous conductive coating on an electrode followed by electropolymerization of an electroactive monomer such as pyrrole. The porous matrix was prepared by polymerizing an SDS microemulsion containing two monomers, acrylamide and styrene [102]. The electropolymerization of pyrrole was performed in an aqueous perchlorate or toluenesulfonate solution. The effects of polymerization potential on the electropolymerization, morphology, and electrochemical properties were reported [101]. The copolymer matrix improves the mechanical behavior of the polypyrrole composite film.

VIII. SUMMARY

Electrochemical techniques are particularly useful for studying diffusion coefficients in microemulsions. Along with the microemulsion structure, the probe partitioning equilibria and kinetics are critically important in interpreting the electrochemical measurements of apparent self-diffusion coefficients. The partitioning of the electroactive solutes into the organic domains depends on hydrophobic interactions as well as electrostatic interactions for ionic surfactants. Explicit accounting of partitioning is necessary in interpreting the measured apparent diffusion coefficient when the probe is soluble in both the continuous and discontinuous phases. However, when the probe is predominantly (more than 95%) present in the dispersed phase, both slow- and fast-kinetics limiting expressions yield diffusion coefficients of comparable magnitude. In conjunction with light scattering, NMR, and other measurements, electrochemical experiments provide qualitative insight into interactions such as electrostatic and hydrophobic effects. Further research is warranted in order to quantify these interparticle and intraparticle interactions.

Microemulsions are attractive as tunable media for fundamental electrochemical studies such as the investigation of electron transfer mechanisms as well as potential practical applications such as electrosynthesis and electrocatalysis. Microemulsions provide both hydrophilic and hydrophobic domains for solubilization of dissimilar reactants, intermediates, or products. The interfacial area per unit volume is enormous, allowing very efficient mass transfer. The literature on microemulsions for the use of electrochemistry is expanding rapidly. Various electrochemical reactions have been investigated in water-continuous, oil-continuous, and bicontinuous microemulsions. Systems studied include methyl viologen, ferrocene and its derivatives, ferricyanide and ferrocyanide, nerve agent simulant, vitamin B_{12}, polyaromatic hydrocarbons, and pyridinium compounds. Encouraging results have been reported on the electrocatalytic reduction of alkyl vicinal dihalides by vitamin B_{12} and that of alkyl halides by metal phthalocyanines and cobalt complexes. Rate enhancements, reaction control, and stereoselectivity have been achieved in microemulsion media. Microemulsions have been used for electropolymerization also. The future of microemulsions as media for electrochemistry is promising. However, numerous challenges exist in understanding electrochemistry in microemulsions. For instance, questions exist about the structure and dynamics of adsorbed species on the electrode surface as well as on the mechanism and kinetics of electron transfer across the surfactant film. Thus, there are abundant opportunities for further fundamental research and innovative applications involving electrochemistry in microemulsions.

REFERENCES

1. G. E. O. Proske, Anal. Chem. *24*:1834 (1952).
2. L. Meites, *Polarographic Techniques*, Interscience, New York, 1955.
3. R. A. Mackay and J. Texter (eds.), *Electrochemistry in Colloids and Dispersions*, VCH, New York, 1992.
4. R. A. Mackay, Colloids Surf. A: *82*:1 (1994).
5. J. F. Rusling, in Electroanalytical Chemistry, Vol. 18 (A. J. Bard, ed.), Marcel Dekker, New York, 1994, pp. 1–88.
6. J. F. Rusling, Electrochemistry and electrochemical catalysis in microemulsions, in *Modern Aspects of Electrochemistry*, No. 26 (B. E. Conway and J. O'M. Bockris, eds.), Plenum, New York, 1994, pp. 49–104.
7. J. P. Hoar and J. H. Schulman, Nature *152*:102 (1943).
8. J. H. Schulman and E. G. Cockbain, Trans. Faraday Soc. **36**:651 (1940).
9. S. E. Friberg, J. Dispersion Sci. Technol. *6*:317 (1985).
10. S. Qutubuddin, Microemulsions, in *Encyclopedia of Chemical Processing and Design*, Vol. 30, Marcel Dekker, New York, 1989, p. 149.
11. L. M. Prince, *Microemulsions*, Academic, New York, 1977.
12. D. O. Shah (ed), *Surface Phenomena in Enhanced Oil Recovery*, Plenum, New York, 1984.
13. C. A. Miller and S. Qutubuddin, in *Interfacial Phenomena in Non-Aqueous Media* (H. F. Eicke and G. D. Parfitt, eds.), Marcel Dekker, New York, 1987, pp. 117–185.
14. P. A. Winsor, *Solvent Properties of Amphiphilic Compounds*, Butterworths, London, 1954.
15. R. L. Reed and R. M. Healy, in *Improved Oil Recovery by Surfactant and Polymer Flooding* (D. O. Shah and R. S. Schecter, eds.), Academic, New York, 1997, p. 383.
16. K. Shinoda and S. Friberg, Adv. Colloid Interface Sci. *91*:223 (1983).
17. L. E. Scriven, Nature *263*:123 (1976).
18. S. Qutubuddin, C. A. Miller, W. J. Benton, and T. Fort, Jr., in *Macro- and Microemulsions: Theory and Applications* (D. O. Shah, ed.) ACS Symp. Ser. Vol. *272*, American Chemical Society, Washington, DC, 1985, pp. 223–251.
19. S. Qutubuddin, C. A. Miller, and T. Fort, Jr., J. Colloid Interface Sci. *101*:46 (1984).
20. J. H. Fendler and E. J. Fendler, *Catalysis in Micellar and Macromolecular Systems*, Academic, New York, 1975.
21. J. H. Fendler, *Membrane Mimetic Chemistry*, Wiley New York, 1982.
22. M. Zulauf and H. F. Eicke, J. Phys. Chem. *84*:1503 (1980).
23. P. E. Luisi and B. E. Straub (eds.), *Reverse Micelles*, Plenum, New York, 1984.
24. K. Chokshi, S. Qutubuddin, and A. Hussam, J. Colloid Interface Sci. *129*:315 (1989).
25. R. A. Mackay, N. S. Dixit, R. Agarwal, and R. P. Seiders, J. Dispersion Sci. Technol. *4*:397 (1983).
26. J. Georges and A. Berthod, Electrochim. Acta *28*:735 (1983).
27. R. Zana and R. A. Mackay, Langmuir *2*:109 (1986).
28. A. J. Bard and L. R. Faulkner (eds.), *Electrochemical Methods: Fundamentals and Applications*, Wiley, New York, 1980.
29. R. S. Nicholson and I. Shain, Anal. Chem. *36*:706 (1964).
30. R. S. Nicholson and I. Shain, Anal. Chem. *37*:178 (1965).
31. R. S. Nicholson and I Shain, Anal. Chem. *37*:190 (1965).
32. R. H. Wopshall and I. Shain, Anal. Chem. *39*:1514 (1967).
33. R. M. Wightman and D. O. Wipf in *Electroanalytical Chemistry* Vol. 16 (A. J. Bard, ed.), Marcel Dekker, New York, (1988), p. 267.
34. J.-W. Chen and J. Georges, J. Electroanal. Chem. 210:205 (1986).
35. Z. Wang, A. Owlia, and J. F. Rusling, J. Electroanal. Chem. 270:407 (1989).
36. A. Owlia, Z. Wang, and J. F. Rusling, J. Am. Chem. Soc. 111:5091 (1989).
37. J. F. Rusling, Z. Wang, and A. Owlia, Colloids Surf. *54*:1 (1990).
38. R. A. Mackay, C. Hermansky, and A. Agarwal, in *Colloids and Interface Science*, Vol. II (M. Kerker, ed.), Academic, New York, 1976, pp. 289–303.

39. R. A. Mackay and R. Agarwal, J. Colloid Interface Sci. 65:225 (1978).
40. R. A. Mackay, Adv. Colloid Interface Sci. 15:131 (1981).
41. R. A. Mackay, in *Microemulsions* (I. D. Robb ed.), Plenum, New York, 1982, pp. 207–219.
42. T. Hanai, Kolloid Z. 171:23 (1960).
43. K. R. Foster, B. R. Epstein, P. C. Jenin, and R. A. Mackay, J. Colloid Interface Sci. 68:233 (1982).
44. B. R. Epstein, K. R. Foster, and R. A. Mackay, J. Colloid Interface Sci. 95:218 (1983).
45. K. R. Foster, E. Cheever, J. B. Leonard, F. D. Blum, and R. A. Mackay, in *Macro and Microemulsions: Theory and Applications* (D. O. Shah, ed.), ACS Symp. Ser. Vol. 272. Am. Chem. Soc., Washington, DC, 1985, pp. 275–286.
46. R. A. Mackay, N. S. Dixit, C. Hermansky, and A. S. Kertes, Colloids Surf. 21:27 (1986).
47. R. A. Mackay, N. S. Dixit, and R. Agarwal, ACS Symp. Ser. 177:179 (1982).
48. A. Berthod and J. Georges, J. Colloid Interface Sci. 106:194 (1985).
49. J. Georges and J.-W. Chen, Colloid Polym. Sci. 264:896 (1986).
50. J. Georges, J.-W. Chen, and N. Arnaud, Colloid Polym. Sci. 265:45 (1987).
51. E. Dayalan, S. Qutubuddin, and J. Texter, in *Electrochemistry in Colloids and Dispersions* (R. A. Mackay and J. Texter, eds.), VCH, New York, 1992, pp. 119–135.
52. P. Guering and B. Lindman, Langmuir 1:464 (1985).
53. M. T. Clarkson, D. Beaglehole, and P. T. Callaghan, Phys. Rev. Lett. 54:1722 (1985).
54. E. Dayalan, S. Qutubuddin, and J. Texter, J. Colloid Interface Sci. 143:423 (1991).
55. J. Texter, J. Electroanal. Chem. 304:257 (1991).
56. D. M. Evans, J. Electroanal. Chem. 258:451 (1989).
57. J. Texter, F. R. Horch, S. Qutubuddin, and E. Dayalan, J. Colloid Interface Sci. 135:263 (1990).
58. J. F. Rusling, C.-N. Shi, and T. F. Kumosinski, Anal Chem. 60:1260 (1988).
59. E. Dayalan, S. Qutubuddin, and A. Hussam, Langmuir 6:715 (1990).
60. J. Texter, T. Beverly, S. R. Templar, and T. Matsubara, J. Colloid Interface Sci. 120:389 (1987).
61. J. Georges and S. Desmettre, Electrochim. Acta 29:521 (1984).
62. R. A. Mackay, S. A. Myers, L. Bobalbhai, and A. Brajter-Toth, Anal. Chem. 62:1084 (1990).
63. R. A. Mackay, S. A. Myers, L. Bobalbhai, and A. Brajter-Toth, Electroanalysis 8:759 (1996).
64. F. C. Anson, J.-M. Saveant, and K. Shigehara, J. Am. Chem. Soc. 105:1096 (1983).
65. D. M. Shah, K. M. Davies, and A. Hussam, Langmuir 13:4729 (1997).
66. H. M. Cheung, S. Qutubuddin, R. A. Edwards, and J. A. Mann, Jr., Langmuir, 3:744 (1987).
67. J. Santhanalakshmi and G. Vijayalakshmi, Langmuir 13:3915 (1997).
68. J. H. Fendler, F. Nome, and H. C. van Woert, J. Am. Chem. Soc. 96:6745 (1976).
69. R. A. Mackay, S. A. Myers, L. Bodalbhai, and A. Brajter-Toth, in *Electrochemistry in Colloids and Dispersions* (R. A. Mackay and J. Texter, eds.), VCH, New York 1992, p. 163.
70. Y. Ohsawa, Y. Shimajaki, and S. Aoyagui, J. Electroanal. Chem. 114:235 (1980).
71. R. A. Mackay, F. R. Longo, B. L. Knier, and H. D. Durst, J. Phys. Chem. 91:861 (1987).
72. B. L. Knier, H. D. Durst, B. A. Burnside, R. A. Mackay, and F. R. Longo, J. Solution Chem. 17:77 (1988).
73. B. A. Burnside, L. L. Szafraniec, B. L. Knier, H. D. Durst, R. A. Mackay, and F. R. Longo, J. Org. Chem. 53:2009 (1988).
74. B. A. Burnside, B. L. Knier, R. A. Mackay, H. D. Durst, and F. R. Longo, J. Phys. Chem. 92:4505 (1988).
75. R. A. Mackay, B. A. Burnside, S. M. Garlick, B. L. Knier, H. D. Hurst, P. M. Nolan, and F. R. Longo, J. Dispersion Sci. Technol 9:493 (1989).
76. P. L. Cannon, Jr., S. L. Garlick, S. D. Christen, N. M. Wong, A. C. Novelli, F. R. Longo, and R. A. Mackay, in *Electrochemistry in Colloids and Dispersions* (R. A. Mackay and J. Texter, eds.), VCH, New York, 1992, p. 147.
77. E. Garcia and J. Texter, Proc. Electrochem. Soc. 93:2166 (1993).
78. E. Garcia, S. Song, L. E. Oppenheimer, B. Analek, A. J. Williams, and J. Texter, Langmuir 9:2782 (1993).
79. E. Garcia and J. Texter, J. Colloid Interface Sci. 162:262 (1994).
80. B. Antalek, A. J. Williams, E. Garcia, and J. Texter, Langmuir 10:4459 (1994).

81. M. O. Iwunze, A. Sucheta, and J. F. Rusling, Anal. Chem. *62*:644 (1990).

82. M. O. Iwunze and J. F. Rusling, J. Electroanal. Chem. *303*:267 (1991).

83. G. Gounili, J. M. Bobbitt, and J. F. Rusling, Langmuir *11*:2800 (1995).

84. A. P. Abbott, G. Gounili, J. M. Bobbitt, J. F. Rusling, and T. F. Kumosinski, J. Phys. Chem. *96*:11091 (1992).

85. J. F. Rusling, T. F. Connors, and A. Owlia, Anal. Chem. *59*:2123 (1987).

86. J. F. Rusling, Acc. Chem. Res. *24*:75 (1991).

87. M. O. Iwunze and J. F. Rusling, J. Electroanal. Chem *266*:197 (1989).

88. E. C. Couture, J. F. Rusling, and S. Zhang, Inst. Chem. Eng. Symp. Ser. (U.K.) *127*:177 (1992).

89. G. N. Kamau, N. Hu, and J. F. Rusling, Langmuir *8*:1042 (1992).

90. G. N. Kamau and J. F. Rusling, Langmuir *12*:2645 (1996).

91. D.-L. Zhou, J. Gao, and J. F. Rusling, J. Am Chem. Soc. *117*:1127 (1995).

92. D.-L. Zhou, H. Carrero, and J. F. Rusling, Langmuir *12*:3067 (1996).

93. J. Gao, J. F. Rusling, and D.-L. Zhou, J. Org. Chem. *61*:5972 (1996).

94. J. Gao and J. F. Rusling, J. Org. Chem. *63*:218 (1998).

95. Y. Shao, M. V. Mirkin, and J. F. Rusling, J. Phys. Chem. B. *101*:3202 (1997).

96. A. C. Onuoha and J. F. Rusling, Langmuir *11*:3296 (1995).

97. A. C. Onuoha, X. Zu, and J. F. Rusling, J. Am. Chem. Soc. *119*:3979 (1997).

98. E. Garcia, L. E. Oppenheimer, and J. Texter, in *Electrochemistry in Colloids and Disperisions*, (R. A. Mackay and J. texter, eds.), VCH, New York, 1992, p. 257.

99. K. L. N. Phani, S. Pitchumani, S. Ravichandran, S. T. Selvan, and S. Bharathey, J. Chem. Soc., Chem. Commun. 1993:179.

100. K. L. N. Phani, S. T. Selvan, A. Mani, and S. Pitchumani, J. Electroanal. Chem *384*:183 (1995).

101. D. A. Kaplin and S. Qutubuddin, Synth. Met. *63*:187 (1994).

102. S. Qutubuddin, C. S. Lin, and Y. Tajuddin, Polymer *35*:4606 (1994).

22

Polymerization in Microemulsions

Françoise Candau
Institut Charles Sadron (CRM-EAHP), CNRS-ULP, Strasbourg, France

I. INTRODUCTION

Free radical polymerization in colloidal dispersions such as emulsions has become a standard technique for the production of industrial polymers in the form of polymer colloids or latexes that are the bases for paints, adhesives, polishes, and other coatings. Isolation of single free radicals in loci of small dimensions affords a means of attaining simultaneously high molecular weights and high reaction rates. Emulsion polymerization has been extensively studied over the past 50 years, and the kinetics of the reaction and the mechanistic aspects of the process are rather well understood (see, e.g., Gilbert [1]). In contrast, the concept of polymerization in microemulsions appeared only around the 1980s [2], likely as a consequence of the numerous studies performed on these systems after the 1974 oil crisis. Since then, the field has developed rapidly, as attested to by the increasing number of papers devoted to microemulsion polymerization. The following interesting features of microemulsions result in a unique microenvironment that can be harnessed to produce novel materials with interesting morphologies or polymers with specific properties [3–6]:

1. Large overall interfacial area (about 100 m^2/mL)
2. Optical transparency and thermodynamic stability
3. Small size of the domains ($\sim 10^{-2}$ μm)
4. Great variety of structures

In this paper I review the salient features of polymerization in microemulsions at the present state of knowledge. I discuss the formulation of polymerizable microemulsions and show how the incorporation of monomers can modify the initial structure of the systems. The kinetic and mechanistic aspects are given and compared to those experienced in conventional emulsion polymerization. I also describe some recent results obtained on the formation of porous solid materials and functionalized microlatex particles, which seem quite promising for future applications.

II. STRUCTURE AND PHASE DIAGRAMS OF MONOMER-CONTAINING MICROEMULSIONS

The structure and phase diagrams of systems containing oil, water, and surfactant (and possibly a cosurfactant) have been described in detail throughout this book. Here, we consider essentially the monophasic domains of the phase diagram that are referred to as microemulsions. Let us recall that microemulsions are thermodynamically stable and isotropic dispersions containing oil and water, the stability being ensured by a very low interfacial tension capable of compensating for the dispersion entropy, which is very large owing to the small droplet size. In most cases it is necessary to associate a cosurfactant (such as an alcohol) with the surfactant in order to achieve the low interfacial tension required. The incorporation of a monomer in a microemulsion by interchanging either the water with an aqueous monomer solution or the oil with a hydrophobic monomer may result in large changes in the phase diagram due to the possible cosurfactant role of the monomer and/or to the change in solubility of the surfactant in oil or water. This is why it is important to study, prior to undertaking polymerization studies, the phase diagram of the systems in the presence of monomer and to perform a search for the most adequate formulation.

Before examining this point, I present a brief description of microemulsion structures.

A. General Features of the Phase Diagrams

One of the most common structures encountered in microemulsions consists of water or oil droplets dispersed in a continuous phase of oil or water, respectively. The type of dispersion results from the preferred curvature C_0 of the surfactant layer, which is by convention positive for oil-in-water (O/W) systems and negative for water-in-oil (W/O) systems. C_0 can be varied by adjusting the surfactant/cosurfactant ratio, which allows swelling of the droplets until a maximum is reached. When the systems become more concentrated, the micellar swelling is mostly limited by attractive interparticle interactions, as observed, for example, for microemulsions close to a critical point.

By varying several parameters such as the W/O ratio, one can induce an inversion from an O/W to a W/O microemulsion and vice versa. The type of structure in the inversion domain depends essentially on the bending constant K_e, a characteristic of the elasticity of the surfactant layer [7]. If K_e is on the order of kT (where k is the Boltzmann constant and T absolute temperature), the persistence length of the film ξ_k (i.e., the distance over which the film is locally flat) is microscopically small. The interfacial film is flexible and is easily deformed under thermal fluctuations. The phase inversion occurs through a bicontinuous structure formed of water and oil domains randomly interconnected [8,9]. The system is characterized by an average curvature around zero, and the solubilization capacity is maximum. When $K_e \gg kT$, ξ_k is large and the layers are flat over macroscopic distances. The transition occurs through a lamellar phase.

Microemulsions can also coexist in equilibrium with various phases, the most widely studied being the so-called Winsor phase equilibria: Winsor I is a globular O/W microemulsion in equilibrium with excess oil, Winsor II a globular W/O microemulsion in equilibrium with excess water, and Winsor III a middle-phase bicontinuous microemulsion in equilibrium with both oil and water phases [10].

The determination of the different domains in the phase diagram is a tedious and time-consuming task. In particular, bicontinuous and globular microemulsions are both transparent, isotropic, and of low viscosity and therefore are not distinguishable. Clear evidence of the bicontinuous structure can be provided by transmission electron microscopy (TEM) [11].

Several studies have attempted to correlate the characteristics of the final products to the initial structures prior to polymerization. It must be reminded that the accurate determination of a microemulsion structure is rather difficult. In particular, when performing scattering experiments, which in principle provide the droplet size, the system must be diluted. However, the dilution procedure is not trivial because of the partitioning of the components of the microemulsion between continuous and dispersed phases. Experiments performed at finite concentration can suffer by a large error, in particular in the vicinity of a critical point where the radiation scattering probes critical fluctuations with a characteristic length much larger than the droplet radius [4].

Another important characteristic of microemulsions is the transience of the aggregates. Surfactant molecules and other constituents are constantly exchanging. The droplets collide, forming aggregates with lifetimes on the order of microseconds, or longer in the case of attractive interactions [12,13]. Thus, one can speculate that polymerization of these dynamic structures will be accompanied by structural changes.

B. Role of Monomer

Incorporation of monomer into the systems can considerably modify the structure of microemulsions and the phase diagram. This was shown by Candau and coworkers [14–22] for water-soluble monomers, in particular for acrylamide (AM). It was found that water-soluble monomers act as cosurfactants, leading to a considerable extent of the microemulsion domain in the phase diagram (Fig. 1). The cosurfactant role of various hydrophilic monomers was confirmed by surface tension experiments [19,22,23]. These results gave clear evidence that the monomer molecules are partially located at the W/O interface between the surfactant molecules. In the case of the widely studied AOT/water-acrylamide/toluene systems [AOT-sodium 1,4-bis(2-ethylhexyl)sulfosuccinate], the interfacial localization of the AM molecules induces attractive interactions between the droplets, as can be seen by light scattering, small-angle neutron scattering, and viscometric experiments [14,15,24]. It should be noted that without acrylamide the systems display essentially hard-sphere behavior.

Upon further addition of acrylamide, the interaction potential becomes so attractive that transient clusters form. Above a threshold volume fraction, a large increase in the electrical conductivity is observed, which is an indication of a percolation phenomenon [25] (Fig. 2). The percolation threshold decreases with increasing AM/H_2O ratio, i.e., with increasing attractive interactions, in good agreement with theoretical analyses [26] and data obtained for other microemulsions containing alcohols as cosurfactants [27–29]. As shown in Sec. III.C, this percolating structure has an effect on the formation of polymer latex particles and the polymerization mechanism.

Under certain conditions, the radius of curvature can become so large that the globular configuration evolves toward a bicontinuous structure. This transition can be induced by the addition of ionic monomers and/or electrolytes to microemulsions stabilized by nonionic surfactants [21]. In this case, the role of the monomer is twofold: As a cosurfactant, it increases the flexibility and fluidity of the interface, which favors the formation of a bicontinuous microemulsion; as an electrolyte, it decreases the solubility in water of the ethoxylated moiety of the nonionic surfactant (salting-out effect) with its progressive transfer in the oil phase via a bicontinuous phase. Here the role of the monomer is clearly demonstrated, since the corresponding systems in the absence of monomer do not have a bicontinuous character but are indeed coarse emulsions. This is illustrated by the example of the pseudo ternary phase diagram given in Fig. 1.

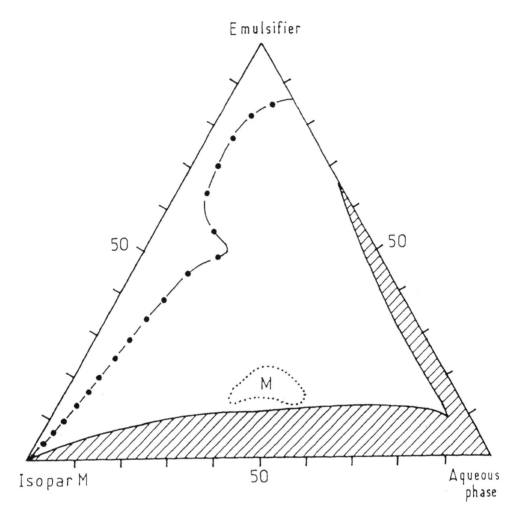

Figure 1 Pseudo ternary phase diagram. The dashed line (–•–) is the boundary between the microemulsion (left) and emulsion (right) domains in the absence of monomers, i.e., in pure water. Addition of monomers (acrylamide + sodium acrylate) to water (1.25 mass ratio) extends the microemulsion domain up to the full line (entire white area). Polymerization reactions have been carried out in the M area. (Isopar M is the oil). (From Ref. 16.)

C. Formulation Rules

First, it must be kept in mind that the formation of a microemulsion requires far more surfactant than that of an emulsion, because of the need to stabilize a large overall interfacial area. This drawback can considerably restrict the potential uses of microemulsion polymerization, since high solids contents and low amounts of surfactant are desirable for most applications. These requirements are far from being achieved at the present state of the art for polymerization in O/W microemulsions, because in most cases the amount of monomer does not exceed a few percent and that of surfactant is around 10% or more.

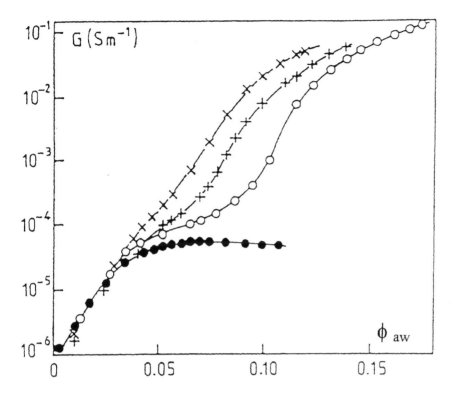

Figure 2 Specific conductivity of inverse microemulsions as a function of the volume fraction Φ_{aw} of AM–H_2O in the disperse phase. AM/H_2O (weight ratio) = ($\bullet$) 0.00; ($\bigcirc$) 0.24; (+) 0.40; (×) 0.67. (AOT/toluene weight ratio = 0.177.) (From Ref. 25.)

In the case of water-soluble monomers, much effort has been devoted to an optimal formulation compatible with an economical process. The reader can refer to Ref. 18 for details of the optimization procedure. Reference 18 concerns acrylamide, but the same procedure was also successfully applied to other water-soluble monomers with small modifications [19,20,22,30].

I summarize briefly below the basic concepts of this approach, which is derived from that developed by Beerbower and Hill [31] for the stability of classical nonionic emulsions, which is referred to as the cohesive energy ratio (CER) concept. The treatment lies in a perfect chemical match between the partial solubility parameters of oil (δ_o^2) and surfactant lipophilic tail (δ_L^2) and of water and hydrophilic head. Under these conditions, one obtains for the optimum HLB (hydrophile–lipophile balance) of the surfactant the relation

$$HLB_o = 20\delta_L^2/(K + \delta_L^2) \tag{1}$$

where K is a constant that depends on the nature of the emulsion [31]. It is thus possible to calculate the required HLB for a given oil. The effect of the type of oil on the formulation of acrylamide-containing microemulsions was investigated [18]. A close correlation was found between the values of δ_o^2 and δ_L^2 and the amount of surfactant needed to form the microemulsion: The lower the amount, the smaller the mismatch between the respective partial solubility parameter values. An additional parameter comes into play when neutral

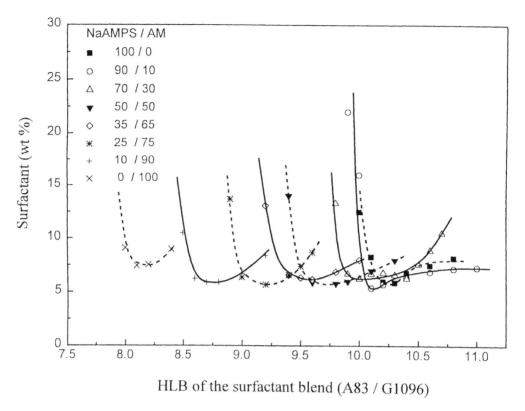

Figure 3 Weight percentage of surfactants required for the emulsion–microemulsion transition versus the HLB value for various compositions of the monomer feed. (From Ref. 30.)

monomers are copolymerized in nonionic micoremulsions with ionic monomers. An illustrative example is provided by the copolymerization of AM with an anionic monomer, sodium 2-acrylamido-2-methylpropanesulfonate (NaAMPS) [30]. The use of Eq. (1) led to the selection of a surfactant mixture made of sorbitan sesquioleate [Arlacel 83; HLB = 3.7, $\delta_L = 7.87$ (cal/cm^3)$^{1/2}$] and a polyoxyethylene sorbitol hexaoleate with 50 ethylene oxide residues [G 1096; HLB = 11.4, $\delta_L = 7.87$ (cal/cm^3)$^{1/2}$ (ICI)] and Isopar M (Esso Chemie) as the oil [$\delta_o = 7.79$ (cal/cm^3)$^{1/2}$]. Note the good agreement between the values of δ_o^2 and δ_L^2 in this case.

Figure 3 represents the percentage of surfactant(s) required for the formation of an AM-NaAMPS microemulsion as a function of the HLB number for different compositions of the monomer feed [30]. The curves delineate the transition between a turbid emulsion and an optically transparent microemulsion. The transition is sharp and can be easily detected by turbidimetry or visually. It can be seen that microemulsions are found in an HLB domain ranging between 8 and 11. The curves exhibit a minimum for an optimum HLB value, which increases as the content of ionic monomer in the feed increases. Note also the low surfactant concentration needed for the formation of clear systems ($5.5\% < S_{\min} < 7.5\%$) in spite of the large proportions of monomers incorporated ($\sim$22%).

Values of HLB of about 8–11 are indicative of systems located in a phase-inversion region, i.e., of microemulsions with a bicontinuous structure [18,31].

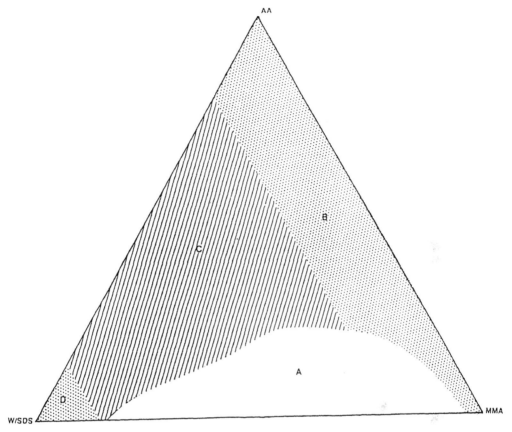

Figure 4 Ternary phase diagram for the system methyl methacrylate (MMA)–acrylic acid (AA)–20 wt% solution of sodium dodecylsulfate in water (w/SDS)–ethylene glycol dimethacrylate (EGDMA) at $25 \pm 0.1°C$ and 1 atm. Compositions are on weight percent basis; EGDMA content is 4% of the combined weight of MMA and AA. Domain A, two-phase region; domain B, W/O microemulsion; domain C, bicontinuous microemulsion; domain D, O/W microemulsion. (From Ref. 50.)

The formation of microemulsions with a bicontinuous structure can be ascribed to the presence of monomers in large proportions in the system, which affects the HLB and interfacial properties as discussed above. However, the cosurfactant effect of the monomer is not sufficient by itself to account for the high HLB values observed. Those values are also partly caused by a salting-out of the ethoxylated surfactants by NaAMPS. Thus, at a given temperature, the HLB of the ethoxylated blend is made more lipophilic upon addition of a salting-out electrolyte in the system. Optimum microemulsification therefore requires a surfactant blend of higher HLB value. This salting-out effect accounts for the increase in the HLB_{opt} values (minima of the curves) observed as greater amounts of NaAMPS are added in the feed as seen in Fig. 3.

Another important aspect of the formulation is illustrated in the studies aimed at preparing porous materials. In this case, one has to formulate systems containing large amounts of hydrophobic monomers (up to 70 wt%) either in the continuous phase of globular microemulsions [32–42] or in the oil domains of bicontinuous microemulsions [43–55]. A typical example of a phase diagram is given in Fig. 4. It shows four detectable regions, three of them being monophasic microemulsions (W/O, O/W, and bicontinuous). The acrylic acid here acts as a cosurfactant, and it was found that surfactant-free polymerizable single-phase microemulsions made from the same components [methyl methacrylate (MMA), acrylic acid, ethylene glycol dimethacrylate (EGDMA), and water] could also form. The morphology of the materials obtained after polymerization is discussed in Sec. IV.A.

There have not been such detailed formulation studies for O/W microemulsions based on hydrophobic monomers. The main reason is that the monomer most investigated so far is styrene trapped in droplets stabilized by aliphatic surfactants. According to the criterion defined above, there is a chemical mismatch between styrene (aromatic) and the hydrophobic tail of the surfactant. In addition, styrene has no amphiphilic character and cannot act as a cosurfactant. As a result, the domain of existence of microemulsions is very limited.

III. POLYMERIZATION IN GLOBULAR MICROEMULSIONS

Most studies have dealt either with the free radical polymerization of hydrophobic monomers—e.g., styrene [56–89], methyl methacrylate (MMA) [68,73,74,84,86,90–93] or derivatives [2,94,97], and butyl acrylate (BA) [98–100]—within the oily core of O/W microemulsions or with the polymerization of water-soluble monomers such as acrylamide (AM) within the aqueous core of W/O microemulsions [101–123]. In the latter case, the monomer is a powder that has to first be dissolved in water (1 : 1 mass ratio) so that the resulting polymer particles are swollen by water, in contrast with O/W latex particles, where the polymer is in the bulk state. The polymerization can be initiated thermally, photochemically, or under γ-radiolysis. The possibility of using a coulometric initiation for acrylamide polymerization in AOT systems was also reported [120]. Besides the conventional dilatometric and gravimetric techniques, the polymerization kinetics was monitored by Raman spectroscopy [73,74], pulsed UV laser source [72,78], the rotating sector technique [105,106], calorimetry, and internal reflectance spectroscopy [95].

For both O/W and W/O systems, the amount of monomer is usually restricted to 5–10 wt% with respect to the overall mass, and that of surfactant(s) lies within the same range or even above. Nevertheless, there have been a few studies in which the formulation deviated from these conditions. For instance, surfactant concentrations of 2 wt% were reported [56–58,69,124,125]. However, in this case the amount of monomer was also very low (< 2 wt%) so that the systems must be considered as micellar solutions rather than true microemulsions. Conversely, a 1994 study of Gan et al. [82] reported the polymerization of styrene up to 15 wt% using only about 1 wt% dodecyltrimethylammonium bromide surfactant (DTAB) in a Winsor I–like system. This system consists of a microemulsion (lower) phase topped off with pure styrene. The polymerization takes place in the microemulsion phase, while the styrene phase acts as a monomer reservoir. Such a polymerization process is novel, but it yields latices of large particle size (~100 nm) that can be more easily obtained by conventional emulsion polymerization.

The main difficulty encountered by most of the authors and one that precludes the use of higher monomer concentrations lies in retaining the optical transparency and stability of the microemulsions upon polymerization. In addition to entropic factors contributing to the destabilization of microemulsions during polymerization, the compatibility between polymer and cosurfactant also influences the system [64]. This is especially true when styrene is polymerized within O/W microemulsions that contain an alcohol because the latter is not a solvent for the polymer. Conversely, the polymerization of acrylamide in alcohol-free W/O microemulsions was reported in 1982 to give transparent microlatices of small particle size ($d\sim$30 nm) [24].

There are some other drawbacks that make alcohols undesirable hosts in monomer-containing microemulsions.

1. They partition between the interfacial film and the other phases, which considerably complicates the dilution procedure required for the determination of the droplet or particle size.
2. They may modify monomer partitioning [126].
3. They act as chain transfer agents, lowering the polymer molecular weight [4].

Despite these difficulties, most of the earliest studies used an alcohol in the formulation of O/W microemulsions, and it was only in 1989 that the polymerization of hydrophobic monomers in a three-component O/W cationic system was reported by Ferrick et al. [124]. This spurred new interest, and systematic studies on ternary microemulsions based on cationic surfactants of different alkyl chain lengths ensued, mainly in the group of Gan and those of Puig and Kaler [67,77,81–87,90–93,127,128]. Nonionic surfactants were also used in ternary O/W microemulsions for the polymerizations of styrene and methyl methacrylate [68] and anionic [AOT] surfactants for the polymerization of tetrahydrofurfuryl methacrylate [95,97].

A. Particle Nucleation: Basic Features

In conventional emulsion polymerization, the polymer content can reach 60% and the amount of surfactant is less than 1%. Particle nucleation occurs at the early stages of the reaction (interval I: degree of conversion about 2–15%). Depending on the water solubility of the monomer, the latex particles are nucleated either by the entry of the free radicals produced by the initiator decomposition in water into the monomer-swollen micelles or by homogeneous nucleation [129]. The latter involves the formation in the aqueous continuous phase of oligomeric radicals that ultimately precipitate upon reaching a critical size (about 65 monomers for MMA and about 8 for styrene) to form primary particles. In interval II, the nucleated polymer particles grow in size at constant number by monomer diffusion from the large reservoir droplets. Inside the particles, the monomer concentration is maintained at equilibrium with the monomer dissolved in the aqueous phase, the particle swelling being limited by the water/particle surface tension. Interval III corresponds to the disappearance of the reservoir droplets and to the consumption of the monomer still contained in the particles. As a result, each final particle usually contains a high number of polymer chains. The nature of the kinetics may be discerned by calculating the average number of radicals, , per particle during any stage of the reaction (Smith–Ewart cases I, II, III) and by the locus of nucleation prevailing in the system [129].

To understand the mechanism occurring in microemulsion polymerization, one has to remember that the concentration of monomer used is low (a few percent) while the concentration of surfactant is much larger than that used in an emulsion. The monomer/surfactant mass ratio is around 0.3–1 compared to 30–60 in an emulsion.

The second difference lies in the structure of the initial systems. In an emulsion the monomer is located in large monomer droplets ($d\sim$1–10 μm) and in small micelles ($d\sim$5–10 nm) and is partially solubilized in the continuous phase. In a globular microemulsion, it is solubilized within swollen micelles of the same size ($d=$ 5–10 nm). These features coupled with the dynamic character of microemulsions are the origin of the difference in the mechanisms observed in the two processes.

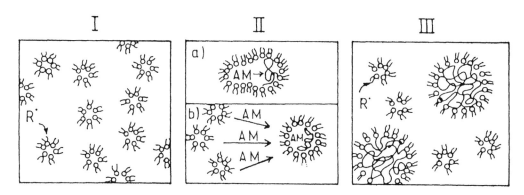

Figure 5 Polymerization mechanism in AOT globular microemulsions. (I) Before polymerization; AOT micelles ($d\sim6$ nm). (II) Polymer particle growth (a) by collisions between particles; (b) by monomer diffusion through the toluene phase. (III) End of polymerization. Polymer particles ($d\sim40$ nm) plus small micelles ($d\sim3$ nm). (From Ref. 23.)

The problem of particle nucleation was first addressed in the 1980s by Candau and coworkers [14,104] for the case of water-in-oil microemulsions. Their studies concerned the polymerization of acrylamide inside water-swollen micelles stabilized by AOT and dispersed in toluene. A thorough investigation of the structures prior to and after polymerization by elastic and quasi-elastic light scattering (QELS), viscometry, and ultracentrifugation experiments yielded two key experimental results:

1. The particle size of the final microlatex ($d\sim20$–40 nm) was larger than that of the initial monomer-swollen micelle. This led to a final number of polymer particles, N, about two or three orders of magnitude smaller than that of the monomer droplets.
2. The average number of polymer chains contained in each particle, n_p, was one.

From these results it was postulated that *particle nucleation was continuous throughout the process* [104], in strong contrast with emulsion polymerization.

To account for the particle growth, it was proposed that only a small fraction of the overall number of droplets are nucleated. The nucleated particles grow, the monomer being supplied from the nonnucleated reservoir droplets either by diffusion through the continuous phase or by "sticky" collisions between droplets. As a result of this growing process and because of the large amount of surfactant involved in the formulation, small micelles are always present in the reaction mixture. Owing to their high overall interfacial area compared to that of the nucleated polymer particles, these micelles capture preferentially the primary radicals generated in the organic phase. Hence, each particle is entered once on average (Fig. 5). Note that these results do not follow the classical Smith–Ewart theory, as each nucleated particle contains only one macromolecule, either growing or in its final form. However, not all particles are active at any given time, and , averaged over the entire population, is less than 1.

A continuous particle nucleation mechanism was further confirmed by transmission electron microscopic (TEM) experiments performed on polyacrylamide samples taken at various degrees of conversion [25]. The number of polymer particles was shown to increase proportionally with conversion (Fig. 6a), whereas the size remained roughly constant.

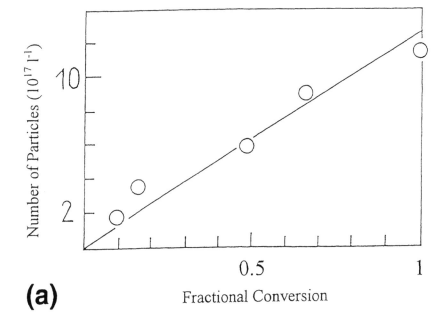

(a)

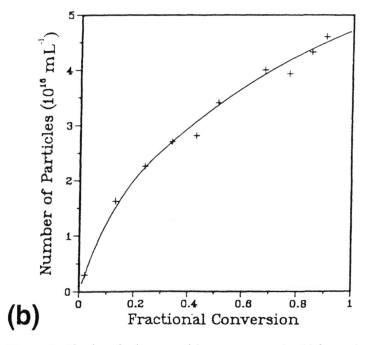

(b)

Figure 6 Number of polymer particles versus conversion (a) for acrylamide polymerization in AOT W/O microemulsion (from Ref. 25); (b) for styrene polymerization in O/W microemulsion (from Ref. 75).

The mechanism of polymerization in ternary and quaternary oil-in-water microemulsions has become understood only in recent years. The onset of turbidity upon polymerization and the lack of stability with time observed by most authors, particularly for MMA monomer, is likely the reason for the slow progress in the comprehension of the mechanism of O/W systems. Only slight changes in the formulation are sufficient to significantly affect the polymerization process and to induce particle coagulation at any stage of the reaction. This may explain the disparity in the kinetic data reported by some authors for very similar systems. With this remark in mind, one can, however, conclude that the scheme that is now well accepted is that of a continuous particle nucleation mechanism as in the case of inverse systems. This view is supported by several features.

First, the number of polymer chains per particle is in general very low, sometimes close to one [4,66,69,70,75,76,83,84,90,91]. As a general trend, n_p increases slightly with the degree of conversion [75,83]. This increase was accounted for by the capture of radicals by pre--existing polymer particles, these competing more effectively with the monomer-swollen micelles as the reaction proceeds. Limited flocculation at later stages of the reaction was also envisioned.

Second, the number of polymer particles, N, was found to increase continuously with the conversion [75,83,93,100] (Fig. 6b). Comparing these results with those obtained for inverse microemulsions (cf. Fig. 6a), one observes a downward curvature of N (conversion) in the case of O/W systems that correlates well with the increase in n_p with time. It must be stressed that the determination of n_p is subject to a large uncertainty, because it is based on the combined knowledge of the polymer molecular weight and the size of the polymer particle. The latter can be determined by either TEM or QELS. At low conversion, the polymer particles are swollen to a certain extent by monomer, and the diameter measured by QELS is not that of the dry particle. This has not always been appreciated by the researchers and can be at the origin of some significant differences between the n_p values reported for similar systems. Another difficulty arises from the increase in both particle size distribution and particle diameter with time [75,93] (Fig. 7). This increase was explained by the same considerations as those given above.

The phase diagrams of the ternary systems allow one to make a direct comparison between emulsion and microemulsion polymerization processes just by varying the surfactant concentration, as shown by Gan and coworkers [84,91,127]. Figure 8 represents the polymerization rate conversion curves for methyl methacrylate polymerization at different surfactant concentrations.

Three distinct intervals are observed in the emulsion regime with a rate plateau (interval II), as commonly observed in emulsion polymerization. This plateau corresponds to the steady-state conditions (constant particle number and monomer concentration in the particles). Only two intervals are observed in the microemulsion regime, with a maximum occurring at around 20% conversion. Two rate intervals with no constant rate have also been observed by several groups for microemulsion polymerization of styrene [66,70,77,79–81,83,84] and butyl acrylate [98,99]. By analogy with emulsion polymerization, the rate decrease observed at high conversion was often misinterpreted as an indication of particle growth, the initial rate being attributed to particle nucleation. This scheme is evidently incompatible with the one to three polymer chains found in each final latex particle. Once the polymer particle is formed in a single nucleation step, its growth is necessarily very limited. In fact, the continuous increase of the particle number with conversion proves unambiguously that particle nucleation occurs all throughout the reaction. It is now generally believed that the rate decrease observed above 20% conversion is due to

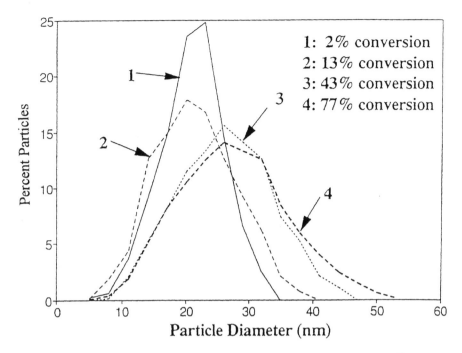

Figure 7 Particle size distribution for polystyrene microlatices at different conversions with 0.27 mM KPS at 70°C. (From Ref. 75.)

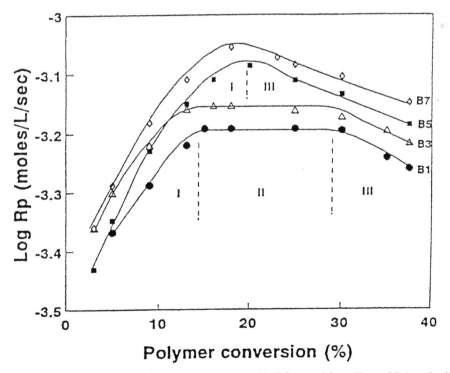

Figure 8 MMA polymerization rate curves at 60°C for emulsions (B_1 and B_3) and microemulsions (B_5 and B_7). (From Ref. 91.)

monomer depletion in the polymer particles, as proposed by El-Aasser and coworkers [75,76]. Termination of chain growth in polymer particles is attributed to chain transfer to monomer [75,76].

Having described the general mechanistic features of microemulsion polymerization, turn now to a more detailed inspection of several parameters.

B. Relevant Parameters in the Mechanism and Kinetics of Oil-in-Water Microemulsions

1. Interfacial Fluidity of Microemulsion Droplets

Styrene has been polymerized in a series of cationic microemulsions containing alcohols of different alkyl chain lengths from 2-ethyoxyethanol to 2-(2-butoxyethoxy)ethanol [80]. The distributions of styrene and cosurfactants in both dispersed and continuous phases were analyzed by dialysis. The concentration dependence of styrene and potassium persulfate initiator (KPS) on the polymerization rates R_p ($R_p \propto$ [styrene] [KPS]$^{0.4}$) was roughly the same in these systems and quite similar to the values found for other ternary or quaternary microemulsions. However, the activation energy E_a decreased steadily as the molar ratio of cosurfactant to surfactant in the dispersed phase increased, i.e., as the interface fluidity increased. It was conjectured that the more fluid the interface, the easier the diffusion of KPS free radicals or oligomeric anion radicals in the microemulsion droplets. The polymer molecular weights were found to depend on which alcohol was used in the formulation. As the chain transfer constants of this series of alcohols were roughly the same, the differences observed in molecular weights were rather attributed to an indirect effect of the interfacial fluidity, affecting the termination rate in the particles.

2. Effect of Monomer

Comparative mechanistic studies on the microemulsion polymerization of styrene and methyl methacrylate were carried out by several groups [67,73–77,79,81,83,90–93,128]. The results could be coherently interpreted in terms of the relative monomer solubilities in water. In the case of styrene, which has a very low solubility in water (0.031%), it was postulated that initiation takes place in the microemulsion droplets. The polymer particles grow by recruiting monomer and surfactant from uninitiated droplets. Homogeneous nucleation in styrene systems may be relatively insignificant due to the large number of microemulsion droplets, which will capture most of the radicals generated in the aqueous phase before they reach a critical size for precipitation.

The experiments performed in MMA systems suggest that homogeneous nucleation can compete with monomer droplet initiation because of the non-negligible solubility (1.56%) and the more polar and cosurfactant character of MMA in water. In a detailed investigation, Bleger et al. [93] proposed that homogeneous nucleation occurred at the early stages of the reaction when the increase in conversion was very slow. In the second stage, which has a much faster polymerization rate, nucleation proceeds via a micellar entry mechanism. An increase in the total number of particles with conversion was also found, providing evidence for continuous generation.

Gan et al. compared the growth of poly(methyl methacrylate) (PMMA) and polystyrene (PS) particles in ternary cationic microemulsions [128]. Different growth patterns were observed, which was attributed to differences in the redistribution of the components during the reaction. The strong interactions between the polar MMA monomer and the cationic surfactant at the W/O interface reduce the swelling of monomer-swollen polymer particles. In contrast, interactions are weak for styrene, and maximum swelling is reached in the early stages of the reaction [75,76].

The dependence of the polymerization rate (R_p) on monomer concentration was found to vary from about 1 [79,80] to 1.3 [92], close to the theoretical value of 1 for emulsion solution or bulk polymerization.

3. Effect of Initiator

The effect of water-soluble (KPS) and oil-soluble (2,2′-azobisisobutyronitrile; AIBN) initiators on the microemulsion polymerization of styrene and methyl methacrylate in ternary and quaternary microemulsions was investigated by several groups [73,77,79,81,90,92]. The polymerization mechanism does not seem to depend on the nature of the initiator used; both KPS- and AIBN-initiated systems produced microlatices with similar hydrodynamic radii, numbers of polymer particles, molecular weights of polymer, and numbers of macromolecules per particle [81]. However, the polymerization rates were generally faster with KPS than with AIBN [73,79,81], and the latex parameters showed stronger dependence on the concentration of KPS than on that of AIBN. This behavior was discussed in terms of the initiators having different efficiencies in producing effective radicals for the polymerization. At an equimolar concentration of initiators, KPS generates more radicals in the aqueous phase than AIBN. These radicals are thus more effective for initiation (in the continuous phase, in the monomer droplets, or at their interfaces) than AIBN radicals due to a significant autotermination of AIBN radical pairs in the small droplets (cage effect) and the low solubility of AIBN in water. In this case, initiation is believed to occur essentially via micellar entry of single radicals arising from the very small portion of AIBN either dissolved in water or desorbed from other swollen micelles.

Bleger et al. [93] used AIBN and a cationic water-soluble initiator, 2,2′-azobis(2-amidinopropane) dihydrochloride (V50) for MMA polymerization in ternary cationic microemulsions. In contrast to the case of the anionic KPS initiator, the free radicals produced by V50 decomposition have the same sign as the microemulsion droplets. The repulsion between these two species favors homogeneous nucleation rather than micellar entry during the first stages of the reaction (see above). This accounts for the longer homogeneous nucleation regime seen with V50. Moreover, homogeneous nucleation is known to yield primary particles prone to only limited flocculation. This is likely the reason for the higher number of polymer chains per particle found at the end of the reaction (three times as many chains as with AIBN initiator).

In the case of thermal initiation of styrene [79,80], the polymerization rate was found to be proportional to $[AIBN]^{0.39}$ and $[KPS]^{0.47}$, in good agreement with other data for three- or four-component microemulsions [66,81]. The dependence on AIBN concentration is consistent with the prediction of 0.40 based on the micellar nucleation theory in emulsion polymerization (Smith–Ewart case 2) (see, e.g., Ref. 129). The dependence on KPS concentration lies between this case and the value of 0.5 for solution or bulk polymerization.

The overall activation energy, E_A, of polymerization was determined for both monomers as a function of the nature of the initiator [66,79,81,90,92]. For styrene polymerization in cationic microemulsions, E_A was found to be much higher for KPS systems ($E_A \sim 95$ kJ/mol) than for AIBN systems (48 kJ/mol) in spite of similar decomposition energies. This difference was attributed to different radical capture efficiencies between the anion radicals of KPS and the uncharged AIBN radicals and the positively charged

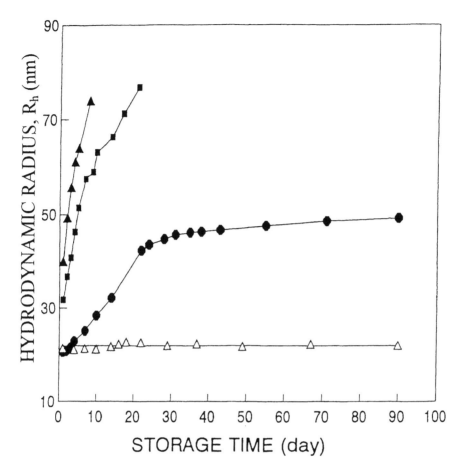

Figure 9 Changes of PMMA particle sizes on long-term storage at 60°C for four microlatices stabilized by different surfactants: (▲) TTAB; (■) TTAC (tetradecyltrimethylammonium chloride); (●) CTAB (cetyltrimethylammonium bromide); (△) STAC. (From Ref. 128.)

interfaces of the droplets [79,81]. However, this effect was not observed for MMA polymerization with the same initiators, which renders the above explanation rather speculative [90,92].

4. Effect of Surfactant

In addition to electrostatic stabilization, steric stabilization was found to be an important factor in the microemulsion polymerization of MMA. A comparative stability study was performed on MMA polymerization in tetradecylammonium bromide (TTAB) or stearyltrimethylammonium chloride (STAC) microemulsions [128]. A strong increase in particle size was observed above 50% conversion for the former system, whereas the size remained rather constant up to 98% conversion for the latter. Similarly, n_p increased from 1.75 to 35.6 when the conversion went from 61% to 93% for TTAB systems but remained around 1.4 for STAC systems. These results give clear evidence of the occurrence of particle coagulation in MMA systems based on TTAB surfactants. Figure 9 shows the change in PMMA particle sizes upon long-term storage with the type of surfactant used. Surfactants with longer carbon chain lengths form thicker interfacial layers, which prevent latex flocculation. The role played by the counterions of the surfactants in the stability of the microlatices was also discussed. Finally, it should be noted that the stability of PS microlatices is not affected by the alkyl chain length of cationic surfactants owing to stronger surfactant adsorption on the surface of PS particles compared to that on PMMA particles [128]. It is indeed known that the surfactant adsorption at a latex/water surface decreases with increases in the polarity of a polymer [130].

5. Emulsion Polymerization Versus Microemulsion Polymerization

The kinetic studies performed on the polymerization of MMA in the emulsion and microemulsion regions of ternary systems based on cetyltrimethylammonium bromide (CTAB) confirmed the difference in mechanisms between the two processes [84,91]. The following kinetic laws were obtained:

Emulsion regime:

$$R_p \propto [\text{CTAB}]^{0.31} \, [\text{KPS}]^{0.82}$$

Microemulsion regime:

$$R_p \propto [\text{CTAB}]^{0.58} \, [\text{KPS}]^{0.33}$$

The value of 0.31 found in emulsion polymerization for the dependence of R_p on the surfactant concentration is much lower than the value of 0.6 based on the Smith–Ewart micellar nucleation mechanism. This low value indicates a significant contribution by homogeneous nucleation as already quoted for this monomer [131].

On the other hand, the micellar nucleation mechanism is likely predominant in microemulsion polymerization due to the large number of micelles and large overall surface area available for capturing free or oligomeric radicals generated in the aqueous phase. This view is supported by the exponent of 0.58 found in microemulsion systems.

Similar studies were conducted for the polymerization of styrene [84]. These led to a single dependence of R_p on surfactant concentration for both emulsion and microemulsion systems ($R_p \propto [\text{TTAB}]^{0.52}$). This confirms that when surfactant concentrations are well above the critical micelle concentration (cmc) (> 3 wt%) in emulsion or microemulsion systems, the micellar nucleation mechanism prevails in both cases because of the very low solubility of this monomer in water.

C. Kinetics in Water-in-Oil Microemulsions

Dilatometric and rotating sector techniques were combined to follow the photopolymerization of acrylamide in AOT reverse micelles with AIBN as the initiator and toluene as the organic phase [105,106]. Very high polymerization rates were observed, with total conversion to polymer achieved in a few minutes and good heat transfer. A monoradical termination reaction was found ($R_p \propto [\text{AIBN}]^1 \, [\text{AM}]^{0.95}$), caused by a degradative chain transfer to toluene. The reaction likely occurs at the water/oil interface by transfer of the growing polymer radical to toluene, followed by the exodiffusion of the new benzylic radical, the latter being too stable to reinitiate polymerization.

A detailed study on the loci of initiation and propagation of AM (co)polymerization in the same AOT systems was performed by Barton's group [35,110–114] using initiators and inhibitors of various solubilities. Initiation with AIBN was shown to take place predominantly in the water/oil interlayer where the encounter with acrylamide cosurfactant was facilitated. With water-soluble ammonium persulfate, initiation occurs, as expected, in the micellar water pools.

Steady-state fluorescence of indolic probes quenched by AM and selectively located at the various locations of the microemulsion (toluene, interfacial layer, and water phase) was used to follow the depletion of this monomer during polymerization [119]. The results show that AM is evenly consumed from all parts of the AOT systems independently of the initial composition and the nature of the initiator.

The mechanistic scheme and kinetics described above for polymerization in globular AOT/water-AM/toluene microemulsions become more complex when these systems are initially percolating (see Sec. II.B). QELS and TEM techniques were used to monitor changes in particle size with conversion for both types of systems [25]. Different behaviors were observed at the early stages of the reaction (< 10–15% conversion). In particular, the large initial rise in size observed for percolating systems was attributed to the formation of a necklace of connecting beads. A comparative study of the kinetics of polymerization of AM is percolating and nonpercolating systems [115,116] yielded different orders of reaction rates with respect to AM ($R_p \propto [AM]^{1.1}$ and $R_p \propto [AM]^{1.8}$, respectively). Also, for a given AM/water mass ratio, the polymerization rate is higher in a nonpercolating system. It was proposed that some local AM concentration gradients could occur in the propagation loci, thus altering the kinetic behavior. However, the exact mechanism requires further studies to be fully elucidated.

IV. POLYMERIZATION IN THE CONTINUOUS OR BICONTINUOUS PHASES OF MICROEMULSIONS

The variety of structures encountered in microemulsions offers great versatility for choosing the locus of polymerization. Besides polymerization in globular microemulsions, several studies have dealt with polymerization of monomers in the other phases of microemulsions. One of the main goals underlying these studies was to use the microstructure of microemulsions as a template to produce solid polymers with similar characteristics. For example, incorporation of large amount of hydrophobic monomers in the continuous phase of W/O microemulsions should yield solid polymers with a Swiss cheese–like structure capable of encapsulating the disperse phase (water). This would allow the inclusion of materials (metallic colloidal particles as catalysts, photochromic compounds, etc.) in the disperse phase that would otherwise be insoluble in the polymer.

In the case of bicontinuous microemulsions, both hydrophobic and hydrophilic monomers have been considered. As seen below, the morphology of the final product depends on the microemulsion composition and on the nature of the monomer. In the studies performed so far, polymerization of hydrophobic monomers led to porous materials, whereas polymerization of water-soluble monomers yielded transparent and stable microlatices. It should be noted that the incentives in these two kinds of studies were different. In particular, the aim of the research on water-soluble monomers was to prepare high molecular weight polymers at large contents that could be used as flocculants [4].

A. Microemulsions Based on Hydrophobic Monomers

In a pioneering work, Stoffer and Bone [32–34] studied the phase behavior of sodium dodecyl sulfate (SDS)–pentanol–methyl methacrylate or methyl acrylate–water systems before and after polymerization. A typical composition of the system was MMA, 41.7% pentanol, 27%; SDS, 14.6%; water, 16.7%. As the monomers forming the continuous phase of the microemulsion polymerize, phase separation occurs.

Polymerization degree (DP) dependences on initiator concentration were shown to follow solution kinetics (DP$\propto$[I]$^{-1/2}$) for water- and oil-soluble initiators. Such behavior could be expected, as the monomer is located in a continuous medium and not in the micellar dispersed phase. The pentanol present in large amounts (27%) acts as a chain transfer agent. The expected linear dependence of 1/DP versus pentanol/monomer ratios gives a value of 5.1×10^{-4} for the chain transfer constant C_s.

The kinetics of the homopolymerization of methyl methacrylate in the continuous phase of AOT reverse micelles was studied by Vaskova et al. [35]. Strong turbidity was observed in the course of polymerization, although MMA was highly diluted with toluene ($C_{MMA} \cong 6\%$). The MMA polymerization in AOT system was compared to that in pure toluene and to that in toluene in the presence of AOT. A very low polymerization rate was found with water-soluble ammonium persulfate. This indicates that only a small amount of MMA is present in the AOT water pools and that the APS radicals remain trapped inside.

Friberg and coworkers examined the problems of stability in a series of papers related to the polymerization of styrene in water-in-oil microemulsions stabilized by SDS and pentanol [64,132–134]. Their conclusion was that entropic conformational factors were not the only ones of importance in the mechanism of destabilization. A correlation was established between polymer solubility in the cosurfactant and that of monomer by using another cosurfactant, butyl cellosolve. The higher solubility of polystyrene in butyl cellosolve gave better stability than pentanol-containing microemulsions.

Gan and Chew [36] reproduced the previous experiments of Stoffer and Bone and confirmed that transparent polymer mixtures at 100% MMA conversion could not be obtained in situ by using pentanol as cosurfactant in which PMMA is insoluble. They subsequently improved the procedure appreciably by replacing 1-pentanol with the polymerizable cosurfactant acrylic acid (AA) [36]. This monomer is soluble in water and dissolves up to 32% PMMA. Transparent solid copolymers could be obtained by fully polymerizing 54% MMA, 34% AA, 10% H$_2$O, and 2% SDS or with other lower water contents. The copolymers were very heterogeneous in composition and had high molecular weight ($\cong 10^6$). Scanning electron microscopic (SEM) observations did not reveal any particular pattern of the structure, indicating that water was isotropically distributed in the polymeric matrix.

Gan and Chew [37,38] extended their studies to microemulsions in which all the components except water were polymerizable. Polymerization in microemulsions containing a polymerizable surfactant (sodium acrylamidoundecanoate [37] or acrylamidostearate [38]), a cosurfactant (acrylic acid), and methyl methacrylate as the continuous phase led, under certain conditions, to transparent solid terpolymers with up to 10–20% water dispersed in the polymer matrices. As in the case of copolymers, no particular structure was shown by SEM for these terpolymers.

Porous solids were obtained by copolymerization of styrene-divinylbenzene [40,42] and cyclohexyl methacrylate-allyl methacrylate [41] in the continuous phase of microemulsions. Menger et al. [40] showed that the pore size of the material, always larger than the initial droplet size, is highly dependent on the water/surfactant ratio in the microemulsion. The same authors prepared porous materials with chemically active groups on the surface [39]. A W/O AOT microemulsion was prepared using styrene/divinylbenzene (6 : 4 w/w) as the oil phase and compound I as the polar comonomer. Polymerization led to porous materials with active surfaces able to complex and remove Cu^{2+} from solution. Such a polymer-Cu^{2+} complex is a good candidate for phosphate ester catalysis.

$$CH_3 \diagdown N \diagup N \diagup CH_3 \diagup (CH_2)_{12}OCH_2 - \text{(styrene)}$$

I

Qutubuddin and coworkers [43,44] were the first to report on the preparation of solid porous materials by polymerization of styrene in Winsor I, II, and III microemulsions stabilized by an anionic surfactant (SDS) and 2-pentanol or by nonionic surfactants. The porosity of materials obtained in the middle phase was greater than that obtained with either oil-continuous or water-continuous microemulsions. This is related to the structure of middle-phase microemulsions, which consist of oily and aqueous bicontinuous interconnected domains. A major difficulty encountered during the thermal polymerization was phase separation. A solid, opaque polymer was obtained in the middle with excess phases at the top (essentially 2-pentanol) and bottom (94% water). The nature of the surfactant had a profound effect on the mechanical properties of polymers. The polymers formed from nonionic microemulsions were ductile and nonconductive and exhibited a glass transition temperature lower than that of normal polystyrene. The polymers formed from anionic microemulsions were brittle and conductive and exhibited a higher T_g. This was attributed to strong ionic interactions between polystyrene and SDS.

Thermal polymerization of styrene was also carried out by Rabagliati et al. [45] in a three-phase Winsor III microemulsion–styrene–brine system. The system exhibited features of a solution polymerization process, and no mention was made of the morphology of the final product.

More recently an important contribution to this field was provided by Cheung and collaborators [46–50], who obtained porous polymeric structures by photopolymerization of monomers in single-phase microemulsions. The system consisted of methyl methacrylate (MMA), acrylic acid (AA), a cross-linking agent, ethylene glycol dimethacrylate (EGDMA), water, and sodium dodecyl sulfate (SDS) as the surfactant. Large amounts of monomers were used in the formulation (up to 70% in some cases). The structures of the initial microemulsions were inferred from conductivity, viscosity, and QELS experiments. An example of the richness of these structures is given in Fig. 4 (Sec. II.C). Interestingly, a close correlation was found between the microstructure of the polymeric material and the nature of the initial microemulsion. The polymeric materials were characterized by SEM, thermogravimetry, adsorption studies, swelling and permeability measurements, and differential scanning calorimetry. The main conclusions reached by these authors were that

Polymerization in microemulsions with a water/oil droplet structure yields closed-cell porous polymeric solids having a morphology characterized by a disjointed cellular structure in which the water pores are distributed as discrete pockets throughout the solid.

Polymerization in microemulsions with a bicontinuous structure results in a polymers with an open-cell structure, i.e., an interconnected porous structure with water channels through the polymer. The surface area increases steadily as water content increases in the precursor microemulsion.

The above morphologies were clearly evidenced by means of scanning electron microscopy and thermogravimetric analysis. The distinction between open-cell and closed-cell porous structures was based on the strong difference in the shapes of the drying rate curves for the two structures. In closed-cell structures, the drying process is diffusion-limited, resulting in an exponential decrease in drying rate with decreasing moisture content. In open-cell structures, the drying process is dominated by the transport of moisture from the interior of the solid to its surface (capillary forces). In this case, a linear decrease in drying rate was observed up to a certain threshold of water content.

These results indicate that the morphology of the polymer retains some memory of the initial structure, as this structure is retained to a certain extent. However, the length scale of the porous structure obtained (1–4 μm) is considerably larger than the length scale characteristic of microemulsions (less than 0.1 μm), owing to phase separation effects or structural changes during polymerization. This instability can be related to the complex effect of the entropic increase of free energy resulting from conformational limitations. The incorporation of the cross-linking agent ethylene glycol dimethacrylate (EGDMA) was found to be quite effective in minimizing the occurrence of phase separation. Upon polymerization, microemulsions with EGDMA concentrations below 10% by weight of monomer content yield transparent polymeric solids [46]. The porous polymers have good mechanical stability, and their rigidity can be varied according to the compositions of the initial microemulsions and, especially, the amount of cross-linking agent used.

In more recent studies performed on very similar systems, Gan and coworkers [51,52] confirmed the findings of the group of Cheung, i.e., that the morphology of the resulting polymeric material depends strongly on the composition of the precursor microemulsions. These authors also attempted to preserve to a greater extent the initial bicontinuous structure of microemulsions by varying the nature of the surfactant and the polymerization conditions [135–137]. The reaction has to be carried out as fast as possible (photoinitiation at 35°C) in order to reduce the structural modifications of the microemulsion at the early stages of the reaction. The use of bicontinuous microemulsions based on a polymerizable surfactant seems to be the key parameter for obtaining transparent solid polymeric materials with an open-cell microstructure [51,135,137]. In particular, the polymerization of a bicontinuous microemulsion based on water, MMA ($<50\%$), EGDMA ($<4\%$), and a polymerizable zwitterionic surfactant, acryloyloxyundecyldimethylammonium acetate (AUDMAA) ($>25\%$) produced transparent solid materials in 10 min [137]. The polymer formation during polymerization was monitored by scanning electron microscopy (SEM) at different time intervals. The widths of the randomly distributed bicontinuous domains were about 50–70 nm, i.e., far below those usually obtained in the previous reports.

The preparation of novel solid materials is a huge field for applications such as microfiltration, separation membranes or their supports, microstructured polymer blends, and porous microcarriers for the culture of living cells and enzymes. The considerable progress accomplished over the last four years makes it possible to envision many future developments. Some attempts for specific applications have already been made as shown below.

The difference in relative permeability of gases through the membranes offers an attractive way for the enrichment of industrial gases. The transport properties of N_2 and O_2 through porous polystyrene materials were found to be much higher than through those prepared by bulk polymerization [43].

Novel conductive composite films were developed using a two-step process. The porous film coated on an electrode was prepared by polymerizing a bicontinuous microemulsion containing acrylamide and styrene monomers followed by electropolymerization of pyrrole [55]. The copolymer matrix was found to improve the mechanical behavior of the pyrrole composite film.

Burban et al. [53] reported the preparation of microporous silica gels by polymerization of partially hydrolyzed tetramethoxysilane gels present in the aqueous phase of bicontinuous microemulsions stabilized with didodecylammonium bromide. When vacuum dried, the gels made in microemulsions had about twice as much specific surface area as conventional vacuum-dried silica gels.

B. Microemulsions Based on Water-Soluble Monomers

The polymerization of water-soluble monomers in the aqueous domains of nonionic bicontinuous microemulsions has been studied by Candau and coworkers over the last decade. By optimizing the procedure (see Sec. II), one can prepare microemulsions containing up to 25 wt% monomers dissolved in the same amount of water and around 8 wt% surfactant [4,30]. Various water-soluble monomers have been investigated: the neutral monomer acrylamide [17,18,138,139], anionic monomers sodium acrylate (NaA) [16,140–143] and sodium 2-acrylamido-2-methylpropanesulfonate (NaAMPS) [22,30,144–150], cationic monomer methacryloyloxyethyltrimethylammonium chloride (MADQUAT) [19,20,22, 144–151], and copolymers of these monomers.

As for polymerization of hydrophobic monomers in the bicontinuous phase of microemulsions, the initial structure is not preserved upon polymerization. However, a notable difference from the former systems is that the final system is a microlatex that is remarkably transparent (100% optical transmission), fluid, and stable, with a particle size remaining unchanged over years even at high volume fractions ($\sim$60%) [20]. The microlatex consists of water-swollen spherical polymer particles with a narrow size distribution according to QELS and TEM experiments. This result is of major importance with regard to inverse emulsion polymerization, which is known to produce unstable latices with a broad particle size distribution [23].

Several factors are responsible for the structural change observed upon polymerization.

Microemulsions have flexible and fluctuating interfaces that undergo easy deformations.

The monomer consumption from the interfacial layer (cosurfactant effect, see Sec. II.B) modifies the film curvature energy. The formation of water-swollen spherical polymer particles dispersed in the oily phase corresponds to the minimum free energy of the system.

The stability of the inverse latex strongly depends on an appropriate formulation. Poor chemical compatibility between oils and emulsifiers produces unstable latices, whereas a good chemical match leads to perfectly transparent and stable latices [17,18]. The latex stability was accounted for by (1) reduced gravity forces ($\sim d^3$, where d is the particle diameter); (2) high entropic contribution from the droplets owing to their large number; and (3) low interfacial tension between polymer droplets and the continuous phase [4].

Candau and Anquetil [30] examined the correlation between the composition of the initial system and the stability of the final microlatex for AM-NaAMPS copolymers. A comparison between the domain of stability of the microlatex and that of the microemulsion

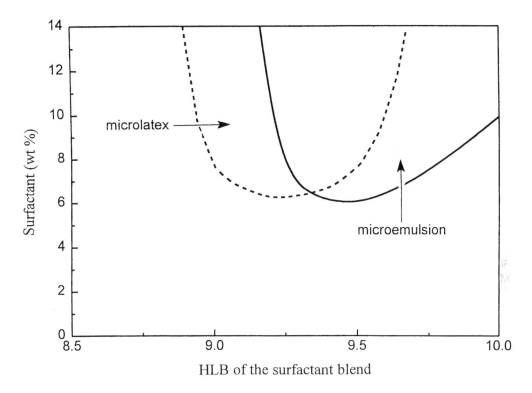

Figure 10 Stability domains for the microlatex (dashed line) and the microemulsion (full line) as a function of the HLB of the surfactant blend in the case of systems based on AM-NaAMPS. (From Ref. 30.)

shows that the former is slightly shifted toward lower HLB values of the nonionic surfactant blend (Fig. 10). The best area for carrying out the polymerization reaction is likely located at the intercept of these two domains. This study demonstrates that it is not necessary to start with a transparent microemulsion to obtain a clear and stable latex. It was proposed that the turbidity observed for some systems was not due to the existence of a polyphasic domain resulting from an insufficient amount of surfactant but rather to the high electrolyte content (ionic monomer), which lowers the cloud point of the ethoxylated surfactants (salting-out). In the course of polymerization, the monomers located partially at the interface diffuse into the particles and the salt-saturated aqueous phase gets progressively used up, accounting for the clarity of the systems at the end of the reaction [16,30].

V. CHARACTERISTICS OF THE FINAL PRODUCTS

The characteristics of both particle latexes and polymers depend critically on the formulation. In fact, the composition chosen for the system depends on whether the ultimate goal of the formulator is to prepare specific polymers or to produce small-sized latex particles. As a general trend, research on O/W systems is focused toward applications based on the properties of latex particles whereas studies on W/O systems are aimed at producing water-soluble polymers of high molecular weights. As a rule, the higher the surfactant-monomer ratio, the smaller the particle size [2,20,85,94,138]. Therefore, high solids contents and small particles can hardly be achieved simultaneously.

A. Polymers

The polymers formed by this process can be recovered from the latexes by precipitation in a large excess of nonsolvent and drying under vacuum. They are further characterized in dilute solutions by means of techniques such as light scattering, gas permeation chromatography (GPC), or viscometry.

The polymer molecular weights are high, usually ranging from 10^6 to 10^7, as expected for polymerization in dispersed media. In some cases, exceptionally high molecular weights have been reported ($M_w \sim 2.5 \times 10^7$–3.3×10^7 [79,81]), but one could question the structure and linearity of the polymers thus formed. When alcohols are used in the formulation, chain transfer reactions can occur that reduce the molecular weight [60,65,101]. The distribution of molecular weights in O/W systems is usually very broad ($M_w/M_n \cong 2$–7, up to 12 in some cases [84]). In the case of water-soluble polymers prepared in W/O or bicontinuous microemulsions, the molecular weight distribution could not be determined because of the difficulties encountered when dealing with such high molecular weight water-soluble polymers [4]. For example, analysis by the classical GPC technique is not possible; high molecular weight monodisperse standards are not available, and there are exclusion phenomena and adsorption tendencies in the pore columns [23]. A decrease in molecular weight is observed upon increasing the initiator concentration. With regard to the effect of monomer/surfactant ratio, a trend similar to that obtained for particle size is observed: The larger the monomer/surfactant ratio, the higher the molecular weight [20,146]. Note that the few polymer chains of high molecular weight (one in the limiting case) confined in the microlatex particle must be strongly collapsed in order to fit such small dimensions ($d < 40$ nm) [4,14].

The large differences observed between the kinetics and mechanism of microemulsion polymerization and those of the other processes (solution, emulsion, and bulk polymerization) can modify some polymer characteristics such as the molecular structure, microstructure, or molecular weight. This effect can be especially important in the case of water-soluble polymers whose efficiency as flocculants depends on their characteristics. Candau and coworkers [140,146] performed a comparative study on the microstructure of copolymers prepared by polymerization in microemulsion, emulsion, or solution. An interesting finding is that microemulsion polymerization seems to improve the structural homogeneity of the copolymers with reactivity ratios close to unity. For example, microemulsion polymerization leads to almost random polyampholytes, whereas those prepared in solution exhibit a strong tendency to alternation [146]. It was shown both theoretically [152] and experimentally that the conformation and solution properties of these ampholytic polymers were directly related to the monomer sequence distribution [149]. At equimolar proportions of anionic and cationic monomers, a random polyampholyte (microemulsion process) is insoluble in water whereas an alternated one (solution process) is soluble. These results are accounted for by the marked differences between the microemulsion process and others, in terms of microenvironment (charge screening and preferential orientation of the monomers at the W/O interface) and mechanism (interparticle collisions with complete mixing) [4,140].

In the above studies, both ionic monomers are located exclusively in the aqueous disperse phase, so monomer partitioning between oil and water can be neglected in the calculation of the reactivity ratios. This no longer holds when styrene is copolymerized with methyl methacrylate [153] or acrylonitrile (AN) [154] in O/W microemulsions. The difference in reactivity ratio values observed between microemulsion and solution polymerizations was attributed to the partitioning of MMA or AN in the droplets and the aqueous phase of the microemulsions. Furthermore, in the case of AN, whose water solubility is higher than that of MMA, the monomer reactivity ratios were found to vary not only with the monomer feed composition but also with the molar ratio of monomers in the microenvironment of the polymerization loci. Obviously, in this case reactivity ratios obtained by using the droplet concentrations as those of reacting monomers lead to apparent rather than true values.

Finally, Puig et al. [155] copolymerized styrene with a water-soluble monomer, acrylic acid, in dodecyltrimethylammonium bromide (DTAB) microemulsions initiated by KPS. In this system, the styrene is solubilized within the droplets while acrylic acid is distributed between the aqueous phase and the W/O interphase. This resulted in copolymers with acrylate units randomly distributed among PS blocks. Initiation was assumed to take place at the micelle surface followed by polymerization in the micellar core.

The applications of the polymers formed by microemulsion polymerization concern essentially porous polymers (see Sec. IV.A) and water-soluble polymers. In the latter case, the microemulsion polymerization can compete with the more classical inverse emulsion polymerization process, and several patents have been issued [156–162]. More details are given in Ref. 4. We recall that the use of a microemulsion rather than an emulsion makes it possible to overcome some of the problems encountered in the latter process such as poor colloidal stability, broad particle size distribution, and excessive amount of coagulum formed during the reaction. It is worthwhile to note that in spite of their stability, inverse microlatices are self-inverting and can be used as such in oil recovery processes and as flocculants in paper manufacture and mining field and water treatment. The very low number of polymer chains per particle achieved in a microemulsion polymerization (one on average) was taken advantage of to produce Mannich polyacrylamides (i.e., polyacrylamides substituted with tertiary aminoethyl groups) without the significant interpolymer cross-linking seen in those prepared in inverse emulsions. The Mannich acrylamide polymers were found to provide superior dewatering characteristics in comparison with those obtained conventionally [162].

Let us also mention two reports on the preparation of polypyrrole [163] and polyaniline [164] conducting polymers in inverse microemulsions.

B. Microlatices

The size of the microlatex particles has usually been determined by QELS and electron microscopy (EM). As a general rule, the diameters of particles prepared in microemulsions are much smaller than those of particles obtained by emulsion polymerization, although they still significantly exceed those of the parent microemulsion droplets due to the particular mechanism occurring during polymerization. They are around 20–60 nm when the starting microemulsions are globular (O/W or W/O). They are bigger, around 50–150 nm, if the microemulsions are initially bicontinuous, simply because of the greater incorporations of monomer (∼25%). As can be expected, the particle size increases with increases in monomer content or decreases in surfactant [2,66,67,80,84,85,91,94,138] and/or initiator [66,75,77] concentration.

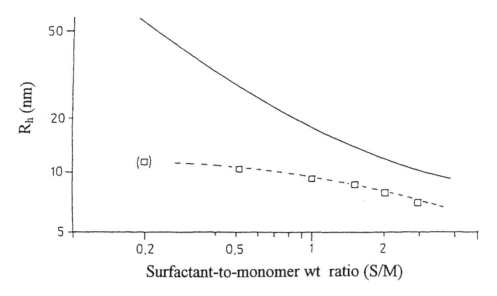

Figure 11 Hydrodynamic radius of polystyrene microlatex particles as a function of the weight ratio of surfactant to monomer. Full line, CTAB; dashed line, TDEA-Cu. (From Ref. 88.)

There have not been detailed studies on the size distribution of the microlatex particles. In general, the polydispersity is characterized by the analysis of the monoexpoentiality of the autocorrelation function of the scattered intensity (QELS), which gives only a rough estimate of this parameter. The variance is generally not very different from that of the starting microemulsion and is in general on the order of 0.03–0.1 [14,16,20,22,25,67,68, 71,85–89,93,94,138]. This corresponds to an index of polydispersity of the particle diameters $d_w/d_n{\sim}1.05$–1.15. Some higher values ($d_w/d_n{\sim}1.40$) have also been reported [70,80]. If we consider W/O microlatices, their polydispersity is thus considerably lower than that of inverse latices prepared by the conventional emulsion process, which commonly have a polydispersity index around 2 [23].

Some efforts were made to control particle size by using appropriate formulations. This is illustrated by the two following examples dealing with direct and inverse microlatices, respectively.

Antonietti and Nestl [88] reported a study using a new class of metallosurfactants that allowed them to reduce both particle size and surfactant concentration. Figure 11 shows the variation of the hydrodynamic radius of polystyrene particles as a function of the weight ratio of surfactant to monomer (S/M) for microemulsions based on a classical surfactant, cetyltrimethylammonium chloride (CTAB) and the metallosurfactant tetradecyl-diethanolamine copper (TDEA-Cu). With this class of surfactants, the authors succeeded in getting a particle diameter as low as 14 nm (width of the distribution = 0.38), with an S/M value of 3. This results in a considerable surface area (${\sim}500$ m^2/g), which renders these systems of interest for subsequent functionalization.

Candau and Anquetil [30] investigated the effect of the type of surfactant on particle size for poly(AM-NaAMPS) microlatices obtained by polymerization in bicontinuous microemulsions. By using different nonionic surfactant blends at the optimal HLB conditions (see Sec. II.C) they showed a significant effect of this parameter on particle size (Fig. 12). The results were accounted for by more or less pronounced salting-out effects

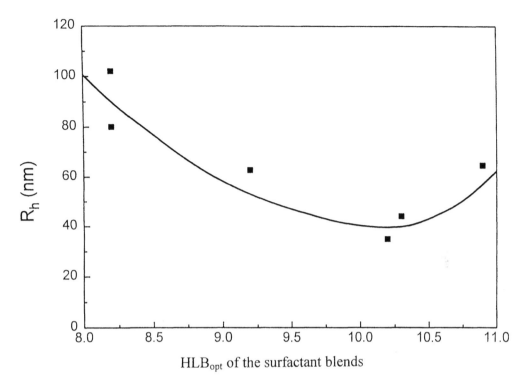

Figure 12 Variation of the hydrodynamic radius of the latex particles versus optimal hydrophilic–hypophilic balance of the surfactant blends. (From Ref. 30.)

of the ethoxylated surfactants by the anionic monomer (NaAMPS). In this respect, the relative proportions of ionic monomers in copolymerization reactions also have a direct effect on the particle size. The dimensions of poly(acrylamide-*co*-sodium acrylate) microlatex particles were shown to decrease linearly with the acrylate content in the comonomer feed [16]. This result was attributed to the combined effects of dehydration of the ethoxylated emulsifier, together with a lowering of the water activity in the particles caused by the sodium acrylate salt.

C. Functionalized Microlatex Particles

The large inner surface area of microemulsions can be easily modified and functionalized by simple copolymerization reactions or by embedding reactions as recently shown by Antonietti et al. [6,86,87,89,96]. Microemulsion copolymerization of styrene with functional monomers acting as cosurfactants and in the presence of a cross-linker resulted in spherical microgels in which most of the functional groups are located at the particle surface [86,87]. The functional additives were shown to stabilize or destroy the initial microemulsions, depending on their nature. Successful additives were based on methacrylate derivatives such as dimethylaminoethyl methacrylate (DAMA).

For most applications, it is desirable to produce particles as small as possible, but, as pointed out earlier, this requires a large amount of surfactant. A description of some procedures that allow the synthesis of functionalized particles of small size follows.

1. Metal-Complexing Microlatices

Metal-complexing microlatices were synthesized via copolymerization of styrene in microemulsions using two comonomers, coupling a 2′,2-bipyridine with or without a spacer to a methacrylic acid unit [6′-methyl-2,2′-bipyridin-6-ylmethyl methacrylate (MBM), structure (1), and 4-(6′methyl-2,2′-bipyridin-6-ylmethoxy)butyl methacrylate (BMBM), structure (2) [96].

In contrast with the case of other pyridine-containing monomers that destabilize the microemulsions [87], these two comonomers led to stable microemulsions and microlatices. Investigation of the size and polydispersity of the microlatex particles revealed that the presence of a C_4 spacer in the comonomer improves the matching between the cationic surfactant (CTAB) and the comonomer. In this case, the average particle size is $d{\sim}30$ nm (width of the distribution $\sigma{\sim}0.3$). Binding experiments with Ni(II), Co(II), Cr(II), and Cu(II) ions in the aqueous disperse phase showed that most of the bipyridine units are located at the latex surface and are thus accessible for the binding process. The binding occurs mostly between two neighboring bipyridine units of the same latex particle. However, there is some pairing between particles, as indicated by an increase in both average particle size and polydispersity.

Another way to functionalize the surface of microlatex particles is to incorporate amphiphilic block copolymers (for example, polystyrene/polyvinylpyridine) as cosurfactants together with the classical surfactants used in the formulation [86,87]. The protruding polyvinylpyridinium chains are anchored to the glassy core through the polystyrene blocks. These blocks copolymers were shown to stabilize the oil/water interface and to effectively bind ions of transition and heavy metals via complexation.

2. Functionalized Microlatices for Biological Applications

The use of microlatices for biological applications is also very attractive. Let us recall that conventional latices prepared from emulsion polymerization are already used for such purposes, for example, in immunoassays, as adsorbents for proteins, for immobilization of enzymes and antibodies, and for controlled release in drug delivery [165]. The latex particle size is in the range 0.1–10 μm. Stable microlatices in the nanosize range (20–30 nm) may be preferred, and some procedures based on inverse microemulsion polymerization have been proposed for the preparation of nanocapsules [4,166–169].

In immunoassay experiments, the size of particles is obviously a critical parameter for the detection sensitivity. A 1992 patent [170] reported the synthesis of nanoparticles with good characteristics with respect to size, polydispersity, and functionalization. The examples reported in this study concerned the copolymerization of styrene with different polymerizable surfactants bearing various functional groups (OH, SO_3H, COOH, etc.). Typical particle sizes were around 20–30 nm. It must be noted that owing to a careful formulation study, the authors succeeded in obtaining microlatices containing about 10 wt% solids stabilized by low surfactant concentration, the excess being eliminated by dialysis.

For medical or pharmaceutical applications, attention must be paid to the problems that can be caused by the possible toxicity of the surfactant remaining in the final product. Antonietti et al. [89] proposed the use of natural, nontoxic, and nondenaturing surfactants based on mixtures of lecithin and sodium chlolate for the formation of globular microemulsions. Pure lecithin is known to form bilayers or liposomes. The role of sodium cholate is to increase the curvature and flexibility of the interfacial layer, allowing the formation of small droplets. The final microlatex particles have a size ranging from 22 to 40 nm, depending on surfactant composition and concentration. The ability to functionalize the surface of these particles was demonstrated by the incorporation of protein molecules.

Daubresse et al [121,171] used an inverse microemulsion polymerization process to prepare a system such as an enzyme nanoparticle recognition molecule that can serve as a diagnostic tool for hybridization of nucleotidic probes. Two functionalized cross-linked microlatices were prepared by copolymerization in AOT micelles of N-6-aminohexyl acrylamide or acrylic acid with AM and N,N-methylenebisacrylamide in the presence of an enzyme (alkaline phosphatase). This process allowed the immobilization of about 50% of the initial enzyme within the latex with retention of most of its original properties. Preliminary results show that nucleotidic probes (29 bases) can be subsequently grafted onto the functionalized latex particles without affecting the catalytic activity of the immobilized enzyme. The process has been extended to the preparation of biodegradable microlatices based on a polymerizable derivative of dextran, i.e., a -glucose polysaccharide.

VI. CONCLUDING REMARKS AND PROSPECTS

The field of polymerization in microemulsions has now reached maturity after a "retardation" period due mainly to the difficulty in preparing stable systems. The intensive efforts of the last few years now allow a good understanding of the mechanisms occurring in this type of polymerization, although some unsolved issues remain. Concerning the preparation of polymers, attention ought to be directed toward polymers and copolymers exhibiting specific properties and microstructures resulting from the microenvironment effect, such as polyampholytes or associating polymers. From the point of view of industrial production, the main limitation arises from the rather high surfactant content used. Nevertheless, thanks to careful optimization of the formulation, industrial water-soluble polymers are now commercially available as advanced flocculants for wastewater treatment [161,162].

Currently, two research areas are in full evolution. One concerns the formation of porous materials, keeping in mind the backdrop of various applications, as briefly pointed out in this chapter. Of particular interest is the recent study of Gan et al. [137] showing the possibility of producing materials with a pore size (50–70 nm) on the order of that of the precursor microemulsion. The second very promising domain for which one can envision many forthcoming developments is related to the functionalization of particles. It can be foreseen that the efforts in this field will be mainly directed toward biological and medical applications.

REFERENCES

1. R. G. Gilbert, *Emulsion Polymerization: A Mechanistic Approach*, Academic, London, 1995.
2. C. Schauber, Thèse Docteur-Ingénieur, Université de Mulhouse, France, 1979.
3. F. Candau, in *Encyclopedia of Polymer Science and Engineering*, 2nd ed., Vol. 9 (H. F. Mark, N. M. Bikales, C. G. Overberger, and G. Menges, eds.), Wiley, 1987, pp. 718–724.
4. F. Candau, in *Polymerization in Organized Media* (C. M. Paleos, ed.), Gordon and Breach, Philadelphia, 1992, Chap. 4, pp. 215–282.
5. B. Gupta and H. Singh, Polym. Plast. Technol. Eng. *31*:635 (1992).
6. M. Antionetti, R. Basten, and S. Lohmann, Macromol. Chem. Phys. *196*:441 (1995).
7. P. G. de Gennes and C. Taupin, J. Phys. Chem. *92*:2294 (1982).
8. L. E. Scriven, Nature (Lond.) *263*:123 (1976).
9. S. Friberg, I. Lapczynska, and G. Gillberg, J. Colloid Interface Sci *56*:19 (1976).
10. P. Winsor, *Solvent Properties of Amphiphilic Compounds*, Butterworth, London, 1954.
11. W. Jahn and R. Strey, J. Phys. Chem *92*:2294 (1988).
12. D. Langevin, Annu. Rev. Phys. Chem. *43*:341 (1992).
13. P. Guering, A. M. Cazabat, and M. Paillette, Europhys. Lett. *2*:953 (1986).
14. F. Candau, Y. S. Leong, G. Pouyet, and S. J. Candau, J. Colloid Interface Sci. *101*:167 (1984).
15. C. Holtzscherer, F. Candau, and R. H. Ottewill, Prog. Colloid Polym. Sci. *81*:81 (1990).
16. F. Candau, Z. Zekhnini, and J. P. Durand, J. Colloid Interface Sci. *114*:398 (1986).
17. C. Holtzscherer and F. Candau, J. Colloid Interface Sci. *125*:97 (1988).
18. C. Holtzscherer and F. Candau, Colloids Surf. *29*:411 (1988).
19. P. Buchert and F. Candau, J. Colloid Interface Sci. *136*:527 (1990).
20. F. Candau and P. Buchert, Colloids Surf. *48*:107 (1990).
21. F. Candau, in *Polymer Association Structures: Microemulsions and Liquids Crystals* (M. El-Nokaly, ed.), ACS Symp. Ser. N. 384, Am. Chem. Soc., Washington, DC, 1989, Chap. 4, pp. 48–61.
22. J. M. Corpart and F. Candau, Colloid Polym. Sci. *271*:1055 (1993).
23. F. Candau, in *An Introduction to Polymer Colloids* (F. Candau and R. Ottewill, eds.), NATO ASI Ser. C No. 303, Klüwer, Dordrecht, 1990, Chap. 3, pp. 73–96.
24. Y. S. Leong and F. Candau, J. Phys. Chem. *86*:2269 (1982).
25. M. T. Carver, E. Hirsch, J. C. Wittmann, R. M. Fitch, and F. Candau, J. Phys. Chem. *93*:4867 (1989).
26. S. A. Safran, G. S. Grest, A. L. R. Bug, and I. Webman, in *Microemulsion Systems* (H. Rosano, and M. Clausse, eds.), Surfact. Sci. Ser. 24, Marcel Dekker, New York, 1987, pp. 235–243.
27. M. Lagues, R. Ober, and C. Taupin, J. Phys. Lett. *39*:487 (1978).
28. A. M. Cazabat, D. Chatenay, D. Langevin, and J. Meunier, Faraday. Disc. Chem. Soc. *76*:291 (1983).
29. B. Lagourette, J. Peyrelasse, C. Boned, and M. Clausse, Nature *5276*:60 (1979).
30. F. Candau and J. Y. Anquetil, in *Micelles, Microemulsions, and Monolayers* (D. O. Shah, ed.) Marcel Dekker, New York, 1998, pp. 193–213.
31. A. Beerbower and M. W. Hill, in *McCutcheon's Detergents and Emulsifier Annual*, Allured Publish. Corp., Ridgewood, NJ, 1971, pp. 223–235.
32. J. O. Stoffer and T. Bone, J. Polym. Sci. Polym. Chem. Ed. *18*:2641 (1980).
33. J. O. Stoffer and T. Bone, J. Dispersion Sci. Technol. *1*:37 (1980).
34. J. O. Stoffer and T. Bone, J. Dispersion Sci. Technol. *4*:393 (1980).
35. V. Vaskova, V. Juranicova, and J. Barton, in *Radical Copolymerization in Heterogeneous Systems*, (J. Barton ed.), Makromol. Chem. Macromol. Symp., Hüthig Wepf Verlag, Basel, 1990, Vol. 31, pp. 201–212.
36. L. M. Gan and C. H. Chew, J. Dispersion Sci. Technol. *4*:291 (1983).
37. L. M. Gan and C. H. Chew, J. Dispersion Sci. Technol. *5*:179 (1984).
38. C. H. Chew and L. M. Gan, J. Polym. Sci. Polym. Chem. Ed. *23*:2225 (1985).
39. F. M. Menger and T. Tsuno, J. Am. Chem. Soc. *112*:6723 (1990).

40. F. M. Menger, T. Tsuno, and G. S. Hammond, J. Am. Chem. Soc. *112*:1263 (1990).
41. B. Gupta and H. F. Eicke, Polym. Sci. Symp. Proc. Polym. 91 *2*:681–684 (1991).
42. B. Gupta and H. Singh, Indian J. Technol. *31*(11):777 (1993).
43. S. Qutubuddin, E. Haque, W. J. Benton, and E. J. Fendler, in *Polymer Association Structures: Microemulsions and Liquid Crystals* (M. El-Nokaly, ed.), ACS Symp. Ser. No. 384, Am. Chem. Soc., Washington, DC, 1989, Chap. 5, pp. 65–83.
44. E. Haque and S. Qutubuddin, J. Polym. Sci. Polym. Lett. Ed. *26*:429 (1988).
45. F. M. Rabagliati, A. C. Falcon, D. A. Gonzales, C. Martin, R. E. Anton, and J. Salager, J. Dispersion Sci. Technol. *7*:245 (1986).
46. W. R. P. Raj, M. Sasthav, and H. M. Cheung, Langmuir *7*:2586 (1991).
47. M. Sasthav and H. M. Cheung, Langmuir *7*:1378 (1991).
48. M. Sasthav, W. R. P. Raj, and H. M. Cheung, J. Colloid Interface Sci. *152*:376 (1992).
49. W. R. P. Raj, M. Sasthav, and H. M. Cheung, Langmuir *8*:1931 (1992).
50. W. R. P. Raj, M. Sasthav, and H. M. Cheung, J. Appl. Polym. Sci. *47*:499 (1993).
51. L. M. Gan, T. H. Chieng, C. H. Chew, and S. C. Ng, Langmuir *10*:4022 (1994).
52. T. H. Chieng, L. M. Gan, C. H. Chew, and S. C. Ng, Polymer *36*:1941 (1995).
53. J. H. Burban, H. E. Mengtao, and E. L. Cussler, AIChE J. *41*(1):159 (1995).
54. J. H. Burban, H. E. Mengtao, and E. L. Cussler, AIChE J. *41*(4):907 (1995).
55. D. A. Kaplin and S. Qutubuddin, Synth. Met. *63*(3):187 (1994).
56. S. S. Atik and K. J. Thomas, J. Am. Chem. Soc. *103*:4279 (1981).
57. S. S. Atik and K. J. Thomas, J. Am. Chem. Soc. *104*:5868 (1982).
58. S. S. Atik and K. J. Thomas, J. Am. Chem. Soc. *105*:4515 (1983).
59. P. Lianos, J. Phys. Chem. *86*:1935 (1982).
60. P. L. Johnson and E. Gulari, J. Polym. Sci. Polym. Chem. Ed. *22*:3967 (1984).
61. A. Jayakrishnan and D. O. Shah, J. Polym. Sci. Polym. Lett. Ed. *22*:31 (1984).
62. C. K. Grätzel, M. Jirousek, and M. Grätzel, Langmuir *2*:292 (1986).
63. H. I. Tang, P. L. Johnson, and E. Gulari, Polymer *25*:1357 (1984).
64. L. M. Gan, C. H. Chew, and S. E. Friberg, J. Macromol. Sci. Chem. *A19*:739 (1983).
65. P. L. Kuo, N. J. Turro, C. M. Tseng, M. S. El-Aasser, and J. W. Vanderhoff, Macromolecules *20*:1216 (1987).
66. J. S. Guo, M. S. El-Aasser, and J. W. Vanderhoff, J. Polym. Sci. Polym. Chem. *27*:691 (1989).
67. V. H. Perez-Luna, J.-E. Puig, V. M. Castano, B. E. Rodriguez, A. K. Murthy, and E. Kaler, Langmuir *6*:1040 (1990).
68. C. Larpent and Th. F. Tadros, Colloid Polym. Sci. *269*:1171 (1991).
69. R. A. Mann, R. G. Gilbert, D. H. Napper, and D. F. Sangster, Preprint of IPUAC Int. Symp. Polymer 91, Melbourne, 1991.
70. L. M. Gan, C. H. Chew, I. Lye, and T. Imae, Polym. Bull. *25*:193 (1991).
71. W. M. Brouwer, J. Appl. Polym. Sci. *38*:1335 (1989).
72. S. Holdcroft and J. E. Guillet, J. Polym. Sci. Polym. Chem. *28*:1823 (1990).
73. L. Feng and K. Y. S. Ng, Macromolecules *23*:1048 (1990).
74. L. Feng and K. Y. S. Ng, Colloids Surf. *53*:349 (1991).
75. J. S. Guo, E. D. Sudol, J. W. Vanderhoff, and M. S. El-Aasser, J. Polym. Sci. Polym. Chem. Ed. *30*:691 (1992).
76. J. S. Guo, E. D. Sudol, J. W. Vanderhoff, and M. S. El-Aasser, J. Polym. Sci. Polym. Chem. Ed. *30*:703 (1992).
77. J. E. Puig, V. H. Perez-Luna, M. Perez-Gonzales, E. R. Macias, B. E. Rodriguez, and E. W. Kaler, Colloid Polym. Sci. *271*:114 (1993).
78. B. G. Manders, A. M. van Herk, A. L. German, J. Sarnecki, R. Schomäcker, and J. Schweer, Macromol. Chem. Rapid Commun. *14*:693 (1993).
79. L. M. Gan, C. H. Chew, and I. Lye, Makromol. Chem. *193*:1249 (1992).
80. L. M. Gan, C. H. Chew, I. Lye, L. Ma, and G. Li, Polymer *34*:3860 (1993).
81. L. M. Gan, C. H. Chew, J. H. Lim, K. C. Lee, and L. H. Gan, Colloid Polym. Sci. *272*:1082 (1994).
82. L. M. Gan, N. Lian, C. H. Chew, and G. Z. Li, Langmuir *10*:2197 (1994).

83. L. M. Gan, C. H. Chew, K. C. Lee, and S. C. Ng, Polymer *35*:2659 (1994).

84. L. M. Gan, K. C. Lee, C. H. Chew, and S. C. Ng, Langmuir *11*:449 (1995).

85. M. Antonietti, W. Bremser, D. Müschenborn, C. Rosenauer, B. Schuppe, and M. Schmidt, Macromolecules *24*:6636 (1991).

86. M. Antonietti, S. Lohmann, and C. van Niel, Macromolecules *25*:1139 (1992).

87. M. Antonietti, S. Lohmann, and W. Bremser, Prog. Colloid Polym. Sci. *89*:62 (1992).

88. M. Antonietti and T. Nestl, Macromol. Chem. Rapid Commun. *15*:111 (1994).

89. M. Antonietti, R. Basten, and F. Groehn, Langmuir *10*:2498 (1994).

90. L. M. Gan, C. H. Chew, K. C. Lee, and S. C. Ng, Polymer *34*:3064 (1993).

91. L. M. Gan, C. H. Chew, S. C. Ng, and S. E. Loh, Langmuir *9*:2799 (1993).

92. L. A. Rodriguez-Guadarrama, E. Mendizabal, J. E. Puig, and E. W. Kaler, J. Appl. Polym. Sci. *48*:775 (1993).

93. F. Bleger, A. K. Murthy, F. Pla, and E. W. Kaler, Macromolecules *27*:2559 (1994).

94. C. Schauber and G. Riess, Makromol. Chem. *190*:725 (1989).

95. A. P. Full, J. E. Puig, L. U. Gron, E. W. Kaler, J. R. Minter, T. H. Mourey, and J. Texter, Macromolecules *25*:5157 (1992).

96. M. Antonietti, S. Lohmann, C. D. Eisenbach, and U. S. Schubert, Macromol. Chem. Rapid Commun. *16*:283 (1995).

97. J. Texter, L. E. Oppenheimer, and J. R. Minter, Polym. Bull. *27*:487 (1992).

98. P. Potisk and I. Capek, Angew. Makromol. Chem. *222*:125 (1994).

99. I. Capek and P. Potisk, Makromol. Chem. Phys. *196*:723 (1995).

100. I. Capek and P. Potisk, J. Polym. Sci. Polym. Chem. Ed. *33*:1675 (1995).

101. Y. S. Leong, G. Riess, and F. Candau, J. Chim. Phys. *78*:279 (1981).

102. Y. S. Leong, S. J. Candau, and F. Candau, in *Surfactants in Solution*, Vol. 3 (K. L. Mittal and B. Lindman, eds.), Plenum, New York, 1984, pp. 1897–1910.

103. F. Candau, in *Comprehensive Polymer Science: Chain Polymerization II*, Vol. 4 (G. G. Eastmond, A. Ledwith, and P. Sigwalt, eds), Pergamon, Oxford, 1988, Chap. 13, pp. 225–229.

104. F. Candau, Y. S. Leong, and R. M. Fitch, J. Polym. Sci. Polym. Chem. Ed. *23*:193 (1985).

105. M. T. Carver, U. Dreyer, R. Knoesel, F. Candau, and R. M. Fitch, J. Polym. Sci. Polym. Chem. Ed. *27*:2161 (1989).

106. M. T. Carver, F. Candau, and R. M. Fitch, J. Polym. Sci. Polym. Chem. Ed. *27*:2179 (1989).

107. J. P. Fouassier, D. J. Lougnot, and I. Zuchowicz, Eur. Polym. J. *11*:933 (1986).

108. N. Girard, Diplôme d'Etudes Approfondis de l'Université de Rennes, 1989.

109. C. Holtzscherer, S. J. Candau, and F. Candau, in *Surfactants in Solution*, Vol. 6 (K. L. Mittal and P. Bothorel, eds.), Plenum, New York, 1986, pp. 1473–1481.

110. V. Vaskova, V. Juranicova, and J. Barton, Angew. Makromol. Chem. *191*:717 (1990).

111. V. Vaskova, V. Juranicova, and J. Barton, Angew. Makromol. Chem. *192*:989 (1991).

112. V. Vaskova, V. Juranicova, and J. Barton, Makromol. Chem. *192*:1339 (1991).

113. J. Barton, Makromol. Chem. Rapid Commun. *12*:675 (1991).

114. V. Vaskova, Z. Hlouskova, J. Barton, and V. Juranicova, Makromol. Chem. *193*:627 (1992).

115. J. Barton, Polym. Int. *30*:151 (1993).

116. J. Barton, J. Tino, Z. Hlouskova, and M. Stillhammerova, Polym. Int. *34*:89 (1994).

117. I. Janigova, K. Csomorova, M. Stillhammerova, and J. Barton, Makromol. Chem. Phys. *195*:3609 (1994).

118. V. Vaskova, M. Stillhammerova, and J. Barton, Chem. Pap. *48*:355 (1994).

119. I. Lacik, J. Barton, and G. G. Warr, Makromol. Chem. Phys. *196*:2223 (1995).

120. E. D. Garcia, L. E. Oppenheimer, and J. Texter, in *Electrochem. Colloids Dispersions* (R. A. Mackay and J. Texter, eds.), VCH, New York, 1992, pp. 257–270.

121. C. Daubresse, C. Grandfils, R. Jerome, and P. Teyssie, J. Colloid Interface Sci. *168*:222 (1994).

122. E. J. Beckman and R. D. Smith, J. Supercrit. Fluids *3*(4):205 (1990).

123. E. J. Beckman and R. D. Smith, J. Phys. Chem. *94*:345 (1990).

124. M. R. Ferrick, J. Murtagh, and J. K. Thomas, Macromolecules *22*:1515 (1989).

125. Y. J. Yang and J. B. F. N. Engberts, Eur. Polym. J. *28*:881 (1992).

126. J. S. Guo, M. S. El-Aasser, E. D. Sudol, H. J. Yue, and J. W. Vanderhoff, J. Colloid Interface Sci. *140*:175 (1990).

127. S. E. Loh, L. M. Gan, C. H. Chew, and S. C. Ng, J. Macromol. Sci. Pure Appl. Chem. *A32*(10):1681 (1995).

128. L. M. Gan, K. C. Lee, C. H. Chew, E. S. Tok, and S. C. Ng, J. Polym. Sci. Polym. Chem. *33*:1161 (1995).

129. F. Candau and R. H. Ottewill (eds.), *An Introduction to Polymer Colloids and Scientific Methods for the Study of Polymer Colloids and their Applications*, Vols. 1 and 2, NATO ASI Ser. C No. 303, Kluwer, Dordrecht, 1990.

130. N. Sutterline, H. J. Kurth, and G. Markett, Angew. Makromol. Chem. *177*:1549 (1976).

131. J. H. Bayendale, M. G. Evans, and S. K. Kilham, J. Polym. Sci. *1*:466 (1946).

132. L. M. Gan, C. H. Chew, S. Friberg, and T. Higashimura, J. Polym. Sci. Polym. Chem. Ed. *19*:1585 (1981).

133. L. M. Gan, C. H. Chew, and S. E. Friberg, J. Polym. Sci. Polym. Chem. Ed. *21*:513 (1983).

134. S. Friberg and P. Liang, J. Polym. Sci. Polym. Chem. Ed. *22*:1699 (1984).

135. T. D. Li, C. H. Chew, S. C. Ng, L. M. Gan, W. K. Teo, J. Y. Gu, and G. Y. Zhang, J. Macromol. Sci. Pure Appl. Chem. *A32*:969 (1995).

136. T. H. Chieng, L. M. Gan, C. H. Chew, L. Lee, S. C. Ng, K. L. Pey, and D. Grant, Langmuir *11*:3321 (1995).

137. L. M. Gan, T. D. Li, C. H. Chew, W. K. Teo, and L. H. Gan, Langmuir *11*:3316 (1995).

138. C. Holtzscherer, J. P. Durand, and F. Candau, Colloid Polym. Sci. *265*:1067 (1987).

139. C. Holtzscherer and F. Candau, in *Surfactants in Solution*, Vol. 10 (K. L. Mittal, ed.), Plenum, New York, 1989, pp. 223–231.

140. F. Candau, Z. Zekhnini, and F. Heatley, Macromolecules *19*:1895 (1986).

141. F. Candau, Z. Zekhnini, F. Heatley, and E. Franta, Colloid Polym. Sci. *264*:676 (1986).

142. F. Candau, Z. Zekhnini, and J. P. Durand, Prog. Colloid Polym. Sci. *73*:33 (1987).

143. F. Candau, D. Collin, and F. Kern, in *Copolymerization in Dispersed Media* (J. Guillot and C. Pichot, eds.), Makromol. Chem. Macromol. Symp., Vol. 35 Hüthig Wepf Verlag, Basel, 1990, pp. 105–119.

144. J. M. Corpart, J. Selb, and F. Candau, in *Polymer 91: International Symposium on Polymer Materials* (K. P. Ghiggino ed.), Makromol. Chem. Macromol. Symp., Vol. 53, Hüthig Wepf Verlag, Basel, 1992, pp. 253–265.

145. J. M. Corpart and F. Candau, Macromolecules *26*:1333 (1993).

146. J. M. Corpart, J. Selb, and F. Candau, Polymer *34*:3873 (1993).

147. M. Skouri, J. P. Munch, S. J. Candau, and F. Candau, Macromolecules *27*:69 (1994).

148. F. Candau, In *Radical Copolymers in Dispersed Media* (J. Guillot, A. Guyot, and C. Pichot, eds.), Makromol. Chem. Macromol. Symp., Vol. 92, Hüthig Wepf Verlag, Basel, 1995, pp. 169–178.

149. S. Neyret, A. Baudouin, J. M. Corpart, and F. Candau, Nuevo Cim. *16*:669 (1995).

150. S. Neyret, L. Ouali, F. Candau, and E. Pefferkorn, J. Colloid Interface Sci. *176*:86 (1995).

151. F. Candau, P. Buchert, and I. Krieger, J. Colloid Interface Sci. *140*:466 (1990).

152. J. Wittmer, A. Johner, and J. F. Joanny, Europhys. Lett. *24*:263 (1993).

153. L. M. Gan, K. C. Lee, C. H. Chew, S. C. Ng, and L. H. Gan, Macromolecules *27*:6335 (1994).

154. K. C. Lee, L. M. Gan, C. H. Chew, and S. C. Ng, Polymer *36*:3719 (1995).

155. J. E. Puig, S. Corona Galvan, A. Maldonado, P. C. Schulz, B. E. Rodriguez, and E. Kaler, J. Colloid Interface Sci. *137*:308 (1990).

156. F. Candau, Y. S. Leong, N. Kohler, and F. Dawans, Fr. Patent 2,254,895, to CNRS-IFP (1984).

157. J. P. Durand, D. Nicolas, N. Kohler, F. Dawans, and F. Candau, Fr. Patents 2,565,623 and 2,565,592, to IFP (1987).

158. J. P. Durand, D. Nicolas, and F. Candau, Fr. Patent 2,567,525, to IFP (1987).

159. F. Candau and P. Buchert, Fr. Patent 87,08295, to Norsolor (1987).

160. F. Candau, P. Buchert, and M. Esch, Fr. Patent 88,17306, to Norsolor-Orkem (1988).

161. J. J. Kozakiewicz and S. Y. Yuang, U.S. Patent 4,956,399, to American Cyanamid (1990).

162. J. J. Kozakiewicz and S. Y. Huang, U.S. Patent 5,132,023, to American Cyanamid (1992).

163. G. Markham, T. M. Obey, and B. Vincent, Colloids Surf. *51*:239 (1990).
164. L. M. Gan, C. H. Chew, H. S. O. Chan, and L. Ma, Polym. Bull. *31*:347 (1993).
165. K. Nustad, S. Funderud, T. Ellingsen, A. Berge, and J. Ugelstad, in *Scientific Methods for the Study of Polymer Colloids and Their Applications* (F. Candau, and R. Ottewill, eds.), NATO ASI Ser. C No. 303, Kluwer, Dordrecht, 1990, pp. 517–565.
166. C. Vauthier-Holtzscherer, S. Benabbou, G. Spenlehauer, M. Veillard, and P. Couvreur, S. T. P. Pharm. Sci. *1*(2):109 (1991).
167. G. Birrenbach and P. P. Speiser, J. Pharm. Sci. *65*:1763 (1976).
168. M. R. Gasco and M. Trotta, Int. J. Pharm. *29*:267 (1986).
169. Ch. Cadic, B. Dupuy, Ch. Basquez, and D. Ducassou, Innov. Tech. Biol. Med. *11*:412 (1990).
170. C. Larpent, J. Richard, and S. Vaslin, Fr. Patent 92,06759, to Prolabo (1992).
171. C. Daubresse, Thèse de Doctorat, Université de Liège, Belgium, 1993.

23

Enzymatic Reactions in Microemulsions

Krister Holmberg
Chalmers University of Technology, Göteborg, Sweden

I. INTRODUCTION

Enzymes have traditionally been used in an aqueous medium, and the vast majority of studies on kinetics and stereoselectivity of bio-organic reactions have been performed on water-based systems. However, in recent years, much interest has been devoted to the use of enzymes in media of low water content. In comparison to an aqueous reaction medium, solvents of low polarity

>   Are better solvents for many lipophilic substrates
>   Can shift thermodynamic reaction equilibria toward condensation
>   May improve thermal stability of the enzymes, enabling reactions to be carried out at higher temperatures
>   May lead to simpler workup, as nonpolar solvents are more easily removed by evaporation under reduced pressure

On the negative side, as always with nonaqueous formulations, are the environmental problems associated with organic solvents.

From a practical point of view the improved solubility of nonpolar reactants is very important. Low reactant solubility in water is a frequently encountered problem in enzyme-catalyzed reactions, leading to low production capacity per vessel volume. Also, the possibility of using hydrolytic enzymes such as lipases, esterases, peptidases, and amylases to catalyze condensation reactions instead of bond cleavage is of considerable practical importance because it opens new ground for enzyme-catalyzed processes.

The characteristic features of enzymes in media of low water content are ideally suited for lipases, whose natural substrates usually are of very low water solubility. Consequently, lipase function in nonpolar media and the use of lipase to catalyze various types of reactions such as ester synthesis, transesterification, and ester hydrolysis have been extensively investigated.

Some enzymes have been found to function in essentially water-free systems. For instance, Klibanov [1] showed that both pancreatic and microbial lipases catalyze transesterification in nearly anhydrous organic solvents. However, it seems that the majority of enzymes lose their activity more rapidly in water-free systems than in media containing a small amount of water. The interest in microemulsions as reaction media stems from the fact that in such systems the amount of water in the formulation can be varied almost

713

at will and that the microheterogeneity of the systems provides a good environment for the enzyme. In the majority of cases, L_2 microemulsions, i.e., systems with a water-in-oil (W/O) structure, have been used, but the literature also contains examples of enzymatic reactions being conducted in microemulsions of higher water content having either a bicontinuous or oil-in water (O/W) structure.

In the literature there is often no clear distinction between microemulsions and micellar systems. For instance, a system containing a small amount of water solubilized in hydrocarbon may be referred to as a W/O microemulsion (or an L_2 microemulsion) or as a system of reverse micelles (or swollen reverse micelles). It has been suggested [2] that the borderline between reverse micelles and microemulsions droplets should be defined by the water-to-surfactant ratio; above a molar ratio of 15, the system should be referred to as a microemulsion. In this chapter no such distinction is made. All systems containing oil and water together with surfactant are termed microemulsions, regardless of the relative component proportions.

Wells and his coworkers seem to have been the first to work systematically with proteins in microemulsions [3–5], although the concept of solubilizing a protein in a hydrocarbon solvent had been described earlier [6]. The use of microemulsions as a medium for bio-organic reactions was pioneered by the groups of Martinek [7], Luisi [8], and Menger [9] toward the end of the 1970s. Interest in the area grew rapidly, and the field is today subject to considerable activity in academia and in industry.

Martinek introduced the term "micellar enzymology" to cover the area of enzymatic catalysis in these systems [10]. The phrase deliberately alludes to molecular biology. The use of enzymes in water-poor media is common in biological systems. Many enzymes, including lipases, esterases, dehydrogenases, and oxidoreductive enzymes, often function in the cells in microenvironments that are hydrophobic in nature. Also, the use of enzymes in microemulsions is not an artificial approach per se. In biological systems many enzymes operate at the interface between hydrophobic and hydrophilic domains, and these interfaces are often stabilized by polar lipids and other natural amphiphiles [11]. It is an established fact today that biological membranes need not be composed of flat bilayers of lipid molecules. Non-bilayer lipid structures seem to be essential for many processes occurring in the living cell, such as fusion and compartmentalization of membranes [12]. So-called lipid particles can be seen as reverse micelles sandwiched between monolayers of polar lipids. It is also known that many enzymes induce the formation of such bilayer structures upon incorporation into both model and biological membranes. Hence, studies of enzymes in W/O microemulsions are of relevance to biology in a wider sense than biocatalysis. However, this review chapter does not attempt to cover work specifically oriented to mimic the function of biological membranes ("membrane mimetics").

II. ENZYME ACTIVITY IN MICROEMULSIONS

A. Structure of the Enzyme-Containing Droplet

In W/O microemulsions the enzyme molecules are invariably located (entrapped) in the water droplet region, but the exact position may vary depending on the hydrophilic–lipophilic balance of the enzyme (and to some extent on the nature of the solvent). A truly hydrophilic enzyme will be located in the water core of the droplet, surrounded by a water layer. α-Chymotrypsin is an example of such a protein. A surface-active enzyme, such as lipase, has a strong driving force for the oil/water interface, where it competes with the surfactant [13]. Access to the interface is probably necessary for the enzyme to catalyze reactions with hydrophobic substances situated in the continuous hydrocarbon domain. (In fact, lipases need a hydrophobic surface in order to open the lid covering the active site [14].) Strongly hydrophobic enzymes will be partially located in the hydrocarbon region. The different locations of the enzyme in microemulsion droplets are schematically illustrated in Fig. 1 [15].

Certain large, hydrophobic proteins, such as bacteriorhodopsin, seem to induce a more complicated microstructure in W/O microemulsions. It has been claimed that the molecule extends over several water droplets, adapting a conformation such that polar parts of the molecule are incorporated into the water droplets and nonpolar parts are exposed to the continuous hydrocarbon domain [16].

Luisi and coworkers [17] studied how size and structure of reversed micelles change upon uptake of a large guest molecule such as a protein. The basis for the discussion is that two different water pools of different radii, R, are being spontaneously created: protein-containing micelles, R_f, and unfilled micelles, R_e (f stands for filled and e for empty). It was demonstrated that the initial micelles of radius R_0 split after addition of enzyme into a larger one of radius R_f and a smaller one of radius R_e. Thus, the micelles accommodating the protein grow in size at the expense of the unfilled micelles. The size difference between filled and unfilled micelles was claimed to be approximately 2 : 1, but it varied with the enzyme.

An alternative model, suggested by Levashov et al. [18], assumes that such an increase in size occurs only when the inner cavity of the initial empty micelle is smaller than the protein molecule. If the size of the initial water cavity is equal to or exceeds that of the protein molecule, then protein entrapment will not lead to any substantial increase in the size of the water droplet. The validity of the so-called fixed-size model was supported by ultracentrifugation measurements [19]. The protein-containing micelles were found to contain practically the same numbers of both surfactant and water molecules as the unfilled ones. A variety of other analytical techniques have been used in an attempt to estimate the sizes of filled and empty droplets; these include quasi-elastic light scattering [20], fluorescence recovery after fringe pattern photobleaching [21], and small-angle neutron scattering [22,23]. With light scattering close to the optical match point of the microemulsion particles, the size distribution of the droplets has also been measured [24]. In general, the results are more in favor of the fixed-size model than of that suggesting large R_f and small R_e. However, there is a level of ambiguity in the reported results, either because of limitations used for data analysis or because of artifacts introduced by the experimental approach [25].

Attempts have been made to construct a thermodynamic model to predict the sizes of protein-containing and unfilled droplets as a function of system parameters such as ionic strength, water content, and protein net charge, size, and concentration [25,26]. In the more elaborate of these methods [25] excellent agreement was obtained between model predictions and small-angle scattering data on micellar sizes for the system α-chymotrypsin in a sodium bis(2-ethylhexyl)sulfosuccinate (AOT)–water–isooctane microemulsion. Both the model and experimental results indicate that filled droplets may be bigger or smaller than empty droplets, depending on conditions. Salt concentration is found to be an important factor governing droplet size, and this effect can be attributed to different degrees of screening of the electrostatic charges within the droplets, filled or empty. For empty droplets there is a repulsive charge interaction from the surfactant headgroups across the water pool. With an increase in salt concentration, screening of the electrostatic interaction is increased, leading to a reduction in droplet

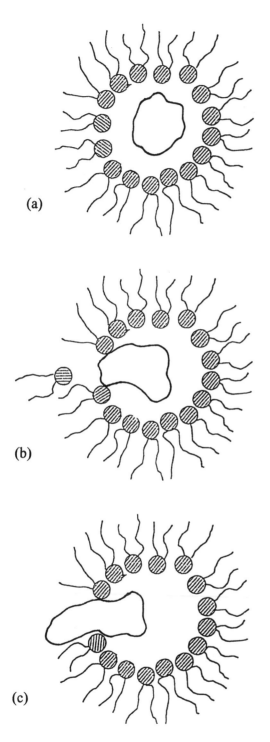

Figure 1 Illustration of an enzyme residing (a) in the droplet core, (b) partly in the interfacial region, and (c) partly in the continuous hydrocarbon domain.

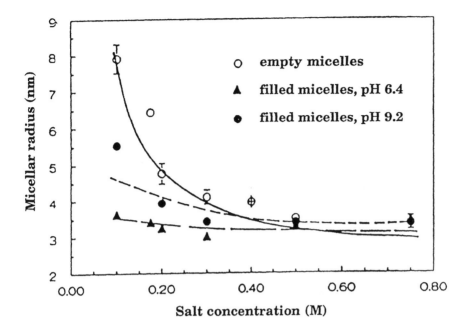

Figure 2 Micellar radius as a function of salt concentration for the system α-chymotrypsin in 100 mM AOT–isooctane–water. Buffers of sodium phosphates and carbonate were used as salt. Enzyme pI is 8.2–8.6. (Redrawn from Ref. 25.)

size. For the enzyme-filled droplet the protein–surfactant charge interactions will be important, and these will depend on the net charge of the protein. At pH<pI of the protein, the across-droplet repulsive interactions are to a large extent screened by the presence of the protein in the droplet core. (The surfactant used, AOT, is negatively charged.) Instead, attractive interactions between protein and surfactant headgroups will determine droplet size. With increasing salt concentration these interactions become less pronounced, but the droplet size does not vary much with ionic strength. At pH>pI, on the other hand, protein–surfactant interaction is repulsive, and the unfavorable interaction becomes less pronounced as the salt concentration is increased, leading to the formation of smaller micelles. The net result of having the protein, a large low dielectric body, in the center of the droplet is that droplets smaller than the empty ones are produced unless the salt concentration is high, as is shown in Fig. 2.

The effect of the enzyme on droplet size is, of course, very dependent on protein net charge, as shown in Fig. 3. Considering that the surfactant AOT is an anionic species, it is interesting to note that at the electrolyte concentration used, filled droplets are smaller than empty ones even when the protein has a slightly negative net charge. The droplet size does not vary with protein concentration, however.

It must be stressed that the above findings concerning relative sizes of filled and empty droplets may not be true for other systems. When the protein is surface-active, which is the case for the most frequently employed enzyme, lipase, the model cannot readily be used, because the protein does not reside in the droplet core. Furthermore, the model treats all ions as point charges; hence, specific ion effects cannot be accounted for.

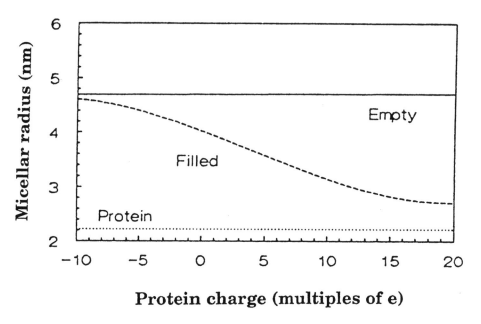

Figure 3 Model prediction of micellar radius as a function of protein charge for a system consisting of a protein of radius 2.2 nm in 100 mM AOT–isooctane–water using 0.2 M sodium buffers. (From Ref. 25.)

B. Relation Between Enzyme Activity and Size of the Enzyme-Containing Droplets

Many investigations have been carried out regarding the effect of water content on the catalytic activity of enzymes in water-poor media. The role of the enzyme-bound water has not been fully clarified; however, the water dependence clearly differs from one enzyme to another. For instance, some lipases exhibit high activity and good stability in organic solvents containing only traces of water, whereas other lipases reach optimal activity at a relatively high water content [27]. A related enzyme, α-chymotrypsin, has been studied in detail with regard to the effect of hydration on activity [28]. Circular dichroism measurements indicated that at very low water content the protein was essentially frozen. The first molecules of water added interacted predominantly with ionizable groups. This led to an even higher degree of rigidity of the protein molecule. A minimum in activity was displayed at a water-to-surfactant ratio of 5. (The molar ratio of water to surfactant is a commonly used parameter and is often referred to as W_0). Further addition of water first led to hydration of hydrogen-binding sites, and subsequently nonpolar regions were covered by a water monolayer. The catalytic activity increased with increasing W_0 up to a value of around 10. Electron paramagnetic resonance (EPR) spectroscopy is also a useful technique to study enzyme active site conformations in organic solvents and in microheterogeneous systems [29]. EPR has been used to characterize the active site dynamics of α-chymotrypsin in W/O microemulsions [30].

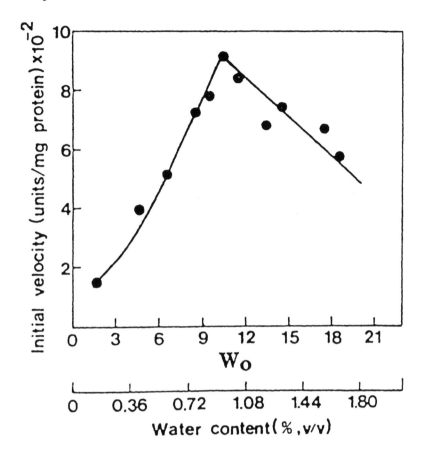

Figure 4   Initial velocity of lipase in W/O microemulsions (50 mM AOT in isooctane) of varying W_0 (surfactant-to-water ratio). The overall enzyme concentration is the same at all W_0 values. (From Ref. 32.)

It has been demonstrated by several groups that the catalytic activity of enzymes confined in microemulsion droplets varies with overall water content. For a variety of enzymes it has been found that there is a bell-shaped dependence of activity on W_0 [31–35]. Figure 4 shows a representative example.

For constant surfactant concentration there is a linear correlation between water concentration and the size of the water droplet in a W/O microemulsion [36]. Thus, droplet size is proportional to W_0. [For the most commonly used surfactant, AOT, the droplet radius, R_d, can be directly obtained from the W_0 value from the relationship R_d (nm)=0.175 W_0 [37].] In general, it seems that maximum activity occurs around a value of W_0 at which the size of the droplet is equivalent to or slightly larger than that of the entrapped enzyme. Figure 5 shows data compiled for 17 different enzymes on the relationship between the hydrodynamic radius of the protein and droplet radius at optimum activity [38]. With only one exception (lipoxygenase), the correlation between the two radii is excellent.

An illuminating example of how droplet size can determine enzymatic activity is given in Fig. 6 [39]. The heterodimeric enzyme γ-glutamyltransferase was studied in a microemulsion based on AOT–isooctane–aqueous buffer. The enzyme consists of one light

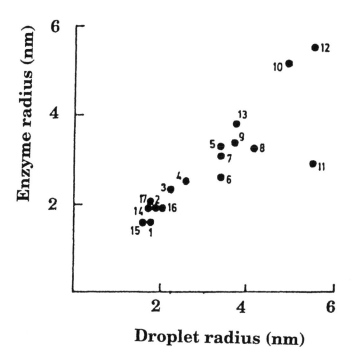

Figure 5 Correlation between radius of entrapped enzyme and that of a microemulsion droplet at maximum activity. The following enzymes were investigated: 1, lysozyme; 2, cytochrome c; 3, trypsin; 4, α-chymotrypsin; 5, ribonuclease; 6, pepsin; 7, lipase; 8, peroxidase; 9, acid phosphatase; 10, alcohol dehydrogenase; 11, lipoxygenase; 12, PGH synthetase; 13, laccase; 14, alkaline phosphatase; 15, lactate dehydrogenase; 16, catalase; 17, alcohol dehydrogenase. (From Ref. 38.)

and one heavy chain that are noncovalently linked. γ-Glutamyltransferase catalyzes γ-glutamyl residue transfer from a donor molecule to an acceptor molecule. Depending on the nature of the acceptor, three types of reactions may occur, all catalyzed by the enzyme:

1. Transpeptidation, i.e., transfer of the γ-glutamyl residue from one peptide to a different peptide
2. Autotranspeptidation, i.e., transfer of the γ-glutamyl residue between two identical peptides
3. Hydrolysis, i.e., splitting off of the glutamyl residue

As is seen from Fig. 6, each of the three reactions exhibits three maxima with respect to W_0 (although no peaks are very distinct). As is also shown in the figure, the radii of the water droplets at the maxima correspond well with the radii of the light chain, heavy chain, and heterodimer, which have molecular masses of 21, 54, and 75 kDa, respectively. This is a strong indication that in the microemulsion with its limited water domain size the two chains that make up the enzyme dissociate and the individual chains become entrapped in droplets. The experiment also shows that both chains as well as the heterodimer are effective catalysts for transpeptidation, that the heavy chain and the dimer are effective for the autotranspeptidation, and that the light chain is most effective for the hydrolysis reaction.

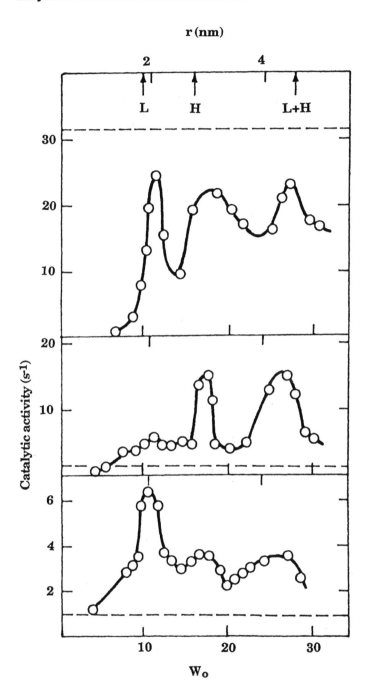

Figure 6 Dependence of γ-glutamyl transferase activity on W_0 in an AOT–octane–water microemulsion. The figure gives activity data for three reactions catalyzed by the enzyme: transpeptidation (top), autotranspeptidation (middle), and hydrolysis (bottom). Values of catalytic activity of the enzyme in aqueous solution, measured under conditions of pH optima, are shown by the dashed lines. A scale of droplet radii corresponding to the W_0 scale as well as indications of the radii of the light (L) and heavy (H) subunits and of the heterodimer (L+H) are given on the upper extension line. (From Ref. 39.)

Micelles containing heavy chains could be separated from those containing light chains by centrifugation, and experiments conducted on the individual chains confirmed the above-described pattern of catalytic efficiency [39].

It was later demonstrated that W_0 is not the only factor governing enzymatic activity in W/O microemulsions. Both a hydrophilic enzyme, α-chymotrypsin, and a lipophilic enzyme, hydroxysteroid dehydrogenase, varied in activity with surfactant concentrations at constant W_0 [36]. Evidently, at least two parameters, molar water-to-surfactant ratio (W_0) and surfactant concentration, are decisive for enzymatic activity in these systems.

There have been many attempts to explain the bell-shaped curve of enzyme activity versus W_0. It is likely that several factors contribute and that the relative importance of different parameters varies with the type of enzyme studied [40,41]. However, it seems probable that diffusion effects play a major role, and a diffusion model applicable to a hydrophilic enzyme located in the core of the water droplet and hydrophilic substrates also situated in the droplets was worked out by Walde and coworkers [42,43]. Before the enzyme-catalyzed reaction can take place, two different diffusion processes must occur. In the first of these, an interdroplet diffusion step, drops containing the substrate and drops containing the enzyme must collide. In the second process, an intradroplet diffusion step, the substrate reaches the enzyme's active site. Whereas the rate of the first process increases with droplet radius, the reverse is true for the second process. These two counteracting dependencies of reaction rate on droplet size (and thus on W_0 at constant surfactant concentration) may lead to a bell-shaped activity versus W_0 curve.

For two out of the three reactions illustrated in Fig. 6, enzyme activity is higher in the microemulsion than in bulk water. (The dashed lines represent activity in aqueous solution.) This phenomenon, called "superactivity," has been noted by many groups [44–46]. In general, enzymatic reactions in microemulsions seem to obey Michaelis–Menten kinetics, but enzymes in such microheterogeneous media often show different specificity than that observed in aqueous solution. For instance, alcohol dehydrogenase–catalyzed oxidation of aliphatic alcohols gives a maximum in reaction rate for octane in microemulsion and for butane in bulk solution [47]. The regiospecificity of lipases with respect to acyl group exchange in triglyceride can be different in microemulsion than in aqueous buffer [48]. It has also been found that the turnover number, k_{cat}, can be orders of magnitude higher in microemulsion than in aqueous solution. However, the Michaelis constant, K_M, is often also large, leading to an overall reaction rate in the same range as in aqueous solution. (K_M appears in the denominator and k_{cat} in the numerator in the Michaelis–Menten equation.) Figure 7 shows an illustrative example of superactivity. The figure also illustrates that the dramatic increase in k_{cat} occurs only at a specific optimum microemulsion composition.

The anomalous activity characteristics have been attributed to conformational changes of the solubilized enzyme [49], but more recent spectroscopic studies seem to indicate that this is not the main cause. Solubilization of an enzyme into microemulsion droplets does not normally lead to major conformational alterations, as indicated, e.g., by fluorescence and phosphorescence spectral investigations [28,50]. The situation is complex, however, and it has been shown by circular dichroism (CD) measurements that the influence of the oil/water interface on enzyme conformation may vary even between enzymes belonging to the same class [51]. In the case of human pancreatic lipase, the conformation of the polypeptide chain is hardly altered after the enzyme is transferred from a bulk aqueous solution to the microenvironment of reverse micelles. Conversely, the CD spectra of the lipases from

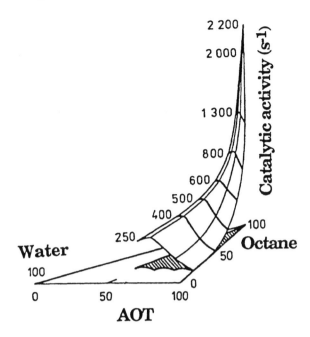

Figure 7 Influence of composition (in weight percent) on rate of pyrocatechol oxidation. The enzyme laccase was added to a microemulsion based on AOT–octane–aqueous buffer. The value of k_{cat} in bulk water is 28 s^{-1}. (From Ref. 44.)

Candida rugosa and *Pseudomonas* sp. are considerably different in reverse micelles from those in aqueous solution, indicating that both enzymes lose their native structure in the microemulsion environment [51].

It has been proposed that repulsive interactions between charges at the droplet surface and charged substrate ions are the main reason for the anomalous behavior of solubilized enzymes. (The majority of studies of enzyme activity in W/O microemulsions have been performed with the anionic surfactant AOT.) Both the superactivity and the bell-shaped curve of activity versus W_0 have been theoretically explained in this way [52,53]. However, later investigations with α-chymotrypsin in AOT-based microemulsions showed that superactivity could be obtained with uncharged as well as with negatively charged substrates [54]. Hence, the superactivity cannot be due solely to a high local substrate concentration arising from electrostatic repulsion between the negatively charged substrate and the negatively charged surfactant palisade layer.

Michaelis–Menten kinetics have been derived for the case of hydrophilic enzyme and hydrophilic substrate both entrapped in water droplets [55]. The normal kinetic scheme for enzymatic reactions in aqueous solution, shown in Fig. 8 (top) is the basis for the treatment, and with consideration taken of droplet collision and disintegration the kinetic scheme becomes more involved (Fig. 8, bottom). Collision of droplets containing enzyme and substrate is followed by decomposition of the transient dimer formed, leading to a droplet containing both enzyme and substrate. The enzyme–substrate complex, E · S, will form within the droplet, and the complex will subsequently decompose into enzyme and product. Finally, the droplet containing both E and P will decompose into droplets containing E and droplets containing P. Reversibility of all processes except the last step is taken into consideration. The treatment leads to the prediction that hyperbolic (Michaelis–Menten)

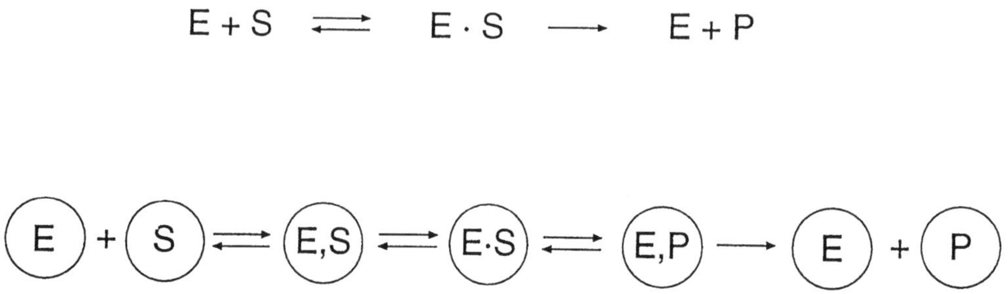

$$E + S \ \rightleftharpoons \ E \cdot S \ \longrightarrow \ E + P$$

$$\boxed{E} + \boxed{S} \rightleftharpoons \boxed{E,S} \rightleftharpoons \boxed{E \cdot S} \rightleftharpoons \boxed{E,P} \longrightarrow \boxed{E} + \boxed{P}$$

Figure 8 Schemes for the reaction between enzyme and substrate in aqueous solution (top) and between enzyme and water-soluble substrate in W/O microemulsion (bottom). E, S, and P stand for enzyme, substrate, and product, respectively. (From Ref. 55.)

kinetics is obeyed when the substrate concentration is varied in microemulsions of fixed composition and that both k_{cat} and K_M are dependent on the size and concentration of the water droplets in the system.

Corresponding treatment of the kinetics for hydrophobic compounds residing in the continuous hydrocarbon domain predicts that both k_{cat} and K_M are independent of the volume fraction of water in the microemulsion [55]. The K_M value in microemulsion is predicted to be larger than the value in aqueous solution by a factor approximately equal to the oil–water partition coefficient of the substrate. The predicted kinetics seems to be supported by previously published experimental data.

C. Influence of the Solvent

The organic solvent used in the microemulsion formulation should be nonpolar. The hydrophobicity of the solvent seems to be a key factor for the catalytic activity of the enzyme. A good correlation between hydrophobicity of the organic solvent and biocatalytic activity is obtained by the use of log P values of solvents [56,57]. (P is the partition coefficient of the solvent in the water–octanol system.) In general, enzyme stability and activity in microemulsions are poor with relatively hydrophilic solvents, for with log P is <2, moderate in solvents where log P is between 2 and 4, and high in hydrophobic solvents with log P>4. Even relatively small changes in solvents, such as going from cyclohexane to nonane, can give rise to large improvements in enzyme stability, as has been demonstrated for lipase from *Chromobacterium viscosum* [58]. The rationale behind this division is that very hydrophobic solvents do not distort the essential water layer around the enzyme, thereby leaving the catalyst in an active state. This limits the choice of organic solvent to aliphatic hydrocarbons with seven or more carbon atoms. On the other hand, lower hydrocarbons are preferred from a workup point of view, since they can be readily removed by evaporation after reaction. As a compromise, heptane, octane (in particular, isooctane), and nonane are the solvents of choice, and these hydrocarbons have been used in almost all enzymatic reactions in microemulsion media. Log P values of some organic solvents are listed in Table 1.

D. Influence of the Surfactant

The choice of surfactant is of importance for the rate of many enzymatic reactions in microemulsions. For instance, it has been found that whereas lipase-catalyzed hydrolysis of triglycerides is rapid in microemulsions based on AOT, it is extremely sluggish when

Table 1 Log P Values of Common
Organic Solvents[a]

Solvent	log P
Dimethyl sulfoxide	−1.3
Ethanol	−0.24
Acetone	−0.23
Ethylacetate	0.68
Diethyl ether	0.85
Pentanol	1.3
Chloroform	2.0
Toluene	2.5
Hexane	3.5
Heptane	4.0
Octane	4.5
Dodecane	6.6

[a] P is the partition coefficient between octanol and water.

normal nonionic surfactants are used [59]. The difference in behavior between microemulsions based on anionic and nonionic surfactants was attributed to differences in accessibility of the triglyceride to the enzyme. In the AOT system the lipase was anticipated to have good access to the interface between the oil and water domains. On the other hand, in the systems based on nonionic surfactants the poly(ethylene glycol) (PEG) chains stretching out from the interface into the droplets were believed to make it difficult for the enzyme to gain access into the interfacial region. The steric repulsion between the protein and the PEG layer was paralleled by the well-known protein-repelling effect of PEG chains grafted on solid surfaces. A later work, using penta(ethylene glycol) monododecyl ether ($C_{12}E_5$) as surfactant, confirmed the view that straight-chain alcohol ethoxylates are unsuitable for use in lipase-catalyzed hydrolysis reactions in microemulsions. At all compositions tested, only partial hydrolysis of triglyceride was obtained [60]. Contrary to what would be expected, an increase in the overall water content of the microemulsion formulation led to a decreased rate of lipase-catalyzed triglyceride hydrolysis, as can be seen from Fig. 9.

NMR self-diffusion measurements were performed on microemulsions containing from 1% up to 40% water while keeping the surfactant concentration constant. As shown in Table 2, all self-diffusion coefficients, including that of water, D_w, decrease with increasing water content, although $D_{palm\,oil}$ seems to have reached a plateau at about 5% water. From the D_w/D_w^0 and D_h/D_h^0 values, with D_w^0 and D_h^0 being the self-diffusion coefficients for neat water and neat hydrocarbon, respectively, it is obvious that whereas the diffusion of water is highly restricted throughout the series, that of isooctane is of the same magnitude as that of neat solvent. Furthermore, diffusion constants of the surfactant closely follow that of water. Evidently, the system is a W/O type of microemulsion, and the structure changes in the direction of more closed water domains when water is added. The conclusion that can be drawn from this study is that in the systems with better defined water-in-oil spheres, i.e., those of high water content, hydrolysis of the triglyceride was inhibited compared with reaction in the more bicontinuous structures. This inhibition may be related to the incorporation of surface-active reaction products in the palisade layer between the water and oil microdomains. It may also be due to the fact that $C_{12}E_5$ seems to be a good substrate for lipases. It was shown that in the presence of lipase the surfactant readily formed esters

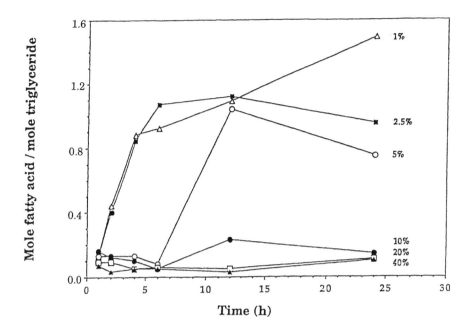

Figure 9 Effect of water content on the formation of fatty acids from palm oil. Reactions were run at 37°C, and the surfactant concentration was kept constant at 17%. (From Ref. 60.)

Table 2 Self-Diffusion Coefficients D (10^{-11} m^2/s) at 37°C for the Components of a Microemulsion Containing 17% $C_{12}E_5$, 5% Palm Oil, and 1–20% 0.1 M Phosphate Buffer at pH 8, with Isooctane Constituting the Balance[a]

Percent water	W_0	D_{water}	D_w/D_w^0	$D_{hydrocarbon}$	D_h/D_h^0	$D_{surfactant}$	$D_{palmoil}$
1.05	1.4	—	—	168	0.71	36	37
2.09	2.8	23	0.085	169	0.72	26	28
5.01	6.7	13	0.048	—	—	18	21
10.0	13.4	16	0.059	—	—	14	17
19.7	26.1	16	0.059	142	0.60	12	20

[a] W_0 is the molar ratio of water to surfactant. Self-diffusion coefficient values are for neat water $D_w^0 = 271$ and for neat isooctane $D_h^0 = 235$. $D_{palmoil}$ for 5% palm oil in isoctane is 60.3. Confidence interval 80%.

with fatty acids generated from triglyceride breakdown [61,62]. Since triglyceride hydrolysis is normally monitored by measuring fatty acid produced during the course of the reaction, this side reaction, which consumes fatty acids, is likely to contribute to the apparent low reaction rate and yield of lipase-catalyzed triglyceride hydrolysis in the presence of normal alcohol ethoxylates, as discussed above.

In a 1994 work, the effect of the surfactant on lipase-catalyzed hydrolysis of palm oil in microemulsion was further investigated [62]. Three surfactants were used: one anionic, one nonionic, and one cationic. As shown in Fig. 10, all three compounds were double-tailed, with similar hydrophilic–lipophilic balance, giving large regions of L2 microemulsions with isooctane and water at 37°C.

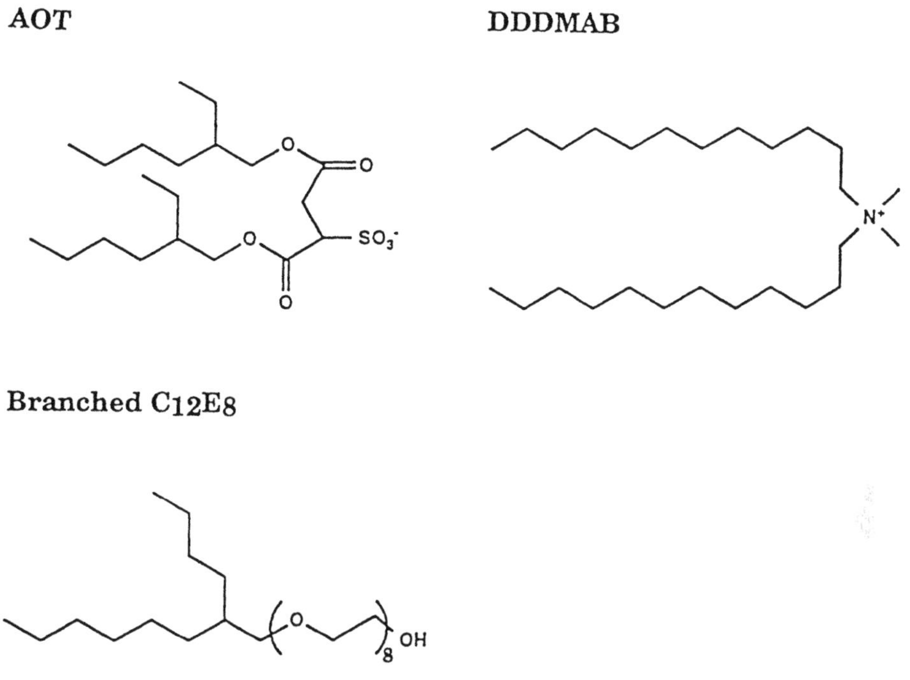

AOT   DDDMAB

Branched $C_{12}E_8$

Figure 10 Structures of branched anionic (AOT), cationic (didodecyldimethylammonium bromide, DDDMAB), and nonionic [octa(ethylene glycol)mono-2-butyloctyl ether, branched $C_{12}E_8$] surfactants.

NMR self-diffusion measurements indicated that all microemulsions consisted of closed water droplets and that the structure did not change much during the course of reaction. Hydrolysis was fast in microemulsions based on branched-chain anionic and nonionic surfactants but very slow when a branched cationic or a linear nonionic surfactant was employed (Fig. 11). The cationic surfactant was found to form aggregates with the enzyme. No such interactions were detected with the other surfactants. The straight-chain, but not the branched-chain, alcohol ethoxylate was a substrate for the enzyme. A slow rate of triglyceride hydrolysis for a $C_{12}E_4$-based microemulsion compared with formulations based on the anionic surfactant AOT [61,63] and the cationic surfactant cetyltrimethylammonium bromide (CTAB) [63] was observed in other cases also. Evidently, this type of lipase-catalyzed reaction should preferably be performed in a microemulsion based on an anionic or branched nonionic surfactant. Nonlipolytic enzymes such as cholesterol oxidase seem to function well in microemulsions based on straight-chain nonionic surfactants, however [64]. CTAB was reported to cause slow inactivation of different types of enzymes [62,64,65] and also, in the case of *Chromobacterium viscosum* lipase [66], to provide excellent stability.

E. Influence of pH

The pH dependence of enzymatic activity in microemulsions sometimes differs from that in aqueous solution. For instance, several groups found that for α-chymotrypsin-catalyzed hydrolysis the optimum pH is shifted to considerably more alkaline conditions in the

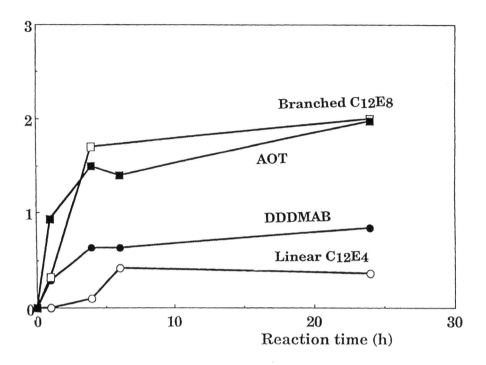

Figure 11 Degree of palm oil hydrolysis at 37°C as a function of time in lipase-containing microemulsions based on isooctane, palm oil, distilled water, and surfactant in weight proportions 50 : 5 : 8 : 37. The double-tailed surfactants of Fig. 10 as well as the straight-chain surfactant tetra(ethylene glycol)monododecyl ether ($C_{12}E_4$) were used. (From Ref. 62.)

microheterogeneous solution than in bulk [36,67,68]. However, other workers reported little or no change in the pH optimum between bulk water and W/O microemulsion with this enzyme [69,70].

The somewhat diverging observations may be due to differences in experimental conditions. An enzyme dispersed in small water pools may experience a local pH different from that recorded in the water phase of the microemulsion before mixing. In fact, it has been seen that optimum pH in a given system may vary with W_0, i.e., with the size of the droplets [36]. Ideally, pH should be recorded on the entire formulation, not on the aqueous phase before mixing. However, direct measurement of pH in the dispersed phase of W/O microemulsions is not straightforward. The measurement is probably best done with some kind of probe technique, and an NMR method based on the degree of protonation of phosphate ions was developed for this purpose [71].

The pH dependence of these systems is complex and not fully understood, however. For instance, the same lipase that when used for hydrolysis showed a typical sigmoidal pH–activity profile exhibited practically no pH dependence when used for catalysis of ester synthesis [66]. Evidently, in the very water poor medium of the esterification, the enzyme does not experience a changed environment.

III. REACTIONS IN MICROEMULSIONS

A. Ester Hydrolysis

A considerable number of mechanistic studies have been made with α-chymotrypsin as hydrolytic enzyme, and many of these were discussed above [35,36,46,67–70,72–74]. From an application point of view, lipases are probably the enzyme of most interest both for hydrolysis and for the reverse reaction, condensation.

Monoglycerides can be obtained in high yields by lipase-catalyzed hydrolysis of the corresponding triglyceride oil in microemulsions of the W/O type. When a 1,3-specific enzyme is used, the hydrolysis takes place in a fairly regioselective manner, conversion into 2-monoglyceride being completed in 2–3 h at a temperature of 37°C [75]. Prolonged reaction time results in a decrease in monoglyceride yield, a process that is not due to lack of regioselectivity of the enzyme but to an acyl group migration in the 2-monoglyceride yielding 1-monoglyceride. The latter compound is a good substrate for the regiospecific enzyme, and complete hydrolysis to fatty acids and glycerol will eventually take place (Fig. 12).

Kinetic studies indicate that effects of temperature and pH, as well as reaction constants, resemble those of aqueous systems. The activity versus W_0 curve shows a typical maximum at W_0 of between 10 and 15, the value varying slightly with the type of lipase used [75,76]. In the AOT–water–isooctane system, which is commonly used for hydrolysis of triglycerides, $W_0 = 11$ was found to correspond to a radius of around 2 nm [37].

Lipase activity is generally higher in AOT-based systems than in microemulsions based on conventional nonionic or cationic surfactants with both triglycerides and nitrophenyl alkanoate esters as substrates. AOT also gave excellent enzyme stability [37,61,64].

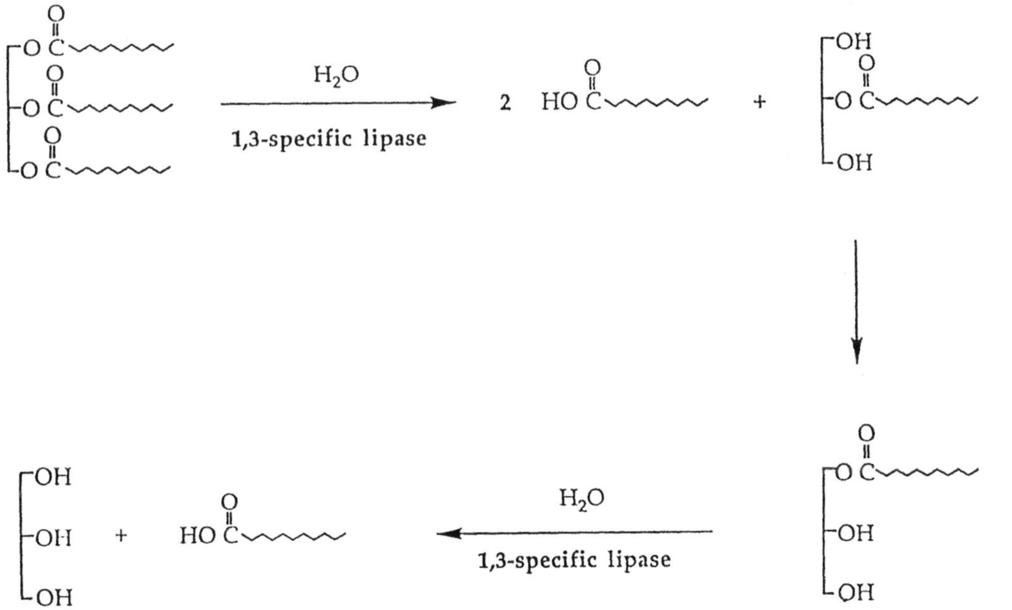

Figure 12 Enzymatic hydrolysis of a triglyceride in combination with migration of the acyl group from the 2- to the 1-position.

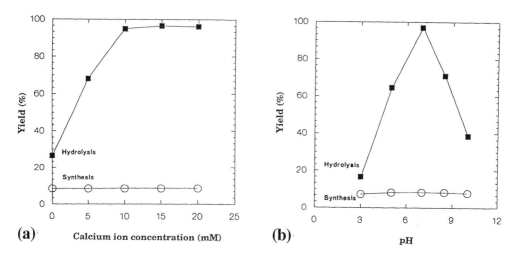

Figure 13 Effect of (a) calcium ions and (b) buffer pH on yield of hydrolysis and synthesis of phosphatidylcholine. The reactions were carried out in W/O microemulsions based on AOT–isooctane–aqueous buffer and run at optimum conditions except for the varied parameters. (Redrawn from Ref. 78.)

A hydrophilic substrate, acetylsalicylic acid, was subjected to lipase catalyzed hydrolysis in a W/O microemulsion [77]. For comparison, the reaction was also carried out in aqueous buffer. Since hydrolysis of acetylsalicylic acid proceeds spontaneously without added catalyst (intramolecular catalysis), reactions without lipase were performed as controls. It was found that addition of lipase did not affect the rate of reaction in aqueous buffer. However, the reaction in miroemulsion was catalyzed by the lipase, and the rate was linearly dependent on lipase concentration. This is a further illustration of the fact that microemulsions, with their large oil/water interfaces, are suitable media for lipase-catalyzed reactions. The same reactions were also performed using α-chymotrypsin as catalyst. This enzyme, which also catalyzes ester hydrolysis but which, unlike lipase, functions independently of a hydrophobic surface, was not more active in microemulsion than in the buffer solution.

Hydrolysis of a phospholipid, phosphatidylcholine, was performed in microemulsions using phospholipase A2 as catalyst [78]. The reaction is fast and gives the lysophospholipid in quantitative yield when the reaction is performed in the presence of at least 10 mM calcium ions in the dispersed aqueous phase, as can be seen in Fig. 13a. The Ca dependence of phospholipase A2 is well known from reactions in aqueous solution. As can be seen from Fig. 13b, the hydrolysis reaction is very pH-sensitive, much more so than is normally seen for lipases.

Enzymatic hydrolysis of lipophilic organophosphorus compounds was performed in microemulsions using phosphotriesterase as catalyst [79]. Destruction of strongly hydrophobic, environmentally harmful substances in general is a type of application where microemulsions have great potential, using either enzymatic [79,80] or nonenzymatic [81] processes.

B. Ester Synthesis

There is considerable current interest in the use of microemulsions as media for lipase-catalyzed stereo- and regiospecific ester synthesis. In some cases lipases seem to display a higher degree of regioselectivity in microemulsions than in aqueous solution.

Lipase-catalyzed synthesis of esters of monofunctional alcohols proceeds in good yield in microemulsions [82–84]. In a systematic study on esterification of different alcohols (primary, secondary, tertiary) with fatty acids of various chain lengths, a pronounced selectivity was obtained [85]. The three different lipases used—*Penicillium simplicissimum, Rhizopus delemar,* and *Rhizopus arrhizus*—exhibited different preferences with regard to acid chain length and type of alcohol. Studies using spectroscopic techniques indicated that the selectivity was related to localization of the enzyme molecule within the droplet. Hence, the hydrophilic–lipophilic character of the protein, not its specificity as expressed in aqueous solution, is responsible for the selectivity. This illustrates the important point that regioselectivity of bio-organic (and organic) reactions may differ between homogeneous and microheterogeneous media.

Esterification of racemic ibuprofen using *Candida cylindracea* lipase with alcohols of varying chain lengths gave the $S(+)$-ibuprofen ester in very high enantiomeric excess [86]. The reverse reaction, hydrolysis, was also carried out with a high degree of specificity, but the reaction was much slower. It is interesting that the high enantioselectivity displayed by the enzyme in the W/O microemulsion was seen neither in esterification in pure organic solvent nor in hydrolysis in aqueous buffer. It is suggested that the large oil/water interfacial area in microemulsions favors enantioselectivity. Excellent enantioselectivity in lipase-catalyzed esterifications of a range of primary and secondary alcohols was reported by several groups [66,87–89].

Surprisingly, esterification of fatty acids with simple sugars, such as glucose and mannitol, in AOT-based microemulsions did not take place at all [82]. No reaction was seen with either of two different lipases. This is probably due to poor phase contact between the very hydrophilic sugar molecule in the water pool and the fatty acid that resides in the hydrocarbon domain. Sugar monoesters can be produced in high yields by lipase-catalyzed esterification in a water-free medium [90].

Synthesis of triglycerides from glycerol and fatty acids was attempted in microemulsions of very low water content [91–95]. The main reaction product is monoglyceride, with smaller amounts of diglyceride formed as well. The yield of triglyceride is negligible. The same reaction proceeds well in monolayer, giving triglyceride in fair yield [94]. It is believed that the reason triglycerides are not readily formed in microemulsions is that the intermediate diglyceride is too lipophilic and has too low a surface activity to stay at the hydrocarbon/water interface. Once formed, it rapidly partitions into the continuous hydrocarbon domain, leaving behind the enzyme, which is located in or at the surface of the water pools. The monolayer situation is very different, since the hydrophobic diglyceride will have no other choice than to stay at the surface. The static air/water interface should constitute an ideal environment for the lipase-catalyzed reaction, as schematically illustrated in Fig. 14.

In a recent work on lipase-catalyzed esterification of dodecanoic acid with glycerol in microemulsions based on different hydrocarbons as oil component, it was demonstrated that there was a clear trend toward more triglyceride and less monoglyceride with increasing hydrocarbon chain length [96]. Whereas shorter chain hydrocarbons, octane and decane,

The monolayer situation:

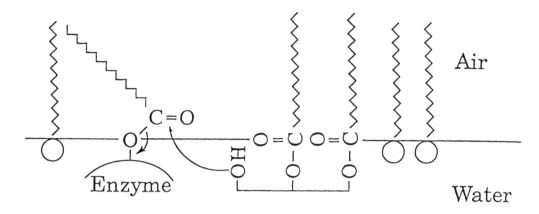

The microemulsion situation:

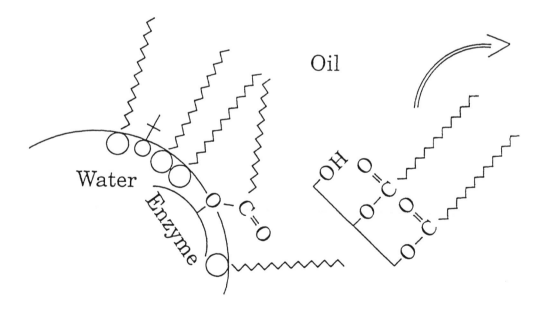

Figure 14 Arrangement of substrates and surfactants at the interfaces of (top) monolayers and (bottom) microemulsions. (From Ref. 94.)

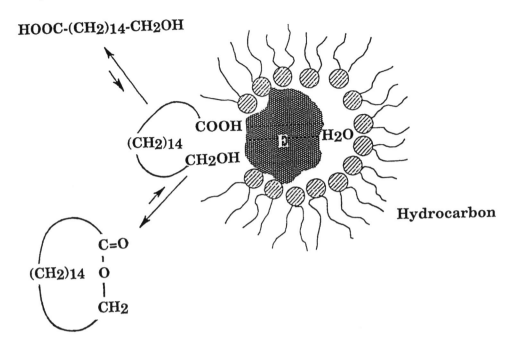

HOOC-(CH2)14-CH2OH

(CH2)14 — COOH / CH2OH

(CH2)14 — C=O / O / CH2

E   H2O

Hydrocarbon

Figure 15 Lactone formation through cyclization of a long-chain ω-hydroxy acid catalyzed by lipase at the interface between the droplet and the continuous domain. (From Ref. 97.)

gave predominantly monoglyceride, the use of a long hydrocarbon, hexadecane, gave triglyceride as the main reaction product. The relative yield of diglyceride was fairly independent of the choice of hydrocarbon.

The strong dependence of hydrocarbon chain length on monoglyceride-to-triglyceride ratio seems to be a good illustration of the effect of partitioning of the reaction intermediate on product composition. Short-chain hydrocarbons are better solvents for mono- and diglycerides than longer chain homologs. The difference in solvency is likely to be particularly pronounced for the diglyceride, as only hydrocarbon molecules smaller than the acyl group should be able to interpenetrate the individual acyl chains. When the solvency is good, as is the case for octane and decane, the product formed at the oil/water interface will be rapidly transported into the continuous oil domain. Since the enzyme is located in the water droplets, such a partitioning will disfavor formation of di- and, particularly, triglyceride, as is illustrated in Fig. 14.

A microemulsion of low water content has been found to be an excellent medium for synthesis of long-chain lactones [97]. These compounds, which are important perfume ingredients, are not easily made by conventional organic synthesis because intermolecular esterification dominates over intramolecular esterification. In the microemulsion, the molecular arrangement at the oil/water interface seems to favor the cyclization reaction (Fig. 15).

α-Chymotrypsin was also used to catalyze condensation reactions in AOT-based microemulsions [98,99]. Amino acid and peptide derivatives were obtained in good yield.

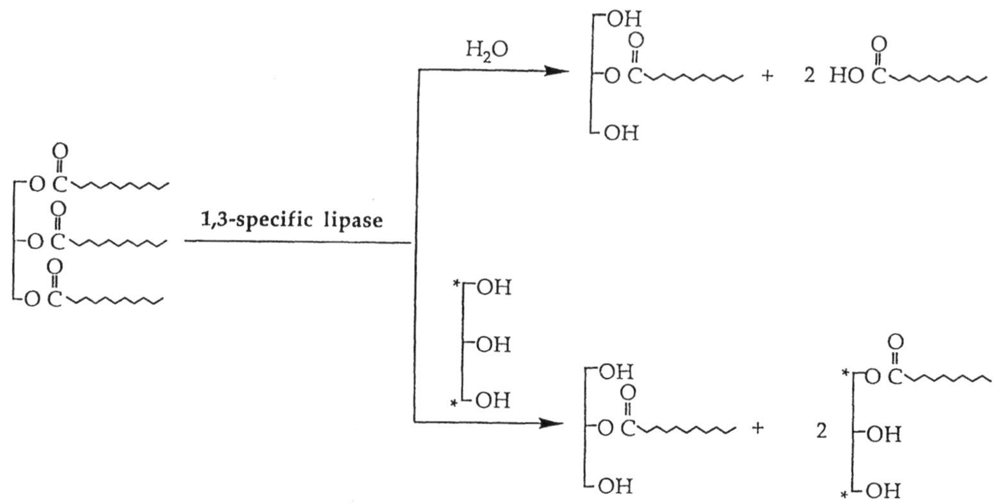

Figure 16 Enzymatic glycerolysis using ^{3}H-labeled glycerol. Hydrolysis occurs in parallel. (From Ref. 101.)

C. Glycerolysis

Water-free microemulsions can be designed with glycerol as the polar component. The solubility of glycerol in surfactant–hydrocarbon systems is normally smaller than in water, however. With AOT as surfactant and heptane as oil component, about 5 mol of glycerol can be solubilized per mole of surfactant at 0°C, this amount decreasing with increasing temperature [100].

Glycerol-containing microemulsions are of interest for lipid conversions, because they open the possibility of performing glycerolysis instead of hydrolysis, provided that glycerol can replace water as activator and stabilizer of the enzyme. Lipase-catalyzed glycerolysis of triglycerides would yield a mixture of 1- and 2-monoglycerides. Since monoglycerides are important emulsifiers in the food industry, this synthesis is of potential industrial interest. Attempts to perform the reaction failed, however, because very little enzyme activity was obtained in the completely nonaqueous system [101]. The use of a combination of water and glycerol as the polar component of an AOT-based microemulsion led to simultaneous hydrolysis and glycerolysis [101,102]. Using ^{3}H-labeled material, monoglyceride originating from added glycerol could be distinguished from that originating from the starting triglyceride, as shown in Fig. 16. It was found that glycerolysis and hydrolysis occur at approximately the same rate. The two reactions probably go through the same intermediate. This work shows that although water can be substituted by glycerol both as the polar component of the microemulsion and as the solvolytic agent in the reaction, it cannot be fully replaced in its role as activator of the enzyme.

D. Transesterification

Lipase-catalyzed transesterification of vegetable oils, i.e., replacement of one acyl group in a triglyceride by another acyl group, can be performed in microemulsions of low water content [91,103,104]. Special attention has been directed toward production from inexpensive

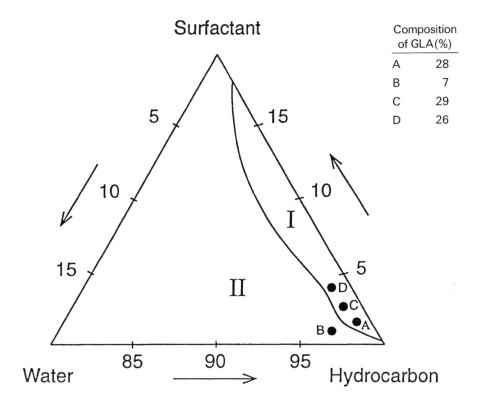

Figure 17 Transesterification of tristearin with γ-linolenic acid using reaction media of slightly different compositions. The table shows the incorporation of γ-linolenic acid in the triglyceride formed after 24 h reaction time.

starting material (such as palm oil) of a triglyceride mixture that corresponds to natural cocoa butter. This reaction requires partial replacement of palmitoyl groups by stearoyl groups in the 1(3)-position while leaving the 2-position essentially unaffected. A high degree of conversion was obtained in lipase-containing microemulsions based on either an anionic or a nonionic surfactant [103].

In a work aimed at incorporating γ-linolenic acid (GLA) into a saturated triglyceride, tristearin was transesterified with GLA using *Rhizopus delemar* lipase as catalyst [105]. Reactions were carried out at four different compositions, all situated in the hydrocarbon-rich corner of the ternary phase diagram. As can be seen in Fig. 17, three samples lie in the isotropic L_2 region, whereas one sample falls within the two-phase region. The reactions were carried out under stirring. Also shown in Fig. 17, the reaction in the two-phase region (emulsion) is very sluggish compared to the reactions in microemulsions. This set of experiments is a good illustration of the benefits of the much larger oil/water interfacial area of microemulsions compared to emulsions.

Attempts have been made to use lipases in microemulsions to introduce polyunsaturated fatty acids such as eicosapentaenoic acid (EPA, C20 : 5) and docosahexaenoic acid (DHA, C22 : 6) into phosphatidylcholine by transesterification of the phospholipid. Phopholipids carrying long polyunsaturated acyl groups are of nutritional as well as pharmaceutical interest. A fair yield of transesterification, i.e., incorporation of the new fatty acid in the 1- or 2-position, was obtained both with sardine oil [106] and with free fatty acids [107] using lipase in W/O microemulsion, but transesterification carried out with phospholipase A2 as catalyst gave a negligible yield [78]. The product, phosphatidylcholine-containing polyunsaturated fatty acids, could be obtained, however, although in low yield, by two consecutive phospholipase A2–catalyzed reactions in microemulsion: hydrolysis of the phospholipid to lysophospholipid and esterification of the latter with a mixture of EPA and DHA [78].

E. Oxidation and Reduction

A number of oxidations and reductions have been carried out in microemulsions using enzymes such as cholesterol oxidase [64,108,109], bilirubin oxidase [110], horseradish peroxidase [111,112], and horse liver alcohol dehydrogenase (HLADH) [73,112–118]. It is noteworthy that cofactor-dependent enzymes also work well in different types of microemulsions. In fact, kinetic studies on the HLDAH-NADH (NADH stands for the reduced form of nicotinamide adenine dinucleotide) system showed that the presence of coenzyme was essential for the long-term stability of the enzyme in a microemulsion [113,115].

HLADH was used as redox catalyst in a coupled substrate–coenzyme regenerating cycle, and the enzymatic activity was studied as a function of oil fraction of the microemulsion [118]. The oil fraction was varied while the surfactant (AOT) concentration was kept constant, leading to a change in the microstructure of the solution from an O/W to a W/O microemulsion via a bicontinuous structure, as determined by self-diffusion NMR. The enzyme exhibited good stability in the various types of structures. Variation of the initial reaction rate could be described by modifying the rate equation, valid in pure buffer, taking into account partitioning of the substrate between the water and hydrocarbon domains.

HLADH has also been used for the preparative reduction of water-insoluble substrates such as cinnamaldehyde [113], cyclohexanone [112], and the steroid eticholane-3β-ol-17-one [112] with simultaneous oxidation of low molecular weight alcohols. With sodium dodecyl sulfate (SDS) or AOT as surfactant, enzyme activity was very dependent on the amount of water added. Relatively high water content gave the highest reaction rate and also the best enzyme stability. A properly formulated microemulsion gave a higher reaction rate than an aqueous buffer [112].

F. Polymerization

The use of W/O microemulsions as a medium for enzyme-catalyzed polymerizations has attracted attention in recent years [119–122]. A nice illustration of the concept is the horseradish peroxidase–catalyzed oxidation of phenols with hydrogen peroxide [119–121]. Using p-ethylphenol as substrate and AOT as surfactant, the polymerization was very fast, and rapid precipitation was seen within a few minutes after addition of H_2O_2. The slightly surface-active alkylphenol is believed to partition at the oil/water interface, which can be regarded as a template for the polymer synthesis. The polymer particles obtained are spherical and typically in the size range 0.3–0.8 μm. It is interesting that when the same polymerization is carried out in surfactant-free solvent systems, spherical particles are not obtained [119]. Even if the particles obtained are 10–100 times larger than the starting microemulsion droplets, it seems that the presence of an interface in the reaction medium governs the morphology of the product.

Alkyl phenol and hydrogen peroxide react in a 1 : 1 stoichiometry, and the resulting polymer is linked by carbon–carbon bonds. Although the phenolic hydroxyl groups are involved in the formation of free radicals leading to onset of polymerization, they do not appear to be involved in bond formation. No ether linkages are detected in the polymer. NMR studies on poly-*p*-ethylphenol indicate that the phenyl rings are connected at the ortho positions [120]. This type of phenol polymerization in microheterogeneous medium is biomimetic in the sense that it resembles lignin synthesis. Lignin is synthesized in the plant by peroxidase-catalyzed oxidation of substituted phenols [123].

G. Workup Aspects

The main practical problem in large-scale use of microemulsions as a medium for biocatalysis is that of workup. Separating surfactant from product is not a trivial issue, since normal purification procedures, such as extraction and distillation, tend to be troublesome due to the well-known problems of emulsion formation and foaming caused by the surfactant.

When a microemulsion is used as reaction medium for synthesis of a surfactant, an obvious way to circumvent the workup problem is to use the reaction product as the microemulsion surfactant. This approach has been used in the synthesis of alkyl glucoside esters [124,125].

A more general approach to facilitate workup of bio-organic reactions in microemulsions, developed by Robinson and coworkers [87,88,126–128], is to use a microemulsion-based gel (MBG) as a reaction medium. An MBG is made by mixing a normal W/O microemulsion with an aqueous gelatin solution above the gelling temperature and then allowing the mixture to cool, for instance in a column, so that a stiff gel is formed. Various spectroscopic techniques have revealed that the microemulsion structure is retained in the gel. The gel, which may contain entrapped enzyme, can be seen as an immobilized enzyme-containing microemulsion. It is resistant to hydrocarbon solvents. The MBG concept is particularly suitable for ester synthesis. By charging a hydrocarbon solution of an acid and an alcohol at the top of the column, the corresponding ester can be recovered from the eluant. This is an elegant way of avoiding the problem of separation of product and surfactant from the reaction mixture. Lipase-containing MBGs have been used to synthesize, on a preparative scale, a variety of esters under mild conditions, and both regio- and stereoselectivity have been demonstrated. The MBGs have not yet been applied to triglycerides.

Another workup approach has been to use the inherent phase behavior of oil–water–surfactant systems to separate product from remaining reactants and from surfactant. A Winsor III system made with a branched-tail phosphonate surfactant was used as reaction medium for lipase-catalyzed hydrolysis of trimyristin. The enzyme resided almost exclusively in the middle-phase microemulsion together with the surfactant. The products formed, 2-myristoylglycerol and sodium myristate, partitioned into the excess hydrocarbon and water phases, respectively, and could easily be recovered [129]. A similar procedure was used for cholesterol oxidation using cholesterol oxidase as catalyst [130].

In a related approach, a Winsor II system was used to effect hydrolysis of D,L-phenylalanine methyl ester using α-chymotrypsin as catalyst. By choosing conditions such that the ester partitioned preferentially into the microemulsion phase along with surfactant and enzyme and the product, L-phenylalanine, into the aqueous phase, workup was smooth [131].

Yet another related approach is to use the strong temperature dependence of microemulsions based on nonionic surfactants. After the reaction is completed in a one-phase microemulsion, the temperature is raised (or lowered) so that a two-phase system forms, consisting of microemulsion in equilibrium with excess water (or oil) phase. If the reaction product is hydrophilic, a temperature increase is chosen and the product is recovered from the aqueous phase. If the product is lipophilic, the temperature is instead decreased, and the product is recovered from the excess hydrocarbon phase. The principle has been applied successfully to an HLADH-catalyzed reaction in microemulsion based on $C_{12}E_5$ [132].

A continuous-flow centrifugal reactor has been designed and used for enzymatic hydrolysis. The reaction is performed in a two-phase system consisting of a W/O microemulsion containing the enzyme in equilibrium with an organic phase containing the substrate. The organic phase is under continuous flow, enabling the reaction and product separation to be carried out in a single-stage unit [79].

H. Enzyme-Catalyzed Reactions in Related Systems

As mentioned above, lipase-catalyzed synthesis of triglycerides from glycerol and fatty acids can be performed in fair yield in monolayer experiments [93,133]. Another air/water interfacial system, foam, also gave di- and triglycerides in higher yields than were obtained in microemulsions [134]. Reactions in foams may be of more practical interest than reactions in monolayers due to the much larger interfacial area in the former type of system.

Lyotropic liquid crystals have been investigated as vehicles for a variety of enzymatic reactions [135,136]. Of particular interest are biphasic systems consisting of a liquid crystalline phase coexisting with an organic solvent phase. The enzyme is confined in the liquid crystals, and the solvent phase can be regarded as a reservoir for the hydrophobic substrate. The enzymes studied, as well as the coenzymes, show a remarkably high stability in these systems.

Lipase-catalyzed hydrolysis has also been performed in O/W emulsions [137,138]. Reverse vesicles have been employed for polyphenol oxidase–catalyzed ortho-hydroxylation of phenols. A higher turnover was reported in this system than in a reverse micellar system based on the same surfactant [139].

IV. CONCLUSIONS

A broad variety of enzymes have been used to catalyze organic reactions in microemulsions. In the majority of cases the enzyme retains both activity and stability in a satisfactory way. Special attention has been given to the use of lipases in W/O microemulsions where the enzyme is located in water droplets of a size not much larger than the hydrodynamic diameter of the protein. Such systems are biomimetic in the sense that lipases in biological systems operate at the interface between hydrophobic and hydrophilic domains, with these interfaces being stabilized by polar lipids and other natural amphiphiles.

Microemulsion media of low water activity enable hydrolytic enzymes to catalyze condensation as opposed to hydrolysis. This has opened the possibility to perform lipase-catalyzed ester synthesis, an area of considerable practical interest. Lipase-catalyzed transesterification in microemulsion of low water content is another area of industrial relevance, in particular for synthesis of triglycerides with unusual fatty acid composition.

ACKNOWLEDGMENTS

I am grateful to Dr. C. P. Singh for help in selecting the material used in this review and to Dr. Norman Burns for linguistic assistance.

REFERENCES

1. A. M. Klibanov, CHEMTECH *16*:354 (June 1986).
2. M.-P. Pileni, J. Phys. Chem. *97*:6961 (1993).
3. M. A. Wells, Biochemistry *13*:4937 (1974).
4. P. H. Poon and M. A. Wells, Biochemistry *13*:4928 (1974).
5. R. L. Misiorowsky and M. A. Wells, Biochemistry *13*:4921 (1974).
6. C. Gitler and M. Montal, FEBS Lett. *28*:329 (1972).
7. K. Martinek, A.V. Levashov, N. L. Klyachko, and I.V. Berezin, Dokl. Akad. Nauk SSSR *236*:920 (1977).
8. P. L. Luisi, F. Henninger, and M. Joppich, Biochem. Biophys. Res. Commun. *74*:1384 (1977).
9. F. M. Menger and K. Yamada, J. Am. Chem. Soc. *101*:6731 (1979).
10. K. Martinek, A. V. Levashov, Yu. L. Khmelnitski, N. L. Klyachko, and I. V. Berezin, Science *218*:889 (1982).
11. K. Larsson, *Lipids—Molecular Organization, Physical Functions and Technical Applications*, Oily Press, Dundee, Scotland, 1994, Chap. 9.
12. B. De Kruijff, Nature *329*:587 (1987).
13. H. Stamatis, A. Xenakis, F. N. Kolisis, and A. Malliaris, Prog. Colloid Polym. Sci. *97*:253 (1994).
14. A. M. Brzozowski, U. Derewenda, Z. S. Derewenda, G. G. Dodsson, D. M. Lawson, J. P. Turkenburg, F. Björkling, B. Huge-Jensen, S. A. Patkar, and L. Thim, Nature *351*:491 (1991).
15. K. Martinek, N. L. Klyachko, A. V. Kabanov, Yu. L. Khmelnitsky, and A. V. Levashov, Biochim. Biophys. Acta *981*:161 (1989).
16. V. Ramakrishnan, A. Darszon, and M. Montal, J. Biol. Chem. *258*:4857 (1983).
17. G. G. Zampieri, H. Jäckle, and P. L. Luisi, J. Phys. Chem. *90*:1849 (1986).
18. A.V. Levashov, Yu. L. Khmelnitsky, N. L. Klyachko, Ya. Chernyak, and K. Martinek, J. Colloid Interface Sci. *88*:444 (1982).
19. D. Chateney, W. Urbach, A. M. Cazabat, M.Vacher, and M. Waks, Biophys. J. *48*:893 (1985).
20. D. Chateney, W. Urbach, C. Nicot, M. Vacher, and M. Waks, J. Phys. Chem. *91*:2198 (1987).
21. P. D. I. Fletcher, B. H. Robinson, and J. Tabony, J. Chem. Soc., Faraday Trans. 1 *82*:2311 (1986).
22. P. D. I. Fletcher, A. M. Howe, N. M. Perrins, B. H. Robinson, C. Toprakcioglu, and J. C. Dore, in *Surfactants in Solution*, Vol. 3 (K. L. Mittal and B. Lindman, eds.), Plenum, New York, 1984, p. 1745.
23. E. Sheu, K. E. Goklen, T. A. Hatton, and S.-H. Chen, Biotechnol. Prog. *2*:175 (1986).
24. S. Christ and P. Schurtenberger, J. Phys. Chem. *98*:12708 (1994).
25. R. S. Rahaman and T. A. Hatton, J. Phys. Chem. *95*:1799 (1991).
26. M. Caselli, P. L. Luisi, M. Maestro, and R. Roselli, J. Phys. Chem. *92*:3899 (1988).
27. R. H. Valivety, P. J. Halling, and A. R. Macrae, in *Biocatalysis in Non-Conventional Media* (J. Tramper, M. H.Vermüe, H. H. Beeftink, and U. von Stockar, eds.), Elsevier, Amsterdam, 1992, p. 549.
28. V. Dorovska-Taran, C.Veeger, and A. J.W. G.Visser, in *Biocatalysis in Non-Conventional Media* (J. Tramper, M. H.Vermüe, H. H. Beeftink, and U. von Stockar, eds.), Elsevier, Amsterdam, 1992, p. 697.
29. D. S. Clark, L. Creagh, P. Skerker, M. Guinn, J. Prausnitz, and H. Blanch, ACS Symp. Ser. *392*:104 (1989).
30. N. S. Kommareddi, K. C. O'Connor, and V. T. John, Biotechnol. Bioeng. *43*:215 (1994).
31. P. L. Luisi, Angew. Chem. *97*:449 (1985).

32. D. Han and J. S. Rhee, Biotechnol. Bioeng. *28*:1250 (1986).
33. K. Martinek, Biochem. Int. *18*:871 (1989).
34. R. Bru, A. Sanchez-Ferrer, and F. Garcia-Carmona, Biotechnol. Bioeng. *34*:304 (1989).
35. V. Papadimitriou, A. Xenakis, and A. E. Evangelopoulos, Colloids Surf. B *1*:295 (1993).
36. H. Ishikawa, K. Noda, and T. Oka, J. Ferment. Bioeng. *70*:381 (1990).
37. P. D. I. Fletcher, B. H. Robinson, R. B. Freedman, and C. Oldfield, J. Chem. Soc., Faraday Trans. 1 *81*:2667 (1985).
38. Yu. L. Khmelnitsky, A.V. Kabanov, N. L. Klyachko, A.V. Levashov, and K. Martinek, in *Structure and Reactivity in Reverse Micelles* (M. P. Pileni and C. Troyanowsky, eds.), Elsevier, Amsterdam, 1989, p. 230.
39. A.V. Kabanov, S. N. Nametkin, G. N. Evtushenko, N. N. Chernov, N. L. Klyachko, A.V. Levashov, and K. Martinek, Biochim. Biophys. Acta *996*:147 (1989).
40. R. M. D. Verhaert and R. Hilhorst, Rec. Trav. Chim. Pays-Bas *110*:236 (1991).
41. C. Oldfield, C. Otero, M. L. Rua, and A. Ballesteros, in *Biocatalysis in Non-Conventional Media* (J. Tramper, M. H. Vermüe, H. H. Beeftink, and U. von Stockar, eds.), Elsevier, Amsterdam, 1992, p. 189.
42. M. Bianucci, M. Maestro, and P. Walde, Chem. Phys. *141*:273 (1990).
43. M. Maestro and P. Walde, J. Colloid Interface Sci. *154*:298 (1992).
44. A.V. Pshezhetsky, N. L. Klyachko, G. S. Pepaniyan, S. Mercker, and K. Martinek, Biokhimiya *53*:1013 (1985).
45. N. L. Klyachko, A. V. Levashov, and K. Martinek, Mol. Biol. Engl. Ed. *18*:830 (1984).
46. Y. Miyake, T. Owari, F. Ishiga, and M. Teramoto, J. Chem. Soc., Faraday Trans. *90*:979 (1994).
47. K. Martinek, Yu. L. Khmelnitsky, A. V. Levashov, and I. V. Berezin, Dokl. Akad. Nauk SSSR *263*:737 (1982).
48. E. Österberg and K. Holmberg, unpublished results.
49. K. Martinek, A.V. Levashov, N. Klyachko, Yu. L. Khmelnitski, and I.V. Berezin, Eur. J. Biochem. *155*:453 (1986).
50. M. Gonnelli and G. B. Strambini, J. Phys. Chem. *92*:2854 (1988).
51. P. Walde, D. Han, and P. L. Luisi, Biochemistry *32*:4029 (1993).
52. E. Ruckenstein and P. Karpe, Biotechnol. Lett. *12*:241 (1990).
53. P. Karpe and E. Ruckenstein, J. Colloid Interface Sci. *141*:534 (1991).
54. Q. Mao and P. Walde, Biochem. Biophys. Res. Commun. *178*:1105 (1991).
55. C. Oldfield, Biochem. J. *272*:15 (1990).
56. C. Laane, S. Borren, R. Hilhorst, and C. Veeger, in *Biocatalysis in Organic Media* (C. Laane, J. Tramper, and M. D. Lilly, eds.), Elsevier, Amsterdam, 1987, p. 65.
57. R. H. Valivety, G. A. Johnston, C. J. Suckling, and P. J. Halling, Biotechnol. Bioeng. *38*:1137 (1991).
58. E. Skrika-Alexopoulos, J. Muir, and R. B. Freedman, in *Biocatalysis in Non-Conventional Media* (J. Tramper, M. H. Vermüe, H. H. Beeftink, and U. von Stockar, eds.), Elsevier, Amsterdam, 1992, p. 705.
59. E. Österberg, C. Ristoff, and K. Holmberg, Tenside *25*:293 (1988).
60. M.-B. Stark, P. Skagerlind, K. Holmberg, and J. Carlfors, Colloid Polym. Sci. *268*:384 (1990).
61. P. Skagerlind, M. Jansson, B. Bergenståhl, and K. Hult, J. Chem. Technol. Biotechnol. *54*:277 (1992).
62. P. Skagerlind and K. Holmberg, J. Dispersion Sci. Technol. *15*:317 (1994).
63. T. P. Valis, A. Xenakis, and F. N. Kolisis, Biocatalysis *6*:267 (1992).
64. K. M. Lee and J. F. Biellmann, Bioorg. Chem. *14*:262 (1986).
65. E. Skrika-Alexopoulos, J. Muir, and R. B. Freedman, Biotechnol. Bioeng. *41*:894 (1993).
66. G. D. Rees and B. Robinson, Biotechnol. Bioeng. *45*:344 (1995).
67. K. Martinek, A.V. Levashov, N. L. Klyachko, V. I. Pantin, and I.V. Berezin, Biochim. Biophys. Acta *657*:277 (1981).
68. S. Barbaric and P. L. Luisi, J. Am. Chem. Soc. *103*:4239 (1981).
69. P. D. I. Fletcher, G. D. Rees, B. H. Robinson, and R. B. Freedman, Biochim. Biophys. Acta *832*:204 (1985).

70. P. D. I. Fletcher, R. B. Freedman, J. Mead, C. Oldfield, and B. H. Robinson, Colloids Surf. *10*:193 (1984).
71. R. E. Smith and P. L. Luisi, Helv. Chim. Acta *63*:2302 (1980).
72. P. L. Luisi, M. Giomini, M.-P. Pileni, and B. H. Robinson, Biochim. Biophys. Acta *947*:209 (1988).
73. K. M. Larsson, A. Janssen, P. Adlercreutz, and B. Mattiasson, Biocatalysis *4*:163 (1990).
74. V. V. Mozhaev, N. Bec, and C. Balny, Biochem. Mol. Biol. Int. *34*:191 (1994).
75. K. Holmberg and E. Österberg, J. Am. Oil Chem. Soc. *65*:1544 (1988).
76. D. Han, J. S. Rhee, and S. B. Lee, Biotechnol. Bioeng. *30*:381 (1987).
77. Y. Miyake, T. Owari, K. Matsuura, and M. Teramoto, J. Chem. Soc., Faraday Trans. *89*:1993 (1993).
78. A. Na, C. Eriksson, S. G. Eriksson, E. Österberg, and K. Holmberg, J. Am. Oil Chem. Soc. *67*:766 (1990).
79. C. Komives, D. Osborne, and A. J. Russel, Biotechnol. Prog. *10*:340 (1994).
80. A. J. Russel and C. Komives, CHEMTECH *24*:26 (January 1994).
81. F. M. Menger and A. R. Elrington, J. Am. Chem. Soc. *113*:9621 (1991).
82. D. G. Hayes and E. Gulari, Biotechnol. Bioeng. *40*:110 (1992).
83. A. Xenakis, T. P. Valis, and N. Kolisis, Prog. Colloid Polym. Sci. *84*:508 (1991).
84. H. Stamatis, A. Xenakis, U. Menge, and F. N. Kolisis, Biotechnol. Bioeng. *42*:931 (1993).
85. H. Stamatis, A. Xenakis, M. Provelegiou, and F. N. Kolisis, Biotechnol. Bioeng. *42*:103 (1993).
86. G. Hedström, M. Backlund, and J. P. Slotte, Biotechnol. Bioeng. *42*:618 (1993).
87. G. D. Rees, T. R. J. Jenta, M. G. Nascimento, M. Catauro, B. H. Robinson, G. R. Stephenson, and R. D. G. Olphert, Indian J. Chem. *32 B*:30 (1993).
88. M. G. Nascimento, M. C. Rezende, R. D. Vecchia, P. C. Jesus, and L. M. Z. Aguiar, Tetrahedron Lett. *33*:5891 (1992).
89. H. Stamatis, A. Xenakis, and F. N. Kolisis, Biotechnol. Lett. *15*:471 (1993).
90. S. E. Godtfredsen, O. Kirk, F. Björkling, and L. Bjerre Christensen, Paper presented at the IUPAC-NOST International Symposium on Enzymes in Organic Solvents, New Delhi, 1992.
91. M. Bello, D. Thomas, and M. D. Legoy, Biochem. Biophys. Res. Commun. *146*:361 (1987).
92. P. D. I. Fletcher, R. B. Freedman, B. H. Robinson, G. D. Rees, and R. Schomäcker, Biochim. Biophys. Acta *912*:278 (1987).
93. C. P. Singh, D. O. Shah, and K. Holmberg, J. Am. Oil Chem. Soc. *71*:583 (1994).
94. C. P. Singh, P. Skagerlind, K. Holmberg, and D. O. Shah, J. Am. Oil Chem. Soc. *71*:1405 (1994).
95. D. G. Hayes and E. Gulari, Biotechnol. Bioeng. *35*:793 (1990).
96. S.-G. Oh, K. Holmberg, and B. W. Ninham, J. Colloid Interface Sci. *181*:341 (1996).
97. G. D. Rees, B. H. Robinson, and G. R. Stephenson, Biochim. Biophys. Acta *1257*:239 (1995).
98. E. Testet, N. Rachdi, M. Baboulene, V. Speziale, and A. Lattes, C. R. Acad. Sci., Ser. 2 *306*:347 (1988).
99. X. Jorba, P. Clapes, J. L. Torres, G. Valencia, and J. Mata-Alvarez, Colloids Surf. A *96*:47 (1995).
100. P. D. I. Fletcher, M. F. Galal, and B. H. Robinson, J. Chem. Soc., Faraday Trans. 1 *80*:3307 (1984).
101. K. Holmberg, B. Lassen, and M.-B. Stark, J. Am. Oil Chem. Soc. *66*:1796 (1989).
102. P. S. Chang, J. S. Rhee, and J. J. Kim, Biotechnol. Bioeng. *38*:1159 (1991).
103. K. Holmberg and E. Österberg, Prog. Colloid Polym. Sci. *74*:98 (1987).
104. E. Österberg, A.-C. Blomström, and K. Holmberg, J. Am. Oil Chem. Soc. *66*:1330 (1989).
105. A.-C. Blomström, E. Österberg, and K. Holmberg, unpublished results.
106. Y. Totani and S. Hara, J. Am. Oil Chem. Soc. *68*:848 (1991).
107. K. Holmberg and C. Eriksson, Indian J. Chem. *31B*:886 (1992).
108. K. M. Min and J. F. Biellmann, Tetrahedron *44*:1135 (1988).
109. G. Hedström, J. P. Slotte, O. Molander, and J. B. Rosenholm, Biotechnol. Bioeng. *39*:218 (1992).
110. C. Oldfield and R. B. Freedman, Eur. J. Biochem. *183*:347 (1989).
111. P. Pietikaeinen and P. Adlercreutz, Appl. Microbiol. Biotechnol. *33*:455 (1990).
112. K. M. Larsson, P. Adlercreutz, and B. Mattiasson, Eur. J. Biochem. *166*:157 (1987).
113. J. P. Samama, K. M. Lee, and J. F. Biellmann, Eur. J. Biochem. *163*:609 (1987).
114. K. M. Lee and J. F. Biellmann, FEBS Lett. *223*:33 (1987).

115. K. M. Larsson, C. Oldfield, and R. B. Freedman, Eur. J. Biochem. *183*:357 (1989).
116. K. M. Lee, D. Martina, C. U. Park, and J. F. Biellmann, Bull. Korean Chem. Soc. *11*:472 (1990).
117. K. M. Larsson, P. Adlercreutz, and B. Mattiasson, Ann. N.Y. Acad. Sci. *613*:791 (1990).
118. K. M. Larsson, P. Adlercreutz, B. Mattiasson, and U. Olsson, J. Chem. Soc., Faraday Trans. *87*:465 (1991).
119. N. S. Kommareddi, M. Tata, C. Karayigitoglu, V. T. John, G. L. McPherson, M. F. Herman, C. J. O'Connor, Y.-S. Lee, J. A. Akkara, and D. L. Kaplan, Appl. Biochem. Biotechnol. *51/52*:241 (1995).
120. M. S. Ayyagari, K. A. Marx, S. K. Tripathy, J. A. Akkara, and D. L. Kaplan, Macromolecules *28*5192 (1995).
121. J. A. Akkara, M. Ayyagari, F. Bruno, L. Samuelson, V. T. John, C. Karayigitoglu, S. Tripathy, K. A. Marx, and D. V. G. L. N. Rao, Biomimetics *2*:331 (1994).
122. H. S. O. Chan, L. M. Gan, H. Chi, and C. S. Toh, J. Electroanal. Chem. *379*:293 (1994).
123. B. Halliwell and J. M. C. Gutteridge, *Free Radicals in Biology and Medicine*, Clarendon Press, Oxford, England, 1989.
124. T. Brenkman, A. R. Macrae, and R. E. Moss, Eur. Patent 92-306 224, to Unilever (1992).
125. P. Skagerlind, K. Larsson, M. Barfoed, and K. Hult, J. Am. Oil Chem. Soc. *74*:39 (1997).
126. G. D. Rees, M. G. Nascimento, T. R. Jenta, and B. H. Robinson, Biochim. Biophys. Acta *1073*:493 (1991).
127. T. R. Jenta, B. H. Robinson, G. Batts, and A. R. Thomson, Prog. Colloid Polym. Sci. *84*:334 (1991).
128. P. C. de Jesus, M. C. Rezende, and M. G. Nascimento, Tetrahedron: Asymmetry *6*:63 (1995).
129. C. Sonesson and K. Holmberg, J. Colloid Interface Sci. *141*:239 (1991).
130. S. Backlund, M. Rantala, and O. Molander, Colloid Polym. Sci. *272*:1098 (1994).
131. T. F. Towey, G. D. Rees, D. C. Steytler, A. L. Price, and B. H. Robinson, Bioseparation *4*:139 (1994).
132. K. M. Larsson, P. Adlercreutz, and B. Mattiasson, Biotechnol. Bioeng. *36*:135 (1990).
133. C. P. Singh and D. O. Shah, Colloids Surf. A *77*:219 (1993).
134. S.-G. Oh, C. P. Singh, and D. O. Shah, Langmuir *8*:2846 (1992).
135. N. L. Klyachko, A. V. Levashov, A. V. Pshezhetsky, N. G. Bogdanova, I. V. Berezin, and K. Martinek, Eur. J. Biochem. *161*:149 (1986).
136. P. Miethe, R. Gruber, and H. Voss, Biotechnol. Lett. *11*:449 (1989).
137. H. Voss and P. Miethe, in *Biocatalysis in Non-Conventional Media (J. Tramper, M. H. Vermüe, H. H. Beeftink, and U. von Stockar, eds.), Elsevier, Amsterdam, 1992, p. 739.*
138. P. Skagerlind, M. Jansson, B. Bergenståhl, and K. Hult, Colloids Surf. B *4*:129 (1995).
139. J. G. T. Kierkels, L. F. W. Vleugels, J. H. A. Kern, E. M. Meijer, and M. Kloosterman, Enzyme Microb. Technol. *12*:760 (1990).

24

Applications of Microemulsions in Enhanced Oil Recovery

Vinod Pillai,† James R. Kanicky, and Dinesh O. Shah
University of Florida, Gainesville, Florida

I. INTRODUCTION

The recovery of oil from a reservoir can be divided into three stages. In the primary oil recovery process, oil is recovered due to the pressure of natural gases, which forces the oil out through production wells. When this pressure decreases to a point where it is no longer capable of expelling the oil, water is injected to repressurize the reservoir. This is generally called secondary oil recovery or water flooding. The average oil recovery during the primary and secondary stages is nearly 35% of oil-in-place. The purpose of the tertiary (enhanced) oil recovery process is to recover at least part of the remaining oil-in-place. Enhanced oil recovery (EOR) methods can be divided into two major groups: thermal processes and chemical flooding processes. In situ combustion, steam injection, and wet combustion methods fall into the first category, whereas caustic flooding, surfactant flooding, micellar polymer flooding, and CO_2 flooding fall into the second category of processes [1–5].

After water flooding, residual oil is believed to be in the form of discontinuous oil ganglia trapped in the pores of rocks in the reservoir. The two major forces acting on an oil ganglion are viscous forces and capillary forces, the ratio of which is represented by the capillary number. At the end of the secondary oil recovery stage, the capillary number is around 10^{-6}. To recover additional oil, the capillary number has to be increased to around $10^{-3} - 10^{-2}$, which can be achieved by decreasing the interfacial tension at the oil/brine interface. Surfactants are used for this purpose.

II. SURFACTANTS IN OIL RECOVERY

A microemulsion is generally defined as a thermodynamically stable, isotropic dispersion of two relatively immiscible liquids, consisting of microdomains of one or both liquids stabilized by an interfacial film of surface-active molecules [1]. Microemulsions may be classified as water-in-oil (W/O) or oil-in-water (O/W) depending on the dispersed and con-

---

† Deceased.

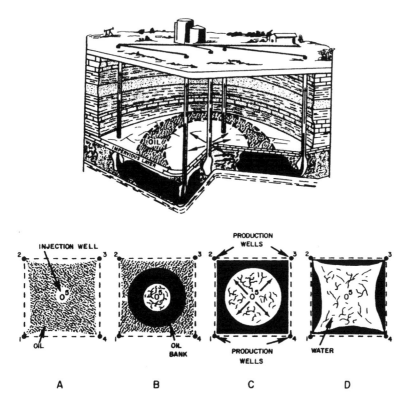

Figure 1 Three-dimensional view of a petroleum reservoir and the displacement of oil by water or surfactant solutions. Prior to EOR, oil in the reservoir is trapped within the rock in the form of oil ganglia (A). Injection of a surfactant solution mobilizes the oil ganglia (B) and forms an oil bank. The oil bank approaches (C) and subsequently reaches (D) the production wells.

tinuous phases present. In both cases, the dispersed phase consists of monodispersed droplets containing comparable amounts of oil and water. Some of these systems may show a bicontinuous, or even a cubic, structure [3,4,6].

Microemulsions were first introduced by Schulman and his coworkers in 1943 [7]. They explained that microemulsions were spontaneously formed with the uptake of water or oil due to a negative transient interfacial tension, which allows the free energy to decrease as the total oil/water interfacial area increases [7,8]. At equilibrium, the oil/water interfacial tension becomes zero or a very small positive number on the order of 10^{-2}–10^{-3} mN/m.

Surfactant solutions for use in improved oil recovery can be of high (2.0–10.0%) or low (0.1–0.2%) surfactant concentration. In the low concentration systems, the ultralow interfacial tension occurs when the aqueous phase of the surfactant solution is at about the apparent critical micelle concentration (cmc).

In the high surfactant concentration systems, a middle-phase microemulsion forms that is in equilibrium with excess oil and brine. The basic components of this microemulsion are surfactant, water, oil, alcohol, and salt. High surfactant concentrations in the injected plug result in a relatively small pore volume (about 3–20%) compared to micellar solutions (15–60%). Figure 1 schematically shows a three-dimensional view of a petroleum reservoir. At the end of water flooding, the oil that remains in the reservoir is believed to be in the form of oil ganglia trapped in the pore structure of the rock as shown in Fig. 1A. These

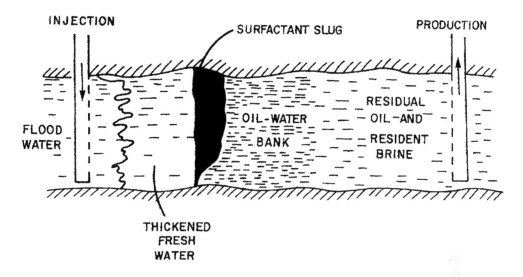

Figure 2 Two-dimensional view of the surfactant–polymer flooding process. Injection of a surfactant solution to coalesce the oil ganglia is followed by injection of a polymer slug to push the oil to production wells.

oil ganglia are entrapped because of capillary forces. However, if a surfactant solution is injected to lower the interfacial tension of the oil ganglia from its value of 20–30 mN/m to 10^{-3} mN/m, the oil ganglia can be mobilized and can move through narrow necks of the pores. Such mobilized oil ganglia form an oil bank as shown in Fig. 1B. Figures 1C and 1D diagrammatically illustrate the oil bank approaching the production well and the subsequent breakthrough of the drive water.

Wagner and Leach [9], Taber [10], and Melrose and Brader [11] suggested that capillary forces are responsible for entrapping a large amount of oil in the form of oil ganglia within the porous rocks of petroleum reservoirs. Foster [12] also showed that interfacial tension at the crude oil/brine interface, which plays a dominant role in controlling capillary forces, should be reduced by a factor of 10,000 to a value of 10^{-3}–10^{-4} mN/m to achieve efficient displacement of crude oil. Such low interfacial tensions can be achieved by appropriate surfactant formulations. Figure 2 schematically illustrates a two-dimensional view of the surfactant–polymer flooding process. A polymer slug, which is used for mobility control (i.e., to make the water more viscous), immediately follows after injection of the surfactant [13]. During this process, the displaced oil droplets coalesce and form an oil bank (see Fig. 3).

Once an oil bank is formed in the reservoir, it has to be propagated through the porous medium with minimum entrapment of oil at the trailing edge of the oil bank. The maintenance of ultralow interfacial tension is necessary to minimize the entrapment of oil in the porous medium. The leading edge of the oil bank coalesces with additional oil ganglia. Besides interfacial tension and interfacial viscosity, another parameter that influences the oil recovery is the surface charge at the oil/brine and rock/brine interfaces [14,15]. It has been shown that a high surface charge density leads to lower interfacial tension, lower interfacial viscosity, and consequently higher recovery of oil as shown in Fig. 4.

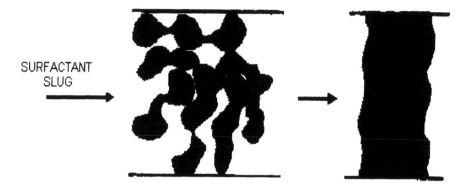

Figure 3 Schematic diagram of coalescence of oil ganglia due to low interfacial viscosity during the surfactant–polymer flooding process. Displaced oil ganglia must coalesce to form a continuous oil bank. For this, a very low interfacial viscosity is necessary.

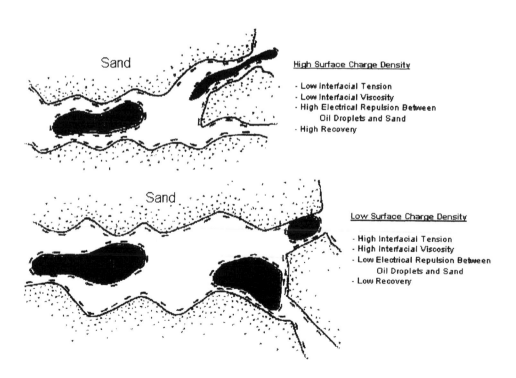

Figure 4 Schematic diagram of the role of surface charge in the oil displacement process. High surface charge density results in high oil recovery, while low surface charge density results in low oil yields.

III. ENHANCED OIL RECOVERY BY MICROEMULSION FLOODING

The success of microemulsion flooding for improving oil recovery depends on the proper selection of chemicals in formulating the surfactant slug. During the past 30 years, it has been reported that many surfactant formulations for enhanced oil recovery generally form multiphase microemulsions [16–19]. From these studies, it is evident that a variety of phases can exist in equilibrium with each other. Figure 5 shows the effect of salinity on the phase behavior of oil–brine–surfactant–alcohol systems. The microemulsion slug partitions into three phases: a surfactant-rich middle phase and surfactant-lean brine and oil phases [20–22] in the intermediate salinity range. The surfactant-rich phase is the middle-phase microemulsion [22]. The middle-phase microemulsion consists of solubilized oil, brine, surfactant, and alcohol. The lower to middle to upper phase ($l \rightarrow m \rightarrow u$) transition of the microemulsion phase can be obtained by varying any of the eight variables listed in Fig. 5.

A. Interfacial Tension

It is well established that ultralow interfacial tension plays an important role in oil displacement processes [16,18]. The magnitude of interfacial tension can be affected by the surface concentration of surfactant, surface charge density, and solubilization of oil or brine. Experimentally, Shah et al. [23] demonstrated a direct correlation between interfacial tension and interfacial charge in various oil–water systems. Interfacial charge density is an important factor in lowering the interfacial tension. Figure 6 shows the interfacial tension and partition coefficient of surfactant as functions of salinity. The minimum interfacial tension occurs at the same salinity where the partition coefficient is near unity. The same correlation between interfacial tension and partition coefficient was observed by Baviere [24] for the paraffin oil–sodium alkylbenzene sulfonate–isopropyl alcohol–brine system.

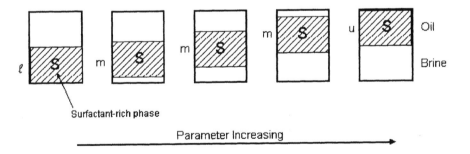

Figure 5 Illustration of the factors influencing the transition of a microemulsion from lower to middle to upper phase. The transition from lower to middle to upper phase ($l \rightarrow m \rightarrow u$) occurs by (1) increasing salinity, (2) decreasing oil chain length, (3) increasing alcohol concentration (C_4, C_5, C_6), (4) decreasing temperature, (5) increasing total surfactant concentration, (6) increasing brine/oil ratio, (7) increasing surfactant solution/oil ratio, and (8) increasing molecular weight of surfactant.

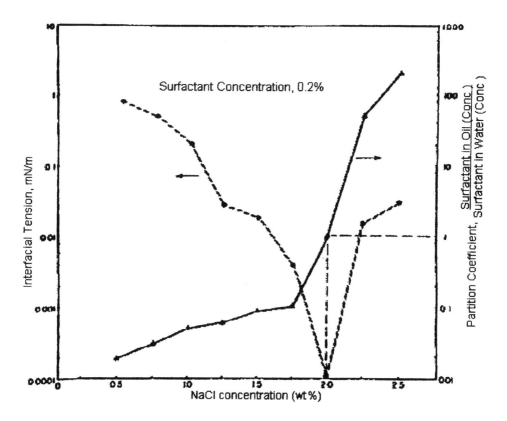

Figure 6 Effect of salinity on interfacial tension and surfactant partitioning in 0.2% TRS 10-80–brine–octane system.

Chan and Shah [25] proposed a unified theory to explain the ultralow interfacial tension minimum observed in dilute petroleum sulfonate solution–oil systems encountered in tertiary oil recovery processes. For several variables such as salinity, oil chain length, and surfactant concentration, a minimum in interfacial tension was found to occur when the equilibrated aqueous phase was at the cmc. This interfacial minimum also corresponded to the partition coefficient near unity for surfactant distribution in oil and brine. It was observed that the minimum in ultralow interfacial tension occurred when the concentration of the surfactant monomers in the aqueous phase was at a maximum.

B. Formation and Structure of Middle-Phase Microemulsion

The $l \rightarrow m \rightarrow u$ transitions of the microemulsion phase as a function of various parameters are shown in Fig. 5. Chan and Shah [26] compared the phenomenon of the formation of middle-phase microemulsions with that of the coacervation of micelles from the aqueous phase. They concluded that the repulsive forces between the micelles decreased due to the neutralization of the surface charge of micelles by counterions. The reduction in repulsive forces enhanced the aggregation of micelles, as the attractive forces between the micelles became predominant. This theory was verified by measuring the surface charge density of the equilibrated oil droplets in the middle phase [14].

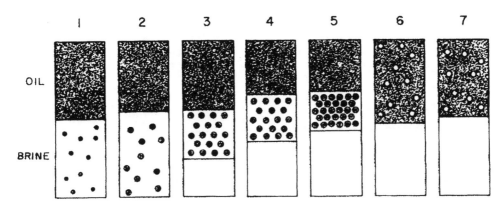

Figure 7 Schematic illustration of middle-phase microemulsion formation in surfactant–brine–oil systems. (●) Oil-swollen micelles (microdroplets of oil); (○) reverse micelles (microdroplets of water).

It was observed that the surface charge density increased to a maximum at the salinity at which the middle phase began to form. Beyond this salinity, the surface charge density decreased in the three-phase region. Based on several observations of different surfactant–brine–oil systems, Chan and Shah [26] proposed the mechanism of middle-phase microemulsion formation shown in Fig. 7. In general, the higher the solubilization of brine or oil in the middle-phase microemulsion, the lower the interfacial tension with the excess phases. The salinity at which equal volumes of brine and oil are solubilized in the middle-phase microemulsion is referred to as the optimal salinity for the surfactant–oil–brine systems under given physicochemical conditions [20,22]. Some investigators [21,27] showed that oil recovery is maximum near the optimal salinity of the system. Therefore, one can conclude that the middle-phase microemulsion plays a major role in enhanced oil recovery processes.

Using various physicochemical techniques such as high resolution NMR, viscosity, and electrical resistivity measurements, Chan and Shah [26] proposed that the middle-phase microemulsion in three-phase systems at or near optimal salinity is a water-external microemulsion of spherical droplets of oil. Extended studies to characterize the middle-phase microemulsions by several techniques including freeze-fracture electron microscopy revealed the structure to be a water-external microemulsion [26]. The droplet size in the middle-phase microemulsion decreases with increasing salinity. A freeze-fracture electron micrograph of a middle-phase microemulsion is shown in Fig. 8. It clearly indicates that the discrete spherical structure of the oil droplets in a continuous aqueous phase is consistent with the mechanism proposed in Fig. 7. This system was extensively studied by Reed and coworkers [20–22].

C. Solubilization

The effectiveness of surfactant formulations for enhanced oil recovery depends on the magnitude of solubilization. By injecting a chemical slug of complete miscibility with both oil and brine present in the reservoir, 100% recovery of oil should be possible.

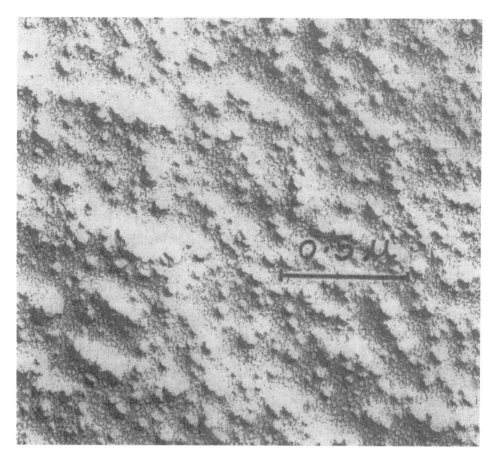

Figure 8 Freeze-fracture scanning electron micrograph of a middle-phase microemulsion. The spherical shapes are oil droplets suspended within the continuous aqueous phase. (The black bar represents 0.5 μm.)

The effect of hydrated radii, valence, and concentration of counterions on oil-external and middle-phase microemulsions was investigated by Chou and Shah [28]. It was observed that 1 mol of $CaCl_2$ was equivalent to 16–19 mol of NaCl for solubilization in middle-phase microemulsions, whereas for solubilization in oil-external microemulsions, 1 mol of $CaCl_2$ was equivalent to only 4 mol of NaCl. For monovalent electrolytes, the values for optimal salinity of solubilization in oil-external and middle-phase microemulsions are in the order LiCl > NaCl > KCl > NH_4Cl, which correlates with the Stokes radii of hydrated counterions. The optimal salinity for middle-phase microemulsions and critical electrolyte concentration varied in a similar fashion with Stokes radii of counterions, which was distinctly different for the solubilization in oil-external miroemulsions. Based on these findings, it was concluded that the middle-phase microemulsion behaved like a water-continuous system with respect to the effect of counterions [28].

The effect of alcohol concentration on the solubilization of brine was studied by Hsieh and Shah [29]. They observed that there was an optimal alcohol concentration that could solubilize a maximum amount of brine and also produce ultralow interfacial tension.

The optimal alcohol concentration depends on the brine concentration of the system. The effect of different alcohols on the equilibrium properties and dynamics of micellar solutions was studied by Zana [30].

D. Phase Behavior

The surfactant formulations for enhanced oil recovery consist of surfactant, alcohol, and brine with or without added oil. As the alcohol and surfactant are added to equal volumes of oil and brine, the surfactant partitioning between oil and brine phases depends on the relative solubilities of the surfactant in each phase. If most of the surfactant remains in the brine phase, the system splits into two phases, and the aqueous phase consists of micelles or oil-in-water microemulsions, depending on the amount of oil solubilized. If most of the surfactant remains in the oil phase, a two-phase system is formed with reverse micelles or the water-in-oil microemulsions in equilibrium with an aqueous phase.

The phase behavior of surfactant formulations for enhanced oil recovery is also affected by the oil solubilization capacity of the mixed micelles of surfactant and alcohol. For low concentration surfactant systems, the surfactant concentration in the oil phase changes considerably near the phase inversion point.

In summary, several phenomena occurring at optimal salinity in relation to enhanced oil recovery by microemulsion flooding are shown in Fig. 9. It is evident that the maximum in oil recovery efficiency correlates well with various transient and equilibrium properties of microemulsion systems. We have observed that surfactant loss in porous media is minimum at optimal salinity, presumably due to reduction in the entrapment process for the surfactant phase. Therefore, the maximum oil recovery may be due to the combined effect of all these processes occurring at optimal salinity.

IV. CURRENT PROGRESS

Until about 1980, the use of surfactants in enhanced oil recovery was mainly in the area of microemulsion flooding. Current low oil prices, however, have not provided adequate financial incentives for continued use of such methods [31], and the oil industry has turned to "gas" flooding, especially with carbon dioxide, as the main enhanced oil recovery tool. With the cheaper production of many of the commonly used surfactants, polymers, and alkalies, enhanced oil recovery with surfactants can once again become an affordable method of industrial oil production [32]. Additional advances in technology and equipment are making possible studies of interfacial systems at high temperatures and pressures [33]. The results of these studies will greatly expand the viability of oil recovery by microemulsion flooding techniques.

One recent attempt to decrease the costs associated with surfactant flooding has been to inject surfactant-producing bacteria into oil reservoirs. This technique involves the injection of selected microorganisms into the reservoir and the subsequent stimulation and transportation of their growth products in order to recover more of the oil-in-place [34]. Some of the mechanisms proposed by which these microbes can stimulate oil production include reservoir repressurization, modification of reservoir rock, degradation and alteration of oil, decrease of viscosity, and increase in emulsification [35].

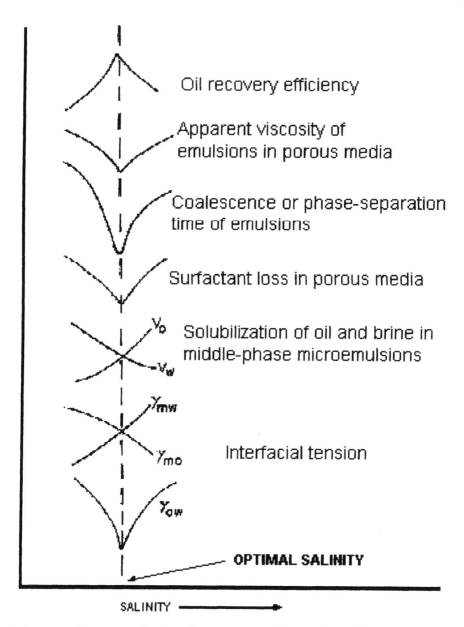

Figure 9   A summary of various phenomena occurring at optimal salinity in relation to enhanced oil recovery.

V.   CONCLUSIONS

Microemulsions are clear, isotropic dispersions of water or oil droplets 10–100 nm in diameter dispersed in a continuous oil or water phase and stabilized by an interfacial film of surfactants. Due to these unique properties, microemulsions are relevant in a variety of technological processes, including enhanced oil recovery (EOR). Several concepts and processes involving EOR are described in a keynote paper presented by Shah at the First European Symposium on Enhanced Oil Recovery [36]. Although oil prices are currently too low to justify the use of surfactants in enhanced oil recovery, advances in technology and lower surfactant costs may one day make the use of surfactants a viable approach in industrial oil recovery.

ACKNOWLEDGMENTS

We thank the industrial consortium of the Alberta Research Council, American Cyanamid Co., Amoco Production Co., Atlantic-Richfield Co., BASF-Wyandotte Co., British Petroleum Co., Calgon Corp., Cities Service Oil Co., Continental Oil Co., Ethyl Corp., Exxon Production Research Co., Getty Oil Co., Gulf Research and Development Co., Marathon Oil Co., Mobil Research and Development Co., Nalco Chemical Co., Phillips Petroleum Co., Shell Development Co., Standard Oil of Ohio Co., Stepan Chemical Co., Sun Oil Chemical Co., Texaco Inc., Union Carbide Corp., Union Oil Co., Westvaco, Inc., and Witco Chemical Co. for supporting the University of Florida Enhanced Oil Recovery Research Program. We also gratefully acknowledge the support of the Department of Energy (grant DE-AC1979BC10075), Defense Advanced Research Project Agency (DARPA), National Science Foundation (grants CBT-8807321 and CTS-8922574), and Electric Power Research Institute (ERPI grant RP800922).

REFERENCES

1. R. Leung, M. J. Hou, C. Manohar, D. O. Shah, and P. W. Chun, in *Macro-and Microemulsions* (D.O. Shah, ed.), Am. Chem. Soc., Washington, DC, 1985, p. 325.
2. M. K. Sharma and D. O. Shah, in *Macro- and Microemulsions* (D. O. Shah, ed.), Am. Chem. Soc., Washington, DC, 1985, p. 1.
3. L. Auvray, J. P. Cotton, R. Ober, and C. Taupin, J. Phys. *45*:913 (1984).
4. L. E. Scriven, Nature *263*:123 (1976).
5. G. Shutang, L. Huabin, and L. Hongfu, SPE Reservoir Eng., August 1995, p. 194.
6. J. Tabony, Nature *319*:400 (1986).
7. T. P. Hoar and J. H. Schulman, Nature *152*:102 (1943).
8. J. H. Schulman, W. Staeckenius, and L. M. Prince, J Phys. Chem. *63*:7716 (1959).
9. O. R. Wagner and R. O. Leach, Soc. Pet. Eng. J. *6*:335 (1966).
10. J. J. Taber, Soc. Pet. Eng. J. *9*:3 (1969).
11. J. C. Melrose and C. F. Brader, J. Can. Pet. Technol. *13*:54 (1974).
12. W. R. Foster, J. Pet. Technol. *25*:205 (1973).
13. S. Nilsson, A. Lohne, and K. Veggeland, Colloids Surf. A *127*:241 (1997).
14. M. Y. Chiang, K. S. Chan, and D. O. Shah, J. Can. Pet. Technol. *17*:1 (1978).
15. F. H. L. Wang, SPE Reservoir Eng., May 1993, p. 108.
16. M. Y. Chiang and D. O. Shah, Paper No. SPE 0988, presented at the Soc. Pet. Eng. 5th Int. Symposium on Oilfield and Geothermal Chemistry, Stanford, CA, 1980.
17. J. L. Cayias, R. S. Schechter, and W. H. Wade, J. Colloid Interface Sci. *59*:31 (1977).
18. P. M. Wilson, C. L. Murphy, and W. R. Foster, Paper No. SPE 5812, presented at the Soc. Pet. Eng. Improved Oil Recovery Symposium, Tulsa, OK, 1976.
19. M. J. Schwuger, K. Stickdorn, and R. Schömacker, Chem Rev. *95*:849 (1995).
20. R. N. Healy and R. L. Reed, Soc. Pet. Eng. J. *14*:451 (1974).
21. R. L. Reed and R. N. Healy, in *Improved Oil Recovery by Surfactant and Polymer Flooding* (D. O. Shah and R. S. Schechter, eds.), Academic, New York, 1977, p. 383.

22. R. N. Healy, R. L. Reed, and D. G. Stenmark, Soc. Pet. Eng. J. *16*:147 (1976).

23. D. O. Shah, K. S. Chan, and V. K. Bansal, Paper presented at the 83rd Natl. Meeting of AIChE, Houston, TX, 1977.

24. M. Baviere, Paper No. SPE 6000, presented at the 51st Annual Fall Technical Conference and Exhibition of the Society of Petroleum Engineers of AIME, New Orleans, 1976.

25. K. S. Chan and D. O. Shah, J. Dispersion. Sci Technol. *1*:55 (1980).

26. K. S. Chan and D. O. Shah, Paper No. SPE 7869, presented at the SPE-AIME Int. Symposium on Oilfield and Geothermal Chemistry, Houston, 1979.

27. D. F. Boneau and R. L. Clampitt, J. Pet. Technol. *29*:501 (1977).

28. S. I. Chou and D. O. Shah, J. Colloid Interface Sci. *80*:311 (1981).

29. W. C. Hsieh and D. O. Shah, Paper No. SPE 6594, presented at the SPE-AIME Int. Symposium on Oilfield and Geothermal Chemistry, La Jolla, CA, 1977.

30. R. Zana, in *Surface Phenomena in Enhanced Oil Recovery* (D. O. Shah, ed.), Plenum, New York, 1981, p. 521.

31. J. P. Brashear, A. Becker, K. Biglarbigi, and R. M. Ray, J. Pet. Technol. *41*:164 (1989).

32. M. Baviere, P. Glenat, N. Plazanet, and J. Labrid, SPE Reservoir Eng. *10*:187 (1995).

33. L. L. Schramm, D. B. Fisher, S. Schurch, and A. Cameron, Colloids Surf. A *94*:145 (1995).

34. J. D. Desai and I. M. Banat, Microbiol. Mol. Biol. Rev. *61*:47 (1997).

35. E. C. Donaldson, G. V. Chilingarian, and T. F. Yen, *Microbial Enhanced Oil Recovery*, Elsevier, New York, 1989, p. 9.

36. D. O. Shah, Fundamental aspects of surfactant–polymer flooding process, Keynote paper presented at the European Symposium on Enhanced Oil Recovery, Bournemouth, England, 1981.

25

Microemulsions in Pharmaceuticals

Martin Malmsten
Institute for Surface Chemistry, Stockholm, Sweden

I. INTRODUCTION

The ever-increasing demands on the performance of pharmaceutical formulations with respect to, e.g., storage stability, increased dosage levels, greater bioavailability, fewer side effects, controlled release, and biological response (e.g., tissue distribution) constitute the main motivation for drug delivery research. In the last few decades, this research has resulted in the development of, e.g., parenteral emulsions, liposomes with improved circulation in the bloodstream, cyclodextrins, and lipoprotein mimics for cancer therapeutics.

Surfactants play a key role in many of the novel drug delivery systems developed, and a wide range of surfactant-containing systems, including emulsions, liposomes, liquid crystalline phases (e.g., lamellar, hexagonal, or cubic), and microemulsions, are being extensively investigated in relation to drug delivery [1–5].

Microemulsions are systems consisting of water, oil, and amphiphile(s) that constitute a single optically isotropic and thermodynamically stable liquid solution [6]. Using this definition of microemulsions, it follows that solutions of micelles or reverse micelles with solubilized oil and water, respectively, should also be referred to as microemulsions, and these systems are therefore included in the present chapter.

Microemulsions are distinctly different from emulsions in that the former are thermodynamically stable one-phase systems whereas the latter are kinetically stabilized dispersions. Thus, microemulsions require no work for their formation, and once formed they are "infinitely" stable (cf. previous chapters). Emulsions, on the other hand, require work for their formation and display a kinetically controlled instability. This fundamental difference between emulsions and microemulsions is not always appreciated within the pharmaceutical literature.

Naturally, the main advantage of microemulsions compared to emulsions is their thermodynamic stability, resulting in ease of preparation and excellent long-term stability. Furthermore, microemulsions are capable of solubilizing large amounts of both water-soluble and oil-soluble drugs, and the microemulsions formed can be used as sustained release formulations. Moreover, microemulsions have been found to improve the drug bioavailability, e.g., in topical administration and in oral administration of peptide and protein drugs, sparingly soluble lipophilic drugs, and drugs labile at the conditions in the stomach. There are also other advantages with microemulsions compared to other drug

Table 1   Pharmaceutical Advantages of Microemulsions

General advantages
  Ease of preparation
  Clarity
  Stability
  Ability to be filtered
  Vehicle for drugs of different lipophilicities in the same system
  Low viscosity (no pain on injection)
Specific advantages
  Water-in-oil (W/O)
    Protection of water-soluble drugs
    Sustained release of water-soluble material
    Increased bioavailability
  Oil-in-water (O/W)
    Increased solubility of lipophilic drugs
    Sustained release of oil-soluble material
    Increased bioavailability
  Bicontinuous
    Concentrated formulation of both oil- and water-soluble drugs

Source: Ref. 9.

delivery systems, including ease of filtration (sterilization) and low viscosity (reducing pain on injection) (Table 1). Drawbacks with microemulsions as drug delivery systems include high surfactant concentration (and concomitant toxicity; see below) and limited solubilizing capacity for high-melting substances.

II. SOLUBILIZATION IN MICROEMULSIONS

Owing to their frequently high content of both water and oil, as well as of surfactant, microemulsions are usually efficient solubilizers of substances of a wide range of lipophilicity. The solubilizing power of microemulsions depends on their structure. Thus, the solubilizing capacity of a W/O microemulsion for water-soluble drugs is typically smaller than that of an O/W microemulsion, while the reverse is true for oil-soluble drugs. Furthermore, the solubilization depends on the microemulsion composition. For example, Jayakrishnan et al. [7] studied the solubilization of hydrocortisone by W/O microemulsions and found that the amount that could be incorporated in the microemulsion depended on the concentration of both the surfactant (Brij 35/Arlacel 186) and the cosurfactant (short-chain alcohols). Furthermore, the solubility of a drug in a microemulsion system depends on the molecular weight of the drug (and oil). For high molecular weight drugs (e.g., proteins), phase separation results only in a limited entropy loss, and consequently the solubilization of such drugs is typically more limited than that of low molecular weight drugs [8].

    It is important to note that the structure and stability of a microemulsion system may be affected by the solubilized drug. The effect of the drug on the microemulsion stability and structure depends on the properties of the drug, notably its lipophilicity. For example, it has been found that it is possible to incorporate lipophilic drugs, e.g., testosterone ethanate, at levels higher than 6 wt% in a 2 wt% soybean oil–20 wt% Brij 96 microemulsion without affecting the stability of the microemulsion [9]. On the other hand, the stability is generally strongly affected by surface-active drugs. For example, sodium salicylate has been found to significantly alter the stability region of microemulsions prepared from lecithin (Fig. 1). Furthermore, Carlfors et al. [10] studied the microemulsions formed by water, isopropyl myristate, and nonionic surfactant mixtures and their solubilization of lidocaine, a local anesthetic. By using NMR self-diffusion measurements, it was found that the surface-active but lipophilic lidocaine lowered the phase inversion temperature (PIT). This is what would be expected from simple packing considerations, since increasing the effective oil volume favors a decrease in the curvature toward the oil as well as the formation of reversed structures [11].

The influence of drugs on the stability and structure of microemulsions has important implications for drug delivery from these systems, since the phase behavior may change dramatically on release, particularly of surface-active drugs, which in turn may lead to changes in the release rate or even cause instability of the microemulsion system as such. Naturally, this effect may be used also as an advantage, e.g., in the in situ transition from an O/W microemulsion to a hexagonal, cubic, or lamellar phase on release of the drug, with obvious consequences in terms of viscosity, transport, etc.

III. SUSTAINED RELEASE FROM MICROEMULSIONS

Due to the wide range of structures occurring in them, microemulsions display a rich behavior regarding the release of solubilized material. Thus, in an O/W microemulsion, hydrophobic drugs, solubilized mainly in the oil droplets, experience hindered diffusion and are therefore released rather slowly (depending on the oil/water partitioning of the substance). Water-soluble drugs, on the other hand, diffuse essentially without obstruction (depending on the volume fraction of the dispersed phase) and are released fast [12,13]. The reverse behavior is expected for W/O microemulsions. For balanced microemulsions, relatively fast diffusion and release occur for both water-soluble and oil-soluble drugs due to the bicontinuous nature of the microemulsion "structure" [1,12,13].

Apart from the microemulsion structure, the microemulsion composition is important for the drug release rate. For example, Osborne et al. [14] studied the in vitro transdermal penetration of radiolabeled water from W/O microemulsions formed by water, octanol, and dioctylsodium sulfosuccinate. It was found that the delivery of the polar water portion in this microemulsion system was highly dependent on the microemulsion composition and, as expected, both the water self-diffusion and the transdermal flux increased on increasing the microemulsion water content. Furthermore, Gasco et al. [15] studied the release of (lipophilic) prednisone from O/W microemulsions formed by lecithin, butanol, and isopropyl myristate and found that the amount of cosurfactant in the microemulsion affected the drug release rate. More precisely, increasing the butanol concentration resulted in a decreased permeability constant for both hydrophilic and hydrophobic membranes. Similar results were obtained by Trotta et al. [16] on the release of steroid hormones of varying lipophilicity from microemulsions prepared from Aerosol OT, isopropyl myristate, water, and varying amounts of butanol.

Naturally, the microemulsion structure and composition are highly interrelated, and solubilized drugs generally affect both at the same time. Thus, for microemulsions formed by lecithin, oil, water, and short-chain alcohols (cf. the study by Gasco et al. [15]), there

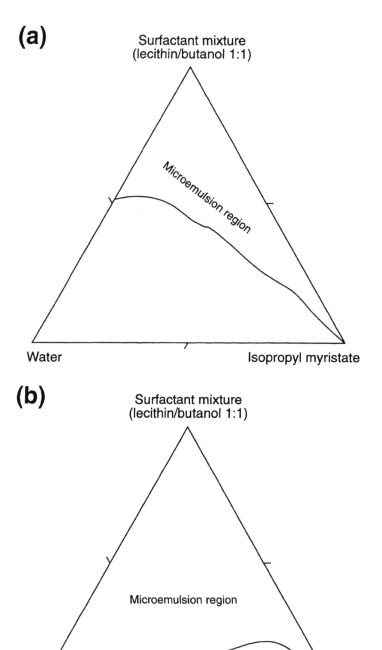

Figure 1 Pseudo ternary phase diagrams at room temperature of (a) quaternary systems containing lecithin, butanol, isopropyl myristate, and water or (b) a 10 wt% aqueous solution of sodium salicylate, at a lecithin/butanol ratio of 1 : 1. (Adapted from Ref. 9.)

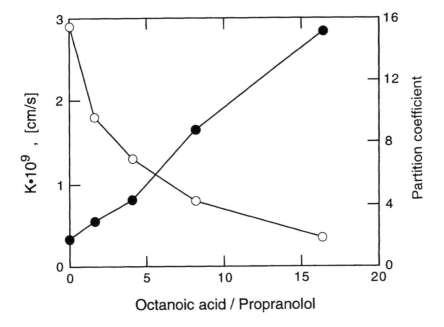

Figure 2 (●) Isopropyl myristate/buffer partition coefficient and (○) permeability coefficient K of propranolol over a hydrophilic membrane from Tween 60–isopropyl myristate–butanol–water microemulsions with varying octanoic acid/propranolol ratios. (Data from Ref. 18.)

is a gradual transition toward structures more curved toward the oil (i.e., smaller droplets) on increasing the alcohol concentration [13]. The smaller oil droplets formed in the presence of butanol hinder the prednisone diffusion more efficiently than larger droplets. Similarly, butanol resulted in a decreased droplet size for the system investigated by Trotta et al. [16]. Quantitatively, however, the experimentally observed decrease in the droplet size on increasing the butanol concentration is rather marginal. On the other hand, butanol was also found to cause an effective enhancement of the drug partitioning to the oil phase [16], which also contributes to the decrease in the release rate.

Apart from the microemulsion structure and composition, the drug release rate from a microemulsion is expected to depend on the oil/water partitioning of the drug. In the study by Trotta et al. [16] on the release of steroid hormones of different lipophilicities from O/W microemulsions, a correlation was found between the partition coefficients and the release rates of the hormones.

Considering the dependence of the release rate on the drug partitioning, it is clear that the rate of release from a microemulsion formulation can be altered by changing the oil/water partitioning of the drug. For example, the effects on the release rate of increasing the effective lipophilicity of timolol by ion-pair formation with octanoic acid was studied by Gallarate et al. [17]. These authors found that on increasing the octanoic acid concentration (i.e., increasing the ion-pair formation), the timolol partitioned more to the oil phase, resulting in an increased penetration of a lipophilic membrane. Similarly, Gasco et al. [18] used ion-pair formation with octanoic acid in O/W microemulsions to increase the lipophilicity of propranolol in order to obtain a disperse phase acting as a reservoir. As shown in Fig. 2, it was found that octanoic acid increased the propranolol partitioning to the oil, resulting in a decrease in the drug release rate.

Table 2 Half-Life Values (h) of Technetium Activity in Rabbits Injected with Technetium in W/O Microemulsion or Aqueous Solution

Rabbit	Microemulsion	Aqueous solution
1	132	10.6
2	151	12.5
3	57[a]	17.9
4	46[a]	12.1
5	69[a]	9.8
6	122	8.2
7	189	9.2

[a] These animals showed biexponential decay; half-lives reported are those of monoexponential functions with the same intercept and the same blood concentration versus time integrals as those of the biexponential fit.

Source: Ref. 20.

Also, strong interactions (complexation) between the drug and the surfactant have been found to strongly affect the drug release rate in microemulsion formulations. For example, Gasco et al. [19] studied the release of doxorubicin, a hydrophilic antitumor drug, from a W/O microemulsion formed by lecithin, water, hexanol, and ethyl oleate and observed a reservoir effect due to complexation between doxorubicin and lecithin.

The possibility of controlling the drug release rate by the microemulsion structure and composition as well as by drug partitioning makes microemulsions of interest for controlled-release applications, and several studies on this aspect of the pharmaceutical use of microemulsions, apart from those mentioned above, have been reported. For example, Bello et al. [20] compared the release of pertechnetate from a lecithin-containing W/O microemulsion and an aqueous solution after subcutaneous administration in rabbits and found that pertechnetate carried by W/O microemulsions was released from the site of administration at a slower rate than that administered from an aqueous solution (Table 2). Along similar lines, Nastruzzi and Gambari [21] performed an antitumor evaluation of lecithin-based microemulsion gels containing an aromatic tetrabenzamidine for topical administration in comparison to intraperitoneal injections of the corresponding aqueous solution and found the former formulation to be quite effective in inhibiting the proliferation at the cutaneous and subcutaneous levels.

IV. TOPICAL ADMINISTRATION FROM MICROEMULSIONS

Transdermal administration of drugs exhibits several advantages over other administration forms, notably oral administration, through the avoidance of systemic side effects. However, the drug efficacy is limited by the transdermal penetration rate, which severely limits the drugs that can be successfully administered by this route. In particular, transdermal penetration is limited by the penetration barrier exerted by the stratum corneum, the outermost layer of the skin, which comprises keratin-rich dead cells embedded in a lipid matrix. In fact, the main function of the stratum corneum is to limit the transdermal penetration in order to prevent the body from dehydrating as well as to protect it against chemical and biological attack.

The lipids present in the stratum corneum, amounting to about 10% of its dry mass, seem to be essential for this function. Friberg and others [22,23] studied the composition and structure of the stratum corneum, and from these studies a layered structure has been

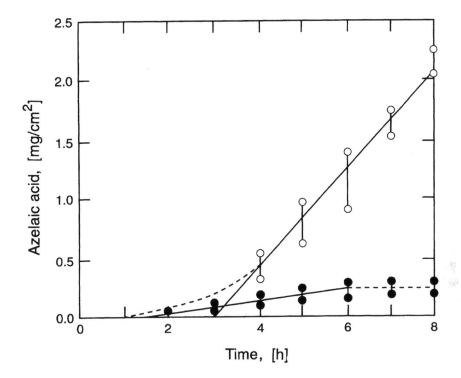

Figure 3 Permeation profiles of azelaic acid over full-thickness abdominal skin from a (○) water-propylene glycol, decanol-dodecanol, Tween 20, 1-butanol, Carbopol 934, water microemulsion and (●) the corresponding polymer solution. (Adapted from Ref. 25.)

inferred. Since no specific lipids have been found to be responsible for the barrier to water transport over the stratum corneum, and since "regenerated" stratum corneum, in which the natural lipids have all been replaced by model lipids, displays similar barrier properties to water transport, it has been inferred that it is the structure of the lipid self-assemblies that governs the barrier properties. This is in analogy to recent findings by Engblom and Engström [24], who showed that Azone, a well-known penetration enhancer used frequently for topical administration of drugs, increased the curvature toward water in lipid systems, exemplified, e.g., by a transition from lamellar to bicontinuous cubic structures. Thus, disruption of the layered structures and generation of water and oil channels correlate with an increased penetration of the stratum corneum.

Considering the solubilizing capacity of microemulsions, these are expected to significantly affect the structure of the stratum corneum lipid self-assemblies, with obvious consequences for drug penetration. There have been several studies of the penetration enhancement due to microemulsion drug carriers. For example, Gasco et al. [25] studied the transport of azelaic acid, a bioactive substance used for treating a number of skin disorders, from a viscosified microemulsion and from the corresponding polymer solution through full-thickness abdominal skin. (The viscosity of microemulsions frequently has to be increased to make them suitable for topical application.) It was found that an O/W microemulsion formed by water–propylene glycol, decanol–dodecanol, Tween 20, 1-butanol, and Carbopol 934 gave significantly better penetration than the corresponding water–propylene glycol–Carbopol "gel" (Fig. 3).

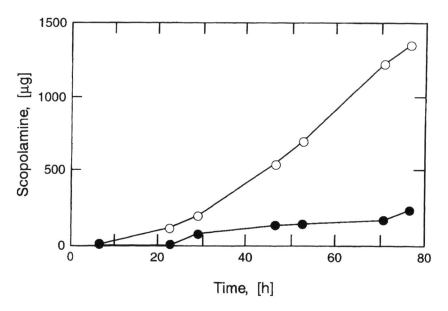

Figure 4 Transport of scopolamine through human skin from (○) a lecithin–isopropyl palmitate–water microemulsion ($[H_2O]/[\text{lecithin}]=3$) and from (●) an aqueous buffer solution. (Adapted from Ref. 28.)

Analogously, Ziegenmeyer and Führer [26] compared the transdermal penetration of tetracycline hydrochloride from a W/O microemulsion prepared from dodecane, decanol, water, and an ethoxylated alkyl ether surfactant with that from conventional formulations and found enhanced absorption from the microemulsion formulation. Similarly, Bhatnagar and Vyas [27] investigated the bioavailability of transdermally administered propranolol, a β-receptor blocking drug, which normally undergoes extensive first hepatic pass effects. It was found that the bioavailability of this drug could be extensively improved by transdermal application from a lecithin-based W/O microemulsion. Moreover, Willimann et al. [28] employed lecithin-containing (W/O) microemulsions for the transdermal administration of scopolamine and broxaterol and found that the transport rate obtained with the lecithin microemulsion gels was much higher than that obtained with an aqueous solution at the same concentration (Fig. 4).

Since transdermal penetration from microemulsion systems is likely to depend on the degree of perturbation of the layered structures of the stratum corneum and the generation of "channels," it could be expected to depend on the microemulsion composition and structure (see above). For example, Osborne et al. [29] studied the transdermal penetration of glucose from W/O microemulsions prepared from octanol, dioctylsodium sulfosuccinate, and water and found that an increased microemulsion water content caused enhanced water penetration. Differences in percutaneous glucose transport were further shown to parallel differences in the diffusion of water within the microemulsion vehicles prior to application to the skin.

An important aspect of enhanced transdermal transport is that it is often accompanied by irritation. This irritation is likely to be due to the disruption of the stratum corneum structure. It is important to note, however, that the composition of an applied multicomponent formulation, specifically that of a microemulsion, changes over time due to evaporation of the most volatile component, usually water. If, for example, what remains on the skin after water evaporation is an oil solution of surfactants, a disordering of the stratum corneum lipid structure may occur, with skin irritation as a consequence. If, on the other hand, the remains form a liquid crystalline phase (notably a lamellar phase), the inherent risk for irritation is reduced [22,23]. Note, however, that the detailed mechanisms of both transdermal penetration enhancement and the occasionally occurring skin irritation are unclear at present despite the rather extensive research in this field.

V. ORAL ADMINISTRATION FROM MICROEMULSIONS

Apart from those involving penetration enhancement in topical administration, the most interesting drug delivery applications of microemulsions lie in the oral administration of peptide and protein drugs. These relatively new types of drugs are becoming increasingly important as therapeutic agents as a consequence of the extensive peptide research over the last few decades, which has resulted in a wide range of biomedical peptide hormones, synthetic peptides, enzyme substrates, and inhibitors [30,31]. Ailments that may be treated more efficiently with peptide drugs include hypertension, mental disorders, autoimmune diseases, cancer, and metabolic and cardiovascular diseases [31].

Despite the many promising features of peptides in drug delivery, many of them are difficult to administer orally without a loss in activity. (Typically, the activity of peptides following oral administration is less than 10% [32].) The generally low bioavailability of peptides administered orally means that the intra- and intersubject variability is magnified, and a major reason for trying to maximize the oral peptide bioavailability is in fact to be able to control the in vivo drug concentration.

As with any substance to be absorbed orally, peptides must pass the physical absorption barrier of the gastrointestinal tract, consisting of mucous, apical, and basal cell membranes and cell content, tight junctions, basement membrane, and the wall of the lymph and blood capillaries [33]. The transport from the lumen to the bloodstream may proceed either transcellularly (i.e., transport through the epithelial cells) or paracellularly (i.e., transport via the tight junctions). The transport via these routes is mainly determined by the lipophilicity, charge, and size of the drug. In particular, the size of protein and polypeptide drugs may cause problems regarding their oral absorption.

A more serious problem than the physical absorption barrier, however, is that of enzymatic activity in the intestinal tract. This enzymatic barrier consists of exo- and endopeptidases and is exceptionally well designed to digest peptides and proteins. The susceptibility of peptides to enzymatic degradation is enhanced by the fact that the peptides contain several linkages, each of which may be susceptible to hydrolysis mediated by one or several peptidases. As an example, the undecapeptide substance P is susceptible to degradation by at least five different enzymes [34].

There have been several approaches reported in the literature to improve the oral bioavailability of peptide drugs, including chemical modification (e.g., PEGylation [35], coadministration of absorption enhancers (e.g., EDTA, salicylates, surfactants, bile salts, and fatty acids [36]), and coadministration of inhibitors to the peptide metabolism [37].

Another important approach for improving the oral biovailability of peptide drugs is the development of drug delivery systems such as liposomes, emulsions, nanocapsules, and nanospheres [38]. Considering their thermodynamic stability, resulting in ease of preparation and excellent long-term storage properties, microemulsions have also been considered as delivery systems for peptide and protein drugs.

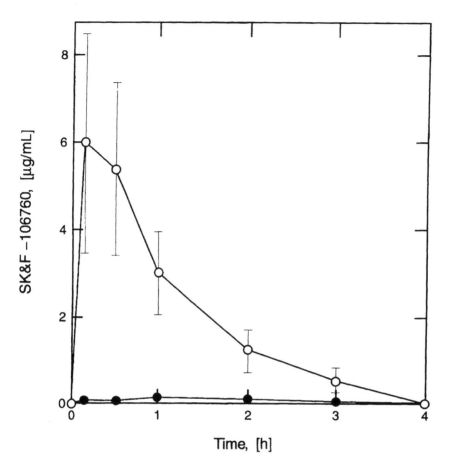

Figure 5   Plasma concentration of SK&F-106760 as a function of time after intraduodenal administration from (●) an aqueous solution or (○) a microemulsion. (Adapted from Ref. 39.)

For example, SK&F-106760 and SK&F-110679, both water-woluble RGD fibrinogen receptor antagonists, were formulated in microemulsions at pharmaceutically relevant levels, and their uptake after intraduodenal administration was investigated [39–41]. It was found that the presence of the SK&F-110679 peptide in varying concentrations did not influence the structure of the W/O microemulsion studied. Furthermore, the bioavailability of SK&F-106760 was increased dramatically for the microemulsion formulations compared to that of the aqueous solutions, as can be seen in Fig. 5.

The peptide drug by far most frequently studied in relation to oral bioavailability is cyclosporine, a potent immunosuppressive agent that prolongs allograft survival in organ transplantation and is also used in the treatment of patients with certain autoimmune diseases. When given orally, the absorption of cyclosporine is incomplete, amounting to about 30% of the dose or less [42]. Furthermore, the absorption of cyclosporine is variable and is affected by physiological and pharmaceutical factors such as bile, food, and drug delivery vehicle. This results in a high intra- and intersubject variability in pharmacokinetic parameters [43].

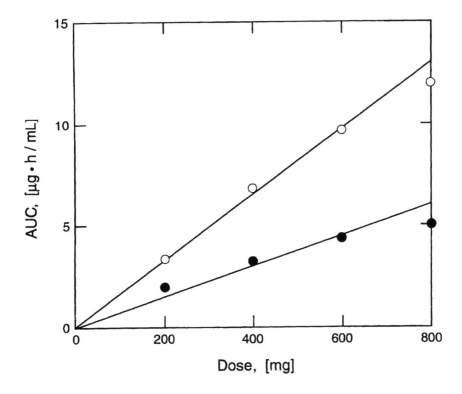

Figure 6 Relationship between cyclosporine bioavailability, given as AUC (integral of the blood concentration versus time curve), and dose after oral administration of (●) a crude emulsion or (○) an O/W microemulsion to healthy volunteers. (Adapted from Ref. 47.)

Since cyclosporine is not subject to extensive enzymatic degradation [30,32], a major reason for the limited bioavailability of this substance appears to be its poor absorption, originating from its high molecular weight and high lipophilicity [44,45]. Considering this, it is not unexpected that the bioavailability of cyclosporine may be improved by coadministration of absorption enhancers, and the traditional administration for cyclosporine has been based on a crude emulsion (Sandimmune). As could be expected, this has been found not only to improve the cyclosporine bioavailability but also to decrease the pharmacokinetic variability. By reducing the droplet size, i.e., increasing the oil/water interfacial area, it is possible to further increase the intestinal absorption [46].

Once more considering the thermodynamic stability of microemulsions, as well as the fact that the droplet size in microemulsions is frequently much smaller than that of the corresponding emulsion, microemulsions have also been applied as drug vehicles for cyclosporine oral administration (Sandimmune Neoral). In a number of investigations, it was found that the bioavailability of cyclosporine may be further improved with this for-mulation at the same time as the pharmacokinetic variability is reduced [43–45,47–49]. An example of this type of results is given in Fig. 6. In line with the findings of Tarr and Yalkowsky [46] on the effect of droplet size on the absorption from emulsions, these results seem to indicate that one contribution to the enhanced absorption from the microemulsion formulation is the small droplet size in this system. However, as shown by Ritschel [49], droplet size is not the only determinant of oral absorption from emulsion and microemulsion systems. Other factors, such as the nature of the oil phase, also have a major influence.

An essential issue for formulation of peptide and protein drugs in microemulsions and other surfactant-containing drug delivery systems is the potential alteration of the peptide and protein biological activity by the surfactant(s). It has been found that particularly anionic, but also cationic and nonionic, surfactants interact to form complexes with a variety of peptides and proteins [50]. More often than not, this complex formation results in conformational changes of the polypeptides (denaturation) and in a loss of biological activity. Just to mention one example of pharmaceutically related studies in this field, Patel et al. [51] studied the formulation of insulin in a microemulsion formed by an ethoxylated fatty acid and found a significant loss of insulin activity on storage of this microemulsion formulation. Care must therefore be taken to avoid biological activity loss upon formulation of peptide and protein drugs in microemulsion systems.

Although the main emphasis of microemulsions for oral administration so far has been on peptide drugs, microemulsions have a general potential for oral delivery of sparingly soluble lipophilic drugs with poor bioavailability as well as of drugs unstable at the conditions present in the stomach. For example, Novelli et al. [52] used a microemulsion to formulate WR2721, which is employed in the radio- and chemotherapy of cancer and needs to be protected from acid hydrolysis in the stomach in order to retain its biological activity. Using a W/O microemulsion of CTAB, isooctane, and butanol, these authors found that the hydrolysis was slowed down considerably in the microemulsion compared to the aqueous solution.

VI. OTHER PHARMACEUTICAL APPLICATIONS OF MICROEMULSIONS

Although sustained release, penetration enhancement in topical administration, improved bioavailability of peptide and lipophilic drugs in oral administration, and protection of drugs against degradation in aqueous systems seem to be the most important applications of microemulsions in drug delivery, these systems have also been explored for other uses. For example, microemulsions containing a polymerizable or crystallizable component have been used for the preparation of solid nanoparticles through polymerization [53] or crystallization [54], respectively. Provided that the microemulsion structure can be retained during the preparation step, extremely small solid particles may be obtained in this manner. In a study by Gasco et al. [53], doxorubicin was incorporated in a matrix through polymerization from a microemulsion consisting of Aerosol OT, butanol, hexane, ethylene glycol, and methylcyan acrylate. Although the particle size increased from 32 nm to 146 nm during the polymerization, the nanoparticles obtained were quite effective in incorporating doxorubicin (Table 3). The doxorubicin incorporated in these particles was released quite slowly, although the total amount released even after several hours was quite low (Fig. 7).

Microemulsions containing fluorocarbons have been investigated as blood substitutes. For example, Cecutti et al. [55] studied these aspects of microemulsions and found that although the inherent incompatibility between hydrocarbons and fluorocarbons puts some demands on the surfactant to be used, the resulting microemulsion (prepared from fluorinated oil, water, and an nonionic surfactant) displayed oxygen absorption similar to that of blood, and at the same time the toxicity was limited and the microemulsions appeared to be well tolerated.

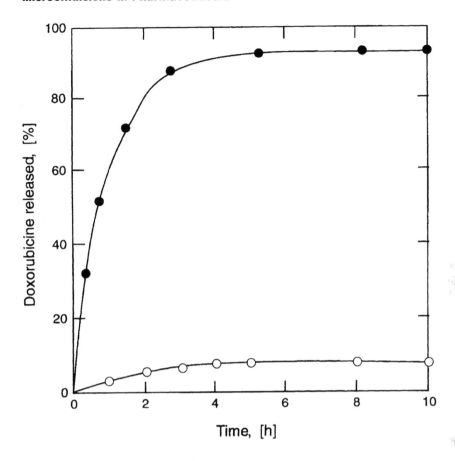

Figure 7 Release profiles for doxorubicin (●) diffusing in solution and (○) encapsulated in solid nanoparticles obtained from polymerization of methylcyan acrylate in a microemulsion consisting of Aerosol OT, butanol, hexane, ethylene glycol, and methylcyan acrylate. (Adapted from Ref. 53.)

Table 3 Amounts of Doxorubicin Incorporated into Nanoparticles Using Different Concentrations of the Drug in Ethylene Glycol as the Disperse Phase of the Microemulsion

Concentration of doxorubicin in ethylene glycol (mg/mL)	Amount of doxorubicin in nanoparticles (mg/100 mg)
0.50	0.3
1.75	0.8
2.50	1.9
5.00	2.5
10.0	4.8
20.0	4.7

Source: Ref. 53.

VII. TOXICITY ASPECTS

An important aspect in all drug delivery is the toxicity of the drug as well as that of the drug carrier. Therefore, toxicity has to be assessed also for microemulsion formulations. In microemulsion systems, the main concern regarding toxicity has to do with the cosurfactants used. For example, the majority of the work on the pharmaceutical application of microemulsions has involved the use of short- or medium-chain alcohols, e.g., butanol. In a range of studies it has been shown that these cause toxic side effects. For example, inhalation studies of the toxicity of 1-butanol, 2-butanol, and *tert*-butanol in rats showed a dose-dependent reduction in fetal weight [56]. Furthermore, aqueous solutions of ethanol, propanol, and butanol were shown to result in elongated mitochondria in hepatocytes after 1 month of exposure [57]. (In addition to the toxicity aspects of these alcohols, microemulsions formed in their presence are often destabilized on dilution of the continuous phase.) Furthermore, many studies so far have involved aliphatic or aromatic oils, such as hexane or benzene, which obviously are unsuitable for pharmaceutical use. Moreover, ionic surfactants could in themselves be toxic and irritant [58].

Nonionic surfactants, such as ethoxylated alkyl ethers and sorbitan esters, as well as nonionic block copolymers [e.g., poly(ethylene oxide)-*block*-poly(propylene oxide)] are generally less irritant and toxic than ionic surfactants [58–60]. Moreover, many nonionic surfactants have the advantage over charged surfactants in that they can form microemulsions even without cosurfactants. For example, this was used by Siebenbrodt and Keipert [61], who investigated the ophthalmic application of microemulsions containing poloxamer L64 [a poly(ethylene oxide)–poly(propylene oxide) block copolymer], propylene glycol, water, and triacetine and found an acceptable tolerance of this model formulation. The comparatively good biological acceptance of nonionic surfactants and block copolymers and the fact that cosurfactants may not be needed for the microemulsion formation constitute the two main motives for the rather extensive use of nonionic surfactants, particularly for topical applications of microemulsions.

Despite the reasonable tolerance of nonionic surfactants, particularly in topical applications, microemulsions prepared from (phospho)lipids seem to be preferred over those prepared by synthetic surfactants from a toxicity point of view. As discussed by Shinoda et al. [13], lecithin in water–oil systems does not spontaneously form the zero mean curvature amphiphile layers required for the formation of balanced microemulsions but rather forms reverse structures. On decreasing the "polarity" of the aqueous phase by addition of a short-chain alcohol, e.g., propanol, lecithin was found to form microemulsions at low amphiphile concentrations over wide ranges of solvent composition. The structure of the microemulsions formed was investigated by NMR self-diffusion measurements, and it was found that with a decreasing propanol concentration there was a gradual transition from oil droplets in water, over a bicontinuous structure, to water droplets in oil [13].

Another way of forming balanced or O/W microemulsions in phospholipid-based systems is through the addition of surfactants that favor structures with a curvature toward the oil. Examples of such substances include nonionic surfactants and block copolymers with long oligo(ethylene glycol) chains, as well as single-chain (an)ionic surfactants [11]. Although this packing concept has not been extensively used for microemulsions so far, it is well known for other surfactant systems [11,62,63]

Despite some rather promising studies of the use of phospholipid-based microemulsions in drug delivery [1], the extensive use of cosurfactants (particularly medium-chain alcohols) means that toxicity still remains a problem for many of these for-

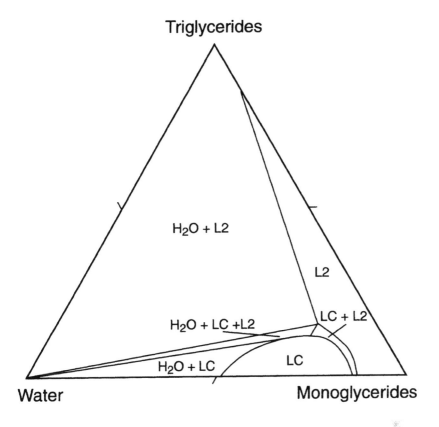

Figure 8 Phase diagram at 40°C of the ternary system water–triglycerides–monoglycerides. LC refers to liquid crystalline phases. (Adapted from Ref. 65.)

mulations. Thus, there is a need to develop new phospholipids and phospholipid-base microemulsions that do not require cosurfactants in order for this type of system to gain wide applicability in drug delivery.

Another type of "biocompatible" microemulsion, formed by water, triglycerides, and monoglycerides [64] has been studied, e.g., by Engström [65] (Fig. 8). It was found that this system formed a rather extensive L_2 phase at 40°C. Based on X-ray diffraction, this phase was proposed to have a lipid bilayer structure even at high oil content. Considering the biocompatibility and the ease of biodegradation of these components, this type of microemulsion is particularly attractive for oral delivery of drugs, where toxicity otherwise limits the applicability of microemulsions.

VIII. SUMMARY

Microemulsions offer an interesting and potentially quite powerful way for drug delivery through a host of administration routes. This is likely to yield formulations with advantageous behavior regarding, e.g., drug release rate, drug stability, drug bioavailability, and formulation storage stability. However, to become a widespread technology base, future microemulsions require development of new surfactants of low toxicity. This is the case particularly for parenteral drug delivery systems.

ACKNOWLEDGMENT

The comments on this chapter tendered by Professor Krister Holmberg, Dr. Björn Bergenståhl, and Dr. Sven Engström are gratefully acknowledged. This work was financed by the Foundation for Surface Chemistry, Sweden.

REFERENCES

1. D. Attwood, in *Colloidal Drug Delivery Systems*, Drugs Pharm. Sci. Ser., Vol. 66 (J. Kreuter, ed.), Marcel Dekker, New York, 1994.
2. A. T. Florence and D. Attwood, *Physicochemical Principles of Pharmacy*, Macmillan, London, 1988.
3. S. Engström, Lipid Technol. *2*:42 (1990).
4. P. Becher (ed.), *Encyclopedia of Emulsion Technology*, Vol. 1, Basic Theory, Marcel Dekker, New York, 1983.
5. P. Becher (ed.), *Encyclopedia of Emulsion Technology*, Vol. 2, Applications, Marcel Dekker, New York, 1983.
6. I. Danielsson and B. Lindman, Colloids Surf. *3*:391 (1981).
7. A. Jayakrishnan, K. Kalaiarasi, and D. O. Shah, J. Soc. Cosmet. Chem. *34*:335 (1983).
8. G. J. Fleer, M. A. Cohen Stuart, J. M. H. M. Scheutjens, T. Cosgrove, and B. Vincent, *Polymers at Interfaces*, Chapman & Hall, London, 1993.
9. M. J. Lawrence, Eur. J. Drug Metab. Pharmacokinet. *3*:257 (1994).
10. J. Carlfors, I. Blute, and V. Schmidt, J. Dispersion Sci. Technol. *12*:467 (1991).
11. J. Israelachvili, *Intermolecular and Surface Forces*, Academic, London, 1992.
12. U. Olsson and H. Wennerström, Adv. Colloid Interface Sci. *49*:113 (1994).
13. K. Shinoda, M. Araki, A. Sadaghiani, A. Khan, and B. Lindman, J. Phys. Chem. *95*:989 (1991).
14. D. W. Osborne, A. J. I. Ward, and K. J. O'Neill, Drug Dev. Ind. Pharm. *14*:1203 (1988).
15. M. R. Gasco, M. Gallarate, and F. Pattarino, Pharmaco *43*:325 (1988).
16. M. Trotta, M. R. Gasco, and F. Pattarino, Acta Pharm. Technol. *36*:226 (1990).
17. M. Gallarate, M. R. Gasco, and M. Trotta, Acta Pharm. Technol. *34*:102 (1988).
18. M. R. Gasco, M. E. Carlotti, and M. Trotta, Int. J. Cosmet. Sci. *10*:263 (1988).
19. M. R. Gasco, F. Pattarino, and I. Voltani, Farmaco *43*:3 (1988).
20. M. Bello, D. Colangelo, M. R. Gasco, F. Maranetto, S. Morel, V. Podio, G. L. Turco, and I. Viano, J. Pharm. Pharmacol. *46*:508 (1994).
21. C. Nastruzzi and R. Gambari, J. Controlled Release *29*:53 (1994).
22. S. E. Friberg, I. Kayali, W. Beckerman, L. D. Rhein, and A. Simion, J. Invest. Dermatol. *94*:377 (1990).
23. S. E. Friberg, J. Soc. Cosmet. Chem. *41*:155 (1990).
24. J. Engblom and S. Engström, Int. J. Pharm. *98*:173 (1993).
25. M. R. Gasco, M. Gallarate, and F. Pattarino, Int. J. Pharm. *69*:193 (1991).
26. J. Ziegenmeyer and C. Führer, Acta Pharm. Technol. *26*:273 (1980).
27. S. Bhatnagar and S. P. Vyas, J. Microencapsulation *4*:431 (1994).
28. H. Willimann, P. Walde, P. L. Luisi, A. Gazzaniga, and F. Stroppolo, J. Pharm. Sci. *81*:871 (1992).
29. D. W. Osborne, A. J. I. Ward, and K. J. O'Neill, J. Pharm. Pharmacol. *43*:451 (1991).
30. J. M. Sarciaux, L. Acar, and P. A. Sado, Int. J. Pharm. *120*:127 (1995).
31. V. H. L. Lee, Pharm. Int. *7*:208 (1986).

32. V. H. L. Lee, Crit. Rev. Ther. Drug Carrier Syst. *5*:69 (1988).
33. E. J. van Hoogdalem, A. G. de Boer, and D. D. Breimer, Pharm. Ther. *44*:407 (1989).
34. N. W. Bunnett, M. S. Orloff, and A. J. Turner, Life Sci. *37*:599 (1985).
35. C. Delgado, G. E. Francis, and D. Fisher, Crit. Rev. Ther. Drug Carrier Syst. *9*:249 (1992).
36. E. S. Swenson and W. J. Curatolo, Adv. Drug. Del. Rev. *8*:39 (1992).
37. S. Fujii, T. Yokohama, K. Ikegaya, F. Sato, and N. Yohoo, J. Pharm. Pharmacol. *37*:545 (1985).
38. P. Covreur and F. Puisieux, Adv. Drug. Del. Rev. *10*:141 (1993).
39. P. P. Constantinides, J.-P. Scalart, C. Lancaster, J. Marcello, G. Marks, H. Ellens, and P. L. Smith, Pharm. Res. *11*:1385 (1994).
40. P. P. Constantinides, C. M. Lancaster, J. Marcello, D. C. Chiossone, D. Orner, I. Hidalgo, P. L. Smith, A. B. Sarkahian, S. H. Yiv, and A. J. Owen, J. Controlled Release *34*:109 (1995).
41. P. P. Constantinides and S. H. Yiv, Int. J. Pharm. *115*:225 (1995).
42. B. D. Kahan, M. Ried, and J. Newburger, Transplant. Proc. *15*:446 (1983).
43. J. Drewe, R. Meier, J. Vonderscher, D. Kiss, U. Posanski, T. Kissel, and K. Gyr, Br. J. Clin. Pharmacol. *34*:60 (1992).
44. J. M. Kovarik, E. A. Mueller, J. B. van Bree, W. Tetzloff, and K. Kutz, J. Pharm. Sci. *83*:444 (1994).
45. J. M. Kovarik, E. A. Mueller, J. B. van Bree, S. S. Flückiger, H. Lange, B. Schmidt, W. H. Boesken, A. E. Lison, and K. Kutz, Transplantation *58*:658 (1994).
46. B. D. Tarr and S. H. Yalkowsky, Pharm. Res. *6*:40 (1989).
47. E. A. Mueller, J. M. Kovarik, J. B. van Bree, W. Tetzloff, J. Grevel, and K. Kutz, Pharm. Res. *11*:301 (1994).
48. E. A. Mueller, J. M. Kovarik, J. B. van Bree, J. Grevel, P. W. Lücker, and K. Kutz, Pharm. Res. *11*:151 (1994).
49. W. A. Ritschel, Meth. Find. Exp. Clin. Pharmacol. *13*:205 (1991).
50. E. D. Goddard and K. P. Ananthapadmanabhan (eds.), *Interactions of Surfactants with Polymers and Proteins*, CRC, Boca Raton, FL, 1993.
51. D. G. Patel, W. A. Ritschel, P. Chalasani, and S. Rao, J. Pharm. Sci. *80*:613 (1991).
52. A. Novelli, I. Rico, and A. Lattes, New J. Chem. *16*:395 (1992).
53. M. R. Gasco, S. Morel, and R. Manzoni, Farmaco *43*:373 (1988).
54. D. Aquilano, R. Cavalli, and M. R. Gasco, Thermochim. Acta *230*:29 (1993).
55. C. Cecutti, I. Rico, A. Lattes, A. Novelli, A. Rico, G. Marion, A. Graciaa, and J. Lachaise, Eur. J. Med. Chem. *24*:485 (1989).
56. B. K. Nelson, W. S. Brightwell, A. Khan, J. R. Burg, and P. T. Goad, Fundam. Appl. Toxicol. *12*:469 (1989).
57. T. Wakabayashi, M. Horiuchi, M. Sakaguchi, H. Onda, and M. Iijima, Acta Pathol. Jpn. *34*:471 (1984).
58. C. Gloxhuber (ed.), *Anionic Surfactants—Biochemistry, Toxicology, Dermatology*, Surfact. Sci. Ser. Vol. 10, Marcel Dekker, New York, 1980.
59. M. J. Schick and F. M. Fowkes (eds.), *Nonionic Surfactants*, Surfact. Sci. Ser. Vol. 1, Marcel Dekker, New York, 1967.
60. P. J. Tarcha (ed.), *Polymers for Controlled Drug Delivery*, CRC, Boca Raton, FL, 1991.
61. I. Siebenbrodt and S. Keipert, Eur. J. Pharm. Biopharm. *39*:25 (1993).
62. A. S. Sadaghiani, A. Khan, and B. Lindman, J. Colloid Interface Sci. *132*:352 (1989).
63. A. S. Sadaghiani and A. Khan, Langmuir *7*:898 (1991).
64. S. Engström, K. Larsson, and B. Lindman, U.S. Patent 5,371,109, to Drilletten AB (1993).
65. L. Engström, J. Dispersion Sci. Technol. *11*:479 (1990).

26
Microemulsions in Cosmetics

Patricia A. Aikens
Uniqema, Wilmington, Delaware

Stig E. Friberg
Clarkson University, Potsdam, New York

I. INTRODUCTION

The properties of microemulsions make them attractive for cosmetic formulations from several points of view. First, the transparent appearance gives a perception of a "clean" system; for a large majority of potential customers this is an essential property. The small droplet size and the transparency make them look like solutions, but the fact that they contain colloidal size droplets of oil in water, or vice versa, significantly enhances their use in cosmetics.

Because of the droplets, microemulsions possess the properties of both water and oil, in contrast with a solution, the properties of which are intermediate between those of oil and water. The consequence is of decisive importance for the formulator. Now both water-soluble compounds such as salts and oil-soluble additives can be freely combined into one transparent liquid.

This advantage from the formulation point of view is further enhanced by the fact that microemulsions are thermodynamically stable. No problems are encountered with changes during long-term storage, the shelf life from the nonchemical aspect is in principle infinite, limited only by the chemical stability of the components.

Finally, the microemulsions form spontaneously, and only mild stirring is necessary to bring about the final state more rapidly. The savings compared to the investment for emulsion production with expensive intensive-mixing equipment usually more than compensates for the increase in cost for the higher concentrations of surfactants and cosurfactants. These reach combined concentrations of 10–15% by weight, about four times that needed for an emulsion, increasing the cost per pound of the product. However, it should be borne in mind that the surfactant–cosurfactant compounds allowed in cosmetics often are chosen for their beneficial effects, thus alleviating the negative economic aspects.

Microemulsions have found a special application as precursors to cosmetic formulations because of the legal restrictions on the extent of volatile organic compounds. Until recently, the fragrance packages used in cosmetics are formulated as alcohol solutions. Environmental legal aspects have stimulated the fragrance industry to reformulate their products using the microemulsion concept in the process and actually achieving a cost reduction.

Some cosmetic products are in the form of "microemulsion gels." This name has emanated from the cosmetics industry; from a colloidal viewpoint these gels are liquid crystals with cubic or hexagonal symmetry. The liquid crystalline structures per se are not the subject of this chapter; reviews in that area are readily available [1–4].

The cosmetics industry has already formulated a long series of microemulsion products, in the following section illustrative examples of these are given within different product areas, including some of the liquid crystalline microemulsion gels.

II. APPLICATIONS

The fact that microemulsions possess the properties of both oil and water makes them ideal for cleaning purposes because the water present retains its capacity to dissolve salts.

A. Cleaners

The cosurfactants for skin cleansers are chosen such that they are nonirritating; hence, pentanol is avoided.

Glycerol, ethylene glycol, or propylene glycol can be used as cosurfactants with oleth-10 to solubilize mineral oil where the proper ratio of glycerol to oleth-10 is crucial for maximum solubilization [5]. Propylene glycol is used in many formulations, and glycerin is used in the clear skin cleanser from Kao [6]. (See Table 1.)

Water-in-oil (W/O) microemulsions have been investigated for topically applied products using hexadecane or decane as oil with a surfactant blend of glycerol monooleate and laureth-23, with and isopropanol as a cosurfactant [7]. Up to 29% water could be solubilized in a system of 35% decane, 21% glycerol monooleate/laureth-23 (5 : 1), and 14% isopropanol. With hexadecane, 25% water was solubilized using 25% glycerol monooleate/laureth-23 (2 : 1) and 12% isopropanol.

B. Hair Products

Microemulsions have been used for hair treatment formulations that contain protein hydrolyzing agents (KOH, NaOH, LiOH) and reducing agents (ammonium thioglycolates, metal or ammonium sulfites or bisulfates) [8]. The combined amount of these agents used for hair relaxers is 0.4–5%; for permanent wave lotions, 2–10%; and for depilatories, 3–15%. It is claimed that the high level of surfactant does not make this product any more irritating to skin than a conventional emulsion. Table 2 shows a typical formulation with anionic

Table 1 Composition of Microemulsion Skin Cleanser

Component	Wt%
PEG-hydrogenated castor oil	6.0
PEG octyl decyl ether	15.0
Glyceryl tri-(2-ethylhexanoate)	27.5
Polyisobutene	27.5
Glycerin	17.5
Water	6.5

Source: Ref. 6.

and nonionic surfactants. Glycerin, sorbitol, or 1,4 butanediol may be substituted for the propylene glycol, and PPG-15 stearyl ether may be included in the oil phase of the formulation in Table 2. The formulation may also be prepared using only nonionic surfactants, a ceteth-20–oleth-3 blend, at a ratio of 3.2 surfactant to 1 oil.

Dimethicone is effective as a hair conditioner, particularly in the two-in-one conditioning shampoos. High molecular weight dimethicone itself is difficult to microemulsify to a clear solution. The Toray Silicone Company (Japan) has patented compositions for clear hair conditioners and conditioning shampoos using a microemulsion of 20% dimethicone with an average particle size of 0.05 μm prepared by emulsion polymerization of dimethylsiloxane tetramer stabilized with dicocoyldimethyl ammonium chloride and tallow trimethylammonium chloride or dodecylbenzene sulfonic acid and heated to 85°C [9]. An example is shown in Table 3.

Microemulsions of functionalized silicone oils are also used to make clear conditioning shampoos in patents by General Electric [10], Dow Corning [11,12], and Unilever [13]. A microemulsified silicone oil solution with a particle size less than 0.15 μm is prepared by emulsion polymerization of the cyclomethicone precursor. A formulation from Unilever is given in Table 4 in which a guar-based deposited aid makes the microemulsified conditioner adhere to the hair during washing.

Table 2 Formulation for a Microemulsion Hair Treatment[a]

	Component	Wt%
A	Light mineral oil	22.1
	Oleth-3 phosphate	4.5
	Oleth-10 phosphate	3.3
	Oleth-3	3.3
	Oleth-10	3.3
B	Water	42.1
	Propylene glycol	8.0
C	NaOH (25%)	13.4

[a] Procedure: Heat A and B separately to 80°C; add B to A slowly with stirring. Remove from heat and add C.
Source: Ref. 8.

Table 3 Composition of Clear Hair Conditioner

Component	Wt%
Stearyl dimethylammonium chloride	1.0
Cetanol	1.0
Stearyl alcohol	1.0
Liquid paraffin	2.0
POE-6 stearyl ether	1.0
Propylene glycol	5.0
Dimethicone microemulsion (20% aq.)	5.0
Water	84.0

Source: Ref. 9.

A latex of polystyrene is prepared using a microemulsion stabilized with nonionic surfactants, and this is then incorporated as an ingredient in hairspray and setting compositions [14]. The very fine particulate size (0.01–0.1 μm) improves the style-forming and retaining capability of the product as well as the feel of the resin on the hair. A example is given in Table 5.

Three microemulsions liquid crystalline gel formulations from Croda, Inc. based on lower EO nonionic surfactants and ethoxylated phosphates for clear hair gel products are shown in Table 6 [15]. One contains cyclomethicone for reduced oily feeling, and another contains a high level of polyol for humectant effect, classifying it as a curl activator.

A very mild hair gel can be made without an ethoxylated ether surfactant by using phosphate and sucrose esters as shown in Table 7. However, long-term stability for this formulation suffers [15].

Table 4 Clear Conditioning Shampoo

Component	Wt%
Sodium lauryl ether sulfate-2EO	16.0
Cocoamidopropyl betaine	2.0
Dimethiconol, 15,000 cS (in 0.036 μm microemulsion)	1.0
Guar hydroxypropyl trimonium chloride (Jaguar C13S)	0.1
Sodium chloride	1.0
Formalin	0.1
Water	79.8

Source: Ref. 13.

Table 5 Blow-Dry Finishing Aid Using a Mircoemulsion Polymer Latex

Ingredient	Wt%
Microemulsion polystyrene latex[a]	
Water	78.0
POE-30 nonylphenyl ether	2.3
Ammonium persulfate	0.2
Styrene	19.5
Blow-dry finishing aid[b]	
Stearyl trimethylammonium chloride	0.4
PEG	0.1
Perfume	0.3
Water	98.7
Microemulsion PS latex[c]	0.5

[a] Styrene is added dropwise for 2 h to an aqueous solution of surfactant and persulfate at 62°C under N_2 and stirred 6 h.
[b] Stearyl trimethylammonium chloride, PEG, and perfume are added to a solution of water and microemulsion polystyrene latex with stirring.
[c] As above.
Source: Ref. 14.

Table 6 Microemulsion Gel Formulations for Hair Care[a]

		Curl activator gel (wt%)	Emollient gel (wt%)	Volatile silicone gel (wt%)
A	Mineral oil	16.05	17.0	10.0
	Volatile silicone 345	—	—	10.0
	DEA oleth-3 phosphate	7.48	2.0	7.0
	DEA oleth-10 phosphate	—	4.0	—
	Oleth-3	—	7.0	7.0
	Oleth-5	10.88	4.0	4.0
	Acetamide MEA	0.2	—	—
B	Water	45.39	54.0	52.0
	Propylene glycol	—	12.0	10.0
	Sorbitol	15.00	—	—
	Hexylene glycol	5.0	—	—
	Preservative	q.s.	q.s.	q.s.

[a] Procedure: Add B (at 80–85°C) to A (at 80–85°C) with stirring.
Source: Ref. 15.

Table 7 Mild Clear Surfactant Microemulsion Gel[a]

Ingredients		Wt%
A	Mineral oil	16.0
	Sucrose stearate and distearate (sold as Crodesta F-10)	5.0
	Sucrose stearate	5.0
	DEA oleth-10 phosphate	2.0
	DEA oleth-3 phosphate	5.0
	1,3-Ethylhexanediol	2.5
B	Poropylene glycol	5.0
	Water	59.5

[a] Procedure: Add B (at 80–85°C) to A (at 80–85°C) with stirring.
Source: Ref. 15.

C. Fragrances and Perfumes

Solubilization of perfume oil in water produces a clear solution without the use of volatile or drying solvents such as ethanol. Generally, fatty alcohol or sorbitan ester ethoxylates are the most suitable surfactants. Isoceteth-20 is a very efficient solubilizer with a branched-chain hydrophobe. It is also available in liquid form, so minimum heat is needed. Two other effective solubilizers are polysorbate 20 and polysorbate 80. Table 8 shows the weight of surfactant needed per gram of fragrance oil [16].

A 1992 patent by Yves Saint Laurent Parfumes (France) describes the solubilization of perfume oils in combination with an emollient using PEG ester surfactants and polyglycerol ester cosurfactants [17]. In the examples in Table 9, the ingredients are stirred slowly at room temperature for 15 min.

Table 8 Solubilization of Fragrance Oils Using Nonionic Surfactants[a]

Fragrance oil (1 g)	Isoceteth-20 (g)	Polysorbate 20 (g)
Pikaki 40-R-15845 (Fritzche Dodge and Olcott, Inc.)	1.8	3.0
Jasmine 42789 (Fritzche Dodge and Olcott, Inc.	0.4	0.8
Dermodor Dulchena 2025 (P. Robertet, Inc.)	2.4	4.8
Iso Rose 11419 (P. Robertet, Inc.)	1.8	3.4
Arpa Fla (Polak's Frutal Works, Inc.)	2.2	4.8
Bath Bouquet 42454 (Fritzsche Dodge and Olcott, Inc.)	0.7	0.8
Fragrance 40081H (Haarmann & Reimer Corp.)	3.0	7.0
Blue Mist W-2597 (Givaudan Corp.)	3.0	7.0

[a] Procedure: Mix surfactant and fragrance oil with enough heat to make a clear solution. Add water slowly with stirring. This procedure can also be used to solubilize the flavor oils methyl salicylate, peppermint oil, or spearmint oil in water (5, 7, 5 g, respectively, of polysorbate-80 per gram of oil) for mouthwash.
Source: Ref. 16.

Table 9 Solubilization of Perfume Concentrate Oil

Ingredient	Emollient perfume water (g)	Perfume essence (g)
Polyglycerol ester of C_{8-10} glycerides	10–15	10–15
Polyglycerol dioleate	1–3	—
Polyglycerol isostearate	—	12–20
DEA-PEA oleyl ether phosphate	0.5–2	8–15
Vegetable oil (macadamia, almond, or borage)	10.0	15.0
Perfume concentrate	4–5	40.0
Preservative (parabens)	0.2	0.2
Demineralized water	to 100 g	to 100 g

Source: Ref. 17.

D. Gels and Antiperspirants

Microemulsion gels are usually based on nonionic ethoxylated ether surfactants based on oleyl, lauryl, or isocetyl alcohol either alone or in combination with ethoxylated phosphate esters. Long hexagonal liquid crystalline structures present in the gel can vibrate when the container is tapped, resulting in a ringing gel [18]. Table 10 presents a versatile ringing gel microemulsion base formula from ICI [16] for three clear cosmetic products.

Revlon has developed a clear silicone microemulsion composition tolerant of high salt content that is suitable for moisturizers and antiperspirants [19]. One example is shown in Table 11.

Clear antiperspirants have become popular because they leave no visible residue. They are not always formulated through a microemulsion route. Clear sticks can be based on dibenzylidene sorbitol acetal (DBSA) gelling agent and zirconium or glycine-complexed aluminum chlorhydrate antiperspirant actives that are soluble in propylene glycol without water. An alternative route is to closely match the refractive indices of the oil and water phases so a clear product (not a microemulsion) results for an antiperspirant gel [20,23]. Clear deodorant sticks with the active ingredient triclosan can be solidified with sodium stearate, which is incompatible with antiperspirant active ingredients (aluminium chlorhydrates).

Table 10 Clear Gel Systems[a]

Ingredient	Hair conditioning setting gel	Hydroxy acid gel	Sunscreen clear gel
A Light mineral oil	11.0/6.0	11.0	10.0
PPPG-15 stearyl ether	—/5.0	—	—
Isoceth-20 (72% active)	27.8/27.8	27.8	26.0
Oleth-2	6.0/6.0	6.0	6.25
Octyldimethyl *p*-aminobenzoic acid	—/—	—	7.0
Benzophenone-3	—/—	—	3.0
B Water	41.8/43.2		35.7
Sorbitol solution (70%)	7.0/7.0		7.0
Propylene glycol	5.0/5.0		5.0
Soyaethyl morpholinium sulfate (35% aq.)	1.4/—	—	—
Lactic, citric, malic acids/tea extract	—/—	12.5	—

[a] Procedure: Add B (at 90°C) to A (at 90°C) with stirring, cool to 80°C, make up lost water, stir until uniform. For sunscreen gel, add sunscreens (premixed at 50°), followed by mineral oil. Heat to 80°C and add B (at 85°C).
Source: Ref. 16.

Table 11 Clear Moisturizing Formula with Sunscreen

Ingredient	Wt%
Urea	12.0
Alkyl methicone (Abil B 9806/9808, Goldschmidt)	12.0
Cyclomethicone	20.0
Magnesium sulfate	11.0
Water	24.0
Propylene glycol	8.0
Alcohol SD-40	10.0
Aluminum chloride	1.0
Octyldimethyl *p*-aminobenzoic acid	2.0

Source: Ref. 19.

Table 12 Flowing Antiperspirant Gel[a]

Ingredient	Wt%
A Cyclomethicone (tetramer)	15.4
Propylene glycol	13.4
Lauryl/myristamide DEA	9.6
PPG-3 myristyl ether	13.5
PEG-150 distearate	9.6
B Aluminum chlorhydrate (50% aq)	38.0

[a] Procedure: Add B (at 40°C) to A (at 70°C) with side scraper agitation.
Source: Ref. 21.

Microemulsion formulations for antiperspirants must be able to tolerate high salt content because of the active ingredient. Water-in-oil formulations are shown in Tables 12, 13, and 14 that incorporate silicone emollients and antiperspirant active components.

Table 13 Firm Antiperspirant Gel

Ingredient		Wt%
A	Cyclomethicone (tetramer)	15.4
	PPG-10 cetyl ether propylene glycol	4.0
	PPG-5–ceteth-20	16.0
	Isostearyl benzoate	20.0
B	Aluminum chlorhydrate (50% aq)	40.0

Source: Ref. 21.

Table 14 W/O Microemulsion Roll-On Antiperspirant[a]

		Wt%
A	Deionized water	8.8
	50% Aluminum sesquichlorhydrate (REACH 301)	40.0
	Dipropylene glycol	3.0
B	PEG-7 glyceryl cocoate	18.2
	Cyclomethicone (and) dimethicone copolyol	20.0
	Cetearyl octanoate	3.2
C	Polysorbate 20	1.0
	Water	4.1
D	Isopropyl myristate	1.0
	Fragrance (99-895, drom International)	

[a] Procedure: Mix A, slowly add B ingredients one at a time, and mix well in between. Slowly add C and mix. Slowly add D and mix until clear.
Source: Ref. 22.

E. Skin Care Products

Sucrose- and glucose-based consurfactants have been used to reduce the temperature sensitivity of ethoxylated nonionic surfactants as the hydroxyl groups of the sugar counter the negative temperature coefficient of the polyethylene oxide [24]. Ethoxylated laury alcohol and ethoxylated Guerbet alcohols (C_{14}, C_{16}), along with sucrose laurate and ethoxylated Guerbet (C_{16}) glucoside can be used to form microemulsions with hexadecane or dodecane that are not as sensitive to temperature as a nonionic surfactant alone would be [24]. The ethoxylate is more hydrophobic (less than 7.5 units EO) than the sugar-based surfactant in all cases.

Silicone emollients are often used in skin products such as moisturizers and antiperspirants because of their ease of spreading and nonoily feel. In addition, the volatile cyclomethicones can provide a cooling sensation as they evaporate after application. Microemulsions can be formed by using nonionic alkoxylated surfactants or dimethicone copolyol surfactants as shown in Table 15.

Table 15 W/O Silicone Emollient Microemulsion[a]

Ingredient	Wt%
A PPG-5-ceteth-20	25.0
PPG-10 cetyl ether propylene glycol	5.0
Cyclomethicone (tetramer)	25.0
Isostearyl benzoate	20.0
B Water	25.0

[a] Procedure: Add B (at 40°C) to A (at 70°C) with side scraper agitation.
Source: Ref. 21.

III. STRUCTURAL CHANGES DURING EVAPORATION

In a large number of applications the change during evaporation is of decisive importance for the function of the personal care product. The example that comes most directly to mind is a fragrance compound in the formulation. The perception of fragrance is one of the deciding features of a personal care product for customer acceptance. The vapor pressure of the fragrance is reduced with time according to Fig. 1, and this variation is essential from an economic point of view because of the perception threshold.

The amount of fragrance represented by the shaded area is wasted, and it may be a considerable part of the total added amount. This is a significant economic factor because the fragrance is one of the more expensive components.

An illustration of the change in vapor pressure with respect to time is given by the following example in which the vapor pressure varies exponentially versus time, reaching 70% of its original value in 30 min.

$$p/p_o = \exp(-\tau t) \tag{1}$$

where

$$t = (\text{time}) = 0.5\text{h}; \quad p/p_o \text{ (vapor pressure)} = 0.7 \text{ (after 0.5 h)} \tag{2}$$

$$\tau = -(\ln 0.7)/0.5 = 0.71 \tag{3}$$

The time to reach the perception limit of $p/p_o = 0.3$ is evaluated from Eq. (1):

$$0.3 = \exp(-0.71t) \tag{4}$$

$$t = 1.70 \text{ h} \tag{5}$$

If the vapor pressure were constant at $p/p_o = 0.35$, the time to use all the fragrance would be calculated as

$$0.35 \, p_o t_{0.35} = p_o \int_o^\infty e^{-0.71t} dt = \frac{p_o}{0.71} \tag{6}$$

$$t_{0.35} = 4.0 \text{ h} \tag{7}$$

Since vapor pressure is constant,

$$\frac{p}{p_o} = \exp(-\tau t) = 0.35; \quad \int p \, dt = \int p_o \exp(-\tau t) \, dt = p_o 0.35t$$

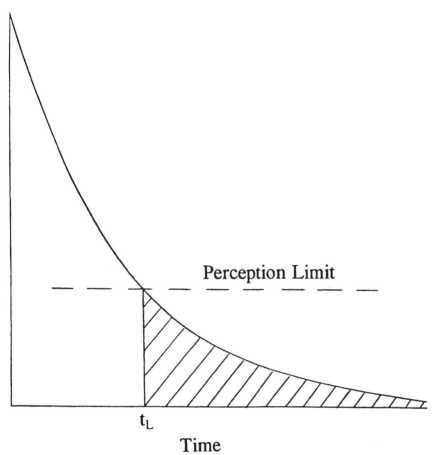

Figure 1 The vapor pressure of a fragrance in a cosmetic product is exponentially reduced with time (the solid line). When the vapor pressure becomes less than the perception limit at time t_L, the useful action of the fragrance is lost, and the amount of fragrance represented by the shaded area is wasted.

This is compared to using all the fragrance if vapor pressure were not constant with time,

$$\int p \, dt = \int p_0 \, \exp(-0.71t) \, dt = \frac{p_0(1-0)}{0.71}$$

This shows that the time that perception lasts more than doubles from 1.7 to 4.0 h if the vapor pressure is kept constant at 0.35 rather than being allowed to decrease exponentially. In addition, and more important, lower levels of vapor pressure at short times after application make it possible to avoid an overbearing and sometimes even unpleasant perception.

The solution to the problem of keeping the vapor pressure constant is found in the use of phase diagrams. During evaporation the vapor pressure will vary within each phase but will remain constant within a three-phase region. This means that the vapor pressure of the fragrance remains constant the entire time the total composition remains within this area during evaporation.

A simple but illustrative example is the combination of phenethyl alcohol with water and a nonionic surfactant polyoxyethylene (4) lauryl ether (BRIJ® 30). Figure 2 shows the W/O microemulsion region to be in equilibrium with a lamellar liquid crystal and a dilute aqueous solution of the fragrance for a large part of the entire phase diagram. This means that the vapor pressure of the phenethyl alcohol remains constant during evaporation as long as the total composition remains within the area ABC of Fig. 2.

The vapor pressure of phenethyl alcohol is small compared to that of water, and the evaporation trajectory will, with good approximation, follow a straight line such as the one shown as the dashed line in Fig. 2. The experimental results of vapor pressures determined for an evaporating composition from Fig. 2 are given in Fig. 3. First, the fact that the vapor pressure is almost constant while all the water evaporates should be noted. Second, and more important, it should be observed that the values of vapor pressures obtained from evaporating compositions agree well with those predicted from static measurements of pressures from different phases. This means that a single determination of vapor pressures from the W/O microemulsion in the diagram makes it possible to predict the variation in fragrance vapor pressure during evaporation from any microemulsion in the system.

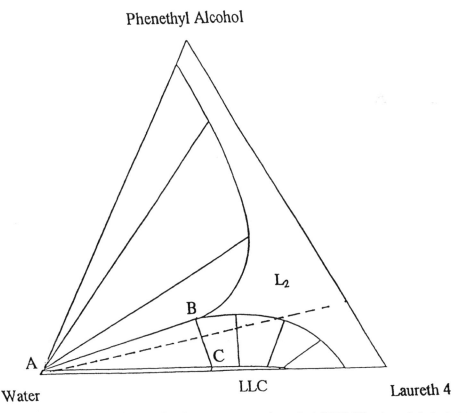

Figure 2 The phase diagram for the system water–laureth-4 (BRIJ 30)–phenethyl alcohol. L_2 is a water-in-oil microemulsion; LLC is the lamellar liquid crystal; and A, B, and C define the three-phase region. The dashed line represents the evaporation path through the three-phase region into the LLC and W/O microemulsion region.

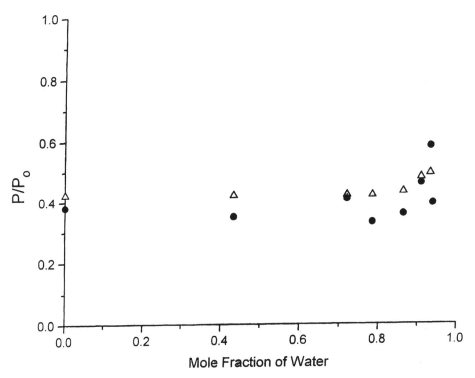

Figure 3 A comparison of ($\triangle$) the estimated vapor pressure of phenethyl alcohol during evaporation of a system containing phenethyl alcohol, water, and laureth-4 and ($\bullet$) actually measured values.

Although fragrance may be the most important ingredient in the cosmetic formulation, the evaporation of water is also essential, and knowledge about the relative rates of water and oil evaporation is necessary to find the final structure of the remaining part of a microemulsion formulation after topical application.

This problem is fundamentally different from the question of fragrance vapor pressure (see above). The relative vapor pressures of water and oil are now only one parameter in the evaporation process; equally important are the transport rates within the microemulsion and across its surface to air. This means that the literature on vapor pressure in microemulsions [25–34] is of limited value and that the evaporation per se must be studied [35,36].

Investigations have been reported in the form of composition trajectories in the condensed phase during evaporation, and a typical result is given in Fig. 4. The figure shows several trends of importance for a cosmetic microemulsion, which will be discussed after the information from the trajectories has been evaluated.

In the colloidal system of water, surfactant, and volatile oil, only the surfactant will not evaporate. Hence, the relative amounts of the remaining two components in the vapor leaving at each point of the trajectory are directly obtained by the points at which the tangents to the evaporation curves intersect the water/oil axis [36].

Points on a trajectory in a system of three components (A, B, C), two of which are volatile (A, B are water and oil) and the third is not (C, surfactant), have weight fractions a, b, c ($a+b+c=1$) of the components A, B, C, respectively, and a line described by $a=a(c)$. The total weight is g, and the amount of each component is $g_A = ag$, $g_B = bg$, $g_C = cg$.

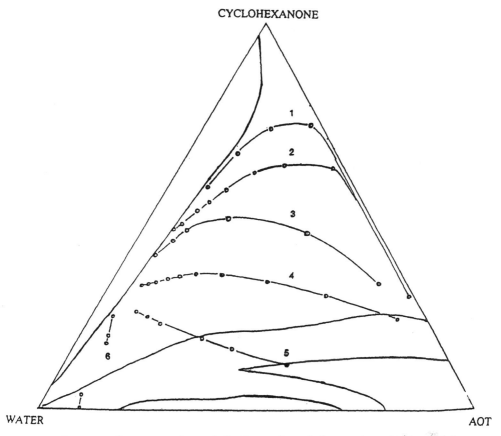

Figure 4 Corresponding evaporation path for the condensed phase system of water, cyclohexanone, and Aerosol OT. Curves 1–6 represent various compositions along the evaporation pathway.

The change in the amount of a single component is

$$dg_A = g\,da = a\,dg \tag{8}$$

$$dg_B = g\,db + b\,dg \tag{9}$$

$$dg_C = 0 \tag{10}$$

The weight fraction of A in the vapor is

$$W_A^V = dg_A/(dg_A + dg_B) \tag{11}$$

since $dg = dg_A + dg_B + dg_C$, and $dg_C = 0$ [Eq. (10)], $W_A^V = dg_A/dg = g(da/dg) + a$ [from Eq. (8)], $dc = (dg_C - c\,dg)/g$, and $dg_C = 0$, $dc = -c(dg/g)$. Rearranging and substituting for dg gives

$$W_A^V = a - c\,(da/dc) \tag{12}$$

which, applied to a point (a_1, b_1, c_1) gives $(W_A^V) = a_1 - c_1(da_1/dc_1)$. This is the value of g_A on the AB axis.

The change in composition with evaporation can easily be followed in Fig. 4. First, the evaporation at the beginning of the process results in a surprisingly similar vapor composition irrespective of the water/oil ratio in the microemulsion. Second, at low water concentrations the relative concentration of oil in the vapor is strongly increased. The final result is that all compositions tend toward the liquid crystalline phase of the surfactant.

This result has two implications from the point of view of skin care. First, the evaporation rate is significantly reduced when liquid crystals appear in the formulation during evaporation, as demonstrated by Moaddel and Friberg [37] and Langlois and Friberg [38]. The importance of slow evaporation to retain as much water as possible is obvious for skin care products. In addition, as recently demonstrated by Bodde et al. [39], the liquid crystals serve to reduce the evaporative water loss from the skin, an essential function. Finally, the publication from the Elias group [40] demonstrates that a combination of compounds that produce a lamellar structure actually will heal damaged stratum corneum structures.

In conclusion, it can be seen that valuable information about the relative amounts of the components can be obtained through examination of the phase diagrams and insight gained pertaining to the behavior of a formulation during evaporation of water and volatile oils. It is obvious that research into the evaporation of microemulsions will provide important information for improving the properties and efficacy of future cosmetic formulations.

ACKNOWLEDGMENT

This research was supported in part by ICI Surfactants, Wilmington, Delaware and by the New York State Commission of Science and Technology at the Center for Advanced Materials Processing, Clarkson University, Potsdam, New York.

REFERENCES

1. B. Lindman and H. Wennerström, Phys. Rep. *52*:1 (1979).
2. G. J. T. Tiddy, Phys. Rep. *57*:1 (1980).
3. A. M. E.-I. Nokaly (ed.), *Polymer Association Structures: Microemulsions and Liquid Crystals*, ACS Symp. Ser. No. 384, Am Cher Soc, Washington, DC, 1989.
4. S. E. Friberg, in *Liquid Crystals—Applications and Uses* (B. Bahadur, ed.), World Scientific, Singapore, 1991, p. 157.
5. N. J. Kale and L. V. Allen, Int. J. Pharm. *57*:87 (1989).
6. C. Fox, Cosmet. Toiletries, *110*(9):59 (1995) (referenced as Jpn Patent 0405,213).
7. A. Jayakrishnan, K. Kalairasi, and D. O. Shah, J. Soc. Cosmet. Chem. *34*:335 (1983).
8. T. W. Clifton and P. H. Cade, World Patent 94/29487, to Croda, Inc. (1994).
9. A. Harashima, O. Tanake, T. Maruyama, and Y. Ohta, Eur. Patent 0 268 982A2, to Toray Silicone Co. (1987) (also Jpn. Patent 274799/86).
10. D. Riccio and J. Merrifield, Br. Patent 2,288,183, to General Electric Company (1994).
11. D. J. Halloran, Eur. Patent 0 514 934A1, to Dow Corning (1992).
12. D. Graiver and O. Tanaka, U.S. Patent 4,999,398, to Dow Corning (1992).
13. D. H. Birtwistle, Eur. Patent 0 529 883A1, to Unilever (1992).
14. K. Yahagi and T. Suzuki, Jpn. Patent 2028898, to Kao Corp. (1985) (also Eur. Patent 0214 626B1, 1986).
15. K. F. Gallagher, Happi, February 1993, p. 58.
16. *Cosmetic and Personal Care Formulary* 'Solubilization', distributed by ICI Surfactants, Wilmington, DE, 1985.

17. N. Dartel and B. Breda, U.S. Patent 5,252,555, to Yves Saint Laurent Parfumes (France) (1993).
18. H. Hoffmann and G. Ebert, Angew. Chem. Int. Ed. Engl. 27:902 (1988).
19. B. Guthauser, U.S. Patent 5,162,378, to Revlon, Inc. (1992).
20. Gillette, World Patent 92/05767 (1992).
21. R. L. Goldemberg, AJ. A. Tassoff, and A. J. DiSapio, Drug Cosmet. Ind., 34 (Feb. 1986).
22. Rehies Technical Literature, Rehies Inc., Berkeley Heights, NJ.
23. E. Abrutyn, (SCC Continuing Ed. Prog., Antiperspirant & Deodorant Technology, Feb. 24, 1993.)
24. K.-H. Oh, J. R. Baran, and W. H. Wade, J. Dispersion Sci. Technol. 16:165 (1995).
25. W. D. Weatherford, Jr. and D. W. Naegeli, J. Dispersion Sci. Technol. 5:159 (1984).
26. C. H. Chew and M. K. Wong, J. Dispersion Sci. Technol. 12:495 (1991).
27. M. Ueda and Z. A. Schelly, J. Colloid Interface Sci. 124:573 (1988).
28. M. Zulauf and H. F. Eicke, J. Phys. Chem. 83:486 (1979).
29. W. D. Weatherford, Jr., J. Dispersion Sci. Technol. 6:467 (1985).
30. J. Biais, P. Bothorel, B. Clin, and P. Lalanne, J. Colloid Interface Sci. 80:136 (1981).
31. J. Biais, L Ödberg, and P. Stenius, J. Colloid Interface Sci. 86:350 (1982).
32. E. Sjöblom, B. Johnsson, A. Johnsson, P. Stenius, P. Saris and L. Ödberg, J. Phys. Chem. 90:119 (1986).
33. L. Damaszewski and R. A. Mackay, J. Colloid Interface Sci. 46:417 (1974).
34. L. Damaszewski and R. A. Mackay, J. Colloid Interface Sci. 48:381 (1974).
35. S. E. Friberg, B. Yu, J. Lin, E. Barni, and T. Young, Colloid Polym. Sci. 271:152 (1993).
36. S. E. Friberg, T. Young, R. A. Mackay, J. Oliver, and M. Breton, Colloids Surf. 100:83 (1995).
37. T. Moaddel and S. E. Friberg, J. Dispersion Sci. Technol. 16:69 (1995).
38. B. Langlois and S. E. Friberg, J. Soc. Cosmet. Chem. 44:23 (1993).
39. H. E. Bodde, H. D. Junginger, and H. I. Maibach, J. Soc. Cosmet. Chem. (in press).
40. G. Grubauer, K. R. Feingold, R. M. Harris, and P. M. Elias, J. Lipid Res. 30:89 (1989).

27

Microemulsions in Foods

Sven Engström and Kåre Larsson
Lund University, Lund, Sweden

I. INTRODUCTION

Certain foods contain microemulsions naturally, and microemulsions as a functional state of lipids have therefore been used in the preparation of foods since ancient times. Microemulsions can also be formed in the intestine during the digestion and absorption of fat. The possibility to produce microemulsions on purpose and use them as tools (i.e., not only as vehicles) in food production, however, is a neglected field in food technology.

This review is focused on lipid systems that form microemulsions and have the potential to be used in foods. Drug delivery systems that have incorporated microemulsions can provide valuable experience for applying microemulsions in foods, and such applications are summarized here. Finally, other applications presently existing only on the laboratory scale are presented as illustrations of microemulsion functionality.

II. MICROEMULSIONS AND SPONGE PHASES FORMED BY FOOD LIPIDS

A. Cereal Lipids

Aqueous solutions of lipids extracted from different cereals have been examined in our laboratory [1–3]. When we first analyzed wheat lipids it was possible to identify four phases at equilibrium: an oil phase free of water, a liquid crystalline phase, a liquid phase consisting of both lipids (mainly the polar fraction) and water, and a water phase. It was possible to describe the phase equilibria in a ternary phase diagram as shown in Fig. 1. It is, in fact, remarkable that this is possible, and the reason probably is that all lipid components (characterized by thin-layer chromatography) tend to behave like two components only: a nonpolar fraction and a polar fraction. As the liquid lipid–water phase could coexist with a water phase, it was at that time logical to term it L_2, and we also pointed out that it could be regarded as a microemulsion as it was thermodynamically stable and contained triglyceride oil, surfactant-like polar lipids, and water. Since that time, however, L_3 phases occurring in lipid–water systems have been described [4], and today it is natural to term this liquid phase L_3 as it exists between a lamellar liquid crystalline phase and the water corner in the phase diagram (see Fig. 1). The relations between the L_2, L_3, L_α, and cubic phases are summarized below and described in detail in Ref. 4.

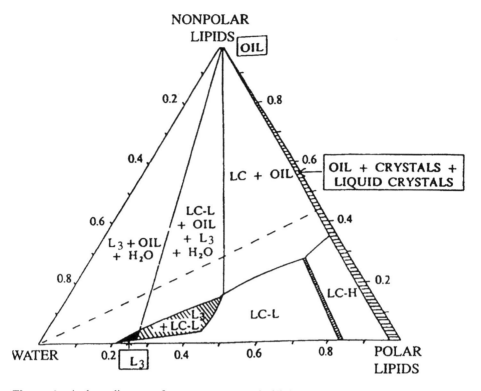

Figure 1 A phase diagram of a system composed of lipids extracted from cereal lipids and water. The weight fraction of each component is given in the diagram. L_3, sponge phase; LC, liquid crystal; LC-L, lamellar liquid crystal; and LC-H, reverse hexagonal liquid crystal.

The same liquid phase (L_3) as is formed by wheat lipids can be obtained in aqueous systems of rye lipids [2] and of oat lipids [3]. Oats are probably the food material richest in polar lipids, and it is therefore realistic to expect industrial production of oat lipids in the future in order to formulate microemulsions for foods. We have found (unpublished observations) that even large protein molecules can be encapsulated into this kind of L_3 phase without being denatured. The incorporation of enzymes in L_2 and L_3 phases, for example, can provide protection against proteolytic degradation.

B. L_2 Phases or Microemulsions in Ternary Systems of Edible Lipids

Phase equilibria in a ternary system consisting of industrially distilled monoglycerides from sunflower oil, water, and soybean oil have been reported [5,6], and the phase diagram is shown in Fig. 2. A large region of the phase diagram consists of a liquid phase with encapsulated reverse micelles: an L_2 phase. A critical ratio of monoglycerides to triglycerides of about 1 : 9 represents a kind of critical micelle concentration. The formation of discrete water domains takes place only above this ratio. It is also natural to term this liquid phase a microemulsion as it consists of oil, water, and an emulsifier, which form a thermodynamically stable liquid phase.

Monoglycerides and triglycerides of saturated chains that are liquid at room temperature have become commercially available—the so-called medium-chain triglycerides. In the

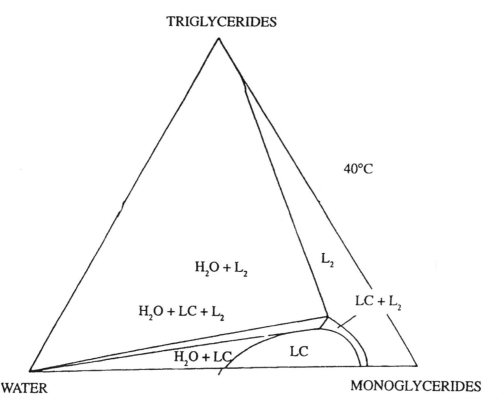

TRIGLYCERIDES

40°C

$H_2O + L_2$

L_2

$LC + L_2$

$H_2O + LC + L_2$

$H_2O + LC$

LC

WATER

MONOGLYCERIDES

Figure 2 A phase diagram of a system composed of edible lipids—monoglycerides from sunflower oil, soybean oil, and water. L_2, reverse micellar or microemulsion phase. The LC (liquid crystalline) region consists of several phases.

corresponding ternary systems, in contrast to the long-chain homologs shown in Fig. 1, the L_2 region is even larger, as no liquid crystalline phase exists at or above room temperature.

The structure of this type of L_2 phase has been analyzed by electron microscopy and by X-ray diffraction [6]. The results indicate that flexible disk-shaped water micelles occur, separated by lipid bilayers. The X-ray data are in agreement with electron micrographs of freeze-etched freeze-fractured samples. An analysis of the X-ray scattering results at a weight ratio of monoglyceride to water of $8:2$ gave these results:

L_2 composition as wt% oil	Water disk diameter (nm)	Water + bilayer thickness (nm)
16.1	72	3.9
18.0	39	4.0
38.7	24	4.1
51.9	13	4.3
69.8	9	4.7

Another type of edible lipid L_2 phase that also might have food applications is shown in the ternary phase diagram in Fig. 3 [7]. By adding a few percent of ethanol to a phospholipid

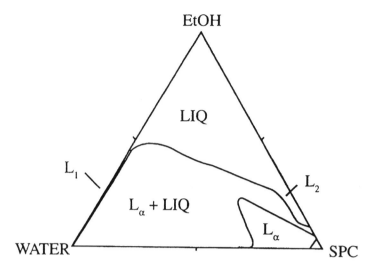

Figure 3   A phase diagram of a system composed of soybean phosphatidylcholine (SPC), ethanol, and water. The phospholipid-rich liquid region (LIQ) is a reverse micellar phase (L$_2$).

such as soybean phosphatidylcholine (SPC), a liquid phase is formed. As the existence range of this phase is connected with the ethanol corner and extends toward the lipid corner in the corresponding ternary phase diagram phospholipid–water–ethanol, it is an L$_2$ phase. Thus it can incorporate water and ethanol as reverse micelles. When this phase is diluted with water it spontaneously forms a liposomal dispersion. Incorporation of additives, such as antioxidants or flavors, will result in liposomal encapsulation. It seems likely that this type of microemulsion will be useful in foods.

C.   Fat Digestion and Absorption

Monoolein, a common food emulsifier, gives rise to a bicontinuous cubic liquid crystalline phase when added to water, as illustrated in Fig. 4. If a triglyceride oil is introduced into the monoolein–water system, a microemulsion (L$_2$) phase is formed above about 10 wt% oil as discussed in Sect. II.B. If lecithin is added, the cubic phase is preserved up to about 30 wt% lecithin, beyond which a lamellar (L$_\alpha$) phase is formed [8]. Adding a bile salt, e.g., sodium taurocholate, in sufficient amount will convert the cubic phase to a micellar solution (L$_1$) [9].

It is interesting to note that during digestion of triglyceride oils, all the substances discussed above are present as well as the phases mentioned. This was nicely illustrated by Patton and Carey [10] in an in vitro study where the fate of a drop of soybean oil in simulated intestinal fluid was monitored under the microscope. The cubic phase is formed at the oil/water interface, and due to its bicontinuous structure it is capable of both delivering the water molecules necessary for hydrolysis and taking care of the resulting fatty acids. The role of lamellar and micellar phases is to act as carriers of lipids to the intestinal wall [11,12].

It turns out that absorption through the intestinal wall is a very effective process, since about 80–85% of the digestible fats and oils are absorbed during half an hour in 20–30 cm of small intestine. This means that roughly 100 g of lipids may be present in the blood after a fatty meal, which is also evident from its milky appearance. Even

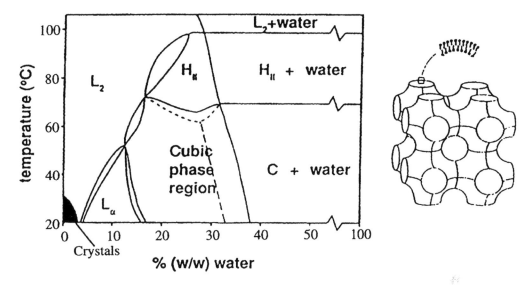

Figure 4 A phase diagram of a system composed of monoolein and water. One representative structure of a bicontinuous cubic liquid crystalline phase is shown to the right. L_α, lamellar liquid crystal; H_{II}, reverse hexagonal liquid crystal; and L_2, reverse micellar phase.

during bile salt–deficient conditions, the absorption effectiveness is high, and the transport of fat in the intestine is then carried out by vesicles rather than by mixed micelles as when bile salts are present [11,12]. Using this ability of microemulsions to promote efficient fat uptake as a method to deliver other bioactive substances such as drugs and vitamins is an area of considerable interest, as discussed below. Food microemulsions play an important role in these studies.

D. The Sponge (L_3) Phase

The effect of adding various lipids to monoolein was discussed above. These lipids are soluble in either water (bile salt) or oil (triglyceride) or hardly soluble at all (lecithin). If a substance that is soluble in both water and oil, e.g., propylene glycol, is added to the monoolein–water system, the cubic liquid crystal undergoes a transition to a sponge or L_3 phase [13], as shown in Fig. 5. The structure of the sponge phase has been described as a "melted" bicontinuous cubic phase [14].

The sponge phase has the same visual appearance as the microemulsion phase, i.e., an isotropic liquid, but it differs from the latter in the way the emulsifier molecules are organized in the system. In a microemulsion the emulsifier forms a monolayer at the oil/water interface, which makes it possible to create systems with high amounts of oil and water if the emulsifier system is well balanced. In a sponge phase, the emulsifier forms a bilayer (normal or reversed), which will limit the incorporation of both oil and water. Although propylene glycol is not a food additive (however, it is used in oral pharmaceutical formulations), we find it relevant to mention it in this context, since it aids in the formation of many different types of phases [14].

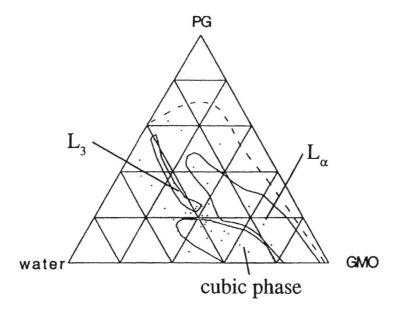

Figure 5 A phase diagram of a system composed of monoolein (GMO), propylene glycol (PG), and water, showing the location of the so-called L_3 or "sponge" phase.

III. MICROEMULSIONS FOR CONTROLLED RELEASE OF FOOD INGREDIENTS

As mentioned above, microemulsions are present during fat digestion, either as an initial formulation of the oil or as an intermediate phase. Therefore, a microemulsion should be considered as a potential candidate to deliver food ingredients, and much can be learned from the drug delivery research field.

Microemulsions have been given much attention as vehicles for drug delivery for several reasons: (1) They are thermodynamically stable, (2) they are easy to produce and handle, (3) they have the capacity to solubilize (and protect) drugs of different polarity, especially poorly water-soluble drugs, (4) they may enhance the absorption of drugs, and (5) they may be formed from food ingredients (water, emulsifiers, and oils). Microemulsions used in pharmaceutical applications often also include non-food ingredients such as propylene glycol and poly(ethylene glycol) [15,16].

There are a few marketed drug delivery systems containing microemulsions that are related to the lipid systems we describe here. Sandimmune Neoral (Novartis, Switzerland) [17], which delivers the lipophilic peptide cyclosporin A, is based on medium-chain triglycerides obtained from coconut oil, a semisynthetic emulsifier (ricinoleate), and propylene glycol. The absorption of the drug taken orally as a microemulsion increases dramatically compared to that of an aqueous suspension [17]. Another marketed example, in which C_8/C_{10} triglycerides from coconut oil are used, is a soft gelatin capsule of vitamin D_3 [18].

One food-grade system that has received much attention as a microemulsion vehicle is the lecithin–triglyceride oil–water system. It is difficult, however, to make a microemulsion with high amounts of oil and water and a small amount of lecithin or any other type of emulsifier [19]. This is not the case if the vegetable oil is replaced by hexadecane, which implies that the slight polarity of the triglyceride oil is sufficient to reduce its flexibility in the oil domain of the microemulsion [4].

IV. APPLICATION ASPECTS

As early as in 1971 Friberg and Rydhag [20] demonstrated the remarkable properties of microemulsions. However, there are still rather few reports in the literature on microemulsions in foods (see Refs. 21–25 in addition to references given above).

As mentioned above, it is possible to solubilize protein molecules into L_2 phases or microemulsions. It is easy to demonstrate this by dissolving a colored protein such as cytochrome c in water and then using this water to prepare a microemulsion, for example, by adding a liquid consisting of sunflower or soybean oil monoglycerides and triglycerides in the weight ratio 7 : 3 [6]. A transparent colored L_2 phase is then obtained. The protein release from such a phase is also easy to follow visually.

If we want to incorporate a certain enzyme into a food product and there is a need to protect that enzyme against proteolytic digestion, the dispersion of such an L_2 phase encapsulating this enzyme might be a good approach. To obtain larger water regions than allowed by unsaturated glycerides with chain lengths around C_{18}, saturated medium-chain-length glycerides can be used. The disadvantage, however, is hydrolytic degradation, which may impart a soapy taste to the product.

We believe that an important application of microemulsions is to provide improved antioxidation effectiveness because of the possibility of a synergistic effect between hydrophilic and lipophilic antioxidants. This is one way of simulating Nature's protection of cell membrane oxidation by tocopherols within the bilayer, which are reactivated by ascorbic acid in the water phase. We describe here how unsaturated oils can be protected, as an illustrative example.

An earlier work [26] showed that soybean oil was effectively protected when contained within an L_2 phase produced by the addition of monoglycerides (sunflower oil monoglycerides) to water. An approximately 1 : 5 ratio of monoglycerides to triglycerides is needed to get enough water into the L_2 phase (about 5 wt%). In such a system 200 ppm of tocopherol in the oil and 5% of ascorbic acid in the reverse micelles gave a dramatic antioxidant effect compared to conventional methods of dissolving or dispersing antioxidants in oils. In fish oils, the same microemulsion-based method to achieve an antioxidant protection effect has also been used [27]. To improve the protection effect even further, glycerol was used instead of water. The existence range of the L_2 phase is almost identical for water and glycerol. The final composition used was 77% fish oil, 20% sunflower oil monoglycerides (Dimodan LS from Grindsted Products, Danisco, Denmark), and 3% glycerol. The concentration of ascorbic acid was 5% in the glycerol phase, and 600 ppm α-tocopherol was added to the oil phase.

A major difference between food microemulsions and most other microemulsions is in the composition of the oil component. In foods the oil is a triglyceride, whereas in other microemulsions the oil is a hydrocarbon, often a mineral oil. The triglyceride molecule itself is surface-active, which in turn implies that triglycerides are not capable of forming separate oil domains in an amphiphile–water system in the same way as mineral oils. Therefore, the composition range in oil–surfactant–water systems that allows microemulsions to form when the oil is a triglyceride is much smaller than the range allowing microemulsion formation when the oil is a hydrocarbon.

V. CONCLUSIONS

It is possible to prepare well-defined food grade microemulsions from extracted cereal lipids. Their structure is similar to that of the sponge or L_3 phase. The other main type of microemulsion formed by food lipids has the L_2 structure, and it can be prepared from liquid mixtures of mono- and triglycerides. An important application of microemulsions is to provide improved antioxidation effectiveness because of the possibility of a synergistic effect between hydrophilic and lipophilic antioxidants, as was illustrated in systems with soybean oil and fish oils.

REFERENCES

1. T. Carlsson, K. Larsson, and Y. Miezis, Cereal Chem. *55*:168 (1978).
2. T. Carlsson, K. Larsson, and Y. Miezis, J. Dispersion Sci. Technol. *1*:417 (1980).
3. G. Jayasinghe, K. Larsson, Y. Miezis, and B. Sivik, J. Dispersion Sci. Technol. *12*:443 (1991).
4. K. Larsson, in *Lipids—Molecular Organization, Physical Functions, and Technical Applications*, Oily Press, Dundee, Scotland, 1994.
5. M. Lindström, H. Ljusberg-Wahren, K. Larsson, and B. Borgström, Lipids *16*:749 (1981).
6. L. Engström, J. Dispersion Sci. Technol. *11*:479 (1990).
7. I. Söderberg, Structural properties of monoglycerides, phospholipids and fats in aqueous systems, Ph.D. Thesis, University of Lund, Sweden, 1990.
8. H. Gutman, G. Arvidsson, K. Fontell, and G. Lindblom, in *Surfactants in Solution*, Vol. 1 (K. L. Mittal and B. Lindman, eds.), Plenum, New York, 1984, pp. 143–152.
9. M. Svärd, P. Schurtenberger, K. Fontell, B. Jönsson, and B. Lindman, J. Phys. Chem. *92*:2261 (1988).
10. J. S. Patton and M. C. Carey, Science *204*:145 (1979).
11. J. E. Staggers, O. Hernell, R. J. Stafford, and M. C. Carey, Biochemistry *29*:2028 (1990).
12. O. Hernell, J. E. Staggers, and M. C. Carey, Biochemistry *29*:2041 (1990).
13. S. Engström, K. Alfons, M. Rasmusson, and H. Ljusberg-Wahren, Progr. Colloid Polym. Sci. 108:93 (1998).
14. U. Olsson and H. Wennerström, Adv. Colloid Interface Sci. *49*:113 (1994).
15. D. Attwood, in *Drugs and Pharmaceutical Science, Vol. 66* (J. Kreuter, ed.), Marcel Dekker, New York, 1994, pp. 31–71.
16. P. P. Constantinides, Pharm. Res. *12*:1561 (1995).
17. J. M. Kovarik, E. A. Mueller, J. B. van Bree, W. Tetzloff, and K. Kutz, J. Pharm. Sci. *83*:444 (1994).
18. I. Holmberg, L. Aksnes, T. Berlin, B. Lindback, J. Zemgals, and B. Lindeke, Biopharm. Drug Disposition *11*:807 (1990).
19. J. Alander and T. Wärnheim, J. Am. Oil Chem. Soc. *66*:1161, 1656 (1989).
20. S. E. Friberg and L. Rydhag, Kolloid-Z. Z. Polym. *244*:233 (1971).
21. R. A. Burns and M. F. Roberts, J. Biol. Chem. *256*:2716 (1981).
22. N. Krog, N. M. Barfod, and R. M. Sanchez, J. Dispersion Sci. Technol. *10*:483 (1989).
23. D. Han, O. S. Yi, and H. K. Shin, J. Food Sci. *55*:247 (1990).
24. T.-L. Lin, S.-H. Chen, N. E. Gabriel, and M. F. Roberts, J. Phys. Chem. *94*:855 (1990).
25. M. El-Nokaly, G. Hiler, and J. McGrady, in *Microemulsions and Emulsions in Foods* (M. El-Nokaly and D. Cornell, eds.), ACS Symp. Ser. No. 448, Am. Chem. Soc., Washington, DC, 1991, pp. 26–43.
26. L. Moberger, K. Larsson, W. Buchheim, and H. Timmen, J. Dispersion Sci. Technol. *8*:207 (1985).
27. M. Jacobsson and B. Sivik, J. Dispersion Sci. Technol. *15*:611 (1994).

28

Application of Microemulsions as Liquid Membranes

John M. Wiencek
The University of Iowa, Iowa City, Iowa

I. INTRODUCTION

The development of efficient separation processes is important when attempting to increase the quantity and/or improve the quality of a wide variety of products. The executive summary of the 1987 National Research Council's report on separation science [1] set several goals for the next few decades. These included:

1. Generating improved selectivity among solutes
2. Concentrating solutes from dilute solutions
3. Understanding and controlling interfacial phenomena
4. Increasing the rate and capacity of separation
5. Developing improved process configurations
6. Improving energy efficiency in separation systems

The general separation class of liquid membranes goes far in addressing many of these issues. Microemulsion liquid membranes have further improved and exploited these advantages. This chapter is the first overall summary of research in the area of microemulsion-based liquid membrane systems. Most of the discussion focuses on emulsified liquid membranes (ELMs), as first invented by Li [2]. Monographs by Noble and Way [3] and Bartsch and Way [4] discuss the state of the art and are highly recommended as additional background in the area of liquid membranes. However, these reviews have very scant (if any) information on the use of microemulsions as liquid membranes; thus, this chapter is intended to bring such information together within a single document.

II. BACKGROUND

A. Liquid Membranes

The term "liquid membrane" has a variety of meanings in the literature. In its simplest form, one can envision a liquid membrane as a fluid phase that is immiscible with both the feed and the receiving phase (see Fig. 1) The configuration is inherently unstable, as depicted in Fig. 1, because density differences between the various fluids will caused rapid and irreversible

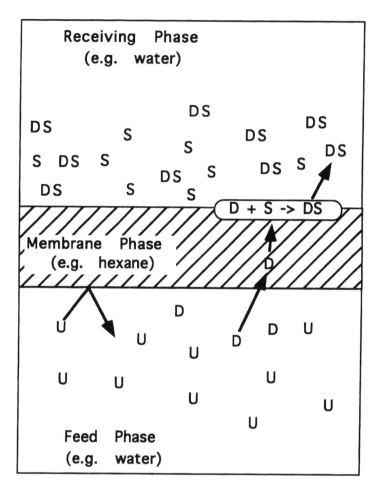

Figure 1   A conceptual liquid membrane. A liquid membrane is a fluid phase that separates two other fluid phases and allows selective transfer of a desired solute (D) while rejecting undesired solute (U). Often, a reaction is used on the receiving phase side to strip the solute (D). The reagent involved in this stripping reaction (S) is preloaded into the receiving phase. DS denotes the stripped solute.

phase separation. To circumvent this problem, a variety of configurations have been employed to stabilize the liquid membranes. This background section briefly reviews these various configurations and attempts to show where microemulsions have been used to enhance a particular separation device.

Liquid membranes are most useful where there is a low driving force for mass transfer. In this case, the fluid liquid membrane can serve as an extracting phase for a desired solute. The solute partitions to satisfy thermodynamic equilibrium constraints. Since the liquid membrane is usually very thin, this partitioning will be completed in a relatively short time and with minimal concentrative effect. In standard liquid–liquid extraction processes, one would employ a stripping step to replenish the extractant and concentrate the extracted solute. For liquid membranes, such a stripping step may be carried out on the opposite side of the liquid membrane (i.e. in the receiving phase). Thus, liquid membrane separations are often called liquid membrane extraction processes in view of the analogy to traditional liquid–liquid extraction. Since extraction and stripping occur simultaneously, the amount of extractant (i.e., liquid membrane) required is greatly reduced, and rates of separation are often dramatically increased. The technique is especially attractive when specialty extractants are required to remove the solute of interest from a mixture. For metal ions, much recent work has been directed to finding novel extractants that solubilize the metal into a variety of organic (i.e., water-immiscible) phases [4]. Thus, much of the work on liquid membranes is directed toward the separation of a particular metal ion from an aqueous mixture of ions. The selectivity is provided by the specialty extractant, and the concentrative effect is provided by the stripping reaction that occurs in the receiving phase. A specific example for mercury is presented in Sec. III.B.

1. U-Tubes

U-tubes are a very simple implementation of the liquid membrane concept. Two miscible fluids, usually aqueous phases, are placed in separate containers. One container is the feed solution, and the other is the receiving solution. These two containers are connected via an inverted or normal U-shaped tube containing an immiscible solvent (e.g., hexane or chloroform). An inverted U-tube is used if the solvent is less dense than the receiving and feed solutions (e.g., hexane vs. water), whereas a normal U-tube is used if the solvent is more dense than the receiving and feed solutions (e.g., chloroform vs. water). Mixing bars and sampling draw-off ports are usually placed in the feed and receiving phase containers (see Fig. 2 for an example). The U-tube is a useful device for preliminary experimental tests on new extractants because it allows for the experimental measurement of the solute concentration in all three phases (feed phase, membrane phase, and receiving phase). U-tubes are not a practical contacting device if the separation is to be used on a large scale. The limited surface area for mass transfer at both the feed phase/solvent and the solvent/receiving phase interfaces as well as the macroscopic diffusion distances in the solvent phase within the U-tube lead to very slow rates of separation.

2. Immobilized on Microporous Supports

Another means of implementing a liquid membrane uses a microporous solid film that maintains the liquid membrane within its pores by capillary action. Such liquid impregnated films are called supported liquid membranes (SLMs). A typical SLM configuration has the organic phase liquid (often containing some sort of complexing reagent to facilitate transport) immobilized in the pores of a microporous membrane with the aqueous feed phase flowing on one side and the aqueous receiving phase on the other side (see Fig. 3). A major disadvantage of SLMs is their instability, which is mainly due to (1) loss of extractant and/or organic solvent into the flowing aqueous phase because of solubility and (2) short-circuiting of the two aqueous phases if the pressure differential across the membrane exceeds the capillary forces that hold the organic liquid in the pores. The operation must be interrupted to replenish lost extractant [5]. In addition, SLMs are known to form emulsions at the membrane interface, leading to contamination of the feed solution with extractant.

3. Emulsion Liquid Membranes

Emulsion liquid membranes (ELMs), first invented by Li [2], are made by forming a surfactant-stabilized emulsion between two immiscible phases. The solute (i.e, the chemical species that is desired) is selectively transported from a feed phase across a thin liquid film

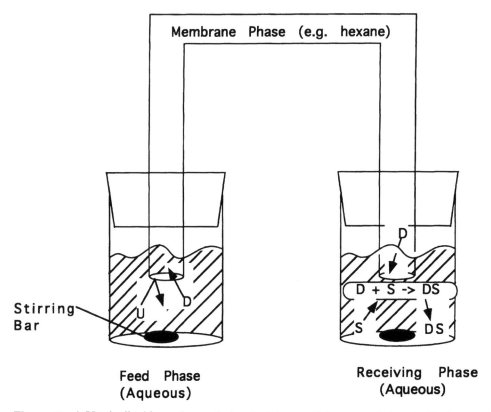

Figure 2 A U-tube liquid membrane device. In this case, it is assumed that the liquid membrane phase is less dense than the feed and receiving phases. Using the same notation as in Fig. 1, the desired solute (D) partitions from the feed phase beaker into the membrane phase. The undesired solute (U) is rejected at this interface. The solute (D) then diffuses the length of the U-tube until it encounters the beaker containing the receiving phase. If a stripping reagent (S) is present, the solute is stripped from the membrane phase at this interface.

of immiscible phase and enriched in the receiving phase. The phases involved are stabilized by forming an emulsion of the membrane and one of the other phases using appropriate surfactants. Usually a water-in-oil emulsion, consisting of an oil phase with appropriate extractants and an aqueous stripping reagent as an internal phase, is dispersed into an aqueous stream containing the solute. A simple example of acetic acid removal from water is shown in Fig. 4. Other separation mechanisms in ELMs are outlined by Li et al [6]. After extraction, the emulsion can be demulsified to recover the enriched stripping phase. Demulsification (emulsion breaking) by application of high voltage electric fields has proven to be most successful [7].

The main advantage of ELMs is the large surface area available for mass transfer. Two disadvantages of ELMs are (1) the lack of stability of the emulsion, which allows leakage of the solute and unreacted stripping reagent into the feed phase, and (2) swelling of the internal microdrops with water from the feed phase. These problems are illustrated conceptually in Fig. 4. Both of these effects reduce separation efficiency. Swelling and leakage of the ELM can occur after prolonged contact with the feed phase [8]. Swelling (commonly called swell in the literature) occurs when the feed stream gets into the internal phase by either

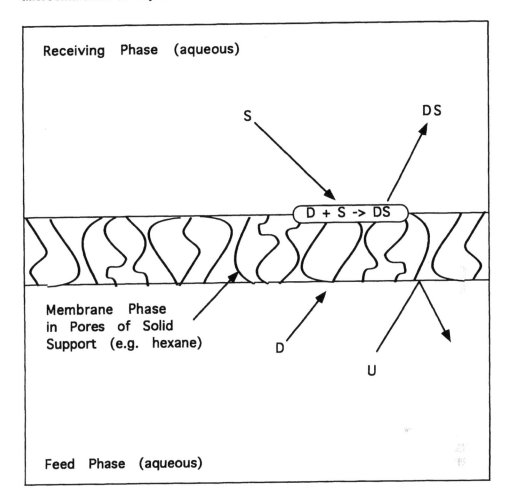

Figure 3 A supported liquid membrane device. A solid microporous sheet (e.g., polypropylene) is soaked in the membrane phase until its pores are saturated with the membrane phase. Usually, the membrane phase is an organic solvent and the microporous sheet is a hydrophobic substance, so the membrane fluid is readily retained in the pores by capillary action. Now this supported liquid membrane can be placed between the feed and receiving phases, and the same mechanism of transport as presented in Figs. 1 and 2 applies.

osmotic pressure or physical breakage and subsequent re-formation of the membranes. The water content in the emulsion can thus increase from 10–20 wt% to 30–50 wt%. Swelling reduces the stripping reagent concentration in the internal phase, which in turn lowers its stripping efficiency. Dilution of the solute that is to be concentrated in the internal phase also results in a less efficient separation. Leakage of the internal phase contents into the feed stream, because of membrane rupture, not only releases extracted solute back into the feed stream but further contaminates the feed stream with stripping reagents. Leakage can be minimized by making a more stable emulsion with a higher concentration of surfactant, but this makes the downstream demulsification and product recovery steps more difficult as well as increasing swell. Lower shear rates would also minimize leakage, but mass transfer resistance could then become very significant. Current research in the area is

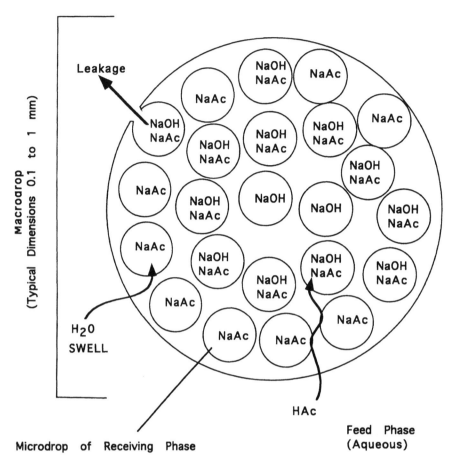

Figure 4 An emulsion liquid membrane (ELM). Again, it is assumed that the membrane phase is an organic solvent. This organic solvent is emulsified with the receiving phase, sodium hydroxide in this example, to yield a water-in-oil emulsion. This emulsion is then dispersed into the feed aqueous phase containing the desired solute, acetic acid (HAc) in this case. A multiple emulsion (W/O/W) forms. The HAc in the feed phase partitions into the membrane portion of the emulsion liquid membrane, diffuses inward until a microdrop containing NaOH is encountered, and then reacts to form sodium acetate. Once in the form of sodium acetate, the acetate cannot partition back into the membrane phase and is trapped. The larger globule of emulsion liquid membrane is called a macrodrop to distinguish it from the microdrops dispersed within the emulsion. For coarse emulsions, the microdrop is roughly 1 μm compared with 0.01 μm for a microemulsion. Note that water can also permeate along with the solute, leading to swelling, and the microdrops can burst, releasing their contents into the feed phase, leading to leakage.

attempting to address these conflicting requirements by use of hybrid techniques (see Sec. II.A.4).

4. Hybrids

Other configurations reported for simultaneous extraction and stripping include hollow fiber–contained liquid membrane (HFCLM) and ELMs in hollow fiber contactors. Neither technique has used microemulsion phases as the liquid membrane. The HFCLM configuration has two sets of microporous hollow fiber membranes, one carrying the feed phase and the other the stripping phase [9,10]. The liquid membrane phase is contained between these two sets of fibers by maintaining the aqueous phase at a pressure higher than the organic phase but lower than its breakthrough value. The HFCLM offers long-term stability because the membrane liquid is connected to a reservoir and is continuously replenished to make up for any loss by solubility. The major disadvantage is the difficulty in mixing the two sets of fibers to achieve a low contained liquid membrane thickness. Presently reported thicknesses are large (approximately 500 μm) and would yield low fluxes if diffusion through the membrane is rate-limiting.

Raghuraman and Wiencek [11] developed a hybrid technique where an emulsion is fed into a hollow fiber contactor on the tube side. Since the solid membrane support is hydrophobic, the continuous phase of the water-in-oil emulsion easily wets the pores of the tube wall and permeates to the shell side. On the shell side of the hollow fiber, the aqueous feed phase is exposed and held at an elevated pressure that prevents the permeating liquid membrane phase from exiting the pores. Thus, extraction occurs on the shell side, and stripping on the tube side of the hollow fiber membrane module. This methodology is closely related to SLMs, but the key difference is the presence of the emulsion on the tube side, which allows for long-term stability because the membrane liquid is continuously replenished to make up for any loss by solubility.

B. Microemulsions as Liquid Membranes

1. History

To the best of my knowledge, Xenakis and Tondre [12] were the first to use the term "microemulsion liquid membrane" with reference to a system using an oil-in-water microemulsion to separate oil-soluble components in a U-tube configuration. In a closely related publication [13], the same authors showed the generality of the idea by reversing the transport, using water-in-oil microemulsions to separate and to concentrate water-soluble solutes.

Microemulsions have not been employed to any significant extent as liquid membranes in SLMs. Although it is not absolutely clear whether microemulsions have been placed within the pores of a solid microporous membrane support to serve as a liquid membrane, a study reported in one publication came very close to doing so. Osseo-Asare and Chaiko [14] used an extractant, dinonylnaphthalenesulfonic acid (HDNNS), to separate cobalt selectively from a multicomponent aqueous stream. In the presence of water, the HDNNS forms a microemulsion. Thus, the authors concluded that their supported liquid membrane, which was initially impregnated with HDNNS, was indeed a microemulsion (supported) liquid membrane once it contacted the aqueous feed and receiving phases.

The great majority of the published work on microemulsions used as liquid membranes is based on the ELM technique. Qutubuddin appears to be the first person to have suggested the technique, and he ultimately published a series of papers with me, his doctoral student at the time [15–18]. I have continued this work with others focusing on a variety of metal ion extractions (especially mercury) as well as protein separations [19–24]. Harada and coworkers [25] claim to have used microemulsions as ELMs to separate a variety of metal ions from solution; however, their work is more likely an equilibrium extraction process as described in Sec. II.B.2. Rautenbach and Machhammer (see Ref. 26) also claimed to have used microemulsions as ELMs, but it was clearly pointed out by Wiencek and

Qutubuddin [26] that their imprecise terminology caused confusion and that their work actually employed macroemulsions.

2. Closely Related Separation Techniques

There are a variety of references to microemulsions within the context of separations, especially as related to extraction. The most common type of confusion is between microemulsion liquid membranes and equilibrium extraction using microemulsions. Equilibrium extraction does not employ a stripping reagent to drive the separation and usually involves a strong interaction between the stabilizing surfactant and the solute. Such methods have been known for a long time [27] and have been patented as an industrial method of purification [28]. More recent work by Vijayalakshmi et al. [29,30] extended these ideas, originally developed for organic molecules, to the separation of multivalent metal ions from water.

Microemulsion liquid extraction is often confused with microemulsion liquid membrane research. Microemulsion liquid extraction has been studied by many groups, for example, Bauer et al. [31], Tondre and Boumezioud [32], and Paatero et al. [33]. The basic idea relies on the formation of a microemulsion phase during a normal equilibrium extraction process. The feed phase is then incorporated as dispersed droplets within the newly formed microemulsion, resulting in extremely high interfacial areas per unit volume. Although similar to microemulsion liquid membranes in its use of the increased surface area, the microemulsion liquid extraction technique incorporates a forward extraction step only and is inherently equilibrium-limited. Thus, the ability of a microemulsion liquid membrane to simultaneously extract and strip a solute is its distinguishing feature in comparison with other, related separation techniques.

3. Advantages and Disadvantages

The use of microemulsions as emulsion liquid membranes offers many potential advantages. Of course, it possesses the same advantages as all liquid membrane systems with respect to its efficiency in selectively separating a single dilute solute from a mixture (see Sec. II.A.3). Microemulsions show added advantages over coarse emulsions when they are employed as emulsion liquid membranes. The low interfacial tensions that are characteristic of microemulsions will lead to smaller macrodrops, which implies faster mass transfer rates due to increased surface area per unit volume. Over time, microdrops dispersed in coarse emulsions will coalesce and phase separate, resulting in leakage. Microemulsions are thermodynamically stable and do not show such phase separation and may therefore be less apt to leak. Another potential advantage is the reversible phase behavior of microemulsions, which can be used to easily form and break the emulsion phase. For example, a simple temperature shift can cause spontaneous emulsification or demulsification of a nonionic microemulsion phase. Wiencek and Qutubuddin [15] were able to show that such advantages do exist in some separation systems such as acetic acid removal from water. However, these advantages are not always realized, as discussed in Sec. III.

III. APPLICATIONS OF MICROEMULSIONS AS ELMs

This section summarizes the use of microemulsions as liquid membranes to separate organic molecules, metal ions, and proteins from dilute aqueous streams. The work summarized is work I conducted within my laboratory or while I was a doctoral student within the laboratory of Professor Qutubuddin. My intent here is to give a glimpse into what can be done with these systems. As shown in Sec. II, the only published work in this area is that of Qutubuddin and myself.

A. Extraction of Organic Substances from Dilute Aqueous Streams

The separation of an organic compound from an aqueous stream typically relies on a reversible aqueous reaction within the receiving phase to strip the compound from the membrane phase. Thus, for organic substances that are water-soluble, the separation technique is most applicable to organic acids and bases (e.g., carboxylic acids, ammonium, and amines). The first reported study of the use of microemulsions as ELMs investigated the separation of acetic acid from a dilute aqueous feed phase [17]. The mechanism of separation is shown in Fig. 4. This mechanism was previously studied by Li, Terry, and Ho [34] using coarse emulsions. Acetic acid (HAc) exists in water in both ionized and un-ionized forms. The un-ionized species shows a slight partitioning into the oil phase of the microemulsion liquid membrane. Since alkanes (e.g., tetradecane) are commonly used to form these microemulsions, the partition coefficients into such phases are very small (<0.1), which suggests that such systems are not effective means to separate acetic acid from water. However, once the HAc is within the microemulsion it rapidly diffuses inward toward the macrodrop center and encounters a dispersed microdrop. At this microdrop, the acetic acid partitions into the aqueous phase of the microdrop, called the receiving phase. The microemulsion's receiving phase contains NaOH, which quickly ionizes the HAc to sodium acetate and prevents it from repartitioning into the membrane phase. Thus, a unidirectional mass transfer of HAc from the aqueous feed phase to the receiving phase is effected. Upon completion of the separation, the microemulsion liquid membrane must be decanted off the feed phase and demulsified to recover the acetate. If HAc is desired, a strong acid must be added back to the mixture to protonate the acetate, which can then be easily recovered by distillation if desired. This method was successfully used to separate HAc under a variety of experimental conditions as presented by Wiencek and Qutubuddin [15,17].

   These studies outlined the criteria for formulating a microemulsion suitable for use as an ELM. First, the microemulsion should be immiscible with the feed phase. Since the feed phase is typically aqueous, this criterion implies a water-in-oil microemulsion. Also, the receiving phase should constitute a high volume fraction of the microemulsion in order to maximize the separation capacity (i.e., amount of stripping reagent incorporated into the microemulsion). The microemulsion should be insensitive to pH, because it is formulated at one pH extreme (e.g., >13 for above example) and dispersed into a system at the other extreme (<4 for the above example). Thus, nonionic surfactants are the ideal candidates for such microemulsion liquid membranes. Finally, the microemulsion should not contaminate the original feed phase. Thus, the use of water-miscible cosurfactants or cosolvents is not feasible for these systems. In view of these constraints, it was determined by trial and error that the use of a twin-tailed ethoxylated phenol gave acceptable performance [15,17]. Further study showed that this system did not leak and had negligible swell over the 5 min required to complete the separation [15]. In addition, placing this microemulsion at 40°C resulted in the spontaneous demulsification of the receiving phase containing concentrated acetate from the microemulsion liquid membrane. Thus, many of the potential advantages of microemulsions over coarse emulsions were illustrated in these studies [15,17].

B. Metal Extraction from Dilute Aqueous Streams

Initial studies on the removal of metal ions from an aqueous stream by microemulsion liquid membranes were focused on copper [16]. The copper was successfully separated from both buffered and unbuffered media by using a microemulsion liquid membrane similar to the one employed for acetic acid but modified to incorporate a liquid metal ion exchanger. This work was very preliminary in nature but did show the capability of microemulsions to separate metal ions from solution.

The most comprehensive work in the area of metal separations using microemulsion liquid membranes is that of Larson and coworkers [19–22], which focused primarily on the removal of mercury as Hg(II) from water. Mercury is a good candidate metal in view of its potential application to environmental remediation as well as the fact that it displays solution chemistry similar to that of many of the transition metal ions. Larson and Wiencek

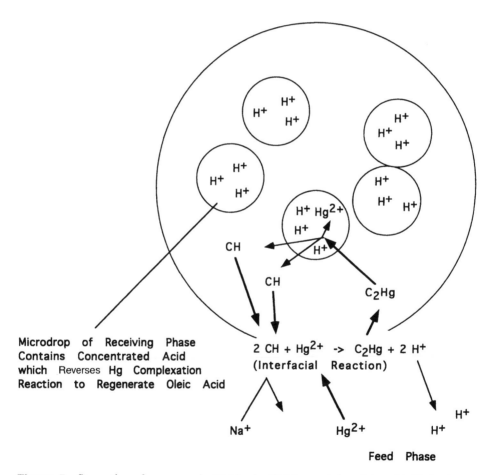

Figure 5   Separation of mercury via ELMs. An ELM containing oleic acid (C) in the membrane phase (i.e., tetradecane) and concentrated acid in the receiving phase is dispersed in a feed phase containing Hg(II). At the macrodrop/feed phase interface, a complexation reaction between two oleic acid molecules and one mercury ion solubilizes the mercury into the hydrophobic membrane phase. This complex then diffuses to the receiving phase, where the low pH affords many protons that can exchange with the mercury and reprotonate the two oleic acid molecules.

[19] showed that mercury can complex with halides to yield a distribution of charged species in water ranging from +2 to −2 in electric charge. Thus, to yield an effective separation strategy, significant effort must be expended on modeling and understanding the distribution of charged species and their propensity to bind with the complexing reagent that is incorporated into the microemulsion liquid membrane [19,35,36]. For the sake of detailed study, Larson and Wiencek [21] used a simple chemistry of oleic acid binding of Hg(II) from acidic aqueous feed phases. A schematic of the separation system is depicted in Fig. 5. Hg(II) is complexed at the macrodrop interface with the feed phase to produce an oil-soluble metallo-organic compound. This complexed mercury diffuses to the center of the macrodrop until (as with acetic acid, above) it encounters a microdroplet containing the stripping reagent. The stripping reagent in this case is a concentrated sulfuric acid solution. The protons react with the oleic acid, releasing the Hg(II) into the microdrop interior. The high proton concentration (i.e., low pH) within the microdrops ensures that the mercury cannot leave the microdrop, since the oleic acid prefers protons to mercury ions at such low pH. The net effect is unidirectional mass transfer of the mercury ions from the feed phase to the receiving phase with a counter transfer of protons, ensuring electroneutrality in both the feed and receiving phases. Once the mercury is separated from the feed, the microemulsion macrodrops are allowed to phase separate and are then decanted off. The feed phase is now free of the toxic mercury ions, and the microemulsion can then be demulsified to recover the mercury in concentrated form.

Larson and Wiencek [21] found that the forward extraction step was easily effected. A noninonic microemulsion enriched with oleic acid as a complexing reagent for Hg(II) was formulated and used in these studies. This microemulsion liquid membrane contained sulfuric acid as the receiving phase and was able to reduce mercury from 460 ppm to 0.8 ppm within 20 min, with most of the separation completed in less than 1 min. This compares favorably to an equilibrium extraction control that resulted in final concentrations of 20 ppm. Thus, some of the advantages of microemulsions were realized. However, the system was difficult of demulsify, and this problem resulted in a separate study focused on the recovery step [8]. The microemulsion swelled with a significant amount of water over the course of the mercury extraction (from 10% to over 50% water by weight), resulting in a white cream at the end of the extraction that more closely resembled a coarse emulsion. This emulsion was very difficult to break and was relatively insensitive to temperature and applied electric fields. Ultimately, butanol was added to break the emulsion, but that resulted in reduced separation efficiency (i.e., a dramatic increase in leakage) upon recycling of the membrane components [8].

C. Extraction of Proteins from Dilute Aqueous Streams

This section is misplaced in some respects. Several efforts have been put forth to separate proteins from aqueous solutions into nonionic microemulsion phases. After several years of investigation, I am convinced that the observed separations rely on an equilibrium partitioning and not on a combined extraction–stripping phenomenon. Thus, such studies should really be incorporated into the discussion in Sec. II.B.2.

The studies employing nonionic microemulsion liquid membranes to separate protein from aqueous solution started with my Ph.D. work [37]. This work was reproduced and summarized in a communication by Qutubuddin et al. [18]. The communication included only a few experiments that yielded a significant partitioning of protein (hemoglobin in this case) into the microemulsion phase. The initial results were confusing and did not display any discernible logic.

In subsequent work, Vasudevan et al. [23] conclusively showed that the hemoglobin was fully dissociated under the conditions presented in Ref. 18 and that only the heme (i.e., iron within porphyrin) was extracted into the microemulsion. Vasudevan and Wiencek [24] went on to prove that under very limiting circumstances, protein will partition into nonionic microemulsion liquid membranes. The underlying extraction mechanism is a weak electrostatic interaction between the trace impurities in the surfactant and the protein. Since the separation is based on an interaction between the surfactant (or some other interfacially active compound) and the solute, no stripping reaction is required, and the system is really an equilibrium microemulsion extraction system as described in Sec. II.B.2.

IV. FUTURE DIRECTIONS

Although microemulsions offer many potential advantages when used as emulsion liquid membranes, their effective usefulness in an industrial setting is questionable. Studies on a variety of systems have shown some disadvantages of microemulsions as opposed to coarse emulsions:

1. During formulation of a suitable microemulsion liquid membrane, the researcher must be able to incorporate a certain amount of additives such as the liquid ion exchanger for metal ion separations. This additive will affect the microemulsion phase behavior and require a screening of surfactant types and concentrations to obtain the desired microemulsion properties discussed in Sec. III.A. In some cases (e.g., quaternary amines), the additive itself is so interfacially active that a microemulsion cannot form.

2. If a suitable microemulsion is formulated, there is no guarantee that it will not leak. In general, only twin-tailed surfactants displayed acceptable results, and often these systems were near phase transitions to liquid crystalline phases. However, experience has shown that if a microemulsion can be formed using a twin-tailed nonionic surfactant, chances are good that it will not leak provided no cosurfactant or cosolvent is used.

3. Once the microemulsion liquid membrane has completed the separation at hand, it must be decanted from the feed phase. The low interfacial tension microemulsion system results in macrodrops an order of magnitude smaller in radius than coarse emulsions, with a concomitant increase in mass transfer area per unit volume. This effect results in very fast mass transfer rates but also very slow disengagement kinetics. A typical batch coarse emulsion system may take 30 s to disengage, whereas a comparable microemulsion system will take almost a full hour.

4. Both coarse emulsion and microemulsion systems can suffer from swelling and leakage. Although such effects were absent from the acetic acid separation system employing microemulsions, they were clearly evident in the mercury separation work, indicating that the supposed advantage of no leakage and no swelling for microemulsions is not universally true but, rather, is very system-dependent.

5. The downfall of microemulsion liquid membranes is the recovery step. Initially, the attractive option of thermally induced phase separation for the recovery of the receiving phase appeared promising based on the results of the acetic acid experiments. The mercury system again showed that this "advantage" is system-dependent and is therefore not really an advantage. In fact, coarse

emulsions are universally amenable to a simple electrostatic demulsification step, and for this reason alone they are currently more attractive than microemulsions as ELMs.

In conclusion, although microemulsion liquid membranes show many potential advantages, such advantages are often system-dependent. In the search for robust technology, this state of affairs is not appealing. I believe that many of the shortcomings of both microemulsions and coarse emulsions as ELMs can be eliminated by controlling the means of contacting the two phases in a nondispersive fashion [11]. If microemulsions with very sensitive phase behavior switches (e.g., temperature) can be routinely developed and formulated, they may have distinct advantages in such a contacting device.

ACKNOWLEDGMENTS

I acknowledge the patience and understanding the editors displayed during the composition of this chapter. Much of this work was supported by the New Jersey Hazardous Management Research Center and the National Science Foundation.

REFERENCES

1. National Research Council, *Separation & Purification—Critical Needs and Opportunities*, NRC Report, National Academy Press, Washington, DC, 1987. p. 21.
2. N. N. Li, U.S. Patent 3,410,794 (1968).
3. R. D. Noble and J. D. Way (eds.), *Liquid Membranes: Theory and Applications*, ACS Symp. Ser. 347, Am. Chem. Soc., Washington, DC, 1987.
4. R. A. Bartsch and J. D. Way (eds.), *Chemical Separations with Liquid Membranes*, ACS Symp. Ser. 642, Am. Chem. Soc., Washington, D.C., 1996.
5. P. R. Danesi, R. Yinger, and P. G. Rickert, J. Membrane Sci. *31*:117 (1987).
6. N. N. Li, J. W. Frankenfeld, and R. P. Cahn, Sep. Sci. Technol. *15*:385 (1981).
7. J. Draxler, W. Furst, and R. Marr, J. Membrane Sci. *38*:281 (1988).
8. K. A. Larson, B. Raghuraman, and J. M. Wiencek, J. Membrane Sci. *91*:231 (1994).
9. S. Majumdar, K. K. Sirkar, and A. Sengupta, in *Membrane Handbook* (W. S. W. Ho and K. K. Sirkar, eds.), Van Nostrand Reinhold, New York, 1992, Chap. 42.
10. A. Sengupta, R. Basu, and K. K. Sirkar, AIChE J. *34*:1698 (1988).
11. B. Raghuraman and J. M. Wiencek, AIChE J. *39*:1885 (1993).
12. A. Xenakis and C. Tondre, J. Phys. Chem. *87*:4737 (1983).
13. A. Xenakis and C. Tondre, in *Surfactants in Solution*, Vol. 3 (K. L. Mittal and B. Lindman, eds.), Plenum, New York, 1984, p. 1881.
14. K. Osseo-Asare and D. J. Chaiko, J. Membrane Sci. *42*: 215 (1989).
15. J. M. Wiencek and S. Qutubuddin, Sep. Sci. Technol. *27*:1211 (1992).
16. J. M. Wiencek and S. Qutubuddin, Sep. Sci. Technol. *27*:1407 (1992).
17. J. M. Wiencek and S. Qutubuddin, Colloids Surf. *29*:119 (1988).
18. S. Qutubuddin, J. M. Wiencek, A. Nabi, and J. Y. Boo, Sep. Sci. Technol. *29*:923 (1994).
19. K. A. Larson and J. M. Wiencek, I&EC Res. *31*2714 (1992).
20. K. Larson, B. Raghuraman, and J. Wiencek, I&EC Res. *33*1612 (1994).
21. K. A. Larson and J. M. Wiencek, Environ. Prog. *13*:253 (1994).
22. K. A. Larson and J. M. Wiencek, in *Emerging Technologies in Hazardous Waste Management IV* (D. W. Tedder and F. G. Pohland, eds.), ACS Symp. Ser. 554, Am. Chem. Soc. Washington, DC, 1994, p. 124.

23. M. Vasudevan, K. Tahan, and J. M. Wiencek, Biotechnol. Bioeng. *46*:99 (1995).
24. M. Vasudevan and J. M. Wiencek, I&EC Res. *35*:1085 (1996).
25. M. Harada, N. Shinbara, M. Adachi, and Y. Miyake, J. Chem. Eng. Jpn. *23*:50 (1990).
26. J. Wiencek and S. Qutubuddin, J. Membrane Sci. *45*:311 (1989).
27. K. Shinoda and T. Ogawa, J. Colloid Interface Sci. *24*:56 (1967).
28. M. L. Robbins U.S. Patent 3,641,181 (1972).
29. C. S. Vijayalakshmi, A. V. Annapragada, and E. Gulari, Sep. Sci. Technol. *25*:711 (1990).
30. C. S. Vijayalakshmi and E. Gulari, Sep. Sci. Technol. *26*:291 (1991).
31. D. Bauer, J. Komornicki, and J. Tellier, U.S. Patent 4,555,343 (1985).
32. C. Tondre and M. Boumezioud, J. Phys. Chem. *93*:846 (1989).
33. E. Paatero, J. Sjöblom, and S. K. Datta, J. Colloid Interface Sci. *138*:388 (1990).
34. N. N. Li, R. E. Terry and W. S. Ho, J. Membrane Sci. *10*:305 (1982).
35. B. J. Raghuraman, N. Tirmizi, and J. Wiencek, Environ. Sci. Technol. *28*:1090 (1994).
36. B. J. Raghuraman, N. P. Tirmizi, B.-S. Kim, and J. M. Wiencek, Environ. Sci. Technol. *29*:979 (1995).
37. J. M. Wiencek, *Liquid membrane separations employing nonionic microemulsions*, Ph.D. Dissertation, Case Western Reserve University, 1989.

29

Application of Microemulsions in Textile Cleaning Using Model Detergency Tests

Hans-Dieter Dörfler
Technical University of Dresden, Dresden, Germany

I. INTRODUCTION

Microemulsions exist in well-defined phase regions of multicomponent systems consisting of water, oil, surfactant, cosurfactant, and electrolyte. They show ultralow interfacial tensions and a high solubilization power toward both hydrophilic and lipophilic substances. Because of these properties, microemulsions are interesting media in textile detergency.

From the literature [1–10] it is clear that multicomponent systems of the water–surfactant–cosurfactant–electrolyte type have already been used in model detergency tests. The fabrics were soiled with different, mostly chemically pure oils. So far only a few investigations [11–14] have been published on the action of microemulsions that are already formulated as cleaning media. Solans et al. [11] carried out model detergency tests with water-rich homogeneous microemulsions consisting of water, n-hexadecane, and dodecyl tetraethylene glycol ether for wool fabrics at washing temperatures between $T = 296$ K and $T = 307$ K. They found that microemulsions showed a much better detergency than macroemulsions consisting of the same components or commercially available liquid detergents. Experiments using commercial grade nonionic surfactants [12] showed the same result. Krüssmann et al. [13,14] investigated two homogeneous microemulsions consisting of water, commercial grade C_{12-14} alkyl polyglycol ethers (Marlipal 24/60 and Marlipal 24/100), and the oil Shellsol D 60 S and obtained good detergency for heavily oil- and pigment-soiled fabrics.

For application in textile cleaning it is useful to compare the detergency of microemulsions with that of a standard detergent solution. In our investigation the influence of the mean ethoxylation number j of the C_{12-14} alkyl polyglycol ethers C_iE_j and the addition of n-pentanol on the detergency was of special interest. Furthermore, for a set of samples from different phase regions of these systems, the relations between the composition of the cleaning agent and the detergency results were also of special interest. Phase diagrams (see Sec. III.A) and properties of multicomponent systems of the water–oil–surfactant–cosurfactant type are discussed in Ref. 17.

The chemically pure surfactants and the oily components used in the works reported in Refs. 15–17 are, however, not suitable for practical applications. Thus in our investigations commercial grade alkyl polyglycol ether mixtures, the so-called Marlipals, and the hydrocarbon n-undecane (Halpaclean: >95% n-undecane) were used. In our previous paper

[18] the phases of quaternary water–n-undecane–alkyl polyglycol ether–n-pentanol systems were determined in dependence on the ethoxylation number j of the alkyl polyglycol ether and on n-pentanol. We used selected samples of these systems for cleaning tests in our model detergency investigations using cotton, polyester, and polyester-cotton mixture test fabrics soiled with sebum, wool fat, pigments, mineral oil, iron oxide, and zinc sulfate.

II. MATERIALS AND METHODS

A. Chemicals

The microemulsions were composed of distilled water, the Halpaclean (which contains >95 wt% n-undecane; Haltermann GmbH, Hamburg), the cosurfactant n-pentanol (Merck, >99%), and commercial grade C_{12-14} alkyl polyglycol ethers [Marlipal 24/40 ($C_{12-14}E_{(4)}$), Marlipal 24/50 ($C_{12-14}E_{(5)}$), Marlipal 24/60 ($C_{12-14}E_{(6)}$), Marlipal 24/70 ($C_{12-14}E_{(7)}$), and Marlipal 24/80 ($C_{12-14}E_{(8)}$); Hüls AG, Marl]. These nonionic are mixtures of alkylpolyglycol ethers C_iE_j with hydrocarbon chain lengths of $i = 12$ (70 wt%) and $i = 14$ (30 wt%) and mean ethoxylation numbers $j = 4, 5, 6, 7, 8$ that are given by the penultimate number in the product labeling [19]. All chemicals except water were used without further purification.

As standard detergent solution [20] the washing machine test detergent was used in the concentration 19.25 g basic powder, 5 g perborate-tetrahydrate, and 0.75 g activator tetraacetylethylenediamine per liter of distilled water. This corresponds to an overall surfactant concentration of 2.7 g/L.

B. Test Fabrics

For the evaluation of the detergency of the microemulsions, three slightly oil- and pigment-soiled test fabrics were used: the cotton fabric *wfk* 10 C; the polyester-cotton mixture in a weight ratio of 65 : 35, *wfk* 20 C; and the polyester fabric *wfk* 30 C. These fabrics are products of the *wfk*-Testgewebe GmbH, Brüggen. They were identically soiled with about 6 mg of pigment mixture, 28 mg of wool fat, and 14 mg of synthetic sebum per gram of fabric. In addition, the heavily oil- and pigment-soiled GFS test fabric (polyester-cotton mixture in a weight ratio of 65 : 35, *wfk*-Forschungsinstitut für Reinigungstechnologie, Krefeld) was employed. Its soiling consisted of 258 ± 8 mg of mineral oil (HD-Mehrbereichsöl SAE 15W-40, Abfüllwerk Haag/Obb), 15 ± 3 mg of $ZnSO_4$ (heptahydrate, purum, Fluka), and 8 ± 1.5 mg of Fe_2O_3 (purum, Merck) per gram of fabric. In order to detect the mineral oil photometrically or by reflectance measurements, it was dyed with 1 wt% Solvent Blue 35 [13].

C. Model Detergency Tests

The model detergency tests were carried out in a laboratory laundrometer LINITEST (Quarzlampen-Gesellschaft mbH, Hanau). The bath ratio of the washing system was 5 : 1, i.e., 20 g of test fabric per 100 g of microemulsion. In comparative detergency tests with the standard detergent solution, the laundrometer box contained 10 g of test fabric and 200 g of detergent solution (bath ratio 20 : 1). The test fabrics were submitted for 20 min to the main washing process at constant temperature. Afterwards they were rinsed three times for 10 min at the same temperature in 200 g of distilled water and dried for at least several hours at room temperature.

The detergency results were evaluated by means of the reflectance values of the cleaned fabrics. Details of the method are described in Ref. 20. As a standard for the soil solubilization, the difference ΔR between the reflectance values of cleaned and uncleaned test fabrics was used. The GFS test fabric was so heavily soiled that reflectance measurements of the uncleaned material were impossible. Thus the detergency for the GFS fabric was evaluated by absolute reflectance values of the cleaned fabrics.

For some samples of the GFS fabric the residual amount of the soil components was analyzed. The mineral oil dyed with Solvent Blue was extracted from the fabric by tetrachloroethylene. The dye content of the solution was measured quantitatively by means of a UV/Vis spectrometer Lambda 15 (Perkin-Elmer). The original mixing ratio of oil and dye was used to calculate the residual mass of the mineral oil [13].

To determine the quantity of metal ions contained by the GFS fabric after the washing process, a piece of fabric of a defined mass was dissolved in sulfuric acid. Using an atomic absorption spectrometer 1100 B (Perkin-Elmer), the contents of iron and zinc ions in the resulting solution were quantified. From these values, the residual amounts of $ZnSO_4$ and Fe_2O_3 on the fabric were determined [13].

III.　RESULTS OF CLEANING TESTS

A.　Summary of the Phase Behavior of the Ternary Systems Water–n-Undecane (Halpaclean)–C_{12-14} Alkyl Polyglycol Ethers (Marlipals)–n-Pentanol as a Basis of the Recipes Used to Prepare Microemulsions for Model Detergency Tests

It is well known [15–18,21] that in a quaternary system of the water–oil–surfactant–cosurfactant type, microemulsions form only in a limited concentration and temperature interval. Thus, an application of microemulsions in textile detergency assumes precise knowledge of the phase diagrams of these systems. The phase diagrams of the ternary and quaternary systems composed of water–n-undecane–C_{12-14} alkyl polyglycol ether–n-pentanol we investigated are presented in Ref 18. The information resulting from the phase diagrams shown in Figs. 1–3 that is relevant to the following model detergency tests is summarized as follows:

In the quaternary phase diagrams of the systems water–n-undecane–C_{12-14} alkyl polyglycol ether for alkyl polyglycol ether concentrations $c_s < 10$ wt%, a three-phase area is found that is surrounded by two-phase regions. As is characteristic for systems with surfactant mixtures, the three-phase area is shifted toward higher temperatures for low surfactant concentrations and high oil concentrations. The same trends are described in Ref. 21. For surfactant concentrations $c_s > 20$ wt%, the lamellar L_α phase is formed.

Near the oil-rich edge of the phase diagram, homogeneous water-in-oil (W/O) microemulsions exist in the entire temperature range 273 K $< T <$ 373 K. In a limited temperature interval around the phase inversion temperature of the system, there is another microemulsion region in the center of the phase diagram at comparable concentrations of water and n-undecane and alkyl polyglycol ether concentrations of about 10 wt% $< c_s <$ 20 wt%. Samples from this region have

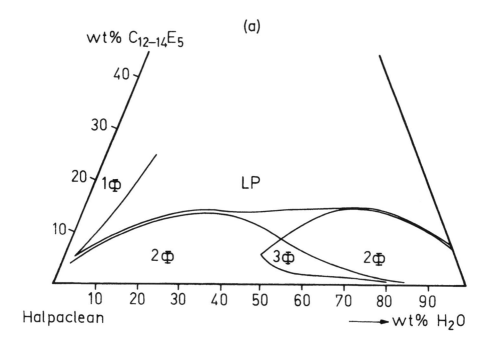

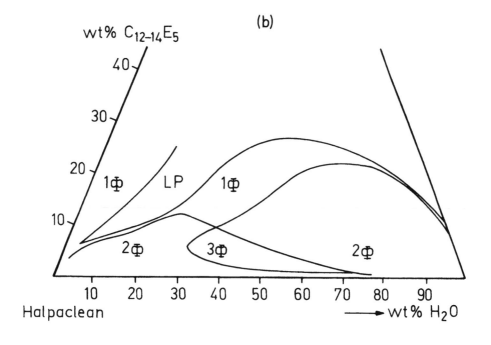

Figure 1 Phase diagrams of the water–oil (Halpaclean)–surfactants ($C_{12-14}E_5$; Marlipal) system with 3.5 wt% n-pentanol at different temperatures. (a) $T = 303$ K; (b) $T = 313$ K; (c) $T = 323$ K; (d) $T = 333$ K. Symbols: 1Φ = homogeneous mixture, 2Φ = two-phase region, 3Φ = three-phase region, LP = lamellar phase L_α.

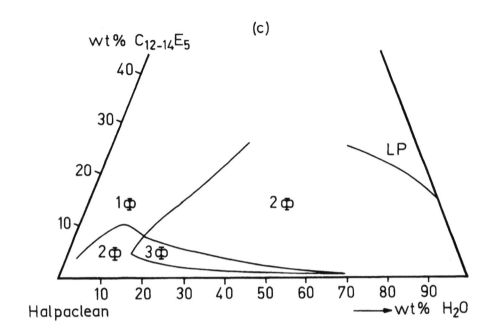

(c)

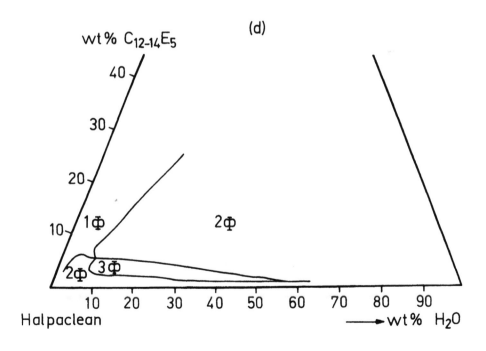

(d)

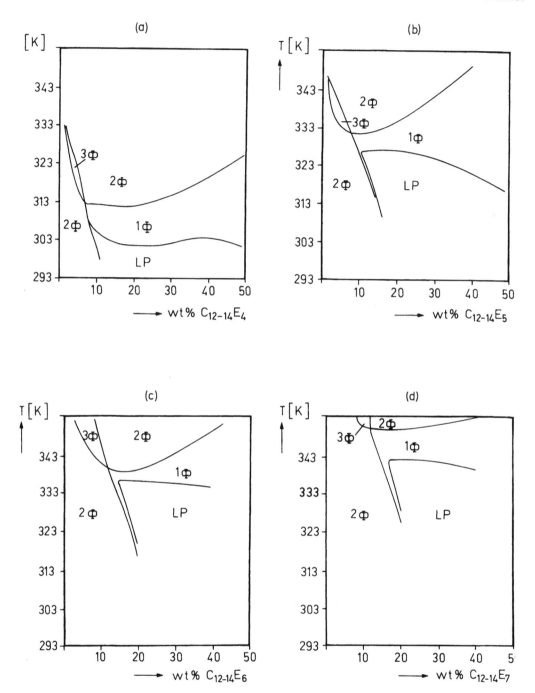

Figure 2 Pseudo-binary concentration–temperature diagrams of the systems (a) water–oil (Halpaclean)–surfactant ($C_{12-14}E_4$; Marlipal); (b) water–oil (Halpaclean)–surfactant ($C_{12-14}E_5$; Marlipal); (c) water–oil (Halpaclean)–surfactant ($C_{12-14}E_6$; Marlipal); and (d) water–oil (Halpaclean)–surfactant ($C_{12-14}E_7$; Marlipal) at constant water and oil (ratio 1 : 1) content. For symbols see legend to Fig. 1.

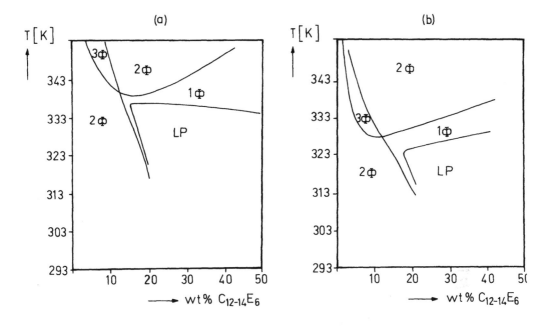

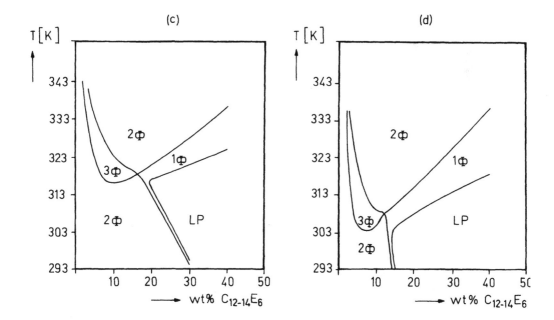

Figure 3 Pseudo-binary concentration–temperature diagrams of the system water–oil (Halpaclean)–surfactant ($C_{12-14}E_6$; Marlipal) at constant water and oil (ratio 1 : 1) content (a) without addition of *n*-pentanol, (b) with addition of 2 wt% *n*-pentanol, (c) with addition of 4 wt% *n*-pentanol, and (d) with addition of 6 wt% *n*-pentanol. For symbols see Fig. 1.

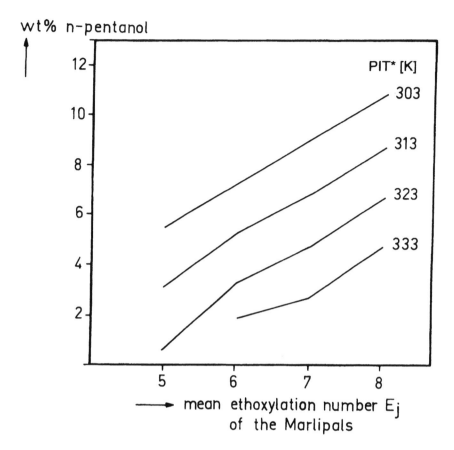

Figure 4   Influence of *n*-pentanol and E_j of surfactant at five different PIT* values (PIT* = phase inversion temperature) in the systems water–oil (Halpaclean)–surfactant ($C_{12–14}E_j$; $j = 5$, 6, 7, 8; Marlipals).

a bicontinuous spongelike structure and show ultralow interfacial tensions on the order of $\gamma \approx 10^{-3}$ mN/m. The solubilization properties of these bicontinuous microemulsions are of special interest for model detergency tests.

The temperature interval in which a central microemulsion region exists is shifted to higher temperatures by increasing the mean ethoxylation number of the $C_{12–14}$ alkyl polyglycol ether, and to lower temperatures by the addition of *n*-pentanol. The variation of the ethoxylation number and a defined addition of *n*-pentanol are thus tools to prepare bicontinuous microemulsions at a given washing temperature.

Based on knowledge of the phase regions in the diagrams, a set of selected samples from the four-component system water–*n*-undecane–$C_{12–14}$ alkyl polyglycol ether–*n*-pentanol were used as cleaning media in model detergency tests. In Table 1 the recipes for these samples, i.e., compositions of the three- or four-component systems, are listed. Furthermore, in Table 1 the mean ethoxylation number *j* of the alkyl polyglycol ethers is given. The last column of Table 1 describes the temperature interval ΔT in which each sample forms homogeneous, bicontinuous microemulsions. As can be seen, the samples contain equal amounts of water and n-undecane.

Samples 1, 2, and 3 are ternary mixtures of water, n-undecane, and C_{12-14} alkyl polyglycol ethers with mean ethoxylation numbers $j = 4$, 5, and 6. These numbers determine the temperature region in which microemulsions are formed. The surfactant concentration was kept as low as possible, i.e., near the transition to the three-phase area in the phase diagram. Samples 4–6 are quaternary mixtures of water–n-undecane–C_{12-14} alkyl polyglycol ether–n-pentanol. The weight ratio of water, n-undecane, and alkyl polyglycol ether and the mean ethoxylation number $j = 6$ of the alkyl polyglycol ether are constant. By addition of different amounts of the cosurfactant n-pentanol, the temperature interval in which microemulsions form is moved to lower temperatures. In samples 7–9, both the ethoxylation number of the alkyl polyglycol ether and the added quantity of n-pentanol are varied. The of water/n-undecane/C_{12-14} alkyl polyglycol ether weight ratio for these samples is kept constant. The shift in the existence temperature of bicontinuous microemulsions due to increasing the mean ethoxylation numbers $j = 5, 6, 8$ is compensated for by a defined addition of n-pentanol. So all three samples form microemulsions at the washing temperature of $T = 313$ K.

In a further test series, the influence of the composition of quaternary systems consisting of water, n-undecane, C_{12-14} alkyl polyglycol ether and n-pentanol on their detergency was investigated. Therefore, 19 samples from different phase regions of the above-mentioned quaternary systems (see Figs. 1–3) were chosen. In Table 2 the composition of the samples and their essential features concerning the phase regions in the phase diagrams at $T = 313$ K are summarized. In the ternary mixture, $c = 3.5$ wt% n-pentanol was added in order to have a large central microemulsion region in the phase diagram at the washing temperature of $T = 313$ K. The alkyl polyglycol ether concentration was varied from $c = 5$ wt% to $c = 20$ wt% in steps of 5 wt%. This specification makes clear that the model detergency tests were carried out not only with bicontinuous or W/O microemulsions but also with heterogeneous regions consisting of microemulsion and water, oil, or the lamellar L_α phase and with a three-phase system of microemulsion, water, and oil phase.

Table 1 Composition of the Ternary and Quaternary Systems of Water, n-Undecane, C_{12-14} Alkyl Polyglycol Ethers and n-Pentanol Used in the Model Detergency Tests[a]

Sample	Water (wt%)	n-Undecane (wt%)	C_{12-14} Alkyl polyglycol ethers (wt%)	j	n-Pentanol (wt%)	ΔT (K)
1	45.0	45.0	10	4	—	306–312
2	44.5	44.5	11	5	—	327–332
3	43.5	43.5	13	6	—	337–340
4	42.7	42.7	12.7	6	1.9	324–328
5	41.8	41.8	12.5	6	3.9	308–313
6	41.0	41.0	12.3	6	5.7	297–302
7	38.6	38.6	19.3	5	3.5	310–316
8	37.5	37.5	18.8	6	6.2	309–316
9	36.4	36.4	18.2	8	9.0	306–319

[a] j-the mean ethoxylation number of the alkyl polyglycol ethers; ΔT-the temperature interval in which the samples form a homogeneous bicontinuous microemulsion.

Table 2 Composition and Phase Regions of the Quaternary Systems Consisting of Water,
n-Undecane, C$_{12-14}$ Alkyl Polyglycol Ether (C$_{12-14}$E$_{(5)}$), and *n*-Pentanol Used in the Model Detergency
Tests[a]

Sample	Water (wt%)	*n*-Undecane (wt%)	C$_{12-14}$ Alkyl polyglycol ethers (wt%)	Phase regions of the sample at $T = 313$ K[b]
10	2	93	5	W/O microemulsion (1Φ)
11	18	77	5	Microemulsion+oil phase (2Φ)
12	38	57	5	Microemulsion+water+oil phase (3Φ)
13	60	35	5	Microemulsion+water phase (2Φ)
14	80	15	5	Microemulsion+water phase (2Φ)
15	4	86	10	W/O microemulsion (1Φ)
16	11	79	10	Microemulsion+L$_\alpha$ phase (2Φ)
17	26	64	10	Microemulsion+oil phase (2Φ)
18	50	40	10	Microemulsion+water phase (2Φ)
19	70	20	10	Microemulsion+water phase (2Φ)
20	6	79	15	W/O microemulsion (1Φ)
21	20	65	15	Microemulsion+L$_\alpha$ phase (2Φ)
22	34	51	15	Bicontinuous microemulsion (1Φ)
23	60	25	15	Microemulsion+water phase (2Φ)
24	8	72	20	W/O microemulsion (1Φ)
25	20	60	20	Microemulsion+L$_\alpha$ phase (2Φ)
26	40	40	20	Bicontinuous microemulsion (1Φ)
27	60	20	20	Microemulsion+water phase (2Φ)
28	76	4	20	Microemulsion+L$_\alpha$ phase (2Φ)

[a] All samples also contained 3.5 wt% *n*-pentanol for 100 wt% of the ternary mixture.
[b] 1Φ=homogeneous mixture; 2Φ=two-phase region; 3Φ=three-phase region.

B. Model Detergency Tests

1. Comparison of the Detergency by Microemulsions with That of a Standard Detergent Solution for Different Fabrics and Soils

The recipes of microemulsions described in Table 1 were tested in laboratory detergency experiments for their suitability as washing media for textile detergency. In the first detergency test series, samples 1 at $T = 308$ K, sample 2 at $T = 329$ K, and sample 3 at $T = 338$ K, i.e., at temperatures at which the mixtures form a bicontinuous microemulsion, were compared with the detergency of the standard detergent solution at the same temperatures. As test fabrics, slightly soiled hydrophilic cotton fabric (*wfk* 10 C), lipophilic polyester fabric (*wfk* 30 C), and polyester-cotton fabric (*wfk* 20 C) were used. They were cleaned independently with the microemulsion and with the standard detergent solution.

The reflectance differences as a measure of detergency are depicted in Figs. 5–7. According to these figures, the microemulsions show, independently of the type of fabric, better detergency than the standard detergent solution. Even at the lowest washing temperature, $T = 308$ K, the solubilization and extraction of oil or pigment soil by the microemulsion of sample 1 is higher than that of the standard washing solution at $T = 348$ K. For samples 2 or 3 at higher washing temperatures, the detergency for the cotton fabric and the polyester-cotton mixture increases slightly (see Figs. 5 and 6). This fact can be explained by the lower viscosity and easier removal of oily soils at higher temperatures since the interfacial tensions of the three microemulsions against oil at each washing temperature

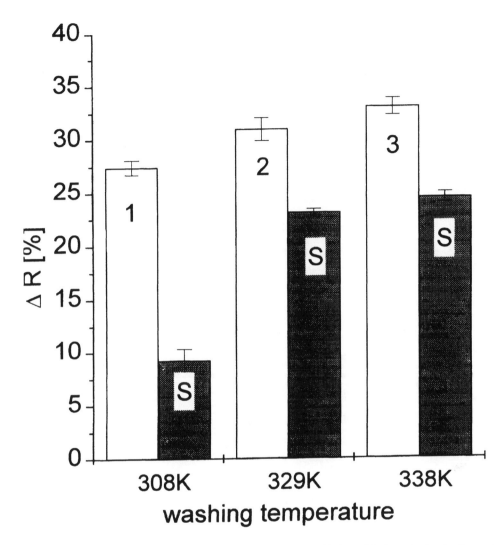

Figure 5 Reflectance differences ΔR of different slightly soiled test fabrics after having been cleaned with the microemulsions formed by (1) sample 1 at $T = 308$ K, (2) sample 2 at $T = 329$ K, and (3) sample 3 at $T = 338$ K (composition of samples according to Table 1) in comparison to the reflectance differences after cleaning with the standard detergent solution (S) at the same temperatures using cotton fabric *wfk* 10 C.

are of the same order, $\gamma \approx 10^{-3}$ mN/m. In contrast, the detergency of the standard detergent solution increased rapidly with temperature, mainly because of the action of the bleaching agent.

In cleaning the lipophilic polyester fabric *wfk* 30 C, the microemulsions do not excel the standard detergent solution as much as in cleaning cotton fabrics (see Fig. 7). Since the detergency of microemulsions is based mainly on the solubilization of oily soil, the higher adsorption of oil on lipophilic polyester fabrics affects the detergency negatively. For samples 2 or 3 at higher washing temperatures, a decrease in the detergency of the microemulsions is observed while that of the standard detergent solution remains nearly constant. Possibly this

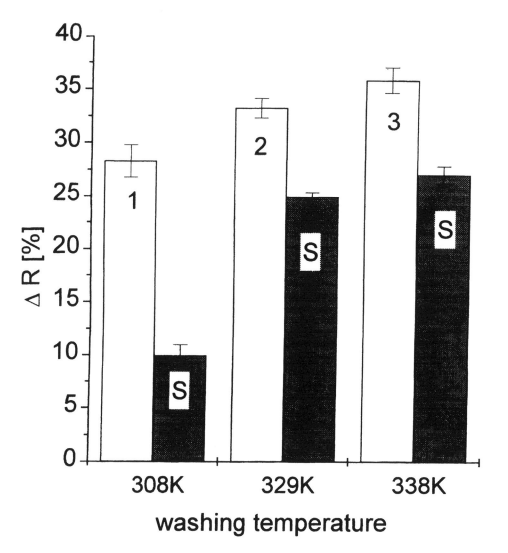

Figure 6 Reflectance differences ΔR of different slightly soiled test fabrics after having been cleaned with the microemulsions formed by (1) sample 1 at $T = 308$ K, (2) sample 2 at $T = 329$ K, and (3) sample 3 at $T = 338$ K (composition of samples according to Table 1) in comparison to the reflectance differences after cleaning with the standard detergent solution (S) at the same temperatures using polyester-cotton mixture fabric *wfk* 20 C.

is due to the higher hydrophilicity of the C_{12-14} alkyl polyglycol ethers contained in samples 2 and 3.

In a further test series we investigated the cleaning of the test fabric *wfk* 20 C with the microemulsions formed by samples 4, 5, and 6 (compositions according to Table 1) at the washing temperatures $T = 298$, 311, 325, and 338 K. The aim of these tests was to find out how the tuning of the existence temperature of homogeneous microemulsions by the addition of *n*-pentanol affected their detergency.

Figure 8 shows the reflectance differences of the cleaned fabrics as a standard for the detergency of the microemulsions in comparison with that of the standard detergent

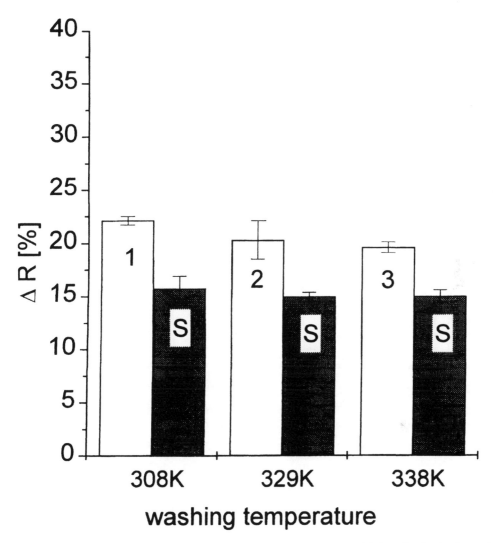

Figure 7 Reflectance differences ΔR of different slightly soiled test fabrics after having been cleaned with the microemulsions formed by (1) sample 1 at $T = 308$ K, (2) sample 2 at $T = 329$ K, and (3) sample 3 at $T = 338$ K (composition of samples according to Table 1) in comparison to the reflectance differences after cleaning with the standard detergent solution (S) at the same temperatures using polyester fabric *wfk* 30 C.

solution. The same trend as in the detergency tests described above was observed: The microemulsion of sample 6 employed at the lowest washing temperature, $T = 298$ K, shows the highest cleaning effect compared with the standard detergent solution. With increasing temperature and using microemulsions with lower quantities of *n*-pentanol, their detergency increases, but much less than that of the standard detergent solution. Figure 8 shows that at nearly the same washing temperature the samples containing *n*-pentanol have the same detergency as systems with lower ethoxylated C_{12-14} alkyl polyglycol ethers. To keep the quantity of surfactant low, it is recommended to use low ethoxylated C_{12-14} alkyl polyglycol ethers, because in these systems a lower surfactant concentration is needed to form

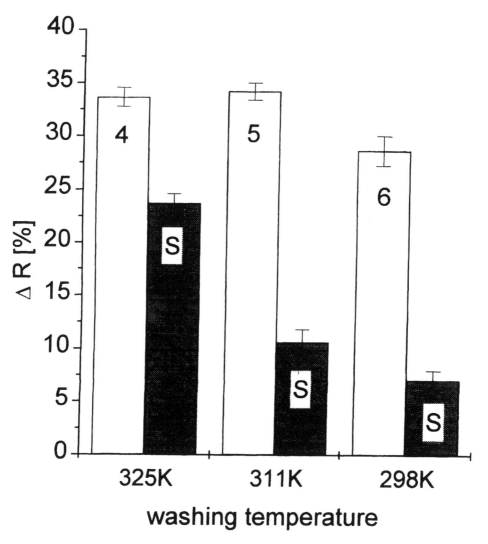

Figure 8 Reflectance differences ΔR of the slightly soiled test fabric *wfk* 20 C after having been cleaned with the microemulsions formed by (6) sample 6 at $T = 298$ K, (5) sample 5 at $T = 311$ K, (4) sample 4 at $T = 325$ K (composition of samples according to Table 1) in comparison to the reflectance differences after cleaning with the standard detergent solution (S) at the same temperatures.

microemulsions at low temperature [18]. In conclusion, we can say that the main advantage of microemulsions in the detergency process is the possibility of making defined changes in their cleaning properties by varying their composition.

In the following washing tests we compared the influence of surfactant concentration in microemulsions and in cleaning with standard detergent solution. We investigated how a variation of the bath ratio (weight ratio of cleaning medium to test fabric) in cleaning with either a microemulsion (sample 1, see Table 1) or the standard detergent solution affected the cleaning results. The test fabric used was *wfk* 20 C at the washing temperature $T = 308$ K.

In Fig. 9 the reflectance differences in the cleaned fabrics can be seen. For better comparison the abscissa is scaled by the surfactant concentration per gram of fabric in the washing bath. In these detergency tests, the use of a microemulsion gave much better results than cleaning with standard detergent solution. The better detergency of the microemulsion at low temperature originates in its ultralow interfacial tensions ($\gamma \approx 10^{-3}$ mN/m) against water and oil and in its high solubilization power. The relatively worse result at a bath ratio of 1 : 1 and the large differences in the reflectance values are probably caused by the fact that the fabric was only partially wetted with the microemulsion by the primitive washing mechanics of the LINITEST laundrometer. But at a bath ratio of 2 : 1 the fabric was totally wetted. The use of higher quantities of microemulsion in the model detergency tests does not notably increase its detergency. Thus micromemulsions can be used in very small amounts without losing their outstanding detergency. This means also that microemulsions can be reused multiple times. How often they can be reused depends on the composition of the soil and microemulsions.

For practical application of microemulsions, the solubilization of heavy oily soil from workwear at low washing temperatures was studied. The detergency of samples 7, 8, and 9 (see Table 1), the pure n-undecane fraction (>95% n-undecane), and the standard detergent solution for the heavily oil-, pigment-, and electrolyte-soiled GFS test fabric was tested at $T = 313$ K. A bath ratio of 5 : 1 was used for the pure n-undecane fraction

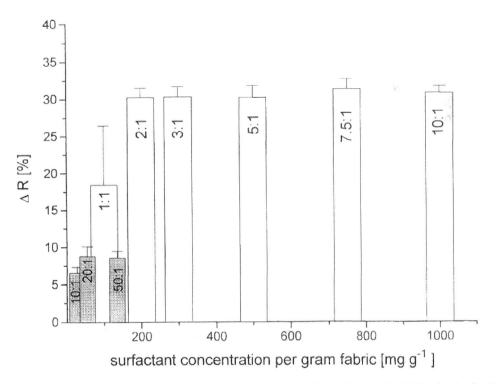

Figure 9 Reflectance differences ΔR of the slightly soiled test fabric *wfk* 20 C after having been cleaned with the microemulsion formed by sample 1 (composition of samples according to Table 1) at the bath ratios (weight ratio of cleaning medium to fabric) 1 : 1, 2 : 1, 3 : 1, 5 : 1, 7.5 : 1, and 10 : 1 (white columns) and with the standard detergent solution (S) at the bath ratios 10 : 1, 20 : 1, and 50 : 1 (screened columns) at a washing temperature of $T = 308$ K.

as for the microemulsions. The residual quantities of each soil component were determined after the washing process either photometrically or by atomic absorption spectroscopy and compared to the initial quantities on the uncleaned GFS fabric.

Figure 10 shows the amount of mineral oil remaining on the fabric. While the standard washing solution extracted only 64 wt% of the oily soil, most of it except for a very small residual content of 2.8–3.6 wt% of the initial weight could be removed by the microemulsions. After cleaning tests with the pure n-undecane fraction, the GFS fabric still contained 4.6 wt% of the initial quantity of oily soil. The composition of the microemulsions affects their detergency (see Fig. 10). The residual amount of oily soil increases with increasing mean ethoxylation number of the C_{12-14} alkyl polyglycol ether and with higher content of n-pentanol. The reason for this could be that although all samples contain the same surfactant concentration, the solubilization power of surfactants decreases with increasing ethoxylation number of the C_{12-14} alkyl polyglycol ether [21].

The application of microemulsions in textile detergency maintains a very high solubilization of oily soil. Concerning electrolytes and pigments, the action of microemulsions as cleaning media is limited. Figure 11 shows the residual quantities of iron oxide and zinc sulfate on the GFS test fabric after the washing process. If microemulsions or pure n-undecane were used as washing media, between 45 and 50 wt% of the initial quantity of these compounds remained on the fabric. The differences between the microemulsions and n-undecane after the washing tests do not exceed the limits of

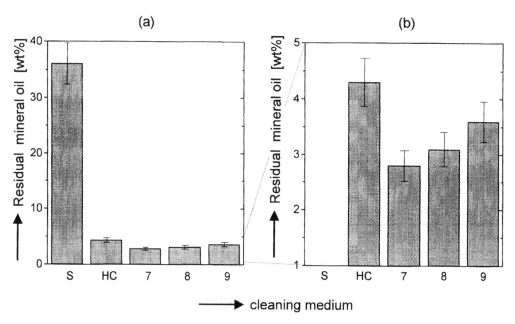

Figure 10 Photometrically determined residual mineral oil on GFS test fabric after cleaning with the standard detergent solution (S), n-undecane fraction ($>$95% n-undecane) (HC), and the microemulsions formed by samples 7, 8, and 9 (composition of samples according to Table 1) at a washing temperature of $T = 313$ K, in weight percent of the initial quantity before washing. (a) Comparison between microemulsions (7, 8, 9), n-undecane fraction ($>$95% n-undecane) (H), and standard detergent solution (S). (b) Comparison between microemulsions (7, 8, 9) and n-undecane fraction ($>$95% n-undecane) (H).

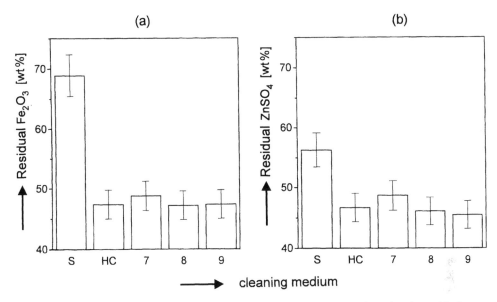

Figure 11 Residual Fe_2O_3 (a) and $ZnSO_4$ (b) on GFS test fabric after cleaning with the standard detergent solution (S), n-undecane fraction ($>95\%$ n-undecane) (HC), and the microemulsions formed by samples 7, 8, and 9 (composition of samples according to Table 1) at a washing temperature of $T = 313$ K, in weight percent of the initial quantity before washing, determined by atomic absorption spectroscopy.

experimental error. It has to be taken into account that in practical washing processes, other detergent components such as builders and bleaching agents play an important role in the removal of hydrophilic pigment soils. The microemulsions used in our model detergency tests, however, do not contain these substances. After cleaning the GFS fabric with standard detergent solution, the residual contents or iron oxide and zinc sulfate were even higher than after cleaning with microemulsions or n-undecane.

2. Relation Between the Composition of the Cleaning Medium and Its Detergency

In all experiments described above, homogeneous, bicontinuous microemulsions were used as cleaning media. To test how the detergency of homogeneous and heterogeneous quaternary systems of microemulsions depends on their composition, we carried out model detergency tests with the quaternary mixtures using microemulsions with the compositions of samples 10–28 (see Table 2) on the test fabrics wfk 10 C and wfk 30 C and GFS fabric at a washing temperature $T = 313$ K. In Fig. 12 the compositions of the samples and the phase regions in the ternary phases diagram of the system water–n-undecane–C_{12-14} alkyl polyglycol ether containing 3.5 wt% n-pentanol are summarized. Figures 13 and 14 show the reflectance differences of various wfk fabrics after cleaning. Figure 15 indicates the absolute reflectance values of the cleaned GFS fabrics.

The best detergency was reached for the hydrophilic cotton fabric wfk 10 C (see Fig. 13). Most of the pieces of fabric cleaned with the different quaternary mixtures show reflectance differences of $\Delta R = 25$–28%. Only the very oil rich W/O microemulsions with the composition of samples 10, 15, and 20 (see Table 2) and the heterogeneous sample 16, which consists of a heterogeneous system containing a W/O microemulsion and the lamellar

Marlipal 24/50

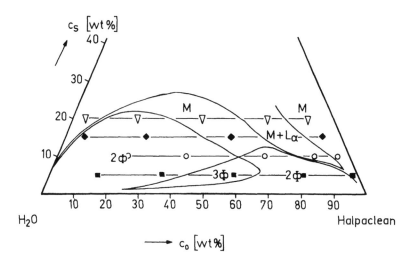

Figure 12 Position of the phase regions and composition of samples 10–28 (composition of samples according to Table 2) in the ternary phase diagram of the system water–n-undecane–C_{12-14} alkyl polyglycol ether $C_{12-14}E_{(5)}$ with 3.5 wt% n-pentanol at $T = 313$ K. M = homogeneous microemulsion; 2Φ = two-phase region consisting of microemulsion and oil or water phase; $M + L_\alpha$ = two-phase region consisting of microemulsion and lamellar phase; 3Φ = three-phase region consisting of microemulsion, oil, and water phase: c_s = alkyl polyglycol ether concentration; c_o = n-undecane concentration. (■) $c_s = 5$ wt%; (○) $c_s = 10$ wt%; (◆) $c_s = 15$ wt%; (∇) $c_s = 20$ wt%. The points for constant c_s are connected.

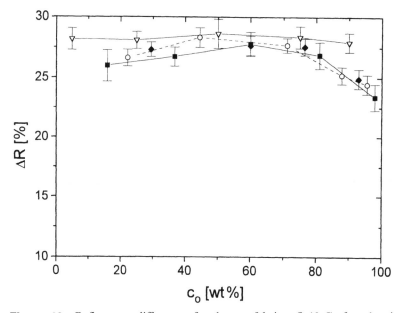

Figure 13 Reflectance differences for the test fabric *wfk* 10 C after cleaning with samples 10–28 (composition of samples according to Table 2) at a washing temperature of $T = 313$ K in dependence on the n-undecane concentration c_o with the alkyl polyglycol ether concentration c_s as the variable parameter. (■) $c_s = 5$ wt%; (○) $c_s = 10$ wt%; (◆) $c_s = 15$ wt%; (∇) $c_s = 20$ wt%. The points for constant c_s are connected.

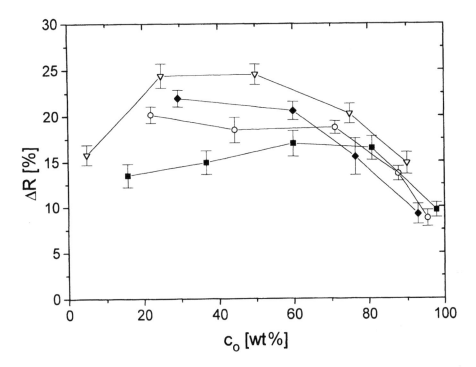

Figure 14   Reflectance differences for the test fabric *wfk* 30 C. For details see Fig. 13.

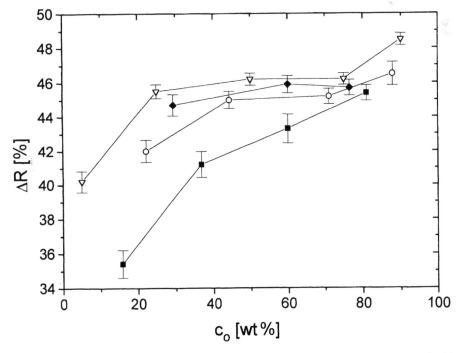

Figure 15   Absolute reflectance values for the GFS test fabric. For details see Fig. 13.

L_α phase, show a slightly lower detergency ($\Delta R = 23$–25%). After cleaning with standard detergent solution, however, a fairly low reflectance difference of $\Delta R = 11\%$ is obtained (see Fig. 5). For the detergency, it is irrelevant whether the samples used in the detergency test form homogeneous microemulsions or heterogeneous systems.

As noted already, the detergency of the quaternary mixtures for lipophilic polyester fabric *wfk* 30 C is lower than for cotton fabrics. Besides, Fig. 14 shows that it is affected much more by the composition of the cleaning medium. After being cleaned with bicontinuous microemulsions or two-phase systems with an alkyl polyglycol ether content $c_s = 20$ wt% and n-undecane concentrations of 30 wt%$<c_o<60$ wt%, the test fabrics show a maximum in the reflectance difference of $\Delta R \approx 11\%$. If the alkyl polyglycol ether content is decreased or the n-undecane concentration is changed to $c_o<20$ wt% or $c_o>60$ wt%, the detergency decreases. The lowest values are found for oil-rich W/O microemulsions of samples 10, 15 and 20 (see Table 2).

For polyester fabric, too, the detergency of the mixtures is not affected whether the cleaning media are homogeneous microemulsions or the lamellar L_α phase. The influence of the composition can be explained by the higher adhesion of the oily soil to lipophilic polyester fabric. So the interfacial tension of the cleaning medium against soil and fabric plays an important role in rolling up and solubilizing the oily soil. In the central isotropic region of the ternary phase diagram, the interfacial tension between the microemulsion and the water or oil phase shows a minimum on the order of $\gamma \approx 10^{-3}$ mN/m, which has a positive effect on the cleaning results for mixed hydrophilic and hydrophobic soils. Only farther away from the isotropic region of the microemulsions in the phase diagram, the surface activity of the surfactant molecules decreases in such a way that the detergency is worse.

Figure 15 shows the detergency of samples 10–28 (see Table 2) for the heavily oil soiled GFS test fabric. Because of the dyeing of the oily soil, the absolute reflectance values of the cleaned fabrics describe the solubilization of the mineral oil. The detergency of the mixtures increases with increasing n-undecane and alkyl polyglycol ether content of the cleaning medium. For all samples with alkyl polyglycol ether contents $c_s>10$ wt% and n-undecane concentrations 30 wt%$<c_o<70$ wt%, the reflectance values of the GFS fabrics cleaned with these samples are on the order of $R \approx 45\%$. This effect correlates with ultralow interfacial tensions of the microemulsions against the water or oil phase of $\gamma<5\times10^{-3}$ mN/m. The best detergency is, however, reached by a W/O microemulsion (sample 24, see Table 2) that contains $c_s = 20$ wt% C_{12-14} alkyl polyglycol ether and $c_o = 72$ wt% n-undecane. This sample shows an extremely low interfacial tension against oil ($\gamma<10^{-4}$ mN/m). From the results of these experiments it follows that the detergency of the investigated quaternary water–n-undecane–C_{12-14} alkyl polyglycol ethers–n-pentanol mixtures correlates directly with their interfacial tension against the oil phase.

C. Changes in the Solubilization Power and the Existence Region of Microemulsions During the Washing Process

The phase behavior of ternary and quaternary systems of the type water–oil–surfactant–cosurfactant is affected strongly by the addition of other components. Therefore, it is questioned how the solubilization of soil during the use of microemulsions as cleaning media in the washing process influences their existence region in the phase diagram and their solubilization power. To test this effect, the temperature dependence of the phase behavior of samples 10–28 (see Table 2) after their use in model detergency tests with the test fabrics *wfk* 10 C, *wfk* 30 C, and GFS were compared with those before use. The following results were obtained:

The cleaning of the slightly soiled test fabrics *wfk* 10 C and *wfk* 30 C does not affect the phase behavior of the samples. Microemulsions exist in the same temperature region within the limits of experimental error ($\Delta T = \pm 2$ K); the alkyl polyglycol ether concentration needed for the formation of homogeneous microemulsions, however, is $\Delta c_s \approx 2$ wt% higher. This might be due to the adsorption of surfactant molecules on the fabric and to the solubilization of oily soil. In the cleaning of light soils, microemulsions are therefore multiply reusable. Their solubilization power can be maintained if, after each washing process, $c_s \approx 2$ wt% alkyl polyglycol ether is added. The reusability is limited by the increasing redeposition of pigment soil. In practical applications pigments could be separated by suitable filter sets.

Model detergency tests with the heavily soiled GFS test fabric make the temperature region in which microemulsions are formed shift by about $\Delta T \approx 5$ K toward lower temperatures. The minimum alkyl polyglycol ether concentration for the formation of microemulsion shifts by solubilization of mineral oil from $c_s = 9$ wt% to $c_s = 16$ wt%. In the cleaning of heavily oil soiled textiles, the reusability is limited by oil solubilization. It has to be taken into account, however, that soiled fabrics in the cleaning practice are not, in large part, saturated with oily soil like the GFS test fabric was. Thus an effective change in the properties of the cleaning medium is expected only after a long series of washing cycles.

IV. EVALUATION OF THE MODEL DETERGENCY TESTS

Summarizing the results of the model detergency tests, the following conclusions can be drawn from this study:

1. Microemulsions of the composition water–*n*-undecane–C$_{12-14}$ alkyl polyglycol ether–*n*-pentanol show, even at a washing temperature of $T = 298$ K, an outstanding detergency, especially for oily soil. They clean much more effectively than a standard detergent solution at $T = 338$ K, although the standard solution contains additional builders and bleaching agents. The most suitable ratio between the concentration of surfactant and cosurfactant is maintained by using microemulsions with the low ethoxylated C$_{12-14}$ alkyl polyglycol ether (Marlipal 24/40) without addition of *n*-pentanol.

2. The outstanding detergency of microemulsions is related to the ultralow interfacial tension ($\gamma \approx 10^{-3}$ mN/m), especially against the oil phase, which indicates low interfacial tensions against oily soil also. An advantage of microemulsions over conventional detergent solutions is that the minimum in the interfacial tension can be turned to a given temperature by varying their components.

3. The solubilization of soil has little influence on the existence regions and properties of the microemulsions investigated here. Thus the microemulsions can be reused several times without an essential change in interfacial tensions or in their detergency.

4. Microemulsions show their full detergency when used even in small quantities (<2 g microemulsion per gram of fabric). An increase in the bath ratio does not improve the detergency in a significant way.

5. The detergency of microemulsions is affected by the properties of the fabric. Whereas hydrophilic cotton fabric is generally cleaned very well, the detergency for lipophilic polyester clearly decreases with decreasing interfacial tension between the microemulsion and the water or soil phase.

6. For practical purposes, in the detergency process, the use of microemulsions should be considered only for special problems such as for the pretreatment of heavily oil soiled workwear. In these processes especially the reusability in the cleaning process is an essential advantage of microemulsions as cleaning media.

7. The composition of microemulsions can be tuned to suit the particular type of soil and fabric. If the amounts of surfactant and the cosurfactant are chosen suitably, the cleaning with microemulsions can be carried out at room temperature.

8. Microemulsions are not only useful in textile detergency, they are also applicable as media for the cleaning of other heavily oil soiled surfaces such as metals, polymers, ceramics, and other materials. Under these conditions, the composition has to be chosen so that it forms a homogeneous microemulsion only during the cleaning process by solubilizing the oily soil.

REFERENCES

1. K. H. Raney, W. J. Benton, and C. A. Miller, J. Colloid Interface Sci. *117*:282 (1987).
2. F. Schambil and M. J. Schwuger, Colloid Polym. Sci. *565*:1009 (1987).
3. Henkel KGaA, Eur. Patent EP 0288858 A1 (1988).
4. F. Mori, J. C. Lim, O. G. Raney, C. M. Elsik, and C. A. Miller, Colloids Surf. *40*:323 (1989).
5. K. Knopf and E. Schollmeyer, Tenside Surfact. Deterg. *27*:96 (1990).
6. J. C. Lim and C. A. Miller, Langmuir *7*:2021 (1991).
7. H. Krüssmann and R. Bercovici, J. Chem. Tech. Biotechnol. *50*:399 (1991).
8. N. Azemar, J. Carrera, and C. Solans, J. Dispersion Sci. Technol. *14*:654 (1993).
9. K. Stickdorn, M. J. Schwuger, and R. Schomäcker, Tenside Surfact. Deterg. *31*:218 (1994).
10. M. J. Schwuger, R. Schomäcker, and K. Stickdorn, Chem. Rev. *95*:849 (1995).
11. C. Solans, J. García Domínguez, and S. E. Friberg, J. Dispersion Sci. Technol. *6*:523 (1985).
12. F. Comelles, C. Solans, N. Azemar, J. Sánchez Leal, and J. L. Parra, Tenside Surfact. Deterg. *22*:323 (1985).
13. H. Krüssmann, T. Bluhm, and B. Föllner, AIF-Forschungsbericht No. 8421, 1993.
14. B. Föllner, T. Bluhm, and H. Krüssmann, Reiniger Wäscher *47*:13 (1994).
15. M. Kahlweit and R. Strey, Angew. Chem. Int. Ed. Engl. *24*:654 (1985).
16. M. Kahlweit, R. Strey, and G. Busse, J. Phys. Chem. *94*:3881 (1990).
17. R. Strey and M. Jonströmer, J. Phys. Chem. *96*:4537 (1992).
18. H.-D. Dörfler, A. Grosse, and H. Krüssmann, Tenside Surfact. Deterg. *32*:484 (1995).
19. Hüls-AG, Product information on surfactants of the type C_{12-14} alkyl polyglycol ether 24, 1995.
20. *wfk*-Testgewebe GmbH Brüggen, Informationen über Testgewebe und deren Einsatz bei Prüfungen des Wascheffekts, 1994.
21. K. R. Wormuth and P. R. Geissler, J. Colloid Interface Sci. *146*:320 (1991).

30

The Future of Microemulsions

Johan Sjöblom
University of Bergen, Bergen, Norway

Stig E. Friberg
Clarkson University, Potsdam, New York

I. INTRODUCTION

To describe the future of microemulsions is a most thrilling task, because the future of microemulsions may be twofold. First the application of these systems in academic research as model systems for intriguing surfactant association structures should be considered, and second, the direct industrial or semi-industrial use of these systems in technological processes. Depending on which future is considered, the conclusions are rather different. In this contribution we try to view both aspects of microemulsions. Academic research in the field of microemulsions in the future will most likely be concerned with highly optimized microemulsion systems and highly advanced instrumentation. In this way one can extract more detailed information on the systems and create the foundation for the development of sophisticated theoretical work. A combination of theoreticians and skilled experimentalists with access to advanced instrumentation and computational facilities is needed to bring forward high-level microemulsion research [1].

Most likely there will be a considerable mismatch between academic research with its well-defined, highly optimized systems and the need for chemical systems in process industry. For example, the highly optimized bicontinuous microemulsion systems seem to be far too sensitive to contamination, process settings, temperature, and retention effects to become an industrial success. Hence in the future one can expect the industrial benefit from these systems to include solubilized W/O or O/W systems. These systems exist, as we know, over rather broad compositional intervals and also for large temperature intervals.

II. FUNDAMENTAL ASPECTS

The fundamental approach to microemulsions has involved determination of appropriate phase diagrams (at least the extension of the solution phases) together with determination of the corresponding solution structures. In this respect, world-leading contributions have been made by groups in Lund, Sweden and in France.

The Lund group successfully combined phase equilibria and NMR measurements to come up with detailed information about different microemulsion systems [2–5]. They have also expanded more-or-less traditional microemulsions to involve polymers and have done careful phase equilibria and structural mappings in the presence of these macromolecules [6]. Pioneering work to directly measure forces in bicontinuous microemulsions has also been reported [7].

The fundamental work in the field of microemulsions in France has been a successful combination of very careful phase equilibria mapping and scattering techniques [8–11].

The theoretical work in Lund and France with the aim to explain fundamental properties of microemulsion membranes and curvatures will proceed in parallel with the experimental impacts.

III. APPLIED ASPECTS

In this prospect of microemulsion applications we have chosen some crucial areas where microemulsions are in use.

A. Microemulsion-Based Washing and Cleaning

The use of microemulsions in the context of washing and cleaning was recently reviewed [1]. There seem to be no reasons to believe that any fundamental new impact is needed in this area from a physicochemical point of view. Large-scale applications in the area of soil remediation can be expected in the near future. In this context it will be essential to estimate microemulsion formation, price, chemical performance, and mechanisms of retention (adsorption) on the solid material when designing these kinds of washing systems. Microemulsions for use in soil remediation have been summarized by Miller and coworkers [12,13] and Schwuger and coworkers [14,15].

B. Microemulsions in Crude Oil Exploitation

By now it is rather obvious that the technological applications (for instance, enhanced oil recovery) within the field of crude oil exploitation have been highly overestimated. In onshore applications, the obvious choice is to drill new holes instead of starting expensive and spectacular chemical operations involving microemulsions. Offshore operations today have greatly benefited from developments in drilling technology, well treatment technology, etc. As a consequence, we have reached (or expect to reach) an exploitation rate in excess of 40% in fields in the North Sea. In future offshore crude oil exploitation, the main topics will include deep sea drilling, high pressure chemistry, and, above all, the chemical or mechanical treatment of extremely stable water-in-crude oil emulsions. In this respect, the chemistry of indigenous components such as asphaltenes and resins will be central but will most likely not involve microemulsions in a traditional sense [16].

C. Microemulsions in Materials Science

Most likely, materials science is an area where sophisticated microemulsion structures will be used in the near future. The use of microemulsion systems for the preparation of fine particles (catalysts, semiconductors, etc.), polymeric particles, latices, and advanced mesoporous inorganic structures for specific catalytic purposes has been reported and will be more extensively investigated in the future.

The systematic utilization of different microemulsion structures, including a variety of particle shapes and functionalities, in fine particle preparation has been performed, for instance, by Pileni and collaborators [17–19].

1. Polymerization

Microemulsions with their microcompartmentalization of hydrophilic and hydrophobic spaces have for a long time been of interest as polymerization media as demonstrated by Antonietti's extensive review [20]. The early publications in the area [21,22] were hampered by problems such as, inadequate stability and incomplete solubility of oligomers [23], and successful polymerization was first reported by Candau and coworkers [24,25].

This polymerization research was concerned with the preparation of ultrasmall latex particles, resulting ultimately [26] in well-defined particles with an average size of 7–8 nm. It is obvious that with the introduction of metallosurfactants [26] and the use of surfactants with an organic counterion [27], the main problem encountered in their application to the preparation of polymer particles, that of high surfactant concentrations, is now alleviated to such a level that serious applications may have a realistic potential.

However, the preparation of latex particles may be perceived as having reached a level at which the potential for a fundamental breakthrough in the final materials per se is rather limited. Pioneering efforts may instead be expected in the development of polymeric microcompartmentalized materials. This development, in a limited form, may be exemplified by the work of Gan and colleagues [28], who polymerized organic monomers solubilized in bicontinuous microemulsions and obtained microporous organic polymers. This area is, of course, of future interest, but the problem of lack of correlation between the microemulsion colloidal structure and the microstructure of the final material may result in a focus on the polymerization of liquid crystalline material where even complex systems [29,30] have been shown to retain their microstructure after polymerization. This area of polymerization has been further developed and systematized by Antonietti [31,32], Antonietti et al. [33], and Fendler [34].

It seems reasonable to expect that microemulsions with their capability for extremely high content of inorganic and organic materials [35] should offer a potential to create organic–inorganic polymer combination materials possessing interesting properties for use in process industry. We are aware of the use of amphiphilic association structures to prepare inorganic condensation polymers [36–40] and even the first results to form organic–inorganic combination materials [35,41], but it seems obvious that efforts to prepare highly structured microcompartmentalized combination materials of this kind should have the potential to become a significant part of the microemulsion-based materials research in the future.

2. Chemical Reactions

The special structure of microemulsions in which both hydrophobic and hydrophilic groups are amenable to reaction per se leads to interesting avenues for specific reactions, as amply demonstrated in other chapters of this book.

In this small section we draw attention to a special form of reaction in which the path of the subsequent reaction is dramatically altered by the microemulsion combination of an O/W interface and relatively high mobility of the microemulsion droplets.

The example is the reaction between tetraalkoxysilanes and water. This reaction in solution takes place in consecutive steps of hydrolysis,

$$Si(OR)_4 + H_2O \rightarrow SiOH(OR)_3 + ROH \tag{1}$$

and condensation,

$$2Si(OH)_{4-x}(OR)_x \rightarrow Si_2O(OH)_{6-2x}(OR)_{2x} + H_2O \tag{2}$$

The continuing condensation gives rise to a complex mixture, the total of which may be described as

$$Si_uO_y(OH)_z(OR)_x \tag{3}$$

in which $x+z = 4u-2y$.

The resulting complex mixture is well illustrated by the silicon NMR spectrum (Fig. 1, bottom).

If the reaction is carried out in a microemulsion under controlled conditions [42], the result is entirely different. As shown in Fig. 1 (top), only two species are found,

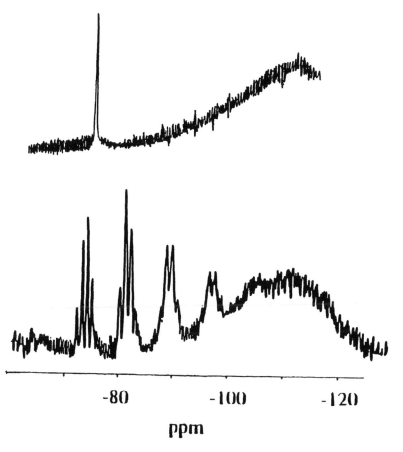

Figure 1 The reaction of tetraethoxysilane and water in a solution gives rise to a large number of hydrolyzed and condensed silicon compounds as illustrated by the bottom spectrum, while the corresponding reaction in microemulsion gave only SiO_2 (top spectrum) in addition to the original compound.

the original $Si(OR)_4$ and the silica of the complete reaction. The reason for this result is, of course, the fact that the initial hydrolysis giving $Si(OH)(OR)_3$ alters the property of the silane from oil-soluble to amphiphilic. Once the molecule has been partially hydrolyzed it remains at the interface, and the hydrolysis and condensation continue unabated to form SiO_2.

Such modification of a reaction path has not been developed into technological applications, but it certainly constitutes a great potential for new processes.

D. Microemulsions in Pharmaceuticals and Cosmetics

Recent findings combining liposomal structures and surfactant formulations in pharmaceuticals and cosmetics seem to open thrilling new possibilities.

1. Liposomal Solutions

The potential of microemulsions as a means of preparing liposomal solutions has not been realized in spite of the fact that the method offers significant advantages over existing methods. Instead, the publications so far have been concerned with simple surfactant solutions, with the process as follows. A micelle-forming surfactant (S_M) and a liposome-forming one (S_L) are combined at a concentration $C_{S_M} > cmc_{S_M}$ (where cmc = critical micelle concentration) [43–45]. The liposome/micelle fraction $S_L/(S_L + S_M)$ has attracted some attention [46].

The liposomal solution can be formed from the micellar solution by either of two mechanisms: dilution or osmotic extraction. In the first mechanism dilution takes place to bring $C_{S_M} > cmc_{S_M}$ where cmc_{S_M} now denotes the cmc for S_M when it is combined with S_L. The principle is given in Fig. 2, which shows the phase equilibria in a system containing S_M with a short hydrocarbon chain, which is of obvious advantage. The second mechanism subjects the system to osmotic separation. The osmosis removes only monomers of S_M, and since the monomer concentration of S_L is extremely low, 10^{-5}–10^{-10} M, the reduction in S_M concentration leaves the concentration of S_L approximately intact.

The principles of the two methods are demonstrated in Fig. 3, which also illustrates the disadvantages of the methods. The first method results in a low concentration of liposomes unless concentrated by osmosis. In short, the micellar system has too high an S_M/S_L ratio to be practical. The osmotic separation results in a higher concentration of S_M, because there is no dilution, but the process is very time-consuming.

These disadvantages have recently been alleviated by the introduction of a method to prepare liposomes from microemulsions formed by a combination of S_L with a hydrotrope (H) [47]. The hydrotropes are short-chain molecules, usually with an ionic group [48,49]. The combination is highly advantageous as a precursor to the preparation of liposomes, as demonstrated in Fig. 4, which illustrates the pronounced difference from micellarly solubilized species as precursors (Fig. 2). The S_M/S_L ratio is typically approximately 2, and the H/S_L ratio is on the order of 0.25, which is one order of magnitude improvement. In addition, it should be realized that the surface activity of the hydrotrope molecule is far less than that of the surfactant, which is another advantage.

There is no doubt that microemulsions of this kind will find use in the future development of liposome technology.

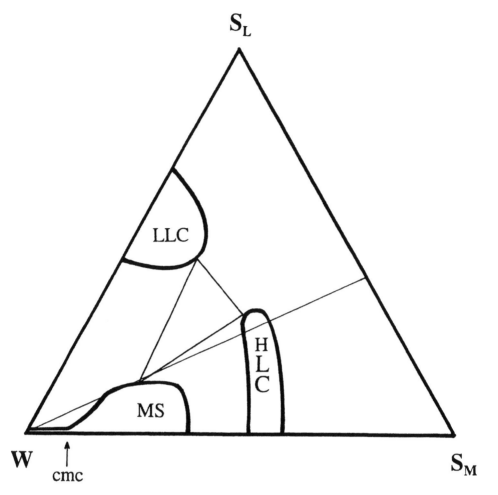

Figure 2 In the system of water (W), a micelle-forming surfactant (S_M), and a liposome-forming surfactant (S_L), the micellar solution (MS) is in equilibrium with the lamellar (LLC) and hexagonal (HLC) liquid crystals.

2. Skin Care Products

The present daily approach to skin care formulations is based on an assumed relation between the original composition and the final activity of the product. So a significant number of authors have tied to relate the size of liposomes in a solution to the final action of an active substance. The results [48,49] have given ample proof that such a relation does not exist. The reason for this lack of success is obvious. Within 25 min after application, 90–95% of the water in a formulation operates, and the new structures appearing during this process cannot a priori be related to the original structures. The biological activity of the formulation depends on the final structure and its interaction with the stratum corneum structural units. Hence, the activity of the formulation depends on the structures found in the water-poor parts of the system as shown in Fig. 5.

This concept presents a significant opportunity for the use of microemulsions in skin care technology, because their advantages in such applications are significant. They are

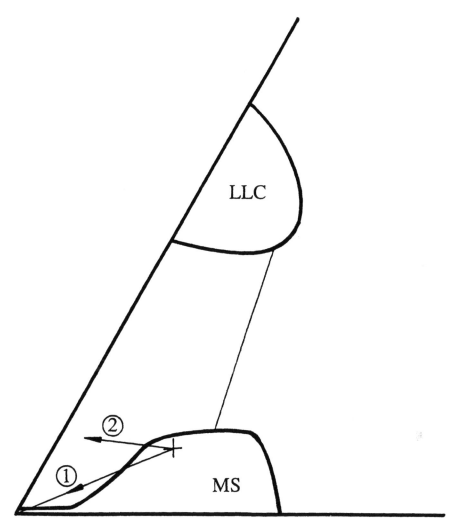

Water

Figure 3 Simple dilution of the micellar solution of S_L (Fig. 2) along line 1 gives a liposomal solution with low concentration of S_L. Removing S_M by osmotic separation (line 2) results in higher concentration of S_L.

transparent, they are infinitely stable, and their preparation requires only the mildest of mixing to form the final and stable state.

With knowledge of the total phase diagram for the system, and realizing that volatile components will evaporate, skin care formulations can be adjusted to give a final product in the form of a lamellar structure compatible with the lamellar structure of the stratum corneum. In addition, the approach makes it possible to formulate systems with an extremely high tendency to transfer active substances from the formulation layer to the skin. An example was presented in 1995 [50] in which a compound initially at 2% by weight in the original formulation exhibited the same or greater tendency to enter the skin as in pure form.

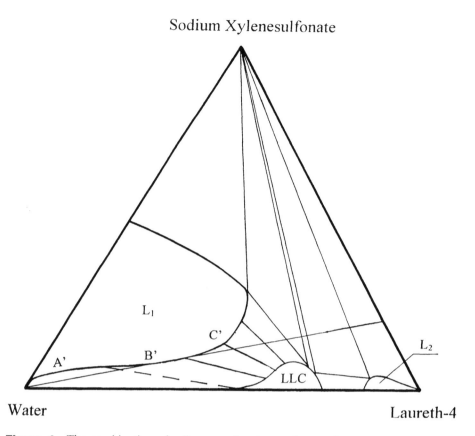

Figure 4 The combination of a liposome-forming surfactant (Laureth 4, L4) and a hydrotrope (sodium xylenesulfonate, SXS) with water gives a huge microemulsion region (L_1), from which a liposomal solution with high concentration of liposome may be made with a high ratio of liposomal surfactants to hydrotropes. The line from the water corner shows the maximum L4/SXS weight ratio.

Once these concepts have been adopted by the personal care community, new and novel applications of microemulsions will emerge.

IV. CONCLUSIONS

From the various equilibrium structures of microemulsions outlined as well as their kinetics, it appears that the areas of highest fundamental future interest and most likely of the greatest potential for microemulsion applications are systems in which structural changes are observed when a substrate is encountered. In the present chapter the change in substrate–microemulsion interaction when the volatile components evaporate in topical applications has been mentioned. For that specific case the interaction between the stratum corneum corneocytes and lipids and the microemulsion time-dependent structures during partial volatilization is an area that would richly reward research efforts. In the same vein, knowledge of the structural changes when microemulsions are exposed to biological plasma is essential to an understanding of medicinal action in general and offers the potential for directing the action of drugs. In this context it should be mentioned that microemulsions consisting of substances active

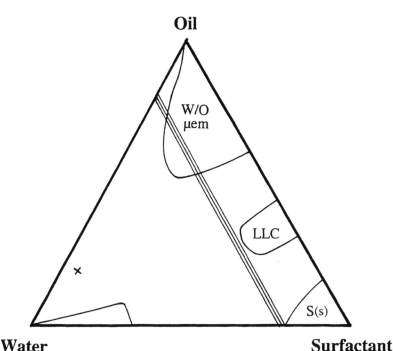

Figure 5 The original skin care formulation (×) is an emulsion, but the final state (area to the right of the triple lines) strongly depends on the surfactant/oil ratio. μem = microemulsion.

in the olfactory and vomero-nasal systems for which the solubilization of components in aqueous microemulsions is followed by aerosol dispersal is an area where little or nothing is known about the biological effect of such particles after evaporation.

Another sophisticated and technologically very promising area is the formation of inorganic-organic materials from suspensions of solid materials in microemulsions followed by evaporation-polymerization processes. Such materials offer a potential for developing easily applied liquids with time-dependent structural and rheological changes in materials properties.

ACKNOWLEDGMENT

We thank the technology program FLUCHA financed by the Norwegian Research Council (NFR) and the oil industry for its financial support.

REFERENCES

1. J. Sjöblom, R. Lindberg, and S. E. Friberg, Adv. Colloid Interface Sci. *65*:125 (1996).
2. B. Lindman, K. Shinoda, U. Olsson, D. Andersen, G. Karlström, and H. Wennerström, Colloids Surf. *38*:205 (1989).
3. U. Olsson and H. Wennerström, Adv. Colloid Interface Sci. *49*:113 (1994).
4. U. Olsson, K. Shinoda, and B. Lindman, J. Phys. Chem. *90*:4083 (1986).
5. H. Wennerström, O. Söderman, U. Olsson, and B. Lindman, Colloids Surf. A *123–124*:13 (1997).

6. H. Bagger-Jörgensen, Thesis, University of Lund, 1997.
7. P. Petrov, Thesis, University of Lund, 1996.
8. A. M. Bellocq and D. Roux, in *Microemulsions: Structure and Dynamics* (S. E. Friberg and P. Bothorel, eds.), CRC Press, Boca Raton, FL, 1987, p. 33.
9. A. M. Bellocq, J. Biais, P. Bothorel, B. Clin, G. Fourche, P. Lalanne, B. Lemaire, B. Lemanceau, and D. Roux, Adv. Colloid Interface Sci. *20*:167 (1984).
10. K. Kekicheff and B. Cabane, J. Phys. *48*:1571 (1987).
11. A. M. Bellocq, in *Emulsions and Emulsion Stability* (J. Sjöblom, ed.), Marcel Dekker, New York, 1996, p. 181.
12. K. H. Raney, W. J. Benton, and C. A. Miller, J. Colloid Interface Sci. *117*:282 (1987).
13. K. H. Raney and C. A. Miller, J. Colloid Interface Sci. *119*:539 (1987).
14. A. Hild, J.-M. Sequaris, H.-D. Narres, and M. J. Schwuger, Colloids Surf. A *123–124*:515 (1997).
15. E. Koglin, A. Tarazona, S. Kreisig, and M. J. Schwuger, Colloids Surf. A *123–124*:523 (1997).
16. J. Sjöblom, T. Skodvin, Ø. Holt, and F. P. Nilsen, Colloids Surf. A *123–124*:593 (1997).
17. I. Lisieki and M. P. Pileni, J. Am. Chem. Soc. *115*:3887 (1993).
18. I. Lisieki and M. P. Pileni, J. Phys. Chem. *99*:5077 (1995).
19. M. P. Pileni, J. Tanori, and A. Filankembo, Colloids Surf. A *123–124*:561 (1997).
20. M. Antonietti, Macromol. Chem. Phys. *196*:441 (1995).
21. J. O. Stoffer and T. Bone, J. Dispersion. Sci. Technol. *1*:37 (1980).
22. S. S. Atik and K. J. Thomas, J. Am. Chem. Soc. *103*:4279 (1981).
23. L. M. Gan, C. H. Chew, and S. E. Friberg, J. Macromol. Chem. *A19*:739 (1983).
24. Y. S. Leong and F. Candau, J. Phys. Chem. *86*:2269 (1982).
25. F. Candau, Y. S. Leong, G. Pouyet, and S. J. Candau, J. Colloid Interface. Sci. *101*:167 (1984).
26. M. Antonietti and T. Nestl, Macromol. Rapid Commun. *15*:111 (1994).
27. M. Antonietti and H. P. Hentze, Adv. Mater. *8*:840 (1996).
28. T. H. Chieng, L. M. Gan, W. K. Teo, and K. L. Pey, Polymer *37*:5917 (1996).
29. S. E. Friberg, B. Yu, and G. A. Campbell, J. Polym. Sci. *28*:3575 (1990).
30. S. E. Friberg, B. Yu, and G. A. Campbell, J. Dispersion Sci. Technol. *14*:205 (1993).
31. M. Antonietti, Chim. Ing. Technol. *68*:518 (1996).
32. M. Antonietti, Macromol. Chem. Phys. *197*:2713 (1996).
33. M. Antonietti, Ch. Benger, J. Conrad, and A. Kaul, Macromol. Symp. *106*: (1996).
34. J. H. Fendler, Chem. Mater. *8*:1616 (1996).
35. S. E. Friberg, C. C. Yang, M. B. Biscoglio, and H. F. Hellbig, J. Mater. Sci. *11*:1373 (1992).
36. S. E. Friberg and C. C. Yang, in *Innovations in Materials Processing Using Aqueous, Colloid, and Surface Chemistry* (F. M. Doyle, S. Raghavan, P. Somasundaran, and S. W. Warren, eds.), Minerals, Metals, and Materials Soc., Warrendale, PA, 1988, pp. 181–191.
37. S. E. Friberg, C. C. Yang, and J. Sjöblom, Langmuir *8*:372 (1992).
38. J. Sjöblom, M. Selle, S. E. Friberg, T. Moaddel, and C. Brancewicz, Colloids Surf. *88*:235 (1994).
39. B. Ammundsen, G. R. Burns, A. Amran, and S. E. Friberg, J. Sol/Gel Sci. Technol. (in press).
40. J. Sjöblom, H. Ebeltoft, A. Bjorseth, S. E. Friberg, and C. Brancewicz, J. Dispersion Sci. Technol. *15*:21 (1994).
41. S. M. Jones, S. E. Friberg, and J. Sjöblom, J. Mater. Sci. *29*:4075 (1994).
42. S. E. Friberg and Z. Ma, J. Non-Cryst. Solids *147*:30 (1992).
43. D. Lichtenberg, R. J. Robson, and E. A. Dennis, Biochim. Biophys. Acta *415*:29 (1975).
44. D. Levy, A. Gulik, M. Seigneureth, and J. L. Rigaud, Biochemistry *29*:9480 (1990).
45. K. Edwards and M. Almgren, J. Colloid Interface. Sci. *147*:1 (1991).
46. A. de la Maza and J. L. Parra, Colloid Polym. Sci. *272*:721 (1994).
47. S. E. Friberg, H. Yang, L. Fei, S. Sadasivan, D. H. Rasmusen, and P. A. Aikens, J. Dispersion. Sci. Technol. (in press).
48. D. Balusubramanian and S. E. Friberg, in *Surface and Colloid Science*, Vol. 15 (E. Matijevic, ed.), Plenum, New York, 1993, pp. 197–220.
49. P. A. Aikens and S. E. Friberg, Opinion Colloid Interface Sci. *1*(5):672 (1996).
50. S. E. Friberg, T. Moaddel, and A. J. Brin, J. Soc. Cosmet. Chem. *46*:255 (1995).

Index

Printed and bound by CPI Group (UK) Ltd, Croydon, CR0 4YY

29/10/2024

01780648-0001